ENERGY
HANDBOOK

ENERGY
HANDBOOK

Robert L. Loftness, D. Sc.

VNR **VAN NOSTRAND REINHOLD COMPANY**
NEW YORK   CINCINNATI   ATLANTA   DALLAS   SAN FRANCISCO
LONDON   TORONTO   MELBOURNE

Van Nostrand Reinhold Company Regional Offices:
New York   Cincinnati   Atlanta   Dallas   San Francisco

Van Nostrand Reinhold Company International Offices:
London   Toronto   Melbourne

Manufactured in the United States of America

Published by Van Nostrand Reinhold Company
135 West 50th Street, New York. N.Y. 10020

Published simultaneously in Canada by Van Nostrand Reinhold Ltd.

15 14 13 12 11 10 9 8 7 6 5 4 3

Library of Congress Cataloging in Publication Data

Loftness, Robert L
    Energy handbook.

    Includes index.
    1. Power resources—Handbooks, manuals, etc.
2. Power (Mechanics)—Handbooks, manuals, etc.
3. Environmental protection—Handbooks, manuals, etc.
I. Title.
TJ163.2.L63   333.7   77-18190
ISBN 0-442-24836-9

Introduction

This handbook has been prepared in recognition that information on energy, once of interest only to a limited group of engineers and scientists, now has far broader significance to the public in general. This sharp change in public interest was occasioned by the quadrupling of the price of oil in 1973 by the cartel of oil-producing nations comprising the Organization of Petroleum Exporting Countries. The selective embargo of October, 1973, by the OPEC made this point even more clear. These actions highlighted the degree to which many nations had become dependent on the importation of a single commodity— a commodity constituting the lifeblood of their economies.

Warnings of the risks inherent in the impending depletion of domestic petroleum resources and in overdependence on foreign sources had been voiced by several individuals as well as expert committees for some time, not only in the United States but also in other industrialized countries. These few now have been joined by many who have a growing premonition that the era of cheap and abundant energy has come to an end and that the distortions now apparent in national economic affairs and international relations may take many years to bring back into balance. The information in this handbook may help to clarify where we have been, where we are, and where we might possibly be going in our energy economies.

The data and information are drawn from many sources. The selection of data has been, of course, a matter of personal choice—a choice based upon an objective to illustrate, in a relatively concise manner, comparisons among alternatives, basic trends, and general principles. The data represent only a small fraction of that contained in the documents to which references are given. It will be noted that seemingly equally qualified authorities often have different conclusions, that much of the data is still imprecise and many of the projections are controversial. These uncertainties appear in many areas—the assessment of resources, the projections of population and energy demand, the relationship of energy to economic welfare, the feasibility of new technologies, the degree to which conservation efforts can be implemented, and the long-term balance between environmental and economic objectives. At the same time, these uncertainties, in themselves, suggest certain courses of action as being most prudent—actions which, in their motivation, lean more toward the less optimistic assessments and projections.

The data are presented in sixteen chapters beginning with a brief chapter on the importance of energy to man. Energy resources are covered in two chapters; one on depletable fossil and mineral resources, and another on renewable resources such as water power and solar energy. Energy consumption patterns, both historical and projected, are presented in separate chapters. Techniques for the recovery of fossil fuel resources and the conversion of these resources to usable fuel forms are discussed in the chapter following. Three relatively new forms of energy—nuclear, geothermal, and solar—are described separately in terms of resource extraction and utilization. Methods for the utilization of basic fuel resources are described in three chapters covering energy conversion and storage, energy efficiency and conservation, and energy transport. The environmental aspects of energy use and the environmental control of energy production are each represented in separate chapters. The concluding two chapters relate to the costs of energy and to the anticipated futures for energy as outlined in various studies. In addition, two appendices are included—one providing a glossary of energy terms and the other a set of conversion tables to assist in the translation of the many different units used to report energy information.

One of the more formidable aspects in the presentation of energy data is the variety of units in which these data are reported. This variety is the result not only of the existence of two measuring systems, the English and the metric systems, but also of the historical development of units which have become common terminology. In the English system, petroleum is measured in 42-gallon barrels—a barrel of uncommon size used to transport oil in 1859 from the first discovery oil well in Pennsylvania to the nearest railroad. The metric system measures petroleum in metric tons. Gas production is described in trillions of cubic feet or in cubic kilometers. Coal production is given in millions of short tons (2000 pounds) or metric tons (1000 kilograms). Overall energy production and consumption is variously reported in joules, Q's (10^{18} Btu), Quads (10^{15} Btu), or calories.

While all of these units, and more, will be found in this handbook, two units are used as primary reference points. These two units have come into common use in the United States and will probably persist for a number of years to the confusion and consternation of everyone trying to make comparisons among data from various countries. These units are expressed in terms of British Thermal Units, the amount of heat required to raise one pound of water one degree Fahrenheit. In discussing energy resources, a unit representing one quintillion (10^{18}) Btu is used—a unit is designated as Q. This unit is used since it is of a size such that resource data can be presented in units generally from 1 to 1000. In discussing energy consumption data, a unit one thousand times smaller and representing one quadrillion (10^{15}) Btu is used. This unit is designated as a Quad or, by some, as a small q. The use of this unit for energy consumption data also provides numbers that are generally between 1 and 1000 (for example, world energy consumption is about 250 Quad). For those who wish to convert these units into the metric unit of joules, one Btu is equivalent to about 1000 joules.

There is today an impending break in the continuity of the development of man— a break which could be occasioned by the depletion of inexpensive fossil fuel resources. One suggestion for accommodation to this growing shortage of fossil fuels is to change wasteful life styles. Some view such changes as a social good—a more restricted energy diet will lead to a less frantic, more congenial and more healthful life. There are visions of improved family life on gasless Sundays, of the conviviality of car-pools and mass transit, of the beneficial effects of walking or bicycling, and of the general strengthening of moral fiber as a result of belt-tightening in the use of energy. In some degree, these are all truths—at least for those who have enjoyed the fruits of an energy-intensive society. But these truths are only a part of a larger truth, a truth relating to the value of energy in the fundamental support of social systems and in the quality of life for all men. There is waste in the use of energy, waste that should be eliminated; but there is also much muscle, muscle that has made man human rather than slave.

In itself, the suggestion that life styles be changed should not be a matter of alarm. Society has adapted to many changes during the past two centuries and, indeed, has welcomed most of them. These changes have come largely as a result of the introduction of new technologies, each technology reducing man's burden but each more energy-intensive than the one it replaced. The steam engine replaced the water wheel and wind-mill; the locomotive, truck, and automobile replaced the horse-drawn cart; the electric light replaced the kerosene lamp. The jet-propelled airplane has given man a new dimension in time and space; the electric motor has made it possible to develop automatic washers, dryers, refrigerators, and air conditioners. Electricity is used to power industrial machines once dependent upon water or hand power, and is essential to the operation of the telegraph, telephone, radio, television, and the computer.

All of these technologies, each requiring energy, have led to a new status for man. Most importantly, these energy-intensive technologies have given man an unmatched productivity and wealth—he works 40 rather than 80 hours per week, he can afford an education, his diet is of his own choosing, he has the convenience of a personal water supply and sanitation, he can control the climate within which he works and lives, he has a mobility that encompasses the world, and he can view events occurring thousands of miles distant.

These all comprise the basic elements of a "life style" for man—a life style that is real for many who live in industrialized, energy-intensive societies, and a life style that is fervently sought by the millions living in the lesser developed nations. Yet without sufficient energy, the creation or maintenance of these elements of a decent human condition are threatened. A return to more labor-intensive economies will not provide the necessary productivity and wealth. Human labor is one of the least efficient and most costly forms of energy. The cost of food necessary to produce a given output of human muscle energy is 25-30 times more expensive than the cost of producing that same energy from electric motors or fossil-fueled engines. In addition, there are many tasks which only machines, powered by synthetic energy, can perform.

The changes in the condition of human life brought about by man's use of new forms of energy in machines that multiplied his productivity, broke the patterns of life that had existed for milleniums. Man began to move away from a condition of economic serfdom toward new levels of humanity. This process has been proceeding at an ever-accelerating pace—a pace dependent upon the consumption of resources which may be sufficient only for a few decades more. The search for new, inexpensive, energy sources is the challenge of modern man.

Contents

ENERGY
HANDBOOK

Energy and Man

Land use in Blythe, California.

EVOLUTION OF ENERGY USE

Cultural Energy. Man's survival has always depended upon his ability to derive adequate energy, in the form of food calories, from his environment and to protect himself from the extremes of heat and cold in that environment. Primitive man, and most animals, were able to adapt to changing conditions in their environment—migrating from one hunting ground to another in search of food and living in natural shelters. Only man, however, has been able to effectively alter his environment to suit himself—and he has done this through his discovery of sources of energy to supplement the energy of his muscles. His first discovery, that of using fire to cook food and keep himself warm, began the long history of man's use of "cultural" energy to increase his supply of food, improve his physical comfort, and to expand the quality of his life.

In terms of the energy he uses today to fulfill his needs, man's available muscular energy is meager indeed. Modern technological man uses 200 times as much cultural energy as he expends in muscular activity. Muscular energy is derived from the calories supplied in food and from a diet of 2,000,000 calories per day; about half the calories are consumed in the operation of the internal organs of the body and the other half are available for useful work (the diet or food calorie is 1000 times larger than the ordinary calorie—a daily consumption of 2000 food calories is perhaps more familiar than that of 2,000,000 ordinary calories). The energy available in 1000 food calories is equivalent to 1.2 kilowatt hours of electricity—in a 16-hour waking day, man expends muscular energy at a rate equal to that of a 75 watt electric light bulb.

Value of Human Muscular Energy. The daily muscular energy available from 1000 food calories (one million ordinary calories) is equivalent to 1.2 kilowatt hours of electricity or 4094 British Thermal Units (Btus).

In terms of other forms of energy 1000 food calories[1] are worth:

3.6¢ of electricity[2]
6.8¢ of gasoline[3]
3.2¢ of coal[4]

[1] Primitive man lived, and the people of some underdeveloped countries now live, on a diet of about 2000 food calories per day with about half of these calories available for muscular activity. The diet in developed countries is nearer 3000 food calories per day and the values for muscular energy given above should be doubled. Even at doubled values, however, the wages for workers are many times the value of their muscular energy.
[2] Electricity at 3¢ per kilowatt hour.
[3] Gasoline at 50¢ per gallon used in an engine of 25% efficiency.
[4] Coal at $20.00 a short ton used in a steam engine of 10% efficiency.

Human Energy Consumption Throughout History

| | Daily Per Capita Consumption (1,000 kilocalories) | | | | |
	Food	Residential and Commercial	Industry and Agriculture	Transportation	Total
Primitive man	2	—	—	—	2
Hunting man	3	2	—	—	5
Primitive agricultural man	4	4	4	—	12
Advanced agricultural man	6	12	7	1	26
Industrial man	7	32	24	14	77
Technological man	10	66	91	63	230

Energy consumption represents six stages of human development. Primitive man, about one million years ago, had only the energy content in the food he ate. Hunting man, who lived about 100,000 years ago, had more food and also used wood for heat and for cooking. About 5000 B.C., primitive agricultural man grew crops and used animals for cultivation. By A.D. 1400, advanced agricultural man in Europe used coal for heating and also used water and wind for power. In the late 19th century, industrial man had added the steam engine as a source of mechanical energy. Modern technological man utilizes the internal combustion engine, steam turbines, gas turbines, and electricity as his sources of energy.

Source: Data taken from Earl Cook, "The Flow of Energy in an Industrial Society," *Scientific American*, September, 1971.

Sprinkler irrigation rigs that "walk" the land near Abbott, New Mexico. USDA Photo by Bill Marr.

ENERGY AND CULTURE

Energy Use and Cultural Change. Earl Cook[1] has prepared the accompanying chart relating the cultural changes in China with changes in the use of energy. Until the 19th century and the advent of the use of fossil fuels, the changes in energy use were primarily in the more efficient use of solar energy in agriculture. In a series of stages, more efficient crops were introduced—Champs rice, sorghum, maize, sweet potatoes, peanuts, and Irish potatoes. Increased food production was also achieved by extending crop land to hillsides and by double and triple cropping in a single year. In the 20th century, the use of synthetic fertilizers once again permitted an increase in food production. Over the centuries, periods of increased food production were matched by increasing population followed by famine, social unrest, and rebellion. Each period of peak population was typified by overplanting and soil depletion as the people struggled to meet their food requirements. As food production declined, so did populations.

In any society, the ability of that society to engage in cultural activities or, indeed, to live in a "civilized" fashion, requires that at least some of the members of that society need not devote their entire time to grubbing for the basic necessities of life. From time to time in the history of China, the more efficient use of solar energy permitted such civilization. In modern times, the use of the energy from wood, oil, and coal has raised the level of productivity to the point where entire nations live as comfortably and are as well educated as were only members of royalty a few centuries ago.

[1] Earl Cook, "Energy for millenium three," *Technol. Rev.*, December 1972.

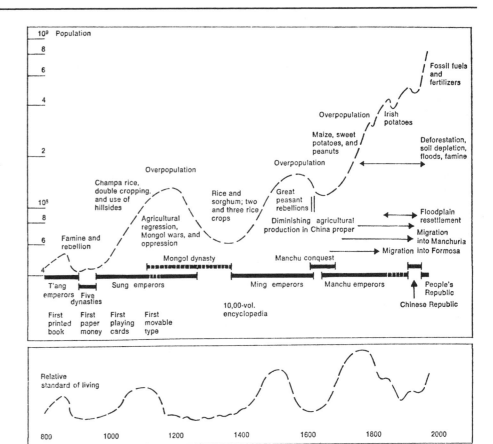

China's history demonstrates the interrelationship of energy and culture (expressed here as standard of living). The introduction of new food plants allowed cultivation of previously unusable land; new varieties of rice allowed double-cropping and planting on land above flood plains; maize and sweet potatoes could be planted on mountain sides, and the Irish potato on mountain tops. In each case, energy supply increased. But erosion, soil depletion, flooding and famine were inevitable, as was, apparently, a change in government, when the energy supply receded.

Source: Earl Cook, "Energy for Millenium Three," *Technology Review*, December, 1972.

Cotton harvesters pick as much in an hour as a man could pick in 72 hours. USDA Photo by Larry Rana.

DEVELOPMENT OF MACHINES

The Industrial Revolution. The dramatic increase in energy consumption during the past two centuries has been a result of the invention of machines for the efficient conversion of energy from one form to another, in particular, the conversion of the heat energy in fuels into mechanical energy. Prior to 1700, the heat from fuels was used for cooking, for the manufacture of glass and metals, and for heating dwellings. Man had also invented various forms of levers, pulleys, and gears that could concentrate muscular energy but not replace it, and he had machines of relatively low power that utilized the energy of falling water and the wind. The horizontal waterwheel was developed in the first century B.C. with a power output of 0.3 kilowatt. By the fourth century A.D. vertical waterwheels of 2 kilowatts were in use. The power output of waterwheels increased steadily and by the 16th century they formed the basis for the industrialization of western Europe. The famous waterwheel at Marly-la-Machine produced 56 kilowatts of power to operate the fountains of Versailles.

The windmill, which first appeared in Europe in the 12th century, was used primarily for the pumping of water and the grinding of grain—activities that could be accommodated to the intermittent nature of wind power. Although power levels of 12 kilowatts were achieved, the sporadic production of power from windmills precluded their use for widespread industrial activities.

The major industrial development in Europe came after the invention of the steam engine which freed industry from the geographical limitations imposed by water power resources and permitted location of industry either near other primary resources or near convenient transportation. In addition, the relatively small size of the steam engine and the portability of fuel necessary to power such engines, gave rise to the development of the railroad.

The use of heated steam to provide mechanical energy was demonstrated by Hero of Alexandria in the first century A.D. with his Sphere of Aeolus—a tiny hollow metal sphere that turned as steam was ejected through two small curved vent tubes mounted at the circumference. The principle employed in Hero's device, that of a steam reaction turbine, was not further developed until the late 19th century.

The first steam "engine" was invented by Thomas Savery in England in the 17th century. His invention utilized the vacuum created by condensing steam to pump water and did not have a moving piston. The development of moving piston steam engines quickly followed and by the early 18th century, steam engines of several kilowatts were being used.

During the past century, the development of water turbines, steam turbines, internal combustion engines and gas turbines, has provided man with high-powered, inexpensive and mobile power sources which have added greatly to his productivity and his personal convenience.

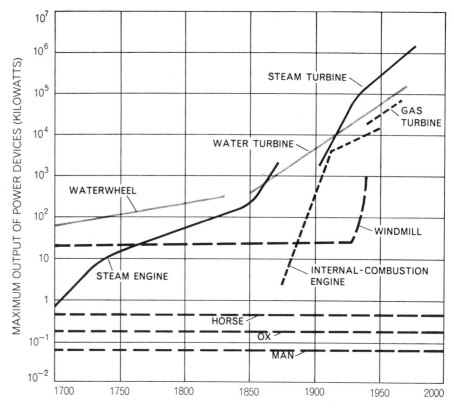

POWER OUTPUT OF BASIC MACHINES has climbed more than five orders of magnitude since the start of the Industrial Revolution (*ca.* 1750). For the steam engine and its successor, the steam turbine, the total improvement has been more than six orders, from less than a kilowatt to more than a million. All are surpassed by the largest liquid-fuel rockets (*not shown*), which for brief periods can deliver more than 16 million kilowatts.

Chauncey Starr, Energy and Power, *Scientific American*, September 1971.

ENERGY USE AND PRODUCTIVITY

Energy and National Productivity. Prior to the 17th century, the productivity of man was determined by his own labor, that of domesticated animals, and the relatively low power available from waterwheels and windmills. As new power sources have been developed, there has been both a rapid increase in population and an increase in average individual affluence. These power sources—machines that can transform the heat energy of wood, coal, and oil to mechanical energy—represent "cultural energy" since they are the product of the culture and civilization of man.

Since the use of cultural energy provides man with "energy" slaves at a cost far below that of human slaves, there is, undoubtedly, some correlation between the use of energy and economic productivity. The correlation between the Gross National Product—a monetary measure of the goods and services produced—and the level of energy consumption is, however, somewhat ambiguous, at least at high levels of energy consumption. For example, New Zealand has half the per capita energy consumption as the United Kingdom but has a higher GNP. As the accompanying figure indicates, there are many countries where the standard of living is considered to be high, yet where the energy consumption/GNP ratio is relatively low. The differences may be due to a lopsided distribution of income between the wealthy and the poor, to the nature of the economy—whether rural or industrial—or to social values that emphasize activities that are inherently not energy-intensive in nature. At the same time, there is little question that the use of energy has raised the average standard of living in those countries where such energy has been available.

The relationship between energy consumption and GNP needs further clarification. In the wealthier countries there is obviously some wasteful use of energy—waste occasioned by the low cost of energy. Yet there must also be a level of per capita energy use below which there would be a true reduction in the standard of living. As the cost of energy rises due to depletion of easily won resources, this balance in energy use will acquire additional meaning.

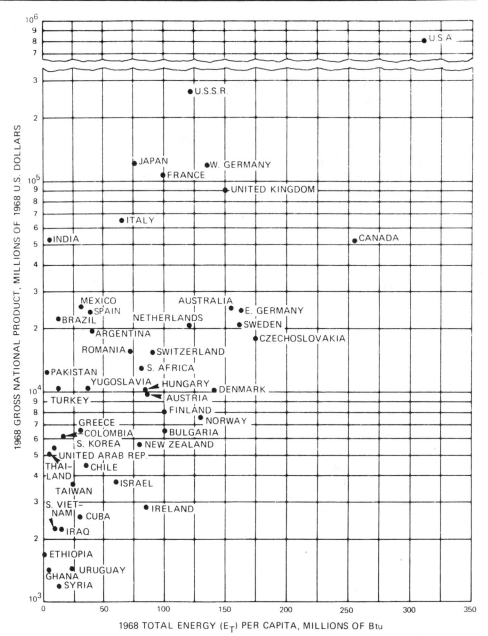

Source: The U.S. Energy Problem, ITC Report C 645 to the National Science Foundation, Volume 1: Summary, InterTechnology Corp., November, 1971.

ENERGY USE AND PRODUCTIVITY—AMERICAS AND AFRICA

Population, Gross National Product, and Energy Consumption—Americas

Country	Population[a] (millions) 1971	Energy Consumption[b] 10^{12} Btus 1971	Annual GNP per Capita[a] $U.S. 1971	Annual Energy Consumption per Capita[c] 10^6 Btus
Northern America				
Canada	21.8	5498.6	2460	252.2
United States	207.1	63535.9	3980	306.8
Middle America				
Costa Rica	1.9	21.8	450	11.5
El Salvador	3.6	20.3	280	5.6
Guatamala	5.3	33.8	320	6.4
Honduras	2.8	15.8	260	5.7
Mexico	52.5	1762.5	530	33.6
Nicaragua	2.1	21.3	370	10.2
Panama	1.5	31.7	580	21.1
Caribbean				
Barbados	0.3	—	440	—
Cuba	8.6	265.2	310	30.8
Dominican Rep.	4.4	29.1	290	6.6
Guadeloupe	0.4	4.1	510	10.1
Haiti	5.4	3.8	70	0.7
Jamaica	2.0	65.6	460	32.8
Martinique	0.4	6.3	610	15.6
Puerto Rico	2.9	289.4	1340	99.8
Trinidad & Tobago	1.1	111.5	870	101.3
Tropical South America				
Bolivia	4.8	28.5	150	5.9
Brazil	95.7	1351.8	250	14.1
Columbia	22.1	393.6	310	17.8
Ecuador	6.3	53.6	220	8.5
Guyana	0.8	19.8	340	24.8
Peru	14.0	235.4	380	16.8
Surinam	0.4	24.7	430	61.8
Venezuela	11.1	734.0	950	66.1
Temperate South America				
Argentina	24.7	1135.0	820	46.0
Chile	10.0	403.8	480	40.4
Paraguay	2.5	9.5	230	3.8
Uruguay	2.9	73.3	520	25.3

Population, Gross National Product, and Energy Consumption—Africa

Country	Population[a] (millions) 1971	Energy Consumption[b] 10^{12} Btus 1971	Annual GNP per Capita[a] $U.S. 1971	Annual Energy Consumption per Capita[c] 10^6 Btus
Northern Africa				
Algeria	14.5	201.0	220	13.9
Libya	1.9	31.1	1020	16.4
Morocco	16.3	84.4	190	5.2
Sudan	16.3	52.2	100	3.2
Tunisia	5.3	36.9	220	7.0
UAR	34.9	264.0	170	7.6
Western Africa				
Dahomey	2.8	2.8	80	1.0
Gambia	0.4	0.7	100	1.8
Ghana	9.3	45.0	170	4.8
Guinea	4.0	11.0	90	2.8
Ivory Coast	4.4	34.0	260	7.7
Liberia	1.2	15.9	210	13.3
Mali	5.2	3.5	90	0.7
Mauritania	1.2	4.7	180	3.9
Niger	4.0	2.9	70	0.7
Nigeria	56.5	91.0	70	1.6
Senegal	4.0	15.0	170	3.8
Sierra Leone	2.7	7.5	150	2.8
Togo	1.9	4.0	100	2.1
Upper Volta	5.5	2.0	50	0.4
Eastern Africa				
Burundi	3.7	1.1	50	0.3
Ethiopia	25.6	27.2	70	1.1
Kenya	11.2	54.8	130	4.9
Malagasy, Republic of	7.1	13.6	100	1.9
Malawi	4.6	6.1	50	1.3
Mauritius	0.9	4.1	230	4.5
Mozambique	7.9	36.1	200	4.6
Reunion	0.5	4.1	610	8.2
Rwanda	3.7	1.1	70	0.3
Somalia	2.9	2.5	60	0.8
Tanzania (United Rep.)	13.6	25.8	80	1.9
Uganda	8.8	19.9	110	2.3
Zambia	4.4	54.9	220	12.5

Population, Gross National Product, and Energy Consumption—Africa (*Continued*)

Country	Population[a] (millions) 1971	Energy Consumption[b] 10^{12} Btus 1971	Annual GNP per Capita[a] $U.S. 1971	Annual Energy Consumption per Capita[c] 10^6 Btus
Middle Africa				
Angola	5.8	23.9	190	4.1
Cameroon (West)	5.9	15.7	140	2.7
Central African Rep.	1.6	2.7	120	1.7
Chad	3.8	2.8	60	0.8
Congo (Dem. Rep.)	17.8	6.6	90	0.4
Congo (Republic of)	1.0	2.7	230	2.7
Equatorial Guinea	0.3	1.5	260	4.8
Gabon	0.5	14.2	310	28.4
Southern Africa				
Botswana	0.6		100	
Lesotho	1.1		80	
South Africa	20.6		650	
Namibia (Southwest Afr.)	0.6		650	
Swaziland	0.4		200	

[a]Population and per capita gross national product data from Paul R. Ehrlich and Anne H. Ehrlich, "Population, resources, environment," W. H. Freeman and Company, San Francisco, 1972.

[b]Energy consumption data from World Energy Conference Survey of Energy Resources, The United States National Committee of the World Energy Conference, 345 East 47th Street, New York, N.Y. 10017 (1974).

[c]In addition to the per capita energy consumption listed in this column, an individual will expend about 2×10^6 Btus of human energy per year.

This cotton gin in Tunica County, Mississippi, bales as much cotton in an hour as a mechanical cotton picker can harvest in a day. It separates cotton fiber from the cotton seed, then cleans it for textile processing. USDA Photo by George Robinson.

ENERGY USE AND PRODUCTIVITY—EUROPE AND ASIA

Population, Gross National Product, and Energy Consumption—Europe

Country	Population[a] (millions) 1971	Energy Consumption[b] 10^{12} Btus 1971	Annual GNP per Capita[a] $U.S. 1971	Annual Energy Consumption per Capita[c] 10^6 Btus
Northern Europe				
Denmark	5.0	722.7	2070	144.5
Finland	4.7	554.2	1720	117.9
Iceland	0.2	24.3	1680	121.3
Ireland	3.0	271.8	980	90.6
Norway	3.9	556.8	2000	142.8
Sweden	8.1	1347.6	2620	166.3
United Kingdom	56.3	8871.9	1790	148.7
Western Europe				
Austria	7.5	697.1	1320	92.9
Belgium	9.7	1680.4	1810	173.2
France	51.5	5500.0	2130	106.8
Germany (FRG)	58.9	8742.8	1970	148.4
Luxembourg	0.4	—	2170	—
Netherlands	13.1	1845.5	1620	140.9
Switzerland	6.4	618.4	2490	96.6
Eastern Europe				
Bulgaria	8.6	963.5	770	112.0
Czechoslovakia	14.8	2621.0	1240	177.0
Germany (DRG)	16.2	2936.0	1430	181.2
Hungary	10.3	930.6	980	90.4
Poland	33.3	3910.0	880	117.4
Romania	20.6	1662.0	780	80.7
Southern Europe				
Albania	2.2	39.6	400	17.1
Greece	9.0	335.1	740	39.5
Italy	54.1	3926.9	1230	72.6
Malta	0.3	—	640	—
Portugal	9.6	196.6	460	20.5
Spain	33.6	1552.0	730	46.2
Yugoslavia	20.8	903.1	510	43.4
USSR	245.0	30349.4	1110	123.9

Population, Gross National Product, and Energy Consumption—Asia

Country	Population[a] (millions) 1971	Energy Consumption[b] 10^{12} Btus 1971	Annual GNP per Capita[a] $U.S. 1971	Annual Energy Consumption per Capita[c] 10^6 Btus
Southwest Asia				
Cyprus	0.6	25.2	830	42.0
Iraq	10.0	169.7	260	17.0
Israel	3.0	218.8	1360	72.9
Jordan	2.4	20.7	260	8.6
Kuwait	0.8	231.5	3540	289.3
Lebanon	2.9	66.0	560	22.7
Muscat & Oman	0.7	0.9	250	1.3
Saudi Arabia	8.0	209.4	360	26.2
Southern Yemen	1.3	23.8	120	18.3
Syria	6.4	83.7	210	13.1
Turkey	36.5	415.0	310	11.4
Yemen (Arab Rep.)	5.9	2.1	70	0.4
Middle South Asia				
Afghanistan	17.4	13.1	80	0.8
Bhutan	0.9	—	60	—
Sri Lanka	12.9	—	180	—
India	569.5	2804.0	100	4.9
Iran	29.2	834	310	28.6
Nepal	11.5	2.7	80	0.2
Pakistan	141.6	294.4	100	2.1
Southeast Asia				
Burma	28.4	52.5	70	1.9
Cambodia	7.3	—	120	—
Indonesia	124.9	428.0	100	3.4
Laos	3.1	7.0	100	2.2
Malaysia	11.1	136.1	330	12.3
Philippines	39.4	301.8	180	7.7
Singapore	2.2	90.0	700	40.9
Thailand	37.4	301.0	150	80.5
Vietnam (Dem. Rep.)	21.6	96.9	90	4.5
Vietnam (Rep.)	18.3	165.3	130	9.0

Population, Gross National Product, and Energy Consumption—Asia (*Continued*)

Country	Population[a] (millions) 1971	Energy Consumption[b] 10^{12} Btus 1971	Annual GNP per Capita[a] $U.S. 1971	Annual Energy Consumption per Capita[c] 10^6 Btus
East Asia				
China (Peoples Rep.)	772.9	12024.2	90	155.6
China (Taiwan)	14.3	—	270	—
Hong Kong	4.3	112.8	710	26.2
Japan	104.7	9349.7	1190	89.3
Korea (Dem. Peo. Rep.)	14.3	894.9	250	62.6
Korean (Rep.)	32.9	762.6	180	23.2
Mongolia	1.3	31.3	430	24.1
Ryukyu Islands	1.0	—	580	—
Oceania				
Australia	12.8	1893.0	2070	147.9
Fiji	0.5	6.1	330	12.3
New Zealang	2.9	219.5	2000	75.7

[a]Population and per capita gross national product data from Paul R. Ehrlich and Anne H. Ehrlich, "Population, resources, environment," W. H. Freeman and Company, San Francisco, 1972.

[b]Energy consumption data from World Energy Conference Survey of Energy Resources, The United States National Committee of the World Energy Conference, 345 East 47th Street, New York, N.Y. 10017 (1974).

[c]In addition to the per capita energy consumption listed in this column, an individual will expend about 2×10^6 Btus of human energy per year.

In Massachusetts, fields of cranberries are flooded and a beater used to separate the berries from the bushes. The floating cranberries are then gathered and transported to market. USDA Photo by George Robinson.

TRENDS IN U.S. ENERGY USE

Historical Trends in U.S. Energy Consumption. As the accompanying charts indicate, the consumption of energy in the United States increased in proportion to the increase in population until about 1900. During the 19th century, the dependence on wood as the primary fuel gave way to coal. The surge in energy consumption after 1900 can be ascribed to a number of factors: the discovery and use of petroleum and gas, both easily transportable and convenient to use; the advent of the internal combustion engine for vehicles; and the discovery of practical systems for the generation and utilization of electric energy—systems deriving their energy from water power or coal-fired steam boilers. Between 1900 and 1960 the average compound energy growth rate has been about 3%.

In the decade 1960-70 the overall growth rate was 4.2% while the growth rate in the electrical sector was 9%. Further details on energy consumption are given in Chapter 4 and on energy consumption projections in Chapter 5.

Future trends in energy consumption are difficult to predict—for example, during 1974 energy consumption did not increase, partly because of the quadrupling in the prices of imported petroleum and partly because of a downturn in the economy. The general increase in the cost of all forms of energy will undoubtedly stimulate efforts to produce and use energy more efficiently with a consequent shift in historical growth patterns.

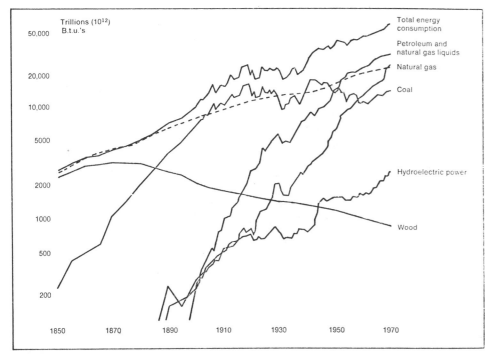

The history of U.S. energy sources and consumption since 1850 is replete with drama—which is often concealed by rounding curves to concentrate on the familiar, inexorable rise in energy demand. Look, for example, for the drop in energy consumption corresponding to economic depressions in the early 1920's and most noticeably in the early 1930's; the increasing consumption of coal in war-time, both in 1918 and 1941-46; and the gradual advance, since late in the 19th century, of energy consumption in relation to population. The logarithmic *vertical scale helps to suggest how similar to those of recent decades have been past patterns of energy growth and replacement. The dotted line shows population growth, plotted on a scale of the ordinate numbers times 10¹.*

Source: Hoyt C. Hottel and Jack B. Howard, "An Agenda for Energy," edited at the Massachusetts Institute of Technology, *Technology Review*, January, 1972.

Winchester, Virginia, November, 1919. (*Courtesy U.S. Department of Agriculture*)

MECHANIZATION OF U.S. AGRICULTURE

Energy and Agricultural Productivity. The importance of the availability of energy to the reduction of human labor and the improvement of crop productivity is illustrated by the impact of the introduction of the tractor and the widespread use of fertilizers, herbicides and insecticides. Petroleum forms the basis for these technologies—in the form of gasoline to operate internal combustion engines and as the raw material for the production of farm chemicals. Between 1935 and 1960 farm output doubled while man-hours of labor were cut in half. As described on page 13, this increase in output was achieved, however, at a decreasing efficiency in total energy use.

The change in agricultural productivity was described as follows in the 1960 U.S. Yearbook of Agriculture:

Horse and mule numbers at that time (1918) were the highest in our history—more than 25 million—but the rate of technological progress had slowed down. The availability of good new land had dwindled to insignificance. One-fourth of the harvested crop acreage was being used to produce feed for power animals.

If methods had not been changed, many more horses, more men to work them, and much more land to grow feed for them would be required for today's net agricultural output. The American economy of the 1960's could not be supported by an animal-powered agriculture on our essentially fixed—in fact, slowly shrinking—land base. National progress on all fronts would have been retarded seriously had not agriculture received new forms of power and sources of energy not restricted by biological limitations.

With the adoption of mechanical forms of power in engines, tractors, and electric motors and development of more and more types of adapted equipment to use that power, American agriculture entered a new era of sharply rising productivity.

Farm efficiency has lifted our living standard . . .

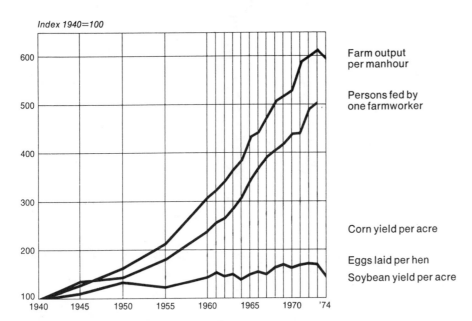

The U.S. is endowed with more good farmland than any other nation, and over the decades each cultivated acre has provided more food and fiber with less and less human toil. Agriculture's prize exhibit has been the rise in the corn yield, but, because of bad weather, it fell in 1973 and, along with total farm output per manhour, in 1974. Many economists fear that the next decade or so will bring slower gains on the farm.

Tom Cardamone for *Fortune Magazine*, April, 1975.

Mechanical lettuce harvester. (*Courtesy U.S. Department of Agriculture*)

ENERGY INPUT AND AGRICULTURAL PRODUCTIVITY

Corn Production and Energy Input. Although corn is only one of many grain crops, it is the most important grain crop in the United States and it ranks third in world production. In the United States, the yield of corn per acre (one acre–0.405 hectare) has increased from 26 bushels per acre in 1909 to 87 bushels per acre in 1971. This sharp rise in productivity has been the result of the increased use of fertilizers, insecticides, and herbicides, through irrigation, and most recently, through the use of high-yield hybrid grains. The increased use of machinery has been responsible for a decrease from 23 to 9 in the number of man-hours of labor per crop acre during the period 1945–70. This increase in productivity has been accompanied, however, by a decreasing efficiency in terms of energy—for every calorie of energy input the calories produced in the form of grain has dropped from 3.70 to 2.82.

As the costs of energy for fertilizers, fuel, and machinery increase, the optimum economic cropping patterns tend to shift back toward traditional, less energy-intensive methods—the rotation of crops with nitrogen-rich legumes, the use of livestock manure, and the increased use of man power. The more traditional farming methods also provide weed and pest control without the extensive use of herbicides and pesticides but at the expense of lower productivity per acre.

American farmers grow half the corn produced in the world. This field near Kearney, Nebraska, yields about 150 bushels per acre. The average U.S. yield in 1975 was 86.2 bushels per acre. USDA Photo by Jim Pickerell.

Average Energy Inputs in Corn Production During Different Years
(All figures per acre)

Inputs	1945	1950	1954	1959	1964	1970
Labor*	23	18	17	14	11	9
Machinery (kcal × 10³)†	180	250	300	350	420	420
Gasoline (gallons)‡	15	17	19	20	21	22
Nitrogen (pounds)§	7	15	27	41	58	112
Phosphorus (pounds)§	7	10	12	16	18	31
Potassium (pounds)§	5	10	18	30	29	60
Seeds for planting (bushels)‖	0.17	0.20	0.25	0.30	0.33	0.33
Irrigation (kcal × 10³)¶	19	23	27	31	34	34
Insecticides (pounds)#	0	0.10	0.30	0.70	1.00	1.00
Herbicides (pounds)**	0	0.05	0.10	0.25	0.38	1.00
Drying (kcal × 10³)††	10	30	60	100	120	120
Electricity (kcal × 10³)‡‡	32	54	100	140	203	310
Transportation (kcal × 10³)§§	20	30	45	60	70	70
Corn yields (bushel) ‖‖	34	38	41	54	68	81

* Mean hours of labor per crop acre in United States (6, 25).   † An estimate of the energy inputs for the construction and repair of tractors, trucks, and other farm machinery was obtained from the data of Berry and Fels (63), who calculated that about 31,968,000 kcal of energy was necessary to construct an average automobile weighing about 3400 pounds. In our calculations we assumed that 244,555,000 kcal (an equivalent of 13 tons of machinery) were used for the production of all machinery (tractors, trucks, and miscellaneous) to farm 62 acres of corn. This machinery was assumed to function for 10 years. Repairs were assumed to be 6 percent of total machinery production or about 15,000,000 kcal. Hence, a conservative estimate for the production and repair of farm machinery per corn acre per year for 1970 was 420,000 kcal. A high for the number of tractors and other farm machinery on farms was reached in 1964 and continues (64, 65). The number of tractors and other types of machinery in 1945 were about half what they are now. ‡ DeGraff and Washbon (66) reported that corn production required about 15 gallons of fuel per acre for tractor use—intermediate between fruit and small grain production. Because corn appeared to be intermediate, the estimated mean fuel (gallons) burned in farm machinery per harvested acre was based on U.S. Department of Agriculture (22, 64) and U.S. Bureau of the Census (65) data. § Fertilizers (N, P, K) applied to corn are based on USDA (25, 26, 61, 62) estimates.   ‖ During 1970, relatively dense corn planting required about one-third of a bushel of corn (25,000 kernels or 34,000 kcal) per acre; the less dense plantings in 1945 were estimated to use about one-sixth of a bushel of seed. Because hybrid seed has to be produced with special care, the input for 1970 was estimated to be 68,000 kcal.   ¶ Only about 3.8 percent of the corn grain acres in the United States were irrigated in 1964 (67), and this is not expected to change much in the near future (68). Although a small percentage, irrigation is costly in terms of energy demand. On the basis of the data of Epp (69) and Thorfinnson et al. (70), an estimated 905,600 kcal is required to irrigate an acre of corn with an acre-foot of water for one season. Higher energy costs for irrigation water are given by *The Report on the World Food Problem* (2). Since only 3.8 percent of the corn acres are irrigated (1964–1970), it was estimated that only 34,000 kcal were used per acre for corn irrigation. The percentage of acres irrigated in 1945 was based on trends in irrigated acres in agriculture (55, 67).   # Estimates of insecticides applied per acre of corn are based on the fact that little or no insecticide was used on corn in 1945, and this reached a high in 1964 (28, 51).   ** Estimates of herbicides applied per acre of corn are based on the fact that little or no herbicides were used on corn in 1945 and that this use continues to increase (28, 51).   †† When it is dried for storage to reduce the moisture from about 26.5 percent to 13 percent, about 408,204 kcal are needed to dry bushels (71). About 30 percent of the corn was estimated to have been dried in 1970 as compared to an estimated 10 percent in 1945.   ‡‡ Agriculture consumed about 2.5 percent of all electricity produced in 1970 (24) and an estimated 424.2 trillion British thermal units of fossil fuel were used to produce this power (72); on croplands this divides to 310,000 kcal per acre for 1970 (6, 51). The fuel used to produce the electrical energy for earlier periods was estimated from data reported in *Statistical Abstracts* (73).   §§ Estimates of the number of calories burned to transport machinery and supplies to corn acres and to transport corn to the site of use is based on data from U.S. Department of Commerce (74), U.S. Bureau of the Census (65, 67, 72), Interstate Commerce Commission (75), and U.S. Department of Transportation (76). For 1964 and 1970 this was estimated to be about 70,000 kcal per acre, it was about 20,000 kcal per acre in 1945.   ‖‖ Corn yield is expressed as a mean of 3 years, 1 year previous and 1 year past (55, 59, 60).

Source: David Pimentel, L. E. Hurd, A. C. Bellotti, M. J. Forster, I. N. Oka, O. D. Sholes, R. J. Whitman, "Food Production and the Energy Crisis," *Science*, Vol. 182, 2 November 1973.
Note: References cited in the table appear in the original article.

Energy Inputs (kilocalories) in Corn Production

Input	1945	1950	1954	1959	1964	1970
Labor*	12,500	9,800	9,300	7,600	6,000	4,900
Machinery†	180,000	250,000	300,000	350,000	420,000	420,000
Gasoline‡	543,400	615,800	688,300	724,500	760,700	797,000
Nitrogen§	58,800	126,000	226,800	344,400	487,200	940,800
Phosphorus ‖	10,600	15,200	18,200	24,300	27,400	47,100
Potassium¶	5,200	10,500	18,900	36,500	30,400	63,000
Seeds for planting#	34,000	40,400	50,400	60,400	68,000	68,000
Irrigation†	19,000	23,000	27,000	31,000	34,000	34,000
Insecticides**	0	1,100	3,300	7,700	11,000	11,000
Herbicides††	0	600	1,100	2,800	4,200	11,000
Drying†	10,000	30,000	60,000	100,000	120,000	120,000
Electricity†	32,000	54,000	100,000	140,000	203,000	310,000
Transportation†	20,000	30,000	45,000	60,000	70,000	70,000
Total inputs	925,500	1,206,400	1,548,300	1,889,200	2,241,900	2,896,800
Corn yield (output)‡‡	3,427,200	3,830,400	4,132,800	5,443,200	6,854,400	8,164,800
Kcal return/input kcal	3.70	3.18	2.67	2.88	3.06	2.82

* It is assumed that a farm laborer consumes 21,770 kcal per week and works a 40-hour week. For 1970: (9 hours/40 hours) $\times$ 21,770 kcal = 4,900 kcal.
† See Table 1.   ‡ Gasoline, 1 gallon = 36,225 kcal (77).   § Nitrogen, 1 pound = 8,400 kcal, including production and processing (78).   ‖ Phosphorus, 1 pound = 1,520 kcal, including mining and processing (79).   ¶ Potassium, 1 pound = 1,050 kcal, including mining and processing (79).   # Corn seed, 1 pound = 1,800 kcal (33). This energy input was doubled because of the effort employed in producing hybrid seed corn.   ** Insecticides, 1 pound = 11,000 kcal including production and processing (similar to herbicide; see ††).   †† Herbicides, 1 pound = 11,000 kcal including production and processing (31).   ‡‡ Each pound of corn was assumed to contain 1,800 kcal (33) and a bushel of corn was considered to be 56 pounds.

Source: David Pimentel, L. E. Hurd, A. C. Bellotti, M. J. Forster, I. N. Oka, O. D. Sholes, R. J. Whitman, "Food Production and the Energy Crisis," *Science,* Vol. 182, 2 November 1973.

Grass drills spread seeding.

ENERGY USE IN THE U.S. FOOD SYSTEM—1

Food System Energy Requirements. Each method of raising, gathering, and distributing food has its own relative efficiency in energy costs. Primitive shifting agriculture—the cultivation of a new plot of ground each year with human labor—is the least energy-intensive food system. Machinery and fertilizers are not used, food is eaten as it is produced with minimum storage and distribution. Productivity per acre is, of course, low both in terms of yield per crop acre and in utilization of the land which is allowed to lie fallow nine years out of ten.

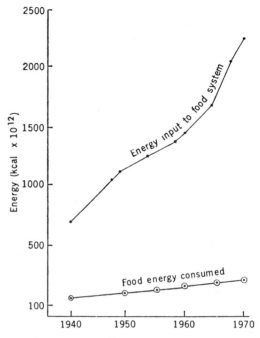

Energy use in the food system, 1940 through 1970, compared to the caloric content of food consumed.

Source: John S. Steinhart and Carol E. Steinhart, "Energy Use in the U.S. Food System," *Energy: Use Conservation and Supply*, American Association for the Advancement of Science, Washington, D.C., 1974.

Energy Use in the U.S. Food System, 1940–70

(In trillions of Btu)

Component	1940	1950	1960	1970
On farm				
Fuel (direct use)	277.8	626.9	746.0	920.6
Electricity	2.8	130.5	182.9	253.2
Fertilizer	49.2	95.2	162.7	373.0
Agricultural steel	6.3	10.7	6.7	7.9
Farm machinery	35.7	119.0	206.3	317.4
Tractors	50.8	122.2	46.8	76.6
Irrigation	71.4	99.2	132.1	138.9
Subtotal	493.6	1,203.9	1,484.0	2,087.6
Processing industry				
Food processing industry	583.3	761.9	888.8	1,222.1
Food processing machinery	2.8	19.8	19.8	23.8
Paper packaging	33.7	67.5	111.1	150.8
Glass containers	55.6	103.2	123.0	186.5
Steel cans and aluminum	150.8	246.0	341.2	484.1
Transport (fuel)	196.8	404.7	608.3	979.7
Trucks and trailers (manufacture)	111.1	196.4	175.4	293.6
Subtotal	1,133.7	1,799.5	2,267.7	3,340.7
Commercial and home				
Commercial refrigeration and cooking	480.1	595.2	738.8	1,043.6
Refrigeration machinery (home and commercial)	39.7	99.2	127.0	242.0
Home refrigeration and cooking	572.2	802.7	1,097.5	1,904.6
Subtotal	1,092.0	1,497.1	1,963.4	3,190.3
Grand total [1]	2,719.3	4,500.5	5,715.1	8,618.4

[1] Totals may not add because of rounding.

Source: U.S. Congress, Senate, Committee on Agriculture and Forestry, *The U.S. Food and Fiber Sector: Energy Use and Outlook*, Committee Print, 93d Cong., 2d sess., prepared by U.S. Department of Agriculture (Washington, D.C.: Government Printing Office, 1974); J. S. Steinhart and C. E. Steinhart, "Energy Use in the U.S. Food System," *Science*, 184 (4134) (April 19, 1974), pp. 307–316.

Source: The Sixth Annual Report of the Council on Environmental Quality, December, 1975.

In the accompanying illustration, John S. Steinhart and Carol E. Steinhart[1] show the energy "subsidies" associated with various food crops and methods of producing them. Cattle raised in feedlots require an energy subsidy 20 times that of cattle fed on the range. Corn requires more energy than does rice but at the same time yields are higher per acre (see page 00 for the relative efficiency with which plants convert solar energy).

[1] John S. Steinhart and Carol E. Steinhart, "Energy use in the U.S. food system," *Energy: Use Conservation and Supply*, American Association for the Advancement of Science, Washington, D.C. 20005, 1974.

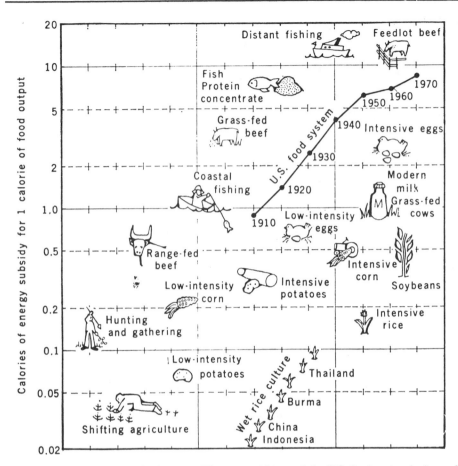

Energy subsidies for various food crops. The energy history of the U.S. food system is shown for comparison.

From *The Fires of Culture: Energy Yesterday and Tomorrow* by Carol Steinhart and John Steinhart. © 1974 by Wadsworth Publishing Company, Inc., Belmont, California 94002. Reprinted by permission of the publisher, Duxbury Press.

Wheat is harvested at Palouse, Washington, using combines, tractors, and trucks representing an investment of over $150,000. USDA Photo by Doug Wilson.

ENERGY USE IN THE U.S. FOOD SYSTEM—2

Food System Energy Requirements—1963. In a study of 1963 data, Eric Hirst[1] found that of the total energy requirement involved in the entire food chain, only 16% represented the energy used on the farm. The remainder of the energy requirement involved food processing, transportation, marketing, and home food preparation. In that year, the energy subsidy, the ratio of energy input to food energy consumed, was 6.41. The dollar cost for all food-related activities was $94 million, corresponding to 23% of the disposable personal income. The energy input into the food system represented 12% of the total national energy requirement and 22% of the electricity consumption.

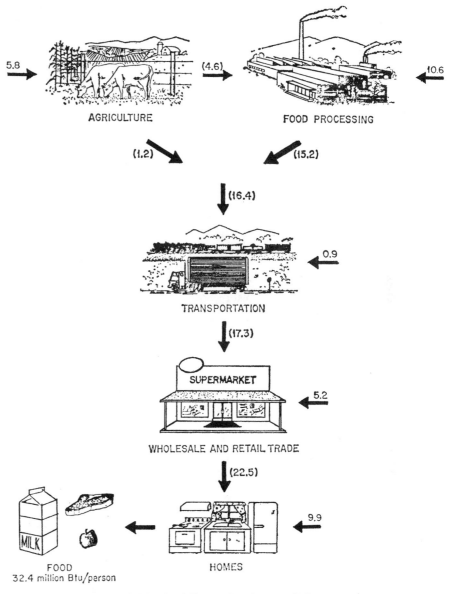

The Total Costs of Food in the United States, in Terms of Dollars, Energy, and Electricity in 1963

Sector	Cost (billion $)	Energy Use (trillion Btu)	Electricity Use (billion kw-hr)
Agriculture			
To PCE	4.5	231	4.1
To processing*	(15.8)	(862)	(15.6)
Food processing*	51.5	2868	51.3
Transportation	2.0	170	1.0
Trade	29.9	982	26.0
Households	6.1	1868	104.4
Total	94.0	6119	186.8

*Agricultural output delivered to processing (shown in parentheses) also appears in the figures for food processing.

Hirsh, E. and Shimotake, H., "Food-related energy requirements," *Science*, vol. 184, pp. 134–138, April 12, 1974. © 1974 by the American Association for the Advancement of Science.

Source: Eric Hirst, "Food-Related Energy Requirements," *Science*, Vol. 184, 12 April, 1974.

Hirst, E., Shimotake, H., "Food-Related Energy Requirements," *Science*, Vol. 184, pp. 134–138, April 12, 1974. © American Association for the Advancement of Science.

[1]Eric Hirst, "Food-related energy requirements," *Science*, vol. 184, 12 April 1974.

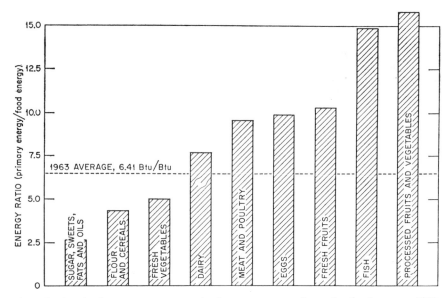

Fig. 3. Ratio of primary energy use to food energy content for major food groups, 1963.

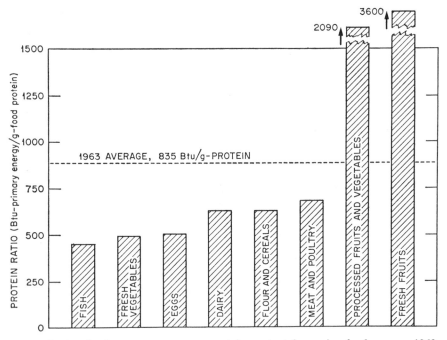

Fig. 4. Ratio of primary energy use to protein content for major food groups, 1963.

Hirsh, E., Shimotake, H., "Food-Related Energy Requirements," *Science*, Vol. 184, pp. 134–138, April 12, 1974. © American Association for the Advancement of Science.

Automatic waterers and a moving feed chain are used for the feeding of chickens at El Dorado, Arkansas. USDA Photo by Michelle Bogre.

ENERGY BALANCE OF THE EARTH

The Global Energy Balance. The dominant factor in the earth's energy balance is radiation from the sun. In comparison, the contributions made by tidal, geothermal, natural radioactivity, and gravitational energy are minor as is the contribution made by the combustion of fossil fuels. The radiant flux from the sun outside the earth's atmosphere is 0.139 watts/cm². Given a radius of the earth of 6371 kilometers, the energy intercepted across the diametral plane of the earth is 1.78×10^{17} watts (5.3×10^{21} Btus/year or 5300 Q/year). Current world energy consumption, derived primarily from fossil fuels, is about 0.25 Q/year, less than 1/20,000 of the solar energy input.

About 35% of the incoming solar energy is directly reflected back into space while about 65% is absorbed and then reradiated at a lower temperature, that is, at longer wave-

Earth Energy Flows	Q/Year
Solar input	5300
Photosynthesis	1.2
Evaporation of water	980
Useful hydropower	0.086
World geothermal heat flow	0.8
Useful geothermal heat flow	0.004
Total tidal energy	0.09
Useful tidal energy	0.0019
Present world energy use	0.25

lengths of radiant energy. Since the temperature of the earth has remained relatively constant over geologic time (temperature changes of only a few degrees were sufficient to cause the ice ages) essentially all the energy received from the sun is reradiated with only a very minor portion of the incoming energy, one part in five thousand, being stored in the form of plant material through photosynthetic conversion of the sun's energy. World photosynthetic production corresponds to 40×10^{12} watts (1.2 Q/year).

About 20% of the solar flux (980 Q/year) is involved in the evaporation of water from the land and oceans. This water, of course, is returned to the earth as rainfall but only about 0.01% (0.086 Q/year) represents potential water power.

Geothermal Energy. The crust of the earth is a poor conductor of heat and the heat flow from the earth's center through the surface is only about 40 calories per cm² per year. The geothermal heat flux over the entire surface of the earth is about 0.8 Q/year. The heat stored in the rock near the surface of the earth can be "mined" and in the known geothermal regions of the earth this represents about 40 Q if the heat can be recovered to a depth of 10 kilometers.

Tidal Energy. The tidal energy generated from the gravitational interaction of the earth and the moon represents about 0.09 Q/year. The usable tidal energy in promising tidal areas where flows are sufficiently large to make power generation practical, is only about 2% of the total tidal energy or about 0.0019 Q/year.

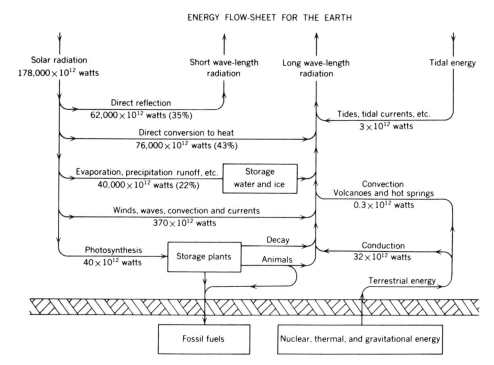

ENERGY FLOW-SHEET FOR THE EARTH

Source: M. King Hubbert, *Energy Resources*, National Academy of Sciences–National Research Council, Washington, 1962.

REFERENCE

M. R. Gustavson, "Dimensions of world energy," M71-71, the MITRE Corporation, Washington, D.C., November 1971.

WORLD PRODUCTION OF ORGANIC MATTER

Annual World Production of Organic Matter. Each year about 80–200 billion metric tons of organic matter are produced throughout the world by photosynthetic processes in plants. The energy for these processes is derived from solar radiation with a net utilization of about one-tenth of 1% of the total energy received from the sun. This organic matter provides food for man and is the basis for all the life-support systems of the biosphere.

In the photosynthetic process in plants, molecules of carbon dioxide and water, drawn from the atmosphere, are converted into complex carbohydrate molecules and oxygen. The rate at which plant growth takes place is dependent upon a number of factors—the amount of incident sunlight, temperature, availability of water and nutrients, as well as the presence of a suitable stable environment for the growth of living things. These factors vary from region to region in the world. The accompanying tables[1] and chart[2] provide two separate estimates of the annual production of organic matter in various environments. The lower of these two estimates indicates the productivity of all the terrestrial environments combined is 373 metric tons/km^2.

This organic matter not only provides the basis for the maintenance of life processes but also can be used as a source of heat energy. Wood was the major fuel in all societies until about a century ago when coal, oil, and gas began to be more widely used. The prospects for commercial photosynthetic production of organic matter as a source of fuel are discussed in Chapter 9 on Solar Energy. As a heat resource, the organic matter produced in plants has a heat content of approximately 8000 Btus per pound. Translating the various estimates of the annual production of organic matter into Q units (10^{18} Btus) gives the following units:

$$78 \times 10^9 \text{ metric tons}—1.37 \text{ Q}$$
$$164 \times 10^9 \text{ metric tons}—2.9 \text{ Q}$$

These estimates might be compared with the 1.2 Q estimated by Gustavson (page 19), and with the current annual world energy consumption, 0.25 Q.

[1] "Man's impact on the global environment," *Report of the Study of Critical Environmental Problems*, The MIT Press, 1970.
[2] George M. Woodwell, "The energy cycle of the biosphere," *Sci. Amer.*, September 1970.

Organic Carbon and Its Rates of Production[a]
(Living and dead, excluding incipient fossil deposits)

Reservoir	Area[b] 10^6 km^2	Organic Carbon Pool[b] 10^9 metric tons	Production/Year[b] 10^9 metric tons
Forest and woodland	48	1012	36
Grassland and tundra	38	314	9
Desert and semidesert	32	59	3
Wetlands	2	30	2
Glaciers and barren	15	0	0
Agricultural	15	165	6
Total terrestrial	150	1580	56
Oceanic	361	703	22
Burning fossil fuel (1970)			4
Atmospheric pool		683	

[a] Resistant humus and other material with decay rates of 0.001 per year or less have been omitted. Production is "net primary production," that is, production from photosynthesis minus plant respiration; it represents the yield to animals and decomposers.

[b] The numbers shown here are intermediate values from the several sources listed in the text. Although these sources present similar estimates, their combined accuracy is not regarded as high; procedures for obtaining global estimates for characteristics of vegetation are still primitive.

Subdivision of Organic Carbon into Materials with Short or Long Residence Times

Reservoir	Organic Carbon Pool		Mean Residence Time (years)	
	Short	Long	Short	Long
Forest and woodland	62	950	3.1	59
Grassland and tundra	11	303	2.6	63
Desert and semidesert	4	55	2.2	39
Wetlands	1	29	1	36
Agricultural	4	161	1	80
Total terrestrial	82	1498	2.3	62
Oceanic	2	701	0.1	701

Mean residence of carbon in the atmosphere: 8.8 years.

Note: The numbers shown here are not data but intelligent guesses of the way the estimates from the above table can be subdivided. Definitions of "short" and "long" residence times are given in the text.

Reprinted from *Man's Impact on the Global Environment: Assessments and Recommendations for Action* by William H. Mathews by permission of the MIT Press, Cambridge, Massachusetts, 1970.

PRODUCTIVITY OF ECOSYSTEMS

Relative Productivity of Ecosystems. Within the figures given on page 20 for the gross productivity of land lies a large variability in the productivity of various individual species. The accompanying figure reflects the differences in productivity due to climate—rainfall, sunlight, and growing season, the presence of nutrients in the soil, and the relative efficiency with which individual plants utilize sunlight. It will be noted that the shorter growing season for forests in temperate zones results in a lower annual productivity than for forests in tropical zones. The higher nutrient content in sewage ponds in contrast to freshwater ponds results in a higher rate of plant growth in sewage ponds. The nutrients brought up from the bottom of the ocean in areas of upwelling of the ocean currents provides the basis for a productivity some six times higher in such areas than in the open ocean where the nutrient content is low.

Among the agricultural ecosystems there is a considerable variation in the productivity of corn and rice culture reflecting climatic conditions, cropping methods, and the intensity of fertilizer use. The productivity of sugarcane is not as high relatively as the figure indicates since the productivity of sugarcane is measured by the weight of the entire stalk while that of the grains is for the grain yield only.

To conserve water and prevent soil erosion, farmers in dry areas employ strip farming. USDA Photo by Lowell Georgia.

Net Biomass Production Levels for Natural and Agricultural Ecosystems

	Net Production (grams per square meter per year)
Natural Ecosystems	
Temperate terrestrial zone	
Oak-Pine Forest (New York)	1195
Beech Forest (Denmark)	1350
Spruce Forest (Germany)	1450
Scotch Pine (England)	1600
Grassland (New Zealand)	3200
Tropical Terrestial Zone	
Forest (West Indies)	6000
Oil-Palm Plantation (Congo)	3700
Forest (Ivory Coast)	1340
Freshwater	
Freshwater Pond (Denmark)	950–1500
Sewage Ponds (California)	5600
Cattail Swamp (Minnesota)	2500
Marine	
Algae (Denmark)	260–340
Seaweed (Nova Scotia)	2000–2600
Algae on Coral Reef (Marshall Islands)	4900
Open Ocean (average)	100
Coastal Zone (average)	200
Upwelling Areas (average)	600
Agricultural Ecosystems	
Temperate Zone	
Corn (Minnesota)	1390
Corn (Israel)	3600
Corn (U.S. average)	2500–4000
Rice (Japan average)	1000–1200
Tropical Zone	
Sugarcane (Hawaii)	7200–7800
Sugarcane (Java)	9400
Rice (Ceylon average)	340–550
Rice (West Pakistan average)	560–700

Source: Data from George M. Woodwell, "The Energy Cycle of the Biosphere," *Scientific American*, September, 1970.

NATURAL AND AGRICULTURAL ECOSYSTEMS

Man and Nature. Man, along with all other living things, is a part of the global ecology—an ecology deriving its energy from the sun and with each living species dependent on others for its survival. Our current understanding of the relationships of various species in ecosystems is based on the work of Raymond L. Lindeman who published a paper in *Ecology* in 1942 which outlined the quantitative relationships among the various users of energy in nature. He postulated that some fraction of the energy entering any animal or plant population is available to another species without serious disruption of either population. Thus, one species can feed on another and he suggested a rough "10% rule" which suggested that grazing animals could consume about 10% of the energy fixed in plants and that carnivores could consume about 10% of the energy consumed by the herbivores while still maintaining a stable ecological relationship. Equally important to the living balance in nature is the chain of decay of dead organic matter—plants, fecal matter, animals, and other debris—that provide the nutrients and growth environment for living matter.

Of the approximately 160 billion metric tons of dry organic matter produced annually through photosynthesis, about two-thirds on land and one-third in the oceans, virtually all is consumed in the respiration of organisms other than green plants, releasing water and carbon dioxide as well as heat. A small fraction of the dry organic matter is stored as humus in the ground, in tissues of living organisms, or as organic sediments that, in the past, have created deposits of petroleum, gas, and coal.

The most stable ecosystems are those in regions where rainfall is regular and the climate benign, such that a great diversity of plant and animal life can be sustained, each species providing an inherent regulation of the population of other species. Regions of variable annual rainfall—the lands bordering between grassland and desert, exhibit great fluctuations in the populations of individual species from year to year with relatively few species evident. Man has created, on the 11% of the land of the earth under cultivation, yet another ecosystem. These agricultural ecosystems are not basically stable and require constant maintenance through the use of fertilizers, pesticides, and cultivation—all of which are dependent upon inputs of energy. Yet without agriculture, the numbers of humans that could be maintained in a natural stable ecosystem of the world might number only in the tens of millions.

Pesticide Usage and Agricultural Yields in Selected World Areas

| Area or Nation | Pesticide Use | | Yield | |
	Grams per Hectare	Rank	Kilograms per Hectare	Rank
Japan	10,790	1	5,480	1
Europe	1,870	2	3,430	2
United States	1,490	3	2,600	3
Latin America	220	4	1,970	4
Oceania	198	5	1,570	5
India	149	6	820	7
Africa	127	7	1,210	6

Source: FAO, *Production Yearbook*, 1963.

Reprinted from *Man's Impact on the Global Environment: Assessments and Recommendations for Action* by William H. Mathews by permission of The MIT Press, Cambridge, Massachusetts, 1970.

Volcanic soil near Pearl Harbor in the Hawaiian Islands is largely devoted to the production of sugar cane. USDA Photo by Charles O'Rear.

WORLD DIETARY STANDARDS

Diet and Human Energy. As the accompanying figures indicate, there is a marked difference in calorie and protein content in the diets of peoples of various countries of the world. A diet of high caloric intake leads in two directions—either to a population with considerable muscular energy to expend or to a population tending toward overweight. A diet high in protein provides the amino acids essential for normal and vigorous growth. The difference in stature of humans from different countries is a reflection largely of the protein content of their diet. While it is possible to obtain the necessary protein from vegetables for an adequate diet, it is generally far simpler to obtain the necessary protein from fish, meat, and fowl. Although man can live as a herbivore, he has found a diet combining that of the herbivore with that of the carnivore is not only more pleasing but more readily provides him the essential foods he needs. This preference for a combination vegetable-meat diet is of particular importance in the determination of the optimum equilibrium population of the world since the world can support more humans on a plant diet than on a diet comprised of meat from animals who, in turn, have derived their support from plants.

Estimated Calories of Food and Grams of Protein in the Average Daily Diets for Various Nations

Country	Prewar Calories	1951/53 Calories	1957/59 Calories	1963/64 Calories	Protein, Grams
U.S.A.	3280	3130	3110	3110	92
France	2880	2840	2940	3070	100
Sweden	3120	3020	2930	2980	84
Spain		2490	2590	2850	78
Brazil	2190	2380	2590	2850	69
Japan	2020	1930	2170	2280	73
India	1950	1750	1910	1990	50
Philippines		1690	1760	1990	46

Source: United Nations, 1966.
From "Resources and Man: A Study and Recommendations," by the Committee on Resources and Man of the Division of Earth Sciences, National Academy of Sciences—National Research Council with the cooperation of the Division of Biology and Agriculture, W. H. Freeman and Company. Copyright © 1969.

Comparison of U.S. and Indian Food Consumption

Food Type	Consumption (kilocalories per capita/year) United States	India
Rice	11,060	317,400
Wheat	186,280	77,650
Corn and other cereals	34,790	134,820
Pulses and nuts	24,240	108,490
Potatoes and other starches	35,280	10,680
Sugar and syrup	206,730	56,520
Vegetables	25,260	11,170
Fruits	32,210	7,170
Fish	6,020	5,730
Meat	285,690	3,000
Eggs	26,320	350
Milk products	122,190	26,980
Fats and oils	216,730	27,190
Totals	1,212,900	787,150

Cereals, pulses and nuts provide about 81 percent of the energy in the average Indian diet. These same foods provide about 21 percent of the average Indian diet. Meat, eggs and milk products supply about 36 percent of the food energy in the U.S. diet, and sugar, syrup, fats and oils another 35 percent (vs 4 and 11 percent in the Indian diet). On the average, Indians have a daily diet of about 2150 kilocalories per person while Americans consume about 3300 kilocalories per person per day, although about 600 of these kilocalories are discarded.

Source: Data from Roger Revelle, "Food and Population," *Scientific American*, September, 1974.

WORLD PROTEIN AND CALORIC INTAKE

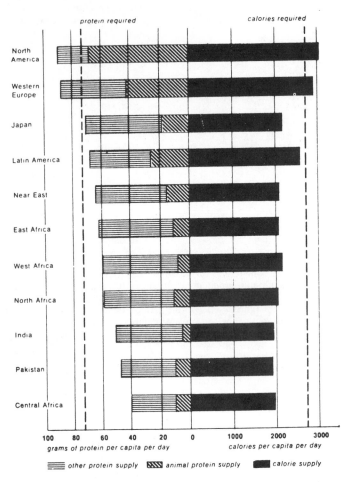

other protein supply   animal protein supply   calorie supply

Dash-lines indicate estimated North American protein and caloric requirements, based on diets sufficient to enable people to attain full body weight.

Source: UN Food and Agriculture Organization Provisional Indicative World Plan for Agricultural Development (Rome: UN Food and Agriculture Organization, 1970).

Source: Economic Growth in the Future, Report of the EEI Committee on Economic Growth, Pricing and Energy Use, Edison Electric Institute, 90 Park Avenue, New York, N.Y. 10016, June 1975.

Harvesting of catfish at the C. H. Block farm near Tunica, Mississippi. U.S. commercial production of catfish is about 50 million pounds each year. USDA Photo by David Warren.

POTENTIAL WORLD FOOD PRODUCTION

The Limits to Human Population. The level at which the earth can maintain the human race is dependent upon a number of factors: the extent to which potentially arable land can be brought into use for the production of food crops, the productivity of that land—in itself dependent upon cropping methods, irrigation, grain types, use of fertilizers, herbicides and pesticides, the nature of the diet deemed acceptable, and the ability to store and transport food. Forecasts of imminent starvation in some regions of the world would tend to indicate the world is already overpopulated at four billion people. Yet Roger Revelle has estimated that a population of 40 to 50 billion could be supported assuming the proper combination of technology, investment, and cooperation, although he does not suggest that a world population of that size is desirable.

While the momentum of world population growth will undoubtedly continue for the next century, a number of factors suggest that ultimately world population levels will stabilize below present levels. It is not likely that much additional land will be brought into production; estimates that the 11% of the land surface now under cultivation can be extended to 24% would require extensive capital investment for reclamation of desert and tropical land in countries with negligible capital resources. Fertilizers, pesticides, and herbicides—currently produced from petroleum—will become increasingly expensive as petroleum resources are depleted. It seems logical to assume that all nations will begin to restrict their numbers to more nearly match their resources and aspirations. Most developed countries are approaching a zero population growth today and other countries are beginning efforts to moderate population growth rates. Finally, there is a growing recognition that man himself is but a part of a somewhat fragile world ecology and if that ecology is to remain in reasonable balance, man cannot become an overly dominant species.

Present and Potential Uses of the Land Surface of the Planet
(Percent of total area)

Use	Present	Potential
Croplands	11	24
Rangelands	20	28
Managed forests	10	15
Reserves (80% forest)	26	0
Not usable	33	33
Total Land	100	100

Source: President's Science Advisory Committee (PSAC), 1967.

Past and Projected World Land-Use Changes
(Percent of total land area)

Use	1950–1968	1968–2000
Cropland	+1.5	+0.4
Rangeland	+5.2	−1.3
Total forest	−0.2	−1.3

Sources: Food and Agriculture Organization (FAO), 1951, 1958, 1969.

Reprinted from *Man's Impact on the Global Environment: Assessments and Recommendations for Action* by William H. Mathews by permission of the MIT Press, Cambridge, Massachusetts, 1970.

Global Land Use
(10⁶ hectares)

	Total	Cropland	Pasture Range	Forest	Other
1950	13,509	1,230	2,187	4,024	6,068
1957	13,670	1,384	2,407	3,839	6,040
1968	13,395	1,447	2,892	3,994	5,062

Sources: Food and Agriculture Organization (FAO) 1951, 1958, 1969. *Yearbook on Food and Agricultural Statistics*, 1950, 1957, 1967.

Percentage of Potentially Arable Land Now Cultivated, and Acres Cultivated Per Person, on Different Continents

Continent	Percent Cultivated	Acres Cultivated per Person
Asia	83	0.7
Europe	88	0.9
South America	11[a]	1.0
Africa	22[b]	1.3
North America	51	2.3
U.S.S.R. (Europe-Asia)	64	2.4
Australasia	2[c]	2.9

[a]Tropical limitation.
[b]Desert and tropical limitation.
[c]Desert limitation.
From "Resources and Man: A Study and Recommendations," by the Committeee on Resources and Man of the Division of Earth Sciences, National Academy of Sciences–National Research Council with the cooperation of the Division of Biology and Agriculture, W. H. Freeman and Company. Copyright © 1969.

REFERENCES

Charles E. Kellogg and Arnold C. Orvedal, "Potentially arable soils of the world and critical measures for their use," *Advan. Agron.*, vol. 21, pp. 109–170, 1969.

Gary H. Heichel, "Energy needs and food yields," *Technol. Rev.*, July–August, 1974.

Nevin S. Scrimshaw, "The world-wide confrontation of population and food supply," *Technol. Rev.*, December 1974.

Tomatoes developed to withstand rough mechanical handling are harvested in this field in Yolo County, California. A second crop of sugar beets or barley is planted after the tomato plants are plowed under. USDA Photo by Jack Clark.

THE FOSSIL FUEL AGE

The Age of Fossil Fuels. Over a period of hundreds of millions of years in geological history there has been a slow accumulation of organic matter in some regions of the earth. This organic matter, derived from growing plants and marine organisms, was subsequently buried under silt and sand and converted by heat and pressure into fossil fuels—coal, oil, and gas. The energy in this organic matter, which came from the photosynthetic energy of the sun, represents a limited inheritance of man. Since about 1850, man has been using these resources at an ever increasing rate. M. King Hubbert[1] has estimated the cycles for

[1]M. King Hubbert, Statement to the subcommittee on the Environment, Committee on Interior and Insular Affairs, House of Representatives, June 4, 1974.

world oil and coal production should consumption continue at its present increasing rate. His estimates indicate a peaking of world oil production in 1990–2000 with essential depletion of oil resources by the year 2050. Even if the consumption of oil remains at its current level of 16 billion barrels per year, oil resources would be depleted during the next century. If coal production could be limited to its current level of three billion tons per year, coal reserves would last about 1000 years. In any event, the world's most inexpensive and convenient fuels, oil and gas, will be gone during the lifetime of children born today. In the panorama of man's history, all of his fossil fuels will have been used in a fraction of recorded history—the fossil fuel age.

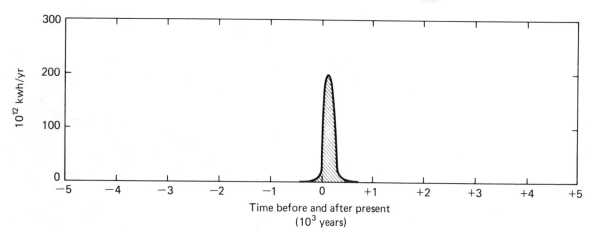

Epoch of fossil fuel exploitation in the longer span of human history.

Source: Hubbert, M. King, 1972, Man's conquest of energy: its ecological and human consequences, on the environmental and ecological forum 1970–1971: U.S. Atomic Energy Commission; available as T 1D-25857 from Natl. Tech. Inf. Service, Springfield, Virginia 22151.

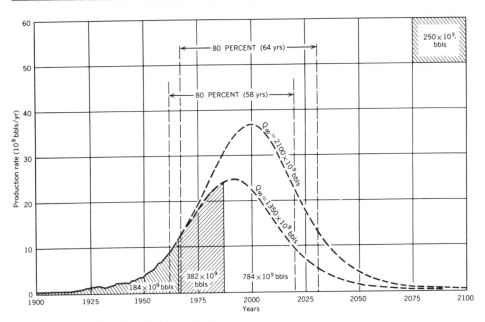

Complete cycles of world crude oil production.

Source: M. King Hubbert, "Energy Resource," *Resources and Man*, Committee on Resources and Man, National Academy of Sciences–National Research Council, W. H. Freeman and Company, San Francisco, 1969. Reproduced with permission of the National Academy of Sciences.

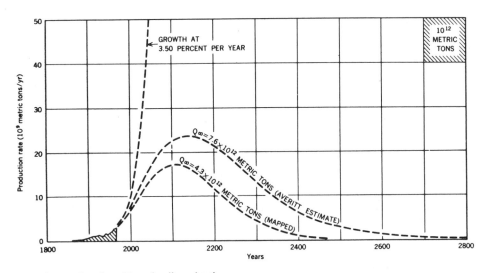

Complete cycles of world crude oil production.

Source: M. King Hubbert, "Energy Resource," *Resources and Man*, Committee on Resources and Man, National Academy of Sciences–National Research Council, W. H. Freeman and Company, San Francisco, 1969. Reproduced with permission of the National Academy of Sciences.

Windmills such as this one near Guymon, Oklahoma, have long been used to pump water for cattle. The depletion of fossil fuels may signal a return to wind power. USDA Photo by Bill Marr.

ENERGY RESOURCES OF MAN

History and Future of Energy Use. Earl Cook has drawn the accompanying chart illustrating past sources of energy and future potential sources. It will be noted that various sources have made, or will make, substantial contributions to man's energy needs for a time and then assume a status of being only of local importance or insignificant. These transitions represent several phenomena. Renewable resources such as wild food, wood, cultivated food, animal power, water, and windpower have become inadequate to meet the energy needs of a burgeoning world population. Fossil fuel resources—coal, oil, and gas—are the significant resources presently but, as indicated on page 27, will be depleted in the next several centuries. Nuclear fission power resources, uranium and thorium, will assume significance for perhaps the next thousand years at which time they too will be depleted. Two energy sources will then remain—nuclear fusion and solar energy—of sufficient magnitude to support the present world population at its current economic level. Nuclear fusion has not yet been established as a practical energy technology and solar energy, which effectively provided all of man's energy resources one hundred years ago, may one day again be the dominant energy resource. The transition of the societies of man to an equilibrium population with energy derived from renewable energy resources rather than fossil resources will provide an interesting chapter in history.

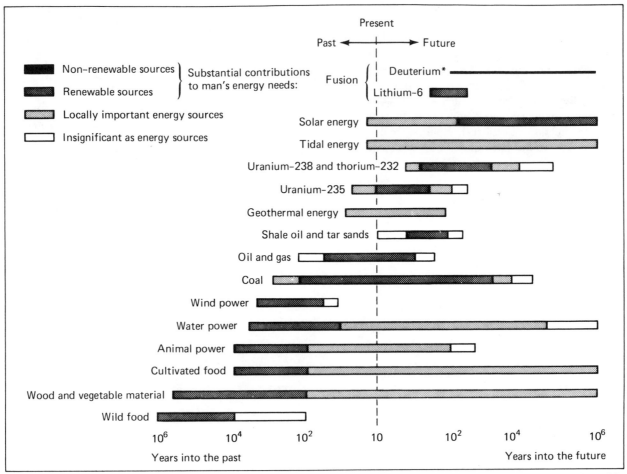

It is only in the last century that man has become dependent on fossil fuels rather than on tapping non-exhaustible natural forces. The author's projected future includes three ages: a "mining phase" for the next few decades, then an age in which reactors provide most of man's energy, and finally — if man is to survive — a renewed reliance on natural forces rather than exhaustible fuels — a "solar phase." *Deuterium will be a renewable source only if technology can make it so.

Source: Earl Cook, "Energy for Millenium Three," *Technology Review*, edited at the Massachusetts Institute of Technology, December, 1972.

2

Fossil and Mineral
Energy Resources

World's first commercial well. This photo, taken near Titusville, Pa., shows in top hat and frock coat, Edwin L. Drake, the man who conceived the idea of drilling for oil. On August 27, 1859, he proved his theory with the primitive rig pictured in the background. With Drake, in this photograph taken in 1861, is Peter Wilson, a Titusville druggist who encouraged him in the venture.

CLASSIFICATION OF MINERAL RESOURCES

Resource Classification. The accompanying table and figure, representing two recent classifications of minerals, demonstrate the differences in terminology with which mineral resources are identified. Additional terms will be found in this handbook. The differences are of both a historical and an economic nature. Resources that have been in long use in different countries, such as coal, tend to have the greatest differences in resource terminology while resources such as uranium and thorium, which have come into use only recently, are generally identified with common terminology.

In addition, the use of a term such as "proved recoverable," for example, will have a different context for a geologist than for a businessman with the former estimating the technically recoverable resources while the latter will estimate the resources recoverable under current market conditions. Finally, there is considerable confusion resulting from the use of different measuring units in the metric and English systems, as well as differences in the meaning of numbers such as billion and trillion.

The common denominator of all fuel resources is, of course, the heat content of the

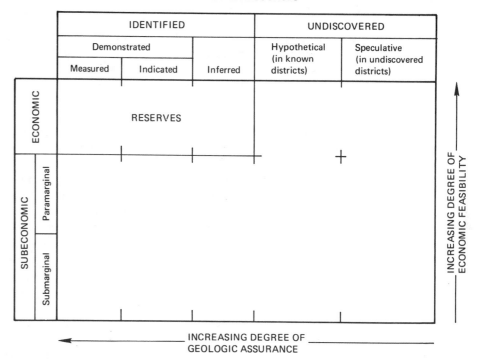

SOURCE: After USBM (1973).

Classification of Mineral Resources

Source: Mineral Resources and the Environment, National Academy of Sciences, Washington, D.C., 1974.

Summary of Reserve and Resource Terminology Used in Present Survey

Type Resource	Reserves		Other Resources[1]
	Total	Recoverable	
Solid Fuels[2]	Known Reserves-in-place	Known Recoverable Reserves	Additional Resources
Oil and Natural Gas	Original Reserves-in-place[3]	Proved Recoverable Reserves	Additional Resources
Natural Gas Liquids	—	Proved Recoverable Reserves	Additional Resources
Oil Shale and Bituminous Sands	—	Potential Total Known Recoverable Resources	
Uranium and Thorium	—	Known Recoverable Reserves[4]	Additional Resources

Note:

Terminology for Hydraulic Resources includes installed and installable capacity (power in MW) and probable annual generation (energy in GWh/year). Similar terminology applies, in general, to other renewable resources.

1. Includes indicated (probable) and inferred (possible) reserves as normally defined.
2. Total resources are also given for solid fuels.
3. Includes past cumulative production.
4. Alternative terminology (OECD) is *reasonably assured resources* (recoverable at costs up to approximately $26 per kilogram of U or Th). Reasonably assured resources recoverable at costs above $26 per kilogram are regarded as part of additional resources.

Source: Survey of Energy Resources 1974, The United States National Committee of the World Energy Conference, New York, 1974.

fuel. In the metric system, heat content is expressed in joules, while in the English system the British Thermal Unit (Btu) is most commonly used. The unit used in this handbook for energy resources is labeled a Q or 10^{18} Btu (equivalent to 1.05×10^{21} joules). This is a convenient resource unit since world energy consumption is now about 0.25 Q and is expected to be near 1.0 Q by the year 2000. Energy consumption figures given in Chapters 4 and 5 are generally given in quadrillions of Btus or Quads (10^{15} Btu). The Quad unit is, therefore, a thousand times smaller than the Q unit, but is more convenient for energy consumption data since its use provides whole rather than fractional numbers.

ESTIMATES OF WORLD FOSSIL AND MINERAL ENERGY RESOURCE BASE

The Fossil and Mineral Resource Base. The difference in estimates of the various fossil and mineral energy resources displayed in the tables is a measure of the difficulty in the assessment of these resources. Proved-recoverable resources of fossil and uranium resources are about 30 Q (30×10^{18} Btu) assuming nonbreeder reactor nuclear plants. If uranium were used in breeder reactors, the world energy resource from proved-recoverable reserves would be more than doubled. Development of nuclear reactors using thorium would add an energy resource about equivalent to that provided by uranium breeder reactors. Nuclear fusion, if successfully developed, would provide an energy resource of 200–10,000 Q.

The resource base for these various fossil and mineral fuels, that is, resources not yet discovered but assumed to be present on the basis of past exploration, is estimated at five to ten times larger than the proved-recoverable reserves. At current levels of world consumption of energy (0.25 Q/year), the proved-recoverable reserves are sufficient for about 120 years and the entire resource base will be consumed in 500–1000 years.

Estimates of Energy Content of World Fossil and Mineral Resources Q (10^{18} Btu)

	Gustavson[a]	Linden[b] Proved-Recoverable	Linden[b] Resource Base	EPP-EEC[c] Reserves	EPP-EEC[c] Recoverable Resources	EPP-EEC[c] Remaining Resource Base	WEC[d] Proved-Recoverable	WEC[d] Resource Base
Coal	190.5	23.8	168.2	—	—	340	16.3	295
Petroleum	11.4	3.9	16.7	3.6	14.4	60	3.9	—
Natural gas	10.5	1.8	10.1	1.9	15.8	32	1.9	5.3–5.4
Tar-sand oil	1.7	—	—	1.0	2.2	—	—	—
Shale oil	1.1	1.6	14.0	1.1	—	12,000	4–20	2,300
Peat	—	—	—	—	—	—	—	1,680
Uranium								
nonbreeding	4.1[e]	0.8	1.6	0.5	1.0	650,000	0.7	—
breeding	580.	61.0	120.0	40.0	77.0	600 million	46.0	—
Thorium (breeding)	—	—	—	22.0	66.0	—	—	—
Fusion	9,500.0[f]	—	—	—	—	—	—	—

[a]M. R. Gustavson, "Dimensions of world energy," M71-71, The MITRE Corporation, Washington, D.C., November 1971.
[b]Henry R. Linden and J. D. Parent, "Analysis of world energy supplies," Institute of Gas Technology, Chicago, Ill., 35 pp.
[c]*Exploring Energy Choices*, Energy Policy Project of the Ford Foundation, Washington, D.C., 1974.
[d]*Survey of Energy Resources 1974*, The United States National Committee of the World Energy Conference, 345 East 47th Street, New York, N.Y. 10017, 1974.
[e]Uranium figures based on an available supply of 10 million short tons of U_3O_8.
[f]Fusion figures based on deuterium present as 1 ppm in oceans.

Estimates of World Fossil and Mineral Energy Resource Base (In Traditional Units)

	Coal Billion (10^9) Short Tons Proved-Recoverable	Coal Resource Base	Petroleum Billion (10^9) bbls Proved-Recoverable	Petroleum Resource Base	Gas Trillion (10^{12}) cu ft Proved-Recoverable	Gas Resource Base	Tar-Sand Oil Billion (10^9) bbls Proved-Recoverable	Tar-Sand Oil Resource Base	Shale Oil Billion (10^9) bbls Proved-Recoverable	Shale Oil Resource Base	Uranium Thousands Short Tons Proved-Recoverable	Uranium Resource Base
Linden[a]	1,158	8,410	633	2,685	1,735	9,740	—	—	270	2,415	2,020	3,995
WEC[b]	650	11,800	674	—	1,518	—	—	—	1,685	—	—	—
Hubbert[c]	—	6,900	2,000	—	10,000	—	300	—	190	—	—	—
NAS[d]	—	10,000	600	1,130	2,000	4,900	750	—	12,000	330,000	—	—

[a]Henry R. Linden and J. D. Parent, *Analysis of World Energy Supplies*, Institute of Gas Technology, Chicago, Ill., 35 pp.
[b]*Survey of Energy Resources 1974*, The United States National Committee of the World Energy Conference, 345 East 47th Street, New York, N.Y. 10017, 1974.
[c]M. King Hubbert, "The energy resources of the earth," *Sci. Amer.*, September 1971.
[d]*Mineral Resources and the Environment*, National Academy of Sciences, Washington, D.C. 1975.

Summary of World Fossil Fuel Resources

Fuel	Total Resources (millions of (megatonnes)	World Use, 1972 (megatonnes)	Resources to Demand Ratio
Solid Fuels Resources*	11.0	2984	3686
Recoverable Reserves*	0.5		172
Petroleum			
Recoverable Reserves*	0.092	2408	38
Natural Gas			
Recoverable Reserves*	0.037	940	40
Oil Shale			
Recoverable Resources*	0.230		
Total Resources (Hubbert)			
≥ 10% Kerogen	2.3		
≥ 4% Kerogen	44.3		
≥ 2% Kerogen	240.0		

*Data from current survey.

Source: Survey of Energy Resources 1974, The United States National Committee of the World Energy Conference, New York, 1974.

World Reserves and Resources of Mineral Fuels

Resource	Units*	Reserves	Total Recoverable Resources**	Sources
Coal	10^9 tons	665	6,000–8,000	(1)(2)
Crude Oil	10^9 bbls.	700	1,300–1,880	(3)(4)(5)(6)
Natural Gas	10^{12} cu. ft.	2,300	3,000–6,000	(3)(4)(8)
Oil Shale	10^9 bbls.	500	1,125–16,000	(1)(7)
Bitumen Rocks	10^9 bbls.	350	1,000–2,500	(1)(8)
Uranium (U_3O_8)	10^3 tons	3,500	5,000–6,500	(1)(9)
Thorium (ThO_2)	10^3 tons	400	2,500–3,450	(1)(9)

Note: Comparisons between and among the resource categories on a common-unit basis such as quads are not advisable because of the widely varying methods of estimating and reporting data and the fundamental differences as to the technical feasibility of exploiting and using the several mineral fuels.

* Conventional U.S. units
** Including reserves
(1) World Energy Conference, **Survey of World Energy Resources**, 1974.
(2) Averitt, U.S.G.S. Bulletin 1412.
(3) "Changes Restructuring World Oil," **The Oil and Gas Journal**, December 30, 1974.
(4) "Productive Capacity Grows as World Demand Falters," **World Oil**, August 15, 1975.
(5) Lawrence and Farrar, 1975, p. 63.
(6) Linden, Institute of Gas Technology, 1975.
(7) Culbertson and Pitman, U.S.G.S. Professional Paper 820.
(8) McKelvey, "World Energy Reserves and Resources," **Public Utility Fortnightly**, September 25, 1975.
(9) Energy Research and Development Administration estimates.

Source: A National Plan for Energy Research, Development and Demonstration: Creating Energy Choices for the Future, ERDA 76-1, Energy Research and Development Administration, April 15, 1976.

Estimated Recoverable Reserves of Coal, Petroleum,* and Natural Gas

Region	Coal	Percent of Total Coal	Petroleum	(In Quads) Natural Gas	Total Oil & Gas	Percent of Total Oil & Gas	Total Coal, Oil & Gas	Percent of Total Coal, Oil & Gas
United States	4,900	34.3%	246	244	490	7.6%	5,390	25.9%
Other North America	142	1.0%	71	80	151	2.3%	293	1.4%
South America	50	0.3%	157	65	222	3.4%	272	1.3%
Subtotal, Western Hemisphere	5,092	35.6%	474	389	863	13.3%	5,955	28.6%
Middle East	35	0.2%	2,343	700	3,043	46.8%	3,078	14.8%
North Africa	neg.	0.0%	227	260	487	7.5%	487	2.4%
Middle Africa	69	0.5%	169	65	234	3.6%	303	1.5%
South Africa	292	2.0%	neg.	neg.	neg.	0.0%	292	1.4%
Western Europe	1,374	9.6%	150	210	360	5.5%	1,734	8.3%
Eastern Europe	1,073	7.5%	17	20	37	0.6%	1,110	5.3%
U.S.S.R.	3,325	23.3%	480	580	1,060	16.3%	4,385	21.1%
China	2,222	15.6%	145	25	170	2.6%	2,392	11.5%
South & East Asia	351	2.5%	110	75	185	2.9%	536	2.6%
Oceania	460	3.2%	13	45	58	0.9%	518	2.5%
Subtotal, Eastern Hemisphere	9,201	64.4%	3,654	1,980	5,634	86.7%	14,835	71.4%
World Total	14,293	100.0%	4,128	2,369	6,497	100.0%	20,790	100.0%
Percentage of Total	68.7%		19.9%	11.4%			100.0%	

* Includes natural gas liquids when data were available.
Tabular data derived from the following sources:
(1) All data for the United States are from the U.S. Geological Survey.
(2) World Energy Conference, 1974.
(3) "Changes Restructuring World Oil," **The Oil and Gas Journal**, December 30, 1974.
(4) "Productive Capacity Grows as World Demand Falters," **World Oil**, August 15, 1975.

Note: Data generally include measured and indicated reserves as of January 1, 1975, although in some regions data are not sufficiently well defined to assure the intended comparability. Coal data are reported in energy units. Where other reserve data were not reported in energy units, conversion from physical units was based on standard conversion factors per barrel of oil or cubic foot of gas.

Source: A National Plan for Energy Research, Development and Demonstration: Creating Energy Choices for the Future, ERDA 76-1, Energy Research and Development Administration, April 15, 1976.

Boulders of oil shale from Wyoming. (ERDA Photo)

WORLD DISTRIBUTION OF RECOVERABLE FOSSIL AND MINERAL ENERGY RESERVES

World Recoverable Energy Reserves. The distribution of the various types of energy reserves among the countries of the world has become an important economic and political factor. In particular, the energy economies of most countries of the world have become dependent upon petroleum which is an easily transported fuel and which was one of the least expensive fuels until 1973 when world prices quadrupled. Currently, the bulk of the proved-recoverable petroleum reserves are in countries bordering the Persian Gulf.

Although North America has some 49% of the world's reserves of fossil fuel, these reserves are largely in coal and oil shale and are not available immediately as substitutes for the growing requirements for petroleum and natural gas.

Measured World Nonrenewable Energy Reserves Q = 10^{18} Btu

Area	Fossil Fuels			Oil Shale and Tar Sands	Uranium (nonbreeders)	Total
	Solid Fuels	Crude Oil	Natural Gas			
Africa	0.361	0.526	0.201	0.081	0.198	1.369
Asia (less USSR)	2.608	2.211	0.432	0.870	0.003	6.126
Europe (less USSR)	2.581	0.057	0.153	0.117	0.046	2.955
USSR	3.325	0.333	0.577	0.139	unknown	>4.376
North America[a]	5.070	0.301	0.380	9.111	0.422	15.286
South America	0.049	0.311	0.060	0.023	0.011	0.457
Oceania	0.459	0.009	0.024	0.009	0.099	0.602
Total	14.457	3.741	1.831	10.351	>10.781	>31.173

[a]According to the Bureau of Mines, North American tar sands and oil shale reserves may be severely overstated. Development of most of these reserves is not economic at the present time.
Source: Energy Perspectives, U.S. Department of the Interior, February 1975.

TOTAL: 31,173

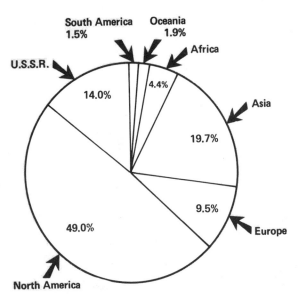

World Recoverable Energy Reserves
(Quadrillion Btu)

Source: Survey of Energy Resources, World Energy Conference, 1974.
Source: Energy Perspectives, U.S. Department of the Interior, February, 1975.
Note: A quadrillion BTU equals 10^{15} BTU or 0.001 (10^{18} BTU).

Specimen of antunite from the Dawn Mine, Spokane, Washington, containing nearly 50 percent uranium by weight. In contrast, the average uranium ore mined in the United States contains less than a half of one percent uranium. (Westcott, USAEC)

Energy Content of Recoverable Reserves of Fossil Fuels by World Regions

Region	Millions of Terajoules	Energy		Major Nations and Resources
		Q $(10^{18} \text{ Btu})^a$	% of World Total	
Northern America	15,681	14.9	48.8	United States, all fossil fuels, 78%
USSR	4,617	4.4	14.5	Coal, lignite and peat, 76%
East Asia	3,380	3.2	10.6	China, P.R. of, coal, 96%
Southwest Asia	1,978	1.9	6.2	Saudi Arabia and Kuwait, oil and gas 68%
Eastern Europe	1,166	1.1	3.7	Poland and Germany, D.R. of, coal 77%
Western Europe	1,160	1.1	3.6	Germany, F.R. of, coal, 83%
Middle South Asia	938	0.89	2.9	Iran, oil and gas, 63%; India, coal, 29%
Northern Africa	615	0.58	1.9	Algeria and Libya, oil and gas, 92%
Oceania	531	0.50	1.7	Australia, coal, 91%
Tropical South America	433	0.41	1.4	Venezuela, oil and gas, 69%
Northern Europe	309	0.28	1.0	U.K., all fossil fuels, 59%
Southern Africa	309	0.28	1.0	South Africa and Swaziland, coal, 98%
Southern Europe	292	0.28	0.9	Yugoslavia, coal 62%; Italy, Oil Shale, 15%
Middle Africa	148	0.14	0.5	Zaire, coal and oil shale, 57%
Southeast Asia	132	0.13	0.4	Indonesia, Malaysia and Brunei, oil and gas, 79%
Western Africa	102	0.10	0.3	Nigeria, oil and gas, 95%
Eastern Africa	61	0.06	0.2	Rhodesia, coal, 70%
Middle America	46	0.04	0.1	Mexico, coal, oil and gas-100%
Temperate South America	37	0.04	0.1	Argentina, all fossil fuels, 77%
Japan	33	0.03	0.1	Coal, 98%
Caribbean	21	0.02	0.1	Trinidad and Tobago, oil and gas, 99%
World total	31,922	30.3	100.0	

[a]Author's note.

Source: Survey of Energy Resources 1974, The United States National Committee of the World Energy Conference, 345 East 47th Street, New York, N.Y. 10017, 1974.

Drilling rig in the Rub' al-Khaki in Saudi Arabia. Courtesy of Arabian American Oil Co. and the American Petroleum Institute Photographic and Film Service.

WORLD COAL RESERVES

World Coal Resources and Reserves. Although estimates of the extent of coal resources are better documented than any other fossil fuel resource, it will be noted from the tables that there is still considerable variation in these estimates and no clear definition between resources, reserves, and recoverable reserves. Generally, reserves are considered to be about half of the estimated resources since past experience indicates that only about one-half of the coal can be retrieved in underground mining. At the same time, the recoverable reserves estimated in the World Energy Conference table are only about 6% of the total resource estimate. This estimate of recoverable reserves, 600 billion tonnes, can be compared to the current annual coal consumption of 3 billion tonnes.

Nearly 90% of the world resources of coal exist in three countries above 30° N latitude—the USSR, the United States, and the Peoples Republic of China. More than half the remaining resources are in Europe.

The Geology of Coal. The widespread occurrence of coal throughout the world reflects the mechanism by which coal is formed—essentially the accumulation of decaying plant debris in low-lying areas with subsequent alternation and solidification of that debris into various grades of coal.

The optimum conditions for coal formation have been in temperate and subtropical regions where high rainfall provided an environment for prolific growth of forests. At the same time, such growth occurred in swampy areas where the debris decayed in an aqueous environment which prevented oxidation in air. Coastal swamps and basin lowlands historically provided such conditions in almost every geologic age.

The conversion of plant debris to coal involves the loss of oxygen, nitrogen, and hydrogen through bacterial action with an attendant increase in carbon content. Substances resistant to bacterial action, such as waxes and resins, remain in the coal. Various levels of coal formation are represented by peat, lignite, bituminous, and anthracite coals.

Coal occurs in successive layers of varying thickness interspersed with layers of sandstone or shale that signal the shifting of land levels with geologic time—plant debris accumulating during periods when the land was essentially at sea level and sediment accumulating during periods when the land was submerged. Coal seams vary in thickness from a few millimeters to many meters. The thickest coal seam that has been found is some 130 meters thick and is in the Peoples Republic of China. Seams less than 0.3 meters in thickness are not considered to be economically recoverable.

Bituminous coal represents the final stage of coal formation in undisturbed flat land. In regions where there has been extensive folding of the earth's crust, the Appalachian Mountains, for example, additional heat and pressure have caused further devolatilization of the coal, producing a product especially high in carbon content—anthracite coal.

	Recoverable Reserves[2]	% World Total	Total Reserves[3]	% World Total	Total Resources[4]	% World Total
U.S.S.R.	150,576	23.1	301,151	19.5	6,298,158	53.1
China, Peoples Rep. of . .	88,185	13.5	330,693	21.4	1,102,310	9.3
Rest of Asia	19,344	3.0	44,620	2.9	119,108	1.0
United States	200,379	30.8	400,758	25.9	3,223,709	27.2
Canada	6,103	0.9	9,958	0.6	119,906	1.0
Latin America	3,090	0.5	10,142	0.7	36,297	0.3
Europe	139,745	21.4	352,526	22.8	669,676	5.6
Africa	17,227	2.6	33,390	2.2	64,864	0.6
Oceania	27,026	4.2	82,341	5.3	220,081	1.9
World Total	651,676	100.0	1,545,741	100.0	11,854,109	100.0

[1]Excludes peat.

[2]Amount of reserves in place that can be recovered under present local economic conditions using available technology.

[3]The porton of total resources which have been carefully measured and assessed as being exploitable under local economic conditions and available technology.

[4]Total amount available in the earth that can be successfully exploited and used by man within the foreseeable future.

Source: World Energy Conference Survey of Energy Resources, 1974.

Note: Estimates of U.S. coal reserves in this survey do not agree with other domestic data; among other reasons, criteria used for thickness of seam and depth of overburden are not those employed by the U.S. Bureau of Mines or U.S. Geological Survey.

Source: Coal Facts 1974–1975, National Coal Association, Washington, D.C.

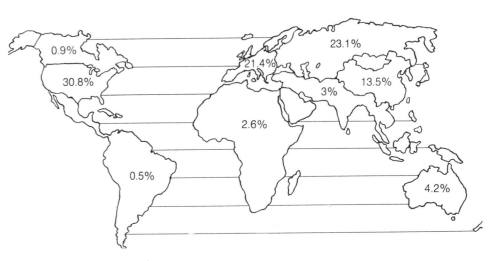

Canada 0.9% · United States 30.8% · Latin America 0.5%

Europe 21.4% · Africa 2.6% · USSR 23.1% · China 13.5%

Rest of Asia 3% · Oceania 4.2%

Recoverable coal reserves of the world.

Source: World Energy Conference Survey of Energy Resources, 1974.

Estimated World Coal Resources, by Continent[1]
[In billion tons]

Continent	Resources determined by mapping and exploration	Probable additional resources in unmapped and unexplored areas	Estimated total resources
Asia and European U.S.S.R.	[2] 7,000 ([3] 140Q)	4,000 (80Q)	11,000 (220Q)
North America	1,720 (34.4Q)	2,880 (57.6Q)	4,600 (92.0Q)
Europe	620 (12.4Q)	210 (4.2Q)	830 (17.6Q)
Africa	80 (1.6Q)	160 (3.2Q)	240 (4.8Q)
Oceania	60 (1.2Q)	70 (1.4Q)	130 (2.6Q)
South and Central America	20 (0.4Q)	10 (0.2Q)	30 (0.6Q)
Total	[2] 9,500 (190.0Q)	7,330 (146.6Q)	[4] 16,830 (337.6Q)

[1] Original resources in the ground in beds 12 inches or more thick and generally less than 4,000 feet below the surface, but includes small amounts between 4,000 and 6,000 feet.
[2] Includes about 6,500,000,000,000 short tons in the U.S.S.R.
[3] Assumes an average of 20,000,000 Btu per ton.
[4] Includes about 9,500,000,000,000 short tons in the U.S.S.R.

Source: Paul Averitt, "Coal Resources of the United States, Jan. 1, 1967," U.S. Geological Survey Bulletin 1275 (Washington: GPO, 1969), p. 82.

Walking dragline with 120 cubic yard bucket removes earth and rock in 200-ton bites to expose coal seam. (*Courtesy Marion Power Shovel Co.*)

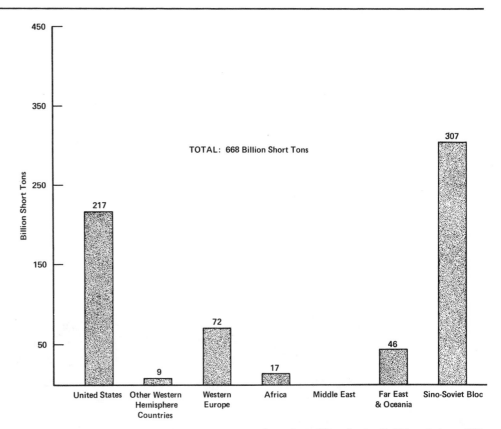

TOTAL: 668 Billion Short Tons

Source: Survey of Energy Resources, World Energy Conference, 1974; U.S. Bureau of Mines, 1974.

World Recoverable Coal Reserves, 1973.

Source: Energy Perspectives, U.S. Department of the Interior, February, 1975.

WORLD PEAT RESERVES

Peat as a Fuel. About one-third of the world's production of peat is used as fuel, primarily in two countries—Ireland and the USSR. Peat is formed in low-lying bogs in regions where the climate is cool and the humidity high throughout the year. Upon drying in air, peat still retains about 25% moisture. The content of organic matter is usually above 50%. Peat is porous and contains about 2% nitrogen and is, therefore, a good soil conditioner and a low-grade fertilizer. The heat content of peat is about 7,200,000 Btu per short ton (compared to 26,000,000 Btu/ton for bituminous coal). The tonnage of peat recovered as fuel is about 2% of the coal tonnage mined.

The total energy content of the peat reserves of the world (211,301 metric megatonnes) is about 1.7 Q (10^{18} Btu).

Peat production in Ireland. Three drum mill raises peat for drying to about 50 percent moisture content. (Consulate General of Ireland)

Reserves and Annual Production of Peat

Name of Continent, Region and Country	Ref. No.	Year of Ref.	Recoverable Reserves Amount (mega-tonnes)	Recoverable Reserves Min thickness (m)	Recoverable Reserves Max Depth (m)	Total Reserves Amount (mega-tonnes)	Total Reserves Total Energy (millions of Gcal)	Total Reserves Total Energy (millions of GJ)	Recent Annual Production Year of Ref.	Recent Annual Production Amount (megatonnes) Fuel	Recent Annual Production Amount (megatonnes) Agriculture	Recent Annual Production Amount (megatonnes) Total
Asia												
Korea, Rep. of	2,3	–	U	0.50	–	163	326	1 365	1971	–	0.009	0.009
Japan	2,3	1963	U	0.50	–	500	1 000	4 186	1971	U	U	0.073
Pakistan	2,3	1966	U	0.50	–	100[1]	200	837	–	–	–	–
Bangladesh	1	1962	35	0.08	4	392[2]	78	327	–	–	–	–
Israel	2,3	–	U	0.50	–	U	U	U	1971	–	0.020	0.020
Total Asia		–	–	–	–	802	1 604	6 715	–	–	0.029	0.102
Europe												
Western Europe												
Germany, F.R. of	1,3	1972	900	2.00	12	900	1 800	7 535	1972	0.400	1.300	1.700
France	2,3	1967	N	0.50	–	N	–	–	1971	–	0.073	0.073
Netherlands	2,3	1955	U	0.50	–	13	26	109	1971	U	U	0.400
Austria	2	1966	U	0.50	–	20	40	167	–	–	–	–
Total, Western Europe		–	–	–	–	933	1 866	7 811	–	0.400	1.373	2.173
Northern Europe												
United Kingdom	2	1968	U	–	–	1 610	3 220	13 478	–	–	–	–
Sweden	1,3	1952	U	–	–	9 400	18 800	78 697	1971	0.023	0.082	0.105
Denmark	1,3	–	0	0.30	4	561	1 122	4 697	1971	0.005	–	0.005
Finland	1,3	1972	4 290	1.00	3	33 000	66 000	276 276	1972	0.170	0.145	0.315
Norway	2,3	1966	U	0.50	–	85	170	712	1971	0.004	0.012	0.016
Ireland	2,3	1967	U	0.50	–	400	800	3 349	1971	4.900	0.050	4.950
Iceland	2	1967	U	0.50	–	2 000	4 000	16 744	–	–	–	–
Total, Northern Europe		–	–	–	–	47 056	94 112	393 953	–	5.102	0.289	5.390
Eastern Europe												
Poland	2,3	1967	U	0.50	–	24 000	48 000	200 928	1971	0.050	–	0.050
Romania	2	1932	U	0.50	–	68	136	569	–	–	–	–
Hungary	2,3	1966	N	0.50	–	N	N	N	1971	–	0.065	0.065
Total, Eastern Europe		–	–	–	–	24 068	48 136	201 479	–	0.050	0.065	0.115
Total Europe		–	–	–	–	72 057	144 114	603 243	–	5.552	1.727	7.678
USSR	2,3	1967	U	0.50	–	125 592	251 194	1 051 500	1971	57.000	130.000	187.000
North America												
United States	2,3	1967	U	0.50	–	12 540	25 080	104 985	1971	–	0.550	0.550
Canada	2,3	1926	U	0.50	–	200	400	1 675	1971	–	0.280	0.280
Total North America		–	–	–	–	12 740	25 480	106 660	–	–	0.830	0.830
South America												
Argentina	1,3	1972	U	U	10	90[3]	180	753	1971	–	0.003	0.003
Uruguay	1	1972	10	0.30	6	10	20	84	–	–	0.003	0.003
Total South America		–	10	–	–	100	200	837	–	–	0.003	0.003
Oceania												
Papua-New Guinea	2	1967	U	0.50	–	10	20	84	–	–	–	–
World Total		–	–	–	–	211 301	422 612	1 769 039	–	62.552	132.589	195.613

Notes:

In addition to the countries listed in the tabulations above, Lesotho (1960), Guyana (1961), Philippines (1966), Bulgaria (1965), Portugal (–), Switzerland (1966) and Australia (1967) have reported negligible resources. Mexico (1967) and Thailand (1967) have reported that resource data have been recorded but are not yet available. Rhodesia (1961), Angola (1965), South Africa (1959), P.R. of China (–), India (–), Indonesia (1962), D.R. of Germany (1973), Spain (–), Cuba (1963), Chile (1967), Venezuela (–), and New Zealand (1967) have indicated that data are unavailable.

Total energy content of peat reserves in each country are calculated values based on the use of 2000 kcal (8.37 million joules)/kg.

1. Excludes reserves in Bangladesh.   2. Excludes 119 million tons of additional resources.   3. Excludes 50 million tons of additional resources.

References

1. Data from current survey; see Appendix 2.
2. "World Power Conference Survey of Energy Resources, 1968," A. Parker, 1968 (Resource data only).
3. "Mineral Yearbook, 1971," Vol. I-II, U.S. Bureau of Mines, 1972 (Production data only)

Source: Survey of Energy Resources 1974, The United States National Committee of the World Energy Conference, New York, 1974.

WORLD OIL RESOURCE ESTIMATES

Petroleum Resources. Petroleum resources, both liquid and gaseous, have become the major sources of energy in many countries because of the availability and convenience of these fuels for both transportation engines and stationary power plants. Although petroleum resources occur in almost every part of the world to some degree, the major commercially valuable resources occur in relatively few locations where geological conditions were appropriate for the formation and storage of these fuels underground.

It is generally believed that petroleum is derived from deposits of plant and animal remains since petroleum deposits are found almost exclusively in sedimentary rock formations which were laid down during geologic periods when plant life flourished. Petroleum is rarely found in igneous or metamorphic rocks. The deposits of oil which have been found were largely formed in marine environments—the deposits now on land were previously also sites of ancient seas.

The creation of sizable reservoirs of petroleum required the deposition of large quantities of organic matter along with rapid burial under sediments to prevent oxidation of the organic matter. There is not a generally accepted theory on the mechanisms by which this organic matter was converted to petroleum but bacterial action, chemical reactions, the effect of radiation from rocks, and heat and pressure probably all played a part in breaking down the complex molecules of organic living matter and subsequent formation of the hydrocarbons typical of petroleum.

The accumulation of petroleum in quantities sufficiently large to be commercially attractive required that the rock in which the petroleum was formed be porous and covered with a layer of impermeable cap rock to prevent the upward flow of the petroleum to the surface. Geologic folding of these rock strata into inverted-trough anticlines provided traps into which the petroleum would migrate and collect in underground oil "pools" in the pore space of the sedimentary rock. Several other types of traps are also found which are associated with fault structures, salt domes, reefs, and lithologic changes in sandstone structure.

The occurrence of petroleum throughout the world is associated with marine sedimentary deposits in past geologic time although the existence of such deposits does not necessarily mean that oil will be found. Regions which have only igneous, metaphoric, or nonmarine sedimentary rock are highly unfavorable for the presence of petroleum. Such unfavorable areas include the land areas in northeastern United States and Canada, eastern Brazil, central and southern Africa, western Australia, Scandinavia, the Indian subcontinent, and eastern Siberia.

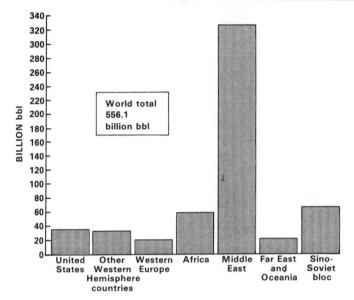

World estimated crude oil reserves, January 1, 1975.

Source: Energy Perspectives **2**, U.S. Department of the Interior, June 1976.

Oil Wells at Texas City. (Marathon Oil Company)

WORLD "PUBLISHED PROVED" OIL RESERVES AT END 1974

COUNTRY/AREA	Thousand Million Tonnes	Share of Total	Thousand Million Barrels
U.S.A.	5·3	5·4%	40·6
Canada	1·1	1·2%	8·8
Caribbean	2·6	2·7%	18·4
Other Western Hemisphere	3·1	3·1%	22·2
TOTAL WESTERN HEMISPHERE	**12·1**	**12·4%**	**90·0**
Western Europe	3·5	3·6%	26·3
Middle East	55·0	56·3%	403·4
Africa	9·1	9·3%	68·3
U.S.S.R.	11·4	11·6%	83·4
Eastern Europe	0·4	0·4%	3·0
China	3·4	3·5%	25·0
Other Eastern Hemisphere	2·8	2·9%	21·0
TOTAL EASTERN HEMISPHERE	**85·6**	**87·6%**	**630·4**
WORLD (excl. U.S.S.R., E. Europe & China)	82·5	84·5%	609·0
WORLD	**97·7**	**100·0%**	**720·4**

SOURCE OF DATA

U.S.A. American Petroleum Institute.

Canada. Canadian Petroleum Association.

All other areas. Estimates published by the "Oil & Gas Journal"
(Worldwide Oil issue 30th December, 1974)

NOTES

1 Proved reserves are generally taken to be the volume of oil remaining in the ground which geological and engineering information indicate with reasonable certainty to be recoverable in the future from known reservoirs under existing economic and operating conditions.

2 The recovery factor, i.e. the relationship between proved reserves and total oil in place, varies according to local conditions and can vary in time with economic and technological changes.

3 For the U.S.A. and Canada the data include oil which it is estimated can be recovered from proved natural gas reserves.

4 The data exclude the oil content of shales and tar sands.

Source: The British Petroleum Company Limited, Britannic House, Moor Lane, London.

Oil Wells in West Texas. (Gulf Oil Corp.)

Estimates of world crude oil resources and reserves
(In megatonnes)

Continent, geographical region, and country	Resources						Reserves			
	Original oil in place	Possible ultimate discoveries	Remaining recoverable resources	Ultimately recoverable resources			Probable recoverable reserves	Proved recoverable reserves	Total reserves	Proved recoverable reserves
	Torrey[1] Jan. 1962	Hendricks[2] 1965	Linden[3] Dec. 1971	Weeks[4] 1962	Ryman[4] 1967	USGS[5] 1973	World Petroleum[4] 1966	World Oil[4] Dec. 1966	1968 WEC study[6]	1974 WEC study
(1)	(2)	(3)	(4)	(5)	(6)	(7)	(8)	(9)	(10)	(11)
Africa	7 356	144 740	65 789	13 157	32 893	38 372–383 720	6 447	4 197	3 441–5 588	12 848
Asia			182 432							
China (Peoples Republic)	220	a				1 500–15 000			44–130	1 726
Far East (exc. Peo. Rep. China)	3 784	13 600		11 642	27 392	11 430–114 300	2 342	2 068	259–262	2 009
Middle East	107 993	122 950		106 540	81 956	51 900–519 000	41 538	37 386	31 247–33 629	50 237
U.S.S.R.	12 116	244 900	b			15 000–150 000			4 700–5 200	8 138
Europe		40 540	19 014							
Eastern Europe	1 629					750–7 500			146	374
Remaining Europe	1 211			2 603	2 740	7 575–75 750	548	493	886–893	1 020
North America										
Caribbean	875	c				482–4 820			128	287
Canada	3 428	40 540		11 445	12 792	1 500–15 000	1 535	1 467	1 293	1 075
Mexico	2 667	c				15 000–150 000			389	414
Middle America (exc. Mexico)						630–6 300				
United States	46 670	134 800	45 160	36 398	26 961	15 000–150 000	17 336	15 287	55 000	5 569
South America		71 430								
Venezuela	24 515					15 000–150 000			3 013	6 093
Remaining South America	6 054					8 100–81 000			1 161–1 193	1 506
Oceania	18	27 210	b			1 530–15 300			68	229
China, Eastern Europe, U.S.S.R.				59 864	68 028		11 796	8 912		
Caribbean, Mexico, South America				31 547	32 118		9 193	8 123		
Western Hemisphere (exc. U.S.A.)			52 143							
Total world (values rounded)	218 540	840 710	364 540	273 200	284 880	183 769–1 837 690	90 735	77 933	101 780–106 930	91 525

[a]Included in U.S.S.R. total.
[b]Included in Asia total.
[c]Included in Canada total.

1. P. D. Torrey, C. L. Moore, and G. H. Weber, "World Oil Resources," *Sixth World Petroleum Congress Proceedings,* Section VIII, paper 9, pages 83–114, Frankfurt-am-Main (June 19–26, 1963), published by Verein zur Förderung des 6 Welt-Erdöl-Kongresses, Hamburg.

2. T. A. Hendricks, "Resources of Oil, Gas, and Natural Gas Liquids in the United States and the World," Geological Survey Circular 522, U.S. Department of the Interior, Washington, D.C. (1965).

3. H. R. Linden, "The Future Development of Energy Supply Systems," a paper prepared for the Fuel Conference in Commemoration of the Golden Jubilee of the Fuel Society of Japan, Tokyo (Oct. 31–Nov. 2, 1972).

4. Data in columns 5, 6, 8, and 9 are the works of other investigators reported by M. K. Hubbert, "Energy Resources," *Resources and Man,* Chap. 8, page 194, W. H. Freeman and Co., San Francisco (1969).

5. J. P. Albers, M. D. Carter, A. L. Clark, A. B. Coury, and S. P. Schweinfurth, *Summary Petroleum and Selected Mineral Statistics for 120 Countries, Including Offshore Areas,* Geological Survey Professional Paper 817, U.S. Department of the Interior, Washington, D.C. (1973).

6. *World Power Conference Survey of Energy Resources, 1968,* Table 5, pages 34–38, published by the Central Office of the World Power Conference, London, W.C.1.

Source: Survey of Energy Resources 1974, The United States National Committee of the World Energy Conference, New York, 1974.
Note: The last figure in the table of 91,525 megatonnes for the total world proved recoverable reserves of crude oil corresponds to 670 billion barrels.

(Y-16) Off-shore gas and oil drilling rigs loom impressively from the gleaming waters of the Gulf of Mexico. (*Courtesy Shell Oil Co.*)

WORLD NATURAL GAS RESOURCES

Natural Gas Resources. The formation of natural gas is probably due to processes similar to those for the formation of petroleum—natural gas is found along with oil in almost all oil fields but, at the same time, there are many gas fields where little or no oil is found. The reasons for these differences in origins for gas and oil are not understood.

Natural gas is composed of the lightest, most volatile hydrocarbons—methane, ethane, propane, and butane. As marketed, natural gas consists largely of methane, since the ethane, propane, and butane are usually removed and marketed separately as special fuels or as feedstocks for the manufacture of petrochemicals.

As it exists in nature, natural gas is found either in solution in oil or as gas trapped above an oil pool (associated gas) or in underground reservoirs with little or no oil present (nonassociated gas). As the gas is withdrawn from underground reservoirs, droplets of hydrocarbons of slightly higher molecular weight are carried along with the gas. These less volatile hydrocarbons condense as liquids and are called natural gas liquids. On the average, about 170 cubic feet of natural gas liquids are produced for every million cubic feet of gas. A number of other elements and compounds are also found in natural gas including nitrogen, hydrogen, hydrogen sulfide, carbon dioxide, and helium. The non-combustible impurities are removed prior to marketing.

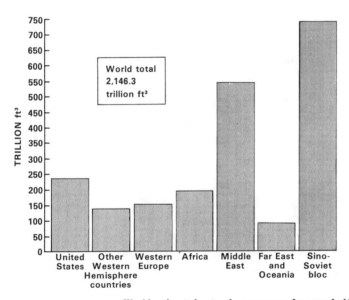

World total 2,146.3 trillion ft³

World estimated natural gas reserves, January 1, 1975.

Source: Energy Perspectives 2, U.S. Department of the Interior, June 1976.

Thirty-four percent of world estimated reserves of natural gas was located in the Sino-Soviet bloc nations at the beginning of 1975, mostly in the U.S.S.R. (95 percent). The Middle East accounted for 28 percent of world reserves, and the United States 11 percent.

The distribution of natural gas production has been distinctly different from the distribution of reserves because of the difficulty and costs of transportation. In 1974, the United States accounted for almost half and the Sino-Soviet bloc more than one-fourth of world production.

The natural gas liquids fractionating plant with a capacity of 138,000 barrels per day at Ras Tanura, Saudi Arabia. (*Courtesy of Arabian American Oil Co. and the American Petroleum Institute Photographic and Film Services*)

Estimates of World Resources of Natural Gas
(In cubic kilometers)

Continent, geographical region, and country	Estimates by other			WEC studies	
	Possible ultimate discoveries	Remaining recoverable resources		Total reserves	Proved recoverable reserves
	Hendricks[1] 1965	Linden[2] Dec. 1971	USGS[3] 1973	1968 study[4]	1974 study
(1)	(2)	(3)	(4)	(5)	(6)
Africa	96 280	48 140	9 648–96 480	2 680–4 745	5 709
Asia					
Far East (inc. Peo. Rep. China)	11 330		8 796–87 960	979–1 024	2 357
Middle East	62 300		10 650–106 500	6 074–6 607	9 884
U.S.S.R., Peo. Rep. China, Mongolia	150 100				
Asia-Pacific (inc. European U.S.S.R.)		121 760			
U.S.S.R.			30 000–300 000	70 404	17 136
Europe (exc. U.S.S.R.)	22 650	11 330	5 736–57 360	4 255	4 513
North America					
United States	70 790	41 260	30 000–300 000	56 640	7 556
Remaining North America	62 300		6 519–65 190	2 137	3 093[a]
Western Hemisphere (exc. U.S.A.)		53 600			
South America	45 300		10 560–105 600	1 557–1 635	1 591
Oceania			3 303–33 030	208	693
Australia, East Indies, Pacific islands	19 820				
Total world	540 870	276 090	115 212–1 152 120	144 934–147 655	52 532

[a]Canada reported natural gas as marketable gas.

1. T. A. Hendricks, "Resources of Oil, Gas, and Natural Gas Liquids in the United States and the World," Geological Survey Circular 522, U.S. Department of the Interior, Washington, D.C. (1965).

2. H. R. Linden, "The Future Development of Energy Supply Systems," a paper prepared for the Fuel Conference in Commemoration of the Golden Jubilee of the Fuel Society of Japan, Tokyo (October 31–November 2, 1972).

3. J. P. Albers, et al, "Summary Petroleum and Selected Mineral Statistics for 120 Countries, Including Offshore Areas," Geological Survey Professional Paper 817, U.S. Department of the Interior, Washington, D.C. (1973).

4. *World Power Conference Survey of Energy Resources, 1968;* Table 7, page 41–44, published by the Central Office of the World Power Conference, London, W.C.1.

Notes: One trillion cubic feet equals 28.3 cubic kilometers. The last figure in the table of 52,532 cubic kilometers for the total world proved recoverable reserves corresponds to 1855 trillion cubic feet with a heat content of 1.90 (10^{18} Btu).

Source: Survey of Energy Resources 1974, The United States National Committee of the World Energy Conference, New York, 1974.

WORLD OIL SHALE RESOURCES

Shale Oil Resources. In addition to the deposits of organic material that have produced petroleum and natural gas, there are deposits that are solid or semisolid and which cannot be recovered by conventional methods of petroleum production. These deposits are found in organic-rich shale and sands and are referred to as oil shales or oil or tar sands. The organic solids in oil shale are a waxlike material, kerogen, and those in tar sands are a heavy bitumen of very high viscosity. Both are extracted by heating in retorts.

The geologic origin of oil shale is from deposits in large shallow lakes and seas where subtropical, stagnant conditions allowed an accumulation of algae, spores, pollen, and other organic material. The most extensive deposits, some nearly 2000 feet thick, occurred in ancient freshwater lakes where conditions were favorable for the growth of algae.

Although the oil available from shale exceeds that available from natural petroleum deposits, extraction of oil from shale has not been widely adopted since petroleum can be produced at far lower cost. Nonetheless, shale oil has been extracted in Scotland, Estoria, France, Sweden, Germany, Spain, South Africa, Australia, and China.

Estimates of Shale Oil Resources of World Land Areas
(In Billions of Barrels)

Continent	Known Resources — Recoverable under Present Conditions 10–100	Known Resources 25–100	Known Resources 10–25	Known Resources 5–10	Possible Extensions of Known Resources (Marginal and Submarginal, Oil Equivalent in Deposits; Range in Grade, Oil Yield in Gallons per Ton of Shale) 25–100	10–25	5–10	Undiscovered and Unappraised Resources 25–100	10–25	5–10	Order of Magnitude of Total Resources[a] (Oil Equivalent in Deposits) 25–100	10–25	5–10
Africa	10	90	b	b	c	c	c	4,000	80,000	450,000	4,000	80,000	450,000
Asia	20	70	14	c	2	3,700	c	5,400	106,000	586,000	5,500	110,000	590,000
Australia and New Zealand	b	b	1	c	c	c	c	1,000	20,000	100,000	1,000	20,000	100,000
Europe	30	40	6	c	100	200	c	1,200	26,000	150,000	1,400	26,000	140,000
North America	80	520	1,600	2,200	900	2,500	4,000	1,500	45,000	254,000	3,000	50,000	260,000
South America	50	b	750	c	c	3,200	4,000	2,000	36,000	206,000	2,000	40,000	210,000
Total	190	720	2,400	2,200	1,000	9,600	8,000	15,000	313,000	1,740,000	17,000	325,000	1,750,000
Total Q (10^{18} Btu)	1.1	4.2	13.9	12.8	5.8	55.7	46.4	87.0	1,815	10,090	98.6	1885	10,150

[a]Of the approximately 2 by 10^{15} bbl here indicated, 190 by 10^9 were considered recoverable under 1965 conditions.
[b]Small.
[c]No estimate.
Source: Duncan and Swanson, 1965, table 3, p. 18; NAS/NRS, "Resources and man," op. cit., p. 200.
Source: Energy Research and Development—Problems and Prospects, Serial No. 93-21 (92-56), A National Fuels and Energy Policy Study, Committee on Interior and Insular Affairs, United States Senate, 1973.

Estimated Shale Oil Resources of the World[a]
(In Megatonnes: Values are Rounded)

Continent	Identified Resources[c] Oil Yield		Hypothetical Resources[d] Oil Yield		Speculative Resources[e] Oil Yield	
	0.1–0.4 (liters/kg)	0.04–1 (liters/kg)	0.1–0.4 (liters/kg)	0.04–1 (liters/kg)	0.–0.4 (liters/kg)	0.04–0.1 (liters/kg)
	24–96 (gals/ton)	9.6–24 (gals/ton)	24–96 (gals/ton)	9.6–24 (gals/ton)	24–96 (gals/ton)	9.6–24 (gals/ton)
(1)	(2)	(3)	(4)	(5)	(6)	(7)
Africa	15,500	small	[b]	[b]	621,000	12,400,000
Asia	14,000	2,170	310	575,000	839,000	17,100,000
Oceania	small	160	[b]	[b]	155,000	3,100,000
Europe	10,900	930	15,500	31,000	186,000	4,040,000
North America	60,600	232,000	50,800	246,000	232,000	6,670,000
South America	small	124,000	[b]	497,000	311,000	5,590,000
Total (megatonnes)	101,000	359,300	66,600	1,349,000	2,344,000	48,900,000
Total (billions of barrels)	728	2,640	490	9,915	17,230	359,000
Total Q (10^{18} Btu)[f]	4.2	15.3	2.8	57.5	99.9	2,082

[a]Data from W. C. Culbertson and Janet K. Pitman, "Oil shale," U.S. Mineral Resources, U.S. Geological Survey Professional Paper 820, pp. 497–503, United States Government Printing Office, Washington (1973).

[b]No estimate.

[c]Identified resources: specific, identified mineral deposits that may or may not be evaluated as to extent and grade, and whose contained minerals may or may not be profitably recoverable with existing technology and economic conditions.

[d]Hypothetical resources: undiscovered mineral deposits, whether of recoverable or subeconomic grade, that are geologically predictable as existing in known district.

[e]Speculative resources: undiscovered mineral deposits, whether of recoverable or subeconomic grade, that may exist in unknown districts or in unrecognized or unconventional form.

[f]Author's note.

Source: Survey of Energy Resources 1974, The United States National Committee of the World Energy Conference, 345 East 47th Street, New York, N.Y. 10017, 1974.

Experimental oil shale facility near Rifle, Colorado. The oil shale mine is near the top of the escarpment. The plant and housing areas are below and are served by the access road running to the highway in the foreground. (ERDA Photo)

WORLD URANIUM RESOURCES

Uranium Geology. Uranium is a common constituent of continental rock with an average crustal concentration of about 2 parts per million (ppm). Concentration of ore quality is generally in the 5000–10,000 ppm range (0.5 to 1.0%). Minerals containing uranium are found in several types of deposits: sandstone, quartz pebble conglomerates, veins in metamorphic rocks, phosphate rocks, lignites, and black shales. Deposits in the United States are mainly found in sandstone and are probably a result of precipitation from hydrogen-sulfide bearing ground waters. Conglomerate ore bodies are found in Canada and South Africa and the principal vein deposits are located in Australia and Canada. Of the total world low-cost reserves (<$10 per pound), about 37% is found in sandstone, 35% in conglomerates, and the remainder in vein and other type deposits.

Three estimates of reasonably assured, low-cost world uranium reserves are given in the accompanying tables: OECD—1,226,000 short tons U_3O_8; L.G Poole—1,290,000 short tons U_3O_8 (995,000 tonnes U); World Energy Conference—1,276,000 short tons U_3O_8 (985,000 tonnes U). These resources of low-cost, reasonably assured uranium represent in heat content only about 0.8 Q when used in thermal reactors and about 46 Q if used in breeder reactors. Estimates of total resources of 4,000,000 tonnes of uranium correspond to 190 Q in a breeder reactor economy.

The 1975 world demand for uranium is expected to be about 19,000 tonnes; the demand in 1985, based on nuclear plants under construction and on order, is expected to be near 98,000 tonnes annually. The present assured supply of uranium will, therefore, be committed by the mid 1980s for use in the current thermal reactors. Further expansion of nuclear power will depend either upon discovery of additional uranium resources or by a shift to breeder reactors which, by transmuting the uranium 238 portion of natural uranium into plutonium 239, extends the supply of fissionable material by a factor of at least 50.

Nations with Major Uranium Resources

Nation	Reasonably Assured Resources up to $26/KgU (tonnes)	Energy Content Nonbreeders Millions of Gwh	Nonbreeders Q (10^{18} Btu)[a]	Breeders Millions of Gwh	Breeders Q (10^{18} Btu)[a]	Total Uranium Resources (tonnes)
United States	329,267	79	0.270	4741	16.18	2,041,156
Canada	185,799	46	0.152	2675	9.13	716,984
Sweden	—	—	—	—	—	308,381
South Africa	202,000	41	0.141	2473	8.44	298,004
Australia	120,949	29	0.10	1742	5.94	160,049
France	34,850	8.4	0.029	502	1.71	85,000
Niger	40,000	9.6	0.033	575	1.96	80,800
India	—	—	—	—	—	61,862
Colombia	—	—	—	—	—	51,000
Argentina	12,665	3.0	0.010	182	0.62	38,590
Gabon	20,400	4.9	0.017	294	1.00	30,240
Rest of Europe	21,834	5.2	0.018	315	1.07	73,863
Rest of World	16,710	4.0	0.013	240	0.81	78,023
Total	984,474	229	0.781	13739	46.46	4,023,948[b]

[a] Author's note.

[b] Total uranium resource of 4 million tonnes is equivalent to 190 Q when used in breeder reactors. Total world energy consumption in 1972 was 0.27 Q.

Source: Survey of Energy Resources 1974, The United States National Committee of the World Energy Conference, 345 East 47th Street, New York, N.Y. 10017.

Uranium: Reasonably Assured Resources
(1,000 tonnes U)
Data available 1st January 1975

COST RANGE	< 15 $/lb U_3O_8 RESERVES	15-30 $/lb U_3O_8
Algeria	28	–
Argentina	9.3	11.3
Australia	243	–[a]
Brazil	9.7	0.7
Canada[b]	144	22[c]
Central African Republic	8	–
Denmark (Greenland)	–	6
Finland	–	1.9
France	37	18
Gabon	20	–
Germany	0.5	0.5
India	3.4	25.8
Italy	–	1.2
Japan	1.1	6.6
Korea	–	2.4
Mexico	5	1
Niger	40	10
Portugal	6.9	–[a]
South Africa[f]	186	90
Spain[e]	10	93.5
Sweden	–	300
Turkey	2.6	0.5
United Kingdom	–	1.8
United States[d]	320	134
Yugoslavia	4.2	2.3
Zaire	1.8	–
Total (rounded)	1 080	730

a) Estimates of resources in this range have not been made and are therefore unknown. Exploration to date has concentrated on proving high grade resources.

b) Categories are by reference to price.

c) Estimates in this price range are preliminary, restricted only to principal deposits, and thus very conservative.

d) Does not include 54,000 tonnes U as a by-product from phosphates or 15,000 tonnes U as a by-product from copper production which might be recovered in the period to the year 2000.

e) Includes some 80,800 tonnes U reasonably assured resources in lignites in the cost range $15-30/lb U_3O_8 for which the availability is uncertain.

f) The 350,000 tonnes U total uranium resource for South Africa as given in Part II has also been supplied apportioned as a best estimate to the various resource categories although reservations have been expressed concerning the accuracy of the split figures.

Source: Uranium Resources, Production and Demand, Organization for Economic Cooperation and Development, Paris, 1975.

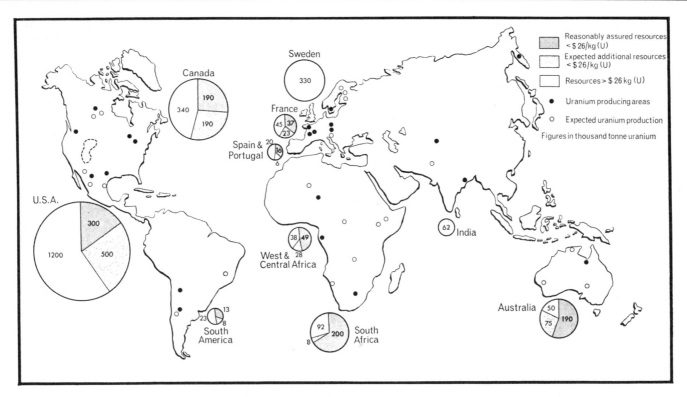

Main uranium producing regions of the world.

Source: Poole, L. G., "World Uranium Resources," *Nuclear Engineering International*, February 1975.

Weight and Price Equivalents

1 short ton U_3O_8 = 0.91 tonnes U_3O_8
= 0.77 tonnes U equivalent
1 lb U_3O_8 = 0.45 kg U_3O_8
= 0.39 kg U equivalent
1 tonne U = 1.3 short tons U_3O_8 equivalent
1 kg U = 2.2 lb U_3O_8 equivalent
$10/lb U_3O_8 = $26/kgU
$10–15/lb U_3O_8 = $26–39/kgU
$15–30/lb U_3O_8 = $39–78/kgU

WORLD THORIUM RESOURCES

Geologic Occurrence of Thorium. Thorium is widely found in low concentrations (8 ppm) in continental rocks, as is uranium. Commercial sources of thorium are derived principally from the mineral, monazite, a heavy, hard mineral of rare earth and thorium phosphates. Deposits of this mineral containing 3 to 9% thorium (30,000–90,000 ppm) are found in placer deposits where the heavy ore has been concentrated by river and ocean wave action from weathered continental rock.

Large placer deposits of monazite occur along the coast of India at Travancore and on the coasts of Brazil and Australia. Stream-deposited placers are found in Idaho, North Carolina, and South Carolina in the United States. Vein deposits of monazite have been found in South Africa and veins of thorite (a silicate ore) are found in Idaho and Montana in the United States. Thorium ores occur in the uranium deposits at Blind River, Canada.

Estimated U.S. Thorium Resources

Cutoff Cost Dollars per Pound ThO$_2$	Short Tons of ThO$_2$		
	Reasonably Assured	Estimated Additional	Total
10	65,000	35,000	400,000
30[a]	200,000	400,000	600,000
50[a]	3,200,000	7,400,000	10,600,000

[a]Includes lower cost resources.
Source: WASH-1535, Environmental Statement Liquid Metal Fast Breeder Reactor Program, Volume IV, Alternative Technology Options, U.S. Atomic Energy Commission, March 1974.

World Thorium Resources—$10 per Pound ThO$_2$
(Tons ThO$_2$)

	Reasonably Assured	Possible Additional	Total
India[a]	300,000	250,000	550,000
United States[b]	65,000	335,000	400,000
Canada	100,000[c]	155,000	255,000
Africa[a]	50,000	50,000	100,000
Australia,[a] S.E. Asia	10,000	—	10,000
Brazil[a]	10,000	20,000	30,000
Total	535,000	810,000	1,345,000

[a]ENEA 1965. Africa includes Central Africa, South Africa, and Madagascar.
[b]USAEC Division of Production and Materials Management, 1973.
[c]Canada, Mineral Bulletin 117, 1971. Mostly by-product of uranium mining. Reasonably assured given as over 100,000; possible additional not given.
Source: WASH-1535, Environmental Statement Liquid Metal Fast Breeder Reactor Program, Volume IV, Alternative Technology Options, U.S. Atomic Energy Commission, March 1974.

World Thorium Resources
(All Values in Terms of Elemental Thorium)

Name of Continent and Country	Lower Cost Resources					Higher Cost Resources	
	Reasonably Assured				Additional Resources (tonnes)	Reasonably Assured (tonnes)	Additional Resources (tonnes)
	Amount (tonnes)	Energy Content, Breeders					
		Millions of GWh	Millions of terajoules	Q (10^{18} Btu)[b]			
Africa							
Malagasy						1,960	
Malawi							8,800
Egypt						14,700	279,800
South Africa						20,434	N
Other[a]						~17,200	30,800
Total Africa						54,294	319,400
Asia							
India						302,370	
Turkey						3,960	
Total Asia						306,330	
Europe							
Norway	132,000	1,890	6,804	6.45			132,000
USSR						~80,000	~80,000
North America							
United States	51,573	744	2,678	2.54	267,443		
Canada	79,832	1,146	4,128	3.91	79,832		
Greenland						26,400	
Total North America	131,405	1,890	6,806	6.45	347,275	26,400	
South America							
Brazil	58,388	836	3,010	2.05	1,059,080		
Uruguay						748	1,496
Total South America	58,388	836	3,010	2.05		748	1,496
Oceania							
Australia						6,160	10,560
World total	321,793	4,616	16,620	15.8	1,406,355	473,932	543,456

[a]Resources for Africa recoverable at costs up to $22/kg ThO$_2$ include 91,000 metric tons in central and southern Africa (including Malagasy) divided equally between reasonably assured and additional resources (Ref. 4).
[b]Author's note.

References
1. Data from current survey; see also Appendix 7/8.
2. "World Power Conference Survey of Energy Resources, 1968." A. Parker. 1968.
3. "Uranium Resources, Production and Demand," ENEA/IAEA, August 1973.
4. "Mineral Facts and Problems, 1970," U.S. Bureau of Mines Bulletin 650 (Resource data only).

Source: Survey of Energy Resources 1974, The United States National Committee of the World Energy Conference, 345 East 47th Street, New York, N.Y. 10017, 1974.

Most of the limited world production of thorium (1000 tonnes/year) has been as a by-product of recovery of rare earth metals from monazite. Thorium is used in special alloys and refractories, as a catalyst and in gas mantles.

The two accompanying estimates of world thorium resources, reasonably assured and lower cost (<$10 per pound), are in reasonable agreement. The World Energy Conference estimate of 321,793 metric tonnes of elemental thorium is equivalent to 403,000 short tons of ThO$_2$ (compared to a 535,000 ton USAEC figure). The heat content represented by the 403,000 ton figure is about 16 Q.

Identified Thorium Resources

[Identified resources are specific, identified mineral deposits that may or may not be evaluated as to extent and grade, and whose contained minerals may or may not be profitably recoverable with existing technology and economic conditions]

Locality [1]	Type of deposit	Thousands of short tons ThO$_2$			Principal published reference
		Recoverable primarily as byproduct or coproduct	Recoverable primarily for ThO$_2$ of grade—		
			>0.1 percent	<0.1 percent	
United States:					
Atlantic coast	Beach placer	16			McCauley (1960, p. 1).
North and South Carolina	Fluviatile placer			56	Overstreet, Theobald, & Whitlow (1959, p. 713–714).
Idaho and Montana	do	2		38	
Lemhi Pass district, Idaho and Montana	Veins		100.0		U.S. Atomic Energy Commission (1969, p. 127).
Wet Mountains, Colo	do		4.5		
Powderhorn district, Colo	do		1.5		
Mountain Pass district, California	do		.5		
Do	Carbonatite	28			
Palmer area, Michigan	Conglomerate			46	Olson & Overstreet (1964, p. 37).
Bald Mountain, Wyo	do			2	Borrowman & Rosenbaum (1962, p. 2).
Total, United States		46	106.5	142	
Australia	Beach placers	50			Griffith (1964, p. 196).
Brazil:					
Atlantic Coast	do	20			Stamper & Chin (1970, p. 685).
Morro de Ferro, Minas Gerais	Alkalic rock complex		14.0		Wedow (1967, p. 30).
Araxa, Minas Gerais	Carbonatite	130			Guimaes (1957).
Canada: Elliot Lake, Ontario	Conglomerate	[2]580			Griffith & Roscoe (1974, p. 11).
Greenland: southwest part	Alkalic rock complex			750	European Nuclear Energy Agency (1967, p. 13).
India; various areas, including Kerala, Bihar, West Bengal	Placers	450			Bhola & others (1965, p. 93).
Kenya: Mrima Hill, Mombasa area	Carbonatite	21			Davidson (1956, p. 206).
Korea	Fluviatile and beach placers	6			Olson & Overstreet (1964, p. 32–33).
Malagasy Republic [1]: Fort Dauphin area	Beach placers	10			Murdock (1963).
Malaysia	Placers	20			Shortt (1970, p. 207).
Malawi	Fluviatile and beach placers			10	Bowie (1959).
South Africa [1]: Palabora, Transvaal	Carbonatite	75			Pike (1958, p. 94).
Uganda: Sukulu, Eastern Province	do	2			Davidson (1956, p. 206).
United Arab Republic: Nile delta	Fluviatile and beach placer	10			Higazy & Naguib (1958).
West Africa: Nigeria, Sierra Leone	Placers	15			Bowie (1959).
Total, world identified resources		1,435	120.5	902	

[1] Not shown, because of inadequate data, are resources in Argentina, Ceylon, parts of Malagasy Republic, Norway, South Africa (Steenkampskraal monazite veins), Uruguay, U.S.S.R., and several other countries.

[2] Includes some hypothetical resources, which are undiscovered mineral deposits, whether of recoverable or subeconomic grade, that are geologically predictable as existing in known districts

Note: Extracted from "United States Mineral Resources" Geological Survey Professional Paper 820, December 1973—P. 474.

Source: Issues for Consideration—Review of National Breeder Reactor Program, Joint Committee on Atomic Energy, Congress of the United States, August 1975.

WORLD NUCLEAR FUSION RESOURCES

Fusion Resource Materials. The generation of power from nuclear fusion is theoretically possible through two reactions: deuterium with tritium, and deuterium with deuterium. Controlled fusion reactions of either type have not yet been achieved but since the deuterium-tritium reaction occurs at a lower temperature than the deuterium-deuterium reaction, experimental development is presently concentrated on the D-T possibility.

Tritium itself does not occur in nature and must be manufactured by nuclear transmutation of one of the isotopes of lithium, lithium 6, which represents 7.42% of natural lithium. Fusion energy derived from the D-T reaction is, therefore, dependent upon the resources of lithium. Lithium occurs in igneous rocks known as pegmatites which are geologically rare, and in the salts of saline lakes. The total resources of lithium, measured, indicated and inferred, are about 9.1 million tons. Of this elemental lithium, lithium 6 totals about 67,500 metric tons. The energy available for this amount of lithium 6 is approximately 204 Q, an amount about equivalent to that available from fossil fuels.

If the deuterium-deuterium fusion reaction can be accomplished, the fusion energy resource becomes vastly larger. Deuterium is abundant, about one atom of deuterium to each 6700 atoms of hydrogen in water. In a cubic meter of water there are 10^{25} atoms of deuterium which would provide, in a fusion reaction, about 7.50×10^9 Btu (7.94×10^{12} joules). The same energy would be available from 300 metric tons of coal or 1500 barrels of oil. The energy available from 1% of the deuterium in the oceans (total volume of the oceans is about 1.5 billion cubic kilometers) would be about 110 million Q (1.2×10^{29} joules), about 500,000 times the energy available from the world's fossil fuels.

Estimated Lithium Reserves of the United States, Canada, and Africa

Location	Li_2O Measured, Indicated, and Inferred (10^6 metric tons)	Lithium Metal (10^6 metric tons)	Lithium 6 (10^4 metric tons)	Number of Lithium 6 Atoms (10^{33} atoms)	Equivalent Fusion Energy (10^{21} joules)	Q (10^{18} Btu)
United States	19.0	8.8	65.4	65.4	234	221
Canada	0.4	0.2	1.4	1.4	5	4.74
Africa	0.2	0.1	0.7	0.7	2.5	2.37
Total	19.6	9.1	67.5	67.5	241.5	228.9

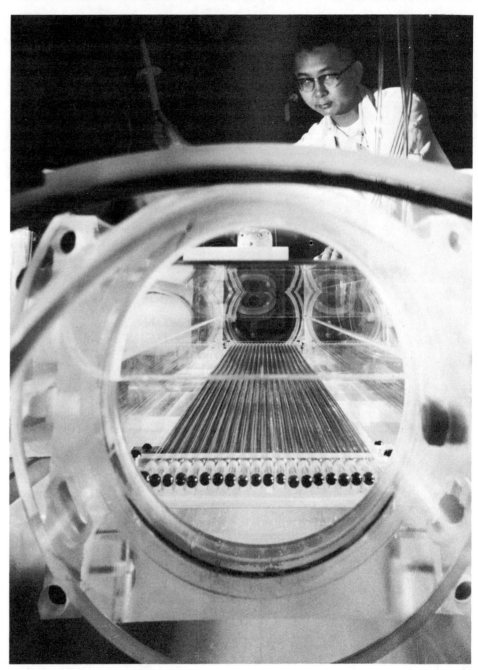

Deuterium and tritium resources are used in carbon dioxide laser fusion research at Lawrence Livermore Laboratory. (Lawrence Livermore Laboratory)

U.S. RECOVERABLE FOSSIL AND MINERAL ENERGY RESOURCES

Estimates of U.S. Depletable Energy Resources. The estimates of recoverable energy resources given in the table represent the data appearing in the pages following. Of these estimates those for coal, petroleum, and natural gas are perhaps the most accurate since these resources have been more extensively surveyed than those for oil shale and uranium. The extent of the recoverable resource from oil shale is difficult to ascertain; the quantity of oil in shale formation is much larger than the 2000 billion barrels given in the table but there is serious question that it can be economically recovered. The estimate of 3500 thousand tons of uranium oxide as the limit of the recoverable ore is also ill-defined for lack of exploration and may well be much larger.

The estimates of recoverable energy resources given in the chart represent those made by the U.S. Geological Survey and the Atomic Energy Commission in 1974. The chart presents data in quadrillion Btu (10^{15} Btu), and the numbers should be divided by 1000 to be comparable to those in the table. The numbers in the table and in the chart are in relatively good agreement considering the difficulty in measuring resources located below ground and in estimating the resources that will be economically recoverable at some future date.

Resource	Amount	Units	Equivalent Quads** (10^{15} Btu)
Coal	600	Billions of Tons	12,000
Natural Gas	750	Trillions of Cubic Feet	775
Petroleum***	142	Billions of Barrels	800
Shale Oil	200	Billions of Barrels	1,200
Uranium	3.6	Millions of Tons of U_3O_8	1,800

*Recoverable resources include both already identified resources and estimated undiscovered resources that are considered to be economically recoverable with existing technology. Resource estimates are subject to large uncertainties because they include resources which have not yet been found. Point estimates have been used for R,D&D planning, but the high degree of uncertainty should be recognized.

**For comparison, total U.S. energy consumption was 73 Quads in 1974.

***Includes Natural Gas Liquids.

Note: Appendix A presents information on conversion factors.

Selected Recoverable Domestic Energy Resources*

Recent Estimates of U.S. Energy Resources

Resource	Recoverable Reserves	Total Recoverable Resources
Coal (billions of tons)	430	1,500
heat content Q = 10^{18} Btu	(11.5 Q)	(36.5 Q)
Petroleum (billions of barrels)	48	132–242
heat content Q = 10^{18} Btu	(0.27 Q)	(0.78–1.4 Q)
Natural Gas (trillions of cubic feet)	266	718–1171
heat content Q = 10^{18} Btu	(0.27 Q)	(0.74–1.14 Q)
Oil Shale (billions of barrels)	80	2,000
heat content Q = 10^{18} Btu	(0.46 Q)	(11.6 Q)
Uranium (thousands of short tons)		
U_3O_8	600	3,500
heat content Q = 10^{18} Btu		
nonbreeder reactors	(0.24 Q)	(1.40 Q)
breeder reactors	(12.0 Q)	(70.0 Q)

Notes:
U.S. Energy Consumption 1974 = 0.072 Q.
Coal = 24,100,000 Btu/ton.
Petroleum = 5,800,000 Btu/barrel.
Natural Gas = 1030 Btu/cubic foot.
Uranium = 400 × 10^9 Btu/ton in nonbreeder reactors.
     = 20,000 × 10^9 Btu/ton in breeder reactors.
1 Q = 10^{18} Btu.
1 Q coal = 41.5 billion short tons.
1 Q oil = 173 billion barrels.
1 Q natural gas = 970 trillion cubic feet.

AVAILABLE ENERGY IN QUADS (10^{15} BTU) SHOWN GRAPHICALLY BY AREA.

TOTAL U.S. ENERGY CONSUMPTION IN 1974 WAS 73 QUADS

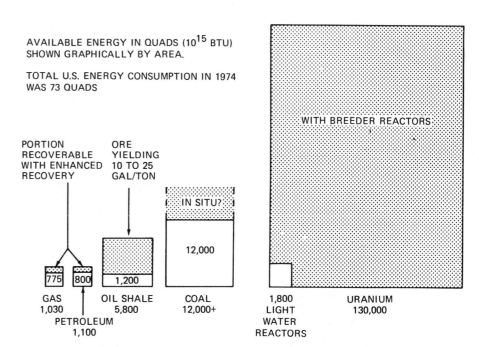

Source: A National Plan for Energy Research, Development and Demonstration: Creating Energy Choices for the Future, ERDA-48, Vol. 1: The Plan, Energy Research and Development Administration, June 1975.

Recoverable Energy Resources of the United States
(In Thermal Equivalents)

Resource	Units	Data In ERDA–48 Report	Data from Recent Official Reports			
			Demonstrated Reserves	Additional Resources	Total	Source and Explanation
Coal	Quads	12,000	4,900	16,500	21,400	ERDA–48 excluded hypothetical resources and those in areas likely to be closed to mining
Natural Gas	Quads	775*	244	706	950	Estimates based on "Mean Values" of USGS range of undiscovered resources
Petroleum**	Quads	800*	246	704	950	Estimates based on "Mean Values" of USGS range of undiscovered resources
Shale Oil	Quads	1,200	727	473	1,200	Western and Alaskan shales yielding 25–100 gal/ton
Uranium***	Quads	1,800	900	900	1,800	Utilized in LWRs
Geothermal	Quads	400	102	3,332	3,434	Recoverable heat with present or near-term technology

* ERDA–48 data based on the USGS "High Probability" estimates but excluding resources that may be produced through techniques to enhance recovery rates.
** Crude oil and natural gas liquids.
*** The use of uranium in breeder reactors could expand the resource base to about 130,000 quads.
Note: See discussions in text and bibliography for detailed references.

Source: A National Plan for Energy Research, Development and Demonstration: Creating Energy Choices for the Future, ERDA 76-1, Energy Research and Development Administration, April 15, 1976.

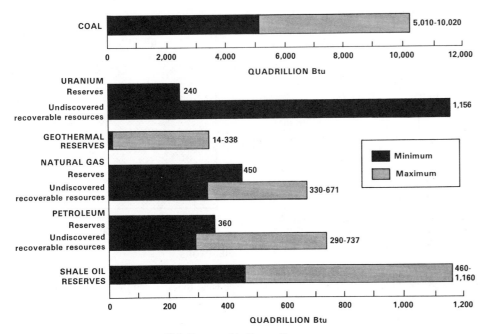

U.S. Recoverable Energy Resources

Source: Energy Perspectives **2**, U.S. Department of the Interior, June 1976.

Steam from the Nero geothermal well in Hawaii. Water from this geothermal source is low in dissolved solids and has the highest temperature, 357°C, of any well in the United States. (DOE Photo)

U.S. COAL PRODUCTION AND SUPPLY

The industrial development of the United States during the late 1800s was largely based on coal as fuel. Demands by the steel industry, railroads, and electric utilities reached 100 million tons per year in the 1880s and coal production exceeded 200 million tons per year by 1900. By 1918 coal production stood at 579 million tons—a level matching current production. During the period from the turn of the century to 1968 the number of coal workers decreased from 622,000 to 130,000 as a result of mechanization of mining operations and improved operating efficiencies. Over the years, the increasing availability of oil and gas and their convenience of use have resulted in a sharp drop in coal use in residences and commercial establishments and a virtual disappearance of coal-fired steam locomotives in favor of diesel locomotives. The decrease in the use of coal for these purposes has, however, been compensated by the growth in the use of coal for the generation of electricity.

U.S. Coal Production and Exports, 1960-74 (Million Short Tons)

ITEM	YEAR							
	1960	1965	1969	1970	1971	1972	1973	1974
NET EXPORTS	37.7	52.0	57.8	72.4	57.9	57.1	53.9	60.0
DOMESTIC PRODUCTION	434.3	527.0	571.0	612.7	560.9	602.5	598.6	596.0

Source: Energy Perspectives, U.S. Department of the Interior, February 1975.

U.S. Coal Production and Exports, 1960-74

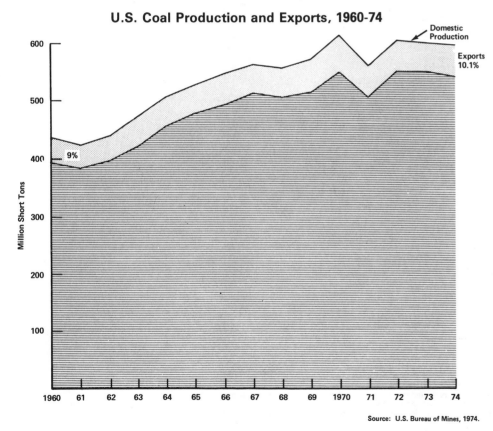

Source: Energy Perspectives, U.S. Department of the Interior, February 1975.

U.S. Low-Sulfur Coal Reserves, Production, Distribution, 1973

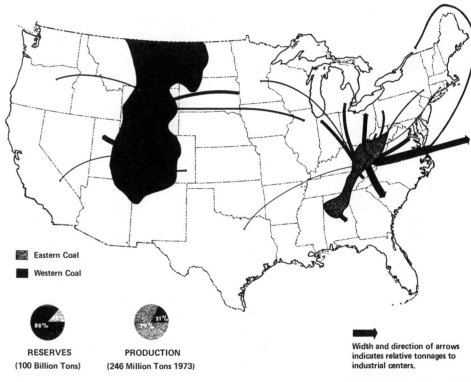

Source: Energy Perspectives, U.S. Department of the Interior, February 1975.

REFERENCE

"Bituminous coal and lignite," from *Mineral Facts and Problems, 1970 Edition*, U.S. Department of the Interior, Bureau of Mines Bulletin 650.

U.S. COAL RESOURCES

Coal Reserves in the United States
(Million short tons)

	Identified resources[1]					Hypothetical resources[2]			Total estimated identified and hypothetical resources remaining in the ground Jan. 1, 1972
	Overburden 0-3,000 ft.					0-3,000 ft.	Overburden 3,000-6,000 ft.	0-6,000 ft.	
	Estimated identified resources remaining in the ground, Jan. 1, 1972					Estimated hypothetical resources in unmapped and un-explored area	Estimated hypothetical resources in deeper structural basins	Total estimated hypothetical resources	
	Bituminous Coal	Sub-bituminous Coal	Lignite	Anthracite and semi-anthracite	Total				
Alabama	13,342	†	2,000	†	15,342	20,000	6,000	26,000	41,342
Alaska	19,413	110,668	†	†	130,081	130,000	5,000	135,000	265,081
Arizona	21,246	†	†	†	21,246	†	†	†	21,246
Arkansas	1,638	†	350	430	2,418	4,000	†	4,000	6,418
Colorado	62,339	18,242	†	78	80,659	146,000	145,000	291,000	371,659
Georgia	24	†	†	†	24	60	†	60	84
Illinois	139,124	†	†	†	139,124	100,000	†	100,000	239,124
Indiana	34,573	†	†	†	34,573	22,000	†	22,000	56,573
Iowa	6,509	†	†	†	6,509	14,000	†	14,000	20,509
Kansas	18,674	†	†	†	18,674	4,000	†	4,000	22,674
Kentucky . . .	64,842	†	†	†	64,842	52,000	†	52,000	116,842
Maryland	1,158	†	†	†	1,158	400	†	400	1,558
Michigan	205	†	†	†	205	500	†	500	705
Missouri	31,014	†	†	†	31,014	18,200	†	18,200	49,214
Montana	2,299	131,855	87,521	†	221,675	157,000	†	157,000	378,675
New Mexico . . .	10,752	50,671	†	4	61,427	27,000	21,000	48,000	109,427
North Carolina . .	110	†	†	†	110	20	5	25	135
North Dakota . .	†	†	350,630	†	350,630	180,000	†	180,000	530,630
Ohio	41,358	†	†	†	41,358	2,000	†	2,000	43,358
Oklahoma	3,281	†	†	†	3,281	20,000	10,000	30,000	33,281
Oregon	50	284	†	†	334	100	†	100	434
Pennsylvania . . .	56,759	†	†	20,510	77,269	10,000	†	10,000	87,269
South Dakota . .	†	†	2,031	†	2,031	1,000	†	1,000	3,031
Tennessee	2,572	†	†	†	2,572	2,000	†	2,000	4,572
Texas	6,048	†	6,824	†	12,872	14,000	†	14,000	26,872
Utah	23,541[3]	180[3]	†	†	23,721[3]	21,000[4]	35,000	56,000	79,721
Virginia	9,352	†	†	335	9,687	5,000	100	5,100	14,787
Washington . . .	1,867	4,190	117	5	6,179	30,000	15,000	45,000	51,179
West Virginia . .	100,628	†	†	†	100,628	†	†	†	100,628
Wyoming	12,705	107,951	†	†	120,656	325,000	100,000	425,000	545,656
Other States . .	610[5]	32[6]	46[7]	†	688	1,000	†	1,000	1,688
Total	686,033	424,073	449,519	21,362	1,580,987	1,306,280	337,105	1,643,385	3,224,372

(Figures are for resources in the ground, about half of which may be considered recoverable. Includes beds of bituminous coal and anthracite 14 in. or more thick and beds of subbituminous coal and lignite 2½ ft. or more thick.)

†A few small resources are listed under other types of coal; a few deposits are omitted because they are not believed to be economically mineable.

[1] Identified resources: Specific, identified mineral deposits that may or may not be evaluated as to extent and grade, and whose contained minerals may or may not be profitably recoverable with existing, technology and economic conditions.

[2] Hypothetical resources: Undiscovered mineral deposits, whether of recoverable or subeconomic grade, that are geologically predictable as existing in known district.

[3] Excludes coal in beds less than 4 ft. thick.

[4] Includes coal in beds 14 in. or more thick, of which 14,000 million tons is in beds 4 ft. or more thick.

[5] California, Idaho, Nebraska, and Nevada.

[6] California and Idaho.

[7] California, Idaho, Louisiana, and Mississippi.

Source: U.S. Geological Survey

Source: Coal Facts 1974-1975, National Coal Association, Washington, D.C.

Estimated Remaining Coal Reserves of the United States by Ranks, Sulfur Content and State in January 1, 1965.

(Million short tons)

Coal rank and State	Sulfur content, percent									Total
	0.7 or less	0.8 - 1.0	1.1 - 1.5	1.6 - 2.0	2.1 - 2.5	2.6 - 3.0	3.1 - 3.5	3.6 - 4.0	Over 4.0	
Bituminous coal:										
Alabama	889.2	1,189.3	5,421.7	5,182.8	458.8	417.4	---	---	18.6	13,577.8
Alaska	20,287.4	1,100.0	---	---	---	---	---	---	---	21,387.4
Arkansas	---	---	1,128.4	293.1	154.0	---	40.3	---	---	1,615.8
Colorado	25,178.3	37,237.2	---	---	---	---	---	---	---	62,415.5
Georgia	---	76.0	---	---	---	---	---	---	---	76.0
Illinois	---	573.7	4,942.4	2,615.1	809.6	16,583.8	33,650.4	57,652.2	19,062.0	135,889.2
Indiana	197.5	173.0	3,645.2	4,248.8	3,543.4	4,110.5	10,872.8	5,105.9	2,944.0	34,841.1
Iowa	---	---	---	---	---	---	117.1	---	6,405.4	6,522.5
Kansas	---	---	519.9	519.7	1,038.7	2,070.6	4,148.0	8,287.3	4,153.8	20,738.0
Kentucky:										
West	---	---	1,119.6	162.0	336.3	3,793.6	12,759.3	13,643.3	5,081.3	36,895.4
East	13,639.4	8,491.9	2,286.8	1,658.8	1,158.3	2,154.4	24.7	---	---	29,414.8
Maryland	---	---	---	124.6	191.8	208.2	378.6	56.4	220.4	1,180.0
Michigan	---	---	---	---	---	---	---	205.0	---	205.0
Missouri	---	---	---	---	---	---	6,456.7	20,669.2	51,634.1	78,760.0
Montana	51.2	218.2	205.0	397.2	400.0	175.0	40.0	27.0	591.0	2,104.6
New Mexico	5,212.0	5,474.0	---	---	---	---	---	---	---	10,686.0
North Carolina	---	---	---	---	---	110.0	---	---	---	110.0
Ohio	---	611.0	369.0	2,110.2	2,750.4	7,810.5	9,785.3	10,148.2	8,439.4	41,024.0
Oklahoma	250.6	772.2	825.0	368.1	---	---	577.2	19.1	490.6	3,302.8
Oregon	---	14.0	---	---	---	---	---	---	---	14.0
Pennsylvania	44.0	1,154.4	7,624.4	12,424.9	19,689.5	9,995.6	5,287.6	1,150.5	580.6	57,951.5
Tennessee	3.3	160.9	715.9	258.7	178.2	190.5	219.7	43.8	68.5	1,839.5
Texas	---	---	---	---	7,978.0	---	---	---	---	7,978.0
Utah	8,551.4	13,584.0	---	1,524.9	---	---	---	---	3,997.7	27,658.0
Virginia	1,981.5	6,077.5	1,637.1	---	123.9	---	---	---	---	9,820.0
Washington	898.9	672.1	---	---	---	---	---	---	---	1,571.0
West Virginia	20,761.0	26,710.6	21,819.7	13,290.6	8,496.1	2,491.8	3,147.4	5,949.2	---	102,666.4
Wyoming	6,222.2	6,596.6	---	---	---	---	---	---	1.1	12,819.9
Other States[1]	---	616.0	---	---	---	---	---	---	---	616.0
Total	104,168.4	111,502.6	52,260.1	45,179.5	47,307.0	50,111.9	87,505.1	122,957.1	103,688.5	724,680.2
Percent of total	14.4	15.4	7.2	6.2	6.5	6.9	12.1	17.0	14.3	100.0
Subbituminous coal:										
Alaska	71,115.6	---	---	---	---	---	---	---	---	71,115.6
Colorado	13,320.8	4,908.7	---	---	---	---	---	---	---	18,229.5
Montana	94,084.4	36,728.0	0.5	1,303.7	---	---	---	---	---	132,116.6
New Mexico	38,735.0	12,000.0	---	---	---	---	---	---	---	50,735.0
Oregon	87.0	87.0	---	---	---	---	---	---	---	174.0
Utah	---	---	150.0	---	---	---	---	---	---	150.0
Washington	3,693.8	500.0	---	---	---	---	---	---	---	4,193.8
Wyoming	35,579.7	72,315.6	---	---	---	---	---	---	8.6	107,903.9
Other States[2]	---	4,047.0	---	---	---	---	---	---	---	4,047.0
Total	256,616.3	130,586.3	150.5	1,303.7	---	---	---	---	8.6	388,665.4
Percent of total	66.0	33.6	0.1	0.3	---	---	---	---	(3)	100.0
Lignite:										
Alabama	---	---	20.0	---	---	---	---	---	---	20.0
Arkansas	280.0	70.0	---	---	---	---	---	---	---	350.0
Montana	60,214.5	24,141.6	2,660.9	---	---	464.7	---	---	---	87,481.7
North Dakota	284,129.1	34,987.3	31,581.6	---	---	---	---	---	---	350,698.0
South Dakota	---	2,031.0	---	---	---	---	---	---	---	2,031.0
Texas	---	---	6,902.0	---	---	---	---	---	---	6,902.0
Washington	---	116.6	---	---	---	---	---	---	---	116.6
Other States[4]	---	42.0	---	---	---	---	---	---	---	42.0
Total	344,623.6	61,388.5	41,164.5	---	---	464.7	---	---	---	447,641.3
Percent of total	77.0	13.7	9.2	---	---	0.1	---	---	---	100.0
Anthracite:										
Alaska	2,101.0	---	---	---	---	---	---	---	---	2,101.0
Arkansas	---	---	---	145.5	286.3	---	---	---	---	431.8
Colorado	---	90.0	---	---	---	---	---	---	---	90.0
New Mexico	---	6.0	---	---	---	---	---	---	---	6.0
Pennsylvania	12,211.0	---	---	---	---	---	---	---	---	12,211.0
Virginia	335.0	---	---	---	---	---	---	---	---	335.0
Washington	5.0	---	---	---	---	---	---	---	---	5.0
Total	14,652.0	96.0	---	145.5	286.3	---	---	---	---	15,179.8
Percent of total	96.5	0.6	---	0.9	2.0	---	---	---	---	100.0
Grand total	720,060.3	303,573.4	93,575.1	46,628.7	47,593.3	50,576.6	87,505.1	122,957.1	103,697.1	1,576,166.7
Percent of total	45.7	19.3	5.9	3.0	3.0	3.2	5.5	7.8	6.6	100.0

[1] Arizona, California, Idaho, Nebraska, Nevada.
[2] Arizona, California, Idaho.
[3] Less than 0.1 percent.
[4] California, Idaho, Louisiana, Nevada.

NOTE: Air quality standards for new plants without sulfur dioxide removal equipment require bituminous coal of no more than 0.7% sulfur. However, Western coals, because of lower heat content, are limited by these standards to less than 0.5% sulfur.

Source: U.S. Bureau of Mines.

Source: Coal Facts 1974-1975, National Coal Association, Washington, D.C.

Strip mining, general. The coal seam at this Western coal mine is 32 feet thick. In the foreground, a shovel loads the coal into a truck. On the bench above, bulldozers and scrapers remove the overburden atop the coal seam. This is the Dave Johnston mine of Pacific Power & Light Co. at Glen Rock, Wyoming. (*Courtesy National Coal Association, October, 1968.*)

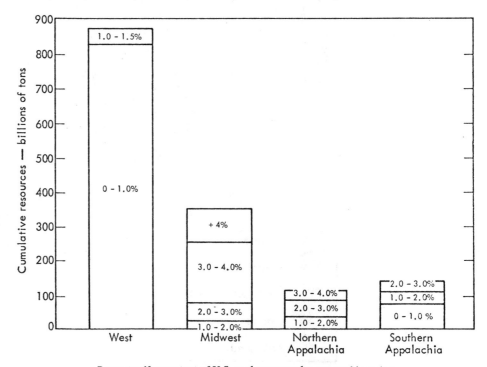

Percent sulfur content of U.S. coal resources by geographic region.

Source: "Symposium on Energy and Resources," The Mitre Corporation, July 30, 1971.

U.S. COAL RESERVES

U.S. Coal Data

Recoverable resources	1429 billion short tons
Demonstrated reserve base	434 billion short tons
Low sulfur resource base	100 billion short tons
Coal productions–1974	0.596 billion short tons
Annual increase in domestic production 1960–74	2.3%
U.S. coal consumption–1974	0.014 Q (10^{18} Btu)
Total U.S. energy consumption–1974	0.072 Q (10^{18} Btu)
Heat content of coal resources	
total resources	36.5 Q (10^{18} Btu)
demonstrated resource base	11.5 Q (10^{18} Btu)
low sulfur resource base	2.7 Q (10^{18} Btu)

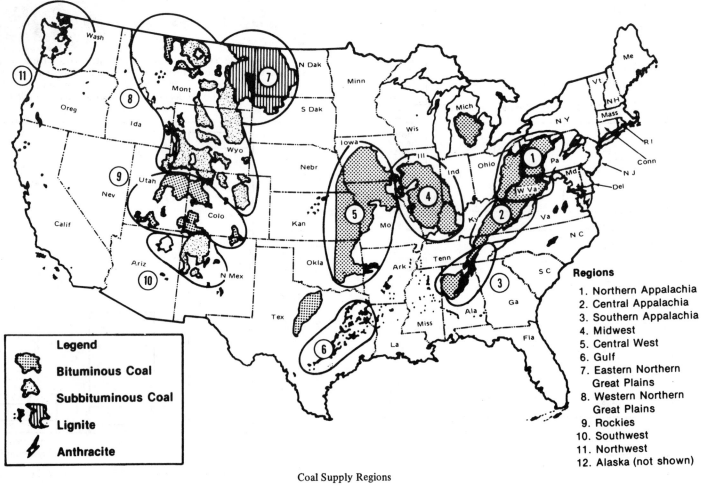

Legend

Bituminous Coal

Subbituminous Coal

Lignite

Anthracite

Regions

1. Northern Appalachia
2. Central Appalachia
3. Southern Appalachia
4. Midwest
5. Central West
6. Gulf
7. Eastern Northern Great Plains
8. Western Northern Great Plains
9. Rockies
10. Southwest
11. Northwest
12. Alaska (not shown)

Coal Supply Regions

Source: National Energy Outlook, FEA-N-75/713, Federal Energy Administration, February 1976.

	Potential Underground Method of Mining				Potential Surface Method of Mining					Total All Methods
	Anthracite	Bituminous	Sub-bituminous	Total Under-ground	Anthracite	Bituminous	Sub-bituminous	Lignite[2]	Total Surface	
Alabama	....	1,798	....	1,798	....	157	....	1,027	1,184	2,982
Alaska	....	....	4,246	4,246	....	1,201	5,902	296	7,399	11,645
Arizona	....	....	....	....	....	.....	350	....	350	350
Arkansas	96	306	....	402	....	231	....	32	263	665
Colorado	28	9,227	4,745	14,000	....	870	....	....	870	14,870
Georgia	....	1	....	1	....	....	....	....	....	1
Illinois	....	53,442	....	53,442	....	12,223	....	....	12,223	65,665
Indiana	....	8,949	....	8,949	....	1,674	....	....	1,674	10,623
Iowa	....	2,885	....	2,885	....	....	....	....	....	2,885
Kansas	....	....	....	....	....	1,388	....	....	1,388	1,388
Kentucky, East	....	9,467	....	9,467	....	3,450	....	....	3,450	12,917
Kentucky, West	....	8,720	....	8,720	....	3,904	....	....	3,904	12,624
Maryland	....	902	....	902	....	146	....	....	146	1,048
Michigan	....	118	....	118	....	1	....	....	1	119
Missouri	....	6,074	....	6,074	....	3,414	....	....	3,414	9,488
Montana	....	1,384	63,781	65,165	....	....	35,431	7,131	42,562	107,727
New Mexico	2	1,527	607	2,136	....	250	2,008	....	2,258	4,394
North Carolina	....	31	....	31	....	....	....	....	....	31
North Dakota	....	....	....	....	....	....	....	16,003	16,003	16,003
Ohio	....	17,423	....	17,423	....	3,654	....	....	3,654	21,077
Oklahoma	....	860	....	860	....	434	....	....	434	1,294
Oregon	....	....	1	1	....	....	....	....	....	1
Pennsylvania	7,030	22,789	....	29,819	90	1,091	....	....	1,181	31,000
South Dakota	....	....	....	....	....	....	....	428	428	428
Tennessee	....	667	....	667	....	320	....	....	320	987
Texas	....	....	....	....	....	....	....	3,272	3,272	3,272
Utah	....	3,780	....	3,780	....	262	....	....	262	4,042
Virginia	138	2,833	....	2,971	....	679	....	....	679	3,650
Washington	....	251	1,195	1,446	....	....	500	8	508	1,954
West Virginia	....	34,378	....	34,378	....	5,212	....	....	5,212	39,590
Wyoming	.....	4,524	23,030	27,554	....	....	23,674	....	23,674	51,228
Total	7,294	192,336	97,605	297,235	90	40,561	67,865	28,197	136,713	433,948

Note: The U.S. Geological Survey has identified 1.6 trillion tons of coal deposits at depths of less than 3,000 feet; it estimates about an equal amount exists in deeper deposits or in unexplored areas. The table above represents the amount of these deposits in relatively thick beds and near enough to the surface to be mined by conventional surface or underground methods. The standard for thickness is 28 inches or more for bituminous coal and anthracite; 60 inches or more for subbituminous and lignite. Maximum depth for all ranks is 1,000 feet, except for lignite; this table includes only lignite that can be surface mined, generally at depths no greater than 120 feet. Within these limits, this table represents the amount of coal in place. The portion that can be recovered ranges from 40 to 90 per cent depending on geology, law and the mining method; a rule of thumb is that at least half the coal in place, nationwide, can be recovered.

[1] Includes measured and indicated categories as defined by the USBM and USGS and represents 100% of the coal inplace.

[2] Only lignite beds that can be mined from surface are included in this table.

Source: U.S. Bureau of Mines

Source: *Coal Facts 1974-1975*, National Coal Association, Washington, D.C.

Strippable Reserves of Coal
(Millions of tons)

Rank	Remaining * Strippable Resources	Strippable ** Reserves	Strippable Reserves		
			Low Sulfur (less than 1%)	Medium Sulfur (1 - 2%)	High Sulfur (over 2%)
BITUMINOUS COAL					
Eastern province—Appalachian region:					
Alabama	667	134	33	74	27
Kentucky—east	4,609	781	532	189	60
Maryland	150	21	0	8	13
Ohio	5,566	1,033	0	126	907
Pennsylvania	2,272	752	0	225	527
Tennessee	483	74	5	43	26
Virginia	1,556	258	154	99	5
West Virginia	11,230	2,118	1,138	669	311
Subtotal	26,533	5,171	1,862	1,433	1,876
Interior and Gulf provinces:[1]					
Arkansas	200	149	3	118	28
Illinois	18,845	3,247	0	80	3,167
Indiana	2,741	1,096	0	293	803
Iowa	1,000	180	0	0	180
Kansas	1,388	375	0	0	375
Kentucky—west	4,746	977	0	0	977
Michigan	6	1	0	0	1[2]
Missouri	3,425	1,160	0	0	1,160
Oklahoma	434	111	10	44	57
Subtotal	32,785	7,296	13	535	6,748
Rocky Mountain and Northern Great Plains provinces:[3]					
Colorado	870	500	476	24	0
Utah	252	150	6	136	8
Subtotal	1,122	650	482	160	8
Alaska	1,201	480	480[4]	0	0
Total bituminous	62,826	13,597	2,837	2,128	8,632
SUBBITUMINOUS COAL					
Rocky Mountain and Northern Great Plains provinces:[5]					
Arizona	400	387	387	0	0
Montana	7,813	3,400	3,176	224	0
New Mexico	3,307	2,474	2,474	0	0
Wyoming	22,028	13,971	13,377	65	529
Subtotal	33,548	20,232	19,414	289	529
Pacific Coast province:[6]					
California	100	25	0	0	25
Oregon	0	0	0	0	0
Washington	500	135	135	0	0
Subtotal	600	160	135	0	25
Alaska	6,190	3,926[4,7]	3,926[4,7]	0	0
Total subbituminous	40,338	24,318	23,475	289	554
LIGNITE					
Interior and Gulf provinces:[8]					
Arkansas	32	25	25	0	0
Texas	3,272	1,309	625	684	0
Subtotal	3,304	1,334	650	684	0
Rocky Mountain and Northern Great Plains provinces:					
Montana	7,058	3,497	2,957	540	0
North Dakota	5,239	2,075	1,678	397	0
South Dakota	399	160	160	0	0
Subtotal	12,696	5,732	4,795	937	0

Strippable Reserves of Coal (*Continued*)

(Millions of tons)

Rank	Remaining* Strippable Resources	Strippable** Reserves	Strippable Reserves		
			Low Sulfur (less than 1%)	Medium Sulfur (1 - 2%)	High Sulfur (over 2%)
Alaska	8	5	5	0	0
Total lignite	16,008	7,071	5,450	1,621	0
Grand total United States	117,987	44,986	31,762	4,038	9,186

*"Remaining strippable resources" means coal still in the ground in beds of at least minimum thickness and under less than maximum overburden. "Resource" figures do not consider whether the coal can be mined economically. Seam thicknesses are generally at least 28 in. for bituminous coal (12 to 18 in the Midwest) and 60 in. for Texas, the Rocky Mountain, Northern Great Plains and Pacific Coast provinces. Overburden is limited to 120 ft. in the East, 90-150 ft. in the Midwest and (depending on the coalbed) 50-250 ft. in the rest of the country.
** "Strippable reserves" means that part of the strippable resources which can be economically recovered with existing strip-mining technology. This is generally determined by the "stripping ratio," which is the number of feet of overburden per foot of coalbed thickness. The strippable reserve figures exclude coal which cannot be mined because of topography, natural and manmade features such as rivers and towns, and other limitations.
1. Bituminous coal resource and reserve not estimated for Texas and Nebraska. 2. There may be isolated areas of some seams which might be classed in the medium-sulfur category.
3. Bituminous coal resource and reserve not estimated for Montana, New Mexico, Idaho, and Wyoming. 4. 478 million tons of bituminous and 3,387 million tons of subbituminous coal reserves in the northern Alaska fields (North Slope) are included in the estimates even though an economic export market, which is essential for exploitation, does not currently exist. 5. Subbituminous coal resource and reserve not estimated for Colorado. 6. Bituminous coal resource and reserve not estimated for Washington; the subbituminous coal resource and reserve estimated for Oregon is less than 1 million tons. 7. Includes 179 million tons of undifferentiated subbituminous coal and lignite. 8. Lignite resource and reserve not estimated for Kansas, Mississippi, Louisiana, and Alabama.
Source: Bureau of Mines, Information Circular 8531

Source: Coal Facts 1974-1975, National Coal Association, Washington, D.C.

Coal cutting machine cuts a slot 6 inches high and 11 feet deep in preparation for blasting. (*Courtesy Mining Equipment Division, Westinghouse Air Brake Co.*)

Aerial view of old spoil piles left 20–30 years ago after strip mining by Burlington-Northern near Colstrip, Montana. Power plant under construction.

Source: EPA–Documerica–Boyd Norton.

ESTIMATES OF UNDISCOVERED U.S. OIL AND GAS RESOURCES

Estimation of Oil and Gas Resources. The range of estimates for undiscovered oil and gas resources shown in the accompanying tables illustrates the difficulty in such predictions. The latest estimates of May 1975 of the U.S. Geological Survey (item 10 in the table) do not appear in the charts and indicate a reduction by a factor of three of estimates made by the USGS the previous year.

Estimates of Undiscovered Recoverable Resources of Oil and Natural Gas of the United States

	Oil and Natural Gas Liquids		Natural Gas	
	Billion (10^9) Tonnes	Billion (10^9) Barrels	Trillion (10^{12}) Cubic Meters	Trillion (10^{12}) Cubic Feet
Oil Companies				
1. Company A. (Weeks, 1960)	22.3	168	—	—
2. Company B. (Hubbert, 1967)	3.2–8.5[a]	24–64[a]	5–14[a]	180–500[a]
3. Company C (1973)	7.3[b]	55	—	—
4. Company D (1974)	11.9	89	12.6	450
5. Company E	12	90	—	—
U.S. Geological Survey				
6. Hendricks (1965)	46	346	37	1,300
7. Theobald et al. (1972)	61	458	56	1,980
8. McKelvey (1974)	27–54	200–400	28–57	990–2,000
9. Hubbert (1974)	9.6	72	15.3	540
10. USGS (1975)	8.1–20	61–149	9.1–18.6	322–655
National Academy of Sciences				
11. National Research Council (1975)	15	113	15	530

[a]Exclusive of Alaska.
[b]Estimated discoverable between 1973 and 1985.
Reproduced with permission of the National Academy of Sciences.

Asphalto Oil Field, Texas. (*Courtesy American Petroleum Institute Photo Library*)

Alternative Estimates of U.S. Undiscovered Natural Gas Resources

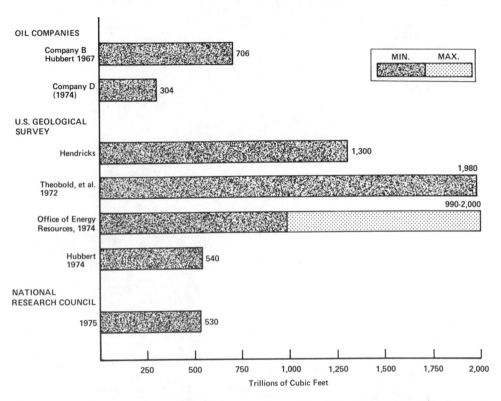

Source: _Mineral Resources and The Environment_, National
Academy of Sciences, 1975, page 89.

Source: Energy Perspectives, U.S. Department of the Interior, February 1975.

Alternative Estimates of U.S. Undiscovered Recoverable Petroleum Resources

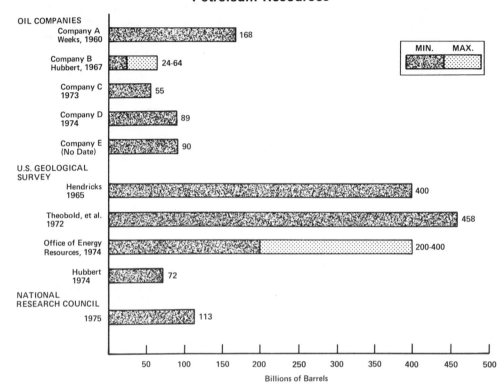

Source: _Mineral Resources and The Environment_, National
Academy of Sciences, 1975, page 89.

Source: Energy Perspectives, U.S. Department of the Interior, February 1975.

U.S. PETROLEUM RESOURCES

U.S. Petroleum Resources, December 31, 1974
[Billion bbl]

| Region | Reserves | | | | Undiscovered recoverable resources | |
| | Demonstrated | | Inferred | Total | Statistical mean | Estimated range[1] |
	Measured	Indicated				
Onshore crude oil:						
Alaska	9.9	N	6.1	16.0	12	6-19
Lower 48	21.1	4.3	14.3	39.7	44	29-64
Total	31.0	4.3	20.4	55.7	56	37-81
Offshore crude oil:						
Alaska	.2	N	.1	.3	15	3-31
Lower 48	3.1	.3	2.6	6.0	11	5-18
Total	3.3	.3	2.7	6.3	26	10-49
Total, United States	[2]34.3	4.6	23.1	62.0	82	50-127
Total natural gas liquids	6.4	—	6.0	12.4		11-22

[1] The low value of the range is the quantity associated with a 95-percent probability that there is at least this amount. The high value is the quantity with a 5-percent probability that there is at least this amount. Totals for the low and high values must be derived by statistical methods; they are *not* obtained by arithmetic summation.

[2] The American Petroleum Institute estimated U.S. measured crude oil reserves at approximately 32.7 billion barrels as of December 31, 1975.

Note.—As of December 31, 1974. N = negligible.

Source: Miller, Betty M., et al. *Geological Estimates of Undiscovered Recoverable Oil and Gas Resources in the United States*, U.S. Geological Survey Circular 725, National Center, Reston, Va., 1975.

Source: Energy Perspectives **2**, U.S. Department of the Interior, June 1976.

U.S. Petroleum Data

Energy content	5,800,000 Btu per barrel
	138,000 Btu per gallon
U.S. Petroleum resources	
measured reserves	48.3 billion barrels
indicated reserves	25–45 billion barrels
undiscovered resources	61–149 billion barrels
total resources	134–242 billion barrels
total energy content	0.8–1.4 Q (10^{18} Btu)
Total U.S. energy consumption (1974)	0.072 Q (10^{18} Btu)
U.S. Petroleum consumption (1974)	6.04 billion barrels
	(0.035 Q)
U.S. Petroleum consumption by sector (1973)	
electric power generation	9.0%
transportation	53%
industrial	10%
household and commercial	16%
nonfuel use	12%

Oil Well in West Texas. (*Courtesy of Amoco Oil Co. and the American Petroleum Institute Photo Library*)

(LL-14) Engineers brave cold to check line of "Christmas Trees" on Prudhoe Bay. These control values mark the vast stores of energy on Alaska's North Slope. (available in color) (*Courtesy Michigan Consolidated Gas Company*)

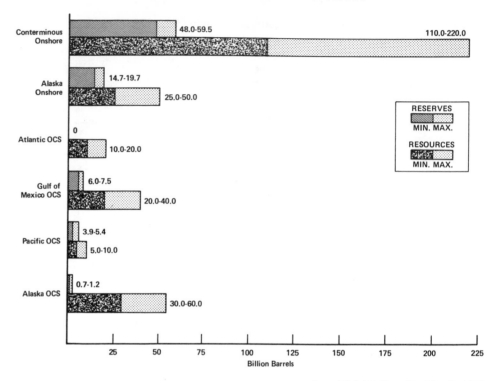

U.S. Recoverable Petroleum Resources

Conterminous Onshore 48.0-59.5   110.0-220.0

Alaska Onshore 14.7-19.7   25.0-50.0

Atlantic OCS 0   10.0-20.0

Gulf of Mexico OCS 6.0-7.5   20.0-40.0

Pacific OCS 3.9-5.4   5.0-10.0

Alaska OCS 0.7-1.2   30.0-60.0

RESERVES
MIN. MAX.
RESOURCES
MIN. MAX.

25 50 75 100 125 150 175 200 225
Billion Barrels

Source: U.S. Geological Survey, News Release, March 1974.

Source: Energy Perspectives, U.S. Department of the Interior, February 1975.

U.S. NAVAL PETROLEUM RESERVES

Naval Petroleum Reserves. During the period 1912 to 1924 four areas were set aside by Presidential Executive Orders to maintain petroleum resources in a standby production status until needed for national defense requirements. Naval Petroleum Reserves (NPR) Nos. 1 and 2 are located west of Bakersfield, California. NPR-1 is presently producing about 4000 barrels per day and production could be increased to about 160,000 barrels per day within 90 days. NPR-2 has been under lease to private operators for about 50 years and is currently producing about 7900 barrels per day. Production could be increased to 50,000 barrels per day within 90 days. NPR-3, located northeast of Casper, Wyoming, is producing 500 barrels per day and could produce 5000 barrels per day within 90 days. NPR-4 is located on the north slope of Alaska near the Prudhoe Bay field. NPR-4 has not been fully explored or developed and no oil is being produced and NPR-4 does not have, therefore, a standby producing capacity.

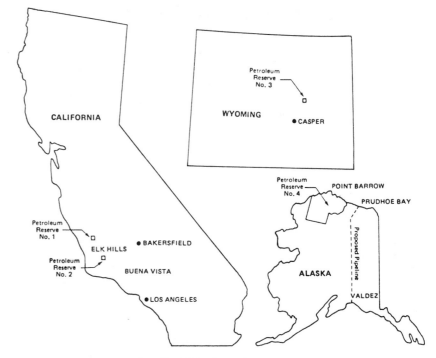

Location of naval petroleum reserves.

Source: Strategic Petroleum Reserves, Hearing before the Committee on Interior and Insular Affairs, United States Senate, May 30 and July 26, 1973. Serial No. 93-11 (92-46).

Proven and Estimated Reserves of the Naval Petroleum Reserves

	Oil		Gas	
	Proved Reserves (billion bbls)	Estimated Additional Reserves (billions bbls)	Proved Reserves (trillions cubic feet)	Estimated Additional Reserves (trillions cubic feet)
NPR-1, Bakersfield	1.300	0.482	1,000	300
NPR-2, Bakersfield	0.051	—	50	—
NPR-3, Wyoming	0.050	—	8	—
NPR-4, Alaska	0.100	15.000	1,000	80,000

Source: Strategic Petroleum Reserves, Hearings before the Committee on Interior and Insular Affairs, U.S. Senate, May 30 and July 26, 1973, Serial No. 93-11 (92-46).

U.S. Petroleum Production, 1960-74
(Millions of Barrels)

	Onshore	Federal OCS	Production Total
1960	2,862.9	52.2	2,915.1
1961	2,916.2	67.3	2,983.5
1962	2,954.7	94.2	3,048.9
1963	3,043.5	110.1	3,153.6
1964	3,080.8	128.5	3,209.3
1965	3,138.6	151.5	3,290.1
1966	3,297.7	198.7	3,496.4
1967	3,497.3	232.9	3,730.2
1968	3,595.3	284.0	3,879.3
1969	3,619.1	332.9	3,952.0
1970	3,737.1	386.2	4,123.3
1971	3,616.3	455.4	4,071.7
1972	3,640.3	453.3	4,093.6
1973	3,562.0	433.6	3,995.3
1974	3,425.8	424.2	3,850.0

Cumulative production of petroleum in the United States between 1960 and 1974 was 53.8 billion barrels, an amount roughly equal to the remaining proved measured reserves (48.3 billion barrels).

Comestic petroleum production increased at an annual rate of 3.5 percent between 1960 and 1970. Since 1970, domestic production has declined at an annual rate of 0.6 percent.

Reference: "Petroleum," from *Mineral Facts and Problems, 1970 Edition*, Bureau of Mines Bulletin 650.

Source: Energy Perspectives, U.S. Department of the Interior, February 1975.

U.S. Petroleum Supply, 1960-74
(Billions of Barrels)

YEAR	DOMESTIC PRODUCTION[1]	NET IMPORTS PRODUCTS	NET IMPORTS CRUDE OIL	TOTAL
1960	2.92	.22	.37	3.51
1965	3.29	.38	.45	4.12
1970	4.12	.68	.48	5.28
1971	4.07	.74	.61	5.42
1972	4.09	.84	.81	5.74
1973	4.00	1.00	1.18	6.18
1974	3.85	.94	1.25	6.04

[1] Includes natural gas liquids.

Net imports of crude oil and petroleum products increased at a 9.6 percent annual rate between 1960 and 1970 and at a 17.2 percent annual rate since 1970.

Source: Energy Perspectives, U.S. Department of the Interior, February 1975, page 86.

The 270,000 deadweight-ton **CONOCO EUROPE** carries over 2 million barrels of oil from the Middle East to various points in northwest Europe. The ship, manned by a crew of 36, travels at 16 knots. (Continental Oil Co.)

U.S. Petroleum Production, 1960-74
(Millions of Barrels)

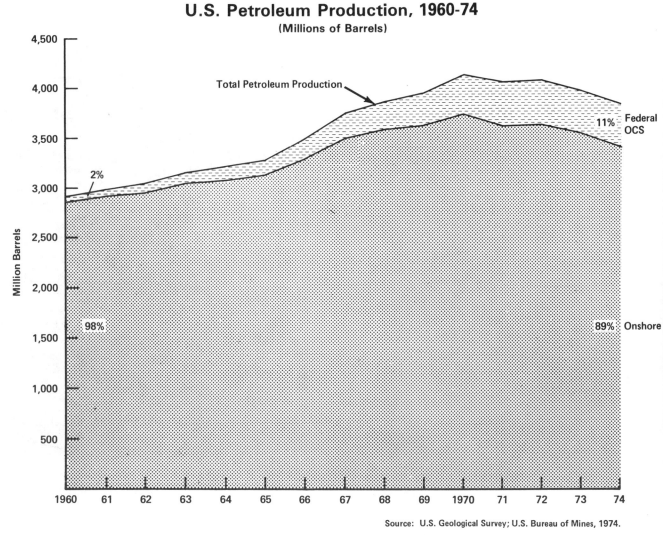

Total Petroleum Production

2%

98%

11% Federal OCS

89% Onshore

Million Barrels

4,500 — 4,000 — 3,500 — 3,000 — 2,500 — 2,000 — 1,500 — 1,000 — 500

1960 61 62 63 64 65 66 67 68 69 1970 71 72 73 74

Source: U.S. Geological Survey; U.S. Bureau of Mines, 1974.

Source: Energy Perspectives, U.S. Department of the Interior, February 1975.

Production facilities at Yates Field, Texas, one of the largest oil reserves in the United States. (Marathon Oil Company.)

U.S. Petroleum Supply, 1960-74

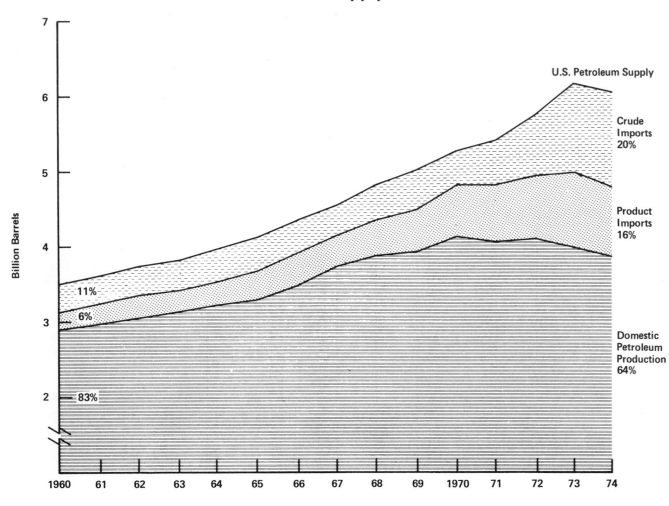

U.S. Petroleum Supply

Crude
Imports
20%

Product
Imports
16%

Domestic
Petroleum
Production
64%

Billion Barrels

11%

6%

83%

1960 61 62 63 64 65 66 67 68 69 1970 71 72 73 74

Source: U.S. Bureau of Mines, 1974.

Source: Energy Perspectives, U.S. Department of the Interior, February 1975.

Horsehead pump at one of the 200 wells requiring pumping at Yates Field, Texas. (Marathon Oil Company)

U.S. NATURAL GAS RESOURCES 1974

U.S. Gas Resources. The significant proved reserves of natural gas in the United States are found in Alaska, Louisiana, and Texas, both onshore and offshore. Sizable undiscovered resources are expected to be found in these same areas as well as the outer continental shelf along the Atlantic seacoast. Estimates of measured and indicated-inferred reserves are in the range of 400-500 trillion cubic feet indicating that present reserves will be depleted in 17 to 22 years at current levels of consumption unless additional gas is discovered.

Estimates of undiscovered recoverable resources of gas vary from 180 to 2000 trillion cubic feet. The latest estimate of the National Academy of Science is 530 trillion cubic feet and of the U.S. Geological Service is 322-655 trillion cubic feet. Total gas resources will therefore be depleted in 30-50 years at current levels of consumption if these estimates prove to be correct.

U.S. Recoverable Natural Gas Resources as of January 1, 1975
[Trillion ft^3]

Region	Reserves			Undiscovered recoverable resources	
	Measured	Inferred	Total reserves	Statistical mean	Estimated range[1]
Onshore:					
Alaska	31.7	14.7	46.4	32	16-57
Lower 48	169.5	119.4	288.9	345	246-453
Total	201.2	134.1	335.3	377	264-506
Offshore:					
Alaska	.1	.1	.2	44	8-80
Lower 48	35.8	67.4	103.2	63	26-111
Total	[2]35.9	67.5	103.4	107	42-181
Total, United States	237.1	201.6	438.7	484	322-655

[1] The low value of the range is the quantity associated with a 95-percent probability that there is at least this amount. The high value is the quantity with a 5-percent probability that there is at least this amount. Totals for the low and high values must be derived by statistical methods; they are *not* obtained by arithmetic summation.

[2] The American Gas Association estimated measured, natural gas reserves at 228.2 trillion ft^3 as of December 31, 1975.

Note.—As of December 31, 1974.

Source: Miller, Betty, M., et al. *Geological Estimates of Undiscovered Recoverable Oil and Gas Resources in the United States*, U.S. Geological Survey Circular 725, National Center, Reston, Va., 1975.

Source: Energy Perspectives **2,** U.S. Department of the Interior, June 1976.

Offshore rigs such as this one in the Gulf of Mexico provide 22 percent of domestic natural gas production. (*Courtesy American Petroleum Institute Photo Library*)

U.S. Recoverable Natural Gas Resources

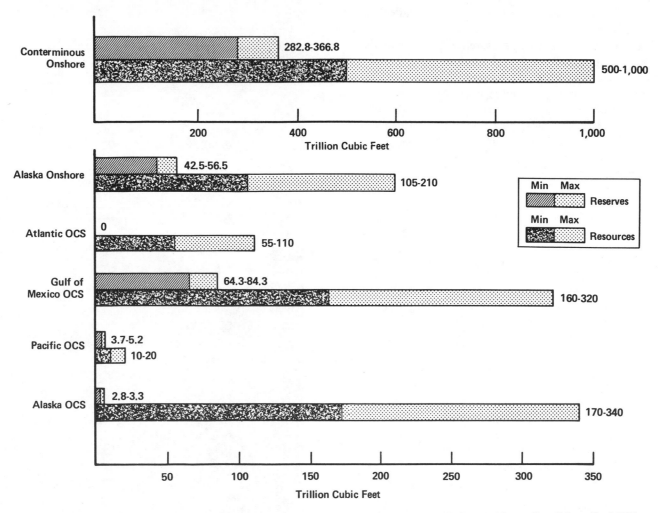

Source: U.S. Geological Survey, News Release, March 1974.

Source: Energy Perspectives, U.S. Department of the Interior, February 1975.

U.S. NATURAL GAS RESERVES 1947-73

Proved Gas Reserves. The extent of proved reserves of natural gas in the United States increased steadily until 1967 to a level of about 290 trillion cubic feet. Except for a one-year increase in 1970 due to the addition of Alaskan reserves, the total reserves have since declined to a level near 265 trillion cubic feet. This decline has occurred since annual additions of newly discovered gas have been only a small fraction of the annual consumption (in 1973 some 23 trillion cubic feet were consumed and only 4 trillion cubic feet were found). Although drilling rates for exploration wells increased 50% between 1966 and 1973, the "finding rate" for new gas fell from 662 thousand cubic feet per foot of exploratory drilling in 1966 to 104 thousand cubic feet in 1973.

U.S. NATURAL GAS RESERVES - ADDITIONS - PRODUCTION*

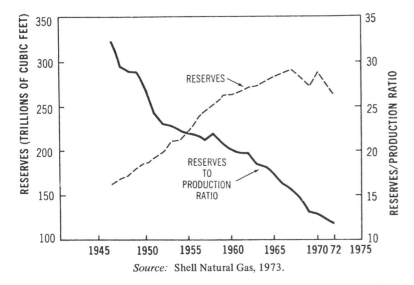

Source: Shell Natural Gas, 1973.

Estimated Total Proved Reserves of Natural Gas in the United States
(Millions of Cubic Feet—14.73 psia, at 60°F.)

State	Production	Reserves as of December 31, 1973			
		Total Gas	Nonassociated	Associated-Dissolved	Underground Storage
Alabama	7,892	327,375	309,338	18,037	0
Alaska	130,815	31,642,626	5,210,166	26,432,460	0
Arkansas	163,586	2,269,353	2,082,558	152,294	34,501
California	478,356	5,199,837	2,380,028	2,540,137	279,672
Colorado	134,276	1,868,299	1,592,362	250,904	25,033
Florida	33,660	148,914	0	148,914	0
Illinois	2,840	380,525	1,103	24,195	355,227
Indiana	1,364	66,682	2,178	3,789	60,715
Kansas	899,460	11,722,395	11,411,765	203,533	107,097
Kentucky	62,396	864,921	709,030	43,187	112,704
Louisiana	8,457,596	69,151,613	57,239,668	11,732,832	179,113
Michigan	47,842	1,548,508	492,825	431,086	624,597
Mississippi	96,657	1,178,218	968,414	120,711	89,093
Montana	60,209	1,092,449	821,513	85,625	185,311
Nebraska	4,446	48,816	13,779	8,795	26,242
New Mexico	1,194,706	12,488,363	9,814,816	2,657,246	16,301
New York	4,283	136,842	31,972	70	104,800
North Dakota	37,099	448,184	6,372	441,812	0
Ohio	89,527	1,179,391	649,260	159,383	370,748
Oklahoma	1,777,787	14,098,735	11,183,035	2,675,836	239,864
Pennsylvania	78,514	1,494,381	876,818	12,050	605,513
Texas	8,240,478	84,936,502	60,530,423	24,268,979	137,100
Utah	51,029	1,024,723	541,376	481,697	1,650
Virginia	5,048	37,273	37,273	0	0
West Virginia	168,023	2,319,828	1,912,318	52,319	355,191
Wyoming	376,758	4,109,523	3,413,115	641,032	55,376
Miscellaneous	759	165,931	14,433	837	150,661
Total United States	22,605,406	249,950,207	172,245,938	73,587,760	4,116,509
Gulf of Mexico	4,164,421	36,785,308	32,092,008	4,693,300	0

Source: American Gas Association Monthly, May 1974.

U.S. NATURAL GAS RESERVES

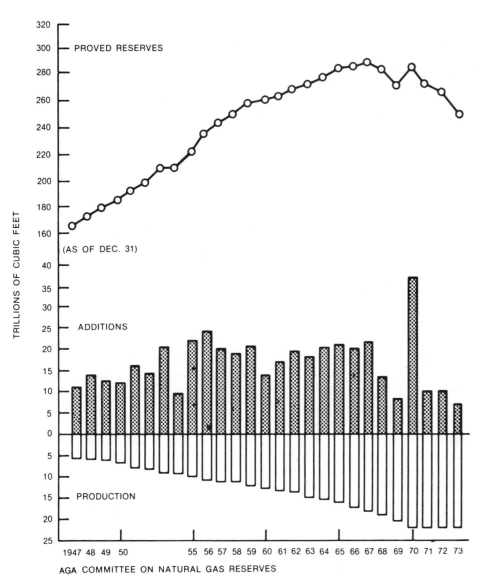

AGA COMMITTEE ON NATURAL GAS RESERVES

Source: Project Independence Report, United States Federal Energy Administration, November 1974.

Proved Reserves of Natural Gas in the United States, 1959-76
(Trillions of cubic feet; 14.73 PSIA at 60°F)

Year	Proved Reserves at Beginning of Year	Proved Reserves at End of Year	Net Change From Previous Year
1959	252.8	261.2	+8.4
1960	261.2	262.3	+1.1
1961	262.3	266.3	+4.0
1962	266.3	272.3	+6.0
1963	272.3	276.2	+3.9
1964	276.2	281.3	+5.1
1965	281.3	286.5	+5.2
1966	286.5	289.3	+2.8
1967	289.3	292.9	+3.6
1968	292.9	287.4	-5.5
1969	287.4	275.1	-12.3
1970	275.1	290.8	+15.7
1971	290.8	278.8	-12.0
1972	278.8	266.1	-12.7
1973	266.1	250.0	-16.1
1974	250.0	237.1	-12.9
1975	237.1	228.2	-8.9
1976	228.2	216.0	-12.2

Note: 1970 figures reflect the addition of Prudhoe Bay, Alaska reserves.

Source: Reserves of crude oil, natural gas liquids, and natural gas in the United States and Canada as of December 31, 1976. Joint publication by the American Gas Association, American Petroleum Institute, and Canadian Petroleum Association. Vol. 31, May 1977.

From *Status Report on the Gas Potential from Devonian Shales of the Appalachian Basin*, Office of Technology Assessment, Congress of the United States, Washington, D.C., November 1977.

U.S. NATURAL GAS SUPPLY 1960-74

U.S. Gas Data

Energy content	1030 Btu per cubic foot
Gas resources	
measured reserves	266.1 trillion cubic feet
indicated reserves	130-250 trillion cubic feet
undiscovered resources	322-655 trillion cubic feet
total gas resources	718-1171 trillion cubic feet
total energy content	0.74-1.14 Q (10^{18} Btu)
U.S. Gas consumption (1974)	22.8 trillion cubic feet
Total U.S. energy consumption (1974)	0.072 Q (10^{18} Btu)
U.S. Gas consumption (1973)	
household and commercial	34%
industrial	43%
transportation	3.5%
electric generation	16.5%
nonfuel use	3.0%
Average annual increase in gas consumption 1960-74	4.1%

U.S. Natural Gas Supply, 1960-74
(Trillion Cubic Feet)

	YEAR				ESTIMATED 1974
	1960	1965	1970	1973	
DOMESTIC PRODUCTION [1]	12.77	16.04	21.92	22.65	21.90
NET IMPORTS	.14	.43	.75	.96	.86
NATURAL GAS SUPPLY	12.91	16.47	22.67	23.61	22.76

[1] Natural gas production refers to marketed production, namely gross withdrawals less gas used for reprocessing and quantities vented and flared.

Source: Energy Perspectives, U.S. Department of the Interior, February 1975.

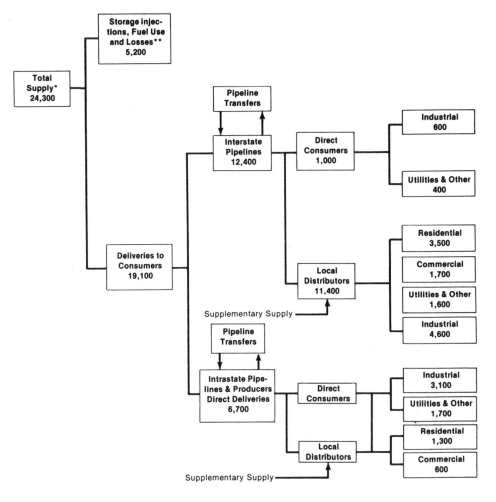

*Supply includes U.S. marketed production, withdrawals from storage, and imports.
**Gas for such purposes as lease and plant fuel, pipeline compressor fuel, extraction loss, and transmission losses.

Note Divisions between interstate and intrastate volumes are estimated.

Source: Based primarily on data from "Natural Gas Production and Consumption: 1974" (Washington, DC: Bureau of Mines, Mineral Industry Surveys, 1975).

Overview—U.S. Natural Gas System 1974 (Bcf)

Source: National Energy Outlook, FEA-N-75/713, Federal Energy Administration, February 1976.

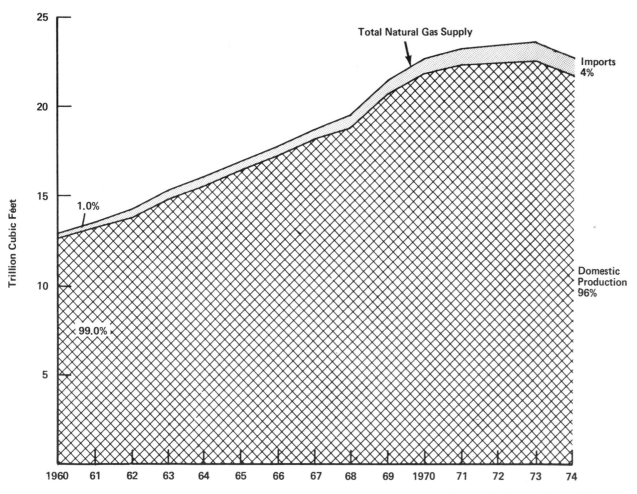

U.S. Natural Gas Supply, 1960-74

Total Natural Gas Supply

Imports 4%

1.0%

Domestic Production 96%

99.0%

Trillion Cubic Feet

1960 61 62 63 64 65 66 67 68 69 1970 71 72 73 74

Source: U.S. Bureau of Mines, 1975;
Project Independence Report, 1974.

Source: Energy Perspectives, U.S. Department of the Interior, February 1975.

Facility for off-loading of liquified natural gas. (*Courtesy of American Gas Association*)

U.S. SHALE OIL RESOURCES

U.S. Shale Oil Deposits. Shales containing oil occur in at least 30 states, including Alaska. The most important deposits occur in the Green River Formation in the states of Colorado, Utah, and Wyoming, where oil content measures 25 gallons per ton in many deposits. The Devonian "black" shales of the central United States extending from Texas to New York and Alabama to Michigan, are a potentially large energy resource with oil content in the 10 gallon per ton range.

The Green River oil shale formation extends over some 16,500 square miles with the richest and thickest deposits being found in the Piceance basin in Colorado. Although the shales at the center of this basin are buried under 1000 feet of overburden, the shale deposits are almost 2000 feet thick and have an average assay of 25 gallons per ton. The outcrop of this deposit along the Colorado River is, however, only 100 feet thick.

The shales in this formation were laid down during the Eocene period about 50 million years ago. Deposits of marine organisms and sediments at the bottoms of lakes were subsequently compacted into finely structured shales which have little pore volume or permeability.

REFERENCES

Noel de Nevers, "Tar sands and oil shales," *Sci. Amer.*, vol. 214, no. 2, February 1966.

Gerald V. Dinneen and Glenn L. Cook, "Oil shale and the energy crisis," *Technol. Rev.*, January 1974, pp. 27–33.

Shale Oil, from *Mineral Facts and Problems, 1970 Edition*, Bureau of Mines Bulletin 650, U.S. Department of the Interior.

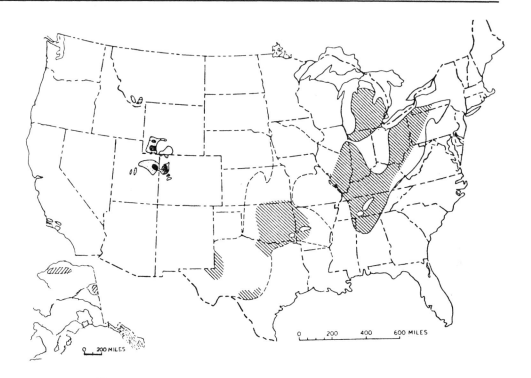

Tertiary deposits Green River Formation in Colorado, Utah, and Wyoming; Monterey Formation, California; middle Tertiary deposits in Montana. Black areas are known high-grade deposits

Mesozoic deposits Marine shale in Alaska

Permian deposits Phosphoria Formation, Montana

Devonian and Mississippian deposits (resource estimates included for hachured areas only in Geological Survey Circular 523). Boundary dashed where concealed or where location is uncertain.

Source: Final Environmental Statement for the Prototype Oil Shale Leasing Program (USDI), 1973, p. 11-5.

Energy Content of Oil Shale Resources of the United States
[Quadrillion Btu]

Deposit	Identified 25-100 gal/ton	Identified 10-25 gal/ton	Hypothetical 25-100 gal/ton	Hypothetical 10-25 gal/ton	Speculative 25-100 gal/ton	Speculative 10-25 gal/ton
Green River formation: Colorado, Utah and Wyoming	2,424	8,120	290	3,480	–	–
Chattanooga shale and equivalent formation: central and eastern United States	–	1,160	–	4,640	–	–
Marine shale: Alaska	Small	Small	1,450	1,160	–	–
Other shale oil deposits	–	Small	–	–	3,480	133,400
Total	2,424	9,280	1,740	9,280	3,480	133,400

Note.—The energy content of the discovered recoverable resource range would lie between 464 million and 1,160 million Btu. Energy content based on conversion factor of 1 bbl of shale oil = 5,800,000 Btu.

Source: Energy Perspectives **2**, U.S. Department of the Interior, June 1976.

U.S. Oil Shale Deposits, Green River Formation

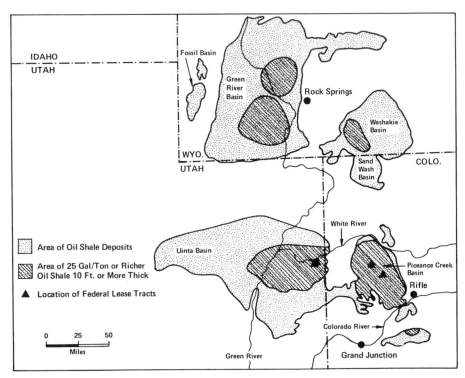

Area of Oil Shale Deposits

Area of 25 Gal/Ton or Richer
Oil Shale 10 Ft. or More Thick

▲ Location of Federal Lease Tracts

Source: Energy Perspectives, U.S. Department of the Interior, February 1975.

Exploration for oil shale in Colorado. (*Courtesy of American Petroleum Institute Photo Library*)

U.S. URANIUM RESOURCES

U.S. Uranium Resources. Uranium is one of the less common elements, comprising about 4 ppm of the earth's crust. It is more abundant than silver or gold but less abundant than nickel, copper, lead, and zinc. Uranium is widely distributed throughout the world but even major deposits are generally of low concentration (100–2000 ppm). It does not occur in the elemental state as do gold, silver, and copper, but as a mineral, most commonly as uranium dioxide (UO_2).

The major sources of uranium ore in the United States are found in sedimentary strata, especially in the Colorado plateau. Uranium is also found in igneous rocks, although the concentration is much lower. Granite deposits in New Hampshire average about 12 ppm uranium. The Chattanooga shales, covering some 40,000 square miles in Tennessee, Kentucky, and Alabama, average about 60 ppm uranium.

Estimates of the extent of recoverable uranium resources in the United States are dependent upon the extraction costs deemed to be economically acceptable. At current prices for uranium in yellowcake form (U_3O_8) of $15 per pound, the cost of uranium represents less than 10% of the cost of producing electric power and less than 5% of the delivered cost of electricity. As of January 1, 1975, reserves of U_3O_8 costing $30 per pound or less are estimated at 600,000 tons while additional potential resources at the same price level are estimated to be about 2,900,000 tons by the U.S. Energy Research and Development Administration. Estimates of uranium resources available at $100 per pound are from 13,200,000 to 28,900,000 pounds at 50% and 5% probability factors.[1]

---

[1] *Uranium Resources to Meet Long Term Uranium Requirements*, EPRI-SR-5, Electric Power Research Institute, Palo Alto, California, November 1974.

Prospector at work collecting uranium ore samples. (Anaconda, New Mexico Operations)

U.S. Uranium Resources
[Thousand tons of U_3O_8]

Operating and future capital cost level per pound of U_3O_8 (excluding profits)	Reserves	Potential resources		
		Probable	Possible	Speculative
$10	270	440	420	145
$15[1]	430	655	675	290
$30[1]	640	1,060	1,270	590
Byproduct 1975-2000[2]	140	—	—	—
Total	780	1,060	1,270	590

[1] Includes lower cost resources.
[2] Byproduct of phosphate and copper production.
Note.—Data as of January 1, 1976. Estimated operating and future capital costs used in calculating resource costs. **Profits and "sunk" costs not included.**
Source: U.S. Energy Research and Development Administration, *ERDA Issues Latest Estimate of U.S. Uranium Resources*, News Release No. 76-26, Apr. 2, 1976.

Source: Energy Perspectives 2, U.S. Department of the Interior, June 1976.

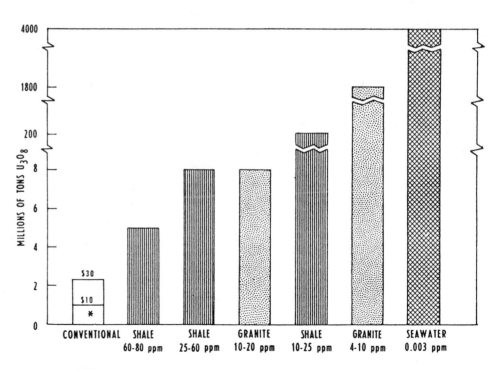

* 700 – 2100 ppm

U. S. URANIUM RESOURCES[6] AT $10 TO $? PER LB U_3O_8

Source: WASH 1243, "Nuclear Fuel Resources and Requirements," United States Atomic Energy Commision, April 1973.

Principal Prospecting Areas for Uranium and Thorium

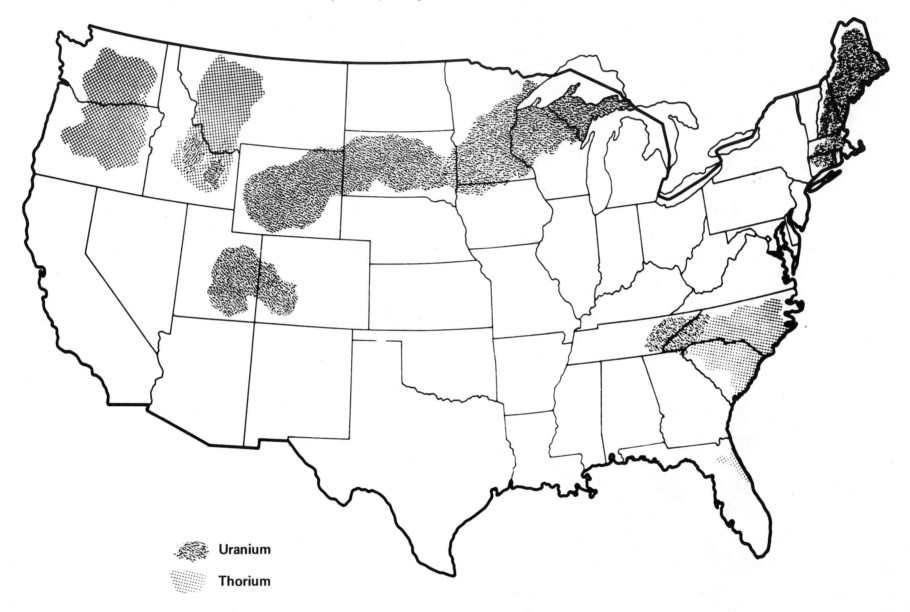

Projected Uranium Reserves and Resources—United States Only

[Thousands of short tons U_3O_8 [1] as of September 1974]

| | Identified [2][6] | | | Potential (undiscovered) [3] | | | |
| | Demonstrated | | | Probable known districts, productive formations | Possible productive provinces, productive formations | Speculative new provinces, new formations | Total, by index cost, identified and potential |
Measured	Indicated	Inferred					
Index costs, dollar per pound, U_3O_8: [4]							
Economic (1974): $8	Reserves, 280			300	210	30	820
Not yet economic:							
$8 to $10	60			160	190	80	490
$10 to $15	Reasonably, 180.			230	250	100	760
$15 to $30	Assured, 180			280	710	210	1,380
Total	700			970	1,360	420	[5]3,450
Decreasing degree of assurance:							
Unknown: $30 to $100: Host formations, other than sandstone (see table 4), 100–500 ppm U_3O_8.							??
Unlikely:							
$100 plus: Chattanooga shales, 60–80 ppm U_3O_8 (approximately).							5,000
$150 plus: Chattanooga shale, 25–60 ppm U_3O_8 (approximately).							8,000

[1] 1.3 short tons U_3O_8 = 1.18 metric tons U_3O_8 = 1,000 kilograms uranium.

[2] Ores to grades down to approximately 0.12 percent U_3O_8; approximately 95 percent from sandstone host rocks.

[3] Ores to grades down to approximately 0.10 percent U_3O_8, primarily from sandstone host, but including small contributions from other host formations such as veins, conglomerates and tuffaceous material at grades down to approximately 0.025 percent U_3O_8, where there are sufficient data to judge the possible quantity of uranium.

[4] The index cost is not the average cost of production. And more importantly, it is not the price at which uranium will be sold. See text for a discussion of index costs, projected actual costs and prices.

[5] This is a new total, approximately 1,200,000 short tons of U_3O_8 higher than Jan. 1, 1974 estimates, a result of the preliminary national uranium resource evaluation program (PNURE), started approximately 18 mos.

[6] There are other small domestic sources of uranium: minus 200,000 metric tons of depleted uranium tails, available to stock LMFBR's (sufficient for at least 2000—000 Mwe LMFBRs); minus 20,000 short tons of U_3O_8 recoverable from copper ore leach solutions between now and year 2000; minus 70,000 short tons of U_3O_8 recoverable from phosphoric acid made from Florida phosphate rock between now and year 2000; minus 2,000–3,000 short tons U_3O_8 per year by year 2000 from lignite gasification assuming 75 percent recovery of U_3O_8 and 20 percent of natural gas demand supplied from lignite (No production now planned.)

Source: Issues for Consideration—Review of National Breeder Program, Joint Committee on Atomic Energy; Congress of the United States, August 1975.

Uranium is found mostly in deposits of uraninite, coffinite, and brilliantly colored minerals irregularly disseminated in sandstones of the Western States. About 4 pounds of uranium oxide is contained in an average ton of ore mined. This photo shows a large backhoe excavating uranium ore from the open pit mine of the Petrotomics Company in Shirley Basin, Wyoming. The backhoe is better adapted to selective excavation of the ore than are larger excavators commonly used for removal of overburden. (Courtesy Petrotomics Company)

U.S. URANIUM PRODUCTION AND RESERVES 1947-74

U.S. Uranium Reserves. The figure of 277,000 tons of uranium oxide (U_3O_8) reserves indicated in the table of reserves 1947-73, was lowered to 200,000 tons in January 1975 by the Energy Research and Development Administration. This change reflected a shift of 77,000 tons from the category of uranium oxide costing $8 per pound or less to the category of higher cost reserves.

Over the years, the finding rate, or gross additions to reserves, has corresponded to the intensity of exploratory drilling. Drilling activity increased steadily during the early 1950s to meet defense requirements and then dropped as these requirements were filled and incentives were dropped. The second peak in drilling activity occurred as commercial nuclear power plants began to be ordered in substantial numbers in the late 1960s. Drilling effort dropped again, however, because of market uncertainties and a generally depressed price level for uranium oxide in the $5-6 per pound range.

Uranium Production—1974

Uranium ore production (ore weighed and sampled by mills) during 1974 by states was:

Uranium Ore Production—1974

State	Tons of Ore (Millions)	Contained Tons of U_3O_8	% of Total (U_3O_8)
New Mexico	3.0	5,400	43
Wyoming	2.5	4,000	32
Others (Colo., Texas, Utah, and Wash.)	1.6	3,200	25
Total	7.1	12,600	100

Concentrate production was 11,500 tons of U_3O_8 in 1974 compared to 13,200 tons in 1973. Concentrate production by state was:

Uranium Concentrate Production—1974

State	Tons of U_3O_8
New Mexico	4,900
Wyoming	3,800
Others (Colo., Texas, Utah, and Wash.)	2,800
Total	11,500

Uranium Ore Reserves and Production
1947 Through 1972—Tons U_3O_8 in Ore

Year End	Shipment To Mills[a]	Cumulative Production	Reserve Estimation[b]	Sum of Cumulative Production and Reserves
1947	—	—	2,200	2,200
1948	83	83	2,200	2,283
1949	502	585	2,200	2,785
1950	810	1,395	3,000	4,395
1951	1,088	2,483	5,800	8,283
1952	1,288	3,771	7,346	11,117
1953	2,315	6,086	15,203	21,289
1954	3,539	9,625	27,582	37,207
1955	4,425	14,050	67,595	81,645
1956	8,434	22,484	120,240	142,724
1957	9,837	32,321	166,300	198,621
1958	14,003	46,324	181,800	228,124
1959	17,377	63,701	197,100	260,801
1960	18,842	82,543	187,100	270,443
1961	18,513	101,056	174,200	275,256
1962	17,085	118,141	166,200	284,341
1963	14,721	132,862	160,231	293,093
1964	13,888	146,750	150,927	297,677
1965	10,578	157,328	144,702	302,030
1966	10,051	167,379	140,835	308,214
1967	10,866	178,245	147,741	325,986
1968	12,850	191,095	160,819	351,914
1969	12,595	203,690	204,080	407,770
1970	13,073	216,763	246,100	462,863
1971	13,089	229,852	273,200	503,052
1972	13,863	243,715	273,200	516,915

[a]Includes miscellaneous U_3O_8 receipts from mine waters, heap leach, in situ, and refining residues.

[b]The reserve estimates since 1961 are based on a maximum forward cost of $8 per pound of U_3O_8 in concentrate. Estimates for the period 1952-61, inclusive, are based on the AEC Domestic Uranium Program Circular 5 (Revised). For the period prior to 1952, the basis is arbitrary thickness and grade cutoffs. The reserve estimates for 1956-62 do not coincide with estimates published in previous years. During 1972 the reserve estimates for 1956-62 were adjusted to more nearly reflect yearly additions during this period based on the redefinition of reserves in 1962.

Source: Statistical Data of the Uranium Industry, January 1, 1973, GJO-100 (73), U.S. Atomic Energy Commission, Grand Junction, Colorado.

U.S. Mill Production, Uranium Oxide (U$_3$O$_8$), 1960-73

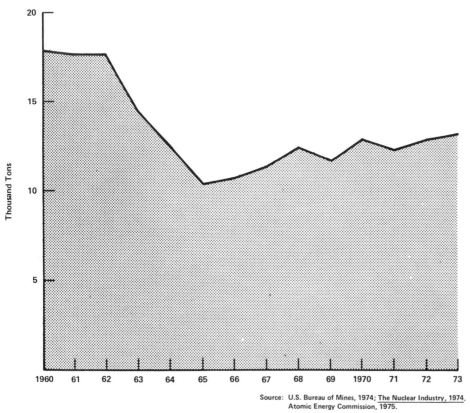

Source: U.S. Bureau of Mines, 1974; The Nuclear Industry, 1974,
Atomic Energy Commission, 1975.

Source: Energy Perspectives, U.S. Department of the Interior, February 1975.

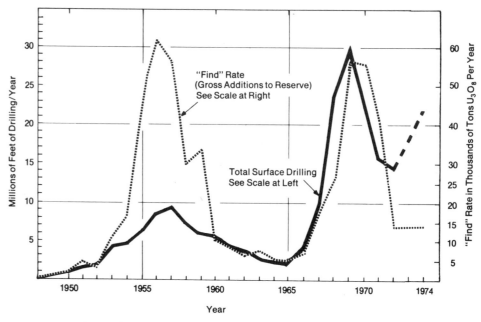

"Find" Rate
(Gross Additions to Reserve)
See Scale at Right

Total Surface Drilling
See Scale at Left

History of surface drilling in U.S. and rate of finding new uranium reserves

Source: Nuclear News, May 1975.

Mining of uranium ore. (United States Vanadium Company)

3

Renewable Energy Resources

Credit: Atomics International Division, Rockwell International.

WORLD RENEWABLE ENERGY RESOURCE BASE

The renewable energy resources of the world are derived from a number of forces: from the gravitational forces of the moon and sun which create the tides; from the decay of radioactive minerals and the heat of the earth which provide geothermal energy; from the rotation of the earth and solar energy which generate ocean currents and the winds and, in turn, waves; from solar energy and the gravitational force of the earth which create the cycle of evaporation and precipitation of water; from the photosynthetic conversion by solar energy of carbon dioxide, water and other elements into organic matter; and finally, from the direct heat of the sun itself. All of these sources are inexhaustible, at least in the sense that little change in their energy output is expected over a span of millions of years.

Until about 1850, almost all of the energy used by man was derived from renewable resources: wood, formed by photosynthesis, was the primary fuel; industrial plants were powered by the flowing water of streams and rivers; ships sailed before the wind; and the human and animal energy, which played a much larger role then in industry and agriculture, was derived from plants grown in the sun. Since that time, man has made increasing use of stored fossil fuels—coal, oil and gas—themselves derived from organic material created by the sun over periods of millions of years. Most recently, man has begun to use nuclear fuels whose origins rest in the early formation of the solar system. The fossil fuels, which have been inexpensive and convenient, have permitted the widespread use of mechanical engines to replace human and animal labor. These fossil resources are, however, but a limited treasure of man—a treasure that is fast disappearing.

Man is now faced with a transition, over the next 50 to 200 years, from the age of fossil fuels to a new age based upon nuclear energy and upon the renewable energy resources he once used. Some of the renewable resources are large in extant but generally diffuse in form and more difficult and expensive to convert to usable mechanical and thermal energy than are the fossil fuels. If it had been otherwise, man would not have shifted from his prior use of renewable resources to the fossil fuels. A return to renewable energy resources will not be a simple matter. Populations have grown and per capita energy consumption has increased many fold. Some renewable resources—the tides and hydro resources—are plainly insufficient to meet current and projected levels of energy consumption. Other competing uses have been found for the wood from the forests, the energy from the sun and winds varies each day, and the heat locked in the earth may not be readily available. The accompanying tables provide some of the estimates of the extent to which the renewable energy resources will be able to meet energy requirements.

Renewable Energy Resource Base

Continuous Supply (10^{12} watts)	WORLD		U.S.	
	Maximum	Possible by 2000	Maximum	Possible by 2000
Solar radiation	28,000		1,600	
Fuel wood	3	1.3	.1	.05
Farm waste	2	.6	.2	.00
Photosynthesis fuel	8	.01	.5	.001
Hydropower	3	1.	.3	.1
Wind power	.1	.01	.01	.001
Direct conversion	?	.01	?	.001
Space heating	.6	.006	.01	.001
Nonsolar				
Tidal	1.	.06	.1	.06
Geothermal	.06	.006	.01	.006
Total	18+	3	1.2	.2
Annual demand year 2000 (10^{12} watts)	~15		~5–6	

CONTINUOUS, OR RENEWABLE, ENERGY SUPPLY can be divided into two categories: solar and nonsolar. Two sets of estimates are again presented, one for the world and one for the U.S. alone. The figure for total solar radiation includes only the fraction (about 30 percent) falling on land areas. If an efficient solar cell existed to convert sunlight directly to electric power, one could think of utilizing solar energy on a large scale. The sunlight that falls on a few percent of the land area of the U.S. would satisfy most of the energy needs of the country in the year 2000 if converted to electricity at an efficiency of 12 percent.

Note: A continuous supply of energy at 1×10^{12} watts is equivalent to 8.76×10^{12} kilowatt hours per year or 29.9×10^{15} BTU/year or 0.0299 Q (10^{18} BTU)/year. U.S. energy consumption in 1974 was 74×10^{15} BTU.

Earth Energy Resources

Energy Stored in the Earth	SERPYs[a]
Heat content of the earth	5,000,000
Solar heat stored in the earth	32,000
Solar heat stored in the oceans	146
Rotational energy of the earth	40,000

Annual Renewed Energy	Kilowatt Hours per Year	Q (10^{18} Btu) per Year	SERPYs per Year
Solar energy received at outer atmosphere	$1,550,000 \times 10^{12}$	5,300	1
Solar energy received at earth's surface	$730,000 \times 10^{12}$	2,500	0.47
Water power	2.5×10^{12}	0.009	0.0000016
Geothermal heat flow	230.0×10^{12}	0.78	0.00015
Heat from radioactive decay	155.0×10^{12}	0.55	0.0001
Solar energy converted to wind	$26,000 \times 10^{12}$	89	0.017

[a]A SERPY is a unit equivalent to the Solar Energy Received Per Year by the earth.
Source: Data from Vincent E. McKelvey, "Solar energy in earth processes," *Technol. Rev.*, as edited at The Massachusetts Institute of Technology, March/April 1975.

Renewable Energy Resources[a]
(Data in Units of Q (10^{18} Btu) per Year)

Energy Sources	World Theoretical Maximum	World Possible Potential	United States Theoretical Maximum	United States Possible Potential
Hydropower	0.275	0.02		0.0024
Solar	2490.0		50	
Photosynthesis, (1% efficiency, 10% of land area.)				0.05
Thermal electric, (10% efficiency, 10% of land area)				0.5
Ocean thermal, ($\frac{1}{2}$°F drop in temp. of Gulf Stream)				0.05
Tidal power		0.002		0.001
Waste materials			0.011	0.0017
Geothermal heat flow (entire earth surface)	0.77		0.017	
Geothermal stores energy			20,000.	0.01
Wind power	88.0	0.6	0.015	0.005

[a]Data from various sources in this chapter.
Note: U.S. energy consumption in 1974 was 0.074 Q.

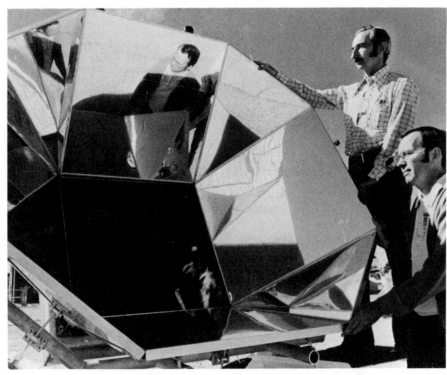

Augmented flat-plate energy collector. (Sandia Laboratory)

WORLD SOLAR ENERGY RESOURCES

Approximately 5300 Q (10^{18} Btu) or 1.55×10^{18} kilowatt hours of solar energy reach the earth's outer atmosphere each year. Of this energy, about 35% is reflected back into space, 18% is absorbed by the atmosphere and 47% reaches the earth's surface. Man's use of energy in 1974 was about 1/13,000 of the solar energy reaching the earth and about 1/28,000 of the energy reaching the outer atmosphere.

Of the various forms of solar energy, man must accept the wind, waves, and ocean thermal gradients as they exist in nature. The two forms that are subject to some modification in use are those related to the direct use of the heat from the sun and the use of the photosynthetic radiation from the sun. Both of these forms are dependent in their use upon the area of the globe man wishes to devote to their exploitation, and in a geographical sense, upon the intensity of solar radiation in a particular region. The accompanying map indicates solar radiation received throughout the world and the chart indicates the land area required to generate usable energy assuming various efficiencies for the conversion of incoming solar energy. For example, assuming an average annual solar energy input of 1500 Btu/ft^2/day (the case in the United States), solar electric power plants of 10% overall efficiency would require about 2% of the area of the United States to provide energy equivalent to that used in 1974 and about 4% of the land area to meet the total energy demand expected in the year 2000. Photosynthetic conversion efficiencies are usually less than 0.5% and the production of organic matter to be subsequently burned for its thermal energy content, would require about 40% of the land area of the U.S. to match 1974 energy demand and 80% of the land area to match energy demand in the year 2000. These land area requirements for production of organic matter could be reduced if the higher photosynthetic conversion efficiencies, up to 3%, achieved experimentally can be matched in practice.

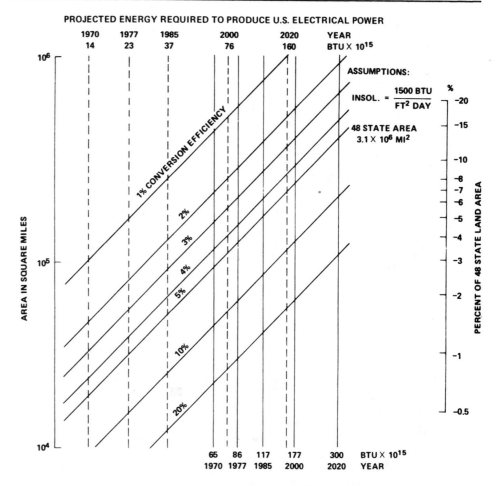

Availability of solar energy vs. land area.

Source: Solar Energy as a National Energy Resource, NSF/NASA Solar Energy Panel, National Science Foundation, December 1972.

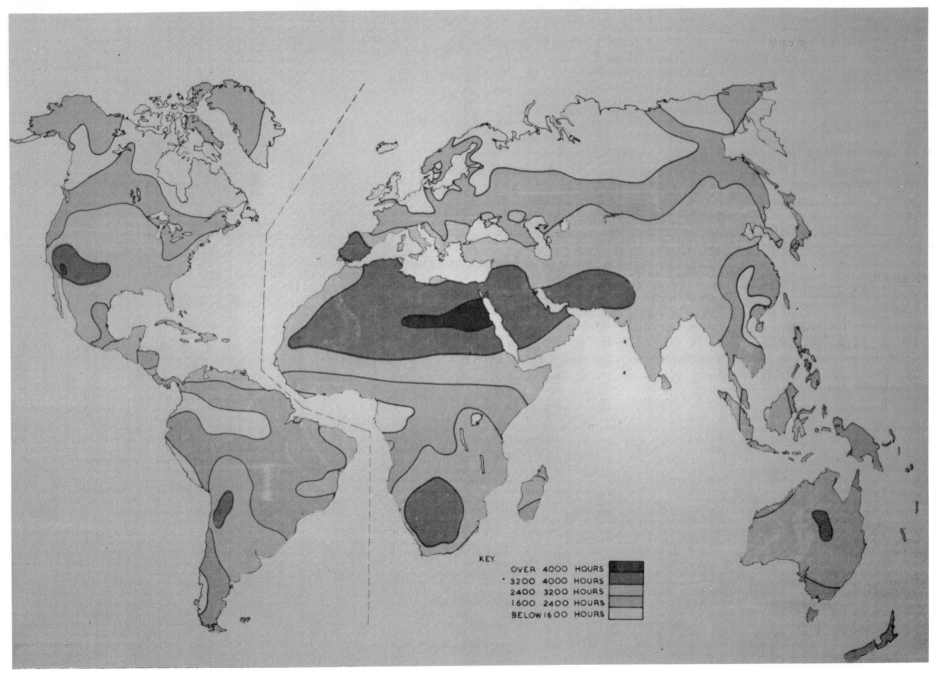

Average annual duration of sunshine.

Source: Arthur H. Stodhart.

Solar furnace at Odeillo, France.

Credit: Creamer Colarossi Basford, Inc.

ORGANIC WASTE RESOURCES

One of the energy resources derived from the photosynthetic energy of the sun is the organic waste material representing the residue of human and animal use of products made from living plants and fossil fuels. These residues include paper, plastics, wood, animal dung, plant stalks and fibers, and food residues. These residues have a heating value and in some countries, such as India, residual organic materials are not wasted. In the United States, however, much of this organic material is not used.

In a study[1] prepared for the Bureau of Mines of the U.S. Department of the Interior, L. L. Anderson has estimated that the total amount of organic waste produced in the United States in 1971 was 880 million tons on a dry, ash-free basis. Of this amount of industrial, urban, and agricultural waste, less than 20%, or 136 million tons, was considered to be collectable for conversion to oil or gas or to be burned directly.

At an average energy value for waste materials of 6500 Btu per pound, the energy content of 136 million tons is 0.0017 Q (10^{18} Btu) per year. This corresponds to about 2.3% of the total United States energy consumption in 1971. Conversion of the organic wastes to synthetic fuels through hydrogenation, pyrolysis or bioconversion would generate

Comparative Power Potential of Waste Materials in the United States

Total thermal energy used 1970	65×10^{15} Btu
Total thermal energy used for the generation of electricity 1970	17×10^{15} Btu
Total thermal energy potential of waste materials per year	11×10^{15} Btu or 0.011 Q (10^{18} Btu)
Total energy potential of readily collectable waste materials per year	1.7×10^{15} Btu or 0.0017 Q (10^{18} Btu)

Amounts (In Millions of Tons) of Dry, Ash-free Organic Solid Wastes Produced in the United States in 1971

Source	Wastes Generated	Readily Collectable
Manure	200	26.0
Urban refuse	129	71.0
Logging and wood manufacturing residues	55	5.0
Agricultural crops and food wastes	390	22.6
Industrial wastes	44	5.2
Municipal sewage solids	12	1.5
Miscellaneous	50	5.0
Total	880	136.3
Net oil potential (10^6 barrels)	1098	170
Net methane potential (10^9 cubic feet)	8.8	1.36

Source: L. L. Anderson, Bureau of Mines.
Hammond, A. L. et al., "Energy and the future," 1973, p. 74.
© American Association for the Advancement of Science.

about 170 million barrels of oil per year (3% of the 1971 crude oil consumption) or 1.36 trillion cubic feet of methane (about 6% of the 1971 consumption of natural gas). These conversion processes are discussed further in Chapter 9 on Solar Energy.

REFERENCE

"Fuel from wastes: A minor energy source," *Science*, 10 November 1972.

[1] L. L. Anderson, *Bureau of Mines Information Circular 8549*, U.S. Department of the Interior, 1972.

OCEAN THERMAL GRADIENT ENERGY RESOURCES

In 1881, the French physicist, Jacques D'Arsonval, suggested that the temperature differentials in the ocean could be used to generate power. Several experimental power plants of this type have been built and are discussed in Chapter 9 on Solar Energy.

In the regions between the Tropics of Cancer and Capricorn, about 90% of the earth's surface is water with surface temperatures above 82°F. Below this surface layer of warm water, some 100-150 feet thick, the water temperature drops rapidly. At depths of about 2000 feet, the temperature is below 40°F due to the movement of cold water from the polar regions toward the equator. This temperature differential between surface and depth provides the source of energy for ocean thermal gradient power plants.

The solar energy input into the region between the Tropics of Cancer and Capricorn is about 1600 Q (10^{18} Btu) annually. This heat creates ocean currents and evaporates sea water and is ultimately radiated back into space so that the temperature equilibrium of the earth is maintained. It is possible, however, that a small portion of the ocean's thermal energy could be extracted without upsetting oceanic or atmospheric conditions.

Advantageous sites for the location of ocean thermal power plants occur within a few miles of shore along both coasts of Africa, the west coasts of the Americas, and near many islands, particularly in the Caribbean. In the United States, the Gulf Stream, which flows at 30 million cubic meters per second in a path 15 miles across and 550 miles long between the tip of Florida and South Carolina, has a temperature differential of at least 30°F throughout the entire year and is considered a logical location for the installation of ocean thermal power plants.

Extraction of the heat in the Gulf Stream with an attendent temperature drop of $\frac{1}{2}$°F would provide sufficient heat to operate 1,600,000 MWe of electrical generating capacity and represent an energy resource of about 0.05 Q (10^{18} Btu) per year. A detailed study[1] indicates that the practical power available from the Gulf Stream would be considerably lower. Assuming a temperature drop of 1.8°F in the warm water passing through the power plants and a cold water intake depth of 400 meters, the available power is 320,000 MWe.

[1]W. P. Goss, W. E. Heronemus, P. A. Mangarella, and J. G. McGown, "Summary of University of Massachusetts research on Gulf Stream based ocean thermal power plants," in *Proc. Third Workshop on Ocean Thermal Energy Conversion (OTEC)*, Houston, Texas, May 8010, 1975, The Johns Hopkins University, Applied Physics Laboratory, August 1975.

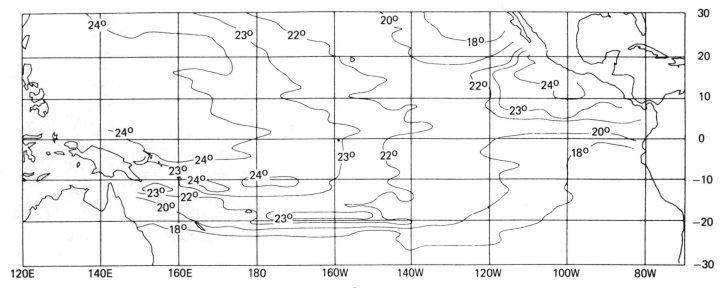

Pacific Ocean area temperature difference (°C) between surface and 1000 m, summer season.

Source: Robert H. Douglass, "Ocean Thermal Energy Conversion: An Engineering Evaluation," in *Proceedings, Third Workshop on Ocean Thermal Energy Conversion (OTEC)*, Houston, Texas, May 8-10, 1975, The Johns Hopkins University, Applied Physics Laboratory, August 1975.

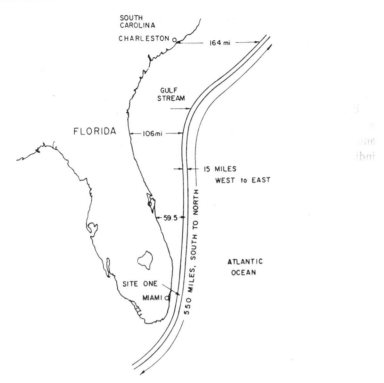

Total length of Gulf Stream sites.

Source: W. P. Goss, W. E. Heronemus, P. A. Mangarella, and J. G. McGowan, "Summary of University of Massachusetts Research on Gulf Stream Based Ocean Thermal Power Plants," in *Proceedings*, *Third Workshop on Ocean Thermal Energy Conversion (OTEC)*, Houston, Texas, May 8–10, 1975, The Johns Hopkins University, Applied Physics Laboratory, August 1975.

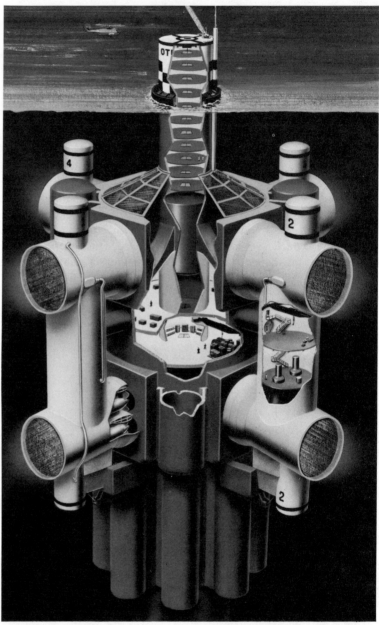

Electricity generator immersed in ocean. A Lockheed engineering study for the National Science Foundation resulted in this artist's concept of a structure that would use the 40°F, ocean-temperature differences to produce electricity. The structure is a platform with crew quarters and maintenance facilities. Attached around the outside are turbine-generators and pumps. It is 250 ft in diameter and 1600 ft long and weighs about 300,000 tons. It would produce enough electricity to fulfill the needs of a city with 100,000 residents.

Courtesy: Lockheed Missiles & Space Company, Inc., A Subsidiary of Lockheed Aircraft Corp., Sunnyvale, Calif.

WORLD WIND ENERGY RESOURCES

Of the total solar energy falling on the earth each year, 5300 Q (10^{18} Btu), about 88 Q or 25,800 kilowatt hours is converted into air in motion. Some fraction of this energy is available in the first 100 meters or so above the ground surface where wind power plants might be installed. The World Meteorological Organization has estimated[1] that a little less than 1% of the wind energy, that is, 0.6 Q (10^{18} Btu) or 175×10^{12} kilowatt hours is available at selected sites throughout the world.

The distribution of wind is not uniform over the earth—wind velocities and frequencies

[1]"Energy from the wind," Tech. Note 4, World Meterological Organization, Geneva, Switzerland, 1954.

are higher in polar and temperate zones than in tropical zones and are generally higher in coastal areas than inland. The attached generalized map of wind energy availability gives data in kilowatt hours of output per year per kilowatt of rated plant capacity. If winds blew constantly throughout the year the maximum output would be 8760 kilowatt hours/kilowatt of installed capacity.

As described further in Chapter 9 on Solar Energy, the output of wind power machines increases as the cube of the wind velocity (for example, a doubling of the wind velocity increases the power output eight times). Because of friction losses, wind power machines usually do not operate at wind velocities much less than 10 miles per hour, and if winds are of gale force, the rotors of wind machines are usually feathered to prevent damage. Wind energy resources do not include, therefore, the entire range of wind velocities.

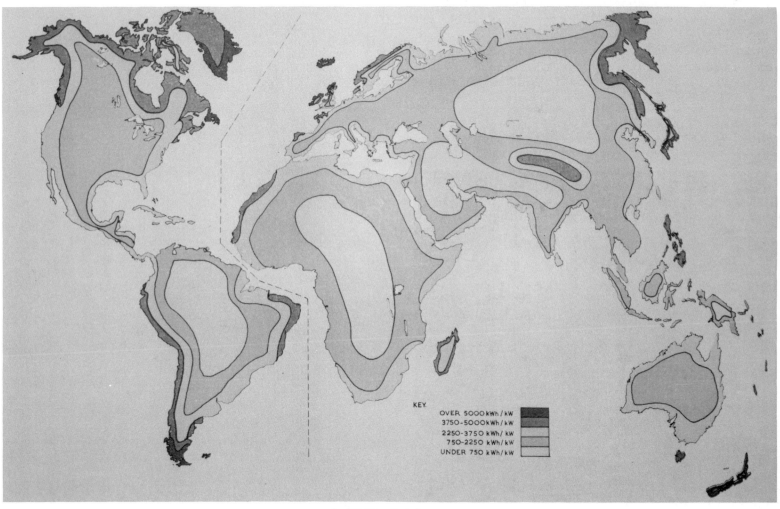

KEY

OVER 5000 kWh/kW	
3750-5000 kWh/kW	
2250-3750 kWh/kW	
750-2250 kWh/kW	
UNDER 750 kWh/kW	

Availability of wind energy.
Annual specific output of windmills rated at 25 mile/h

Source: Arthur H. Stodhart, "Wind Data for Wind Driven Plant," *Wind Energy Conversion Systems Workshop Proceedings*, NSF/RA/W-73-006, National Science Foundation, December 1973.

U.S. WIND ENERGY RESOURCES

Although the energy in the winds is theoretically quite large, this energy is relatively diffuse and intermittent in nature. Practical limitations on windmill height and rotor diameter result in individual wind mills having power capabilities in the range of 1000–2000 kilowatts. The theoretical power that can be extracted by a rotor is 59.3% of the energy in the wind passing through the area swept by the rotor. Well-designed practical rotors can extract 40–45% of the energy from the wind. When rotors are coupled to electrical generators, frictional losses lower the overall efficiency to 30–35%. The wind power plant at Grandpa's Knob[1] in Vermont had a rated capacity of 1250 kilowatts and a rotor diameter of 175 feet. Test operations of this plant were conducted from October 1941 to March 1945 when operations were terminated after failure of one of the rotor blades.

An estimate[2] has been made that by the year 2000, wind power plants, placed at suitable locations throughout the United States and its territorial waters, could generate 1.536×10^{12} kilowatt hours annually—about 19% of the projected electrical energy requirement at that time. About 62% of this wind energy would be generated from plants located offshore along the Atlantic and Gulf Coasts and about 26% from plants located along the chain of Aleutian Islands.

[1]P. C. Putnam, *Power From the Wind.* New York, Van Nostrand Reinhold Company, 1948.
[2]*Solar Energy as a National Energy Resource: NSF/NASA Solar Energy Panel*, December 1972.

Windmill and power lines near the Dave Johnston Power Plant, 20 miles east of Casper, Wyoming.

Credit: EPA-Documerica—Boyd Norton.

Maximum Electrical Energy Production From Wind Power

Site	Annual power production	Maximum possible by year
(1) Offshore, New England	159×10^9 kWh	1990
(2) Offshore, New England	318×10^9 kWh	2000
(3) Offshore, Eastern Seaboard, along the 100 meter contour, Ambrose shipping channel south to Charleston, S.C.	283×10^9 kWh	2000
(4) Along the E-W Axis, Lake Superior (320 m)	35×10^9 kWh	2000
(5) Along the N-S Axis, Lake Michigan (220 m)	29×10^9 kWh	2000
(6) Along the N-S Axis, Lake Huron (160 m)	23×10^9 kWh	2000
(7) Along the W-E Axis, Lake Erie (200 m)	23×10^9 kWh	2000
(8) Along the W-E Axis, Lake Ontario (160 m)	23×10^9 kWh	2000
(9) Through the Great Plains from Dallas, Texas, North in a path 300 miles wide W-E, and 1300 miles long, S to N. Wind Stations to be clustered in groups of 165, at least 60 miles between groups (sparse coverage)	210×10^9 kWh	2000
(10) Offshore the Texas Gulf Coast, along a length of 400 miles from the Mexican border, eastward, along the 100 meter contour	190×10^9 kWh	2000
(11) Along the Aleutian Chain, 1260 miles, on transects each 35 miles long, spaced at 60-mile intervals, between 100 meter contours. Hydrogen is to be liquefied and transported to California by tanker.	402×10^9 kWh	2000

Estimated Total Production Possible: 1.536×10^{12} kWh by year 2000

"Solar Energy as a National Energy Resource: NSF/NASA Solar Energy Panel," December 1972.

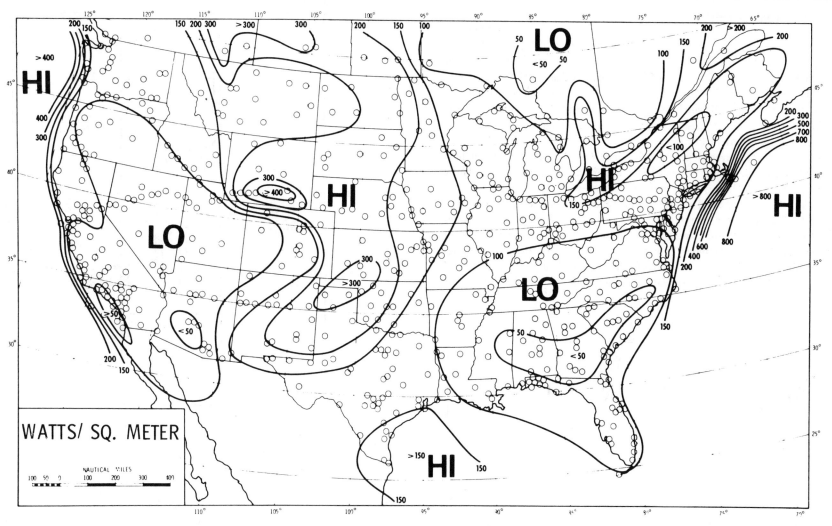

Available wind power, annual average.

Source: Jack W. Reed, *Wind Power Climatology of the United States*, SAND74-0348, Sandia Laboratories, Albuquerque, New Mexico, June 1975.

WAVE ENERGY RESOURCES

The exploitation of the energy in waves represents another technique for tapping the large solar energy input to the earth. The differential heat of the atmosphere and the rotational forces of the earth create the winds which, in turn, interact with the surface of the oceans to create waves. A number of mechanical devices—rocking booms, articulated floats, and buoys—have been studied which are capable of extracting the energy of mechanical oscillation in waves. All of these devices translate mechanical motion into hydraulic pressure which is used to drive a turbine generator.

The energy available from waves depends, of course, on their size and frequency. Storm waves of 600 meters length and with a 20 second period, such as occur in Atlantic storms, contain as much as 1500 kilowatts per meter. Storm waves in the North Sea can be 400 meters in length with a period of 16 seconds and an energy content of about 700 kilowatts/meter. Practical wave energy generators are not expected to operate under storm conditions but rather under those wave conditions that prevail for perhaps 70% of the year. Average energy available under "normal" wave conditions might be 10 kilowatts/meter. If such wave energy could be extracted with an overall plant efficiency of 25%, a facility to generate 1000 megawatts of electric power would require 400 kilometers of coastline.

The amount of power available in the waves around Britain. The most powerful waves are the least frequent. Thus a device designed to make the most of these conditions (typified by line A) would be under-utilised most of the time. A wavepower device designed to more modest levels (as in line B) would generate its maximum output for most of the time.

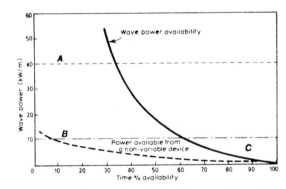

Source: Malcolm Woolley and Jim Platts, "Energy on the Crest of a Wave," *New Scientist*, 1 May 1975. This article first appeared in *New Scientist*, London, the weekly review of *Science and Technology*.

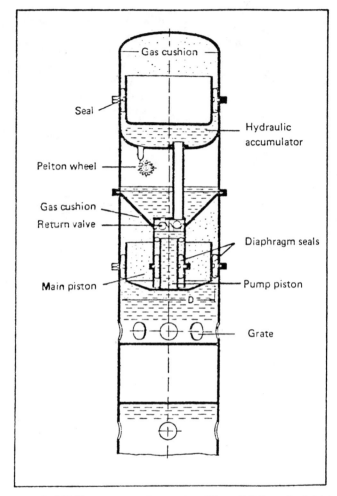

General arrangement of 1 kW wave-powered generator with small Pelton wheel driving generator.

Source: "Wave Motion can be used to Tap Wind Energy," *Energy International*, Vol. 12, No. 4, April 1975.

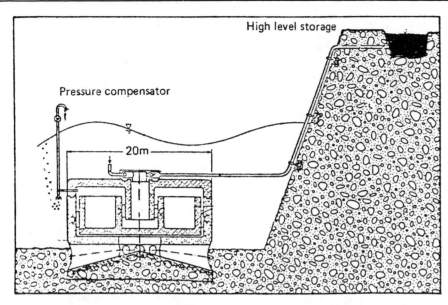

Scheme for 1 Mwe wave generator supplying water of Cliff-top Reservoir.

Source: "Wave Motion can be used to Tap Wind Energy," *Energy International*, Vol. 12, No. 4, April 1975.

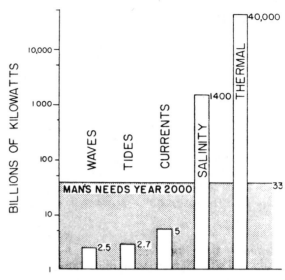

Estimates of power dissipation rates in the oceans.

Note: A power dissipation rate for wave energy of 2.5 billion kilowatts would produce 2.2×10^{12} kilowatt hours/year and is equivalent to 0.0075 Q (10^{18} Btu).

Source: Isaacs, John D. and Seymour, Richard J; "The Ocean as a Power Resource," *International Journal of Environmental Studies*, 1973, Vol. 4, pp. 201–205.

REFERENCE

J. LaStella, "Tapping ocean wave energy for powering gas turbine plants," *Gas Turbine World*, May 1975.

WORLD HYDROELECTRIC RESOURCES

The use of the energy of water flowing in streams to turn water wheels for the grinding of grain dates from Roman times. The conversion of hydraulic energy to mechanical energy for the operation of factories reached its peak in the 1700s, at which time the steam engine came into use and permitted the location of factories elsewhere than on river banks. Hydraulic resources became important once again with the development of efficient electric generators and transmission technology that permitted location of hydroelectric plants several hundred miles from the point of energy consumption. By 1925 about 40% of the electricity generated in the world was being produced in hydroelectric power plants.

The energy that can be derived from flowing water is a function of the quantity of water and the height through which the water can be made to fall from reservoir to the hydraulic turbine. The construction of diversion and storage dams for hydroelectric power plants requires suitable topography, river banks that provide a firm foundation for the dam structure, and a relatively steep drop in the elevation of the river to provide sufficient head for operation of the turbines. Practical considerations of dam construction limit the maximum potential hydroelectric generation to some 16% of the energy flowing in the streams and rivers of the world.

World Hydroelectric Resource Data

Annual evaporation of water from the oceans	400,000 cubic kilometers
Annual rainfall on land areas	100,000 cubic kilometers
Annual rainfall reevaporated from the land	63,000 cubic kilometers
Annual river flow to the oceans	37,000 cubic kilometers
Average elevation of the continents	800 meters
Average annual energy flowing in rivers	290×10^6 terajoules/year or 0.275 Q (10^{18} Btu)/year[a]
Percentage of river flow usable for power generation under average flow conditions	16%
Percentage of river flow usable for power generation at maximum flow available 95% of the time	6%
Potential annual usable energy recovery	
at average flow	17×10^6 terajoules/year or 0.016 Q (10^{18} Btu)/year[a]
at maximum flow (95% ofthe time)	45×10^6 terajoules/year or 0.043 Q (10^{18} Btu)/year[a]
Total installed capacity potential	
at 87% load factor	600,000 MWe
at 48% load factor	3,000,000 MWe
Present world installed capacity	329,000 MWe
Present world hydroelectric output	4.7×10^6 terajoules/year or 0.0045 Q (10^{18} Btu)/year[a] or 1.3×10^{12} kilowatt hours/year

[a]Author's note.

Note: Figures for annual energy available from hydroelectric plants should be multiplied by about 3 to obtain equivalent fossil fuel resources required to generate the same power.

Source of Data: Survey of Energy Resources 1974, The U.S. National Committee of the World Energy Conference, 345 East 47th Street, New York, N.Y. 10017, 1974.

Georgia Power Company's 65,000-kilowatt hydroelectric generating station at Barlett's Ferry Dam. (Georgia Power Company)

WORLD REGIONAL AND NATIONAL HYDROELECTRIC RESOURCES

World Hydroelectric Power Capacity and Production Capability, 1974.

Continent or area	Total capacity (MW)	Percent of total world capacity	Developed capacity (MW)	Percent capacity developed	Average annual production capability (GWh)	Percent of total world production capability	Developed average annual production capability (GWh)	Percent production capability developed
Africa	437,104	19.3	8,154	1.9	2,019,934	20.6	30,168	1.5
Asia (less U.S.S.R.)	684,337	30.3	47,118	6.9	2,638,169	26.9	198,433	7.5
Europe (less U.S.S.R.)	215,407	9.5	103,998	48.3	772,368	7.4	382,317	52.9
U.S.S.R.	269,000	11.9	31,500	11.7	1,095,000	11.2	123,000	11.2
North America	330,455	14.7	90,210	27.3	1,487,847	15.2	453,334	30.5
South America	288,289	12.7	18,773	6.5	1,637,031	16.7	91,415	5.6
Oceania	36,515	1.6	7,609	20.8	202,071	2.0	28,897	14.3
Total	2,261,107	100.0	307,362	13.6	9,802,420	100.0	1,307,564	13.3

Note.—Based on capacity under average annual stream flow conditions.

Source: World Energy Conference, *Survey of Energy Resources,* New York, 1974.

Source: Energy Perspectives 2, U.S. Department of the Interior, June 1976.

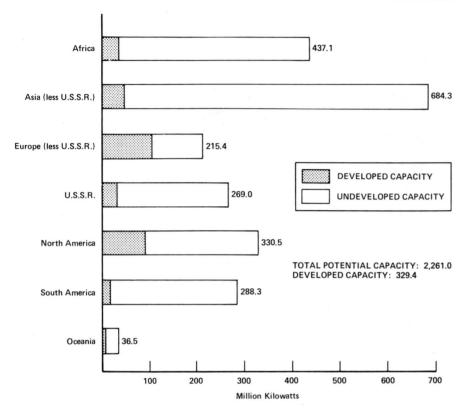

TOTAL POTENTIAL CAPACITY: 2,261.0
DEVELOPED CAPACITY: 329.4

Source: Survey of Energy Resources,
World Energy Conference, 1974.

World hydroelectric power, developed and undeveloped, 1973.

Source: Energy Perspectives, U.S. Department of the Interior, February 1975.

Comparison of Installed Hydroelectric Capacity and Annual Generation in 1925 and 1971.

Area or Nation	1925[1]			1971[2]		
	Installed Capacity	Annual Generation		Installed Capacity	Annual Generation	
	(MW)	(GWh)	(TJ)	(MW)	(GWh)	(TJ)
Western Europe	~10 000	30 833	111 000	97 822	366 533	1 319 518
United States	8 800	25 530	91 909	53 404	256 781	924 412
Canada	3 600	9 949	35 816	32 501	178 169	641 408
Japan	3 000	7 610	27 396	19 897	82 270	296 172
USSR		268	965	31 500	123 000	442 800
Eastern Europe		268	965	6 176	15 784	56 821
Other East Asia				14 521	54 954	197 834
South Asia	~1 000	368	1 325	12 469	60 902	219 247
Africa		65	234	8 154	30 168	108 594
Australia		343	1 235	4 310	12 353	44 471
New Zealand		306	1 102	3 200	16 000	57 600
Other Oceania				99	544	1 959
Latin America		3 155	11 358	23 078	109 799	395 277
World Total	26 400	78 678	283 241	307 131	1 307 257	4 706 113

1. From Darmstadter et al, *Energy in the World Economy*
2. From United Nations, J Series *World Energy Supplies, 1968–1971* Includes very small amounts of geothermal energy

Source: Survey of Energy Resources 1974, The United States National Committee of the World Energy Conference, 345 East 47th Street, New York, N.Y. 10017. 1974.

25 Nations with Largest Hydraulic Resources.

Nation Pays	Hydraulic Resources at:					
	G_{av}			G_{95}		
	GWh/yr	MW	% of World	GWh/yr	MW	Rank
China, P.R. of	1 320 000	330 000	13.5	480 000	60 000	2
USSR	1 095 000	269 000	11.1	400 000	50 000	3
United States	701 500	186 700	7.2	216 100	27 000	6
Zaire	660 000	132 000	6.7	620 800	77 600	1
Canada	535 200	94 500	5.5	240 900	30 100	5
Brazil	519 300	90 200	5.3	382 500	47 800	4
Malagasy	320 000	64 000	3.3	91 500	11 400	10
Colombia	300 000	50 000	3.1	25 900	3 200	21
India	280 000	70 000	2.9	216 600	27 100	7
Burma	225 000	75 000	2.3	24 000	3 000	23
N. Vietnam and Laos	192 000	48 000	2.0	24 000	3 000	24
Argentina	191 000	48 100	1.9	25 900	3 200	22
Indonesia	150 000	30 000	1.5	128 000	16 000	8
Japan	130 000	49 600	1.3	63 200	7 900	13
Ecuador	126 000	21 000	1.3	9 600	1 200	25
Papua–New Guinea	121 700	17 800	1.2	32 600	4 100	17
Norway	121 000	29 600	1.2	30 700	3 800	19
Cameroon	114 800	23 000	1.2	115 000	14 400	9
Peru	109 200	12 500	1.1	30 700	3 800	20
Pakistan	105 000	20 000	1.1	42 000	5 200	15
Sweden	100 300	20 100	1.0	32 000	4 000	18
Mexico	99 400	20 300	1.0	85 100	10 600	11
Venezuela	98 000	11 600	1.0	60 100	7 500	14
Chile	88 600	15 800	0.9	64 300	8 000	12
Gabon	87 600	17 500	0.9	38 400	4 800	16
All Other Nations	2 011 800	514 800	20.5	954 600	119 100	
World Total	9 802 400	2 261 100	100.0	4 434 500	553 800	

Source: Survey of Energy Resources 1974, The United States National Committee of the World Energy Conference, 345 East 47th Street, New York, N.Y. 10017. 1974.

Index of Hydroelectric Projects with Capacities of 1000 MW or More

Hydroelectric Project, River Basin and County	Ultimate Capacity (MW) *Puissance finale (M-W)*	Hydroelectric Project, River Basin and Country	Ultimate Capacity (MW)
1 Grand Coulee, Columbia, U.S.	9771	33 Talbingo, Murray, Australia (UC)	1500
2 Guri, Orinoco, Venezuela	6500	34 McNary, Columbia, U.S.	1406
3 Sayansk, Yenisei, USSR (UC)	6400	35 Marimbondo, Parana, Brazil (UC)	1400
4 Krasnoyarsk, Yenisei, USSR	6096	36 Agua Vermelha, Parana, Brazil (UC)	1380
5 Bratsk, Yenisei, USSR	4600	37 Sir Adam Beck-2, St. Lawrence, Canada	1370
6 Sukhovo, USSR (UC)	5225	38 Johnson, St. Lawrence, Canada (UC)	1353
7 Churchill Falls, Churchill, Canada (UC)	5225	39 Hoover, Colorado, U.S.	1345
8 Ust-Illinsk, Yenisei, USSR (UC)	4300	40 Wanapum, Columbia, U.S.	1330
9 Cabora Basa, Zambesi, Mozambique (UC)	4000	41 Saratov, Volga, USSR	1290
10 Ilha Solteira, Parana, Brazil (UC)	3200	42 Priest Rapids, Columbia, U.S.	1262
11 Paulo Afonso, Sao Francisco, Brazil	2939	43 Castaio, Castaio, U.S. (UC)	1250
12 John Day, Columbia, U.S. (UC)	2700	44 Keban, Euphrates, Turkey (UC)	1240
13 Nurek, Amv Darya, USSR (UC)	2700	45 Kettle Rapids, Nelson, Canada (UC)	1224
14 22nd Congress, Volga, USSR	2560	46 Jupia, Parana, Brazil	1222
15 Mica, Columbia, Canada (UC)	2500	47 Rocky Reach, Columbia, U.S.	1215
16 Sao Simao, Parana, Brazil (UC)	2500	48 Furnas, Parana, Brazil	1200
17 V. I. Lenin, Volga, USSR	2300	49 Toktogul, Syr Darya, USSR (UC)	1200
18 Bennett, Mackenzie, Canada (UC)	2270	50 El Chocan, Negro, Argentina (UC)	1200
19 Iron Gate, Danube, Romania-Yugoslavia (UC)	2160	51 Manic-3, St. Lawrence, Canada (UC)	1176
20 High Aswan, Nile, Egypt	2100	52 Sanmen Hsia, Hwang Ho, P.R. China	1100
21 Tarbella, Indus, Pakistan (UC)	2100	53 Nizhne-kamskaya, Volga, USSR (UC)	1090
22 Itumbiara, Parana, Brazil (UC)	2080	54 Dworshak, Columbia, U.S. (UC)	1060
23 Chief Joseph, Columbia, U.S.	2073	55 Bersimis-1, St. Lawrence, Canada	1050
24 Robert Moses, St. Lawrence, U.S.	1950	56 Satto Osorio, Parana, Brazil (UC)	1050
25 St. Lawrence, St. Lawrence, U.S.-Canada	1880	57 Estreito, Parana, Brazil	1050
26 The Dalles, Columbia, U.S.	1813	58 Bhakra, Indus, India	1050
27 Kemano, Fraser, Canada	1670	59 Zeya, Zeya, USSR (UC)	1020
28 Beauharnois, St. Lawrence, Canada	1641	60 Manic-2, St. Lawrence, Canada	1016
29 Cheboksary, Volga, USSR (UC)	1632	61 Votkinsk, Volga, USSR	1000
30 Inguri, Inguri, USSR (UC)	1600	62 Mangla, Indus, Pakistan (UC)	1000
31 Kariba, Zambesi, Rhodesia-Zambia	1500	63 Chirkey, Sulak, USSR (UC)	1000
32 Tumut-3, Murray, Australia (UC)	1500	64 Kaniji, Niger, Nigeria (UC)	1000

Source of Data: 1. U.S. Geological Survey

2. Brazilian National Committee, WEC

Source: Survey of Energy Resources 1974, The United States National Committee of the World Energy Conference, 345 East 47th Street, New York, N.Y. 10017. 1974.

U.S. HYDROELECTRIC RESOURCES

By 1973, 31% of the potential hydroelectric power capacity of the United States had been developed, compared to a world average of 15%. Of the total capacity of the electric utility industry in the United States, about 15% was represented by the contribution from hydroelectric facilities. Plans for new hydroelectric power plants include the addition of some 11,000 MWe by 1990, increasing the developed capacity from 31 to 37% of the total potential capacity. The ultimate potential hydroelectric capacity in the United States corresponds to about 0.0024 Q (10^{18} Btu) at average flow conditions. This hydroelectric capacity would replace about 0.0072 Q of fossil fuels since thermal power stations have an efficiency of only 30-40%.

Planned New Hydroelectric Facilities in 1978-1990

Census Division	New Projects	Total Capacity (MW)	Average Size (MW)
New England (St. John, Maine)	1	830	830
Middle Atlantic	1	180	180
South Atlantic	10	2,168	217
East South Central	6	1,124	187
West South Central	4	480	120
Mountain	14	5,027	359
Pacific	5	1,333	267
Total	**41**	**11,142**	**272**

Source: Federal Power Commission, Bureau of Power, April 1971.

U.S. Hydroelectric Power by Region, 1975.

Region[1]	Developed			Undeveloped			Total	
	Number of sites	Capacity (MW)	Average annual generation (GWh)	Number of sites	Capacity (MW)	Average annual generation (GWh)	Capacity (MW)	Average annual generation (GWh)
New England	229	1,508.9	6,061	97	3,204.7	7,395	4,713.6	13,456
Middle Atlantic	159	4,235.5	25,491	82	4,093.1	12,167	8,328.6	37,658
East North Central	213	927.1	4,386	71	1,356.3	5,377	2,283.4	9,763
West North Central	78	2,769.3	10,269	78	5,950.7	17,687	8,720.0	27,956
South Atlantic	159	5,604.5	15,833	139	8,278.7	17,250	13,883.2	33,082
East South Central	54	5,386.0	21,408	51	3,237.1	6,524	8,623.0	27,932
West South Central	41	2,257.4	6,431	66	2,417.5	4,261	4,674.9	10,691
Mountain	178	6,656.8	29,560	335	21,319.5	71,816	27,976.3	101,375
Pacific, contiguous	289	25,775.2	141,915	418	34,051.5	95,937	59,826.7	237,852
Total contiguous United States	1,400	55,120.7	261,354	1,337	83,909.1	238,413	139,029.7	499,766
Pacific, noncontiguous	52	140.9	643	84	32,523.4	172,371	32,664.2	173,014
Total United States	1,452	55,261.6	261,997	1,421	116,432.5	410,784	171,693.9	672,780

[1] Census. Pacific region is divided between the contiguous and noncontiguous States (Alaska and Hawaii). Regions are defined in the glossary.
Note.—As of January 1, 1975.
Source: Federal Power Commission.

Source: Energy Perspectives 2, U.S. Department of the Interior, June 1976.

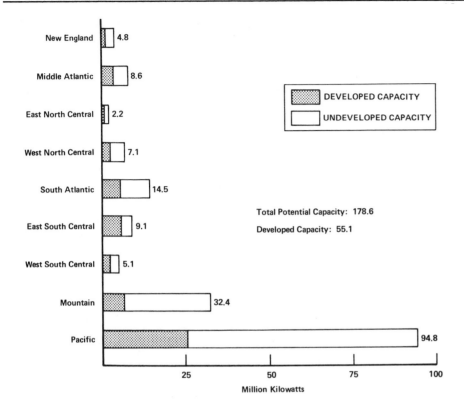

DEVELOPED CAPACITY
UNDEVELOPED CAPACITY

New England 4.8
Middle Atlantic 8.6
East North Central 2.2
West North Central 7.1
South Atlantic 14.5
East South Central 9.1
West South Central 5.1
Mountain 32.4
Pacific 94.8

Total Potential Capacity: 178.6
Developed Capacity: 55.1

25   50   75   100
Million Kilowatts

Source: Federal Power Commission, 1974.

U.S. hydroelectric power by region, 1973.

Source: Energy Perspectives, U.S. Department of the Interior, February, 1975.

Noxon Dam.

Credit: The Washington Water Power Company.

U.S. HYDROELECTRIC RESOURCES BY STATES

Hydroelectric Power Resources of the United States—Conventional Hydro Plants

	Developed		Undeveloped		Total Potential	
	Installed Capacity (KW)	Av. Annual Generation (GWH)*	Capacity (KW)	Av. Annual Generation (GWH)	Capacity (KW)	Av. Annual Generation (GWH)
New England						
Maine	509,470	2,528	1,714,074	4,289	2,223,544	6,817
New Hampshire	428,942	1,306	802,100	983	1,231,042	3,289
Vermont	199,742	833	337,701	701	537,443	1,534
Massachusetts	219,001	869	266,728	742	485,729	1,611
Rhode Island	2,860	8	—	—	2,860	8
Connecticut	131,115	390	183,200	552	314,315	942
Total	**1,491,130**	**5,934**	**3,303,803**	**7,267**	**4,794,933**	**13,201**
Middle Atlantic						
New York	3,809,050	23,777	1,292,175	3,293	5,101,225	27,070
New Jersey	8,294	48	241,000	1,001	249,294	1,049
Pennsylvania	429,520	1,816	2,980,630	7,057	3,410,150	8,873
Total	**4,246,864**	**25,641**	**4,513,805**	**11,351**	**8,760,669**	**36,992**
East North Central						
Ohio	2,399	12	249,200	1,206	251,599	1,218
Indiana	109,987	590	315,000	1,085	424,987	1,675
Illinois	42,931	196	206,200	1,101	249,131	1,297
Michigan	394,538	1,734	272,200	782	666,738	2,516
Wisconsin	418,668	1,922	213,200	949	631,868	2,871
Total	**968,523**	**4,454**	**1,255,800**	**5,123**	**2,224,323**	**9,577**
West North Central						
Minnesota	170,295	906	157,210	723	327,505	1,629
Iowa	135,675	811	345,200	1,632	480,875	2,443
Missouri	392,600	1,009	2,024,600	9,269	2,417,200	10,278
North Dakota	400,000	1,886	195,000	840	595,000	2,726
South Dakota	1,392,196	4,437	303,000	884	1,695,196	5,321
Nebraska	238,236	1,177	1,035,710	3,388	1,273,946	4,565
Kansas	5,150	13	303,000	1,460	308,150	1,473
Total	**2,734,152**	**10,239**	**4,363,720**	**18,196**	**7,097,872**	**28,435**
South Atlantic						
Maryland	494,280	1,750	163,000	497	657,280	2,247
District of Columbia	3,000	5	—	—	3,000	5
Virginia	735,981	1,144	1,276,231	2,962	2,012,212	4,106
West Virginia	208,010	1,015	1,994,040	6,793	2,202,050	7,808
North Carolina	1,765,525	5,196	919,350	2,000	2,684,875	7,196
South Carolina	1,034,153	2,788	1,834,530	2,356	2,868,683	5,144
Georgia	1,069,136	3,259	3,197,766	4,646	4,266,902	7,905
Florida	38,968	240	83,500	69	122,468	309
Delaware	—	—	—	—	—	—
Total	**5,349,043**	**15,397**	**9,468,417**	**19,323**	**14,817,470**	**34,720**

Source: U.S. Energy Outlook, New Energy Forms, National Petroleum Council, Washington, D.C., 1973.

Hydroelectric Power Resources of the United States—Conventional Hydro Plants

	Developed		Undeveloped		Total Potential	
	Installed Capacity (KW)	Av. Annual Generation (GWH)*	Capacity (KW)	Av. Annual Generation (GWH)	Capacity (KW)	Av. Annual Generation (GWH)
East South Central						
Kentucky	670,617	3,053	1,484,800	3,300	2,155,417	6,353
Tennessee	1,893,500	8,044	688,100	2,189	2,581,600	10,233
Alabama	2,267,485	8,223	1,629,520	2,939	3,897,005	11,162
Mississippi	—	—	140,300	381	140,300	381
Total	**4,831,602**	**19,320**	**3,942,720**	**8,809**	**8,774,322**	**28,129**
West South Central						
Arkansas	900,340	2,285	915,100	1,903	1,815,440	4,188
Louisiana	—	—	76,000	461	76,000	461
Oklahoma	363,400	1,142	913,600	2,043	1,277,000	3,185
Texas	434,180	1,278	1,160,035	1,837	1,594,215	3,115
Total	**1,697,920**	**4,705**	**3,064,735**	**6,244**	**4,762,665**	**10,949**
Mountain						
Montana	1,511,848	8,219	6,269,220	21,876	7,781,068	30,095
Idaho	1,250,839	7,588	12,391,631	41,100	13,642,470	48,688
Wyoming	212,140	940	1,286,290	6,117	1,498,430	7,057
Colorado	314,270	1,279	1,784,600	7,123	2,098,870	8,402
New Mexico	24,300	96	154,200	595	178,500	691
Arizona	1,879,190	8,102	3,676,000	15,749	5,555,190	23,851
Utah	207,915	957	1,320,050	4,917	1,527,965	5,874
Nevada	682,120	2,126	8,800	35	690,920	2,161
Total	**6,082,622**	**29,307**	**26,890,791**	**97,512**	**32,973,413**	**126,819**
Pacific						
Washington	9,549,389	62,516	23,498,864	74,075	33,048,253	136,591
Oregon	3,449,205	20,847	5,686,190	21,141	9,135,395	41,988
California	5,323,697	24,506	11,908,750	35,367	17,232,447	59,873
Total	**18,322,291**	**107,569**	**41,093,804**	**130,583**	**59,416,095**	**238,452**
Alaska	83,636	338	32,511,100	172,496	32,594,736	172,834
Hawaii	18,698	107	35,000	229	53,698	336
United States Total	**45,826,491**	**223,311**	**130,443,695**	**477,133**	**176,270,186**	**700,144**

* GWH = 1,000,000 KWH.

Source: Federal Power Commission, "Hydroelectric Power Resources of the United States, Developed and Undeveloped, January 1, 1968," Report P-36.

Source: U.S. Energy Outlook, New Energy Forms, National Petroleum Council, Washington, D.C., 1973.

WORLD TIDAL RESOURCES

Mills utilizing the kinetic energy of tidal currents date from the 11th century along the Atlantic Coast of Europe. Waterwheels 20 feet in diameter were installed in 1580 under the London Bridge to pump water for the city. A mill constructed in the 18th century in Rhode Island employed wheels 26 feet in width and 11 feet in diameter. With the advent of steam power and hydroelectric plants, the use of tide mills gradually disappeared because of the lower cost and operating flexibility of these new sources of power.

Extensive studies have been made of large tidal electric installations in a number of countries; the Severn estuary in England, the Carlingford and Strangford estuaries in Northern Ireland, the San Jose Gulf in Argentina, various sites in western Australia, sites along the Arctic coast of the USSR near Murmansk, Cook Inlet in Alaska, and the Bay of Fundy on the Atlantic coast between Maine and New Brunswick. There are however, only two tidal power plants in operation; a small 400 kilowatt experimental plant in Kislaya Bay in the USSR and a 240 megawatt plant on the Rance River on the north coast of France near St. Malo. The Rance plant is discussed in the page following.

Although the technology for the construction of tidal power plants is available, the relatively better economics of other power sources and the intermittent nature of tidal power generation have deterred investment in tidal projects. Plans have recently been announced[1] for the construction of a 6000 megawatt tidal power station at Mezenskaya on the White Sea along the Arctic coast of the USSR. This project would require a dam over 58 kilometers long.

In terms of world energy requirements, tidal power represents a minor energy source. Development of the total tidal potential would represent 0.002 Q (10^{18} Btu)/year in hydraulic energy (about 0.006 Q if the same energy were to be provided by fossil fuels).

REFERENCE

T. J. Gray and O. K. Gashus, *Tidal Power*.
   New York-London: Plenum Press, 1972.

Tidal Power Sites and Maximum Potential Power

Location	Average Range R (meters)	R^2 (m^2)	Basin Area S (km^2)	$R^2 S$ (m^2)(km^2)	Average Potential Power P (10^3 kw)	Potential Annual Energy E (10^6 kwh)
North America						
Bay of Fundy						
Passamaquoddy	5.52	30.5	262	7,930	1,800	15,800
Cobscook	5.5	30.3	106	3,210	722	6,330
Annapolis	6.4	41.0	83	3,440	765	6,710
Minas-Cobequid	10.7	114	777	88,600	19,900	175,000
Amherst Point	10.7	114	10	1,140	256	2,250
Shepody	9.8	96	117	11,200	2,520	22,100
Cumberland	10.1	102	73	7,450	1,680	14,700
Peutcodiac	10.7	114	31	3,530	794	6,960
Memramcook	10.7	114	23	2,620	590	5,170
Subtotal					29,027	255,020
South America						
Argentina						
San José	5.9	34.8	750	26,100	5,870	51,500
Europe						
England						
Severn	9.8	96.0	70	7,460	1,680	14,700
France						
Aber-Benoit	5.2	27.0	2.9	78	18	158
Aber-Wrac'h	5.0	25.0	1.1	28	6	53
Arguenon & Lancieux	8.4	70.6	28.0	1,980	446	3,910
Frênaye	7.4	54.8	12.0	658	148	1,300
La Rance	8.4	70.6	22.0	1,550	349	3,060
Rotheneuf	8.0	64.0	1.1	70	16	140
Mont Saint-Michel	8.4	70.6	610	43,100	9,700	85,100
Somme	6.5	42.3	49	2,070	466	4,090
Subtotal					11,149	97,811
USSR						
Kislaya Inlet	2.37	5.62	2.0	11	2	22
Lumbovskii Bay	4.20	17.6	70	1,230	277	2,430
White Sea	5.65	31.9	2,000	63,800	14,400	126,000
Mezen Estuary	6.60	43.6	140	6,100	1,370	12,000
Subtotal					16,049	140,452
Grand Total					63,775	559,483

Sources: N. W. Trenholm, 1961: L. B. Bernshtein, 1965 (1961), Table 5-5, p. 173.
Source: Resources & Man, W. H. Freeman & Co., pp. 212–213.
Reproduced with permission of the National Academy of Sciences.

[1] Eric Jeffs (Ed.), "Is tidal power showing signs of revival," *Energy Int.*, vol. 11, no. 12, December 1974.

Credit: U.S. Corps of Engineers.

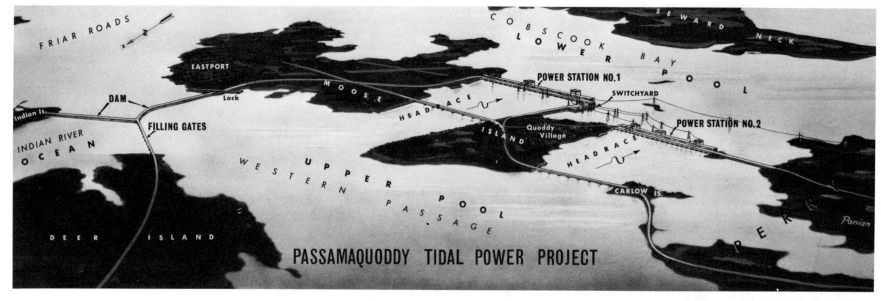

Credit: U.S. Corps of Engineers.

LA RANCE TIDAL POWER STATION

The Marémotrice tidal power station is located near St. Malo on the Rance estuary in northwestern France. The tidal range is a maximum of 13.5 meters at the equinoctal spring tide. The plant, completed in 1967, utilizes 24 horizontal turbine/pump units, each with a rating of 10 megawatts. The annual energy production is about 550 billion kilowatt hours representing a load factor of about 25%.

The Marémotrice generating units are capable of operating either as turbines or pumps. The pumping capability is used to boost the level of the water in the storage basin behind the dam during periods when demand for power is low. This pumped storage capability permits a greater overall utilization of the plant capacity.

REFERENCE

"French stem the tides," *Electrical World*, November 7, 1966.

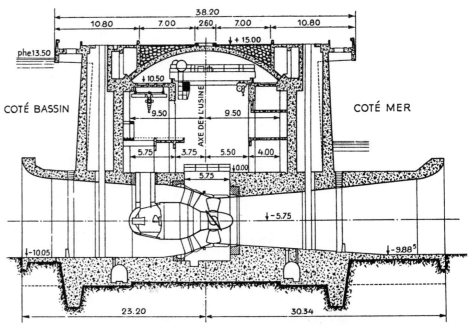

Coupe transversale de l'usine. (Document E.D.F.)
Source: Électricité de France.

OPERATION DATA
Électricité-de France
Maremotrice tidal power plant at la Rance

Gross production

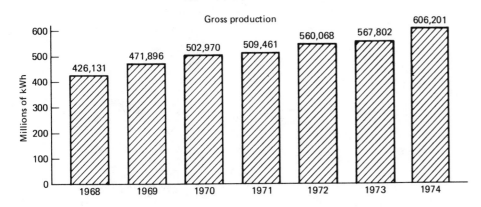

Group availability

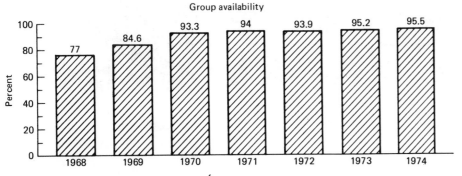

Source: Électricité de France.

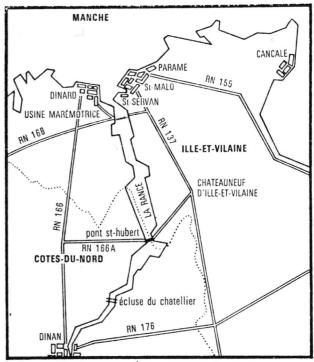

Source: Électricité de France.

Marémotrice Tidal Power Plant under construction in January 1965 at La Rance, France.

Credit: Électricité de France.

Marémotrice Tidal Power Plant at La Rance, France. The last of the 24 turbine units (10,000 kilowatts each) was put into service on December 4, 1967. The total capacity of the plant is 240,000 kilowatts and annual production of electricity is about 560 million kilowatt hours.

Credit: Électricité de France.

WORLD GEOTHERMAL ENERGY REGIONS

The exploitation of geothermal energy has been in regions throughout the world where the presence of geysers, hot springs, and fumaroles has indicated relatively high temperature geothermal reservoirs near the earth's surface. All of these geothermal reservoirs are located in regions at the edge of crustal plate boundaries which are tectonically active and which exhibit recent volcanism and regional or local high levels of heat flow. Further development of geothermal energy is expected to occur in regions with these characteristics.

Fewer than 100 potential geothermal areas have been explored and less than 1000 geothermal wells have been drilled since 1920. Generally, geothermal resources are expected to provide energy to meet local requirements on a scale comparable to the energy provided by hydroresources. The characteristics of various geothermal systems are described further in Chapter 12 on Geothermal Energy.

Status of Geothermal Exploration and Development, 1972

Nation	Electric-Power Generation/ Construction	Experimental Power Stations	Significant Direct Utilization	Other Geothermal-Field Discoveries	Additional Exploration Underway[c]
Chile				x	
China		x			
Ethiopia					x
Guadeloupe (Fr. W. Indies)				x	
Hungary			x		
Iceland	x		x	x	
Indonesia					x
Italy	x				x
Japan	x	x[b]	x	x	x
Kenya			x		x
Mexico	x[a]	x	x	x	x
New Zealand	x		x	x	
Nicaragua				x	
Philippines		x		x	
El Salvador	x[a]			x	
Taiwan				x	
Turkey				x	
USSR	x	x	x	x	x
United States	x	x[b]	x	x	x
Zaire		x[b]			

[a]Under construction.
[b]Inactive.
[c]Other geothermal exploration/interest: Algeria, Argentina, Bulgaria, Burundi, Colombia, Costa Rica, Czechoslovakia, Ecuador, Fiji Islands, Greece, Guatemala, India, Israel, Malawi, Mali, Morocco, New Britain, New Hebrides, Peru, Poland, Portugal (Azores Is.), Rwanda, Spain (Canary Is.), Tanzania, Tunisia, TFAI (French Somaliland), Uganda, Venezuela, Yugoslavia, Zambia.

"Worldwide Status of Geothermal Resources Development," by James B. Koenig in Geothermal Energy: Resources Production, Stimulation, edited by Paul Kruger and Carel Otte with permission of the publishers. Stanford University Press, © 1973 by the Board of Trustees of the Leland Stanford Junior University.

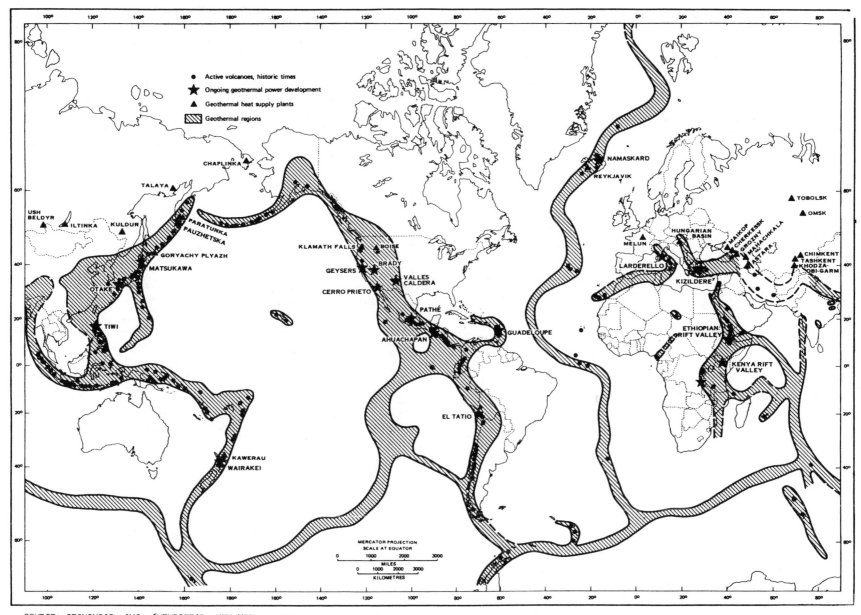

Active volcanoes, historic times
Ongoing geothermal power development
Geothermal heat supply plants
Geothermal regions

NAMASKARD
REYKJAVIK

TOBOLSK
OMSK

CHAPLINKA
TALAYA
USH BELDYR   ILTINKA   KULDUR
PARATUNKA
PAUZHETSKA
GORYACHY PLYAZH
MATSUKAWA
OTAKE
TIWI

KLAMATH FALLS   BOISE
GEYSERS   BRADY
VALLES CALDERA
CERRO PRIETO
PATHÉ
GUADELOUPE
AHUACHAPAN

EL TATIO

KAWERAU
WAIRAKEI

MELUN
HUNGARIAN BASIN
LARDERELLO
KIZILDERE
MAIKOP   CHERKESSK
GROZNY   MAHACHKALA
ARTARA
CHIMKENT
TASHKENT
KHODZA-
OBI-GARM

ETHIOPIAN RIFT VALLEY
KENYA RIFT VALLEY

MERCATOR PROJECTION
SCALE AT EQUATOR
0   1000   2000   3000
MILES
0   1000   2000   3000
KILOMETRES

SOURCE: GEONOMICS, INC. (HEMPSTEAD, NEW YORK, 1974).

Geothermal regions of the world.

Source: A Technology Assessment of Geothermal Energy Resource Development, prepared for the National Science Foundation by The Futures Group, U.S. Government Printing Office, Washington, D.C. April 1975.

U.S. GEOTHERMAL RESOURCES

Estimates of the extent of recoverable geothermal energy in the United States vary widely and depend upon the extent to which various types of geothermal resources are believed to be amenable to recovery. Only one geothermal area producing dry steam, that at The Geysers in California, has been discovered in the United States. Hydrothermal resources of various qualities of temperature and mineral content have been identified but have not been exploited although several experimental programs are underway.

The largest geothermal potential is represented by the heat stored in dry hot rock, heat that could possibly be "mined" if suitable techniques for underground rock fracturing can be developed. Estimates of United States geothermal resources are given in the accompanying tables.

In an assessment of geothermal resources prepared by the U. S. Geological Survey in 1975 and published in *USGS Circular 726*, the geothermal reserves of the United States, that is, the resources recoverable at present prices and with current technology, were estimated to be 3500 megawatt-centuries (or 11,700 megawatts for 30 years) in high temperature hot water and steam systems. Paramarginal resources were estimated at 125,000 megawatts for 30 years and total resources at 500,000 megawatts for 30 years.

Estimates of Potential Electricity Produced from Geothermal Energy
(MWe)

Source	1972	1974	1985	1990	2000
Bureau of Mines (1972)	192	—	4,000	—	40,000
Dept. of Interior (1972)	192	—	19,000	—	75,000
National Petroleum Council (1972) Case I	—	—	19,000	—	—
National Petroleum Council (1972) Case IV	—	—	3,500	—	—
Geothermal Energy, W. Hickel (1972)	—	—	132,000	—	395,000
Rex and Howell (1973)	—	—	—	400,000	—
Calif. Div. of Oil & Gas 1972 (in SRI 1973)	—	—	—	—	7,500 (in California)
SRI (1973)	—	—	11,800	—	4,400 (in California)
Futures Group "Normal Program" (1974)	—	396	9–11,000	—	55–200,000
Futures Group "Crash Program" (1974)	—	396	27–40,000	—	270–800,000

Source: Task Force Report Geothermal Energy, Project Independence, U.S. Government Printing Office, November 1974.

Geothermal Resources—Estimated Recoverable Heat with Present or Near-Term Technology without Regard to Cost†; (In Quads).*

RESOURCE TYPE	KNOWN	INFERRED	STATE OF TECHNOLOGY
HYDROTHERMAL CONVECTIVE**			
VAPOR DOMINATED (>150°C)	2	2	COMMERCIAL
LIQUID DOMINATED			
HIGH TEMPERATURE (>150°C)	20	110	TEST PHASE
LOW TEMPERATURE (90°-150°C)	80	250	TEST PHASE
GEOPRESSURED			
ELECTRICAL UTILIZATION	100	230	EXPERIMENTAL
METHANE PRODUCTION	500	1500	
HOT DRY ROCK	80	240	EXPERIMENTAL
MAGMA ++	80	240	UNEXPLORED
TOTAL	~900	~2500	
GRAND TOTAL (KNOWN PLUS INFERRED)	3400 QUADS		

* NORMAL GRADIENTS ARE NOT INCLUDED AT THIS TIME AS THEY ARE NOT PRESENTLY CONSIDERED RECOVERABLE. 1 QUAD=10^{15} Btu's.
** DOES NOT INCLUDE LESS THAN 90°C SYSTEMS, ALTHOUGH SUCH SYSTEMS MAY BE ECONOMICALLY EXPLOITABLE ESPECIALLY FOR NON-ELECTRIC APPLICATIONS.
+ ASSUMING 2% EXTRACTION RECOVERY, 8% CONVERSION EFFICIENCY.
++ MAGMA RESOURCES MAY BE RENEWED BY NATURAL RESUPPLY FROM THE INTERIOR OF THE EARTH; THEREFORE, THIS ESTIMATE MAY BE CONSERVATIVE.

SOURCE: DEFINITION REPORT: GEOTHERMAL ENERGY RESEARCH, DEVELOPMENT AND DEMONSTRATION PROGRAM (ERDA-86), OCTOBER 1975.

Source: A National Plan for Energy Research, Development and Demonstration: Creating Energy Choices for the Future, ERDA 76-1, Energy Research and Development Administration, April 15, 1976.

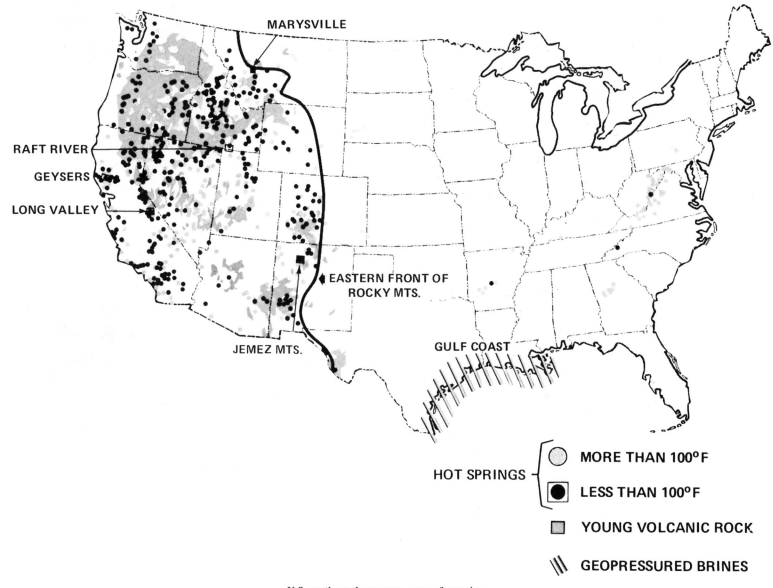

U.S. geothermal resources, areas of promise.

Source: U.S. Geological Survey, 1974, *Energy Perspectives*, U.S. Department of the Interior, February 1975.

A commercial geothermal plant at "The Geysers" located in Sonoma County, Ca., about 75 miles north of San Francisco.

4

Energy Consumption Trends

WORLD ENERGY CONSUMPTION 1900–73

Since 1900, the consumption of energy throughout the world has increased by a factor of ten, the world population has more than doubled, and per capita use of energy has quadrupled. For the past 50 years, world energy consumption has increased at an average rate of $3\frac{1}{2}\%$ per year and per capita consumption has increased at a rate of 2.0% per year.

The growth in energy consumption has not been uniform during these years—energy consumption remained relatively constant during the depression years of the 1930s while in the post-World War II period, energy consumption increased at a rate of 4.8% per year and per capita energy consumption at a rate of 2.8% per year.

World Production of Energy

Year	Annual Production Q (10^{18} Btu)[a]	Growth Rate %/year	World Population (millions)	Per Capita Energy Production (10^6 Btu/year)
1900	21	—	1571	13
1925	43	2.9	1965	22
1950	76	2.2	2486	30
1960	112	4.0	2982	38
1965	147	5.5	3289	45
1970	191	5.4	3632	52
1973 (est.)	218	4.9	3855	57

[a]Author's note.

Source: Survey of Energy Resources 1974, The U.S. National Committee of the World Energy Conference, 345 East 47th Street, New York, N.Y. 10017, 1974.

Physical Quantities of Fuels Consumed in 1970

Energy Source	World Quantity Millions of Tons	World Quantity Cubic Miles	United States Quantity Millions of Tons	United States Quantity Cubic Miles	United States Per Capita Tons
Coal (crushed)	2657	0.72	525	0.14	2.6
Petroleum	2350	0.59	800	0.20	4.0
Natural gas	820		460		2.3
One atmosphere		260		145	
65 atmospheres		4.00		2.25	
Liquified		0.43		0.24	
Falling water					
Falling 100 feet		3400		730	

Source: Fisher, John C.; *Energy Crises in Perspective*, John Wiley & Sons, New York, 1974.

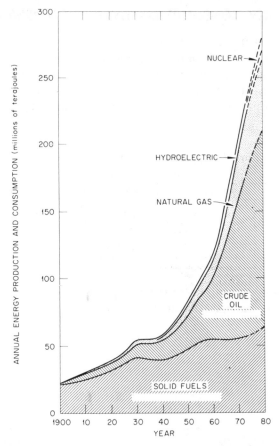

Changing use of energy resources in the twentieth century.

Source: Survey of Energy Resources 1974; in United States National Committee of the World Energy Conference, 325 East 47th Street, New York, N.Y. 10017, 1974.

WORLD ENERGY CONSUMPTION BY REGION 1950-90

During the period 1960-72, the rate of increase of energy consumption on a worldwide basis was 4.9% per year while in the United States the growth rate was 4.1% per year. The U.S. Department of the Interior has projected an annual rate of increase in world energy consumption of 3.3% during the period 1972-90. The corresponding rate of increase in energy consumption in the United States during that period is projected to be 3.0% per year.

World Energy Consumption, by Region, 1960-90
(Quadrillion Btu)[1]

REGION	1960	1965	1970	1972	1980	1985	1990
UNITED STATES	44.6	53.3	67.0	72.0	86.3	102.9	121.9
WESTERN EUROPE	26.4	34.4	46.0	49.1	62.6	75.2	87.2
JAPAN	3.7	6.2	12.0	13.4	20.4	26.7	34.0
SINO-SOVIET BLOC	39.0	45.2	58.3	63.7	82.0	94.0	109.0
REST OF WORLD	18.0	24.3	33.6	35.7	45.0	52.1	60.4
TOTAL	131.7	163.4	216.9	233.9	296.3	350.9	416.5

[1] 1 Quadrillion Btu = 500,000 barrels petroleum per day for a year
 = 40 million tons of bituminous coal
 = 1 trillion cubic feet of natural gas
 = 100 billion kWh (based on a 10,000-Btu/kWh heat rate)

Source: Energy Perspectives, U.S. Department of the Interior, February 1975.

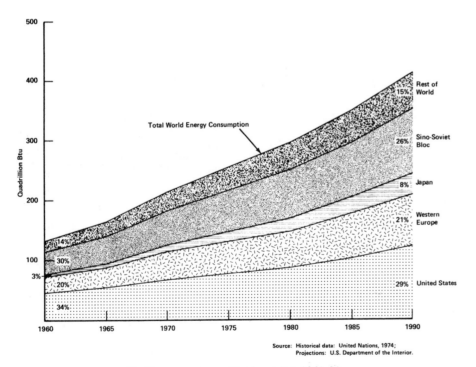

Source: Historical data: United Nations, 1974;
Projections: U.S. Department of the Interior.

World energy consumption by region, 1960-90.

Source: Energy Perspectives, U.S. Department of the Interior, February 1975.

World Primary Energy Consumption

COUNTRY/AREA	1974 Oil	1974 Natural Gas	1974 Solid Fuels	1974 Water Power	1974 Nuclear	1974 TOTAL	1973 Oil	1973 Natural Gas	1973 Solid Fuels	1973 Water Power	1973 Nuclear	1973 TOTAL
					MILLION TONNES OIL EQUIVALENT							
U.S.A.	785·4	560·4	331·9	76·9	28·8	1 783·4	818·0	572·3	335·0	75·6	21·8	1 822·7
Canada	88·1	64·1	15·0	59·4	4·9	231·5	83·7	63·8	16·4	55·5	4·4	223·8
Other Western Hemisphere	178·9	48·0	14·0	31·3	0·3	272·5	169·9	42·5	16·4	28·5	—	257·3
TOTAL WESTERN HEMISPHERE	1 052·4	672·5	360·9	167·6	34·0	2 287·4	1 071·6	678·6	367·8	159·6	26·2	2 303·8
Belgium & Luxembourg	27·5	10·0	11·6	0·1	—	49·2	31·5	8·2	11·0	0·1	—	50·8
Netherlands	35·4	32·9	3·3	—	0·8	72·4	41·3	32·2	3·1	—	0·3	76·9
France	120·1	17·0	27·0	12·6	3·0	179·7	127·3	15·7	26·3	10·6	3·0	182·9
W. Germany	134·4	32·1	82·9	5·0	2·7	257·1	149·7	27·0	87·9	3·0	2·9	270·5
Italy	100·7	15·9	9·2	11·0	0·9	137·7	103·6	14·4	7·6	10·1	0·8	136·5
U.K.	105·8	30·8	68·9	1·3	7·3	214·1	113·4	26·1	78·5	1·2	5·9	225·1
Scandinavia	51·1	—	7·3	36·7	0·4	95·5	55·9	—	7·0	32·6	0·5	96·0
Spain	38·2	1·0	12·3	8·2	1·9	61·6	36·3	1·0	11·7	7·5	1·7	58·2
Other Western Europe	86·0	6·4	39·4	25·4	1·5	158·7	90·0	5·2	38·7	23·6	1·6	159·1
TOTAL WESTERN EUROPE	699·2	146·1	261·9	100·3	18·5	1 226·0	749·0	129·8	271·8	88·7	16·7	1 256·0
Japan	261·1	5·1	58·8	19·0	4·1	348·1	268·3	4·8	59·8	15·8	2·1	350·8
Australasia	33·9	4·1	26·1	2·4	—	66·5	32·8	3·9	29·4	2·2	—	68·3
U.S.S.R.	341·8	216·6	369·1	38·7	4·0	970·2	317·7	200·4	361·8	36·9	3·0	919·8
Eastern Europe	78·3	43·5	225·1	5·7	0·3	352·9	74·8	39·4	218·7	5·3	0·1	338·3
China*	48·8	4·4	331·5	9·5	—	394·2	41·7	4·0	316·3	9·0	—	371·0
Other Eastern Hemisphere	227·4	42·0	134·8	21·1	1·0	426·3	220·0	37·9	127·5	19·5	1·0	405·9
TOTAL EASTERN HEMISPHERE	1 690·5	461·8	1 407·3	196·7	27·9	3 784·2	1 704·3	420·2	1 385·3	177·4	22·9	3 710·1
WORLD	2 742·9	1 134·3	1 768·2	364·3	61·9	6 071·6	2 775·9	1 098·8	1 753·1	337·0	49·1	6 013·9

The above data are not comparable with previous years, see page 24.   *Includes Albania, N. Korea and N. Vietnam.

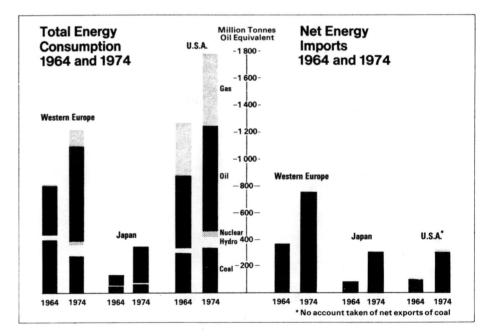

Source: The British Petroleum Company Limited, Britannic House, Moor Lane, London EC2Y 9BU.

SOURCES OF WORLD ENERGY SUPPLY 1950-68

Although there has been an increase in the production of all basic energy forms—coal, petroleum, natural gas, hydropower, and nuclear power—during the period 1950-68, there has been a marked shift in the relative production and consumption of these energy forms. Coal has dropped from 56% to 34% of the total energy supply while petroleum has increased from 29% to 43% and gas from 9% to 17%. The contribution of hydropower and nuclear power combined remained constant at 6.5% during that period.

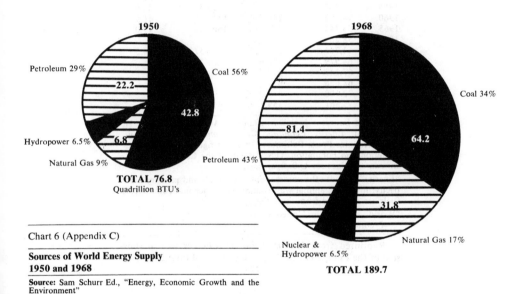

Chart 6 (Appendix C)

**Sources of World Energy Supply
1950 and 1968**

Source: Sam Schurr Ed., "Energy, Economic Growth and the Environment"

Source: Exploring Energy Choices, Energy Policy Project of The Ford Foundation, Washington D.C. 1974.

World Energy Consumption, by Source and Major Region, 1950 and 1968

	1950					1968				
	Coal[a]	Oil[b]	Natural Gas	Hydro[c]	Total	Coal[a]	Oil[b]	Natural Gas	Hydro and Nuclear[c]	Total
					(trillion BTU)					
North America	14,013	14,264	6,226	2,357	36,860	13,968	29,734	21,032	3,860	68,594
Canada	1,100	775	76	756	2,707	639	2,682	1,468	1,373	6,162
United States	12,913	13,489	6,150	1,601	34,153	13,329	27,052	19,564	2,487	62,432
Western Europe	13,533	2,506	49	1,395	17,483	13,541	21,535	1,566	4,942	41,584
Oceania	581	243	0	66	890	958	1,086	0	196	2,240
Latin America	235	1,747	199	216	2,397	414	5,501	1,475	644	8,034
Asia (excl. Communist)	2,026	1,084	52	642	3,804	4,475	10,428	714	1,140	16,757
Japan	1,076	87	3	573	1,739	2,259	5,615	88	729	8,691
Other Asia	950	996	49	68	2,063	2,216	4,813	626	411	8,066
Africa	797	478	0	22	1,297	1,531	1,583	46	183	3,343
U.S.S.R. and Communist										
Eastern Europe	10,453	1,870	298	221	12,842	20,840	10,876	6,977	1,150	39,843
U.S.S.R.	6,369	1,660	207	191	8,427	12,294	9,167	6,140	1,027	28,628
Eastern Europe	4,084	210	87	33	4,414	8,546	1,709	837	123	11,215
Communist Asia	1,160	11	0	79	1,250	8,443	669	—	230	9,342
World	42,798	22,203	6,824	4,998	76,823	64,170	81,412	31,810	12,345	189,737

					(percent of each region's total energy consumption)					
North America	38.0	38.7	16.7	6.4	100.0	20.4	43.3	30.7	5.6	100.0
Canada	40.6	28.6	2.8	27.9	100.0	10.4	43.5	23.8	22.3	100.0
United States	37.8	39.5	18.0	4.7	100.0	21.3	43.3	31.3	4.0	100.0
Western Europe	77.4	14.3	0.3	8.0	100.0	32.6	51.8	3.8	11.9	100.0
Oceania	65.3	27.3	0	7.4	100.0	42.8	48.5	0	8.7	100.0
Latin America	9.8	72.9	8.3	9.0	100.0	5.2	68.5	18.4	8.0	100.0
Asia (excl. Communist)	53.3	28.5	1.4	16.9	100.0	26.7	62.2	4.3	6.8	100.0
Japan	61.9	5.0	0.2	32.9	100.0	26.0	64.6	1.0	8.4	100.0
Other Asia	46.0	48.3	2.4	3.3	100.0	27.5	59.7	7.8	5.1	100.0
Africa	61.4	36.9	0	1.7	100.0	45.8	47.4	1.4	5.5	100.0
U.S.S.R. and Communist										
Eastern Europe	81.4	14.6	2.3	1.7	100.0	52.3	27.3	17.5	2.9	100.0
U.S.S.R.	75.6	19.7	2.5	2.3	100.0	42.9	32.0	21.4	3.6	100.0
Eastern Europe	92.5	4.8	2.0	0.7	100.0	76.2	15.2	7.5	1.1	100.0
Communist Asia	92.8	0.9	0	6.3	100.0	90.4	7.2	—	2.5	100.0
World	55.7	28.9	8.9	6.5	100.0	33.8	42.9	16.8	6.5	100.0

Source: Reference 2.

Note: Dashes indicate "not available."

[a]Principally bituminous coal, but also includes anthracite, a variety of low-quality coals, and lignite.

[b]Including, where known, natural gas liquids.

[c]Also includes small quantities of geothermal electricity. The nuclear portion of this column for 1968 was (in trillion BTU): Canada, 8; United States, 130; Western Europe, 355; Japan, 11; Eastern Europe, less than 0.5 trillion; and U.S.S.R., unavailable. This column also includes net regional imports of electricity.

Source: Finkel, Asher J., ed., *Energy, The Environment and Human Health,* Publishing Sciences Group, Inc., Acton, Massachusetts, 1974, from Sam Schurr, Ed., *Energy, Economic Growth, and the Environment*, Johns Hopkins University Press, Baltimore, 1972.

WORLD PER CAPITA ENERGY CONSUMPTION

Per capita energy consumption in the United States is more than twice that of the industrialized nations of Europe and of Japan, more than five times the world average per capita energy consumption, and more than ten times that of the developing countries. The rate of increase in per capita energy consumption during the period 1950-68 has been higher, however, in almost all countries of the world compared to the United States.

Differences in energy consumption per capita among the countries of the world reflect differences in general economic level, in the type of economy, and patterns of living. Additional information on energy consumption in selected countries is given on page 122, on United States per capita energy consumption on page 137, and the relationship of gross national product to energy consumption in Chapter 15 on Energy Costs.

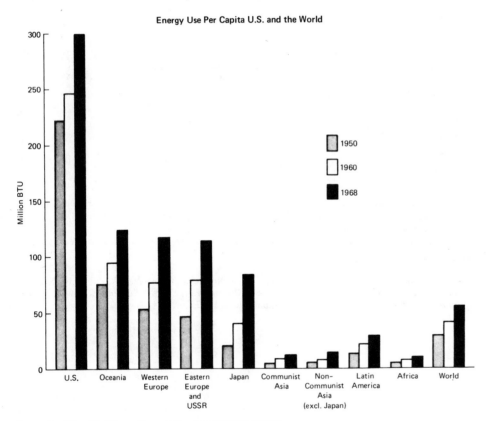

Energy Use Per Capita U.S. and the World

☐ 1950
☐ 1960
■ 1968

Source: Sam Schurr Ed., "Energy, Economic Growth and the Environment"

Source: Exploring Energy Choices, Energy Policy Project of The Ford Foundation, Washington, D.C. 1974.

Per Capita Energy Consumption, U.S. and World Selected Years, 1920–1973

Year	Total U.S. (Trillion BTU)	Per Capita U.S. (Million BTU)	Total World (Trillion BTU)	Per Capita World (Million BTU)
1920	19,768	185.7		
1925	20,879	180.2	44,249	23.4
1930	22,253	180.6		
1940	23,877	180.1		
1950	33,992	223.2	76,823	30.7
1955	39,703	239.3	99,658	36.6
1960	44,569	246.7	124,046	41.5
1965	53,343	274.5	160,722	49.0
1967	58,265	293.2		
1968	61,763	307.7	189,737	54.5
1969	64,979	320.6		
1970	67,143	327.7		
1971	68,698	331.8		
1972	72,108	345.3		
1973P	75,561	359.1		

Notes and Sources: Joel Darmstadter, "Appendix," in Sam H. Schurr, ed., *Energy, Economic Growth, and the Environment*, Johns Hopkins Press, Baltimore, 1972, for figures on world consumption and population. U.S. Bureau of the Census, and U.S. Bureau of Mines for U.S. figures.

"Energy" includes the commercial fossil fuels and primary electricity: firewood, animal wastes and most other non-commercial fuels are excluded from these figures.

"World" includes the United States.

P—Preliminary

Reprinted from *Exploring Energy Choices* with permission of the Energy Policy Project of The Ford Foundation.

World Energy Consumption and Population, by Major Regions, 1950, 1960, 1968

	1950			1960			1968		
Region	Total Consumption (trillion BTU)	Population (million)	Consumption per Capita (million BTU)	Total Consumption (trillion BTU)	Population (million)	Consumption per Capita (million BTU)	Total Consumption (trillion BTU)	Population (million)	Consumption per Capita (million BTU)
North America	36,860	166.1	221.9	48,701	198.7	245.1	68,594	222.0	309.9
Canada	2,707	13.7	197.6	3,885	17.9	217.0	6,162	20.8	296.3
United States	34,153	152.3	224.3	44,816	180.7	248.0	62,432	201.2	310.3
Western Europe	17,483	302.4	57.8	26,066	326.5	79.8	41,584	350.6	118.6
Oceania	890	12.2	73.0	1,398	15.4	90.8	2,240	18.3	122.4
Latin America	2,397	161.9	14.8	4,939	212.4	23.3	8,034	267.4	30.0
Asia (excl. Communist)	3,804	805.4	4.7	8,228	970.6	8.5	16,757	1,182.7	14.2
Japan	1,739	82.9	21.0	3,672	93.2	39.4	8,691	101.1	86.0
Other Asia	2,063	722.5	2.9	4,556	877.4	5.2	8,066	1,081.6	7.5
Africa	1,297	217.0	6.0	2,162	276.0	7.8	3,343	336.5	9.9
U.S.S.R. & Com. East Europe	12,842	269.8	47.6	25,973	312.9	83.0	39,843	341.9	116.5
U.S.S.R.	8,427	180.0	46.8	17,898	214.4	83.5	28,628	237.8	120.4
Eastern Europe	4,414	89.7	49.2	8,075	98.5	82.0	11,215	104.1	107.7
Communist Asia	1,250	569.8	2.2	6,579	677.5	9.7	9,342	765.2	12.2
World	76,823	2,504.5	30.7	124,046	2,989.9	41.5	189,737	3,484.5	54.5

Source: Reference 3.

Source: Finkel, Asher J., ed., *Energy, The Environment, and Human Health*, Publishing Sciences Group, Inc., Acton, Massachusetts, 1974, from R. G. Ridker, ed., *Population, Resources, and the Environment*, Vol. 3 of *Research Reports of The Commission on Population Growth and The American Future*, U.S. Government Printing Office, Washington, D.C., 1972.

ENERGY CONSUMPTION—SELECTED COUNTRIES 1970

The accompanying tables provide comparisons among the countries of the European Communities,[1] the United States, the USSR and Japan and a tabulation of the energy resources consumed in India. Energy production is expressed in terms of tonnes of coal equivalent.

It can be noted that the United States produces more than twice the electrical energy per capita than the other countries, that the number of passenger vehicles per capita in the United States is twice that of western European countries, and five times that of Japan. On the other hand, passenger travel by railroad is 50–30 times greater, on a per capita basis, in these countries than it is in the United States.

In contrast to the industrialized countries, where energy is derived principally from oil, natural gas, and coal; the Indian economy in 1969 obtained more than half its energy from dung, firewood, and vegetable waste.

Energy Consumption, India, 1953–69

	Commercial Energy				Noncommercial Energy		
	Coal	Lignite	Oil	Hydropower	Dung	Firewood	Vegetable Waste
	(ton × 10^{-6})			(KWH × 10^{-10})	(ton × 10^{-6})		
1953–54	34.1	—	3.5	2.9	46.4	86.3	26.4
1955–56	35.1	—	4.4	3.7	48.8	88.8	27.7
1960–61	49.9	—	6.7	7.8	54.0	100.0	30.7
1962–63	59.6	—	8.5	11.8	54.9	101.6	31.1
1965–66	67.7	2.5	10.7	15.2	44.06	126.4	32.5
1968–69	69.5	3.4	14.8	20.8	49.40	142.0	34.2

Forms of Energy in Total Commercial and Noncommercial Consumption, India (Percent)

	Commercial					Noncommercial			
	Coal	Lignite	Oil	Hydropower	Total	Dung	Firewood	Vegetable Waste	Total
1953–54	18.5	—	11.9	1.4	31.8	—	—	—	68.2
1955–56	18.1	—	13.2	1.7	33.0	—	—	—	67.0
1960–61	20.6	—	16.2	2.6	39.6	—	—	—	60.4
1962–63	22.2	—	18.9	3.9	45.0	—	—	—	53.0
1965–66	21.6	0.5	20.0	4.1	46.2	5.6	38.3	9.9	53.9
1968–69	19.4	0.6	23.5	4.2	47.7	5.5	37.7	9.1	52.3[a]

[a]The coal equivalent for the last of above-indicated years is 358.65 million tons of coal.
Source: Datta, R. L., *Solar Energy in Developing Countries: Perspectives and Prospects*, National Academy of Sciences, Washington, D.C., March 1972.

[1]The European Communities include the following countries in the Six: Belgium, France, Federal Republic of Germany, Italy, Netherlands, and Luxembourg; and the following additional countries in the Nine: Denmark, Ireland, and the United Kingdom.

Comparative Data—European Communities and Other Countries 1972-73

	Six[a]	Nine[b]	U.S.A.	USSR	Japan
Population (millions)	192.5	256.6	210.4	247.5	108.4
Production of primary energy (millions of tons of coal equivalent)	346.4	503.8	2205.2	1411.6	65.6
Internal consumption of primary energy (millions of tons of coal equivalent)	915.3	1260.4	2390.0	1208.6	384.5
Crude petroleum production (millions of tons of coal equivalent)	16.3	16.8	755.7	572.6	1.0
Production of electricity (millions of tons of coal equivalent)	42.3	54.8	119.7	46.9	33.7
Consumption of electricity per capita (tons of coal equivalent	4.8	4.9	11.4	5.0	3.9
Consumption of electricity per capita (kilowatt hours)					
industrial	1740.0	1807.0	3600.0	2000.0	2150.0
other	1310.0	1565.0	4607.0	930.0	1445.0
Steel production (millions of tonnes)	122.9	150.1	139.9	131.5	119.3
Passenger car production (millions)	8.6	10.5	8.8	0.7	4.0
Dwellings completed (per 1000 inhabitants)	8.7	8.1	9.7	9.0	17.9
Rail transport (billions of passenger-kilometers)	132.9	165.6	19.6	285.8	197.8
Passenger motor vehicles (per 1000 inhabitants)	251.0	246.0	460.0	7.0	117.0
Television sets (per 1000 inhabitants)	232.0	247.0	449.0	160.0	222.0
Telephones (per 1000 inhabitants)	216.0	234.0	604.0	49.0	282.0

[a]Six: Federal Republic of Germany, France, Italy, Netherlands, Belgium, Luxembourg.
[b]Nine: The six countries above plus the United Kingdom, Ireland, Denmark.
Source: Eurostat: *Basic Statistics of the Community 1973-1974*, Statistical Office of the European Communities, 1049 Brussels, Belgium.

WORLD COAL PRODUCTION

Historical Trends in Coal Use. Although coal was used as a fuel at least 2000 years ago in China and in the Roman Empire, wood was the primary fuel throughout the world until about the year 1200. The mining and use of coal increased rapidly after that time, especially in England, Scotland, and northern Europe where coal was used in industry for the manufacture of glass, brick, and iron. By 1550, England was the largest producer of coal in the world with production at about 200,000 metric tonnes per year. Deforestation and urbanization contributed to an increased use of coal in homes as well.

World consumption of coal increased at about 2% per year during the period 1000 A.D. to 1860; at 4% per year during the period 1860 to 1914, at 0.75% per year from 1916 to 1946, and at an average of 3.6% per year from 1946 to 1974. During recent years, however, world coal production has been declining with estimates of a 2.0% annual growth between 1961 and 1970 and 1.2% from 1965 to 1970.

In 1925, about 80% of the coal being mined was in three countries—the United States, the United Kingdom, and Germany. During the period 1925-71, production in the United States remained relatively constant while production in the United Kingdom and France and Belgium declined markedly. Major increases in coal production have occurred in China, the USSR, Poland, Czechoslovakia, India, Australia, and South Africa. These shifts in coal production have been caused by two major factors—depletion of domestic resources and a greater use of less expensive and more convenient oil and gas either from domestic production or from imports.

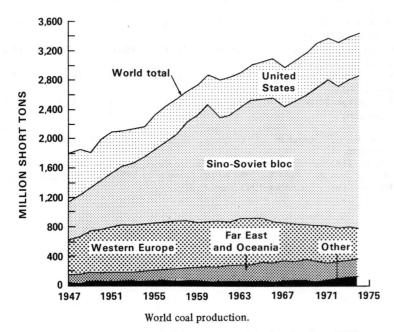

World coal production.

Source: Energy Perspectives 2, U.S. Department of the Interior, June 1976.

Comparison of Coal and Lignite Production in 1925 and 1971

Nation	Annual Production (Megatonnes)	
	1925[1]	1971[2]
United States	528	510
Germany	286	482*
United Kingdom	247	147
France	48	35
Japan	32	34
Czechoslovakia	31	113
Poland	29	180
Belgium	23	12
China	24	410
USSR	22	620
India	21	73
Australia	15	73
South Africa	12	61
Canada	12	18
Spain	7	14
Hungary	6	28
Yugoslavia	4	31
Romania	3	22
Bulgaria	1	28
Turkey	<1	32
Korea	<1	32**
Greece	<1	11
Total	1348	3029
World Total	1381	3069

*Includes East and West Germany
**Includes North and South Korea

[1] J. Darmstadter, et al, *Energy in the World Economy*
[2] *Mineral Yearbook, 1971*, U.S. Bureau of Mines

Source: Survey of Energy Resources 1974, The United States National Committee of the World Energy Conference, 345 East 47th Street, New York, N.Y. 10017. 1974.

World Coal¹ Production, 1947–74
[Million short tons]

Year	United States	Other Western Hemisphere countries	Western Europe	Africa	Middle East	Far East and Oceania	Sino-Soviet bloc	World total
1947	688	22	493	29	4	100	479	1,815
1948	657	26	533	30	3	107	531	1,886
1949	481	26	578	32	6	116	587	1,825
1950	560	26	592	33	6	117	664	1,998
1951	576	26	627	34	7	126	722	2,118
1952	507	25	641	36	7	132	785	2,133
1953	488	24	637	37	8	134	835	2,163
1954	421	23	646	38	9	134	902	2,173
1955	491	24	652	41	9	138	998	2,353
1956	530	24	663	43	10	147	1,069	2,486
1957	518	22	669	45	11	160	1,150	2,575
1958	432	21	655	47	12	164	1,357	2,688
1959	433	20	630	46	11	170	1,465	2,777
1960	434	21	618	48	11	189	1,578	2,899
1961	420	21	613	49	11	202	1,420	2,736
1962	439	21	580	50	12	213	1,453	2,768
1963	477	22	622	51	13	222	1,521	2,928
1964	504	24	628	55	9	222	1,590	3,031
1965	527	22	601	59	10	237	1,622	3,079
1966	547	21	567	58	11	247	1,664	3,114
1967	565	22	540	59	11	248	1,552	2,997
1968	557	22	525	62	11	257	1,652	3,086
1969	571	22	506	63	12	253	1,761	3,188
1970	613	28	500	66	10	240	1,861	3,317
1971	561	30	496	70	10	226	1,991	3,385
1972	602	33	458	70	11	234	1,930	3,339
1973	599	35	466	75	12	233	1,984	3,404
1974²	608	38	438	78	13	247	2,029	3,450

¹ Anthracite, bituminous coal, and lignite.
² Preliminary.
Source: U.S. Department of the Interior, Bureau of Mines, *Minerals Yearbook 1950-74*.

Source: Energy Perspectives 2, U.S. Department of the Interior, June 1976.

WORLD PETROLEUM AND GAS PRODUCTION 1973

The exploitation of petroleum and gas resources began in 1859 with the drilling of the first oil well by Drake in Pennsylvania. For a number of years, production of petroleum was relatively modest with the major use in heating. The advent of the internal combustion engine in the late 19th century led to an ever-increasing demand for refined petroleum products. By 1945, petroleum was supplying as much energy as coal in the United States, and by 1974, oil and gas were providing about 77% of the energy consumed.

Normally, most oil wells produce both oil and natural gas. As a liquid, oil is readily transported from the well to the point of consumption. Natural gas, on the other hand, can be transported only through pipelines or as a liquid produced by refrigerating the gas with subsequent transport in insulated tanks. For many years, the lack of extensive pipeline distribution systems or refrigeration technology limited the utilization of natural gas to local regions. In many cases, natural gas produced as a by-product from oil wells has been, and still is, burned or "flared" at the well head. The difficulty in transport of natural gas except by pipeline, has meant there has been relatively little international trade in natural gas and domestic consumption has been restricted to domestic production capability. International shipments of natural gas, in the form of liquified natural gas (LNG), are steadily increasing, however.

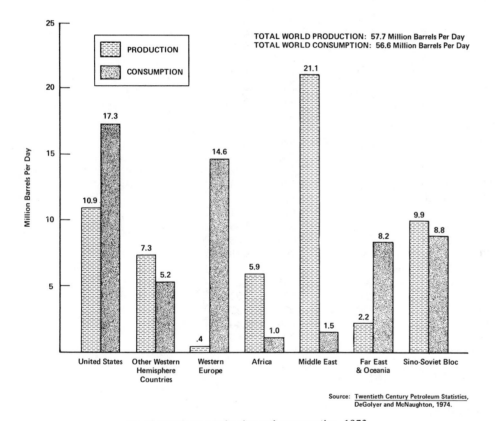

World petroleum production and consumption, 1973.

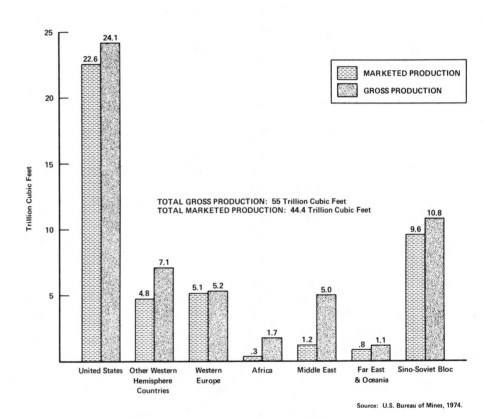

World natural gas production, 1973.

WORLD OIL PRODUCTION AND CONSUMPTION 1964-74

World Oil Production 1964-1974

COUNTRY/AREA	1964	1965	1966	1967	1968	1969	1970	1971	1972	1973	1974	Yearly Change† 1974 over 1964	1974 over 1969
NORTH AMERICA U.S.A.					THOUSAND BARRELS DAILY								
Crude Oil	7 615	7 805	8 295	8 810	9 095	9 240	9 635	9 465	9 440	9 210	8 795	+ 1·4%	− 1·0%
Natural Gas Liquids	1 155	1 210	1 285	1 410	1 505	1 590	1 660	1 695	1 745	1 740	1 690	+ 3·7%	+ 1·2%
	8 770	9 015	9 580	10 220	10 600	10 830	11 295	11 160	11 185	10 950	10 485	+ 1·7%	− 0·7%
Canada	850	935	1 015	1 110	1 195	1'310	1 475	1 585	1 830	2 115	2 000	+ 9·0%	+ 9·2%
Mexico	355	360	370	410	440	460	485	485	505	550	625	+ 5·7%	+ 6·0%
TOTAL NORTH AMERICA	9 975	10 310	10 965	11 740	12 235	12 600	13 255	13 230	13 520	13 615	13 110	+ 2·7%	+ 0·8%
CARIBBEAN Venezuela	3 395	3 505	3 405	3 580	3 645	3 630	3 760	3 620	3 305	3 460	3 060	− 1·1%	− 3·5%
Colombia	170	200	195	190	175	210	220	215	195	185	170	—	− 4·2%
Trinidad	135	135	150	180	185	160	140	130	140	165	180	+ 2·8%	+ 2·8%
TOTAL CARIBBEAN	3 700	3 840	3 750	3 950	4 005	4 000	4 120	3 965	3 640	3 810	3 410	− 0·9%	− 3·2%
SOUTH AMERICA Argentina	275	270	285	315	345	355	390	425	435	420	415	+ 4·2%	+ 3·1%
Brazil	90	95	115	145	165	175	165	170	165	165	175	+ 6·7%	—
Other South America	115	115	120	150	165	· 150	135	135	230	380	325	+ 10·8%	+ 16·6%
TOTAL SOUTH AMERICA	480	480	520	610	675	680	690	730	830	965	915	+ 6·5%	+ 5·9%
TOTAL WESTERN HEMISPHERE	**14 155**	**14 630**	**15 235**	**16 300**	**16 915**	**17 280**	**18 065**	**17 925**	**17 990**	**18 390**	**17 435**	**+ 2·0%**	**+ 0·1%**
WESTERN EUROPE France	55	60	60	55	55	50	45	35	30	25	20	− 9·2%	−15·4%
W. Germany	150	155	155	155	155	155	150	145	140	130	120	− 2·1%	− 4·7%
Austria	50	55	55	55	55	55	55	50	50	50	45	− 1·7%	− 4·1%
Turkey	20	30	40	55	60	70	70	70	65	70	65	+13·6%	− 1·7%
Other Western Europe	140	135	125	125	125	130	130	125	150	170	195	+ 2·7%	+ 7·8%
TOTAL WESTERN EUROPE	415	435	435	445	450	460	450	425	435	445	445	+ 0·4%	− 0·7%
MIDDLE EAST Iran	1 710	1 910	2 110	2 600	2 840	3 375	3 845	4 565	5 050	5 895	6 060	+13·4%	+12·4%
Iraq	1 255	1 315	1 390	1 230	1 505	1 525	1 565	1 700	1 465	2 020	1 935	+ 4·4%	+ 4·8%
Kuwait	2 115	2 170	2 275	2 290	2 420	2 575	2 735	2 925	3 000	2 755	2 275	+ 0·7%	− 2·4%
Neutral Zone	360	370	420	415	405	420	505	545	565	535	545	+ 4·1%	+ 5·3%
Qatar	215	235	290	325	340	355	370	430	485	570	520	+ 9·3%	+ 7·9%
Saudi Arabia	1 730	2 025	2 395	2 600	2 830	2 995	3 550	4 500	5 730	7 345	8 210	+16·8%	+22·4%
Abu Dhabi	185	280	360	380	495	600	695	935	1 050	1 305	1 415	+22·4%	+18·7%
Oman	—	—	—	55	240	330	330	285	280	295	295	*	− 2·1%
Other Middle East	50	60	65	75	105	185	245	320	345	390	465	+25·1%	+19·7%
TOTAL MIDDLE EAST	7 620	8 365	9 305	9 970	11 180	12 360	13 840	16 205	17 970	21 110	21 720	+11·0%	+11·9%
AFRICA Algeria	565	575	730	835	915	955	1 040	780	1 070	1 095	1 040	+ 6·2%	+ 1·7%
Libya	860	1 225	1 505	1 745	2 605	3 110	3 320	2 765	2 240	2 180	1 525	+ 5·9%	−13·3%
Other North Africa	130	130	140	170	290	420	555	505	435	335	325	+ 9·4%	− 5·2%
Nigeria	120	275	420	320	145	540	1 085	1 530	1 820	2 055	2 330	+34·2%	+33·9%
Other West Africa	50	40	45	85	120	155	215	230	270	350	420	+24·0%	+22·9%
TOTAL AFRICA	1 725	2 245	2 840	3 155	4 075	5 180	6 215	5 810	5 835	6 015	5 640	+12·6%	+ 1·8%
SOUTH EAST ASIA Indonesia	470	485	475	510	600	750	855	890	1 080	1 335	1 395	+11·5%	+13·2%
Other South East Asia	75	80	95	110	125	140	160	220	250	325	305	+15·2%	+17·1%
TOTAL SOUTH EAST ASIA	545	565	570	620	725	890	1 015	1 110	1 330	1 660	1 700	+12·0%	+13·8%
U.S.S.R.	4 485	4 885	5 335	5 795	6 190	6 595	7 090	7 470	7 890	8 455	9 055	+ 7·3%	+ 6·6%
Eastern Europe	300	305	315	325	330	325	325	340	340	345	350	+ 1·4%	+ 1·5%
China**	190	215	265	240	280	315	430	540	630	845	1 050	+18·9%	+27·3%
Other Eastern Hemisphere	85	105	140	175	195	210	360	500	555	580	575	+20·8%	+22·1%
WORLD (exl. U.S.S.R., E. Europe & China)	24 545	26 345	28 525	30 665	33 540	36 380	39 945	41 975	44 115	48 200	47 515	+ 6·7%	+ 5·5%

*Greater than 300%   †Based on weight   **Includes Albania

Source: The British Petroleum Company Limited, Britannic House, Moor Lane, London EC2Y 9BU

World Oil Production

COUNTRY/AREA	1964	1965	1966	1967	1968	1969	1970	1971	1972	1973	1974	Yearly Change 1974 over 1964	1974 over 1969
NORTH AMERICA U.S.A.					MILLION TONNES								
Crude Oil	379·2	387·6	412·0	437·5	452·9	458·8	478·6	469·9	470·1	457·3	436·8	+ 1·4%	− 1·0%
Natural Gas Liquids	41·7	43·6	46·1	50·5	53·8	56·4	58·9	60·1	62·1	61·7	59·9	+ 3·7%	+ 1·2%
	420·9	431·2	458·1	488·0	506·7	515·2	537·5	530·0	532·2	519·0	496·7	+ 1·7%	− 0·7%
Canada	40·8	44·6	49·2	53·8	58·2	62·2	71·5	76·6	88·8	102·3	96·5	+ 9·0%	+ 9·2%
Mexico	17·6	17·8	18·3	20·2	21·7	22·8	23·9	23·9	24·8	26·9	30·6	+ 5·7%	+ 6·0%
TOTAL NORTH AMERICA	479·3	493·6	525·6	562·0	586·6	600·2	632·9	630·5	645·8	648·2	623·8	+ 2·7%	+ 0·8%
CARIBBEAN Venezuela	177·4	182·1	177·0	186·1	189·9	188·7	195·2	187·7	171·5	179·0	158·2	− 1·1%	− 3·5%
Colombia	8·7	10·2	10·0	9·6	8·8	10·7	11·2	11·0	10·0	9·4	8·7	—	− 4·2%
Trinidad	7·1	7·0	7·9	9·5	9·3	8·2	7·3	6·7	7·3	8·6	9·4	+ 2·8%	+ 2·8%
TOTAL CARIBBEAN	193·2	199·3	194·9	205·0	208·2	207·6	213·7	205·4	188·8	197·0	176·3	− 0·9%	− 3·2%
SOUTH AMERICA Argentina	14·3	14·0	15·0	16·4	17·9	18·6	20·4	22·1	22·6	22·0	21·6	+ 4·2%	+ 3·1%
Brazil	4·5	4·6	5·6	7·1	8·0	8·5	8·0	8·3	8·1	8·1	8·5	+ 6·7%	—
Other South America	5·5	5·6	5·8	7·1	7·8	7·2	6·4	6·4	10·8	18·0	15·5	+ 10·8%	+ 16·6%
TOTAL SOUTH AMERICA	24·3	24·2	26·4	30·6	33·7	34·3	34·8	36·8	41·5	48·1	45·6	+ 6·5%	+ 5·9%
TOTAL WESTERN HEMISPHERE	**696·8**	**717·1**	**746·9**	**797·6**	**828·5**	**842·1**	**861·4**	**872·7**	**876·1**	**893·3**	**845·7**	**+ 2·0%**	**+ 0·1%**
WESTERN EUROPE France	2·8	3·0	2·9	2·8	2·7	2·5	2·3	1·9	1·5	1·3	1·1	− 9·2%	−15·4%
W. Germany	7·7	7·9	7·9	7·9	8·0	7·9	7·5	7·4	7·1	6·6	6·2	− 2·1%	− 4·7%
Austria	2·7	2·8	2·7	2·7	2·7	2·7	2·8	2·5	2·5	2·6	2·2	− 1·7%	− 4·1%
Turkey	0·9	1·6	2·1	2·8	3·1	3·6	3·5	3·5	3·4	3·5	3·3	+13·6%	− 1·7%
Other Western Europe	7·5	6·8	6·5	6·5	6·5	6·7	6·7	6·5	7·7	8·6	9·8	+ 2·7%	+ 7·8%
TOTAL WESTERN EUROPE	21·6	22·1	22·1	22·7	23·0	23·4	22·8	21·8	22·2	22·6	22·6	+ 0·4%	− 0·7%
MIDDLE EAST Iran	85·4	95·0	105·2	129·6	141·8	168·1	191·3	227·0	251·9	293·1	301·2	+13·4%	+12·4%
Iraq	61·7	64·4	68·1	60·3	73·9	74·9	76·9	83·5	72·1	99·0	94·9	+ 4·4%	+ 4·8%
Kuwait	106·7	109·1	114·4	115·2	122·1	129·5	137·5	147·1	151·2	138·4	114·4	+ 0·7%	− 2·4%
Neutral Zone	18·8	19·4	21·7	21·5	21·0	21·7	26·0	28·3	29·3	27·6	28·1	+ 4·1%	+ 5·3%
Qatar	10·2	11·1	13·8	15·5	16·3	17·0	17·7	20·5	23·2	27·3	24·9	+ 9·3%	+ 7·9%
Saudi Arabia	86·2	100·6	118·8	129·0	140·9	148·6	176·2	223·4	285·4	364·7	407·8	+16·8%	+22·4%
Abu Dhabi	9·0	13·5	17·3	18·3	23·9	28·9	33·4	44·9	50·6	62·6	67·9	+22·4%	+18·7%
Oman	—	—	—	2·9	12·1	16·4	16·6	14·4	14·2	14·7	14·8	*	− 2·1%
Other Middle East	2·5	3·0	3·3	3·6	5·3	9·5	12·4	16·2	17·5	19·6	23·5	+25·1%	+19·7%
TOTAL MIDDLE EAST	380·5	416·1	462·6	495·9	557·3	614·6	688·2	805·3	895·4	1 047·0	1 077·5	+11·0%	+11·9%
AFRICA Algeria	26·5	26·6	34·2	39·1	42·9	44·5	48·5	36·5	50·1	51·2	48·5	+ 6·2%	+ 1·7%
Libya	41·4	58·9	72·3	84·0	125·7	149·8	159·8	133·1	108·2	104·9	73·5	+ 5·9%	−13·3%
Other North Africa	6·6	6·7	7·0	8·6	14·5	20·9	27·7	25·2	21·6	16·7	16·0	+ 9·4%	− 5·2%
Nigeria	6·0	13·5	20·4	15·6	7·2	26·4	52·9	74·7	88·9	100·1	113·5	+34·2%	+33·9%
Other West Africa	2·4	2·0	2·2	4·1	5·8	7·5	10·9	11·5	13·6	17·6	21·0	+24·0%	+22·9%
TOTAL AFRICA	82·9	107·7	136·1	151·4	196·1	249·1	299·8	281·0	282·4	290·5	272·5	+12·6%	+ 1·8%
SOUTH EAST ASIA Indonesia	23·3	24·0	23·5	25·2	29·7	37·1	42·2	44·1	53·4	66·0	69·0	+11·5%	+13·2%
Other South East Asia	3·6	4·0	4·7	5·4	6·2	6·8	7·8	11·0	12·3	16·0	15·0	+15·2%	+17·1%
TOTAL SOUTH EAST ASIA	26·9	28·0	28·2	30·6	35·9	43·9	50·0	55·1	65·7	82·0	84·0	+12·0%	+13·8%
U.S.S.R.	223·6	242·9	265·1	288·1	309·2	328·3	353·0	372·0	394·0	421·0	451·0	+ 7·3%	+ 6·6%
Eastern Europe	15·0	15·2	15·6	16·2	16·4	16·1	16·3	16·8	17·3	17·1	17·1	+ 1·4%	+ 1·5%
China**	9·3	10·7	13·3	11·9	14·0	15·7	21·3	26·9	31·5	42·1	52·4	+18·9%	+27·3%
Other Eastern Hemisphere	4·2	5·1	7·0	8·5	9·5	10·3	17·6	24·3	26·7	28·2	27·9	+20·8%	+22·1%
TOTAL EASTERN HEMISPHERE	**764·0**	**847·8**	**950·0**	**1 025·3**	**1 161·4**	**1 301·4**	**1 469·0**	**1 603·2**	**1 735·2**	**1 950·5**	**2 005·2**	**+10·1%**	**+ 9·0%**
WORLD (excl. U.S.S.R., E. Europe & China)	1 212·9	1 296·1	1 402·9	1 506·7	1 650·3	1 783·4	1 959·8	2 060·2	2 168·5	2 363·6	2 330·2	+ 6·7%	+ 5·5%
WORLD	**1 460·8**	**1 564·9**	**1 696·9**	**1 822·9**	**1 989·9**	**2 143·5**	**2 360·4**	**2 475·9**	**2 611·3**	**2 843·8**	**2 850·9**	**+ 8·9%**	**+ 5·9%**

*Greater than 300%.   ** Includes Albania.

Source: The British Petroleum Company Limited, Britannic House, Moor Lane, London EC2Y 9BU.

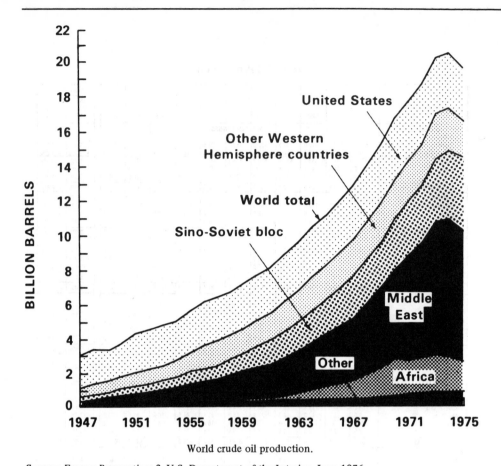

World crude oil production.

Source: Energy Perspectives 2, U.S. Department of the Interior, June 1976.

Orthoflow fluid catalytic cracking refinery of the Getty Oil Company. This refinery processes 102,000 barrels of oil daily. (*Courtesy Getty Oil*)

WORLD PETROLEUM REFINING CAPACITY

Although the refining capacity of the United States had doubled from 7 to 14 million barrels per day in the period 1950-73, the percentage of world refining capacity represented by the United States has dropped from 59% to 23% during that same period. This shift in location of refineries represents both a generally greater increase in oil consumption abroad as well as economic conditions which have made construction of refineries abroad more attractive than in the United States.

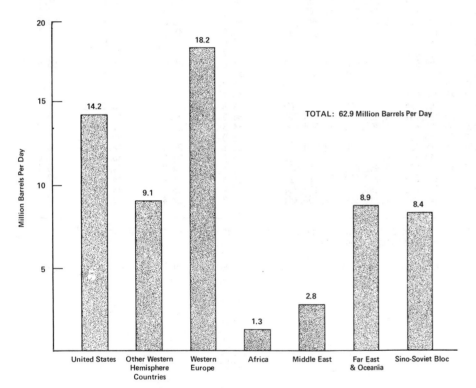

TOTAL: 62.9 Million Barrels Per Day

Source: Oil and Gas Journal, 1974;
U.S. Bureau of Mines, 1974.

World petroleum refining capacity, 1973.

Table 2-9. Number and Capacity of U.S. Refineries, 1918 - 1974

	Number of Refineries As of January 1					Capacity as of January 1 (Barrels Per Day)					
Year	Operating	Shut-down	Total Operating and Shut-down	Build-ing[1]	Total Operating, Shut-down and Building	Operating	Shutdown[2]	Total Operating and Shutdown	Building[3]	Total Operating, Shutdown and Building	Daily Average Crude Oil Runs To Stills (barrels)
1974	257	27	284	5	289	14,220,316	269,120	14,489,436	1,319,490	15,808,926	12,044,000*
1973	252	25	277	0	277	13,454,471	320,175	13,774,646	100,500	13,875,146	12,431,000
1972	250	32	282	2	284	13,034,818	402,650	13,437,468	223,470	13,660,938	11,696,000
1971	253	26	279	3	282	12,658,248	361,830	13,019,978	621,300	13,641,278	11,199,000
1970	262	19	281	4	285	11,882,393	191,930	12,074,323	386,700	12,461,023	10,870,000
1969	264	20	284	2	287	11,575,829	163,680	11,739,509	362,500	12,102,009	10,630,000
1968	270	21	291	1	292	11,172,694	360,160	11,532,854	751,550	12,284,404	10,312,000
1967	260	21	281	4	285	10,412,447	347,160	10,759,607	391,700	11,151,307	9,815,000
1966	267	19	286	3	289	10,171,159	321,580	10,492,739	148,300	10,641,039	9,444,364
1965	273	27	300	1	301	10,161,311	613,284	10,774,595	74,960	10,849,555	9,043,403
1964	282	22	304	1	305	10,063,164	322,210	10,385,374	54,700	10,440,074	8,806,910
1963	287	21	308	2	310	9,814,791	303,530	10,118,321	178,300	10,296,621	8,686,718
1962	287	24	311	1	312	9,812,248	292,899	10,105,147	110,350	10,215,497	8,409,947
1961	289	22	311	0	311	9,629,685	380,388	10,010,073	36,500	10,046,573	8,183,994
1960	290	20	310	2	312	9,543,329	358,095	9,901,424	70,947	9,972,371	8,067,032
1959	291	22	313	0	313	9,450,741	369,105	9,819,846	108,400	9,928,246	7,993,591
1958	289	29	318	2	320	8,939,907	467,800	9,407,707	185,265	9,592,972	7,605,737
1957	298	21	319	3	322	8,808,841	314,833	9,123,674	256,350	9,380,024	7,919,003
1956	294	24	318	2	320	8,380,801	251,589	8,632,390	267,000	8,899,390	7,937,448
1955	296	30	326	4	330	8,069,154	351,476	8,420,630	146,800	8,567,430	7,480,049
1954	308	29	337	7	344	7,782,103	224,794	8,006,897	397,500	8,404,397	6,957,710
1953	315	28	343	4	347	7,481,701	156,960	7,638,661	509,721	8,148,382	6,999,630
1952	327	23	350	0	350	7,161,366	171,519	7,332,885	282,680	7,615,565	6,670,106
1951	325	32	357	1	358	6,701,815	261,829	6,963,644	160,100	7,123,744	6,494,258
1950	320	47	367	2	369	6,222,998	473,302	6,696,300	145,600	6,841,900	5,739,362
1949	336	39	375	3	378	6,230,505	208,490	6,438,995	341,500	6,780,495	5,236,633
1948	352	38	390	2	392	5,825,566	208,686	6,034,252	367,250	6,401,502	5,596,583
1947	361	38	399	0	399	5,336,399	233,083	5,569,482	162,200	5,731,682	5,074,646
1946	364	29	393	1	394	5,086,165	229,691	5,315,856	53,100	5,368,956	4,740,266
1945	380	33	413	1	414	5,077,690	223,463	5,301,153	36,075	5,337,228	4,711,052
1944	384	68	452	0	452	4,709,382	383,641	5,093,023	118,270	5,211,293	4,551,049
1943	386	85	471	1	472	4,409,013	492,998	4,902,011	195,100	5,097,111	3,917,090
1942	430	92	522	1	523	4,496,843	459,756	4,956,599	43,400	4,999,999	3,655,077
1941	420	136	556	6	562	4,180,588	538,381	4,718,969	141,225	4,860,194	3,860,800
1940	461	86	547	10	557	4,196,694	431,952	4,628,646	92,567	4,721,213	3,535,962
1939	435	103	538	7	545	3,933,785	574,770	4,508,555	142,250	4,650,805	3,391,342
1938[4]	431	120	551	10	561	3,970,196	380,955	4,351,151	283,020	4,634,171	3,191,822
1938[5]	431	120	551	10	561	4,151,276	199,875	4,351,151	22,550	4,373,701	—
1937	423	149	572	11	583	3,966,616	328,265	4,294,881	81,200	4,376,081	3,242,301
1936	422	210	632	15	647	3,749,835	367,212	4,117,047	46,899	4,163,946	2,919,590
1935	435	196	631	7	638	3,614,749	443,751	4,058,500	13,900	4,072,400	2,646,000
1934	454	137	591	13	604	3,553,569	364,648	3,918,217	44,450	3,962,667	2,453,797
1933	372	133	505	18	523	3,445,118	444,118	3,889,510	31,545	3,921,055	2,359,600
1932	365	108	473	6	479	3,624,992	389,616	4,014,608	8,720	4,023,328	2,240,429
1931	346	89	435	10	445	3,706,610	236,075	3,942,685	45,000	3,987,685	2,450,981
1930	358	54	412	8	420	3,634,825	130,760	3,765,585	37,200	3,802,785	2,540,951
1929	341	72	413	14	427	3,325,890	183,650	3,509,540	99,000	3,608,540	2,706,049
1928	326	97	423	5	428	3,036,125	214,255	3,250,380	22,000	3,272,380	2,495,342
1927	327	138	465	7	472	2,834,282	226,725	3,061,007	61,000	3,122,007	2,270,781
1926	352	158	510	2	512	2,562,357	290,610	2,852,967	5,500	2,858,467	2,134,970
1925 (May 1)	365	185	550	4	554	2,511,817	342,025	2,853,842	11,000	2,864,842	2,027,178
1924 (November 1)	357	190	547	8	555	2,480,922	333,410	2,814,332	18,200	2,832,532	1,592,433
1922	325	154	479	30	509	1,854,590	254,610	2,109,200	59,950	2,169,150	1,371,797
1921	350	65	415	44	459	1,794,395	94,405	1,888,800	76,600	1,965,400	1,214,693
1920	373[6]	(6)	373	99	472	1,530,565[6]	(6)	1,530,565	263,500	1,794,065	1,185,560
1919	—	—	—	—	289	1,295,115	0	1,295,115	0	1,295,115	990,466
1918	—	—	—	—	267	1,186,155	0	1,186,155	0	1,186,155	893,219

[1] New plants.
[2] Beginning in 1938, shutdown capacity includes inoperative portions of operating refineries as well as plants completely shutdown. In previous years, shutdown capacity represented only plants completely shutdown.
[3] Beginning in 1938, building capacity includes additional capacity being built at existing refineries as well as new plants under construction. In previous years, building capacity included only new plants being built.
[4] New basis; see footnotes 2 and 3.
[5] Old basis; see footnotes 2 and 3.
[6] Shutdown facilities included with operating facilities.
*DOT — TSC computation from weekly averages in American Petroleum Industry, Weekly Statistical Bulletin, Vol. 55, 1974.

Source: U.S. Department of the Interior, Bureau of Mines, "Petroleum Refineries in the United States and Puerto Rico: January 1, 1974," July 22, 1974, p. 3 for 1974 and equivalent tables in earlier editions; and the "Annual Petroleum Statement."

Source: Energy Statistics: A Supplement to the Summary of National Transportation Statistics, Report No. DOT-TSC-OST-75-33, U.S. Dept. of Transportation, August 1975.

Estimated Worldwide Crude Oil Refining Capacity by Area (As of January 1)
(Barrels per day)

Year	United States [1]	Other Western Hemisphere	Total Western Hemisphere	Middle East	Africa	Asia	Europe	Total free world	Sino-Soviet bloc	Total world	United States as a percent of free world	United States as a percent of total world	Western Hemisphere as a percent of free world	Western Hemisphere as a percent of total world
1950	6,696,300	1,450,500	8,146,800	916,500	39,000	266,000	859,200	10,227,500	1,134,000	11,361,500	65.5	58.9	79.6	71.7
1951	6,963,644	1,770,400	8,734,044	940,090	41,000	282,900	1,014,900	11,013,744	(2)	(2)	63.2	----------	79.3	----------
1952	7,332,885	1,700,100	9,032,985	950,000	49,000	344,000	1,495,000	11,870,985	(2)	(2)	61.8	----------	76.1	----------
1953	7,619,720	2,180,200	9,790,920	1,093,900	52,100	379,200	1,835,400	13,151,520	1,087,000	14,238,520	57.9	53.5	74.5	68.8
1954	7,983,977	2,392,700	10,376,677	1,159,700	72,400	459,800	2,077,500	14,140,077	1,246,100	15,392,177	56.5	51.9	73.4	67,4
1955	8,363,044	2,566,500	10,929,544	1,184,300	68,400	613,700	2,365,900	15,161,844	1,784,000	16,945,844	55.2	49.4	72.1	64.5
1956	8,582,636	2,776,800	11,359,436	1,272,800	72,400	731,000	2,373,700	15,809,336	1,481,000	17,290,336	54.3	49.6	71.9	65.7
1957	9,071,697	3,087,400	12,159,097	1,235,200	93,400	876,000	2,762,700	17,126,397	1,550,000	18,676,397	53.0	48.6	71.0	65.1
1958	9,358,307	3,438,600	12,796,907	1,253,700	98,400	1,088,500	2,902,900	18,140,407	(2)	(2)	51.6	----------	70.5	----------
1959	9,761,446	3,680,500	13,441,946	1,432,800	100,100	1,229,600	3,633,200	19,837,646	(2)	(2)	49.2	----------	67.8	----------
1960	9,543,329	3,762,300	13,305,629	1,434,800	116,500	1,374,600	3,977,500	20,209,029	3,322,000	23,531,029	47.2	40.6	65.8	56.5
1961	9,998,573	4,188,000	14,186,573	1,481,800	122,500	1,431,700	4,476,100	21,698,673	(2)	(2)	46.1	----------	65.4	----------
1962	10,033,047	4,442,900	14,475,947	1,514,100	140,300	1,913,000	4,918,100	22,961,447	(2)	(2)	43.7	----------	63.0	----------
1963	10,010,921	4,687,200	14,698,121	1,754,900	196,800	2,129,800	5,149,400	23,929,021	(2)	(2)	41.8	----------	61.4	----------
1964	10,305,774	4,828,400	15,134,174	1,801,300	365,200	2,537,000	6,263,500	26,101,174	(2)	(2)	39.5	----------	58.0	----------
1965	10,419,851	5,021,500	15,441,351	1,801,400	502,900	3,176,600	7,510,500	28,522,751	4,656,000	33,178,751	36.5	31.4	54.1	46.5
1966	10,393,839	5,344,500	15,738,339	1,910,200	613,300	3,516,700	8,541,900	30,320,439	6,305,000	36,625,439	34.3	28.4	51.9	43.0
1967	10,658,407	5,544,500	16,202,907	1,960,200	704,500	3,955,700	9,526,900	32,350,207	(2)	(2)	32.9	----------	50.1	----------
1968	11,353,404	5,883,100	17,236,504	2,052,700	720,100	4,323,900	11,085,000	35,418,204	6,842,000	42,260,204	32.1	26.9	48.7	40.8
1969	11,702,309	6,290,200	17,993,509	2,296,500	776,100	4,919,800	12,884,300	38,849,209	6,375,000	45,224,209	30.1	25.9	46.3	39.8
1970	12,021,273	6,512,500	18,533,733	2,437,900	784,900	5,565,900	13,941,400	41,263,833	6,952,000	48,215,833	29.1	24.9	44.9	38.4
1971	12,860,228	7,128,600	19,988,828	3,171,700	924,900	6,060,900	15,177,200	45,323,528	7,388,000	52,711,528	28.4	24.4	44.1	37.9
1972	13,292,468	7,578,600	20,871,068	2,851,300	902,400	6,817,700	16,982,800	48,425,268	7,690,000	56,115,268	27.4	23.7	43.1	37.2
1973	13,642,446	8,175,800	21,818,246	2,757,700	825,100	7,915,700	16,826,700	50,143,446	8,110,000	58,253,441	27.2	23.4	43.5	37.5

[1] Includes operating capacity plus operable shutdown but excludes shutdown inoperable capacity.
[2] Data unavailable.

Sources: United States: U.S. Bureau of Mines; rest of the world: "Oil & Gas Journal," "World Wide Issues," and "International Petroleum Encyclopedia".

Source: *Fiscal Policy and the Energy Crisis*, hearings before the Subcommittee on Energy of the Committee on Finance, United States Senate, January 25, 28, and 29, 1974, U.S. Government Printing Office, Washington, D.C. 1974.

WORLD PETROLEUM TRADE

The international production and distribution of petroleum was dominated for many years by a relatively few large multinational companies. Although this petroleum trade had a monopoly character, international prices tended to be in "reasonable" relationship to development, production, and distribution costs. In 1948, the control of these companies over the markets and distribution in Europe and North America as well as the oil reserves of the Mideast, enabled them to negotiate arrangements whereby they received 82% of the monopoly rents derived from oil produced in the Mideast.

After the oil fields in the Mideast had been developed, the bargaining position of the producing countries improved and they demanded an increase in their share of the monopoly rents. Accommodation to these demands was made by United States companies through agreement to the concept of "posted" prices that were far higher than actual prices but which could be used as a basis for the calculation of United States foreign tax credits. United States companies were able, thereby, to increase payments without decreasing their revenues.

As the dependence on imported oil grew in Europe and the United States, the producing countries made further demands for greater participation in ownership and profits from oil-producing properties in their countries. By 1970, the OPEC countries had been able to increase their share of monopoly rents to 70% and had obtained the right to set their own posted prices. By 1973, control had passed from the multinational companies to the OPEC countries. In 1973, posted prices for oil from OPEC countries were quadrupled and in October of that year, an embargo was imposed on export of petroleum to the United States in protest to United States political positions in the Mideast. Although the embargo was lifted in the spring of 1974, world oil prices have remained at the high posted levels.

The crux of the immediate energy crisis stems from at least three factors: the dependence of European countries, Japan and the United States on imported oil; the adverse impact of high world oil prices on the trade balances of the oil-consuming countries; and the uncertainty with regard to investment in alternate energy sources. This uncertainty for investments is caused by the large spread between selling price and production costs

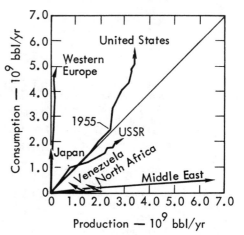

World Petroleum flow, 1940–1972. All arrows except the USSR have tails at 1940 and heads at 1972. USSR arrow has tail at 1950 and head at 1972.

Source: Carl J. Anderson, et al, *An Assessment of U.S. Energy Options for Project Independence*, UCRL-51638, Lawrence Livermore Laboratory, September 1974.

for Mideast oil which makes it possible for these countries to reduce the prices for oil at will to meet competition from alternate energy sources. These factors are destabilizing in both a political and an economic sense and have complicated planning efforts.

In the long run, the energy crisis has a different character. The world supplies of oil and gas, including those of the Mideast, are limited and within a period of 25–50 years these supplies will have been consumed. Development of alternate sources of energy—coal, nuclear, and renewable resources derived from the sun—will be essential to replace the liquid and gaseous fossil fuels as they are depleted.

To meet its needs for energy, Western Europe depends heavily on imports from the Middle East and North Africa—much more than the U.S. and Canada do. This dependence has made Europe more eager than the U.S. to favor the Arab side in the Arab-Israeli dispute. But its policy tilt did not save Europe from severe fuel shortages this winter. Because of the sharply rising price of petroleum, industrialized countries with their own fuel resources—among them, the U.S. and Canada—will enjoy an economic advantage in the years ahead. In the Eighties, Britain should get a tremendous boost from its North Sea oil and natural-gas deposits. Norway could become a Scandinavian Abu Dhabi.

The figures are for 1972. The table was compiled by FORTUNE from data supplied by the Organization for Economic Cooperation and Development, the Common Market, and the United Nations.

	BELGIUM	DENMARK	FRANCE	ITALY	NETHERLANDS	NORWAY	UNITED KINGDOM	WEST GERMANY	CANADA	UNITED STATES
TOTAL ENERGY USED (millions of metric tons in coal equivalents)	63.2	27.9	235.2	173.1	82.5	19.5*	306.6	355.1	213*	2,600.7
From Crude Oil (%)	60	95	67	74	50	53*	50	55	47*	46
From Natural Gas (%)	14	0	7	11	46	0	11	9	32*	32
CRUDE IMPORTED (%)	100	99	99	99	98	100	100	94	62	20
Source: North Africa & Arab Countries	75.5	66.6	79.5	75.2	70.8	46.2	70	72.9	23.1	22.3
Iran	14.2	19.5	6.0	14.3	11.6	39.6	13.4	9.6	12.4	6.1
West Africa	4.2	13.9	11.3	2.7	16.9	8.7	10.4	10.7	12.2	11.7
Western Hemisphere	2.9	0	1.5	1.2	0.7	5.5	5.1	3.6	52.3	52.3
Far East	1.7	0	0	0.4	0	0	0.2	0	0	7.5
Eastern Europe	1.3	0	1.5	6.1	0	0	0.2	2.8	0	0
NATURAL GAS IMPORTED (%)	100	0	50	9	0	0	3	52	1	5
Source: Algeria			10				100			0.2
Canada										99.1
Libya				100						
Mexico										0.7
Netherlands	100		90					100		
United States									100	

*FORTUNE estimate

Where NATO nations get their fuel.

Source: Fortune, February 1974.

WORLD ELECTRIC POWER CAPACITY AND GENERATION

Growth of Electric Power. The rate of growth in the consumption of electric power has been considerably higher than the rate of growth of overall energy consumption. Over the past half century, world consumption of electricity has grown at a rate of about 7.0% per year, representing a doubling of electrical energy consumption every ten years. The rate of increase in overall energy consumption during this same period has been 4.8% per year.

The higher rate of growth for electric energy consumption has resulted from the shift away from other forms of energy to electricity—hydroelectric power has replaced the direct use of water power in factories and electricity is supplanting coal, oil, and gas for the heating of residential and commercial buildings. The installation of electric air conditioning has resulted in considerable growth of electric power since 1945.

By 1970, about 27% of the energy consumed in the United States was in the form of electricity. It is expected that this figure will rise to about 50% by the end of the century as electricity generated from coal and nuclear fuel is substituted for the dwindling domestic supplies of gas and oil.

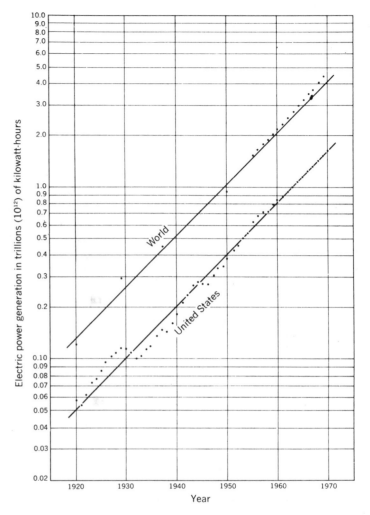

Generation of electric power over the past half century, for selected years worldwide (11.1) and annually for the United States (11.2). The trend lines correspond to uniform growth at 7.2 percent per year (100 percent per decade). World electric power generation has been growing at about 8 percent per year over the past decade.

Source: Fisher, John C., *Energy Crises in Perspective*, John Wiley & Sons, New York, 1974.

Surry Nuclear Power Station of Virginia Electric and Power Company. Each of the two generating units is designed to produce 788,000 electrical kilowatts. (*Courtesy Virginia Electric and Power Company*)

World Generation of Electricity
Kilowatt-Hours in Billions

Year	Total World	United States*	All Other Countries
1975	6,439p	2,003	4,436p
1974	6,253r	1,968	4,285r
1973	6,075r	1,959	4,116r
1972	5,654r	1,853p	3,801p
1971	5,224	1,716r	3,508r
1970	4,904p	1,640	3,264p
1969	4,443	1,553	2,890
1968	4,083	1,436	2,647
1967	3,755	1,317	2,438
1966	3,515	1,249	2,266
1965	3,260	1,158	2,102
1964	3,024	1,084	1,940
1963	2,783	1,011	1,772
1962	2,568	946	1,622
1961	2,372	881	1,491
1960	2,236	844	1,392
1959	2,058	795	1,263
1958	1,890	725	1,165
1957	1,804	716	1,088
1956	1,678	685	993
1955	1,539	629	910
1954	1,360	545	815
1952	1,151	463	688
1950	965	389	576
1948	797	337	460
1944	660	280	380
1940	505	180	325
1935	375	119	256
1930	310	115	195
1925	180	85	95
1920	126	57	69

Sources: 1972-75 United Nations' Statistical Yearbook, Federal Power Commission; 1955-71 "World Power Data," Federal Power Commission; 1948-54 United Nations' Statistical Yearbook; 1940-44 Estimated from United Nations' Monthly Bulletin of Statistics; 1920-35 Statistischer Bericht fur die Elektroindustrie International.

* Alaska and Hawaii included since 1960.
p—preliminary.   r—revised.

Source: EEI Pocketbook of Electric Utility Industry Statistics, 23rd Edition, Edison Electric Institute, 90 Park Avenue, New York, N.Y. 10016, 1978.

WORLD POWER DATA
Ten Countries with Greatest Installed Capacity
1972*

	Country	Installed Capacity (Kw in Thousands)			Energy Production (Kwhr in Millions)			Population†† (Thousands)	Kwhr per Capita
		Hydro	Thermal	Total	Hydro	Thermal	Total		
1.	United States**	56 566	343 040	399 606	272 734	1 474 589	1 747 323†	208 842‡	8 367
2.	USSR	34 846	151 393	186 239	122 899	734 536	857 435	247 459	3 465
3.	Japan	20 765	64 531	85 296	88 658	325 633	414 291	106 958	3 873
4.	United Kingdom	2 158	73 288	75 446	4 305	259 376	263 681	55 788	4 726
5.	Germany (West)	4 839	52 778	57 617	13 689	261 080	274 769	61 674	4 455
6.	Canada	32 500	17 444	49 944	178 169	59 458	237 627	21 848	10 876
7.	France	15 800	28 200	44 000	48 417	114 995	163 412	51 700	3 161
8.	Italy	15 945	21 099	37 044	45 402	89 528	134 930	54 345	2 483
9.	Spain	11 450	8 550	20 000	36 520	32 390	68 910	34 494	1 998
10.	China (Mainland)	4 600	12 900	17 500	39 400	66 600	106 000	800 721	132

* Preliminary.
** Including Alaska and Hawaii.
† Net imports not included.
‡ U. S. population includes armed forces abroad.
†† Population for Japan as of March 31, 1972; for all other countries as of July 1, 1972.
Source: Federal Power Commission.

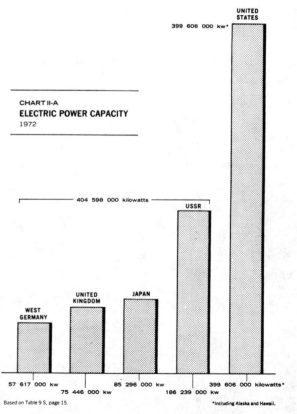

CHART II-A
ELECTRIC POWER CAPACITY
1972

404 598 000 kilowatts

UNITED STATES — 399 606 000 kw*

USSR

JAPAN

UNITED KINGDOM

WEST GERMANY

57 617 000 kw   75 446 000 kw   85 296 000 kw   186 239 000 kw   399 606 000 kilowatts*

Based on Table 9 S, page 15.   *Including Alaska and Hawaii.

Source: Statistical Year Book of the Electric Utility Industry for 1973, Edison Electric Institute, 90 Park Avenue, New York, N.Y., Nov. 1974.

U.S. ENERGY CONSUMPTION 1850-75

The sources of energy in the United States have shifted, since 1850, from wood to coal and then to oil and gas. Wood constituted 90.7% of United States energy sources in 1850 but had essentially disappeared as an energy source by 1960. Coal became the major fuel in 1910 and it, in turn, has been replaced by oil and gas which by 1974 were supplying 46.2% and 30.4% of the nation's energy supply. The contribution of hydropower has remained relatively constant at about 4% of the United States energy supply since 1920.

Total consumption of energy in the United States reached at least a temporary peak in 1973 with consumption dropping about 2% in 1974 from the 1973 level as a result of a general drop in economic activity and conservation measures instituted after the embargo on export of petroleum from the Middle East to the United States. Energy consumption in the first quarter of 1975 was about 1% below the same period in 1974.

Total U.S. Energy Consumption, 1947-75

Year	Coal[1] Trillion Btu	Coal[1] Million short tons	Natural gas Trillion Btu	Natural gas Billion ft[3]	Petroleum[2] Trillion Btu	Petroleum[2] Million bbl	Hydropower and geothermal Trillion Btu	Hydropower and geothermal Billion kWh	Nuclear power Trillion Btu	Nuclear power Billion kWh	Total gross energy consumption (trillion Btu)
1947	15,824	605	4,518	4,366	11,367	1,990	1,326	85.0	—	—	33,035
1948	14,897	570	5,033	4,862	12,558	2,020	1,393	88.5	—	—	33,881
1949	12,631	483	5,289	5,110	12,120	2,128	1,449	96.4	—	—	31,489
1950	12,913	494	6,150	5,942	13,489	2,375	1,440	102.7	—	—	33,992
1951	13,225	506	7,248	7,003	14,848	2,584	1,454	106.6	—	—	36,775
1952	11,868	454	7,760	7,498	15,334	2,671	1,496	112.0	—	—	36,458
1953	11,893	455	8,156	7,870	16,098	2,775	1,439	111.6	—	—	37,586
1954	10,195	390	8,548	8,259	16,132	2,849	1,388	112.0	—	—	36,263
1955	11,540	447	9,232	8,920	17,524	3,100	1,407	120.3	—	—	39,703
1956	11,752	457	9,834	9,502	18,627	3,233	1,487	129.8	—	—	41,700
1957	11,168	434	10,416	10,064	18,570	3,234	1,551	137.0	1	([3])	41,706
1958	9,849	386	10,995	10,623	19,214	3,371	1,636	147.6	2	0.2	41,696
1959	9,810	385	11,990	11,585	19,747	3,481	1,591	145.0	2	.2	43,140
1960	10,140	398	12,699	12,269	20,067	3,611	1,657	154.0	6	.5	44,569
1961	9,906	390	13,228	12,750	20,487	3,641	1,680	157.8	18	1.7	45,319
1962	10,189	403	14,121	13,612	21,267	3,796	1,821	172.5	24	2.3	47,422
1963	10,714	423	14,843	14,341	21,950	3,925	1,767	168.6	34	3.2	49,308
1964	11,264	446	15,648	15,118	22,386	4,034	1,907	182.3	35	3.3	51,240
1965	11,908	472	16,098	15,598	23,241	4,202	2,058	196.8	38	3.7	53,343
1966	12,495	498	17,393	16,854	24,394	4,111	2,073	199.0	57	5.5	56,412
1967	12,256	491	18,250	17,685	25,335	4,585	2,344	224.7	80	7.7	58,265
1968	12,659	509	19,580	18,973	27,052	4,902	2,342	225.2	130	12.5	61,763
1969	12,733	516	21,020	20,388	28,421	5,160	2,659	254.5	146	13.9	64,979
1970	12,698	524	22,029	21,367	29,537	5,365	2,650	252.6	229	21.8	67,143
1971	12,043	502	22,819	22,132	30,570	5,553	2,862	273.1	404	37.9	68,698
1972	12,423	523	23,035	22,429	32,966	5,990	2,946	283.9	576	54.0	71,946
1973	13,294	562	22,712	22,245	34,851	6,317	2,998	288.6	888	83.3	74,743
1974	[4]13,103	[4]553	22,028	21,512	33,414	6,070	3,290	316.7	1,202	112.7	73,037
1975[5]	[4]13,092	[4]554	20,399	19,985	32,701	5,954	3,158	304.0	1,652	155.0	71,002

[1] Includes anthracite, bituminous coal, and lignite.
[2] Includes domestically produced crude oil, natural gas liquids, and condensate, plus imported crude oil and products.
[3] Less than 0.1.
[4] Includes coke net imports.
[5] Preliminary.
Source: U.S. Department of the Interior, Bureau of Mines, Division of Interfuels Studies.

Source: Energy Perspectives 2, U.S. Department of the Interior, June 1976.

U.S. Energy Consumption Trends, 1850-1974
(Quadrillion Btu)[1]

YEAR	COAL	PETROLEUM	NATURAL GAS	HYDROPOWER	NUCLEAR	FUEL WOOD	TOTAL
1850	.2	—	—	—	—	2.1	2.3
1860	.5	—	—	—	—	2.6	3.1
1870	1.0	—	—	—	—	2.9	4.0
1880	2.0	.1	—	—	—	2.9	5.0
1890	4.1	.2	.3	—	—	2.5	7.1
1900	6.8	.2	.3	.3	—	2.0	9.6
1910	12.7	1.0	.5	.5	—	1.9	16.6
1920	15.5	2.6	.8	.8	—	1.6	21.3
1930	13.6	5.4	2.0	.8	—	1.5	23.3
1940	12.5	7.5	2.7	.9	—	1.4	25.0
1950	12.9	13.5	6.2	1.4	—	1.2	35.2
1960	10.1	20.1	12.7	1.7	—	—	44.6
1970	12.7	29.5	22.0	2.7	.2	—	67.1
1971	12.0	30.6	22.8	2.9	.4	—	68.7
1972	12.4	33.0	23.0	2.9	.6	—	71.9
1973	13.4	34.7	22.8	2.9	.9	—	74.7
1974	13.0	33.8	22.3	2.9	1.2	—	73.2

[1] 1 Quadrillion Btu = 500,000 barrels petroleum per day for a year
= 40 million tons of bituminous coal
= 1 trillion cubic feet of natural gas
= 100 billion kWh (based on a 10,000-Btu/kWh heat rate)

Source: Energy Perspectives, U.S. Department of the Interior, February 1975.

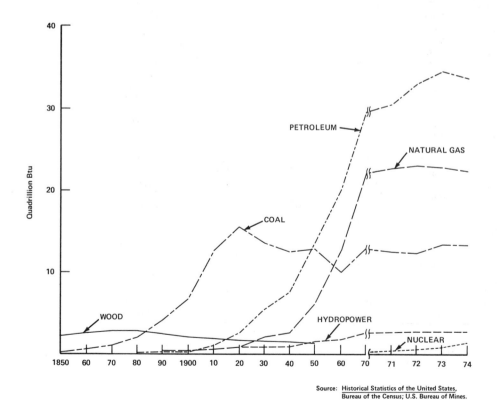

U.S. energy consumption trends, 1850-1974.

Source: Historical Statistics of the United States, Bureau of the Census; U.S. Bureau of Mines.

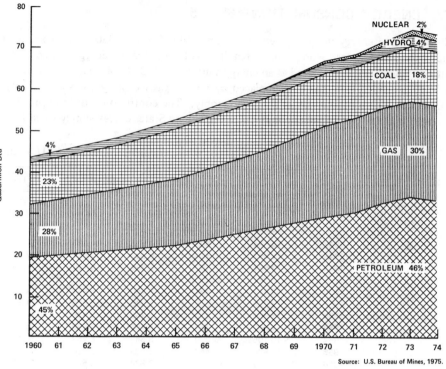

U.S. gross energy consumption patterns by source, 1960-74.

Source: Energy Perspectives, U.S. Department of the Interior, February 1975.

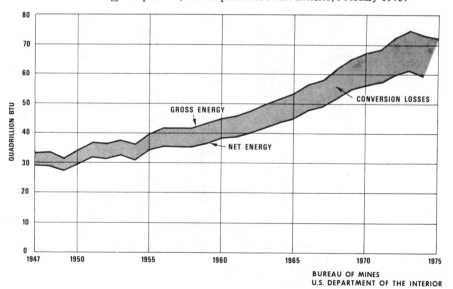

United States gross and net energy consumption, 1947-1975.

BUREAU OF MINES
U.S. DEPARTMENT OF THE INTERIOR

Source: Walter G. Dupree, Jr. and John S. Corsentino, *United States Energy Through the Year 2000 (Revised)*, Bureau of Mines, U.S. Department of the Interior, December 1975.

U.S. ENERGY CONSUMPTION–RESOURCE PERCENTAGE

U.S. Energy Consumption Patterns, 1850-1974
(Percent)

YEAR	COAL	PETROLEUM	NATURAL GAS	HYDROPOWER	NUCLEAR	FUEL WOOD
1850	9.3	–	–	–	–	90.7
1860	16.4	.1	–	–	–	83.5
1870	26.5	.2	–	–	–	73.3
1880	41.1	1.9	–	–	–	57.0
1890	57.9	2.2	3.7	.3	–	35.9
1900	71.3	2.4	2.6	2.6	–	21.1
1910	76.8	6.1	3.3	3.2	–	10.6
1920	72.8	12.2	3.9	3.6	–	7.5
1930	58.7	23.2	8.5	3.4	–	6.2
1940	50.1	30.0	10.9	3.7	–	5.3
1950	36.7	38.4	17.5	4.1	–	3.3
1960	22.8	45.0	28.5	3.7	–	–
1970	19.2	43.9	32.7	3.9	.3	–
1971	18.2	44.2	32.9	4.1	.6	–
1972	17.3	45.8	32.0	4.1	.8	–
1973	17.8	46.5	30.5	4.0	1.2	–
1974	17.8	46.2	30.4	4.0	1.9	–

Source: Energy Perspectives, U.S. Department of the Interior, February 1975.

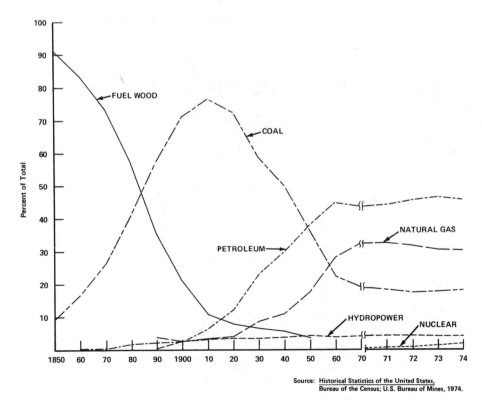

U.S. energy consumption patterns, 1850-1974.

Source: Energy Perspectives, U.S. Department of the Interior, February 1975.

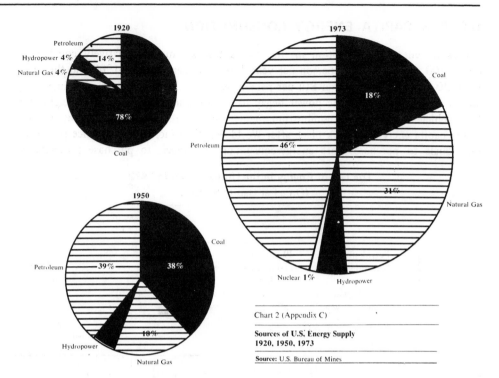

Chart 2 (Appendix C)

**Sources of U.S. Energy Supply
1920, 1950, 1973**

Source: U.S. Bureau of Mines

Sources of U.S. energy supply—1920, 1950, 1973.

Source: Exploring Energy Choices, Energy Policy Project of The Ford Foundation, Washington, D.C. 1974.

U.S. PER CAPITA ENERGY CONSUMPTION

Per capita energy consumption in the United States remained relatively constant at about 100 million Btu per person per year during the period 1850–1900 and then rose by 1920 to about 200 million Btu per person per year. During this period of increasing energy consumption, electric power began to be introduced and passenger, freight, and farm vehicles using internal combustion engines rapidly grew in number. Another surge in per capita energy use took place after World War II as increasing affluence permitted individuals to purchase more energy-consuming appliances, to purchase products re-

quiring energy to produce, and to travel more widely. In 1973, per capita energy consumption had reached 355 million Btu/year.

During the past century, growth in the per capita gross national product has paralleled the growth in per capita energy consumption at roughly 100,000 Btu per dollar of gross national product. Since 1880, the Btu/GNP dollar ratio has fallen from 125,000 to about 94,000 Btu/GNP dollar indicating some improvement in the efficiency of energy use in the generation of the gross national product. This improvement may also reflect the shift to less energy-intensive service industry from industrial and agricultural activities.

The Use of Energy in the United States in 1972
(On a Per Capita Basis)

		Percentage of total (%)
Petroleum	29 barrels	46
Natural gas	110,000 cubic feet	32
Coal	2.5 short tons	17
Hydropower	1,350 kilowatt hours	4
Nuclear power	280 kilowatt hours	1

Source: Bureau of Mines.

U.S. Energy Consumption Per Capita, 1850–1973
(Millions of Btu)[1]

YEAR	AMOUNT	YEAR	AMOUNT
1850	101.2	1930	188.6
1860	100.4	1940	189.5
1870	99.0	1950	231.7
1880	99.4	1960	246.6
1890	111.1	1970	329.3
1900	125.9	1971	333.4
1910	174.3	1972	344.6
1920	200.1	1973	355.1

[1] 100 million Btu = 18 barrels of petroleum equivalent.

Source: Energy Perspectives, U.S. Department of the Interior, February 1975.

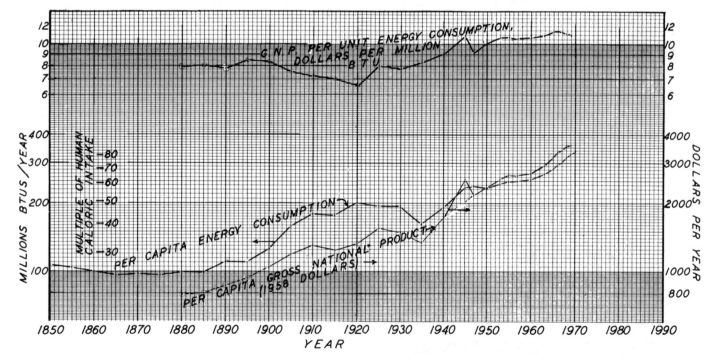

Per capita energy consumption, per capita gross national product, and their ratio, as functions of time.

Source: H. C. Hottel and J. B. Howard, *New Energy Technology—Some Facts and Assessments*, The MIT Press, Cambridge, Mass., 1971.

Selected U.S. Economic, Demographic, and Energy Indicators, 1947-75

Year	Gross national product (billion 1972 dollars)	Population (millions)	Gross energy input[1] (quadrillion Btu)	Net energy input[2] (quadrillion Btu)	Gross energy per dollar of 1972 GNP (Thousand Btu)	Gross energy per capita (million Btu)	Net energy per capita (million Btu)	Conversion efficiency (percent)[3]
1947	468.3	144.1	33.0	29.2	70.5	229.0	202.8	88.5
1948	487.7	146.6	33.9	29.1	69.5	231.2	198.6	85.8
1949	490.7	149.2	31.5	27.3	64.2	211.1	182.7	86.5
1950	533.5	152.3	34.0	29.7	63.7	223.2	194.8	87.3
1951	576.5	154.9	36.8	32.1	63.8	237.6	206.9	87.1
1952	598.5	157.6	36.5	31.6	61.0	231.7	200.6	86.6
1953	621.8	160.2	37.6	32.6	60.5	234.7	201.1	85.7
1954	613.7	163.0	36.3	31.2	59.1	222.7	191.5	86.0
1955	654.8	165.9	39.7	34.3	60.6	239.3	206.7	86.4
1956	668.8	168.9	41.7	35.8	62.4	246.9	211.7	85.7
1957	680.9	172.0	41.7	35.6	61.2	242.2	206.9	85.4
1958	679.5	174.9	41.7	35.5	61.4	238.4	202.8	85.1
1959	720.4	177.8	43.1	36.4	59.8	242.4	205.0	84.6
1960	736.8	180.7	44.6	38.2	60.5	246.8	211.5	85.7
1961	755.3	183.8	45.3	38.7	60.0	246.5	210.6	85.8
1962	799.1	186.5	47.4	40.5	59.3	254.1	217.2	85.5
1963	830.7	189.2	49.3	42.0	59.3	260.5	222.0	85.2
1964	874.4	191.8	51.2	43.6	58.6	266.9	227.3	85.5
1965	925.9	194.3	53.3	45.3	57.6	274.3	233.1	85.0
1966	981.0	196.6	56.4	47.6	57.5	286.9	242.1	84.4
1967	1,007.7	198.7	58.3	49.4	57.9	293.4	248.6	84.7
1968	1,051.8	200.7	61.7	52.2	58.7	307.4	260.1	84.6
1969	1,078.8	202.7	65.0	54.4	60.3	320.7	268.4	83.7
1970	1,075.3	204.9	67.1	56.0	62.4	327.5	273.3	83.6
1971	1,107.5	207.0	68.7	57.0	62.0	331.9	275.4	83.0
1972	1,171.1	208.8	71.9	59.5	61.6	345.3	285.0	82.6
1973	1,233.4	210.4	74.7	61.3	60.6	355.0	291.3	82.1
1974	1,210.7	211.9	72.9	59.3	60.2	345.0	279.8	81.3
1975[4]	1,186.4	213.4	71.1	59.3	59.9	338.0	269.4	80.9

[1] Gross energy is the total of inputs into the economy of the primary fuels (petroleum, natural gas, and coal, including imports) or their derivatives, plus the generation of hydro and nuclear power converted to equivalent energy inputs.
[2] Net energy is the sector inputs (household and commercial, transportation, and industrial), and consists of direct fuels and purchased electricity.
[3] The conversion factor is the percent of total gross energy going into the final consuming sector.
[4] Estimated.

Source: *Energy Perspectives 2*, U.S. Department of the Interior, June 1976.

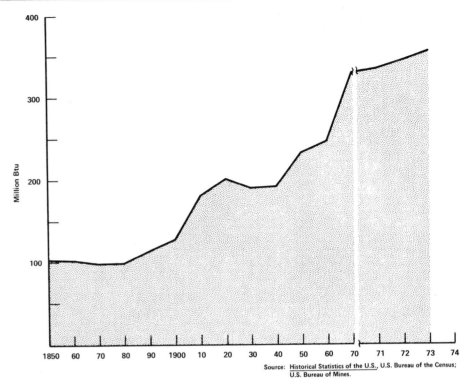

Source: Historical Statistics of the U.S., U.S. Bureau of the Census; U.S. Bureau of Mines.

U.S. energy consumption per capita, 1850-1973.

Source: *Energy Perspectives*, U.S. Department of the Interior, February 1975.

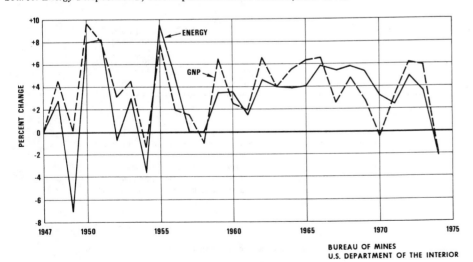

BUREAU OF MINES
U.S. DEPARTMENT OF THE INTERIOR

Changes in energy and GNP, 1947-1974.

Source: Walter G. Dupree, Jr. and John S. Corsentino, *United States Energy Through the Year 2000* (*Revised*), Bureau of Mines, U.S. Department of the Interior, December 1975.

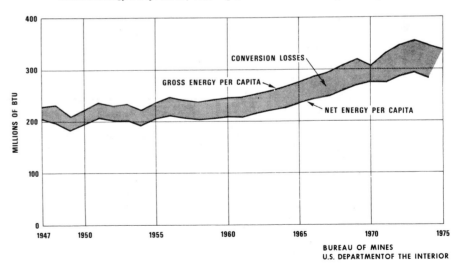

BUREAU OF MINES
U.S. DEPARTMENT OF THE INTERIOR

United States net and gross energy per capita, 1947-1975

Source: Walter G. Dupree, Jr. and John S. Corsentino, *United States Energy Through the Year 2000* (*Revised*), Bureau of Mines, U.S. Department of the Interior, December 1975.

U.S. ENERGY RESOURCE PRODUCTION AND CONSUMPTION

The United States has been consuming more energy than it has produced from domestic energy resources since 1955. The deficit in domestic energy resources has been filled largely through the importation of oil and, to a minor extent, the importation of natural gas. In 1974, the United States imported 17% of its energy resource requirements, 40% of its petroleum needs, and 3% of its natural gas needs. In that same year, the United States exported 10% of its coal production.

U.S. Energy Balances, 1974 and 1975
[Trillion Btu]

Sector and year	Anthracite	Bituminous coal and lignite	Coke[1]	Natural gas, dry[2]	Petroleum[3]	Hydropower[4]	Nuclear power[5]	Utility electricity[6]	Total
Final consuming sectors:									
Household and commercial:									
1974	66	248	—	7,518	6,061	—	—	3,723	17,616
1975 (estimated)	56	226	—	7,373	5,829	—	—	3,783	17,267
Industrial:									
1974	35	4,312	59	10,018	5,907	37	—	2,665	23,033
1975 (estimated)	32	4,237	18	8,991	5,635	36	—	2,708	21,657
Transportation:[7]									
1974	—	2	—	685	17,720	—	—	19	18,426
1975 (estimated)	—	1	—	635	17,857	—	—	19	18,512
Miscellaneous and unaccounted for:									
1974	—	—	—	—	246	—	—	—	246
1975	—	—	—	—	68	—	—	—	68
Total final consumption:									
1974	100	4,562	59	18,221	29,934	37	—	[8]6,407	[9]59,320
1975 (estimated)	88	4,464	18	16,998	29,389	36	—	[8]6,510	[9]57,504
Energy conversion sector:									
Electricity generation, utilities:									
1974	38	8,482	—	3,512	3,480	3,253	1,202	[8]−6,407	[10]13,560
1975 (estimated)	39	8,785	—	3,174	3,312	3,122	1,652	[8]−6,510	[10]13,574
Total resources consumed:									
1974	138	13,044	59	21,733	33,414	3,290	1,202	0	[11]72,880
1975 (estimated)	128	13,248	18	20,173	32,701	3,158	1,652	0	[11]71,078

[1] Net imports of coke. The rest of domestic coke consumption is accounted for in the Btu value of coal carbonized for coke.

[2] Excludes natural gas liquids.

[3] Petroleum products, including still gas, liquefied refinery gas, and natural gas liquids.

[4] Outputs of hydropower (adjusted for net imports) are converted to theoretical energy inputs using 10,389 Btu/kWh, the national average heat rate for fossil-fueled steam electric plants provided by the Federal Power Commission.

[5] Energy inputs for nuclear powerplants are calculated using the average heat rate of 10,660 Btu/kWh, based on information from the Energy Research and Development Administration.

[6] Excludes electricity generated by nonutility plants, the energy inputs to which are included in the other inputs of the final consuming sectors. Distribution to sectors is based on sales reported in the Edison Electric Institute, *Statistical Yearbook of the Electric Utility Industry for 1974.* Energy converted to energy equivalent at the theoretical value of contained energy of 3,412 Btu/kWh.

[7] Includes bunkers and military uses.

[8] Utility electricity, generated and imported, distributed to the final consuming sectors.

[9] Also termed "net energy inputs."

[10] Conversion losses in the electricity generation utilities sector. Total energy inputs to this sector is the sum of conversion losses and utility electricity distributed (footnote 8).

[11] Also termed "gross energy inputs."

Note.—Data may not add to totals because of rounding.

Source: U.S. Department of the Interior, Bureau of Mines, Division of Interfuels Studies

Source: Energy Perspectives 2, U.S. Department of the Interior, June 1976.

Apparent Consumption of Primary Energy Resources and Selected Related Products

Commodity	1974	1975 (Estimated)	Percentage Change from 1974
Primary energy sources:			
Bituminous coal and lignite (million short tons)	552.7	562.0	+1.7
Anthracite (do)	5.4	5.0	−7.8
Crude petroleum, runs to stills (million barrels)	4,428.7	4,517.6	+2.0
Natural gas, dry[1] (billion cubic feet)	21,223.1	19,700.0	−7.2
Natural gas liquids[2] (million barrels)	616.1	594.2	−3.6
Hydropower, utility[3] (million kilowatt hours)	313,141.4	300,550.0	−4.0
Hydropower, industrial[4] (do)	3,537.1	3,460.0	−2.2
Products:			
All oils, domestic product demand[5] (million barrels)	6,069.5	5,954.0	−1.9
Coke (million short tons)	64.1	60.0	−6.4
Electricity from fuel burning plants:[4]			
Nuclear power, utility (million kilowatt hours)	112,739.9	155,000.0	+37.5
Conventional plants, utility (do)	1,451,773.7	1,452,450.0	Negligible
Industrial (do)	98,904.5	83,000.0	−16.1

[1] Excludes shrinkage resulting from extraction of natural gas liquids, transmission losses, and gas unaccounted for.

[2] Liquids recovered from natural gas processing plants.

[3] Net generation, adjusted to include net imports. The net trade is hydropower with an undetermined portion of steam plant power.

[4] Net generation.

[5] Includes natural gas liquids.

Source: Division of Interfuels Studies, Office of Assistant Director—Fuels, Bureau of Mines, U.S. Department of the Interior.

Source: Department of the Interior News Release, April 5, 1976.

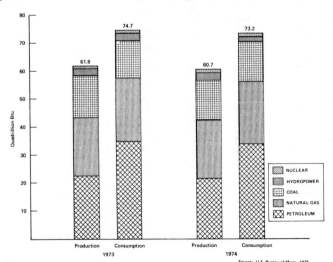

U.S. energy production and gross consumption, 1973 and 1974.

Source: Energy Perspectives, U.S. Department of the Interior, February 1975.

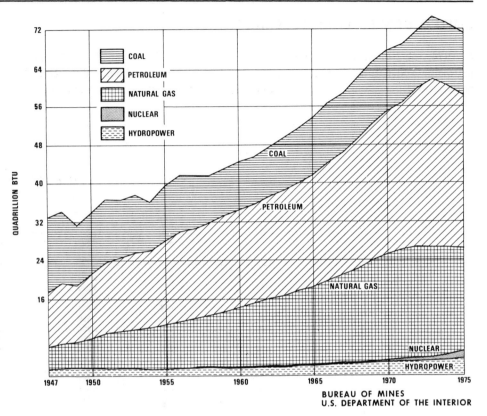

United States gross energy consumption by source, 1947–1975.

Source: Walter G. Dupree, Jr. and John S. Corsentino, *United States Energy Through the Year 2000 (Revised),* Bureau of Mines, U.S. Department of the Interior, December 1975.

U.S. Energy Production and Gross Consumption, 1973 and 1974
(Quadrillion Btu)[1]

SOURCE	1973		1974	
	PRODUCTION	CONSUMPTION	PRODUCTION	CONSUMPTION
COAL	14.4	13.4	14.4	13.0
PETROLEUM[2]	21.4	34.7	20.7	33.8
NATURAL GAS[3]	22.3	22.8	21.6	22.3
HYDROPOWER	2.8	2.9	2.8	2.9
NUCLEAR POWER	.9	.9	1.2	1.2
TOTAL	61.8	74.7	60.7	73.2

[1] 1 Quadrillion Btu = 500,000 barrels petroleum per day for a year
= 40 million tons of bituminous coal
= 1 trillion cubic feet of natural gas
= 100 billion kWh (based on a 10,000-Btu/kWh heat rate)

[2] Includes natural gas liquids.

[3] Excludes natural gas liquids.

Source: Energy Perspectives, U.S. Department of the Interior, February 1975.

UNITED STATES ENERGY TRADE

In 1974, the United States exported coal equivalent to 1645 trillion Btu and imported petroleum and natural gas with a combined energy content of 13,552 trillion Btu. Imports represented about 17% of the total United States energy consumption. In 1973, crude oil from the Middle East represented about 25% of the total crude oil imports. About 2% of the imports of refined petroleum products were obtained directly from refineries in the Middle East and an undetermined percentage from Middle East crude oil refined at locations outside the United States. About 74% of the imported petroleum products came from refineries in the Caribbean and Central and South America.

U.S. Trade in Fossil Fuels, 1947–75
[Trillion Btu]

Year	Exports				Imports				Net trade[1]			
	Coal	Petroleum	Natural gas	Total	Coal	Petroleum	Natural gas	Total	Coal	Petroleum	Natural gas	Total
1947	2,015	944	19	2,971	8	954	–	962	2,007	-10	19	2,016
1948	1,373	786	19	2,178	8	1,151	–	1,159	1,365	-365	19	1,019
1949	856	683	21	1,560	8	1,399	–	1,407	848	-716	21	153
1950	766	639	27	1,432	10	1,862	–	1,872	756	-1,223	27	-440
1951	1,637	888	25	2,550	8	1,857	–	1,865	1,629	-969	25	685
1952	1,365	911	28	2,304	8	2,097	8	2,113	1,357	-1,186	20	191
1953	954	853	29	1,836	7	2,285	–	2,292	947	-1,432	29	-456
1954	885	736	30	1,651	5	2,289	7	2,301	880	-1,553	23	-650
1955	1,493	772	32	2,297	9	2,728	11	2,748	1,484	-1,956	21	-451
1956	2,022	910	37	2,969	9	3,160	11	3,180	2,013	-2,250	26	-211
1957	2,218	1,202	43	3,463	10	3,455	39	3,504	2,208	-2,253	4	-41
1958	1,444	573	40	2,057	8	3,725	141	3,874	1,436	-3,152	-101	-1,871
1959	1,071	439	19	1,529	10	3,875	139	4,024	1,061	-3,436	-120	-2,495
1960	1,044	411	12	1,467	7	3,955	161	4,123	1,037	-3,544	-149	-2,656
1961	1,001	363	11	1,375	4	4,209	227	4,440	997	-3,846	-216	-3,065
1962	1,103	350	16	1,469	6	4,572	416	4,994	1,097	-4,222	-400	-3,525
1963	1,376	427	18	1,821	5	4,654	420	5,079	1,371	-4,227	-402	-3,258
1964	1,351	412	17	1,780	8	4,909	457	5,374	1,343	-4,497	-440	-3,594
1965	1,418	378	27	1,823	5	5,349	471	5,825	1,413	-4,971	-444	-4,002
1966	1,381	400	25	1,806	5	5,592	495	6,092	1,376	-5,192	-470	-4,286
1967	1,382	643	84	2,109	6	5,484	582	6,072	1,376	-4,841	-498	-3,963
1968	1,407	483	97	1,987	6	6,106	673	6,785	1,401	-5,623	-576	-4,798
1969	1,564	482	53	2,099	3	6,805	750	7,558	1,561	-6,323	-697	-5,459
1970	1,991	547	72	2,610	1	7,388	846	8,235	1,990	-6,841	-774	-5,625
1971	1,569	463	83	2,115	3	8,406	964	9,373	1,566	-7,943	-881	-7,258
1972	1,545	464	80	2,089	1	10,112	1,047	11,160	1,544	-9,648	-967	-9,071
1973	1,497	483	79	2,059	[2]31	13,466	1,055	14,552	1,466	-12,983	-976	-12,493
1974	1,685	467	79	2,230	[2]141	13,052	982	14,175	1,544	-12,585	-903	-11,944
1975[3]	1,850	447	77	2,374	[2]79	12,746	972	13,797	1,771	-12,299	-895	-11,423

[1] Net trade = exports - imports.
[2] Includes imports of coke, not significant before 1973.
[3] Preliminary.
Source: U.S. Department of the Interior, Bureau of Mines, Division of Interfuels Studies.

Source: Energy Perspectives 2, U.S. Department of the Interior, June 1976.

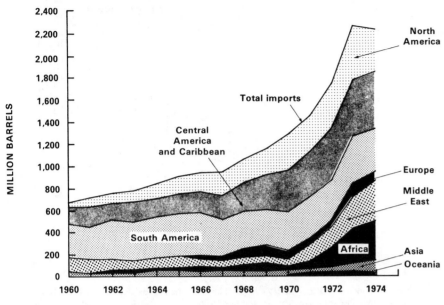

U.S. petroleum imports by region, 1960–74.

Source: Energy Perspectives 2, U.S. Department of the Interior, June 1976.

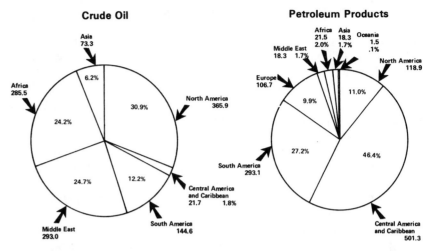

Crude Oil

Asia 73.3
Africa 285.5
6.2%
30.9%
24.2%
North America 365.9
24.7%
12.2%
Central America and Caribbean 21.7   1.8%
Middle East 293.0
South America 144.6

TOTAL IMPORTS: 1,184.0

Petroleum Products

Africa 21.5   2.0%
Asia 18.3   1.7%
Oceania 1.5   .1%
Middle East 18.3   1.7%
North America 118.9
Europe 106.7
9.9%
11.0%
27.2%
46.4%
South America 293.1
Central America and Caribbean 501.3

TOTAL IMPORTS: 1,079.6

Source: U.S. Bureau of Mines, 1974.

U.S. petroleum imports by region, 1973 (millions of barrels).

Source: Energy Perspectives, U.S. Department of the Interior, February 1975.

U.S. Energy Trade
(Trillion Btu)[1]

SOURCE	1973	1974
COAL:		
IMPORTS	3	24
EXPORTS	1,460	1,669
NET TRADE	1,457	1,645
PETROLEUM:		
IMPORTS	13,078	13,130
EXPORTS	484	462
NET TRADE	12,594	12,668
NATURAL GAS:		
IMPORTS	1,061	966
EXPORTS	79	82
NET TRADE	982	884

[1] 1 Quadrillion Btu = 1,000 trillion
 = 500,000 barrels petroleum per day for a year
 = 40 million tons of bituminous coal
 = 1 trillion cubic feet of natural gas
 = 100 billion kWh (based on a 10,000-Btu/kWh heat rate)

Source: Energy Perspectives, U.S. Department of the Interior, February 1975.

U.S. Net Trade,[1] Mineral Fuels

Mineral fuels	Exports Quantity 2/	Exports Trillion Btu	Imports Quantity 2/	Imports Trillion Btu	Net trade Quantity 2/	Net trade Trillion Btu	Percentage change from 1974
Anthracite:							
1974	-1,169	-29.7	--	--	-1,169	-29.7	
1975 (estimated)	-1,175	-29.8	--	--	-1,175	-29.8	+0.5
Bituminous coal and lignite:							
1974	-59,926	-1,621.6	2,080	49.1	-57,846	-1,572.5	
1975 (estimated)	-66,000	-1,786.0	1,150	27.1	-64,850	-1,758.9	+12.1
Coke:							
1974	-1,278	-33.2	3,540	92.0	2,262	58.8	
1975 (estimated)	-1,300	-33.8	2,000	52.0	700	18.2	-69.1
Natural gas, dry: 3/							
1974	-76,789	-78.6	959,284	982.3	882,495	903.7	
1975 (estimated)	-75,000	-76.8	950,000	972.8	875,000	896.0	-0.8
Petroleum (crude):							
1974	-1.1	-6.4	1,269.2	7,361.4	1,268.1	7,355.0	
1975 (estimated)	-2.1	-12.2	1,461.1	8,474.4	1,459.0	8,462.2	+15.1
Petroleum products: 4/							
1974	-79.4	-460.8	952.4	5,690.8	873.0	5,230.0	
1975 (estimated)	-75.0	-435.0	715.0	4,272.0	640.0	3,837.0	-26.7
Total mineral fuels:							
1974	XX	-2,230.3	XX	14,175.6	XX	11,945.3	
1975 (estimated)	XX	-2,373.6	XX	13,798.3	XX	11,424.7	-4.4

XX Not applicable.
1/ Minus sign indicates exports.
2/ Quantities used are: Anthracite, bituminous coal, lignite, and coke: thousand short tons; natural gas, million cubic feet; and petroleum (crude and products), million barrels.
3/ Excludes natural gas liquids.
4/ Includes natural gas liquids.

Source: Division of Interfuel Studies, Office of Assistant Director-Fuels, Bureau of Mines, U.S. Department of the Interior.

Source: **Division of Interfuel Studies, Office of Assistant Director-Fuels, Bureau of Mines, U.S. Department of the Interior.**

Source: **Department of the Interior News Release, April 5, 1976.**

U.S. DEMAND FOR PETROLEUM PRODUCTS

Since 1950, the United States demand for petroleum has nearly tripled from 6.4 million barrels/day to about 18 million barrels/day. Total consumption in 1973 was 18.1 million barrels/day while in 1974, consumption dropped slightly to 17.7 million barrels/day. Domestic production of petroleum has been declining. Production was 9.5 million barrels/day in 1972, 9.1 million barrels/day in 1973, and 8.6 million barrels/day in 1974. Over the same three-year period, imports rose from 4.7. million barrels/day in 1972 to 6.9 million barrels/day in 1974. The cost of petroleum imports in 1974 was more than $27 billion. By 1976, oil imports were at a level of 7.3 million barrels a day or 42% of consumption requirements. The cost of this imported oil was $35 billion.

Of the demand for petroleum in 1973, transportation was responsible for 53%, electrical generation for 9%, household and commercial use for 20%, and industrial use for 18%.

Consumption of gasoline rose steadily from 4.5 million barrels/day in 1963 to 6.7 million barrels/day in 1973. Consumption of gasoline declined slightly in 1974 to 6.5 million barrels/day.

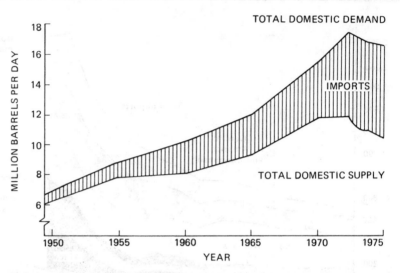

U.S. dependence on imported oil (Source: U.S. Federal Energy Administration).

Source: Environmental Quality–1976, The Seventh Annual Report of the Council on Environmental Quality, September 1976.

Domestic Gasoline Consumption

Year	Consumption (million barrels per day)
1963	4.479
1964	4.530
1965	4.713
1966	4.913
1967	5.048
1968	5.344
1969	5.596
1970	5.839
1971	6.064
1972	6.374
1973	6.671
1974	6.542

Source: Bureau of Mines.
Source: Energy Statistics, Committee on Finance, U.S. Senate, July 1975.

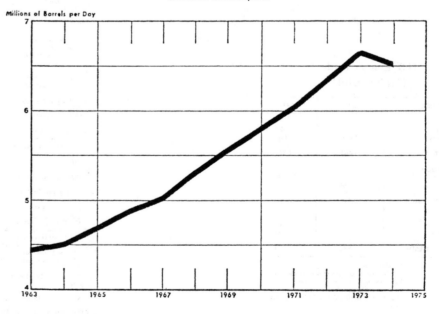

Gasoline Consumption

Source: Energy Statistics, Committee on Finance, U.S. Senate, July 1975.

TOTAL: 6,298

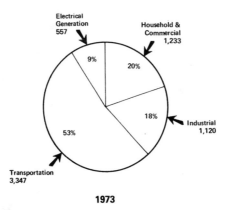

1973

U.S. petroleum consumption, by sector, 1973 (millions of barrels).

Source: Energy Perspectives, U.S. Department of the Interior, February 1975.

Petroleum Consumption, by Major Products[1] and Major Consuming Sectors, 1975 (Estimated)

	Household and commercial		Industrial		Transportation 2/		Electricity generation, utilities		Miscellaneous and unaccounted for		Total domestic product demand	
	Million barrels	Trillion Btu	Million barrels	Trillion Btu	Million barrels	Trillion Btu	Million barrels	Trillion Btu	Million barrels	Trillion Btu	Million barrels	Trillion Btu
Fuel and power:												
Liquefied gases	171.0	685.9	3/53.6	215.0	31.0	124.3	--	--	--	--	255.6	1,025.2
Jet fuels:												
Naphtha type	--	--	--	--	82.0	439.1	--	--	--	--	82.0	439.1
Kerosine type	--	--	--	--	279.1	1,582.5	5.0	28.4	--	--	284.1	1,610.9
Total	--	--	--	--	361.1	2,021.6	5.0	28.4	--	--	366.1	2,050.0
Gasoline	--	--	--	--	2,452.1	12,868.6	--	--	--	--	2,452.1	12,868.6
Kerosine	43.0	243.8	12.1	68.6	--	--	--	--	--	--	55.1	312.4
Distillate fuel	480.0	2,796.0	122.5	713.6	353.0	2,056.2	77.0	448.5	7.0	40.8	1,039.5	6,055.1
Residual fuel	153.0	961.9	175.0	1,100.2	100.0	628.7	451.0	2,835.4	4.3	27.1	883.3	5,553.3
Still gas	--	--	176.0	1,056.0	--	--	--	--	--	--	176.0	1,056.0
Petroleum coke	--	--	61.0	367.5	--	--	--	--	--	--	61.0	367.5
Total	847.0	4,687.6	600.2	3,520.9	3,297.2	17,699.4	533.0	3,312.3	11.3	67.9	5,288.7	29,288.1
Raw material: 4/												
Plant condensate	--	--	5.0	27.1	--	--	--	--	--	--	5.0	27.1
Special naphthas	--	--	29.0	152.2	--	--	--	--	--	--	29.0	152.2
Lubes 5/ and waxes	--	--	36.0	215.0	26.0	157.7	--	--	--	--	62.0	372.7
Petroleum coke 6/	--	--	24.0	144.5	--	--	--	--	--	--	24.0	144.5
Asphalt and road oil	172.0	1,141.4	--	--	--	--	--	--	--	--	172.0	1,141.4
Petrochemical feedstock offtake:												
Liquefied refinery gas 7/	--	--	41.3	159.8	--	--	--	--	--	--	41.3	159.8
Liquefied petroleum gas 7/, 8/	--	--	209.0	730.1	--	--	--	--	--	--	209.0	730.1
Naphtha (-400 degrees)	--	--	58.0	304.4	--	--	--	--	--	--	58.0	304.4
Still gas	--	--	12.0	72.0	--	--	--	--	--	--	12.0	72.0
Miscellaneous (+400 degrees)	--	--	53.0	308.7	--	--	--	--	--	--	53.0	308.7
Total	172.0	1,141.4	467.3	2,113.8	26.0	157.7	--	--	--	--	665.3	3,412.9
Miscellaneous and Unaccounted for	--	--	--	--	--	--	--	--	--	--	--	--
Total domestic product demand	1,019.0	5,829.0	1,067.5	5,634.7	3,323.2	17,857.1	533.0	3,312.3	11.3	67.9	5,954.0	32,701.0

1/ Includes liquefied refinery gas and natural gas liquids.
2/ Includes bunkers, military transportation, and all military use of distillate and residual fuel oils.
3/ Includes secondary recovery of petroleum and agriculture uses.
4/ Includes some fuel and power used by raw materials industries.
5/ Lubricants are distributed on basis of data from Bureau of the Census.
6/ Includes portions of petroleum coke estimated to be consumed in nonfuel uses.
7/ Includes ethane.
8/ Includes LP-gas for synthetic rubber.

Source: Division of Interfuels Studies, Office of Assistant Director-Fuels, Bureau of Mines, U.S. Department of the Interior.

Source: Department of the Interior News Release, April 5, 1976.

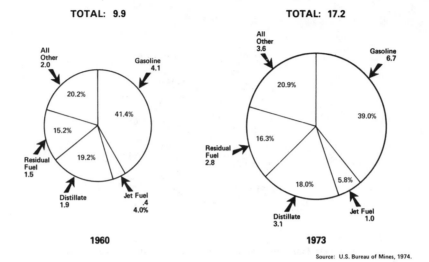

TOTAL: 9.9

All Other 2.0

Gasoline 4.1

20.2%

41.4%

15.2%

19.2%

Residual Fuel 1.5

Distillate 1.9

Jet Fuel .4 4.0%

1960

TOTAL: 17.2

All Other 3.6

Gasoline 6.7

20.9%

39.0%

16.3%

Residual Fuel 2.8

18.0%

5.8%

Distillate 3.1

Jet Fuel 1.0

1973

Source: U.S. Bureau of Mines, 1974.

U.S. petroleum consumption by product, 1960 and 1973 (millions of barrels per day).

Source: Energy Perspectives, U.S. Department of the Interior, February 1975.

Petroleum Consumption, by Major Products[1] and Major Consuming Sectors, 1974

	Household and commercial		Industrial		Transportation [2]		Electricity generation, utilities		Miscellaneous and unaccounted for		Total domestic product demand	
	Million barrels	Trillion Btu	Million barrels	Trillion Btu	Million barrels	Trillion Btu	Million barrels	Trillion Btu	Million barrels	Trillion Btu	Million barrels	Trillion Btu
Fuel and power:												
Liquefied gases	172.2	690.7	[3] 55.8	223.8	31.2	125.1	--	--	--	--	259.2	1,039.6
Jet fuels:												
Naphtha type	--	--	--	--	81.2	434.8	--	--	--	--	81.2	434.8
Kerosine type	--	--	--	--	276.2	1,566.0	5.2	29.5	--	--	281.4	1,595.5
Total	--	--	--	--	357.4	2,000.8	5.2	29.5	--	--	362.6	2,030.3
Gasoline	--	--	--	--	2,402.4	12,607.8	--	--	--	--	2,402.4	12,607.8
Kerosine	49.9	282.9	14.5	82.2	--	--	--	--	--	--	64.4	365.1
Distillate fuel	492.7	2,870.0	126.3	735.7	364.2	2,121.5	79.5	463.1	10.1	58.8	1,072.8	6,249.1
Residual fuel	167.4	1,052.5	194.0	1,219.7	112.7	708.5	475.2	2,987.6	8.5	53.4	957.8	6,021.7
Still gas	--	--	175.7	1,054.2	--	--	--	--	--	--	175.7	1,054.2
Petroleum coke	--	--	62.1	374.1	--	--	--	--	--	--	62.1	374.1
Total	882.2	4,896.1	628.4	3,689.7	3,267.9	17,563.7	559.9	3,480.2	18.6	112.2	5,357.0	29,741.9
Raw material: [4]												
Plant condensate	--	--	6.1	33.0	--	--	--	--	--	--	6.1	33.0
Special naphthas	--	--	32.0	167.9	--	--	--	--	--	--	32.0	167.9
Lubes [5] and waxes	--	--	37.7	225.1	25.8	156.5	--	--	--	--	63.5	381.6
Petroleum coke [6]	--	--	25.0	150.6	--	--	--	--	--	--	25.0	150.6
Asphalt and road oil	175.6	1,165.3	--	--	--	--	--	--	--	--	175.6	1,165.3
Petrochemical feedstock offtake:												
Liquefied refinery gas [7]	--	--	41.9	162.1	--	--	--	--	--	--	41.9	162.1
Liquefied petroleum gas [7], [8]	--	--	211.7	739.5	--	--	--	--	--	--	211.7	739.5
Naphtha (-400 degrees)	--	--	61.9	324.9	--	--	--	--	--	--	61.9	324.9
Still gas	--	--	14.4	86.4	--	--	--	--	--	--	14.4	86.4
Miscellaneous (+400 degrees)	--	--	56.2	327.4	--	--	--	--	--	--	56.2	327.4
Total	175.6	1,165.3	486.9	2,216.9	25.8	156.5	--	--	⊥	--	688.3	3,538.7
Miscellaneous and Unaccounted for	--	--	--	--	--	--	--	--	24.2	133.8	24.2	133.8
Grand total domestic product demand	1,057.8	6,061.4	1,115.3	5,906.6	3,293.7	17,720.2	559.9	3,480.2	42.8	246.0	6,069.5	33,414.4

1/ Includes liquefied refinery gas and natural gas liquids.
2/ Includes bunkers, military transportation, and all military use of distillate and residual fuel oils.
3/ Includes secondary recovery of petroleum and agriculture uses.
4/ Includes some fuel and power used by raw materials industries.
5/ Lubricants are distributed on basis of data from Bureau of the Census.
6/ Includes portions of petroleum coke estimated to be consumed in nonfuel uses.
7/ Includes ethane.
8/ Includes LP-gas for synthetic rubber.

Source: Division of Interfuels Studies, Office of Assistant Director-Fuels, Bureau of Mines, U.S. Department of the Interior.

Source; Department of the Interior News Release, April 5, 1976.

U.S. SECTORIAL DEMAND FOR FOSSIL FUELS 1947-71

Sectorial Demand for Natural Gas, 1947–1971
(In Trillions of Btu)

Year	Household and Commercial	Industrial			Transportation	Electrical Generation	Total Non-energy uses	Total Fuel uses	Total
		Fuel Uses	Non-fuel uses	Total					
1947	1,125	2,505	502	3,007	--	386	502	4,016	4,518
1948	1,262	2,778	498	3,276	--	495	498	4,535	5,033
1949	1,387	2,889	443	3,332	--	569	443	4,846	5,288
1950	1,642	3,302	425	3,727	130	651	425	5,725	6,150
1951	2,007	3,810	441	4,251	199	791	441	6,807	7,248
1952	2,213	4,010	381	4,391	214	942	381	7,379	7,760
1953	2,294	4,242	312	4,554	238	1,070	312	7,845	8,156
1954	2,566	4,187	350	4,537	239	1,206	350	8,198	8,548
1955	2,850	4,573	362	4,935	253	1,194	362	8,870	9,232
1956	3,151	4,709	386	5,095	306	1,283	386	9,449	9,835
1957	3,391	4,933	397	5,330	310	1,385	397	10,019	10,416
1958	3,712	5,162	378	5,540	323	1,421	378	10,618	10,996
1959	4,024	5,533	388	5,921	362	1,684	388	11,603	11,991
1960	4,268	5,915	372	6,287	359	1,785	372	12,327	12,699
1961	4,477	6,133	338	6,471	391	1,889	388	12,890	13,228
1962	4,849	6,531	311	6,842	396	2,034	311	13,810	14,121
1963	5,027	6,862	298	7,160	438	2,218	298	14,545	14,843
1964	5,343	7,154	297	7,451	451	2,403	297	15,351	15,648
1965	5,517	7,379	292	7,671	517	2,392	292	15,805	16,097
1966	5,945	7,908	295	8,203	552	2,692	295	17,097	17,392
1967	6,223	8,138	461	8,599	594	2,834	461	17,789	18,250
1968	6,451	8,819	455	9,274	610	3,245	455	19,125	19,580
1969	6,890	9,171	714	9,885	651	3,594	714	20,306	21,020
1970	7,108	9,475	687	10,162	744	4,015	687	21,342	22,029
1971P	7,346	9,753	686	10,439	825	4,124	686	22,048	22,734

P Preliminary.

Source: Division of Fossil Fuels, Bureau of Mines, U.S. Department of the Interior.

Source: Dupree, Walter G., Jr., and West, James A., *United States Energy Through the Year 2000*, U.S. Department of the Interior, December 1972.

Sectorial Demand for Petroleum, 1947-71[1]
(All Figures in Trillions of Btu)

Year	Household and Commercial			Industrial			Transportation[2]	Electrical Generation	Other Not Specified	Total Non-fuel use	Total Fuel Use	Total Input
	Fuel use	Non-fuel use	Total	Fuel use	Non-fuel use	Total						
1947	1,925	326	2,251	2,085	432	2,517	5,761	468	371	758	10,610	11,368
1948	2,154	385	2,539	2,102	428	2,530	6,157	444	889	813	11,746	12,559
1949	2,093	379	2,472	2,064	402	2,466	6,183	577	422	781	11,339	12,120
1950	2,603	435	3,038	2,213	453	2,666	6,785	662	337	888	12,600	13,488
1951	2,722	480	3,202	2,509	535	3,044	7,482	499	621	1,015	13,833	14,848
1952	2,833	517	3,350	2,514	520	3,034	7,868	492	590	1,037	14,297	15,334
1953	2,869	522	3,391	2,584	560	3,144	8,158	577	829	1,082	15,017	16,099
1954	3,094	556	3,650	2,542	576	3,118	8,358	480	531	1,132	15,005	16,137
1955	3,386	615	4,001	2,754	652	3,406	9,109	512	496	1,267	16,257	17,524
1956	3,523	660	4,183	2,975	713	3,688	9,448	497	809	1,373	17,252	18,625
1957	3,432	637	4,069	2,715	763	3,478	9,649	512	862	1,400	17,170	18,570
1958	3,889	679	4,568	2,519	773	3,292	9,819	515	1,020	1,452	17,762	19,214
1959	3,997	721	4,718	2,674	915	3,589	9,923	546	971	1,636	18,111	19,747
1960	4,189	734	4,923	2,674	1,008	3,682	10,372	564	526	1,742	18,325	20,067
1961	4,275	753	5,028	2,634	1,048	3,682	10,575	577	625	1,801	18,686	20,487
1962	4,423	804	5,227	2,750	1,130	3,880	11,001	579	580	1,934	19,333	21,267
1963	4,434	824	5,258	2,739	1,255	3,994	11,506	600	592	2,079	19,871	21,950
1964	4,350	841	5,191	2,922	1,262	4,184	11,791	636	585	2,103	20,284	22,387
1965	4,744	891	5,635	2,826	1,313	4,139	12,179	744	545	2,204	21,038	23,242
1966	4,830	936	5,766	2,883	1,470	4,353	12,777	905	594	2,406	21,989	24,395
1967	5,289	917	6,206	2,820	1,612	4,432	13,408	1,013	276	2,529	22,806	25,335
1968	5,145	984	6,129	3,186	1,780	4,966	14,535	1,180	242	2,763	24,289	27,052
1969	5,260	1,009	6,269	3,220	1,951	5,171	15,125	1,628	229	2,960	25,462	28,422
1970	5,371	1,082	6,453	3,252	2,015	5,267	15,592	2,087	215	3,097	26,517	29,614
1971P	5,435	1,110	6,545	3,191	2,028	5,219	16,139	2,417	172	3,138	27,354	30,492

P Preliminary.
[1] Petroleum products refined and processed from crude oil, including still gas, liquefied refinery gas and natural gas liquids.
[2] Includes bunkers and military transportation.

Source: Division of Fossil Fuels, Bureau of Mines, U.S. Department of the Interior.

Source: Dupree, Walter G., Jr. and West, James A., *United States Energy Through the Year 2000*, U.S. Department of the Interior, December 1972.

Sectorial Demand for Coal 1947-71[1]
(In Trillions of Btu)

Year	Household and Commercial	Industrial			Transportation	Electrical Generation	Miscellaneous	Total Non-Fuel Uses	Total Fuel Uses	Total
		Fuel Uses	Non-Fuel Uses	Total						
1947	3,399	7,175	123	7,298	3,030	2,084	13	123	15,701	15,824
1948	3,238	6,395	121	6,516	2,624	2,392	127	121	14,776	14,897
1949	3,025	5,462	109	5,571	1,892	2,021	122	109	12,522	12,631
1950	2,913	5,828	129	5,957	1,701	2,228	115	129	12,785	12,914
1951	2,648	6,260	143	6,403	1,525	2,537	112	143	13,082	13,225
1952	2,416	5,548	125	5,673	1,086	2,588	107	125	11,745	11,870
1953	2,072	5,958	147	6,105	809	2,805	102	147	11,746	11,893
1954	1,752	4,737	123	4,860	516	2,866	201	123	10,072	10,195
1955	1,745	5,577	149	5,726	464	3,482	123	149	11,391	11,540
1956	1,629	5,659	146	5,805	378	3,815	124	146	11,605	11,751
1957	1,225	5,546	149	5,695	270	3,881	97	149	11,019	11,168
1958	1,187	4,558	117	4,675	133	3,743	112	117	9,733	9,850
1959	969	4,419	111	4,530	102	4,050	156	111	9,696	9,807
1960	983	4,548	125	4,673	87	4,252	145	125	10,015	10,140
1961	868	4,357	121	4,478	21	4,374	165	121	9,785	9,906
1962	872	4,405	122	4,527	18	4,636	153	122	10,084	10,206
1963	731	4,635	119	4,754	18	5,065	146	119	10,595	10,714
1964	609	4,933	130	5,063	19	5,395	178	130	11,134	11,264
1965	678	5,227	139	5,366	18	5,843	4	139	11,770	11,909
1966	677	5,337	135	5,472	16	6,327	3	135	12,360	12,495
1967	585	5,065	134	5,199	13	6,457	1	134	12,121	12,255
1968	529	4,973	151	5,124	11	6,994	1	151	12,508	12,659
1969	447	4,899	159	5,051	8	7,227	---	152	12,574	12,733
1970	427	4,853	151	5,004	8	7,483	---	151	12,771	12,922
1971P	390	4,332	133	4,465	7	7,698	---	133	12,427	12,560

P Preliminary.
[1] Includes anthracite, bituminous and lignite coals.

Source: Dupree, Walter S., and West, James A., *United States Energy Through the Year 2000*, U.S. Department of the Interior, December 1972.

Domestic Supply and Demand for Coal

	1974 Thousand short tons	1974 Trillion Btu	1975 (estimated) Thousand short tons	1975 (estimated) Trillion Btu	Percentage change from 1974
ANTHRACITE					
Supply:					
Production 1/	6,617	168.1	6,200	157.4	-6.3
Exports 2/	-1,169	-29.7	-1,175	-29.8	+0.5
Imports	---	---	---	---	
Stock change: withdrawals (+), additions(-)	NA	NA	NA	NA	
Losses, gains, and unaccounted for	---	---	---	---	
Total	5,448	138.4	5,025	127.6	-7.8
Demand by major consuming sectors: 3/					
Household and commercial 4/	2,577	65.5	2,210	56.1	-14.2
Industrial 5/	1,373	34.9	1,265	32.1	-7.9
Electricity generation, utilities	1,498	38.0	1,550	39.4	+3.5
Total	5,448	138.4	5,025	127.6	-7.8
BITUMINOUS COAL AND LIGNITE					
Supply:					
Production 1/	603,406	14,481.7	640,000	15,360.0	+6.1
Exports	-59,926	-1,621.6	-66,000	-1,786.0	+10.1
Imports	2,080	49.1	1,150	27.1	-44.7
Stock change: withdrawals (+), addition (-)	+7,265	+136.9	-9,421	-260.5	
Losses, gains, and unaccounted for	-116	-2.2	-3,729	-92.2	
Total	552,709	13,043.9	562,000	13,248.4	+1.7
Demand by major consuming sectors:					
Fuel and power:					
Household and commercial 4/	8,840	248.0	8,000	226.0	-9.5
Industrial 5/	149,619	4,197.0	146,010	4,125.1	-2.4
(Coal carbonized for coke) 6/	(89,747)	(2,517.5)	(85,000)	(2,401.4)	(-5.3)
Transportation 7/	80	2.2	25	0.7	-68.8
Electricity generation, utilities	390,068	8,481.6	404,000	8,784.6	+3.6
Total	548,607	12,928.8	558,035	13,136.4	+1.7
Raw material: Industrial 8/					
Crude light oil	1,087	30.5	1,000	28.2	-8.0
Crude coal tar	3,015	84.6	2,965	83.8	-1.7
Total raw material	4,102	115.1	3,965	112.0	-3.3
Grand total	552,709	13,043.9	562,000	13,248.4	+1.7

NA Not available
1/ Includes use by producers for power and heat.
2/ Includes shipments to U.S. Armed Forces in West Germany.
3/ Except for small quantities used as raw material for coal chemicals, all anthracite is used for fuel and power.
4/ Data represent "retail deliveries to other consumers." These are mainly household and commercial users, with some unknown portion of use by small industries.
5/ Includes consumption by coke plants, steel and rolling mills, and other industrial uses. Adjusted to exclude coal equivalent of raw material use.
6/ Figures in parentheses are not added into totals.
7/ Includes bunkers and military transportation.
8/ Coal equivalent based on British thermal unit value of raw material consumption of coal chemicals listed.

Source: Division of Interfuels Studies, Office of Assistant Director-Fuels, Bureau of Mines, U.S. Department of Interior.

Source: Department of the Interior News Release, April 5, 1976.

Domestic Supply and Demand for Natural Gas

	1974		1975 (estimated)		Percentage change from 1974
	Million cubic feet	Trillion Btu	Million cubic feet	Trillion Btu	
Supply:					
Marketed production 1/	21,600,522	23,689.3	20,100,000	22,186.0	-6.9
Transfers out, extraction loss 2/	-887,490	-2,479.2	-769,000	-2,391.1	-13.4
Domestic production 3/	20,713,032	21,210.1	19,331,000	19,794.9	-6.7
Exports	-76,789	-78.6	-75,000	-76.8	-2.3
Imports	959,284	982.3	950,000	972.8	-1.0
Stock change: Withdrawals (+), additions (-)	-83,663	-85.7	-221,000	-226.3	
Transmission loss and unaccounted for 4/	-288,731	-295.6	-285,000	-291.8	
Total	21,223,133	21,732.5	19,700,000	20,172.8	-7.2
Demand by major consuming sectors:					
Fuel and power:					
Household and commercial 5/	7,341,745	7,517.9	7,200,000	7,372.8	-1.9
Industrial	9,073,193	9,291.0	8,098,000	8,292.3	-10.7
Transportation	668,834	684.9	620,000	634.9	-7.3
Electricity generation, utilities	3,429,231	3,511.5	3,100,000	3,174.4	-9.6
Total	20,513,003	21,005.3	19,018,000	19,474.4	-7.3
Raw Materials: Industrial 6/					
Carbon black	40,130	41.1	32,000	32.8	-20.3
Other chemicals 7/	670,000	686.1	650,000	665.6	-3.0
Total	710,130	727.2	682,000	698.4	-4.0
Grand total	21,223,133	21,732.5	19,700,000	20,172.8	-7.2

1/ Marketed production represents gross withdrawals less the quantities used for repressuring and the amount rented or flared. British thermal unit value of production is for wet gas prior to extraction of natural gas liquids. Higher Btu values assigned to extraction loss represent the Btu value of natural gas liquids production for each year.

2/ Extraction loss from cycling plants represents offtake of natural gas for natural gas liquids as reported to the Bureau of Mines. Energy equivalent of extraction loss is based on annual outputs of natural gasoline and associated products at 110,000 Btu per gallon, annual outputs of LPG at 95,500 Btu per gallon, annual outputs of ethane, since 1967, at 73,390 Btu per gallon, and beginning with 1973, energy equivalent for plant condensate is computed at 129,000 Btu per gallon.

3/ Domestic production is the marketed production less the shrinkage resulting from the extraction of natural gas liquids.

4/ Transmission loss and unaccounted for was formerly included in the industrial sector.

5/ Includes deliveries to municipalities and public authorities for institutional heating, street lighting, etc., formerly included in the industrial consuming sector.

6/ Includes some fuel and power used by raw material industries.

7/ Estimated from partial data.

Source: Division of Interfuels Studies, Office of Assistant Director--Fuels, Bureau of Mines, U.S. Department of the Interior.

Source: Department of the Interior News Release, April 5, 1976.

Domestic Supply and Demand for Petroleum[1]

	1974 Million bbl	1974 Trillion Btu	1975 (estimated) Million bbl	1975 (estimated) Trillion Btu	Percentage change from 1974
Supply, crude oil:					
Production (including lease condensate)------------------	3,199.3	18,556.0	3,056.1	17,725.4	-4.5
Exports---	-1.1	-6.4	-2.1	-12.2	
Imports---	1,269.2	7,361.4	1,461.1	8,474.4	+15.1
Stock change: withdrawals (+), addition, (-)----------	-22.5	-130.5	+11.0	+63.8	
Losses, transfers for use as fuel, and unaccounted for--	-16.2	-94.0	-8.5	-49.3	
Total--	4,428.7	25,686.5	4,517.6	26,202.1	+2.0
Refinery input:					
Crude oil---	4,428.7	25,686.5	4,517.6	26,202.1	+2.0
Transfers in, natural gas liquids 2/-------------------	272.4	1,245.3	278.0	1,270.0	-2.1
Other hydrocarbons-------------------------------------	13.1	45.9	13.1	46.4	--
Total--	4,714.2	26.977.7	4,808.7	27,518.5	+2.0
Supply, refined products:					
Refinery output---------------------------------------	4,714.2	26,977.7	4,808.7	27,518.5	+2.0
Unfinished oil reruns, net-----------------------------	37.4	217.9	15.0	42.4	-59.9
Processing gain, net-----------------------------------	175.2	---	159.5	---	
Total--	4,926.8	27,195.6	4,983.2	27,560.9	+1.1
Exports 3/--	-79.4	-460.8	-75.0	-435.0	-5.5
Imports 3/--	952.4	5,690.8	715.0	4,272.0	-24.9
Stock change, including natural gas liquids------------	-42.8	-231.6	-34.0	-170.0	
Transfers in, natural gas liquids 2/, 4/---------------	343.7	1,233.9	316.2	1,121.1	-8.0
Losses, gains, and unaccounted for---------------------	-31.2	-13.5	48.6	352.0	
Total--	6,069.5	33,414.4	5,954.0	32,701.0	-1.9
Demand by major consuming sectors:					
Fuel and power:					
Household and commercial------------------------------	882.2	4,896.1	847.0	4,687.6	-4.0
Industrial--	628.4	3,689.7	600.2	3,520.9	-4.5
Transportation 5/-------------------------------------	3,267.9	17,563.7	3,297.2	17,699.4	+ .9
Electricity generation, utilities---------------------	559.9	3,480.2	533.0	3,312.3	-4.8
Other, not specified----------------------------------	18.6	112.2	11.3	67.9	-39.2
Total--	5,357.0	29,741.9	5,288.7	29,288.1	-1.3
Raw material 6/:					
Petrochemical feedstock offtake-----------------------	386.1	1,640.3	373.3	1,575.0	-3.3
Other nonfuel use-------------------------------------	302.2	1,898.4	292.0	1,837.9	-3.4
Total--	688.3	3,538.7	665.3	3,412.9	-3.3
Miscellaneous and unaccounted for---------------------	24.2	133.8	---	---	
Total domestic product demand-------------------------	6,069.5	33,414.4	5,954.0	32,701.0	-1.9

1/ Supply and demand for crude oil and petroleum products. Petroleum products include products refined and processed from crude oil, including still gas and liquefied refinery gas; also natural gas liquids transferred from natural gas.

2/ Btu values for natural gas liquids for each year shown are implicitly derived from weighted averages of major natural gas liquids, with natural gasoline and other products at 110,000 Btu per gallon, liquefied petroleum gases at 95,500 Btu per gallon, ethane at 73,390 Btu per gallon, and plant condensate at 129,000 Btu per gallon.

3/ Btu values for imported and exported refined products for 1974 are totals of the Btu values of the respective products imported and exported. The 1974 average Btu value is applied to 1975 estimates.

4/ Includes natural gas liquids other than those channeled into refinery input as follows: Petrochemical feedstocks, direct uses for fuel and power, and other uses.

5/ Includes bunkers and military fuel uses.

6/ Includes some fuel and power use by raw materials industries.

Source: Division of Interfuels Studies, Office of Assistant Director--Fuels, Bureau of Mines, U.S. Department of the Interior.

Source: Department of the Interior News Release, April 5, 1976.

U.S. GROSS AND NET ENERGY CONSUMPTION BY SECTOR

The gross energy consumption in the transportation, industrial, and household-commercial sectors increased at annual rates of 4.3%, 3.1%, and 2.9% during the period 1960–73. Aggregate consumption in these sectors dropped about 2% during 1974.

Gross energy consumption in the electrical sector increased at a somewhat higher rate, 6.9% per year, during the 1960–73 period. During 1974, gross energy consumption in the electrical sector increased at a rate of 0.4% per year. Consumption in the early months of 1975 indicated a return to considerable growth in electrical energy at a rate of 4.3% per year.

U.S. Gross Energy Consumption, By Sector, 1960–73
(Quadrillion Btu)[1]

YEAR	HOUSEHOLD & COMMERCIAL	INDUSTRIAL	TRANSPORTATION	ELECTRICAL	TOTAL
1960	10.2	16.6	10.8	8.3	44.6
1963	11.0	17.7	12.0	9.7	49.3
1965	13.1	19.5	12.7	11.1	53.3
1968	12.9	19.1	15.3	13.9	61.8
1970	14.0	20.4	16.5	16.2	67.1
1971	14.2	20.2	17.0	17.3	68.7
1972	14.6	20.7	18.0	18.6	71.9
1973	14.7	21.7	18.6	19.7	74.7

[1] 1 Quadrillion Btu = 500,000 barrels petroleum per day for a year
= 40 million tons of bituminous coal
= 1 trillion cubic feet of natural gas
= 100 billion kWh (based on a 10,000-Btu/kWh heat rate)

Note.—Gross consumption of energy refers to the total energy consumption of the economy; net energy refers to the energy used by the final consuming sectors, and does not include the losses experienced in converting primary sources to secondary sources.

Source: Energy Perspectives, U.S. Department of the Interior, February 1975.

U.S. Net Energy Consumption,
By Final Consuming Sector, 1960–73
(Quadrillion Btu)[1]

YEAR	HOUSEHOLD & COMMERCIAL	INDUSTRIAL	TRANSPORTATION	TOTAL
1960	11.4	15.9	10.8	38.1
1961	11.8	15.9	11.0	38.7
1962	12.4	16.7	11.4	40.5
1963	12.7	17.4	12.0	41.0
1964	12.9	18.2	12.3	43.4
1965	13.8	18.8	12.7	45.3
1966	14.5	19.8	13.4	47.7
1967	15.3	20.1	14.0	49.4
1968	15.6	21.4	15.2	52.2
1969	16.4	22.3	15.8	54.5
1970	17.0	22.6	16.5	56.1
1971	17.4	22.5	17.1	57.0
1972	18.1	23.3	18.0	59.4
1973	18.4	24.2	18.8	61.4

[1] 1 Quadrillion Btu = 500,000 barrels petroleum per day for a year
= 40 million tons of bituminous coal
= 1 trillion cubic feet of natural gas
= 100 billion kWh (based on a 10,000-Btu/kWh heat rate)

[1] Net energy refers to the energy used by the final consuming sectors, and does not include the losses experienced in converting primary to secondary sources.

Source: Energy Perspectives, U.S. Department of the Interior, February 1975.

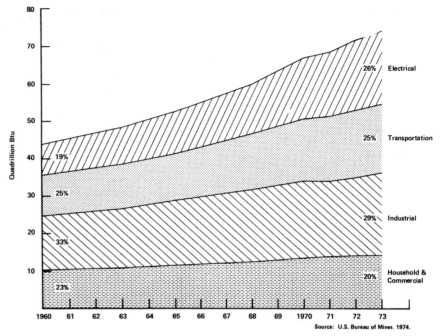

U.S. gross energy consumption, by sector, 1960–73.

Source: Energy Perspectives, U.S. Department of the Interior, February 1975.

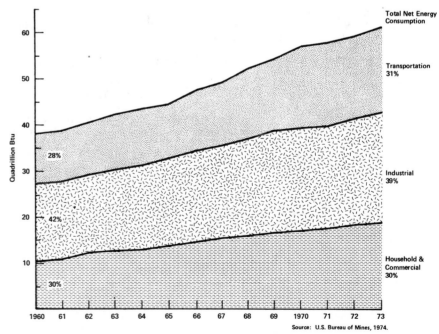

U.S. net energy consumption, by final consuming sector, 1960–73.

Source: Energy Perspectives, U.S. Department of the Interior, February 1975.

U.S. Consumption of Energy Resources by Major Sources and Consuming Sectors
(Trillion Btu)

Consuming sectors	Anthracite	Bituminous coal and lignite	Coke 1/	Natural gas, dry 2/	Petroleum 3/	Hydro-power 4/	Nuclear power 4/	Total gross energy inputs 5/	Utility electricity distributed 6/	Total net energy inputs 7/	Percentage change from 1974
Household and commercial:											
1974 -------------------------	66	248	--	7,518	6,061	--	--	13,893	3,723	17,616	
1975 (estimated)---------------	56	226	--	7,373	5,829	--	--	13,484	3,783	17,267	-2.0
Industrial:											
1974 -------------------------	35	4,312	59	10,018	5,907	37	--	20,368	2,665	23,033	
1975 (estimated)---------------	32	4,237	18	8,991	5,635	36	--	18,949	2,708	21,657	-6.0
Transportation: 8/											
1974 -------------------------	XX	2	--	685	17,720	--	--	18,407	19	18,426	
1975 (estimated)---------------	XX	1	--	635	17,857	--	--	18,493	19	18,512	+0.5
Electricity generation, utilities: 4/											
1974 -------------------------	38	8,482	--	3,512	3,480	3,253	1,202	19,967	6,407	XX	
1975 (estimated)---------------	39	8,785	--	3,174	3,312	3,122	1,652	20,084	6,510	XX	+0.6
Miscellaneous and unaccounted for:											
1974 -------------------------	--	--	--	--	246	--	--	246	XX	246	
1975 (estimated)---------------	--	--	--	--	68	--	--	68	XX	68	
Total energy inputs: 9/											
1974 ---------------------	138	13,044	59	21,733	33,414	3,290	1,202	72,880	XX	59,320	(Gross)
1975 (estimated)----------	128	13,248	18	20,173	32,701	3,158	1,652	71,078	XX	57,504	-2.5

XX Not applicable

1/ Net imports of coke (2,262,000 tons in 1974 and 700,000 tons in 1975). The remainder of domestic coke consumption is accounted for in utilizing the gross total Btu value of coal carbonized for coke.

2/ Excludes natural gas liquids.

3/ Petroleum products including still gas, liquefied refinery gas, and natural gas liquids.

4/ Outputs of hydropower (adjusted for net imports or net exports) are converted to theoretical energy inputs calculated from national average heat rates for fossil-fueled steam-electric plants provided by the Federal Power Commission using 10,389 Btu per net kilowatt-hour. Energy inputs for nuclear power are converted at an average heat rate of 10,660 Btu per kilowatt-hour based on information from the Energy Research and Development Administration. Excludes inputs for power generated by nonutility plants which are included within the other consuming sectors.

5/ Gross energy is that contained in all types of commerical energy at the time it is incorporated into the economy, whether energy is produced domestically or imported. Gross energy comprises inputs of primary fuels (or their derivatives) and outputs of hydropower and nuclear power converted to theoretical fuel inputs. Gross energy includes energy used for production, processing, and transportation of energy proper.

6/ Utility electricity, generated and imported, distributed to the other consuming sectors as energy resource inputs. Distribution to sectors is based on sales reported in the Edison Electric Institute "Statistical Yearbook of the Electric Utility Industry for 1974." Conversion of electricity to energy equivalent by sectors was made at the value of contained energy corresponding to 100 percent efficiency using a theoretical rate of 3,412 Btu per kilowatt-hour.

7/ Energy inputs into the final consuming sectors consisting of direct fuels and electricity distributed. Conversion losses in the electric sector constitute the difference between net and gross energy totals.

8/ Includes bunkers and military uses.

9/ Data may not add to totals shown because of independent rounding.

Source: Division of Interfuels Studies, Office of Assistant Director - Fuels, Bureau of Mines, U.S. Department of the Interior.

Source: Department of the Interior News Release, April 5, 1976.

1974 Energy Consumption by Fuel by Sector
(10^{12} Btus)

	Household/ Commercial	Industrial (Fuel & Power)	Raw Materials	Transportation	Electric Utilities	Misc.	Fuel Total
Coal							
Bituminous	229	4,167		2	8,630		13,028
Anthracite	62	32			38		132
Natural Gas	7,116	10,394	689	664	3,328		22,191
Carbon Black			47				47
Petroleum Products							
Gasoline				12,596			12,596
Jet Fuel				2,006			2,006
Distillate	3,000	769		2,027	396	50	6,242
Residual	1,125	1,182		685	3,018	62	6,072
Kerosene	289	83					372
Liquid Gas	734	186		133			1,053
Liquid Refinery Gas			176				176
Liquid Petroleum Gas		740					740
Still Gas		1,038	61				1,099
Naphthas			291				291
Special Naphthas			165				165
Lubes and Waxes			228	160			388
Asphalt	1,241						1,241
Petroleum Coke		391	170				561
Miscellaneous			346			107	453
Fossil Fuels	13,796	18,982	2,173	18,273	15,410	219	68,853
Electricity*	3,372	2,416		16			5,804
Sector Totals	17,168	21,398	2,173	18,289	15,410	219	

*Nine percent has been deducted from electricity distributed for transmission loss.

Source: Oil, Coal, and Gas from Bureau of Mines. (1974 preliminary) Electricity from Edison Electric Institute

Source: As given in *National Energy Outlook*, FEA-N-75/713, Federal Energy Administration, February 1976.

ENERGY FLOW PATTERNS

In the assessment of energy consumption trends and projections, frequent use is made of flow diagrams of energy systems which detail all of the steps in energy utilization from resource extraction to end use. These flow diagrams provide detailed estimates of energy consumption by sector and demand category, of the efficiency of energy conversion systems, of transportation modes, of fuel extraction and refining requirements, and of energy resource requirements. The construction of flow diagrams for energy consumption in future years involves many judgmental decisions: expected consumption, relative availability and cost of various energy resources, improvements in equipment efficiencies, and the probable availability of new technology.

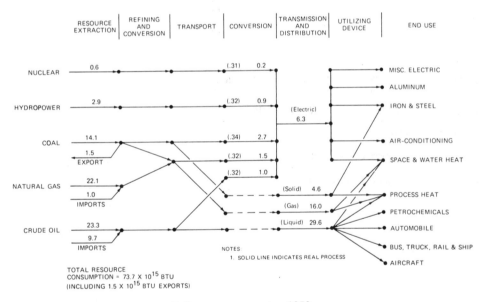

Reference energy system 1972.

Source: A National Plan for Energy Research, Development and Demonstration: Creating Energy Choices for the Future, Volume 1: The Plan, ERDA-48, U.S. Energy Research and Development Administration, June 1975.

Part of the pattern of energy flow is the movement of coal along the nation's railroads. (*Courtesy Dept. of Transportation*)

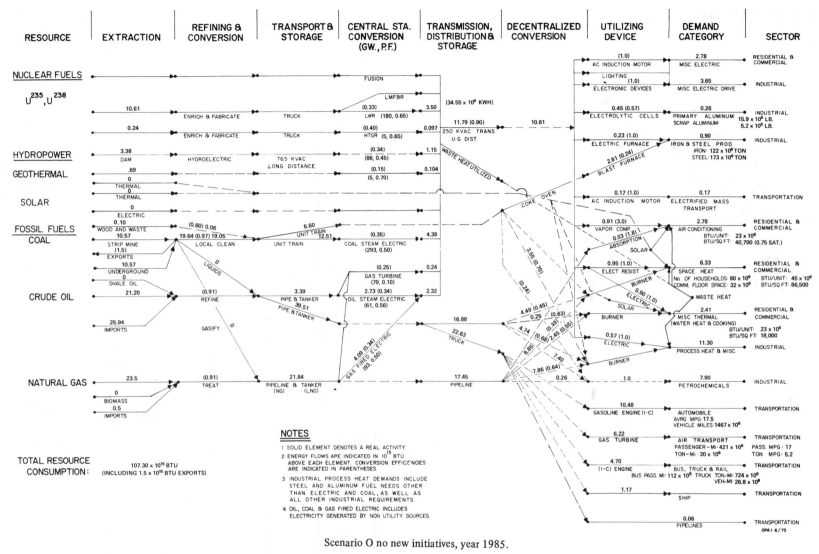

| RESOURCE | EXTRACTION | REFINING & CONVERSION | TRANSPORT & STORAGE | CENTRAL STA. CONVERSION (GW., P.F.) | TRANSMISSION, DISTRIBUTION & STORAGE | DECENTRALIZED CONVERSION | UTILIZING DEVICE | DEMAND CATEGORY | SECTOR |

Scenario O no new initiatives, year 1985.

NOTES

1. SOLID ELEMENT DENOTES A REAL ACTIVITY.

2. ENERGY FLOWS ARE INDICATED IN 10^{15} BTU ABOVE EACH ELEMENT. CONVERSION EFFICIENCIES ARE INDICATED IN PARENTHESES.

3. INDUSTRIAL PROCESS HEAT DEMANDS INCLUDE STEEL AND ALUMINUM FUEL NEEDS OTHER THAN ELECTRIC AND COAL, AS WELL AS ALL OTHER INDUSTRIAL REQUIREMENTS.

4. OIL, COAL & GAS FIRED ELECTRIC INCLUDES ELECTRICITY GENERATED BY NON UTILITY SOURCES.

TOTAL RESOURCE CONSUMPTION: 107.30 x 10^{15} BTU (INCLUDING 1.5 x 10^{15} BTU EXPORTS)

Source: A National Plan for Energy Research, Development and Demonstration: Creating Energy Choices for the Future, Volume 1: The Plan, ERDA-48, U.S. Energy Research and Development Administration, June 1975.

U.S. END USES OF ENERGY

TOTAL GROSS ENERGY USE: 74.7 Quadrillion Btu

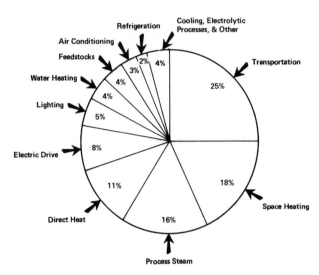

Source: U.S. Energy Prospects: An Engineering Viewpoint,
National Academy of Engineering, 1974, page 26.

U.S. gross energy end uses, 1973.

Source: Energy Perspectives, U.S. Department of the Interior, February 1975.

**End Uses of Energy
United States, 1968**

Source: Stanford Research Institute, "Patterns of Energy Consumption in U.S."

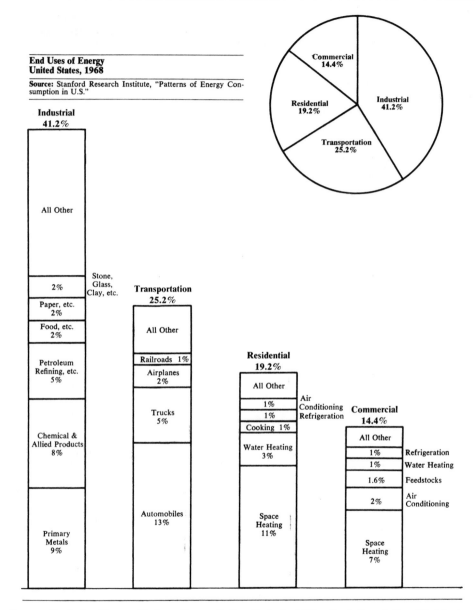

Source: Exploring Energy Choices, Energy Policy Project of The Ford Foundation, Washington, D.C. 1974.

Energy Consumption in the United States by End Use 1960-1968
(Trillions of Btu and Percent per Year)

Sector and End Use	Consumption 1960	Consumption 1968	Annual Rate of Growth	Percent of National Total 1960	Percent of National Total 1968
Residential					
Space heating	4,848	6,675	4.1%	11.3%	11.0%
Water heating	1,159	1,736	5.2	2.7	2.9
Cooking	556	637	1.7	1.3	1.1
Clothes drying	93	208	10.6	0.2	0.3
Refrigeration	369	692	8.2	0.9	1.1
Air conditioning	134	427	15.6	0.3	0.7
Other	809	1,241	5.5	1.9	2.1
Total	7,968	11,616	4.8	18.6	19.2
Commercial					
Space heating	3,111	4,182	3.8	7.2	6.9
Water heating	544	653	2.3	1.3	1.1
Cooking	98	139	4.5	0.2	0.2
Refrigeration	534	670	2.9	1.2	1.1
Air conditioning	576	1,113	8.6	1.3	1.8
Feedstock	734	984	3.7	1.7	1.6
Other	145	1,025	28.0	0.3	1.7
Total	5,742	8,766	5.4	13.2	14.4
Industrial					
Process steam	7,646	10,132	3.6	17.8	16.7
Electric drive	3,170	4,794	5.3	7.4	7.9
Electrolytic processes	486	705	4.8	1.1	1.2
Direct heat	5,550	6,929	2.8	12.9	11.5
Feed stock	1,370	2,202	6.1	3.2	3.6
Other	118	198	6.7	0.3	0.3
Total	18,340	24,960	3.9	42.7	41.2
Transportation					
Fuel	10,873	15,038	4.1	25.2	24.9
Raw materials	141	146	0.4	0.3	0.3
Total	11,014	15,184	4.1	25.5	25.2
National total	43,064	60,526	4.3	100.0%	100.0%

Note: Electric utility consumption has been allocated to each end use.

Source: Patterns of Energy Consumption in the United States, prepared by the Stanford Research Institute for the Office of Science and Technology, Executive Office of the President, Washington, D.C., January 1972.

U.S. ENERGY RESOURCE INPUTS—INDUSTRIAL

During the period 1947-71, energy consumption in the industrial sector increased from 13,300 to 22,500 trillion Btu. During that period, gas consumption increased by a factor of over three, petroleum consumption by a factor of two, electricity consumption by a factor of five, while coal consumption dropped by about 37%.

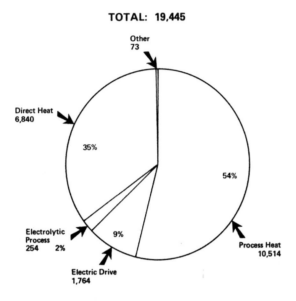

TOTAL: 19,445

Other 73

Direct Heat 6,840

35%

54%

Electrolytic Process 254   2%

9%

Process Heat 10,514

Electric Drive 1,764

Note.—Nonenergy uses not shown.
[1] 1 Quadrillion Btu = 1,000 trillion Btu.

Source: U.S. Bureau of Mines, 1973.

U.S. net energy consumption, industrial sector, 1970 (trillion Btu)[1].

Source: Energy Perspectives, U.S. Department of the Interior, February 197?.

Energy Consumption by the Industrial Sector, 1947-75
[Quadrillion Btu]

Year	Coal[1]			Natural gas			Petroleum[2]			Electricity purchased[3]	Total fuel uses	Total nonfuel uses	Total energy inputs
	Fuel uses	Nonfuel uses	Total	Fuel uses[4]	Nonfuel uses	Total	Fuel uses	Nonfuel uses	Total				
1947	7.2	0.1	7.3	2.5	0.5	3.0	2.1	0.4	2.5	0.5	12.2	1.1	13.3
1948	6.4	.1	6.5	2.8	.5	3.3	2.1	.4	2.5	.5	11.8	1.0	12.8
1949	5.5	.1	5.6	2.9	.4	3.3	2.1	.4	2.5	.5	10.9	1.0	11.9
1950	5.8	.1	6.0	3.3	.4	3.7	2.2	.5	2.7	.6	11.9	1.0	12.9
1951	6.3	.1	6.4	3.8	.4	4.3	2.5	.5	3.0	.7	13.2	1.1	14.4
1952	5.5	.1	5.7	4.0	.4	4.4	2.5	.5	3.0	.7	12.8	1.0	13.8
1953	6.0	.1	6.1	4.2	.3	4.6	2.6	.6	3.1	.8	13.5	1.0	14.6
1954	4.7	.1	4.9	4.2	.4	4.5	2.5	.6	3.1	.8	12.3	1.0	13.3
1955	5.6	.1	5.7	4.6	.4	4.9	2.8	.7	3.4	1.0	13.9	1.2	15.1
1956	5.7	.1	5.8	4.7	.4	5.1	3.0	.7	3.7	1.1	14.5	1.2	15.7
1957	5.5	.1	5.7	4.9	.4	5.3	2.7	.8	3.5	1.1	14.3	1.3	15.6
1958	4.6	.1	4.7	5.2	.4	5.5	2.5	.8	3.3	1.1	13.3	1.3	14.6
1959	4.4	.1	4.5	5.5	.4	5.9	2.7	.9	3.6	1.2	13.8	1.4	15.3
1960	4.5	.1	4.7	5.9	.4	6.3	2.7	1.0	3.7	1.3	14.4	1.5	15.9
1961	4.4	.1	4.5	6.1	.3	6.5	2.6	1.0	3.7	1.3	14.4	1.5	15.9
1962	4.4	.1	4.5	6.5	.3	6.8	2.8	1.1	3.9	1.4	15.1	1.6	16.7
1963	4.6	.1	4.8	6.9	.3	7.2	2.7	1.3	4.0	1.5	15.7	1.7	17.4
1964	4.9	.1	5.1	7.2	.3	7.5	2.9	1.3	4.2	1.5	16.5	1.7	18.2
1965	5.2	.1	5.4	7.4	.3	7.7	2.8	1.3	4.1	1.6	17.1	1.7	18.8
1966	5.3	.1	5.5	7.9	.3	8.2	2.9	1.5	4.4	1.8	17.9	1.9	19.8
1967	5.1	.1	5.2	8.1	.5	8.6	2.8	1.6	4.4	1.9	17.9	2.2	20.1
1968	5.0	.2	5.1	8.8	.5	9.3	3.2	1.8	5.0	2.0	19.0	2.4	21.4
1969	4.9	.2	5.1	9.2	.7	9.9	3.2	2.0	5.2	2.2	19.4	2.9	22.3
1970	4.9	.2	5.0	9.5	.7	10.2	3.3	2.0	5.3	2.2	19.8	2.9	22.6
1971	4.2	.1	4.3	9.9	.7	10.6	3.2	2.0	5.2	2.3	19.6	2.8	22.4
1972	4.1	.1	4.2	9.9	.7	10.6	3.5	2.3	5.8	2.5	20.0	3.1	23.1
1973	4.2	.1	4.3	9.3	.7	10.1	3.8	2.4	6.2	2.6	20.0	3.2	23.3
1974[6]	[5]4.2	.1	[5]4.3	[5]9.6	.7	[5]10.3	3.7	2.4	6.1	2.7	20.2	3.2	[5]23.4
1975[6]	[5]4.2	.1	[5]4.3	[5]8.6	.7	[5] 9.3	3.5	2.3	5.8	2.7	19.0	3.1	[5]22.1

[1] Includes anthracite, bituminous, and lignite coals.
[2] Petroleum refined and processed from crude oil, including still gas, liquefied refinery gas, and natural gas liquids.
[3] Utility electricity, generated and imported, distributed on basis of historical series in the *Edison Electric Institute Yearbook*. Conversion of electricity to energy equivalent was made at the value of contained energy corresponding to 100-percent efficiency using a theoretical rate of 3,412 Btu/kWh.
[4] Includes transmission and other unaccounted for losses.
[5] Includes net imports of coke of 59 × 10¹² in 1974 and 18 × 10¹² in 1975
[6] Preliminary
Note.—Data may not add to totals shown because of rounding.
Source: U.S. Department of the Interior, Bureau of Mines, Division of Interfuels Studies.

Source: Energy Perspectives 2, U.S. Department of the Interior, June 1976.

Total Consumption of Energy by the Industrial Sector, 1947-75
[Physical units]

Year	Natural gas (million cubic ft³) Fuel uses	Nonfuel uses	Total	Petroleum[1] (million bbl) Fuel uses	Nonfuel uses	Total	Coal[2] (thousand short tons) Fuel uses	Nonfuel uses	Total	Electricity purchased (billion kWh)
1947	2,420,689	484,882	2,905,571	344.7	78.3	423.0	268,790	4,613	273,403	135
1948	2,684,282	480,646	3,164,928	348.6	78.4	427.0	239,678	4,636	244,314	147
1949	2,791,305	427,892	3,219,197	341.7	73.9	415.6	204,845	4,074	208,919	142
1950	3,190,905	410,852	3,601,757	363.3	83.5	446.8	218,683	4,824	223,507	164
1951	3,680,694	426,423	4,107,117	415.9	100.2	516.1	233,439	5,337	238,776	192
1952	3,874,587	368,399	4,242,986	417.0	98.0	515.0	206,502	4,644	211,146	200
1953	4,088,186	300,942	4,389,128	427.9	105.7	533.6	221,415	5,450	226,865	224
1954	4,045,096	338,101	4,383,197	421.3	108.5	529.8	174,626	4,522	179,148	235
1955	4,418,768	349,794	4,768,562	455.6	124.2	579.8	209,353	5,593	214,946	295
1956	4,549,541	372,598	4,922,139	495.0	136.6	631.6	212,318	5,492	217,810	326
1957	4,766,625	383,788	5,150,413	449.5	146.3	595.8	207,839	5,565	213,404	332
1958	4,987,057	365,237	5,352,294	421.0	149.3	570.3	171,206	4,394	175,600	323
1959	5,346,144	374,612	5,720,756	445.6	179.5	625.1	165,732	4,192	169,924	356
1960	5,714,486	359,628	6,074,114	444.3	199.6	643.9	170,537	4,688	175,225	383
1961	5,895,291	326,377	6,221,668	436.0	206.6	642.6	163,532	4,558	168,090	383
1962	6,279,008	300,000	6,579,008	455.9	222.9	678.8	165,414	4,581	169,995	411
1963	6,630,360	287,378	6,917,738	455.9	246.1	702.0	173,759	4,442	178,201	429
1964	6,911,947	286,759	7,198,706	491.5	253.1	744.6	184,681	4,880	189,561	453
1965	7,149,904	283,296	7,433,200	475.0	265.4	740.4	195,496	5,192	200,688	479
1966	7,662,332	286,154	7,948,486	484.6	296.4	781.0	199,798	5,147	204,945	524
1967	7,885,653	446,961	8,332,614	470.9	327.5	798.4	189,592	5,003	194,595	547
1968	8,546,122	440,973	8,987,095	535.2	362.0	897.2	186,061	5,541	191,002	599
1969	8,895,419	692,251	9,587,670	541.9	401.7	943.6	182,900	5,686	188,586	632
1970	9,190,960	665,884	9,856,844	546.5	414.9	961.4	181,027	5,610	186,637	648
1971	9,603,790	648,699	10,252,489	537.7	432.3	970.0	154,589	4,277	158,866	672
1972	9,618,143	653,939	10,272,082	595.5	510.0	1,105.5	154,658	4,595	159,253	731
1973	9,144,606	650,000	9,794,606	647.2	518.4	1,165.6	156,448	4,379	160,827	772
1974	9,073,193	710,130	9,783,323	628.4	512.7	1,141.1	150,992	4,102	155,094	781
1975[3]	8,098,000	682,000	8,780,000	600.2	493.3	1,093.5	147,275	3,965	151,240	794

[1] Petroleum products refined and processed from crude oil, including still gas, liquefied refinery gas, and natural gas liquids.
[2] Includes anthracite and bituminous coals and lignite.
[3] Preliminary.
Source: U.S. Department of the Interior, Bureau of Mines, Division of Interfuels Studies.

Source: Energy Perspectives 2, U.S. Department of the Interior, June 1976.

Use of Electric Energy by Industry Groups in the United States[a,b]

SIC		1966 (Million kWh) Purchased (1)	Generated Less Sold (2)	Total (3)	Percent (4)	1967 Cost of Electricity Purchased as a Percent of Value of Shipments (5)
20	Food and kindred products	23,574	2,573	26,147	5.4	0.4
21	Tobacco manufactures	675	—	675	0.1	0.01
22	Textile mill products	18,426	701	19,127	3.9	0.9
23	Apparel and related products	3,167	19	3,186	0.7	0.3
24	Lumber and wood products	6,485	944	7,429	1.5	0.8
25	Furniture and fixtures	2,260	80	2,340	0.5	0.5
26	Paper and allied products	24,166	23,025	47,191	9.8	1.0
27	Printing and publishing	5,056	—	5,056	1.0	0.4
28	Chemicals and allied products	92,155	21,016	113,171	23.4	1.4
29	Petroleum and coal products	15,756	4,282	20,038	4.1	0.6
30	Rubber and plastics products, n.e.c.	10,688	614	11,302	2.4	0.9
31	Leather and leather products	1,195	85	1,280	0.3	0.4
32	Stone, clay, and glass products	22,793	1,662	24,455	5.0	1.4
33	Primary metal industries	99,700	23,460	123,160	25.4	1.5
33 1	Steel rolling and finishing	(41,911)	(10,676)	(52,587)	(10.9)	—
33 2	Iron and steel foundries	(5,703)	(46)	(5,749)	(1.2)	—
33 3	Primary nonferrous metals	(40,767)	(12,603)	(53,370)	(11.0)	—
34	Fabricated metal products	12,222	117	12,339	2.5	0.6
35	Machinery, except electrical	14,939	649	15,588	3.2	0.5
36	Electrical equipment and supplies	16,903	300	17,203	3.6	0.5
37	Transportation equipment	23,077	278	23,355	4.8	0.4
38	Instruments and related products	2,223	578	2,801	0.6	0.4
39 & 19	Miscellaneous manufacturing, including ordnance	5,092	41	5,133	1.1	0.5
	Withheld	1,739	1,623	3,362	0.7	
	Total	402,291	82,047	484,358	100.0%	0.7

[a] Source:
Cols. (1), (2), and (3): U.S. Department of Commerce, Bureau of the Census M66(AS)-4, *Annual Survey of Manufacturers—1966.* "Fuels and Electric Energy Used, by Major Industry Groups, and by Divisions and States."
Col. (4): Calculated from U.S. Department of Commerce, "Fuels and Electric Energy Used in Manufacturing," Preliminary Report, *1967 Census of Manufacturers*, Series MC67(P)-7, July 1969, and U.S. Department of Commerce, "General Statistics for Industry Groups and Industries," *Annual Survey of Manufacturers—1968*, M58(AS)-1, June 1970.
[b] ( ) Do not add to total SIC 33 due to disclosure problems.

Source: Engineering for Resolution of the Energy-Environment Dilemma, Committee on Power Plant Siting, National Academy of Engineering, Washington, D.C. 1972.

U.S. ENERGY RESOURCE INPUTS—HOUSEHOLD AND COMMERCIAL

During the period 1947-71, energy inputs into the household and commercial sectors increased from about 7100 Btu to 17,400 trillion Btu. Consumption of natural gas increased by over a factor of six, petroleum by a factor of three, electricity by a factor of eight, while coal consumption decreased by a factor of nine.

Of the total energy use in the household and commercial sectors, between 67% and 68% was used in space heating.

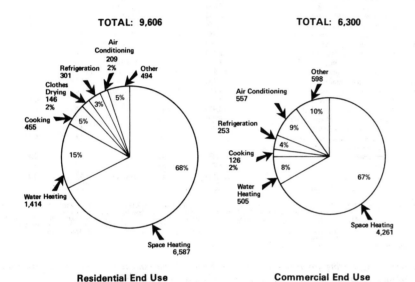

Residential End Use

Commercial End Use

Note.—Does not include nonenergy uses.

[1] 1 Quadrillion Btu = 1,000 trillion Btu.

Source: U.S. Bureau of Mines, 1973.

U.S. net energy consumption, residential and commercial sectors, 1970 (trillion Btu)[1].

Source: Energy Perspectives, U.S. Department of the Interior, February 1975.

Sources of Residential Average Annual Use[a,b]

Electric Appliance	1949 Saturation (Percent)	1949 Appliance Annual Use (kWh)	1949 Contribution to Annual Use (kWh)	1959 Saturation (Percent)	1959 Appliance Annual Use (kWh)	1959 Contribution to Annual Use (kWh)	1969 Saturation (Percent)	1969 Appliance Annual Use (kWh)	1969 Contribution to Annual Use (kWh)
Refrigerator	69	360	248	97	415	403	98	660	647
Range	31	1,350	419	63	1,350	851	80	1,350	1,080
Water heater	16	4,050	648	44	4,490	1,976	71	5,175	3,674
Space heating	2	8,860	177	18	10,710	1,928	31	11,260	3,491
Air conditioner:									
Room [c]	5	1,250	6	15	1,355	203	36	1,680	605
Central	—	—	—	2	3,500	70	9	4,100	369
Television	1	400	4	74	400	296	99	400	396
Washer:									
Automatic	8	100	8	28	100	28	49	100	49
Nonautomatic	38	50	19	45	50	23	40	50	20
Dryer	4	940	38	9	1,130	102	35	1,335	467
Freezer	7	895	63	21	900	189	38	980	372
Dishwasher	1	325	3	5	285	14	16	340	54
Miscellaneous [d]	—	—	1,132	—	—	1,323	—	—	2,376
			2,765			7,406			13,600

[a] Source: *The Methodology of Load Forecasting*, A Report to the Federal Power Commission, prepared by the Technological Advisory Committee on Load Forecasting Methodology for the National Power Survey (1969), p. III-12.
[b] Covers a large utility system in the Southeast.
[c] Saturation is defined as the percentage of residential customers having one or more room conditioners.
[d] Lighting, small appliances, supplemental heat, and other uses.

Source: Engineering for Resolution of the Energy-Environment Dilemma, National Academy of Engineering, Washington, D.C. 1972.

Energy Consumption by the Household and Commercial Sector, 1947-75
[Trillion Btu]

Year	Petroleum products and natural gas liquids						Coal[2]	Natural gas	Total fossil fuels	Electricity distributed[3]	Total energy consumption
	Kerosine	Distillate fuel oil	Residual fuel oil	Liquefied gases	Asphalt	Total[1]					
1947	354	1,106	355	110	326	2,251	3,399	1,125	6,775	391	7,166
1948	401	1,244	368	141	385	2,539	3,238	1,262	7,039	442	7,481
1949	375	1,183	380	156	379	2,472	3,025	1,387	6,884	488	7,372
1950	555	1,373	457	218	435	3,038	2,913	1,642	7,593	546	8,139
1951	491	1,547	478	207	480	3,202	2,648	2,007	7,857	615	8,472
1952	493	1,626	497	217	517	3,350	2,416	2,213	7,979	666	8,645
1953	469	1,649	514	237	522	3,391	2,072	2,294	7,757	733	8,490
1954	483	1,865	495	251	556	3,650	1,752	2,566	7,968	797	8,765
1955	478	2,077	543	288	615	4,001	1,745	2,849	8,595	854	9,449
1956	488	2,198	551	287	660	4,183	1,629	3,151	8,963	935	9,898
1957	431	2,196	512	293	637	4,069	1,225	3,391	8,685	1,019	9,704
1958	506	2,404	664	314	679	4,568	1,187	3,712	9,467	1,095	10,562
1959	491	2,428	704	375	721	4,718	969	4,024	9,711	1,203	10,914
1960	448	2,551	786	404	734	4,923	983	4,268	10,174	1,262	11,436
1961	445	2,657	761	412	753	5,028	868	4,477	10,373	1,385	11,758
1962	467	2,719	787	450	804	5,227	872	4,849	10,948	1,490	12,438
1963	453	2,711	787	483	824	5,258	731	5,027	11,016	1,645	12,661
1964	404	2,658	793	495	841	5,191	609	5,343	11,143	1,792	12,935
1965	449	2,802	983	511	891	5,635	678	5,517	11,830	1,948	13,778
1966	433	2,723	1,132	543	936	5,766	677	5,945	12,388	2,101	14,489
1967	388	3,099	1,208	594	917	6,206	585	6,223	13,014	2,257	15,271
1968	433	2,975	1,096	642	984	6,129	529	6,451	13,109	2,467	15,576
1969	428	2,981	1,120	731	1,009	6,269	447	6,890	13,606	2,752	16,358
1970	415	3,060	1,173	723	1,082	6,453	427	7,108	13,988	3,000	16,988
1971	401	3,050	1,148	732	1,108	6,440	406	7,366	14,212	3,209	17,421
1972	375	3,191	1,176	788	1,137	6,667	308	7,613	14,588	3,478	18,066
1973	348	3,122	1,206	749	1,264	6,689	296	7,318	14,303	3,709	18,012
1974	283	2,870	1,053	691	1,165	6,061	314	7,518	13,893	3,723	17,616
1975[4]	244	2,796	962	686	1,141	5,829	282	7,373	13,484	3,783	17,267

[1] Data may not add to totals shown due to rounding.
[2] Includes anthracite, bituminous coal and lignite.
[3] Utility electricity, generated and imported, distributed on basis of sales reported in the *Edison Electric Institute Statistical Yearbook*. Conversion of electricity to energy equivalent was made at the value of contained energy corresponding to 100 percent efficiency using a theoretical rate of 3,412 Btu/kWh.
[4] Preliminary.
Source: U.S. Department of the Interior, Bureau of Mines, Division of Interfuels Studies, 1976.

Source: Energy Perspectives 2, U.S. Department of the Interior, June 1976.

Consumption of Energy by the Household and Commercial Sector, 1947-75
[Physical units]

Year	Coal[1] (thousand short tons)	Natural gas (million ft[3])	Petroleum products and natural gas liquids (million bbl)						Electricity purchased[2] (million kWh)
			Kerosine	Distillate fuel oil	Residual fuel oil	Liquefied gases	Asphalt	Total	
1947	128,657	1,087,000	62.5	189.9	56.4	27.4	49.1	385.3	115
1948	122,994	1,219,402	70.7	213.6	58.6	35.1	58.0	436.0	130
1949	114,639	1,340,362	66.2	203.0	60.4	38.8	57.1	425.5	143
1950	110,422	1,586,207	97.8	235.7	72.7	54.4	65.6	526.2	160
1951	100,057	1,939,034	86.6	265.5	76.0	51.6	72.3	552.0	180
1952	91,212	2,137,635	86.9	279.2	79.1	54.0	77.9	577.1	195
1953	77,976	2,216,153	82.7	283.1	81.8	59.0	78.8	585.4	215
1954	65,425	2,479,205	85.2	320.1	78.8	62.5	83.8	630.4	234
1955	66,039	2,753,171	84.3	356.6	86.3	71.8	92.7	691.7	250
1956	61,685	3,044,435	86.0	377.3	87.6	71.5	99.4	721.8	274
1957	46,382	3,276,185	76.0	377.0	81.4	73.0	96.0	703.4	299
1958	45,005	3,586,025	89.3	412.7	105.6	78.4	102.3	788.3	321
1959	36,700	3,887,708	86.5	416.8	111.9	93.6	108.7	817.5	353
1960	37,180	4,123,389	79.0	438.0	125.1	100.6	110.6	853.3	370
1961	32,805	4,325,427	78.4	456.1	121.1	102.8	113.5	871.9	406
1962	32,955	4,685,231	82.4	466.8	125.2	112.2	121.1	907.7	437
1963	27,603	4,856,804	79.9	465.4	125.2	120.3	124.2	915.0	482
1964	22,949	5,162,009	71.2	456.3	126.2	123.3	126.7	903.7	525
1965	25,676	5,346,450	79.2	481.0	156.3	127.3	134.2	978.0	571
1966	25,587	5,760,999	76.4	467.4	180.0	135.3	141.0	1,000.1	616
1967	22,134	6,029,855	68.5	532.0	192.1	148.1	138.2	1,078.9	661
1968	19,983	6,250,997	76.3	510.7	174.3	160.0	148.2	1,069.5	723
1969	16,875	6,682,804	75.4	511.8	178.1	182.3	152.0	1,099.6	807
1970	16,114	6,894,007	73.2	525.3	186.6	180.2	163.1	1,128.4	879
1971	15,201	7,144,389	70.7	523.6	182.6	182.6	167.0	1,126.5	941
1972	11,708	7,412,543	66.2	547.8	187.1	196.5	171.3	1,168.9	1,019
1973	11,117	7,167,428	61.3	536.0	191.8	186.8	190.4	1,166.3	1,087
1974	11,417	7,341,745	49.9	492.7	167.4	172.2	175.6	1,057.8	1,091
1975[3]	10,210	7,200,000	43.0	480.0	153.0	171.0	172.0	1,019.0	1,109

[1] Includes anthracite and bituminous coals and lignite.
[2] Utility electricity, generated and imported, distributed on basis of sales reported in the *Edison Electric Institute Statistical Yearbook*.
[3] Preliminary.
Source: U.S. Department of the Interior, Bureau of Mines, Division of Interfuels Studies.

Source: Energy Perspectives 2, U.S. Department of the Interior, June 1976.

U.S. ENERGY RESOURCE INPUTS—TRANSPORTATION

In 1971, automobiles accounted for about 49% of the energy use in the transportation sector while trucks accounted for 20% of the energy use. Since 1955, relative energy consumption has decreased in transport by water, rail and bus and increased in transport by air and pipeline.

Total energy inputs into the transportation sector doubled between 1947 and 1971 from 8820 trillion Btu to 16,988 trillion Btu.

Average Travel and Mileage for Passenger Cars and Trucks, 1950–69

	Cars		Trucks	
Year	Average Travel per Vehicle (miles)	Average Miles Traveled per Gallon (miles/gallon)	Average Travel per Vehicle (miles)	Average Miles Traveled per Gallon (miles/gallon)
1950	9,015	14.40	10,524	9.51
1951	9,187	14.54	10,537	9.31
1952	9,361	14.27	10,759	9.13
1953	9,370	14.39	10,926	9.17
1954	9,308	14.57	10,883	8.44
1955	9,359	14.53	10,697	8.37
1956	9,348	14.36	10,813	8.31
1957	9,391	14.40	10,328	7.93
1958	9,494	14.30	10,348	7.96
1959	9,529	14.30	10,552	7.97
1960	9,446	14.28	10,583	7.96
1961	9,465	14.38	10,461	7.82
1962	9,441	14.37	10,406	7.80
1963	9,240	14.26	11,644	8.44
1964	9,286	14.25	11,723	8.43
1965	9,286	14.15	11,587	8.60
1966	9,384	14.10	11,207	8.52
1967	9,399	14.05	11,268	8.42
1968	9,488	13.91	11,571	8.37
1969	9,633	13.75	11,565	8.36

RCE: *Estimated Motor Vehicle Travel in the United States and Related Data*, Bureau of Public Roads, American Petroleum Institute.
Source: The U.S. Energy Problem, Volume II (NTIS), 1971, p. N-13.

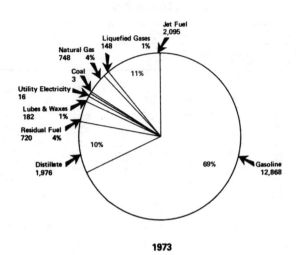

TOTAL: 18,756

1973

[1] 1 Quadrillion Btu = 1,000 trillion.   Source: U.S. Bureau of Mines, 1974.

U.S. net energy consumption, transportation sector, 1973 (trillion Btu)[1].

Source: Energy Perspectives, U.S. Department of the Interior, February 1975.

Total Energy Requirements for Automobiles in the United States

	1960	1970
Gasoline consumption (10^{15} Btu)	5.60	8.94
Gasoline refining and retail sales (10^{15} Btu)	1.28	2.07
Oil consumption, refining, retail sales (10^{15} Btu)	0.07	0.11
Automobile manufacturing (10^{15} Btu)	0.86	0.80
Automobile retail sales (10^{15} Btu)	0.17	0.21
Repairs, maintenance, parts (10^{15} Btu)	0.35	0.37
Parking, garaging (10^{15} Btu)	0.22	0.44
Tire manufacturing and retail sales (10^{15} Btu)	0.09	0.23
Insurance (10^{15} Btu)	0.19	0.31
Taxes (highway construction) (10^{15} Btu)	0.73	1.00
Total (10^{15} Btu)	9.56	14.48
Total automobile mileage (10^9 miles)	590	901
Total energy required (Btu/mile)	16,200	16,100
Total United States energy consumption (10^{15} Btu)	44.57	67.44
Percent of total energy consumption devoted to automobiles	21	21

Energy Consumption by the Transportation[1] Sector, 1947-75
[Trillion Btu]

| Year | Petroleum products and natural gas liquids | | | | | | Coal[2] | Natural gas | Total fossil fuels | Electricity distributed[4] | Total energy consumption |
	Gasoline	Jet fuel	Liquefied gases	Distillate fuel oil	Residual fuel oil	Total[1]					
1947	4,131	–	9	246	1,374	5,761	3,030	–	8,791	29	8,820
1948	4,527	–	9	301	1,320	6,157	2,624	–	8,781	27	8,808
1949	4,747	–	8	325	1,103	6,183	1,892	–	8,075	25	8,100
1950	5,166	NA	12	463	1,145	6,785	1,701	130	8,616	24	8,640
1951	5,666	NA	28	519	1,269	7,482	1,525	199	9,206	23	9,229
1952	5,941	114	35	599	1,179	7,868	1,086	214	9,168	22	9,190
1953	6,265	185	48	571	1,089	8,158	809	238	9,205	20	9,225
1954	6,398	246	52	708	954	8,358	516	239	9,113	18	9,131
1955	6,944	282	62	824	997	9,109	464	253	9,826	19	9,845
1956	7,143	409	74	827	997	9,448	378	306	10,132	17	10,149
1957	7,303	425	76	841	1,003	9,649	270	310	10,229	15	10,244
1958	7,462	524	81	811	939	9,819	133	323	10,275	16	10,291
1959	7,717	580	85	667	874	9,923	102	362	10,387	17	10,404
1960	7,854	738	86	867	826	10,372	87	359	10,818	18	10,836
1961	7,967	839	84	872	814	10,575	21	390	10,986	19	11,005
1962	8,219	974	89	929	790	11,001	18	396	11,415	18	11,433
1963	8,580	1,050	95	1,038	743	11,506	18	438	11,962	19	11,981
1964	8,701	1,096	112	1,103	779	11,791	19	451	12,261	20	12,281
1965	9,028	1,214	114	1,082	742	12,179	18	517	12,714	18	12,732
1966	9,412	1,354	116	1,145	750	12,777	16	552	13,345	16	13,361
1967	9,671	1,670	111	1,175	781	13,408	13	594	14,015	17	14,032
1968	10,265	1,944	118	1,411	798	14,535	11	610	15,156	18	15,174
1969	10,719	2,017	140	1,504	746	15,125	8	651	15,784	17	15,801
1970	11,185	1,973	128	1,571	736	15,592	8	745	16,345	16	16,361
1971	11,615	2,051	126	1,681	687	16,160	6	766	16,932	17	16,949
1972	12,337	2,091	141	1,887	652	17,107	4	787	17,898	17	17,915
1973	12,872	2,131	135	2,132	731	18,001	3	744	18,748	15	18,763
1974	12,608	2,001	125	2,122	709	17,564	2	685	18,251	19	18,270
1975[5]	12,869	2,022	124	2,056	629	17,699	1	635	18,335	19	18,354

[1] Includes bunkers and military transportation.
[2] Includes anthracite, bituminous coal and lignite.
[3] Data may not add to totals shown because of rounding.
[4] Utility electricity, generated and imported, distributed on basis of sales reported in the *Edison Electric Institute Statistic Yearbook*. Conversion of electricity to energy equivalent was made at the value of contained energy to 100 percent efficiency using a theoretical rate of 3,412 Btu/kWh.
[5] Preliminary.
Note.–NA = not available.
Source: U.S. Department of the Interior, Bureau of Mines, Division of Interfuels Studies, 1976.

Source: Energy Perspectives 2, U.S. Department of the Interior, June 1976.

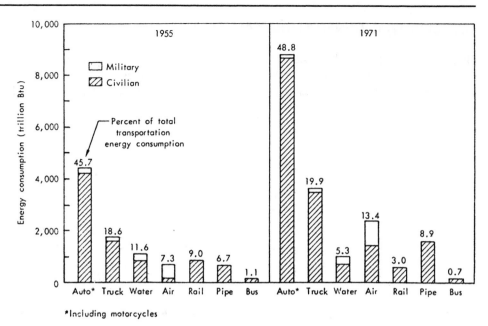

Modal distribution of transportation energy consumption, 1955 and 1971.

Source: Transportation Energy Use in the United States, R-1391-NSF, Rand Corporation.

U.S. ELECTRIC GENERATING CAPACITY

During 1974, the total installed electric generating capacity in the United States increased from 459 million kilowatts to 495 million kilowatts. Additions during the year included 10 million kilowatts of nuclear capacity, 24 million kilowatts of fossil fuel capacity, and 2 million kilowatts of hydroelectric capacity. Of the total installed capacity, about 80% was represented by fossil fuel plants, 14% by hydroelectric plants, and 6% by nuclear plants.

During 1975, the installed generating capacity in the United States reached 524 million kilowatts.

Installed Generating Capacity of the U.S. Electric Utility Industry, 1947–75, and Projected 1985 Capacity

Year	Hydropower (MW)	Conventional steam (MW)	Nuclear steam (MW)	Internal combustion (MW)	Total (MW)	U.S. population (millions)	U.S. generating capacity per capita (kW)	GNP (billion 1972 dollars)	MW capacity ÷ billion dollars of GNP (1972 dollars)
1947	14,971	36,034	—	1,317	52,322	144.1	0.36	468.3	111.7
1948	15,652	39,304	—	1,604	56,560	146.6	.39	487.7	116.0
1949	16,654	44,640	—	1,806	63,100	149.2	.42	490.7	128.6
1950	17,675	49,333	—	1,911	68,919	152.3	.45	533.5	129.2
1951	18,868	54,865	—	2,042	75,775	154.9	.49	576.5	131.4
1952	20,419	59,679	—	2,128	82,226	157.6	.52	598.5	137.4
1953	22,045	67,235	—	2,222	91,502	160.2	.57	621.8	147.2
1954	23,211	77,102	—	2,279	102,692	163.0	.63	613.7	167.2
1955	25,005	87,112	—	2,355	114,472	165.9	.69	654.8	174.8
1956	25,654	92,591	—	2,452	120,697	168.9	.72	668.8	180.5
1957	27,036	99,437	105	2,545	129,123	172.0	.75	680.9	189.6
1958	29,359	110,528	105	2,605	142,597	174.9	.82	679.5	209.9
1959	31,074	122,980	105	2,682	156,841	177.8	.88	720.4	217.8
1960	32,376	132,521	297	2,817	168,002	180.7	.93	736.8	228.0
1961	35,481	141,784	442	2,961	180,668	183.7	.98	755.3	239.2
1962	37,342	149,943	734	3,048	191,067	186.5	1.02	799.1	239.1
1963	40,214	166,229	861	3,245	210,549	189.2	1.11	830.7	253.5
1964	42,188	175,871	906	3,320	222,285	191.9	1.16	874.4	254.2
1965	43,782	188,054	926	3,365	236,127	194.3	1.22	925.9	255.0
1966	44,977	197,415	1,942	3,509	247,843	196.6	1.26	981.0	252.6
1967	48,112	214,435	2,887	3,818	269,252	198.7	1.36	1,007.7	267.2
1968	51,168	233,095	2,817	3,978	291,058	200.7	1.45	1,051.8	276.7
1969	52,753	252,411	3,980	4,205	313,349	202.7	1.55	1,078.8	290.5
1970	55,056	275,191	6,493	4,350	341,090	204.9	1.67	1,075.3	317.2
1971	55,898	298,345	8,687	4,466	367,396	207.0	1.78	1,107.5	331.7
1972	56,566	322,944	15,300	4,796	399,606	208.8	1.91	1,171.1	341.2
1973	61,782	352,072	21,118	4,903	439,875	210.4	2.09	1,233.4	356.6
1974[1]	63,589	375,736	31,662	5,001	475,988	211.9	2.25	1,210.7	393.2
1975[2]	65,601	395,397	38,943	5,028	504,969	213.4	2.37	1,186.4	425.6
1985	[3]97,000	594,162	200,000	8,838	900,000	235.7	3.82	1,908.0	471.7

[1] Preliminary.

[2] Estimated.

[3] Includes geothermal, 3,000.

Sources: Edison Electric Institute, *Historical Statistics of the Electric Utility Industry*, New York, 1947-70; Federal Power Commission; U.S. Department of the Interior, Bureau of Mines, *U.S. Energy Through the Year 2000*, Revised, Washington, D.C., 1975.

Source: Energy Perspectives 2, U.S. Department of the Interior, June 1976.

The source for installed generating capacity in the United States is the Federal Power Commission. Data beginning with the year 1963 reflect the change to the maximum generator name-plate rating made by the Federal Power Commission.

Pumped-storage capacity is included with hydro-electric capacity, and gas turbine capacity is included with steam.

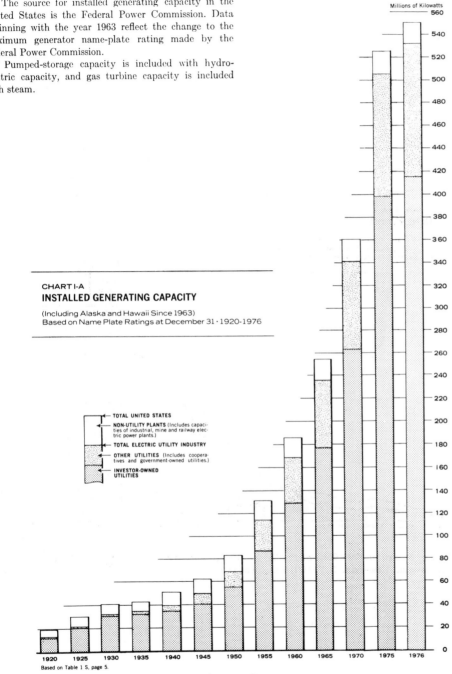

Millions of Kilowatts

CHART I-A
INSTALLED GENERATING CAPACITY

(Including Alaska and Hawaii Since 1963)
Based on Name Plate Ratings at December 31 · 1920-1976

← TOTAL UNITED STATES
← NON-UTILITY PLANTS (Includes capacities of industrial, mine and railway electric power plants.)
← TOTAL ELECTRIC UTILITY INDUSTRY
← OTHER UTILITIES (Includes cooperatives and government-owned utilities.)
← INVESTOR-OWNED UTILITIES

Based on Table 1 S, page 5.

Generating capacity.

Source: *Statistical Year Book of the Electric Utility Industry for 1976*, Edison Electric Institute, 90 Park Avenue, New York, N.Y. 10016, October 1977.

Installed Generating Capacity Total U.S.*
Kilowatts in Millions (Name Plate)

December 31st	Total Capacity	Total Electric Utility Industry				Industrial Plants†
		Total	Investor-Owned	Cooper-atives**	Govern-ment-Owned	
1976p	550.6	531.4	415.8	9.9	105.7	19.2
1975r	527.6	508.4	399.0	9.1	100.3	19.2
1974	495.5	476.2	376.1	7.5	92.6	19.3
1973	459.3	439.9	346.5	7.3	86.1	19.4
1972p	418.5	399.6	314.9	6.7	78.0	18.9
1971r	388.4	369.1	288.3	5.5	75.3	19.3
1970	360.3	341.1	262.7	5.2	73.2	19.2
1969	332.6	313.3	240.1	4.3	68.9	19.3
1968	310.2	291.1	220.8	3.4	66.9	19.1
1967	288.2	269.3	203.6	3.0	62.7	18.9
1966	266.8	247.8	185.7	2.7	59.4	19.0
1965	254.5	236.1	177.6	2.3	56.2	18.4
1964	228.7	222.3	167.7	2.0	52.6	18.2
1963	240.5	210.5	158.4	1.9	50.2	18.2
1962	208.7	191.1	144.6	1.5	45.0	17.6
1961	198.4	180.7	136.7	1.5	42.5	17.7
1960	185.8	168.0	128.5	1.4	38.1	17.8
1959	174.3	156.8	119.0	1.1	36.7	17.5
1958	160.7	142.6	108.2	1.0	33.4	18.1
1957	146.2	129.1	97.4	0.9	30.8	17.1
1956	137.3	120.7	91.1	0.8	28.8	16.6
1955	130.9	114.5	86.9	0.8	26.8	16.4
1954	118.9	102.6	79.1	0.8	22.7	16.3
1953	107.4	91.5	71.2	0.6	19.7	15.9
1952	97.3	82.2	64.4	0.5	17.3	15.1
1950	82.8	68.9	55.2	0.3	13.4	13.9
1947	65.1	52.3	42.0	0.1	10.2	12.8
1945	62.9	50.1	40.3	—	9.8	12.8
1940	50.9	39.9	34.4	—	5.5	11.0
1937	44.4	35.6	31.9	—	3.7	8.8
1932	42.9	34.4	32.0	—	2.4	8.5
1927	34.6	25.1	23.4	—	1.7	9.5
1922	20.5	14.2	13.4	—	0.8	6.3
1917	15.5	9.0	8.4	—	0.6	6.5
1912	11.0	5.2	4.8	—	0.4	5.8
1907	6.8	2.7	2.5	—	0.2	4.1
1902	3.0	1.2	1.1	—	0.1	1.8

Sources: 1940-76 Federal Power Commission; 1902-37 U. S. Census of Central Electric Light & Power Stations.

* Alaska and Hawaii included since 1963.

** Included in Government-Owned prior to 1947.

† Excludes small isolated plants in hotels, laundries, institutions, and so on, for which information is not available.

p—preliminary.   r—revised.

Source: *EEI Pocketbook of Electric Utility Industry Statistics*, 23rd Edition, Edison Electric Institute, 90 Park Avenue, New York, N.Y. 10016, 1978.

U.S. GENERATION OF ELECTRICITY

During the period 1947-73, the generation of electricity grew at an annual rate near 7%. During 1974, however, the increase in electrical generation was at a rate of only 0.4% per year. There were significant regional differences in the composite growth rate in 1974. Consumption of electricity dropped between 0.5% and 3.3% in the New England, Mid-Atlantic, Central Industrial, Southeast and Pacific Southwest regions while consumption increased between 1.7% and 7.0% in the West Central, South Central, Rocky Mountain and Pacific Northwest regions.

Total consumption of electricity in 1974 was 1982 billion kilowatt hours. This represented an average national per capita consumption of about 9376 kilowatt hours per person. In 1975, preliminary data indicate a very slight increase in the consumption of electricity over the 1974 year with consumption reaching 2008 billion kilowatt hours representing a per capita consumption of 9421 kilowatt hours per person.

U.S. Power Generation by Type of Capacity, 1947-75, and Projected 1985 Power Generation

Year	Hydropower and geothermal (million kWh)	Conventional steam (million kWh)	Nuclear steam (million kWh)	Internal combustion (million kWh)	Total (million kWh)	U.S. population (millions)	kWh per capita	GNP (billion 1972 dollars)	kWh per dollar GNP (1972 dollars)
1947	78,426	174,500	—	2,813	255,739	144.1	1,774.7	468.3	0.546
1948	82,470	196,928	—	3,300	282,698	146.6	1,928.4	487.7	.580
1949	89,748	197,879	—	3,473	291,100	149.2	1,951.1	490.7	.593
1950	95,938	229,543	—	3,660	329,141	152.3	2,161.1	533.5	.617
1951	99,750	267,252	—	3,671	370,673	154.9	2,393.0	576.5	.643
1952	105,103	290,385	—	3,736	399,224	157.6	2,533.2	598.5	.667
1953	105,233	333,542	—	3,890	442,665	160.2	2,763.2	621.8	.712
1954	107,069	360,834	—	3,783	471,686	163.0	2,893.8	613.7	.769
1955	112,975	430,119	—	3,944	547,038	165.9	3,297.4	654.8	.835
1956	122,029	474,552	—	4,087	600,668	168.9	3,556.4	668.8	.898
1957	130,232	497,203	10	4,062	631,507	172.0	3,671.6	680.9	.927
1958	140,262	500,599	165	4,072	645,098	174.9	3,688.4	679.5	.949
1959	137,782	567,651	188	4,385	710,006	177.8	3,993.3	720.4	.986
1960	145,516	602,824	518	4,492	753,350	180.7	4,169.1	736.8	1.022
1961	151,850	633,872	1,692	4,625	792,039	183.7	4,311.6	755.3	1.049
1962	168,283	676,956	2,270	4,805	852,314	186.5	4,570.1	799.1	1.067
1963	165,755	742,780	3,212	5,046	916,793	189.2	4,845.6	830.7	1.104
1964	177,073	798,564	3,343	5,010	983,990	191.9	5,127.6	874.4	1.125
1965	193,851	852,655	5,657	5,089	1,055,252	194.3	5,431.0	925.9	1.140
1966	194,756	938,910	3,520	5,164	1,144,350	196.6	5,820.7	981.0	1.167
1967	221,518	980,336	7,655	4,856	1,214,365	198.7	6,111.6	1,007.7	1.205
1968	222,491	1,089,239	12,528	5,185	1,329,443	200.7	6,624.0	1,051.8	1.264
1969	250,193	1,172,481	13,928	5,580	1,442,182	202.7	7,114.9	1,078.8	1.337
1970	247,456	1,256,294	21,797	6,062	1,531,609	204.9	7,474.9	1,075.3	1.424
1971	266,320	1,303,465	37,899	6,252	1,613,936	207.0	7,796.8	1,107.5	1.457
1972	272,734	1,413,882	54,031	6,676	1,747,323	208.8	8,368.4	1,171.1	1.492
1973	271,634	1,494,901	83,334	6,347	1,856,216	210.4	8,822.3	1,233.4	1.505
1974[1]	300,447	1,445,785	112,740	5,989	1,864,961	211.9	8,801.1	1,210.7	1.540
1975[2]	291,000	1,443,200	155,000	5,800	1,895,000	213.4	8,880.0	1,186.4	1.597
1985	391,000	2,419,975	1,139,000	10,025	3,960,000	235.7	16,801.0	1,908.0	2.075

[1] Preliminary.
[2] Estimated.
Sources: Edison Electric Institute, *Historical Statistics of the Electric Utility Industry*, New York, various years; Federal Power Commission; U.S. Department of the Interior, Bureau of Mines, *U.S. Energy Through the Year 2000*, Revised, Washington, D.C., 1975.

Source: Energy Perspectives 2, U.S. Department of the Interior, June 1976.

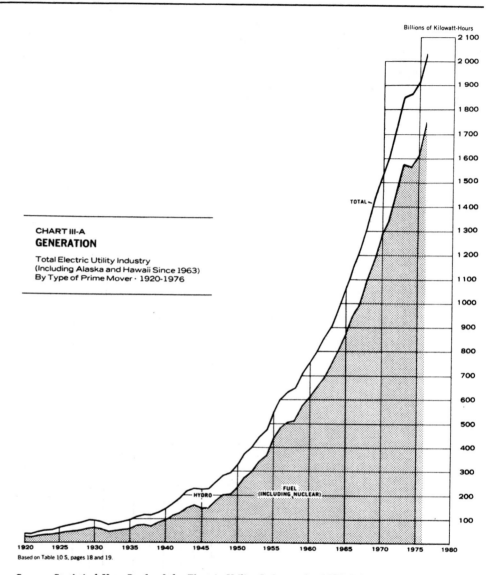

Billions of Kilowatt-Hours

CHART III-A
GENERATION

Total Electric Utility Industry
(Including Alaska and Hawaii Since 1963)
By Type of Prime Mover · 1920-1976

TOTAL

FUEL (INCLUDING NUCLEAR)

HYDRO

Based on Table 10 S, pages 18 and 19.

Source: Statistical Year Book of the Electric Utility Industry for 1976, Edison Electric Institute, 90 Park Avenue, New York, N.Y. 10016, October 1977.

Generation of Electricity—Total U.S.*
Kilowatt-Hours in Billions

| | | Total Electric Utility Industry | | | | |
Year	Total Generation	Total	Investor-Owned	Cooper-atives**	Govern-ment-Owned	Industrial Plants †
1976p	2,123.4	2,036.5	1,582.0	40.6	413.9	86.9
1975r	2,003.0	1,917.6	1,486.8	35.3	395.5	85.4
1974	1,968.0	1,866.4	1,442.1	33.8	390.5	101.6
1973	1,958.7	1,856.2	1,448.9	35.1	372.2	102.5
1972p	1,853.4	1,747.3	1,356.7	31.6	359.0	106.1
1971r	1,715.8	1,612.6	1,248.6	27.2	336.8	103.2
1970	1,639.8	1,531.6	1,183.2	23.4	325.0	108.2
1969	1,552.8	1,442.2	1,102.2	17.5	322.5	110.6
1968	1,436.0	1,329.4	1,019.3	14.1	296.0	106.6
1967	1,317.3	1,214.4	928.4	12.4	273.6	102.9
1966	1,249.4	1,144.3	880.8	11.2	252.3	105.1
1965	1,157.6	1,055.3	809.5	8.6	237.2	102.3
1964	1,083.7	984.0	756.2	7.9	219.9	99.7
1963	1,011.4	916.8	701.3	6.9	208.6	94.6
1962	943.5	852.3	651.0	6.1	195.2	91.2
1960	841.6	753.4	578.6	5.0	169.8	88.2
1959	795.3	710.0	544.2	4.4	161.4	85.3
1958	724.8	645.1	490.4	3.4	151.3	79.7
1957	716.4	631.5	480.9	3.0	147.6	84.9
1956	684.8	600.7	459.0	3.4	138.3	84.1
1955	629.0	547.0	420.9	3.0	123.1	82.0
1954	544.6	471.7	371.0	2.5	98.2	72.9
1953	514.2	442.7	354.3	1.9	86.5	71.5
1950	388.7	329.2	266.9	1.0	61.3	59.5
1947	307.4	255.7	208.1	0.4	47.2	51.7
1945	271.3	222.5	180.9	—	41.6	48.8
1940	179.9	141.8	125.4	—	16.4	38.1
1937	146.5	118.9	110.5	—	8.4	27.6
1932	99.4	79.4	74.5	—	4.9	20.0
1927	101.4	75.4	70.9	—	4.5	26.0
1922	61.2	43.6	41.6	—	2.0	17.6
1917	43.4	25.4	24.4	—	1.0	18.0
1912	24.8	11.6	11.0	—	0.6	13.2
1907	14.1	5.9	5.6	—	0.3	8.2
1902	6.0	2.5	2.3	—	0.2	3.5

Sources: 1940-76 Federal Power Commission; 1902-37 U.S. Census of Central Electric Light & Power Stations.

* Alaska and Hawaii included since 1963.

** Included in Government-Owned prior to 1947.

† Excludes small isolated plants in hotels, laundries, institutions, and so on, for which information is not available.

p—preliminary.   r—revised.

Source: EEI Pocketbook of Electric Utility Industry Statistics, 23rd Edition, Edison Electric Institute, 90 Park Avenue, New York, N.Y. 10016, 1978.

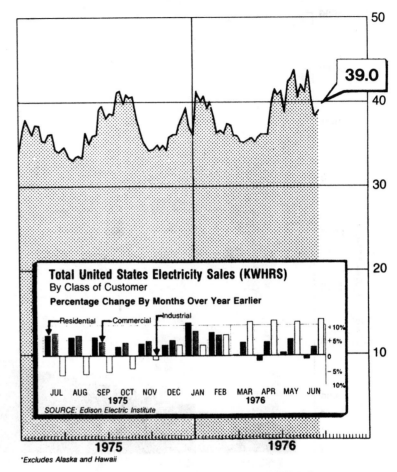

U.S. weekly electric output* (billions of KWHRS).

*Excludes Alaska and Hawaii

Source: Edison Electric Institute, 90 Park Avenue, New York, N.Y. 10016

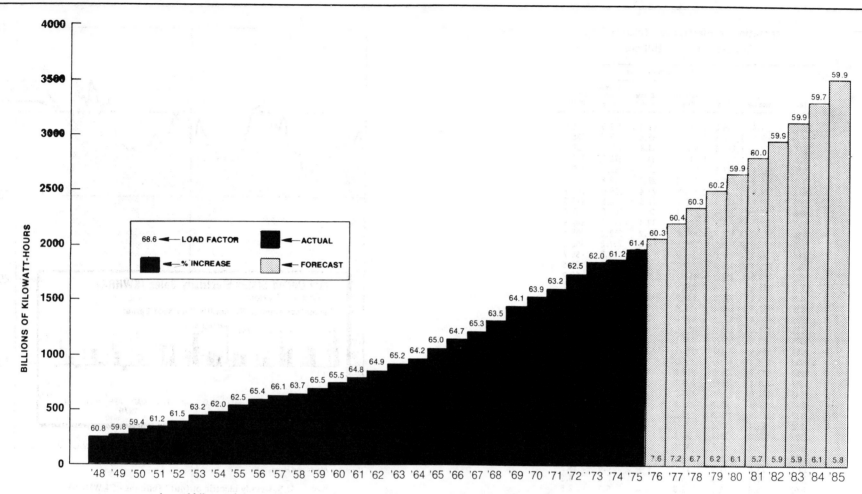

Annual kilowatt-hour requirements and annual load factors—total electric utility industry of the contiguous United States.

Source: *59th Electric Power Survey*, A Report of the Electric Power Survey Committee of the Edison Electric Institute, 90 Park Avenue, New York, N.Y. 10016 (April 1976).

U.S. ELECTRIC GENERATION—SEASONAL VARIATIONS

The average annual load factors achievable in the generation of electric power, normally in the 60–65% range, depend largely upon the seasonal and daily variations in the demand for electricity. Seasonal peaks in demand occur in both summer and winter as a result of air-conditioning and heating and lighting loads. Daily peaks in the demand for electricity reflect the normal cycle of human activity throughout the day.

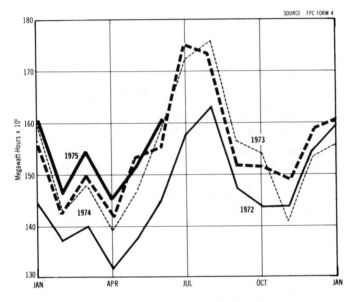

The total monthly consumption is given as a point for the month indicated.
The lines merely connect sets of points to identify the year.

Monthly net generation of electric power by U.S. electric utility plants, 1972, 1973, 1974, 1975.

Source: FPC News, Federal Power Commission, January 2, 1976.

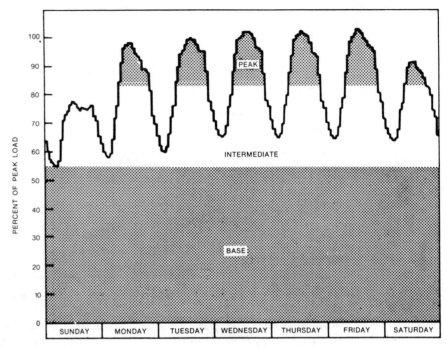

Weekly load curve.

Source: Project Independence, Project Independence Report, Federal Energy Administration, November 1974.

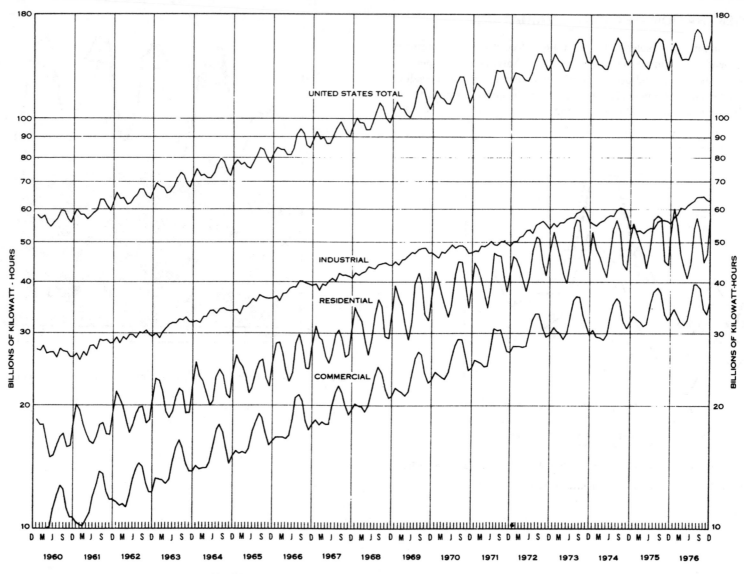

Monthly residential, commercial, and industrial sales (United States).

Source: FPC News, Federal Power Commission, Volume 10, No. 34-B, Aug. 26, 1977.

U.S. SECTOR CONSUMPTION OF ELECTRIC ENERGY

In 1973, residential consumption of electrical energy accounted for 32%, commercial consumption for 23%, and industrial consumption for 40%. Regional differences in consumption of electricity reflect both differences in industrialization and differences in fuel resource availability. Per capita energy consumption ranges from 5000 kilowatt hours per person in New England to 12,000 kilowatt hours per person in the East South Central region.

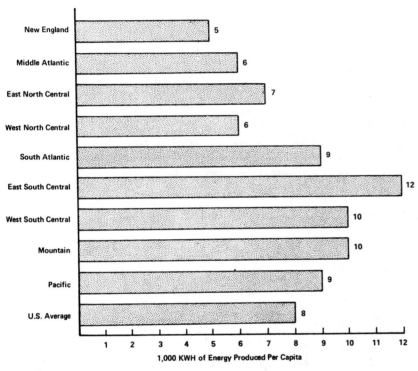

Source: Fuel and Energy Data, United States by States and Regions, 1972, U.S. Bureau of Mines Information Circular 8647, 1974.

U.S. electricity production per capita, by census region, 1972.

Source: *Energy Perspectives*, U.S. Department of the Interior, February 1975.

Electrification of the United States 1960–1968

End Uses	1968 Percent Electric	1968 Percent of United States		Percent Annual Growth	
		Energy	Electricity	Total Energy	Energy for Electricity
Electrified					
Industrial drive	~100	10.3	39.7	4.4	4.4
Refrigeration	~100	3.0	11.6	5.6	6.0
Electrolytic processes	100	1.5	5.8	3.7	3.7
Air conditioning	96	1.5	5.6	11.3	11.1
Television	100	0.7	2.6	9.6	9.6
Lighting and other	~100	5.0	19.1	9.4	9.4
	~100	22.0	84.4	6.0	6.0
Partially electrified					
Clothes drying	70	0.4	1.0	10.2	9.9
Cooking	40	1.4	2.1	2.0	3.0
Water heating	38	4.2	6.2	4.1	3.4
Direct heat	6	11.0	2.6	2.8	5.8
Space heating	5	20.8	3.3	4.1	24.
	11	37.8	15.2	3.6	6.6
Unelectrified					
Transportation	—	25.6	0.4	4.1	—
Process steam	—	14.6	—	3.8	—
	—	40.2	0.4	4.0	—
All end uses	25	100	100	4.3	6.1

~ indicates approximate percentage.

Source: Fisher, John C.; *Energy Crises in Perspective*, John Wiley & Sons, New York, 1974.

Energy Sales—Total Electric Utility Industry
(By Years and Classes of Service)
Kilowatt-Hours in Millions

Year	Total Sales	Exports to Canada and Mexico	Total Sales to Ultimate Customers	Residential	Commercial and Industrial		Street and Highway Lighting	Other Public Authorities	Railroads and Railways	Interdepartmental
					Small Light and Power*	Large Light and Power*				
1976	1 852 003	2 378	1 849 625	613 072	440 625	725 169	14 413	45 625	4 338	6 383
1975	1 738 107	5 083	1 733 024	586 149	418 069	661 558	13 907	43 625	4 273	5 443
1974	1 703 495	2 726	1 700 769	554 960	392 716	689 435	13 313	40 721	4 258	5 366
1973	1 705 773	2 570	1 703 203	554 171	396 903	687 235	12 836	42 340	4 186	5 532
1972	1 580 466	2 752	1 577 714	511 423	361 859	639 467	12 193	43 190	4 440	5 142
1971	1 469 955	3 514	1 466 441	479 080	333 752	592 700	11 673	39 819	4 537	4 880
1970	1 395 568	4 209	1 391 359	447 795	312 750	572 522	11 183	37 816	4 633	4 660
1969	1 311 008	3 830	1 307 178	407 922	286 686	557 220	10 772	35 861	4 531	4 186
1968	1 206 606	4 285	1 202 321	367 692	265 151	518 834	10 302	32 162	4 540	3 640
1967	1 111 373	4 350	1 107 023	331 525	242 492	486 043	9 863	29 426	4 572	3 102
1966	1 042 158	3 176	1 038 982	306 572	225 878	465 077	9 240	25 922	4 514	1 779
1965	957 113	3 699	953 414	280 970	202 112	433 365	8 782	21 675	4 652	1 858
1964	894 609	4 253	890 356	262 010	183 539	409 356	8 290	20 651	4 721	1 789
1963	832 796	1 985	830 811	241 692	166 516	388 399	7 748	20 194	4 667	1 595
1962	777 749	1 661	776 088	226 414	144 095	373 916	7 350	18 349	4 717	1 247
1961	721 645	917	720 728	209 021	134 864	347 427	6 762	16 438	4 685	1 531

* Small Light and Power and Large Light and Power are not wholly comparable on a year-to-year basis due to changes from one classification to another.

Energy Sales—Investor-Owned Electric Utilities
(By Years and Classes of Service)
Kilowatt-Hours in Millions

Year	Total Sales	Sales for Resale†	Total Sales to Ultimate Customers	Residential	Commercial and Industrial		Street and Highway Lighting	Other Public Authorities	Railroads and Railways	Interdepartmental
					Small Light and Power*	Large Light and Power*				
1976	1 530 286	99 496	1 430 790	453 774	355 555	569 134	11 276	34 810	3 930	2 311
1975	1 447 147	101 273	1 345 874	439 496	340 082	515 798	10 868	33 528	4 158	1 944
1974	1 418 958	93 201	1 325 757	416 333	319 706	542 234	10 409	30 876	4 161	2 038
1973	1 417 386	82 990	1 334 396	416 851	323 018	545 912	10 034	32 452	4 086	2 043
1972	1 316 379	76 572	1 239 807	386 561	297 028	507 080	9 581	33 491	4 333	1 733
1971	1 215 614	70 009	1 145 605	358 564	272 703	468 512	9 206	30 618	4 424	1 578
1970	1 142 777	63 032	1 079 745	333 895	254 684	447 664	8 814	28 721	4 520	1 447
1969	1 070 721	58 641	1 012 080	304 497	232 965	433 422	8 441	26 928	4 417	1 410
1968	984 542	55 572	928 970	272 792	215 090	403 468	8 089	23 908	4 424	1 199
1967	890 755	45 334	845 421	244 416	194 772	371 319	7 695	21 726	4 445	1 048
1966	840 093	43 776	796 317	226 727	182 923	354 759	7 299	19 427	4 403	779
1965	770 804	39 471	731 333	207 580	164 354	329 541	6 943	17 683	4 543	689
1964	717 951	37 980	679 971	192 855	148 050	310 678	6 592	16 580	4 606	610
1963	665 808	35 160	630 648	178 157	133 679	291 542	6 251	15 859	4 567	593
1962	621 040	32 492	588 548	165 598	115 792	281 701	5 947	14 464	4 674	372
1961	575 573	28 817	546 756	153 541	107 522	261 441	5 596	13 481	4 639	536

* Small Light and Power and Large Light and Power are not wholly comparable on a year-to-year basis due to changes from one classification to another.
† To non-investor-owned electric utilities and exports.

Source: Statistical Year Book of the Electric Utility Industry for 1976, Edison Electric Institute, 90 Park Avenue, New York, N.Y. 10016, October 1977.

U.S. ELECTRIC UTILITY FUEL USE

Of the total energy inputs into the electrical sector in 1971, about 44% was in the form of coal, 14% as petroleum, 24% as natural gas, 16% as hydropower, and 2% as nuclear energy. In 1974, these percentages had shifted to 47% coal, 16% petroleum, 17% gas, 14% hydropower, and 6% nuclear energy.

In terms of all the energy resources consumed in 1973, consumption in the electrical sector accounted for 69% of the coal, 9% of the petroleum, and 16% of the natural gas.

Energy Consumption by the Electrical Sector, 1947–75
[Trillion Btu]

Year	Petroleum products				Coal[1]	Natural gas	Total fossil fuels	Hydropower[2]	Nuclear[3]	Total energy consumption[4]
	Jet fuel	Distillate fuel	Residual fuel	Total						
1947	–	84	384	468	2,084	386	2,938	1,326	–	4,264
1948	–	87	357	444	2,392	495	3,331	1,393	–	4,724
1949	–	73	504	577	2,021	569	3,167	1,449	–	4,616
1950	–	77	586	662	2,228	651	3,541	1,440	–	4,981
1951	–	56	443	499	2,537	791	3,827	1,454	–	5,281
1952	–	48	443	492	2,588	942	4,022	1,496	–	5,518
1953	–	40	537	577	2,805	1,070	4,452	1,439	–	5,891
1954	–	36	445	480	2,866	1,206	4,552	1,388	–	5,940
1955	–	34	478	512	3,482	1,194	5,188	1,407	–	6,595
1956	–	31	465	497	3,815	1,283	5,595	1,487	–	7,082
1957	–	31	482	512	3,881	1,385	5,778	1,551	1	7,330
1958	–	31	484	515	3,743	1,421	5,679	1,636	2	7,317
1959	–	29	517	546	4,050	1,684	6,280	1,591	2	7,873
1960	–	27	537	564	4,251	1,785	6,600	1,657	6	8,263
1961	–	24	553	577	4,374	1,889	6,840	1,680	17	8,537
1962	–	24	555	579	4,636	2,034	7,249	1,821	23	9,093
1963	–	24	576	600	5,065	2,218	7,883	1,767	33	9,683
1964	–	22	614	636	5,395	2,403	8,434	1,907	34	10,375
1965	–	22	722	744	5,843	2,392	8,979	2,058	38	11,075
1966	–	21	884	905	6,327	2,692	9,924	2,073	57	12,054
1967	–	17	996	1,013	6,457	2,834	10,304	2,344	80	12,728
1968	–	18	1,163	1,181	6,994	3,245	11,420	2,342	130	13,892
1969	–	71	1,557	1,628	7,227	3,594	12,449	2,659	146	15,254
1970	–	144	1,943	2,087	7,261	4,015	13,363	2,650	229	16,242
1971	–	206	2,338	2,543	7,330	4,117	13,990	2,862	404	17,256
1972	50	347	2,737	3,134	7,836	4,086	15,056	2,946	576	18,578
1973	36	417	3,203	3,656	8,618	3,681	15,955	2,998	888	19,841
1974	30	463	2,988	3,480	8,520	3,512	15,512	3,253	1,202	19,967
1975[5]	28	449	2,835	3,312	8,824	3,174	15,310	3,122	1,652	20,084

[1] Includes anthracite, bituminous coal and lignite.
[2] Includes net imports. Outputs converted to theoretical inputs using the national average heat rate for fossil-fueled steam electric plants provided by the Federal Power Commission.
[3] Energy inputs calculated as for hydropower through 1970 (see footnote 2). Beginning in 1971, an average heat rate of 10,660 Btu/kWh is used based on information from the Energy Reserch and Development Administration.
[4] Data may not add to totals because of rounding.
[5] Preliminary.
Source: U.S. Department of the Interior, Bureau of Mines, Division of Interfuels Studies.

Source: *Energy Perspectives 2*, U.S. Department of the Interior, June 1976.

Consumption of Energy by the Electrical Sector, 1947–75
[Physical units]

Year	Coal[1] (thousand short tons)	Petroleum products (million bbl)				Natural gas (million ft³)	Hydropower[2] (million kWh)	Nuclear (million kWh)
		Jet fuel	Distillate fuel	Residual fuel	Total			
1947	89,531	–	14.4	61.1	75.5	373,037	84,981	–
1948	99,586	–	14.9	56.8	71.7	478,097	88,535	–
1949	83,964	–	12.6	80.1	92.7	550,121	96,361	–
1950	91,871	–	13.2	93.1	106.3	628,919	102,671	–
1951	105,768	–	9.6	70.5	80.1	763,898	106,554	–
1952	107,071	–	8.3	70.5	78.8	910,117	111,977	–
1953	115,883	–	6.9	85.4	92.3	1,034,272	111,625	–
1954	118,401	–	6.1	70.7	76.8	1,165,498	113,980	–
1955	143,759	–	5.9	76.0	81.9	1,153,280	120,304	–
1956	158,279	–	5.4	74.0	79.4	1,239,311	129,775	–
1957	160,761	–	5.3	76.6	81.9	1,338,079	136,959	10
1958	155,714	–	5.4	77.0	82.4	1,372,853	147,581	165
1959	168,427	–	5.0	82.2	87.2	1,627,097	145,002	188
1960	176,633	–	4.7	85.4	90.1	1,724,763	153,958	518
1961	182,138	–	4.2	87.9	92.1	1,825,341	157,754	1,692
1962	193,128	–	4.1	88.3	92.4	1,965,590	172,458	2,270
1963	211,193	–	4.1	91.6	95.7	2,142,930	168,573	3,212
1964	225,271	–	3.8	97.6	101.4	2,321,889	182,258	3,341
1965	244,887	–	3.7	114.9	118.6	2,318,253	196,843	3,657
1966	266,394	–	3.6	140.6	144.2	2,608,768	199,030	5,520
1967	273,970	–	2.9	158.4	161.3	2,746,352	224,650	7,655
1968	296,942	–	3.0	185.0	188.0	3,143,858	225,242	12,528
1969	310,310	–	12.2	247.6	259.8	3,486,391	254,540	13,928
1970	322,357	–	24.8	309.0	333.8	3,894,019	252,571	21,801
1971	327,926	–	35.3	371.8	407.1	3,992,983	273,110	37,899
1972	348,612	8.8	59.6	435.3	503.7	3,978,673	283,866	54,032
1973	386,879	6.3	71.6	509.5	587.4	3,605,333	289,596	83,334
1974	391,566	5.2	79.5	475.2	559.9	3,429,231	316,678	112,740
1975[3]	405,550	5.0	77.0	451.0	533.0	3,100,000	304,000	155,000

[1] Includes anthracite and bituminous coals and lignite.
[2] Includes net imports and negligible amount of hydropower generated by industrial establishments.
[3] Preliminary.
Source: U.S. Department of the Interior, Bureau of Mines, Division of Interfuels Studies.

Source: *Energy Perspectives 2*, U.S. Department of the Interior, June, 1976.

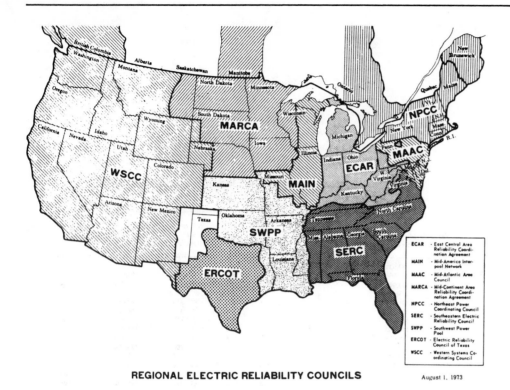

REGIONAL ELECTRIC RELIABILITY COUNCILS

August 1, 1973

ECAR	East Central Area Reliability Coordination Agreement
MAIN	Mid-America Inter-pool Network
MAAC	Mid-Atlantic Area Council
MARCA	Mid-Continent Area Reliability Coordination Agreement
NPCC	Northeast Power Coordinating Council
SERC	Southeastern Electric Reliability Council
SWPP	Southwest Power Pool
ERCOT	Electric Reliability Council of Texas
WSCC	Western Systems Co-ordinating Council

Fossil Steam Generation Capacity and Fuel Consumption of Units Reporting on FPC Form 36, by Reliability Council Area, 1972

Reliability Council	Nameplate Rating MW	Nameplate Capacity (MW) Currently Burning			Fuel Burned in 1972*		
		Coal	Oil	Gas	Coal 1,000 Tons	Oil** 1,000 BBLS	Gas 1,000 Mcf
ECAR	49,109.8	45,508.9	2,567.8	1,033.1	106,446.6	6,411.7	43,970.6
MAIN	23,535.7	20,956.6	1,230.1	1,349.0	42,494.3	8,231.7	73,308.6
MAAC	27,715.3	15,207.3	12,508.0	0.0	30,907.5	92,887.2	22,778.4
MARCA	8,184.5	7,501.4	211.8	471.3	17,910.2	688.2	118,664.1
NPCC	21,917.7	3,000.3	18,417.4	500.0	7,334.9	143,834.4	54,807.6
SERC	57,267.1	42,840.0	12,234.6	2,192.5	91,467.4	90,972.2	241,475.2
SWPP	25,454.4	2,542.8	216.5	22,695.1	4,855.1	4,647.8	1,068,902.2
ERCOT	26,053.0	1,186.8	104.0	24,762.2	1,789.9	26.3	991,235.4
WSCC	32,780.7	9,263.0	3,602.7	19,915.0	21,627.0	44,983.5	785,847.6
Unaffiliated	14,848.2	8,368.5	2,828.1	3,651.6	19,229.4	20,117.3	189,466.5
TOTAL	286,866.4	156,375.6	53,921.0	76,569.8	344,062.3	412,800.3	3,590,456.2
% of Capacity	100.0	54.5	18.8	26.7			
% of Energy***					55.2	18.3	26.5

* FPC News Release No. 19370 of June 27, 1973, reports fuel consumption under steam-electric boilers in the contiguous United States as:
    Coal -- 350.78 million tons
    Oil -- 432.32 million barrels (including distillate oil used under steam-electric boilers)
    Gas -- 3,763.74 million Mcf
** Includes distillate oil used under steam-electric boilers
*** Based on average 1971 heating values of fuels burned by electric utilities:
    Coal: 11,169 BTU/LB
    Oil: 147,017 BTU/GAL
    Gas: 1,030 BTU/Cu. Ft.

Source: Energy Emergency Legislation, Hearing before the Committee on Interior and Regular Affairs, United States Senate, Serial No. 93–22, November, 1975, U.S. Government Printing Office, Washington, 1973.

5

Energy Consumption Projections

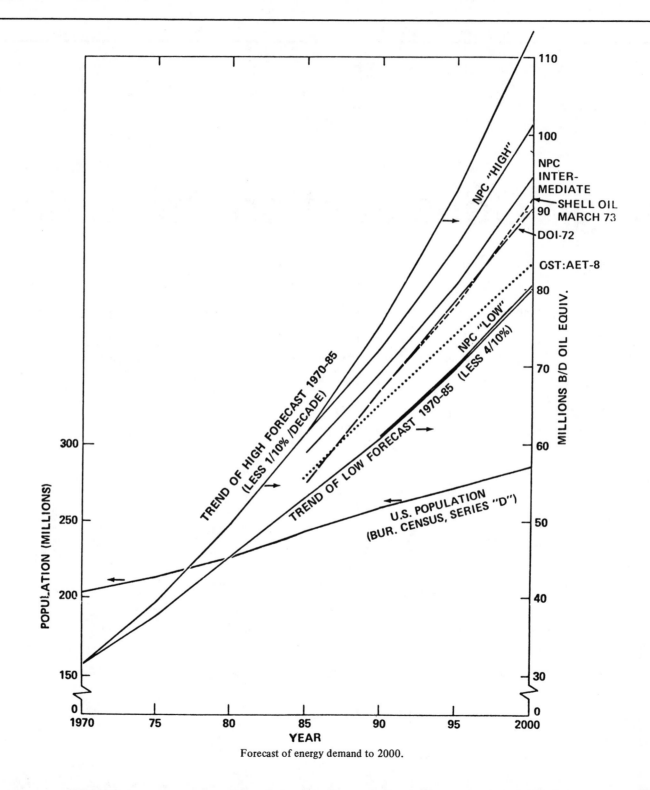

Forecast of energy demand to 2000.

WORLD POPULATION TRENDS

One of the factors in assessing future energy requirements is the rate at which populations are increasing. During the period 1950-72, the world average annual increase in population has been 2.36% with annual increases as high as 3.86% in Latin America and as low as 0.90% in Europe. (A 3% annual rate of increase corresponds to a doubling of the population in 25 years.) Several countries of Western Europe–East Germany, West Germany, and Luxembourg–have reached the state of zero population growth and a number–Sweden, United Kingdom, Finland, Austria, Belgium, and Hungary–have a growth rate of only 0.3%.

At the start of the 19th century in the developed countries of Western Europe, the birthrate was 37 per 1000 population and the death rate was about 32 per 1000, producing an overall growth rate of about 0.5% annually. During the years following, the death rate was gradually reduced to 10 per 1000 as improved public sanitation, medical practice, and nutrition were introduced. For a time, birthrates remained high with the consequence that population growth was greatly accelerated. By 1900, however, birthrates also began to drop as a result of changing economic and social conditions–movement of families from rural to urban communities, deferred marriages as education periods lengthened, improved literacy, and better understanding of birth control methods. Birthrates have now reached a level of 15 to 18 per 1000 population.

The situation now in developing countries is much the same as the situation was in Europe in 1800. Birthrates are near 40 per 1000 population while death rates have dropped from 35 to 18 per 1000. As a consequence, population growth rates average 2.9% per year in the developing countries.

Another contrast between the populations of developed and developing countries is provided by the age profile of these populations. The age distribution in developed countries is fairly uniform while in developing countries the age distribution profile has a pyramidal shape. In developing countries, 40% of the population is below 15 years of age and, on the average, there is less than one working adult for each child. In developed countries about 27% of the population is below age 15 and the ratio of working adults to children is double that of the developing countries.

Age Distribution of the Population of the Philippines and Sweden, 1965

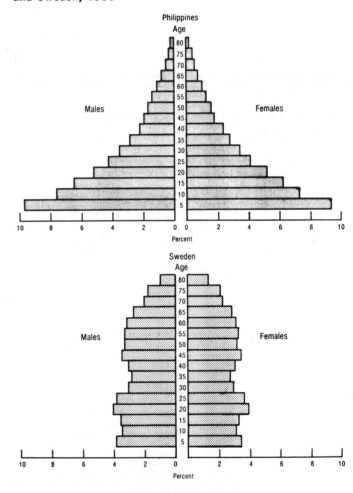

Source: UN Department of Economic and Social Affairs, *The Determinants and Consequences of Population Trends*, Population Studies No. 50 (1970), Vol. I, p. 267.

Age Distribution of the Population of the Phillipines and Sweden, 1965.

Source: Environmental Quality–1974, the Fifth Annual Report of the Council on Environmental Quality, U.S. Government Printing Office, December 1974.

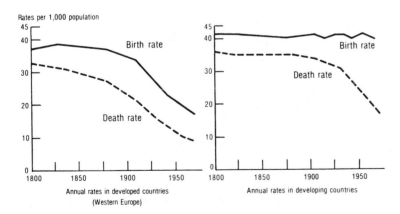

Source: U.S. Department of State, based on UN data.

Annual birth and death rates in developed and developing countries, 1800-1968.

Source: Environmental Quality, the Fifth Annual Report of the Council on Environmental Quality, U.S. Government Printing Office, December 1974.

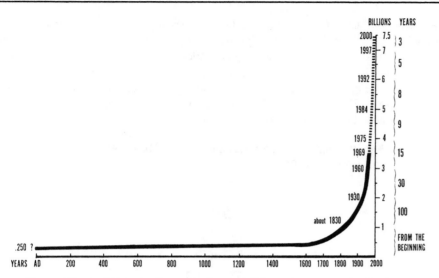

The world's population growth—past and projected (assuming constant fertility levels).

Source: U.S. Department of State

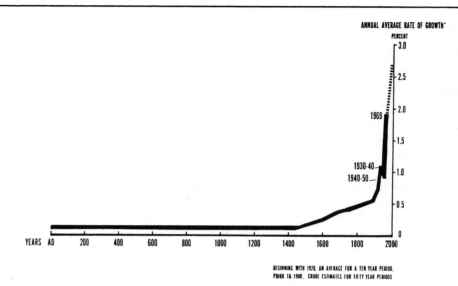

Rates of growth of world population—past and projected.

Source: U.S. Department of State

World Trends in Population and Vital Statistics, by Geographical Region and Development Status, 1950-72

Topic	World total	Development status[1]		Geographical region						
		More developed	Less developed	Africa	Northern America	Latin America	Asia	Europe	Oceania	Soviet Union
Population (millions)										
1950	2,486	852	1,634	217	166	162	1,355	392	13	180
1960	2,982	970	2,012	270	199	213	1,645	425	16	214
1970	3,632	1,078	2,554	344	228	283	2,056	462	19	243
1972[2]	3,780	1,102	2,678	363	231	300	2,149	469	20	248
Average annual increase, 1950-72 (percent)[3]	23.6	1.31	2.90	3.04	1.77	3.86	2.68	0.90	2.45	1.72
Average annual rates per 1,000 population, 1965-71										
Births	34	19	41	47	19	38	[4](29)(43)	18	25	18
Deaths	14	9	16	21	9	10	[4](12)(15)	10	10	8
Natural Increase	20	10	25	26	10	28	[4](17)(28)	8	15	10
Life expectancy at birth, 1965-70[5]	53.1	70.4	49.6	43.3	70.5	60.2	(52.2)(48.8)	70.9	64.8	70.3

[1] Development status is based on level of economic development. The United Nations recognizes two categories, less developed and more developed (Generally, less developed countries have high fertility and the more developed countries have low fertility.)

[2] Population Council estimate based on the 1971 population for each region and on the rates of natural increase shown.

[3] Births minus deaths.

[4] Figures are for East Asia and balance of Asia.

[5] Life expectancy at birth is defined by the United Nations as "the average number of years of life to which a group of newborn infants could look forward if they were subjected to risks of death at each age according to the mortality rates observed at each level of age, in the area and during the period to which the measure refers."

NA=not available.

Sources: UN, *Demographic Yearbook, 1971*, 23rd issue (New York: UN Department of Economic and Social Affairs, 1972); UN Population Division, Working paper No. 37, December 17, 1970 (mimeo).

Source: Environmental Quality, the Fifth Annual Report of the Council on Environmental Quality, U.S. Government Printing Office, December 1974.

WORLD POPULATION PROJECTIONS

Population growth is one of the basic factors in determining the quality of life for humanity. Demands for food, energy, and other resources become increasingly difficult to meet as population increases.

The rate of population growth, the sum of birth and death rates, has increased dramatically in the last century. In 1830, the world population was one billion; by 1930, the population was two billion; and by 1960, the world population had reached three billion. In 1972, the total world population stood at 3.78 billion.

Estimates of future world population growth vary widely. Bogue has estimated that by the year 2000, world population will be 4.5 billion. The United Nations' high estimate for the same year is 7 billion, the U.N. medium estimate is 6.1 billion, and the U.N. low estimate is 5.4 billion. An increase in population at a rate of 2% per year would result in a world population of 6.5 billion in the year 2000. (During the 1960-72 period, world population grew at a rate of 2.36% per year.)

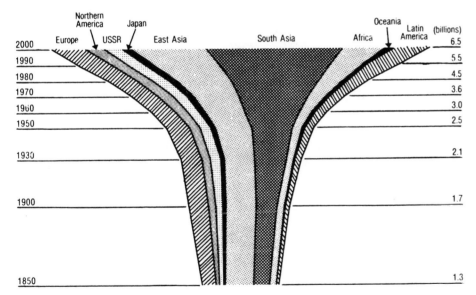

[1]Assumes continuation of 1963 fertility levels.

Source: UN Secretariat, Population Division, and J.A. Durand, "The Modern Expansion of World Population," *Proceedings of the American Philosophical Society*, Vol. III, No. 3, June 1967, cited in UN Fund for Population Activities, *WPY Bulletin*, No. 1, May 1973.

Estimated growth and regional distribution of the world's population, 1850-2000[1].

Source: Environmental Quality—1974, the Fifth Annual Report of the Council on Environmental Quality, U.S. Government Printing Office, December 1974.

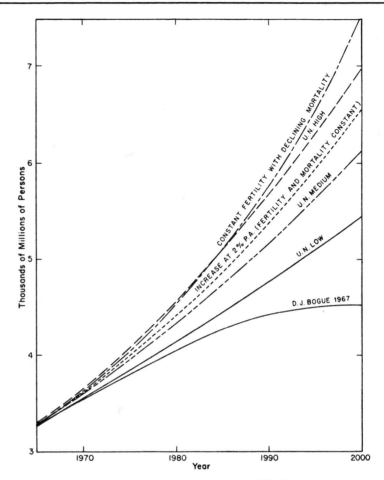

Six estimates of world population, 1965-2000.

Source: Resources and Man, Committee on Resources and Man, National Academy of Sciences–National Research Council, W. H. Freeman and Company, San Francisco, 1969.

Although birthrates have been falling in many countries, there is a built-in momentum in population growth that assures an increase in population even if the fertility rate drops to the zero population growth level. The high birthrates in recent years have resulted in a large number of young people who are potential parents and even if these young people have children only at the replacement level (2.1 children per couple), the population will still continue to grow for a time before population growth levels off. If birthrates could be reduced to the replacement level by 1985, the world population would stabilize at 6.4 billion in the year 2075. If replacement levels are reached by 2005, world population would stabilize at 8.5 billion in the year 2095; and if replacement levels are not reached until the year 2052, world population would increase to 11.2 billion sometime after the year 2100.

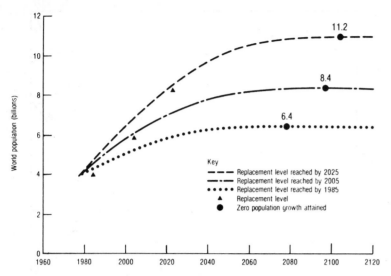

Explanation   The table shows world population growth assuming replacement levels are achieved by and maintained after the dates indicated.

Source: U.S. Department of State, *The Population Explosion: A Present Danger* (1974).

Momentum of population growth.

Source: Environmental Quality—1974, the Fifth Annual Report of the Council on Environmental Quality, U.S. Government Printing Office, December 1974.

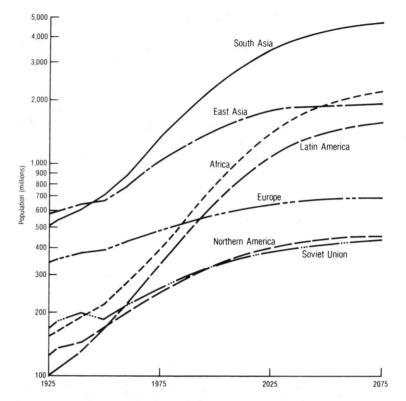

[1]According to the UN medium variant of long-range projections (charted on a logarithmic scale)

Source: UN Population Commission, *Preliminary Version of Concise Report on the World Population Situation*, E/CN.9/IIISS/CRP.1 (14 February 1974), Figure 2, p. 50.

Population of major regions of the world, 1925-2075[1]

Source: Environmental Quality—1974, the Fifth Annual Report of the Council on Environmental Quality, U.S. Government Printing Office, December 1974.

Time Required to *Double* a Population

Annual Rate of Population Growth	Number of Years To Double Population
4.0	17.3
3.5	20.1
3.0	23.1
2.5	27.6
2.0	34.6
1.5	46.2
1.0	69.3

Note: To maintain the *same* standard of living for its people a country must *double its* output of goods and services (*GNP*) in the *same time* that population doubles to improve standards of living it must *more than double* its GNP in the *same time*.

Source: U.S. Department of State

WORLD ENERGY CONSUMPTION PROJECTIONS

The world energy consumption data in the accompanying tables and graphs were prepared by the U.S. Department of the Interior. During 1960-72, world energy consumption grew at a rate of 4.9%/year and the projected growth rate from 1972-90 is 3.3%/year. Annual rates of increase in the consumption of various resources during 1972-90 are projected to be: coal, 1.8%/year; petroleum, 2.4%/year; natural gas, 2.9%/year; hydropower and geothermal, 2.1%/year; and nuclear power, 23.6%/year. Of the total world consumption in the year 1990, coal is projected to constitute 22%; petroleum, 40%; natural gas, 19%; hydropower and geothernal, 5%; and nuclear power, 15%.

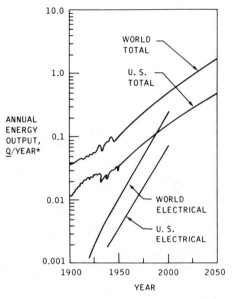

$$*\underline{Q} = 10^{18} \text{ Btu} = 2.93 \times 10^{14} \text{ kWh.}$$

Annual energy output, U.S. total and world total.

World Energy Consumption by Source, 1960-90
(Quadrillion Btu)[1]

ENERGY SOURCE	1960	1965	1970	1972	1980	1985	1990
COAL	61.5	62.6	66.8	66.3	79.2	85.7	92.0
PETROLEUM	45.3	64.4	96.9	107.5	132.3	147.2	165.0
NATURAL GAS	18.0	26.6	40.6	45.8	56.8	67.3	77.1
HYDROPOWER & GEOTHERMAL	6.9	9.6	11.8	12.9	15.4	17.1	18.8
NUCLEAR		.2	.8	1.4	12.6	33.6	63.6
TOTAL	131.7	163.4	216.9	233.9	296.3	350.9	416.6

[1] 1 Quadrillion Btu = 500,000 barrels petroleum per day for a year
    = 40 million tons of bituminous coal
    = 1 trillion cubic feet of natural gas
    = 100 billion kWh (based on a 10,000-Btu/kWh heat rate)

Source: Energy Perspectives, U.S. Department of the Interior, February 1975.

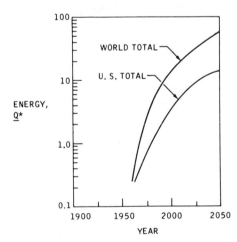

$$*\underline{Q} = 10^{18} \text{ Btu} = 2.93 \times 10^{14} \text{ kWh.}$$

Cumulative demand for energy, U.S. total and world total (from 1960).

World Energy Consumption, by Region, 1960-90
(Quadrillion Btu)[1]

REGION	1960	1965	1970	1972	1980	1985	1990
UNITED STATES	44.6	53 3	67.0	72.0	86.3	102.9	121.9
WESTERN EUROPE	26.4	34.4	46.0	49.1	62.6	75.2	87.2
JAPAN	3.7	6.2	12.0	13.4	20.4	26.7	34.0
SINO-SOVIET BLOC	39.0	45.2	58.3	63.7	82.0	94.0	109.0
REST OF WORLD	18.0	24.3	33.6	35.7	45.0	52.1	60.4
TOTAL	131.7	163.4	216.9	233.9	296.3	350.9	416.5

[1] 1 Quadrillion Btu = 500,000 barrels petroleum per day for a year
    = 40 million tons of bituminous coal
    = 1 trillion cubic feet of natural gas
    = 100 billion kWh (based on a 10,000-Btu/kWh heat rate)

Source: Energy Perspectives, U.S. Department of the Interior, February 1975.

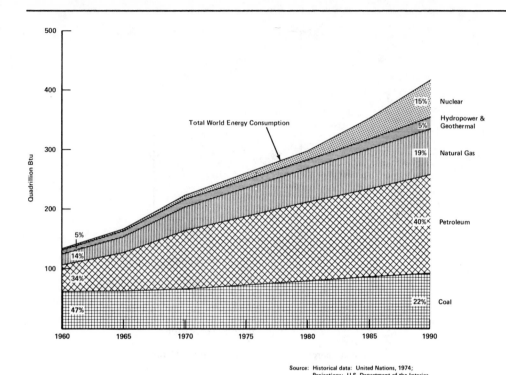

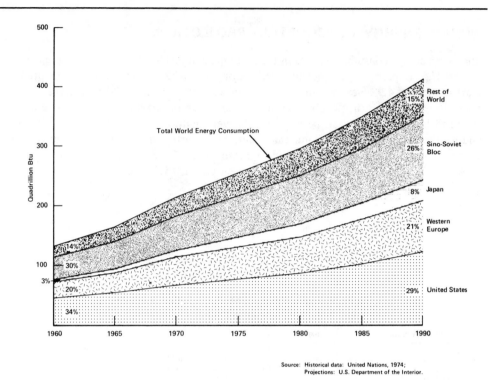

World energy consumption by source, 1960–90.

Source: Energy Perspectives, U.S. Department of the Interior, February 1975.

World energy consumption by region, 1960–90.

Source: Energy Perspectives, U.S. Department of the Interior, February 1975.

WORLD ELECTRICITY CONSUMPTION PROJECTIONS

World energy inputs into the electrical sector increased at a rate of 7.6%/year during 1960–72 and are projected to increase at 6.4%/year between 1972 and 1990. Although the inputs of fossil fuels—coal, petroleum, and gas—will continue to increase steadily in the electrical sector, the relative percentage contribution of fossil fuels will decrease with an attendant increase in the contribution of nuclear power from a few percent in 1975 to 37% in 1990.

World Energy Inputs, Electrical Sector, 1960–90
(Quadrillion Btu)[1]

YEAR	FOSSIL FUELS	HYDROPOWER & GEOTHERMAL	NUCLEAR	TOTAL
1960	16.1	6.9	---	23.0
1965	24.4	9.6	0.2	34.2
1970	36.6	11.8	.8	49.2
1972	41.6	12.8	1.4	55.8
1980	63.0	15.4	12.6	91.0
1985	73.3	17.1	33.6	124.0
1990	87.6	18.8	63.6	170.0

[1] 1 Quadrillion Btu = 500,000 barrels petroleum per day for a year
= 40 million tons of bituminous coal
= 1 trillion cubic feet of natural gas
= 100 billion kWh (based on a 10,000-Btu/kWh heat rate)

Source: Energy Perspectives, U.S. Department of the Interior, February 1975.

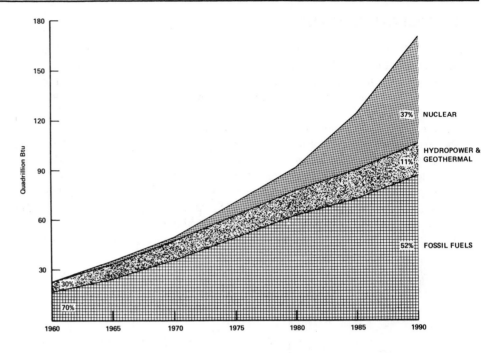

Source: Historical data: United Nations, 1974;
Projections: U.S. Department of the Interior.

World energy inputs, electrical sector, 1960–90.

Source: Energy Perspectives, U.S. Department of the Interior, February 1975.

U.S. POPULATION TRENDS

Between 1800 and 1972, the population of the United States increased from 5 to 210 million. During this period a considerable contribution to the growth of the population was made through immigration. There was a steady, long-term decline in the birthrate from 55 per 1000 population in 1820, to 44 per 1000 in 1880, 32 per 1000 in 1900, and 18 per 1000 in 1936. The birthrate rose again after World War II to 27 per 1000 population and then declined to about 15 per 1000 population in 1973 and 1974.

The fertility rate (the average number of births per woman) was 2.03 in 1972 and 1.90 in 1973, well below the 2.1 necessary for the United States population to reproduce itself, assuming no immigration. Zero population growth has not been reached, however, because the population has a disproportionate number of young females—37% in the child-bearing age (18-44) and 26% under the age of 15. Assuming continuation of the present birthrate, zero population growth will be achieved in about 70 years. Immigration currently adds about 400,000 persons per year to the United States population, and contributes about 0.2%/year to the growth rate.

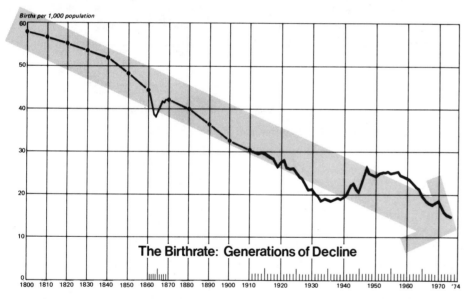

The Birthrate: Generations of Decline

Viewed in historical perspective, the high birthrates experienced by the U.S. after World War II were just a temporary aberration in a protracted downward movement. The 1974 rate of about 14.8 per thousand population seems more in step with the trend. In only two earlier recorded periods, however, has the U.S. birthrate fallen with anything like the rapidity of the last decade or so. The first was during the Civil War, when large numbers of men were mobilized and separated from their wives. In order to show the extraordinary Civil War experience, the chart reflects yearly data between 1860 and 1870; otherwise the data for years prior to 1910 are charted at intervals of a decade. The birthrate again dropped steeply during the 1920's. A lot of that decline can be accounted for by the fact that immigration was curbed and many Americans moved from rural to urban places. Both immigrants and rural residents tend to have higher birthrates than do native-born urban dwellers. During the Depression, the birthrate bottomed out at 18.4.

The birthrate: generations of decline.

Source: Tom Cardamone for *Fortune Magazine*, December, 1974.

Trends in U.S. Population and Vital Statistics, 1940-72

Category	1940	1950	1960	1970	1972
Total U.S. population (1,000)	132,165	151,326	179,975	203,810	209,851
Total births (1,000)	[1] 2,559	[1] 3,632	[1] 4,258	3,718	[2] 3,256
Crude birth rate (per 1,000 of population)	19.4	24.1	23.7	18.2	15.6
General fertility rate (births per 1,000 females, 15-44 years of age)	79.9	106.1	119.1	87.4	73.4
Total immigration (1,000)	52.8	103.5	251.5	373	385
Immigration rate (per 1,000 of population) [3]	[4] 0.4	[4] 0.7	[4] 1.5	1.8	1.8
Total deaths (1,000)	1,417	1,452	1,712	1,921	1,962
Crude death rate (deaths per 1,000 population)	10.8	9.6	9.5	9.4	9.4
Life expectation at birth (years)	62.9	68.2	69.7	70.8	72.2
Life expectation at 5 years (years)	62.5	65.5	67.0	NA	67.8
Rate of natural increase (births-deaths, per 1,000 population)	8.6	14.5	14.2	8.4	[2] 6.2

[1] Figures adjusted for underregistration.

[2] Preliminary figures.

[3] Immigration rates for 1940, 1950, and 1960 are averages for 1931-40, 1941-50. 1951-60 respectively.

[4] Excludes Alaska and Hawaii.

Sources: National Center for Health Statistics, *Vital Statistics of the United States*, annual reports; U.S. Immigration and Naturalization Service, annual reports; and unpublished data, as cited in U.S. Bureau of the Census, *Statistical Abstract of the United States, 1973* (1973), Tables 65, 78, 80, and 141.

Source: Environmental Quality, the Fifth Annual Report of the Council on Environmental Quality, U.S. Government Printing Office, December 1974.

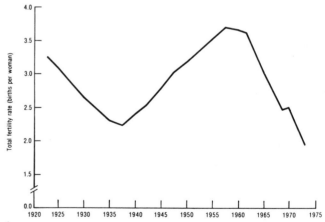

[1] 5-year data for 1920-59, single-year data for 1960-73.

Source: U.S. Bureau of the Census, *Special Studies, Fertility Indicators: 1970*, Series P-23, No. 36, April 6, 1971, p.9; 1971, 1972, and 1973 figures are unpublished Bureau of Census data.

U.S. total fertility rate, 1920-73.[1]

Source: Environmental Quality—1974, the Fifth Annual Report of the Council on Environmental Quality, U.S. Government Printing Office, December 1974.

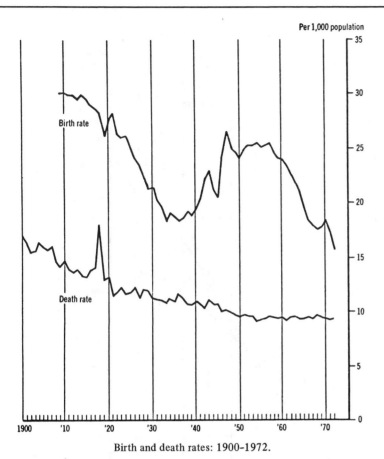

Birth and death rates: 1900-1972.

Source: Social Indicators 1973, Executive Office of the President, Office of Management and Budget, U.S. Government Printing Office, Washington, D.C. 1973.

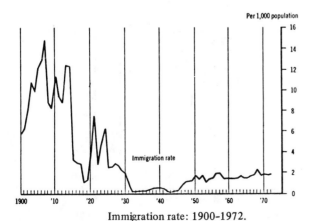

Immigration rate: 1900-1972.

Source: Social Indicators 1973, Executive Office of the President, Office of Management and Budget, U.S. Government Printing Office, Washington, D.C., 1973.

Rate of population growth: 1910-1972

Source: Social Indicators 1973, Executive Office of the President, Office of Management and Budget, U.S. Government Printing Office, 1973.

U.S. POPULATION PROJECTIONS

The present rate of growth of the United States population, including immigration, is 0.72%/year. At this rate, the population will be about 250 million in the year 2000 and would double in 97 years. Should the present low fertility rate of 1.8 births per woman continue, however, then the population growth rate will continue to decline as the proportion of childbearing women declines. In this event, zero population growth would be achieved in about 70 years.

To reflect the trend in declining birthrates, in 1972 the Bureau of the Census revised its projections of United States population growth, dropping the high growth Series A and B, and adding a new low projection, Series F, based on a fertility rate of 1.8 births per woman per year.

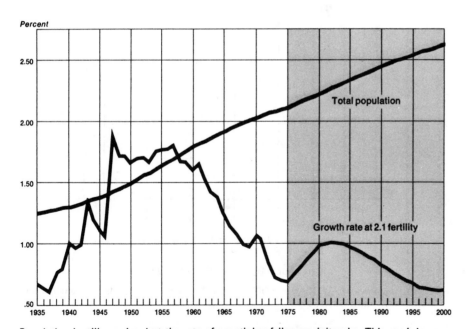

Population is still growing, but the rate of growth has fallen precipitously. This year's increase of a mere 0.7 percent is the lowest since the 1930's. The lines on the chart represent actual experience for the years through 1974 and projections thereafter. The projections assume a net gain of 400,000 people a year through immigration and a fertility rate of 2.1 children per woman. Such a fertility rate would speed up growth for a time because so many women are entering the childbearing ages. But without immigration the population would level off at about 275 million before the middle of the next century.

Source: Tom Cardamone for *Fortune Magazine*, December, 1974.

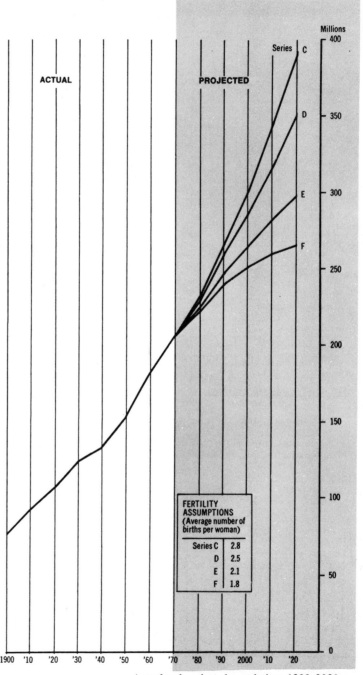

FERTILITY ASSUMPTIONS
(Average number of births per woman)

Series C	2.8
D	2.5
E	2.1
F	1.8

Actual and projected population: 1900-2020.

Source: Social Indicators 1973, Executive Office of the President: Office of Management and Budget, U.S. Government Printing Office, 1973.

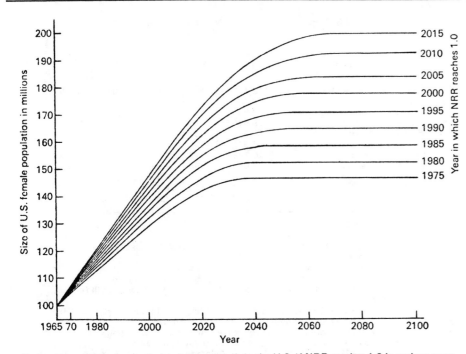

Projections of the course of population growth in the U.S. if NRR reaches 1.0 in various years. Total population size would be slightly less than twice the size of the female population, since there are slightly fewer men than women (after Frejka *Population Studies*, Nov. 1968).

From "Population, Resources, Environment: Issues in Human Ecology," Second Edition, by Paul R. Ehrlich and Anne H. Ehrlich. W. H. Freeman and Company. Copyright © 1972.

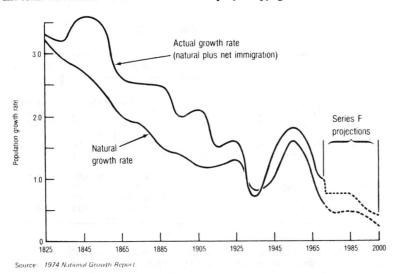

Source: *1974 National Growth Report.*

U.S. population growth, actual 1825–1970 and estimated 2000.

Source: Environmental Quality–1974, the Fifth Annual Report of the Council on Environmental Quality, U.S. Government Printing Office, December, 1974.

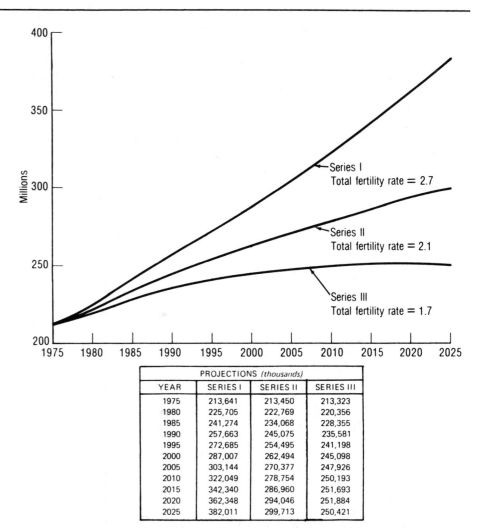

PROJECTIONS *(thousands)*			
YEAR	SERIES I	SERIES II	SERIES III
1975	213,641	213,450	213,323
1980	225,705	222,769	220,356
1985	241,274	234,068	228,355
1990	257,663	245,075	235,581
1995	272,685	254,495	241,198
2000	287,007	262,494	245,098
2005	303,144	270,377	247,926
2010	322,049	278,754	250,193
2015	342,340	286,960	251,693
2020	362,348	294,046	251,884
2025	382,011	299,713	250,421

Source: U.S. Bureau of the Census, *Current Population Reports*, Series P–25, No. 541, "Projections of the Population of the United States, by Age and Sex, 1975 to 2000, with Extensions of Total Population to 2025, advance report, (Washington, D.C.: Government Printing Office, 1975), p. 1.

U.S. population projections, 1975–2025.

Source: The Sixth Annual, Report of the Council on Environmental Quality, December 1975.

U.S. REGIONAL POPULATION TRENDS

One of the factors in determining future energy consumption patterns is the distribution of the population between urban and rural areas. Since 1900, the rural population of the United States has dropped from about 60% to 25% of the total population. During that period, the percentage of the population living in cities of one million or more has remained about the same while the percentage living in smaller cities and towns has increased.

This movement from the country to towns and cities has resulted in a proportionately larger number of smaller households and has occurred during a period when energy demand for household heating and cooling, and for the operation of appliances, has increased substantially.

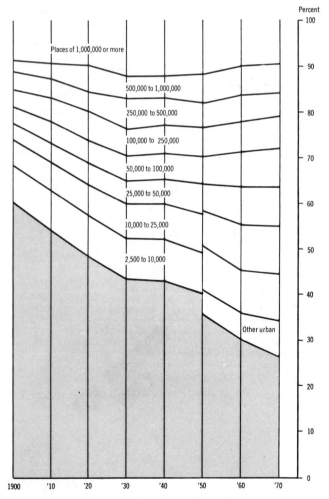

Population in urban and rural areas: 1900-1970, by size of place.

Source: Social Indicators 1973, Executive Office of the President, Office of Management and Budget, U.S. Government Printing Office, 1973.

U.S. Population Trends, by Region, 1940-73
[In thousands]

Region [1]	1940	1950	1960	1970	1973
Total United States	**132,165**	**151,326**	**179,975**	**203,810**	**209,851**
New England	8,437	9,314	10,532	11,880	12,151
Middle Atlantic	27,539	30,164	34,270	37,272	37,528
East North Central	26,626	30,399	36,291	40,322	30,896
West North Central	13,517	14,061	15,424	16,359	16,704
South Atlantic	17,823	21,182	26,091	30,791	32,459
East South Central	10,778	11,477	12,073	12,844	13,289
West South Central	13,065	14,538	17,010	19,389	20,257
Mountain	4,150	5,075	6,916	8,344	9,149
Pacific	10,229	15,115	21,368	26,609	27,417
Puerto Rico	1,869	2,211	2,350	2,712	NA

NA—Not available.

[1] Regions:

New England—Maine, New Hampshire, Vermont, Massachusetts, Rhode Island, Connecticut.

Middle Atlantic—New York, New Jersey, Pennsylvania.

East North Central—Ohio, Indiana, Illinois, Michigan, Wisconsin.

West North Central—Minnesota, Iowa, Missouri, North Dakota, South Dakota, Nebraska, Kansas.

South Atlantic—Delaware, Maryland, District of Columbia, Virginia, West Virginia, North Carolina, South Carolina, Georgia, Florida.

East South Central—Kentucky, Tennessee, Alabama, Mississippi.

West South Central—Arkansas, Louisiana, Oklahoma, Texas.

Mountain—Montana, Idaho, Wyoming, Colorado, New Mexico, Arizona, Utah, Nevada.

Pacific—Washington, Oregon, California, Alaska, Hawaii.

Sources: U.S. Bureau of the Census, *Statistical Abstract of the United States, 1973* (1973), Table 14, and *Current Population Reports*, Series P–25, Nos. 500 and 508 (1974).

U.S Urban and Rural Population, 1950-70
[In thousands]

Class and size	1950	1960	1970
Urban			
Total	**96,469**	**125,269**	**149,326**
Places of 500,000 or more	26,591	28,595	31,736
Places of 100,000–500,000	17,721	22,418	24,728
Places of 10,000–100,000	29,606	46,355	55,987
Places of less than 10,000	15,207	18,050	21,689
Unincorporated parts of urbanized areas	7,344	9,851	15,186
Rural			
Total	**54,230**	**54,054**	**53,887**
Places under 2,500	10,054	10,391	10,508
Other rural	43,725	43,664	43,379
Total U.S. population	**150,699**	**179,323**	**[1]203,213**

[1] Excludes Hawaii and Alaska.

Source: U.S. Bureau of the Census, *U.S. Census of Population, 1970*, Vol. I, Part A, as cited in U.S. Bureau of the Census, *Statistical Abstract of the United States, 1973* (1973), Table 18.

Source: Environment Quality–1974, the Fifth Annual Report of the Council on Environmental Quality, U.S. Government Printing Office, December 1974.

U.S. POPULATION AGE GROUP TRENDS

The trends in age group composition of the United States population over the past 75 years have shown a transition from a population of predominantly children and young people to a population with the majority of working age and with 10% above the age of 65. Projections of age distributions for the year 2000 show a further decline (from 36.8% to 27.5%) in the percentage of young (under 20) in the population, a growing percentage (53.2% to 60.9%) of those of working age, and a slight increase (10.0 to 11.5%) of those over the age 65. These projections are based on a United States population of 251 million in the year 2000. (This population projection should be compared to that of 279 million used by the U.S. Department of the Interior in its energy consumption projections.)

U.S. Population Trends, by Age, 1950–73
[In Thousands]

Year	Total All Ages	Under 5 Years	5–14 Years	15–34 Years	35–64 Years	65 Years and Over	Median Age
1950	152,271	16,410	24,588	46,391	52,486	12,397	30.2
1960	180,671	20,337	35,735	47,497	60,422	16,679	29.4
1970	204,879	17,167	40,688	61,750	65,094	20,177	28.0
1973	210,404	16,714	38,983	67,487	65,891	21,133	NA

NA—Not available.
Sources: U.S. Bureau of the Census, *Current Population Reports*, Series P-25, Nos. 311, 483, 493, as cited in U.S. Department of Commerce, *Statistical Abstract of the United States, 1973* (1973), Table 3, and Series P-25, No. 519.
Source: Environmental Quality–1974, the Fifth Annual Report of the Council on Environmental Quality, U.S. Government Printing Office, December 1974.

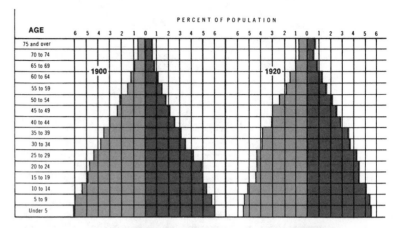

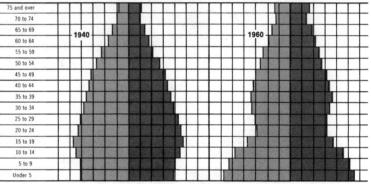

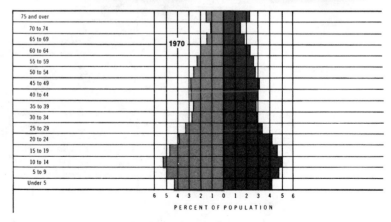

Population by age and sex: 1900–1970.

Source: Social Indicators 1973, Executive Office of the President, Office of Management and Budget, U.S. Government Printing Office, 1973.

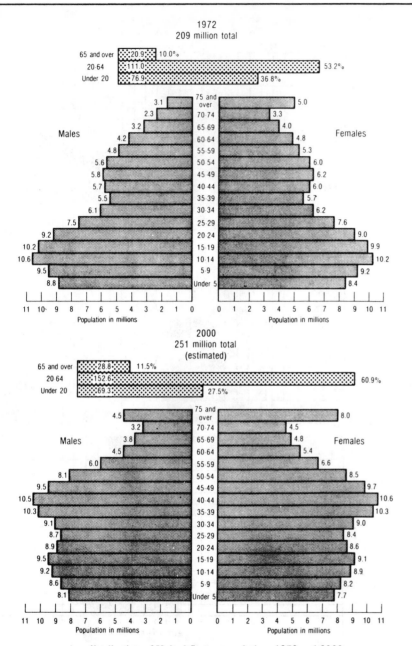

Age distribution of United States population, 1972 and 2000.

Source: Environmental Quality—1974, the Fifth Annual Report of the Council on Environmental Quality, U.S. Government Printing Office, December 1974.

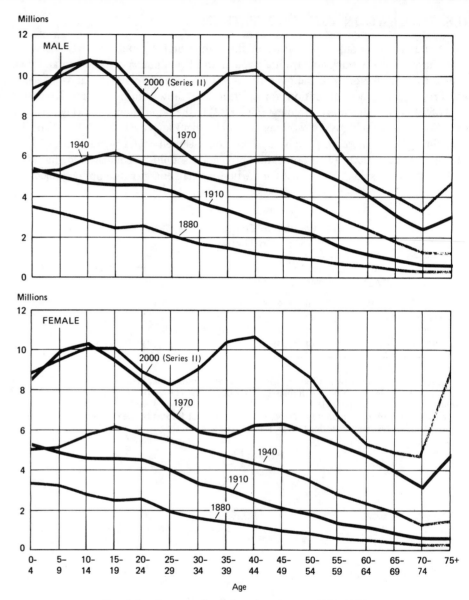

Population, by sex and age, for selected years: 1880–2000.

Source: Social Indicators 1976, U.S. Department of Commerce, Office of Federal Statistical Policy and Standards, Bureau of the Census, December 1977.

U.S. ENERGY DEMAND FORECASTS

Projections of energy consumption in the United States in the year 2000 vary from a low of 120 quadrillion Btu/year to a high of 220 quadrillion Btu/year (consumption in 1974 was about 74 quadrillion Btu/year).

The high energy consumption growth estimates are usually based on the historic growth rate of 4.1%/year for the period 1960-72, and assume that energy prices will return to a relatively low level compared to other commodity prices and that there will be little incentive for conservation.

A moderate growth rate in energy consumption is usually assumed to be about 3%/year with some improvement in the efficiency of energy use but with the standard of living and economic development at historic levels.

The low growth rate projections of about 2%/year are based on energy prices remaining high with considerable incentive for conservation, and upon some reduction in the production of goods with life styles somewhat more austere than before.

An additional factor contributing to the differences in energy growth projections stems from differences in population growth projections. The Council on Environmental Quality projections are based on a population of 251 million in the year 2000; the Energy Research and Development Administration estimates are based on a population of 265 million in that year; and Department of the Interior estimates are based on a population of 279 million in the year 2000.

Additional detail on estimates of future energy consumption is given in Chapter 16, Energy Futures.

Level of Demand for Energy Services

	1972	1985	2000
RESIDENTIAL			
Heating, cooling, & electricity	66.7x10⁶ households	80.0x10⁶ households	99x10⁶ households
COMMERCIAL			
Heating, cooling, & electricity	23.5x10⁹ ft² floorspace	32.0x10⁹ ft²	42.0x10⁹ ft²
INDUSTRIAL			
Process & direct heat	7.84x10¹⁵ Btu	3% growth/yr	3% growth
Petrochemicals	4.19x10¹⁵ Btu	5% growth/yr	5% growth
Electricity	2.57x10¹⁵ Btu	4% growth/yr	4% growth
Iron	84.5x10⁶ ton	122x10⁶ ton	153x10⁶ ton
Aluminum	4x10⁶ ton	8x10⁶ ton	14x10⁶ ton
TRANSPORTATION			
Private Auto			
Air—passenger	992x10⁹ vehicle-mile	1467x10⁹ vehicle-mile	2050x10⁹ vehicle-mile
—freight	153x10⁹ passenger-mile	421x10⁹ passenger-mile	874x10⁹ passenger-mile
Bus, truck & rail	4x10⁶ ton-mile	30x10⁶ ton-mile	99x10⁶ ton-mile
—passenger	89x10⁹ passenger-mile	11x10⁹ passenger-mile	161x10⁹ passenger-mile
—freight	461x10⁹ ton-mile	72x10⁹ ton-mile	1040x10⁹ ton-mile
Ship	0.7x10¹⁵ Btu	4%/yr growth	3%/yr growth

Source: A National Plan for Energy Research, Development and Demonstration: Creating Energy Choices for the Future, Volume 1: The Plan, ERDA-48, Energy Research and Development Administration, June 1975.

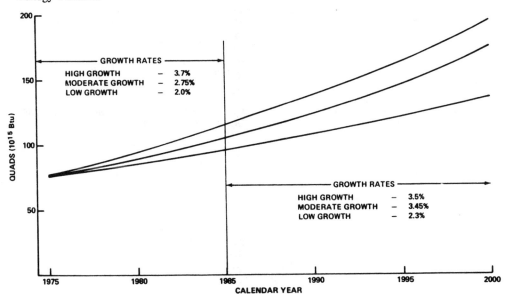

Total United States energy consumption (1975 interim projections).

Source: Roger W. A. Legassie, Testimony at Hearings on Growth Rates of Electricity and the Role of; Nuclear Energy, United States Committee on Interior and Insular Affairs, Subcommittee on Energy and the Environment, April 28, 1975.

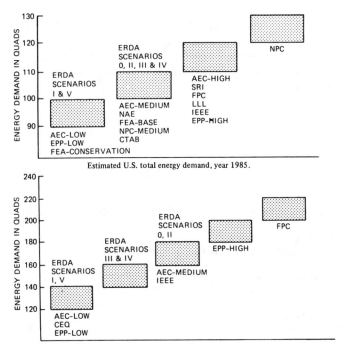

Estimated U.S. total energy demand, year 1985.

Estimated U.S. energy demand, year 2000.

Source: A National Plan for Energy Research, Development and Demonstration: Creating Energy Choices for the Future, Volume 1: The Plan, ERDA-48, Energy Research and Development Administration, June 1975.

U.S. ECONOMIC AND ENERGY INDICATORS

Projections of economic and energy indicators made by the U.S. Department of the Interior in 1972 assume a population of 279 million in the year 2000 (average of the Series D and E projections of the Bureau of the Census), an increase in the gross per capita energy consumption at a rate of 2.5%/year–from 333 million Btu per person in 1971 to 686 million Btu per person in the year 2000, and a decline in the energy/GNP ratio from 93,000 Btu per 1958 dollar to 78,700 Btu per dollar. Net per capita energy growth is projected to be at a lower rate, 2.1%/year, than gross per capita energy growth as the econonmy shifts toward secondary energy forms, principally electricity.

During the 1971–2000 period, coal consumption is projected to increase from 511 million tons/year to 1310 million tons/year; petroleum consumption will increase from 15.1 million barrels/day to 35.6 million barrels/day; natural gas usage is projected to increase from 22 trillion cubic feet/year to 33 trillion cubic feet/year; hydropower inputs will more than double from 266 to 700 billion kilowatt hours/year and energy inputs from nuclear power will increase from 38 billion kilowatt hours/year to 5500 billion kilowatt hours/year.

In terms of the contribution of the various energy forms to the total energy inputs during the period 1971-2000, the share for coal will decline from 18.2% to 16.3%, the share for petroleum will decrease from 44.1% to 37.2%, the share for natural gas will decrease from 33.0% to 17.7%, that for hydropower will decline from 4.1% to 3.1%, while the share from nuclear power will increase from 0.6 to 25.7%.

ASSUMPTIONS USED IN THE FORECAST

Gross National Product (GNP)

This is a comprehensive measure of aggregate economic activity. GNP measures final goods and services, eliminating intermediate products consumed in the production of end products. Because the use of energy is so intimately bound to economic activity, a forecast of GNP is necessary for forecasting energy consumption. For purposes of this study GNP was assumed to be the following:

Year	GNP (billion 1958 dollars)
1974	891.2
1980	1,092
1985	1,294
2000	2,105

Population

The forecast was based on the following forecast of population:

Year	Population (millions)
1974	211.9
1980	224
1985	236
2000	264

Source: Walter G. Dupree, Jr. and John S. Corsentino, *United States Energy Through the Year 2000 (Revised)*, Bureau of Mines, U.S. Department of the Interior, December 1975.

Selected United States Economic, Energy, and Demographic Indicators: Actual for 1974 and Projected to the Year 2000

	1974	1980	1985	2000
Gross Energy Inputs (Quadrillion Btu)	73.1	87.1	103.5	163.4
Net Energy Inputs (Quadrillion Btu)	59.9	68.9	77.5	110.2
Population (Millions)	212	224	236	264
Gross National Product (Million 1958 dollars)	821	1,092	1,294	2,105
Energy/GNP Ratio (Thousand Btu per 1958 dollar)	89	80	80	77.6
Gross Energy/Capita Ratio (Million Btu per person)	345	389	439	619
Net Energy/Capital Ratio (Million Btu per person)	283	308	328	417
Net Energy/Gross Energy Ratio	81.9	79.1	74.9	67.5

Source: Walter G. Dupree, Jr., and John S. Corsentino, *United States Energy Through the Year 2000 (Revised)*, Bureau of Mines, U.S. Department of the Interior, December 1975.

U.S. ENERGY FORECASTS—SOURCE AND SECTOR

The data given in the accompanying tables are based on the economic and energy consumption projections given in 5-10. The data represent one of the higher projections of growth in energy consumption as given in 5-9.

Projections of U.S. Gross Energy Consumption by Source, 1975-2000

Fuel source	1975[1]	1980		1985		2000[3]
		Bureau of Mines	Federal Energy Administration[2]	Bureau of Mines	Federal Energy Administration[2]	Bureau of Mines
Coal:						
Quadrillion Btu	13.4	17.2	15.7	21.3	20.6	34.7
Million tons	567	736	719	923	962	1,560
Petroleum:						
Quadrillion Btu	32.7	41.0	35.6	45.6	40.9	51.2
Billion bbl	6.0	7.4	6.5	8.3	7.5	9.3
Natural gas:						
Quadrillion Btu	20.2	20.6	22.7	20.1	24.2	19.6
Trillion ft³	19.7	20.0	22.0	19.5	23.4	19.0
Oil shale:						
Quadrillion Btu	—	—	—	.9	.6	5.7
Billion bbl	—	—	—	.2	.1	1.0
Nuclear power:						
Quadrillion Btu	1.7	4.5	3.9	11.8	8.67	46.1
Billion kWh	155	427	388	1,139	867	5,120
Hydropower and geothermal:						
Quadrillion Btu	3.1	3.8	3.7	3.8	3.9	6.1
Billion kWh	304	343	NA	391	NA	605
Total (quadrillion Btu)	71.1	87.1	81.6	103.5	98.9	163.4

[1] Actual.
[2] Reference case, with oil imports at $13 per barrel.
[3] FEA forecast does not extend to 2000.
Sources: Dupree, Walter G., and John S. Corsentino, *U.S. Energy Through the Year 2000*, Revised, Washington, D.C., U.S. Department of the Interior, Bureau of Mines, Dec. 1975; Federal Energy Administration, *National Energy Outlook*, Washington, D.C., 1976.

Source: Energy Perspectives 2, U.S. Department of the Interior, June 1976.

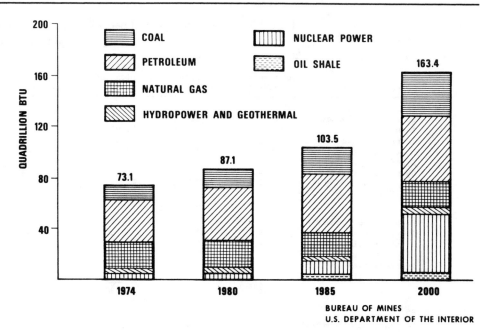

United States energy consumption by source, 1974–2000.

Source: Walter G. Dupree, Jr. and John S. Corsentino, *United States Energy Through the Year 2000 (Revised)*, Bureau of Mines, U.S. Department of the Interior, December 1975.

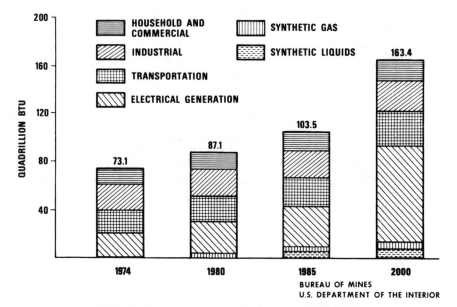

United States energy consumption by sector, 1974–2000.

Source: Walter G. Dupree, Jr. and John S. Corsentino, *United States Energy Through the Year 2000, (Revised)*, Bureau of Mines, U.S. Department of the Interior, December, 1975.

United States Consumption of Energy Resources by Major Sources, 1974 Preliminary and Projected to the Year 2000
(in standard physical units)

Consuming Sector	Coal million short tons 1/	Petroleum million barrels 2/	Natural Gas billion cubic feet	Oil Shale million barrels equivalent	Nuclear Power billion kWh	Hydropower & Geothermal billion kWh	Synthetic Liquids distributed million barrels	Synthetic Gas distributed billion cu. ft.	Utility Electric distributed billion kWh
1974									
Household & Commercial	10.9	1,115.0	6,970	--	--	--	--	--	1,080
Industrial 3/	154.9	1,137.1	10,900	--	--	3	--	--	781
Transportation	.1	3,274.7	650	--	--	--	--	--	3
Electrical generation	390.6	554.0	3,260	--	110	291	--	--	--
Synthetic gas	--	--	--	--	--	--	--	--	--
Synthetic liquids	--	--	--	--	--	--	--	--	--
Total	556.5	6,080.8	21,780	--	110	294	--	--	1,864
1980									
Household & Commercial	4	1,376	7,770	--	--	--	--	120	1,695
Industrial	185	1,358	9,700	--	--	--	--	--	1,057
Transportation	--	3,750	580	--	--	--	--	--	17
Electrical generation	547	924	1,940	--	427	343	--	--	--
Synthetic gas	--	25	--	--	--	--	--	--	--
Synthetic liquids	--	--	--	--	--	--	--	--	--
Total	736	7,433	19,990	--	427	343	--	120	2,769
1985									
Household & Commercial	3	1,427	8,250	--	--	--	22	230	2,290
Industrial	190	1,516	9,220	--	--	--	23	270	1,647
Transportation	--	4,174	580	--	--	--	65	--	23
Electrical generation	704	1,123	1,460	--	1,139	391	--	--	--
Synthetic gas	26	25	--	--	--	--	--	--	--
Synthetic liquids	--	--	--	157	--	--	--	--	--
Total	923	8,265	19,510	157	1,139	391	110	500	3,960
2000									
Household & Commercial	--	1,442	8,740	--	--	--	170	2,160	4,320
Industrial	228	1,878	8,740	--	--	--	225	2,510	4,302
Transportation	--	5,103	580	--	--	--	605	--	28
Electrical generation	941	851	970	--	5,120	605	--	--	--
Synthetic gas	300	--	--	--	--	--	--	--	--
Synthetic liquids	91	--	--	1,040	--	--	--	--	--
Total	1,560	9,274	19,030	1,040	5,120	605	1,000	4,670	8,650

1/ Includes anthracite, bituminous, and lignite.
2/ Petroleum products refined and processed from crude.
3/ Miscellaneous and unaccounted energy assigned to this sector. For 1974 this was 37.9 million barrels of petroleum product.

Source: Walter G. Dupree, Jr. and John S. Corsentino, *United States Energy Through the Year 2000 (Revised)*, Bureau of Mines, U.S. Department of the Interior, December 1975.

United States Consumption of Energy Resources by Major Sources and Consuming Sectors, 1974 Preliminary and Projected to the Year 2000
(Trillion Btu)

Consuming Sector	Coal	Petroleum	Natural Gas	Oil Shale	Nuclear Power	Hydropower and Geothermal	Total Gross Energy Inputs	Synthetic Liquids Distributed	Synthetic Gas Distributed	Utility Electricity Distributed	Total Net Energy Inputs
1974											
Household & Commercial	291	6,390	7,116	--	--	--	13,797	--	--	3,687	17,484
Industrial 1/	4,208	6,044	11,129	--	--	34	21,415	--	--	2,665	24,081
Transportation	2	17,608	664	--	--	--	18,274	--	--	16	18,290
Electrical generation	8,668	3,448	3,328	--	1,173	3,018	19,635	--	--	--	--
Synthetic gas	--	--	--	--	--	--	--	--	--	--	--
Synthetic liquids	--	--	--	--	--	--	--	--	--	--	--
Total	13,169	33,490	22,237	--	1,173	3,052	73,121	--	--	6,368	59,855
1980											
Household & Commercial	100	7,600	8,000	--	--	--	15,700	--	110	5,790	21,600
Industrial	4,800	7,500	10,000	--	--	--	22,300	--	--	3,600	25,900
Transportation	--	20,700	600	--	--	--	21,300	--	--	60	21,360
Electrical generation	12,250	5,100	2,000	--	4,550	3,800	27,700	--	--	--	--
Synthetic gas	--	140	--	--	--	--	140	--	--	--	--
Synthetic liquids	--	--	--	--	--	--	--	--	--	--	--
Total	17,150	41,040	20,600	--	4,550	3,800	87,140	--	110	9,450	68,860
1985											
Household & Commercial	100	7,880	8,500	--	--	--	16,480	120	210	7,810	24,620
Industrial	4,930	8,370	9,500	--	--	--	22,800	130	240	5,620	28,790
Transportation	--	23,040	600	--	--	--	23,640	360	--	80	24,080
Electrical generation	15,700	6,200	1,500	--	11,840	3,850	39,090	--	--	--	--
Synthetic gas	520	140	--	--	--	--	660	--	--	--	--
Synthetic liquids	--	--	--	870	--	--	870	--	--	--	--
Total	21,250	45,630	20,100	870	11,840	3,850	103,540	610	450	13,510	77,490
2000											
Household & Commercial	--	7,960	9,000	--	--	--	16,960	940	1,940	14,740	34,580
Industrial	5,910	10,370	9,000	--	--	--	25,280	1,230	2,260	14,680	43,450
Transportation	--	28,170	600	--	--	--	28,770	3,330	--	100	32,200
Electrical generation	20,700	4,700	1,000	--	46,080	6,070	78,550	--	--	--	--
Synthetic gas	6,000	--	--	--	--	--	6,000	--	--	--	--
Synthetic liquids	2,140	--	--	5,730	--	--	7,870	--	--	--	--
Total	34,750	51,200	19,600	5,730	46,080	6,070	163,430	5,500	4,200	29,520	110,230

1/ Miscellaneous and unaccounted energy assigned to this sector. For 1974 this was 219 trillion Btu of petroleum products.

Source: Walter G. Dupree, Jr. and John S. Corsentino, *United States Energy Through the Year 2000* (*Revised*), Bureau of Mines, U.S. Department of the Interior, December 1975.

U.S. ELECTRICITY DEMAND FORECASTS

Projections of the consumption of electricity in the year 2000 vary from 3000 billion kilowatt hours/year to 10,000 billion kilowatt hours/year (consumption in 1971 was 1600 billion kilowatt hours). The growth rate for electricity consumption in recent years has been near 7%/year. High growth projections assume a continuation of this historic trend while low growth projections assume growth rates nearer 5%/year.

The data given in the tables are based on a growth rate of 6.1% for net electricity generation but a lower growth rate, 5.6%, for the energy inputs required to generate this electricity based on an assumed improvement of 17.5% in the efficiency of generation.

Nuclear power is projected to provide 51.1% of the installed electrical generation capacity in the year 2000.

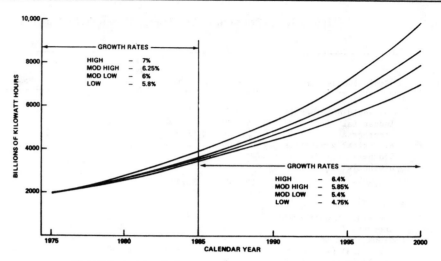

United States electricity consumption (1975 interim projections).

Source: Roger W. A. Legassie, Testimony at Hearings on Growth Rates of Electricity and the Role of Nuclear Energy, United States Committee on Interior and Insular Affairs, Subcommittee on Energy and the Environment, April 28, 1975.

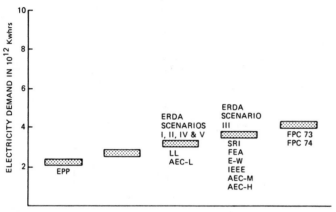

Estimated U.S. electricity demand, year 1985.

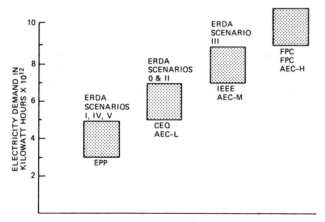

Estimated U.S. electricity demand year 2000.

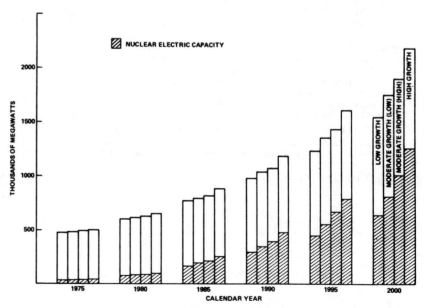

United States electrical system capacity (1975 interim projections).

Source: A National Plan for Energy Research, Development and Demonstration: Creating Energy Choices for the Future, Volume 1: The Plan, ERDA-48, Energy Research and Development Administration, June 1975.

Source: Roger W. A. Legassie, Testimony at Hearings on Growth Rates of Electricity and the Role of Nuclear Energy, United States Committee on Interior and Insular Affairs, Subcommittee on Energy and the Environment, April 28, 1975.

Electric Utility Industry Projections to the Year 2000

Year and type of plant	Installed generating capacity (MW)	Apparent capacity factor	Net generation (billion kWh)	Heat rate (Btu/kWh)	Energy resource inputs (trillion Btu)
1974:[1]					
Fuel-burning plants[2]	379,744	0.44	1,466	10,535	15,444
Nuclear plants	31,662	.39	110	10,660	1,173
Hydropower and geothermal plants	63,168	.53	291	10,389	3,018
All plants	474,574	.45	1,867	10,517	19,635
1980:					
Fuel-burning plants	484,000	.46	1,950	9,930	19,350
Nuclear plants	75,000	.65	427	10,660	4,550
Hydropower plants[3]	88,000	.50	385	9,500	3,650
Geothermal plants	1,000	.80	7	21,690	150
All plants	648,000	.49	2,769	10,000	27,700
1985:					
Fuel-burning plants	603,000	.46	2,430	9,640	23,400
Nuclear plants	200,000	.65	1,139	10,400	11,840
Hydropower plants[3]	94,000	.45	370	9,200	3,400
Geothermal plants	3,000	.80	21	21,690	450
All plants	900,000	.50	3,960	9,870	39,090
2000:					
Fuel-burning plants	824,000	.41	2,960	8,930	26,400
Nuclear plants	900,000	.65	5,085	9,060	46,080
Hydropower plants[3]	153,000	.40	535	8,500	4,550
Geothermal plants	10,000	.80	70	21,690	1,520
All plants	1,887,000	.52	8,650	9,080	78,550

[1] Preliminary.
[2] Includes steam, internal combustion, and gas turbine plants. Heat rate based on energy inputs to all fuel-burning plants.
[3] Includes pumped storage.
Source: Dupree, Walter G., and John S. Corsentino, *United States Energy Through the Year 2000*, Revised, Washington, D.C., U.S. Department of the Interior, Bureau of Mines, 1975.

Source: Energy Perspectives 2, U.S. Department of the Interior, June 1976.

Fossil Fuel Inputs to the Electrical Sector 1974[p] and Projected to the Year 2000

	1974	1980	1985	2000
Coal				
Million short tons	390.6	547	704	941
Trillion Btu	8,668	12,250	15,700	20,700
Percent of total [1]	56.1	63.3	67.1	78.4
Natural Gas				
Billion cubic feet	3,260	1,940	1,460	970
Trillion Btu	3,328	2,000	1,500	1,000
Percent of total [1]	21.5	10.3	6.4	3.8
Petroleum				
Million barrels	554	830	1,000	750
Trillion Btu	3,448	5,100	6,200	4,700
Percent of total [1]	22.4	26.4	26.5	17.8
Total				
Trillion Btu	15,444	19,350	23,400	26,400

[p] Preliminary.
[1] Percentages may not add to 100 due to independent rounding. Based on Btu input.

Source: Walter G. Dupree, Jr. and John S. Corsentino, *United States Energy Through the Year 2000* (*Revised*), Bureau of Mines, U.S. Department of the Interior, December 1975.

Comparative Analysis of Electrical Sector Energy Inputs, 1975–2000, for the 1972 and 1975 Versions of *United States Energy Through the Year 2000* (Trillion Btu)

	1980	1985	2000
Coal			
1972 Study	10,660	14,220	17,520
1975 Study	12,250	15,700	20,700
Difference [1]	-1,590	-1,480	-3,180
Percent decline [1]	-14.9	-10.4	-18.2
Liquid Hydrocarbons			
1972 Study	5,000	6,650	5,040
1975 Study	5,100	6,200	4,700
Difference [1]	-100	450	340
Percent decline [1]	-2.0	6.8	6.8
Gaseous Fuels			
1972 Study	3,600	3,450	2,640
1975 Study	2,000	1,500	1,000
Difference [1]	1,600	1,950	1,640
Percent decline [1]	44.4	56.5	62.1
Nuclear Energy			
1972 Study	6,720	11,750	49,230
1975 Study	4,550	11,840	46,080
Difference [1]	2,170	-90	3,150
Percent decline [1]	32.3	.7	6.4
Hydropower [2]			
1972 Study	3,990	4,320	5,950
1975 Study	3,650	3,400	4,550
Difference [1]	340	920	1,400
Percent decline [1]	8.5	21.3	23.5
Geothermal [3]			
1972 Study	-	-	-
1975 Study	150	450	1,520
Difference [1]	-	-	-
Percent decline [1]	-	-	-
Total Energy Inputs			
1972 Study	29,970	40,390	80,380
1975 Study	27,700	39,090	78,550
Difference [1]	2,270	1,300	1,830
Percent decline [1]	7.6	3.2	2.3

[1] Negative figure indicates increase.
[2] 1972 version included pumped storage.
[3] Not specifically broken out in 1972 version of United States Energy Through the Year 2000.

Source: Walter G. Dupree, Jr. and John S. Corsentino, *United States Energy Through the Year 2000* (*Revised*), Bureau of Mines, U.S. Department of the Interior, December 1975.

United States Electric Utility Power Statistics (on population, capacity, and consumption basis) 1974 Preliminary and Projected to the Year 2000

	1974	1980	1985	2000
Generating capacity (Thousand MW)	474.6	648.0	900.0	1,887.0
Population (Millions)	211.9	224.1	235.7	264.4
Kilowatt capacity per capita	2.2	2.9	3.8	7.1
Power consumption (Billion kWh)	1,867	2,769	3,960	8,650
Annual per capita consumption (kWh)	8,811	12,356	16,800	32,715
Nuclear power capacity (Percent of total)	6.7	11.6	22.2	47.7

Source: Walter G. Dupree, Jr. and John S. Corsentino, *United States Energy Through the Year 2000* (*Revised*), Bureau of Mines, U.S. Department of the Interior, December 1975.

U.S. ENERGY DEMAND FLOW CHARTS

In a booklet published by the Joint Committee on Atomic Energy, U.S. Congress, changes in energy flow patterns have been graphically compared. Two charts from that booklet, illustrating energy flows in 1970 and 1985, project an increase in imported oil from 3.5 million barrels/day to 14.6 million barrels/day, a drop in the efficiency of total energy use, and relatively little increase in the consumption of natural gas. Increased energy demand in 1985 is met by coal, imported oil, and nuclear power.

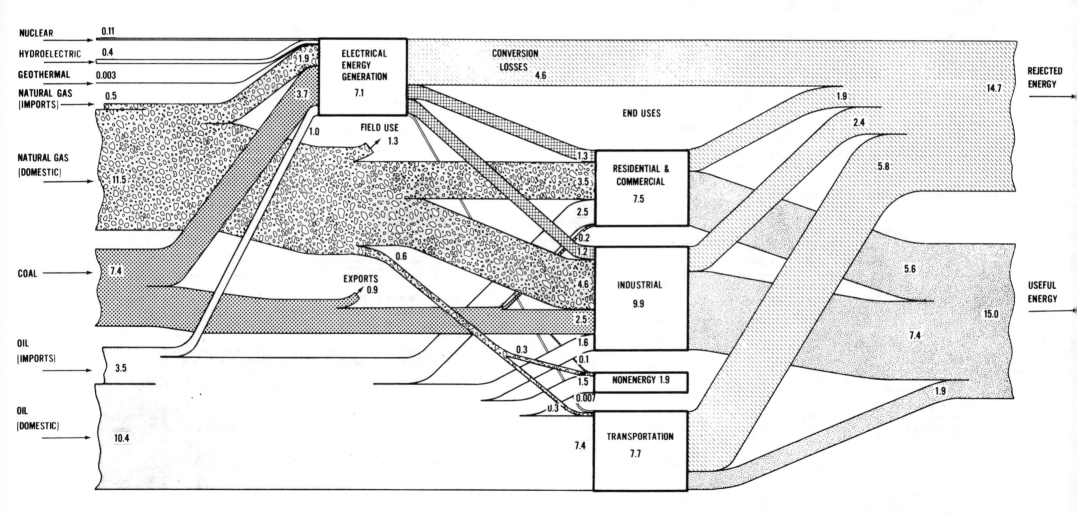

(UNITS: MILLION BBLS./DAY OIL EQUIVALENT)

Source: Certain Background Information for Consideration When Evaluating the "National Energy Dilemma," Joint Committee on Atomic Energy, U.S. Government Printing Office, Washington, 1973.

1985

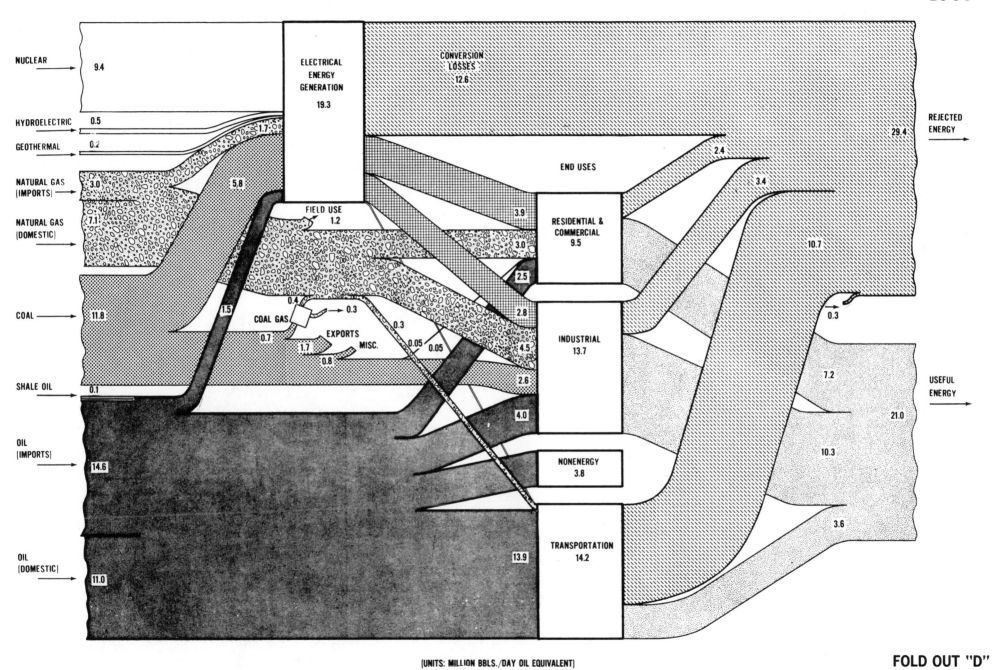

(UNITS: MILLION BBLS./DAY OIL EQUIVALENT)

FOLD OUT "D"

Source: Certain Background Information for Consideration When Evaluating the "National Energy Dilemma," Joint Committee on Atomic Energy, U.S. Government Printing Office, Washington, 1973.

Recovery of Fossil Fuels

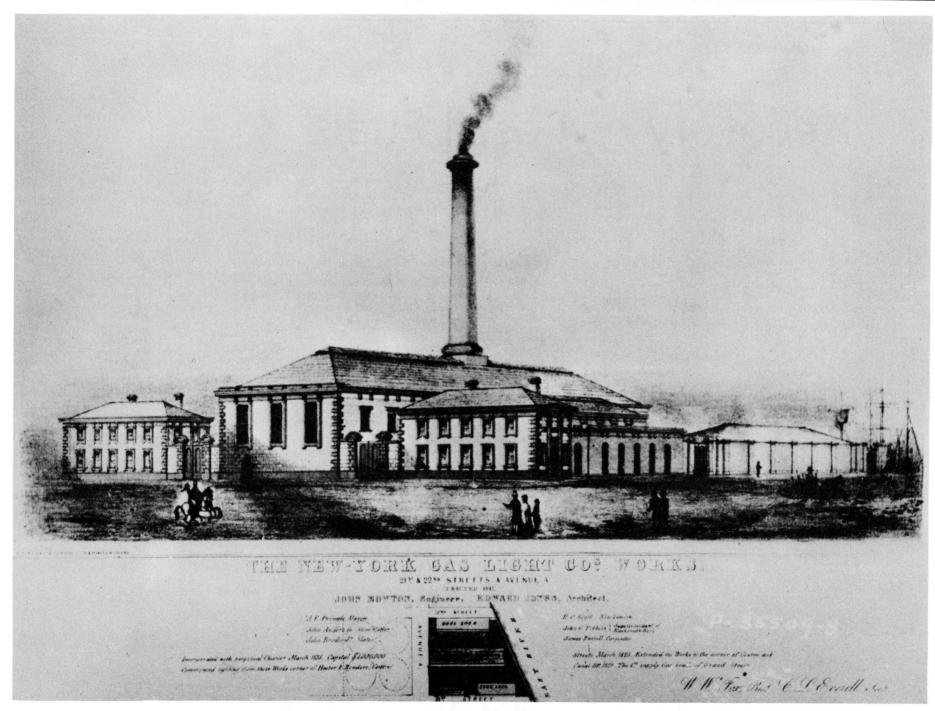

This early gas manufacturing plant was erected in New York in 1847. (*Photo courtesy Consolidated Edison Co. of N.Y.*)

COAL MINING METHODS

Depending on the orientation of coal seams relative to the surface terrain, access to deep underground coal beds may be made by slope, drift, or vertical shafts. Mining techniques include room and pillar, longwall, and shortwall. In room and pillar mining, pillars of coal are left in place in the mined areas to provide support for the overlying rock. The average national recovery rate using the room and pillar technique is 50 to 60% of the coal in place. Longwall mining involves the use of a coal cutter and conveyor which continuously traverses a long block of coal. Roof support is provided by self-advancing hydraulic props which protect the machine and the operators in the working location. The mine roof is permitted to cave, however, immediately after mining. Coal recovery with this method is 80 to 95%, production rates are high but so are capital costs. Not all deposits are suited for longwall mining. The shortwall mining method is similar to longwall mining except that the longwall coal cutter and conveyor are replaced by conventional continuous-mining machines and shuttle cars.

Surface mining methods include area stripping, contour stripping, and auger mining. Area stripping is used in flat or slightly rolling terrain where the coal beds are continuous over large areas and are near the surface of the ground. Recovery rates average about 80% but can be as high as 90%. Contour stripping is used in mountainous regions such as the narrow valleys of Appalachia. Bench cuts are made in the sloping hillsides at the level of the coal seam. Overburden is removed and dumped on the downslope. The depth of the bench cut which can be made into the hillside is limited by the height of the highwall on the upslope side. As a result, the extent of recovery of coal is influenced by the steepness of the slopes and varies over a wide range. Contour strip mines are usually small and short-lived.

Auger mining is used to remove coal from coal seams that are exposed along steep slopes. Rotating, spiral cutting tools, 18 inches to 7 feet in diameter, are employed which can penetrate to depths of about 200 feet. The rate of recovery from auger mining is variable, depending largely on the ground slope and the size of the coal seams. Recovery rates are usually 20 to 25%. About 3% of United States coal production is from auger mining.

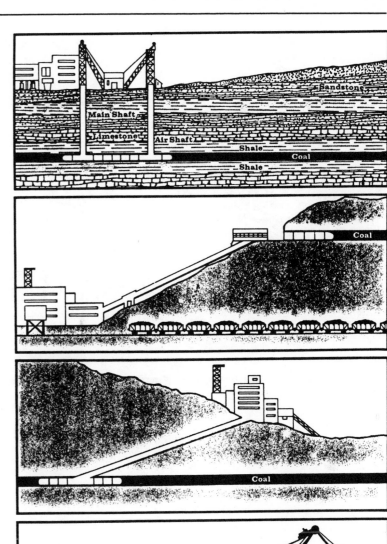

Shaft Mine

Drift Mine

Slope Mine

Surface Mine

Four types of bituminous coal mines.

Source: Coal Facts, 1974-1975, National Coal Association, Washington, D.C.

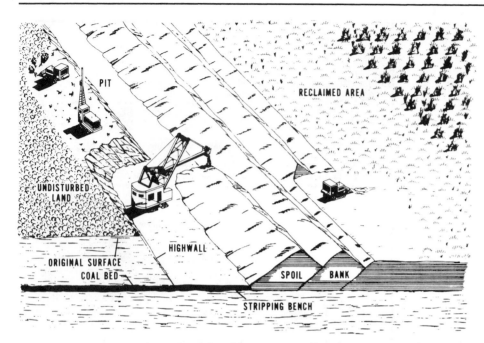

Area strip mining with concurrent reclamation.

Source: Synthetic Liquid Fuels Development: Assessment of Critical Factors, ERDA 76-129/1, Energy Research & Development Administration, 1976.

Strip mining, shovels. Dipper of a surface coal mining shovel dwarfs a mechanic, left, as it lifts 140 cubic yards—about 210 tons—of earth and rock. Once every minute the shovel removes this much overburden, about three railroad carloads, from above a coal seam near Marissa, Ill. Smaller shovels load the coal into trucks. (*Courtesy Bucyrus-Erie Co., November, 1964*)

Loading, mines. The steel arms of this high capacity loading machine sweep the coal onto a conveyor belt and into a shuttle car (seen at left) for delivery to the mine transportation system. Only 33 inches high, this machine works seams 38 inches or higher. It has a gathering arm reach of 106 inches and a rated loading speed of 15 to 30 tons a minute. The loader is a Joy model 14BU10-31B and the scene is at Consolidation Coal Co.'s Montour #4 Mine in Pennsylvania. (*Courtesy Joy Manufacturing Co., Dec., 1967*)

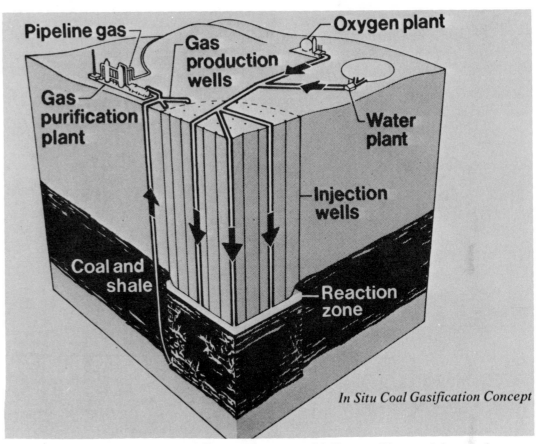

In Situ Coal Gasification Concept

In situ coal gasification. This sketch illustrates the concept of the Lawrence Livermore Laboratory for the gasification of deep coal deposits. A chemical explosive, aluminized ammonium nitrate-fuel oil, would be used to create a zone of fractured coal. An oxygen-steam mixture would be forced through the fractured zone, partially burning the coal and converting the gases to methane, hydrogen, and carbon monoxide.

Source: U.S. Atomic Energy Commission.

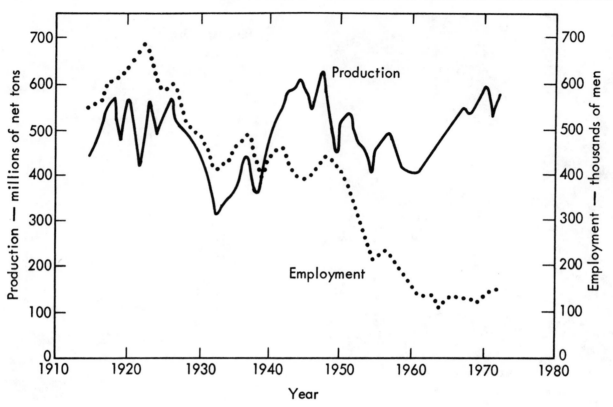

History of soft coal employment and production.

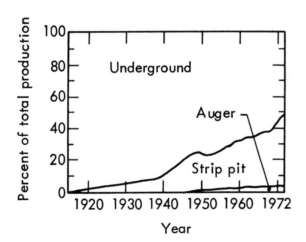

Production by mine type of U.S. soft coal.

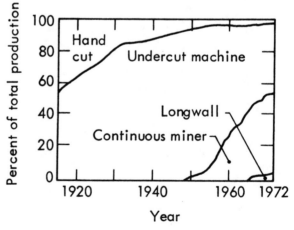

Underground production of U.S. soft coal by mining method.

Source: An Assessment of U.S. Energy Options for Project Independence, UCRL-51638, prepared for the U.S. Atomic Energy Commission by the Lawrence Livermore Laboratory, September 1, 1974.

COAL MINING PRODUCTIVITY

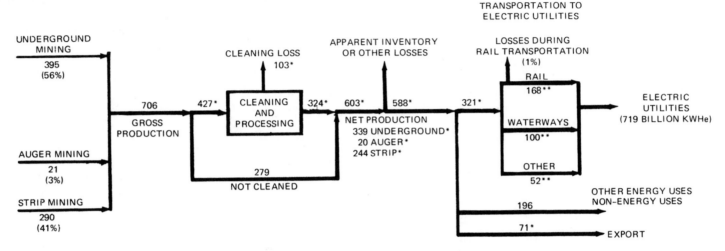

*NOTE: NUMBERS ARE IN UNITS OF MILLIONS OF TONS OF BITUMINOUS COAL.
NUMBERS WITH ONE ASTERISK (*) WERE OBTAINED FROM NATIONAL COAL ASS'N. (NCA),
BITUMINOUS COAL DATA, 1971 ED. TRANSPORTATION FIGURES (**) WERE ESTIMATED
FROM NCA, COAL TRAFFIC ANNUAL, 1970 ED. ALL OTHER FIGURES WERE INFERRED
FROM BITUMINOUS COAL DATA, 1971 ED. SEE FOOTNOTE (45)

U.S. production and disposition of bituminous coal, 1970.

Source: Comparative Risk-Cost-Benefit Study of Alternative Sources of Electrical Energy, WASH-1224, United States Atomic Energy Commission, December 1974.

Output Per Man Per Day at U.S. and European Coal Mines
(Net Tons of 2,000 Pounds)[1]

	United States Bituminous Mines				Anthracite Mines	United Kingdom	Belgium[2]	France	Netherlands	West Germany	Poland
	Underground	Strip	Auger	Average All Mines							
1920	4.50	9.91	....	4.56	2.00	1.24	0.74	0.87	1.13	....	1.20
1936	4.42	13.91	....	4.62	2.79	1.71	1.29	1.43	2.94	....	3.09
1937	n.a.	n.a.	....	4.69	2.77	1.69	1.26	1.36	2.81	....	2.99
1938	4.60	15.00	....	4.89	2.79	1.67	1.20	1.35	2.61	....	n.a.
1939	4.92	14.68	....	5.25	3.02	1.67	1.23	1.40	2.62	....	n.a.
1940	4.86	15.63	....	5.19	3.02	1.62	1.22	1.28	2.24	....	n.a.
1941	4.83	15.59	....	5.20	3.04	1.61	1.16	1.17	2.18	....	n.a.
1942	4.74	15.52	....	5.12	2.95	1.57	1.07	1.15	1.91	....	n.a.
1943	4.89	15.15	....	5.38	2.78	1.54	0.97	1.06	1.92	....	n.a.
1944	5.04	15.89	....	5.67	2.79	1.50	0.86	0.94	1.77	n.a.	n.a.
1945	5.04	15.46	....	5.78	2.79	1.49	0.92	0.97	1.34	n.a.	n.a.
1946	5.43	15.73	....	6.30	2.84	1.55	0.90	1.02	1.74	n.a.	1.77
1947	5.49	15.93	....	6.42	2.78	1.61	0.95	1.05	1.81	1.32	1.86
1948	5.31	15.28	....	6.26	2.81	1.64	0.97	1.07	1.85	1.40	1.99
1949	5.42	15.33	....	6.43	2.87	1.72	1.02	1.20	1.91	1.50	2.02
1950	5.75	15.66	....	6.77	2.83	1.78	1.02	1.32	1.93	1.54	n.a.
1951	6.08	16.02	....	7.04	2.97	1.80	1.07	1.43	1.91	1.61	n.a.
1952	6.37	16.77	20.07	7.47	3.06	1.77	1.15	1.49	1.77	1.63	n.a.
1953	7.01	17.62	25.30	8.17	3.28	1.78	1.18	1.66	1.73	1.61	1.95
1954	7.99	19.64	24.12	9.47	4.02	1.78	1.21	1.66	1.65	1.64	1.91
1955	8.28	21.12	22.22	9.84	3.96	1.76	1.27	1.74	1.64	1.70	1.86
1956	8.62	21.18	24.85	10.28	4.25	1.76	1.28	1.81	1.65	1.72	1.81
1957	8.91	21.64	26.19	10.59	4.18	1.76	1.27	1.85	1.65	1.75	1.78
1958	9.38	21.54	28.15	11.33	4.36	1.81	1.27	1.85	1.68	1.81	1.85
1959	10.08	22.65	28.77	12.22	5.12	1.90	1.39	1.89	1.78	2.03	1.91
1960	10.64	22.93	31.36	12.83	5.60	2.17	1.58	1.98	1.97	2.27	1.98
1961	11.41	25.00	30.61	13.87	5.63	2.27	1.70	2.07	2.27	2.43	2.06
1962	11.97	26.76	36.51	14.72	5.92	2.24	2.00	2.12	2.28	2.61	2.13
1963	12.78	28.69	38.87	15.83	6.27	2.36	2.01	2.16	2.30	2.78	2.21
1964	13.74	29.29	42.63	16.84	6.11	2.46	1.94	2.25	2.36	2.88	2.29
1965	14.00	31.98	45.85	17.52	6.55	2.58	2.07	2.25	2.42	2.98	2.38
1966	14.64	33.57	44.43	18.52	6.60	2.62	2.20	2.32	2.48	3.22	2.46
1967	15.07	35.17	46.48	19.17	7.21	3.15	2.32	2.47	2.61	3.60	2.63
1968	15.40	34.24	40.46	19.37	7.62	3.46	2.46	2.59	2.72	3.89	2.67
1969	15.61	35.71	39.88	19.90	7.45	3.64	2.65	2.91	3.11	4.04	2.75
1970	13.76	35.96	34.26	18.84	7.10	3.63	2.93	2.78	3.43	4.14	2.81
1971	12.03	35.69	39.00	18.02	6.30	3.72	2.89	2.89	3.53	4.22	2.91
1972	11.91	35.95	43.00	17.74	6.88	3.81	2.91	2.99	3.61	4.68	3.68
1973	11.20	34.60	41.10	16.76	n.a.	3.87	2.81	3.95	3.02	4.76	n.a.

[1] Table does not include lignite mines except for the United States where such data are included in bituminous. The European data do not include above-ground workers at underground mines or any workers at surface mines. The United States underground figures include above-ground workers. Data for the United States relate to man-days worked; those for other countreis relate to man-shifts worked. Although conditions in each country are different, the data represent trends.

[2] The calculation of output per man shift in Belgium takes into account the shifts worked by under-officials, except for 1966-1973. This is not the practice in other countries.

n.a. — Not available.

Source: U.S. Bureau of Mines and Annual Bulletin of Coal Statistics for Europe, ECE.

Source: Coal Facts 1974-1975, National Coal Association, Washington, D.C.

50 Biggest Bituminous Mines in 1973
The 50 biggest bituminous coal mines in 1973 produced 146,453,223 tons or 24.8% of the estimated total production. 43 were commercial operations; 7 captive, 25 increased their output over the previous year.

	COMPANY	NAME OF MINE	STATE	PRODUCTION 1973	1972	1950
1.	Utah International Inc.	Navajo (s)	N.M.	7,389,321	6,898,262	New 1963
2.	Peabody Coal Co.	River King (s)	Ill.	6,526,267	6,775,551	New 1957
3.	Peabody Coal Co.	Sinclair (s)	Ky. W.	5,290,991	5,476,921	New 1962
4.	Southwestern Illinois Coal Corp.	Captain (s)	Ill.	4,451,313	4,481,000	New 1964
5.	Consolidation Coal Co. Central Div.	Egypt Valley (s)	Ohio	4,256,821	3,822,173	New 1967
6.	Western Energy Co.	Colstrip (s) (c)	Mont.	4,253,681	5,500,700	New 1968
7.	Peabody Coal Co.	River Queen (d&s)	Ky. W.	4,172,223	4,660,542	New 1957
8.	Peabody Coal Co.	No. 10	Ill.	4,147,069	4,693,393	New 1952
9.	Peabody Coal Co.	Lynnville (s)	Ind.	4,064,910	4,173,500	New 1955
10.	Clinchfield Div., Pittston Co.	Moss No. 3	Va.	3,902,707	4,582,152	New 1958
11.	Central Ohio Coal Co.	Muskingum (s) (c)	Ohio	3,667,844	4,309,953	New 1952
12.	Peabody Coal Co.	Black Mesa (s)	Ariz.	3,246,500	2,953,654	New 1970
13.	Washington Irrigation Dist.	Centralia (s) (c)	Wash.	3,229,176	2,596,729	New 1970
14.	Amax Coal Co.	Ayrgem (s)	Ky. W.	3,206,242	3,183,229	New 1969
15.	Peabody Coal Co.	Ken (d & s)	Ky. W.	3,202,350	2,771,077	672,357
16.	Peabody Coal Co.	Universal (s)	Ind.	3,043,781	2,253,106	New 1970
17.	Amax Coal Co.	Leahy (s)	Ill.	2,942,035	2,356,216	New 1971
18.	Pacific Power & Light Co.	Dave Johnston (c)	Wyo.	2,897,383	2,618,137	New 1958
19.	United States Steel Corp.	Robena (c)	Pa.	2,871,000	3,008,000	3,137,832
20.	Arch Minerals Corp.	Seminoe No. 1 (s)	Wyo.	2,865,100	2,060,506	New 1971
21.	Monterey Coal Co.	Monterey No. 1 (c)	Ill.	2,694,505	1,974,000	New 1970
22.	Consolidation Coal Co. Christopher Div.	Humphrey No. 7	W. Va.	2,691,889	3,007,689	New 1956
23.	C&K Coal Co.	Fox (s)	Pa.	2,620,000	2,300,000	New 1962
24.	Peabody Coal Co.	Camp No. 1	Ky. W.	2,619,513	1,185,289	New 1971
25.	Inland Steel Co.	Inland (c)	Ill.	2,588,482	2,126,257	New 1968
26.	Kemmerer Coal Co.	Sorensen (s)	Wyo.	2,546,435	1,843,596	New 1963
27.	Peabody Coal Co.	Homestead (s)	Ky. W.	2,448,696	2,468,618	New 1963
28.	Peabody Coal Co.	Vogue (s)	Ky. W.	2,412,445	2,810,869	589,608
29.	Consolidation Coal Co. Mountaineer Div.	Robinson Run	W. Va.	2,401,441	1,836,821	New 1968
30.	Old Ben Coal, Inc.	Enos (s)	Ind.	2,396,288	2,137,383	1,224,190
31.	Old Ben Coal Corp.	No. 24	Ill.	2,377,482	2,506,740	New 1965
32.	Florence Mining Co.	No. 1	Pa.	2,353,371	2,326,630	New 1967
33.	Consolidation Coal Co. Ohio Valley Div.	Ireland	W. Va.	2,342,829	2,300,291	New 1956
34.	Freeman Coal Mining Corp.	Orient No. 3	Ill.	2,207,429	2,443,969	New 1950
35.	Consolidation Coal Co. Mountaineer Div.	Loveridge	W. Va.	2,184,618	2,438,908	New 1958
36.	Eastern Associated Coal Corp.	Kopperston No. 1 & 2	W. Va.	2,167,186	2,521,472	1,463,827
37.	Peabody Coal Co.	Sunnyhill	Ohio	2,141,238	2,170,603	New 1960
38.	Old Ben Coal Corp.	No. 26	Ill.	2,100,316	2,041,351	New 1968
39.	Amax Coal Co.	Wright (s)	Ind.	2,096,526	2,444,367	New 1955
40.	Mathies Coal Co.	Mathies	Pa.	2,035,869	2,205,283	886,811
41.	Amax Coal Co.	Minnehaha (s)	Ind.	2,011,682	1,808,297	419,770
42.	Old Ben Coal Corp.	No. 21	Ill.	2,004,279	1,732,530	New 1960
43.	Peabody Coal Co.	Star	Ky. W.	1,998,867	1,493,537	New 1971
44.	Itmann Coal Co.	Itmann	W. Va.	1,986,996	1,895,631	103,543
45.	Peabody Coal Co.	Big Sky (s)	Mont.	1,971,643	1,601,181	New 1970
46.	Peabody Coal Co.	Squaw Creek (s)	Ind.	1,961,214	2,086,731	New 1958
47.	Consolidation Coal Co. Midwest Div.	Hillsboro	Ill.	1,887,638	1,990,686	New 1965
48.	Consolidation Coal Co. Christopher Div.	Osage No. 3	W. Va.	1,861,562	2,127,831	877,807
49.	Pittsburg & Midway Coal Mining Co.	Colonial (s)	Ky. W.	1,846,098	1,991,303	678,058
50.	Freeman Coal Mining Corp.	Orient No. 6	Ill.	1,829,970	1,843,546	New 1968
	TOTAL PRODUCTION, 50 MINES			146,453,223	142,835,710	10,053,803
	U.S. TOTAL PRODUCTION, Bituminous and Lignite			591,000,000	595,386,000	516,311,000
	% LISTED MINES TO NATIONAL TOTAL			24.8%	23.9%	1.9%

(c) Captive Mines   (s) Strip Mines   (d) Deep Mines—tonnage not coded is deep.

Source: Keystone Coal Industry Manual

Source: Coal Facts 1974-1975, National Coal Association, Washington, D.C.

COAL CLEANING

Coal cleaning processes are designed to convert raw coal from the mines into a coal product which is of uniform quality. In particular, these processes are used to remove some of the sulfur in the coal. The introduction of mechanized mining has given special impetus to coal cleaning and preparation since such mining methods do not differentiate well between coal and impurities.

The physical cleaning of coal involves crushing, grinding, and washing of coal so that materials of different density than coal can be removed. The sulfur in coal occurs in three forms: as organic, sulfate, or pyritic sulfur. Organic sulfur is chemically bound in the coal and generally cannot be removed by physical separation. Organic sulfur represents from 30 to 70% of the total sulfur in most coals. Sulfates in coal are normally water-soluble oxidation products and are, therefore, readily removed in coal washing operations. Sulfate content in coal is usually less than 0.05%. Pyritic sulfur in coal is in the form of the min-

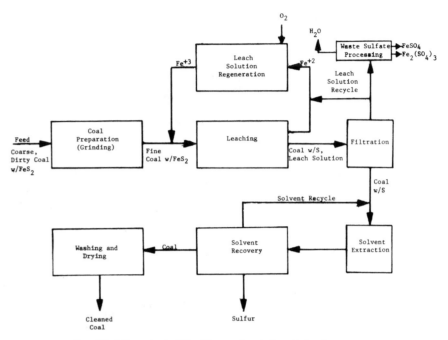

Simplified flow chart of the Meyers chemical coal cleaning process.

Source: L. Lorenzi, Jr., "Chemical Coal Cleaning," ENV/SEM. 4/Aa.6, Economic Commission for Europe, Second Seminar on Desulphurization of Fuels and Combustion Gases, Washington, D.C. November 1975.

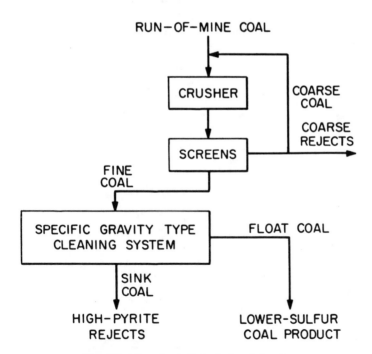

Simplified flow chart of physical coal cleaning.

Source: Proposed Final Environmental Statement Liquid Metal Fast Breeder Reactor Program, Volume III, WASH-1535, U.S. Atomic Energy Commission, December 1974.

eral pyrite which occurs as discrete, although often small, particles. Pyrite is a heavy mineral with a specific gravity of about 5.0 in contrast to the specific gravity of 1.6 for coal. The pyrite content of coal can be appreciably reduced by gravity separation based on these differences in specific gravity. A number of physical cleaning devices are under development: wet concentrating tables, concentric spirals, hydrocyclones, electrokinetic separators, and froth flotation. Several processes for removal of pyritic sulfur by chemical means are also being developed. The Meyers process involves the treatment of crushed coal with a warm ferric sulfate solution in a large reactor with the introduction of air or oxygen. The chemical reaction transforms the pyrite (FeS_2) and ferric sulfate into three products: ferrous sulfate, sulfuric acid, and elemental sulfur. The liberated sulfur is removed by dissolution in a warm naphtha bath. The Battelle hydrothermal process involves the treatment of the coal with sodium hydroxide which reacts with the pyrite to produce soluble sodium sulfide (Na_2S).

COAL CHARACTERISTICS

ASTM Specifications Classify Coals According to Rank

Class	Group		Fixed carbon limits, % (Dry, mineral-matter-free basis)		Volatile matter limits, % (Dry, mineral-matter-free basis)		Calorific value limits, Btu/lb (Moist,[1] mineral-matter free basis)		Agglomerating character
			Equal or greater than	Less than	Greater than	Equal or less than	Equal or greater than	— Less than	
I	Anthracitic	1 Meta-anthracite	98	—	2	—	—	—	
		2 Anthracite	92	982	8	—	—	—	Nonagglomerating
		3 Semianthracite[2]	86	928	14		—	—	
II	Bituminous	1 Low volatile bituminous coal	78	86	14	22	—	—	
		2 Medium volatile bituminous coal	69	78	22	31	—	—	
		3 High volatile A bituminous coal	—	69	31	—	14,000[3]	—	Commonly agglomerating[4]
		4 High volatile B bituminous coal	—	—	—	—	13,000[3]	14,000	
		5 High volatile C bituminous coal	—	—	—	—	11,500	13,000	
							10,500	11,500	Agglomerating
III	Subbituminous	1 Subbituminous A coal	—	—	—	—	10,500	11,500	
		2 Subbituminous B coal	—	—	—	—	9,500	10,500	
		3 Subbituminous C coal	—	—	—	—	8,300	9,500	Nonagglomerating
IV	Lignitic	1 Lignite A	—	—	—	—	6,300	8,300	
		2 Lignite B	—	—	—	—	—	6,300	

The American Society for Testing & Materials classification index set forth in "Specification for classification of coals by rank" (ASTM D 388-66 [1972]) does not include a few coals, principally nonbanded varieties, which have unusual physical and chemical properties and which come within the limits of fixed carbon or calorific value of the high-volatile bituminous and subbituminous ranks. All of these coals either contain less than 48% dry, mineral-matter-free fixed carbon or have more than 15,500 Btu/lb on a moist, mineral-matter-free basis.

[1] Moist refers to coal containing its natural inherent moisture but ,not including visible water on the surface of the coal.
[2] If agglomerating, classify in the low-volatile group of the bituminous class.
[3] Coals having 69% or more fixed carbon on the dry, mineral-matter-free basis shall be classified according to fixed carbon, regardless of calorific value.
[4] It is recognized that there may be nonagglomerating varieties in these groups of the bituminous class, and there are notable exceptions in the high volatile C bituminous group.

Reprinted with permission from *POWER*, February, 1974.

Analysis of Typical Coals Shows Wide Variations in Composition

Coal rank[1]				Coal analysis, Bed-moisture basis						Rank FC	Rank HV
Class	Group	State	County	M[2]	VM	FC	A	S	HV		
I	1	PA	Schuylkill	4.5	1.7	84.1	9.7	0.77	12,745	99.2	14,280
I	2	PA	Lackawanna	2.5	6.2	79.4	11.9	0.60	12,925	94.1	14,880
I	3	VA	Montgomery	2.0	10.6	67.2	20.2	0.62	11,925	88.7	15,340
II	1	WV	McDowell	1.0	16.6	77.3	5.1	0.74	14,715	82.8	15,600
II	1	PA	Cambria	1.3	17.5	70.9	10.3	1.68	13,800	81.3	15,595
II	2	PA	Somerset	1.5	20.8	67.5	10.2	1.68	13,720	77.5	15,485
II	2	PA	Indiana	1.5	23.4	64.9	10.2	2.20	13,800	74.5	15,580
II	3	PA	Westmoreland	1.5	30.7	56.6	11.2	1.82	13,325	65.8	15,230
II	3	KY	Pike	2.5	36.7	57.5	3.3	0.70	14,480	61.3	15,040
II	3	OH	Belmont	3.6	40.0	47.3	9.1	4.00	12,850	55.4	14,380
II	4	IL	Williamson	5.8	36.2	46.3	11.7	2.70	11,910	57.3	13,710
II	4	UT	Emery	5.2	38.2	50.2	6.4	0.90	12,600	57.3	13,560
II	5	IL	Vermilion	12.2	38.8	40.0	9.0	3.20	11,340	51.8	12,630
III	1	MT	Musselshell	14.1	32.2	46.7	7.0	0.43	11,140	59.0	12,075
III	2	WY	Sheridan	25.0	30.5	40.8	3.7	0.30	9,345	57.5	9,745
III	3	WY	Campbell	31.0	31.4	32.8	4.8	0.55	8,320	51.5	8,790
IV	1	ND	Mercer	37.0	26.6	32.2	4.2	0.40	7,255	55.2	7,610

[1] Coal rank, as defined by ASTM D 388-66 [1972], is outlined in tabular form on p S.7 [2] Column headings are defined as follows: M is percent equilibrium moisture, VM percent volatile matter, FC percent fixed carbon, A percent ash, S percent sulfur, HV higher heating value in Btu/lb, Rank FC percent dry, mineral-matter-free fixed carbon, and Rank HV moist, mineral-matter-free heating value in Btu/lb.

Reprinted with permission from *POWER*, February, 1974.

Average Trace Elements Content of Coal and Rocks

Element	Coal, ppm	Shale, ppm	Igneous Rock ppm
Arsenic	25	13	1.8
Barium	1–3000	580	425
Beryllium	0.1–1000	3	2.8
Boron	100–1000	100	10
Copper	300	45	55
Germanium	25–3000	1.6	5.4
Gold	≤0.125	0.005	0.004
Iodine	6	2.2	0.5
Lanthanum	≤1000	20	30
Molybdenum	10	2.6	1.5
Platinum	≤0.035		0.005
Selenium	≤7	0.6	0.05
Strontium	1000	300	375
Tin	10	6	2
Uranium	1–200	3.7	2.7
Zirconium	≤250	160	165

Source of Data: H. J. M. Bowen, *Trace Elements in Biochemistry*

Average Analysis of Typical Plant Material, and Bituminous Coal Containing 7.3% Ash

Element	Original Plant Material %	Average Bituminous Coal %
Carbon	45.0	80.0
Oxygen	43.0	5.0
Hydrogen	5.5	5.0
Nitrogen	2.1	1.5
Sulfur	0.1	1.0
Phosphorus	0.2	0.2
Subtotal	95.9	92.7
Potassium	1.80	3.20
Chlorine	0.60	0.30
Silicon	0.55	1.00
Calcium	0.38	0.67
Magnesium	0.18	0.20
Sodium	0.14	0.25
Iron	0.03	0.05
Aluminum	0.02	0.03
Other	0.40	1.50
Subtotal	4.10	7.30

Source of data: H. J. M. Bowen, *Trace Elements in Biochemistry*

Source: *Survey of Energy Resources 1974*, The United States National Committee of the World Energy Conference, 345 East 47th Street, New York, N.Y. 10016, 1974.

COAL CONVERSION PROCESSES

The conversion of coal to other forms of useful fuels has long been a part of industrial technology. In the early 19th century coal was converted to gas—water gas, city gas, town gas—for street lighting and home use. The production of liquids from coal was initiated in the 1920s in Germany and became the primary source of liquid fuels in that country during World War II. With the advent of inexpensive natural gas, the use of manufactured gas made from coal rapidly declined. The availability, until recently, of cheap oil, also made the production of synthetic oil from coal economically unattractive. Several factors have contributed to a renewed interest in gaseous and liquid products made from coal: the increased cost of natural gas and petroleum, regional shortages of these natural fuels, and a need to produce environmentally acceptable fuels from coal.

Historical Perspective. The basic science on which coal gasification technology rests dates to 1670 when Reverend John Clayton, a Yorkshire clergyman, reported the generation of a luminous gas when coal was heated in a chemical retort. A century later, in 1792, William Murdoch, a Scotsman, illuminated his home with gas obtained by distilling coal in an iron retort. Lowe, in 1875, developed the cyclic air/coke–steam/coke process for producing water gas.[1]

Larger scale production of synthetic fuels from coal began about 1910 with the Fischer-Tropsch Process in Germany and is still used in the SASOL plant in South Africa to produce a variety of fuels from coal. Another process, the Lurgi[2] process, was developed in Germany in the 1930s to gasify noncaking coal with oxygen and steam to produce a gas of about 400 Btu/scf. One of the major uses for this gas was as a town gas for distribution in European cities. Other uses are to produce synthetic gas for production of ammonia, methanol, and a variety of liquid fuels and petrochemical products. During the past 50 years, some 50 Lurgi units have been put into operation.

Koppers, Inc., in cooperation with engineers of Heinrich Koppers, GmbH, Essen, Germany, designed and built the first suspension demonstration unit for gasifying coal in 1948. It was based on concepts of Dr. Friedrich Totzek. The first commercial Kopper-Totzek unit was installed in Finland in 1952.

In 1922, Dr. Fritz Winkler discovered a fluidized bed reaction process while working on a process for preparing activated carbon. A gasifier based on his work was put into operation at Leuna in present-day East Germany. Since that time, 36 gasifier units have been designed, constructed and commissioned by Bamag Verfahrenstechnik, GmbH, Cologne, Germany. Locations of Winkler units are given in 6-9.

Coal gasification operations survived the worldwide shift to natural gas in nations that wished to produce ammonia, methanol, hydrogen, urea, or other fuels or chemicals from native coals. Coal is gasified today in Turkey, India, South Africa, Scotland, Morocco, Yugoslavia, Korea,[3] Spain, and China.

The various new processes for conversion of coal to synthetic fuels under development by the Department of Energy and by private industry are designed to improve existing processes, and remove existing disadvantages. Most important among the disadvantages is the relatively high costs in comparison to the cost of fuel from natural gas during the past 25 years.

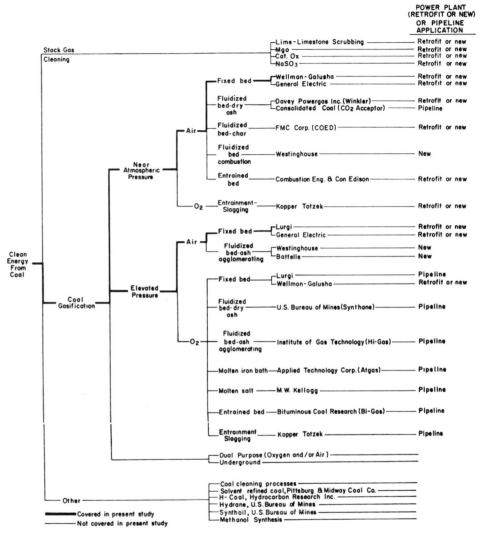

Potential routes for obtaining clean energy from coal.

Source: Evaluation of Fixed-Bed, Low-BTU Coal Gasification Systems for Retrofitting Power Plants, EPRI 203-1, prepared by the Tennessee Valley Authority for the Electric Power Research Institute, February 1975.

---

[1] Air is first blown into coke to produce combustion and raise the coke bed temperature to incandescence. The air supply is shut off and steam is introduced to produce H_2 and CO. Air is again introduced when the bed temperature falls to levels too low for H_2 and CO production.

[2] Lurgi-Gesellschaft für Warme und Chemotechnik, GmbH, Frankfurt am Main, Germany.

[3] Squires, A.M, "Clean fuels from coal gasification," *Science* 184, 340 (1974).

Source: Synthetic Fuels from Coal Task Force Report, Project Independence Blueprint, Federal Energy Administration, November 1974.

Commercial Ventures in Synthetic Fuels

	Company	Site	Process	Approximate Coal Feed (tons/day)	Plant Output (10^6 ft^3/day)	Status
1.	El Paso Natural Gas Company	Four Corners Area, New Mexico	Lurgi with Methanation	28,250	288	El Paso Natural Gas Co. plans to construct and operate the Burnham Coal Gasification Complex on the Navajo Indian Reservation. Capital costs (1973 basis) are estimated at $114 million for mining and $491 million for gasification. The average unit synthetic fuel cost over the 25-year plant life is projected to be $1.17 per 1000 std ft^3. Gas heating value is expected to be 930 Btu/ft^3.
2.	WESCO — Texas Eastern Transmission Corp. Pacific Lighting Corp. Utah International Corp. Fluor Corp. (Engr.)	Four Corners Area, San Juan Co. New Mexico	Lurgi with Methanation	22,000	250 (1000 - 4 plants)	Plans are to construct and operate 4 plants near Farmington, New Mexico. Plant cost is near $450 million. Utah Int. will supply coal and water. Site development is planned for spring 1975 and component assembly should be complete in late 1977. 1974 activities involve engineering design, preparation of FPC application, and EPA environmental impact statements.
3.	Panhandle Eastern Pipeline Co; Peabody Coal Co.	Eastern Wyoming	Lurgi with Methanation	25,000	250	Plant operation is anticipated in 1978-1980 assuming timely receipt of government authorizations. Plant cost will be near $500 million.
4.	Panhandle Eastern Pipeline Co.; Peabody Coal Co.	Southern Illinois	Lurgi with Methanation	–	–	Feasibility studies are underway.
5.	Natural Gas Pipeline Company of America	Dunn County North Dakota	Lurgi with Methanation	108,500 (4 mines)	1000 (4 plants)	Plans are for four plants and possibly eight. Mining will be from 110,000 acres of leased land in N. Dakota. Each mine will cost about $100 million and the associated gasification plant will be $370 million. The first plant is scheduled to go on line in 1982.
6.	American Natural Gas Company (North American Coal Co.)	Beulah-Hazen Area North Dakota	(High Btu)	–	1000 (4 plants)	Plans are for four plants. The state has allocated water for first plant. Each plant will cost about $450 million. First is scheduled for operation in 1980 with succeeding plants following at 3-4 year intervals.
7.	Northern Natural Gas Co.; Cities Service Gas Co.	Powder River Basin Montana	(High Btu)	–	1000 (4 plants)	Northern Natural and Cities Service plan to construct four $250 million ft^3/day coal gasification plants. Peabody Coal has agreed to supply ~800 million tons of coal and the gas companies are negotiating for another like amount. Through 1975, $10-11 million will be spent on initial development; construction could start in 1976-77 with operation in 1979-1980.
8.	Colorado Interstate Gas Corp.; Westmorland Coal Co.	Southeast Montana	(High Btu)	25,000	250	Colorado Interstate has an option on 300 million tons of coal and 10,000 acre-ft per year of water to be supplied by Westmoreland for development of a gasification project.
9.	The Columbia Gas System, Inc.	Illinois	(High Btu)	–	300	Columbia will exchange 50% of its interest in 43,400 acres of its 300,000 acres in W. Virginia for 50% interest in 35,000 acres in Illinois held by Exxon's Carter Oil Co. The Illinois coal will be held by Columbia for gasification pending development of technology in gasification. Plant cost is estimated at $635 million and three mines will cost $75 million.
10.	Consolidated Natural Gas Co.	Southwest Pennsylvania	(High Btu)	–	–	Consolidated has 70% interest in SW Pennsylvania acreage containing over 300 million tons of coal and is negotiating for more. A commitment on first gasifier is expected in 1976.

(continued)

Commercial Ventures in Synthetic Fuels (*continued*)

	Company	Site	Process	Approximate Coal Feed (tons/day)	Plant Output (10^6 ft^3/day)	Status
11.	Pennsylvania Gas and Water Co.	Pennsylvania	Hygas or similar	5,000	80	The company has proposed to Office of Coal Research a plan for financing and operating a demonstration plant.
12.	Texas Gas Transmission Co.; Consolidation Coal Co.	–	–	–	–	Feasibility studies are underway for a plant in Central United States, perhaps near Evansville, Indiana or Ohio River.
13.	Southern Natural Gas	Illinois	(High Btu)	–	250	Southern has option for coal reserves in Illinois Basin from Consolidation Coal. The option will be exercised if the FPC allows cost of coal reserves (~$28 million) in Southerns rate base.
14.	Conoco	Westfield, Scotland	Lurgi – Methanation	–	2.6	Operational; variety of coals.
15.	Northern Illionis Gas Co.; State of Illinois	Illinois	COED plus char gasification with Methanation	10,000	80-90 (plus 10,000 bbl/day syn crude)	Demonstration plant with 1980 completion goal. Office of Coal Research is being asked to provide 50% of funding of projected $250 million plant.
16.	Consolidation Coal Co.	Cresap, W. Virginia	Project Gasoline	–	–	Reactivated in 1974.
17.	Commonwealth Edison Co.; Electric Power Research Institute	Pelein, Illinois	Lurgi – no Methanation	1,400	192 (low-Btu gas)	Demonstration plant will cost $19 million; EPRI will contribute $11 million and Commonwealth Edison $8 million. 1974 construction start; plant testing in 1976.
18.	Northern States Power Co.; Pittsburg-Midway Coal Co.; Foster Wheeler Corp. (Engr.)	S. Dakota (Sioux Falls)	Low-Btu gas (Elevated Pressure Entrained Bed)	1,200	–	Northern States, P-M, and Foster Wheeler have proposed that OCR partly finance a pilot plant to first establish gasifier operation and eventually produce 60 megawatts of electric power in combined cycle.
19.	Exxon	Baytown, Texas		0.5	–	Proprietary liquefaction process in operation since 1967.
20.	Panhandle Eastern; Tennessee Gas; Consolidated Natural Gas; Republic Steel; Rocky Mountain Energy and FMC	Princeton, New Jersey	COGAS (oil)	~36	240,000 BPD	Expected to be on stream in late 70's.
21.	Same as 20	Leatherhead, England	COGAS (oil)	~36	240,000 BPD	Expected to be on stream in late 70's.
22.	Michigan Wisconsin Pipeline Co.	–	–	–	–	Leased 3 billion tons of lignite in N. Dakota from North American Coal Corp.
23.	Standard Oil of Ohio	Ohio	COED	900	–	Sohio is arranging for a privately-sponsored large-scale demonstration of a liquefaction process and a test utilization of the product in a large electric power plant.

Source: Synthetic Fuels from Coal Task Force Report, Project Independence Blueprint, Federal Energy Administration, November 1974.

COAL GASIFICATION

Coal gasification processes involve the reaction of coal at high temperature with steam and air or oxygen to produce a mixture of gases, typically carbon monoxide, carbon dioxide, hydrogen, and methane, as well as hydrogen sulfide from the reactions with the sulfur in the coal. The gases from the gasifier are further upgraded in several steps. By reaction with water, the carbon monoxide in the gas is converted to hydrogen and carbon dioxide. Carbon dioxide and hydrogen sulfide are removed in a purification system, and the hydrogen is converted to methane by reaction with carbon monoxide.

The composition of the synthesis gas from the gasifier is a function of the feed rates of

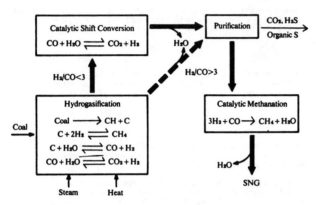

A generalized flow chart for the gasification of coal. The scheme is similar for naphtha gasification, except that desulfurization occurs first.

Source: Hammond, A. L., et al, "Energy and the Future," 1973. © American Association for the Advancement of Science.

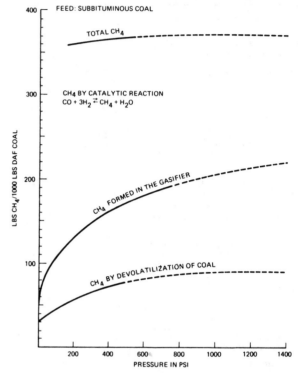

Methane Formation as a Function of Pressure. The total CH_4 generation is due to (1) devolatilization (curve 1), (2) hydrogenation (curve 2), and (3) subsequent reaction between CO and H_2 (not shown).

Source: Synthetic Fuels from Coal Task Force Report, Project Independence Blueprint, Federal Energy Administration, November 1974.

Selected Analyses of Gaseous Fuels Derived from Coal

Column No.		1	2	3	4
Analyses, % by vol					
H_2	Hydrogen	47.9	2.4	34.0	14.0
CH_4	Methane	33.9	0.1	15.5	3.0
C_2H_4	Ethylene	5.2	—	4.7	—
CO	Carbon monoxide	6.1	23.3	32.0	27.0
CO_2	Carbon dioxide	2.6	14.4	4.3	4.5
N_2	Nitrogen	3.7	56.4	6.5	50.9
O_2	Oxygen	0.6	—	0.7	0.6
C_6H_6	Benzene	—	—	2.3	—
H_2O	Water	—	3.4	—	—
Specific gravity (relative to air)		0.413	1.015	0.666	0.857
Higher heat value—Btu/cu ft					
@ 60F & 30 in. Hg		590	—	534	163
@ 80F & 30 in. Hg		—	83.8	—	—

Col. No.	Kind of Gas	Col. No.	Kind of Gas
1	Coke-oven gas	3	Carbureted water gas
2	Blast-furnace gas (lean)	4	Producer gas

Analyses of gaseous fuels derived from coal by processes in use since the early 1800's.

Source: Steam; Its Generation and Use, 38th Edition, Babcock and Wilcox, 161 East 42nd St., New York, N.Y., 10017.

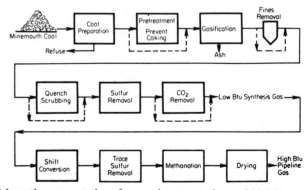

Schematic representation of processing steps to low and high Btu gases.

Source: Evaluation of Coal Conversion Processes to Provide Clean Fuels, EPRI 206-0-0, prepared by the University of Michigan, College of Engineering, for the Electric Power Research Institute, Palo Alto, California, February 1974.

oxygen (or air) and steam relative to the feed rate for coal, the gasification pressure and temperature, and the flow of the feed and product gases relative to the feed coal. High gasifier pressures promote the production of methane—an important factor if the objective is to produce a synthetic natural gas.

Differences in coal gasification systems relate to methods for handling operations at each stage: the admission of coal into pressurized systems, the handling of agglomerating coals, the removal of sulfur, the removal of ash, and the catalytic shift conversion and methanation.

Summary Review of Typical Product Compositions and Heating Values for Coal Gasifiers

GASIFICATION PROCESSES

TYPICAL PRODUCT COMPOSITIONS AND HEATING VALUES

	KOPPERS TOTZEK	LURGI Oxygen	LURGI Air	WINKLER Oxygen	WINKLER Air	BI-GAS	SYN-THANE	ATGAS	HYGAS Electro-thermal	HYGAS Oxygen	HYGAS Steam-iron	HYGAS Air	HYDRANE	CO_2 ACCEPTOR	KELLOGG SALT	U GAS	HANNA U-GROUND GASIF. Dry Basin	BYU
Carbon Monoxide, CO	50.4	9.2	13.3	25.7	19.0	22.9	10.5	69.7	21.3	18.0	7.4	13.5	3.9	14.1	26.0	17.0	9.0	37.5
Carbon Dioxide, CO_2	5.6	14.7	13.3	15.8	6.2	7.3	18.2	–	14.4	18.5	7.1	12.7	–	5.5	10.3	8.8	17.1	5
Hydrogen, H_2	33.1	20.1	19.6	32.2	11.7	12.7	17.5	9.6	24.2	22.8	22.5	16.6	22.9	44.6	34.8	11.6	21.4	39
Methane, CH_4	0	4.7	5.5	2.4	0.5	8.1	15.4	20.0	19.9	14.1	26.2	8.4	73.2	17.3	5.8	4.1	4.5	2
Ethane C_2H_4		0.5					0.6		0.8	0.5	1.0	0.6		0.37				1.5
Other Hydrocarbons																		
Water, H_2O	9.6	50.2	10.1	23.1	11.5	48.0	37.1		17.1	24.4	32.9	18.3		17.1	22.6	12.0		15
Hydrogen Sulfide, H_2S	0.3	0.6	0.6	0.25	0.13	0.7	0.3		1.3	0.9	1.5	0.8		0.03	0.2	0.6		
Carbonyl Sulfide, COS			0.1	0.04	0.02												0.11	
Nitrogen, N_2	1.0		37.5	0.8	51.1	0.3	0.5	0.7				28.9		0.2	0.3	45.4	46.4	
Other									1.0	0.8	1.4	0.2					1.5	
Ammonia, NH_3														0.8				
Heating Value Btu/scf Water-free basis																		
High	298	302	180	275	118	378	405	457	437	374	565	236	826	440	329	150	160	340
Low	279	270	163	250	111	349	368	432	396	338	507	214	740	391	298	139	145	322

1. Coal less than 1% sulfur

Source: Evaluation of Coal Conversion Processes to Provide Clean Fuels, EPRI 206-0-0, prepared by the University of Michigan, College of Engineering, for the Electric Power Research Institute, Palo Alto, California, February 1974.

COAL CONVERSION RESEARCH AND DEVELOPMENT REQUIREMENTS

Areas of Research and Development. There are potential problems and improvements needed in both the processing scheme and the equipment used for the coal gasification and liquefaction processes. A greater part of the identified problem areas appear to be common to all processes, although there are a number of technical uncertainties that are specific to the individual processes. The major areas of concern are as follows:

General:

Development of high-capacity coal mining methods. It is apparent that the larger plants discussed in this report require coal feed rates equivalent to multiples of the output of the largest currently operating coal mines.

Improved feed equipment for injecting pulverized coal into a pressurized system. Injecting coal into high pressure reactors presents serious mechanical problems. Lock hoppers for feeding coal as a dry solid have not been proven at or above pressures of 400 psig. Techniques for feeding coal in the form of a slurry have not been demonstrated on a commercial scale. In the low Btu coal gasification processes, the coal must be fed at the gasification rate otherwise complete combustion of the coal could result; also there is the danger of the combustion zone being discharged through the grate.

Development of an efficient hot char transport and injection system. Char recycle is one of the most important steps in a number of coal gasification and liquefaction process schemes. A hot char transport and injection system which will minimize erosion problems, allow operation at high temperature (above 1100°F) and high pressures, and be capable of metering precise quantities of char for recycle feed has not yet been demonstrated at pilot scale.

Development of satisfactory control equipment. For coal gasification and liquefaction plants, reliable and durable high temperature valves suitable for operation with solids or gases at temperatures up to 2100°F and pressures as high as 1500 psig in pipelines of large diameters have to be developed since completely satisfactory valves for these conditions of service are not known to exist. For coal liquefaction plants, there is the added problem of handling coal-oil slurry which causes extreme erosion of pump casings and impellers, pipe bends, and valves. Handling these slurries leads to deposition of solids in the low velocity portion (or at cold spots) in pumps, lines shutoff valves, automatic control valves or instruments resulting in sticking, plugging, and failures. Again, satisfactory equipment to handle the combination of slurry service temperatures between 600° and 875°F and pressures of 100 to 3500 psig (as are the conditions in coal liquefaction plants) do not currently exist. For coal liquefaction plants to operate reliably, improvements are required in heat exchangers for slurries, pump and valve design, metallurgy; and solids separation.

Quantitative definition of pollutants. For most coal gasification and liquefaction processes, there is lack of information on the amount of pollutants present in the waste streams. Such data are necessary for the design and evaluation of waste treatment systems.

Improved methods for conservation of water. Coal gasification and liquefaction plants are large consumers of water; the feasibility of treating process waste water for reuse within the plant facility should be explored.

Applicability of the process to other types of coal. Tests on some of the coal conversion processes have been conducted with only selected types of coal, and it is not known whether other types of coal may also be used with these processes. For example, IGT has so far only used lignite in its HYGAS pilot plant tests.

Scale-up of process equipment. Research on some of the coal conversion processes has been carried out only on bench-scale equipment to date. For example, the Synthoil process was investigated in a reactor of $\frac{5}{16}$-inch I.D. by 68 feet long. It is not known whether these bench-scale test data may be successfully extrapolated for the design of scale-up equipment.

Investigation of personnel health and safety aspects. As the synthetic fuel industry is presently nonexistent, a complete evaluation of all aspects affecting working personnel and local resident health must be accomplished. In addition, the new features of each conversion process equipment train must be evaluated in terms of personnel safety.

Gasification:

Development of methods for slag and ash removal. Plugging of slag tap holes has been experienced in a number of slagging gasifiers. The withdrawal of slag and ash from the bottom of a high temperature, elevated pressure coal gasifier or producer gas unit is a major design problem. One method proposed is to drop the slag into a water-filled quench tank and then pass the slag-water slurry through a pressurized lock hopper to the ash disposal area. There is a need to develop and test such systems.

Improvements in gas turbine design. One of the most efficient uses for low Btu gas is the generation of electricity in combined cycle power plants. However, the low Btu gas has a low heat value to mass ratio which affects the gas turbine design. Turbines presently available cannot necessarily handle low Btu gases. Also, the current turbine technology limits the inlet turbine temperature to about 1800°F, and significant increases in the efficiencies of combined cycle power generation may be realized if higher inlet temperatures are used. Turbine blade coatings may also be necessary to reduce the corrosion and erosion caused by the presence of sulfur compounds and particulates in the low Btu gas. The upper concentration limits of the sulfur compounds and the particulates in the low Btu gas, which may be tolerated in the currently available gas turbines, have not yet been determined.

Retrofit of boilers for low Btu gas. There are considerable differences of opinion on this subject. Some claim that existing boilers designed to fire a conventional fuel (gas, oil, or coal) can be readily modified to burn other, including synthetic fuels. The amount of derating a given boiler would experience depends on its original design and its vintage derating. The Electric Power Research Institute (EPRI) has initiated studies to identify gas- and oil-fired utility boilers that are candidates for retrofit. Similar studies should be conducted for industrial boilers.

Development of high temperature particulate removal system. For high Btu coal gasification, incomplete char recovery from the gasifier effluent by cyclones may lead to possible shift catalyst poisoning or char accumulation downstream. For low Btu coal gasification used for combined cycle power generation, incomplete char and particulate removal from the gasifier effluent may cause severe turbine blade corrosion and erosion. The capability of currently available cyclone systems to effect almost complete particulate removal under high temperature conditions has not yet been demonstrated.

Development of high temperature acid gas removal systems. For low Btu gas utilized in a combined cycle power generation system or as an on-site fuel gas, it will be necessary to remove the hydrogen and carbonyl sulfides at temperatures above 1100°F. Methods for removing hydrogen sulfide at such high temperatures have not yet been developed.

Investigation of material of construction problems in quenching. Water quenching is often used to scrub the heavy hydrocarbons, particulate matter, and other residue trace components from coal gasifier effluents before the acid gas removal step. In view of the high pressure and temperature and the corrosive atmosphere of the hydrogen sulfide, considerable pilot plant testing is required for investigating the problems of construction materials in quenching.

Improved refractories for coal gasifiers. The elevated pressures and temperatures of the gasifiers environment present a severe refractory service environment, especially in regions where the molten slag (3000°F) is in contact with the refractory and causes fluxing and erosion problems.

Development of methanation reactors for high Btu coal gasification. The methanation step is of considerable importance in the production of high Btu gas from coal. The methanation reaction is very exothermic, and careful temperature control is required to prevent catalyst deactivation and temperature runaway. A number of methanation reactor designs such as fixed bed, fluidized bed, tubewall reactors, packed-bed reactors, and ebullated reactors have been proposed. To date they have not been proven on a commercial scale at the high pressures and high carbon oxide contents typical of product gases from coal gasification. Although the Westfield, Scotland, coal gasification plant has claimed recent success in the operation of its demonstration methanator, the details of the testing conditions, such as composition of the influent stream, the methanation pressure and temperature, the type of catalyst used, and the reactor design, have not yet been disclosed for technical evaluation.

Improved method for pretreatment of caking coals. Present methods of pretreatment for caking coals, such as utilizing air to induce mild surface oxidation of coal particles, often lead to the generation of large quantities of pretreatment off-gas and pretreatment fines, both of which may contain relatively high levels of sulfur and are of limited use. Recycling of process char residues to the reactor as part of the coal feed may eliminate the requirement for coal pretreatment, but has yet to be demonstrated on pilot scale.

Liquefaction:

Development of improved methods of solid/liquid separation. Current approaches, derived from available chemical process industry techniques, are subject to temperature and pressure limitations and as a result require considerable steam processing in use.

Development of improved catalysts. Existing coal conversion catalysts can be improved in terms of high temperature stability, resistance to sulfur, lifetime, activity, selectivity, and costs. The achievement of these improved characteristics will permit lower temperature and pressure operation of liquefaction processes and reductions in energy consumption.

Development of materials to withstand liquefaction environments. Currently available materials are subject to degradation through exposure to sulfur compounds, with the exception of several alloys which are prohibitively expensive in the quantities needed to support a liquefaction industry. New, inexpensive alloys having improved sulfur resistance *and* good mechanical properties over a wide temperature range are required.

An examination of the state of the art of synthetic fuels technology and the technical problems and uncertainties leads to the conclusion that there are a number of R&D needs that apply generally to each type of fuel. R&D programs oriented to these needs, *in addition to the process development programs*, should lead to a more rapid introduction of synthetic fuels into the United States marketplace. A description of these needs is given below.

Basic technology. There is a need to study better caking prevention methods, effect of time, temperature, and pressure on coal gasification with steam, air, or oxygen, and particularly the testing of a wide variety of coals.

Solids handling. Technology and equipment are needed to move coal and/or coal slurries continuously into and out of pressure vessels at high temperatures and pressures.

Field erection and testing of large pressure vessels. Techniques and procedures are required to build large gasifiers and other equipment for synthetic fuel production and to develop testing methods to insure safe operation.

Removal of pollutants from hot gases. Process economics would be improved if procedures were available to remove pollutants such as H_2S from process gas streams at processing temperatures so that the gases would not have to be cooled for treatment and then reheated.

Equipment for combined cycles. Turbine and boiler research and development are needed to speed the development of the combined cycle approach to electric power generation.

Environmental problems. Because synthetic fuel plants will use large quantities of coal and other raw materials, a constant examination of the environmental situation and related R&D is needed to assure reliable procedures.

Methanation. Research on new catalysts for the methanation reaction is needed along with studies on catalyst overheating and poisoning and the cause and prevention of carbon deposits.

One step catalytic reaction for high Btu gas. Process economics should be improved if a catalyst is developed that could perform the shift reaction and the methanation reaction both in one step.

Low cost hydrogen production. All synthetic fuel processes require the manufacture of hydrogen in the process. R&D is needed to decrease the cost of hydrogen production.

Packaged low Btu unit. A smaller "packaged" unit for manufacturing low Btu gas could find application at individual plants such as glass or steel manufacturing.

Solid-liquid separations. Improved methods are needed for separating solids from liquids at elevated temperatures.

Char utilization. More effective usage of the char from synthetic fuel processes, either as a fuel or for hydrogen production, would improve process economics.

Materials of construction. Many materials of construction in synthetic fuel plants will be exposed to both corrosive and erosive conditions, often at elevated temperatures, for a number of years. Better knowledge for predicting equipment life is needed.

Refining of various syncrudes. Studies on the refining and processing of syncrudes is needed to determine what changes in refinery operation might be required.

Source: *Synthetic Fuels from Coal Task Force Report*, Project Independence Blueprint, Federal Energy Administration, November 1974.

Pilot plant near Tacoma, Washington, for processing 50 tons per day of coal by the Solvent Refined Coal (SRC) process. By treating pulverized coal with hydrogen at elevated temperatures and pressures in the presence of a solvent, a low-ash, low-sulfur synthetic fuel is produced. Operation of the plant began in 1974. (*Courtesy Gulf Oil Corp.*)

DEVELOPMENT STATUS OF COAL GASIFIER SYSTEMS

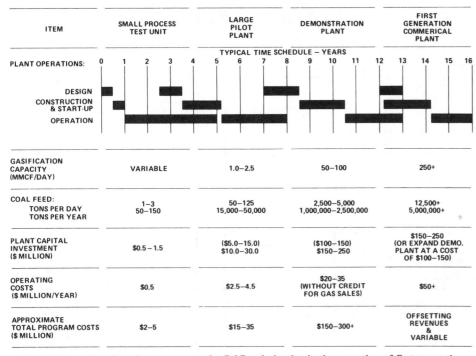

ITEM	SMALL PROCESS TEST UNIT	LARGE PILOT PLANT	DEMONSTRATION PLANT	FIRST GENERATION COMMERICAL PLANT
PLANT OPERATIONS: DESIGN, CONSTRUCTION & START-UP, OPERATION	(TYPICAL TIME SCHEDULE – YEARS 0–16)			
GASIFICATION CAPACITY (MMCF/DAY)	VARIABLE	1.0–2.5	50–100	250+
COAL FEED: TONS PER DAY / TONS PER YEAR	1–3 / 50–150	50–125 / 15,000–50,000	2,500–5,000 / 1,000,000–2,500,000	12,500+ / 5,000,000+
PLANT CAPITAL INVESTMENT ($ MILLION)	$0.5–1.5	($5.0–15.0) $10.0–30.0	($100–150) $150–250	$150–250 (OR EXPAND DEMO. PLANT AT A COST OF $100–150)
OPERATING COSTS ($ MILLION/YEAR)	$0.5	$2.5–4.5	$20–35 (WITHOUT CREDIT FOR GAS SALES)	$50+
APPROXIMATE TOTAL PROGRAM COSTS ($ MILLION)	$2–5	$15–35	$150–300+	OFFSETTING REVENUES & VARIABLE

Typical time schedule and program costs for R&D culminating in the operation of first generation commercial plants to produce synthetic pipeline gas from coal gasification processes.

Source: Clean Energy from Coal Technology, Office of Coal Research, Dept. of the Interior, 1973.

Coal Conversion and Combustion Processes Receiving Federal Support[1]

Process	Development scale	Size	Environmental data
Coal cleaning			
Meyers Chemical Cleaning (TRW/EPA)	Pilot	750 PPH	Some
Physical Cleaning Test Facility (EPA/Bureau of Mines)	Pilot	480 TPD	None
General Public Utilities	Commercial	1200 MW	None
EPA Test Support	Commercial	650 MW	None
Fluidized bed combustion			
Atmospheric bed (PER)	PDU	Unk	Some
	Pilot	30 MW	NA
CPU–400, gas turbine (CPC)	PDU	1 TPD	None
Pressurized FBC (Esso)	Unk	Unk	Some
Pressurized bed (IEA)	Pilot	30 MW	None
Pressurized FBC (NRDC/BCURA)	Pilot	Unk	Unk
High-Btu gasification			
Bi-gas (BCRI)	Bench	1 TPD	Some
	Pilot	120 TPD	NA
Methanator section	PDU	Unk	Unk
Clean fuels (Coalcon)	Demo	2600 TPD	Unk
Hydrocarbonization (Union Carbide)	PDU	20 PPH	Unk
CO₂ acceptor (CONOCO)	Pilot	40 TPD	Some
Hydrane (Bureau of Mines)	Bench	12 PPH	Some
	Pilot	25 TPD	Unk
Hygas (IGT)	Pilot	75 TPD	Unk
Liquid phase methanation (Chemical Systems)	Pilot	2 MM SCFD	Unk
Synthane (Bureau of Mines)	PDU	40 PPH	Some
(Lummus)	Pilot	75 TPD	NA
High- and low-Btu gasification			
Ash agglomerating (Battelle/ Union Carbide)	Pilot	25 TPD	Some
Ash agglomerating (IGT)	PDU	500 PPH	None
	Pilot	250 TPD	NA
Low-Btu gasification			
Boiler fuel gas (Westinghouse)	PDU	12 TPD	Some
	Pilot	120 TPD	NA
Boiler fuel gas (Combustion Engineering)	PDU	120 TPD	NA
Boiler fuel gas (Foster Wheeler)	Pilot	500 TPD	Unk
Flash pyrolysis (Garrett)	Bench	3 PPH	Some
	Pilot	250 TPD	NA
Molten salt (Kellogg)	Bench	Unk	Some
Multiple fluidized bed (BCRI)	PDU	100 PPH	None
Liquefaction			
CSF (Consol, Fluor)	Pilot	20 TPD	Some
	Bench	20 PPH	None
H-Coal (HRI)	PDU	200 PPH	None
SRC (Pittsburgh and Midway)	Pilot	50 TPD	None
Synthoil (Bureau of Mines)	Bench	50 PPH	Some
	PDU	10 TPD	NA

[a] Based on reference 33.

MM SCFD = million standard cubic feet per day
MW = megawatts
NA = not applicable
PDU = process development unit
PPH = pounds per hour
TPD = tons per day
Unk = information was not or could not be obtained during study

Source: Environmental Quality–1976, The Seventh Annual Report of the Council on Environmental Quality, September 1976.

Summary Development Status of Representative Coal Gasifier Systems

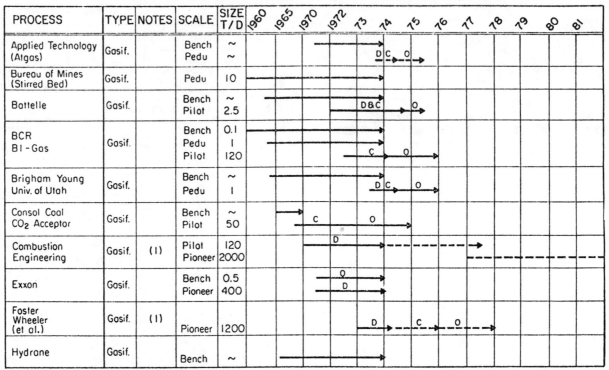

PROCESS	TYPE	NOTES	SCALE	SIZE T/D	Timeline (1960–1981)
Applied Technology (Atgas)	Gasif.		Bench / Pedu	~ / ~	Firm plans ~1972→74; D C O ~74–75
Bureau of Mines (Stirred Bed)	Gasif.		Pedu	10	Firm plans 1960→74
Battelle	Gasif.		Bench / Pilot	~ / 2.5	Firm plans ~1965→74; D&C ~74; O ~75
BCR BI-Gas	Gasif.		Bench / Pedu / Pilot	0.1 / 1 / 120	Firm plans 1960→74; C 73→74; O 74→76
Brigham Young Univ. of Utah	Gasif.		Bench / Pedu	~ / 1	Firm plans 1965→74; D C ~74; O ~75
Consol Coal CO$_2$ Acceptor	Gasif.		Bench / Pilot	~ / 50	Firm plans 1960→~70; C ~72; O ~74–75
Combustion Engineering	Gasif.	(1)	Pilot / Pioneer	120 / 2000	Firm plans 1972→74; D 73; proposal 74→77; proposal 78→81
Exxon	Gasif.		Bench / Pioneer	0.5 / 400	Firm plans 1972→74; O ~73; D ~73
Foster Wheeler (et al.)	Gasif.	(1)	Pioneer	1200	D 73→74; C ~75; O ~76
Hydrane	Gasif.		Bench	~	Firm plans 1965→74

PROCESS	TYPE	NOTES	SCALE	SIZE T/D	Timeline (1960–1981)
IGT Hygas	High Btu Gasif.		Bench / Pilot	~ / 75	Firm plans 1965→72; O ~70; →72
Koppers-Totzek	Gasif.	(2)	Comm.	~	
Lurgi	Gasif.	(2)	Comm.	~	D&C ~74→75
Molten Salt (Atomics International)	Gasif.		Bench / Pilot / Pioneer	~ / 120/240 / 2400	Firm plans 1965→74; D ~75; proposal 75→; proposal 79→81
Molten Salt (M.W. Kellogg)	Gasif.		Bench	~	Firm plans 1965→73
Bureau of Mines Synthane	Gasif.		Bench / Pilot	1< / 75	Firm plans 1965→73; proposal 73→77
Westinghouse Bechtel et al.	Gasif.		Pedu / Pilot / Pioneer	4 / 120 / 1200	D&C 73→74; O ~74; proposal 75→76 D,C&O; proposal 77→81 D C O
IGT, U-Gas	Gasif.	(3)	Pioneer	1000	

——— Firm Plans
---- Proposal
D – Design of Plant
C – Construction of Plant
O – Operation

NOTES: (1) Basic Data from BCR PEDU
(2) Koppers-Totzek and Lurgi Processes are Commercially Available
(3) Conceptual Design Underway Based on IGT DATA
(4) C.F. Braun is Evaluating Processes for High BTU Gas Production

Source: Evaluation of Coal Conversion Processes to Provide Clean Fuels, EPRI 206-0-0, prepared by the University of Michigan, College of Engineering, for the Electric Power Research Institute, Palo Alto, California, February 1974.

COAL GASIFICATION—LOW Btu PROCESSES

A Summary of Some Proposed Processes for the Production of Low-Btu Gas from Coal

Process[a]	Gasifier type	Gasifier pressure (psig)	Oxidizing medium	Comments
Lurgi	Downward-moving stirred bed, nonslagging	300–450	Air	Process is in commercial operation on sized noncaking coal. Plans are under way to test operation on caking bituminous coal
Koppers-Totzek	Cocurrent solid-gas combustion, slagging	1–5	Oxygen or oxygen-enriched air	Process is in commercial operation using oxygen. Tests are planned using enriched air. Can handle any type of coal
Wellman-Galusha	Downward-moving stirred bed, nonslagging	1–300	Air	Process is in commercial operation using coke or noncaking coals, mostly in the steel and ceramics industries. Bureau of Mines has a pilot plant operating on caking coal at pressures up to 125 psig, capacity about 20 tons/day. Tests are planned at 300 psig to increase throughput
Union Carbide	Ash-agglomerating fluidized bed, separate fluidized regenerator	100	Air	Process is in the pre-pilot-plant stage. Plans are proceeding for design and construction of a 25-ton/day pilot plant
ATGAS (Applied Technology Corp.)	Coal is dissolved in molten iron with limestone-air injection	1	Air	Bench-scale unit operating (2.5 ft diam). Development work is in progress. Can handle any type of coal
General Electric	Slow-moving bed with inert bulk diluents	300	Air	Few details available. Process is in bench-scale stage. Uses extrusion feeder instead of lock hoppers; membrane scrubber for H_2S removal. Plans are under way for 6-ton/day pilot plant
Bi-gas	Two-stage entrained bed	300	Air	Pilot plant with 120-ton/day capacity is under construction. Process can handle any type of coal

[a]Processes such as Hygas, Synthane, and CO_2 Acceptor are not included in this table since they are intended primarily for high-Btu gas production.

Source: *Proposed Final Environmental Statement: Liquid Metal Fast Breeder Reactor Program, Volume III*, WASH-1535, U.S. Atomic Energy Commission, December 1974.

Preliminary Comparison of Some Proposed Low-Btu Gas Processes

Process	Claimed or potential advantages	Potential disadvantages or problems
Lurgi	1. Process is in large-scale commercial operation on noncaking coal	1. Possible problem with caking coals 2. Low degree of automation 3. High operating and maintenance requirements
Koppers-Totzek	1. Process is in commercial operation using O_2	1. May have difficulty operating with air instead of oxygen. Oxygen adds appreciably to cost 2. Not readily adaptable to high-pressure operation
Wellman-Galusha	1. Many gasifiers are in commercial operation on coke and noncaking coals, principally in steel and ceramics industries	1. Process is lower pressure than Lurgi, hence probably lower throughput per gasifier 2. Low degree of automation 3. High operating and maintenance requirements 4. Possible difficulty with swelling caking coals
Union Carbide ash-agglomerating fluidized bed	1. Can handle any type of coal 2. Fluidized-bed technique gives high throughput per unit volume 3. Separate regenerator permits use of air without nitrogen dilution of product	1. Process is still in the development stage 2. Possible difficulty in establishing proper conditions for fluidization of ash particles
ATGAS	1. Can handle any type of coal 2. Gas purification greatly simplified 3. Low cost claimed	1. Process is still in the development stage 2. Possible engineering and materials problems
General Electric	1. Can handle swelling caking coals 2. Eliminates lock-hopper feeders 3. Simplified H_2S removal 4. Eliminates stirring devices	1. Process is still in the development stage 2. Use of inert bulk diluents may pose problems
Bi-gas	1. Can handle any type of coal 2. Entrained-bed gasifier is simple and reliable 3. High throughput per unit volume	1. Process is still in the development stage 2. Molten ash may cause deposits in gasifier

Source: *Proposed Final Environmental Statement: Liquid Metal Fast Breeder Reactor Program, Volume III*, WASH-1535, U.S. Atomic Energy Commission, December 1974.

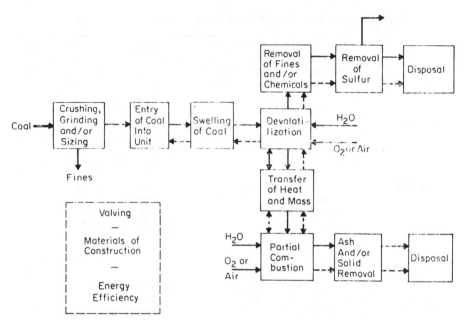

Schematic representation of processing steps and processing requirements for coal gasification.

Source: Evaluation of Coal Conversion Processes to Provide Clean Fuels,; EPRI 206-0-0, prepared by the University of Michigan, College of Engineering, for the Electric Power Research Institute, Palo Alto, California, February 1974.

Coal-to-gas pilot plant near Homer City, Pa. The plant is designed to process 120 tons of coal per day to produce 2.4 million cubic feet of high-Btu gas. This Bi-Gas process employs a two-stage, high-pressure, oxygen-blown system using pulverized coal and steam. Developer of the process is the Bituminous Coal Research, Inc. (*Courtesy Stearns-Roger Corp., 1977.*)

Plant Locations of Winkler Gasifiers

Plant No.	Plant	Year	Product	Capacity Per Generator				No. Gen
				Normal		Maximum		
				1000 NM^3/hr	1000 scfh	1000 NM^3/hr	1000 scfh	
1	Leuna-Werk Leuna, Germany	1926-1930	Fuel Gas Water Gas	60 30	2240 1120	100 50	3730 1870	5
2	Braunkohle-Benzin A.G. Bohlen, Germany	1936	Water Gas	27.6	1030	30	1120	3
3	Braunkohle-Benzin A.G. Magdeburg, Germany	1936	Water Gas	27.6	1030	33	1230	3
4	Yahagi Japan	1937	Water Gas	8.75	330	-	-	1
5	Uraunkohle-Benzin A.G. Zeitz, Germany	1938	Water Gas	22.5	840	-	-	3
6	Dai-Nihonyinzo-Hiryo Japan	1938	Synthesis Gas	14	520	-	-	2
7	Nippon Tar Japan	1938	Water Gas	14	520	-	-	2
8	Toyo-Koatsu Japan	1939	Synthesis Gas	15	560	20	750	2
9	Sudetenlandische Treibstoffwerke Brux, Czechoslovakia	1943	Water Gas	27.6	1030	30	1120	5
*10	Fabrika Azotnih Jendinjenja Gorazde, Yugoslavia	1953	Synthesis Gas	5	190	-	-	1
11	Calvo Sotelo Puertollano, Spain	1954	Water Gas	9.5	350	-	-	1
12	Union Rheinische Braunkohlen Wesseling, Germany	1956	Synthesis Gas	12	450	17	630	1
13	Calvo Sotelo Puertollano, Spain	1957	Synthesis Gas	9.5	350	-	-	1
*14	Azot Sanyyii TAS Kutahya, Turkey	1959	Synthesis Gas	12	450	18	670	2
*15	Neyveli Fignite Corporation Madras, India	1959	Synthesis Gas	41.6	1550	-	-	3
16	Union Rheinische Braunkohlen Wesseling, Germany	1960	Synthesis Gas	12	450	17	630	1

*Presently Operating

Source: *Synthetic Fuels from Coal Task Force Report*, Project Independence Blueprint, Federal Energy Administration, November 1974.

COMMERCIAL GASIFIER FOR LOW Btu PROCESSES

Commercially available technologies for synthetic fuel production include the Lurgi, Fischer-Tropsch, Koppers-Totsek, Winkler, Wellman-Galusha, and Ignifluid processes. These processes are being used throughout the world to produce ammonia, liquid fuel, and petrochemical products.

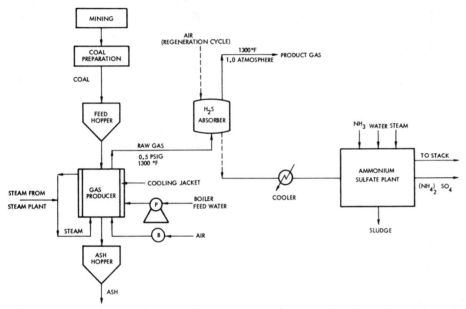

Flow diagram for atmospheric pressured fixed-bed process for production of low Btu gas (fuel gas).

Source: Synthetic Fuels from Coal Task Force Report, Project Independence Blueprint, Federal Energy Administration, November 1974.

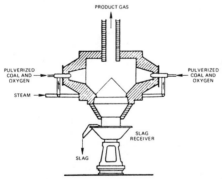

The Koppers-Totzek gasifier operates at near atmospheric pressure. Operating flame temperatures are near 3000°F and at these temperatures the products are CO, H_2, and a small amount of CO_2 with almost no methane, tars or oils. Some ash is removed as liquid and the rest goes out with the product gas. About 52 commercial Koppers-Totzek units are in operation, mainly for production of hydrogen for ammonia synthesis.

Source: Synthetic Fuels from Coal Task Force Report, Project Independence Blueprint, Federal Energy Administration, November 1974.

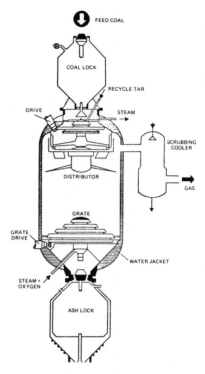

The Lurgi gasifier operates at a pressure of about 400 psi. Lump coal is admitted through the upper coal lock and feeds downward over a distributor and into the gasifier which is a water-jacketed vessel about 12 feet in diameter and 25 feet high. The coal is held on a rotating grate with steam and oxygen coming up through the grate. Ash falls through the grate and is removed the ash lock at the bottom. Caking coals cannot normally be handled in the Lurgi gasifier.

Source: Synthetic Fuels from Coal Task Force Report, Project Independence Blueprint, Federal Energy Administration, November 1974.

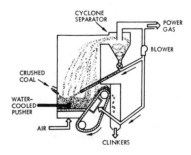

The Mark I Coal Gasifier is similar to the Ignifluid unites operating in Morocco, France, Scotland and Korea. Units operate at atmospheric pressure and the incoming air is used to cool the main grate. Sulfur removal is not provided.

Source: Synthetic Fuels from Coal Task Force Report, Project Independence Blueprint, Federal Energy Administration, November 1974.

COAL GASIFICATION—HIGH Btu PROCESSES

Survey of Representative U.S. High Btu Coal Gasification Processes. The HYGAS[1] process under development by IGT. In this process, ground dried coal at minus 8 mesh size is fed into a pretreatment section where a flow of air heated to about 800°F destroys the caking tendency of coal by mild surface oxidation of the coal particles. The coal then is mixed with an oil by-product of the process and the resultant slurry pumped to 1500 psig and moved to the hydrogasification reactor. In the hydrogasifier, the incoming slurry is sprayed onto a fluidized drying section where oil is flashed upwards with hot gases and dried coal is routed downwards to two hydrogasification stages operating at 1000 psig and between 1200° and 1800°F. Coal reacts with raw hydrogen-rich gas and steam to form methane, CO_2, H_2, CO, H_2S, and other impurities, as well as some light and heavy oils. The gas leaving the hydrogasifier is passed through a quench section for removal of particulates, water soluble trace components, excess steam, and oil vapors. An acid-gas removal system next absorbs the CO_2 and H_2S present with sulfur recovery in a Claus plant. The purified gas then undergoes a shift reaction to provide a stoichiometric balance of CO and H_2 for methane formation. The gas is then catalytically methanated to yield a high Btu gas. Carbon remaining in the char from the bottom of the hydrogasifier is used to generate hydrogen in the steam-iron process.

The steam-iron process appears to be an economical method for hydrogen production. In the oxidizer unit, hydrogen is generated from the reaction of steam and iron at 1500°F and HYGAS pressure. The resulting iron oxide is then returned to the reducer unit where it reacts with hot producer gas from char to make iron for recycle.

The HYGAS plant is being modified to incorporate a steam-iron hydrogen generator system (and thereby eliminate the need for purchased electrical power for electrohydrogasification).

The Synthane Process[2] is a gasification system developed by the Bureau of Mines for converting bituminous coal, subbituminous coal, and lignite into a satisfactory substitute for natural gas.

Crushed coal passes through a lock hopper into a fluidized bed pretreater where the caking properties of raw coal are destroyed by contacting the coal with a steam-oxygen mixture at 800°F. Decaked coal from the pretreater enters the top of the fluidized bed

[1] "Process design and cost estimate for production of 266 million SCF/day of pipeline gas by hydrogasification of bituminous coal-hydrogen by the steam-iron process," R & D Rep. 22, Institute of Gas Technology, February 1967.

[2] *Synthetic Fuels from Coal Task Force Report*, Project Independence Blueprint, Federal Energy Administration, November 1974.

Summary of High-Btu Coal Gasification Processes and Projects Commercial Processes and Ventures

Name of process	Owner(s) or contractor and site	Plant output (million cubic feet per day)	Status and funding
Lurgi Pressure Gasification (Lurgi Gesellschaft fur Warme and Chemotechnik m.b.H.).	El Paso Natural Gas Co. (Four Corners area, New Mexico).	250	El Paso plans to construct and operate the Burnham coal gasification complex Indian on the Navajo Reservation. In the initial announcement, capital costs were estimated at $353,200,000 for the gasification plant and $65,300,000 for the associated mine. Initial gas production was scheduled for June 1976, and the estimated 1977 cost of gas at the plant outlet was $1.20 per 1,000 SCF.
Lurgi pressure gasification	Transwestern Coal Gasification Co., Pacific Coal Gasification Co., and Western Gasification Co. (Four Corners area, New Mexico).	250	The firms plan to construct and operate a $405,900,000 coal gasification plant on the Navajo Indian Reservation near Farmington, N. Mex. Utah International will supply the coal and water. Plant operation is scheduled for 1976.
Do	Panhandle Eastern Pipe Line Co. and Peabody Coal Co. (eastern Wyoming).	250	Plant operation anticipated in the 1978–80 period, assuming timely receipt of all required governmental authorizations. Plant investment will be about $400,000,000.
Do	Michigan Wisconsin Pipe Line Co. (central North Dakota).	250	Michigan Wisconsin has acquired options to 3,500,000,000 tons of North Dakota lignite and has entered into agreements with Lurgi and with Lummus Corporation for preliminary design of a 250,000,000 cubic feet per day plant.
Do	Northern Natural Gas Co. and Cities Service Gas Co. (Powder River Basin, Mont.).	250	Northern Natural and Cities Service are considering construction of four 250,000,000 cubic feet per day coal gasification plants. Peabody Coal has agreed to supply about 500,000,000 tons of coal, and the gas companies are negotiating for another like amount. $10,000,000 will be spent for preliminary development through 1975. Construction of the 1st plant could start in 1976, with operation in 1979.
Do	Natural Gas Pipeline Co. of America (Dunn County, N. Dak. .	250	Rights to 2,000,000,000 tons of lignite have been obtained from Star Drilling Inc. The lignite will be reserved for possible future use in a coal gasification project.
	Colorado Interstate Gas Co. and Westmoreland Resources (southeast Montana).		Colorado Interstate has an option on 300,000,000 tons of coal and 10,000 acre-feet per year of water to be supplied by Westmoreland for development of a coal gasification project.
	Columbia Gas System, Inc		Core drilling program in West Virginia to identify possible sites for coal gasification facilities. 233,000,-000 tons of recoverable coal reserves have been proved on part of land to which the company has coal rights.
	Texas Gas Transmission Corp. and Consolidation Coal Co.		Texas Gas Transmission Corp. has acquired ½ interest in a large block of Illinois Basin coal reserves controlled by Consolidation Coal Co. The largest parcel will be held for 10 years for possible use in a coal gasification project.
	Eastern Gas & Fuel Associates and Texas Eastern Transmission Corp.		The companies have acquired coal mining and prospecting rights to about 40,000 acres of Federal and State leases in northwestern New Mexico. If the reserves are developed, Eastern Gas would conduct the mining and Texas Eastern would build and operate the gasification plant.
Do	Transcontinental Gas Pipe Line Corp. (Powder River Basin, northeast Wyoming).	250	Transco has acquired joint rights from Tipperary Corp. and Stoltz, Wagner & Brown to evaluate coal under more than 20,000 acres in Wyoming for a coal gasification project.
	Consolidated Natural Gas Co. (southwest Pennsylvania)		Consolidated has a 70-percent interest in southwestern Pennsylvania acreage containing over 300,000,-000 tons of coal and is negotiating for more reserves. The company expects to make a commitment on its 1st gasification plant in 18 to 24 months.

reactor which contains two operating zones. Steam and oxygen enter at the bottom to effect fluidization.

The gasifier operates at 600 to 1000 psi. The top portion of the gasifier operates as a dense fluidized bed at 1100° to 1450°F and the lower portion as a dilute fluidized bed at 1750° to 1850°F. As in most gasifiers, devolatilization and limited methanation take place in the top portion and hydrogen formation in the lower. Ash and unconverted coal or char are removed through a lock hopper. The char is burned to generate process steam. The Synthane process requires a minimum of oxygen and approximately one-third of the coal is converted to char. A 75 ton/day pilot plant for producing pipeline quality gas was compiled at Bruceton, Pennsylvania, in 1975.

In addition to the HYGAS process and the Synthane process, two other processes for converting coal to high Btu pipeline gas have reached pilot plant stage—the BI-GAS process of Bituminous Coal Research and the CO_2 Acceptor process of Consolidation Coal Co. The Bi-GAS process features a two-stage reactor, which operates as an entrained bed system to gasify ground, dried coal at a pressure of 1100 psig. A fully integrated 5 tons of coal per hour gasification plant has been constructed by Stearns-Rogers, Inc., at Homer City, Pennsylvania. The CO_2 Acceptor process is a unique fluidized-bed system where gasification is accomplished in a mixture of calcined dolomite and lignite. The calcined dolomite acceptor not only removes both CO_2 and H_2S from the raw product gas to minimize the gas cleanup requirements, but also reacts with CO_2 exothermically to provide

Summary of High-British Thermal Unit Coal Gasification Processes and Projects Principal Research and Development Projects

Name of process	Owner(s) or contractor and site	Plant output (million cubic feet per day)	Status and funding
COGAS	COGAS Development Co. (FMC Corp., Panhandle Eastern Pipe Line Co., Tenneco Inc., Consolidated Natural Gas Service Co., Republic Steel Corp., Rocky Mountain Energy Co.) (Princeton, N.J.)		The 1st stage of the COGAS Process is based on results from the COED pilot plant in Princeton, N.J., which was designed to produce oil, char, and a relatively small amount of gas. The COED pilot plant was funded by OCR and completed in 1970 at a cost of $4,500,000. COGAS Development Co. will invest $7,000,000 in the COGAS pilot plant over an 18- to 24-mo. period. The pilot plant has been in operation for over a year.
HYGAS	Institute of Gas Technology (Chicago, Ill.)	1.5	Pilot plant in operation. Preliminary demonstration, plant design complete. Original cost of pilot plant was approximately $9,500,000.
CO_2 acceptor	Consolidation Coal Co. (Pilot plant constructed and operated by Stearns-Roger Corp.) (Rabid City, S. Dak.)	(1)	Pilot plant in operation. A methanation stage is not incorporated, but may be added at a later date. The original pilot plant cost was about $9,300,000.
BI-GAS	Bituminous Coal Research, Inc. (Homer City, Pa.)	2.3	Construction contract has been awarded to Stearns-Roger Corp. Plant cost will be about $18,000,000, and construction will take 18- to 24-mo. Initial startup is scheduled for late 1974.
Kellogg molten salt process	M. W. Kellogg Co.		OCR funded a bench-scale program from 1964 to 1967. Total expenditures were $1,700,000. Major difficulties were experienced with materials of construction. OCR ceased sponsorship because of this problem, budgetary restrictions, and assignment of higher priorities to other coal gasification processes. M. W. Kellogg has carried out additional development work since 1967, but support has not yet been obtained for construction of a large-scale pilot plant.
SYNTHANE	U.S. Bureau of Mines (Bruceton, Pa.)	1.4	Construction contract has been awarded to Rust Engineering Co. Estimated cost is $10,000,000. Operation is expected by mid-1974.
Hydrane	U.S. Bureau of Mines (Pittsburgh, Pa.)		A 10-lb per hour integrated pilot plant is in operation. Scale-up to a 24-ton per day pilot plant is planned
	Exxon Corp. (Baytown, Tex.)		Exxon has spent $20,000,000 on coal gasification and liquefaction since 1966 and is committing $10,-000,000 for additional R. & D. 1st phase of gasification experiments to be finished in 1974. If justified, next phase will be construction of $75,000,000 to $80,000,000 pilot plant charging 500 tons per day of coal.
	Battelle Columbus Laboratories (West Jefferson, Ohio)	2.8	Battelle has been granted a 30-mo, $4,100,000 contract under the joint OCR/AGA program.

[1] Up to 2 (no heating value specified).
[2] Synthesis gas.

Source: Fiscal Policy and the Energy Crisis, Hearings before the Subcommittee on Energy of the Committee on Finance, United States, Seanate, Part 4 of 4 Parts, January 25, 28, and 29, 1974.

heat for the endothermic steam-carbon reaction. The CO_2 Acceptor process pilot plant at Rapid City, South Dakota, is designed to produce 2 million standard cubic feet per day of 375 Btu per standard cubic foot gas from 45 tons of lignite and 3 tons of dolomite per day.

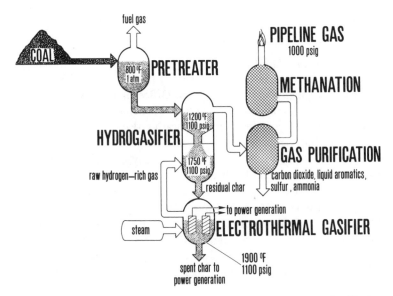

IGT hygas process—hydrogen-rich gas by electrothermal gasification of resid. char.

Source: Office of Coal Research; Annual Report 1972, U.S. Department of the Interior.

World's first large scale coal gasification pilot plant to produce pipeline quality synthetic natural gas. Commercial development of this virtually pollution-free process could substantially increase gas supply in coming years. (*Courtesy of Institute of Gas Technology*)

COAL LIQUEFACTION PROCESSES

Liquid products can be produced from coal by dissolution in gaseous or liquid solvents. The first systematic study of the action of solvents on coal was made in 1862 by de Marsilly. In 1911, gases and oils were made by Berguis by a catalytic hydrogenation process using impure iron oxide in a pasting oil. Solvent extraction of bituminous coal was first studied by Pott and Broche in 1927; Sinnatt made an extensive study of solvent effects in 1933. In 1936, a systematic study of coal hydrogenation and liquefaction was begun by the U.S. Bureau of Mines.

Catalytic hydrogenation of coal was used in 1927 in Germany to produce commercial gasoline and in 1942 some 32 million gallons of aviation gasoline were made from coal in that country.

In general, coal liquefaction processes involve the following steps: the coal is crushed, dried and pulverized, and a solvent added to produce a slurry; the slurry is heated, usually in the presence of hydrogen, to dissolve the coal; the extract is cooled to remove hydrogen, hydrocarbon gases and hydrogen sulfide; and the liquid is flashed at low pressure to separate condensible vapors from the extract. Mineral matter and organic solids are then separated and used to produce hydrogen for the process. Additional steps include separation of the solvent and major products and desulfurization of the extract.

Survey of U.S. Coal Liquefaction Processes. Coal liquefaction in the United States has also been actively under investigation and numerous processes are at either the process development unit (PDU) or pilot plant stage of evaluation. Projects sponsored by the U.S. Department of the Interior have included: (1) the Solvent Refined Coal (SRC) process of the Pittsburgh and Midway Coal Company, (2) the COED (Char-Oil-Energy-

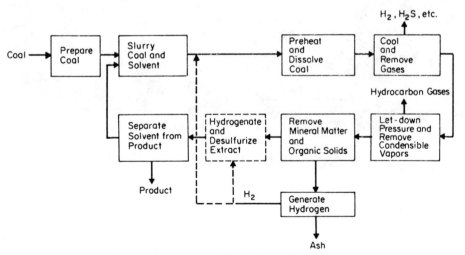

Processing steps in coal dissolution and liquefaction.

Source: *Evaluation of Coal Conversion Processes to Provide Clean Fuels*, EPRI 206-0-0, prepared by the University of Michigan, College of Engineering, for the Electric Power Research Institute, Palo Alto, California, February 1974.

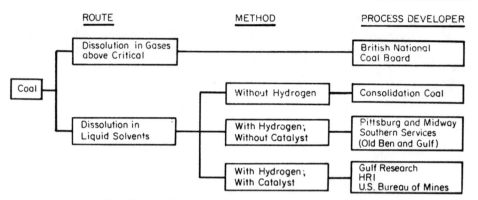

Classification of coal dissolution and liquefaction processes.

Source: *Evaluation of Coal Conversion Processes to Provide Clean Fuels*, EPRI 206-0-0, prepared by the University of Michigan, College of Engineering, for the Electric Power Research Institute, Palo Alto, California, February 1974.

Development) process of FMC Corp., (3) Project Lignite at the University of North Dakota, (4) Clean Coke Process of U.S. Steel, and (5) the Synthoil Process at the U.S. Bureau of Mines. Originally sponsored by the Office of Coal Research, development work on the H-Coal process has been conducted on bench-scale and process development units (2.5 tons per day of coal feed) at Hydrocarbon Research's Trenton, New Jersey, facility for the past eight years. Work on the H-Coal process is underway at several organizations,[1] and a pilot plant (250 to 700 tons per day of coal feed) representing one-tenth the size of a commercial unit has been designed under OCR sponsorship.

The COED process converts coal to a high grade synthetic crude oil, pipeline quality gas, and char by multistage fluidized-bed pyrolysis. The process has been under development by FMC since 1962 under the sponsorship of the Office of Coal Research. The COED pilot plant at Princeton, New Jersey, has a capacity of 35 tons of coal per day yielding about 30 barrels per day of coal-derived oil for hydrotreating. Startup of the pilot plant began in 1970 and operability of the process was demonstrated by long extended runs of over 30 days.

The COGAS process of FMC is in many ways identical to the COED process. The latter produces a solid char; the former gasifies the char and produces gas for synthesis and/or in-plant heating. In general, the COED process is OCR sponsored while COGAS is privately sponsored.[2]

Exxon has a liquefaction pilot plant (0.5 ton/day) operating in Baytown, Texas, with plans for scale-up. The process is proprietary.

The solvent refined coal process (SRC) and closely derived variations are under active development at several locations. This process has good potential for generating a clean boiler fuel (environmentally acceptable) that can be substituted for natural gas in oil/gas electric power plants. A pilot plant at Fort Lewis, Washington, is designed for 50

[1] Arco, Ashland Oil, Sun Oil, Standard Oil of Indiana, EPRI, and HRI.
[2] Partners include Panhandle Eastern, Tennessee Gas, Consolidated Natural Gas, Republic Steel, Rocky Mountain Energy Co, and FMC.

ton/day coal conversion to a product containing ~0.8% sulfur and a heating value of 15,900 Btu/lb. The pilot plant is also designed to: (1) test production scale-up factors; (2) demonstrate high volume solid/liquid separation, and (3) assess the effect of higher hydrogen pressures.

A related project, consisting of a 5 ton/day pilot plant, is located at Wilsonville, Alabama. This project is sponsored by the Electric Power Research Institute (EPRI) and the Southern Company.

Another related project is the Old Ben Project, a 900 ton/day prototype unit near Toledo, Ohio, on a site adjacent to an Edison Co. power plant. Sohio plans to undertake the project with partners from the private sector. The project goal is to: (1) demonstrate clean boiler fuel production at a meaningful (900 ton/day) commercial level and (2) demonstrate the usefulness of the new fuel in a large electric power plant.

Gulf Research and Development has work underway on a small plant (120 lbs/day). A special fixed bed reactor (designed to minimize plugging) is a major feature.

In addition to the coal liquefaction processes mentioned above, Garrett Research and Development Co. has announced plans for the construction of a coal liquefaction pilot plant in LaVerne, California, based on a rapid pyrolysis process. Other types of coal liquefaction processes are also being developed by TOSCO and Union Carbide.

Coal Liquefaction Process Operating Conditions and Typical Products

PROCESS	H-COAL	PARSONS MODIFIED PAMCO	PAMCO (S.SERV.)	BUR. OF MINES	GULF CCL	GULF CCL	CONSOL
Hydrogen used in dissolution?	Yes	Yes	Yes	Yes	Yes	Yes	No
Subsequent Extract Hydrogenation?	No	Yes	No	No	No	No	Yes
Catalytic Dissol.	Yes	No	No	Yes	Yes	Yes	No
Approximate Reactor Temperature.	850°	840°F	850°F	840°F	800°F	800°F	730°F
Reactor Pressure.	3000 psig	1200 psig	1500 psig	4000 psig	3000 psig	3000 psig	400 psig
Coal	Ill. No.6	Ill. No.6	--	Kentucky	Big Horn Subbit.	Pittsburg Seam (Bit.)	Pittsburg Seam Coal
Sulfur, Wt.%	5%	3.38%	5%	4.6%	0.54%	1.49%	3.67%
Solvent to Coal Ratio (to slurry).	1:1(by Wt.)	2.0:1(by Wt.)	2:1(by Wt.)	1.22:1.0(by Wt.)	2.33:1.0(by Wt.)	2.33:1.0(by Wt.)	2:1 (by Wt.)
Percent Coal Dissolved (MAF).	90%+	90%+	90%+	90%+	91%	90%	63%
Hydrogen Consumption Scf/ton Coal (MAF).	15,300	12,600	@2% by Wt. 7600	9000	22,800	17,500	16,300
Solids Separation. Solids Content in Product.	Hydroclones and/or filtration	Filtration	Filtration 0.23 Wt.%	Centrifuge 1.3 Wt.%	Hydroclones & Filt. 0.02 Wt.%	Hydroclones & Filt. 0.03 Wt. %	Hydroclones
Principal Products							
1. Fuel	Fuel Oil	Residual Fuel Oil	Solvent Refined Coal	Fuel Oil	Filtrate Fuel Oil	Filtrate Fuel Oil	Fuel Oil
Yield bbl/ton	1.73 bbl/ton	1.43bbl/ton	1116 lb/ton*	3 bbl/MAF ton*	2.3 bbl/ton	3.6 bbl/ton*	1.52 bbl/ton coal
API gravity	-3.1°API	-9.7°API 60/60		Sp Gr=1.12-1.14	9.0°API	1.2° API	10.3°API
Viscosity				Visc = 75-204 SSF@ 180°F	7.1 CS @100°F	4.3 CS @210°F	
Sulfur, Wt.%	0.5%	<0.5%	<1.2%	0.31%	0.04%	0.11	.128%
Nitrogen, Wt.%				0.9%	0.40%		
2. Fuel	Naphtha	Distillate Fuel Oil			Light Ends	Light Ends	Naphtha
API gravity	38.4°API	13.9°API 60/60			35.3°API	0.45 bbl/ton*	58.0°API
Yield bbl/ton	0.54bbl/ton	0.71 bbl/ton			0.9 bbl/ton*		0.52 bbl/ton
Viscosity					1.2 CS @100°F		
Sulfur, Wt.%	<0.1%	0.2%			0.04%		.056%
Nitrogen, Wt.%					0.19%		

*Exclusive of H_2 Production

Source: Evaluation of Coal Conversion Processes to Provide Clean Fuels, EPRI 206-0-0, prepared by the University of Michigan, College of Engineering, for the Electric Power Research Institute, Palo Alto, California, February 1974.

Technological Status of Liquefaction-Desulfurization Processes

Process	Process developer and sponsor	Pilot plant Location	Pilot plant Capacity (tons coal/day)	Pilot plant Status and cost	Major technological problem areas	Process improvement research and development
Solvent Refined Coal	Pittsburg and Midway Coal Mining Co.; Office of Coal Research (OCR)	Tacoma, Wash.	50	Under construction; startup fall 1974; $17 million	1. Solid separation of unreacted coal 2. Production of H_2 for process 3. Extent of sulfur removal	1. Solids separation techniques 2. Increased sulfur removal 3. Use of $CO + H_2$ instead of H_2 for hydrogenation 4. Gasification of unreacted solids for H_2 production
	Southern Services Co.; Electric Power Research Institute	Wilsonville, Ala.	6	Under construction; startup Jan. 1974; $6 million	Same as above	Same as above
Meyers process	TRW, Inc.; EPA	Redondo Beach, Calif.	Bench scale (12-ton/day pilot plant to be built)		1. Removal of elemental sulfur from the coal after the leach step	1. Studies of organic sulfur removal by solvent extraction
H-Coal process	Hydrocarbon Research, Inc.; OCR and group of oil companies	Trenton, N.J.	3–8 (250-ton/day pilot plant proposed)	Process development unit in operation	1. Solid separation of unreacted coal 2. Production of H_2 for process 3. Catalyst regeneration	1. Solids separation techniques 2. Evaluation of catalysts 3. Scale-up to commercial-sized equipment
Consol process	Consolidation Coal Co; OCR	Cresap, W.Va.	24	Operated 1967–1970; currently shut down; may be used for coal liquefaction demonstrations	1. Solid separation of unreacted coal 2. Equipment mechanical problems 3. Production of H_2 for process	1. Solids separation techniques 2. Mechanical modification of pilot plant to permit continuous operation
COED process	FMC Corp.; OCR Cogas Devel. Corp. (FMC Corp., Panhandle Eastern Pipeline Co., and Tenn. Gas Pipeline Co.)	Princeton, N.J.	36	In operation 1970 to date; $4.5 million	1. Separation of solids from oil produced during oil pyrolysis 2. Gasification of the residual char produced during coal pyrolysis	1. Development of an oil absorption system to eliminate oil filtration 2. Product evaluation studies 3. Development of char gasification techniques
Bureau of Mines Hydrodesulfurization process	U.S. Bureau of Mines	Bruceton, Pa.	0.5 (6-ton/day pilot plant proposed)	Began operation Aug. 1973	1. Solids separation of unreacted coal 2. Production of H_2 for the process 3. Scale-up of reactor to commercial sizes	1. Filter and centrifuge development for solids separation 2. Substitution of $H_2 + CO$ for H_2 used during hydrogenation 3. Determination of catalyst consumption

Source: Proposed Final Environmental Statement: Liquid Metal Fast Breeder Reactor Program, Volume III, WASH-1535, U.S. Atomic Energy Commission, December 1974.

COAL LIQUEFACTION—COMMERCIAL PLANTS

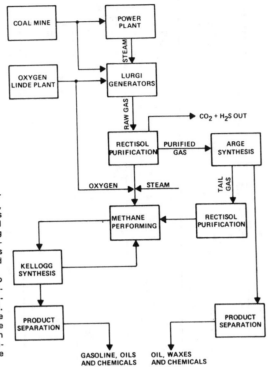

Raw gas from the Lurgi generators, after cooling and particulate and tar removal, goes to Rectisol purification where it is washed with cold methanol. The purified gas is then split into two streams, one going to Arge synthesis and one to methane reforming. The purpose of the Arge plant is to convert H_2 and CO to waxes, oils and chemicals.

Fluidized iron oxide catalyst is fed into the hot gases as they enter the synthesis reactor which operates at about 310°C. Approximate pressure in the system is 350 psi. The gases and oils are separated from the catalyst, which is recycled. The gases are then cooled to separate the oil product with some of the gases being recycled to synthesis and some going back to methane reforming.

Source: Synthetic Fuels from Coal Task Force Report, Project Independence Blueprint, Federal Energy Administration, November 1974.

Products from SASOL Fischer-Tropsch Plant

	Planned Production
Refinery products:	
Gasoline, bbl/day	4,300
Diesel oil, bbl/day	335
Fuel oil, bbl/day	180
Paraffine waxes, 105–240°F mp ton/year	18,000
Liquefied petroleum gas, imp gal/day	720
Pitch and tar road primers, imp gal/day	2,685
Chemical products:	
Ethanol, imp gal/year	4,000,000
Propanol, imp gal/year	2,000,000
Butanol, imp gal/year	525,000
Acetone, imp gal/year	210,000
Methyl ethyl ketone, imp gal/year	260,000
Mixed solvents, imp gal/year	60,000
Benzene, imp gal/year	500,000
Toluene, imp gal/year	280,000
Xylene and solvent naptha, imp gal/year	500,000
Creosote wood preservative, imp gal/year	1,000,000
Crude phenols, ton/year	6,000
Ammonium sulfate, ton/year	35,000

[a]SWAMINATHAN, Petroleum Processing, 10 (7), 987 (July 1955).

bbl = barrel.

imp gal = imperial gallon = 1.2 United States gallons.

Source: Synthetic Fuels from Coal Task Force Report, Project Independence Blueprint, Federal Energy Administration, November 1974.

Conceptual Commercial Size Coal Liquefaction Plants

Process	Engineering Design	Plant Size tons coal/day	Coals	Main Fuel Oil Products		Estimated Overall Thermal Efficiency %	Plant Capital Cost[a] Million $	Date of Study	MW Potential at 35% Efficiency	Cost $/KW
Modified SCR	Ralph M. Parsons (17)	10,000	Illinois No. 6, 3.4% S	0.2% S 0.5% S	13.9° API -9.7° API	63.5	270	1973	620	435
Consol CSF	Foster- Wheeler (18)	20,000 (MF)	Pittsburgh Seam 4.2% S	0.056% S 0.128% S	58° API 10.3° API	71–73.9	230	1972	1530	150
H-Coal	Hydrocarbon Research, Inc. Inc. (10)	25,000 (MF)	Illinois No. 6, 5% S	0.1% S 0.5% S	27° API -3.1° API	69.6	299	1973	1800	166
		35,211	Wyodak 0.7%	<0.2%	39.3° API	b	445	1973	2000	222
Gulf CCL	Gulf R&D (8)	33,000	Big Horn 0.54%	<0.04% S 0.04% S	35.3° SPI 9° API	b	423	1973	2300[b]	184[c]

[a]Does not include interest during construction.

[b]Not given.

[c]Estimated.

(MF) Moisture free.

Source: Evaluation of Coal Conversion Processes to Provide Clean Fuels, EPRI 206-0-0, prepared by the University of Michigan, College of Engineering, for the Electric Power Research Institute, Palo Alto, California, February 1974.

Commercial Plants for Synthesis of Liquid Fuels by Fischer-Tropsch Process

Country	Name of Company	Startup, Year	Starting Material	Normal Pressure Synthesis	Medium Pressure Synthesis Double Tube	Fixed Bed	Fluid Bed	Catalyst Co	Fe	Design Capacity, 1000 Ton/Yr
Germany	Brabag	1936	lignite	X				X		210
Germany	Gewerkschaft Viktor	1936	coke, coke gas	X				X		50
Germany	Ruhrbenzin A.G.	1936 1938	coke	X	X			X		30 42
Germany	Gewerkschaft Rheinpreussen	1936	coke, coke gas	X				X		75
Germany	Krupp-Treibstoffwerke	1937, 1939 1939, 1952	coke	X	X			X X	X	47 13
Germany	Wintershall A.G.	1937, 1939	lignite	X				X		80
Germany	Essener Steinkohle Chemischewerke	1937, 1938	coke, coke gas	X				X		50
France	Harnes	1937	coke	X				X		30
	Society Kuhlmann	1952				X			X	2
Germany	Hoesch-Treibstoffwerke	1939	coke		X			X		60
Germany	Schaffgotsch-Werke	1939	coke, coke gas		X			X		60
Japan	Mike	1940		X				X		40
Japan	Takikawa	1942, 1943	coal	X				X		100
Japan	Rumoi	1939		X				X		50
Japan	Amagasaki	1939		X				X		70
Manchuria	Fushin	1939	coal	X				X		50
U.S.	Hydrocol Co.	1950	natural gas				X		X	360
South Africa	Sasol	1955	coal				X		X	185
South Africa	Sasol	1955	coal			X			X	69

Source: Synthetic Fuels from Coal Task Force Report, Project Independence Blueprint, Federal Energy Administration, November 1974.

OIL EXPLORATION

Deposits of crude oil and natural gas are believed to have been produced in marine environments from the decay of organic plant and animal matter. In historical times, such accumulations could have occurred in shallow seas that are now dry land due to geologic uplifting processes or to changes in the level of the sea. In addition, places where rivers and glaciers dumped soil, silt, and organic matter could also mark potential sites for petroleum. Petroleum deposits occur, therefore, in sedimentary rock of fairly recent geological origin, mainly in the cenozoic period during the past 100 million years although older deposits are also possible. Exploration for oil is conducted in areas where such sedimentary deposits are likely to have occurred.

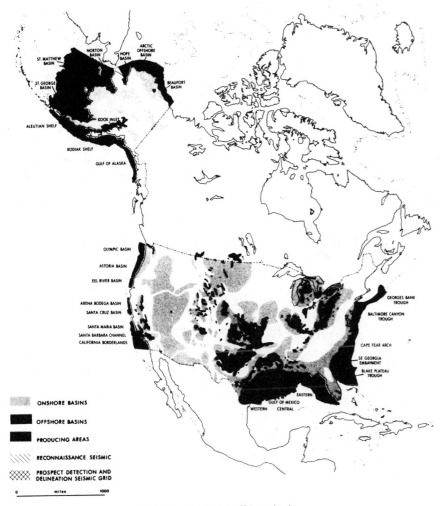

LEGEND
- ONSHORE BASINS
- OFFSHORE BASINS
- PRODUCING AREAS
- RECONNAISSANCE SEISMIC
- PROSPECT DETECTION AND DELINEATION SEISMIC GRID

0 ___ miles ___ 1000

U.S.A. onshore and offshore basins.

Source: "Offshore Oil," *Shell Reports*, July, 1975.

Geological Time Scale

Era	System or Period (rocks) (time)	Series or Epoch (rocks) (time)	Approximate age in millions of years (beginning of unit)
Cenozoic	Quaternary	Recent	0.01
		Pleistocene	2 to 3
	Tertiary	Pliocene	7
		Miocene	25
		Oligocene	40
		Eocene	60
		Paleocene	68 to 70
Mesozoic	Cretaceous (*chalk*)		135
	Jurassic (*Jura Mountains, France*)		180
	Triassic (*from three-fold division in Germany*)		225
Paleozoic	Permian (*Perm, a Russian province*)		270
	Carboniferous (*from abundance of coal*)		325
	Devonian (*Devonshire, England*)		400
	Silurian (*an ancient British tribe, the Silures*)		440
	Ordovician (*an ancient British tribe, the Ordovices*)		500
	Cambrian (*from Cambria, the Roman name for Wales*)		550 to 600
Precambrian	Many local systems and series are recognized, but no well-established worldwide classification has yet been delineated.		3 500 or more

Adapted from J. Gilluly, A. C. Waters, and A. O. Woodford, *Principles of Geology*, 3rd edition, page 106, W. H. Freeman and Co., San Francisco (1968).

Source: Survey of Energy Resources 1974, The United States National Committee of the World Energy Conference, 345 East 47th Street, New York, N.Y., 10017, 1974.

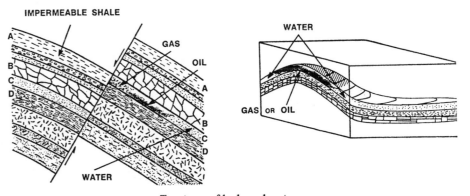

Two types of hydrocarbon traps.

Source: Action in the Atlantic, American Gas Association, 1515 Wilson Boulevard, Arlington, Virginia 22209, 1973.

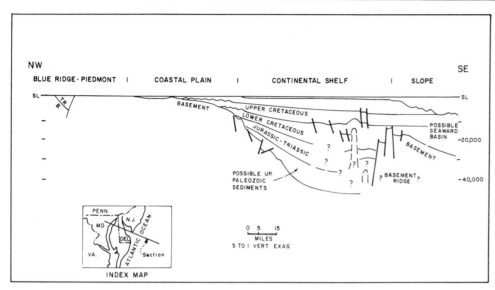

Baltimore Canyon Trough diagrammatic cross section.

Source: Offshore Oil, Shell Reports, July, 1975.

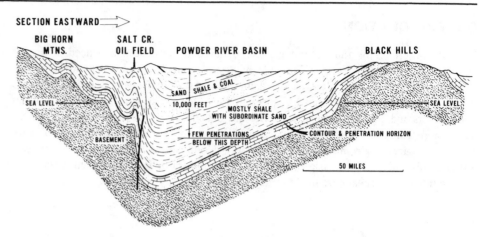

Section of oil-bearing formation in the Powder River Basin.

Source: James E. Wilson, Statement in *Trends in Oil and Gas Exploration,* Hearings before the Committee on Interior and Insular Affairs, United States Senate, Part I, August 8 and 9, 1972.

In addition to the search for suitable sedimentary deposits where heat and pressure might convert the organic matter into oil or gas, the geologist also looks for other conditions which facilitate the formation of a recoverable accumulation of oil or gas. As deposited, the organic matter in sand and silt is but a small fraction of the total material. Accumulations of oil or gas depend upon migration of small quantities of these materials to a trap or pocket in the surrounding rock formation. Since oil and gas are lighter than water, they tend to migrate upwards in a rock formation, if that formation is porous, until they reach a layer of rock that is impermeable. Geological exploration is directed, therefore, to areas of sedimentary rock covered with an impermeable rock layer and where folding or a fault movement has created a suitable trap for the collection of oil or gas.

Source: Synthetic Fuels from Coal Task Force Report, Project Independence Blueprint, Federal Energy Administration, November 1974.

U.S. EXPLORATORY OIL AND GAS ACTIVITY

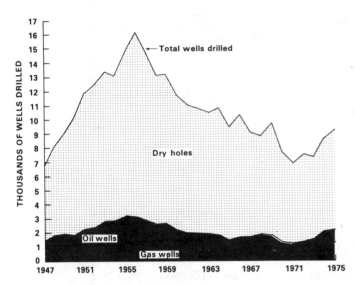

Total exploratory wells drilled for oil and gas.

Source: Energy Perspectives 2, U.S. Department of the Interior, June 1976.

While total exploratory wells drilled for hydrocarbons increased at an annual rate of 10.2 percent from 1947 to 1956, this ratio declined at an annual rate of 2.9 percent from 1956 to 1975.

The average depth of all exploratory wells drilled for hydrocarbons increased approximately 1.5 percent annually since 1947, but this trend was reversed in 1973.

During the 1947-75 period, slightly more than one of five wells drilled proved successful. Since 1973 the success ratio has increased, as a result of a number of factors, including the development of shallower fields, which were previously considered uneconomic.

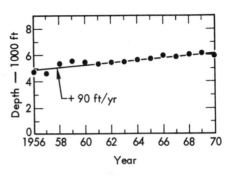

Average gas well depth (including dry holes) between 1956 and 1970.

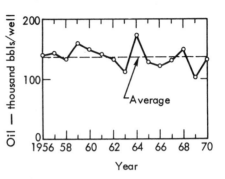

Historic average oil found per well (including dry holes). The average is 137,400 ± 17,900 bbl/well.

Source: An Assessment of U.S. Energy Options for Project Independence, UCRL-51638, prepared for the U.S. Atomic Energy Commission by the Lawrence Livermore Laboratory, September 1, 1974.

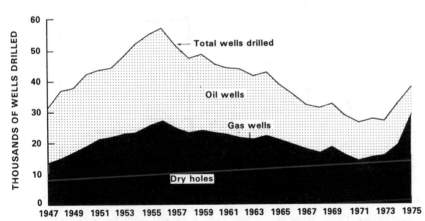

From 1947 to 1956, the total number of wells drilled for hydrocarbons increased at a 7.1-percent annual rate. In 1956, the total number of wells drilled peaked at 57,170, and then declined at an annual rate of 4.4 percent to the 1973 level of 26,592. During the 1947-73 period, the average depth increased at a 1.4-percent annual rate.

In 1975, 40 percent more wells were drilled than in 1973, and the percentage of successful wells exceeded 63 percent for the first time since 1950.

The recent increases in drilling were primarily a response to increased oil prices after the embargo.

Total wells drilled for oil and gas.

Source: Energy Perspectives 2, U.S. Department of the Interior, June 1976.

Rotary bit used in drilling oil and gas wells.

Source: American Gas Association.

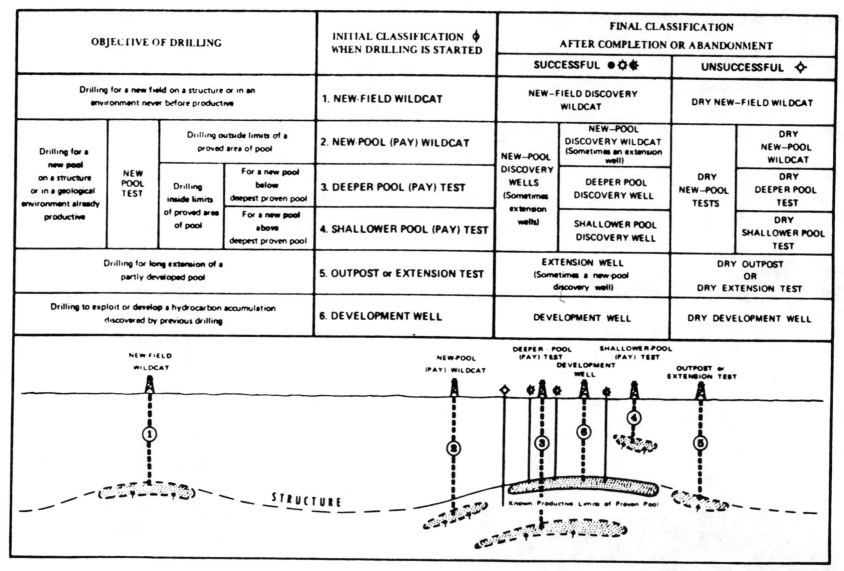

OBJECTIVE OF DRILLING			INITIAL CLASSIFICATION ♦ WHEN DRILLING IS STARTED	FINAL CLASSIFICATION AFTER COMPLETION OR ABANDONMENT			
				SUCCESSFUL ● ✧ ✦		UNSUCCESSFUL ♢	
Drilling for a new field on a structure or in an environment never before productive			1. NEW-FIELD WILDCAT	NEW—FIELD DISCOVERY WILDCAT		DRY NEW-FIELD WILDCAT	
Drilling for a new pool on a structure or in a geological environment already productive	NEW POOL TEST	Drilling outside limits of a proved area of pool	2. NEW-POOL (PAY) WILDCAT	NEW—POOL DISCOVERY WELLS (Sometimes extension wells)	NEW—POOL DISCOVERY WILDCAT (Sometimes an extension well)	DRY NEW—POOL TESTS	DRY NEW—POOL WILDCAT
		Drilling inside limits of proved area of pool — For a new pool below deepest proven pool	3. DEEPER POOL (PAY) TEST		DEEPER POOL DISCOVERY WELL		DRY DEEPER POOL TEST
		For a new pool above deepest proven pool	4. SHALLOWER POOL (PAY) TEST		SHALLOWER POOL DISCOVERY WELL		DRY SHALLOWER POOL TEST
Drilling for long extension of a partly developed pool			5. OUTPOST or EXTENSION TEST	EXTENSION WELL (Sometimes a new-pool discovery well)		DRY OUTPOST OR DRY EXTENSION TEST	
Drilling to exploit or develop a hydrocarbon accumulation discovered by previous drilling			6. DEVELOPMENT WELL	DEVELOPMENT WELL		DRY DEVELOPMENT WELL	

AAPG and API classification of wells.

Source: Fiscal Policy and the Energy Crises, Hearings before the Subcommittee on Energy of the Committee on Finance, United States Senate, Part 3 of 4 parts, January 2, 3, and 24, 1974.

OFFSHORE OIL RECOVERY

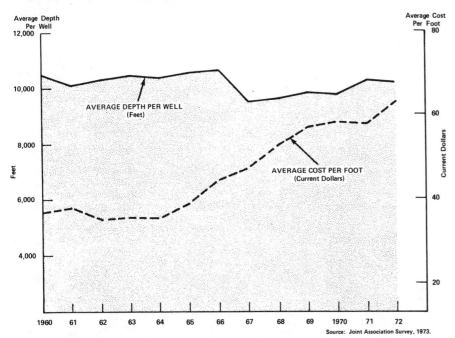

U.S. offshore oil well drilling depths and estimated costs, 1960–72.

Source: Energy Perspectives, U.S. Department of the Interior, February 1975.

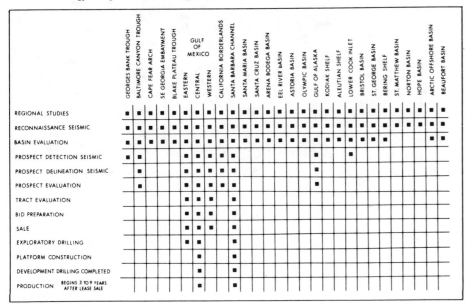

U.S.A. offshore work status.

Source: Offshore Oil, Shell Reports July, 1975.

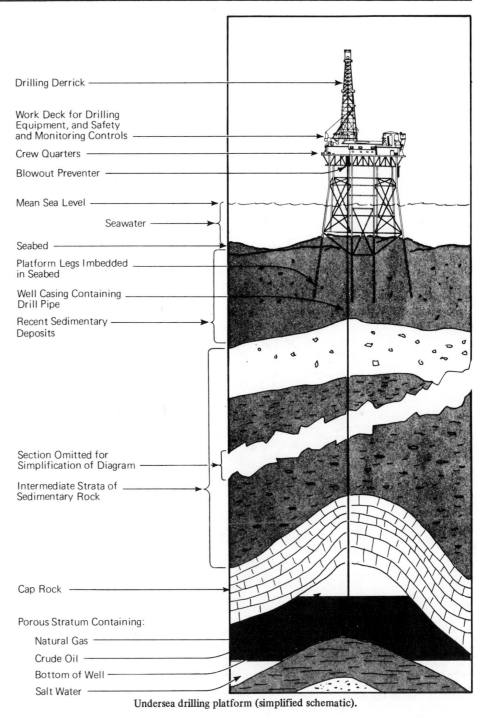

Undersea drilling platform (simplified schematic).

Source: Action in the Atlantic, American Gas Association, 1515 Wilson Boulevard, Arlington, Virginia, 22209, 1973.

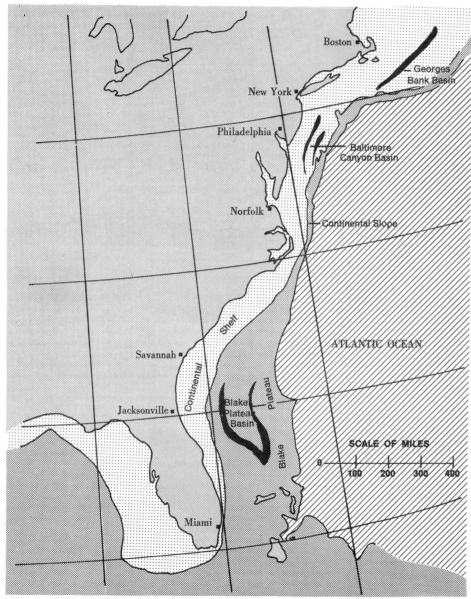

In the search for the needed new supplies of natural gas, rigs are going into more inhospitable areas, and drilling deeper; with costs as high as three million dollars per well. (*Courtesy Panhandle Eastern Pipe Line Co.*)

Map shows the three most promising areas for production of natural gas and oil on the Atlantic Outer Continental Shelf. Northernmost is the Georges Bank Basin, beginning 70 miles south of Cape Cod and 100 miles southeast of Long Island, and extending northeastward toward Nova Scotia. Forty miles off the Middle Atlantic states of Maryland, Delaware, and New Jersey lies the Baltimore Canyon Basin. And beginning 150 miles off the coasts of Florida, Georgia, and the Carolinas is the Blake Plateau Basin. All are thought to contain significant gas and oil bearing sediments, based on preliminary geological seismic studies. Similar estimates have been proven out in Canada, where important new discoveries of gas and oil have been made on the Canadian Atlantic Shelf, in a basin similar to the Georges Bank area off our own Atlantic Coast.

Source: Action in the Atlantic, American Gas Association, 1515 Wilson Boulevard, Arlington, Virginia, 22209, 1973.

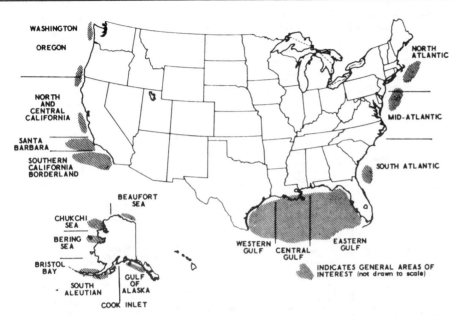

OCS regions of interest for oil/gas exploration.

Source: An Analysis of the Feasibility of Separating Exploration from Production of Oil and Gas on the Outer Continental Shelf, Congress of the United States, Office of Technology Assessment, May 1975.

Drilling rig in Gulf of Mexico. (*Courtesy Panhandle Eastern Pipe Line Co.*)

ENHANCED OIL AND GAS RECOVERY

Estimates indicate there were some 440 billion barrels of oil originally in place in discovered reservoirs in the United States. Of this amount, about 105 billion barrels have already been extracted and an additional 35 billion barrels can be produced by conventional techniques. The recovery of the remaining 305 billion barrels is the object of enhanced recovery techniques. Estimates of the potential for enhanced recovery range from 15 to 50 billion barrels.

The most significant methods of stimulating oil recovery are fluid injection (air, gas, water, steam, miscible fluids), earth fracturing by hydraulic pressure, *in situ* combustion (to reduce oil viscosity and promote flow), and the use of chemical explosives to fracture oil-bearing structures.

Underground reservoirs of oil and gas are not spaces filled with pure oil or gas but rather formations of porous rock with the gas and oil filling the pores in the rock. These reservoirs are under pressure, and gas and oil will flow into a well at a rate depending upon the pressure, the porosity of the rock, and the fluidity of the oil or gas. It was found in some fields that the pressure was maintained by an expanding mass of water below or at the edges of the oil or gas reservoir. The first enhanced recovery technique, that of injection of water, or water flooding, was first discovered by accident in the late 1890s and was widely used in Pennsylvania at that time and later in other states.

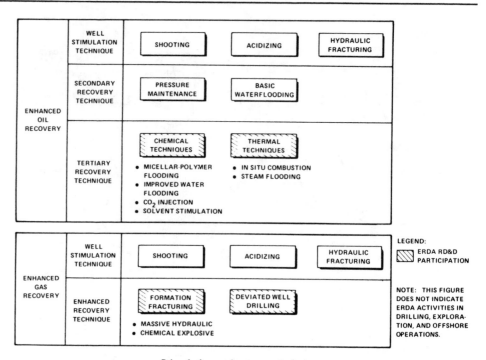

Stimulation and recovery techniques.

Source: Fossil Energy Research Program of the Energy Research and Development Administration, ERDA 77-33, April 1977.

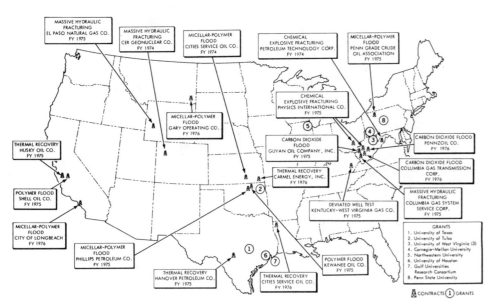

Contracts and grants for cooperative research on enhancement of recovery of oil and gas (Energy Research and Development Administration).

Source: Enhancement of Recovery of Oil and Gas, BERC-76/3, Progress Review No. 7, Energy Research and Development Administration, July 1976.

Enhanced movement of oil through the pore spaces in rock can also be achieved by reducing the viscosity of the oil by mixing it with other materials—gases such as carbon dioxide or propane or aqueous solutions containing surfactants not unlike household detergents. Thermal methods to heat and thin the oil include the injection of steam and underground ignition and burning of part of the oil to heat the oil-bearing rock. The practicality of the more advanced recovery techniques is strongly dependent on the nature of individual oil fields and, of course, on the price at which the oil can be sold.

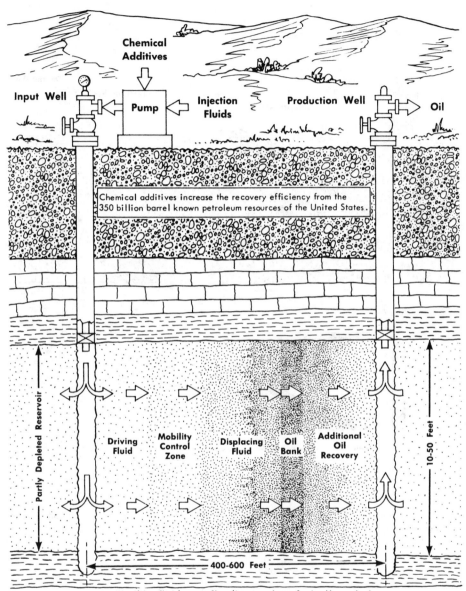

In miscible-fluid waterflooding, a slug of micellar solution is moved through the reservoir by water to displace oil.

Improving secondary and tertiary methods of petroleum recovery.

Source: Enhancement of Recovery of Oil and Gas, Progress Review No. 1, Energy Research and Development Administration, January 1975.

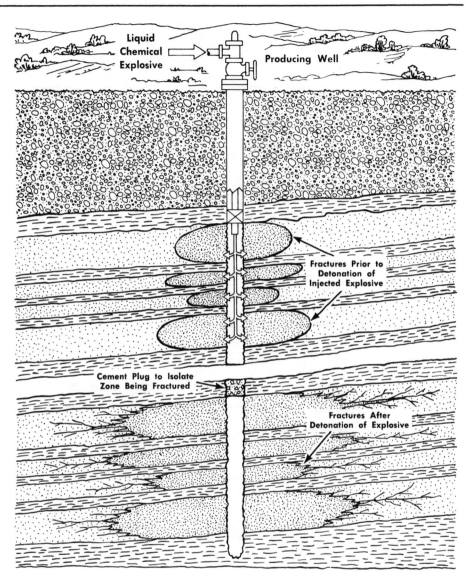

The injection and detonation of liquid chemical explosive increase production of petroleum from tight formations.

Chemical explosive fracturing.

Source: Enhancement of Recovery of Oil and Gas, Progress Review No. 1, Energy Research and Development Administration, January 1975.

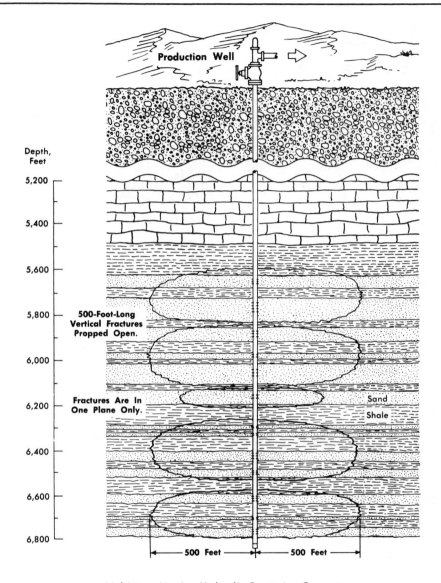

Multistage Massive Hydraulic Fracturing Concept.

Production stimulation by massive hydraulic fracturing.

Source: Enhancement of Recovery of Oil and Gas, Progress Review No. 1., Energy Research and Development Administration, January 1975.

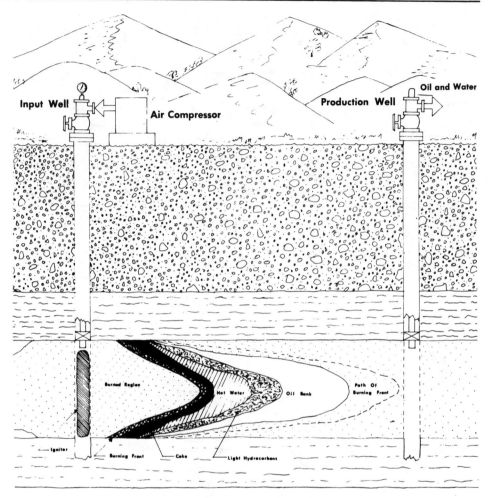

Stimulation by combustion drive.

Source: Enhancement of Recovery of Oil and Gas, BERC-75/3, U.S. Energy Research and Development Administration, July 1975.

ENHANCED GAS RECOVERY—NUCLEAR STIMULATION

About 317 trillion cubic feet of natural gas is estimated to be recoverable from "tight" sandstone formations in western United States by use of nuclear explosives to fracture these formations. Three stimulation tests have been performed. The Gasbuggy test in 1967 in northwestern New Mexico was conducted with a 26-kiloton nuclear explosive; the Rulison test in 1969 was conducted with a 43-kiloton nuclear explosive at a site in western Colorado. Both tests indicated a significant increase in gas production. A third test, the Rio Blanco experiment, was conducted about 35 miles from the Rulison experiment in Colorado. The Rio Blanco experiment involved the detonation of three 30-kiloton nuclear explosives in the same wellbore at about 350-foot intervals more than a mile underground. The Rio Blanco test was intended to stimulate gas flow in the sandstones in the Mesa Verde and Ft. Union formations by interconnection of the gas-bearing sands in a 1400-foot horizon. For reasons still unknown, the fracture zones of the three explosives did not produce interconnecting fractures as anticipated.

A fourth nuclear gas stimulation experiment, the Wagon Wheel program, involving the use of five sequentially detonated explosives of 100 kilotons each, planned for southwestern Wyoming, has been indefinitely deferred because of public opposition.

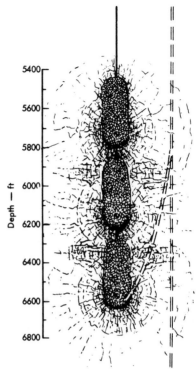

Cross section of the Rio Blanco experiment. The dashed lines show the proposed postshot fracture and pressure monitoring hole.

Source: Trends in Oil and Gas Exploration, Hearings before the Committee on Interior and Insular Affairs, United States Senate, August 8 and 9, 1972.

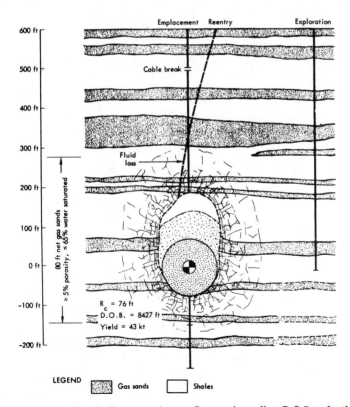

Postshot cross section of the Rulison experiment. R_c = cavity radius; D.O.B. = depth of burial.

Source: Trends in Oil and Gas Exploration, Hearings before the Committee on Interior and Insular Affairs, United States Senate, August 8 and 9, 1972.

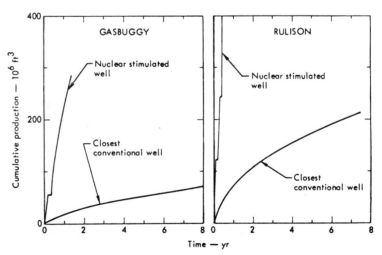

Gas production from the Gasbuggy and Rulison experiments.

Source: Trends in Oil and Gas Exploration, Hearings before the Committee on Interior and Insular Affairs, United States Senate, August 8 and 9, 1972.

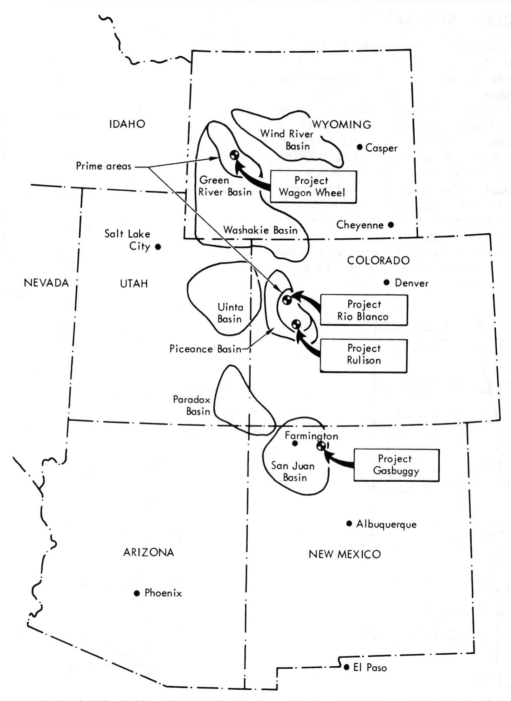

Low-permeability natural gas basins of the Rocky Mountain states. These basins contain reservoirs thick enough to be considered for nuclear stimulation.

Source: Trends in Oil and Gas Exploration, Hearings before the Committee on Interior and Insular Affairs, United States Senate, August 8 and 9, 1972.

CHARACTERISTICS OF PETROLEUM PRODUCTS

Natural petroleum products, gas and oil, vary considerably in composition depending upon the field from which they are taken. Generally, however, natural gas can be used as it is produced with modest separation of gas oil and other impurities while oil is subjected to extensive refining and separation processes. There are two broad classifications of crude oil—the paraffinic type, as produced in the Rostam field in the Arabian Gulf, with high API gravity and low sulfur content as well as other contaminents such as metals and nitrogen; and the naphthenic type, typical of the San Ardo field in California, which has a low API gravity, may be either high or low in sulfur content, and is often high in nitrogen and metals. Crude oils are also classified as sweet or sour—a sweet crude has a sulfur content below 0.5%, does not contain hydrogen sulfide, and has only a minor portion of the sulfur content present as mercaptans.

Each refinery processes a different mixture of crude oils and it is not possible to describe the operation of an "average" refinery. Process stages are similar, however. The crude oil is first distilled into four fractions of different boiling range—a fraction of low boiling point that forms straight-run gasoline; a fraction boiling between 400°F and 650°F that is called the straight-run distillate and is sold primarily as kerosine, jet fuel, Nos. 1 and 2 fuel oil, and diesel fuel; a fraction that boils between 650°F and 850°F called the gas oil fraction that is usually processed further to produce marketable products of lower boiling point; and a fraction that boils at temperatures over 850°F that is also processed further to produce fuels or asphalt.

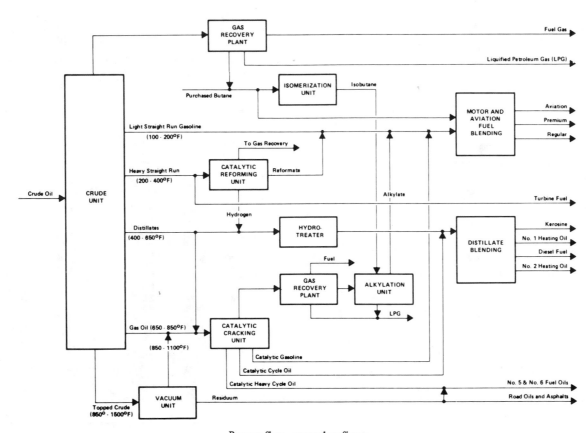

Process flow—example refinery.

Source: Factors Affecting U.S. Petroleum Refining: A Summary, National Petroleum Council, Washington, D.C., May 1973.

Range of Analyses of Fuel Oils

Grade of Fuel Oil	No. 1	No. 2	No. 4	No. 5	No. 6
Weight, percent					
Sulfur	0.01-0.5	0.05-1.0	0.2-2.0	0.5-3.0	0.7-3.5
Hydrogen	13.3-14.1	11.8-13.9	(10.6-13.0)°	(10.5-12.0)°	(9.5-12.0)°
Carbon	85.9-86.7	86.1-88.2	(86.5-89.2)°	(86.5-89.2)°	(86.5-90.2)°
Nitrogen	Nil-0.1	Nil-0.1	—	—	—
Oxygen	—	—	—	—	—
Ash	—	—	0-0.1	0-0.1	0.01-0.5
Gravity					
Deg API	40-44	28-40	15-30	14-22	7-22
Specific	0.825-0.806	0.887-0.825	0.966-0.876	0.972-0.922	1.022-0.922
Lb per gal	6.87-6.71	7.39-6.87	8.04-7.30	8.10-7.68	8.51-7.68
Pour point, F	0 to −50	0 to −40	−10 to +50	−10 to +80	+15 to +85
Viscosity					
Centistokes @ 100F	1.4-2.2	1.9-3.0	10.5-65	65-200	260-750
SUS @ 100F	—	32-38	60-300	—	—
SSF @ 122F	—	—	—	20-40	45-300
Water & sediment, vol %	—	0-0.1	tr to 1.0	0.05-1.0	0.05-2.0
Heating value					
Btu per lb, gross (calculated)	19,670-19,860	19,170-19,750	18,280-19,400	18,100-19,020	17,410-18,990

° Estimated.

Source: Steam: Its Generation and Use, 38th Edition, Babcock and Wilcox, 161 East 42nd Street, New York, N.Y., 10017, 1972.

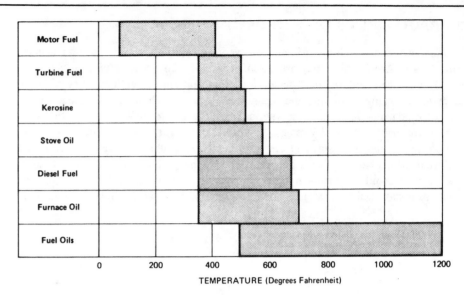

Typical product boiling ranges.

Source: Factors Affecting U.S. Petroleum Refining: A Summary, National Petroleum Council, Washington, D.C., May 1973.

Selected Samples of Natural Gas from United States Fields

	Sample No. Source of Gas	1 Pa.	2 So. Cal.	3 Ohio	4 La.	5 Okla.
Analyses						
Constituents, % by vol						
H_2	Hydrogen	—	—	1.82	—	—
CH_4	Methane	83.40	84.00	93.33	90.00	84.10
C_2H_4	Ethylene	—	—	0.25	—	—
C_2H_6	Ethane	15.80	14.80	—	5.00	6.70
CO	Carbon monoxide	—	—	0.45	—	—
CO_2	Carbon dioxide	—	0.70	0.22	—	0.80
N_2	Nitrogen	0.80	0.50	3.40	5.00	8.40
O_2	Oxygen	—	—	0.35	—	—
H_2S	Hydrogen sulfide	—	—	0.18	—	—
Ultimate, % by wt						
S	Sulfur	—	—	0.34	—	—
H_2	Hydrogen	23.53	23.30	23.20	22.68	20.85
C	Carbon	75.25	74.72	69.12	69.26	64.84
N_2	Nitrogen	1.22	0.76	5.76	8.06	12.90
O_2	Oxygen	—	1.22	1.58	—	1.41
Specific gravity (rel to air)		0.636	0.636	0.567	0.600	0.630
Higher heat value						
Btu/cu ft @ 60F & 30 in. Hg		1,129	1,116	964	1,002	974
Btu/lb of fuel		23,170	22,904	22,077	21,824	20,160

Source: Steam: Its Generation and Use, 38th Edition, Babcock and Wilcox, 161 East 42nd Street, New York, N.Y. 10017, 1972.

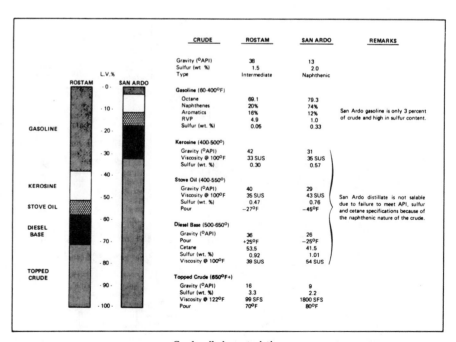

Crude oil characteristics.

Source: Factors Affecting U.S. Petroleum Refining: A Summary, National Petroleum Council, Washington, D.C., May 1973.

OIL SHALE RECOVERY

Oil shale is a finely textured sedimentary rock often with a laminar structure. The deposits in Colorado are highly consolidated and impervious. Mixed with this rock is a solid, insoluble, organic material called kerogen—a mixture of complex compounds, typically macromolecules with a naphthenic structure and a molecular weight over 3000. When the oil shale is heated, the kerogen decomposes and yields a crude oil that can be refined to produce various familiar petroleum products.

Oil shale industries have been operated in eight countries of the world during the past

The Colony Plant (near Grand Valley in the Piceance Basin), after years of research, has taken oil shale from an underground mine and has successfully processed it in an experimental retort. Colony, a joint venture of the Oil Shale Corporation and Atlantic Richfield, is now planning a full scale operation. It will employ 1000 persons and produce between 40,000 and 85,000 barrels of synthetic crude oil a day. Near Rifle, CO. (*Courtesy David Hiser, August, 1973*)

Oil Shale Industries of the Past

Nation	Period of Operation	Peak Shale Throughput (tons/year)
South Africa	1935–62	~250,000
Australia	1940–52	350,000 (1947)
France	until 1957	500,000 (1950)
Spain	until 1966	1,000,000 (late 1950s)
Sweden	1940–66	~2,000,000
Scotland ·	1862–62	3,300,000 (1913)

Source: Proposed Final Environmental Statement: Liquid Metal Fast Breeder Reactor Program, vol. III, WASH-1535, U.S. Atomic Energy Commission, December 1974.

Sample Composition of Oil Shale Section Averaging 25 Gallons of Oil per Ton

	Weight-Percent
Organic matter:	
Content of raw shale	13.8
Ultimate composition:	
Carbon	80.5
Hydrogen	10.3
Nitrogen	2.4
Sulfur	1.0
Oxygen	5.8
Total	100.0
Mineral matter:	
Content of raw shale	86.2
Estimated mineral constituents:	
Carbonates, principally dolomite	48
Feldspars	21
Quartz	13
Clays, principally, illite	13
Analcite	4
Pyrite	1
Total	100

Source: Mineral Facts and Problems. Bureau of Mines Bulletin 650. U.S. Department of the Interior. (Washington, D.C., U.S. Government Printing Office, 1970), p. 185.

Source: Energy from Oil Shale: Technical, Environmental, Economic, Legislative, and Policy Aspects of an Undeveloped Energy Source, report prepared for the Subcommittee on Energy of the Committee on Science and Astronautics, U.S. House of Representatives, Serial I, November 1973.

century. Only the industries in the Soviet Union and mainland China are in commercial operation today although an experimental oil shale plant of 2500 tons/day is operating in Brazil. During the 1945–70 period, shale was mined at an average annual rate of 13 million tons in the Soviet Union, largely from underground mines. New open-pit mines are being opened and production is expected to increase to 30–35 million tons/year.

About 60% of the Soviet shale has been burned directly in pulverized-shale boilers to produce electric power (the kerogen content is two or three times higher than in United States shales and the shale has a heating value of about 5800 Btu lb). The remaining 40% of Soviet oil shale is retorted to produce crude naphtha, petrochemical products, and heating gas.

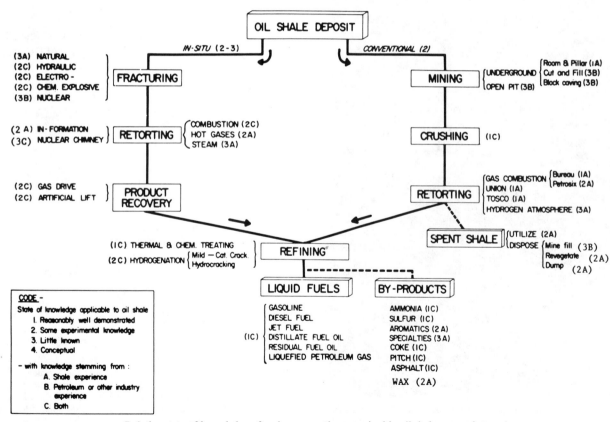

Relative state of knowledge of various operations required in oil shale processing.

Source: Proposed Final Environmental Statement: Liquid Metal Fast Breeder Reactor Program, Volume III, WASH-1535, U.S. Atomic Energy Commission, December 1974.

OIL SHALE SURFACE PROCESSING

Of the many surface retorting processes which have been proposed, only three (Bureau of Mines Gas Combustion, Union Oil, and Tosco) have been developed in the United States to a significant extent. All three systems operate at retorting temperatures near 900°F with local temperatures as high as 2200°F. Extraction efficiencies, based on upgraded oil yield, is usually 85–95% of the crude shale-oil potential.

The kerogen distills into about 66% oil, 25% cokelike solid, and about 9% combustible gas containing some hydrogen sulfide. The retort gas, which has a heating value of about 100 Btu/ft^3, is usually recycled and burned to provide additional heat for retorting. The raw oil is a dark, viscous liquid and is refined to yield low-sulfur liquid fuels and by-products such as ammonia, sulfur, coke, pitch, asphalt, and aromatic chemicals.

Disposal of spent, pulverized shale from the retorts poses a particular problem. A 100,000 barrel/day oil plant would require the disposal of 45 to 60 million tons of spent shale per year either by recycle to the mine, surface layer compaction, or dumping.

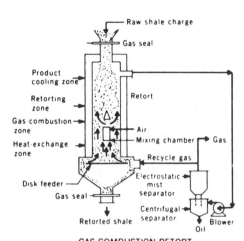

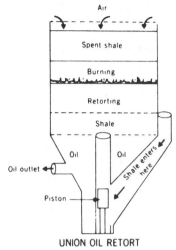

GAS-COMBUSTION RETORT

Recycle gas is mixed with air and burned within the retort. Gases flow upward and shale moves downward.

UNION OIL RETORT

Shale is introduced near bottom of retort and forced upward. Air enters at the top and flows downward.

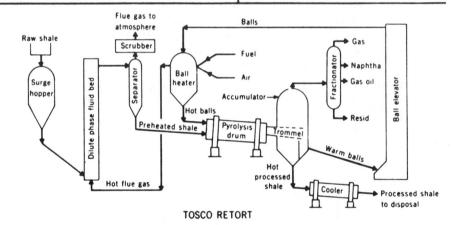

TOSCO RETORT

Ceramic balls transfer heat to the shale. No combustion takes place in retort.

Schematic representations of three oil shale retorting processes.

Source: Proposed Final Environmental Statement: Liquid Metal Fast Breeder Reactor Program, Volume III, WASH-1535, U.S. Atomic Energy Commission, December 1974.

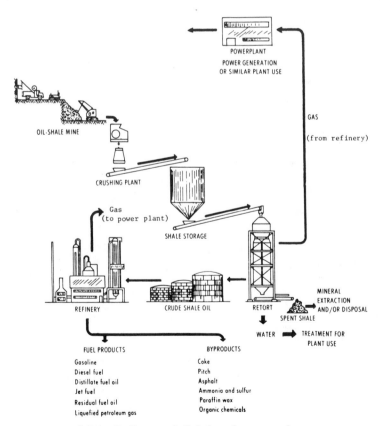

Schematic diagram of oil shale surface processing.

Source: Proposed Final Environmental Statement: Liquid Metal Fast Breeder Reactor Program, Volume III, WASH-1535, U.S. Atomic Energy Commission, December 1974.

OIL SHALE IN SITU PROCESSING

During the past 20 years, a number of projects have been undertaken to recover the oil in shale by *in situ* methods, that is, without removing the shale from the ground. The *in situ* concept presents a number of potential advantages—it might be applied to deposits that cannot be mined easily, it avoids the problems of disposal of spent shale, and avoids some of the costs of mining, crushing, and retorting which represent about 60% of the cost of producing shale by surface mining techniques.

Two general methods for *in situ* recovery have been tried. One method, first tested by Sinclair Oil in 1953, uses a pattern of wells drilled from the surface into the shale formation. The formation is then fractured by hydraulic or explosive techniques, and the shale ignited at one of the wells. Air is injected into this well to support the combustion of the shale and to force the hot combustion gases, oil, and other products to the other wells where they are extracted to the surface. The Sinclair tests confirmed that the fracturing and combustion steps were possible but that the fracturing techniques did not produce sufficient heat transfer surface underground to permit significant extraction of the oil content of the shale. Adequate recovery by underground methods apparently requires a void fraction of 15-35% in the shale.

The other *in situ* recovery method assures the creation of adequate heat transfer surface in the shale by mining techniques. Mine shafts or drifts are cut into the shale, some shale is removed for retorting above ground while the remainder is fragmented underground and retorted in place. The shale rubble is ignited at the top of the underground cavity and the combustion and retorting zones move through the shale rubble from top to bottom. Oil is condensed and collected at the bottom of the chamber. The use of nuclear explosives to create underground chambers filled with shale rubble has also been suggested.

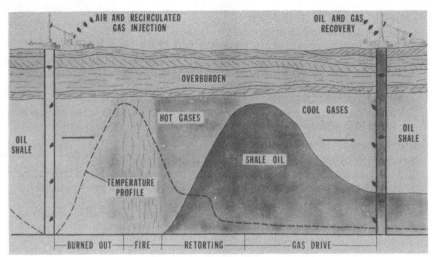

Schematic representation of an *in situ* retorting operation.*

Source: Energy from Oil Shale: Technical, Environmental, Economic, Legislative, and *Policy Aspects of an Undeveloped Energy Source,* report prepared for the Subcommittee on Energy of the Committee on Science and Astronautics, U.S. House of Representatives, Serial I, November 1973.

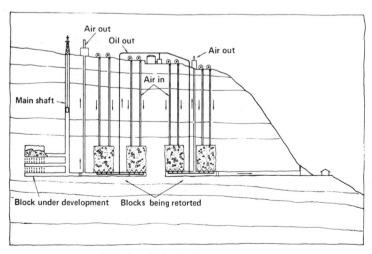

Oil shale rubblization/recovery process.

Source: James S. Kahn, "A Proposal for an AEC *In Situ* Oil Shale Program," in *In Situ Recovery of Shale Oil,* NSF-RA-N-75-001, resulting from the UCSD/NSF(RANN) Workshop held at the University of California, San Diego, September 3 to 7, 1974, sponsored by the National Science Foundation.

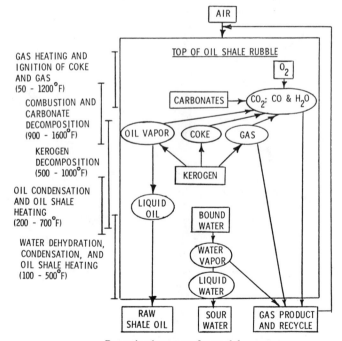

Retorting heat transfer model.

Source: P. W. Snyder, Jr., T. H. Timmins, and W. F. Johnson, "An Evaluation of In Situ Oil Shale Retorting," in *In Situ Recovery of Shale Oil,* NSF-RA-N-75-001, resulting from the UCSD/NSF(RANN) Workshop held at the University of California, San Diego, September 3 to 7, 1974, sponsored by the National Science Foundation.

TAR SANDS RECOVERY

Tar sands, also known as oil sands and bituminous sands, are deposits of sand or rock containing hydrocarbon (bitumen) material that is viscous, semisolid, or solid in form and cannot be recovered by conventional crude-oil production techniques. The bitumen content may be as high as 25% but is usually considerably lower. A deposit containing 14% bitumen is considered rich.

The exploitation of tar sand deposits has been undertaken on a small scale at LaBrea (Trinidad), Selenizza (Albania), Dernea (Romania), and Cheildag (USSR). The first major venture for producing synthetic crude from tar sands was begun in 1967 by the Great Canadian Oil Sands, Ltd., in the Athabasca area in the northeastern part of the Province of Alberta, Canada. The Athabascan deposit contains over 700 billion barrels of bitumen and is the world's largest. The other major tar sands deposits occur in eastern Venezuela in the Orinoco Tar Belt and in the Llanos area of Colombia.

The Canadian process employs bucket-wheel excavators to strip the tar sands and transfer them to a conveyor belt which feeds to a hot water extraction plant. About 3.5 tons of sand and overburden must be moved for each barrel of syncrude produced. The separated crude bitumen is upgraded to synthetic crude oil by coking and hydrogenation of the coker distillates. Major difficulties have been encountered in the equipment used to excavate and transfer the tar sands because of the properties of the tar sands which vary with particle size, oil saturation, moisture content, and temperature. At freezing temperatures the bitumen becomes almost solid and binds the tar sands into a mixture difficult to excavate. When the weather is warm, the tar sands become sticky and cling to everything they touch—requiring continual cleaning of equipment if it is to be kept operating.

Its oil scrubbed out, the moist and rich looking oil sands become the clean, white powdery granules commonly found on ocean beaches. (*Courtesy Sun Oil Co.*)

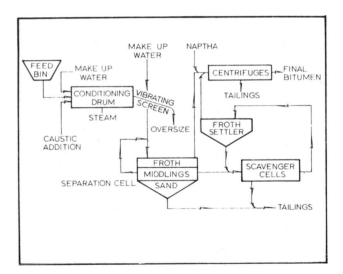

Oil is removed from tar sands by mixing the sands with hot water in a flotation tank. Globules of oil rise to the surface of the tank and the sand sinks to the bottom. The oil-rich froth from the surface of the tank is then processed to remove water and fine sand. The asphaltic crude oil is heated to separate the petroleum fractions and to produce coke.

Source: A. R. Allen, "Coping with the Oil Sands," Great Canadian Oil Sands, Ltd., 1974.

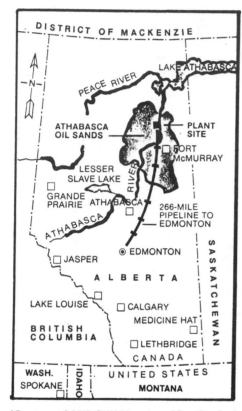

(Courtesy of OUR SUN Magazine of Sun Co., Inc.)

Completed GCOS complex on the Athabasca River marks the world's first commercial venture to produce synthetic crude oil from the tar sands. *(Courtesy Sun Oil Co.)*

Giant bucketwheel excavator *(Courtesy Sun Oil Co.)*

Nuclear Power

Haddam Neck nuclear power plant. The 575 electrical megawatt nuclear power plant of the Connecticut Yankee Atomic Power Company at Haddam Neck, Connecticut, incorporates a pressurized water reactor. The plant began commercial operation in 1968. (*Courtesy Connecticut Yankee Atomic Power Company*)

NUCLEAR POWER FORECASTS

Two studies,[1] one by the Organisation for Economic Cooperation and Development and the other by the U.S. Atomic Energy Commission, provide projections for the growth of nuclear power throughout the world which indicate that by the year 2000, nuclear power may provide nearly 50% of the electrical energy and about 20% of the total energy used in these countries.

Uncertainties in these estimates are highlighted by the assumptions made for the various cases of the AEC report:

Cases A and X represent the lowest forecasts and assume that the long-term growth in power demand will be low and that the trend in construction time slippage will continue.

Cases C and Z are the highest forecasts and assume improvements in the regulatory process, substantial improvement in construction schedules, a relatively high growth in power demand and the predominance of nuclear over fossil power plants for new plant additions.

[1]*Energy Prospects to 1985*, Organisation for Economic Cooperation and Development, Paris, 1974.
Nuclear Power Growth—1974–2000, WASH-1139(74), U.S. Atomic Energy Commission, 1974.

Cases B and Y are based on the continuation of current trends in growth of power demand, and some improvement in construction schedules and the regulatory process.

Case D represents a low long-term growth in the demand for electricity, actions to achieve conservation of energy, and improvements in regulation and construction schedules.

New AEC Forecast of Nuclear Power Capacity

(thousands of megawatts at end of calendar year)

United States Cases	February 1974 Forecast					WASH-1139(72)			
	1980	1985	1990	2000		1980	1985	1990	2000
A	85	231	410	850	Low	127	256	412	825
B	102	260	500	1200	Most Likely	132	280	508	1200
C	112	275	575	1400	High	144	332	602	1500
D	102	250	475	1090	—	—	—	—	—
Foreign Cases									
X	113	290	640	1600	*Low	143	312	600	1635
Y	140	387	780	2130	*Most Likely	161	359	724	2060
Z	157	420	900	2550	*High	173	414	850	2500

* Excludes Peoples Republic of China.

Source: Nuclear Industry, The Monthly Magazine of the Atomic Industrial Forum, Inc., New York, August 1974.

NUCLEAR GENERATING UNIT CAPACITY

	kilowatts
■ Licensed To Operate	
65 licensed by NRC to operate	46,666,000
2 others authorized to operate (ERDA-owned)	940,000
▲ Being Built	
77 construction permits	82,354,500
12 site work authorized	12,954,000
● Planned	
54 reactors ordered	61,254,000
22 reactors not ordered*	26,660,000
232	230,828,500

* Since these units have not been ordered and site information is incomplete, there are no further references to them in this document. There are no symbols for units planned but not sited.

Because of space limitations, symbols do not reflect precise locations. See reverse side for site information.

Energy Research & Development Administration
June 30, 1977

Nuclear power reactors in the United States.

Point Beach nuclear power plant. The two 497 electrical megawatt nuclear power plants at Two Creeks, Wisconsin began operation in 1970 and 1972. The plants are operated by the Wisconsin Electric Power Co. and the Wisconsin Michigan Power Co. (*Courtesy Wisconsin Electric Power Co.*)

Nuclear Power Growth Estimate (GWe)

	1975	1976	1977	1978	1979	1980	1981	1982	1983	1984	1985	1986	1987	1988	1989	1990	1991	1992	2000
Australia														0.5	0.5	1	1	1	6
Austria	–	–	0.7	0.7	0.7	0.7	0.7	2	2	2	3	3	4	4	5	6	7	8	14
Belgium	1.7	1.7	1.7	1.7	1.7	3.5	4.5	5.5	6.5	8.5	9.5	10.5	12.5	13.5	14.5	16.5	17.5	18.5	30
Canada	2.5	3.3	4	4.8	6.1	7.2	8.3	9.5	12.5	14.7	18.4	22	26	30	36	41	46	51	115
Denmark	–	–	–	–	–	–	0.9	0.9	1.8	1.8	2.7	2.7	3.6	3.6	4.9	4.9	6.2	11.4	
Finland	–	–	0.4	0.4	1.5	1.5	1.5	1.5	2.7	2.7	3.9	3.9	3.9	3.9	4.9	4.9	5.9	5.9	13
France	2.3	5	5.9	7.7	13.2	20.4	27.2	33.8	40.8	47.8	56	62	69	76	83	90	97	104.0	170
Germany	3.2	7	9.1	10.6	14.1	19.1	24.6	28.6	32.6	39.6	44.6	51	57	64	70	77	82	87.0	134
Greece	–	–	–	–	–	–	–	–	–	0.6	0.6	0.6	1.2	1.2	1.2	1.2	1.8	1.8	4
Ireland	–	–	–	–	–	–	–	–	–	0.7	0.7	0.7	1.3	1.3	1.3	2	2	2	6
Italy	0.6	1.4	1.4	1.4	1.4	1.4	3.4	7.4	12.4	19.4	26.4	32	38	45	53	62	70	79	140
Japan	6.6	9	11	13	15	17	24	30	36	41	49	54	60	67	73	84	88	95	157
Luxembourg	–	–	–	–	–	–	–	–	–	1.2	1.2	1.2	1.2	1.2	1.2	1.2	1.2	1.2	1.2
Netherlands	0.5	0.5	0.5	0.5	0.5	0.5	1.5	1.5	2.5	2.5	3.5	3.5	4.5	5.5	6.5	7.5	7.5	8.8	16
New Zealand	–	–	–	–	–	–	–	–	–	–	–	–	1.2	1.2	1.2	1.2	1.2	1.2	3
Norway	–	–	–	–	–	–	–	–	–	–	–	–	0.9	0.9	0.9	1.8	1.8	1.8	4
Portugal	–	–	–	–	–	–	–	0.7	0.7	1.4	1.4	2.3	2.3	2.3	3.3	3.3	3.3	4.3	8
Spain	1.1	1.1	3	4.8	6.7	8.7	11.7	14.7	17.7	20.7	23.7	27	30	34	38	42	46	50	80
Sweden	3.2	3.2	4.7	5.6	6.5	7.4	8.3	8.3	9.3	10.3	11.3	12.3	13.3	14.3	15.3	16.3	17.3	17.3	24
Switzerland	1	1	1	1.9	2.9	3.8	4.8	5.8	6.9	6.9	8	8	8	8	8	8	8	8	12
Turkey	–	–	–	–	–	–	–	–	–	0.6	0.6	0.6	1.6	1.6	2.2	2.2	2.2	3.3	16
United Kingdom	4.8	7.2	8.4	10.4	11.1	11.1	11.1	10.6	12.5	14.7	15.4	16.3	19	23	27	31	37	43	115
United States	40.1	47.5	54.6	61.8	68.2	82.2	102	126	152	179	205	234	266	301	341	385	432	481	1 000
OECD, High Estimate	68	88	106	125	150	185	234	287	348	416	484	548	622	703	791	890	981	1 079	2 080
Low Estimate	68	86	101	118	139	171	215	264	324	376	437	491	553	620	692	774	847	926	1 685
African région[1]	–	–	–	–	–	–	0.5	1.4	1.4	2.3	3.1	3.9	3.9	5.4	5.4	6.9	7.7	8.9	29
American région[2]	0.3	0.3	0.9	0.9	2.8	3.6	4.4	7	8.6	12.2	14.4	18	22	26	31	35	41	47	147
Asian région[3]	0.7	0.9	2.3	2.9	4	5	8.3	12.1	15.8	21.7	28.2	35	44	51	61	72	84	94	224
TOTAL (High Estimate)	69	89	110	129	157	194	247	307	374	452	530	605	692	785	888	1 004	1 113	1 229	2 480
Low Estimate	69	87	105	122	146	179	227	285	342	409	479	542	615	693	778	875	961	1 056	2 005

1) Algeria, Egypt, Iraq, Kuwait, Morocco, Saudi Arabia, South Africa, Tunisia.
2) Argentina, Brazil, Chili, Colombia, Cuba, Mexico, Jamaica, Peru, Uruguay, Venezuela.
3) Bangladesh, Hongkong, India, Indonesia, Iran, Israel, Korea, Malaysia, Pakistan, Philippines, Singapore, Taiwan, Thailand.

Reproduced from the 1975 edition of *Uranium Resources; Production and Demand*, jointly prepared by NEA and IAEA and published by the Organization for Economic Cooperation and Development, Paris, France, 1975.

U.S. PROJECTED URANIUM REQUIREMENTS

The lifetime requirement for uranium oxide for a 1000 MWe light or heavy water nuclear power plant is about 4000 short tons of U_3O_8. The fuel requirement for the 235,000 megawatts of nuclear capacity in operation, being built, or planned in the United States (see 7-1) is, therefore, about 940,000 short tons over an operating lifetime of 25 years. Estimates of United States domestic reserves of U_3O_8 include about 600,000 tons in the proven category at $30 per pound and about 2,900,000 tons in the potential resource category (see 2-24).

The accompanying charts indicate the requirements for uranium oxide based on the projected growth in nuclear power over the next half century. The introduction by 1988 of the breeder reactor, which uses uranium some 50 times more efficiently than present reactors, could limit the total uranium requirement to that in the potential resource category.

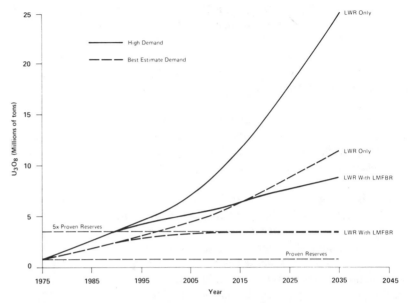

Lifetime U$_3$O$_8$ Commitments Versus Year of Commitment
(0.2% Tails; 8-years Allowed for Plant Construction)

Source: "Liquid Metal Fast Breeder Reactor: Decision Process and Issues," Electric Power Research Institute, Appendix A in *Issues for Consideration–Review of National Breeder Reactor Program,* Joint Committee on Atomic Energy, U.S. Congress, August 1975.

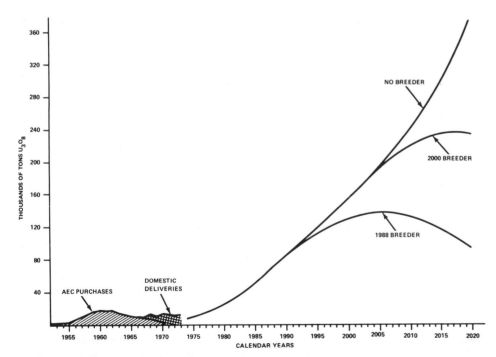

Projected Annual Domestic requirements for 1988, 2000 and no breeder reactors. (0.20% Tails Assay)

Source: The Nuclear Industry 1974, WASH 1174-74, United States Atomic Energy Commission, 1974.

Lifetime U$_3$O$_8$ Commitment Required @ 0.2% Tails

Reactor	ST U$_3$O$_8$/MWe [1]	SWU's/MWe	
		First Core	Annual Reload*
BWR - Pu	4.02	239	105
PWR - Pu	4.10 (Base = 1.0)	222	117
HWR - Nat U$_3$O$_8$ (Throw-away)	4.16 (~1.0)	—	—
ATR	2.60 (~0.63)	—	—
HWR - Th [2]	1.44 (~0.35)	519	19
HTGR (Std)	2.98 (~0.73)	366	85
HTGR (High Gain)	1.92 (~0.47)	336	45

Source: Milton Klein Report, ERDA-1 (Report of the Liquid Metal Fast Breeder Review Group)

(1) 1st Core + 24 FPY's
(2) Requires enriched U_3O_8 plus D_2O enrichment plant
* At 75% capacity factor.

NUCLEAR FUEL CYCLE

The utilization of uranium for the production of power involves a number of steps comprising what is termed the "nuclear fuel cycle."

Mining. Uranium ore is recovered from both open pit and underground mines. Concentrations of uranium in the ore are low, ranging from 2 to 20 pounds of U_3O_8 per ton with an average of 5 pounds per ton.

Milling. Uranium ore is pulverized and the uranium dissolved out of the ore by chemical reagents. The dissolved uranium is recovered from the leach liquor by solvent extraction or ion exchange techniques and, upon removal of excess water, the product is a crude uranium concentrate, containing 70 to 90% U_3O_8, known as "yellowcake."

Conversion. To be used as a reactor fuel, uranium must be refined to remove impurities which, in the reactor, would absorb neutrons unproductively and lower the efficiency of the power plant. The crude concentrates are purified by solvent extraction and the product is calcined to produce uranium trioxide (UO_3), a fine powder known as "orange oxide" because of its brilliant orange color. The orange oxide is converted by hydrogenation to uranium dioxide (UO_2) and this, in turn, is converted to uranium tetrafluoride (UF_4) by reaction with hydrogen fluoride gas, producing a product termed "green salt." This product is further processed to produce uranium hexafluoride (UF_6) by reaction with fluorine gas. This final product is volatile and is used in the next step in the fuel cycle, the enrichment process.

Enrichment. In order to be useful in most nuclear reactors, the concentration of U-235

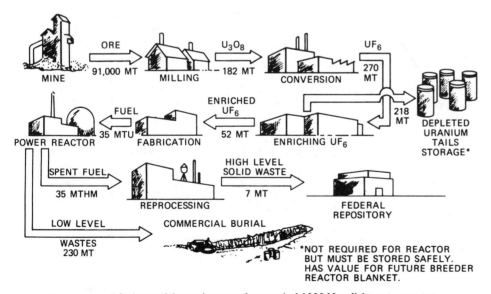

Average annual fuel materials requirements for a typical 1000 Mwe light water reactor.

Material Flow in the Nuclear Fuel Cycle

Material	Point to point movements	Packaging
Uranium ore	Uranium mines to uranium mills	None
Uranium ore concentrates	Uranium mill to feed preparation plant	Drums
Normal UF_6	Feed preparation plant to enriching (gaseous diffusion) plant	Pressurized cylinders
Enriched UF_6	Enriching plant to materials processing and/or fuel fabrication plant	Pressurized cylinders and protective packages
Enriched uranium oxide powder or pellets	Materials processing plant to fuel fabrication plant	Drums and birdcages with protective packages
New fabricated fuel assemblies	Fuel fabrication plant to nuclear power reactors	Special fuel element shipping containers
Low level radioactive wastes	All facilities to commercial burial sites	Drums and shielded protection packages
Irradiated fuel	Nuclear power reactor to chemical reprocessing plant	Shielded casks
Uranyl nitrate	Chemical reprocessing plant to fuel fabrication plant or to UF_6 feed preparation plant	Tank trucks, tank cars and protective packages
Plutonium oxide	Chemical reprocessing plant to fuel fabrication plant	Protective packages and special plutonium containers
Fission product radioisotopes	Chemical reprocessing plant to Government or industry users	Shielded casks
High level radioactive solid wastes	Chemical reprocessing plant to Federal repository	Shielded casks

Source: The Nuclear Industry 1974, WASH 1174-74, United States Atomic Energy Commission, 1974.

in the uranium must be increased from the 0.7% in natural uranium to 2 to 4%. This is accomplished in enrichment plants (see 7-5 to 7-7).

Fuel Fabrication. In the manufacture of fuel elements, the enriched UF_6 from the enrichment plants is first converted into other forms—uranium dioxide, uranium metal, or uranium carbide. These fuel materials are then fabricated into pellets or rods and placed in tubes, typically of stainless steel or zirconium, which are sealed and assembled in bundles to form fuel elements.

Fuel Use. During the approximately two years the fuel elements are in an operating nuclear power plant, some neutrons split the U-235 atoms in the fuel producing heat and radioactive fission fragments, while other neutrons convert some of the U-238 in the fuel into Plutonium-239 which is also fissionable and contributes to the fission chain reaction. After the U-235 is consumed to about $\frac{1}{4}$ of its original concentration, the fuel elements are removed from the reactor for reprocessing to recover the unused U-235 and the Pu-239.

Reprocessing. After removal from the reactor, the fuel elements contain intensely radioactive fission products and they are, therefore, stored under water for several months to allow this radioactivity to decay to a lower level. The elements are then shipped in shielded casks to a reprocessing plant.

Uranium Ore Processing Mills in the United States

Company and location	Capacity TPD. ore	Remarks
AEC licensed active (13):		
1. Anaconda, grants, N. Mex	3,000	
2. Kerr-McGee, Grants, N. Mex	7,000	
3. United Nuclear Homestake, Grants, N. Mex	3,500	
4. Mines Development, Edgemont, S. Dak	650	Currently processing vanadium only.
5. Atlas Minerals, Moab, Utah	1,500	
6. Rio Algom, La Sal, Utah	500	
7. Exxon, Powder River Basin, Wyo	2,000	
8. Federal American Partners, Gas Hills, Wyo	950	
9. Petrotomics, Shirley Basin, Wyo	1,500	
10. Union Carbide, Gas Hills, Wyo	1,000	
11. Utah International, Gas Hills, Wyo	1,200	
12. Utah International, Shirley Basin, Wyo	1,200	
13. Western Nuclear, Jeffrey City, Wyo	1,200	
AEC licensed closed (2):		
1. Foote Mineral, Shiprock, N. Mex		Shutdown 1968.
2. Susquehanna, Riverton, Wyo		Shutdown 1963.
Agreement state licensed active (4):		
1. Conoco & Pioneer Nuclear, Falls City, Tex	1,750	
2. Cotter Corp., Canon City, Colo	450	
3. Union Carbide, Uravan, Colo	1,200	
4. Dawn Mining, Ford, Wash	500	

Source: ERDA Authorizing Legislation Fiscal Year 1976, Hearings before the Subcommittee on Legislation of the Joint Committee on Atomic Energy, Congress of the United States, Part 2, February 18 and 27, 1975.

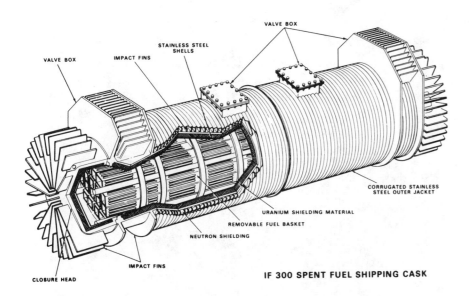

IF 300 SPENT FUEL SHIPPING CASK

Spent Fuel Case. A cutaway view of the largest and first multi-element spent fuel cask (IF-300) licensed by AEC. Weighing approximately 140,000 pounds, it can accommodate 7 pressurized water reactor fuel assemblies or 18 boiling water reactor assemblies, each up to 180 inches long, with a burnup capacity of 35,000 megawatt days per metric ton of uranium. (*Courtesy General Electric*)

Because of the radioactivity of the fuel elements, reprocessing operations are carried out remotely in shielded vaults. The fuel elements are disassembled mechanically and the fuel cladding removed. The fuel is then dissolved in acid and the fission products separated from the uranium and plutonium by a solvent extraction process. The uranium and plutonium are separated in a subsequent step by a similar process. The uranium may, at this point, be returned to an enrichment plant to restore the U-235 content to its former level.

Waste Disposal. The radioactive material remaining from fuel reprocessing is processed further to reduce its volume. Wastes with high levels of radioactivity are converted to ceramic beads for storage while lower level wastes are stored in liquid form (see 7-10).

Characteristics of 1000-MWe LWR Maximum Fuel Cycle Requirements

	Initial Core	Annual Reload	Lifetime Average Annual Fuel Requirement
Irradiation level (MWt-day/MTU)[a] $\times 10^{-3}$	24	33	33
Fresh fuel assay (wt% U-235)	2.26	3.21	3.21
Spent fuel assay (wt% U-235)	0.74	0.90	0.90
Ore supply (ST) $\times 10^{-3}$	237	91	96
Yellowcake U_3O_8 supply (ST)[a]	498	191	201
Natural UF_6 (ST)[a]	624	239	252
Separative work (MT SWU)[a]	174	94	99
Enriched UF_6 (ST)[a]	196	53	56
Enriched UO_2 (ST)[a]	150	41	43
Fuel loading (STU)[a]	132	36	38

[a]*Bases:*
    Reactor plant load factor—75%.
    Enrichment tails assay—0.3%.
    No plutonium recycle.
    Reloads include recovered uranium.
    Losses of 1% each in fuel fabrication and reprocessing.
        ST = short ton = 2000 lb.
        STU = short ton uranium.
       MTU = metric ton uranium.
  MT SWU = separative work (units in metric tons).

Source: Proposed Final Environmental Statement Liquid Metal Fast Breeder Reactor Program, Volume III, WASH-1535, U.S. Atomic Energy Commission, Washington, D.C. December 1974.

Uranium milling plant. The Humble Oil & Refining Company's Highland Uranium Mill near Casper, Wyoming, can handle about 2,000 tons of ore per day. (Norton Pearl Photography. U.S. Atomic Energy Commission)

URANIUM ENRICHMENT PLANT CAPACITY

The accompanying charts illustrate the requirement for additional enrichment plants to supply the fuel needs of nuclear power plants being built throughout the world. The present plants in the United States are in the process of being improved through the Cascade Improvement Program and the Cascade Uprating Program and could reach a level of about 27,000,000 separative work units[1] per year. World requirements in the year 2000 are projected to be 200,000,000 separative work units per year. To meet these requirements would involve the construction of some 18 new plants each with a capacity of 8.75 million SWU/year. United States requirements for enriched uranium will require the construction of six new plants of a similar size by the year 2000. These projections assume that the fast breeder reactor will be introduced in the year 1988.

[1]*Separative work units* are commonly used to express the capacity and production rate of a uranium enrichment plant. A separative work unit is a measure of the effort expended in the plant to separate a quantity of uranium of a given assay into two components, one having a higher percentage of U-235 and one having a lower percentage. Separative work is generally expressed in kilogram units. It is common practice to refer to a kilogram separative work unit simply as a separative work unit or a SWU. Plant capacity is also expressed in terms of *tonnes*, one tonne representing 1000 kilograms.

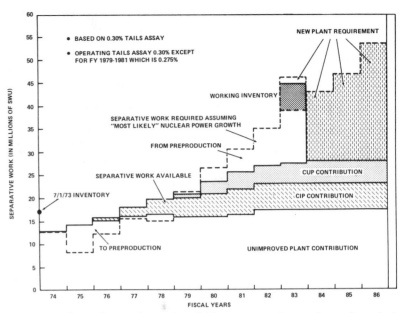

Projections of annual separative work available versus annual separative work required.

Source: The Nuclear Industry 1973, WASH 1174-73, U.S. Atomic Energy Commission 1973.

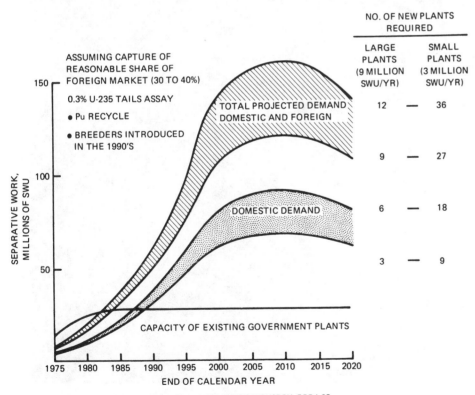

	NO. OF NEW PLANTS REQUIRED	
	LARGE PLANTS (9 MILLION SWU/YR)	SMALL PLANTS (3 MILLION SWU/YR)
	12 —	36
	9 —	27
	6 —	18
	3 —	9

Annual demand for U.S. separative work.

Source: A National Plan for Energy Research, Development and Demonstration: Creating Energy Choices for the Future, ERDA 76-1, Energy Research and Development Administration, April 15, 1976.

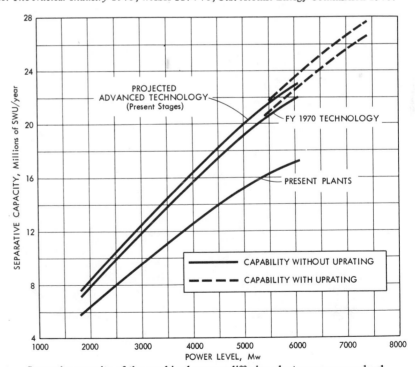

Separative capacity of the combined gaseous diffusion plants versus power level.

Source: AEC Gaseous Diffusion Plant Operations, ORO-684, United States Atomic Energy Commission, Washington, D.C. 20545, January 1972.

Oak Ridge gaseous diffusion plant. The U.S. Atomic Energy Commission's Oak Ridge (Tennessee) Gaseous Diffusion Plant is the first such facility built in the world for production of enriched uranium. The original gaseous diffusion process building "K-25," which is the U-shaped building, was constructed in 1943 and began operating in 1945 producing enriched uranium for the Manhattan Project of World War II. Today the enlarged plant, representing an investment of 841 million dollars produces enriched uranium for nuclear power plants around the world. The plant is operated for the AEC by Union Carbide Corporation. (USAEC photo by F. W. Hoffman)

URANIUM ENRICHMENT PROCESSES

As found in nature, uranium is composed of several isotopes, principally U-235 and U-238 which differ in weight by 3 mass units (the difference between 235 and 238), but which are identical chemically. Separation processes for these isotopes, therefore, are based usually on methods that take advantage of the difference in weight. Of the processes given in the accompanying list, all except two, the photoexcitation and nuclear spin processes, depend upon the weight differential between the two isotopes.

Two processes, the thermal diffusion and electromagnetic separation processes, were developed on a significant scale during World War II, but have since been abandoned. The bulk of the Western world's supply of enriched uranium has been produced by the gaseous diffusion process in plants built by the Atomic Energy Commission at Paducah, Kentucky; Portsmouth, Ohio; and Oak Ridge, Tennessee. Additional gaseous diffusion enrichment plants are under construction in France and are planned in the United States. Enrichment plants employing the centrifuge process are being built by Germany, The Netherlands, and the United Kingdom under a tripartite agreement among these countries. Plans are being considered for the construction of a commercial enrichment plant using the Becker separation nozzle process and a plant is under construction in South Africa which reportedly is based on a variant of this process. Recent attention has been focused on the development of a laser photoexcitation process which may prove to be practical. The enrichment processes of commercial interest are discussed in the pages following.

Isotope Enrichment Processes[a]

1. Phase equilibrium separation processes
   fractional distillation
   gas–liquid absorption
   fractional sublimation
   liquid–liquid extraction
   fractional crystallization
2. Chemical exchange separation processes
   exchange distillation
   gas–liquid chemical exchange
   liquid–liquid chemical exchange
   exchange chromatography
   redox ion exchange
3. Diffusion separation processes
   gaseous diffusion
   sweep and mass diffusion
   carrier diffusion
   thermal diffusion
   standing sound waves
   liquid diffusion
   electrical fields in liquids
4. Molecular flow processes
   molecular distillation
   isotope chopper
5. Aerodynamic processes
   centrifuge
   Becker separation nozzle
   vortex tube
   Fenn shock
6. Photoexcitation processes
7. Electromagnetic process
8. Nuclear spin processes

[a]Descriptions of most of these processes are given in: *Report of Uranium Isotope Separation Review Ad Hoc Committee*, ORO-694, Oak Ridge Operations Office, U.S. Atomic Energy Commission, June 1972.

Summary Comparison of Processes.

Section	Process	Separation Factor, $\alpha - 1$	Specific Energy Consumption, kw/kg USW/yr — Theoretical Minimum	Specific Energy Consumption, kw/kg USW/yr — Practical	Specific Uranium Inventory, kg/kg USW/yr	Unit Capital Cost, $/kg USW/yr
	Gaseous Diffusion	0.0043	0.073	0.266	< 0.1	100
3.1	Distillation of UF_6	< 0.00009	--	> 0.62	> 150	> 126
4.3	Liquid-Liquid Chemical Exchange, $NOUF_6$-UF_6	0.0016	--	--	--	--
4.5	Redox Ion Exchange	0.0005	--	> 0.71	> 22	> 137
5.1	Mass Diffusion with UF_6 and N43	0.00281	0.17	1.0-1.2	--	--
5.3	Thermal Diffusion with UF_6	--	61	--	--	--
5.6	Electromigration in UCl_4	--	--	1817	--	--
6.1	Molecular Distillation	< 0.0043	> 0.073	--	--	--
6.2	Isotope Chopper	< 0.0043	>> 0.073	--	--	--
7.1	Becker Separation Nozzle	0.015	--	0.656	--	< 164

Source: Report of Uranium Isotope Separation Review Ad Hoc Committee, ORO-694, Oak Ridge Operations Office, United States Atomic Energy Commission, June 1972.

THE GASEOUS DIFFUSION ENRICHMENT PROCESS

The gaseous diffusion process is based upon the phenomenon that, in a gas made up of molecules containing two isotopes, the lighter of the molecules will have an average higher velocity and will diffuse at a greater rate through a porous membrane. The maximum theoretical separation that can be achieved is equal to the square root of the ratio of the masses of the gas molecules. In the diffusion process, the process gas is uranium hexafluoride and the square root of the mass ratio is 1.00429. Since this number is close to unity, the degree of enrichment which can be achieved in a single diffusion stage is very small. A number of stages in series is therefore required—enrichment from the 0.711%

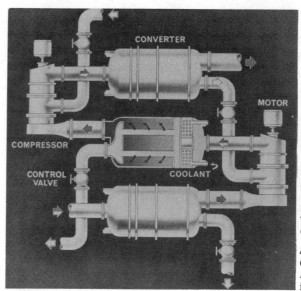

Stage arrangement. UF$_6$ gas is introduced inside the porous barrier tubes, about one-half of the gas diffuses through the barrier and is slightly enriched with respect to U-235 and is fed to the next higher stage; the remaining undiffused portion of the gas is recycled to the next lower stage. Axial flow compressors are used to compress the gas to maintain interstage flow and a gas cooler is provided to remove the heat generated during the compression of the gas.

Source: AEC Gaseous Diffusion Plant Operations, ORO-684, United States Atomic Energy Commission, Washington D.C. 20545, January 1972.

Interior section of an AEC gaseous diffusion plant. The gaseous diffusion process, which enriches uranium for use in atomic energy activities, is done at AEC-owned, contractor-operated plants in Kentucky, Ohio and Tennessee. In the process, gaseous uranium hexafluoride (UF$_6$) is put through a series of barriers, which partially separate the highly fissionable, lighter and faster-moving uranium-235 atoms from the heavier and slower-moving uranium-235 atoms that make up the bulk of the material. The photo shows sections of the basic equipment required for the process. An electric motor drives a compressor which, in turn, compresses the UF$_6$ gas so it will flow through the porous membranes in each converter tank. The product stream from one converter enters a central compressor stage and is mixed with a stream of depleted uranium from another converter stage tank. This mixture is compressed and fed to the central converter. The enriched product from this central converter moves to the next compressor stage upstream, while the depleted stream is sent to the stage below. Groups of stages are coupled in this way to make up operating units and these units, in turn, make up a gaseous diffusion cascade. (Oak Ridge Operations Office, Frank Hoffman)

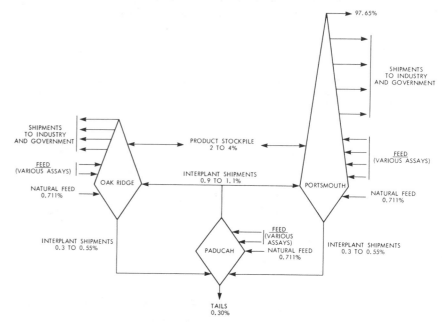

Mode of operation for gaseous diffusion plants (% Values Are Weight % U-235). The three gaseous diffusion plants at Oak Ridge, Paducah, and Portsmouth are operated in an integrated fashion resulting in a gain in separative capacity of about 1% in comparison to independent operation. Although the three plants are at some distance apart geographically, material flow is maintained by shipments of materials in cylinders.

Source: AEC Gaseous Diffusion Plant Operations, ORO-684, United States Atomic Energy Commission, Washington, D.C. 20545, January 1972.

U-235 found in natural uranium to the 4.0% required in a typical light water reactor, requires about 1200 stages, while enrichment to 90% requires about 3000 diffusion stages.

To take advantage of the small separation factor, the pore size in the diffusion membranes must be uniform and quite small—less than two millionths of an inch in diameter—to assure diffusive flow and not just simple gas flow. Because the pore size is so small, very large surface areas of diffusion barrier are required if significant production rates are to be achieved.

The nature of the diffusing gas, uranium hexafluoride, places a number of constraints on plant design. UF_6 is solid at room temperatures (UF_6 becomes a gas at 57°C) and, consequently, diffusion plants must be operated at temperatures and pressures sufficient to keep the UF_6 in gaseous form. In addition, although UF_6 is a stable compound, it is extremely reactive with water, it rapidly corrodes common metals and it is not compatible with organic materials such as lubricating oils. This chemical reactivity requires the use of construction metals such as nickel and aluminum in a leak-tight system.

$$\text{Kinetic Energy} = \tfrac{1}{2} MV^2$$

$$\tfrac{1}{2} M_1 V_1^2 = \tfrac{1}{2} M_2 V_2^2$$

$$\frac{V_1}{V_2} = \sqrt{\frac{M_2}{M_1}} = \alpha^*$$

For Uranium Isotopes In Gaseous UF_6

$$\alpha^* = \sqrt{\frac{M_{U^{238}F_6}}{M_{U^{235}F_6}}} = \sqrt{\frac{352}{349}} = \underline{1.00429}$$

Maximum theoretical separation factor α^*.

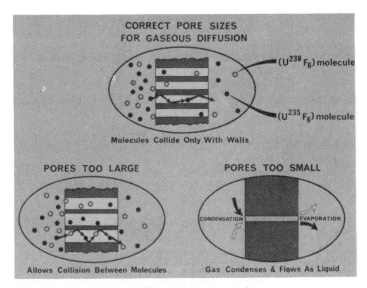

Molecular flow thru porous membrane.

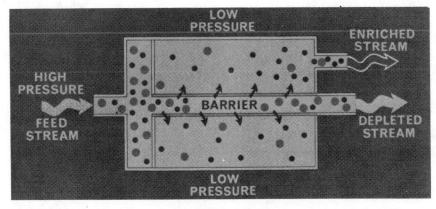

Gaseous diffusion stage.

Source: *AEC Gaseous Diffusion Plant Operations*, ORO-684, United States Atomic Energy Commission, Washington, D.C. 20545, January 1972.

ALTERNATE ENRICHMENT PROCESSES

Plants for the production of enriched uranium by the centrifuge process are being expanded at Capenhurst, England; and Almelo, Holland; under a tripartite agreement between Germany, The Netherlands and the United Kingdom. A capacity of 2,000,000 separative work units per year is planned by 1982.

Uranium hexafluoride gas introduced into a centrifuge rotating at high velocities, distributes itself in the intense gravitational field such that the concentration of the heavier U-238 isotope is slightly higher at the periphery than the concentration of the U-235 isotope. Countercurrent flow also produces an axial concentration difference which is several times larger than the radial concentration difference. Theoretically the *separative power* of a countercurrent centrifuge is:

$$S = \rho D \left(\frac{\Delta M v^2}{2RT} \right)^2 \frac{\pi L}{2}$$

where ρ is the density, D is the diffusion coefficient in the gas, ΔM is the difference in masses of the two isotopes, v is the peripheral velocity, L is the rotor length, R is the gas constant, and T is the temperature.

High production rates can be achieved by increasing the centrifuge rotational velocity and the centrifuge length. High rotational velocities require construction materials of high strength. Maximum operating centrifuge length depends on problems arising from resonance vibrations which develop in long rotating bodies at certain critical speeds determined by the ratio of the length to the diameter of the centrifuge. A centrifuge operating below the first resonance is termed a subcritical machine while a centrifuge operating at speeds above a resonance level is termed a supercritical machine.

The nozzle separation process, developed by Professor E. W. Becker at the Karlsruhe Nuclear Research Center in Germany, utilizes the pressure gradient developed in a curved expanding supersonic jet to effect a separation of uranium isotopes. The feed gas is comprised of about 5 mole % UF_6 and 95% helium, the latter gas being added to increase the velocity of the jet. As the gas traverses the curved wall, the heavier components are enriched at the wall surface. A knife-edge is placed to divide the gas into two fractions, a heavy fraction with an increased concentration of U-238 and a light fraction enriched in U-235. The process requires high-precision fabrication to maintain a uniform spacing between knife-edge and wall of 0.0005 inch. Energy consumption per unit of separative work in the Becker nozzle process is about twice that of the diffusion process.

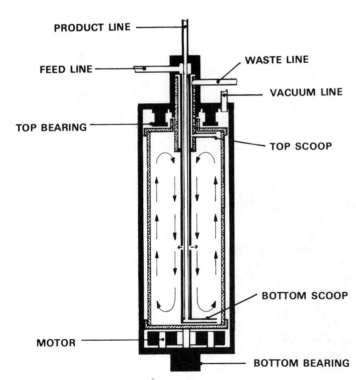

A counter current centrifuge.

Source: Urenco-Centec, 40 West Street, Marlow, Buckinghamshire, England.

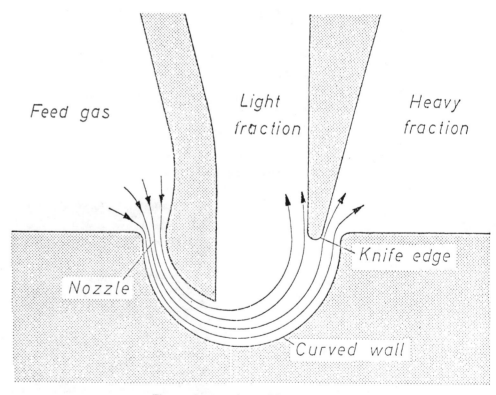

The nozzle separation enrichment process.

Source: Report of Uranium Isotope Separation Review Ad Hoc Committee, ORO-694, United States Atomic Energy Commission, June 1972.

A ten-stage pilot plant was operated at Karlsruhe in 1967 and development has continued since that time.

The laser enrichment process, currently at a laboratory stage of development, involves a two-step ionization process of U-235 with mass spectrographic separation of the uranium isotopes. In the process being developed at Lawrence Livermore Laboratory, uranium metal is ionized in an oven and then illuminated with a xenon laser at a frequency at which U-235 atoms are excited but U-238 atoms are not. The uranium vapor is then irradiated by a krypton laser which ionizes the U-235 atom excited by the xenon laser.

The ionized U-235 atoms are than deflected in an electric field and collected. Enrichments of 3% have been achieved in a single stage and theoretical efficiencies per stage are over 90%. Development problems included the tendency of uranium, at vaporization temperatures of 2100°C, to alloy with other metals, the need to develop high-power lasers that can operate continuously and can be tuned to specific frequencies, and the need to reduce the charge neutralization of the U-235 ions that takes place through collisions in the gas mixture.

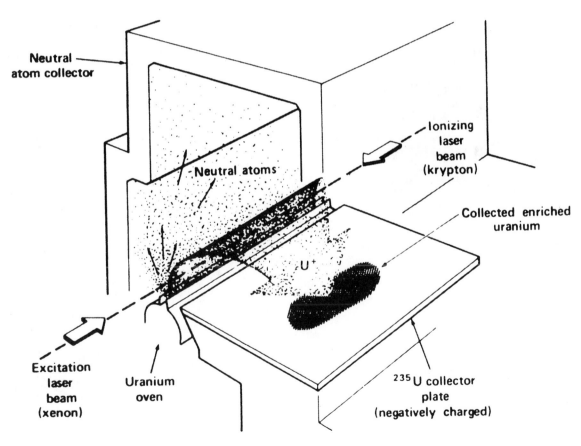

The laser enrichment process.

Source: Nuclear News, July 1975.

U.S. NUCLEAR FUEL FABRICATION CAPACITY

Existing Facilities for Fabrication of UO₂ Fuel Elements

Facility	MTU/Year
General Electric Co., Wilmington, N.C.	1000
Westinghouse Electric Corp., Columbia, S.C.	1000
Babcock & Wilcox Co., Lynchburg, Va.	300
Combustion Engineering, Inc., Windsor, Conn.	300
Gulf United Nuclear Fuels Corp.,[a] New Haven, Conn.	300
Exxon Nuclear Co., Inc., Richland, Wash.	150
	3050
Total "existing" capacity (assuming Gulf United Nuclear shut down)	2750 MTU/year

[a]In accordance with current plans, the Gulf United Nuclear Fuels Corp. facility will be shut down about April 1974—upon completion of existing contractual commitments.

Planned Additions to Existing Capacity

Facility	MTU/Year	Operational
Combustion Engineering, Inc.–New Plant	600	1978
General Electric Co., Wilmington, N.C. (increase capacity to 2700 MTU/year)	1700	1979
Exxon Nuclear Co., Inc.–New Plant	600	1980
Westinghouse Electric Corp., Columbia, S.C. (increase capacity to 2700 MTU/year)	1700	1980
Combustion Engineering, Inc., Windsor, Conn. (increase capacity to 600 MTU/year)	300	1980
Babcock & Wilcox Co., Lynchburg, Va. (increase capacity to 600 MTU/year)	300	1980
Exxon Nuclear Co., Inc., Richland, Wash. (increase capacity to 400 MTU/year	250	1980
Total 1978–80 additions to capacity	5450	

Total U.S. Capacity Projected for 1980–8200 MTU/year

Source: Nuclear Energy, Federal Energy Administration Project Independence Blueprint Final Task Force Report, November 1974.

Fuel tubes containing UO₂ pellets are supported by spacers to assure separation of individual tubes and an adequate flow of cooling water through the fuel assembly. (*Courtesy General Electric*)

BWR/6 FUEL
ASSEMBLIES
& CONTROL
ROD MODULE

1.TOP FUEL GUIDE
2.CHANNEL
FASTENER
3.UPPER TIE
PLATE
4.EXPANSION
SPRING
5.LOCKING TAB
6.CHANNEL
7.CONTROL ROD
8.FUEL ROD
9.SPACER
10.CORE PLATE
ASSEMBLY
11.LOWER
TIE PLATE
12.FUEL SUPPORT
PIECE
13.FUEL PELLETS
14.END PLUG
15.CHANNEL
SPACER
16.PLENUM
SPRING

GENERAL ⓖ ELECTRIC

BWR/6 fuel assemblies and control rod module.

The feed material for the fabrication of fuel for Light Water Reactors is UF_6 enriched from 2 to 4% in the U-235 isotope. The UF_6 is converted to UO_2 and the UO_2 is formed into high-density pellets, sintered, and precision-ground to final dimensions. Finished pellets are loaded into Zircaloy or stainless steel tubes, fitted with end caps and welded. (*Courtesy General Electric*)

U.S. REPROCESSING REQUIREMENTS AND CAPACITY

The accompanying tables represent two estimates of projected fuel reprocessing requirements and one estimate of the reprocessing capacity expected to be available. The estimates made in the National Academy of Engineering study were based on projections of nuclear power growth made in 1968. Delays in construction and deferral of a number of plants since that time have reduced the expected installed capacity in 1980 from 153,000 megawatts to nearer 100,000 megawatts.

The expected shortfall in reprocessing capacity will require the construction of additional spent-fuel storage pools at nuclear plants throughout the country.

Projection of U.S. Reprocessing Capacity
(in metric tonnes of uranium)

Adapted by Nuclear Industry from a July 1974 table by Nuclear Assurance Corp.

	Inventory	1974	1975	1976	1977	1978	1979	1980	1981	1982	1983	1984	1985
Capacity required for reactors firmly committed as of July '74	477	319	725	1162	1512	1701	2079	2212	3004	3829	3496	3840	3489
Planned capacity*													
AGNS		—	—	750	1500	1500	1500	1500	1500	1500	1500	1500	1500
NFS		—	—	—	—	—	375	750	750	750	750	750	750
GE		—	—	—	—	—	—	300	300	300	500	500	500
Total capacity		—	—	750	1500	1500	1875	2550	2550	2550	2750	2750	2750
Probable annual throughput		—	—	750	1500	1500	1875	2550	2550	2550	2750	2750	2750
Unreprocessed inventory	477	796	1521	1933	1945	2146	2350	2012	2466	3745	4491	5581	6320
Additional capacity if Exxon plant of 1,500 tonne/yr. capacity built		—	—	—	—	—	—	—	—	750	1500	1500	1500
Unreprocessed inventory with Exxon	477	796	1521	1933	1945	2146	2350	2012	2466	2995	2241	1831	1070

* Assumes AGNS and NFS come to full load as presently planned, and assumes GE decides to rebuild Morris, can begin operations in 1980 and reach commercial capacity as indicated.

Source: Nuclear Industry, The Monthly Magazine of the Atomic Industrial Forum, Inc., New York, N.Y., July 1974.

Projected Fuel Processing Requirements and High-Level Waste Conditions for the Civilian Nuclear Power Program[a]

	Calendar Year			
	1970	1980	1990	2000
Installed capacity, MW(e) [b]	14,000	153,000	368,000	735,000
Electricity generated, 10^9 kWhr/year [b]	71	1,000	2,410	4,420
Spent fuel shipping				
Number of casks shipped annually	30	1,200	6,800	9,500
Number of loaded casks in transit	1	14	60	85
Spent fuel processed, metric tons/year [b]	94	3,500	13,500	15,000
Volume of high-level liquid waste generated [c, d]				
Annually, 10^6 gal/year	0.017	0.97	2.69	4.60
Accumulated, 10^6 gal	0.017	4.40	23.8	60.1
Volume of high-level waste, if solidified [c, e]				
Annually, 10^3 ft^3/year	0.17	9.73	26.9	46.0
Accumulated, 10^3 ft^3	0.17	44.0	238	601
Solidified waste shipping [f]				
Number of casks shipped annually	0	3	172	477
Number of loaded casks in transit [g]	0	1	4	10
Significant radioisotopes in waste [h, i]				
Total accumulated weight, metric tons	1.8	450	2,400	6,200
Total accumulated beta activity, megacuries	210	18,900	85,000	209,000
Total heat-generation rate, megawatts	0.9	80	340	810
^{90}Sr generated annually, megacuries	4.0	230	560	770
^{90}Sr accumulated, megacuries	4.0	960	4,600	10,000
^{137}Cs generated annually, megacuries	5.6	320	880	1,500
^{137}Cs accumulated, megacuries	5.6	1,300	6,500	15,600
^{129}I generated annually, curies	2.0	110	440	670
^{129}I accumulated, curies	2.0	480	2,700	7,600
^{85}Kr generated annually, megacuries	0.6	33	90	150
^{85}Kr accumulated, megacuries	0.6	124	570	1,200
^{3}H generated annually, megacuries	0.04	2.1	6.2	12
^{3}H accumulated, megacuries	0.04	7.3	36	90
^{238}Pu generated annually, megacuries	0.0007	0.041	0.2	0.6
^{238}Pu accumulated, megacuries	0.0007	1.20	8.3	31
^{239}Pu generated annually, megacuries	0.00009	0.005	0.05	0.2
^{239}Pu accumulated, megacuries	0.00009	0.02	0.24	1.2
^{240}Pu generated annually, megacuries	0.00012	0.007	0.06	0.21
^{240}Pu accumulated, megacuries	0.00012	0.04	0.4	1.9
^{241}Am generated annually, megacuries	0.009	0.5	4.4	15
^{241}Am accumulated, megacuries	0.009	2.3	25	120
^{243}Am generated annually, megacuries	0.00021	0.01	0.1	0.5
^{243}Am accumulated, megacuries	0.00021	0.23	1.3	5.2
^{244}Cm generated annually, megacuries	0.13	7.4	18	23
^{244}Cm accumulated, megacuries	0.13	30	140	260
Volume of cladding hulls generated [j]				
Annually, 10^3 ft^3	0.3	8	40	90
Accumulated, 10^3 ft^3	0.3	40	320	1,030

[a] From ORNL (1970).
[b] Data from Phase 3, Case 42 Systems Analysis Task Force (Apr. 11, 1968).
[c] Based on an average fuel exposure of 33,000 MWd/ton, and a delay of 2 years between power generation and fuel processing.
[d] Assumes wastes concentrated to 100 gal per 10,000 MWd (thermal).
[e] Assumes 1 ft^3 of solidified waste per 10,000 MWd (thermal).
[f] Assumes 10-year-old wastes, shipped in thirty-six 6-in.-diam. cylinders per shipment cask.
[g] One-way transit time is 7 days.
[h] Assumes LWR fuel continuously irradiated at 30 MW/ton to 33,000 MWd/ton, and fuel processing 90 days after discharge from reactor; LMFBR core continuously irradiated to 80,000 MWd/ton at 148 MW/ton, axial blanket to 2,500 MWd/ton at 4.6 MW/ton, radial blanket to 8,100 MWd/ton at 8.4 MW/ton, and fuel processing 30 days after discharge.
[i] Assumes 0.5% of Pu in spent fuel is lost to waste.
[j] Based on 2.1 ft^3 of cladding hulls per ton of LWR fuel processed, and 8.7 ft^3 of cladding hardware per ton of LMFBR mixed core and blankets processed.

Reproduced with permission of the National Academy of Sciences. A report of the National Academy of Engineering. Published by the National Academy of Science.

Shipping cask for spent fuel. This cask for shipping irradiated fuel was built by the General Electric Company and weighs 140,000 pounds when fully loaded. The cask, which uses uranium as a shielding material, can accommodate 7 fuel assemblies from a pressurized water reactor or 18 from a boiling water reactor. (J. E. Westcott, United States Atomic Energy Commission)

RADIOACTIVE WASTE MANAGEMENT

Radioactive wastes of varying degrees of activity are generated in many sectors of the fuel cycle. These wastes are generally placed in three categories; high-level wastes, transuranium-contaminated wastes and low-level wastes.

Comparison of Estimated Annual Quantities of Radioactive Solid Wastes from an LWR and an LMFBR

	1000-MWe LWR with Plutonium Recycle	1000-MWe LMFBR
Produced at Reactor Site		
Other-than-High-Level		
Cubic Feet	8,000-12,000	1,000-2,000
Number of 55-Gal Drums	560-1,120	140-280
Burial Ground Area Used, sq ft	800-1,600	200-400
Noble Gases		
Cylinders	-	5
Produced at Reprocessing Plant Site		
High-Level Solid		
Cubic Feet	55	55
Number of 12-In.-Dia. x 10 Ft. Long Containers	9	9
RSSF Repository Space Required, sq ft	11	11
Cladding Hulls		
Cubic Feet	60	165
Number 3.5 ft^3 containers	17	48
Repository[a] or Burial Space Required, sq ft	12	35
Other Solid Wastes		
Cubic Feet	600-4,000	5,000-10,000
Number of 55-Gal Drums	80-540	700-1,400
Burial Ground, sq ft	100-800	1,000-2,000
Noble Gases		
Cylinders		1
Produced at Fabrication Plant Site		
Other-than-High-Level Including Plutonium Contaminated Wastes		
Cubic Feet	10,000-30,000	10,000-30,000
Number of 55-Gal Drums	1,400-4,200	1,400-4,200
Repository[a] Space, sq ft	2,000-6,000	2,000-6,000

[a]If required by future regulations.

Source: Proposed Final Environmental Statement: Liquid Metal Fast Breeder Reactor Program, Volume II, WASH-1535, U.S. Atomic Energy Commission, December 1974.

Fuel Cycle Wastes Projected for the Year 2000[1]

	Annual generation		Total accumulated inventory	
	Volume (10³ Ft³)	Activity (Mci)	Volume (10³ Ft³)	Activity (Mci)
HLW (solid)	50.0	60,000.0	471	75,000.0
CLAD	54.0	250.0	450	900.0
TRU	700.0	5.5	5,400	24.2
LLW	54,000.0	1.0	330,000	4.0
Gases	2,200.0	23.0	21,000	1,500.0
Tails	2,400,000.0	.8	31,000,000	9.9

[1] Assumes pu recycle.

Types of Wastes:

The staff has adopted for this discussion a set of waste types parallel to that used by Oak Ridge National Laboratory in their inventory projections. These types are:

- high-level wastes (HLW): as defined in 10 CFR 50, Appendix F, those highly radioactive wastes from the first cycle of reprocessing or equivalent or the concentrated wastes from subsequent steps in the reprocessing or equivalent;

- transuranic contaminated wastes (TRU): those wastes which are not high-level wastes but which contain transuranic isotopes in concentrations in excess of 10 nCi/gm (this includes a range of materials from the highly radioactive fuel rod hulls and intermediate level waste streams to slightly contaminated trash such as wipes and gloves);

- low level wastes: waste materials containing sufficient radioactivity to prohibit uncontrolled release but with transuranic content less than 10 nCi/gm;

- mill tailings: sands and slime residues which remain after extraction of the source material (uranium or thorium) from the bulk of the ore;

- "gases": a loose term applied to volatile materials in effluent streams (future capture and containment at such materials has been the subject of considerable discussion);

Status of Waste Disposal/Storage Activities

Waste type source	Industry (or ERDA) strategy established	Major regulations established by NRC		Detailed regulatory criteria established by NRC
		Waste producing facilities	Offsite storage/ disposal	
HLW:				
Commercial	Yes	Yes	Yes	Part
ERDA	Part	(⁴)	(Yes)	No
Hi-γ TRU:				
Commercial	No (in prep.)	Yes	Yes[1]	No[1]
ERDA	Part		(Yes)	No
Lo-γ TRU:				
Commercial	Part	Yes	Yes[1]	Part[1]
ERDA	Part		(Yes)	No
LLW (non-TRU):				
Commercial	Yes	Yes	Yes[3]	Part
ERDA	Yes		(Yes)	No
Gases: Commercial	No	Yes	Yes	No
Mill tailings	No (or part)	Yes	No[3]	Part
Decom. facilities:				
Commercial	No (or part)	Yes	Yes	Part
ERDA	No (or part)			

[1] Proposed regulation on TRU gives ERDA possession; licensing by NRC was not made specific in the Energy Reorganization Act.
[3] Shared with agreement States.
[3] Based on previous legal interpretations of definitions of source and byproduct materials.
[4] Except specific licensed ERDA facilities.

Storage and Disposal of Radioactive Waste, Hearing before the Joint Committee on Atomic Energy, Congress of the United States, November 19, 1975.

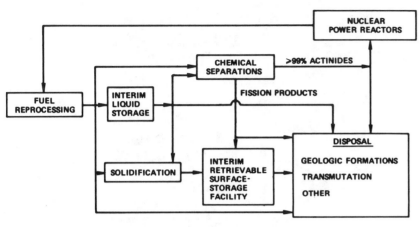

Options for management of high-level wastes.

Characteristics of Typical Solidified High-Level Waste

	Calcine	Borosilicate Glass
Form	Granular	Monolithic
Description	Amorphous	Black glass
Fission product oxide, mol %	Up to 90	Up to 30
Bulk density, g/cm^3	1.0–1.7	3.0–3.4
Thermal conductivity, W/(m^2) ($^\circ$C/m)	0.15–0.4	0.8–1.2
Leachability at 25 $^\circ$C, g/cm^2-day	$1–10^{-1}$	$10^{-5}–10^{-7}$
Relative surface/volume ratio[a]	10^4	1
Hardness	Soft	Very hard
Frangibility	Crumbly	Brittle
Residual nitrate, wt %	<4.0	<0.005
Volume, liter/1,000 MWdth	1.5–5.0	1.5–5.0
Maximum stable temperature, $^\circ$C	600	600[b]

SOURCE: Summary of data presented to NAS–NRC Panel on Engineered Storage by J. L. McElroy, Battelle Memorial Institute, Battelle Pacific Northwest Laboratories, Richland, Washington.
[a]Based on monolith glass samples approximately 25 × 250 cm long and calcine particles averaging between 10 and 100 μ (0.001–0.01 cm) in diameter.
[b]Onset of devitrification.

Source: Interim Storage of Solidified High-Level Radioactive Wastes. Panel on Engineered Storage, Committee on Radioactive Management, National Academy of Sciences, Washington, D.C. 1975.

High-level wastes result from the reprocessing of highly irradiated nuclear fuel. The intensity of the long-lived radioactivity requires storage in isolation and under essentially perpetual surveillance. Nuclear Regulatory Commission regulations require that high-level wastes be converted to solid, inert materials which are noncombustible and cannot turn into gaseous form and become airborne. These solids, in turn, are placed in sealed canisters. Until investigations of permanent storage in geologic formations are completed, plans are being made for surface storage of canisters such that canisters can be later retrieved and transferred to permanent storage facilities.

Transuranium-contaminated wastes consist of solid materials which are contaminated with significant amounts of long-lived alpha emitters such as plutonium. These wastes

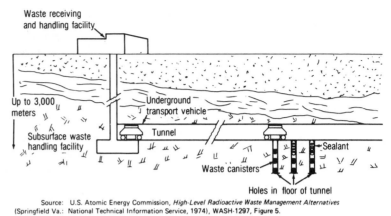

Source: U.S. Atomic Energy Commission, *High-Level Radioactive Waste Management Alternatives* (Springfield Va.: National Technical Information Service, 1974), WASH-1297, Figure 5.

Solid radioactive waste emplacement in a mined cavity.

Source: The Sixth Annual Report of the Council on Environmental Quality, December 1975.

may include rags, paper, plastics, tools, equipment, and fuel cladding. Radiation and heat levels of this category of waste are low compared to those of the high-level wastes, but special handling and storage is required. These wastes must be stored in a readily retrievable manner to facilitate transfer to permanent storage facilities at a later date.

Low-level wastes include those materials of such low concentrations or radioactivity that they do not represent a significant environmental hazard. Low-level wastes are either released into the environment or, if in solid form, are buried in areas where the geologic and hydrologic conditions limit the migration of radioactivity.

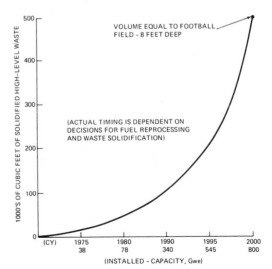

Volume of solidified high-level radioactive waste.

Source: Presentation on the Nuclear Fuel Cycle by the staff of the Energy Research and Development Administration to the EEI Nuclear Fuels Committee, January 27, 1976.

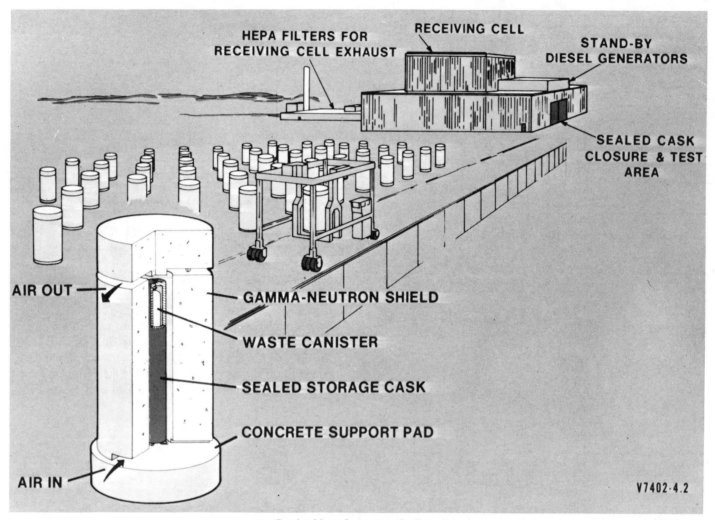

Retrievable surface storage facility.

Sealed cask radioactive waste storage. The sealed cask concept developed by the Atlantic Richfield Hanford Company, Richland, Washington, proposes that high-level radioactive waste be solidifed and placed in steel canisters surrounded by concrete shielding. Natural convection of air would provide cooling for the decay heat of the radioactive material. The 75-ton assembly would be built to withstand any credible natural disaster, such as a tornado or earthquake. (U.S. Atomic Energy Commission)

Rock salt deposits in the United States (after Pierce and Rich, U.S.G.S. Bull. 1148).

Source: Presentation on the Nuclear Fuel Cycle by the staff of the Energy Research and Development Administration to the EEI Nuclear Fuels Committee, January 27, 1976.

NUCLEAR FISSION PRINCIPLES

A nuclear power plant is similar, in operating principle, to power plants employing coal or oil as fuel. Each type generates steam to drive a turbine generator that produces electricity. The heat energy of the steam is converted into mechanical energy in the turbine, and the generator converts the mechanical energy into electricity.

In a conventional power plant, the heat of combustion of coal or oil is used to produce steam. In a nuclear power plant, the heat energy is derived from the fission of uranium or plutonium atoms. Several isotopes (U-235, U-233, and Pu-239) fission readily when struck by a neutron, breaking into two lighter elements that fly apart at high speed. At the same time, two or more new neutrons are released. The kinetic energy of the lighter elements, the fission fragments, is converted into heat as these fragments collide with surrounding atoms, and this heat, in turn, can be used to generate steam. The neutrons produced in one fission can, under appropriate circumstances, cause the fission of other atoms in a chain reaction. The designs of nuclear reactors take various forms to take advantage of the characteristics of various nuclear fuels, coolants, and construction materials.

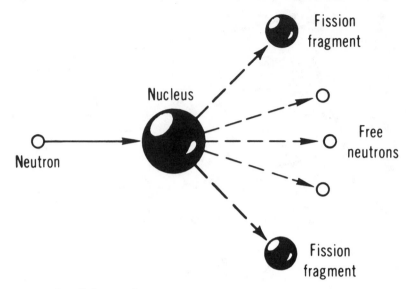

In a fission reaction, a neutron causes the nucleus of an atom to split into two lighter atoms plus two or three free neutrons. If, on the average, one of these free neutrons can be made to cause fission of another atom, a chain reaction can be maintained.

Source: Nuclear Power Plants, Division of Technical Information, U.S. Atomic Energy Commission 1966.

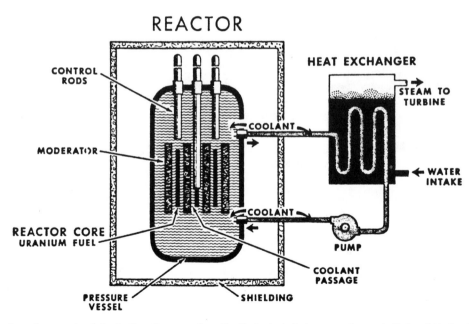

A nuclear reactor is basically a furnace where the fissioning of atoms can be controlled and the heat put to useful work. The diagram shows the location of fuel, moderator, control rods, and coolant in a typical power reactor. These components are enclosed in a pressure vessel. The coolant, heated in passing by the fuel elements, flows through a heat exchanger where it turns water in a secondary circuit into steam. The steam is then used to drive a turbine generator. (Courtesy United States Atomic Energy Commission.)

The maintenance of the nuclear chain reaction requires that at least one neutron from each fission cause a fission in another atom. A number of factors improve the probability of this happening. The number of fissionable atoms in a given space can be increased by enriching the fuel in the fissionable isotope (for example, natural uranium contains only 0.7% of the fissionable isotope U-235—for use in light water cooled reactors, U-235 content is increased to 2 to 4%). There is also a greater probability that a neutron produced in one fission will be "captured" to cause another fission if the speed of the neutron is reduced. This reduction in speed is accomplished by interspersing atoms of light elements, typically carbon, hydrogen, or deuterium (in the form of graphite or light or heavy water) between fuel elements such that the neutrons lose energy and speed through collision with these atoms. These materials which moderate the speed of neutrons are termed "moderators."

A number of factors also interfere with the maintenance of the chain reaction. The materials in the core of a reactor—fuel assemblies, moderator, and coolant—all absorb neutrons to varying degrees. Some neutrons also "leak" out of the outer surface of the core assembly. Generally, attempts are made to minimize parasitic absorption characteristics. Advantage is taken of the high neutron absorption of some elements, however, by using these elements in control rods which can be inserted into the reactor core to regulate the level of the chain reaction.

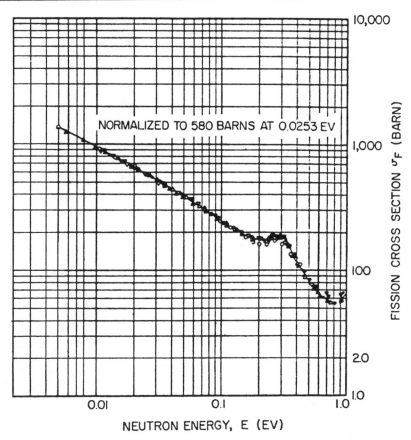

The fission cross section of U-235 is highest at low neutron energies. The high energy neutrons emitted in fission are slowed down through collisions with the atoms of moderators such as carbon, hydrogen, or deuterium.

Source: Robert L. Loftness, *Nuclear Power Plants: Design, Operating Experience and Economics*, D. Van Nostrand Company, Inc., Princeton, N.J., 1974.

Ionizing radiation from spent nuclear fuel creates visible glow in water storage pool.

Source: Department of Energy.

PRESSURIZED WATER REACTORS (PWR)

The Pressurized Water Reactor (PWR) is the most common of all reactor types in use throughout the world, both for commercial power stations and for military applications. PWRs use ordinary water (H_2O) as coolant and moderator and, because of the neutron absorption properties of H_2O, reactors of this type require the use of enriched uranium as fuel. In addition, the efficiency of the PWR, in common with other water-cooled reactors, is limited to about 32% since coolant temperatures cannot be much above 600°F (water cannot be maintained in a liquid phase above the critical point of 705°F, regardless of pressure). Operating pressures are high, near 2000 pounds per square inch, and primary system components are therefore constructed of heavy steel.

On the other hand, PWRs can be relatively small in size compared to a number of other reactor types, and they have been suitable, therefore, for power plants for mobile applications such as submarines. The extensive industrial experience in producing hardware for PWRs for both commercial and military uses, has made this reactor type a leader among the reactors in operation throughout the world. The first commercial PWR, a 60 MWe plant, began operation in 1957 at Shippingport, Pennsylvania.

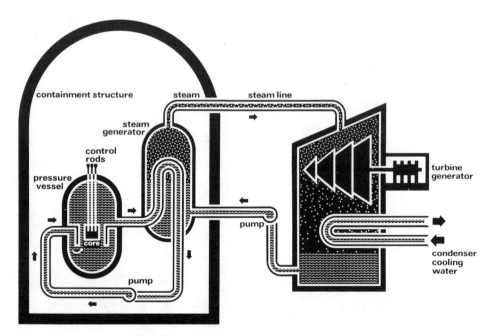

Pressurized water reactor (PWR). As with the boiler in a coal-, oil- or gas-burning power plant, a nuclear power reactor produces steam to drive a turbine which turns an electric generator. Instead of burning fossil fuel, a reactor fissions nuclear fuel to produce heat to make the steam. The PWR shown here is a type of reactor fueled by slightly enriched uranium in the form of uranium oxide pellets held in zirconium alloy tubes in the core. Water is pumped through the core to transfer heat to the steam generator. This coolant water is kept under pressure in the core to prevent boiling and transfers heat to the water in the steam generator to make the steam. Three U.S. companies manufacture PWRs: Westinghouse Electric Corp., Combustion Engineering, Inc., and The Babcock & Wilcox Co.

Point Beach. The 40-ton reactor vessel head of Point Beach nuclear plant, Unit 1, was lowered over the control rod drive shafts after the core was first loaded full with 121 fuel assemblies. The Point Beach plant uses a pressurized water nuclear reactor. It is located at Two Creeks, Wisconsin and operated by Wisconsin Michigan Power Co. and Wisconsin Electric Power Co. It started operation in 1970. (*Wisconsin Electric Power Co.*)

Pressurised Water Reactor (PWR)

Fuel: Uranium dioxide in Zircaloy cans.

Moderator: Light water (ordinary water, H_2O).

Core layout: Fuel pins, arranged in clusters, are placed inside a pressure vessel containing the light water moderator, which also is the coolant.

Heat extraction: The light water in the pressure vessel at high pressure is heated by the core. It is pumped to a steam generator where it boils water in a separate circuit; the steam drives a turbine coupled to an electric generator.

Indicative data for a reactor of 700 MW(E) size:

uranium enrichment (% U-235)	3.2%
coolant outlet temperature	317°C
coolant pressure	2235 psia
steam cycle efficiency	32%
core dimensions	3.0 m dia. × 3.7 m high

Source: Reproduced with permission of the United Kingdom Atomic Energy Authority from the UKAEA Annual Rep. 1973/74.

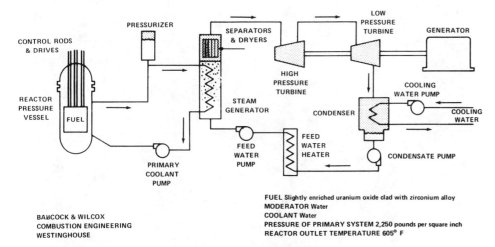

FUEL Slightly enriched uranium oxide clad with zirconium alloy
MODERATOR Water
COOLANT Water
PRESSURE OF PRIMARY SYSTEM 2,250 pounds per square inch
REACTOR OUTLET TEMPERATURE 605° F

BABCOCK & WILCOX
COMBUSTION ENGINEERING
WESTINGHOUSE

Pressurized water reactor power plant.

Source: The Nuclear Industry 1974, WASH 1174-74, U.S. Atomic Energy Commission, Washington, D.C., 1974.

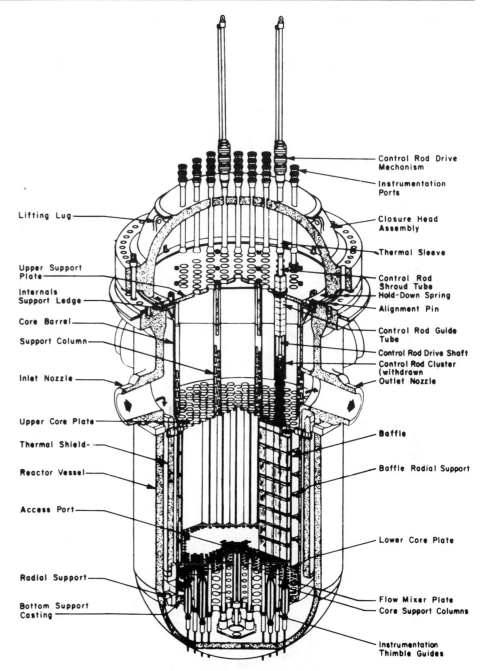

Cutaway view of internals of typical PWR vessel.

Source: Draft Environmental Statement Liquid Metal Fast Breeder Reactor Program, Volume IV: Alternative Technology Options, WASH-1535, United States Atomic Energy Commission, March 1974.

BOILING WATER REACTORS (BWR)

As with pressurized water reactors, the neutron absorption properties of the light water (H_2O) coolant used in boiling water reactors requires the use of slightly enriched uranium as fuel. Of all the reactors currently in use, the boiling water reactor is perhaps simplest in concept. The light water coolant is allowed to boil in the reactor core and the steam produced is used directly to operate a turbine. This simplicity is not without penalty, however. The pressure vessel for the boiling water reactor is about three times larger in volume than a pressurized water reactor of the same power rating since space must be provided in the BWR for water circulation and steam separation. In addition, any radioactivity leaking from fuel elements into the coolant is transferred to the external turbine. The relative merits of boiling and pressurized water reactors are such, however, that both types are quite competitive.

Boiling Water Reactor (BWR)

Fuel: Uranium dioxide in Zircaloy cans.
Moderator: Light water (ordinary water, H_2O).
Core layout: Fuel pins, arranged in clusters, are placed inside a pressure vessel containing the light water moderator, which also is the coolant.
Heat extraction: The light water in the pressure vessel is heated by the core and allowed to boil at pressure. The steam from the boiling coolant drives a turbine coupled to an electric generator.
Indicative data for a reactor of 600 MW(E) size:

uranium enrichment (% U-235)	2.6%
coolant outlet temperature	286°C
coolant pressure	1050 psia
steam cycle efficiency	32%
core dimensions	3.7 m dia. × 3.7 m high

Source: Reproduced with permission of the United Kingdom Atomic Energy Authority from the UKAEA Annual Rep. 1973/74.

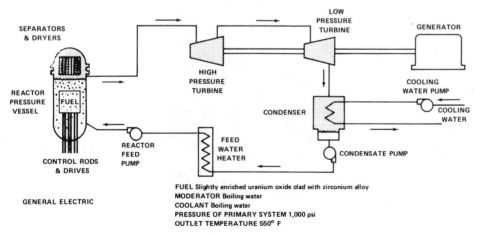

Boiling water reactor power plant.

Source: The Nuclear Industry 1974, WASH 1174-74, U.S. Atomic Energy Commission, Washington, D.C. 1974.

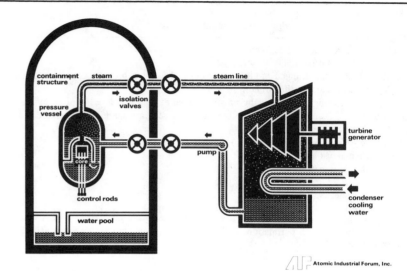

Boiling water reactor (BWR). As with the boiler in a coal-, oil- or gas-burning power plant, a nuclear power reactor produces steam to drive a turbine which turns an electric generator. Instead of burning fossil fuel, a reactor fissions nuclear fuel to produce heat to make the steam. The BWR shown here is a type of reactor fueled by slightly enriched uranium in the form of uranium oxide pellets held in zirconium alloy tubes in the core. Water is pumped through the core, boils, and produces steam that is piped to the turbine. General Electric Co. makes BWRs.

Steam separator assembly for a boiling water reactor. (*Courtesy General Electric*)

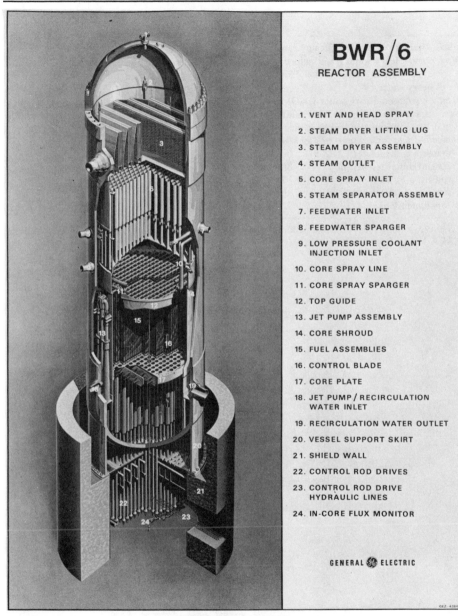

BWR/6
REACTOR ASSEMBLY

1. VENT AND HEAD SPRAY
2. STEAM DRYER LIFTING LUG
3. STEAM DRYER ASSEMBLY
4. STEAM OUTLET
5. CORE SPRAY INLET
6. STEAM SEPARATOR ASSEMBLY
7. FEEDWATER INLET
8. FEEDWATER SPARGER
9. LOW PRESSURE COOLANT INJECTION INLET
10. CORE SPRAY LINE
11. CORE SPRAY SPARGER
12. TOP GUIDE
13. JET PUMP ASSEMBLY
14. CORE SHROUD
15. FUEL ASSEMBLIES
16. CONTROL BLADE
17. CORE PLATE
18. JET PUMP/RECIRCULATION WATER INLET
19. RECIRCULATION WATER OUTLET
20. VESSEL SUPPORT SKIRT
21. SHIELD WALL
22. CONTROL ROD DRIVES
23. CONTROL ROD DRIVE HYDRAULIC LINES
24. IN-CORE FLUX MONITOR

GENERAL ⊕ ELECTRIC

BWR/6 reactor assembly.

Reprinted by permission of General Electric.

Steel pressure vessel for a boiling water reactor being lifted by crane up the side of the reactor building at the Browns Ferry site, Decatur, Alabama. (*Courtesy General Electric*).

CANADIAN DEUTERIUM URANIUM REACTORS (CANDU)

The CANDU nuclear power reactor system has been under development since the 1950s by the Atomic Energy of Canada Limited and Canadian industry. The major development has been concentrated on reactors cooled as well as moderated with heavy water and using natural uranium as fuel. Separate developments are also underway on reactors using heavy water as moderator but light water or organic liquid as coolant. The light water-cooled concept is similar to the steam-generating heavy water reactor recently selected for construction in the United Kingdom. The light water- and organic-cooled concepts use slightly enriched uranium as fuel.

Canadian Natural Uranium Heavy Water Power Reactors in Operation, under Construction, or Committed

NAME[1]	LOCATION	TYPE[2]	POWER MWe NET	NUCLEAR DESIGNER[3]	DATE OF FIRST POWER
NPD	Ontario	PHW	22	AECL & CGE	1962
Douglas Point	Ontario	PHW	208	AECL	1967
Pickering A	Ontario	PHW	514 x 4	AECL	1971-73
Gentilly 1	Québec	BLW	250	AECL	1971
KANUPP	Pakistan	PHW	125	CGE	1971
RAPP 1	India	PHW	203	AECL	1972
RAPP 2	India	PHW	203	AECL	1976
Bruce A	Ontario	PHW	745 x 4	AECL	1976-79
Gentilly 2	Québec	PHW	600	AECL	1979
Point Lepreau	New Brunswick	PHW	600	AECL	1980
Cordoba	Argentina	PHW	600	AECL	1980
Pickering B	Ontario	PHW	514 x 4	AECL	1981-83
Wolsung 1	Korea	PHW	600	AECL	1982
Bruce B	Ontario	PHW	750 x 4	AECL	1983-86
Darlington	Ontario	PHW	800 x 4	AECL	1986-88
		TOTAL	16,703		

[1]NPD   Nuclear Power Demonstration
KANUPP   Karachi Nuclear Power Project
RAPP   Rajasthan Atomic Power Project

[2]PHW   Pressurized Heavy Water Coolant
BLW   Boiling Light Water Coolant

[3]AECL   Atomic Energy of Canada Limited
CGE   Canadian General Electric Company Limited

Source: Atomic Energy of Canada Limited.

CANDU

Fuel: Uranium dioxide in Zircaloy cans.
Moderator: Heavy water (D_2O) which allows unenriched uranium fuel to be used.
Core layout: Each cluster of fuel elements is in a separate pressure tube; the pressure tubes are in a tank of heavy water.
Heat extraction: Heavy water (D_2O) at pressure is heated by passing over the fuel in the pressure tubes. It is pumped to a steam generator where it boils light water (H_2O) in a separate circuit; the steam drives a turbine coupled to an electric generator.
Indicative data for a reactor of 600 MW(E) size:

uranium enrichment (% U-235)	0.7% (natural)
coolant outlet temperature	305°C
coolant pressure	1285 psia
steam cycle efficiency	30%
core dimensions	7.1 m dia. × 5.9 m high

Source: Reproduced with permission of the United Kingdom Atomic Energy Authority from the UKAEA Annual Rep. 1973/74.

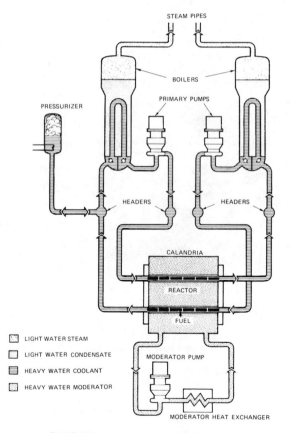

STEAM PIPES
BOILERS
PRIMARY PUMPS
PRESSURIZER
HEADERS   HEADERS
CALANDRIA
REACTOR
FUEL
MODERATOR PUMP
MODERATOR HEAT EXCHANGER

▢ LIGHT WATER STEAM
▢ LIGHT WATER CONDENSATE
▤ HEAVY WATER COOLANT
▦ HEAVY WATER MODERATOR

CANDU reactor simplified flow diagram.

Source: Atomic Energy Canada Limited.

Pickering generating station. Located 20 miles from Toronto on the shore of Lake Ontario, the 2,000,000 kilowatt Pickering Station is one of the largest operating nuclear power stations in the world. It is powered by four CANDU-PHW reactors using natural uranium as fuel and heavy water as moderator and coolant. The station was built by Ontario Hydro, with nuclear design by Atomic Energy of Canada Limited. (Atomic Energy of Canada Limited)

In contrast to the heavy pressure vessels employed in pressurized and boiling water reactors, the CANDU reactor consists of a large cylindrical tank, called the calandria, which contains the cool, low-pressure moderator. This tank is penetrated by a number of horizontal tubes which provide channels in which the fuel elements are placed and through which the pressurized, high temperature heavy water coolant flows. This coolant removes the heat generated in the fuel and this heat is then used to generate steam in an external heat exchanger. Fuel is loaded and unloaded from the reactor during operation.

REFERENCES

L. W. Woodhead, "Performance of Canadian commercial nuclear units and heavy water plants," Atomic Energy of Canada Limited, Sheridan Park Research Community, Mississauga, Ont., Canada L5K 1 B2, 1975.

Hugh C. McIntyre, "Natural uranium heavy-water reactors," *Sci. Amer.*, October 1975.

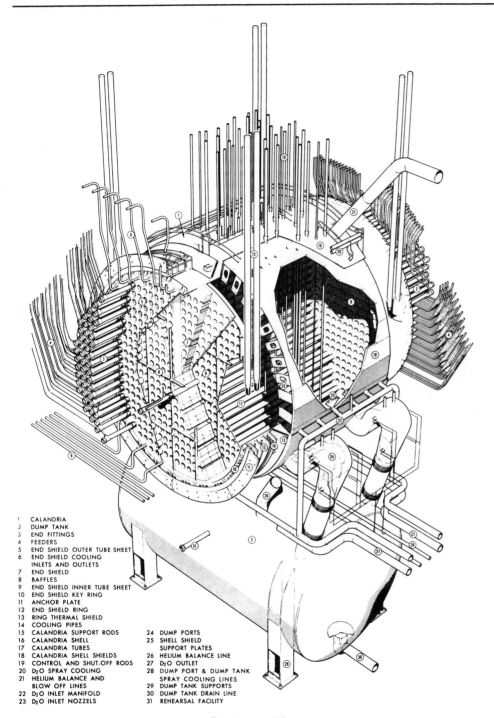

1 CALANDRIA	
2 DUMP TANK	
3 END FITTINGS	
4 FEEDERS	
5 END SHIELD OUTER TUBE SHEET	
6 END SHIELD COOLING INLETS AND OUTLETS	
7 END SHIELD	
8 BAFFLES	
9 END SHIELD INNER TUBE SHEET	
10 END SHIELD KEY RING	
11 ANCHOR PLATE	
12 END SHIELD RING	
13 RING THERMAL SHIELD	
14 COOLING PIPES	
15 CALANDRIA SUPPORT RODS	24 DUMP PORTS
16 CALANDRIA SHELL	25 SHELL SHIELD
17 CALANDRIA TUBES	SUPPORT PLATES
18 CALANDRIA SHELL SHIELDS	26 HELIUM BALANCE LINE
19 CONTROL AND SHUT-OFF RODS	27 D₂O OUTLET
20 D₂O SPRAY COOLING	28 DUMP PORT & DUMP TANK
21 HELIUM BALANCE AND	SPRAY COOLING LINES
BLOW OFF LINES	29 DUMP TANK SUPPORTS
22 D₂O INLET MANIFOLD	30 DUMP TANK DRAIN LINE
23 D₂O INLET NOZZELS	31 REHEARSAL FACILITY

Reactor assembly.

Source: Atomic Energy of Canada Limited.

CANDU reactors are fuelled on load. This eliminates the need for refuelling shutdowns so that the reactor is available 100% of the time. The fuelling machine shown is at the Pickering Generating Station, one of the world's largest nuclear power plants. (*Courtesy Ontario Hydro*)

STEAM GENERATING HEAVY WATER REACTORS (SGHWR)

The steam-generating heavy water reactor (SGHWR) is a heavy water-moderated, light water-cooled concept and represents a variation of the heavy water reactor designs developed in Canada. In contrast to the horizontal calandria pressure tube core assembly used in the CANDU heavy water-cooled reactor, the SGHWR calandria pressure tube assembly is oriented vertically to permit boiling of the light water coolant in the pressure tubes. The steam generated in the core is passed directly to the turbine as in boiling water reactors. Because of the vertical orientation of the calandria, the concept of on-power refueling is more difficult to achieve than in the horizontal CANDU configuration. Off-power refueling, therefore, has been adopted. The use of light-water as a coolant requires that the fuel be slightly enriched.

The United Kingdom has adopted the SGHWR for its next group of nuclear power stations. Design of the 660 MWe power stations is based on the 93 MWe SGHWR prototype which has been in operation at Winfrith since 1967.

Steam Generating Heavy Water Reactor (SGHWR)

Fuel: Uranium dioxide in Zircaloy cans.

Moderator: Heavy water (D_2O).

Core layout: Each cluster of fuel elements is in a separate pressure tube; the pressure tubes are in a tank of heavy water. Heavy water is the most efficient moderator and compensates for the neutron absorption in the pressure tubes.

Heat extraction: Light water (ordinary water, H_2O) at pressure is heated by passing over the fuel in the pressure tubes and allowed to boil; the steam from the boiling coolant drives a turbine coupled to an electric generator.

Indicative data for a reactor of 600 MW(E) size:

uranium enrichment (% U-235)	2.24%
coolant outlet temperature	272°C
coolant pressure	900 psia
steam cycle efficiency	32%
core dimensions	6.5 m dia. × 3.7 m high

Source: Reproduced with permission of the United Kingdom Atomic Energy Authority from the UKAEA Annual Rep. 1973/74.

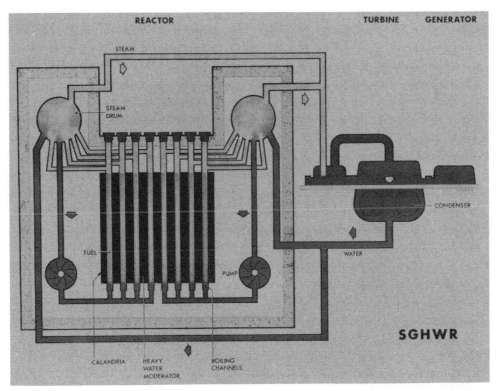

Flow diagram for the Steam Generating Heavy Water Reactor. (United Kingdom Atomic Energy Authority)

Main Characteristics of the 660 MW(e) Commercial SGHWR Design in Comparison with the Winfrith Prototype Reactor

	Commercial design	Winfrith prototype
Net electrical output, MW(e)	625	93
Gross electrical output, MW(e)	**660**	**100**
Reactor heat output, MW(th)	1900	294
Average fuel rating, MW/te(U)	19·77	14·3
Steam pressure at steam drum, Bar a	55	55
Average coolant quality at inlet to steam drum, wt %	11·9	11
Final feed temperature, °C	194	199
Number of pumps per reactor	4	4
Power of pump motors, kW	1864	446
Number of fuel channels	532	104
Lattice pitch, mm	260	260
Pressure tube internal diameter, mm	130	130
Calandria tube internal diameter, mm	180	180
Overall diameter of calandria, mm	7520	3700
Overall height of calandria, mm	4680	3960
Number of shut down tubes	64	12
Weight of uranium in the core, te	100·1	21·41
Heavy water inventory, te	115	32·4
Fuel centre temperature, °C	1910	1905
Fuel can diameter, mm	16	16
Fuel can length, mm	3880	3660
Number of pins per cluster	36	36
Average initial enrichment, % (U-235)	2·139	
Average replacement enrichment, % (U-235)	2·105	2·3
Average discharge irradiation of initial fuel, MWd/te(U)	12 729	
Average discharge irradiation of replacement fuel, MWd/te(U)	20 730	21 000

Source: Rippon, Simon. "The Commercial Steam Generating Heavy Water Reactor," *Nuclear Engineering International,* August 1974.

The 93 MWe prototype Steam Generating Heavy Water Reactor at Winfrith. The plant has been in operation since 1967. (United Kingdom Atomic Energy Authority)

MAGNOX GAS-COOLED REACTORS

The Magnox gas-cooled reactors, which derive their name from the magnesium alloy used to clad the uranium metal fuel elements, represented the initial course of reactor development in both the United Kingdom and France. The Magnox reactors use natural uranium as fuel, carbon dioxide as the primary reactor coolant, and graphite as the moderator. The use of natural uranium fuel and graphite as moderator requires that the reactor core be relatively large, in comparison to other reactor types, in order to provide the conditions necessary for maintenance of the nuclear chain reaction. Coolant temperatures are limited by the creep strength of the magnesium alloy fuel element cladding to about 700°F and as a result, overall steam cycle efficiency is about 31%.

Magnox

Fuel: Uranium metal. To conserve neutrons and allow natural uranium to be used, the fuel is clad in a magnesium alloy (Magnox) with low neutron absorption.

Moderator: Graphite.

Heat extraction: Carbon dioxide gas is heated by passing it over the fuel in the core and transfers its heat to water in a steam generator; the steam drives a turbine coupled to an electric generator.

Indicative data for a reactor of 600 MW(E) size:

uranium enrichment (% U-235)	0.7% (natural)
coolant outlet temperature	400°C
coolant pressure	300 psia
steam cycle efficiency	31%
core dimensions	14 m dia. × 8 m high

Source: Reproduced with permission of the United Kingdom Atomic Energy Authority from the UKAEA Annual Rep. 1973/74.

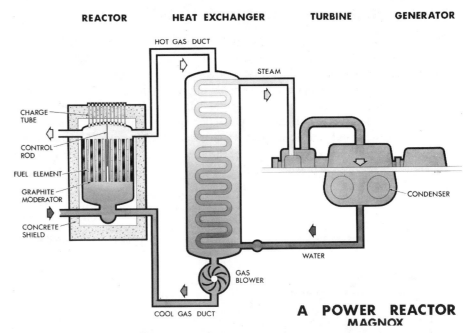

A POWER REACTOR
MAGNOX

Flow diagram for Magnox nuclear power station. Gas coolant is carbon dioxide. Fuel elements are of natural metallic uranium clad with a magnesium alloy. (United Kingdom Atomic Energy Authority)

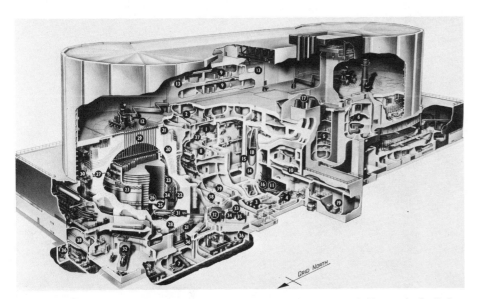

Plant layout for Magnox nuclear power station. (United Kingdom Atomic Energy Authority)

Magnox nuclear power plants at Hunterston, Scotland. (United Kingdom Atomic Energy Authority)

ADVANCED GAS-COOLED REACTORS (AGR)

The development of carbon dioxide-cooled, graphite-moderated reactors of higher thermal efficiency than the Magnox reactors has been a major program in the United Kingdom. Coolant temperatures near 1200°F provide an overall steam cycle efficiency of about 42%. Operation at these temperatures requires the use of fuel elements comprised of uranium oxide clad in stainless steel rather than the uranium metal, magnesium alloy-clad elements used in the Magnox reactors. Operating pressures are also 600 pounds per square inch versus the 300 pounds per square inch used in the Magnox reactors. The reactor core structure and the steam generators of the advanced gas-cooled reactor are housed within a prestressed concrete pressure vessel.

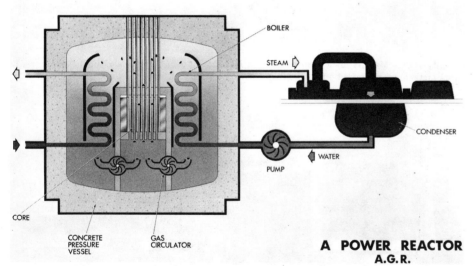

REACTOR   TURBINE   GENERATOR

BOILER

STEAM

CONDENSER

WATER

PUMP

CORE

CONCRETE PRESSURE VESSEL

GAS CIRCULATOR

A POWER REACTOR
A.G.R.

Flow diagram for an advanced gas-cooled reactor (AGR). Reactor core and steam generators are located inside a pre-stressed concrete pressure vessel. (United Kingdom Atomic Energy Authority)

Advanced Gas-Cooled Reactor (AGR)

Fuel: Uranium dioxide in stainless steel cans. The fuel can operate at higher temperatures and heat output rates than Magnox reactor fuel, giving a smaller size of reactor core and a more efficient steam cycle. To achieve these advantages and a greater heat rating, the proportion of U-235 in the fuel has to be increased (enriched uranium).

Moderator: Graphite.

Core layout: Clusters of fuel elements are joined together end-to-end in a stringer, placed in vertical holes in the graphite.

Heat extraction: Carbon dioxide gas is heated by passing over the fuel in the core and transfers its heat to water in a steam generator; the steam drives a turbine coupled to an electric generator.

Indicative data for a reactor of 600 MW(E) size:

uranium enrichment (% U-235)	2.3%
coolant outlet temperature	650°C
coolant pressure	600 psia
steam cycle efficiency	42%
core dimensions	9.1 m dia. × 8.5 m high

Source: Reproduced with permission of the United Kingdom Atomic Energy Authority from the UKAEA Annual Rep. 1973/74.

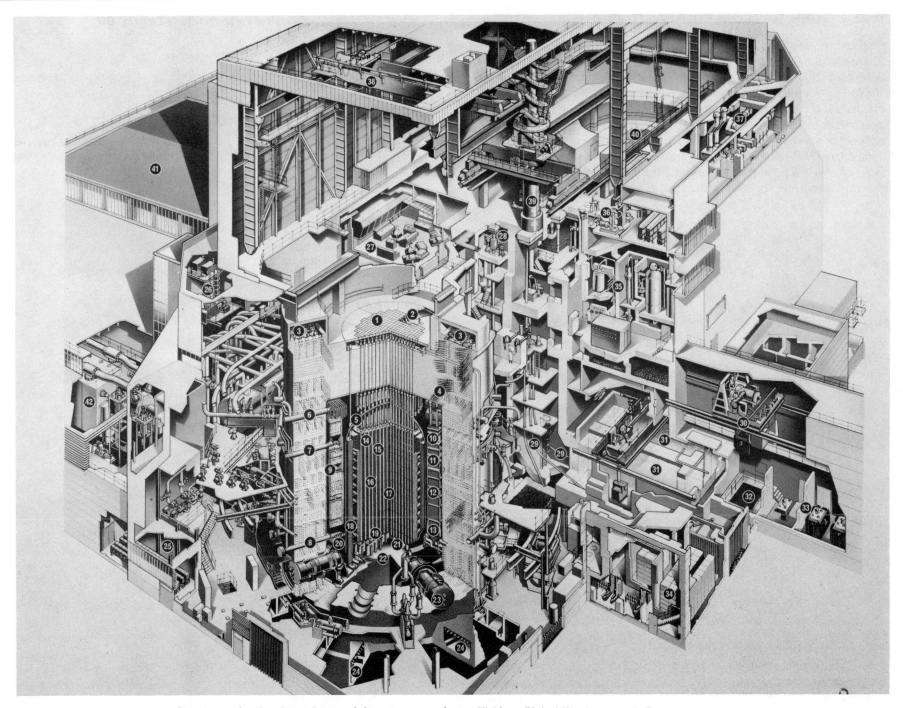

Plant layout for the advanced gas-cooled reactor power plant at Hinkley. (United Kingdom Atomic Energy Authority)

HIGH TEMPERATURE GAS-COOLED REACTORS (HTGR)

The high temperature gas-cooled reactor (HTGR) uses helium as the reactor coolant, graphite as the moderator and core structural material, and a mixture of thorium and uranium particles coated with thin layers of pyrolytic graphite as fuel. During operation the thorium in the fuel is converted to fissionable U-233.

The use of helium as coolant has the advantage that the coolant is always in the gaseous phase and is chemically inert. Operation at temperatures of 1400°F is possible, therefore, and steam cycle efficiencies of 42% are achieved. The relatively poor heat transport properties of helium require that relatively high pressures be employed in the primary coolant circuit. Prestressed concrete pressure vessels are used to house all primary system components including the reactor core, the steam generators, and the helium circulators.

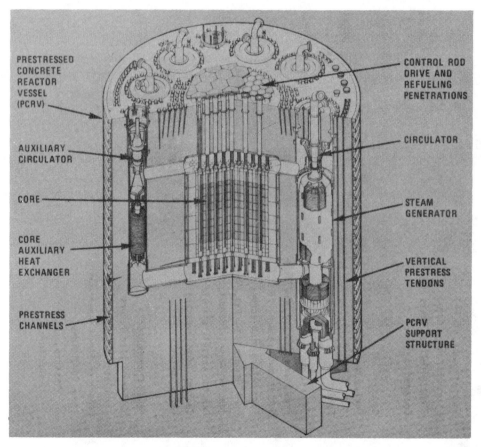

HTGR nuclear Steam System.

Source: Proposed Final Environmental Statement Liquid Metal Fast Breeder Reactor Program, Volume III, WASH-1535, U.S. Atomic Energy Commission, December 1974.

Operating Parameters for a Large HTGR

General	
Thermal power	3000 MWt
Electric power	1160 MWe
Plant lifetime	40 years
Conversion ratio[a]	0.66
Reactor	
Fuel, startup	Th/U-235 (93% enriched)
recycle	Th/U-235 (93% enriched)/U-233 (recycle)
Fuel form	coated particles in cylindrical bonded rods
Moderator	graphite
Avg. power density	8.4 kW/liter
Outlet temperature	1366°F
Temperature rise across core	760°F
Fuel temperature, Avg./Max.	1634/2467°F
Reactor vessel, height	20.8 ft
Reactor vessel, diameter	27.8 ft
Coolant inlet pressure	710 psi
Vessel material	prestressed concrete
Other Components	
Number of circulators	6
Circulator speed	7050 rpm
Number of steam generators	6
Steam conditions	
pressure	2400 psi
temperature	950°F

[a]The conversion ratio of 0.66 shown in the table is the current economic optimum for the fuel cycle. It is adjustable upward as uranium prices escalate.

Source: Proposed Final Environmental Statement Liquid Metal Fast Breeder Reactor Program, vol. III, WASH-1535, U.S. Atomic Energy Commission, December 1974.

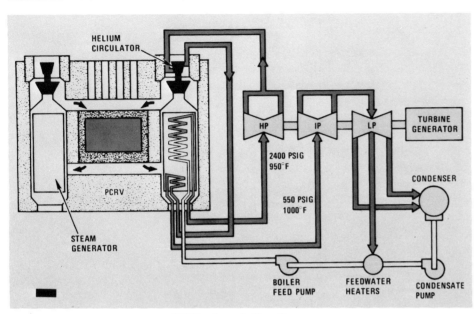

HTGR schematic flow diagram.

Source: General Atomic Company, San Diego, California.

High Temperature Reactor (HTR)

Fuel: Small spheres (about 800 μm dia.) of uranium dioxide with a coating of silicon carbide, bonded in a graphite matrix and assembled with graphite into a fuel element. The fuel can operate at higher temperatures than metal-clad fuel, and a larger proportion can be "burned-up" in each cycle in the reactor.

Moderator: Graphite, replaced with the fuel at each fuel change.

Core layout: Fuel elements are arranged with vertical coolant passages in the graphite moderator.

Heat extraction: Graphite would be corroded by carbon dioxide to an unacceptable extent at HTR temperatures. Helium gas is heated by passing over the fuel in the core and transfers its heat to water in a steam generator; the steam drives a turbine coupled to an electric generator. Potential developments are to replace the steam generator and turbine by a gas turbine; and to use the heat directly in chemical processes.

Indicative data for a reactor of 1300 MW(E) size:

uranium enrichment (% U-235)	10%
coolant outlet temperature	720°C
coolant pressure	715 psia
steam cycle efficiency	39%
core dimensions	9.8 m dia. × 6 m high

Source: Reproduced with permission of the United Kingdom Atomic Energy Authority from the UKAEA Annual Rep. 1973/74.

Three HTGR experimental reactors have been built—the Dragon reactor at Winfrith in England, the AVR reactor at Jülich in Germany, and the Peach Bottom reactor in Pennsylvania. A prototype 330 MWe HTGR commercial plant has been built at Ft. St. Vrain in Colorado and is scheduled to be in operation in early 1977. Plans for the construction of a number of large HTGR commercial plants in the United States were dropped in 1975.

REFERENCE

"Fulton station HTGR," Nuclear Engineering International, August 1974.

BREEDER REACTOR PRINCIPLES

The term "breeder" reactor is derived from the fact that such reactors produce more fissionable material than they consume (including fuel reprocessing and fabrication). Breeding is achieved by combining both fissionable and fertile materials in the reactor core under conditions which provide not only enough neutrons to propagate a chain reaction in the fissionable material but also enough to convert more fertile material into fissionable material than was originally present.

The interest in breeder reactors stems from a number of characteristics of such reactors: generally, breeder reactors operate at higher temperatures than current commercial reactors and therefore have a higher thermal efficiency; power costs are less sensitive to the cost of uranium raw material; and, breeder reactors are able to produce a fissionable material, U-233, from thorium and can convert most of the nonfissionable U-238 in natu-

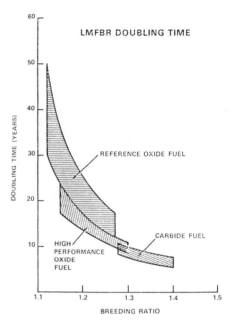

In order to provide sufficient fuel for new power plants, the doubling time for fast breeder reactors should be less than ten years. To achieve these objectives, development of higher performance oxide fuels and carbide fuels is underway. (United States Atomic Energy Commission)

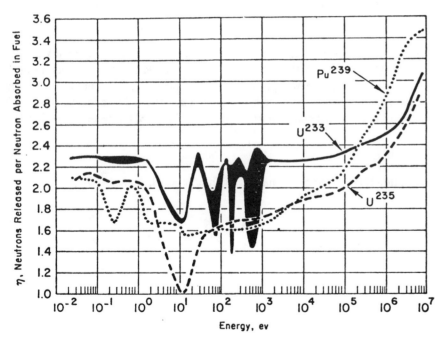

The number of neutrons released per neutron absorbed depends upon the type of fuel as well as the neutron energy. In a breeder reactor at least two neutrons are needed per fission—one to maintain the chain reaction and the other to breed new fuel from a fertile element.

Pu-239 is the best fuel for breeder reactors operating at high of "fast" neutron energies, while U-233 is best for breeder reactors operating at low or "thermal" neutron energies.

Source: Robert L. Loftness, *Nuclear Power Plants: Design, Operating Experience and Economics*, D. Van Nostrand Company Inc., Princeton, N.J., 1964.

Uranium Breeding Cycle

$$_{92}U^{238} + _{0}n' \longrightarrow _{92}U^{239}$$
$$_{92}U^{239} \longrightarrow _{93}Np^{239} + \beta^{-} \text{ (half-life: 24 min.)}$$
$$_{93}Np^{239} \longrightarrow _{94}Pu^{239} + \beta^{-} \text{ (half-life: 2.3 days)}$$
$$_{94}Pu^{239} + _{0}n' \longrightarrow \text{Fission products} + x_{0}n' \quad (x > 2)$$

Thorium Breeding Cycle

$$_{90}Th^{232} + _{0}n' \longrightarrow _{90}Th^{233}$$
$$_{90}Th^{233} \longrightarrow _{91}Pa^{233} + \beta^{-} \text{ (half-life: 22 min.)}$$
$$_{91}Pa^{233} \longrightarrow _{92}U^{233} + \beta^{-} \text{ (half-life: 27 days)}$$
$$_{92}U^{233} + _{0}n' \longrightarrow \text{Fission products} + y_{0}n' \quad (y > 2)$$

The two nuclear fuel breeding cycles involve the transmutation of the fertile isotopes, U-238 and Th-232, into Pu-239 and U-233 by absorption of neutrons and subsequent beta decay.

Source: H. C. Hottel and J. B. Howard, *New Energy Technology: Some Facts and Assessments*, The MIT Press, Cambridge, Massachusetts, 1971.

FFTF . . . A construction photograph made in mid-1973 of the Atomic Energy Commission's Fast Flux Test Facility near Richland, Wash. The 400-thermal megawatt reactor will be used to test fuels and materials scheduled for use in the liquid metal fast breeder demonstration plant planned for early 1980s operation at Oak Ridge, Tennessee. (Credit: Battelle-Northwest Laboratory)

ral uranium (U-238 represents 99.3% of natural uranium) into a new fissionable fuel, Pu-239. Breeder reactors can, therefore, extend the supply of nuclear fuels by a factor of at least 100.

Reactors employing both the U-238 to Pu-239 cycle and the Th-232 to U-233 cycle have been built, although the major effort internationally has been on liquid-metal-cooled reactors employing the U-238/Pu-239 cycle.

These two breeding fuel cycles differ in that Pu-239 produces the most neutrons in reactors designed to operate on neutrons of high velocity, that is, fast neutrons, while U-233 produces the most neutrons during fission caused by neutrons of lower velocity, that is, thermal neutrons. Fast breeder reactors are therefore, constructed so that there is little moderation of neutron velocity while thermal breeder reactors incorporate some moderator material to reduce neutron velocities.

One of the figures of merit used in judging breeder reactors is the efficiency with which they are able to produce more fuel than they consume. This is usually expressed as the "doubling time," or the time required to produce twice as much fuel as is being consumed. Design objectives are usually to set the doubling time at less than ten years.

LIQUID METAL FAST BREEDER REACTORS

The liquid metal fast breeder reactor (LMFBR) takes advantage of the properties of a liquid metal, sodium, as the coolant to transfer heat from the reactor core to the steam generators. Sodium has heat transport properties roughly equivalent to those of water but because of its wide liquid range (melting point—208°F; boiling point—1616°F), reactors using sodium as a coolant can be operated at high temperatures with only modest primary system pressure. Plant thermal efficiencies are therefore higher than for water-cooled reactors (40% versus 30%) and vessels and piping are of much lighter construction. On the other hand, systems containing sodium must be leak-tight since sodium reacts vigorously with water vapor and oxygen. Sodium also becomes radioactive as it passes through the neutron field in the reactor core and all components of the primary coolant system must, therefore, be placed in shielded positions to protect operating personnel. Fuel handling

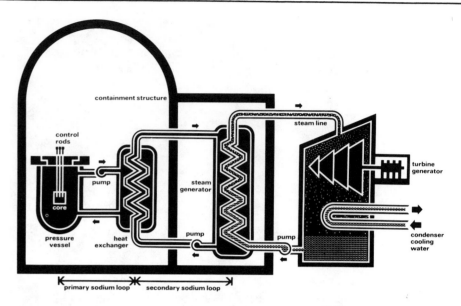

Liquid metal fast breeder reactor (LMFBR). As with the boiler in a coal-, oil- or gas-burning power plant, a nuclear power reactor produces steam to drive a turbine which turns an electric generator. Instead of burning a fossil fuel to make heat, the reactor fissions nuclear fuel to make the steam. In an LMFBR, molten sodium in the primary loop is pumped through the reactor core containing the fuel. This sodium collects the heat and transfers it to a secondary sodium loop in the heat exchanger, from which it is carried to the steam generator.

In addition to producing electricity, this type of reactor also produces more fissionable material than it consumes, which is why it is called a "breeder reactor." When irradiated, certain non-fissionable materials may be transformed into material that is fissionable. An LMFBR begins operation with a core of fissionable uranium-235 surrounded by non-fissionable uranium-238. During operation, the U-238 is bombarded by high-velocity neutrons and transmuted to fissionable plutonium-239. The plutonium is extracted periodically and fabricated into new fuel.

The country's first commercial LMFBR demonstration plant, designed by Westinghouse Electric Corp., is being built in Tennessee by Commonwealth Edison Co., the Tennessee Valley Authority, and the Atomic Energy Commission.

Fast Reactor

Fuel: A mixture of plutonium and uranium dioxides in stainless steel cans.

Moderator: None.

Core layout: Assemblies of fuel elements are placed inside a tank containing the liquid sodium coolant. The core is surrounded by a "blanket" of uranium carbide in stainless steel cans.

Heat extraction: The sodium is heated by the core and pumped through an intermediate heat exchanger where it heats sodium in a separate secondary circuit. The sodium in the secondary circuit transfers its heat to water in a steam generator; the steam drives a turbine coupled to an electric generator.

Indicative data for a reactor of 1300 MW(E) size:

fuel enrichment (% Pu)	20%
coolant outlet temperature	620°C
coolant pressure	5 psia
steam cycle efficiency	44%
core dimensions	2.3 m dia. × 1.1 m high
core and blanket dimensions	3.1 m dia. × 2.1 m high

Source: Reproduced with permission of the United Kingdom Atomic Energy Authority from the UKAEA Annual Rep. 1973/74.

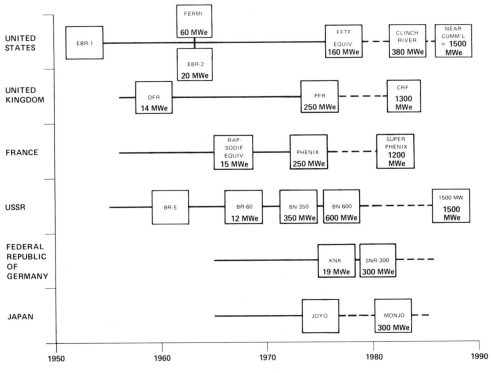

International LMFBR progress.

Source: Liquid Metal Fast Breeder Reactor: Decision Process and Issues, EPRI SR-20, Electric Power Research Institute, Palo Alto, California, November, 1975.

Comparison of Fast Reactor Power Plants Operating and Planned Around the World

Designation	BN 350	PHENIX	PFR DOUNREAY	BN 600	SNR KALKAR	SUPER PHENIX	MONJU FBR	CLINCH RIVER
Operating date	1972	1973	1974	(1977)	(1978-79)	(1980-81)	(1982)	(1982)
Loop or pool/No. of circuits	Loop/6	Pool/3	Pool/3	Pool/3	Loop/3	Pool/4	Loop/3	Loop/3
Electrical power output (MWe)	150 + desal.	250	250	600	282	1200	300	380
Thermal reactor power (MWth)	1000	563	600	1500	736	2900	540	975
Coolant flow (t/h)	14 000	10 000	10 500	24 000	12 200	56 500	15 400	18 800
Reactor inlet temperature (°C)	300	400	400	320	377	380	390	388
Reactor outlet temperature (°C)	500	562	560	520	546	540	540	534
Fuel (driver region)	UO_2	UO_2+PuO_2	UO_2+PuO_2	UO_2+PuO_2	UO_2+PuO_2	UO_2+PuO_2	UO_2+PuO_2	UO_2+PuO_2
Fuel (blanket region)	UO_2	UO_2	UO_2	UO_2	depl. UO_2	depl. UO_2	UO_2	depl. UO_2
Cladding material	ss	ss316	ss316	ss	aust. ss 1-4988	ss316	ss316	20% cw316
Pin diameter (mm)	6·1	6·6	5·84	6·9	6·0	8·65	6·5	5·8
Max. clad mid-wall temp. (°C)	680	693	692	700	620	620	700	663
Power density peak/av. (kW/l)	—/500	646/406	770/500	840/550	510/330	—/275	528/292	740/380
Maximum burnup (MWd/t)	50 000	50 000	61 000	100 000	100 000	70 000	80 000	80 000
Reactor vessel height dia. (m)	13·9/6	12/12	12·8/12·3	13/12·8	14/6·5	18·4/21	17·5/6·5	16·8/6·2
Wall thickness (mm)	30	15	16	—	25-40	25	35	60
Active core height/dia. (m)	1·06/1·5	0·85/1·39	0·9/1·45	0·75/2·05	0·95/1·78	1·0/3·66	0·93/1·49	0·91/1·87
No. of fuel assemblies	200	103	78	370	205	358	198	198
No. of pins per assembly	169	217	325	127	166	271	169	217

Source: Rippon, Simon, "Prototype Fast Breeder Reactors Operating in Europe and the U.S.S.R.," *Nuclear Engineering International,* June/July 1975.

Foreign LMFBR Programs

Country	Project	Power MWe	Criticality Date	Full Power Date	Present Status
United Kingdom	DFR	14	1959	1959	operating
	PFR	250	Mar. 1974	—	~20%
	CRF	1300	—	—	under design, construction start 1977
Fed. Rep. of Germany (+Benelux)	Karlsruhe KNK	19	1977		fast core being loaded
	SNR-300	300	—	—	under construction
France	Rapsodie	37 (thermal)	1967		
	Phenix	250	Aug. 1973	Mar. 1974	
	Superphenix	1200		1982	under design, construction start 1976
Japan	Joyo	100 (thermal)	1975		
	Monju	300	1979 or 1980		under design, construction start 1975 or 1976
	Commercial LMFBR	1500			under design, construction start ~1980
USSR	BR-1, -2, -5				
	BOR-60	12	Dec. 1969		
	BN-350	~305	End of 1972	July 1973	steam generator problems
	BN-600	600	~1976		under construction
	LMFBR	1000–1500	—		under design

Source: Liquid Metal Fast Breeder Reactor: Decision Process and Issues, prepared for the Ad Hoc Subcommittee for the Review of the Liquid Metal Fast Breeder Reactor Program, Joint Committee on Atomic Energy, Electric Power Research Institute, Palo Alto, California, May 9, 1975.

and maintenance procedures are also more difficult with sodium-cooled reactors than they are for water-cooled reactors since the sodium is not only opaque but, because of its chemical reactivity, systems cannot be opened to the air unless the sodium is drained from the systems.

Development of liquid-metal-cooled reactors dates from the early years of nuclear power development. In the United States, the first small fast reactor, Clementine, was operated in 1946. In 1951, the Experimental Breeder Reactor-1 generated the first electricity from nuclear energy. Other liquid-metal-cooled reactors followed: the Sodium Reactor Experiment, the Hallam Nuclear Power Facility, the Experimental Breeder Reactor-2, SEFOR, Fermi, two submarine reactors S1G and S2G, and a series of small liquid-metal-cooled reactors, the SNAP reactors, designed for use in space. In foreign countries, the United Kingdom has two operating LMFBRs, France has two, and the USSR has three. Additional projects are underway in these countries as well as in West Germany and Japan.

REFERENCE

See Robert L. Loftness, *Nuclear Power Plants: Design, Operating Experience and Economics,* D. Van Nostrand Company, Inc., Princeton, N.J., 1964, for details of the early development of liquid-metal-cooled reactors.

LIGHT WATER BREEDER REACTORS (LWBR)

The light water breeder reactor (LWBR) program was initiated in 1965 with the objective of adapting light water-moderated and cooled reactor technology employed in pressurized water reactors, to a thermal breeder using the Th-U-233 fuel cycle. Ordinary pressurized water reactors, using enriched U-235 fuel, have a conversion ratio of about 0.6, that is, during operation, about 0.6 atoms of fissionable Pu-239 are formed for every atom of U-235 consumed. To increase this conversion ratio to 1.0, at which point the reactor makes as much new fuel as it consumes, the LWBR incorporates a number of features. The basic fuel is U-233 rather than U-235 and the fertile material is thorium rather than U-238 to take advantage of the greater number of neutrons produced per fission of U-233 in a reactor in which the water content is minimized by close spacing of the fuel. Zirconium alloy is used instead of stainless steel for fuel cladding and core structure because of the lower neutron absorption of the zirconium alloys. Reactor control is accomplished by use of movable fuel assemblies to avoid the neutron losses associated with neutron-absorbing control rods.

A demonstration core assembly incorporating these features is being tested in the nuclear power plant at Shippingport, Pennsylvania.

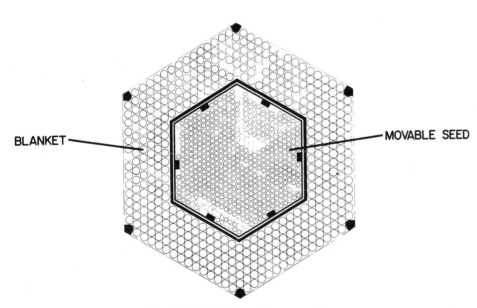

Typical LWBR fuel module cross section.

Source: Naval Nuclear Propulsion Program–1975, Hearing before the Subcommittee on Legislation of the Joint Committee on Atomic Energy, Congress of the United States, March 5, 1975.

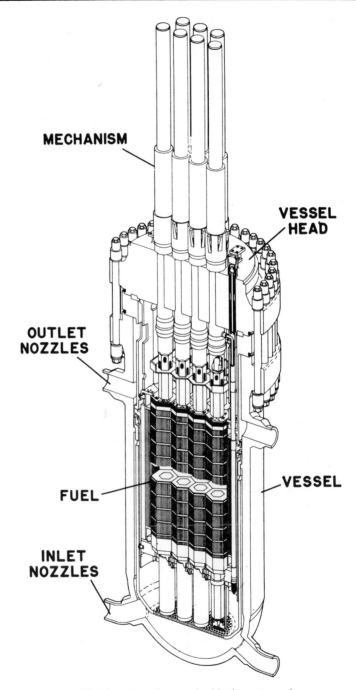

LWBR demonstration core in shippingport vessel.

Source: Naval Nuclear Propulsion Program–1975, Hearing before the Subcommittee on Legislation of the Joint Committee on Atomic Energy, Congress of the United States, March 5, 1975.

MOLTEN SALT BREEDER REACTORS (MSBR)

The molten salt breeder reactor (MSBR) concept is based on the use of a circulating fluid fuel in conjunction with on-line continuous fuel processing. The MSBR employs a graphite moderator structure through which the molten salt fuel circulates. The concept is classed as a thermal breeder and operates on the Th-U-233 fuel cycle. The molten salt fuel and coolant is comprised of fluoride salts of lithium, beryllium, thorium, and uranium. Operating temperature of the molten salt mixture is 1300°F (melting point of this salt mixture is 930°F).

Development of the molten salt reactor concept at Oak Ridge National Laboratory has included two experimental reactors. The aircraft reactor experiment (ARE) was operated for about nine days in 1954 at power levels up to 2.5 thermal megawatts and at fuel tem-

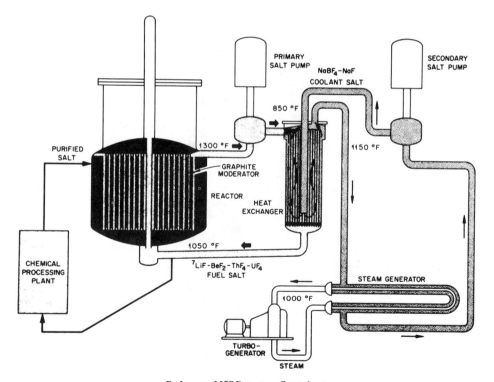

Reference MSBR system flow chart.

Source: Proposed Final Environmental Statement: Liquid Metal Fast Breeder Reactor Program, Volume III, WASH-1535, U.S. Atomic Energy Commission, December 1974.

Principal Operating Parameters of a 1000 MWe MSBR

General	
Thermal power, MW(th)	2,250
Electric power, MW(e)	1,000
Plant lifetime, Years	30
Fuel processing scheme	on-line, continuous processing
Breeding ratio	1.06
Reactor	
Fuel salt	^{7}LiF—BeF$_2$—ThF$_4$—UF$_4$
Moderator	unclad, sealed graphite
Reactor vessel material	modified Hastelloy-N
Power density, KW/liter	22
Exit temperature, °F	1,300
Temperature rise across core, °F	250
Reactor vessel height, ft.	20
Reactor vessel diameter, ft.	22
Vessel design pressure, psia	75
Peak thermal neutron flux, neutrons/cm^2-sec	8.3×10^{14}
Other Components and Systems Data	
Number of primary circuits	4
Fuel salt pump flow, gpm	16,000
Fuel salt pump head, ft.	150
Intermediate heat exchanger capacity, MW(th)	556
Secondary coolant salt	NaF—NaBF$_4$
Number of secondary circuits	4
Secondary salt pump flow, gpm	20,000
Secondary salt pump head, ft.	300
Number of steam generators	16
Steam generator capacity, MW(th)	121

Sources: R. C. Robertson, *A Conceptual Design Study of a Single-Fluid Molten Salt Breeder Reactor*, Rep. ORNL-4541, Oak Ridge National Laboratory, June 1971, and M. W. Rosenthal et al., *The Development Status of Molten Salt Breeder Reactors*, Rep. ORNL-4812, Oak Ridge National Laboratory, August 1972.

peratures up to 1580°F. The ARE had a moderator structure of beryllium oxide. The molten salt reactor experiment (MSRE) was completed in 1965 and during the next four years operated for about 13,000 equivalent full-power hours. The design power level of the MSRE was 8 thermal megawatts.

Although there are no current plans for the construction of a MSBR prototype reactor, research and development work is continuing at Oak Ridge National Laboratory on basic technology including tritium confinement, fuel salt processing, structural materials behavior in the presence of fission products and nuclear radiation, and system components.

REFERENCE

L. E. McNeese and M. W. Rosenthal, "MSBR: A review of its status and future," *Nuclear News*, September 1974.

GAS-COOLED FAST BREEDER REACTORS (GCFR)

Although most of the development work on fast breeder reactors has concentrated on reactors using liquid metal as a coolant, modest design and development work has been undertaken on fast reactors employing a gas, helium, as a coolant. The use of helium as a coolant presents several advantages—its optical clarity permits visual maintenance of components, it does not become radioactive, it is chemically inert and does not undergo

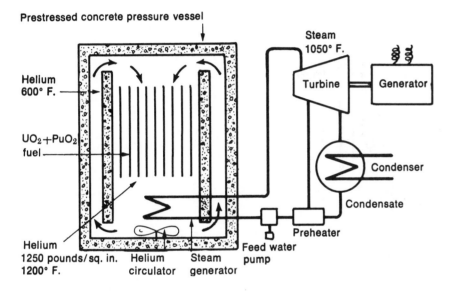

The breeder reactor utilizes fuel composed of plutonium and the abundant 238-isotope of uranium; fission of one plutonium atom by a fast (unmoderated) neutron from the uranium yields 2.5 new neutrons, and one of these continues the fission reaction while the remaining 1.5 are absorbed by the uranium-238 to produce new plutonium to replace that consumed in fission. In the gas-cooled fast reactor, helium gas—inert even at these temperatures—is circulated through the reactor and then across a steam generator to produce steam at over 1,000°F.; this steam in turn drives a turbine connected to a generator, then is cooled in a condenser and returned through a preheater to the steam generator. Because operation is at high temperatures, the overall efficiency may be as high as 40 per cent and the amount of heat to be distributed from condenser to environment per unit of energy produced is relatively low.

Source: Manson Benedict, "Electric Power from Nuclear Fission," *Technology Review*, edited at the Massachusetts Institute of Technology October/November, 1971.

GCFR Demonstration Plant Performance Characteristics

Net electric power	311 MW
Proportions:	
Fuel rod diameter	0.72 cm
Coolant void fraction	0.45
Core diameter	200 cm
Core length	100 cm
Reactor vessel diameter	84 ft
Reactor vessel height	71 ft
Operating Conditions:	
Maximum cladding hot spot	700°C
Pressure	85 atm
Pumping power (% thermal output)	5%
Gas in	595°F
Gas out	1010°F
Performance:	
Fuel rating	0.6 MW/kg
Linear rating, max	12.5 kW/ft
Overall efficiency	36%
Conversion ratio	1.33[a]
Doubling time	21 years[a]

[a]These figures are for the 300-MWe demonstration plant. For the proposed 1000-MWe plant, the breeding ratio has been calculated to be in the range of 1.40 to 1.50, and the corresponding doubling time is 8 to 10 years.

Source: Proposed Final Environmental Statement Liquid Metal Fast Breeder Reactor Program, Volume III, WASH-1535, U.S. Atomic Energy Commission, December 1974.

phase changes. The low neutron absorption characteristics of helium give the GCFR a potentially high breeding ratio. On the other hand, as a gas, helium has relatively poor heat transport properties in contrast to liquid sodium, and GCFR reactors must therefore be operated at high pressures to obtain sufficient heat transfer from the reactor core to the steam generators. As with other high temperature gas-cooled reactors, all major components of the primary coolant system are contained within a prestressed concrete reactor vessel (PCRV).

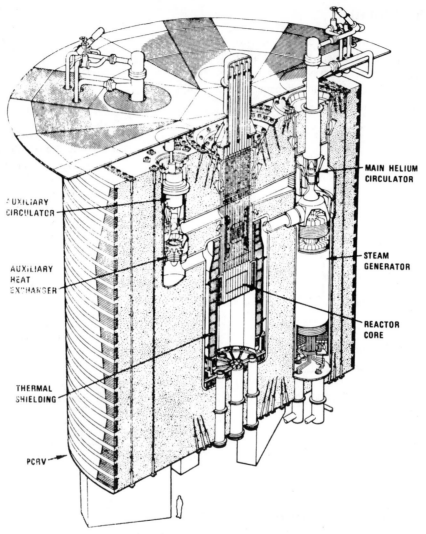

300-MWe GCFR demonstration plant.

Source: Proposed Final Environmental Statement Liquid Metal Fast Breeder Reactor Program, Volume III, WASH-1535, U.S. Atomic Energy Commission, December 1974.

Research and development is continuing on various elements of GCFR technology, particularly fuels and materials development, physics, and safety. Conceptual designs have been completed for a 300 MWe GCFR demonstration plant and design work on a 700 MWe demonstration plant has been initiated. In Europe, core heat transfer studies are being undertaken at the Swiss Federal Institute for Reactor Research, the Nuclear Energy Agency is supporting an effort in GCFR research, and eight companies have formed the Gas Breeder Reactor Association to conduct related research. At present, there are no firm plans to construct a prototype gas-cooled fast reactor.

NUCLEAR POWER PLANT OPERATING EXPERIENCE

Data on power plant reliability and performance are often reported in terms of two factors. "Capacity factor" or "load factor" is the percentage of kilowatt hours produced in a given period compared with the maximum number of kilowatt hours that could have been produced. "Availability factor" is the percentage of time that a plant is available for use during a given period, regardless of its actual use.

The availability factor gives a measure of the technical performance of a plant, whereas the capacity factor or load factor reflects the daily and seasonal variation in load demand below the maximum rating of the plant and the relative economics of operating a particular plant in contrast to other plants on the same system. For example, during the first half of 1975, the availability factors for oil and coal plants was 75.1% and 80.9% while the availability factor for nuclear plants was 71.3% but, since nuclear plants were cheaper to operate, the average capacity factor for nuclear plants was 64.2% as against 34.4% for oil and 55.9% for coal-fired plants.

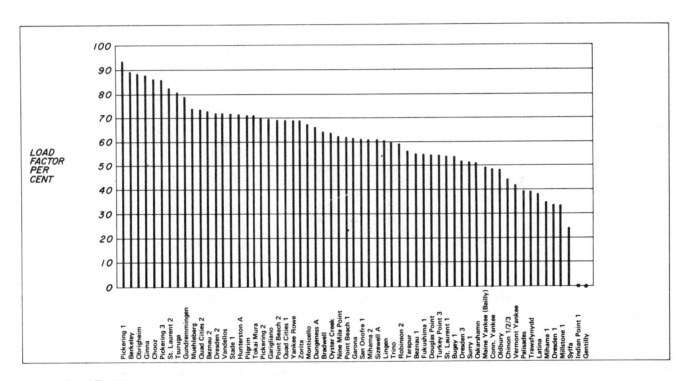

Average Load Factors

56.1% Average of all load factors for 23 U.S. nuclear stations for 1973
62% Average of all load factors for 38 Non-U.S. nuclear stations for 1973

| | Position of U.S. Plants by Groups | | | | | |
	Top 10 Plants	Plants 11-20	Plants 21-30	Plants 31-40	Plants 41-50	Bottom 11 Plants
No. of U.S.	1	3	7	2	5	5

Median position for 23 U.S. nuclear stations is No. 35 in the field of 61.

Source: Power Engineering, October 1974.

Power Plant Reliability in First Half, 1975.

Utility	Capacity[1] Nuclear 1st Quarter	2nd Quarter	Oil 1st Quarter	2nd Quarter	Coal 1st Quarter	2nd Quarter	Availability[2] Nuclear 1st Quarter	2nd Quarter	Oil 1st Quarter	2nd Quarter	Coal 1st Quarter	2nd Quarter
[3]Baltimore G&E	*	69.6	28.3	19.9	68.9	62.4	*	—	—	—	—	—
Boston Edison	49.3	54.2	51.6	45.2	*	*	71.0	80.8	67.5	61.9	—	—
Carolina P&L	95.0	43.0	*	*	49.0	59.0	96.0	45.0	*	*	95.0	88.0
Con. Ed. of NY	51.0	80.0	42.0	43.0	*	*	55.6	88.7	63.5	68.4	*	*
Consumers Power	1.0	28.0	33.0	46.0	66.0	56.0	1.5	55.0	60.8	60.4	73.0	71.7
Dairyland Power Coop.	87.0	31.5	5.4	7.8	68.9	39.6	—	—	—	—	—	—
Duke Power	37.7	69.4	—	—	55.3	42.4	48.2	78.0	—	—	87.1	72.6
[4]Florida P&L	93.7	49.2	41.8	61.9	*	*	94.0	53.0	—	—	*	*
Jersey Central P&L	87.0	34.7	18.4	15.4	73.2	75.6	91.4	42.2	73.6	80.3	76.8	78.7
Maine Yankee	79.9	27.6	*	*	*	*	98.9	36.4	*	*	*	*
Metro. Edison	98.7	60.8	9.3	3.8	58.9	61.9	97.5	66.3	97.0	90.2	72.9	80.3
Niagara Mohawk	72.4	91.0	64.3	53.7	77.2	64.8	80.0	98.0	—	—	—	—
Northeast Util.	82.2	78.1	47.2	42.3	*	*	93.1	83.2	—	—	*	*
[5]Omaha Pub. Pow.	28.0	31.8	*	*	41.9	36.4	39.9	54.4	*	*	96.7	94.4
Pacific G&E	69.2	55.0	36.9	21.4	*	*	93.5	65.7	84.3	79.4	*	*
Phila. Elec.	65.4	65.9	30.5	25.4	59.4	57.1	—	81.5	—	81.5	—	66.6
Rochester G&E	76.0	21.8	*	*	76.2	59.9	76.2	34.2	*	*	79.0	89.7
[6]Sacramento M.U.D.	*	91.2	*	*	*	*	*	—	*	*	*	*
[7]So. Cal. Edison	84.0	83.0	39.0	40.0	60.0	56.0	—	—	—	—	—	—
[8]TVA	73.7	—	*	*	44.6	48.7	86.0	—	*	*	74.2	76.4
Vermont Yankee	86.0	80.4	*	*	*	*	90.8	87.8	*	*	*	*
Va. Elec. & Pow. Co.	66.5	61.4	47.2	43.7	41.5	41.4	72.8	66.2	83.0	75.0	71.0	68.0
Wisc. Elec. Power	68.2	85.5	*	*	47.2	35.4	—	—	*	*	—	—
Wisc. Pub. Serv.	76.1	64.7	—	—	58.1	45.4	93.9	87.2	—	—	91.5	96.1
Yankee Atomic	98.4	96.8	*	*	*	*	98.9	97.8	*	*	*	*
Averages [9]Quarters:	70.7	58.2	35.4	33.5	59.1	52.6	77.9	65.1	75.7	74.6	81.7	80.2
[10]First Half:	64.2		34.4		55.9		71.3		75.1		80.9	

(1) "Capacity Factor" is the percentage of kilowatt hours produced in a given period, compared with the maximum number that could be produced.

(2) "Availability Factor" is the percentage of the time that a plant is available for use during a given period, regardless of its actual use.

(3) Nuclear contribution did not begin until May 8.

(4) One unit was being refueled in second quarter.

(5) Fort Calhoun Unit 1 was out of service for refueling and inspection from February 7 through May 9.

(6) Rancho Seco Unit 1 did not go into service until April 18.

(7) Southern California Edison figures represent the 12-month period through June.

(8) Browns Ferry Units 1 and 2 were not operating in second quarter, but are included in calculation of averages.

(9) Averages for combined fossil (oil and coal) plants for first quarter were 48.0, Capacity Factor, and 79.2, Availability Factor. For second quarter, the averages were 43.7, Capacity Factor, and 77.9, Availability Factor.

(10) Averages for combined fossil (oil and coal) plants for first half were 45.9, Capacity Factor, and 78.5, Availability Factor.

* Not applicable.

Note: Figures for Commonwealth Edison were not ready in time for this survey, but will be included in the survey to be completed after the third quarter.

Source: News Release 9/75/N, Atomic Industrial Forum, Inc., New York, N.Y. September 30, 1975.

U.S. Nuclear Plants: Operating Experience
(Based on NRC "Operating Units Status Report" Gray Book of January, 1978)

Plant	Utility	Mwe net ultimate	Unit Availability[1] Dec. 1977	Unit Capacity Factor[2] Dec. 1977	Cumulative	Forced Outage[3] Rate Dec. 1977	Cumulative
Arkansas One 1	Arkansas P&L	850	74.6	71.4	63.2	25.5	11.3
Beaver Valley	Duquesne Light	852	94.1	91.2	50.0	3.6	18.0
Big Rock Point	Consumers Power	72	100	91.4	55.3	0	17.0
Browns Ferry 1	TVA	1,065	.0F	.0	39.2	.0	50.3
Browns Ferry 2	TVA	1,065	51.4	36.5	31.5	48.6	57.0
Browns Ferry 3	TVA	1,054	96.3	80.5	74.8	3.7	9.9
Brunswick 1	Carolina P&L	821	84.0	65.9	46.0	15.9	44.3
Brunswick 2	Carolina P&L	821	.0	.0	37.3	94.7	18.6
Calvert Cliffs 1	Baltimore G & E	845	94.6	92.5	79.2	5.4	5.9
Calvert Cliffs 2	Baltimore G & E	845	97.4	103.0*	84.9	2.6	4.5
Cook 1	Indiana Michigan Power	1,060	90.6	85.2	68.4	9.4	6.9
Cook 2	*Indiana Michigan Power	1,095	—	—	—	—	—
Cooper	Nebraska Public Power Dist.	778	100	77.9	58.7	.0	6.1
Crystal River 3	Florida Power	825	99.4	87.0	70.0	.6	9.9
Davis-Besse 1	Toledo Edison	906	76.9	40.2	32.6	20.6	17.5
Dresden 1	Commonwealth Edison	200	96.4	64.0	49.2	3.6	11.1
Dresden 2	Commonwealth Edison	800	68.3	49.3	50.8	31.7	16.0
Dresden 3	Commonwealth Edison	800	84.1	57.5	55.7	15.9	14.3
Duane Arnold	Iowa EL&P	550	100	93.9	56.9	.0	5.3
Farley 1	Alabama Power	829	68.8	64.2	64.2	23.0	23.0
FitzPatrick	Power Auth. State of N.Y.	821	100	90.4	71.8	.0	20.3
Ft. Calhoun	Omaha Public Power Dist.	457	73.2F	64.4	60.2	6.6	5.3
Ft. St. Vrain	*P.S. of Colorado	330	—	—	—	—	—
Ginna	Rochester G&E	490	97.6	85.3	66.0	2.4	11.0
Haddam Neck	Conn. Yankee Power	575	82.5	69.0	80.7	1.8	8.6
Hatch 1	Georgia Power	786	78.5 a	70.0	53.8	21.5	18.2
Humboldt Bay	Pacific G&E	68	.0 a	.0	56.4	.0	1.9
Indian Point 1	Con Edison	265	— b	—	—	—	—
Indian Point 2	Con Edison	873	100	91.9	55.7	.0	10.1
Indian Point 3	Con Edison/PASNY	873	46.0 c	38.3	72.1	2.0	4.2
Kewaunee	Wisconsin P.S.	540	98.5	97.6	72.5	1.5	5.0
LaCrosse	Dairyland Power Co-op	50	.0F	.0	48.2	.0	4.0
Maine Yankee	Maine Yankee Power	790	100	99.8	66.9	.0	3.5
Millstone 1	Northeast Nuclear Energy	652	59.9	51.5	60.9	40.1	20.1
Millstone 2	Northeast Nuclear Energy	828	.0F	.0	63.3	.0	23.5
Monticello	Northern States Power	545	95.3	92.6	69.9	.0	8.0
Nine Mile Point 1	Niagara Mohawk Power	610	100	97.6	57.0	.0	11.3
Oconee 1	Duke Power	886	85.1	75.6	56.6	14.9	18.7
Oconee 2	Duke Power	886	73.0	49.8	57.2	27.0	23.7
Oconee 3	Duke Power	886	68.6 F	52.2	65.5	22.9	14.4
Oyster Creek	Jersey Central P&L	640	96.9	97.5	69.2	3.1	5.0
Palisades	Consumers Power	800	95.3	103.8*	43.5	4.7	40.5
Peach Bottom 2	Philadelphia Electric	1,065	93.3	74.8	56.7	6.7	9.2
Peach Bottom 3	Philadelphia Electric	1,065	60.6	38.3	59.3	26.7	9.5
Pilgrim 1	Boston Edison	655	61.9	13.2	47.1	38.0	11.4
Point Beach 1	Wisconsin Electric Power	497	100	96.9	73.6	.0	3.1
Point Beach 2	Wisconsin Electric Power	497	100	99.1	75.5	.0	1.9
Prairie Island 1	Northern States Power	550	100	101.5*	67.4	.0	15.3
Prairie Island 2	Northern States Power	520	51.6F	46.1	72.2	.8	8.5
Quad Cities 1	Commonwealth Edison	800	92.0	75.8	56.8	8.0	9.2
Quad Cities 2	Commonwealth Edison	800	100	74.1	59.1	.0	12.2
Rancho Seco	Sacramento Mun. Util. Dist.	918	98.4	99.1	45.9	1.6	49.0
Robinson 2	Carolina P&L	712	99.7	97.8	72.7	.3	14.9
Salem 1	PSE&G	1,090	86.1	71.6	42.9	13.9	20.3
San Onofre 1	So. Calif. Edison	430	99.6	97.8	71.4	.0	10.0
St. Lucie 1	Florida P&L	810	98.6	99.4	77.6	1.4	10.0
Surry 1	Virginia E&P	820	79.2	77.7	59.3	.3	18.1
Surry 2	Virginia E&P	820	99.8	97.1	58.5	.2	25.5
Three Mi. Island 1	Metropolitan Edison	819	100	101.1*	74.9	.0	5.6
Trojan	Portland General Elec.	1,130	95.7	80.2	48.1	4.6	21.3
Turkey Point 3	Florida P&L	693	.0F	.0	68.7	.0	2.8
Turkey Point 4	Florida P&L	693	96.5	94.9	66.9	3.5	3.3
Vermont Yankee	Vermont Yankee Power	540	95.0	91.9	64.7	.0	5.1
Yankee (Rowe)	Yankee Atomic Power	175	98.7	97.5	72.1	.0	1.6
Zion 1	Commonwealth Edison	1,100	82.8	74.9	51.3	5.3	12.9
Zion 2	Commonwealth Edison	1,100	97.1	84.6	57.5	2.9	20.3

* Plants in power ascension phase.   F Refueling.

[1] NRC definition: the time percentage in the month the unit was available to generate electricity.

[2] NRC definition: net power generated × 100 ÷ max. dependable capacity × gross hrs.

[3] NRC definition: forced outage hrs. × 100 ÷ generator on-line hrs. + forced outage hrs.

* Capacity factors greater than 100% are attributed to low temperature of condenser cooling water decreasing back pressure, increasing unit efficiency.

a Seismic modifications continuing.
b All fuel unloaded, no decision on future operation.
c Unit returned to service on 12-17-77 following 10-week outage for turbine maintenance.

Source: Nuclear Industry, February 1978.

UNIT YEAR AVERAGES
1965–1974

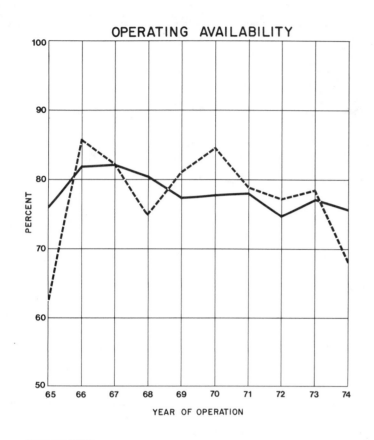

OPERATING AVAILABILITY

SAMPLE TABLE

Year of Operation	Fossil–390 MW and Above No. of Units	Nuclear No. of Units
1965	15	3
1966	21	5
1967	32	6
1968	52	7
1969	76	7
1970	88	9
1971	104	11
1972	128	11
1973	151	22
1974	173	42

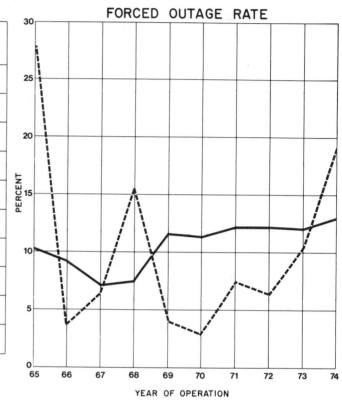

FORCED OUTAGE RATE

———— FOSSIL
-------- NUCLEAR

Comparison of fossil units 390 MW and above to all nuclear units.

Source: Report on Equipment Availability for the Ten-Year Period, 1965–1974, EEI Publication No. 75-50, Edison Electric Institute, 90 Park Avenue, New York, N.Y. 10016, November 1975.

NUCLEAR FUSION PRINCIPLES

In contrast to nuclear fission, in which a heavy element, such as U-235, splits into smaller fragments and releases heat upon impact by a neutron, nuclear fusion involves the joining together or fusing of the nuclei of light elements, typically the deuterium and tritium isotopes of hydrogen, forming a new element of less mass than the original reactants with the mass difference being converted to heat. This heat, in turn, can be used to generate electrical energy.

The fusion of two nuclei can occur if the nuclei collide with sufficient energy. The fundamental problem in causing fusion to occur is that the nuclei are all positively charged and, therefore, repel each other when they come in close proximity. This repelling force can be overcome if the energy or velocity of the nuclei is high enough—a condition which can be achieved by heating a gas of the nuclei to temperatures of the order of 100 million degrees Centigrade. At these temperatures, the mixture of positively charged nuclei of deuterium and tritium and the electrons which have been stripped from the nuclei, form a plasma, somewhat like a gas, but more nearly a fourth state of matter since the plasma has properties unlike those of gases, liquids, or solids at more conventional temperatures.

Since there is no solid material which can exist at fusion temperatures (or contain the equivalent pressure of 1.5 million atmospheres), two separate approaches are being studied for creating the conditions necessary for fusion to occur. One method uses magnetic fields to confine the plasma. A number of experimental devices have been built using magnetic confinement techniques and there has been steady improvement in creating the temperature/density-confinement time conditions necessary for the generation of power from fusion. The immediate goal in the development program is to achieve plasma condi-

U.S. Fusion Power R&D Contractors to ERDA

Name	Purpose	Proposed fiscal year 1976 operating expenses
Principal contractors:		
Argonne National Laboratory	Advanced development and fusion power technology	$2,050
Brookhaven National Lab	do	1,465
General Atomics	Noncircular closed systems research	13,530
Lawrence Livermore Lab	Open systems research	21,865
Los Alamos Scientific Laboratory	High density closed systems research	14,640
Massachusetts Institute of Technology	Experimental research on high field tokamaks	2,470
Holifield National Laboratory	Theoretical and experimental research on closed confinement systems, advanced development	19,195
Pacific Northwest Labs	Fusion power technology	1,820
Princeton Plasma Physics Laboratory	Low density closed systems research	26,460
University of Texas	Turbulent heating studies and theoretical tokamak research	1,450
Other contractors: Contract research primarily at universities		15,055

Source: ERDA Authorizing Legislation Fiscal Year 1976, Hearings before the Subcommittee on Legislation of the Joint Committee on Atomic Energy, Congress of the United States, Part 2, February 18 and 27, 1975.

BREAK-EVEN PLASMA CONDITIONS FOR FUSION POWER

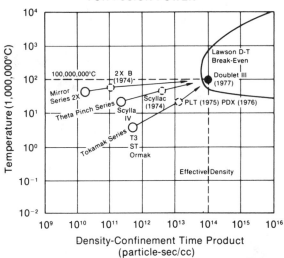

Doublet III, with construction now funded by ERDA for operation in 1977, will be the first fusion experiment whose goal is to achieve break-even, the range of conditions necessary for deuterium-tritium fueled production of a net power output. Operating regimes of other controlled thermonuclear experiments are shown by solid circles (existing devices) and dashed circles (planned). (*Adapted from* High-Level Radioactive Waste Management Alternatives, *WASH 1297, ERDA (formerly AEC), May 1974.*)

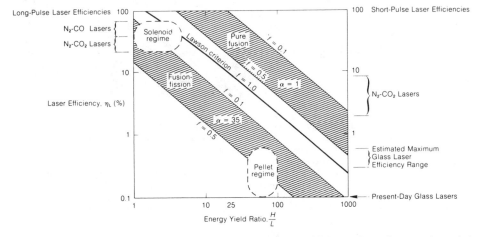

Various laser-fusion yield ratios require different ranges of laser efficiency, depending on values of circulating energy fraction, f, and neutron multiplication factor, α. The Lawson criterion curve represents break-even ($f = 1.0$). Potentially feasible plants fall in the bands bounded by f values of 0.1 and 0.5.

Source: Research Progress Report FF-3, Fossil Fuel and Advanced Systems Division, Electric Power Research Institute, Palo Alto, California, July 1975.

Source: Research Progress Report FF-2, Fossil Fuel and Advanced Systems Division, Electric Power Research Institute, Palo Alto, California, January 1975.

tions such that as much energy is generated in the fusion reaction as is needed to operate the equipment. The conditions for this "break even" are defined by the Lawson Criterion, as shown in the charts.

The other technique for achieving fusion conditions uses high-energy lasers to compress a pellet of the two hydrogen isotopes, deuterium and tritium. Laser fusion attains "break even" conditions under different circumstances than magnetic confinement fusion. Both processes require temperatures of 100 million degrees Centigrade but the nature of the magnetic confinement process is such that fuel densities in the plasma are relatively low and confinement times must, therefore, be relatively long. The impact of convergent laser beams on fuel pellet, however, produces a compaction of the fuel to high densities and confinement times are, therefore, very short. While the major problem with magnetic confinement fusion is the achievement of stable plasma conditions, the major problem with laser fusion is to generate and focus enough energy on a fuel pellet to achieve the required fuel density.

Scyllac is an experiment in the controlled Thermonuclear Research (CTR) Program at the AEC's Los Alamos (N.M.) Scientific Laboratory. The goal of AEC's CTR program is to develop the fusion process (the joining or fusing of light atoms) as a major source of abundant, economical and environmentally attractive energy, particularly for the generation of electrical power. The Scyllac experiment is an important step in perfecting this process. It has an outer circumference of about 85 feet. A quartz tube (located a few feet off the floor) forms a vacuum chamber in which a very hot gas, called a plasma, is formed and confined by an intense magnetic field. The field is created by a coil (the inner-most ring next to the scientists in the photograph) surrounding the quartz tube. The power for the field comes from the capacitor bank, located around the periphery. (Los Alamos Scientific Laboratory)

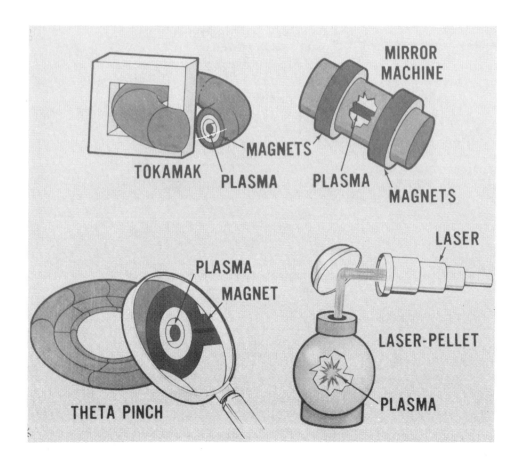

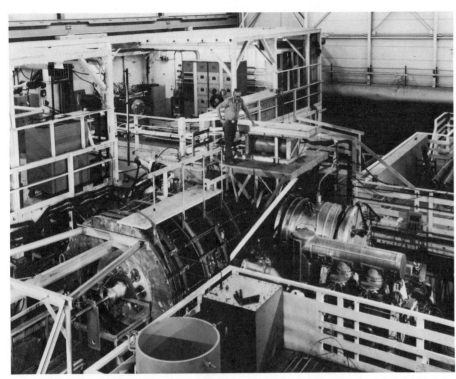

EXIT . . . The "mirror" approach to controlled thermonuclear research, is one of the two basic magnetic confinement fusion concepts. The 2XII, shown here, at the AEC's Lawrence Livermore Laboratory in California is providing important and encouraging results. (Credit: Lawrence Livermore Laboratory)

ORMAK . . . During construction of ORMAK, a thermonuclear fusion research experiment at the AEC's Oak Ridge National Laboratory, a worker inspects the vertical magnetic field coils that surround ORMAK's doughnut-shaped plasma region. The coils, in turn, are surrounded by the inch-thick aluminum shell seen in the photograph.

The ORMAK device is one of a class of Tokamak experiments. The Tokamak concept was invented by the late Soviet Academician Lev Artsimovich. (Credit: Oak Ridge National Laboratory)

MAGNETIC CONFINEMENT FUSION APPROACHES

To achieve the physical conditions necessary for fusion reactions—temperatures high enough to initiate the fusion reaction and confinement times for the reactants that are sufficiently long for enough material to react to yield a net power output—two generic types of magnetic confinement devices are being developed. Both employ intense and specially shaped magnetic fields to hold the hot, ionized plasma in place and prevent leakage to the walls of the fusion equipment. Limits on magnetic field strength require that the density of the plasma gas be kept well below atmospheric density in order to keep gas pressure on the confining magnetic field and surrounding support structures within practical bounds. Plasma densities range from 1/100,000 to 1/1000 of atmospheric density. Corresponding confinement times defined by the Lawson criterion, range from about 1 second to a few hundredths of a second. Power densities vary as the square of the fuel density—at lower density power production is 100,000 to 300,000 kilowatts per cubic meter of reacting plasma; at the higher densities, power production is above a billion kilowatts per cubic meter. The lower range of power production can be handled in a steady fashion since these levels are comparable to those in the furnace of an ordinary steam power plant. The higher values of power density would need to be handled in a "pulsed" mode of operation.

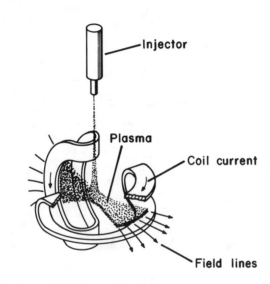

**Minimum-B Magnetic Mirror
(Yin-Yang Coils)**

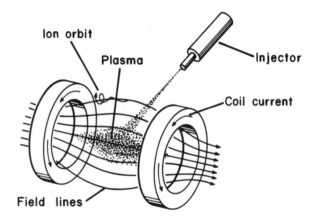

Simple Magnetic Mirror

The minimum-B magnetic mirror configuration has magnets shaped like the threads on a baseball. This configuration provides a shaped magnetic field that is better able to hold the plasma in place than the simple magnetic mirror. Plasma heating is provided by a neutral particle injector. (Los Alamos Photo Laboratory as shown in R.F. Post and F. L. Ribe, "Fusion Reactors as Future Energy Sources," *Science*, 1 November 1974.)

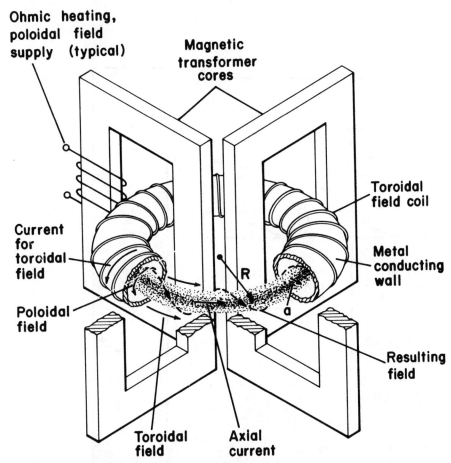

Tokamak magnetic confinement fusion system uses a toroidal magnetic field to confine the plasma inside a torus. Plasma heating is provided by resistance heating of the plasma. (Los Alamos Photo Laboratory, as shown in R. F. Post and F. L. Ribe, "Fusion Reactors as Future Energy Sources," *Science*, 1 November 1974.)

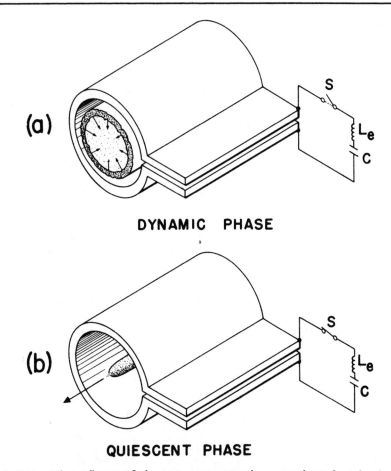

Theta-pinch magnetic confinement fusion system uses magnetic compression to heat the plasma which is confined in a toroidal field. (Los Alamos Photo Laboratory, as shown in R. F. Post and F. L. Ribe, "Fusion Reactors as Future Energy Sources," *Science*, 1 November 1974.)

Basic to all the magnetic confinement systems is the property of charged plasma particles that constrains them to move in a helical orbit that lies along the direction of the lines of force of a magnetic field. The simplest system, a long, straight tube surrounded by a magnetic coil, fails because there is no confinement at the ends of the tube and the plasma is lost too rapidly at the ends to allow a net fusion energy production. The two generic types of fusion systems being developed address this problem of the "ends" in different manners. The "open-ended" system is represented by the magnetic mirror concept in which an extra-strong magnetic field at the ends (the mirrors) of the confinement region repel the charged particles and reflect them back and forth for a long enough period for the fusion reactions to occur.

The "closed" systems solve the end problem by employing a confinement region that has a torus or doughnut shape. One of these closed systems is the "Tokamak" concept,

first developed by Soviet scientists, in which the simple toroidal magnetic field is augmented by inducing a strong electrical current in the plasma itself in order to heat the plasma to reaction temperatures. Another of the closed systems, the theta-pinch magnetic confinement system, employs magnetic compression for plasma heating by pulsing the strength of the magnetic field.

REFERENCES

David J. Rose, "Controlled nuclear fusion: Status and outlook," *Science*, 21 May 1971.
R. F. Post and F. L. Ribe, "Fusion reactors as future energy sources," *Science*, 1 November 1974.

FUSION POWER REACTOR CONCEPTS

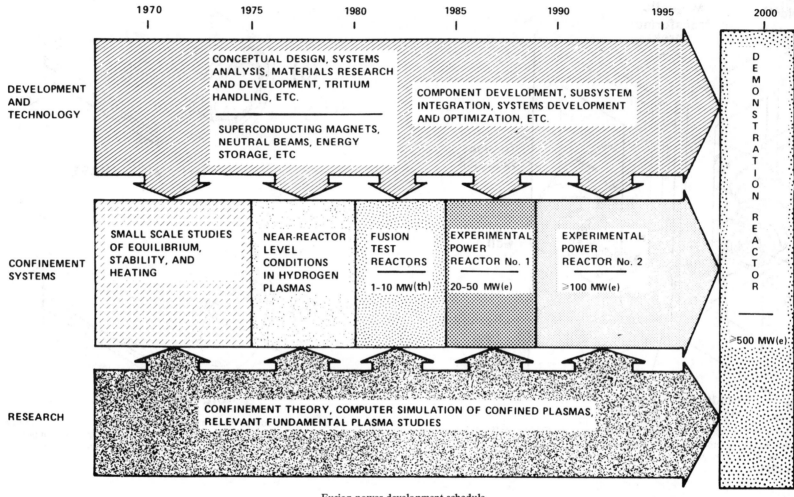

Fusion power development schedule.

Source: Energy Research and Development Administration.

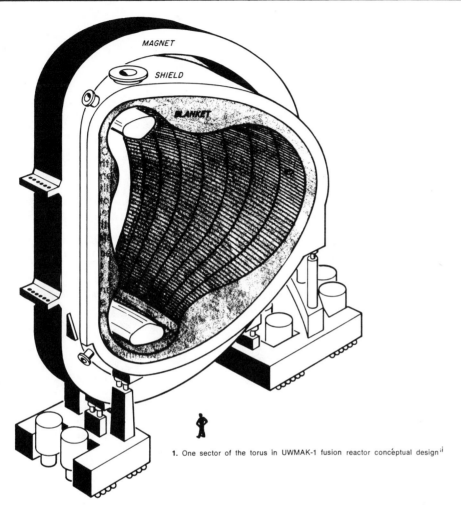

1. One sector of the torus in UWMAK-1 fusion reactor conceptual design[i]

One of twelve sectors of the torus of a 1500 MWe fusion power plant design prepared by the University of Wisconsin Fusion Feasibility Study Group. Energy from the deuterium-tritium plasma within the torus is absorbed in a lithium blanket and removed with lithium coolant to external steam boilers and turbine electric generators. Core structural material is stainless steel. The reactor uses the Tokomak magnetic confinement principle.

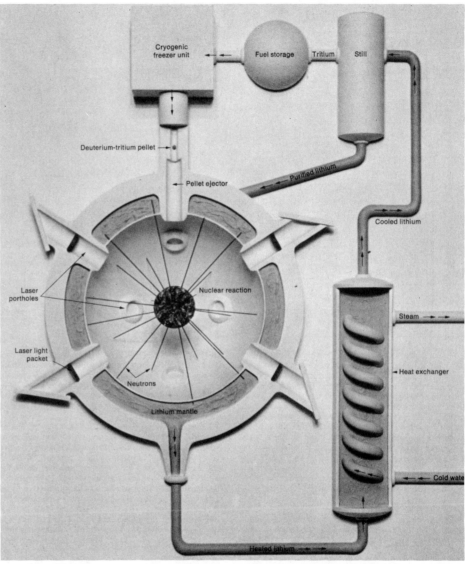

Laser beams converge on a pellet of frozen hydrogen to unlock the immense energies of controlled thermonuclear fusion. This cutaway model shows in principle how a laser fusion reactor would operate; it is based on the latest device developed by scientists. From a cryogenic freezer unit (upper left), tiny pellets of "heavy" hydrogen (deuterium and tritium) are fired at the rate of two or three per second into a vacuum inside a heavy, spherical pressure vessel. Precisely timed lasers, arrayed at portholes around the sphere, simultaneously fire extremely short, sharply focused light pulses at each pellet as it reaches the center of the sphere, raising its temperature to approximately 100,000,000° Centigrade. Thus each pellet turns briefly into a miniature sun in an intense thermonuclear reaction. The process transforms the hydrogen pellet into helium and releases energy in the form of neutrons. As the neutrons streak away from the reaction, their high energy is absorbed as heat in a mantle of molten lithium circulating in a chamber between the inner and outer walls of the sphere. The lithium, heated to as much as 1,000° Fahrenheit, then moves through a heat exchanger (right) to produce steam for the generation of electric power. Before the molten lithium is recycled through the sphere to pick up more heat, a still (top right) separates it from radioactive tritium created by the neutron bombardment. The gaseous tritium is pumped back into the pellet freezer, thus reducing the amount of costly hydrogen pellets that is required as fuel.

8

Geothermal Energy

Geysers (*Courtesy Electric Power Research Institute, Palo Alto, California*).

USES OF GEOTHERMAL ENERGY

Evidence of the heat energy stored in the earth is apparent from the hot springs, geysers, and fumaroles which occur in many countries. The magnitude of this geothermal heat source is particularly evident during volcanic eruptions when material at temperatures up to 800°C is ejected. The extent of the stored energy within the earth has been dramatically manifested in the eruptions of the volcanoes at Santorini about 1500 B.C. and Krakatoa in 1883 A.D., both of which ejected many cubic miles of rock. Except for these isolated signs of the earth's heat, there is generally no physical evidence of that heat at the earth's surface since the rate of heat flow through the insulating layer of the earth's crust is too low to be felt.

Until the beginning of this century, the utilization of geothermal heat has been limited to the use of warm water for baths. Geothermal sources provided heated water for several Roman baths in England and geothermal hot springs began to enjoy wide use throughout the world as therapeutic treatment in the late 19th century when bathing became popular once again. Since that time, more extensive use has been made of geothermal energy for both power and nonpower applications. The use of geothermal energy for the heating of buildings has been made in Iceland, Hungary, the USSR, New Zealand, and the United States. Geothermal heat has been used for the heating of greenhouses and for the processing of paper and minerals. Geothermal waters have also served as sources of boron, calcium chloride, and carbon dioxide.

Utilization of Geothermal Energy

Electric power production
  Direct use of dry steam
  Flashing of hot water to steam
    surface flashing
    in situ flashing
  Binary and hybrid cycles
  Innovative single-well converters
Direct use of thermal waters
  agriculture
  aquaculture
  space heating
  industrial processing
  medical therapy
By-products
  Mineral extraction
  Water resources

Source: Stimulation of Geothermal Energy Resources, ERDA-37, Energy Research & Development Administration, Division of Geothermal Energy, Washington, D.C., March 1975.

Utilization of Geothermal Resources for General Purposes

Use	Country	Localities
Space heating	Iceland	Reykjavik, Hveragerdi, Selfoss, Saudárkrokur, Olafsfjördur, and Dalvik.
	Hungary	various localities
	USSR	Caucausus Mountains, Kazakhstan, Kamchatka
	New Zealand	Rotorua
	United States	Klamath Falls, Boise
Air-conditioning	New Zealand	Rotorua
Greenhouse heating	Iceland	Hveragerdi
	USSR	various localities
	Hungary	various localities
	Japan	various localities
	Italy	Castelnuovo
	United States	Lakeview, Oregon
Product processing:		
paper	New Zealand	Kawerau
diatomite	Iceland	Námafjall
salt	Japan	Shikabe, Hokkaido
By-products:		
dry ice	United States	Imperial Valley, Calif.
boron	Italy	Larderello
calcium chloride	United States	Imperial Valley, Calif.

Source: From L. J. P. Muffler, *Geothermal Resources*, U.S. Geological Survey Professional Paper, 820, 1973.

Engineering Criteria for Thermal Water Utilization

Type of Utilization	Temp (°C) not Less than	Discharge (m³/day) not Less than	Depth of Aquifer (m) not Deeper than	Mineral Content (g/l) not More than
Generation of electric power with direct steam-water cycle	100	10,000	3000	4
Generation of electric power with secondary fluid cycle	70	2500	2500	50
Heat supply of population centers	70	1000	2500	2 (50)[a]
Cooling supply	70	500	1500	50
Hot water supply	40	1000	1500	1 (50)[a]
Greenhousing-hothousing	40	500	1500	10 (50)[a]
Hot water irrigation	25	250	1000	2
Warming soil	25	500	1500	50
Melting permafrost	25	250	3000	50
Swimming pools and bath facilities	25	250	1000	50

[a] Allowable mineral content with secondary fluid heat exchangers.
*From F. Makarenko and V. Kononov (1970).

Source: Stimulation of Geothermal Energy Resources, ERDA-37, Energy Research and Development Administration, Division of Geothermal Energy, Washington, D.C. March 1975.

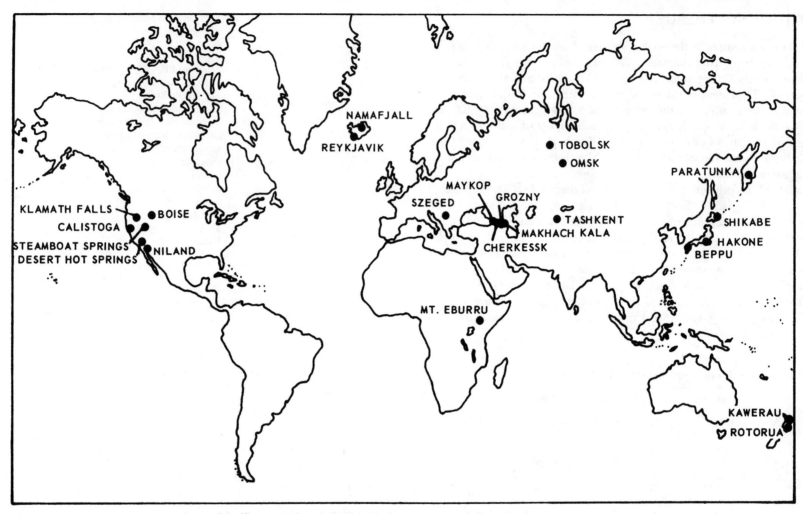

Areas of significant use of geothermal energy for purposes other than generation of electric power.

"Worldwide Status of Geothermal Resources Development," by James B. Koenig in Geothermal Energy: Resources, Production, Stimulation, edited by Paul Kruger and Carel Otte, with permission of the publishers, Stanford University Press. © 1973 by the Board of Trustees of the Leland Stanford Junior University.

GEOTHERMAL POWER PLANT INSTALLATIONS

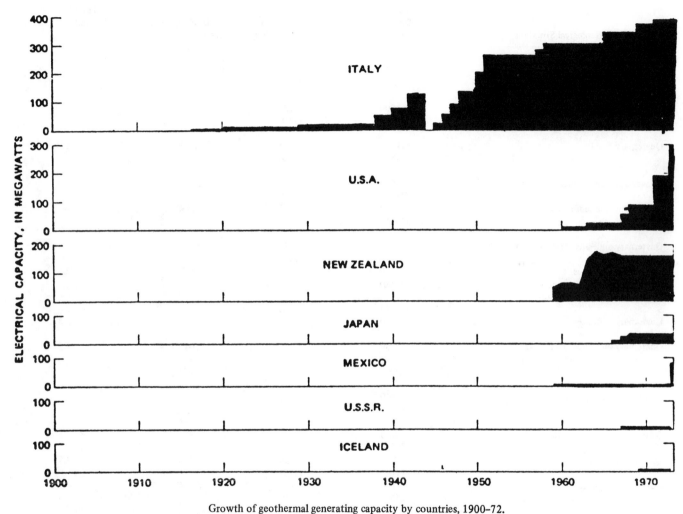

Growth of geothermal generating capacity by countries, 1900–72.

Source: L. J. P. Muffler, *Geothermal Resources*, U.S. Geological Survey Professional Paper 820, 1973.

Geothermal Power Installations in Operation or at an Advanced Stage of Development

	Geological Situation	Average (max.) Drillhole Depth (m)	Average (max.) Temperature (°C)	Discharge Type (S = steam; W = water)	Total Dissolved Solids in Water (g/kg)	Total Generating Capacity (MW)	
						Installed	Planned Addition
Chile							
El Tatio	Quaternary and Tertiary rhyolite, andesite; Mesozoic sediments	650 (900)	230 (260)	S + W	15	—	15
El Salvador							
Ahuachapan	Quaternary andesite	1000 (1400)	230 (250)	S + W	20	30	50
Iceland							
Namafjall	Quaternary basalt	1000 (1400)	250 (280)	S + W	1.0	2.5	—
Italy							
Larderello region	Triassic-Jurassic sediments	600 (1600)	200 (260)	S	—	406	—
Mount Amiata	Triassic-Jurassic sediments; Quaternary volcanics	750 (1500)	170 (190)	S(+W)	—	25	—
Japan							
Matsukawa, N. Honshu	Quaternary andesites; Miocene sandstones	1000 (1500)	220 (270)	S	—	20	—
Otake Kyushu	Quaternary andesites	500 (1500)	230 (250)	S + W	2.5	11	—
N. Hachimantai	Quaternary andesites, dacites	800 (1700)	— (>200)	S + W	—	10	—
Hatchobaru, Kyushu	Quaternary andesites	1000 —	250 (300)	S + W	5.5	—	50
Onikobe, Honshu	Quaternary andesites, dacites, granite	300 (1350)	— (288)	S(shallow) + W (deep)	1.5	—	20
Mexico							
Cerro Prieto	Sandstone, shales, granite	800 (2600)	300 (370)	S + W	17	75	75
New Zealand							
Wairakei	Quaternary rhyolite, andesite	800 (2300)	230 (260)	S + W	4.5	192	—
Kawerau	Quaternary rhyolite, andesite	800 (1100)	250 (285)	S + W	3.5	10	—
Broadlands	Quaternary rhyolite, andesite	1100 (2420)	255 (300)	S + W	4	—	100
Philippines							
Tiwi, S. Luzon	Quaternary andesites	920 (2300)	—	S + W	—	—	10.5
Turkey							
Kizildere	Pliocene-Miocene sandstones limestones; Palaeozoic schists	700 (1000)	190 (220)	S + W	5	—	10
USA							
The Geysers	Jurassic-Cretaceous graywackes and shales, basalt	1500 (2900)	250 (285)	S	—	600	300
USSR							
Pauzhetsk	Quaternary andesite, dacite, rhyolite	— (800)	185 (200)	S + W	3	5	7

Source: A. J. Ellis, "Geothermal systems and power development," *American Scientist*, vol. 63, September–October 1975.

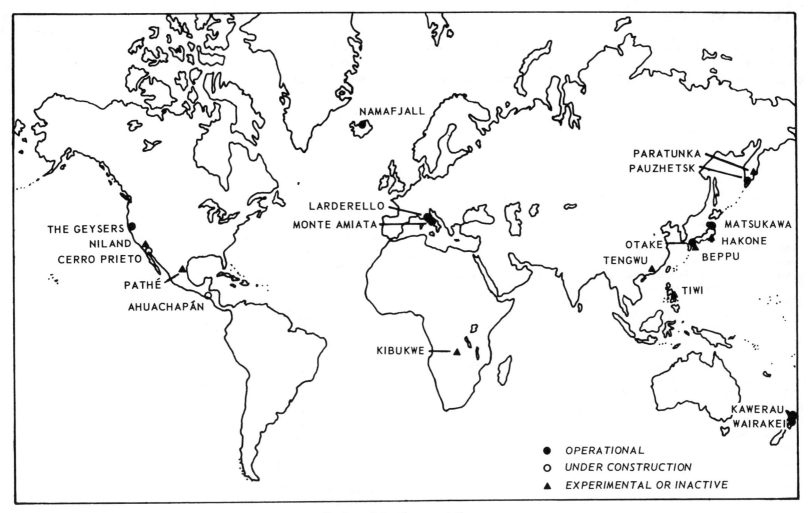

Geothermal electric power stations.

"Worldwide Status of Geothermal Resources Development," by James B. Koenig in Geothermal Energy: Resources, Production, Stimulation, edited by Paul Kruger and Carel Otte, with permission of the publishers, Stanford University Press. © 1973 by the Board of Trustees of the Leland Stanford Junior University.

U.S. GEOTHERMAL DEVELOPMENT

Evidence of geothermal resources in the United States is generally indicated by the presence of hot springs and by heat flow measurements. There are few hot springs with temperatures exceeding 100°F east of the Rocky Mountains and heat flows in the eastern part of the United States are normal. These facts, plus the knowledge that there has been little geological activity in the eastern two-thirds of the United States, indicate that this portion of the country is unlikely to have hidden geothermal systems and hold little promise for geothermal development except in the special case of the geopressured zones along the Gulf Coast.

Location of Selected Nongovernmental Geothermal Developments

California
  The Geysers
    Current power output—396 MWe
    PG&E plants scheduled for operation—total of 908 MWe by late 1976
    Exploration and test wells (Union Oil, Pacific Energy, Magma, Geothermal Kinetics, Burmah Oil and Gas (Signal Co.), Shell Oil, Sun Oil/Aquitaine
  Mono-Lake—Casa Diablo
    10 MWE generating plant planned for 1975 (Southern California Edison & Magma)
    Surprise Valley (northeastern corner of California)
      Exploration (Magma, Gulf Oil, American Thermal Resources, Hunt Oil)
  Imperial Valley
    Production test wells, binary system development (Chevron, Magma, San Diego Gas & Electric)
    25 MWe power plant planned (San Diego Gas & Electric)
    50 MWe power plant planned (San Diego Gas & Electric)
    Tests on brine reinjection systems (Phillips Petroleum, Southern Cal. Edison)
Arizona
    2 test wells (Geothermal Kinetics)
    Test well (Nix Drilling Co.)
Nevada
    Test well (Chevron and American Thermal Resources)
    Test well (Phillips Petroleum and Southern Pacific Land Co.)
    Test well (Union Oil, Magma)
New Mexico
    12 wells drilled since 1960
    Baca 11, planned, power capacity of 6.5 MW (Union Oil)
Oregon
    Test well (Gulf Oil)
    2 greenhouses utilizing geothermal heat under construction (Boise Cascade)
Utah
    Test well (Geothermal Kinetics and Utah Power and Light
    Test well (Nix Drilling Co.)

Location of Government Geothermal Development

U.S. Geological Survey—exploration and research evaluation techniques
    Clear Lake—Geysers, California
    Long Valley, California
    Imperial Valley, California
    Northwest Nevada
    Southwestern Oregon
    Yellowstone National Park
    Jemez Mountain Area, New Mexico
    Southern Idaho
Atomic Energy Commission
    Los Alamos, New Mexico—hot dry rock experiment
    Nevada—field exploration leading to hot water pilot plant (high temperature—low salinity)
    Imperial Valley—equipment design and testing at Lawrence Livermore Lab, Cal. leading to hot water pilot plant (high temperature—high salinity)
    Raft River Valley, Idaho—exploratory drilling leading to hot water pilot plant (moderate temperature—low salinity)
National Science Foundation
    New Mexico—heat flow measurements
    Hawaii—nature and extent of resource
    Imperial Valley, California—nature and extent of resource
    Montana—hot dry rock evaluation
    Nevada—delineation of reservoirs
    (does not include basic research not geographically oriented)
U.S. Bureau of Reclamation
    Imperial Valley, California—pilot desalination plant
U.S. Naval Weapons Center
    China Lake, California—exploration and drilling

Source: A Technology Assessment of Geothermal Energy Resource Development, NSF-RA-X-75-011, prepared for the National Science Foundation by The Futures Group, U.S. Government Printing Office, April 1975.

Known Geothermal Resource Areas within the Conterminous United States as of August 1972.

Name	Location on Map	Name	Location on Map
California		**Oregon**	
The Geysers	1	Breitenbush Hot Springs	1
Salton Sea	2	Crump Geyser	2
Mono-Long Valley	3	Vale Hot Springs	3
Calistoga	4	Mount Hood	4
Lake City	5	Lakeview	5
Wendel-Amedee	6	Carey Hot Springs	6
Cosco Hot Springs	7	Klamath Falls	7
Lassen	8		
Glass Mountain	9	**Washington**	
Sespe Hot Springs	10		
Heber	11	Mount St. Helens	1
Brawley	12		
Dunes	13	**Idaho**	
Glamis	14		
East Mesa	15	Yellowstone	1
		Frazier	2
Nevada			
		Montana	
Beowawe	1		
Fly Ranch	2	Yellowstone	1
Leach Hot Springs	3		
Steamboat Springs	4	**New Mexico**	
Brady Hot Springs	5		
Stillwater-Soda Lake	6	Baca Location No. 1	1
Darrough Hot Springs	7		
Gerlach	8	**Utah**	
Moana Springs	9		
Double Hot Springs	10	Crater Springs	1
Wabuska	11	Roosevelt	2
Monte Neva	12		
Elko Hot Springs	13		

Source: Draft Environmental Statement Liquid Metal Fast Breeder Reactor Program, Volume IV, Alternative Technology Options, WASH-1535, United States Atomic Energy Commission, March 1974.

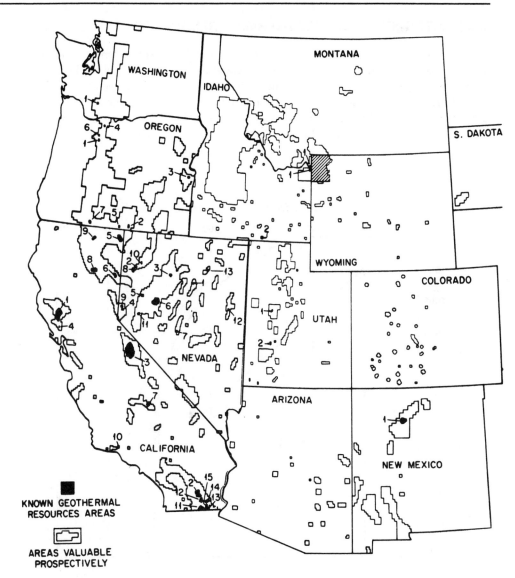

KNOWN GEOTHERMAL RESOURCES AREAS

AREAS VALUABLE PROSPECTIVELY

Map of geothermal resource areas in the Western United States.

Source: Draft Environmental Statement Liquid Metal Fast Breeder Reactor Program, Volume IV, Alternative Technology Options, WASH-1535, United States Atomic Energy Commission, March 1974.

The western one-third of the country, however, shows considerable promise for geothermal development—there is an abundance of hot springs; and volcanic activity and high heat-flow regions are common. About 1.8 million acres of land in the western states have been classified by the U.S. Geological Survey as known geothermal resource areas. Another 96 million acres are listed as having "prospective value" as geothermal resources.

Among the areas of development of the geothermal energy cycle are the location and assessment of geothermal reservoirs, extraction and conversion technology, and environmental control techniques.

CHARACTERISTICS OF GEOTHERMAL RESOURCES

At the center of the earth, the inner core of nickel and iron is at a temperature near 8000°F. Just below the earth's crust, which averages about 50 miles in thickness, the temperature of the molten silicon-iron-magnesium magma is in the range of 2300°F to 2700°F. Heat from the earth's core is constantly being lost through the solid crust and radiated into space. The rock of the crust has, however, low heat conductivity and the rate of heat loss is only about 1.2 microcalories per square centimeter per second (suffi-

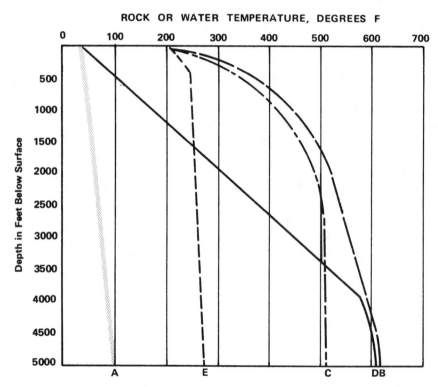

Temperature variation of rock and water in different areas and at increased depths.

Temperature increases about 1°F for each 100 feet in depth as illustrated by Line A. Water boiling temperature of 212°F is reached at a depth of about 16,200 feet. Curve B represents the theoretical increase in the boiling point of water with depth allowing for decreases in the density of water as the temperature increases. Curve C shows the temperatures in an upflowing part of a vigorous hot-spring system such as the one at Wairakei, New Zealand, where water has circulated to great depths. Curve D represents a system in which surface rocks are impermeable and insulate the reservoir and prevent loss of water and steam. Curve E represents a very leaky spring system that discharges large quantities of boiling water. Curves D and E illustrate the difficulty in assessing the potential for geothermal energy from surface indications.

Source: Natural Steam for Power, a pamphlet published by the U.S. Department of the Interior Geological Survey, 1969.

Characteristics of Selected Geothermal Fields

Field	Reservoir Experiment °C	Reservoir Fluid	Enthalpy, cal/g	Average Well Depth, meters	Fluid Salinity, ppm	Mass Flow per Well, kg/hr	Non-condensable Gases, %
Larderello	245	Steam	690	1,000	<1,000	23,000	5
The Geysers	245	Steam	670	2,500	<1,000	70,000	1
Matsukawa	230	Mostly steam	550	1,100	<1,000	50,000	<1
Otake	200+	Water	~400	500	~4,000	100,000	<1
Wairakei	270	Water	280	1,000	12,000	—	<1
Broadlands	280	Water	400+	1,300	—	150,000	~6
Pauzhetsk	200	Water	195	600	3,000	60,000	—
Cerro Prieto	300+	Water	265	1,500	~15,000	230,000	~1
Niland	300+	Brine	240	1,300	260,000	~200,000	<1
Ahuachapán	230	Water	235	1,000	10,000	320,000	~1
Hveragerdi	260	Water	220	800	~1,000	250,000	~1
Reykjanes	280	Brine	275	1,750	~40,000	~400,000	~1
Namafjall	280	Water	260	900	~4,000	400,000	6

"Worldwide Status of Geothermal Resources Development," by James B. Koenig in Geothermal Energy: Resources, Production, Stimulation, edited by Paul Kruger and Carel Otte with permission of the publishers, Stanford University Press © 1973 by the Board of Trustees of the Leland Stanford Junior University.

cient, for example, to melt about a half centimeter of ice in a year). Geothermal heat at the earth's surface is only 1/27,000 of the solar input (2 calories per minute per cm²). In "normal" areas of the earth's crust, therefore, the average temperature increase with depth is about 25°C per kilometer and temperatures at reasonable drilling depths (3–5 kilometers) are too low to serve as practical heat sources for geothermal power plants. This normal heat flow is about 25 kW per square kilometer or 3×10^{10} kW for the entire globe.

The earth's crust is not of uniform thickness, however, and the molten magma of the upper mantle approaches nearer the earth's surface at some points with consequent higher heat flows and thermal gradients. In regions of geologic movement, typically those of strong seismic activity, the molten magma may break through the surface in volcanic eruptions. There are, therefore, a number of areas in the world that represent potential sources of geothermal energy. Warm thermal springs, resulting from the geothermal heating of natural ground water, occur in many regions of the world. Geological formations suitable for the production of steam occur less frequently—the two notable steam sources occur at Lardarello, Italy, and The Geysers in California.

Geothermal systems producing hot water are characterized by the presence of water as the heat transfer medium with relatively small amounts of steam. Rainwater, collected over areas of tens or hundreds of square kilometers, percolates through the ground to depths of 2 to 6 kilometers where it is heated by conduction from the hot rock. The water expands upon heating and moves upward in a relatively restricted area. If the rock above is highly fractured or permeable, the hot water will move upward to the surface with relatively little storage. If, however, the zone of geothermal energy is covered by impermeable rock with few fractures, energy will be stored underground. The energy in such systems can be extracted through holes drilled through the impermeable rock. The largest potential for the extraction of geothermal energy is represented by areas that are "dry," that is, without natural sources of underground water. Extraction of energy from these hot dry rock sources will require the injection of water into artificially created fractured rock zones underground and the subsequent extraction of hot water or steam.

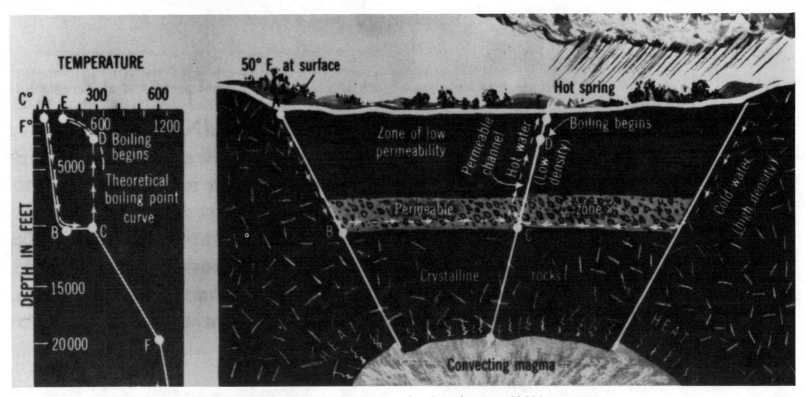

A generalized illustration of a hot spring type of geothermal system, with high rate of upflow.

Source: U.S. Dept. of Interior, Geological Survey.

Tests on sugar beet leaf to determine the potential effect of emissions which might be released from geothermal power facilities in the Imperial Valley of California. (*Credit:* Lawrence Livermore Laboratory.)

Natural steam escaping through fissures in the earth's surface at Steamboat Springs. (*Credit:* Lawrence Berkeley Laboratory.)

GEOTHERMAL HOT DRY ROCK SOURCES

In order to extract the heat from subterranean hot, dry rock, techniques to create passages for the insertion and removal of a heat transfer medium, typically water, must be employed. Methods for producing large volumes of fractured rock underground, into which insertion and extraction wells can be drilled, include hydraulic fracturing, thermal stressing, and chemical and nuclear explosive fracturing. Hydraulic and explosive fracturing methods have already been used to stimulate production from natural gas reservoirs.

Hydraulic fracturing is a relatively simple concept—liquid at high pressure is forced into a rock formation to widen existing fractures and to create new fractures through rock displacement. Successful fracturing depends, however, on the nature of the underground rock formation—leakage may be too rapid into existing fractures to result in pressures high enough to cause additional fracturing, the rock may give elastically rather than by fracture, and the fractures may close again after the pressure is released.

Thermal stress fracturing depends upon the differential contraction, and consequent fracturing, of hot rock as cold fluids are injected into the formation.

Explosive fracturing with chemical explosives can be used to create a fractured zone adjacent to the well bore, to interconnect zones of tight fractures, and to increase permeability around a well through the use of conical-shaped charges. Use of chemical explosives in hot geothermal formations has been limited by the availability of explosives capable of withstanding the temperatures and pressures in the rock formations.

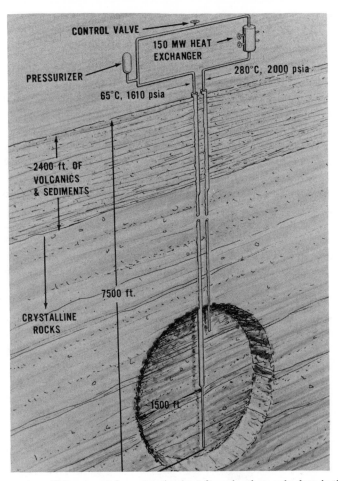

Geothermal concept . . . This concept for extracting heat from dry, hot rocks deep in the earth is currently under investigation by the AEC's Los Alamos Scientific Laboratory. Initial cracks would be accomplished by conventional hydraulic fracturing techniques used in oil field operations. Water would be pumped down one pipe, take the heat from the rocks and return to the surface to run electrical generators. If the cracks propogate due to the heating and cooling process as expected, a pancake-shape fracture zone would be established and the quality of the geothermal source may improve as energy is drawn from it. Such a well could last for 30 years or more.

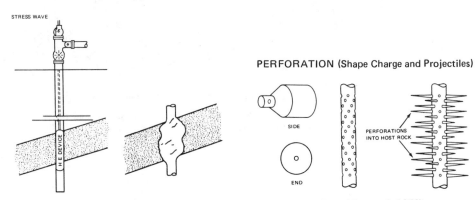

Methods of chemical explosive stimulation. (from Austin and Leonard, 1973)

Source: Stimulation of Geothermal Energy Resources, ERDA-37, Energy Research & Development Administration, Division of Geothermal Energy, Washington, D.C., March 1975.

REFERENCES

Anthony H. Ewing, "Stimulation of geothermal systems;" Henry J. Ramey, Jr., Paul Kruger, and Raj Raghavan, "Explosive stimulation of hydrothermal reservoirs;" Morton Smith, R. Potter, D. Brown, and R. L. Aamodt, "Induction and growth of fractures in hot rock;" and Carl F. Austin and Guy William Leonard, "Chemical explosive stimulation of geothermal wells," in *Geothermal Energy: Resources, Production, Stimulation*, Paul Kruger and Carel Otte, Eds. Stanford, Calif.: Stanford University Press, 1973.

GEOTHERMAL HOT DRY ROCK SOURCES—NUCLEAR STIMULATION

The high yield of nuclear explosives has led to the study of such explosives for creating large volumes of fractured rock in underground formations (one metric ton of chemical explosive, with an explosive yield of 10^9 calories, occupies a volume of about $\frac{3}{4}$ cubic meter—a nuclear explosive occupying the same volume could yield a million times more energy).

In the United States three nuclear experiments have been conducted for the stimulation of natural gas production in deep, low-permeability gas fields with nuclear explosives ranging between 29 and 90 kilotons in yield. Nuclear fracturing for the stimulation of oil production has been demonstrated in the Soviet Union using nuclear explosives of 2.3 to 8 kiloton yield.

Detonation of a nuclear explosive underground produces a chimney, roughly cylindrical in shape, which is nearly filled with broken rock rubble with a void at the top of the chimney and molten rock at the bottom. The volume of the chimney is about 3600 cubic meters per kiloton of explosive. Assuming an extractable energy constant of 45 kilowatt-hours per cubic meter and a conversion efficiency of 21%, a single 1000 kiloton explosive produces a chimney volume with a heat content capable of producing about 60 megawatt-years of electrical energy, that is, enough heat to operate a 2 megawatt plant at 100% load factor for 30 years.

Since the yield of geothermal energy from the cavity created by a single nuclear explosive is too small to be of commercial interest, heat must also be extracted from the rock adjacent to the chimney. The extent to which this is possible depends on the degree of fracturing of the rock adjacent to the chimney which, in turn, is a function of the character of the rock formation. In addition, to obtain liquid flow through this adjacent fractured region, arrays of nuclear explosives are proposed such that the fracture zones overlap and circulation can be established through the fractured rock between chimneys.

Number of Nuclear Explosives in a Dual-Emplacement Array Needed to Support a 200-Mw Plowshare Geothermal Plant

Fracturing Efficiency Coefficient, M	Number of Explosives of Given Size		
	200 kt	500 kt	1000 kt
20	238	96	48
54	106	42	22
129	48	20	10

Note: Conditions: rock temperature, 350°C; placement depth of device, 3 km.

Source: Burnham, J. B. and Stewart, D. H. "Recovery of Geothermal Energy from Hot, Dry Rock with Nuclear Explosives," in *Geothermal Energy: Resources, Production, Stimulation*, edited by Paul Kruger and Carel Otte, with permission of the publishers, Stanford University Press. © 1973 by the Board of Trustees of the Leland Stanford Junior University.

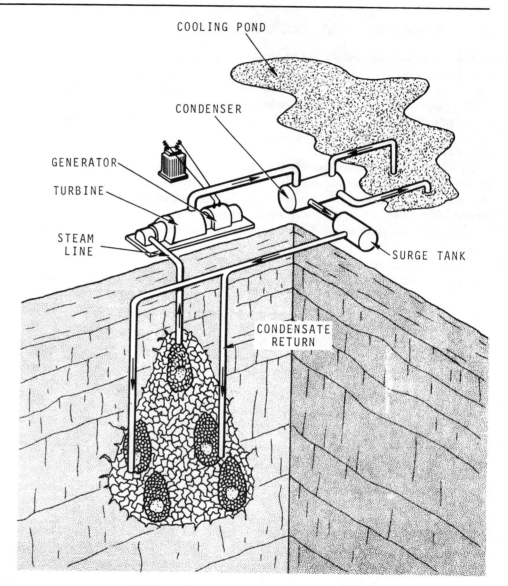

The Plowshare concept of geothermal-heat extraction.

Source: A Feasibility Study of a Plowshare Geothermal Power Plant, American Oil Shale Corporation and the U.S. Atomic Energy Commission, April 1971.

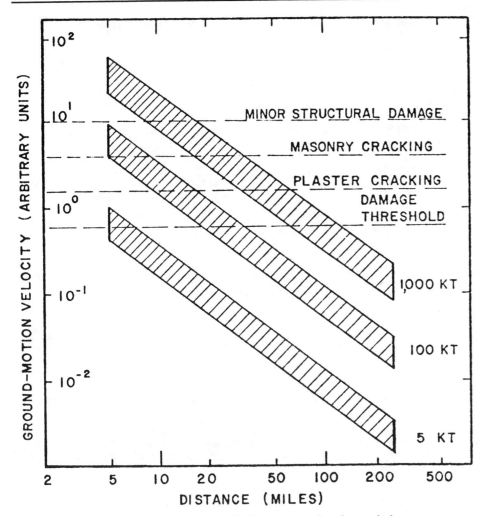

Typical seismic damage associated with underground nuclear explosions.

Source: Sandquist, G. M. and Whan, G. A. "Environmental Aspects of Nuclear Stimulation," in *Geothermal Energy: Resources, Production, Stimulation*, edited by Paul Kruger and Carel Otte, with permission of the publishers Stanford University Press. © 1973 by the Board of Trustees of the Leland Stanford Junior University.

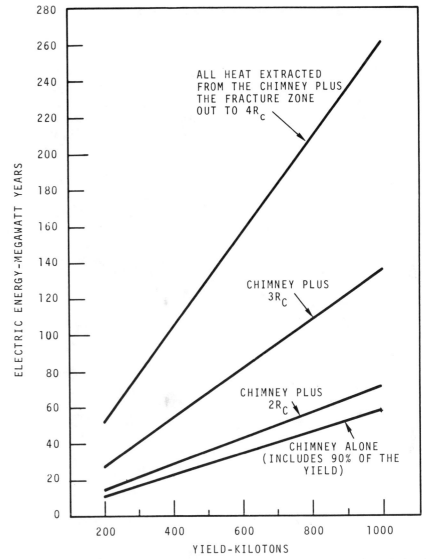

Estimated electric energy available from the fractured volumes produced by single explosives (initial rock temperature = 350°C, final rock temperature = 150°C, conversion efficiency = 21%, extractable-energy constant = 45 kilowatt-hours per cubic meter).

Source: A Feasibility Study of a Plowshare Geothermal Power Plant, American Oil Shale Corporation and the U.S. Atomic Energy Commission, April, 1971.

GEOTHERMAL GEOPRESSURED RESERVOIRS

Large bodies of sand and shale containing saline waters at temperatures up to 350°F, have long been known to exist along the northern shore of the Gulf of Mexico in the region from Brownsville, Texas, to New Orleans, Louisiana. These bodies of water are believed to have been trapped through geologic processes as sediments were deposited over a period of millions of years from the ancestral Mississippi and Rio Grande Rivers. As these deposits grew to massive size, they would slide, from time to time, into the deeper waters of the Gulf, carrying with them trapped water plus the water of hydration of the clay deposits. Continued deposition and subsidence added weight to the earlier layers and prevented the escape of trapped water such that the normal compaction of layers of sand did not occur. The trapped water became the load-bearing element of the formations and the resulting high pressure became known as "geopressure." In addition, much of the bound water of the clay was released as the clay was transformed into shale at the increased temperatures. Another feature of the water in geopressured zones is the pres-

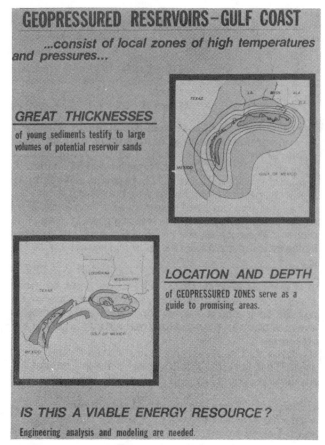

Source: Geothermal Energy; Hearings before the Subcommittee on Energy of the Committee on Science and Astronautics, U.S. House of Representatives, September 11, 13, and 18, 1973.

Well and Aquifer Data

Depth of reservoir	12,000 ft
Initial pore pressure	10,000 psig
Stratum thickness	200 ft
Rock porosity (percent)	20
Rock permeability	100 MD
Rock compressibility (effective)	10^{-5} psi^{-1}
Reservoir temperature	260 °F
Water viscosity (at 260°F)	0.30 cp
Well diameter (pipe I.D.)	9 in
Well radius	0.375 ft
Reservoir radius	10 miles

Source: [10]

Data for Power Production Estimate

Well flow rate	80,000 BWPD
Flowing surface pressure	2,000 psig
Wellhead water temperature	260 °F
Natural gas content of water	30 SCF/bbl
Gas production rate	2,400 MCF/day
Gas heating value	1,000 Btu/SCF
Water temperature at discharge	100 °F

Power Production Estimate

Kind of Energy	Conversion Efficiency (percent)	Electric Power (megawatts)
Natural gas	25	7.5
Geohydraulic	75	1.5
Geothermal	11	6.0
Total		15.0

Source: C. D. Durham, "Proposed Geopressured Energy Investigation, Statement of the Geothermal Energy Research, Development and Commercial Demonstration Act of 1973 H.R. 11212," as given in *A Technology Assessment of Geothermal Energy Resource Development*, NSF-RA-X-75-011, National Science Foundation, April 1975.

ence of dissolved hydrocarbons, predominately natural gas, with 10–40 standard cubic feet of gas per barrel of water.

The geopressured zones extend for some 800 miles along the Gulf coast and lie at depths ranging from 4000 to 25,000 feet. The zones are large individually but are discontinuous and occur at various levels. A typical well into a geopressured zone might be expected to flow at a rate of 80,000 barrels per day with surface pressures over 2000 pounds per square inch for a period of 15 years. After that period, flow would be expected to continue for another 30 years but at pressures below 2000 psi as the subsurface zone gradually loses pressure.

The hot water under pressure from the geopressured zones can be used to generate power by three different mechanisms: 1) from the hydraulic energy of the 2000 psi water, 2) from the natural gas dissolved in the water, and 3) from the geothermal heat of the water. The combination of these energy sources is expected to produce about 15 MWe per well.

TYPES OF GEOTHERMAL POWER PLANTS

The principle characteristics of geothermal electric generating plants that differ from fossil fueled plants are: the much larger flow of steam required by a geothermal plant to produce a kilowatt-hour of electrical power, the relatively small size of an individual geothermal power plant, and the number of different systems employed for the extraction of steam from geothermal sources.

The large steam flow required for geothermal plants is a result of the low temperatures and pressures of the steam from geothermal wells. For example, the steam entering the turbines at The Geysers is about 350°F and the pressure is about 125 psi. In contrast, steam conditions in a modern fossil fuel power plant are about 1000°F and 3600 psi. As a result, the heat rate for the geothermal plant at The Geysers is about 22,000 Btu/kilowatt-hour as compared to 8300 Btu/kilowatt-hour for a fossil-fired power plant. A geothermal plant requires about $2\frac{1}{2}$ times as many pounds of steam per hour per kilowatt-hour generated and, because of the lower pressure, the volume of steam is 50 times greater. The mass flow from geothermal sources must be even larger for hot water systems at lower temperatures and for systems that require the use of intermediate heat exchangers.

Types of Geothermal Power Plants

Heat Source	Generation Mode
Dry steam	Steam turbine
Hot water (T > 180°C)	Steam turbine
Hot water (T < 150°C)	Binary cycle
Hot water (moderate salinity)	Hybrid cycle
Hot brine (pressurized)	Binary cycle
Hot brine (flashed)	Impact turbine
	Helical screw expander
	Bladeless turbine

Source: Stimulation of Geothermal Energy Resources, ERDA-37, Energy Research & Development Administration, Division of Geothermal Energy, Washington, D.C., March 1975.

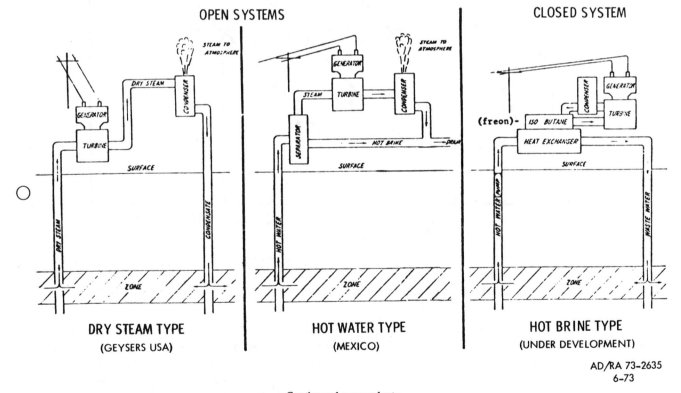

Geothermal power plants.

Source: Geothermal Energy, Hearings before the Subcommittee on Energy of the Committee on Science and Astronautics, U.S. House of Representatives, September 11, 13, and 18, 1973.

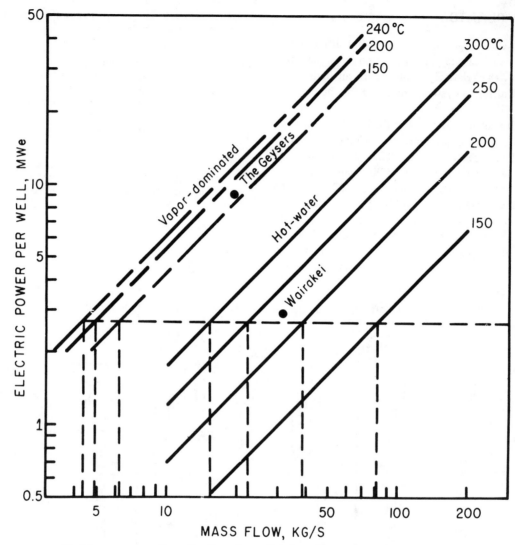

Electric power per well as a function of mass flow for various temperatures of hot-water
and vapor-dominated systems.

Source: D. F. White and D. L. Williams, Eds., *Assessment of Geothermal Resources of the United States—1975*,
Geological Survey Circular 726, 1975.

The size of geothermal power plants is dependent on the economic length of pipelines necessary to transport steam or hot water from wells, each of which may produce the equivalent of only a few megawatts of electric power. Each well at The Geysers produces an equivalent of 7.5 megawatts and the largest single power plant size at The Geysers is 110 MWe (see 8-9).

The production of clean steam suitable for injection into a turbine controls the nature of the equipment necessary for extraction of heat from the steam or hot water coming from a geothermal source. If this geothermal source produces steam which is dry and clean, such as The Geysers or Larderello, it can be introduced directly into the turbine. If the geothermal source produces steam mixed with hot water which is relatively low in salt content, a simple moisture separator is placed before the turbine inlet. On the other hand, geothermal hot water systems which contain high concentrations of salt require the use of binary cycles in which the turbine is operated on a secondary fluid, such as iso-butane, vaporized in a heat exchanger through which the geothermal brine is circulated. Other systems—impact turbines, helical screw expanders, and bladeless turbines—which can directly handle hot brines are under development (see 8-10).

GEOTHERMAL DRY STEAM POWER PLANTS—THE GEYSERS

The Geysers power plant is located in Sonoma County, some 80 miles north of San Francisco. The first development of the use of geothermal energy at The Geysers took place in the 1920s when two small steam-engine generators, utilizing steam from shallow wells, provided power to light The Geysers Resort. Thirty years later, the Magma Power Company and the Thermal Power Company began a more extensive development of the geothermal resource. The first 11 Mwe unit went into operation in 1960 under a contract between Pacific Gas and Electric and the Magma and Thermal Power Companies. In 1967, Union Oil of California joined Magma and Thermal in an expanded joint venture comprising some 15,000 acres. Expansion of power production is expected to proceed at a rate of about 100 megawatts per year until the limits of the field are reached. Exploratory drilling is continuing to extend the present proven productive capacity of 750 MW.

The Geysers is one of a few vapor (or steam) dominated geothermal reservoirs (the other commercial sources are at Larderello in Italy and Matsukawa in Japan). The steam

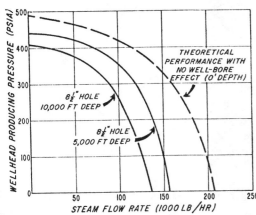

Steam-flow rate vs. wellhead pressure for two well depths for a typical steam well at The Geysers.

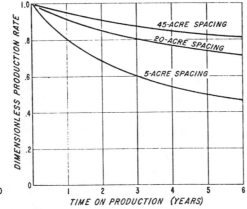

Effect of well density on production rate as determined by a reservoir simulation model.

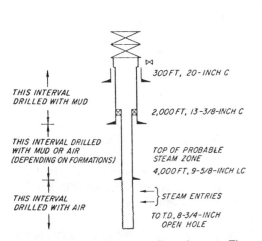

Typical well-completion configuration at The Geysers.

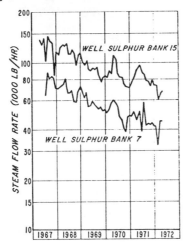

Steam-production rate vs. time for geothermal steam wells at The Geysers.

Source: Budd, Jr., Chester F. "Steam Production at the Geysers Geothermal Field," in *Geothermal Energy: Resources, Production, Stimulation,* Paul Kruger and Carel Otte, Eds., with permission of the publishers, Stanford University Press. © 1973 by the Board of Trustees of the Leland Stanford Junior University.

Venting geothermal wells and gathering pipes at the Geysers. In the Geysers area, which is about 2 by 8 miles, over 100 wells have been drilled; the deepest of these is more than 8,000 feet. Temperatures in the reservoirs from which the heat is drawn are about 255°C (480°F). The basic source appears to be a mass of heated rocks at a depth of 3–5 miles and covering an area of about 100–500 square miles. (*Credit:* Department of Energy.)

A commercial geothermal plant at the Geysers located in Sonoma County, California, about 75 miles north of San Francisco.

Source: U.S. Dept. of Interior Geological Survey.

pressure in the reservoir is below the hydrostatic pressure normally expected at that depth, indicating a layer of tight, impermeable rock over the reservoir. Steam pressures in the reservoir are 450–500 psig. As the steam flows from the reservoir to the surface, it expands and cools so that the pressure at the wellhead is about 125 psig. The turbines are designed to operate at 89–100 psig intake pressure. The average initial production of each well at The Geysers is about 150,000 lb/hr of steam. Seven wells are needed to provide the 1 million lb/hr of steam to operate each 55 MW turbine-generator set. Completed wells are as shallow as 600 feet and as deep as 9000 feet. A typical well is about 6000 feet in depth and consists of 1900 feet of $13\frac{3}{8}$-inch casing, 2000 feet of $9\frac{5}{8}$-inch casing, and 2000 feet of $8\frac{3}{4}$-inch open hole for steam entry.

The average density of wells at The Geysers is about one per 5 acres. Steam production rates from wells at this spacing have shown a decline of about 50% over a six-year period. Calculations indicate that this decline in production rate could be reduced by wider spacing of the wells (for example, spacing at a 45 acre interval per well would limit the decline in steam production to 20% in six years). Optimum economic power production involves, therefore, a compromise among a number of factors—particularly the cost and losses in surface steam lines to the turbine and the cost of new wells at that particular location.

The start-up period for a plant at The Geysers is about 24 hours since initial production of steam from a well causes small rocks and sand to be dislodged from the well bore. Since this debris would damage the turbine, the initial flow of steam is vented until full steam flow is achieved and the steam is clean. The Geysers power plants are therefore used as base-load plants rather than to provide power for the daily peak loads on the PG&E system.

GEOTHERMAL BRINE POWER PLANTS

A number of different power cycles have been considered for extraction of the thermal energy from geothermal waters. Differences in these cycles result from attempts to avoid problems of corrosion, erosion, and solids deposition associated with saline brines. Among these conceptual power cycles are: direct steam cycles, indirect steam cycles, steam-to-secondary fluid cycles, and total flow concepts.

The direct steam cycle operates on steam flashed directly from the geothermal brine. This cycle has been used with geothermal waters of low salinity but the use of this process with high salinity brines will require development of highly efficient separators and steam purifiers to prevent carry-over of brine into the turbine which would result in accelerated corrosion as well as salt deposition on the turbine blades.

In indirect steam cycles, the steam flashed from the brine is passed through a tubular heat exchanger in which pure water is boiled to produce steam for operation of the turbine. Although this cycle keeps the impure, flashed steam from the geothermal brine out of the turbine, the steam must be handled in the heat exchanger which is resistant to corrosion from H_2S, CO_2, ammonia, boric acid, and other corrosive materials. In addition, the heat exchanger must be designed for easy cleaning of the deposits which form on the flashed brine side.

The steam-to-secondary fluid cycles, often called binary cycles, are similar to the indirect steam cycle with the exception that the fluid used to operate the turbine is not water but a fluid selected for its thermodynamic properties to match the temperature interval available. Among these secondary fluids are isobutane, freon, and ammonia. Corrosion and salt deposition problems in the heat exchanger are similar to those in the indirect steam cycle.

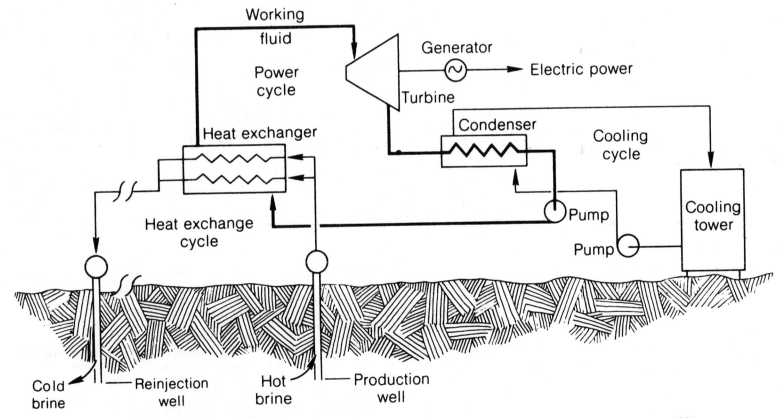

Binary cycle is proposed for low-salinity geothermal demonstration plant where brine resource is low to moderate temperature (300–450°F). Heat exchange cycle isolates hot brine from power cycle, and the thermodynamic properties of the working fluid—probably isobutane—make economical power generation possible.

Source: Research Progress Report FF-3, Fossil Fuel and Advanced Systems Division, Electric Power Research Institute, Palo Alto, California, July 1975.

Comparison of Alternative Geothermal Power Options

Option	Advantages	Disadvantages	Development Potential
1. Steam turbine operating on steam flashed from the brine.	Simplest flow arrangements. High cycle efficiency. Low capital cost. Flash evaporators less susceptible to brine fouling and corrosion than heat exchanger surfaces.	Additional steam purification equipment required to protect turbine from corrosion and deposition. Turbine design modifications may be required to accommodate impurities in the steam. Multiflash staging requires a multistage admission turbine.	High development potential with low development risks.
2. Steam turbine operating on clean steam generated by a secondary brine-to-steam heat transfer system.	Turbine sees only clean steam. Conventional steam clean-up equipment is adequate.	Fouling and corrosion presents a problem in tubular steam generator. Lower turbine throttle temperature and pressure than option 1. Increased capital cost due to secondary steam generator loop. Surface condenser required. Larger capital cost and lower power capability than option 1.	High development potential with low development risk. Development risk is probably lower than option 1, but economics are less favorable.
3. Isobutane turbine, flashed steam to isobutane heat transfer system.	Potentially smaller (and lower cost) turbine than steam cycles.	All of option 2 (except turbine throttle pressure) plus: larger steam generators due to poor thermal conductivity of isobutane. Larger isobutane pumps and pump power. High working pressures increase heat exchanger costs. Development of large secondary fluid turbine required.	Has all of the development problems of Option 2 plus unavailability of isobutane turbines in the desired size range.
4. N-Butane turbine, flashed steam-to-N-butane heat transfer system.	Potentially smaller turbine than steam cycles.	All of option 3 plus lower power capability.	Low.
5. Freon-turbine flashed steam-to-freon heat transfer system.	Potentially smaller turbine than butane or steam cycles.	All of option 3. Freon decomposition may lead to corrosion.	Low.
6. Ammonia turbine, flashed steam-to-ammonia heat transfer system.	Potentially smallest turbine.	All of option 3.	Low.
7. Total flow concept.	Eliminates separators, flash evaporators, and steam generators. Uses rugged turbomachinery with replaceable parts. High efficiency. Potentially a simple, low-cost system.	All of option 3. More severe corrosion/erosion fouling problems than options 1 and 2. Requires development of novel turbomachinery concepts and development of basic analysis methods for high-velocity, two-phase brine flow.	Low. Development potential uncertain. Requires further analytical evaluation.
8. Isobutane turbine, flashed steam-to-isobutane plus brine-to-isobutane heat exchanger.	Relatively simple flow arrangement, single flash vessel and no downhole pump.	All of option 3 plus scale deposition on tubes of brine heat exchanger.	Low development potential due to scale deposition problems.
9. Isobutane turbine and downhole pump, total well flow to isobutane heat transfer system	Simple flow arrangement, steam separator not required. High efficiency possible. Scale deposition in well is minimized.	All of option 3 plus deposition on heat exchanger tubes. Large expensive alloy heat exchanger required due to low coefficients and corrosiveness of brine.	Low development potential due to scale deposition problems.
10. Downhole heat exchanger.	Turbine sees only clean steam. May ease heat exchanger corrosion and fouling problems.	Thermal refluxing problem with concentric tube in well casing, i.e., cold downstream must be thermally insulated from hot upstream. Uncertain knowledge of heat extraction mechanism from geothermal deposits. Outside scaling of casing and plugging of formation may be a problem.	Development potential is uncertain.
11. Injected hydrocarbon into hot brine with vapor passed to mixed vapor turbine.	Simple flow arrangement. Higher hydrocarbon temperature than option 3. May ease well casing corrosion and/or scale deposition problems.	Removal of brine droplets from mixed hydrocarbon-steam vapor presents purification problems. Thermal refluxing in well. Larger fluid flow requirements in well. Requires special turbine development. Loss of hydrocarbon with brine, condensate, and noncondensable gas streams.	Requires further evaluation to establish development potential.

Source: "Power Generation from Imperial Valley Geothermal Brine—Preliminary Cycle Screening," Rockwell International Corp., Atomics International Division, 1974, as given in *A Technology Assessment of Geothermal Energy Resource Development*, NSF-RA-X-75-011, National Science Foundation, April 1975.

In the total flow concept, the entire output of the geothermal well is passed directly through a prime mover which, in turn, drives an electric generator. The development of prime movers—Pelton wheels, radial inflow turbines, bladeless turbines, helical screw expanders—capable of operating in the extremely corrosive and erosive brine solutions represents a distinct challenge.

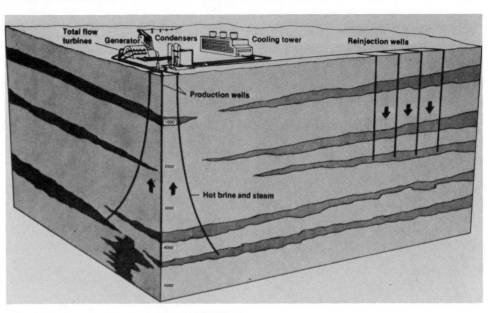

Total flow geothermal energy concept-100 MW.

Geothermal . . . The Atomic Energy Commission's Lawrence Laboratory has developed a new "total flow" concept for producing electrical power from the huge reserves of energy stored in geothermal hot brine deposits. In current technology only the steam phase of geothermal wellhead products is used. In the "total flow" method, depicted in this drawing, energy would be extracted from the entire wellhead product, generally, both steam and hot brine. By producing energy from all the material handled, the power output could be increased by as much as 60 percent.

The sketch shows wells drilled to brine deposits. After initial pumping to the surface, the hot water continues to rise like water in a coffee pot. The mixture of liquid brine and steam is put through a turbine to produce electricity. The residue is reinjected into the ground, thus eliminating the surface environmental problem of disposing of the large amounts of salt in the brine. (Credit: LLL)

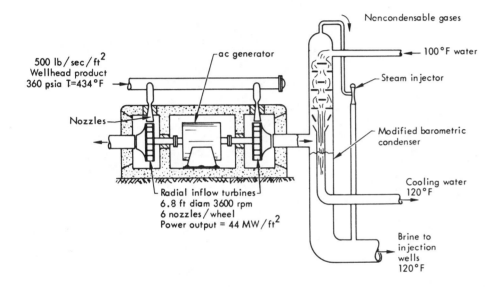

Radial impact turbine for the extraction of geothermal energy from hot brines in the total flow concept.

Source: C. F. Austin, G. H. Higgins, and J. H. Howard, "The Total Flow Concept for Recovery of Energy from Geothermal Hot Brine Deposits," UCRL-51366, United States Atomic Energy Commission, 1973.

REFERENCE

Chester F. Budd, Jr., "Steam production at The Geysers geothermal field," in *Geothermal Energy*, P. Kruger and C. Otte, Eds. Stanford, Calif.: Stanford University Press, 1973.

Solar Energy

Solar collector driving steam engine, exhibited by Mouchot at the World's Fair in Paris, 1878.

ENERGY FROM THE SUN

Efforts to concentrate the energy from the sun to produce heat or power reportedly date from the time of Archimedes who, about 200 B.C., is said to have used reflecting mirrors to set fire to an attacking fleet of ships. In 1878, Mouchot displayed a working solar-powered water pump at the Paris Exposition. In 1913, an American, Frank Schuman, constructed a 13,000 ft^2 cylindrical trough reflector in Egypt. Shuman's collector produced steam which operated a 55-horsepower reciprocating engine. Two solar furnaces capable of producing very high temperatures for materials research have been built in France. F. Tromber built a 50-thermal kilowatt furnace at Montlouis in the 1950s and in 1970, a 1000-thermal kilowatt solar furnace was completed at Odeillo. In 1968, a 100-thermal kilowatt central receiver was constructed at Genoa, Italy, by G. Francis. This receiver produced superheated steam at 950°F and 2000 pounds per square inch.

Systems for the utilization of solar energy may be classified in five categories: low temperature collection systems for space heating and cooling; concentrating collection systems which produce temperatures high enough for the generation of electric power; photoelectric systems for the direct conversion of solar energy to electrical energy; photosynthetic systems based on the production of organic material, including cropping systems and the utilization of organic waste material; and systems that are derived from solar energy, including systems based on energy from the wind, waves, and temperature differences in the ocean. Each of these major systems is in itself characterized by several different technical approaches which are described in more detail in the pages following.

Composition of the Earth's Atmosphere

Total Weight of Atmospheric Materials (99% in lowest 30 kilometers)	5.3×10^{15} tons
Volume concentrations	
nitrogen	78%
oxygen	20%
argon	0.9%
carbon dioxide	0.3%
water vapor	0.3%
water droplets and ice	0.008%
ozone	0.000005%
dust (0.1–1.0 micron)	0.0000001%

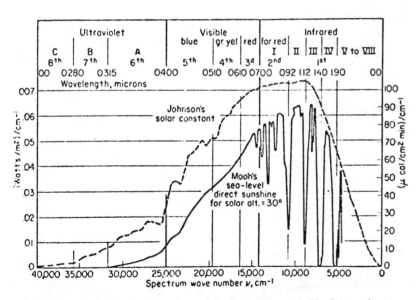

Total solar flux incident outside the atmosphere of the Earth, represented by Johnson's solar constant, is 178×10^{12} kilowatts, continuous for the whole globe, or 1.5×10^{18} kilowatt-hours/year (5.12×10^{21} Btu/year). Depletion in the atmosphere due to absorption by water vapor, carbon dioxide, oxygen, nitrogen, ozone, and particles of dust, reduce the average solar energy available at the bottom of the atmosphere on the land areas of the Earth to 2.16×10^{17} kilowatt-hours/year (7.37×10^{20} Btu/year).

Source: Task Force Report—Solar Energy, Project Independence, under the direction of the National Science Foundation for the Federal Energy Administration, November 1974.

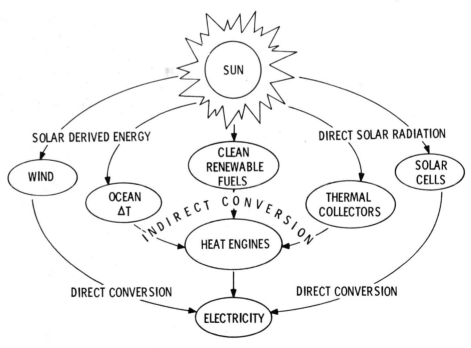

Electricity from the sun.

Source: Solar Energy Research, Staff Report of the Committee of "Science and Astronautics, U.S. House of Representatives, December 1972.

Of the 3.8×10^{26} watts of power radiated from the sun, about 1.7×10^{17} watts are intercepted by the earth. The average flux of solar energy outside the Earth's atmosphere is about 1.35 kW/m^2, varying from 1.43 kW/m^2 in January when the sun is closest to the earth to 1.33 kW/m^2 in July when the sun is farthest away. The term *insolation* is applied to the solar radiation received on a flat, horizontal surface. Insolation values vary with latitude and season.

In passing through the atmosphere of the Earth, the direct radiation from the sun is partially absorbed or scattered by the molecules, water vapor, and dust in the air. Absorption of solar energy at infrared wavelengths (>0.7 microns) is due largely to water vapor and, to a lesser extent, to carbon dioxide. Absorption at ultraviolet wavelengths (<0.3 microns) is principally due to ozone. During the hours around noon on a clear day at midlatitudes, the direct radiation is reduced by some 30% by these processes. Light to

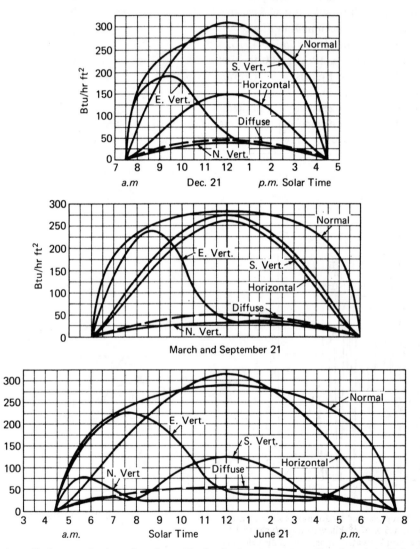

Total cloudless day solar radiation on a horizontal surface. The border between western Canada and the United States is at 49° north latitude, New York is at 41°, Chicago at 42°, San Francisco at 37°, Miami is at 26°, Brownsville at 26°, and Los Angeles at 34°.

Source: Energy Research and Development and Small Business, Hearings before the Select Committee on Small Business, United States Senate, May 13 and 14, 1975.

Values of Perpendicular and Diffuse Solar Radiation for Clear and for Industrial Atmospheres (ASHAE, 1956) and Average Total Insolation through Complete Overcasts of Various Cloud Types (Haurwitz, 1948)

Solar altitude α, degrees	Optical air-mass path* m ~ csc α	Standard; Cloudless Atmosphere			Industrial, Cloudless Atmosphere			Through Complete Overcasts, Blue Hill, Average Total Insolation on Horizon			
		Direct, perpendicular radiation I, B/hr ft²	Diffuse on horizontal I_g difference, B/hr ft²	Total on horizontal W_g B/hr ft²	Direct, perpendicular radiation I, B/hr ft²	Diffuse on horizontal I_g difference, B/hr ft²	Total on horizontal W_g B/hr ft²	Cirro-stratus W_g B/hr ft²	Alto-cumulus W_g B/hr ft²	Strato-cumulus W_g B/hr ft²	Fog W_g B/hr ft²
5	10.39	67	7	13	34	9	12	–	–	–	–
10	5.60	123	14	35	58	18	28	–	–	15	10
15	3.82	166	19	62	80	24	45	50	35	25	15
20	2.90	197	23	90	103	31	64	70	50	35	20
25	2.36	218	26	118	121	38	89	95	65	40	20
30	2.00	235	28	146	136	44	112	120	75	50	25
35	1.74	.248	30	172	148	48	133	145	90	60	30
40	1.55	.258	31	197	158	52	154	165	105	70	35
45	1.41	266	32	220	165	55	172	185	115	80	40
50	1.30	273	33	242	172	58	190	205	130	85	40
60	1.15	283	34	279	181	63	220	235	150	45	45
70	1.06	289	35	307	188	69	246	260	100	110	50
80	1.02	292	(35)	(322)	195	–	–	–	–	–	–
90	1.00	294	(36)	(328)	200†	–	–	–	–	–	–

*Smithsonian Meteorological Tables, 6th rev. ed., 1951, p. 422.
†192 would be more consistent with the curve from 70° down.

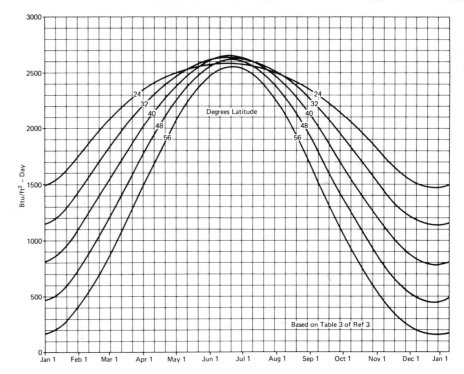

Source: from *Task Force Report—Solar Energy*, Project Independence, under the direction of the National Science Foundation for the Federal Energy Administration, November 1974.

moderate cloud conditions may reduce direct solar radiation to 10–50% of the value outside the atmosphere and very cloudy conditions may reduce direct radiation to less than 1%. Even under cloudy conditions, however, there is appreciable diffuse radiation derived from the scattered direct radiation. Flat plate solar collectors can collect diffuse radiation with fair efficiency whereas the concentrating collectors, which depend upon focusing of direct radiation, have very poor efficiencies under cloudy conditions.

REFERENCE

Arthur D. Watt, "Terrestrial insolation," *Solar Energy for Earth*, American Institute of Aeronautics and Astronautics, 1290 Avenue of the Americas, New York, N.Y. 10019, April 1975.

DISTRIBUTION OF SOLAR ENERGY IN THE UNITED STATES

The extent to which solar energy can be used for the heating and cooling of buildings or for the generation of electrical power, varies considerably throughout the United States. The average solar radiation per year decreases, of course, toward the northern latitudes but, as illustrated in the accompanying chart, the solar radiation reaching the Earth's surface is also determined by the extent of cloud cover. For this reason, solar radiation levels are markedly lower in most of the eastern portion of the United States than they are in the western states.

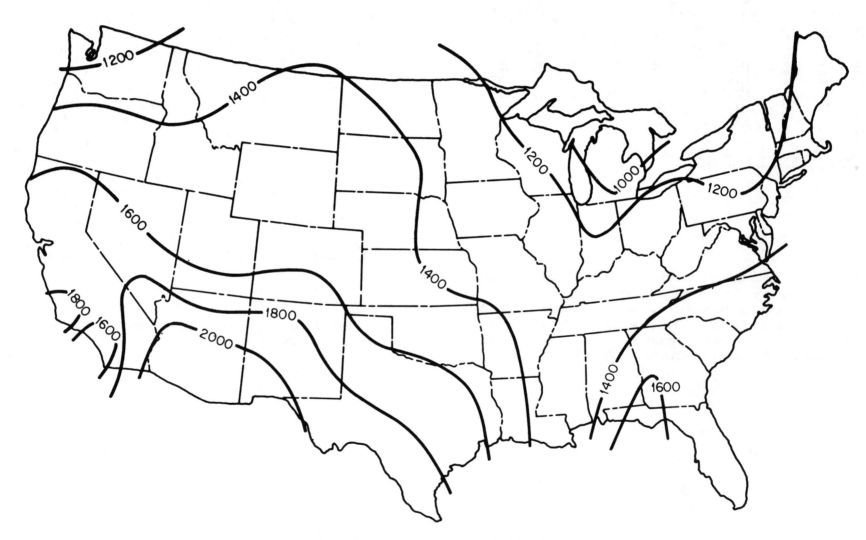

Distribution of solar energy over the United States*

*Figures give solar heat in Btu/ft^2 per average day.

Source: WASH-1535, *Environmental Statement Liquid Metal Fast Breeder Reactor Program, Volume IV, Alternative Technology Options,* United States Atomic Energy Commission, March 1974.

**Quantities of Solar Energy Received by Different Areas when the Average Intensity
of Radiation is 1 cal cm^{-2} min^{-1} = 184/Btu/ft^2/day**

Area	Langleys	kcal min^{-1}	kcal day^{-1a}	Btu hr^{-1}	kw (heat)	hp (heat)
1 cm^2	1.0	0.001	0.500	0.238	7.00×10^{-5}	9.39×10^{-5}
1 ft^2	929	0.929	464	221	0.065	0.087
1 m^2	10^4	10	5.0×10^3	2380	0.700	0.938
100 m^2 (roof)	10^6	10^3	5.0×10^5	2.38×10^6	70.0	93.8
1 acre	4.05×10^7	4.05×10^4	2.02×10^7	9.64×10^6	2.83×10^3	3.79×10^3
1 km^2	10^{10}	10^7	5.0×10^9	2.38×10^9	7.00×10^5	9.38×10^5
1 mile^2	2.59×10^{10}	2.59×10^7	1.3×10^{10}	6.15×10^9	1.81×10^6	2.42×10^6

Conversion factors: 1 kcal = 1000 cal; 1 Btu = 0.252 kcal; 1 kw = 14.3 kcal min^{-1}; 1 hp =
0.742 kw; 1 ft^2 = 929 cm^2; 1 acre = 43,560 ft^2.
[a]Assuming 500 min day^{-1} of solar radiation.

Area	Btu/Average Day*	KWH (Heat)/Day
1 cm^2	0.198	5.8×10^{-5}
1 ft^2	184.1	0.54
1 m^2	1982.5	5.83
100 m^2 (roof)	19.8×10^4	582
1 acre	80.3×10^6	23.6×10^3
1 km^2	19.83×10^9	58.3×10^5
1 mile^2	51.2×10^9	15.5×10^6

Source: Task Force Report—Solar Energy, Project Independence, under the direction of the National Science Foundation for the Federal Energy Administration, November 1974.

Average Solar Insolation and Degree Days by Census Region

Region	Annual		Maximum (July)		Minimum (Jan.)		Average Degree Days
	WH/ft^2/day	Btu/ft^2/day	WH/ft^2/day	Btu/ft^2/day	WH/ft^2/day	Btu/ft^2/day	
New England	346	1181	540	1843	130	444	6500
Middle Atlantic	346	1181	551	1881	140	478	6000
East North Central	356	1215	572	1952	151	515	7000
West North Central	389	1324	659	2249	194	662	6000
South Atlantic	442	1509	572	1452	248	846	4000
East South Central	421	1406	572	1952	205	700	4200
West South Central	464	1584	659	2249	281	959	1800
Mountain	464	1584	702	2396	238	812	6000
Pacific	421	1406	702	2396	162	552	2400

Source: Task Force Report—Solar Energy, Project Independence, under the direction of the National Science Foundation for the Federal Energy Administration, November 1974.

HEATING AND COOLING REQUIREMENTS IN THE UNITED STATES

A measure of the heating and cooling requirements for various regions of the United States is illustrated in the accompanying charts. This measure is given in degree-days, a number representing the difference between the average temperature and 65°F. Thus, if the average temperature is 45°F, there are 65 – 45 = 20 degree-days of heating required for that day. Conversely, if the temperature is 85°F, there are 85 – 65 = 20 degree-days of cooling required. Daily figures are added to obtain composite figures for the heating and cooling seasons.

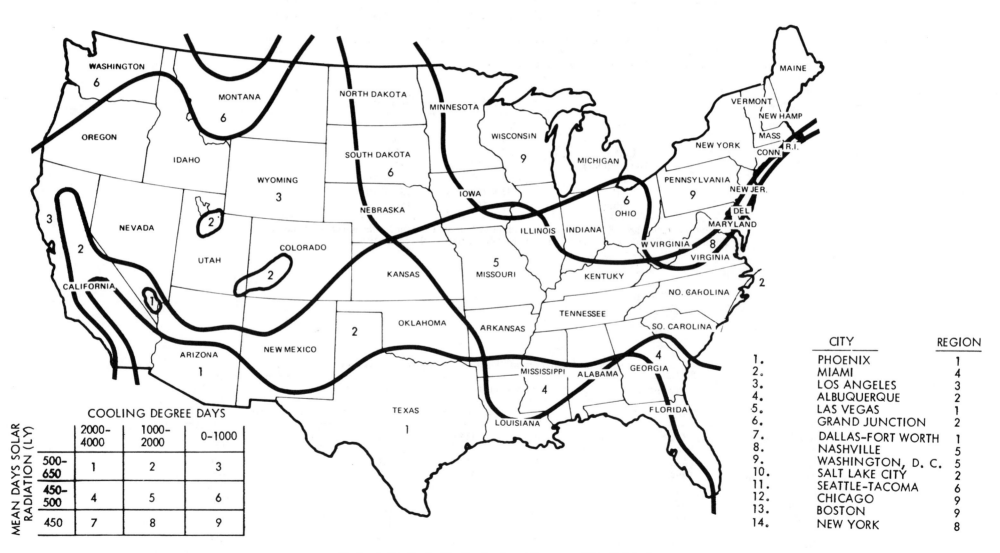

CITY	REGION
1. PHOENIX	1
2. MIAMI	4
3. LOS ANGELES	3
4. ALBUQUERQUE	2
5. LAS VEGAS	1
6. GRAND JUNCTION	2
7. DALLAS–FORT WORTH	1
8. NASHVILLE	5
9. WASHINGTON, D. C.	5
10. SALT LAKE CITY	2
11. SEATTLE–TACOMA	6
12. CHICAGO	9
13. BOSTON	9
14. NEW YORK	8

COOLING DEGREE DAYS

MEAN DAYS SOLAR RADIATION (LY)	2000–4000	1000–2000	0–1000
500–650	1	2	3
450–500	4	5	6
450	7	8	9

Regional climatic classification for the cooling season (May–October).

Source: Solar Energy Task Force Report, Project Independence Blueprint, Federal Energy Administration, November 1974.

The amount of energy required to meet these heating and cooling requirements is, of course, related to the degree-day figures. Requirements for individual buildings depend, however, on building design and construction, the nature of the activities in the building, as well as preferences for temperature levels. Buildings which are well insulated and tightly constructed to prevent infiltration of cold or warm air and which admit the sun in winter and exclude it in summer, have lower energy requirements than those which are not so constructed.

Beyond the design of buildings to take natural advantage of the sun's energy, a number of systems for collection of the sun's energy to provide augmented heating as well as cooling have been developed and are described on the pages following.

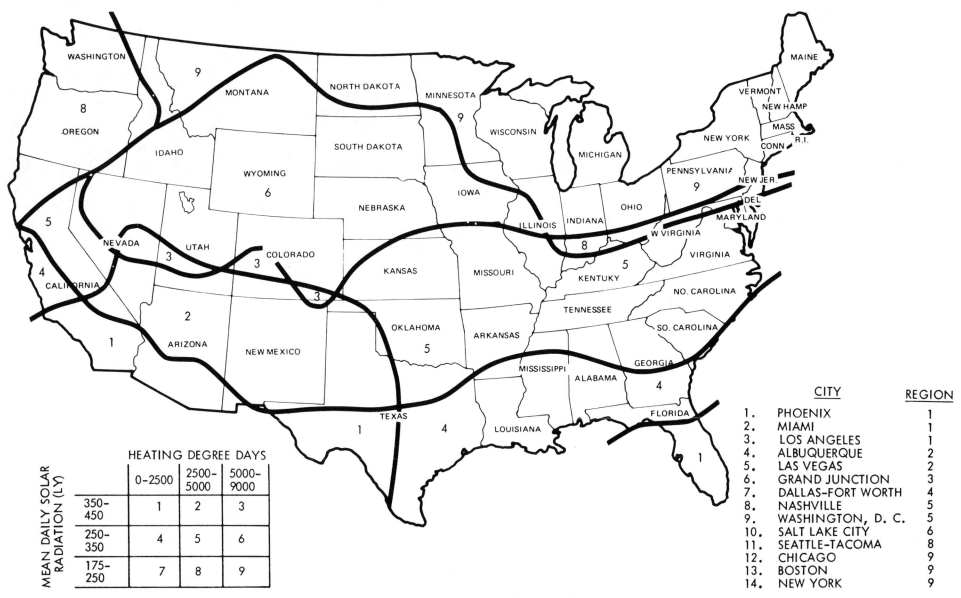

CITY		REGION
1.	PHOENIX	1
2.	MIAMI	1
3.	LOS ANGELES	1
4.	ALBUQUERQUE	2
5.	LAS VEGAS	2
6.	GRAND JUNCTION	3
7.	DALLAS–FORT WORTH	4
8.	NASHVILLE	5
9.	WASHINGTON, D. C.	5
10.	SALT LAKE CITY	6
11.	SEATTLE–TACOMA	8
12.	CHICAGO	9
13.	BOSTON	9
14.	NEW YORK	9

HEATING DEGREE DAYS

MEAN DAILY SOLAR RADIATION (LY)	0–2500	2500–5000	5000–9000
350–450	1	2	3
250–350	4	5	6
175–250	7	8	9

Regional climatic classification for the heating season (November–April).

Source: *Solar Energy Task Force Report*, Project Independence Blueprint, Federal Energy Administration, November 1974.

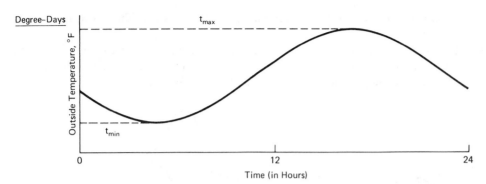

$$t_{average} = \frac{t_{max} + t_{min}}{2} = \text{daily mean temperature}$$

$$\text{degree day} = \text{Daily Mean Temperature} - 65°F.$$

Number of Btu which a heating plant must furnish to a building for a given number of degree days is:

$$\text{Btu required} = \text{Heat rate of building (Btu/hour-degree)} \times 24 \text{ hours/day} \times \text{degree days}$$

where "Btu required" is the heat supplied by the heating system to maintain the desired inside temperature. "Heat rate of building" is the hourly building heat loss divided by the difference between inside and outside design temperatures.

Source: Solar Energy Task Force Report, Project Independence Blueprint, Federal Energy Administration, November 1974.

Solar heating and cooling system installed in the George A. Town Elementary School, Atlanta, Georgia. Reflectors on back slope of collectors permit collection of both direct and reflected sunlight. (*Courtesy Westinghouse Photo*.)

SOLAR THERMAL COLLECTION SYSTEMS

Solar thermal systems have a number of common functional subsystems—collectors, heat transport systems, heat storage systems, and heat utilization equipment. Different subsystem designs are being developed to meet the requirements for different heating, cooling, and power applications.

Collectors for solar thermal systems include flat-plate, parabolic trough, paraboloidal dishes, and central receiver concepts. The simple flat-plate collector is comprised of a coated, flat absorber plate with channels along the plate for passage of a heat transport medium, typically air or water. The flat-plate collectors are suitable for low temperature applications such as the heating and cooling of buildings. Transparent covers of glass or plastic are placed over the absorber plate to reduce convective losses, but below temperatures of 200°F, covers reduce overall efficiency because of multiple reflection and transmission losses. Insulation is placed between the absorber plate and the structure upon which it is mounted to prevent heat losses to the structure. Flat-plate collectors can utilize both direct and diffuse solar radiation and can function, therefore, on hazy days.

Higher temperatures can be achieved by modification of the "flat"-plate collector to include reflective side panels which concentrate the incoming sunlight into smaller areas. These augmented flat-plate designs can produce concentration ratios as high as ten depending upon the angle the collector faces the sun. Temperatures up to 500°F can be achieved.

Parabolic trough reflectors provide an even higher concentration ratio, from 30–50, although they can collect only the direct or near-direct solar radiation and, therefore, can

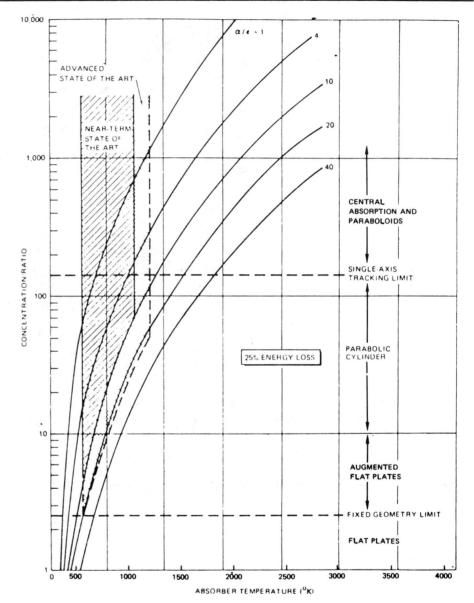

Effect of concentration ratio and absorption/reflection (α/ϵ) ratio on absorber temperature for various collectors.

Reprinted with permission from American Institute of Aeronautics and Astronautics, *Solar Energy for Earth*, April, 1974, pg. 34.

Characteristics of Solar Collector Systems.

Type	Application	Collection Method	Tracking Requirement
Flat Plate	Low Temperature (typically 150°C)	Solar energy is absorbed directly on a surface with no concentration. Both the specular and diffuse components of the solar input are collected.	None
Linear Focus	Low to Moderate Temperature (typically 300°C)	Solar energy is concentrated to a "line" by means of a parabolic trough mirror or a linear fresnel lens. Energy is then absorbed on the surface of a pipe located at the line and transferred to a fluid flowing within the pipe. Only the specular component of the solar input is collected.	Must track the sun in at least one dimension.
Central Focus	Moderate to High Temperature (up to 1000°C or higher)	A large number of nearly flat mirrors reflect solar energy to a central "point" at the top of a tower. The concentrated flux is then absorbed on a surface and transferred through the walls to a working fluid, or absorbed directly in a working fluid. Only the specular component of the solar input is collected.	The mirrors must individually track the sun in two dimensions.

Source: Solar Energy Final Task Force Report, Federal Energy Administration, Project Independence Blueprint, under the direction of the National Science Foundation, November 1974.

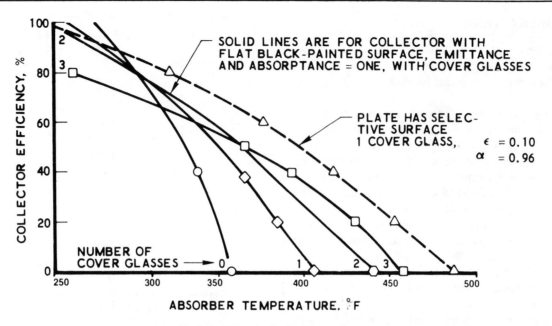

Typical efficiencies for flat-plate collectors.

Reprinted with permission from American Institute of Aeronautics and Astronautics, *Solar Energy for Earth*, April, 1974, pg. 39.

be used only in locations where the sky is generally clear. Parabolic trough collectors are usually mounted on a single axis to permit turning the collector so that the sun's rays can be focused on the heat absorber pipe placed at the focus of the parabola. Selective coatings are used on the heat absorber pipe to maximize absorption and minimize reradiation of heat.

Paraboloidal dishes, with two-axis tracking of the sun, are capable of concentration ratios of over 1000 and, by producing higher temperatures (near 1000°F), provide higher overall cycle efficiencies. Because of the compound curves of paraboloidal dishes, manufacturing is more difficult than it is for parabolic trough or flat-plate collectors.

The central receiver collector concept is basically a modification of the paraboloidal dish with central focusing of sunlight on a receiver from a large number of flat-plate mirrors steered on two axes.

SOLAR THERMAL COLLECTOR DESIGNS

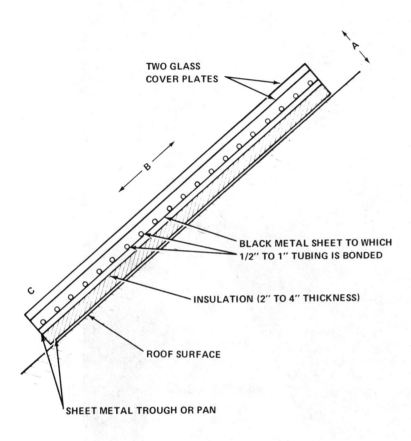

TWO GLASS
COVER PLATES

BLACK METAL SHEET TO WHICH
1/2″ TO 1″ TUBING IS BONDED

INSULATION (2″ TO 4″ THICKNESS)

ROOF SURFACE

SHEET METAL TROUGH OR PAN

NOTES: ENDS OF TUBES MANIFOLDED TOGETHER
ONE TO THREE GLASS COVERS DEPENDING
ON CONDITIONS
DIMENSIONS: THICKNESS (A DIRECTION) 3 INCHES TO 6 INCHES
LENGTH (B DIRECTION) 4 FEET TO 20 FEET
WIDTH (C DIRECTION) 10 FEET TO 50 FEET
SLOPE DEPENDENT ON LOCATION AND ON
WINTER-SUMMER LOAD COMPARISON

Flat-plate collector for residential heating and cooling. Air or water is circulated through the collector and transports heat into the building.

Source: Solar Energy as a National Energy Resource, NSF/NASA Solar Energy Panel, National Science Foundation, December 1972.

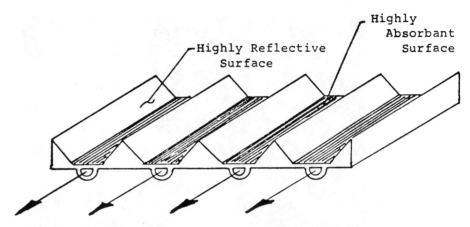

Highly Reflective
Surface

Highly
Absorbant
Surface

Augmented flat-plate collectors produce higher temperatures by employing reflecting surfaces to concentrate sunlight on absorbent surfaces. Optimum groove geometries center around opening angles near 30° and a depth-to-base ratio of 2. Concentration ratios of 3 to 1 can be achieved.

Source: John R. Howell and Richard B. Bannerot, *The Evaluation of Geometry Modification to Improve the Directional Selectivity of Solar Energy Collectors,* NSF/RANN/SE/GI-41003/PR/74/4, National Science Foundation, January 31, 1975.

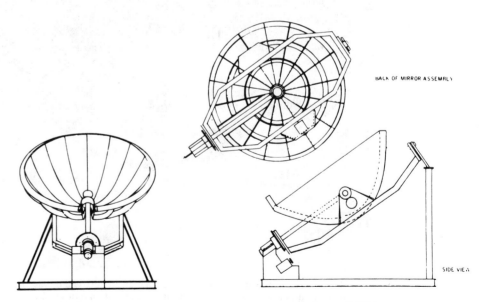

BACK OF MIRROR ASSEMBLY

SIDE VIEW

Paraboloidal dish solar collector is capable of concentration ratios near 1000. Temperatures near 1000°F can be achieved.

Reprinted with permission from American Institute of Aeronautics and Astronautics, *Solar Energy for Earth,* April, 1974, pg. 40.

Parabolic trough solar collector. Rays of the sun are focussed on the glass tube along the axis of the reflector. Water or other suitable liquid is heated and circulated to heat utilization equipment. Concentration ratios of 30–50 can be achieved.

Source: Energy Research and Development Administration.

SOLAR PHOTOELECTRIC COLLECTORS

Photoelectric solar collectors are based on the principle that some solid materials, called semiconductors, generate free electrical charges upon absorption of a photon from sunlight. These charges are collected on electrodes applied to the semiconductor material. Solar cell materials most commonly used are single-crystal silicon, cadmium sulfide/copper sulfide, and polycrystalline silicon. The maximum theoretical conversion efficiency is about 25% for single-crystal silicon when operating at room temperature. Efficiencies of 12–15% have been achieved with single-crystal silicon cells and efficiencies of 4–6% have been achieved with cadmium sulfide/copper sulfide cells in terrestrial applications.

In order to obtain desired voltages and currents, solar cells are connected in series-parallel arrays. Power stations utilizing solar cells would be comprised of large numbers of panels each with hundreds of individual solar cells. The direct current produced by the solar cells is converted into alternating current for use in terrestrial applications.

One of the major advantages of solar cells is that they can utilize the entire solar radiation, both direct and diffuse and, since they convert solar energy directly to electricity, a heat transport medium need not be circulated to dissipate waste heat.

The use of photoelectric cells for power production is presently limited by one or more of the following factors: cell cost, cell lifetime, materials availability, and industrial manufacturing capability. Single-crystal silicon cells currently cost some $20,000–$30,000 per peak kilowatt and present industrial production capacity would support only 100 MWe per year if all refined silicon were used for solar power plants. Cadmium sulfide/copper sulfide cells are less expensive but manufacturing reproducibility is poor and cell performance degrades unpredictably in the presence of oxygen, water vapor, and sunlight, with

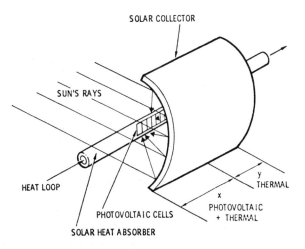

Solar collection system proposed by Sandia Laboratories combines photovoltaic cells with thermal conversion. Silicon solar cells are mounted in good thermal contact with circulating fluid loop. In addition to direct conversion of sunlight to electricity, fluid loop provides additional heat for heating and cooling applications.

Source: Solar Energy: Sandia's Photovoltaic Research Program, SLA-74-0281, Sandia Laboratories, May 1974.

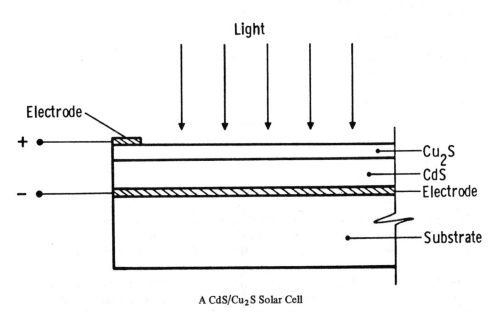

A CdS/Cu$_2$S Solar Cell

Source: Glen E. Brandvold, *Testimony Regarding Solar Research and Development,* presented to the Joint Committee on Atomic Energy, U.S. Congress, May 8, 1974.

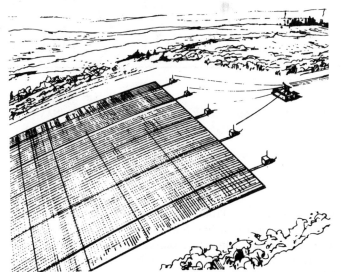

Electricity from solar cells on the earth.

Source: Solar Energy Research, Staff Report of the Committee on Science and Astronautics, U.S. House of Representatives, December 1972.

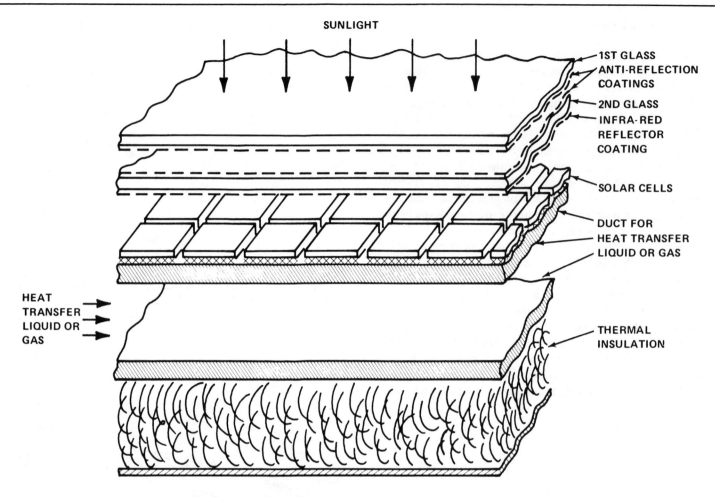

Structure of combination thermal-photovoltaic solar collector.

Source: Solar Energy as a National Energy Resource, NSF/NASA Solar Energy Panel, National Science Foundation, December 1972.

accelerated degradation at higher temperatures. Production costs are potentially low, $15 to $20 per peak kilowatt, for processes in which thin films of cadmium sulfide and copper sulfide are sprayed on glass substrates. Cadmium resources may limit large-scale use of CdS/Cu_2S cells. Present United States cadmium output of 3000 metric tons per year could support about 6000 MWe per year if all cadmium were used for this purpose. Polycrystalline silicon solar cells, made by vapor deposition of silicon on substrate materials, are in the early stages of development. Conversion efficiencies are, as yet, only 1.5% to 2%.

In order to provide a continuous source of power, a terrestrial photoelectric power plant would require some form of energy storage such as batteries, pumped hydro, or flywheels, and a converter system to change the direct current output to alternating current. Assuming an overall system efficiency of 10% for the combined collection—storage—convertor system, a 1000 MWe power station located in the southwestern United States would require about 9500 acres (15 square miles) of cell surface and a total land area of about 19,000 acres (30 square miles). The total station area is larger than the cell surface area because some space must be allowed between solar cell panels to avoid shading and to permit maintenance.

SOLAR RESIDENTIAL SYSTEMS

The basic elements of solar heating and cooling systems include flat-plate collectors, heat transport systems, and storage systems, coupled with conventional heating, cooling, and air-conditioning components.

The solar collectors, designed to produce temperatures in the 200-300°F range, are mounted on the roof at a fixed angle. Air, water, or an organic liquid are used to transport the heat generated by the sun to a storage system which provides heat during the times when the sun is not shining. This heat storage system may be simply an insulated container filled with rocks if the heat transport medium is air, or a tank if the heat transport medium is water. Other heat storage systems utilizing the latent heat of fusion of salts have also been used. Heat from the storage unit is utilized to heat water for domestic use, to provide space heating, and to operate absorption air-conditioning units.

In almost all regions of the United States, some auxiliary heating—gas, oil, or electricity—is desirable to meet requirements during very cold periods or periods of extended cloud cover. A typical house of 1500 ft² in a temperate, sunny location, would require solar collectors of 600-800 ft² to provide about 75% of the heating and cooling requirements.

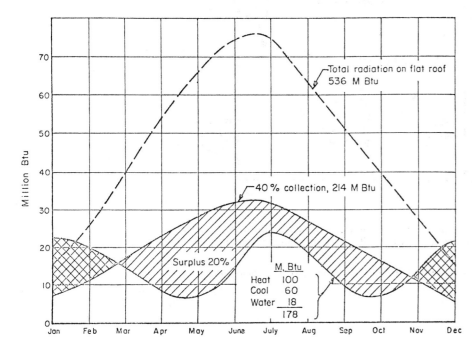

Solar energy supply and requirements for a residence in Columbus, Ohio. Shortfall in solar energy occurs during winter months.

Source: Proceedings of the Solar Heating and Cooling for Buildings Workshop, National Science Foundation, March 1975.

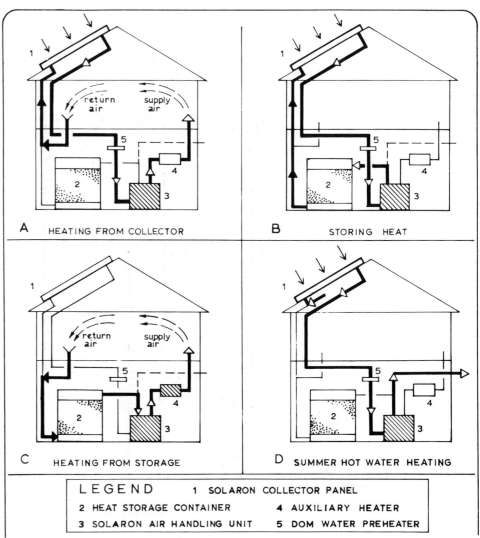

LEGEND	1 SOLARON COLLECTOR PANEL
2 HEAT STORAGE CONTAINER	4 AUXILIARY HEATER
3 SOLARON AIR HANDLING UNIT	5 DOM WATER PREHEATER

Figure A—Air, the circulating heat transfer media is drawn through the flat plate collector where it is normally heated to about 120-150°F. When the space requires heat the solar heated air is drawn through the air handling unit in which motorized dampers are automatically opened to direct the hot air to the space.

Figure B—When the space temperature is satisfied, the automatic control system diverts the air into the heat storage unit where the heat is absorbed by the pebble bed.

Figure C—At night or on cloudy days when solar energy is unavailable and when heat is needed in the space, the automatic control system directs the building return air into the bottom of the heat storage unit, up through the pebbles where the air is heated, through the air handling unit and into the space.

Figure D—In the summer, when space heating is not required, outside air is drawn through the collector where it is heated and then through the water heat exchanger coil. The air is then vented to the outside.

Source: Solaron Corporation, Stapleton Field Industrial Park, 4850 Olive Street, Denver, Colorado 80022.

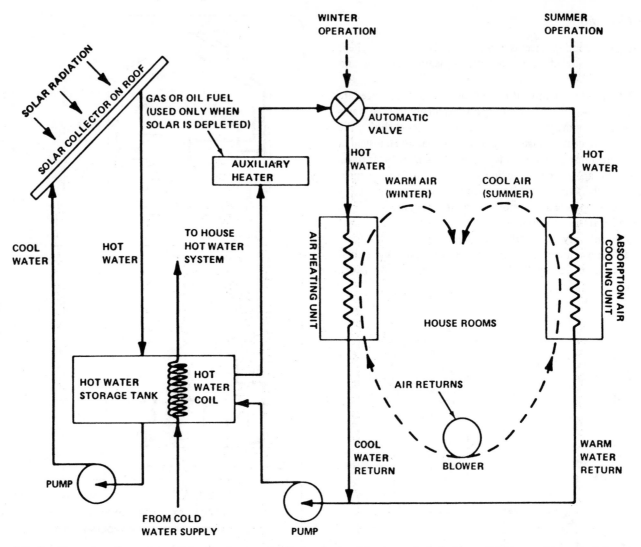

Solar heating and cooling system. Hot water from roof collectors in stored in storage tank during sunlight hours. Hot water is drawn from storage tank as needed to provide heat to air heating unit or absorption refrigeration unit. Auxiliary heater provides heat when solar energy heat is depleted.

Source: Solar Energy as a National Energy Resource: NSF/NASA Solar Energy Panel, National Science Foundation, December 1972.

SOLAR CENTRAL RECEIVER POWER PLANTS

The central receiver solar power plant concept employs mirrors, or heliostats, which can continuously track the sun and maintain a focus on a tower-top receiver where a heat transport fluid is heated. This heat transport fluid is piped to the ground where its heat is used to drive a turbine generator. Solar concentration ratios of about 1000 can be achieved such that temperatures near 900–1000°F are possible, matching the temperature requirements of modern turbines. A number of different heat transport fluids have been studied which are used in either direct or indirect cycles. In the direct cycle concept, the heat transport fluid is vaporized and admitted directly to the turbine. In the indirect cycle, the heat transport fluid is passed through a heat exchanger where its heat is given up to vaporize a second fluid which is then used to drive the turbine.

A typical central receiver solar thermal power plant might consist of a number of modular solar collector-tower units, each having 0.5 square kilometers of reflectors which are focused on a receiver mounted at the top of a tower 260 meters high. A 100 MWe plant would require only one of these solar collector-tower modules if electrical energy is needed only during sunlight hours (this plant would include a one-half hour storage capacity to smooth out transients due to intermittent cloud cover). Two collector-tower modules would be required to maintain power production for three hours beyond daylight hours (with the required heat storage), and three collector-tower modules, with 12 hours of storage, would be required to maintain full power production at 100 MWe for 12 hours beyond daylight hours.

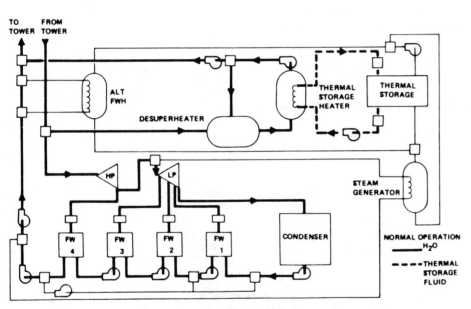

Diagram of central-tower solar-thermal powerplant.

Reprinted with permission from American Institute of Aeronautics and Astronautics, *Solar Energy for Earth*, April, 1974, pg. 35.

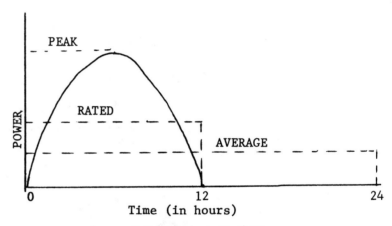

Ratings of solar power plants.

Peak Power is the power output of a solar energy system on a clear day due to direct peak solar radiation at normal incidence. Peak solar radiation is normally assumed to be 1000 watts per square meter, or 93 watts per square foot.

Rated Power for direct solar systems is the power output integrated over time (for the direct radiation) divided by the time (i.e., area under the curve divided by the normal time of operation during a day).

Average Power is the power integrated over time divided by 24 hours (i.e., one complete day).

Source: Solar Energy Task Force Report, Project Independence Blueprint, Federal Energy Administration, November 1974.

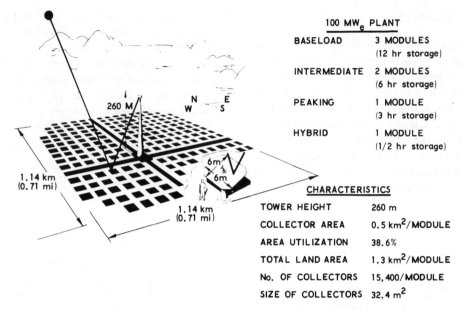

100 MW$_e$ PLANT		
BASELOAD	3 MODULES	(12 hr storage)
INTERMEDIATE	2 MODULES	(6 hr storage)
PEAKING	1 MODULE	(3 hr storage)
HYBRID	1 MODULE	(1/2 hr storage)

CHARACTERISTICS

TOWER HEIGHT	260 m
COLLECTOR AREA	0.5 km^2/MODULE
AREA UTILIZATION	38.6%
TOTAL LAND AREA	1.3 km^2/MODULE
No. OF COLLECTORS	15,400/MODULE
SIZE OF COLLECTORS	32.4 m^2

Modular central receiver concept for a solar electric power plant.

Reprinted with permission from American Institute of Aeronautics and Astronautics, *Solar Energy for Earth*, April, 1974, pg. 33.

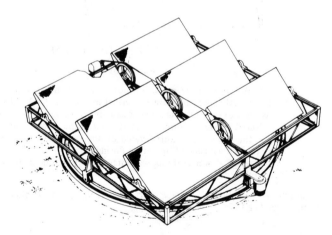

"Ganged" heliostat concept for central-tower solar-electric system.

Reprinted with permission from American Institute of Aeronautics and Astronautics, *Solar Energy for Earth*, April, 1974, pg. 34.

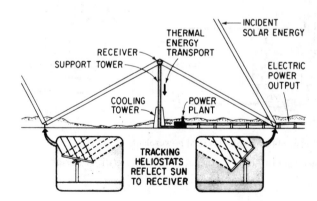

Solar central receiver electric generation system.

Reprinted with permission from American Institute of Aeronautics and Astronautics, *Solar Energy for Earth*, April, 1974, pg. 35.

SOLAR DISTRIBUTED COLLECTOR POWER PLANT

Solar thermal power plants based on the distributed collection concept employ large numbers of individual collectors distributed in a field. The heat generated in each collector is piped to central heat storage and power generation units.

Collectors may be flat-plate units, parabolic troughs, or paraboloidal dishes. The flat-plate collectors have the advantage of utilizing both direct and diffuse sunlight and are less expensive to fabricate, but the temperatures which can be obtained are low, 200-500°F, and cycle efficiency is, therefore, relatively low. In contrast, the parabolic trough and paraboloidal dishes operate only on direct sunlight but can achieve temperatures above 600°F and, as a result, cycle efficiencies are higher. Collectors may be mounted to track the sun in either one or two directions.

A number of different fluids—water, air, sodium—have been studied for transporting the heat from the collectors to the heat storage and turbine-generator units. Since the piping system for the transport of these fluids is a major cost factor, selection of the heat transport medium and operating temperatures represents a balance between higher material costs for systems operating at higher temperatures and the higher efficiencies which can be achieved at the higher temperatures.

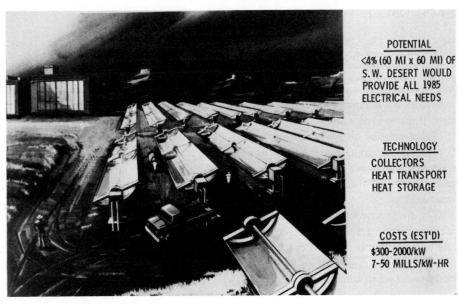

POTENTIAL
<4% (60 MI x 60 MI) OF S.W. DESERT WOULD PROVIDE ALL 1985 ELECTRICAL NEEDS

TECHNOLOGY
COLLECTORS
HEAT TRANSPORT
HEAT STORAGE

COSTS (EST'D)
$300-2000/kW
7-50 MILLS/kW-HR

Electric power from thermal energy.

Solar Thermal Electric Power Plant employing parabolic trough collectors. Heat from the collectors is piped to the heat storage and turbine-generator units. A 100 MWe plant would require a collection area of about 1.5 square kilometers. (*Credit: National Aeronautics and Space Administration.*)

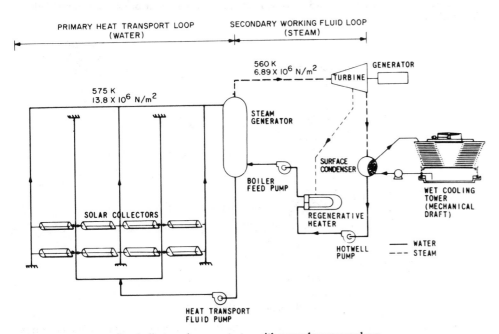

Parabolic trough concentrator with secondary power loop.

Reprinted with permission from American Institute of Aeronautics and Astronautics, *Solar Energy for Earth*, April, 1974, pg. 36.

SPACE SOLAR POWER PLANTS

The average availability of solar energy for a solar power station in geosynchronous orbit, 35,800 kilometers from Earth, is about 15 times that available on Earth. The concept proposed by Glaser[1] for a 5000 MWe system employs concentrators to reflect sunlight onto photoelectric solar cells assembled into two large arrays each 4.33 X 5.2 kilometers in dimension. A microwave transmitting antenna, 1 kilometer in diameter, is placed between the two arrays and transmits electrical energy to a ground-based receiving antenna, 7.12 kilometers in diameter. The total mass of the solar satellite power station would be 10.4 million kilograms (22.9 million pounds). Ion thrusters would be employed to counteract solar pressure, the recoil pressure from the microwave transmission, and to keep the arrays facing toward the sun against the cyclic torques exerted by the Earth's gravitational field.

As with ground-based solar photoelectric power plants, the economic viability of space satellite power stations will depend upon the development of methods for producing solar cells at sharply reduced costs. The development of space shuttles for transport of equipment from the Earth to earth orbit, and space tugs for transfer of equipment to synchronous orbit, will also be required.

[1]Peter E. Glaser, "The satellite solar power station," *Proc. IEEE Int. Microwave Symp.* June 1973.

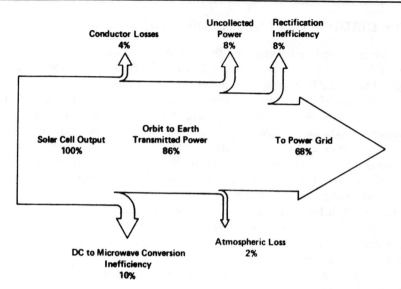

Microwave transmission efficiency.

Source: Vol. II, Briefings before the Task Force on Energy, Subcommittee on Science, Research, and Development of the Committee on Science and Astronautics, U.S. House of Representatives, Serial Q, March 1972.

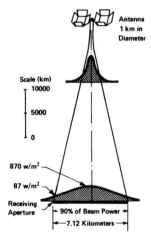

Microwave beam characteristics.

Collection and Conversion of Microwaves. A receiving antenna rectifies and converts the microwave beam into DC, which can then be used directly, be fed into high-voltage DC transmission systems, or be converted into 60-cycle AC. In the receiving antenna, rectifying elements are distributed uniformly so the microwave energy intercepted in a local region is immediately converted into DC. The rectification is accomplished by means of rectifying diodes which already have been demonstrated to have a 75% conversion efficiency. With improved circuits and diodes an efficiency of about 90% should be achievable.

Source: Vol. II, Briefings before the Task Force on Energy; Subcommittee on Science, Research, and Development of the Committee on Science and Astronautics, U.S. House of Representatives, Serial Q, March 1972.

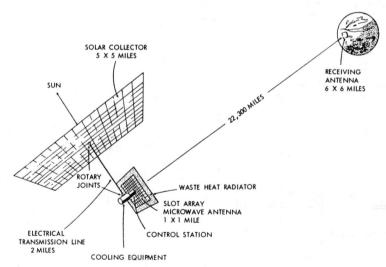

(~60 Satellite Power Stations Needed to Meet One-Half the Projected USA Power Needs in 1990.)

Satellite power station (10^7 kW).

Satellite solar power station. The station would be located in synchronous orbit 22,300 miles from the earth and would remain fixed with regard to the ground receiver. The five-by-five mile square solar collector panel would intercept about 85,000 megawatts of radiant solar energy. Solar cells of 18% efficiency would convert this into 15,000 megawatts of electrical power which would be transmitted to earth by microwave radiation. The receiving antenna would convert the incoming microwave radiation into 10,000 megawatts of electrical power.

Source: National Aeronautics and Space Administration.

WIND ENERGY

The energy in the wind is derived primarily from the solar heating of the atmosphere which, by creating air masses of different densities, results in air movements locally and regionally. These air movements are further modified by the forces imposed by the Earth's rotation and, near the ground, by the resistance to air movement presented by topographical features.

Systems for the conversion of the energy of the wind to useful mechanical energy for grinding of grain and pumping of water date from early in the history of man. The use of wind energy for the generation of electric power began soon after electricity was introduced into commercial use. In 1895, the first wind-electric system was built in

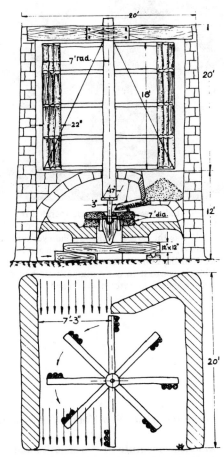

Persian windmill had wind vanes comprised of bundles of reeds mounted on spokes tied to a vertical shaft. Power was used for grinding grain.

Source: Wulff, Hans E. *The Traditional Crafts of Persia*, M.I.T. Press, 1966, as given in a special report by Volta W. Torrey, *Windmills in the History of Technology*, Technology Review, edited by the Massachusetts Institute of Technology, March/April 1975.

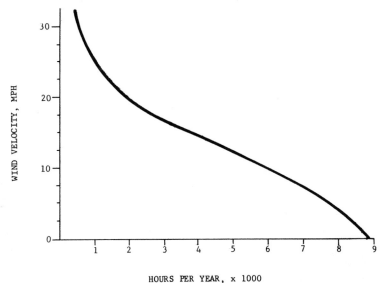

VELOCITY DURATION CURVE FOR AMARILLO, TEXAS

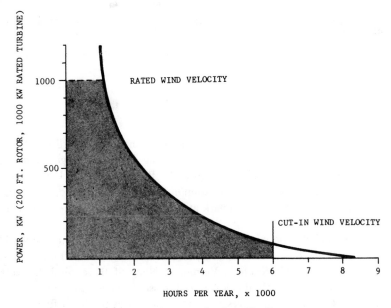

Power duration curve for Amarillo, Texas.

Rated power of wind power plants is the constant output power obtained any time the wind velocity is at or above a *derived* design *rated wind velocity* and below a storm emergency shutdown velocity. Output power varies between zero at the *cut-in wind velocity* to rated power at rated wind velocity. *Equivalent duration* in hours is the total yearly energy output divided by the rated power. When divided by 8760 (hours per year) it reflects the equivalent of a load factor.

Source: Solar Energy Task Force Report, Project Independence Blueprint, Federal Energy Administration, November 1974.

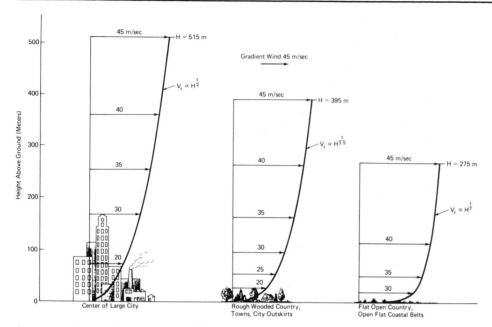

Roughness of terrain lowers wind velocities near the ground surface. Most effective locations for wind power plants are in flat open country or on the crests of hills.

Source: Solar Energy Task Force Report, Project Independence Blueprint, Federal Energy Administration, November 1974.

The 1250 kilowatt Smith-Putnam wind power plant which operated from **1941–1945.** The plant was designed for operation in a 13 meter/second wind and was situated on a 2000-foot hill at Grandpa's Knob near Rutland, Vermont. It was the largest wind turbine ever built with a tower height of 110 feet and a rotor of 175 feet from tip to tip. The rotor drove an ac synchronous generator which fed power into the electrical network of the Central Vermont Power Company. Production models would have cost $191/kilowatt in 1945 in contrast to conventional power plants at $125/kilowatt and the project was therefore dropped after one of the rotor blades failed.

Denmark. This system, designed by Prof. LaCour, consisted of an 80-foot tower supporting a 75-foot diameter, four-bladed rotor driving a mechanical gear-and-shaft drive train which powered a generator on the ground. By 1910, several hundred wind generators, of 5 to 25 kilowatts, were in operation in Denmark. Altogether, there were some 33,000 windmills in operation in Denmark at that time, providing the equivalent of 200 megawatts of power.

A number of wind-electric power plants have been built throughout the world in the ensuing years. The largest of these plants, a 1250 kilowatt installation, operated from 1942 to 1945 at Grandpa's Knob in Vermont. None of these wind-electric systems proved to be competitive with the low-priced fossil fuels which were available until the mid 1970s. Impending shortages and increasing prices for fossil fuels have revived interest in wind-powered electric systems. At the same time, other forms of energy, such as nuclear power, may be more attractive in some countries for the large-scale generation of electricity. For example, in 1974 the Swedish State Power Board concluded that electricity from windmills could cost three to four times as much as electricity from nuclear power, and that some 1500 windmills, mounted on 200-foot towers, would be required to produce as much electricity as a single 1000 megawatt nuclear power station.

REFERENCES

Palmer C. Putnam, *Power from the Wind.* New York: Van Nostrand Reinhold Company, 1948.

Solar Energy for Earth, American Institute of Aeronautics and Astronautics, 1290 Avenue of the Americas, New York, N. Y. 10019, April 1975.

Wind Energy Conversion Systems: Workshop Proceedings, NSF/RA/W-73-006, National Science Foundation, December 1973.

Second Workshop on Wind Energy Conversion Systems, Washington, D. C., June 9–11, 1975, NSF-RA-N-75-050, MTR-6970, The MITRE Corporation.

WIND ENERGY SYSTEMS

The power which can be derived from wind varies as the cube of the wind velocity. The theoretical limit for extraction of the energy of the wind by a propeller-type rotor is 59.3%. In practice, well-designed rotors are able to extract between 40 and 45% and, when the rotor is used to drive an electrical generator with a gear-type transmission between the rotor and the generator, the maximum extractable power is between 30 and 35%.

Propellor-type wind conversion systems are comprised of a support tower, a rotor, a mechanical energy transmission system, and an energy convertor such as an electrical generator or a water pump. A control system is also required to adjust the pitch of the rotor blade and to rotate the rotor assembly to keep the rotor facing the wind. In addition, since the wind is variable, some form of energy storage is also required if power is to be available on demand.

Fluctuating forces imposed by the wind on the rotor and tower place severe mechanical stresses on structural and operating components. Changing wind speeds and gusts cause the tower to sway and the rotor blades to twist and flex and, in addition, change the shaft torque and power output.

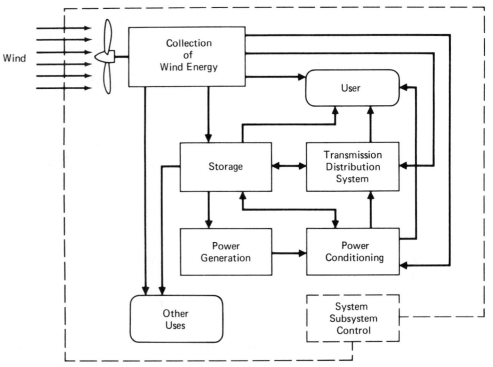

Functional block diagram for wind energy systems.

Shaft Horsepower as a Function of Rotor Diameter and Wind Velocity for Windmill with 75% Actual Efficiency

Rotor Diameter (ft)	Wind Velocity (mph)			
	10	20	30	40
	Shaft Horsepower			
50	6	48	162	384
100	24	190	638	1,520
150	53	424	1,432	3,400
200	96	768	2,550	6,040
300	212	1,700	5,728	13,600

Note: One Horsepower is equal to 0.746 kilowatts.
Source: Johnson, C. C., Smith, R. T. and Swanson, R. K.; "Wind Power Development and Applications," *Power Engineering*, October 1974.

Two types of electrical generators have been studied—synchronous generators which must be rotated at constant speed and require that the pitch of the rotor blades be continuously adjusted as the wind velocity varies; and field-modulated generators which can be operated at variable speeds but require the addition of power-conditioning equipment to produce 60-cycle alternating current.

The development of adequate and cost-effective energy storage systems for use with wind energy systems, is a major requirement for commercial success. Lead-acid batteries, although highly developed, are too expensive for use in large systems. Other batteries under development, such as sodium-sulfur batteries, may provide a less expensive energy storage system. Other storage system possibilities include pumped hydro, compressed air, flywheels, and the electrolytic generation and storage of hydrogen.

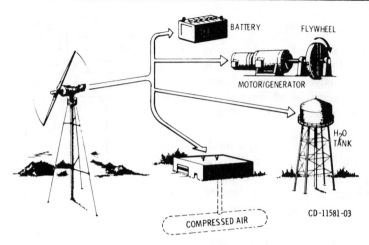

Some storage mechanisms for wind energy.

Source: Wind Energy Developments in the 20th Century, National Aeronautics and Space Administration, Lewis Research Center, Cleveland, Ohio, 1975.

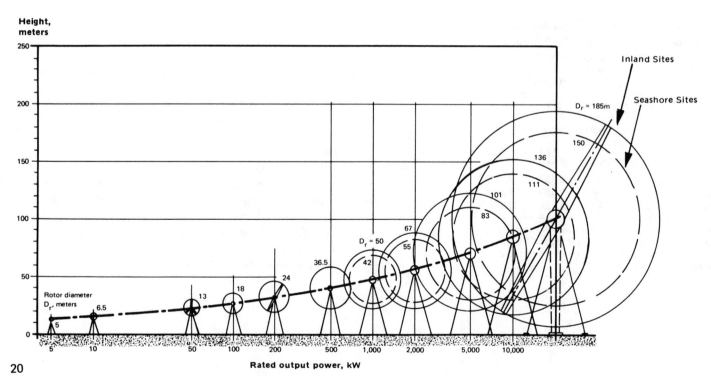

Typical family of horizontal-axis wind turbines for average wind speeds of 17 mph.

Source: Frank R. Eldridge, *Wind Machines*, MTR-6971, NSF-RA-N-75-051, The MITRE Corporation, McLean, Virginia 22101, October 1975. Originally from Swedish project study 1973, by Olle Ljungström, STU.

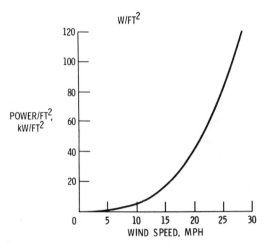

Power density of wind.

Source: Wind Energy Developments in the 20th Century, National Aeronautics and Space Administration, Lewis Research Center, Cleveland, Ohio, 1975.

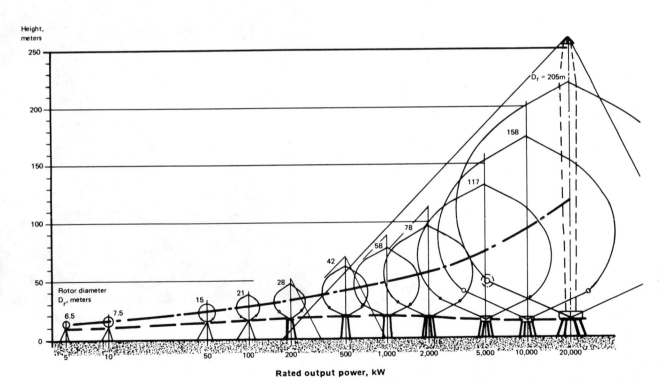

Typical family of vertical-axis, darrieus-type wind turbines for average wind speeds of 17 mph.

Source: Frank R. Eldridge, *Wind Machines*, MTR-6971, NSF-RA-N-75-051, The MITRE Corporation, McLean, Virginia 22101, October 1975. Originally from Swedish project study 1973, by Olle Ljungström, STU.

ERDA/NASA 100 kW WIND TURBINE

The 100 kilowatt experimental wind turbine completed in 1975 at NASAs Plum Brook Station, is the largest wind machine to be built in the United States since the 1250 kilowatt machine at Grandpa's Knob was shut down in 1945. The 100 kilowatt wind turbine is mounted on an open truss tower which stands 100 feet high. The rotor contains two aluminum blades, each measuring 62.5 feet in length and weighing 2000 pounds. The machine starts generating power in an 8 mile per hour wind and reaches its maximum output of 100 kilowatts in a wind of 18 mph. The maximum blade rotational speed is 40 revolutions per minute. When wind speeds exceed 40 mph, the wind turbine is automatically shut down by feathering the rotor blades.

General Specifications of 100 kW Experimental WTG

Power:
Blade power (assuming 7° coning; 0° inclination, kW 133
Generator output, kW 100

Desired rotor power coefficient 0.375

Cut-in wind speed (first load applied), m/sec . . . . 3.52 (8 mph)

Rated wind speed (100 kW bus), m/sec . . . . 7.92 (18 mph)

Feather wind speed, m/sec . . . . 26.4 (60 mph)

Hurricane wind speed, m/sec . . . . 66 (150 mph)

Location to rotor with respect to tower . . . . downwind

Direction of rotation (looking up-wind) . . . . counterclockwise

Blade Specifications

Number of blades 2

Diameter, m . . . . 37.5 (125 ft)

Cong angle - fixed, deg 7

Effective diameter of circle swept by airfoils, m . . . . 37.2 (124 ft)

Inclination of axis of rotation relative to horizontal, deg 0

Effective circular area swept by airfoils, m^2 . . . . 1071.9 11,910 ft^2)

Area of one blade projection on swept circular area, m^2 . . . 16.1 (179 ft^2)

Slenderness ratio relative to blade radius 22

Rotor rpm 3

Maximum thrust from the wind (two blades), newtons . . . 44 482 (10 000 lbf)

Source: Richard L. Puthoff and Paul Sirocky, *Status Report of 100 kW experimental Wind Turbine Generator Project*, NASA Technical Memorandum NASA TM X-71758, June 1975.

ERDA/NASA 100 kW experimental wind turbine, Plum Brook Station, Sandusky, Ohio.

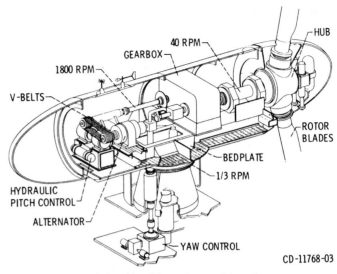

100-kilowatt wind turbine drive train assembly and yaw system.

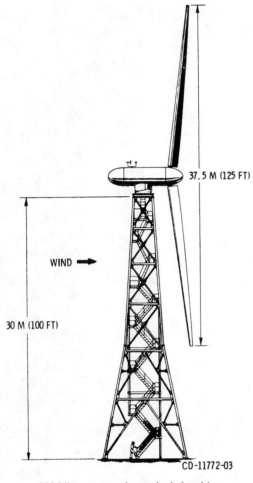

100-kilowatt experimental wind turbine.

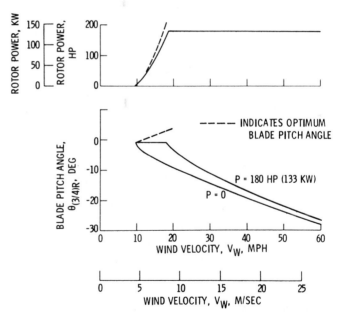

Rotor power and blade angle variation for the 100-kilowatt experimental wind turbine.

100 kilowatt experimental wind turbine at the NASA Plum Brook site near Sandusky, Ohio.

Source: Ronald L. Thomas and John E. Sholes, *Preliminary Results of the Large Experimental Wind Turbine Phase of the National Wind Energy Program*, NASA Technical Memorandum NASA TM X-71796, Oct. 1975.

VERTICAL AXIS WIND TURBINES

Wind turbines mounted with the axis of rotation in a vertical position have the advantage that they are omnidirectional, that is, they need not be turned to face the wind. The vertical orientation eliminates the need for some of the complex mechanical devices and control systems necessary for the horizontally mounted wind machines.

Two types of vertical axis wind machines have received attention. The Darrieus rotor consists of two or three convex metal blades with an airfoil cross section, mounted on a central shaft which is supported by bearings at the top and bottom. The rotor assembly is held in position by guy wires running from the top of the rotor to the ground. The Savonius rotor consists of a long, solid S-shaped surface mounted to turn at the center of the S. The Savonius rotor is self-starting and has an efficiency of about 31% while the Darrieus rotor has a slightly higher efficiency of 35% but is not self-starting.

The vertical axis wind machines have been used in the past only for low-power applications.

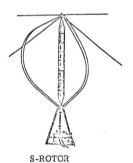

DARRIEUS ROTOR

S-ROTOR

BACKGROUND

    PATENTED IN 1931 (US AND FRANCE)

    CURRENTLY UNDER STUDY AT NATIONAL AERONAUTICAL ESTABLISHMENT, OTTAWA, CANADA

CHARACTERISTICS

    EFFICIENCY ~ 35%

    TIP SPEED TO WIND SPEED ~ 6 TO 8

    POTENTIALLY LOW CAPITAL COST

    CURRENTLY NOT SELF STARTING

BACKGROUND

    PATENTED IN 1929 (US AND FINLAND) BY S. J. SAVONIUS

    CURRENTLY USED AS AN OCEAN CURRENT METER

    OTHER APPLICATIONS SHOWN FEASIBLE

CHARACTERISTICS

    TIP SPEED TO WIND SPEED ~ .8 TO 1.8

    EFFICIENCY ~ 31%

    SELF STARTING

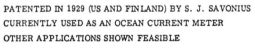

VERTICAL AXIS ROTORS OPERATE INDEPENDENTLY OF WIND DIRECTION AND THUS HAVE A POTENTIAL FOR HIGH EFFICIENCY IN CHANGING WINDS

Vertical axis wind rotors.

Source: W. Vance, "Vertical Axis Wind Rotors—Status and Potential," in *Wind Energy Conversion Systems: Workshop Proceedings*, June 11–13, 1973, NSF/RA/W-73-006, National Science Foundation.

Sandia Laboratories, Albuquerque, N.M. The Atomic Energy Commission's Sandia Laboratories in Albuquerque, New Mexico have filed patent applications for improvements in the vertical axis wind turbine, which has been mounted atop a building at Sandia. The patent application covers design of the starter buckets and the air-foil blades. The 15-foot blades drive into the wind, producing about three horsepower in a 20 mile per hour wind. At left is Lou Feltz, mechanical designer at Sandia, and at right is Ben Blackwell, project leader.

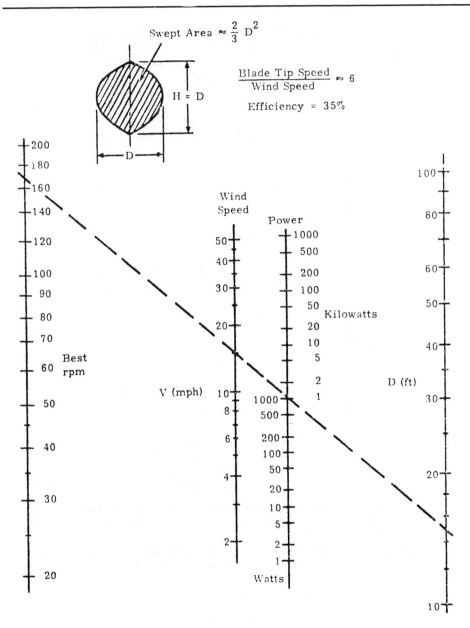

Swept Area $\approx \frac{2}{3} D^2$

$\frac{\text{Blade Tip Speed}}{\text{Wind Speed}} \approx 6$

Efficiency = 35%

H = D

D

Example (broken line): Given wind speed = 15 mph, diameter D = 15 feet, find output shaft power and operating shaft speed for maximum power. Shaft speed = 168 rpm; shaft power = 1080 watts.

Vertical-axis wind turbine performance nomogram.

Source: Reproduced with permission of the National Research Council of Canada.

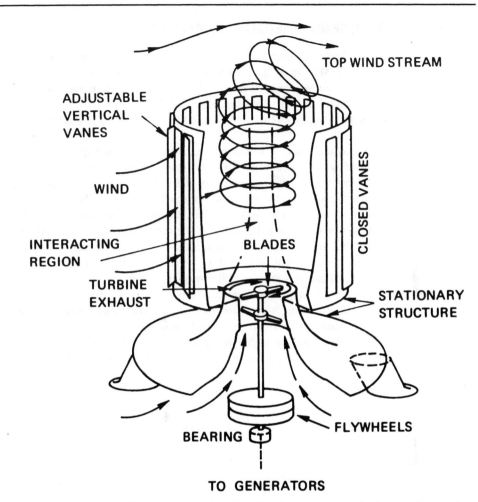

TOP WIND STREAM

ADJUSTABLE VERTICAL VANES

WIND

INTERACTING REGION

TURBINE EXHAUST

CLOSED VANES

BLADES

STATIONARY STRUCTURE

FLYWHEELS

BEARING

TO GENERATORS

Vortex tower concept is independent of wind direction. Low pressure region in core of tower draws in outside air and creates swirling vortex which turns rotor blades. *Credit:* U.S. Department of Energy.

ENERGY FROM OCEAN THERMAL GRADIENTS

The recovery of energy from temperature differentials in an ocean was demonstrated by Georges Claude in 1931 at Mantanzas Bay in Cuba. Operating on a temperature difference of 14°C (25°F), his turbine generator produced 22 kilowatts at an overall efficiency of less than 1%. Two additional experimental plants of 3500 kilowatts were installed off the Ivory Coast in 1956 by the French. Mechanical failures resulted in abandonment of the plants.

In the Claude process, sea water is vaporized and the vapor passed directly to the turbine. Evaporation of water at 80°F in sufficient quantities to operate a turbine requires a vacuum near 0.03 atmospheres. Creation of this vacuum requires pumping large quantities of cold water through the turbine condensor and, as a result, the overall efficiency of the Claude process is low. Attention has, therefore, been directed toward closed Rankine cycle systems in which the heat of the ocean water is transferred to a second working fluid (ammonia, propane, or halogenated hydrocarbons such as Freon). This second fluid is then evaporated and used to drive the turbine. These working fluids afford higher cycle efficiencies and can utilize smaller turbines.

Ocean thermal gradient power plants operate, in any event, at relatively low overall efficiencies. Heat engine efficiency is specified by Carnot's law which states that the maximum theoretical efficiency is defined by the difference between the high and low temperatures available divided by the higher temperature. If, for example, the higher temperature available is 82°F and the lower temperature is 43°F, the ideal efficiency is

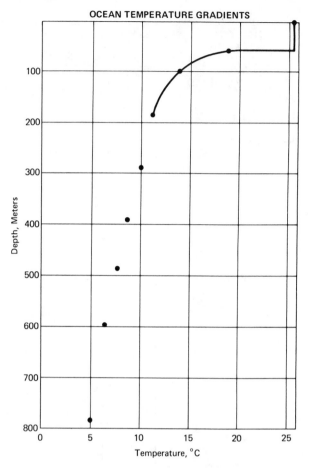

OCEAN TEMPERATURE GRADIENTS

Ocean temperature gradients.

Typical ocean temperature profile as a function of depth. The surface layer, which is 100–150 feet deep, takes its heat from the sun and stays at about 25 °C. The cold underlying water comes from the Arctic region and is usually near 5 °C.

Source: Clarence Zener and Abrahim Lavi, Statement before the Subcommittee on Energy, Committee on Science and Astronautics, House of Representatives, May 23, 1974.

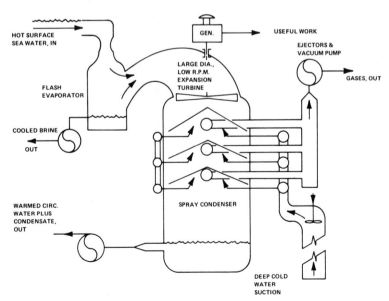

The Claude ocean thermal differences process.

Source: Solar Energy as a National Energy Resource, NSF/NASA Solar Energy Panel, National Science Foundation, December 1972.

7.2%. In contrast, modern fossil fuel steam power plants have efficiencies of about 40%. The significant aspect of ocean thermal gradient plants is, however, the availability of an energy source that is constantly renewed by the sun. Power costs will be a function of plant costs and operating reliability rather than plant efficiency.

Uncertainties with regard to ocean thermal gradient power plants relate to construction costs, station-keeping problems, transmission of power to shore, and systems component life as defined by corrosion or fouling in the marine environment.

Working Fluid Comparison

	Ammonia		Propane		R-12/31	
	Liquid	Vapor	Liquid	Vapor	Liquid	Vapor
Thermal conductivity, 50°-70°F, Btu/hr-ft^2-°F	0.29	0.014	0.07	0.01	0.05	0.006
Heat capacity, Btu/lb-°F	1.13	0.19	0.62	0.24	0.24	0.097
Heat of vaporization, Btu/lb	500		140		70	
Materials compatibility	Wet NH$_3$ not compatible with copper		Excellent, except some plastics		Excellent	
Toxicity	Severe but easily detected		Slight - difficult to detect		Slight - difficult to detect	
Flammability	Moderate		Explosion hazard		Not flammable	
Solubility in water	High		Low		Very low; hydrolysis	
Effect on external environment	Slight		Undesirable local effects		Potentially severe problem	
Problem of contamination as working fluid	Moderate		Negligible		Negligible	
Availability	Good		Good		Potential problem	

Source: Douglass, Robert H; "Ocean Thermal Energy Conversion: An Engineering Evaluation," in *Proceedings, Third Workshop on Ocean Thermal Energy Conversion* (*OTEC*), Houston, Texas, May 8-10, 1975, The Johns Hopkins University, Applied Physics Laboratory, August, 1975.

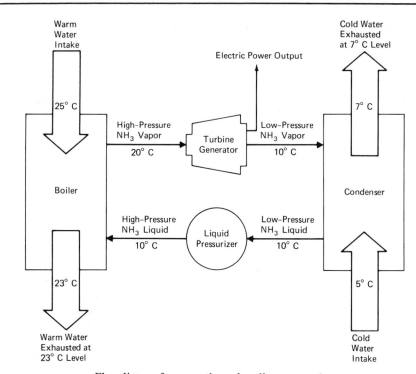

Flow diagram for ocean thermal gradient power plant.

Source: Clarence Zener and Abrahim Lavi, Statement before the Subcommittee on Energy, Committee on Science and Astronautics, House of Representatives, May 23, 1974.

Support barge for the Hughes Glomar Explorer has been designated as a test bed for ocean thermal energy conversion equipment. (*DOE Photo.*)

OCEAN THERMAL GRADIENT POWER PLANTS

Design aspects of a closed Rankine cycle ocean thermal gradient power plant, as conceived by the TRW Systems Group, are shown in the accompanying figures. The 100 megawatt electrical plant is comprised of four 25-MWe power modules with one central cold water supply pipe of fiber-reinforced plastic extending to a depth of 4000 feet. Ammonia is used as the working fluid to drive the turbine generator. The temperature difference available to operate the plant is assumed to be 39.3°F (21.8°C).

Characteristics of Ammonia Vapor Turbine for 25-MWe-Net-Power-Output Module

Shaft power, HP	44,236
Shaft speed, rpm	1,800
Ammonia flow rate, lb/sec	1,933
Overall efficiency	0.896
Inlet total pressure, psia	122.1
Inlet temperature, °F	67
Rotor exit static pressure, psia	87
Inlet duct diameter, in.	64
Diffuser exit diameter, in. (efficiency 75%)	168
Rotor diameter, tip/root, in.	112/73
Rotor speed, tip/root, ft/sec	880/565
Height of assembly (including generator) ft.	23
Turbine assembly weight, lb	50,000
Generator (4 pole, 1800 rpm), lb	250,000
Gross power output, MW_e	32

Source: Douglass, Robert H; "Ocean Thermal Energy Conversion: An Engineering Evaluation," in Proceedings, *Third Workship on Ocean Thermal Energy Conversion*; (OTEC), Houston, Texas, May 8–10, 1975, The Johns Hopkins University, Applied Physics Laboratory, August, 1975.

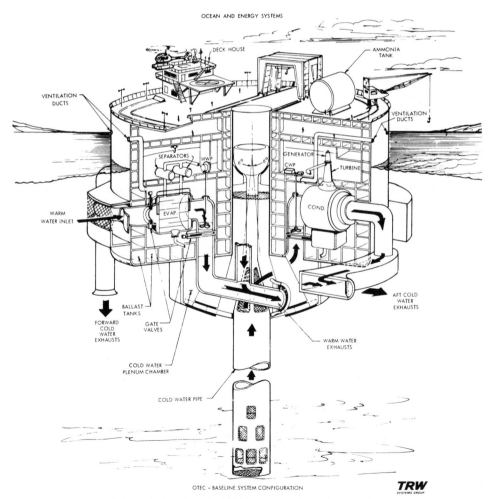

Schematic drawing of a 100 MWe ocean thermal power plant.

Source: TRW Systems Group.

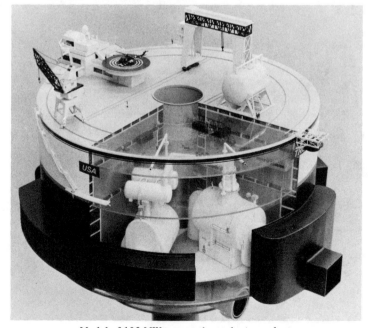

Model of 100 MWe ocean thermal power plant.

Source: TRW Group.

SOLAR BIOMASS ENERGY

The energy from the sun which results in the photosynthetic production of organic material (biomass) can be utilized in two ways. Organic material can be grown specifically as a source of heat energy or the waste organic material derived from the organic materials used in society can be collected and either burned directly or converted to various fuels. Some of these processes are discussed in the pages following.

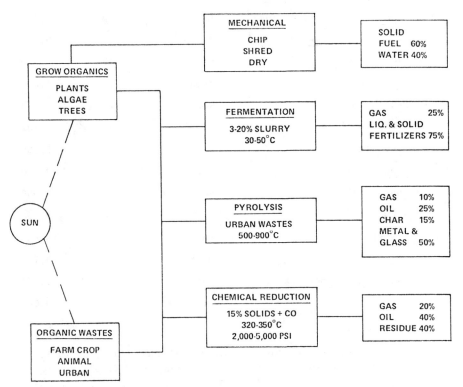

Production of fuels from solar energy.

Source: Solar Energy as a National Energy Resource, NSF/NASA Solar Energy Panel, National Science Foundation, December 1972.

Summary of Organic Material Conversion Processes

PROCESS	FEEDSTOCK	CONVERSION TEMPERATURE	CONVERSION PRESSURE	CHARACTERISTICS OF PROCESS	PRODUCT FORM	PROCESS YIELD (% OF ORIGINAL MASS)	PRODUCT HEATING VALUE	PERCENT OF FEEDSTOCK HEATING VALUE RECOVERED (ASSUMING 8000 BTU/LB.)
Fermentation	Aqueous slurry (3-20% solids)	20° to 50° C	Atmospheric	Fermentation of wastes or algae grown on wastes of energy crops	50 to 70% Methane Remainder CO_2	20 to 26%	Unpurified 600 BTU/SCF Purified 1000 BTU/SCF	60 to 80%
Biophotolysis	Aqueous slurry for algae, bacteria and/or protein-enzyme complexes	20° to 50° C	Atmospheric	Sunlight produces intracellular enzymatic reduction of H_2O ($2H + $ to H_2)	Hydrogen		330 BTU/SCF	10% (Maximum) of total insolation
Acid Hydrolysis	5% acidified slurry (H_2SO_4 with cellulose)	20° to 50° C	Atmospheric	Glucose fermented to ethyl alcohol. Cellulose hydrolized to glucose.	Ethyl alcohol			
Enzyme Hydrolysis	Aqueous slurry (cellulose-rich)	20° to 50° C	Atmospheric	Extracellular enzymatic conversion of cellulose to sugar to alcohol	Ethyl alcohol	90%		
Combustion	Dried feedstock (10% to 25% H_2O)	1200° to 1300° C	Atmospheric	Augments (i.e., 5 to 20%) boiler fuel (i.e., coal, oil or gas).	Heat Steam Can be converted to electricity		3150 BTU/lb.	Heat, 90% Steam, 90% Electricity, 40%
Pyrolysis	Dried feedstock	500° to 1200°	Atmospheric	All of the gas and 1/3 of the char produced is used to supply heat in typical process. Oxygen-free environment used.	Oil Char Gas	40% 20%	12,000 BTU/lb. 9,000 BTU/lb. 770 BTU/SCF	82% (60% for oil only)
Chemical Reduction	Aqueous slurry (15% solids)	250° - 400°C	1500-5000 psi	Uses CO & H_2 3/8 of product oil used by process	Oil	23% (2 barrels/ton)	15,000 BTU/lb.	37% (corrected for CO use) 65% (anticipated, not counting oil used by process)
Hydro-Gasification	Animal manure (other wastes can also be used)	550°C	1500 psi	Hydrogen atmosphere produced from manure. Purification and methanation of product gas required.	C_2H_6 (12%) CH_4 (42%) CO_2 (37%) or CH_4/C_2H_6	40%	1,000 BTU/SCF	60%
Catalytic Gasification	Dried feedstock, mixed with alkali carbonate catalyst. (12-25% by wgt.)	650° - 750°C	30 to 260 psi	Nickel catalysts used for second conversion step. Inert atmosphere required.	CO_2 CH_4 or CH_4, only	90%	500 BTU/SCF 1,000 BTU/SCF	50% (dried feedstock) or 78% (charred animal waste)

Source: Solar Energy Task Force Report, Project Independence Blueprint, Federal Energy Administration, November 1974.

ENERGY FROM CULTIVATED PLANTS

One pound of dry plant material will yield about 7500 Btu when burned directly. When converted to natural gas, the same pound of material will yield about 5 cubic feet of methane. To satisfy the current United States natural gas consumption of 22 trillion cubic feet per year, would require about 2.2 billion tons of dry plant biomass. Assuming a productivity of 30 tons/acre-year, some 730 million acres of land or 1.1 million square miles, would be required. (This is equivalent to the present area under cultivation in the United States.)

As indicated in the accompanying table, the biomass yields under normal conditions are considerably less than 30 tons/acre-year for almost all plants. Achievement of yields of 30 tons/acre-year in an energy plantation is predicated on the use of intensive agricultural techniques.

In terms of the production of electrical energy, a 100 MWe electric power station operating at an annual load factor of 80% would require a land area of 245 square miles to provide a biomass at a rate of 15,070 tons/day, assuming a productivity of 30 tons/acre-year.

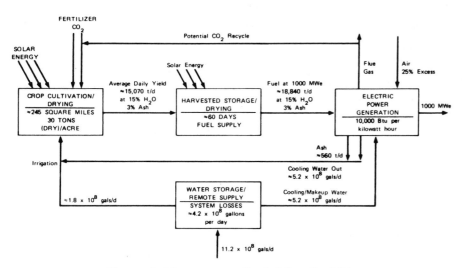

Biomass electric power generation logistics– direct firing.

Source: Effective Utilization of Solar Energy to Produce Clean Fuel, Stanford Research Institute, PB-233 956, prepared for the National Science Foundation, June 1974.

Energy Contact (Btu/pound) of Plant Biomass

Plant	Plant Part	Percent H_2O	Btu/Lb
Sugarcane (<u>Saccharum officinarum</u>)[*]	Bagasse	12	7,281
Sugarcane (<u>Saccharum officinarum</u>)[†]	Bagasse	52	4,000
Bamboo (<u>Phyllostachys</u> spp.)[*]	Cane	10.5	7,398
Buckwheat (<u>Fagopyrum esculentum</u>)[*]	Hulls	10	7,594
Chamise (<u>Adenostoma fasiculatum</u>)[‡]	Leaves	0	9,750
Chamise (<u>Adenostoma fasiculatum</u>)[‡]	Stems	0	9,450
Coconut (<u>Cocos nucifera</u>)[*]	Shells	13	7,560
Beech (<u>Fagus sp.</u>)[*]	Wood	13	7,506
Birch (<u>Betula sp.</u>)[*]	Wood	12	7,578
Oak (<u>Quercus sp.</u>)[*]	Wood	13	7,182
Oak (<u>Quercus sp.</u>)[*]	Bark	7	8,139
Oak (<u>Quercus sp.</u>)[†]	Bark	0	8,370
Pine (<u>Pinus sp.</u>)[*]	Wood	12	7,956
Pine (<u>Pinus sp.</u>)[†]	Bark	0	9,030
Fir (<u>Abies sp.</u>)[†]	Bark	0	8,810
Spruce (<u>Picea sp.</u>)[†]	Bark	0	8,740
Redwood (<u>Sequoia sempervirens</u>)[†]	Bark	0	8,350
Oilseed crop[§]	Seed	--	9,000

[*] <u>Handbook of Chemistry and Physics</u>, C. D. Hodgeman (Ed.), p. 1945. The Chemical Rubber Publishing Co., Cleveland, Ohio (1962).

[†] <u>Steam: Its Generation and Use</u>, Babcock and Wilcox Co., New York, N.Y. (1963).

[‡] R. C. Rothermel and C. W. Philpot, 1973. Predicting changes in chaparral flammability. <u>Journal Forestry</u>, Vol. 71, No. 10 (October 1973).

[§] R. S. Loomis, et al., 1971. Agricultural productivity. <u>Ann. Rev. Plant Physiol.</u> 22:431–468.

Source: Effective Utilization of Solar Energy to Produce Clean Fuel, PB-233956, Stanford Research Institute, prepared for the National Science Foundation, June 1974.

Aboveground, Dry Biomass Yields of Selected Plant Species or Complexes

Species	Location	Yield (ton/acre-year)
Annuals		
Sunflower & Jerusalem artichoke	Russia	13.5
Sunflower hybrids (seeds only)	California	1.5
Exotic forage sorghum	Puerto Rico	30.6
Forage sorghum (irrigated)	New Mexico	7–10
Forage sorghum (irrigated)	Kansas	12
Sweet sorghum	Mississippi	7.5–9
Exotic corn (137-day season)	North Carolina	7.5
Silage corn	Georgia	6–7
Hybrid corn	Mississippi	6
Kenaf	Florida	20
Kenaf	Georgia	8
Perennials		
Water hyacinth	Florida	16
Sugarcane	Mississippi	20
Sugarcane (state average)	Florida	17.5
Sugarcane (best case)	Texas (south)	50
Sugarcane (10-year average)	Hawaii	26
Sugarcane (5-year average)	Louisiana	12.5
Sugarcane (5-year average)	Puerto Rico	15.3
Sugarcane (6-year average)	Philippines	12.1
Sugarcane (experimental)	California	32
Sugarcane (experimental)	California	30.5
Sodangrass	California	15–16
Alfalfa (surface irrigated)	New Mexico	6.5
Alfalfa	New Mexico	8
Bamboo	South East Asia	5
Bamboo (4-year s??nd)	Alabama	7
Abies sacharinensis (dominant species) and other species	Japan	6
Cinnamomum camphora (dominant species) and other species	Japan	6.8
Fagus sylvatica	Switzerland	4.3
Larix decidua	Switzerland	2.2

Aboveground, Dry Biomass Yields of Selected Plant Species or Complexes (*Cont.*)

Species	Location	Yield (ton/acre-year)
Picea abies (dominant species) and other species	Japan	5.5
Picea omorika (dominant species) and other species		6.4
Picea densiflora (dominant species) and other species	Japan	6.1
Castanopsis japonica (dominant species) and other species	Japan	8.3
Betula maximowicziniana (dominant species) and other species	Japan	3
Populus davidiana (dominant species) and other species	Japan	5.5
Hybrid poplar (short-rotation)		
Seedling crop (1 year old)	Pennsylvania	4
Stubble crop (1 year old)	Pennsylvania	8
Stubble crop (2 years old)	Pennsylvania	8
Stubble crop (3 years old)	Pennsylvania	8.7
American sycamore (short rotation)		
Seedlings (2 years old)	Georgia	2.2
Seedlings (2 years old)	Georgia	4.1
Coppice crop (2 years old)	Georgia	3.7
Black cottonwood (2 years old)	Washington	4.5
Red alder (1–14 years old)	Washington	10
Eastern cottonwood (8 years old)		3
Eucalyptus sp.	California	13.4
Eucalyptus sp.	California	24.1
Eucalyptus sp.	Spain	8.9
Eucalyptus sp.	India	17.4
Eucalyptus sp.	Ethiopia	21.4
Eucalyptus sp.	Kenya	8.7
Eucalyptus sp.	South Africa	12.5
Eucalyptus sp.	Portugal	17.9
Miscellaneous		
Algae (fresh-water pond culture)	California	8–39
Tropical rainforest complex (average)		18.3
Subtropical deciduous forest complex (average)		10.9
Puckerbrush complexes (average)	North Carolina	2.2
Puckerbrush complexes (average)	Maine	4.4
World's oceans (primary productivity)		6

Source: Effective Utilization of Solar Energy to Produce Clean Fuel, PB-233956, Stanford Research Institute, prepared for the National Science Foundation, June 1974.

ENERGY FROM CEREAL GRAINS

One of the options for the production of fuels from solar energy is to use idle land for the production of cereal grain and convert the carbohydrates in the grain to ethyl alcohol by the process of fermentation. One hundred million of the unused acres in the United States would produce about 18 billion gallons of alcohol assuming an average yield of 70 bushels of grain per acre. This alcohol production represents about 20% of the approximately 90 billion gallons of motor fuel consumed in the United States in 1971. (The energy equivalent of ethyl alcohol is about 80,000 Btus per gallon while for gasoline the energy equivalent is about 135,000 Btus per gallon.)

$$(C_6H_{10}O_5)_n + nH_2O = nC_6H_{12}O_6$$
Starch .... Dextrose

$$C_{12}H_{22}O_{11} + H_2O \longrightarrow 2\ C_6H_{12}O_6$$
Maltose .... Dextrose

$$C_6H_{12}O_6 \longrightarrow 2\ C_2H_5OH + 2\ CO_2$$
Dextrose .... Ethyl Alcohol Carbon Dioxide

Grain	Starch	Protein	Oil	Fiber	Other Constituents†
Hard Wheat	64	14	2	2	18
Soft Wheat	69	10	2	2	17
Dent Corn	72	10	5	2	11
Sorghum	71	13	3	2	11

* Moisture-free basis.

† Minerals, sugars, pentosans and vitamins.

The theoretical yield from the conversion of starch and sugar into ethyl alcohol is 0.568 pounds of alcohol per pound of starch. In actual practice, yields generally are about 90 to 95% of theoretical.

Source: U.S. Energy Outlook: New Energy Forms, National Petroleum Council, Washington, D.C., 1973.

Fermentative Conversion Cost of 190° and 200° Proof Ethyl Alcohol from Corn
(Exclusive of Cost of Corn)

	Cost/Gallon* (Cents)
190° Proof, Alcohol (2.82 Gallons/Bushel)	
Base Conversion Cost	33.2
Depreciation ($1.45 million/year, 10 years, 17.7 million gallons)	8.2
Total	41.4
By-Product Feed Credit (6.8 lb/gal. alc. at $60/ton)	34.0
Net	7.4
200° Proof, Alcohol (2.7 Gallons/Bushel)	
Alcohol (1.048 gal. at 7.4¢)	7.8
Cost of Dehydration	2.4
Total Cost, Exclusive of Corn, Profit, Packaging and Sales Expenses	10.2

* Costs are in constant 1970 dollars.

Source: Dwight L. Miller, "Corn and Its Uses," National Corn Growers Association, April 5, 1972; updated to May 1973.

Effect of Corn Cost on Ethyl Alcohol Cost
(Basis: 2.7 Gal. 200° Proof Alcohol/Bushel)

Corn Price/Bushel (Dollars)	Alcohol Cost/Gallon, (Cents)		
	Corn	Conversion	Total Base Cost*
1.00	37.0	10.2	47.2
1.25	46.3	10.2	56.5
1.50	55.5	10.2	65.7
1.75	64.8	10.2	75.0
2.00	74.0	10.2	84.2

* These costs do not include profits, packaging and sales expenses.

Source: Dwight L. Miller, "Corn and Its Uses," National Corn Growers Association, April 5, 1972; updated to May 1973.

Source: U.S. Energy Outlook: New Energy Forms, National Petroleum Council, Washington, D.C., 1973.

ENERGY FROM ORGANIC WASTES

A number of processes have been developed for the recovery of the energy in organic wastes. Three processes are illustrated in the accompanying diagrams.

Pyrolysis. Organic materials can be converted into gases, liquids, and solids through pyrolysis at temperatures of 500–900°C by heating in a closed vessel in the absence of oxygen. The pyrolytic destructive distillation of wood has long been used to recover methanol, acetic acid, turpentine, and charcoal. This process can handle all forms of organic materials, including rubber and plastics which are difficult to handle by other processes. The gases produced are a mixture of nitrogen, methane, carbon monoxide, carbon dioxide, and other hydrocarbons. The liquids produced are oil-like materials and the solids are similar to charcoal.

Chemical Reduction. The conversion of organic waste materials into oil can be accomplished by treatment of the waste at temperatures of 300–350°C and pressures of 2000–

Composition of Municipal Trash

	By Weight (Percent)
Paper, paperboard	50
Iron and steel	9
Aluminum	1
Glass, ceramics, rocks	10
Garbage, yard wastes	20
Plastics, textiles, misc.	10

Source: U.S. Energy Outlook: New Energy Forms, National Petroleum Council, Washington, D.C., 1973.

4000 pounds per square inch in the presence of water, carbon monoxide, and catalysts. The process is continuous with the organic material remaining in the reaction vessel for 1 to 2 hours. About 40% of the organic material is converted to oil. The product yield is about 2 barrels of oil per ton of dry waste material. The oil has a heating value of about 15,000 Btu/pound.

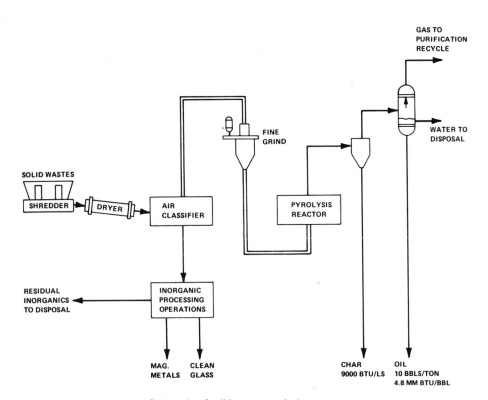

Schematic of solid waste pyrolysis process.

Source: Solar Energy as a National Energy Resource, NSF/NASA Solar Energy Panel, National Science Foundation, December 1972.

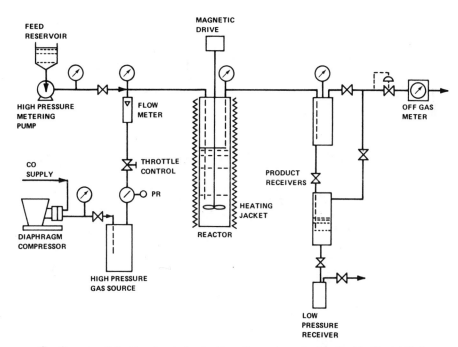

Continuous unit for the chemical reduction of organic wastes to oil with CO and H_2O.

Source: Solar Energy as a National Energy Resource, NSF/NASA Solar Energy Panel, National Science Foundation, December 1972.

Anaerobic Fermentation. In the presence of moisture and the absence of oxygen, most organic materials will undergo natural fermentation in which 60–80% of the carbon in the organic material is converted to a mixture of carbon dioxide, methane, traces of hydrogen sulfide, and nitrogen. The conversion process takes place at temperatures of 15–50°C and near atmospheric pressure. The volatile gases produced have a heating value of between 500–700 Btu/ft^3. The carbon dioxide and hydrogen sulfide can be removed, providing a gas that is essentially pure methane and which has a heating value of 1000 Btu/ft^3 corresponding to that of natural gas.

In general, one pound of dry organic material will produce 4.5–6.5 ft^3 of methane (at standard temperature and pressure) or 9–13 million Btu per ton.

Anaerobic fermentation processes have been used for many years to reduce the volume and weight of domestic sewage solids by conversion to gases and liquids. Generally, the production of combustible gas has been of secondary interest, although some of the larger sewage treatment facilities have used the gas to supply their own power requirements.

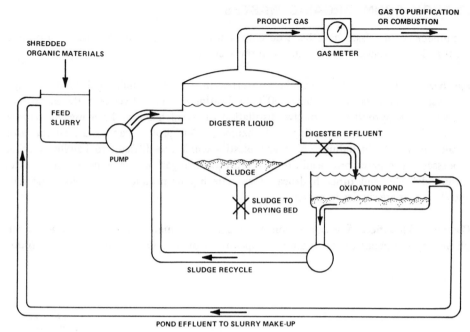

Continuous unit for converting organic material to methane by anaerobic fermentation.

Source: Solar Energy as a National Energy Resource, NSF/NASA Solar Energy Panel, National Science Foundation, December 1972.

U.S. WASTE ENERGY RECOVERY PROJECTS

Projected Implementations of Energy Recovery Systems by 1980

Location	Tons per day	Description	Location	Tons per day	Description
California San Diego County	200	Pyrolysis; EPA is sponsoring project to demonstrate the Garrett Research and Development system; oil produced will be accepted by San Diego Gas and Electric; project in engineering design phase.	**Missouri** St. Louis	8,000	Solid waste as fuel; Union Electric Company plans to implement, by mid-1977, a system to handle the residential, commercial and selected industrial waste from the entire metropolitan area; Union Electric will process raw waste, recover magnetic metal, aluminum, and glass as well as fuel.
Connecticut Bridgeport	1,200	Solid waste as fuel; state-wide resource recovery authority is reviewing proposals, Northeast Utilities will accept the fuel.	**New Jersey** Essex County	1,000	Solid waste as fuel; request for proposals being prepared; supplemental fuel to be accepted by Public Service Gas and Electric or other industrial steam boilers.
District of Columbia	1,000	Solid waste as fuel; D.C., Fairfax County, Arlington County, the City of Alexandria, and the Metropolitan Washington Council of Governments are studying the feasibility of a supplemental fuel system on a region-wide basis. Virginia Electric Power Company and Potomac Electric Power Company are cooperating in the studies.	Hackensack-Meadowlands	2,000	Solid waste as fuel; detailed proposals are currently being reviewed; it is anticipated that the fuel will be accepted by Public Service Gas and Electric or industrial steam boilers.
Illinois Chicago	2,000	Solid waste as fuel; construction started in early March, Commonwealth Edison will accept the fuel.	Union County-Middlesex County	1,000	Solid waste as fuel; feasibility of producing a supplemental fuel for Public Service Gas and Electric is being assessed.
Chicago area excluding the City	1,000	Solid waste as fuel; several suburbs have approached Commonwealth Edison to determine the feasibility of implementing supplemental fuel systems	**New York** Albany area	500	Solid waste as fuel; feasibility of producing supplemental fuel for industrial steam boilers, state-owned heating plant and municipal electric utility is being assessed.
Iowa Ames	200	Solid waste as fuel; construction to begin by June 1974; municipal electric utility will accept the fuel.	Hempstead	1,000	Detailed proposals have been received for design and construction of energy and materials recovery systems.
Maryland Baltimore	1,000	Pyrolysis; EPA is sponsoring project to demonstrate the Monsanto system; pyrolysis gas will be combusted on-site to generate steam for sale to Baltimore Gas and Electric; plant will be operational in early 1975.	Monroe County	500	Solid waste as fuel; feasibility study to produce a supplemental fuel for Rochester Gas and Electric completed; request for proposals being prepared.
Montgomery County	1,200	Solid waste as fuel; County is planning project with Potomac Electric Power Company cooperation; feasibility study has been completed; County Council and County Executive have approved the plan.	New York City	2,000	Solid waste as fuel; City has completed feasibility study of using waste as supplemental fuel in Consolidated Edison's boilers; City writing request for proposals to design and construct supplemental fuel facility; City and Consolidated Edison plan contract to determine feasibility of designing new steam-electric boiler to burn 50 percent solid waste.
Massachusetts Braintree	240	Water wall incineration; plant has been operating since 1972; contract signed early 1974 for sale of steam to Weymouth Art Leather Co.	Westchester County	1,500	Feasibility study completed; County most interested in energy recovery for County-owned industrial park.
East Bridgewater	1,200	Solid waste as fuel; privately financed processing facility; Weyerhauser is accepting the fuel for its industrial steam boilers.	**Ohio** Akron	1,000	Water wall incineration; detailed engineering study is underway; steam product will be used for downtown heat and air conditioning and for B.F. Goodrich process steam.
Saugus (near Boston)	1,200	Water wall incineration; plant under construction; steam product will be sold to General Electric Co. for process steam.	Cleveland	500	City has received bids for a steam generation system; the super-heated steam will be used for electric generation by the municipal utility.
Lawrence	1,000	Solid waste as fuel; Lawrence will be the first implementation under the statewide solid waste master plan approved in early 1974; master plan calls for supplemental fuel production for steam and steam-electric boilers, and materials recovery.			

**Projected Implementations of Energy Recovery Systems
by 1980** (*Continued*)

Location	Tons per day	Description
Oregon Lane County	700	Solid waste as fuel; feasibility study completed to use waste as supplemental fuel in a Eugene municipal steam power plant that currently burns wood waste; additional waste fuel is required because wood wastes are becoming scarce.
Pennsylvania Philadelphia	2,400	Solid waste as fuel; Combustion Equipment Associates has announced plans to construct and operate, with private financing, a facility to produce supplemental fuel for industrial steam boilers.
Puerto Rico San Juan	1,000	San Juan planning to initiate feasibility study for a solid waste as fuel system; supplemental fuel would be used by Commonwealth-owned San Juan steam-electric station.
Tennessee Knoxville	500	Pyrolysis; TVA is studying feasibility of implementing a Torrax gas pyrolysis system to produce gas as supplemental fuel for TVA steam-electric boiler.
Memphis	500	Solid waste as fuel; detailed proposals have been requested to implement a wet processing system to produce supplemental fuel for a TVA steam-electric boiler.
Nashville	750	Water wall incineration; construction is complete; public authority has been formed to construct and operate the facility; steam product will be used for downtown heating and air conditioning.

Source: Environmental Protection Agency. April 1974.

Municipal waste is used as fuel in this power plant which provides electricity for a manufacturing plant. (*DOE Photo.*)

Energy Conversion and Storage

The Old Red Mill, © 1907

ENERGY CONVERSION

All dynamic processes, from the changes of the seasons to the growth and death of living things, involve the conversion of energy from one form to another. Radiant energy from the sun is converted to the mechanical energy of the wind and waves and, through photosynthesis, into the latent stored energy in plants. This stored energy in plants provides food for animals who convert it into mechanical energy to move muscles and heat energy to keep warm. Over the years, the stored energy in plants has been accumulated in deposits of coal, oil and gas, forming the concentrated energy resources which are essential for the operation of the wide variety of devices typical of modern industrial society.

These energy conversion devices are used to convert the stored chemical energy in fossil fuels into three forms of energy particularly useful to man—mechanical energy to operate machines, heat energy to process materials and to heat living spaces, and radiant energy to provide light. Heat energy is derived directly by burning fossil fuels; mechanical energy is derived from heat energy in engines of various types; and radiant energy is provided by the luminosity of the flames of burning fuels. In addition, another form of energy, electrical energy, may be generated from mechanical energy and used to transfer energy over distances to be converted again to mechanical, heat, or radiant energy. Electrical energy also has a unique capability for transferring information over distances by converting radiant light energy and mechanical sound energy into electrical impulses that can be reconstituted in their original energy forms. Electrical energy is, therefore, a particularly useful intermediate form of energy.

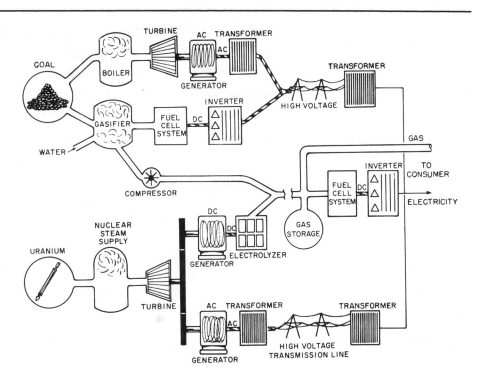

Coal and uranium energy systems.

Source: Proposed Final Environmental Statement: Liquid Metal Fast Breeder Reactor Program, Volume III, WASH-1535, U.S. Atomic Energy Commission, December 1974.

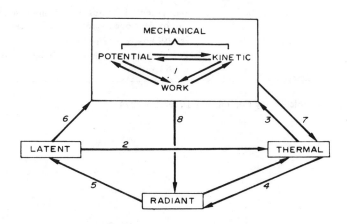

1. Pendulum, flywheel, electrical generator
2. Fire, nuclear pile, respiration
3. Heat engine: power plant, gasoline engine
4. Radiation behaviour
5. Photosynthesis
6. Muscle, battery
7. Friction, electric heater, brake
8. Electric light

Conversions among five basic manifestations of energy.

Source: Luten, Daniel B, "United States Requirements," in *Energy, The Environment, and Human Health*, Proceedings of a Congress on Environmental Health, © American Medical Association, Publishing Sciences Group, Inc., Acton, Massachusetts, 1974.

Energy Conversion Device Summary

Device	Availability	System Efficiency	Application	Remarks
Steam turbines	commercially available	35–41%	Predominantly fossil and nuclear central station-base load.	High reliability systems at modest efficiency. Present efficiency cannot be significantly improved for direct cycle.
Internal combustion engines	commercially available	35%	Small central station and peaking.	Fossil fuel only. Application limited to small plants except for emergency and peaking purposes at large stations.
Gas turbines	commercially available and ASUD[a]	27–38%	Direct cycle for peaking. Topping cycle for fossil, central station.	High temperature turbines under development for better cycle efficiency (up to 45%). Good application to gas-cooled nuclear plants.
Binary cycles	ASUD	Up to 55%	Potassium topping cycle for fossil central station.	Basic technology available. Development of large plants within 5 years possible. Mercury binary might have application to LMFBR.
Fuel cells	ASUD	Up to 70%	Central station or dispersed generation—energy storage features.	Applicable to fossil and nuclear central station power plant and hydrogen economy. Fossil central station may take 5–10 years to develop.
Batteries	ASUD	60% (turnaround)	Energy storage for load leveling and reliability.	Has potential for improving electric power systems. Prospects for development within 5 years are good.
Magnetohydrodynamics	ASUD	Up to 60%	Direct or topping cycle for central station.	Potential for improving efficiency of central station power plants but development may require 10 years or more.
Thermionic devices	ASUD	Up to 50%	Topping cycle on fossil or nuclear central station.	Has potential for fossil application if proper devices can be developed—probably 5–10 years. Little potential for nuclear application.
Thermoelectric devices	suitable materials being sought	6%	Low power only.	Poor efficiency makes unsuitable for central station consideration.

[a] ASUD—Advanced systems under development.
Source: Proposed Final Environmental Statement: Liquid Metal Fast Breeder Reactor Program, Volume III, WASH-1535, U.S. Atomic Energy Commission, December 1974.

ENERGY CONVERSION SYSTEMS

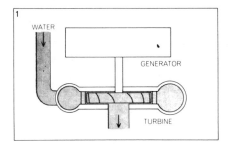

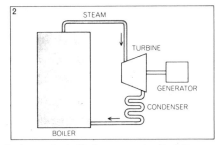

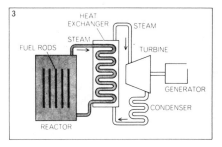

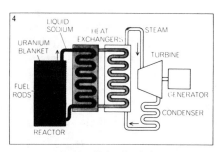

ELECTRIC-POWER GENERATING MACHINERY now in use extracts energy from falling water, fossil fuels or nuclear fuels. The hydroturbine generator (1) converts potential and kinetic energy

into electric power. In a fossil-fuel steam power plant (2) a boiler produces steam; the steam turns a turbine; the turbine turns an electric generator. In a nuclear power plant (3) the fission of ura-

nium 235 releases the energy to make steam, which then goes through the same cycle as in a fossil-fuel power plant. Under development are nuclear breeder reactors (4) in which surplus

neutrons are captured by a blanket of nonfissile atoms of uranium 238 or thorium 232, which are transformed into fissile plutonium 239 or U-233. The heat of the reactor is removed by liquid sodium.

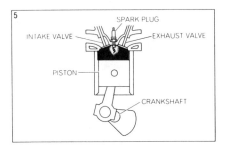

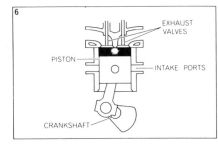

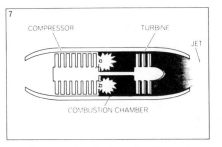

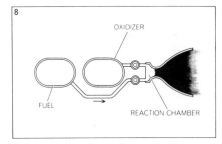

PROPULSION MACHINERY converts the energy in liquid fuels into forms of mechanical or kinetic energy useful for work and transportation. In the piston engine (5) a compressed charge of

fuel and air is exploded by a spark; the expanding gases push against the piston, which is connected to a crankshaft. In a diesel engine (6) the compression alone is sufficient to ignite the charge

of fuel and air. In an aircraft gas turbine (7) the continuous expansion of hot gas from the combustion chamber passes through a turbine that turns a multistage air compressor. Hot gases leaving the

turbine provide the kinetic energy for propulsion. A liquid-fuel rocket (8) carries an oxidizer in addition to fuel so that it is independent of an air supply. Rocket exhaust carries kinetic energy.

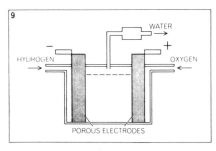

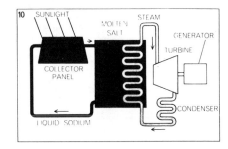

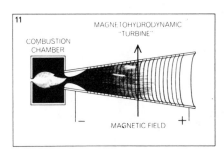

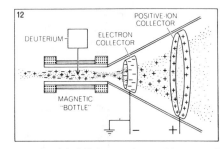

NOVEL ENERGY CONVERTERS are being designed to exploit a variety of energy sources. The fuel cell (9) converts the energy in hydrogen or liquid fuels directly into electricity. The "combustion" of the fuel takes place inside porous electrodes. In a recently proposed solar power plant (10) sunlight falls on specially coated

collectors and raises the temperature of a liquid metal to 1,000 degrees F. A heat exchanger transfers the heat so collected to steam, which then turns a turbogenerator as in a conventional power plant. A salt reservoir holds enough heat to keep generating steam during the night and when the sun is hidden by clouds. In a mag-

netohydrodynamic "turbine" (11) the energy contained in a hot electrically conducting gas is converted directly into electric power. A small amount of "seed" material, such as potassium carbonate, must be injected into the flame to make the hot gas a good conductor. Electricity is generated when the electrically charged particles of

gas cut through the field of an external magnet. A long-range goal is a thermonuclear reactor (12) in which the nuclei of light elements fuse into heavier elements with the release of energy. High-velocity charged particles produced by a thermonuclear reaction might be trapped in such a way as to generate electricity directly.

Source: Claude M. Summers, "The Conversion of Energy," *Scientific American*, September 1971.

TURBINES

Not unlike the ordinary water wheel, all turbines consist of blades mounted on a shaft which is caused to rotate by the transfer of kinetic energy from the working fluid—water, steam, or combustion gas—to the blades of the turbine. The energy available in the working fluid is a function of both the mass flow of the fluid and the head or pressure of that fluid. Water turbines operate at high mass flows and low heads; such turbines are therefore large in size and rotate at relatively slow speeds. Steam and combustion turbines operate at high pressures and proportionately low mass flows. These turbines are relatively smaller and rotate at high speeds.

The power that can be developed by a hydroelectric generating plant is a product of the available hydraulic head and the flow rate. The head, or difference in elevation between the water level upstream of the turbines and the level downstream of their discharge, is usually created by the construction of a dam. The dam provides a reservoir of water which can be used to supply a steady flow to the turbines even though river flow changes with the seasons. Water is carried to the turbines by inlet pipes (penstocks). The low rotational speeds and available hydraulic head make it possible to start hydroelectric turbine units quickly and to change power output easily. Hydroelectric plants are, therefore, well adapted for meeting peak loads, for frequency-control and spinning-reserve duty. Overall efficiency of hydraulic turbines is 90 to 95%.

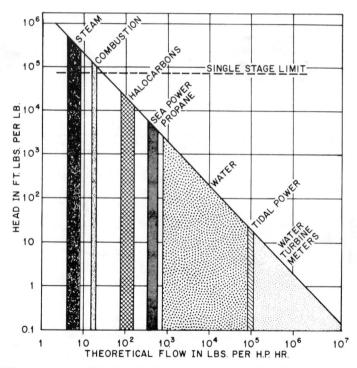

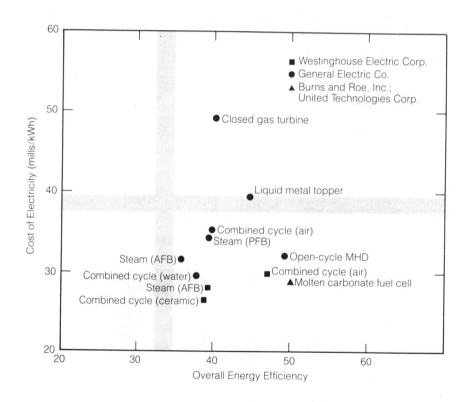

Source: J. Hilbert Anderson, "Turbines for Sea Solar Power Plants," in *Solar Sea Power Plant Conference and Workshop,* June 27-28, 1973, Carnegie-Mellon University, Pittsburgh, Pennsylvania, sponsored by The National Science Foundation.

Relative cost and energy efficiencies of turbine power systems, individually and in combined cycles, with other energy systems including liquid metal topping cycles, atmospheric fluidized bed (AFB), fuel cells, and MHD.

Source: EPRI Journal, April 1977.

TransCanada Pipelines Station 2, Burstall, Saskatchewan. Aft end view of 26,400 bhp Rolls-Royce RB.211 industrial jet engine suspended from small overhead crane just before installation in C-B Coberra 264 gas turbine. In background is C-B RF2BB-30 centrifugal compressor.

Turbine runner at TVA's Fort London, 1948.

A steam turbine is a heat engine—taking heat from a high temperature source, converting part of it into mechanical energy and rejecting the remainder at a low temperature. Typically, heat from an energy source (fossil fuel or nuclear) is transferred to water in a boiler producing high pressure, high temperature steam. The steam is admitted to the turbine where the kinetic energy of the steam molecules is transferred to the turbine blades causing a rotation of the turbine shaft which, in turn, drives an electrical generator. After the steam has expanded to lower pressures and temperatures in passing through the turbine, it is condensed and pumped back to the boiler. The heat removed in the condenser is rejected to the environment. Improvements in the design of steam turbine systems have resulted in a steady increase in system efficiency. In 1903, turbine units were about 5000 kW in size and operated at steam conditions of 175 psi and 375°F with plant efficiencies of 9.2%. Modern steam turbines are rated at 1,000,000 KW and operate at 3500 psi and 1000°F with efficiencies near 39%. Large steam turbine power plants are generally operated as base-load plants, that is, continuous operation at or near full capacity.

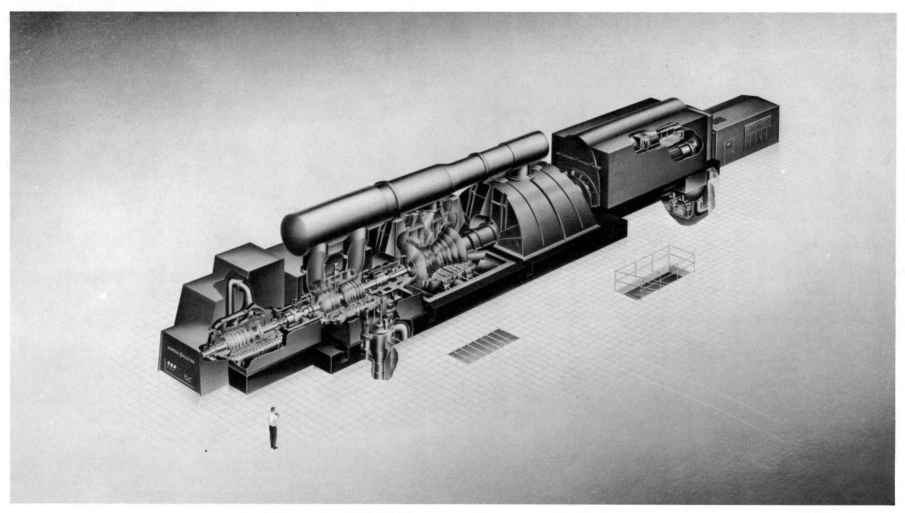

Perspective rendering of TC4F-Fossil, large steam turbine-generator (Code G-3) Detroit Type.

In contrast to the closed Rankine steam cycle used with steam turbines where the working fluid is recycled, gas turbines operate on an open Brayton cycle with fuel and air being admitted at one end of the turbine and the combustion gases ejected at the other. Typically, a gas turbine unit consists of two sections—a forward compressor section in which air is compressed to provide the pressure required for efficient operation, and a turbine section into which the compressed air and fuel are admitted and the fuel is burned to drive the turbine. The compressor and turbine are mounted on the same shaft. Improvements in efficiency have been achieved by raising the turbine inlet temperature and by raising the compression and expansion efficiencies of the compressor and gas turbine. Commercially available simple-cycle gas turbine power plants have an efficiency of about 27%. Continued improvement in efficiency is expected from research efforts on aerodynamic design, metal alloys allowing high turbine inlet temperatures, and improved methods of cooling turbine blades and nozzles. Factory-assembled gas turbine power plants in sizes up 100 MWe are available. Although fuel costs for gas turbine units are high, the ease of installation, short delivery schedules, and quick-start capability have made these units attractive for intermediate and peak-load service. About 8% of the installed generating capacity in the United States is now in gas turbine power plants.

COMBINED-CYCLE POWER PLANTS

Steam turbine systems cannot take full advantage of the high temperatures available from the combustion of fossil fuels because of materials limitations nor of the heat available at the low temperature end of the steam cycle because of economic considerations. The combination of two or more heat engine cycles that cover different parts of the temperature range is referred to as a combined-cycle plant; the combination of two different cycles represents a binary cycle. The addition of a second cycle at the high temperature end of the steam cycle is referred to as a topping cycle; if added at the low temperature end it is a tailing or bottoming cycle.

The most common binary cycle power plants are comprised of a steam turbine cycle topped with a gas turbine cycle. In this combined cycle, the hot exhaust gases from the gas turbines are used to generate steam for the steam turbine generators. Cycle efficiencies near 45% are achieved.

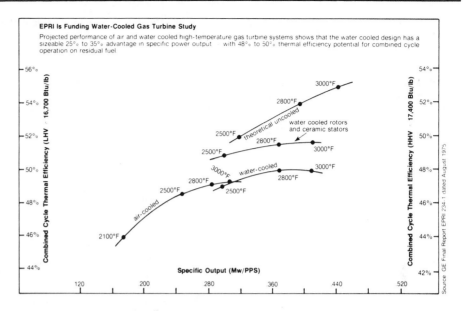

Source: Irwin Stambler, "EPRI Gas Turbine Outlook," Gas Turbine World, November 1975.

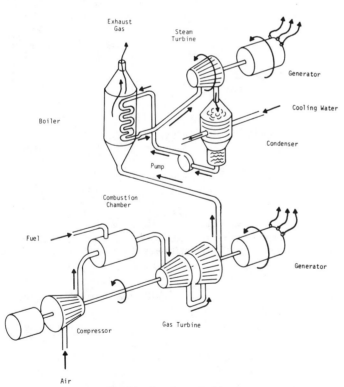

Combined cycle gas turbine.

Source: Proposed Final Environmental Statement: Liquid Metal Fast Breeder Reactor Program, Volume III, WASH-1535, U.S. Atomic Energy Commission, December 1974.

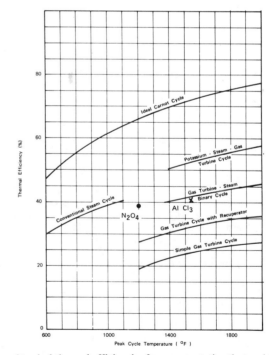

Comparison of typical thermal efficiencies for representative thermodynamic cycles.

Source: Energy Conservation, The Report and Recommendations of the Technical Advisory Committee on Conservation of Energy, National Power Survey, Federal Power Commission, December 1974.

European Combined Cycle Plants in Operation and Under Construction

Plant	User and Location[a]	Type	Fuel[b]		Output MW	Overall Efficiency LCV %	Output to Intake Air Ratio kW/kg	Gas Turbine		In Service
			GT	Steam Generator				Type[c]	Rating MW	
Kornneuburg	NEWAG/Verbundgesellschaft (A)	A/A_3	NG	NG	75	32.6	441	2× B-12/8	2 × 25	1960
Hohe Wand	NEWAG/Niogas (A)	D	NG	C H NG	75	43.7	938	K-VM 51	10.3	1965
Neuchâtel	Neuchâtel (CH)	A/A_1	H	—	26.2	28.9	—	S-N 110 11	19.1	1967
Liège	Socolie (B)	A/A_3	NG	NG	46.3 (34.2)	29.7 (32)	—	S-N 110 11	23.1	1968
Vitry I	EdF (F)	C	NG	NG/CG	322	39.6	1006	B-13 S	42.0	1970
Vitry II		C	NG	NG/CG	322			B-13 S	42.0	
Altbach A Block IV	Neckarwerke (D)	A/A_2	NG/D	NG/H	251	39.9	738	K-V 93	51.0	1971
Kellermann, Lünen	STEAG (D)	E	—	CG	170	36.9	500	K-V 9	74.0	1972
Emden	NWK (D)	C	NG	NG	452	45.5	1329	K-V 93	52.1	1973
Gersteinwerk F	VEW (D)	B	NG	NG/H	417	43.6	1226	K-V 93	52.1	1973
Gersteinwerk G		B	NG	NG/H	417	43.6	1226	K-V 93	52.1	1973
Gersteinwerk H		B	NG	NG/H	417	43.6	1226	K-V 93	52.1	1973
Gersteinwerk J		B	NG	NG/H	417	43.6	1226	K-V 93	52.1	1973
Robert, Frank, Block IV	PREAG (D)	C/C_1	NG	NG	510	—	1417	B-13 C	55.7	1974
Marbach III	EVS (D)	A/A_2	NG	NG	320	—	889	B-13 B	55.4	1974
Emsland B	VEW (D)	B	NG/D	NG/D	417	43.5	1226	K-V 93	51.9	1974
Veltheim, Block IV	GWK (D)		NG/D		374	—	1039	B-13 C	58.7	1974
HKW 3 Duisburg	StwD (D)	C	NG/D	NG/H	172	44.3	860	K-V 82	31.5	1974
Mainzwerk II	KMW AG (D)	C	D	H	338	—	939	B-13 C	67.7	1975
Emsland C	VEW (D)	B	NG/D	NG/D	417	43.5	1226	K-V 93	51.9	1975
Gebersdorf	GFA (D)	C	NG/D	NG/D	403	43.2	1185	K-V 93	51.0	1975
Donge	PNEM (NL)	A/A_1	NG	—	118	44.4	328	B-13 D	75.5	1976
Theiss B	NEWAG (A)	A/A_2	NG	NG/H	310	—	885	SL-GT 120	60.0	1978

[a]Newag, Niederösterreicher Elektrizitätswerke AG; Neuchâtel, Electricité Neuchâteloise SA; EdF, Electricité de France; STEAG, Steinkholen Elektrizitäts AG; NWK, Nordwestdeutsche Kraftwerk AG; PREAG, Preussische Elektrizitäts AG; VEW, Vereinigte Elektrizitätswerke; GWK, Gemeinschafts Kraftwerk Weser GmbH; StwD, Stadtwerke Duisburg; KMWAG, Kraftwerke Mainz Wiesbaden AG; GFA; Grosskraftwerk Franken AG; PNEM, Noordbrabantsche Electriciteits Maatschappij. EVS, Energieversorgung Schwaben.

[b]NG, natural gas; D, distilate oil; H, heavy oil; CG, coal gas.

[c]Kwu, Kraftwerk Union; B. Brown Boveri; S, Sulzer; SL, Stal Laval.

Source: Konrad Goebel, "European approach to combined cycles brings early energy savings," *Energy International*, March 1975.

During the period 1922 to 1950, the General Electric Company constructed a series of six combined-cycle plants employing a mercury topping cycle and a conventional steam cycle. Although these plants performed well, additional plants were not built after 1950 because the efficiency of conventional steam power plants had improved beyond the efficiency of the mercury topping cycle plants.

Binary bottoming cycles employing ammonia or other refrigerants have not been adopted for utility use because cost studies have indicated only marginal gains in plant efficiency and power production costs. The major interest in bottoming cycle technology rests in the application of these cycles to the utilization of the relatively low temperature water available from geothermal sources.

Compact Comanche combined cycle plant producing power.

LESTER, Pa., Aug. 9—Covering less than an acre, this compact combined cycle Comanche Station of Public Service Company of Oklahoma is now producing power, Westinghouse Electric Corporation, the plant's manufacturer, announced here today. The 50-foot-high steel-sided building has a very low profile—taking on the appearance of a small industrial plant instead of a conventional power plant. The two heat recovery steam generators, shown on the left, which produce steam from the exhaust heat of two gas turbines, are less than one-fourth the height of conventional boiler-building structures for comparable steam power plants.

AUTOMOTIVE ENGINE AND FUEL EFFICIENCIES

Several types of combustion engines have been developed for the propulsion of vehicles as well as for stationary use—external combustion engines, internal combustion engines, and gas turbines.

There are two principal external combustion engines—the steam engine and the Stirling engine. The working fluid for both engines operates in a closed Rankine cycle. The steam engine, common for many years on locomotives, generates steam for the power cycle in a boiler externally fired with wood, coal, or oil. The steam is allowed to expand through the drive cylinders to provide propulsive power to the piston rods and drive train. The steam engine is basically a heavy engine with relatively low cycle efficiency.

The Stirling engine uses hydrogen as the working fluid. The hydrogen is heated by an external burner fired with gasoline or diesel fuel and is allowed to expand inside the cylinders of the engine and then is recycled to the burner. The Stirling engine has the advantage of low exhaust emission due to the steady, controlled burning of the fuel, and engine design permits balancing of components such that operation is quiet and vibrationless. On the other hand, the Stirling engine is heavy, requires a large cooling system, and requires expensive nickel-chrome alloys to contain the diffusive high-pressure hydrogen gas. The Stirling engine was invented in 1816 by Robert Stirling, a Scotish inventor, but its development was overshadowed first by the steam engine and later by the internal combustion engine.

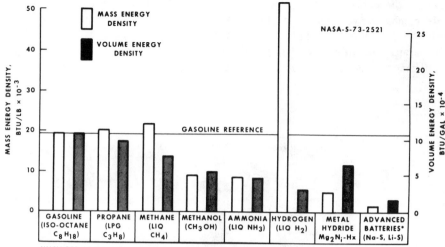

Energy density characteristics of various transportation fuels.

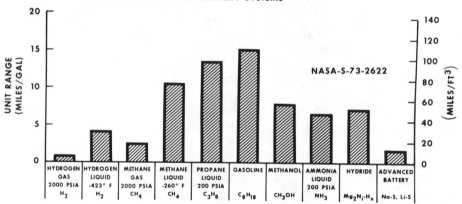

Relative range for selected synthetic fuels.

Source: *A Hydrogenergy Carrier*, Volume II, Systems Analysis, NASA Grant NGT 44-005-114, Systems Design Institute, University of Houston, Johnson Space Center, Rice University, 1973.

The Weight of Various Automobile Energy Storage Methods

	Pounds per 100 miles traveled at 30 h.p.-hr.
Gasoline (Otto cycle)	44
Thermal energy (lithium fluoride)	500-1,000
Flywheel (steel)	3,000
Rubber spring	4,000
Hydraulic	15,000
Batteries (lead-acid) (sodium-sulfur)	4,000 400

Chemical Energy Storage for the Automobile (Equivalents to 20 gallons of gasoline)

	Weight of fuel and container (lbs.)	Volume of fuel (cu. ft.)
Gasoline	142	2.67
Methanol	308	5.4
Liquid hydrogen	182	9.4
Hydrogen from:		
Ammonia	424	6.5
Solid magnesium hydride	661	6.6
Liquid C_6H_{12}	729	12.4
Compressed hydrogen at 3,000 p.s.i.	5,000	46

Gasoline, or other such liquid hydrocarbons, remains the automobile energy storage method of choice, when compared with any other mechanical, electrical, or chemical means. Besides the obvious weight and volume advantages shown here, gasoline is relatively benign in terms of its principal combustion products, carbon dioxide and water.

Source: Kummer, Joseph T. "The Automobile as an Energy Converter," *Technology Review*, edited at the Massachusetts Institute of Technology, February 1975.

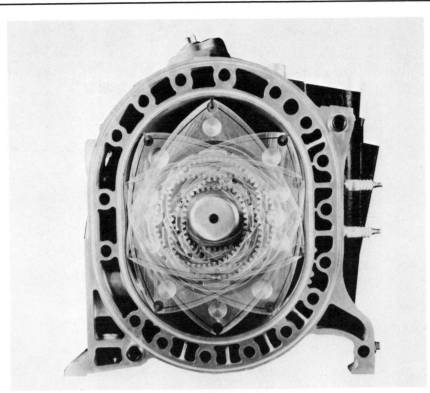

50,000 Mile/three-year rotary engine warranty is one of many standard features on all new Mazda rotary powered cars and trucks currently on sale. The Mazda rotary engine, shown in cutaway above, has far fewer internal moving parts than even a conventional four-cylinder piston engine. In a Mazda rotary power-plant, two revolving rotors perform all the functions that require four, six or eight pistons and a complicated valve train in conventional engines. Nearly three quarters of a million rotary engine Mazda vehicles are on the road today throughout the world.

The internal combustion engine derives propulsive power through the ignition of a fuel–air mixture within a cylinder to generate combustion gases at high pressure that move a piston attached to a drive shaft by a connecting rod. Ignition of the fuel–air mixture is achieved in two different ways: in a Diesel engine the mixture is ignited by compression while in the ordinary gasoline engine the mixture is ignited by an electric spark. In order to achieve ignition in a Diesel engine, the compression ratio must be high—from 14 to 22. In contrast, the compression ratio in the gasoline spark-ignition engine is about 8.5. Because of the high compression ratio, Diesel engines are of much heavier construction than gasoline engines.

A spark-ignition internal combustion engine of design different to the common piston engine, is represented by the Wankel engine. The Wankel engine uses a triangular rotor operating inside an oval-shaped chamber which provides the four stages—compression, ignition, power, and exhaust—of the Otto cycle in various positions in the chamber as the rotor turns. Because of the absence of reciprocating movement, the Wankel engine operates quietly and smoothly.

Test stand for 200 horsepower Stirling engine. (*Credit:* United Stirling (Sweden).)

MAGNETOHYDRODYNAMICS

The magnetohydrodynamic (MHD) generator produces electrical energy directly from thermal energy with a potential conversion efficiency in the range of 50 to 60%. The principle of operation of MHD generators is similar to that of conventional rotating generators except that the rotating wire conductor of the conventional generator is replaced by a moving electrically conductive fluid. As the working fluid flows through the magnetic field, a voltage drop is induced across the stream causing an electrical current to flow between the electrodes. The electrodes of an MHD generator are generally two opposite walls of a rectangular duct to which electrical leads are attached (the adjacent side walls are electrical insulators). The MHD working fluid can be either a plasma or ionized gas in an open or closed cycle system, or a homogeneous mixture of a liquid metal and an inert gas in a closed cycle system.

In the open cycle plasma system, fossil fuel is burned at a sufficiently high temperature to ionize the product gases. Electrical conductivity is further enhanced by "seeding" the gas with a readily ionized material such as salts of potassium or cesium. The conductive gas is expanded through the generator section, producing electricity. The hot gases can be used further to produce steam to generate electricity in a conventional steam turbine-generator system. The seed material is extracted from the gases for reuse.

The closed cycle plasma system utilizes a seeded inert gas which is heated indirectly through a heat exchanger from a nuclear or fossil fuel heat source. After expansion through the MHD generator, the gas is recycled through the heat exchanger.

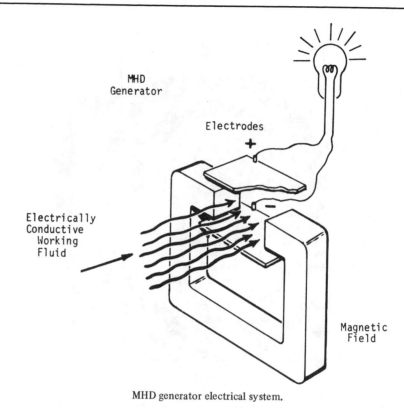

MHD generator electrical system.

Source: Proposed Final Environmental Statement: Liquid Metal Fast Breeder Reactor Program, Volume III, WASH-1535, U.S. Atomic Energy Commission, December 1974.

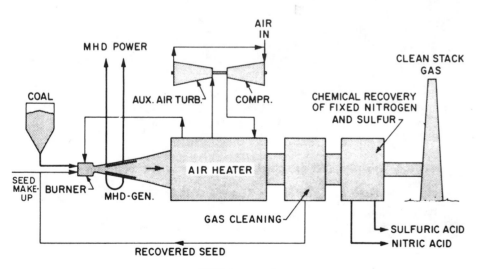

MHD power cycle.

Source: Arthur R. Kantrowitz, Statement included in Hearings on the *Energy Research and Development Policy Act*, before the Committee on Interior and Insular Affairs, United States Senate, June 21, 21; July 11 and 12, 1974.

The liquid metal MHD concept uses two fluid circuits, one for a liquid metal and another for an inert gas. The liquid metal is heated by a fossil or nuclear heat source and the inert gas dispersed into the liquid metal. Heating by the liquid metal causes the gas to expand and accelerates the mixture of the two fluids through the MHD generator. The liquid metal provides the moving conductor in the generator. At the exit of the generator the two fluids are separated, the liquid metal is reheated and the gas is cooled and recompressed and then both are recycled.

All MHD power generator concepts are in the developmental stage although the first patents related to MHD appeared about 1910. The major MHD programs at present are in the United States and the USSR and are centered on open cycle plasma systems. The United States program has focused on the development of system components. Several generators have been built to deliver relatively large amounts of power (18 to 32 MWe) for short periods of time (several minutes), and a variety of electrode configurations have been tested for relatively long periods of time at low power levels. The USSR program has involved the construction of complete power plants (25 MWe). Operation at about half of design output has been achieved in short-term operation.

View of the MKVI MHD generator from exhaust end. The magnets are pulled apart to expose the generator for work. AVCO Everett MHD Research Facility, Everett, Mass.

Research and Development Requirements for Open Cycle MHD Generators.

Materials that will operate for extended periods in the high temperature, erosive environment.

A high efficiency coal-combustor capable of handling coals having 10% or more ash.

The plasma generator must extract 20 to 25% of the total enthalpy of the combustion products; thus far only 8% has been achieved, but scaling laws indicate that increasing the generator size should help achieve this goal. Additional study of the above mentioned generator problems is required.

The scrubber must remove 99.9% of the seed from the spent combustion gases—thus far 99% removal has been achieved.

An overall isentropic generator efficiency of at least 70% is required, while only 40% has been demonstrated. This goal should also be attained by increasing the generator size.

A diffuser efficiency of 70 to 80% is required while 35% has been reported.

The stack gases must be cleaned to acceptable levels; recent results indicate that present-day technology should be able to meet EPA standards.

Long-term tests must be undertaken to demonstrate component longevity once the required performance levels have been demonstrated.

THERMIONIC ENERGY CONVERSION

Interest in thermionic energy conversion systems is based on studies which have indicated that conventional fossil fuel power plant efficiencies can be raised from 41% to about 50% through the use of thermionic topping cycles.

The operation of thermionic devices is based on the emission of electrons by metals at high temperatures. A thermionic converter contains an electron emitter and collector in a sealed envelop at reduced pressure. As the emitter is heated the electrons escape from the surface of the hot emitter and move to the cooler electron collector. External connection of the emitter and collector surfaces results in the flow of an electrical current. In order to achieve a reasonable power density between the emitter and collector, an ionized vapor such as cesium is introduced to neutralize the space charge developed by the flow of electrons.

Thermionic devices have been shown to be technically feasible but further development work is required particularly on problems associated with vaporization of the emitter surfaces, thermal warping, insulator shorting, seal failures, and the development of materials and fabrication methods to reduce costs.

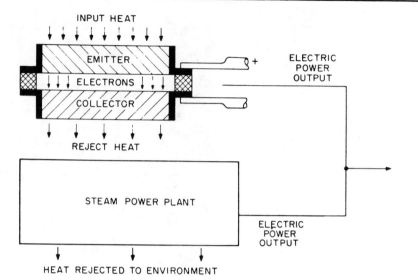

Schematic of thermionic topping cycle for steam power plants.

What is a thermionic diode?

A thermionic diode is a device which directly converts heat into electricity. It consists essentially of an emitter exposed to intense heat (in this case, a pulverized coal flame with highly preheated combustion air) and a collector. A small gap between the emitter and the collector is filled with cesium vapor, which ionizes when heated and carries electrons from the emitter to the collector. Actually, the electrons flow from the emitter, through the gap, to the collector and then return to the emitter through an external load where they do useful work. Large internal surfaces in the diodes allow a comparatively large reserve of precharged cesium, to maintain a satisfactory pressure in the entire system. Another method of controlling cesium vapor pressure is to provide a chamber, separate from diodes, where cesium temperature could be changed to control its vapor pressure in the functioning gap of the diodes to which it is connected. The cesium does not wear out, requires no make-up.

The thermionic diode requires a temperature substantially higher than the thermoelectric generator, another device converting heat directly into electricity, but its efficiency is much higher.

Bear in mind both the thermionic and thermoelectric devices have a dc output. To convert this to ac would call for an inverter, probably solid state.

Source: Joseph F. Engelberger, "Thermionic topping may yield substantial savings," *Power,* November 1967.

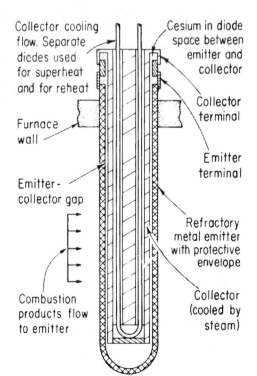

Collector cooling flow. Separate diodes used for superheat and for reheat

Furnace wall

Emitter-collector gap

Combustion products flow to emitter

Cesium in diode space between emitter and collector

Collector terminal

Emitter terminal

Refractory metal emitter with protective envelope

Collector (cooled by steam)

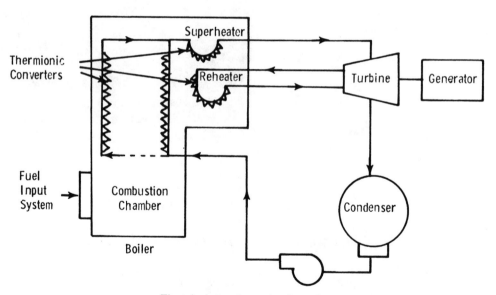

Thermionic topping cycle schematic.

Source: Atomic Energy Commission Authorizing Legislation Fiscal Year 1975, Hearings before the Joint Committee on Atomic Energy, Congress of the United States, March 4 and 5, 1974, Part 4.

FUEL CELLS

A fuel cell is a device that produces electrical energy from the controlled electrochemical oxidation of fuel. The basic components of a fuel cell are the electrodes (anode and cathode) separated by an electrolyte which may be either acidic or basic. The electrodes provide sites for the electrochemical reaction of the fuel and oxidant and also act as conductors for electron flow to the external circuit. The electrical charge is transferred through the electrolyte within the cell by migration of hydroxyl ions from cathode to anode. Continuous operation requires the removal of heat, water, and any inert materials that enter the cell with the reactants. Reaction kinetics are enhanced by the incorporation of a catalyst such as platinum on the electrode surfaces.

The first demonstration of the principle of a fuel cell was conducted by Sir William Grove in 1839. Grove reacted hydrogen and oxygen on platinum electrodes in a dilute sulfuric acid electrolyte to produce electricity, water and heat. In 1959, the Allis Chalmers Manufacturing Company demonstrated a tractor powered by a 20 KWe hydrogen-oxygen fuel cell. The major fuel cell development activity in the United States is being conducted by the Pratt and Whitney Division of United Technologies Corporation.

Past and Projected Fuel Cell Progress

	1964	1968	1972	1976	1980
Application	Space	Military	Residential	Utility	Utility
Programs	Gemini; Apollo	1.5 kW	Target	FCG-1 (Specs)	RP-114 (Goals)
Precious Metal Content (grams/kW)	1,000–2,000	200–800	100	10	0
Fuel/Oxidant	H_2/O_2	CH_4/Air	Nat. Gas/Air	Naphtha/Air	HC[a]/Air
Operating Life (hours)	100[b]	1,000	10,000	40,000	40,000
Heat Rate (Btu/kWh)	N/A	20,000	11,000	9,300	7,500
Capital Cost ($/kW)	100,000	100,000	1,000	200	150

[a]Wide range of available utility fuels.
[b]Life of space systems if operated on air as oxidant.
Source: Electric Power Research Institute, Palo Alto, California.

J. Fuel Cell Parameters

Best performances to date are as follows:

H_2-air*	250°C	1 W/cm^2
		100 W/kg
		20 kW m^{-3}
Hydrocarbon-air*	200°C	0.1 W/cm^2
		25 W/kg
		10 kW m^{-3}
Coal-air*	1,000°C	1.0 W/cm^2
		200 W/kg
		20 kW m^{-3}
Hydrazine-air**	100°C	0.1 W/cm^2
		120 W/kg
		40 kW m^{-3}

* With high catalyst loadings.
** This fuel has carcinogenic properties.

Source: Technology of Efficient Energy Utilization, Report of a NATO Science Committee Conference, Les Arcs, France, October 8-12, 1973, Scientific Affairs Division, North Atlantic Treaty Organization, Brussels, Belgium. Conference report has been reprinted (with NATO's permission) by Pergamon Press, Oxford, U.K., and now available from that publisher.

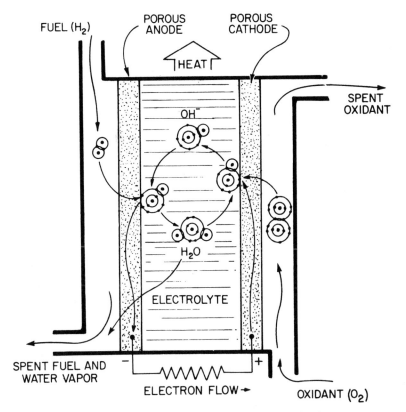

Electrochemical oxidation of hydrogen in a fuel cell.

Source: Proposed Final Environmental Statement: Liquid Metal Fast Breeder Reactor Program, Volume III, WASH-1535, U.S. Atomic Energy Commission, December 1974.

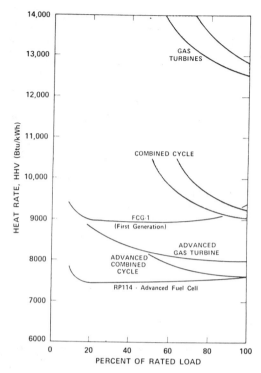

Comparison of heat rates of fuel cells and gas turbine and combined cycle power plants.

Source: Fickett, A. P.; "An Electric Utility Fuel Cell: Dream or Reality?," American Power Conference, April, 1975, as shown in *Assessment of Fuels for Power Generation by Electric Utility Fuel Cells*, EPRI 318, prepared by Arthur D. Little, Inc., for the Electric Power Research Institute, Palo Alto, California, October 1975.

FCG-1–26 megawatt fuel cell powerplant being developed for electric utility applications.

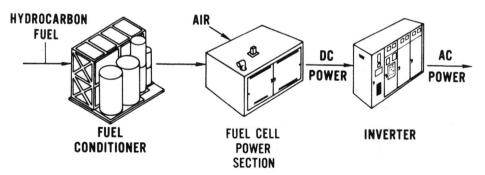

Major components of a fuel cell power plant.

Source: William H. Podolny, Supplemental Information to a Statement before the Energy Research and Water Resources Subcommittee, Committee on Interior and Insular Affairs, United States Senate, March 6, 1975.

A practical fuel cell power plant for use within an electric utility network must utilize available utility fuels and provide alternating current power compatible with other generation in the system. A fuel cell system for utility use must contain, therefore, a fuel processor to convert available fuels to the hydrogen required by the fuel cell, the fuel cell itself, and a power conditioner to convert the direct current produced by the fuel cell to alternating current.

The most highly developed fuel cell power plant uses phosphoric acid, immobilized by an inert matrix, as the electrolyte. The electrodes are porous graphite structures containing small amounts of platinoid catalysts (10 grams/kW). Historically, the major diffuculty encountered with the phosphoric acid cell has been that of reducing the noble metal catalyst requirements while simultaneously preventing the degradation of the catalyst.

Source: Proposed Final Environmental Statement: Liquid Metal Fast Breeder Reactor Program, Volume III, WASH-1535, U.S. Atomic Energy Commission, December 1974.

HEAT PUMPS

A heat pump is basically an air-conditioning system with the flexibility to interchange the functions of the evaporator and the condenser, allowing it to either heat or cool the desired space. During the summer, cool air is delivered indoors and the heated air from the condenser is exhausted outdoors. During the winter, the heat pump flow is reversed and heat is "pumped" from outdoors to inside of the building. The heat delivered is several times the electrical resistance heat equivalent of the input power to the compressor. The coefficient of performance (COP) of the heat pump depends mostly on the temperature of the outside air to which heat is rejected in the summer and from which heat is drawn in the winter.

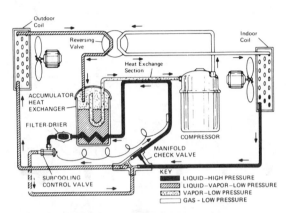

Westinghouse HI/RE/LI heat pump.

Source: Solar Heating and Cooling of Buildings, Phase O, NSF-RA-N-74-023A, Westinghouse Electric Corporation, prepared for the National Science Foundation, May 1974.

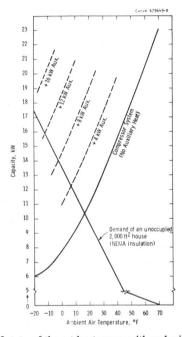

Heating capacity of state-of-the-art heat pump with and without auxiliary heat.

Source: Westinghouse Electric Corporation Research Laboratory, Pittsburgh, Pennsylvania, March, 1975.

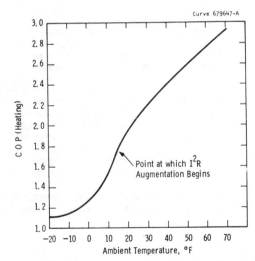

Coefficient of performance (COP) of heat pump as a function of outdoor temperatures. At temperatures below 15 °F, electrical resistance (I^2R) heating or fossil fuel heating is added to augment the heat pump.

Source: Westinghouse Electric Corporation Research Laboratory, Pittsburgh, Pennsylvania, March, 1975.

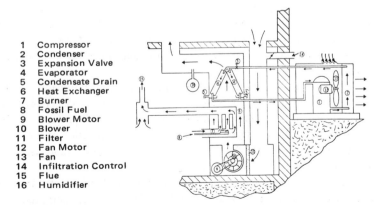

1	Compressor
2	Condenser
3	Expansion Valve
4	Evaporator
5	Condensate Drain
6	Heat Exchanger
7	Burner
8	Fossil Fuel
9	Blower Motor
10	Blower
11	Filter
12	Fan Motor
13	Fan
14	Infiltration Control
15	Flue
16	Humidifier

Fossil fuel system.

Source: Westinghouse Electric Corporation Research Laboratory, Pittsburgh, Pennsylvania, March, 1975.

ENERGY STORAGE

The efficient use of electrical energy requires the use of storage systems that permit a proper matching of demand for electricity with generating capacity without excessive investment in that generating capacity. The basic problem is that the load demand for electricity exhibits marked variations on a daily, weekly, and seasonal basis. Generating capacity must be large enough to meet the peak demand and to afford some margin of reserve capacity to assure system reliability. At the same time, off-peak loads may be less than half of peak loads and, as a result, considerable generating equipment is standing idle part of the time. The lowest cost energy could be provided by large base-load plants operating essentially at continuous full capacity.

In the absence of a practical system for the direct storage of electricity, storage is accomplished by conversion of the electricity to some other form of energy which can be reconverted to electricity as needed. The most widely used storage system is the "pumped hydro" system in which electricity is used to pump water into elevated reservoirs during periods of low electricity demand. The potential energy in the water is recovered and converted to electricity in a hydroelectric plant during peak demand periods. The first pumped storage facility in the United States was built in western Connecticut in the 1930's and had a power capability of 32 megawatts. The largest United States pumped storage plant is located on the shores of Lake Michigan and has a capacity of 1900 megawatts. The upper reservoir is a man-made lake about one mile wide and two miles long.

A number of additional concepts for energy storage are at various stages of development: batteries, chemical energy systems, thermal energy systems, underground pumped hydro, flywheels, compressed air, and superconducting magnets.

Lead-acid batteries represent a well-developed electrochemical system, especially for short-term applications, but use as bulk storage systems for utility applications, where low cost must be combined with long cycle life, has not been established. Performance goals include a specific energy of 220 watt-hours per kilogram, a lifetime of 4 years, a charge-discharge cycle life of 1000, and a storage efficiency of 70%. In addition to improvements in lead-acid batteries, development work is also underway on zinc/chlorine batteries.

Among the promising advanced battery systems are the sodium/sulfur and lithium/iron sulfide concepts. The sodium/sulfur battery uses a solid electrolyte while the lithium/iron sulfide battery uses a fused salt electrolyte. Both batteries operate at elevated temperatures (300°C to 400°C). (See 10-12.)

The best known of the chemical energy storage systems is that related to the concept of a "hydrogen economy." In this system electricity is used to electrolyze water and produce hydrogen which then can be stored either in gaseous or liquid form for later transport to points of use where it can be used for heating, the operation of fuel cells, or processing chemicals. Although the hydrogen system offers considerable versatility, costs are presently higher than for other storage systems.

The thermal storage of energy for utility application was demonstrated in 1929 in Germany where pressurized hot water was stored at the Berlin–Charlottenburg station. Thermal storage does not appear to be cost competitive with pumped hydro or compressed gas systems for power generation stations. Considerable application of thermal storage heaters for residential space heating has taken place, however, in Germany and other countries.

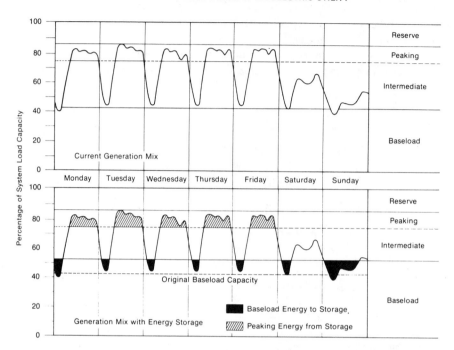

WEEKLY LOAD CURVE OF AN ELECTRIC UTILITY

How energy storage can help

These two figures typify the weekly load curve of a utility with and without energy storage.

As illustrated by the first curve, intermediate and peaking power today involves extensive generating capacity and significant amounts of fossil fuel.

On the other hand, if large-scale energy storage were available (second curve), the relatively efficient and economical baseload generation could be increased and the excess beyond off-peak demand (lower shaded areas) could be employed to charge the storage system.

Discharge of the stored energy (upper shaded areas) during periods of peak power demand would then reduce or replace fuel-burning peaking plant capacity, thus conserving fuel resources. In addition, the higher baseload level would replace part of the intermediate generation. Assuming that new baseload plants use nonfossil fuel, there are further savings of both cost and fossil fuel resources.

Use of energy storage to generate peaking power in this manner is termed "peak shaving." Load leveling describes the more extensive use of storage to eliminate most or all conventional intermediate cycling equipment.

Because of storage system inefficiencies, both peak shaving and load leveling entail losses: 25% was used to develop these illustrated examples. Energy storage on a utility system may result in an overall energy saving, but not necessarily. The outcome depends on whether the higher efficiency of baseload plants, compared with peaking and intermediate equipment, makes up for the storage system inefficiency.

Source: Kalhammer, Fritz R. and Cooper, Vance R.; "Energy Storage," *Research Progress Report FF-2,* Fossil Fuel and Advanced Systems Division, Electric Power Research Institute, Palo Alto, California, January 1975.

The system utilizes ceramic or iron storage elements either in a furnace or in wall heaters that are heated electrically at night during periods when demand for electricity is otherwise low. The storage elements retain sufficient heat to provide adequate heat during the day without additional input of electrical energy.

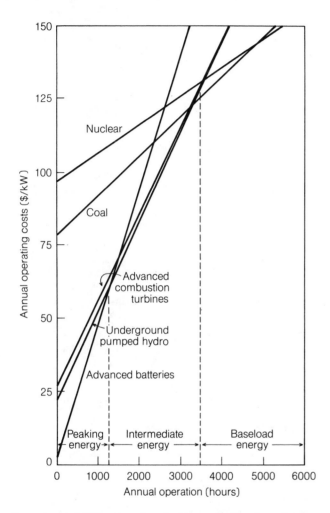

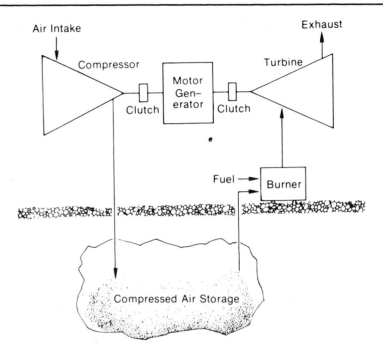

Large reservoirs, such as underground caverns or domed aquifers, are key to use of compressed-air energy. Off-peak power drives motor to compress air for storage. Later expansion of air, along with fuel combustion, drives gas turbine to generate peaking power. (*Adapted from* Exploratory Evaluation of Compressed-Air Storage Peak Power Systems, *United Aircraft Research Lab, September 1973*.)

Simplified compressed-air storage peaking system.

Source: Kalhammer, Fritz R. and Cooper, Vance R.; "Energy Storage," *Research Progress Report FF-2*, Fossil Fuel and Advanced Systems Division, Electric Power Research Institute, Palo Alto, California, January 1975.

The relative overall economy of different modes of power generation depends on annual hours of operation. For intermediate duty, pumped hydro storage is economical. For peaking services battery storage has the advantage of low capital cost.

Source: "Storage Batteries: The Case and the Candidates," *EPRI Journal*, October 1976.

Underground pumped hydro systems are similar in concept to conventional pumped hydro storage systems except that the lower reservoir is placed in an underground cavern.

Flywheels are widely used to smooth the pulsed power derived from engines. They have also been used to provide propulsive power from trolleybuses. Conventional flywheels, however, are too limited in capacity and too expensive for storing energy for utility systems. Recent developments in lightweight composite materials of high strength have led to research on high-speed flywheels which might meet requirements for storage efficiency and cost.

Compressed air storage systems are useful in conjunction with gas turbine power plants.

Since in normal operation about two-thirds of the energy produced in a gas turbine is needed to operate the compressor stage, it is advantageous to use low-cost, off-peak power to compress air for storage. This compressed air is later used, along with fuel combustion, to drive the gas turbine to generate peaking power. A large compressed air storage system is being constructed at Huntorf in Germany. The system will charge salt caverns of 300,000 cubic meters capacity with air at 70 atmospheres. The air will be used to operate a 290 MW gas turbine generator. Completion is expected in 1977.

Superconducting magnets represent the only storage system in which electric energy is stored directly without the conversion and reconversion steps of the other methods. The system depends upon the disappearance of resistance to the flow of electrons in certain materials at very low temperatures. Small superconducting storage magnets have been built for operation of experimental apparatus but the capital cost for utility application appears to be too high to be attractive.

CHARACTERISTICS OF ENERGY STORAGE SYSTEMS

Energy Density

	psi	°F	Btu/Ft^3
Pumped storage (100 ft head)			14
Hot rocks/metal 60–500°F			8,000–12,000
Molten Salts 60–500°F			10,000–20,000
Steam	15	212	40
	130	347	340
	500	467	1,270
Water	15	212	9,000
	130	347	16,000
	500	467	21,000
Hydrogen			
gas	15	60	280
	1,000	60	18,500
liquid	15	–425	200,000
hydride (Mg_2N_i or F_eT_i)			250,000
Ammonia			340,000
Methanol			430,000
Gasoline			830,000
Batteries			10,000–80,000

Source: NSF/RA/W-73-006, National Science Foundation, December 1973.

CHARACTERISTIC / TECHNOLOGY	TYPICAL ECONOMIC MODULE (MWE)	EARLIEST COMMERCIAL AVAILABILITY	STORAGE EFFICIENCY (%)	REMARKS
BATTERIES	1	1975–82	70–80	PROVEN TECHNOLOGY
FLYWHEELS	1	1985	70–90	
HYDROGEN/FUEL CELLS	1	1985	40–60	STORAGE OPTIONS FOR HYDROGEN
COMPRESSED-AIR (ADIABATIC)	10 (30 MWH)	1982	70–80	HIGH-GRADE THERMAL
COMPRESSED-AIR (ISOTHERMAL)	10	1975	NA	REQUIRES FUEL
PUMPED HYDRO	100 (?)	1975	70–75	SPECIAL SITUATIONS
SUPERCONDUCTING MAGNETS	500	1995	90	

Nominal characteristics of storage technology.

Source: Zlotnick, Martin "Energy Storage for Wind Energy Conversion Systems," in *Proceedings of the Second Workshop on Wind Energy Conversion Systems*, Washington, D.C. June 9–11, 1975, sponsored by the Energy Research and Development Administration and the National Science Foundation, NSF-RA-N-75-050, MTR-6970, The MITRE Corporation.

Heat-Storage Materials

Sensible-Heat Storage	Sp. Ht., Btu/lb F	True Density, lb/ft^3	Heat Capacity, Btu/ft^3 F — No Voids	Heat Capacity, Btu/ft^3 F — 30% Voids
Water	1.00	62	62	—
Scrap iron	0.12[a]	490	59	41
Magnetite (Fe_3O_4)	0.18[a]	320	57	40
Scrap aluminum	0.23[a]	170	39	27
Concrete	0.27	140±	38	26
Stone	0.21	170±	36	25
Brick	0.20	140	28	20
Sodium (to 208°F)	0.23	59	14	—

Latent-Heat Storage	Melting Point, °F	Density, lb/ft^3	Heat of Fusion, Btu/lb	Heat of Fusion, Btu/ft^3
Calcium chloride hexahydrate	84–102	102	75	7,900
Sodium carbonate decahydrate	90–97	90	115	10,400
Glauber's salt	90	92	105	9,700
Sodium metal	208	59	42	2,500
Ferric chloride	580	181	114	20,600
Sodium hydroxide	612	133	90	12,000
Lithium nitrate	482	149	158	23,500
Hypophosphoric acid	131	94?	92	8,700
Lithium hydride	1260	51	1800	92,000

[a]Over interval 77°F to 600°F.

Reprinted from *New Energy Technology: Some Facts and Assessments* by H. C. Hottel and J. B. Howard by permission of The MIT Press, Cambridge, Massachusetts, 1972.

STORAGE BATTERY PERFORMANCE

Although a number of rechargeable batteries have been developed which are useful for specialized applications, only the conventional lead-acid battery is of low enough cost to be considered at present for use in vehicles or by electric utilities. Unfortunately, the energy storage capacity and cycle life of lead-acid batteries are insufficient to provide adequate performance characteristics.

The two most promising secondary batteries under development are the sodium/sulfur battery and the lithium/sulfur battery. The sodium/sulfur battery is being developed by several companies including the Ford Motor Company, General Electric, and TRW. The Ford Motor Company has reported operation of cells with an energy storage capacity of about 38 watt-hours/lb, about two or three times that of lead-acid batteries. The sodium/sulfur cells operate at about 350°C with molten sodium and sulfur electrodes separated by a solid ceramic material, beta alumina. Key research and development problems relate to fabrication of the beta alumina electrolyte material, improvement of mechanical characteristics, electrode design, metal corrosion, and the control of impurities.

The development of the lithium/sulfur battery is being pursued by the Argonne National Laboratory and by Atomics International. The positive electrode for this cell is iron sulfide; the negative electrode is solid lithium-aluminum alloy; and the electrolyte is a lithium chloride–potassium chloride molten salt mixture. Operating temperatures are 375 to 400°C. Research and development problems relate to the development of low cost inter-electrode separators and electrical connectors which are resistant to the cell environment.

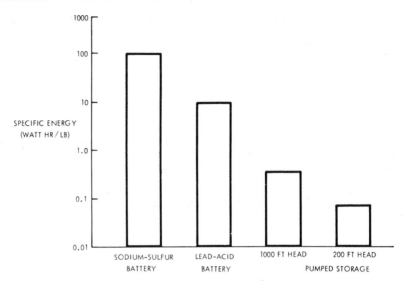

Comparison of energy storage systems specific energy. Batteries are considered an alternate to pumped hydrostorage in utility systems. Higher specific energy of battery systems results in lower land requirements. A 60 megawatt pumped hydroplant with 200-foot head would require 15 acres; a sodium-sulfur battery system of the same capacity would require 0.1 acre.

Source: Statement of G. W. Elverum, Jr., in Briefings before the Task Force on Energy of the Subcommittee on Science, Research and Development of the Committee on Science and Astronautics, U.S. House of Representatives, Vol. III, Serial U, 1972.

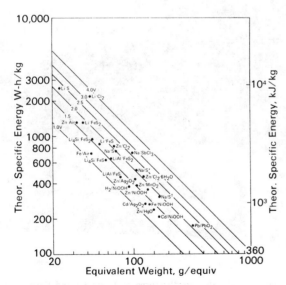

Theoretical Specific Energy of a Number of Electrochemical Couples. * Corresponds to $2Na + Na_2S_{5.2}$, Operation in the Two-phase Region Only. † Corresponds to $4.4Na + 3Na_2S_{5.2} \rightarrow 5.2Na_2S_3$, Operation in the Single-phase Region Only.

"Critical Materials Problems in Energy Production," C. Stein, ed., Academic Press, N.Y.

Battery Parameters

The state of the art (and not the best) is about as follows:

	W/kg	Wh/kg
Zinc-air	100	100
Ni-H$_2$ (compressed)	300	80
Ni-Fe	40	50
Zn-O$_2$ (compressed)	300	160
Aluminium-air	150	240
Sodium-sulphur	100	100
Sodium-air	200	400
Li-Cl$_2$ (liquid storage)	300	500
H$_2$-air	100	2000

It is reasonable to expect that in 10 years the following maximum performances will be attained: 10 W/cm^2; 100 kw m^{-3}; 75% voltage efficiency; 1 kW/kg; 1 kWh/kg.

Source: Technology of Efficient Energy Utilization, Report of a NATO Science Committee Conference, Les Arcs, France, October 8–12, 1973, Scientific Affairs Division, North Atlantic Treaty Organization, Brussels, Belgium. Conference report has been reprinted (with NATO's permission) by Pergamon Press, Oxford, UK, and now available from that publisher.

Energy Density Requirements for Cruise (Heitbrink and Tricklebank, 1970)

| Type of Vehicle | Constant Speed Cruise | | Acceleration | | |
	Range, miles	Velocity mph	Energy Density Whr/lb	Power Density W/lb	Power Density W/lb
Urban car	50	40	25	20	65
Commuter car	100	60	55	33	70–103
Family car	200	70	122	43	73–110
Metro truck	100	40	33	13	40
Urban coach	125	30	42	11	35

Reprinted from *New Energy Technology: Some Facts and Assessments* by H. C. Hottel and J. B. Howard by permission of The MIT Press, Cambridge, Massachusetts, 1972.

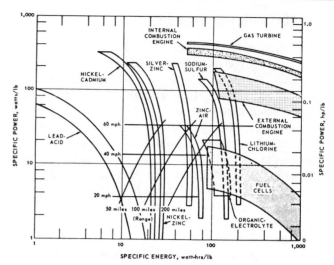

Vehicle Requirements and Motive Power Source Requirements (Assumes 2,000 lb vehicle, 500 lb motive power source and steady driving; power and energy taken at output of conversion device) (NAPCA, 1970).

Reprinted from *New Energy Technology: Some Facts and Assessments* by H. C. Hottel and J. B. Howard by permission of the MIT Press, Cambridge, Massachusetts, 1972.

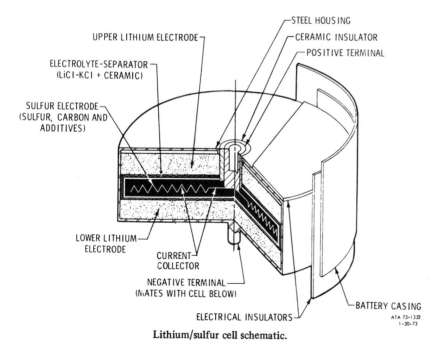

Lithium/sulfur cell schematic.

Source: Energy Research and Development—An Overview of Our National Effort, Hearing before the Subcommittee on Energy of the Committee on Science and Astronautics, U.S. House of Representatives, May 15, 1973.

Lined up at the GM Technical Center are the three 512 vehicles and the 511 car developed by the General Motors Engineering Staff. From left to right are the 511 commuter, a three-wheeled gasoline powered car driven by Edward A. Rishavy and accompanied by James J. Gumbleton; the 512 hybrid gasoline-electric, driven by Daniel L. Frank; the 512 electric driven by Stanley L. Genslak; and the 512 gasoline vehicle driven by Albert G. Lucas.

PETROLEUM STORAGE

Alternative Means of Petroleum Storage
(Costs and Other Characteristics)[a]

Type	Construction Cost $/BBL	Annual Maintenance Cost, $/BBL	Annual Cost $/BBL[b]	Geographical Limitations	Special Considerations
Above ground steel tanks	3.00–5.00	.06–.10	$1.975	Essentially none. Several acres required at each location.	Capacity of steel plate fabrication industry may significantly restrict construction rate.
Underground salt domes (on-shore)	0.50–0.75	.002–.005	$1.242	Gulf Coast locations only.	Costs assume a brine disposal pipeline to the Gulf.
salt domes (off-shore)	0.75–1.00	.005–.01	$1.285	Gulf Coast locations primarily.	Possible technical problems in drilling large diameter (24 to 36 inch) shafts from platforms. Multiple drilling locations required per salt dome.
Abandoned mines	0.20–1.50	0.5% of investment	$1.395	Limited number of fixed locations, not necessarily near refining centers and/or distribution facilities.	Most are probably unsuitable for petroleum storage.
Mined caverns	5.00–1.00	0.5% of investment	$2.713	Must be built in existing, hard-rock formations.	None.

[a]Cost and advantages or disadvantages of shut-in production are discussed elsewhere. Sufficient cost data were unavailable for inclusion of loading-roof pits.

[b]Assumes 15% cost of capital and costs of carrying oil at either $7 or $11 per barrel.

Source: Bass, Essley, and Webster, *Emergency Energy Capacity:* An Interim Report, U.S. Department of Treasury, Washington, D.C. October, 1973, consistent with NPC study, "Emergency Preparedness."

Source: Project Independence, Project Independence Report, Federal Energy Administration, November 1974.

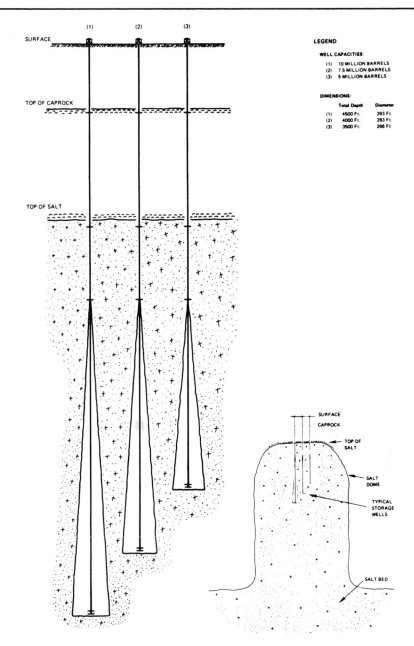

Typical salt dome and storage wells. Relatively low cost of salt dome oil storage facilities is a result of ease with which underground cavities can be mined by leaching with hot water.

Source: Strategic Petroleum Reserves, Hearings before the Committee on Interior and Insular Affairs, United States Senate, May 30 and July 26, 1973.

NET ENERGY ANALYSES

Net energy analyses refer to studies of the amount of electricity generated by a power plant relative to the amount of external energy required to construct, fuel, and operate the plant. The Energy Research and Development Administration has made such studies for light water reactor nuclear power plants.[1] Drawing from studies by the State of Oregon, the Center for Advanced Computation of the University of Illinois (Pilati and Richard), Development Sciences, Inc., and the Institute for Energy Analysis, ERDA has drawn the conclusion that light water nuclear power plants produce substantially more energy than is required to construct, fuel, and operate the plants. The major conclusions are:

The direct and indirect external energy inputs are about 26% of the energy output. This value is within the range of values for electrical generating facilities fueled with conventional fossil fuels.

Uranium enrichment, using the diffusion process, accounts for about 91% of all direct energy inputs and for 72% of all direct and indirect energy inputs. Use of the gas centrifuge process now being developed would significantly reduce the energy required for enrichment. The construction and operation of the power plant account for another 16%.

The internal energy loss due to the thermodynamic efficiency of a nuclear plant is about 68%. This means that the nuclear fuel must generate 3000 Btu of energy in order to produce 1000 Btu of output as electricity.

According to two 1975 studies (Development Sciences, Inc., and Oregon), 6000 to 7000 Btu of resource base must be available in order to have the 3000 Btu energy input to the reactor. This additional resource base is necessary to allow for uranium not recovered during mining or lost during processing as well as the fissionable uranium left in the tails during the enrichment process. However, all the 3000 to 4000 Btu of energy remaining are not irrevocably lost or consumed; much of the energy may be utilized in the future if the economics of processing these resources become attractive.

[1]*Appendix B, A National Plan for Energy Research, Development and Demonstration: Creating Energy Choices for the Future, ERDA 76-1*, Energy Research and Development Administration, April 15, 1976.

Comparison of Net Energy Results

Investigator	Units of External Energy Input Per 1,000 Units of Output
Development Sciences, Inc.	238
State of Oregon Study	194*
University of Illinois, Center for Advanced Computation (Pilati and Richard)	210
Institute for Energy Analysis	248
ERDA–76–1	262

* Adjusted to comparable basis

Source: Appendix B, A National Plan for Energy Research, Development and Demonstration: Creating Energy Choices for the Future, ERDA 76-1, Energy Research and Development Administration, April 15, 1976.

Energy Requirements for a Large Nuclear Power Plant

1000 MWe Pressurized Water Reactor—No Recycle—0.20% Enrichment Tails Assay
30 Year Life—61% Average Capacity Factor—160,300,000 MWh Output (547 Trillion Btu)*

Process	Quantity	Electrical Inputs, in MWh			Thermal Inputs, In Millions Btu			Total Inputs, Billion Btu**
		Direct	Materials	Construction	Direct	Materials	Construction	
Mining	3909 MTU	47,760	22,050	6,210	1,005,000	621,500	441,500	2,935
Milling	3909 MTU	67,430	16,140	2,780	1,310,000	621,500	121,200	3,037
Conversion	3909 MTU	39,830	16,620	620	4,826,000	426,000	29,000	5,334
Enrichment	3124 × 10³ SWU	8,778,000	18,120	24,050	1,048,000	428,000	1,016,400	103,037
Fuel Fabrication	683 MTU		67,750	137,200	600	147,900	1,579,000	26,000	4,096
Power Plant Construction and Operation	30 years	0	256,500	205,000	378,000	8,179,000	9,583,000	23,401
Fuel Storage	683 MTU	7,280	2,430	3,880	4,560	64,300	174,300	240
Waste Storage	30 years	130	4,560	320	2,080	158,200	22,950	398
Transportation								
—Natural	3909 MTU	0	410	0	31,900	24,470	0	61
—Fuel	693 MTU	0	1,546	0	120,300	92,290	0	230
Totals		9,008,000	475,600	243,460	8,874,000	12,195,000	11,414,000	142,769

*Electricity converted to thermal equivalent at 3413 Btu per kilowatt hour
Legend: MTU—Metric tons uranium
        SWU—Separative work units (directly proportional to the energy required in the enrichment process)
**Electricity converted to thermal equivalent at average heat rate of 11,400 Btu per kilowatt hour

Note: The conversion of electrical to thermal units plus the addition of thermal inputs from different energy sources ignores the quality aspect of preferred fuels. This is a serious shortcoming, as there is no single factor that is satisfactory for converting different energy sources to an equivalent base.

Source: Institute for Energy Analysis with adjustments by ERDA.

Source: Appendix B, A National Plan for Energy Research, Development and Demonstration: Creating Energy Choices for the Future, ERDA 76-1, Energy Research and Development Administration, April 15, 1976.

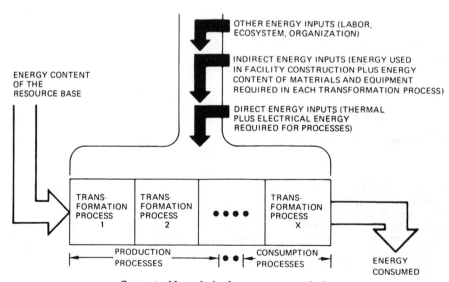

Conceptual boundaries for net energy analysis.

Source: Appendix B, A National Plan for Energy Research, Development and Demonstration: Creating Energy Choices for the Future, ERDA 76-1, Energy Research and Development Administration, April 15, 1976.

Energy Efficiency and Conservation

Waterval–Waterfall–Wasserfall–Cascade

EFFICIENCY OF ENERGY CONVERTERS

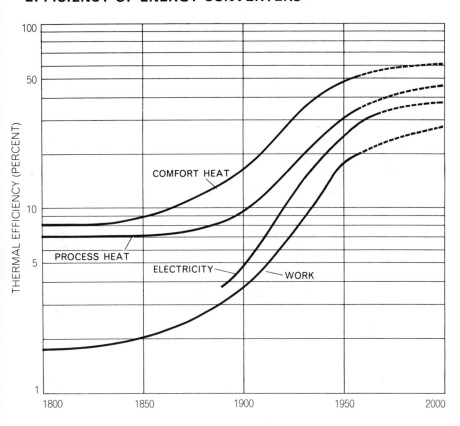

Efficiency of energy converters rose steeply from 1850 to 1950. From here on improvements will be much harder to win, partly because of thermodynamic limitations. A simple unweighted average of efficiencies in four major categories of energy use gives a value of about 8 percent in 1900, 30 percent in 1950 and a projected 45 percent in A.D. 2000.

Source: Chauncey Starr, "Energy and Power," *Scientific American*, September 1971.

Efficiency of energy converters runs from less than 5 percent: for the ordinary incandescent lamp to 99 percent for large electric generators. The efficiencies shown are approximately the best values attainable with present technology. The figure of 47 percent indicated for the liquid-fuel rocket is computed for the liquid-hydrogen engines used in the Saturn moon vehicle. The efficiencies for fluorescent and incandescent lamps assume that the maximum attainable efficiency for an acceptable white light is about 400 lumens per watt rather than the theoretical value of 220 lumens per watt for a perfectly "flat" white light.

Source: Claude M. Summers, "The Conversion of Energy," in *Energy and Power*, American Book, W. H. Freeman and Company, San Francisco, 1971.

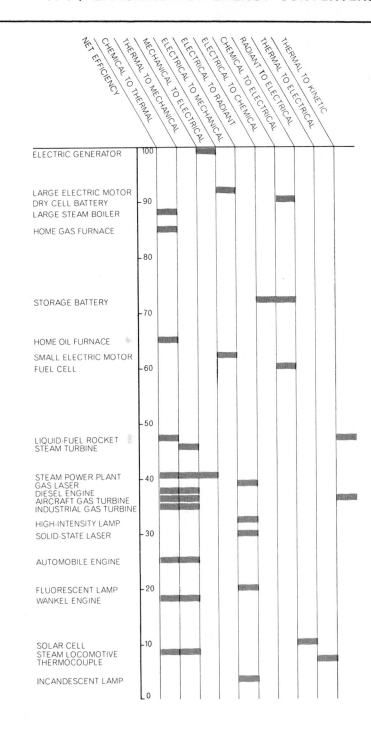

FUEL USE EFFICIENCIES

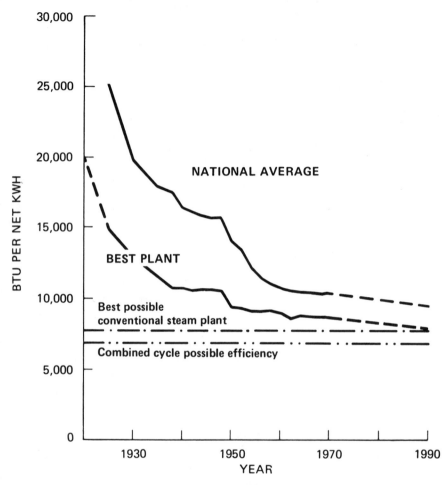

NATIONAL AVERAGE

BEST PLANT

Best possible
conventional steam plant

Combined cycle possible efficiency

NOTE: Based on chart of Federal Power Commission, *The 1970 National Power Survey*, *Part I*, Dec., 1971.

The efficiency of electric power generating plants can be measured by the number of Btu's required to generate a kilowatt-hour of electrical energy. In 1900, power plants required 93,700 Btu's to generate one kilowatt hour. In, 1970, only 9000 Btu's were required to generate the same kilowatt hour of electrical energy.

Source: Potential for Energy Conservation in the United States: 1974-1978, A Report of the National Petroleum Council, Washington, D.C., 1974.

Efficiency of Fuels Delivered to Consumer
[Efficiency of utilization not included]

Fuel	Efficiency of production (percent) ×	Efficiency of generation or refining (percent) ×	Efficiency of transmission (percent) =	Efficiency delivered to consumer (percent)
Coal	96 ×	──────── ×	97 =	93
Petroleum	96 ×	──────── ×	97 =	93
Natural gas	96 ×	──────── ×	97 =	93
Heating oil	96 ×	87 ×	97 =	81
Manufactured gas	31–46 ×	──────── ×	95 =	37
Electricity (coal)	96–97 ×	22 ×	80 =	17

Estimated Average Utilization Efficiency of Domestic Heating and Water Heating Equipment Using Different Fuels
[In percent]

Fuel	Equipment	Estimated efficiency of utilization
Coal (bituminous)	Central heating, hand fired	45.0
	Central heating, stoker fired	55.0
	Water heating, pot stove, 50 gallons per day	14.5
Oil	Central heating (84 percent of gas efficiency)	63.0
	Storage water heater, 50 gallons per day	44.2
	Storage water heater, 100 gallons per day	50.4
Gas	Central heating	75.0
	Room heater, unvented [1]	[2] 91.0
	Room heater, vented [1]	75.0
	Storage water heater, 50 gallons per day	54.1
	Storage water heater, 100 gallons per day	63.7
Electricity	Central heating, resistance [1]	95.0
	Central heating, heat pump (air to air)	226.0
	Room heaters [1]	100.0
	Storage water heater, 50 gallons per day	85.6
	Storage water heater, 100 gallons per day	92.2

[1] Estimated.
[2] This is a minimum figure and assumes adequate building ventilation to prevent condensation of the water vapor in the products of combustion.

Estimated Ultimate Fuel Efficiency [1]
[In percent]

Fuel and use	Delivered efficiency ×	Utilization efficiency =	Ultimate efficiency
Coal (bituminous):			
Central heating, hand fired	93.0 ×	45.0 =	41.8
Central heating, stoker fired	93.0 ×	55.0 =	51.1
Water heating, pot stove	93.0 ×	14.5 =	13.5
Oil:			
Central heating	81.0 ×	63.0 =	51.0
Water heating, 100 gallons per day	81.0 ×	50.4 =	40.8
Natural gas:			
Central heating	93.0 ×	75.0 =	69.7
Water heating, 100 gallons per day	93.0 ×	63.7 =	59.2
Electricity:			
Central heating, resistance	17.0 ×	95.0 =	16.1
Central heating, heat pump	17.0 ×	226.0 =	38.4
Water heating, 100 gallons per day	17.0 ×	92.2 =	15.6

[1] Based on data from tables 12–156a and 12–156b.

Source: Conservation of Energy, Committee on Interior and Insular Affairs, United States Senate,; Serial No. 92-18, 1972.

POWER CONSUMPTION OF APPLIANCES

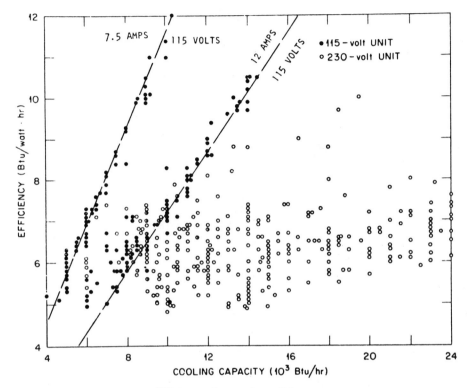

Efficiency of room air conditioners.

Efficiency of room air conditioners operating on standard 115-volt circuits is generally worse than the efficiency of air conditioners rated at higher capacity and requiring special installation of 230-volt wiring.

Source: Proposed Final Environmental Statement Liquid Metal Fast Breeder Reactor Program, Volume III, WASH 1535, U.S. Atomic Energy Commission, December 1974.

Annual Energy Requirements of Electric Household Appliances

	Average Wattage	Estimated kWh Consumed Annually		Average Wattage	Estimated kWh Consumed Annually
Food Preparation			water heater	2,475	4,219
blender	300	1	water heater		
broiler	1,140	85	(quick-recovery)	4,474	4,811
carving knife	92	8			
coffee maker	894	106	*Comfort Conditioning*		
deep fryer	1,448	83	air cleaner	50	216
dishwasher	1,201	363	air-conditioner (room)	860	860*
egg cooker	516	14	bed covering	177	147
frying pan	1,196	100	dehumidifier	257	377
hot plate	1,200	90	fan (attic)	370	291
mixer	127	2	fan (circulating)	88	43
oven, microwave (only)	1,450	190	fan (rollaway)	171	138
range			fan (window)	200	170
with oven	12,200	1,175	heater (portable)	1,322	176
with self-cleaning oven	12,200	1,205	heating pad	65	10
roaster	1,333	60	humidifier	177	163
sandwich grill	1,161	33			
toaster	1,146	39	*Health and Beauty*		
trash compactor	400	50	germicidal lamp	20	141
waffle iron	1,200	20	hair dryer	381	14
waste dispenser	445	7	heat lamp (infrared)	250	13
			shaver	15	0.5
Food Preservation			sun lamp	279	16
			tooth brush	1.1	1.0
freezer (15–21 cu. ft.)			vibrator	40	2
chest-type, manual					
defrost upright-					
type	—	1,320			
manual defrost	—	1,320			
automatic defrost	—	1,985	*Home Entertainment*		
refrigerators/freezers			radio	71	86
manual defrost,			radio/record player	109	109
10–15 cu. ft.	—	700	television		
automatic defrost,			black & white		
16–18 cu. ft.	—	1,795	tube-type	100	220
automatic defrost,			solid-state	45	100
20 cu. ft. and up	—	1,895	color		
			tube-type	240	528
Laundry			solid-state	145	320
clothes dryer	4,856	993	*Housewares*		
iron (hand)	1,100	60			
washing machine			clock	2	17
(automatic)	512	103	floor polisher	305	15
washing machine			sewing machine	75	11
(nonautomatic)	286	76	vacuum cleaner	630	46

The estimated annual kilowatt-hour consumption of the electric appliances listed is based on normal usage. When using these figures for projections, such factors as the size of the specific appliance, the geographical area of use, and individual usage should be taken into consideration. Note that the wattages are not additive since all units are normally not in operation at the same time.

*Based on 1000 hours of operation per year. This figure will vary widely depending on area and specific size of unit. See EEI-Pub #76-2 "Air Conditioning Usage Study" for an estimate for your location.

Source: EEI-Pub #75-61, Edison Electric Institute, 90 Park Avenue, New York, N.Y. 10016.

Properties of Incandescent and Electrical Discharge Systems for Illuminating Purposes

Lamp type	Power (W)	Lamp efficiency lm/W	Lamp efficiency %	Ballast power (W)	Ballast + lamp efficiency lm/W	Ballast + lamp efficiency %	Fitting loss %	Overall efficiency lm/W	Overall efficiency %
Incandescent	100	15	5	—	—	—		7-13	2.5-4.5
Incandescent halogen	100	30	13	—	—	—	10-50 (indoor)	15-27	6.5-12
Fluorescent	40	80	22	13.5	60	16.5		30-54	8.5-15
	65	85	—	11	73	—		36-66	—
Low-pressure Sodium	90	140	—	35	100	—		55-80	—
	180	180	27	30	155	23	20-45 (outdoor)	85-125	13-18.5
High-pressure Sodium	250	100	—	33	—	—		60-86	—
	400	120	29	39	108	24		72-103	14-20
High-pressure mercury	80	44	—	8.5	39	—		22-31	—
	250	54	—	18.5	50	—		28-40	—
	400	57	15	26	53.5	14		30-43	8-11
High-pressure metal-halide	400	85	23	26	75	21.5	30-40 (flood light)	45-52	13-15
	1000	90	—	43	86	—		51-60	—
	2000	100	—	68	97	—		58-68	—

Source: Technology of Efficient Energy Utilization, Report of a NATO Science Committee Conference, Les Arcs, France, October 8–12, 1973, Scientific Affaris Division, North Atlantic Treaty Organization, Brussels, Belgium. Conference report has been reprinted (with NATO's permission) by Pergamon Press, Oxford, UK, and available from that publisher.

TRANSPORT EFFICIENCIES

Energy and Price Data for Transport

Mode	Energy (Btu/ton-mile)	Price (cents/ton-mile)
Intercity Freight Transport		
Pipeline	450	0.27
Railroad	670	1.4
Waterway	680	0.30
Truck	2,800	7.5
Airplane	42,000	21.9
Passenger Transport		
Intercity[a]		
Bus	1,600	3.6
Railroad	2,900	4.0
Automobile	3,400	4.0
Airplane	8,400	6.0
Urban[b]		
Mass transit	3,800	8.3
Automobile	8,100	9.6

[a]Load factors (percentage of transport capacity utilized) for intercity travel are about: bus, 45%; railroad, 35%; automobile, 48%; and airplane, 50%.

[b]Load factors for urban travel are about: mass transit, 20%; automobile, 28%.

Source: Proposed Final Environmental Statement Liquid Metal Fast Breeder Reactor Program, Volume III, WASH-1535, U.S. Atomic Energy Commission, December 1974.

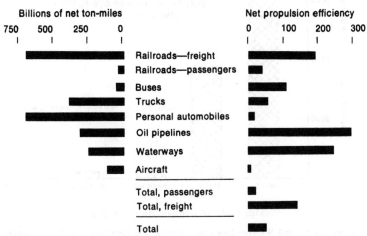

Intercity transport of passengers and freight in the U.S. varies markedly in efficiency, as measured in terms of the net payload (pound or individuals) carried per gallon of fuel consumed. If our goal is to increase a total national propulsion efficiency, we will concentrate on those methods of moving goods and people which show the highest net propulsion efficiency.

Source: Rice, Richard A., "System Energy and Future Transportation," *Technology Review*, edited at the Massachusetts Institute of Technology, January 1972.

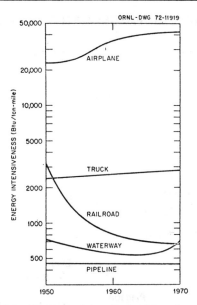

Historical variation in energy intensiveness for inter-city freight modes, plotted semilogarithmically.

Source: Eric Hirst, *Energy Intensiveness of Passenger and Freight Transport Modes*, Oak Ridge National Laboratory, April 1973.

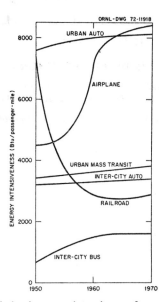

Historical variation in energy intensiveness for passenger modes.

Source: Eric Hirst, *Energy Intensiveness of Passenger and Freight Transport Modes*, Oak Ridge National Laboratory, April 1973.

ENERGY REQUIREMENTS—PASSENGER TRAVEL

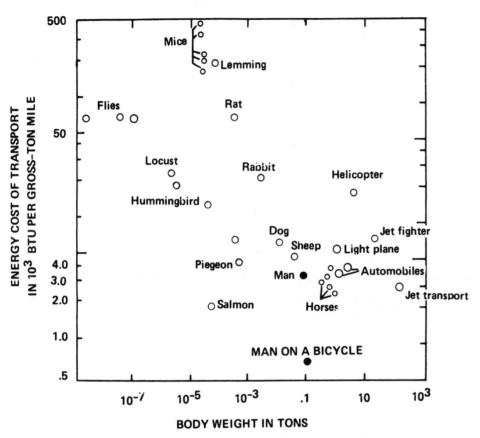

Graph data from article by S. S. Wilson, (*Scientific American*, March 1973)
Energy expended for moving creatures and machines.

Source: Nina Dougherty and William Lawrence, *Bicycle Transportation*, U.S. Environmental Protection Agency, December 1974.

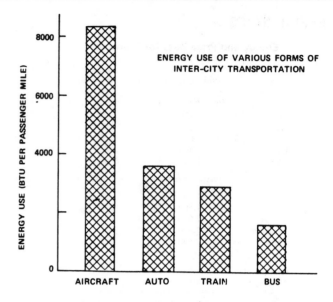

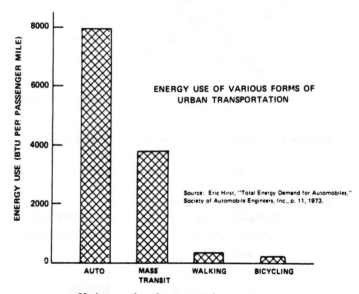

Various modes of transportation energy use.

Source: Nina Dougherty and William Lawrence, *Bicycle Transportation*, U.S. Environmental Protection Agency, December 1974.

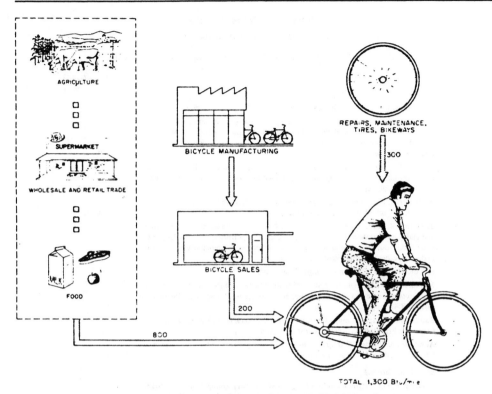

Direct and indirect energy requirements for bicycle travel.

Source: Eric Hirst, "Transportation Energy Research at ORNL," in *Energy Conservation Research*, Proceedings of the NSF/RANN Conference on Energy Conservation Research, Airlie House, Virginia, February 18–20, 1974, National Science Foundation.

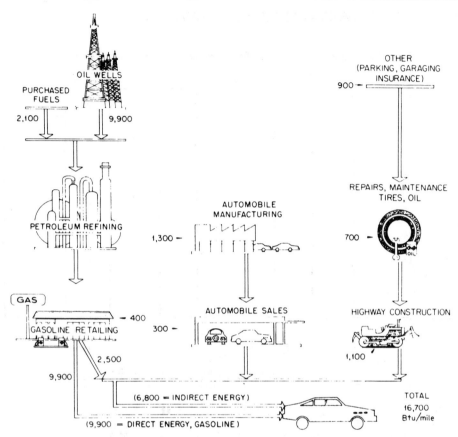

Direct and indirect energy requirements for the automobile in the U.S. 1971.

Source: Eric Hirst, "Transportation Energy Research at ORNL," in *Energy Conservation Research*, Proceedings of the NSF/RANN Conference on Energy Conservation Research, Airlie House, Virginia, February 18–20, 1974, National Science Foundation.

ENERGY EFFICIENCY—PASSENGER TRAVEL

Inter-City Passenger Transport Fuel Efficiencies by Vehicle Type (1)

VEHICLE TYPE	AVERAGE SEATING CAPACITY	Fuel Economy		Average Fuel Efficiencies			
		Average km/litre	Average miles/U.S. gall.	Seat-km litre	Seat-miles gallon	Passenger km/litre (2)	Passenger miles/gallon (2)
AUTOMOBILE							
Luxury 1800kg 4000lb	6	6	14	36	84	18	42
Standard 1800-1125kg 4000-2540lb	5	8	18	40	90	20	45
Compact 1125kg 2500lb	4	11	26	44	104	22	52
BUS							
Highway Diesel	50	3	7	150	350	68	158
RAIL							
Diesel/Electric	100	1	2.5	100	250	40	100
AEROPLANE							
Short-Range (B-737, etc)	130	.18	.43	24	56	12	28
Medium-Range (Airbus, etc)	260	.11	.27	30	70	15	35
Long-Range (B-707)	164	.07	.17	12	28	6	14
Jumbo-Jet (B-747)	435	.06	.14	25	59	13	30

Sources: William P. Goss and Jon G. McGowan, "Energy Requirements for Passenger Ground Transportation Systems", paper presented at the Intersociety Conference on Transportation, Denver, September 1973. Eric Hirst. "Energy Consumption for Transportation in the United States", Oak Ridge National Laboratory Report CRNL-NSF-EP-15, March 1972. Richard A. Rice, "Historical Perspective in Transport System Development", Advanced Urban Transportation Systems, Carnegie-Mellon University, Pittsburg, Pa., 1971. Council on Environmental Quality, "Energy on the Environment", United States Government Printing Office, August 1973. Jane's All the World's Aircraft, 1973.

1) C.K. Orski, "The Potential for Fuel Conservation; The Case of the Automobile", Transportation Research, September 1974.

2) The Average passenger-kilometres/litre (passenger-miles/U.S. gallon) have been computed using load factors encountered under typical urban operating conditions; automobile 50 per cent; bus 45 per cent; railroad 40 per cent and aeroplane 50 per cent.

Source: Energy Prospects to 1985, Volume II, Organization for Economic Co-operation and Development, Paris, 1974.

In passenger transit, the high performers in terms of net passenger miles moved per gallon of fuel are buses and commuter trains; the more exotic, faster means of transport are lower in efficiency, and so are such hard-to-die luxuries as superliners and Pullman (overnight) trains. A trend to the most efficient forms of passenger transport, writes the author, would considerably increase the U.S. national propulsion efficiency. Note that the horizontal scales are logarithmic.

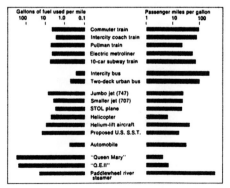

Source: Rice, Richard A. "System Energy and Future Transportation," *Technology Review*, edited at the Massachusetts Institute of Technology, January 1972.

Passenger Transportation
Present Fuel Consumption
(Based on Seating Capacity and Average Fuel Use Rates)

	Seat Miles/Gallon (Representative)
(Rail—(U.S. Current—includes allowances for engine idling between runs.)	
3000 hp locomotive, turbocharged, 0.5 mpg, 9 coaches per locomotive, 60–80 seats each (Amtrak data)—Relatively new Amtrak locomotives. Many Amtrak cars were less than 60 seats when acquired, now being refurbished with higher density seating.	270–360
2250 hp E-8, not turbocharged, 0.63 mpg, 6 coaches per locomotive, 60–80 seats each (Amtrak data)—relatively old Amtrak locomotives, extra engines sometimes added for reliability.	225–300
2250 hp E-8, not turbocharged, 0.63 mpg, 4–5 coaches per locomotive, 60–80 seats each (Amtrak data interpolated).	150–250
2250 hp E-8, not turbocharged, 0.62 mpg, 4 cars per locomotive—sleepers, dining, and club cars (Southern Railroad data)—locomotive performance essentially the same as Amtrak data. For cross-country, added engine for reliability and added luxury cars sharply reduce seat miles per gallon.	50
Rail diesel car (RDC) (Budd), 3 mpg, 75–85 seats.	250
Rail diesel car (RDC) (B&M) 2 mpg, 75 seats—illustrates difference between manufacturers estimate and operating experience. Single B&O demonstration of same cars got 3.48 mpg.	150
Rail turbine train; 0.33 mpg, 320 seats (296 + 24 snack bar) (Amtrak data) (Amtrak's French RTG)—delivery test at 80 mph average.	110
Autotrain, 0.37 mpg, 3600 hp locomotive, 18 cars per locomotive, 30 automobile—miles/gallon, at 5 seats per auto.	150
Bus—(U.S. Current)	
Intercity, 6.0 mpg (Greyhound), 47 seats (TSC Industry average)—over-the-road test of Greyhound and Trailways buses by TSC indicated 8.8 mpg at 50 mph, 8.1 mpg at 60 mph.	282
Urban, 3.6–4.6 mpg, 50 seats (FHWA, UMTA data).	180–230
Automobile—(FHWA Data)	
Urban subcompact — 4 seats, 24 mpg	96
Urban compact — 5 seats, 18 mpg	90
Urban standard — 6 seats, 14.4 mpg	86.4
Urban luxury — 6 seats, 9.0 mpg	54
Intercity subcompact — 4 seats, 30 mpg	120
Intercity compact — 5 seats, 22.5 mpg	112.5
Intercity standard — 6 seats, 18.0 mpg	108
Intercity luxury — 7 seats, 12.0 mpg	72
Air—(NASA Data)	
Twin engine turbofan, 68–106 seats	
short (250 mi. stage) .34–.44 mpg	30–38
medium (500 mi. stage) .44–.54 mpg	37–47
3 & 4 engine turbofan, 131–200 seats	
medium (500 mi. stage) .21–.29 mpg	35–41
long (1,000 mi. stage) .26–.34 mpg	44–51
3 & 4 engine turbofan, widebody, 256–385 seats	
wide body jets use new high by pass turbofan engines with low specific fuel consumption	
medium (500 mi. stage) .11–.19 mpg	44–51
long (1,000 mi. stage) .14–.22 mpg	54–60

Source: Report to Congress on the Rail Passenger Service Act by Transportation Secretary Claude Brinegar, July 22, 1974.

AUTOMOBILE FUEL ECONOMY

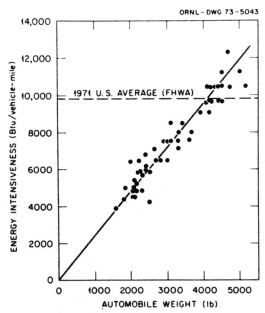

Automobile fuel consumption as a function of vehicle weight for 1971 and 1972 model cars.

Source: Conservation and Efficient Use of Energy, (*Part 2*), Joint Hearings before certain Subcommittees on Government Operations and Science and Astronautics, House of Representatives, July 10, 1973.

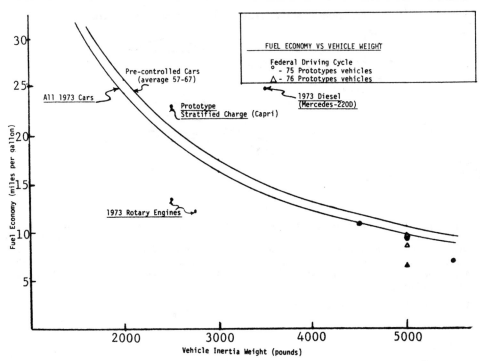

Source: Conservation and Efficient Use of Energy (Part I), Joint Hearings before certain Subcommittees of the Committees on Government Operations and Science and Astronautics, House of Representatives, June 19, 1973.

Fuel Economy in Miles per Gallon for Various Model Years and Inertia Weight Categories

Year	1750	2000	2250	Inertia Weight 2500	2750	3000	3500	4000	(__ indicates no data) 4500	5000	5500
57	--	26.4	--	--	13.2	--	14.7	13.0	--	--	12.5
58	--	25.3	18.2	--	13.2	--	13.6	15.2	12.5	8.6	--
59	--	28.6	--	--	--	15.2	15.0	13.2	12.7	13.8	--
60	--	20.4	--	22.3	24.5	--	15.7	12.4	10.8	10.9	--
61	--	29.4	--	20.9	16.3	17.2	11.4	14.0	10.5	10.6	--
62	--	25.8	--	--	18.0	16.3	13.0	13.8	11.1	10.8	--
63	--	23.2	19.5	--	16.1	14.7	12.6	12.0	11.1	10.6	--
64	--	22.8	--	--	17.3	16.2	13.7	12.9	11.4	11.0	--
65	--	23.8	--	--	18.3	15.2	13.7	12.3	11.7	10.3	--
66	--	20.9	--	12.7	14.9	14.6	13.9	12.3	12.1	11.3	9.3
67	--	22.6	25.7	--	18.7	15.9	13.1	12.1	11.6	11.2	10.3
68	--	19.3	20.5	18.5	19.7	15.6	13.3	12.0	11.3	9.5	--
69	--	22.2	20.3	18.8	--	15.4	13.3	11.9	11.3	9.1	10.8
70	--	23.4	19.3	17.5	18.5	15.9	13.3	12.0	10.7	10.1	9.9
71	27.2	22.6	21.4	19.3	18.3	14.8	12.2	11.7	10.7	9.6	10.9
72	--	23.0	21.9	19.6	20.0	14.4	13.3	11.1	10.7	9.6	9.3
73	24.8	23.8	21.9	19.7	17.5	15.6	13.9	10.8	10.1	9.3	8.6
74	--	24.1	21.4	18.7	17.7	14.8	13.7	10.8	9.6	9.1	8.2
75	--	--	20.1	17.4	16.6	--	14.3	--	10.1	9.6	8.4
57-67 Aver.	--	23.2	21.7	19.1	17.1	15.4	13.5	12.6	11.7	10.9	10.5

Source: A Report on Automotive Fuel Economy, U.S. Environmental Protection Agency, Washington, D.C., 1974.

Effect of Speed on Automobile Fuel Consumption Rates

Vehicle Speed	km/hour	50	65	80	100	115
	miles/hour	30	40	50	60	70
Fuel Economy	km/litre	8.95	8.96	8.28	7.44	6.35
	miles U.S./ gallon	21.05	21.07	19.49	17.51	14.93
Percentage decrease in Fuel Economy	cumulative		0%	8.1%	11.3%	17.3%
			0%	8.1%	19.4%	36.7%

Source: U.S. Department of Transportation, "The Effect of Speed on Automobile Gasoline Consumption Rates," October 1973.

RELATIONSHIP BETWEEN FUEL CONSUMPTION AND ACCELERATION

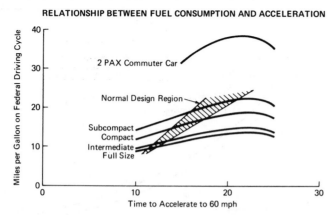

Comparing systems. This graph shows how gas mileage is sensitive to the acceleration design characteristics of cars. Smaller cars, unlike full-size cars, are deliberately designed to operate on the more favorable part of the curve, trading acceleration for energy savings. But if different size cars were designed for the same performance level, the disparity in fuel economy would be reduced substantially. Also shown is the curve for a theoretical two-passenger trunkless car with a number of possible energy-saving adaptations.

Source: Mosaic, National Science Foundation, Vol. 5, No. 2, Spring 1974.

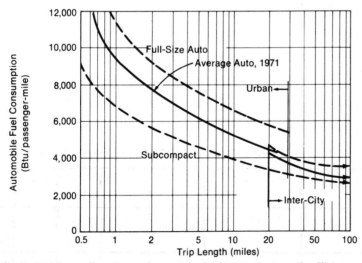

Big vs. small, short vs. long. For all cars, longer trips lead to greater per mile efficiency, and at intercity distances the differences between subcompact, average, and full-size cars grow smaller. But in city driving, and especially for short trips, the subcompact car—even with reduced seating taken into account—has a marked fuel consumption advantage over the full-size car.

Source: Mosaic, National Science Foundation, Vol. 5, No. 2, Spring 1974.

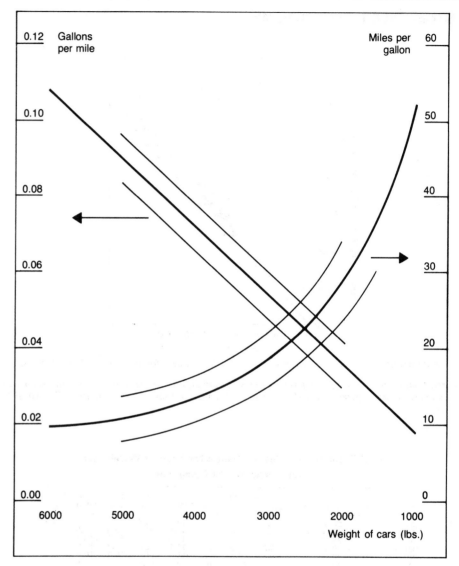

The weight of the car and the gasoline consumption or gasoline efficiency of that car are directly related. Sixty different cars weighing from 1600 to 5100 lbs. were tested for their gasoline consumption over a 300- 340-mile course representing typical driving conditions, with the above results.

Source: Tien, John K., Clark, Roy W. and Malu, Mahendra K., "Reducing the Energy Investment in Automobiles," *Technology Review*, edited at the Massachusetts Institute of Technology, February 1975.

ENERGY CONSERVATION POSSIBILITIES

Sectoral Energy Conservation Possibilities

Type of Conservation	Potential Savings in % of sectoral consumption (year)	Costs/Benefits	Type of policy action available
I. ELECTRICITY CONVERSION Improve efficiency of generation and transmission			Encourage R & D by utilities
(a) Smoothing daily demand cycle and reduction of peak load	Slight increase in generation efficiency	Lower capital requirements for new plant: possibility of earlier retirement for old plant	Revised tariff structures (less favourable rates for heavy users)
(b) Waste heat utilisation (total energy systems and district heating)	10 per cent increase in efficiency (1985) Total energy systems use 15 per cent less primary energy than systems with fuel heating and electrical cooling	Less economies of scale for small plants - less thermal pollution	Removal of legal and institutional obstacles. Encouragement of initiatives by utility companies
II. INDUSTRY Eliminate waste:			
Less non-productive idling of plant	10 per cent fuel	Small cost compared to savings	Information and encouragement
Better regulation and control heat recovery	5 per cent electricity (immediately)		
Replacement of old equipment and processes by a more efficient inventory			Ensure that energy price to industry reflects real long-run marginal costs of production
Improved thermal insulation	15 per cent (1985)		Accelerate efficiency improvements by tax incentives and/or credit facilities
Increased maintenance of energy-using equipment			
More heat recovery systems			
Recycling of selected materials	Scrap processing requires 5 per cent (aluminium and copper) 15 per cent (steel) of energy required to process ore	Scrap collection system Recycle-oriented design of products Less environmental pollution	Subsidy for R & D taxes or lower depletion allowances on virgin material
III. TRANSPORTATION			
Driving at slower speeds	25 per cent less fuel consumption at 80 km/hour than at 115 km/hour	Reduction in accidents involving injury/death	Speed limits
Switch to smaller cars Automobile efficiency improvements:			
Redesign of body	5%	Small cost	
Change to radial tyres	10%	Small cost/longer tread life, greater safety	Graduated tax on fuel economy
Improved load to engine matching	15-20%	$100-$200	
Installation of overdrive	20% during use	$150	
	Switch to smaller cars and efficiency improvements estimated to save 6% (1980) and 22% (1985) of transportation sector fuel		

(bracketed notes in Costs/Benefits column: "Less specific fuel consumption" / "Less engine wear")

Sectoral Energy Conservation Possibilities

Type of Conservation	Potential Savings in % of sectoral consumption (year)	Costs/Benefits	Type of policy action available
Increased commuter car pooling	1-2% (1985)	Reduces congestion	Encouragement reduced road tolls for cars containing 2 or more occupants during commuting hours.
Increased use of public transportation in cities:	Depends on present state of development of public transport system	Reduces traffic congestion and pollution	High city centre parking fees and fines
More priority bus lanes		Low cost/discourages private cars	Subsidies to public transport
Fast commuter bus services		" " "	
Extension of rail and subway systems		High investment	
More cycle paths		Low cost/increased safety encourages switch to bicycle for short trips	
Introduction of fast inter-city train service		High investment/competes with automobile and aeroplane (less than 500 km.)	Subsidy
Switch some inter-city freight to rail	1-3% (1985)	Investment in terminal handling facilities/competes with road haulage	Subsidy
Increased load factors in airplane flights	22% jet fuel savings (1980)	2.5% of transportation sector	Encouragement of co-operation between airplanes
Lower plane cruising speed	3% " "		
Longer Term Aspects:			
New propulsion technology			Support for R & D
Development of urban clusters			
Advanced communications systems to save trips			
IV. RESIDENTIAL/COMMERCIAL			
Adjustment of thermostats by 1°-2°C	6% per degree of space heating/cooling requirements (immediately)		Encouragement
More energy-conscious construction	30-40% of space heating needs in new buildings or 20-25% of sectoral consumption (1985)	2-3% of cost of new building	Revised insulation standards for new buildings; financial incentives for insulation of existing buildings
Improved thermal insulation and temperature control			
Increased use of heat pumps	5-10% space heating needs, depending on degree of insulation		
Heat recovery from ventilation stream			
More efficient appliances	1-2% of sectoral consumption (1980)	Higher initial cost/lower running costs	Efficiency labelling requirement, possibly tax on efficiency
Reduce excess lighting	4% of total electricity consumption		Encouragement
District heating schemes (using waste heat from electricity generation or burning municipal waste)		Reduces thermal pollution	Removal of institutional barriers, grants for pilot schemes
Increase R & D on better design and operation of buildings, and on appliance technology	Increases the probability of achieving maximum potential	Accelerates the impact of new energy-saving technology	Support for R & D

Source: Energy Prospects to 1985, Volume 1, Organization for Economic Co-operation and Development, Paris, France, 1974.

Possible Annual Energy Savings by Sector and End Use

Sector and End-Use	Savings in Gross Energy Input (10^15 Btu)	
	1980	1990
Industrial	4.5 to 6.4	9.0 to 12.0
Process Steam	included above	
Direct Heat	included above	
Electric Drive	included above	
Electrolytic Processes	included above	
Other	included above	
Residential	3.6	15.0
Space Heating	2.2	
Water Heating	0.25	
Air Conditioning	0.50	
Refrigeration	0.10	
Cooling	0.05	
Other, Including Lighting	0.50	
Commercial[a]	1.5	
Transportation	4.8	8.0
Total	14.4 to 16.3 (15-17%)[d]	32.0 to 35.0 (23-25%)
Equivalent Generation[b] (10^9 kWhr)	1440 to 1630	3560 to 3890
Equivalent Baseload Generating Capacity[c] (GW)	220 to 250	540 to 590

[a] Commercial end-uses.

[b] At the electric utility generation bus, assuming average plant heat rates of 10,000 Btu/kWhr in 1980, 9000 Btu/kWhr in 1990, and 8000 Btu/kWhr in 2000.

[c] At 75% generating plant capacity factor.

[d] Percentage savings refer to the total projections of U.S. energy consumption in 1980 of 96 x 10^15 Btu, in 1990 of 140 x 10^15 Btu, and in 2000 of 190 x 10^15 Btu considered in "The Nation's Energy Future," USAEC Report WASH-1281, December 1973.

Source: Proposed Final Environmental Statement Liquid Metal Fast Breeder Reactor Program, Volume III, WASH-1535, U.S. Atomic Energy Commission, December 1974.

Impact of Energy Conservation Actions

	1985 Energy Savings (MMB/D)	1985 Oil Import Reductions (MMB/D)
Transportation Sector:		
Auto efficiency standards	1.0	1.0
National Van Pool program	0.1	0.1
Improved airline load factors	0.1	0.1
Household/Commercial Sector:		
Thermal efficiency standards for new buildings	0.3	0.3
Appliance standards for labeling	0.2	0.1
Insulation tax credit	0.1	0.1
Elimination of gas pilot lights	0.2	0.2
Industrial Sector and Others:		
Industrial energy conservation program	0.6	0.3
Increased dispersed solar equipment	0.1	—
Solid waste energy combustion	0.2	—
Total	2.9	2.2

Source: National Energy Outlook, FEA-N-75/713, Federal Energy Administration, February 1976.

Potential Savings with Improved Use Efficiency

(A) United States Total Energy Savings (1975-2000), in Percentage of Sector Energy Use Compared to Base Case.

Sector	Immediate (0-1 yr) [Operational and housekeeping changes]	Near-Term (2-5 yr) [Some investments and process and equipment changes]	Long-Term (5-25 yr) [Major investments and process and equipment changes]	Annual Savings in the year 2000
Industry (Chapter 2)	10-15	10-15	10-20	30-50
Commerce (Chapter 3)	5-10	5-10	5-10	15-30
Residential (Chapter 5)	5-10	10-15	10-20	25-45
Transportation (Chapter 5)	10-15	10-15	10-20	30-50

(B) United States Electrical Energy Savings (1975-2000) in Percentage of Sector Electricity Use Compared to Base Case

Sector	Immediate (0-1 yr) [Operational and housekeeping changes]	Near-Term (2-5 yr) [Some investments and process and equipment changes]	Long-Term (5-25 yr) [Major investments and process and equipment changes]	Annual Savings in the year 2000
Industry (Chapter 2)	5-10	5-10	5-10	15-30
Commerce (Chapter 3)	5-10	5-10	10-20	20-40
Residential (Chapter 4)	5-10	5-10	15-25	25-45
Other[a]	0-5	5-10	5-10	10-25

[a] Includes estimated savings in agriculture (Chapter 6) and transportation and communication (Chapter 5).

Source: Smith, C. B., Ed., Efficient Electricity Use. New York: Pergamon Press, 1976, p. 810.

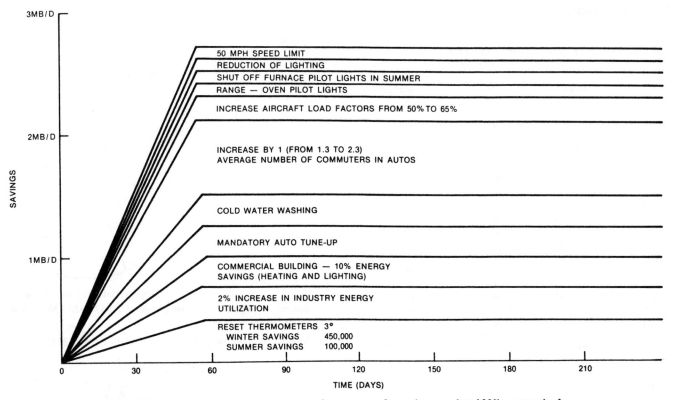

Potential energy saving of emergency conservation programs [assuming complete 100% cooperation].

Source: Project Independence, Project Independence Report, Federal Energy Administration November 1974.

Potential Electricity Savings for the United States

Sector	Potential Savings 1975–2000[a] (in Percent of Sector Electricity Use) %	Sector Importance[b] (Year 2000) (as a Percent of Total Electrical Energy) %	Weighted Annual Savings (as a Percent of Total Electricity Use in Year 2000)[c] %
Industry	15–30	45	7–14
Commerce	20–40	15	3–6
Residential	25–45	20	5–9
Other	10–25	20	2–5
Totals	—	100%	17–34

[a]See text and Table 14.1B.
[b]Projected for year 2000.
[c]Calculated by multiplying sector importance times annual savings. Totals are rounded. These electricity savings (17 to 34%) are equivalent to 10 to 20% of total energy for the base case where electricity accounts for 60% of total energy use.
Source: Smith, C. B., Ed., *Efficient Electricity Use.* New York: Pergamon Press, 1976, p. 813.

Potential Total Energy Savings for the United States

Sector	Annual Savings (in Percent of Sector Energy Use) for the Year 2000[a]	Sector Importance Year 2000[b]	Weighted Annual Savings (in Percent of Total U.S. Energy Use in Year 2000[c]	Potential Savings Year 2000[d] 10⁹ GJ/yr	Mbpd
Industry	30–50	40	12–20	22–36	10–16
Commerce	15–30	15	2–5	4–9	2–4
Residential	25–45	20	5–9	9–16	4–7
Transportation	30–50	25	8–13	14–23	6–10
Totals	—	100	27–47	49–84	22–37

[a]See text and Table 14.1 from reference.
[b]Estimated for year 2000.
[c]Calculated by multiplying sector importance times annual savings.
[d]Based on year 2000 total energy use of 180×10^9 annual GJ/yr (170×10^{15} Btu/year, 80 mbpd). This corresponds to an annual growth rate of 3.25% (22-year doubling time). Numbers are rounded. Table uses the conversion that 1 mbpd = 2.24×10^9 GJ/yr to calculate energy use in equivalent quantities of crude oil.
Source: Smith, C. B., Ed., *Efficient Electricity Use.* New York: Pergamon Press, 1976, p. 812.

FEDERAL ENERGY CONSERVATION—1974

Energy Conservation Performance by Agencies Fiscal Year 1974

Agency	Adjusted FY 73 Baseline[a]	Annual FY 74 Consumed[a]	Percent Saved	Percent of Total Government Use
Defense	1,900,401	1,426,567	25	84.49
NASA	38,118	29,043	24	1.72
GSA	69,252	54,967	21	3.26
Transportation	25,810	20,699	20	1.22
Labor	1,818	1,450	20	.08
Veterans	32,910	26,734	19	1.58
Interior	13,594	11,250	17	.66
Agriculture	11,161	9,427	16	.55
Commerce	3,704	3,225	13	.19
HEW	8,218	7,285	11	.43
Atomic energy	96,895	87,094	10	5.16
EPA	359	323	10	.01
HUD	449	411	9	.02
Treasury	3,319	3,029	8	.17
State	6.2	5.7	7	.001
Justice	6,834	6,414	6	.38
	2,212,848	1,687,923	24	100

[a]In billion Btus (Btus × 10^9)
Source: Federal Energy Administration Release, Jan. 3, 1975.

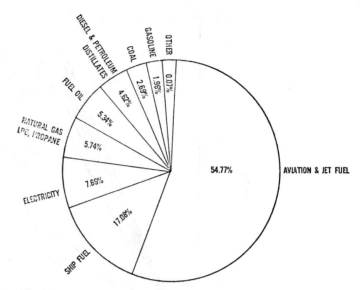

Federal energy savings by source fiscal year 1974.

Savings are expressed as a percentage of the 524,920,330,000,000 [524,920 × 10^9] Btu's saved in total by federal agencies.

Source: Federal Energy Administration Release, Jan. 3, 1975.

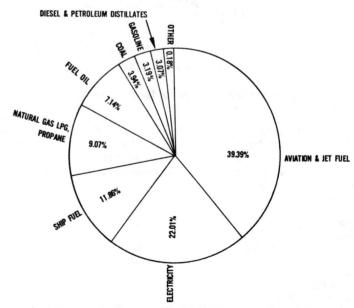

Federal energy consumption by source fiscal year 1974.

Energy use expressed as percentage of the 1,687,926,470,000,000 [1,687,926 × 10^9] total Btus used by federal agencies.

Source: Federal Energy Administration Release, Jan. 3, 1975.

RESIDENTIAL ENERGY REQUIREMENTS

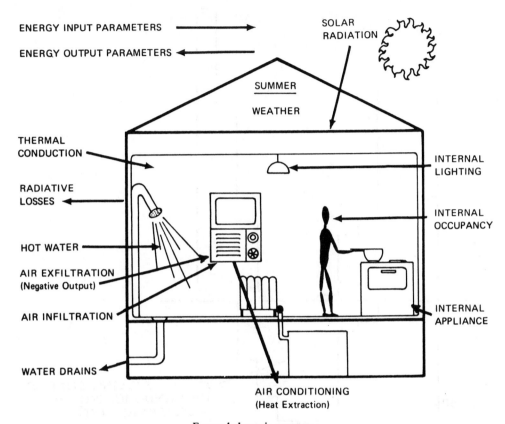

Energy balance in summer.

Source: Residential Energy Conservation, HUD-HAI-8, prepared for the Department of Housing and Urban Development by Hittman Associates, Inc., Columbia, Maryland, July 1974.

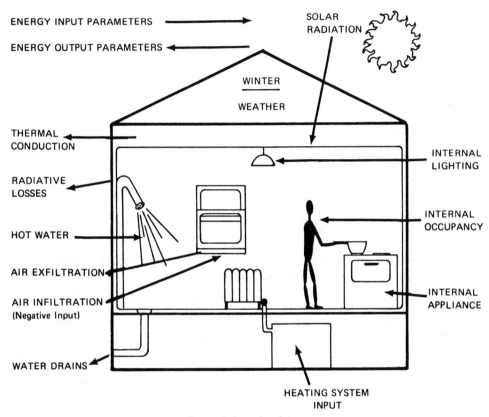

Energy balance in winter.

Source: Residential Energy Conservation, HUD-HAI-8, prepared for the Department of Housing and Urban Development by Hittman Associates, Inc., Columbia, Maryland, July 1974.

Breakdown of Heating and Cooling Loads for a Typical House in the Baltimore-Washington Area*

		% of Heating Load	% of Cooling Load
Ceiling		3.7	2.3
Floor		2.2	2.4
Total Window		13.6	4.1
Total Door		1.4	0.4
Total Wall		23.9	14.2
Infiltration Load		55.2	41.5
Internal Load		—	35.1
	Total	100.0	100.0
Total Load, Therms		710	282
Total Load, kJ		750×10^5	300×10^5

* These loads are for an average house in the Baltimore-Washington area. The values presented will vary significantly for different geographic locations.

Annual Primary Energy Consumption for a Typical Residential Structure in the Baltimore-Washington Area

Component	All electric house, therms*	Minimum energy house, therms
Hot Water Heating	480**	270(G)
Lights	218	218 (E)
Range	120	50(G)
Refrigerator/Freezer	200	200 (E)
Clothes Dryer	108	40(G)
Colour TV	54	54 (E)
Furnace Fan	0	43 (E)
Dishwasher	40	40 (E)
Clothes Washer	11	11 (E)
Iron	16	16 (E)
Coffee Maker	11	11 (E)
Outside Light	90	90 (E)
Miscellaneous	131	131 (E)
	1,487	1,174

(E) Electric
(G) Gas

* One therm = 10^5 Btu = 1.055×10^5 kJ.
** Electric power based on power plant energy consumption of 10,910 Btu/kW-hr.

* These features are for a single family house located in the Baltimore/Washington area of the United States.

[1] ANDERSON, R.W. and HARVEY, D.G., "Energy Consumption in Single Family Residences" 8th Intersociety IECEC Conference, August 1973.

Source: Technology of Efficient Energy Utilization, Report of a NATO Science Committee, Les Arcs, France, October 8–12, 1973, Scientific Affaris Division, North Atlantic Treaty Organization, Brussels, Belgium. Conference report has been reprinted (with NATO's permission by Pergamon Press, Oxford, UK, and available from that publisher.

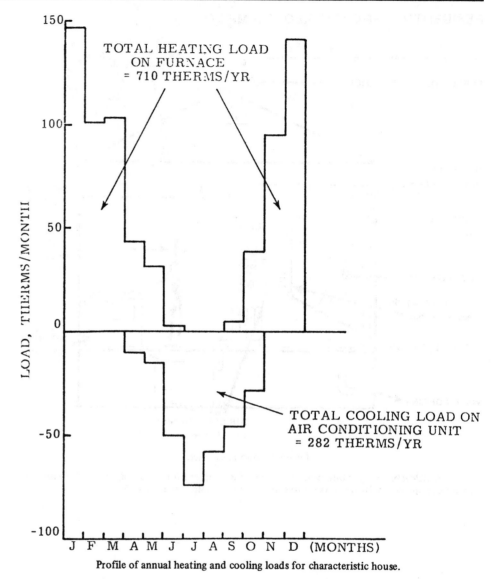

Profile of annual heating and cooling loads for characteristic house.

Source: Residential Energy Consumption: Single-Family Housing Final Report, Department of Housing and Urban Development, March 1973.

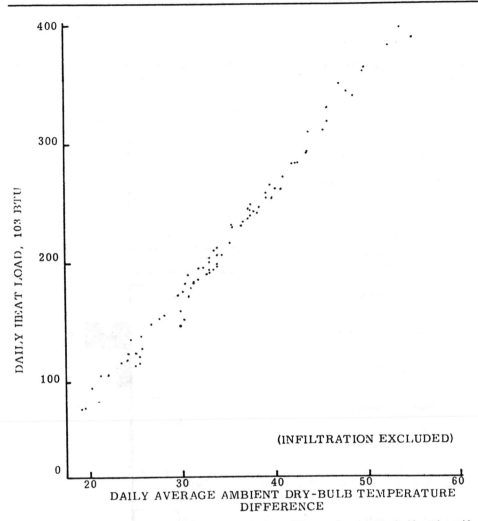

(INFILTRATION EXCLUDED)

Typical heating load correlation with dry-bulb temperature difference between the inside and outside air.

Source: Residential Energy Consumption: Single-Family Housing Final Report, Department of Housing and Urban Development, March 1973.

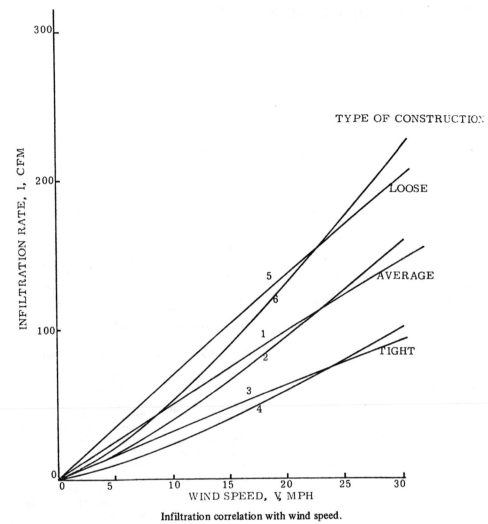

Infiltration correlation with wind speed.

Source: Residential Energy Consumption: Single-Family Housing Final Report, Department of Housing and Urban Development, March 1973.

RESIDENTIAL ENERGY CONSERVATION

Design Parameters for Characteristic Single-Family Detached Dwelling in Baltimore-Washington Area

STRUCTURAL PARAMETERS:

Number of stories	Two
Basement	Full (unfinished)
Garage	Attached, open carport
Floor area, ft^2	1695 finished
Construction type	Wood frame
Exterior walls:	
Outside surface	Wood shiplap
Sheathing	Plywood, 1/2 inch
Insulation	R-7 batting*
Inside surface	Drywall, 1/2 inch
Ceiling insulation	Loose fill blown-in, 5 inches
Attic	Ventilated, unheated
Roof	Asphalt shingles
Windows:	
Type	Aluminum casement
Glazing	Single
Area, ft^2	190
Exterior doors:	
Type	Wood panel
Number	3
Total area, ft^2	60
Patio door:	
Type	Aluminum frame
Glazing	Single
Area, ft^2	40

*R7 equivalent to 2-1/4 in. fiberglass batting.

ENERGY CONSUMPTION PARAMETERS:

Energy consuming equipment:

Heating system	Forced air, gas
Cooling system	Forced air, electric
Hot water heater	Gas
Cooking range	Gas (90 therms/year)*
Clothes dryer	Gas
Refrigerator/Freezer	Electric (1830 Kw-hr/year)*
Lights	Electric (2000 Kw-hr/year)*
Color TV	Electric (500 Kw-hr/year)*
Furnace fan	Electric (394 Kw-hr/year)*
Dishwasher	Electric (363 Kw-hr/year)*
Clothes washer	Electric (103 Kw-hr/year)*
Iron	Electric (144 Kw-hr/year)*
Coffee maker	Electric (106 Kw-hr/year)*
Miscellaneous	Electric (1200 Kw-hr/year)*

*Energy input to structure due to use of item.

Factors affecting heating/cooling load: (Base case)

Exterior glass areas	70% draped
	20% shaded
	10% open
	Patio door on south wall
	No awnings, no storm windows
External landscaping	No shading effect
Dwelling facing	North
External colors	White roof and walls
People	Two adults, two children
Weather	Data for 1954 from Baltimore, Md. weather station
Garage location	West side

Source: Residential Energy Conservation, HUD-HAI-8, prepared for the Department of Housing and Urban Development by Hittman Associates, Inc., Columbia, Maryland, July 1974.

Effects of Structure Related Modifications on the Characteristic Single-Family Residence

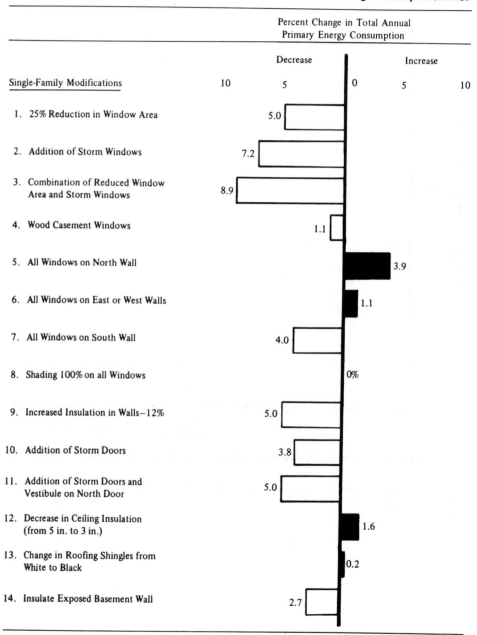

Source: Residential Energy Conservation, HUD-HAI-8, prepared for the Department of Housing and Urban Development by Hittman Associates, Inc., Columbia, Maryland, July 1974.

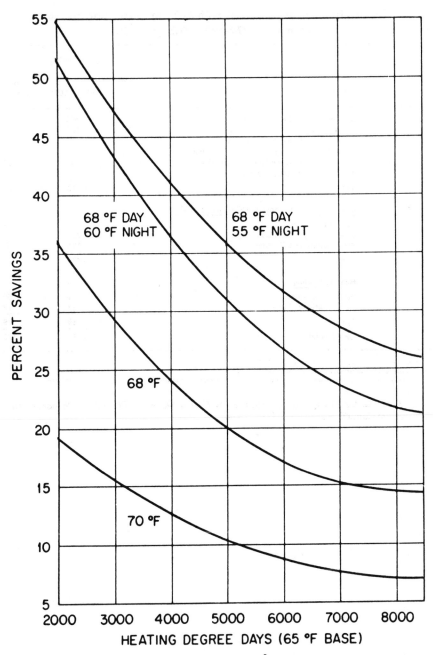

Predicted energy savings for several thermostat settings (72°F is the reference setting, and night setback is from 10 p.m. to 6 a.m.).

Source: David A. Pilati, *The Energy Conservation Potential of Winter Thermostat Reductions and Night Setback*, National Science Foundation, 1975.

Increases in Thermostat "Comfort" Settings During Summer Season Possible Through Clothing Changes

Women	Raise Thermostat Setting (°F)	Men	Raise Thermostat Setting (°F)
Replace light slacks with light skirt	1.5	Replace heavy trousers with light trousers	.6
Replace long-sleeved dress with sleeveless dress	.2	Replace winter-weight jacket with summer jacket	2.5
Replace dress made of tightly woven cloth with one having an open weave	.5	Replace long-sleeved shirt with short-sleeved shirt	.8
Remove stockings	.1	Replace long light trousers with Bermuda shorts	1.0
Replace full slip with half slip	.6	Remove summer weight jacket	2.0
Remove full slip	1.0	Remove undershirt (T-shirt)	.5
Replace pumps with sandals	.2	Remove tie and open collar	.2
Remove hat	.2	Replace knee-length socks with ankle ankle socks	.6
Remove light long-sleeved sweater	1.7	Remove light long-sleeved sweater	2.0
Remove heavy long-sleeved sweater	3.7	Remove heavy long-sleeved sweater	3.7

Chart provided by Dr. Ralph Goldman, U.S. Army Ergonomics Laboratory and the John B. Pierce Foundation, New Haven, Connecticut.

The Federal Energy Administration has estimated that raising the thermostat setting during the summer season by 1°F would save the equivalent of 100,000 barrels of oil per day in the United States.

Source: Energy Reporter, Federal Energy Administration, Citizen Newsletter, August/September 1975.

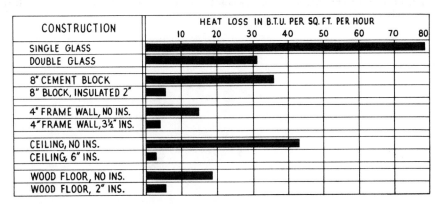

This bar graph shows transfer of heat through common types of material. Next to glass, the big losers are concrete and uninsulated frame construction. Adequate insulation and double glass lower heat loss from 17% to 70%, depending upon materials and location.

Source: Conservation and Efficient Use of Energy, Part 1, Joint Hearings before certain Subcommittees of the Committees on Government Operations and Science and Astronautics, House of Representatives, June 19, 1973.

Estimated Energy Savings Through Implementation of Currently Available Conservation Measures

Sector	Conservation Measure	Type[a]	Estimated Percent Implementation Existing (1973)	Estimated Percent Implementation Projected (Year)	Estimated Reduction in Annual Energy Consumption for Indicated Increase in Implementation, % of 1973 Energy Consumption In Sector[b]
Residential	adequate insulation	TM	30	70(1978)	6
	storm windows, storm doors	TM	30	70(1978)	4
	maximum efficiency heating/cooling and hot water equipment	TM	30	85(1990)	5
	adequate maintenance of heating/cooling and hot water equipment	OP	30	85(1975)	8
	10°F night setback of thermostat	LS	40	85(1975)	3
	other energy conservative habits—control of lights, doors, windows, hot water, shading	LS	20	80(1975)	3

[a]LS—lifestyle, TM—technical measure, OP—operational.

[b]Energy savings are not generally additive because each measure is assumed to be independent of the others.

Source: Conservation and Efficient Use of Energy, Part 4, Joint Hearings before certain Subcommittees of the Committees on Government Operations and Science and Astronautics, House of Representatives, July 12, 1973.

Existing Residential Construction Short-Term Energy Conservation Potentials—1974-1978 (Trillion BTU's)

Rank	Life-Style	Savings Total Potential Percent*	Savings Total Potential BTU x 10^{12}	Maximum Assumed Achievable
1	Set Thermostat Back to 68°F (day)	15	993	745
3	Set Water Heater Back to 120°F	22.5	405	203
7	Set Air Conditioner Thermostat Up to 78°F	15	97	49
8	Set 8-Hour Thermostat Back to 60°F (night)	7	463	46
9	Reduce Bathing Water Consumption One-Third	14	252	25
10	Turn Off Pilots in Gas Furnace†	30	56	17
	Total		2,266	1,085
	Insulation, Storm Doors and Windows, Caulking			
2	Ceiling Insulation	11	733	367
4	Weatherstripping and Caulking	5	325	140
6	Storm Doors and Windows	2	128	64
	Total		1,186	571
	Heating/Cooling			
5	Furnace Tune-up	10	662	66
11	Air Conditioner Tune-up	15	97	13
	Total		759	79
	Grand Total	35.0‡	4,211	1,735

* Percentages applied to 1972 utilization data in Table 10.

† Assumes homeowners will be educated on relighting furnaces.

‡ Total potential savings may not be additive as measures included therein are interdependent and/or mutually exclusive.

Source: Potential for Energy Conservation in the United States: 1974-1978, A Report of the National Petroleum Council, Washington, D.C., 1974.

ENERGY CONSERVATION POTENTIAL—COMMERCIAL

Estimated Energy Savings Through Implementation of Currently Available Conservation Measures

Sector	Conservation Measure	Type[a]	Existing (1973)	Projected (Year)	Estimated Reduction in Annual Energy Consumption for Indicated Increase in Implementation, % of 1973 Energy Consumption In Sector[b]
			Estimated Percent Implementation		
Commercial	adequate insulation	TM	50	70(1990)	3
	storm windows, storm doors, vestibules, revolving doors	TM	40	70(1980)	3
	maximum efficiency heating/cooling, hot water, and ventilation equipment	TM	30	80(1990)	5
	adequate maintenance of heating/cooling, hot water ventilation and other energy-related equipment	OP	30	80(1980)	8
	automatic control systems adjusted to minimize energy consumption	OP	5	20(1980)	3
	manual control to minimize energy consumption—night setback, lights, etc.	OP	40	70(1980)	2
	other energy conservative measures	OP	40	70(1980)	2

[a]LS—lifestyle, TM—technical measure, OP—operational.
[b]Energy savings are not generally additive because each measure is assumed to be independent of the others.
Source: *Conservation and Efficient Use of Energy*, Part 4, Joint Hearings before certain Subcommittees of the Committees on Government Operations and Science Astronautics, House of Representatives, July 12, 1973.

Energy Conservation Measures by Rank, Category and Yield-Commercial Sector

Rank by Potential Yield	Category	Conservation Measure	Total Potential Total (QBTU)	Total Potential Percent of 1972 Commercial Sector	Assumed Achievable Total (QBTU)	Assumed Achievable Percent of 1972 Commercial Sector
1.	I	Establish a 68°F. maximum occupied temperature level in apartments and hotel/motels and 65°F. in commercial establishments; hospitals and nursing homes excepted.	1.084	9.0	.358	3.0
2.	I	Establish a 5°F. night setback below day levels in apartments and 10°F. for commercial buildings during unoccupied hours; hospitals and nursing homes excepted.	.688	5.7	.227	1.9
3.	II	Caulk and weatherstrip around all windows and between building walls and window frames.	.591	4.9	.089	.7
4.	II	Scheduled maintenance on equipment and systems.	.587	4.9	.088	.7
5.	III	Insulate ceiling, above or below roofs, using insulation having an equivalent "R" factor of 19.	.542	4.5	.011	.0
6.	III	Insulate sidewalls using insulation having an equivalent "R" factor of 11.	.513	4.3	.010	.0
7.	III	Install storm sash, or high efficiency glass.	.493	4.1	.010	.0
8.	I	Reduce lighting levels to a minimum acceptable level where possible.	.292	2.4	.096	.7
9.	II	Establish minimum ventilating air requirements for occupancy periods and zero ventilation during unoccupied periods where possible.	.159	1.3	.024	.2
10.	II	Use restricted flow shower heads. (2.5 gallons per minute maximum)	.097	.8	.015	.1
11.*	I	Establish a cooling comfort level of 78°F. if basic energy is necessary.	.096	.8	.032	.2
12.	I	Cease cooling of building at least one hour before termination of occupancy.	.086	.7	.028	.2
13.	II	Use automatic shutoff faucets in lavatories.	.064	.5	.010	.0
14.	II	Reduce water distribution pressure to a maximum of 25 p.s.i.	.063	.5	.009	.0
15.	I	Reduce temperature of general purpose hot water by 20°F. (120°F. minimum) except where dishwashers require otherwise.	.059	.5	.019	.1

*Reheat systems should be considered as separate cases.

Source: *Potential for Energy Conservation in the United States: 1974-1978*, A Report of the National Petroleum Council, Washington, D.C., 1974.

ENERGY CONSUMPTION—INDUSTRIAL

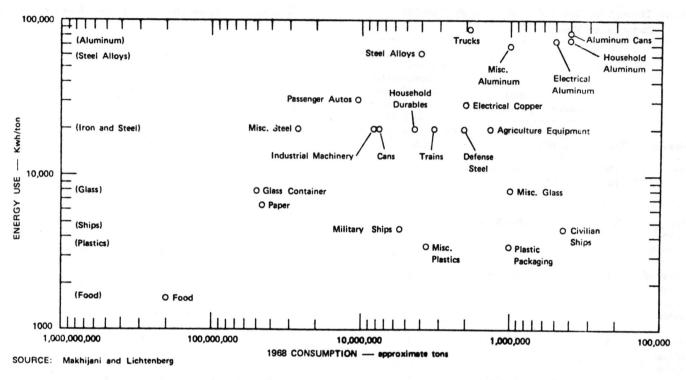

SOURCE: Makhijani and Lichtenberg

Energy use in commodities (U.S.).

Source: Stanford Research Institute, *Support of Energy Program Planning*, sponsored by the Advanced Projects Agency, Menlo Park, California, September 1972, as shown in *Energy Facts*, prepared by the Science Policy Research Division, Congressional Research Service, Library of Congress, for the Subcommittee on Energy of the Committee on Science and Astronautics, U.S. House of Representatives, November 1973.

Typical Energy Contents of Materials and Manufactured Products

	Energy* megajoules/kg	Cost of energy/* Value of product
Metals		
Steel (various forms)	25-50	0.3
Aluminum (various forms)	60-270	0.4
Copper	25-30	0.05
Magnesium	80-100	0.1
Other Products		
Glass (bottles)	30-50	0.3
Plastic	10	0.04
Paper	25	0.3
Inorganic chemicals (average value)	12	0.2
Cement	9	0.5
Lumber	4	0.1

* These are typical values. The actual value depends on the purity, form, manufacturing process and other variables.

Source: Technology of Efficient Energy Utilization, Report of a NATO Science Committee Conference, Les Arcs, France, October 8-12, 1973, Scientific Affairs Division, North Atlantic Treaty Organization, Brussels, Belgium. Conference report has been reprinted (with NATO's permission) by Pergamon Press, Oxford, UK, and available from that publisher.

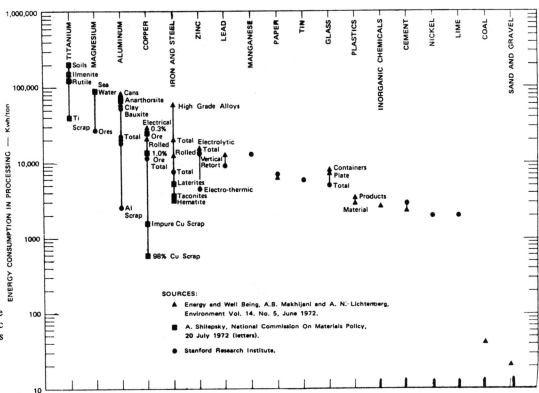

Energy consumption in processing of selected materials.

Source: Stanford Research Institute, *Support of Energy Program Planning*, sponsored by the Advanced Research Projects Agency, Menlo Park, Calif, September 1972, as shown in *Energy Facts*, prepared by the Science Policy Research Division, Congressional Research Service, Library of Congress, for the Subcommittee on Energy of the Committee on Science and Astronautics, U.S. House of Representatives, Nov. 1973.

ENERGY CONSERVATION POTENTIAL—INDUSTRIAL

Comparison of Specific Fuel Consumption of Known Processes With Theoretical Minimum for Selected U.S. Industries

	1968 Specific Fuel Consumption (Btu/ton)	Potential Specific Fuel Consumption Using Technology Existing in 1973 (Btu/ton)	Theoretical Minimum Specific Fuel Consumption Based Upon Thermodynamic Availability Analysis (Btu/ton)
Iron and Steel	26.5×10^6	17.2×10^6	6.0×10^6
Petroleum Refining	4.4×10^6	3.3×10^6	0.4×10^6
Paper	*39.0×10^6	*23.8×10^6	†Greater than -0.2×10^6 Smaller than $+0.1 \times 10^6$
Primary Aluminum Production**	190×10^6	152×10^6	25.2×10^6
Cement	7.9×10^6	4.7×10^6	0.8×10^6

*Includes 14.5×10^6 Btu/ton of paper produced from waste products consumed as fuel by paper industry.

**Does not include effect of scrap recycling.

†Negative value means that no fuel is required.

From "Potential Fuel Effectiveness in Industry," Copyright 1974, the Ford Foundation. Reprinted with permission of the Ballinger Publishing Company.

Industries Ranked by Dollars on Energy Requirements

Sic Code	Description	Value of Shipments (mill $)	Cost of Materials (mill $)	Purchased Fuels and Electric Energy (mill $)	Purchased Fuels and Electric Energy (% of shipment value)
331	Blast furnace & basic steel products	25,790.4	14,677.7	1,250.5	4.8
281	Industrial chemicals	16,524.3	7,545.1	1,119.4	6.8
291	Petroleum refining	24,583.7	20,091.9	585.6	2.4
262	Papermills, except building paper	5,825.7	3,229.6	375.3	6.4
333	Primary nonferrous metals	5,290.6	3,930.0	320.0	6.0
371	Motor vehicles and equipment	58,138.0	37,575.7	305.1	0.5
282	Plastic materials and synthetics	9,345.6	4,569.4	286.5	3.1
263	Paperboard mills	3,536.4	1,885.5	256.2	7.2
324	Cement, hydraulic	1,560.3	564.9	243.1	15.6
332	Iron & steel foundries	4,977.5	1,956.8	207.4	4.2
327	Concrete, gypsum & plaster products	6,545.8	3,291.1	179.7	2.7
335	Nonferrous rolling and drawing	11,316.2	8,062.9	173.4	1.5
322	Glass & glassware, pressed or blown	3,052.0	1,009.3	146.7	4.8
209	Miscellaneous food and kindred products	12,580.7	8,651.8	136.0	1.1

Source: The Costs of Purchased Fuels and Electric Energy by Industry, Economic Stabilization Program, Cost of Living Council, Washington, D.C., December 26, 1973.

Estimated Energy Savings Through Implementation of Currently Available Conservation Measures

Sector	Conservation Measure	Type[a]	Estimated Percent Implementation		Estimated Reduction in Annual Energy Consumption for Indicated Increase in Implementation, % of 1973 Energy Consumption In Sector[b]
			Existing (1973)	Projected (Year)	
Industrial	steel—basic oxygen process	TM	53	80(1980)	8.7
	steel—continuous casting	TM	—	—	—
	aluminum—primary production	TM		30(1980)	2.4
	increased recycling and reuse of materials and products	TM	—	—(1975)	0.8
	effective waste heat utilization	TM	—	—	—
	total energy systems	TM	—	—	—
	industrial practices—education and action eliminating wasteful practices	TM	—	—(1975)	0.2
	industrial practices—rescheduling for off-peak periods	OP	—	—(1975)	2.0

[a]LS—lifestyle, TM—technical measure, OP—operational.
[b]Energy savings are not generally additive because each measure is assumed to be independent of the others.
Source: Conservation and Efficient Use of Energy, Part 4, Joint Hearings before certain Subcommittees of the Committees on Government Operations and Science and Astronautics, House of Representatives, July 12, 1973.

General Principles of Energy Management

Principle	Relative Cost	Relative Time to Implement	Relative Complexity	Relative Benefit (Typical)
1. Review historical energy use (review of historical data	Low	<1 year	Low	5-10%
2. Energy audits (review of current practices)	Low	<1 year	Low	5-10%
3. Analysis of energy use (engineering analysis, computer simulation, availability studies)	Low to Moderate	1-2 years	Moderate to High	10-20%
4. More efficient equipment	Moderate to High	years	Moderate to High	10-30%
5. More efficient processes	Moderate to High	years	Moderate to high	10-30%
6. Energy containment (heat recovery and waste reduction)	Low to Moderate	years	Moderate to high	10-50%
7. Substitute material	Low to Moderate	<1 year	Low	10-20%
8. Material economy (scrap recovery, salvage, and recycle)	Low	1-2 years	Low to High	10-50%
9. Material quality selection (material purity and properties)	Low	1 year	Low	5-15%
10. Aggregation of energy uses	Moderate to High	years	Moderate to High	20-50%
11. Cascade of energy uses	Moderate to High	years	Moderate to High	20-50%
12. Alternative energy sources (energy form or fuel substitution)	Moderate to high	years	Moderate to High	10-30%
13. Energy conversion	Moderate to High	years	Moderate to High	10-30%
14. Energy storage	Moderate to High	years	Moderate to High	10-30%
15. Economic evaluation (cost benefit, rate of return, life-cycle costing)	Low	<1 year	Low	5-15%

Source: Proceedings of an EPRI Workshop on Technologies for Conservation and Efficient Utilization of Electric Energy, EPRI EM-313-SR, July 1976.

ENERGY CONSUMPTION AND CONSERVATION—TRANSPORTATION

Transportation Energy and Related Consumption by Mode and Purpose*—1972

	Thousand Barrels Per Day			Quadrillion BTU's	Percent of Total Transportation Energy
	Gasoline	Distillate†	Other		
Highway					
Passenger Cars					
Private (Personal Use)	3,886.0	—	—	7.46	42.96
Commercial and Other	870.8	—	—	1.67	9.63
Total	4,756.8‡	—	—	9.13	52.59
Single Unit (Light Trucks)					
Private (Personal Use)	669.1			1.29	7.40
Commercial §	705.3	31.9	—	1.42	8.15
Government	32.9	.2	—	.06	.36
Total	1,407.3	32.1	—	2.77	15.91
Combination (Heavy) Trucks					
Commercial §	69.4	483.0	—	1.16	6.11
Government	.9	6.2	—	.01	.07
Total	70.3	489.2	—	1.17	6.18
Buses					
School	20.4	.4	—	.04	.23
Urban	2.0	18.8	1.6‖	.05	.25
Intercity	3.0	12.1	—	.03	.16
Total	24.4	31.3	—	.12	.64
Motorcycles	22.3	—	—	.04	.25
Total Highways	6,281.0	552.6	1.6	13.23	75.57
	6,835.1				
Airways					
Airlines					
Scheduled	—	670.0	—	1.39	7.41
Supplemental	—	6.0	—	.01	.07
Total	—	676.0	—	1.40	7.48
General Aviation	46.0	37.0	—	.17	.92
Military	—	288.0#	—	.57	3.18
Factory and Miscellaneous	—	20.0	—	.04	.22
Total Airways	46.0	1,021.0	—	2.18	11.80
	1,067.0				
Railways	—	247.3	3.1**	.53	2.77
	—	.8††	—	—	.01
Total Railways	—	248.1	3.1	.53	2.78
	251.2				

Transportation Energy and Related Consumption by Mode and Purpose*—1972 (Continued)

	Thousand Barrels Per Day			Quadrillion BTU's/Year	Percent of Total Transportation Energy
	Gasoline	Distillate†	Other		
Waterways					
Private and Commercial	44.8	65.9	—	.23	1.22
Commercial					
At Port	—	—	41.0	.09	.45
At Sea	—	—	196.0	.45	2.18
Total Waterways	44.8	65.9	237.7	—	—
	348.4			.77	3.85
Urban Public Transit (Nonhighway)††					
Rapid Transit	—	—	3.4	.01	.04
Surface Railway	—	—	.2	—	—
Trolley Coach	—	—	.2	—	—
Total UPT	—	—	3.9	.01	.04
Pipeline					
Total Pipeline	—	167.9‡‡	371.0§§	—	
	538.9			1.15	5.96
Total Transportation Energy	6,371.8	2,055.5	617.3	17.86	100.00
	9,044.6				
Miscellaneous					
Farm Equipment‖‖	134.0	144.0	—	.56	—
Construction Equipment	45.0	281.0	—	.69	—
Utility Engines##	22.0	—	—	.04	—
Snowmobiles	5.0	—	—	.01	—
Race Cars	.5	—	—	—	—
Total Miscellaneous	206.5	425.0	—	1.30	—
	631.5				
GRAND TOTAL	6,578.3	2,480.5	617.3	19.16	—
	9,676.1				

*Data may not agree with Bureau of Mines data as some volumes are estimated and some are based on Federal Highway Administration or tax data which could include changes in secondary inventories.
†Distillate as used includes the full range of middle distillate oils including diesel fuels, kerosine jet fuel, marine diesel and also naphtha jet fuel.
‡Due to the necessity of using data as described in footnote (*), this volume and the respective BTU value does not precisely agree with the values shown by the Patterns of Consumption/Energy Demand Task Group.
§Private business and for hire.
‖Propane.
#242,000 barrels per day naphtha jet fuel; 46,000 barrels per day kerosine jet fuel.
**Residual oil.
††Electricity converted to distillate equivalent.
‡‡Liquids pipeline fuels converted to distillate equivalent.
§§Natural gas pipeline fuels converted to distillate equivalent.
‖‖Fuel for motive purposes.
##Small horsepower engines, lawnmowers, tillers, etc.

Source: Potential for Energy Conservation in the United States: 1974-1978, A Report of the National Petroleum Council, Washington, D.C., 1974.

Source: Potential for Energy Conservation in the United States: 1974-1978, A Report of the National Petroleum Council, Washington, D.C., 1974.

Estimated Energy Savings Through Implementation of Currently Available Conservation Measures

| Sector | Conservation Measure | Type[a] | Estimated Percent Implementation | | Estimated Reduction in Annual Energy Consumption for Indicated Increase in Implementation, % of 1973 |
			Existing (1973)	Projected (Year)	Energy Consumption In Sector[b]
Transportation	transition to smaller cars	TM	20	70(1985)	9.0
	use of low loss (radial) tires	TM	5	60(1976)	1.5
	increase in number of passengers per car	LS	2.2/car[c]	2.5/car[c] (1985)	7.0
	50 mph maximum cruising speeds	LS	—	Aver #(1985)	6.0
	growth in railroad's share of ton-miles	TM	—	4% Increase[c] (1980)	1.5
	growth in mass transit, clustering, mode shifts	LS	—	—(1985)	1.5
	growth in intercity railroad passenger-miles	LS	—	—(1985)	3.5

[a]LS–lifestyle, TM–technical measure.
[b]Energy savings are not generally additive because each measure is assumed to be independent of the others.
[c]Note departure from column heading.

Source: *Conservation and Efficient Use of Energy*, Part 4, Joint Hearings before certain Subcommittees of the Committees on Government Operations and Science and Astronautics, House of Representatives, July 12, 1973.

Summary of Measures and Potential for Energy Conservation in Commercial Air Transport

| Conservation Measure | Goal | Effectiveness | | Percent Reduction in Airline Energy Requirements | |
		Short Term	Long Term	1980	2000
Improve load factor	Industry average ton load factor of 59 percent	X	X	21.7	21.5
Reduce cruising speed	Minimum fuel consumption speed for each aircraft	X	X	3.6	3.8
Maintain aircraft delay level	Maintain per-operation delay at 1971 level	--	X	--	1.0
Substitute transport modes	Replace air travel with autos, TACV, and trucks in routes less than 500 mi	--	X	--	5.0[a]
Total reduction as a percent of:					
Base-case airline energy		--	--	25.3	31.3
Total transport energy		--	--	2.4	7.9
Total energy--all uses		--	--	0.6	2.0

[a]Represents net energy conservation of mode substitution, i.e., the difference between the energy required by airlines and that required by automobiles, TACV, and trucks. The gross reduction in airline fuel requirements is 7.8 percent.

Source: *The Potential for Energy Conservation in Commercial Air Transport*, The Rand Corporation, R-1360-NSF, October 1973.

12

Energy Transport

The Chinese were the first to pipe natural gas; they were using bamboo pipes 3000 years ago. This photo, taken in recent years, shows crude gas piping in use in Tseliutsing, Szechwan Province.

MODES OF ENERGY TRANSPORT

The transport of energy, in one form or another, has been a determinant in the creation and the structure of modern society. Initially, industries were established on streams and rivers where water power was available and could be converted directly into mechanical energy. The invention of the steam engine permitted the use of coal resources, but industrial plants were generally located adjacent to coal mines until the development of railroads in the early 1900s made the bulk transport of coal from mines possible. In the late 1900s, the discovery of oil deposits, the development of pipeline technology and tanker technology, and the introduction of the internal combustion engine led to another shift in industrial and social development. Also in the late 1900s, the advent of electrical energy and the transportability of this energy over wires, added yet another aspect to the development of the structure of society. Today, the transport of energy has become essential to the economic welfare of nations. The availability of that energy has created the basic fabric of life in all developed countries.

The cost, feasibility, and convenience of transport of various forms of energy have led to the present pattern of energy resource utilization on a regional, national, and international scale. Petroleum, which can be easily and cheaply transported by tanker, represents the major energy commodity in international trade. Natural gas, which can be transported cheaply by pipeline but is expensive to transport across oceans, is used mainly in land areas where production and consumption are contiguous. Although some coal is traded internationally, most coal is consumed within a few hundred miles of the point where it is mined because of transport costs. Electricity is also generally used within a few hundred miles of its point of generation because of the cost of transmission over long distances.

Primary Resources	Resource Extraction Processes	Supply Transportation	Primary Conversion	Product Transportation	Secondary Conversion	Distribution	End-Use Conversion
Foreign oil	Import	Tankers	Sour-crude refining	Tankers	Combustion power generation	Trucks	Auto engines
Domestic oil	Exploration and production	Crude pipelines	Sweet-crude refining	Liquid pipelines	Combined-cycle power generation	Tank cars	Bus/Truck engines
Oil shale	Mining and extraction	Gas pipelines	Shale refining	Low-Btu gas pipelines	Large fuel cells	Gas lines	Aircraft engines
Tar sand		Unit trains	Power generation	High-Btu gas pipelines		Power lines	Ship engines
High-sulfur coal	Mining and retorting	Slurry pipelines	Hydrogen from coal	Unit trains		Service stations	Combustion
Low-sulfur coal	Surface mining		High-Btu gasification	Power transmission		Local deliveries	Resistance heating
Natural gas	Underground mining		Solvent refining				Heat pumps
Imported gas			Liquefaction				Catalytic burners
Imported methanol	Enrichment		Low-Btu gasification				Electro-mechanical devices
Uranium			Methanol refining				Steam generation
			Nuclear-thermal water splitting				Small fuel cells
			Nuclear power generation				Chemical manufacturing
							General manufacturing

Transportation of energy is an important link in the utilization of energy between the resource extraction, primary conversion, secondary conversion, and end-use stages.

Source: Milton Searl, "Sensing the Energy Future," *EPRI Journal*, October 1976.

			Wood	Coal	Petroleum	Gas	Heat (Steam)	Electricity	Hydro Power
Land	Batch	Armload	▨						
		Pack	▨						
		Basket		▨					
		Pot			▨				
		Wagon	▨	▨	▨				
		Truck	▨	▨					
		Rail	▨	30	>15				
		Vehicle Fuel Tank			▨				
	Continuous	Aqueduct							▨
		Pipeline			10	20			
		Transmission Line						50	
		Slurry Pipeline		30					
Sea	Batch	Cargo Ship	▨	<30					
		Collier		<30					
		Barge		<30					
		Tanker			5				
		LNG Tanker				>20			
		Supertanker			<5				

Transport of basic energy commodities is accomplished either by land or sea. Transport by sea is, by nature, a batch process. Transport over land may be either continuous (pipeline) or by batch process. Shaded areas indicate practical means of transport for the various commodities. The numbers represent the cost of transport, in cents per million BTUs, over a 100-mile distance.

Daniel B. Luten, "The Economic Geography of Energy," in *Energy and Power*, A Scientific American Book, W. H. Freeman and Company, San Francisco, 1971.

Barge transport is one of the least expensive ways to move coal. This is the docking facility at the Bruce Mansfield electric power plant at Shippingport, Pennsylvania, on the Ohio River. (*Courtesy Ohio Edison Company*)

Loaded cars of a unit train travel from the coal mine to the electric utilities or other large customer. When the train is unloaded, it will speed back to the mine to be refilled and then return once again. Unit trains which carry nothing but coal directly from mine to market have helped lower transportation costs, thus making the delivered coal price a bargain. (*Courtesy National Coal Association, 1966*)

A coalburg tug and eight coal barges go downstream through the London locks, one of three lock systems on the Kanawha River. The other locks are at Marmet and Winfield. The coal is being transported to a power plant on the Ohio River, South of Charleston, WV.

WORLD OIL MOVEMENTS

Until 1973, products derived from petroleum provided relatively inexpensive sources of energy. These products are also convenient to transport. In addition, petroleum products have a wide range of uses—as fuel for power plants, for heating, and for internal combustion engines in transport vehicles. While petroleum products have become more expensive, there are no readily available substitutes for many applications and demand is expected to grow.

The world geographic distribution of petroleum resources does not match the geographic consumption needs and, as a result, petroleum products represent a major commodity in international trade.

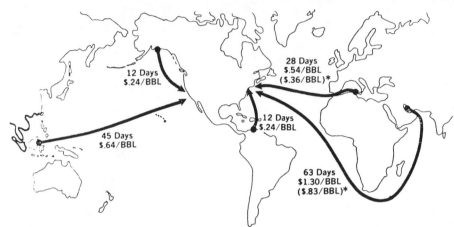

*Costs with supertankers and superports.

Transportation of U.S. petroleum imports 1972–days per voyage and costs per barrel with and without superports.

Costs in 1972 dollars. Costs are for U.S. flag vessels between Alaska and the U.S. and for foreign flag vessels on other routes. Alaskan and Indonesian costs are based on 160,000 DWT tankers. Venezuelan cost is based on 65,000 DWT tankers. Mediterranean and Persian Gulf costs are based on 65,000 DWT tankers without use of superports (and 375,000 DWT tankers with superports).

Source: White House News release (U.S. Treasury Dept.), April 18, 1973, chart 5.

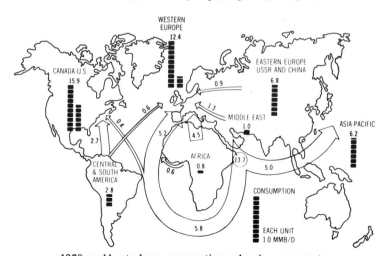

1970 world petroleum consumption and major movements.

Source: The National Energy Outlook, Shell Report, March, 1973.

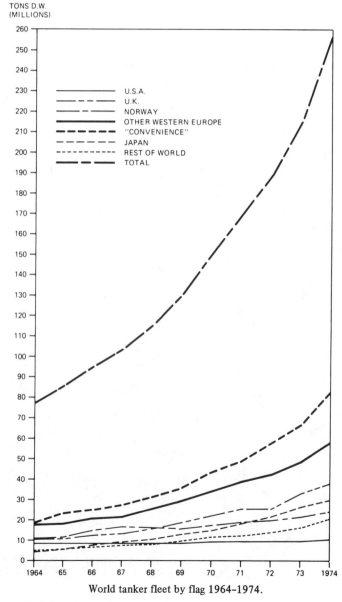

World tanker fleet by flag 1964–1974.

Source: Energy Statistics: A Supplement to the Summary of National Transportation Statistics, DOT-TSC-OST-75-33, U.S. Department of Transportation, August 1975.

World Tanker Fleet at End 1974 (10 000 Long Tons D.W. and over)
(Excluding 41.2 million D.W.T. combined carriers)
By Flag and Ownership

FLAG	OWNERSHIP						Change 1974 over 1973	Share of Total 1974
	Oil Company	Private	Government	Other	Total 1974	Total 1973		
	MILLION LONG TONS DEADWEIGHT							
Liberia	19·8	53·8	—	0·3	73·9	59·2	+14·7	29%
Norway	0·5	23·4	—	—	23·9	21·2	+ 2·7	9%
U.K.	21·0	11·0	0·2	—	32·2	27·8	+ 4·4	13%
Japan	3·8	25·4	—	—	29·2	26·1	+ 3·1	11%
U.S.A.	4·3	4·6	1·5	—	10·4	9·5	+ 0·9	4%
Panama	4·0	4·4	—	—	8·4	7·6	+ 0·8	3%
France	8·4	3·3	0·1	—	11·8	9·6	+ 2·2	5%
Greece	—	13·9	—	—	13·9	12·6	+ 1·3	6%
Other Western Europe	13·0	18·6	0·1	0·1	31·8	26·0	+ 5·8	12%
Other Western Hemisphere	5·7	0·2	0·2	—	6·1	5·2	+ 0·9	2%
U.S.S.R., E. Europe & China	—	—	6·6	—	6·6	5·4	+ 1·2	3%
Other Eastern Hemisphere	2·9	4·6	0·1	—	7·6	5·4	+ 2·2	3%
TOTAL	83·4	163·2	8·8	0·4	255·8	215·6	+40·2	100%
Fleet as at end 1973	66·7	141·1	7·6	0·2	215·6			
Net increase 1974	16·7	22·1	1·2	0·2	40·2			

By Age, Size and Propulsion
Million Long Tons Deadweight

SIZE IN '000 D.W.T.	YEAR OF CONSTRUCTION								PROPULSION		New Building in progress and on order at end 1974*
	Up to end 1945	1946-1950	1951-1955	1956-1960	1961-1965	1966-1970	1971-1974	Total	Motor	Other	
10– 25	2·9	0·6	5·6	6·8	1·9	2·4	1·7	21·9	15·4	6·5	1·1
25– 45	1·0	0·9	4·0	14·6	4·0	1·4	4·6	30·5	11·7	18·8	7·1
45– 65	—	—	0·9	5·1	14·4	1·8	0·4	22·6	7·5	15·1	1·8
65–125	—	—	—	2·2	14·0	21·9	8·3	46·4	30·1	16·3	17·6
125–205	—	—	—	—	0·1	10·2	7·7	18·0	10·2	7·8	21·4
205–285	—	—	—	—	—	26·2	79·3	105·5	7·3	98·2	57·9
285 and over	—	—	—	—	—	1·9	9·0	10·9	—	10·9	57·5
TOTAL	3·9	1·5	10·5	28·7	34·4	65·8	111·0	255·8	82·2	173·6	164·4
MOTOR	0·3	0·5	5·0	8·4	16·6	24·5	26·9	82·2			
OTHER	3·6	1·0	5·5	20·3	17·8	41·3	84·1	173·6			

*Excludes 7·3 million D.W.T. combined carriers

(Source: John I. Jacobs & Co. Ltd.)

Employment of Tankers 1974
Estimated Proportions of World's Active Ocean-Going Fleet on Main Voyages

VOYAGES TO	VOYAGES FROM					TOTAL
	U.S.A.	Caribbean	Middle East	N. Africa	Others	
U.S.A.	3·0%	3·5%	4·5%	0·5%	3·5%	15·0%
Canada	—	0·5%	2·0%	—	—	2·5%
Other Western Hemisphere	—	—	5·5%	0·5%	1·5%	7·5%
Western Europe, N. & W. Africa	—	1·0%	45·5%	2·0%	4·0%	52·5%
E. & S. Africa, S. Asia	—	—	1·5%	—	—	1·5%
Japan	—	—	12·5%	0·5%	2·5%	15·5%
Other Eastern Hemisphere	—	—	4·0%	—	—	4·0%
U.S.S.R., E. Europe & China	—	—	1·5%	—	—	1·5%
TOTAL	3·0%	5·0%	77·0%	3·5%	11·5%	100·0%

Source: The British Petroleum Company Limited, Britannic House, Moor Lane, London EC2Y 9BU

PETROLEUM PIPELINES

The first oil produced in 1859 from Drake's discovery well at Titusville, Pennsylvania, was hauled in wooden barrels on horse-drawn carts to the nearest railroad. The cost was near $5.00 per 42-gallon barrel and about two gallons leaked out along the road. In 1865, Samuel Van Syckel built the first successful pipeline, a 2-inch cast-iron pipe, over a distance of 5 miles to the railroad. He charged 50¢ to $2.00 per barrel for transport. A part of that cost was for armed guards to protect the pipeline from the teamsters who had, two years earlier, torn out an earlier pipeline during its construction.

Although the economic advantages of pipelines resulted in the construction of some 45,000 miles of pipelines in the United States by the year 1907, the stimulus for the construction of large, long-distance lines came during World War II when the submarine threat to Atlantic Coast shipping led the U.S. government to finance the construction of two lines from Texas to New Jersey—the "Big Inch Line," 24 inches in diameter and 1341 miles long, and the "Little Big Inch Line," 20 inches in diameter and 1475 miles long. Since that time, advances in materials and welding technology have permitted a steady increase in the size of pipelines. The pipeline from the North Slope of Alaska to Valdez is 48 inches in diameter.

Characteristics of the proposed Trans-Alaska Pipeline System (T.A.P.S.)

789 miles long, running north and south between Prudhoe Bay and Valdez.

4 feet in diameter.

In geological fault areas, the pipe is designed with flexibility to move 3 feet vertically and 20 feet horizontally without rupture. Can bend 6 feet in a 100-foot span without a wrinkle.

Oil pressure would be regulated by a series of pump stations and temporary storage tanks.

Power-operated, remote-control, cutoff valves would be spaced at 15-mile intervals.

Estimated cost of construction is $2.8 billion, or approximately $3.5 million per mile.

Approximate time of construction is 3 years.

Expected initial oil flow is 600,000 barrels per day at 2 miles per hour.

Anticipated peak oil flow is 2,000,000 barrels per day at 7 miles per hour.

Source: *Energy Statistics*: *A Supplement to the Summary of National Transportation Statistics*, DOT-TSC-OST-7; 5-33, U.S. Department of Transportation, August 1975.

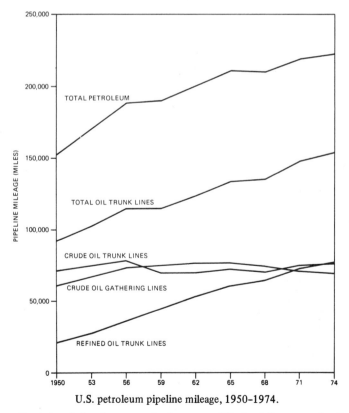

U.S. petroleum pipeline mileage, 1950–1974.

Source: Energy Statistics: A Supplement to the Summary of National Transportation Statistics, DOT-TSC-OST-75-33, U.S. Department of Transportation, August 1975.

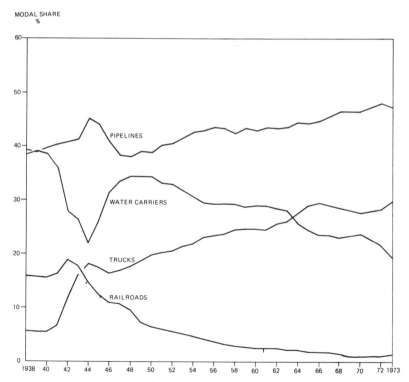

Modal shares for total crude petroleum and petroleum products transported in the U.S., 1938–1973.

Source: Energy Statistics: A Supplement to the Summary of National Transportation Statistics, DOT-TSC-OST-75-33, U.S. Department of Transportation, August 1975.

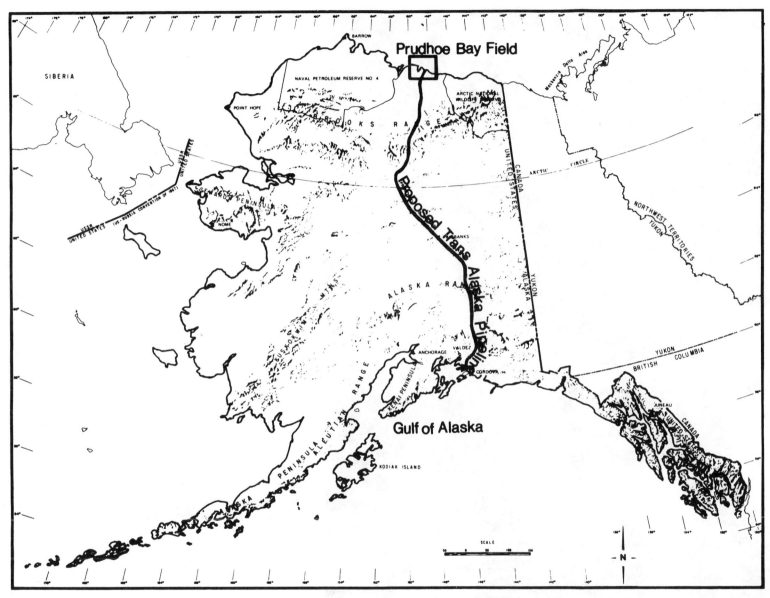

North Alaska production areas.

Source: *National Energy Outlook*, FEA-Np75/713, Federal Energy Administration, February 1976.

This close-up of a pipe wrapping machine shows how protective coatings are applied on a natural gas pipeline. (*Courtesy of Texas Eastern Transmission Corp.*)

Proposed Arctic natural gas transportation routes.

Source: Environmental Quality, The eighth annual report of the Council on Environmental Quality, Executive Office of the President, Dec. 1977.

COAL SLURRY PIPELINES

The use of pipelines to transport slurries of water and granular minerals dates from 1914 when a pipeline was used to pump coal and water into London. In 1957, a pipeline 108 miles long was constructed in Ohio to transport coal at a rate of 1.3 million tons/year and in 1970 another coal slurry pipeline of 273 miles was completed to transport 4.8 million tons of coal per year from Black Mesa, Arizona, to the Mohave power plant in southern Nevada.

For purposes of transport as a slurry, coal is ground to a consistency of table sugar and mixed with an equal volume of water. In the 18-inch Black Mesa pipeline, the coal slurry moves at a rate of $3\frac{1}{2}$ miles/hour. The three pumping stations along the 273-mile route consume energy equivalent to about 4% of the energy content of the coal transported. Upon arrival at the power plant, the slurry is filtered to remove the water, the coal is then fed to the boilers and the water is used to provide about one-eighth of the cooling water requirement of the power plant.

A 25 million ton/year slurry pipeline has a carrying capacity equivalent to seven 100-car unit trains operating on a daily basis.

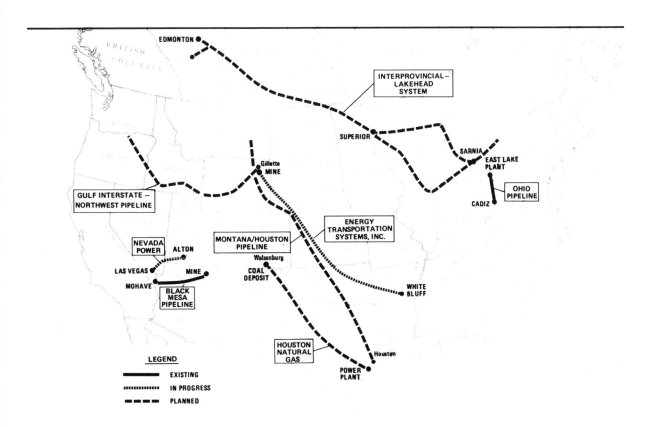

Coal slurry pipelines.

Source: "Innovation in Energy Transportation," by Energy Transportation Systems Inc., October 1975.

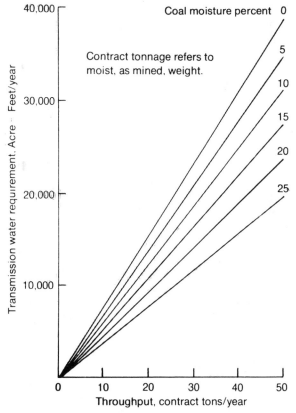

Transmission water requirements as a function of coal throughput moisture content.

Source: A Technology Assessment of Coal Slurry Pipelines, Office of Technology Assessment, Congress of the United States, Jan. 1978.

Summary of Selected Commercial Slurry Pipelines

Slurry material	System or location	Length (miles)	Diameter (inches)	Annual thruput (million tons/year)	Initial operation
Existing					
Coal	Consolidation	108	10	1.3	1957
	Black Mesa	273	18	4.8	1970
Limestone	Calaveras	17	7	1.5	1971
	Rugby	57	10	1.7	1964
	Trinidad	6	8	0.6	1959
Copper concentrate	Bougainville	17	6	1.0	1972
	West Irian	69	4	0.3	1972
	KBI Turkey	38	5	1.0	....
	Pinto Valley	11	4	0.4	1974
Magnetite concentrate	Tasmania	53	9	2.3	1967
	Waipipi (land)	4	8	1.0	1971
	Waipipi (offshore)	1.8	12	1.0	1971
	Pena Colorada	30	8	1.8	1974
Gilsonite	American Gilsonite	72	6	0.4	1957
Tails	Japan	44	12	0.6	1968
Nickel refinery tailings	Western Mining	4.3	4	0.1	1970
In Progress					
Coal	Nevada Power - Utah/Nevada	180	24	10	
	Energy Transportation Systems Inc. - Wyo./Ark.	1,036	38	25	
Magnetite and Hematite	Sierra Grande	20	8	2.1	
	Brazil	250	20	12	
	Mexico	17	10	1.5	
Planned					
Coal	Interprovincial - Lakehead System	2,100	30-48	15	
	Houston Natural Gas - Colorado/Texas	750	22	9	
	Gulf Interstate - Northwest Pipeline	1,100	24-20	10	
	Brown & Root/ Texas Eastern - Montana/Houston	1,260	36-40	22-29	
Phosphate	Australia	200	16-22	4.0-6.0	
Sulfur/hydrocarbon	Canada	800	12-16	....	
Magnetite and Hematite	Africa	350	18	6.6	
	Brazil	240	20	12.0	
	India	36	20-22	10.0	
	Mexico	17	10	1.5	
	Australia	44	8	0.9	

Source: *"Innovation in Energy Transportation,"* by Energy Transportation Systems Inc., October 1975.

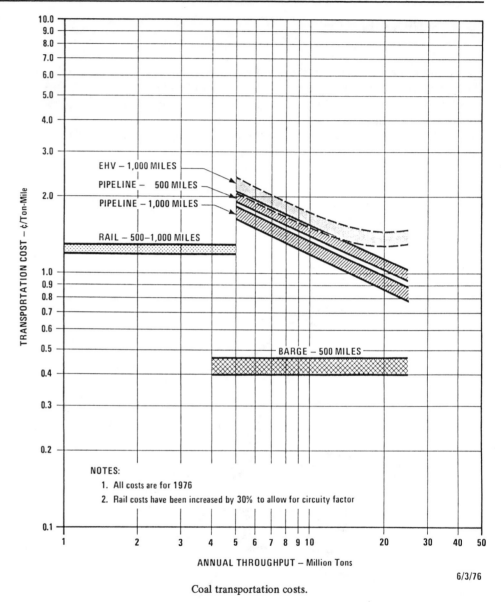

NOTES:
1. All costs are for 1976
2. Rail costs have been increased by 30% to allow for circuity factor

6/3/76

Coal transportation costs.

Source: Wasp, E. J.; Vice President, Energy Transportation Systems Inc.

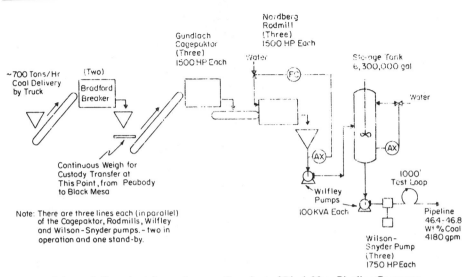

Schematic flow sheet for coal preparation plant of Black Mesa Pipeline Company.

Source: Evaluation of Coal Conversion Processes to Provide Clean Fuels, EPRI 206-0-0, prepared by the University of Michigan, College of Engineering, for the Electric Power Research Institute, Palo Alto, California, February 1974.

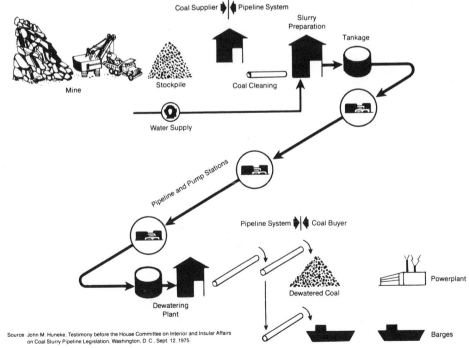

Schematic of slurry pipeline system.

Source: A Technology Assessment of Coal Slurry Pipelines, Office of Technology Assessment, Congress of the United States, Jan. 1978.

Coal, crushed to size of coarse sand in rod mill, emerges, mixed with water, to begin 275-mile journey through coal slurry pipeline in Arizona. (*Courtesy National Coal Association, 1974*)

TRANSPORT OF LIQUEFIED NATURAL GAS

Although gas can be transported efficiently and economically over land by pumping the gas at high pressure through pipelines, the shipment of natural gas by tanker is most efficiently carried out by condensing the gas to a liquid commonly called liquefied natural gas or LNG. As LNG, six hundred times as much gas can be shipped in a tanker of given storage capacity as could be shipped if the gas were in gaseous form.

Natural gas is liquefied by cooling it in stages in tanks of boiling ammonia and boiling ethylene until the liquid temperature of −259°F is reached. The LNG is at atmospheric pressure during shipment and a small portion of the liquid boils off during transit. About 0.25% is lost per day.

Tankers with a capacity of 72,000 cubic meters of LNG are in use and larger tankers of 125,000 cubic meters capacity are under construction. A tanker of 125,000 cubic meters can transport LNG equivalent to 2.6 billion cubic feet of natural gas in gaseous form. Present consumption of natural gas in the United States is 22,000 billion cubic feet per year.

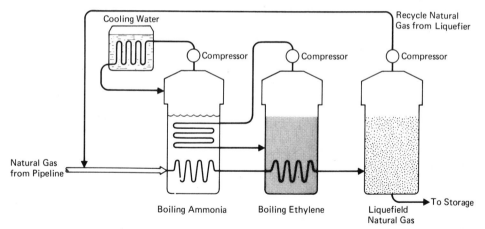

Natural gas is liquefied in a cascade process in which the gas is cooled in stages in tanks of boiling ammonia and boiling ethylene. The liquefied natural gas (LNG), at a temperature of −259°F, is then transferred to insulated storage tanks.

Noel de Nevers, "Liquid Natural Gas," *Scientific American*, Oct. 1967.

Proposed El Paso system.

Source: National Energy Outlook, FEA-N-75/713, Federal Energy Administration, February 1976.

Federal Power Commission Short-Term[a] LNG Import Applications, June 1, 1974

Applicant	Docket No.	Source	Mode of Transport	Point of Delivery	Quantity (MMcf)[b]	Price ($/Mcf)[b]	Date of Authorization
1. Boston Gas Company	CP69-112	Algeria	Ship	Boston, Mass.	200	1.14	10/25/68
2. Lowell Gas Company	CP70-143	Algeria	Ship	Boston, Mass.	374[c]	1.52	12/17/69; 1/16/70
3. Willbros Terminal Company	CP70-194	Canada	Truck	Boston, Mass.	266[d]	2.20	2/13/70; 5/12/70
4. Texas Eastern Transmission Corporation	CP70-208	Algeria	Ship	Staten, Island, N.Y.	2,600	1.37	3/16/70; 4/8/70
5. Boston Gas Company	CP70-291	Algeria	Ship	Boston, Mass.	1,600	1.70	7/14/70; 11/16/70
6. Fall River Gas Company	CP70-305	Canada	Truck	Fall River, Mass.	150[e]	1.12	7/2/70; 7/17/70; 9/22/70
7. Lowell Gas Company	CP71-9	Canada	Truck	Tewksbury, Mass.	630	1.25 before 10/7/70 1.94 10/7/70 to 4/30/71	8/28/70; 3/15/71
8. Boston Gas Company	CP71-61	Canada	Truck	Boston, Mass.	220	1.25 10/19/70 to 11/1/70 2.00 11/1/70 to 4/1/71	11/4/70
9. Boston Gas Company	CP71-247	Canada	Truck	Boston, Mass.	714*	1.30 4/1/71 to 9/1/71* 1.45 9/1/71 to 11/1/71* 1.92 11/1/71 to 4/1/72*	6/4/71
10. Boston Gas Company	CP71-248	Algeria	Ship	Boston, Mass.	1,250	1.66	6/4/71
11. Lowell Gas Company	CP72-10	Canada	Truck	Tewksbury, Mass.	696*	1.30–1.45 4/1/71 to 10/31/71* 1.92 11/1/71 to 4/1/72*	8/5/71
12. Fall River Gas Company	CP72-18	Canada	Truck	Fall River, Mass.	120*	1.30 before 8/31/71* 1.45 8/31/71 to 10/31/71*	9/8/71
13. Texas Eastern Transmission Corporation	CP72-93	Libya	Ship	Staten Island, N.Y.	11,160*	.81*	f
14. Distrigas Corporation	CP72-165 (Terminated)	—	—	—	—	—	—
15. Lowell Gas Company	CP72-301 (Withdrawn)	—	—	—	—	—	—
16. Boston Gas Company	CP73-40	Algeria	Ship	Boston, Mass.	675[g]	1.87	9/25/72
17. Lowell Gas Company	CP73-63	Canada	Truck	Tewksbury, Mass.	539*	1.13 9/1/72 to 11/15/72[h] 1.58 11/15/72 to 4/30/73	11/27/72
18. Distrigas Corporation	CP73-78	Algeria	Ship	Everett, Mass., & Staten Island, N.Y.	59,000*	i	Withdrawn 2-6-74
19. Boston Gas Company	CP73-98	Algeria	Ship	Everett, Mass.	2,380*	1.43*	1/9/73
20. Brockton Taunton Gas Company	CP73-307	Canada	Truck	Easton, Mass.	400*	1.46*[h]	7/31/73
21. Providence Gas Company	CP73-338	Canada	Truck	Exeter, R.I.	120*	1.50*[h]	7/31/73
22. Lowell Gas Company	CP74-3	Canada	Truck	Tewksbury, Mass.	591	1.58* to 1.70[h]	10/10/73

*Volume is billion Btu and price is $/million Btu.

[a]To date, less than 3 years; also typified as not requiring substantial new construction of facilities.

[b]Unless otherwise noted, volumes and prices are as reported in the respective applications filed with the FPC and have been founded to the nearest MMcf and ¢/Mcf where necessary; delivery at shiprail or truckside.

[c]Estimated from the original filing on basis of 52 Mcf per metric ton.

[d]Estimated from reported volumes delivered.

[e]Estimated from the original filing on basis of 83.3 cf per gallon.

[f]Two shiploads authorized 2-15-72; Examiner's Decision approving import issued 5/2/72; withdrawn 9/17/73.

[g]Option for an additional 370 MMcf equivalent, exercisable until 2/1/73.

[h]Not including transportation.

[i]40¢/MMBtu base prices plus a freight rate of 60¢/MMBtu for first 20 trillion Btu; 42¢/MMBtu for the next 20 trillion Btu; and 30¢/MMBtu for remaining Btu.

Source: FPC News, Federal Power Commission, July 19, 1974.

Self-supporting or free standing	Membrane
Conch Methane.	Gaz Transport.
Gaz Transport.	Gazocean/Technigaz.
Esso International.	Conch Ocean.
Kvaerner/Moss.	Bridgestone.
Gazocean/Technigaz.	Ishakawajima Harima
A. G. Weser.	Heavy Industries. (IHI)
Zellentank.	
McMullen.	
Chicago Bridge & Iron (CB&I).	

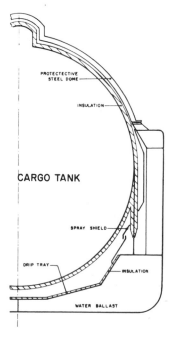

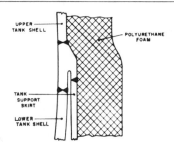

Detail of tank-support skirt connection for Kvaerner-Moss spherical tank design

The Kvaerner/Moss Rosenberg self-supporting tank is a spherical single barrier design made of aluminum or 9 percent nickel alloy and capable of use on ships with capacity up to 125,000 cubic meters. The insulation is closed cell plastic foam. The tank uses a skirt support system holding it tight to the ship's hull.

The Gazocean/Technigaz membrane system was used on the ship *Pythagore.* It is of rectangular design with a primary barrier of corrugated stainless steel. The secondary barrier is also stainless steel with PVC for insulation.

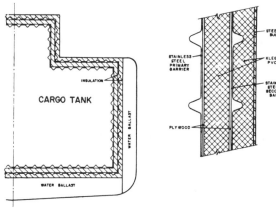

LNG tanker support systems.

Source: National Gas Survey, Volume II, Supply Task Force Reports, U.S. Federal Power Commission, 1973.

A number of these specially designed tankers already are in service transporting natural gas in liquefied form at minus 260 degrees Fahrenheit. Liquefied natural gas takes up only 1/600th the volume compared to its gaseous state. (*Courtesy of Brooklyn Union Gas Company*)

POWER TRANSMISSION—OVERHEAD ALTERNATING CURRENT

The function of transmission systems is to transport bulk electric energy from generation stations to the main substations serving areas of electric load. Transmission systems consist of both overhead transmission lines and underground cables operating at 69,000 volts (69 kV) or higher; terminal equipment including high voltage transformers, converters, switchgear, lightning arrestors, inductive reactors and capacitors; and control and metering equipment, communications equipment and computers. Transmission at voltages between 200 and 1000 kV is referred to as extra high voltage (EHV) transmission while operation in the 1000 to 1500 kV range is referred to as ultra high voltage (UHV).

In addition to providing transmission within individual utility service areas, transmission systems of adjacent electric utilities are generally interconnected to provide reliability of service and better operating economics.

At present, transmission is predominantly overhead in the United States with less than 1% of the electric power transmission lines installed underground. Historically, there has been a steady upward trend in transmission line voltages for two reasons—transmission losses are reduced as the voltage increases and the power carrying capability quadruples with every doubling of the voltage. Each upward step in transmission voltage has required the development and testing of new equipment—transformers, switch gear, and insulators. The highest voltage transmission line in the United States operates at a level of 765 kV.

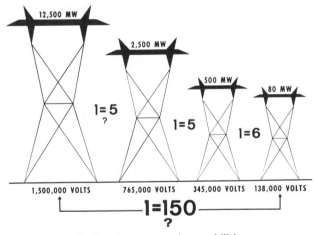

Ratios of power-carrying capabilities

Source: The Utility Industry Today, by the Editors of Transmission & Distribution, Cleworth Publishing Company, Inc, Cos Cob, Connecticut, 1971.

Circuit Miles of Overhead Electric Line of 22,000 Volts and above in Service—Total U.S. Electric Utility Industry.

Nominal Voltage	Circuit Miles		
	1976	1975	1974
Total†	520 635	513 925	504 890
22 000– 25 000	40 663	39 739	39 531
26 000– 30 000	16 023	15 535	13 446
33 000– 36 000	57 536	58 066	56 921
38 000– 42 000	8 233	8 145	8 944
44 000– 50 000	30 681	31 038	30 181
55 000– 60 000	8 495	8 456	7 756
66 000– 75 000	91 472	90 642	92 005
80 000–100 000	8 221	8 204	8 279
110 000–120 000	80 132	79 282	77 678
132 000–140 000	58 534	58 184	57 167
154 000–165 000	20 911	20 960	20 589
220 000–240 000	52 789	51 457	51 730
287 000–288 000	2 198	2 177	760
330 000–345 000	28 188	26 106	24 313
500 000	14 011	13 757	13 412
765 000	1 703	1 332	1 333
800 000	845	845	845

Source: Edison Electric Institute Statistical Year Book for 1976, Edison Electric Institute, 90 Park Avenue, New York, N.Y. 10016, Oct. 1977.

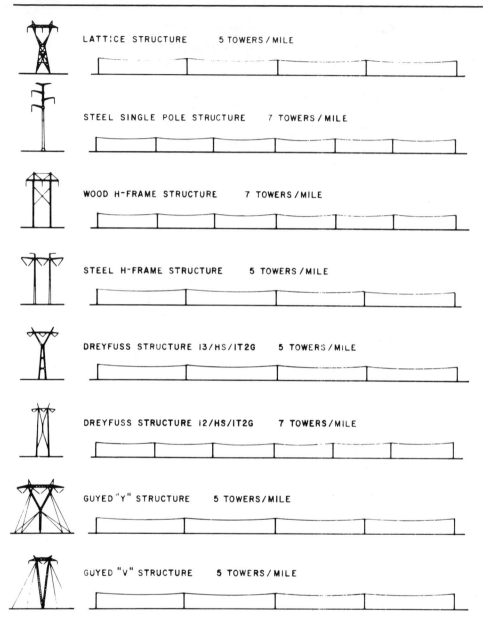

Transmission towers. Functional structures now in use. Number of towers per mile required to maintain the necessary span length follows the tower structure.

Source: J. T. Kitchings, H. H. Shugart, and J. D. Story, *Environmental Impacts Associated with Electric Transmission Lines*, ORNL-TM-4498, Oak Ridge National Laboratory, March 1974.

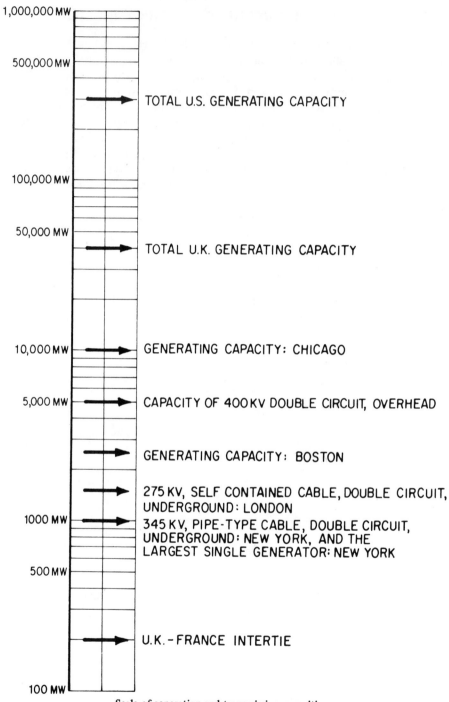

Scale of generation and transmission capacities.

Source: Underground Power Transmission, A Study Prepared by Athur D. Little, Inc., for the Edison Electric Institute, 90 Park Avenue, New York, N. Y., October 1971 (ERC Pub. No. 1–72).

LEGEND

- – – – – 230 KV AC
- ————— 345 KV AC
- •–•–•–• 500 KV AC
- ✚✚✚✚ 765 KV AC
- ○○○○ ± 400 KV DC
- ✚ NUMBER OF CIRCUITS

Transmission system in the United States in 1970.

Source: Transmission Line Reference Book 345 kv and Above, Electric Power Research Institute, Palo Alto, California, 1975.

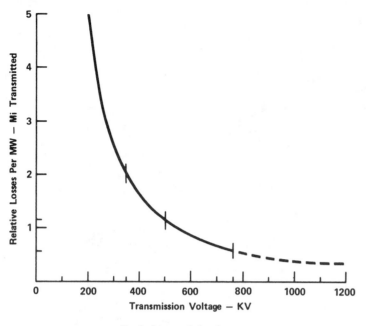

Typical transmission losses.

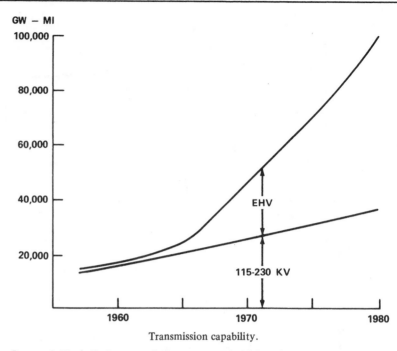

Transmission capability.

Source: Proposed Final Environmental Statement: Liquid Metal Fast Breeder Reactor Program,
Volume III, WASH-1535, U.S. Atomic Energy Commission, December 1974.

Source: Proposed Final Environmental Statement: Liquid Metal Fast Breeder Reactor Program,
Volume III, WASH-1535, U.S. Atomic Energy Commission, December 1974.

Power lines near Searchlight, Nevada. 1972. (*Courtesy U.S. Environmental Protection Agency*)

POWER TRANSMISSION—DIRECT CURRENT

High voltage dc transmission systems are being installed in larger numbers because of the ability of dc systems to transport more electric energy per unit width of right-of-way than an equivalent ac line. Because of the absence of reactive currents and the resistance-increasing skin-effect experienced with ac transmission, both the induced dielectric losses and resistive losses are lower for dc transmission. For dc and ac lines of the same insulation level and conductor size, the dc line losses are about 65% of the ac line losses for transmission of a given amount of power. This advantage of dc transmission is partially offset, however, by the losses at the dc conversion terminals, which are about $4\frac{1}{2}$ times higher than those experienced in ac transformers.

The economics of dc systems favor their use for the transmission of large amounts of power over long distances—distances of 300 to 400 miles for overhead lines and 30 to 50 miles for underground lines. The dc "links" are also being built to interconnect ac transmission systems. Such links, by converting ac to dc and then dc back to ac, essentially "decouple" the two ac systems and remove the necessity for synchronization of the frequencies of the two ac systems.

REFERENCE

Eric Jeffs, "CIGRE reveals mounting confidence in HVDC," *Energy International*, October 1974.

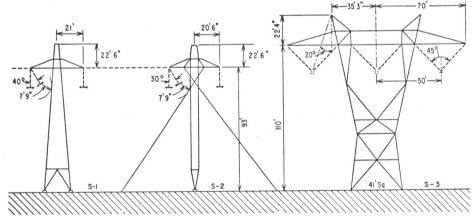

Typical transmission tower structures for 735-800 kV direct current (S-1 and S-2) and alternating current (S-3).

Source: Transmission Line Reference Book 345 kv and Above, Electric Power Research Institute, Palo Alto, California, 1975.

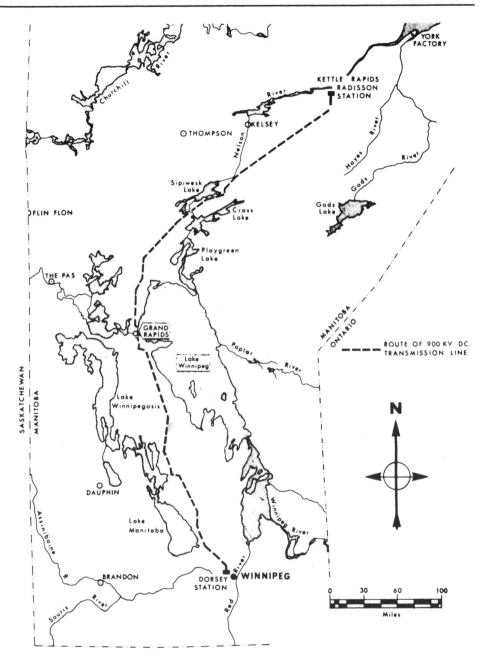

ATOMIC ENERGY OF CANADA LIMITED
NELSON RIVER TRANSMISSION FACILITIES

Route of the 900 kV direct current transmission line from Kettle Rapids generating station to Winnipeg.

Source: The Utility Industry Today, by the Editors of Transmission & Distribution, Cleworth Publishing Company, Inc., Cos Cob, Connecticut, 1971.

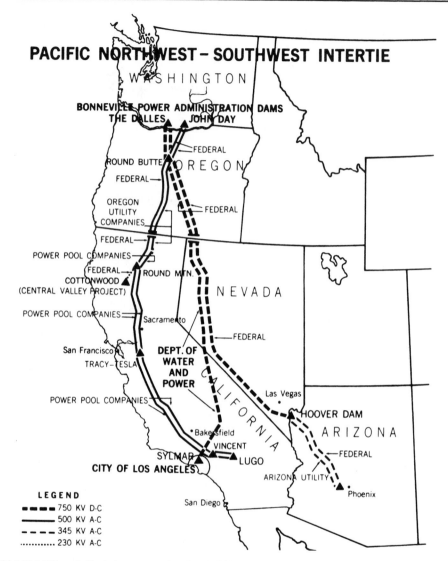

PACIFIC NORTHWEST – SOUTHWEST INTERTIE

WASHINGTON

BONNEVILLE POWER ADMINISTRATION DAMS
THE DALLES   JOHN DAY

FEDERAL

ROUND BUTTE   OREGON
FEDERAL

OREGON
UTILITY
COMPANIES   FEDERAL

FEDERAL

POWER POOL COMPANIES
FEDERAL   ROUND MTN.
COTTONWOOD
(CENTRAL VALLEY PROJECT)   NEVADA

POWER POOL COMPANIES   Sacramento
FEDERAL

San Francisco   CALIFORNIA
TRACY–TESLA   DEPT. OF
WATER
AND
POWER   Las Vegas

POWER POOL COMPANIES   HOOVER DAM
ARIZONA
•Bakersfield   FEDERAL
VINCENT
SYLMAR   LUGO   ARIZONA UTILITY
CITY OF LOS ANGELES   Phoenix

San Diego

LEGEND
▬ ▬ ▬ 750 KV D-C
——— 500 KV A-C
– – – – 345 KV A-C
··········· 230 KV A-C

500 kV AC lines indicated are operational as is 800 kV DC line from The Dalles to Los Angeles. Second DC line to Hoover Dam is planned.

Source: The Utility Industry Today, by the Editors of Transmission & Distribution, Cleworth Publishing Company, Inc., Cos Cob, Connecticut, 1971.

Direct current transmission tower. (*Courtesy Bonneville Power Administration*).

POWER TRANSMISSION—UNDERGROUND

The simplest and least expensive method for providing the necessary insulation between electrical conductors and between these conductors and other objects is to separate the conductors by mounting them on spaced insulators with air as the general insulating medium—the usual practice in conventional overhead transmission lines. In many urban areas, however, it is not possible to construct overhead lines and underground lines must be used. While underground transmission cables are more expensive than overhead transmission lines, they offer the advantage of requiring narrower rights-of-way.

The problem of insulation of underground cables is difficult for two reasons: materials of high dielectric strength must be used which provide a continuous insulation around and along the conductor as well as other system components, and energy losses in the underground conductor must be kept to a minimum since cooling of underground conductors adds to the cost and system complexity. A number of different types of underground cables are in use and under development. The various types are listed in 12-9 and 12-10 along with the comparative advantages and disadvantages.

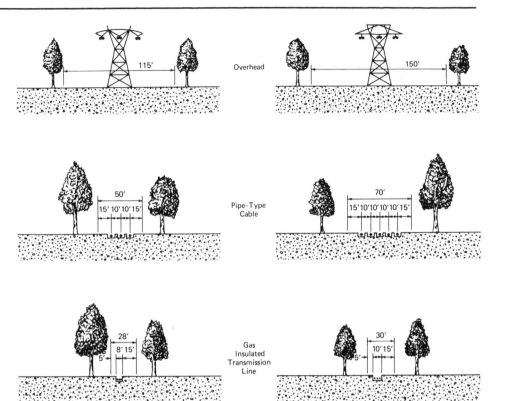

Comparison of rights-of-way required by overhead and underground transmission lines.

Source: Guide to the Use of Gas Cable Systems, EPRI 78-25, ERDA E(49-18)-1615, prepared by the I-T-E Imperial Corporation for the Electric Power Research Institute and the Energy Research and Development Administration, September 1975.

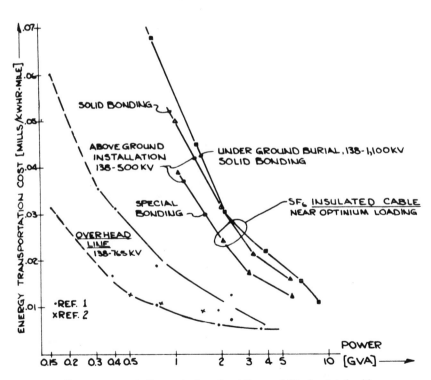

Energy transportation cost of overhead lines and SF_6 insulated cable.

Source: Guide to the Use of Gas Cable Systems, EPRI 78-25, ERDA E(49-18)-1615, prepared by the I-T-E Imperial Corporation for the Electric Power Research Institute and the Energy Research and Development Administration, September 1975.

Total Circuit Miles of Underground and Submarine Transmission Lines in Service in the United States by Nominal Voltage Classes as of June 30, 1974

69 kV	115 kV	138 kV	161 kV	230 kV	345 kV	500 kV	765 kV	±400 kV dc	Total Circuit Miles
850	501	1,121	15	96	101	0	0	0	2,684

Source of Reference: FPC Form 12F.
Source: FPC News, Federal Power Commission, March 14, 1975.

CAPABILITIES OF UNDERGROUND TRANSMISSION SYSTEMS

Range of Application of Various Families of Underground Cable: Summary of Conclusions.

Cable Type	Features Needing Further Development	Best Distance	Best Power Level (MW)	Economic Incentive	Amount R & D Needed	When Available	Further Comments
H.P. oil-filled, pipe-type[1], paper tape	None	$< 100^b$	200-600	—	—	Now	No advantage in voltages > 345 kV
HPOF[a], paper, forced-cooled	Cooling theory & practice, higher volt.	$< 100^b$	500-1500	Moderate	Little D	Soon	Payoff small at lower voltages
HPOF[a], synth. laminate (PPC) tape	Synth. laminates	$< 100^b$	300-700	Moderate	Mod. D	< 1980	No advantage in voltages > 500 kV
HPOF[a], synth. tape	Synth. tapes	$< 100^b$	300-700	Moderate	Mod. D, some R	< 1980	
HPOF[a], PPC or synth., forced-cooled	Cooling & tapes	$< 100^b$	800-2000	Moderate	Mod. D, some R	< 1980	Pays off at higher voltages
Extruded solid dielectric[c], Cu or Al conductor	Dielectric breakdown & reliability	$< 100^b$	200-600	Large	Some D, some R	< 1980	Available now at lower voltages; dielectric breakdown poorly understood
Extruded solid dielectric[c], Na conductor	Dielectric breakdown & Na technology	$< 100^b$	200-400	Large	Mod. D, some R	< 1980	Available now at lower voltages; Na successful at lower voltages
Gas-spacer (SF$_6$), naturally cooled	Scaling to higher voltages	all^d	500-1500	Moderate	Little D	< 1980	Compatible with OH
Gas-spacer, forced-cooled	Scaling to higher voltages	all^d	1000-5000	Large	Mod. D	< 1990	Pays off at very large cap., high volt.
Microwave waveguides	Generators, rectifiers, launchers, precision fabrication & laying	10-100	Enormous	None	Much R & D	> 1990	State of the art not yet ripe
Cryoresistive, vacuum ins.	Termination, splices other comp., inst. tech	> 10	800-2000	Moderate	Mod. D, little R	< 1980	Nearly ready for demonstration of tech. feasibility, but not in economically desirable config.
Cryoresistive, tape ins.	Termination, splices other comp., inst. tech	$10-100^b$	1500-5000	Moderate	Mod D, some R	< 1980	Full-power laboratory model could be designed & tested soon
Superconducting ac	Cryo. dielectrics, low-loss high T_c ac supercon.	> 10	1500-5000	Moderate	Much D, much R	< 1990	Delay syst. design pending AR & ED results
Superconducting dc	Dielectrics, high T_c supercon.	> 100	> 5000	Large	Much D, some R	< 1990	Available sooner if vigorously pursued
Dc HPOF paper or extruded dielectric	Dc behavior of dielectrics	> 100	1000-5000	Large	Little D	Soon	Use limited by converter costs; smaller sizes available almost at once; some R may be needed for higher voltage

a. Can be laid in short lengths   b. Limited by charging current & compensation   c. Most advantageous when laid in long lengths; compatible with integrated laying techniques   d. Especially in Scenario D (1,000 ft.)

Source: *Underground Power Transmission*, A Study Prepared by Arthur D. Little, Inc., for the Edison Electric Institute, 90 Park Ave., New York, N.Y., October 1971 (ERC Pub. No. 1-72).

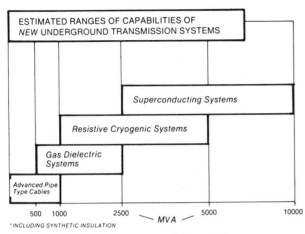

(Courtesy Boston Edison Co.)

Source: The Utility Industry Today, by the Editors of Transmission & Distribution, Cleworth Publishing Company. Inc., Cos Cob, Connecticut, 1971.

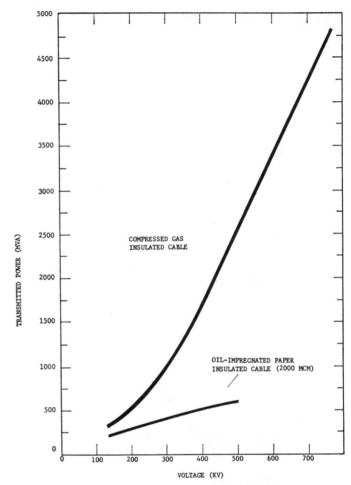

A comparison of power transmission capability of compressed gas insulated cable and oil-impregnated paper insulated cable. Figure courtesy Edison Electric Institute.

Source: The Utility Industry Today, by the Editors of Transmission & Distribution, Cleworth Publishing Company, Inc., Cos Cob, Connecticut, 1971.

Energy Densities and Power Losses in Underground Power Systems

HPOF-PPC-345kV	Max. Power (MW)	Energy Density (MW/cm²)	Power Loss (%/mi)	Attenuation (dB/mi)
Naturally cooled	550	1.75	2.5×10^{-2}	1.1×10^{-3}
Forced cooled	1,020	3.3	4.3×10^{-2}	1.9×10^{-3}
Gas Spacer Cable-500kV				
Naturally cooled	2,200	0.36	1.3×10^{-2}	0.57×10^{-3}
Forced cooled	6,500	1.1	4.5×10^{-2}	2.0×10^{-3}
Superconducting DC	10,000	7.75	4.5×10^{-4}	1.9×10^{-5}
Microwave (waveguide)	10,000	0.38	6×10^{-2}	*2.6×10^{-3}

*For comparison, typical small microwave guides in normal application have attenuations of up to 20 dB/mile.

Source: Underground Power Transmission, A Study Prepared by Arthur D. Little, Inc., for the Edison Electric Institute, 90 Park Avenue, New York, N.Y., October 1971 (ERC Pub. No. 1-72).

COMPONENTS OF UNDERGROUND CABLE SYSTEMS

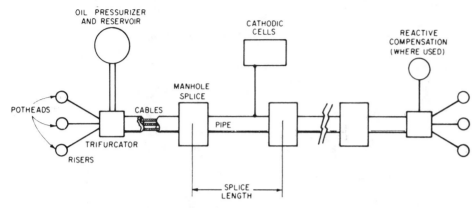

Components of a single-circuit pipe-type cable system.

Source: Underground Power Transmission, A Study Prepared by Arthur D. Little, Inc., for the Edison Electric Institute, 90 Park Ave., New York, N.Y., October 1971 (ERC Pub. No. 1-72).

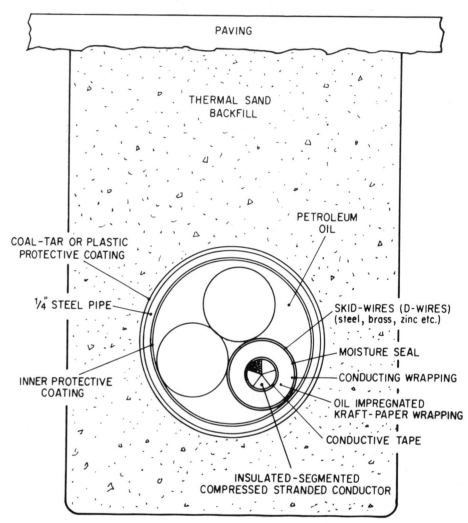

Cross section of a single-circuit pipe-type cable system.

Source: Underground Power Transmission, a Study prepared by Arthur D. Little, Inc., for the Edison Electric Institute, 90 Park Ave., N.Y., N.Y. October 1971 (ERC Pub. No. 1-72)

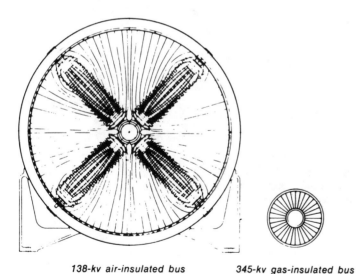

138-kv air-insulated bus   *345-kv gas-insulated bus*

Comparison of air-insulated and gas-insulated bus.

Gas-insulated underground transmission systems consist of a small diameter tube held by insulators along the center line of a larger tube. The use of a gas, sulfur hexafluoride (SF_6) allows the size of the outer tubing to be reduced to one-eighth the diameter required for a system using air as the insulating medium. A 345-kv system of this type using air would be approximately twelve feet in diameter, whereas an SF_6-insulated system is only fifteen inches in diameter.

Source: Research Progress Report, EPRI TD-1, Transmission and Distribution Division, Electric Power Research Institute, Palo Alto, California, October 1974.

POWER TRANSMISSION—SUPERCONDUCTING

Since cooling is required, in any event, for high-capacity underground transmission cables, development is underway on cooled underground cables that take advantage of the lower resistivity of conductors at very low temperatures. At a temperature of 77°K, the boiling point of liquid nitrogen, the resistivity of copper and aluminum is reduced by a factor of ten. At 20°K, the boiling point of liquid hydrogen, the resistance loss is reduced by a factor of 500. These reductions in resistance loss must be balanced against the cost of refrigeration—in the cryoresistive range (77°K), the costs about balance, while in the superconducting range (20°K), there is a net gain in performance. There is, however, an additional advantage of superconducting systems: these systems have a very large power transmission capability. For example, a typical design consisting of coaxial conductors (with inner conductors 6.7 centimeters in diameter) would be capable of transmitting 4710 million volt-amperes at 230,000 volts. This represents more than eight times the capacity of a 345,000 volt overhead transmission line.

Development objectives for superconducting underground transmission systems include work on stable superconducting materials, insulating dielectric materials, long-life refrigerating systems, and terminal equipment.

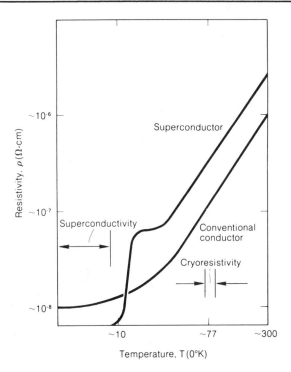

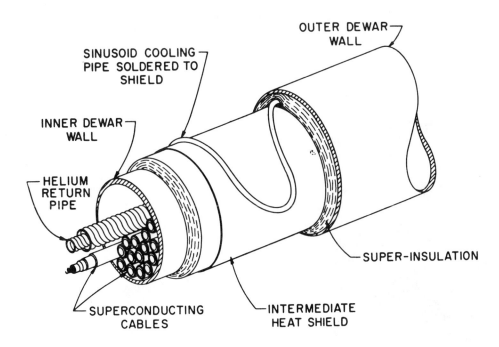

Conceptual 3000 MVA underground superconducting cable.

Source: Energy Research and Development—An Overview of our National Effort, Hearing before the Subcommittee on Energy of the Committee on Science and Astronautics, U.S. House of Representatives, May 15, 1973.

For all metals, resistivity to current flow decreases with temperature and the relationship is essentially linear over a wide range. Cryogenic technology today involves two temperature regimes, one that is within the linear spectrum and one that is below it.

Cryoresistive denotes the regime near 77°K (−196°C) where the resistivity of many conventional conductors, including copper and aluminum, falls by a factor of 10 from its ambient value. Liquid nitrogen is the usual refrigerant.

Superconductive denotes the much lower regime around 10°K (−263°C) where some metals, such as niobium and many alloys, undergo more dramatic change. As the temperature approaches absolute zero, their dc resistivity—ordinarily higher than that of other metals—completely disappears. (The phenomenon is less marked in ac circuitry.) Liquid helium can produce and maintain this effect; thus, the voltage source can be removed from a superconducting dc circuit and its current will continue to flow, undiminished and indefinitely.

The refrigeration energy used for cryoresistive transmission ordinarily exceeds the electrical energy saved in specific performance. Superconducting generation and transmission, on the other hand, have the potential for a net increase in electrical efficiency. The power density is greatly enhanced in both cases. Therefore, the resulting savings in materials and structures, and especially in right-of-way cost, can produce a favorable overall economic benefit-cost ratio.

Source: Rabinowitz, Mario; "Superconductivity: Worldwide Research," in *Research Progress Report TD-3*, Transmission and Distribution Division, Electric Power Research Institute, Palo Alto, California, September 1975.

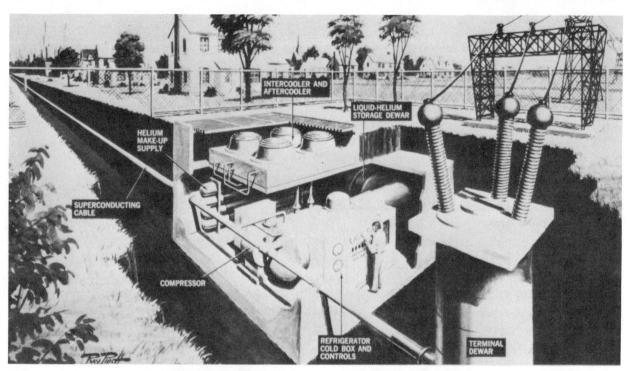

Typical refrigeration system required for a superconducting cryogenic cable.

Source: Research Progress Report; Transmission and Distribution Division, EPRI TD-1, Electric Power Research Institute, Palo Alto, Calif., Oct. 1974.

Model of superconducting cable. (*Courtesy Linde Division, Union Carbide Corporation*)

ALTERNATE CHEMICAL ENERGY TRANSPORT SYSTEMS

Pipelines carrying oil, gas, or coal slurries represent systems by which energy is transported in the form of a chemical which can be used upon delivery for generating electricity, for heating, for the production of other chemicals, or for operating combustion engines. Several synthetic chemicals—hydrogen, methanol, and gas mixtures—have also been suggested as possible alternative energy transport media. All of these synthetic chemicals are derivatives of primary energy sources and, as such, are more expensive than the basic raw energy materials. As long as oil and natural gas are available, their ready transportability and low cost provide difficult competition for the synthetic chemical transport materials. The synthetic chemicals can become attractive, however, when the natural liquid and gaseous fuels are limited.

The hydrogen transport system involves the substitution of hydrogen, produced from coal or by electrolysis, for natural gas. The methanol transport systems would replace other liquid fuels with methanol produced from coal or other organic materials. The Eva-Adam energy transport system employs a two-way flow of gaseous chemicals. In this system, methane and water are reacted in a high temperature energy source (such as a nuclear power plant), to produce hydrogen and carbon monoxide. These two gases are then transported by pipeline to the point of consumption where they are reacted to produce methane and water as well as heat. The methane and water are then recycled back in a separate pipeline to the original heat source.

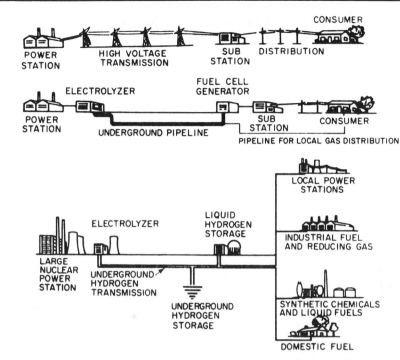

In the hydrogen economy energy is transported in the form of hydrogen produced by electrolysis. Hydrogen is used for heating, production of chemicals, and to generate electricity.

Source: W. A. Sevian, et al, *Analysis of Hydrogen Energy Systems*, BNL-50393, Report to the Subgroup on Synthetic Fuels, Office of Science and Technology, February 1973.

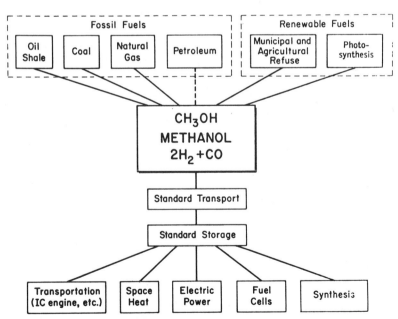

Sources, transport, and possible applications of methanol. Another form of energy transport system can be based on methanol produced from fossil fuels or renewable organic materials.

Reed, T. B., and Lemer, R. M., "Methanol: A Versatile Fuel for Immediate Use," *Science*, Vol. 182 pp. 1299–1304, December 28, 1973. © American Association for the Advancement of Science.

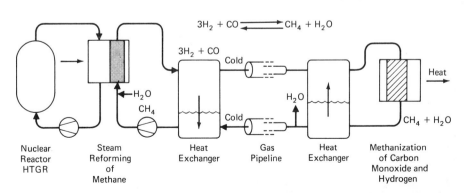

Source: Common project of KFA Juelich and Rheinische Braunkohlenwerke AG, Koeln (Germany).

Summary of Synthetic Fuels Production Cost[a]

Fuel	Fossil-Based Process	Fuel Cost (¢/10^6 Btu)	Electrical- (or other) Based Processes	Fuel Cost (¢/10^6 Btu)
Hydrogen	natural gas, 40¢/10^3 ft^3	97	water electrolysis	
	coal, $7/ton	132	power, 8 mills/kWhr	368
	lignite, $2/ton	78	advanced technology, 8 mills/kWhr	233
	(liquefaction	150)	advanced technology + by-product credits, 8 mills/kWhr	174
			off-peak power, 2.5 mills/kWhr, adv. tech.	195[b]
Ammonia	natural gas, 45¢/10^3 ft^3	157	H_2 via H_2O electrolysis, 8 mills/kWhr	517
			H_2 via H_2O electrolysis, 2.5 mills/kWhr	228
Hydrazine		~2100		
Methanol	natural gas, 40¢/10^3 ft^3	158	H_2 via H_2O electrolysis, 8 mills/kWhr (CO_2 from air)	~550
	coal, $7/ton (~27¢/10^6 Btu)	148		
	lignite, $2/ton (~15¢/10^6 Btu)	~125		
Ethanol	petroleum feed stocks	~460	fermentation from corn, $1.25/bu	880
Methane	well-head gas	15–40	urban and agricultural wastes	~115
	LNG, imported	80–100		
	coal	80–100		
Gasoline	crude oil	105		

[a]Costs are based on 15% fixed charge rate and large plant capacities.

[b]Water electrolysis plant operated at a 0.3 load factor.

Authors Note: These production costs for synthetic fuels should be adjusted to reflect increases in the costs of fossil fuels and electricity upon which the processes are based. During the period 1973–1977, costs for fossil fuels have risen by a factor of two to four. The average cost of electricity for industrial use has risen to 20 mills/kWh.

Source: John W. Michel, *Hydrogen and Synthetic Fuels for the Future*, Oak Ridge National Laboratory, research sponsored by the U.S. Atomic Energy Commission, 1973.

Energy Transmission—Relative Transportation Costs (Near-Urban Environment)

	Volume (10^6 Btu/hr)	Size	Investment per 100 Miles (10^6 $)	Total Cost, Operating Plus Investment[a] per 100 Miles (¢/10^6 Btu)
H_2 gas	21,000	36-in pipe	68[b,c]	5.7[c]
NH_3	21,000	36-in pipe	62[b,c]	5.0[c]
CH_3OH	21,000	30-in pipe	52[b,c]	4.2[c]
Natural gas	21,000	36-in pipe	62[b,c]	5.1[c]
H_2 liquid	550	3-in pipe	79[d]	250.0
H_2 Liquid	1,800	1,100,000-gal barge	2.4	6.9
H_2 Liquid	62	13,000-gal trailer	0.15	34.0
Electric (superconducting line)	13,600	4,000-MVA line	140	19.3
Electric (above ground)	8,500	2500 MW, 750 kV	30	8.3
Electric (below ground)	8,500	2500 MW, 345 kV	~300	~100

[a]Unit cost assumption: 100% utilization of facility, 15% = fixed charge rate (depreciation, interest, return on investment), operating power 6 mills/kWh.

[b]Reported pipeline costs can vary by as much as a factor of four, depending on terrain, congestion, etc.

[c]For cross-country straight lines, 50% reduction is likely.

[d]No reliquefaction included.

Source: John W. Michel, *Hydrogen and Synthetic Fuels for the Future*, Oak Ridge National Laboratory, research sponsored by the U.S. Atomic Energy Commission, 1973.

PROPERTIES OF HYDROGEN

Physical Properties of Hydrogen.

Liquid

Melting point at atmospheric pressure, °F	-434.6
Boiling point at atmospheric pressure, °F	-422.9
Critical temperature, °F	-400
Critical pressure, psia	191
Specific gravity (liquid water = 1.00)	0.07 (4.37 lbs/ft^3)
Density (liquid), lb/gal (at -422.9°F and 30 in Hg)	0.583
Specific heat, Btu/lb °F	0.57
Viscosity (at normal bp), poises	182 × 10^{-6}
Heat of fusion, Btu/lb	25.2
Inversion point, Joule Thomson	-92°F
Heat of vaporization (nearly all para), Btu/lb	190.5

Gas

Specific gravity 68°F (air = 1.00)	0.06953
Density, lb/ft^3 (60°F and 30 in. Hg)	0.00532
Specific volume, ft^3/lb (60°F and 30 in. Hg)	187.9
Gross heat of combustion, Btu/ft^3 (incl. latent heat energy of steam)	325.1
Gross heat of combustion, Btu/lb (incl. latent heat energy of steam)	61,084
Btu/ft^3 of gas/air mixture (F/A ratio 0.420 vol. 0.020 wt.; or 28.7% H_2 by vol.)	81.3
Ft^3 of air required per ft^3 of combustible	2.382
Pound air required per pound combustible	34.226
Flame temperature (F/A ratio 0.462 vol. 0.0313 wt.; or 31.6% H_2 by vol.), °F	3,700
Ignition temperature in air (auto ignition), °F	1065
Ignition temperature in oxygen, °F	1040
Flammability limits, % vol. H_2 in air	4.1-74.2
Flammability limits, % vol. H_2 in oxygen	4.6-93.9
Detonation limits, % vol. H_2 in air	18.3-59
Detonation limits, % vol. H_2 in oxygen	15.0-90
Nonflammable limits, air-hydrogen-carbon dioxide	less than 8% O_2
Nonflammable limits, air-hydrogen-nitrogen	less than 6% O_2
Emissivity of flame (blackbody = 1.00)	0.085

Chemical Properties. Hydrogen in liquid or gaseous form will react violently with strong oxidizers such as oxygen and spontaneously with fluorine and chlorine trifluoride.

Hydrogen gas is colorless, odorless, nontoxic (though asphyxiating), and noncorrosive. When its temperature is that of the ambient air, its density is only about $\frac{1}{14}$ of the air density, and the gas is thus strongly buoyant; however, the vapor at the boiling point is as heavy as air at 70°F.

Liquid hydrogen is a transparent, colorless liquid of low viscosity. It does not form

A Comparison of Some Properties of Three Combustible Gases

	H_2 (Hydrogen)	CH_4 (Methane)	C_3H_8 (Propane)
Boiling Point (°K)	20.3	111.7	230.8
Lower Flammability Limit in Air (Vol %)	4.1	5.3	2.3
Upper Flammability Limit in Air (Vol %)	74.8	15.0	9.5
Lower Detonation Limit in Air (Vol %)	18.0	6.3	
Upper Detonation Limit in Air (Vol %)	59.0	13.5	
Ignition Temperature (°K)	850	807	736
Ignition Energy (milliJoules)	0.02 0.6 (at lower flammability limit)	0.3	0.25
Flame Temperature (°K) Air	2400	2190	2200
Flame Velocity (cm/sec)	275	37	41
Quenching Distance (cm) (one atm)	0.06	0.23	0.20
Air Required for Combustion, lbs of air/lbs of gas (295°K, one atm)	34.2	17.23	15.67

NOTES:
1. Average thermal radiation from hydrogen flame is approximately one-tenth the radiation from hydro-carbon flames.
2. Hydrogen diffuses through air 2.82 times faster than methane.
3. Propane is a 'heavier-than-air' gas.

Source: A Hydrogenergy Carrier, Volume II, Systems Analysis, NGT 44-005-114, NASA-ASEE Systems Design Institute, University of Houston, Johnson Space Center, Rice University, 1973.

solutions with any material except, to a slight extent, with helium. In particular, gases like oxygen and nitrogen condense and freeze to solids in liquid hydrogen without entering into solution. At about 14°K (-435°F) liquid hydrogen freezes to a solid. The temperature and pressure at the triple point (at which solid, liquid, and gaseous hydrogen coexist) are 14.0°K and 0.071 atmospheres for normal hydrogen, and 13.8°K and 0.069 atmospheres for parahydrogen. Solid hydrogen freezes into a white crystalline or snowlike mass.

Hydrogen diffuses approximately 3.8 times faster than air. A spill of 500 gallons of liquid hydrogen on the ground will diffuse to a nonexplosive mixture after about one minute. Air turbulence increases the rate of hydrogen diffusion.

Hydrogen in both the liquid and gaseous states is particularly subject to leakage because of its low viscosity and low molecular weight. Leakage rate is inversely proportional to viscosity. Because of its low viscosity alone, the leakage of liquid hydrogen will be roughly 100 times that of JP-4 fuel, 50 times that of water, and 10 times that of liquid nitrogen. Likewise, the leakage of gaseous hydrogen will be greater than that of air.

REFERENCE

Cloyd, D. R. and N. J. Murphy; Handling Hazardous Materials, Chapter 1, Liquid Hydrogen, NASA SP-5032, September, 1965.

Source: *A Hydrogenergy Carrier*, Volume II, Systems Analysis, NGT 44-005-114, NASA-ASEE Systems Design Institute, University of Houston, Johnson Space Center, Rice University, 1973.

Environmental Aspects of Energy Use

Emissions from stacks of Holmes Road Incinerator in south Houston, Texas during April, 1972. The incinerator was burning old automobile batteries, plastics, and other modern municipal trash which it was not designed to handle. Complaints of air pollution by residents and teachers at nearby schools resulted in an executive order by the mayor of Houston closing the incinerator in January, 1974 (see DOCUMERICA images #'s 14,959/14,960 for documentation of closed incinerator in July, 1975.)

POTENTIAL POLLUTION FROM USE OF ENERGY RESOURCES

The environmental impact of the production and use of energy resources takes many forms. There is a potential for modification of the global climate and there are observed changes in local climates. Recovery of energy resources from the land can despoil the land and waters. Use of energy resources produces direct and indirect effects on health and causes damage to property. The production and use of energy has changed the aesthetics of our environment—we have industrial centers, sprawling suburbs, airports, superhighways and transmission lines. And machines that use energy are noisy.

There is no way, however, for man to live on earth without disturbing the ecological balance that might exist if man were not present. Even without the use of energy resources, man must till the land and domesticate animals if he is to survive in his present numbers. The use of energy resources has greatly assisted in this task of survival and has permitted the luxuries of leisure, education, and mobility. It is not likely that man will willingly give up the advantages offered by the use of energy.

The current concern over environmental issues stems from the realization that the use of energy resources need not necessarily result in the environmental impact that it now has. The setting of standards for each of the activities relating to energy production

Coal, potential pollution

	Water	Air	Land	Solid Waste
Production	Acid mine drainage	Mine fires	Strip mining damage	Waste from underground mining
	Leaching of waste piles	Waste pile fires		
	Erosion and silting of streams			
Upgrading	Preparation plant effluent streams	Particulates from fine coal drying		Waste from coal cleaning
	Leaching of waste piles			
		Nitrogen oxides		
		Waste bank fires		
Transportation				
Utilization	Thermal pollution	Sulfur oxides } Nitrogen oxides } Power Particulates } plants		Disposal of fly ash and slag
		Particulates } Hydrogen sulfide } Coke Carbon monoxide } ovens Hydrocarbons }		

Many different environmental pollutants are associated with the many stages of fuel production and utilization in our complex technological society. Some, such as fly ash and slag from the combustion of coal (above), have obvious—if incompletely realized—uses. Others, notably the waste heat from electric power generation, present problems to which only compromises and partial solutions can be envisioned.

Source: Perry, Harry and Berkson, Harold, "Must fossil fuels pollute?" *Technol. Rev.*, edited at The Massachusetts Institute of Technology, December 1971.

Oil, potential pollution:

	Water	Air	Land	Solid Waste
Production	Disposal of brine			
Upgrading	Thermal pollution	Sulfur oxides		Spent phosphoric acid catalyst
	Sulfuric acid	Hydrocarbons		
	Spent caustic	Nitrogen oxides		Spent clay
Transportation	Tanker accidents		Alaskan pipeline	
Utilization	Thermal pollution	Auto and diesel exhaust emissions		
		Sulfur oxides		
		Nitrogen oxides		

Oil production yields unique fluid pollutants as well as the gaseous by-products of combustion with which we are increasingly familiar. The possibilities of converting these to useful by-products which will substitute for primary resources are limited.

Natural gas, potential pollution:

	Water	Air	Land	Solid Waste
Production				
Upgrading				
Transportation		Nitrogen oxides at compressor stations		
Utilization	Thermal pollution	Nitrogen oxides		

Natural gas, of which—among all energy sources—world supplies are least adequate, presents fewest pollution problems. The authors report that "the most advanced example of commercially practiced conservation of waste material in the fuel industry" is to be found in its treatment.

Source: Perry, Harry and Berkson, Harold, "Must Fossil Fuels Pollute?" *Technology Review*, edited at The Massachusetts Institute of Technology, December, 1971.

and use is complex. In each case, some agreement must be reached on the importance of the environmental impact and a balance achieved between costs and benefits associated with correction of the problem. Resolution of environmental issues is the subject of intense national and international debate. Ultimately, the public will decide, balancing the perceived risks against the apparent benefits in making life more fulfilling; recognizing that life itself is a terminal disease.

Nuclear fuels, potential pollution:

	Water	Air	Land	Solid Waste	Radiation
Production	Leaching of waste banks Uranium mine water		Strip mining damage	Waste from underground mining	Exposure of miners
Upgrading	Leaching of waste banks	Particulate emission and waste banks		Wastes from ore dressing	Exposure of plant workers
Transportation					
Utilization	Thermal pollution			Waste disposal from fuel processing plants	During generation and disposal of waste

Energy from nuclear fuels commits us to disposing of many pollutants familiar from conventional fossil plants and adds a new dimension—radiation—which remains a highly controversial problem for the future.

Source: Perry, Harry and Berkson, Harold, "Must Fossil Fuels Pollute," *Technology Review,* edited at The Massachusetts Institute of Technology, December, 1971.

Environmental Factors for Some Emerging Energy Technologies[1]

	Low Btu gasification		High Btu gasification		Coal liquefaction		Oil shale Colorado
	Western coal	Illinois coal	Western coal	Illinois coal	Western coal	Illinois coal	
Air (lb/h)							
Particulates	0.86	0.86	727	944	633	612	453.6
SO_2	580	2,250	1,800	10,400	1,493	1,957.7	5,324.1
NO_x	1,130	1,130	7,110	7,770	8,507.5	8,507.5	1,966.7
CO	32.3	32.3	377	414	340	340	174.5
HC	32.6	32.5	115	126	2,607.6	2,607.6	2,652.3
NH_3	56.8	45.4	34.7	54.7	—	—	—
Water (lb/h)							
Suspended solids	0	0	0	0	0	0	0
Dissolved solids	0	0	0	0	0	0	0
Organic material	0	0	0	0	0	0	0
Thermal (Btu/h)	Negligible	Negligible	Negligible	Negligible	Negligible	Negligible	Negligible
Solid wastes (ton/day)	5,350	7,320	5,560	7,930	5,519	8,423	164.3×10^3
Land use (acres)	3,190	3,190	1,400	1,400	3,254	3,254	2,000
Water requirements (gal/day)	7.46×10^6	7.46×10^6	64.2×10^6	64.2×10^6	33.3×10^6	33.3×10^6	21.1×10^6
Occupational health (per year)							
Deaths	0.71	0.71	1.8	1.8	0.511	0.511	0.755
Injuries	14.2	14.2	61	61	9.9	9.9	79.2
Man-days lost	7,500	7,500	16,600	16,600	2,372	2,372	77.0
Efficiency (percent)							
Primary product efficiency	75.8	75.8	68.2	67.9	62.5	62.5	66.7
Total products efficiency	86.5	83.9	68.2	67.9	62.5	62.5	79.7
Overall efficiency	86.5	83.9	68.2	67.9	62.5	62.5	76.9
Ancillary energy (Btu/day)	0	0	0	0	0	0	5.59×10^{10}

[1] On the basis of a 10^{12} Btu/day output from each process.

Source: Radian Corporation, *A Western Regional Energy Development Study: Primary Environmental Impacts,* Vol. 2, prepared for the Council on Environmental Quality and the Federal Energy Administration under contract no. EQ4AC037 (Springfield, Va.: National Technical Information Service, 1975).

Source: *The Sixth Annual Report of the Council on Environmental Quality,* December 1975.

Environmental Factors for Some Energy Systems[1]

Line No.	End use	Air emissions (pounds/hour)					Solid waste (tons/day)	Land use (acres)	Water required (gallons/day)	Energy efficiency (percent)
		Partic- ulates	SO₂	NO₂	CO	HC				
	Electricity									
1	Surface coal mine (Montana) Rail to Chicago Coal-fired powerplant (Chicago) Total for scenario	9,919	127,812	106,707	15,527	3,329	10,480	50,311	117x10⁶	36
2	Surface coal mine (Montana) Slurry pipeline to Chicago Coal-fired powerplant (Chicago) Total for scenario	8,290	124,033	84,950	4,973	1,482	9,130	41,992	149x10⁶	36
3	Surface coal mine (Montana) Rail to Chicago Low Btu gasification (Chicago) Low Btu gas powerplant (Chicago) Total for scenario	6,671	16,907	99,912	16,321	2,776	16,230	66,640	137x10⁶	28
4	Surface coal mine (Illinois) Low Btu gasification (mine mouth) Low Btu gas powerplant (mine mouth) Long distance transmission to Chicago Total for scenario	14,416	65,736	72,803	2,532	302	20,203	79,101	140x10⁶	28
5	Oil well (Gulf coast) Pipeline to Chicago Refine (Chicago) Oil-fired powerplant (Chicago) Total for scenario	1,284	24,743	89,129	4,081	11,400	9,940	25,708	161x10⁶	33
	Liquid fuels									
6	Surface oil shale mine (Colorado) Retort (mine mouth) Crude pipeline to Chicago Refine (Chicago) Total for scenario	1,923	6,099	4,541	858	6,016	185,362	20,135	38x10⁶	68
7	Surface coal mine (Montana) Liquefaction (mine mouth) Crude pipeline to Chicago Refine (Chicago) Total for scenario	2,804	3,309	13,549	1,333	7,475	6,792	15,218	57x10⁶	52
8	Oil well (Gulf coast) Crude pipeline to Chicago Refine (Chicago) Total for scenario	427	1,029	2,041	597	3,154	3	7,725	11x10⁶	93
9	Surface coal mine (Illinois) Rail to Chicago Liquefaction (Chicago) Refine (Chicago) Total for scenario	8,359	4,353	15,725	2,316	7,657	11,148	32,368	57x10⁶	47
	Gas									
10	Surface coal mine (Montana) High Btu gasification (mine mouth) Gas pipeline to Chicago Total for scenario	2,780	3,876	19,199	6,272	1,148	5,560	9,827	64x10⁶	65
11	Surface coal mine (Illinois) High Btu gasification (mine mouth) Gas pipeline to Chicago Total for scenario	6,073	10,424	8,017	618	165	7,930	15,252	64x10⁶	61
12	Gas well (Gulf coast) Gas pipeline to Chicago Total for scenario	52	166	2,090	59	81,700	0	16,892	0	96

[1] On the basis of a 10¹³ Btu per day output from the trajectory.

Source: Radian Corporation, *A Western Regional Energy Development Study: Primary Environmental Impacts*, Vol. II, prepared for the Council on Environmental Quality and the Federal Energy Administration under contract no. EQ4AC037 (Springfield, Va.: National Technical Information Service, 1975), PB-246 265.

Source: The Sixth Annual Report of the Council on Environmental Quality, December 1975.

CHANGES IN THE GLOBAL ENVIRONMENT

Long before man appeared on earth, the forces of nature had caused dramatic and continuing changes in the physical environment. Continents have moved, mountains have been formed only to be eroded by wind and water, lakes have filled with silt to become meadows, and volcanoes have erupted to fill the air with dust. Species of plants and animals have come and gone as the environment changed.

During recent geologic time, the last several million years, the earth's climate has been greatly changed by series of warmer and colder periods marked by the formation and retreat of massive continental glaciers. There have been about six of these cold periods during the last 500,000 years. The most recent period began about 70,000 years ago with the peak of glaciation occurring about 20,000 years ago. About 44 million square kilometers were covered with ice in contrast to some 15 million square kilometers today. The 47 million additional cubic kilometers of water drawn from the oceans to form the ice in these glaciers resulted in a lowering of the average level of the sea by some 130 meters (430 feet). The forces that caused the ice ages are not known—it is possible that they were a result of changes in the sun's output of energy, or that they represented an adjustment in the earth's ocean-atmosphere-land energy balance.

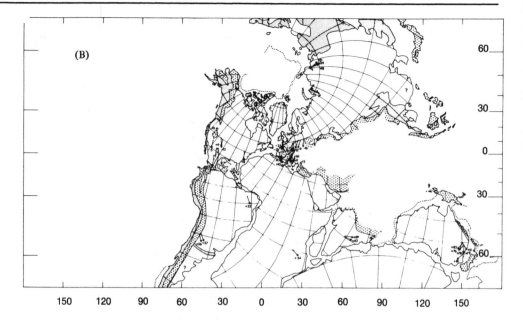

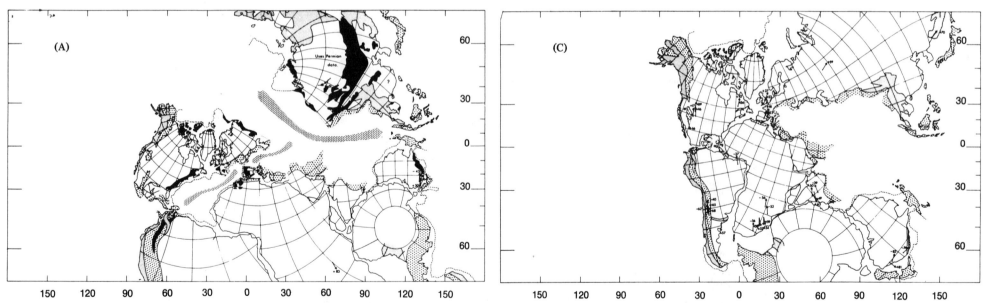

The motions of the continents are shown in these reconstructions, slightly modified from drawings made by A. G. Smith and G. E. Drewry at the Sedgwick Museum, University of Cambridge, and J. C. Briden of Leeds University. The reconstructions rely upon measurements of magnetism and radioactivity in rocks, as described in the text. Map (A) shows the world of about 340 million years ago (±30 million years), during the Lower Carboniferous Period. England and eastern North America are shown to have been in the tropical rain belt near the equator, and this position may have encouraged the rich vegetation that resulted in coal formation in these regions. Map (B) shows the world of 250 million years ago (±25 million years). Northwest Africa has moved into the rain belt; to the east is a large sea called Tethys' Ocean, and the winds probably come from the east, maintaining conditions favorable for eventual coal formation. Thirty million years later (map (C); 220 million years ago, ±20 million years), during the Triassic Period, similar conditions prevail for the central Sahara. Reptiles are evolving toward their domination of the earth. The maps are Mercator projections, in which the meridians of longitude map into parallel vertical lines. The stippled areas denote places where mountain ranges formed.

Source: Newell, Reginald E., "The Earth's Climatic History," *Technology Review*, edited at The Massachusetts Institute of Technology, December, 1974.

While the activities of man have resulted in local climatic changes (see 13-7), the present world energy production is about 1/20,000 that of the energy received on earth from the sun, and it seems unlikely that man's energy use is enough by itself to cause a shift in the global balance.

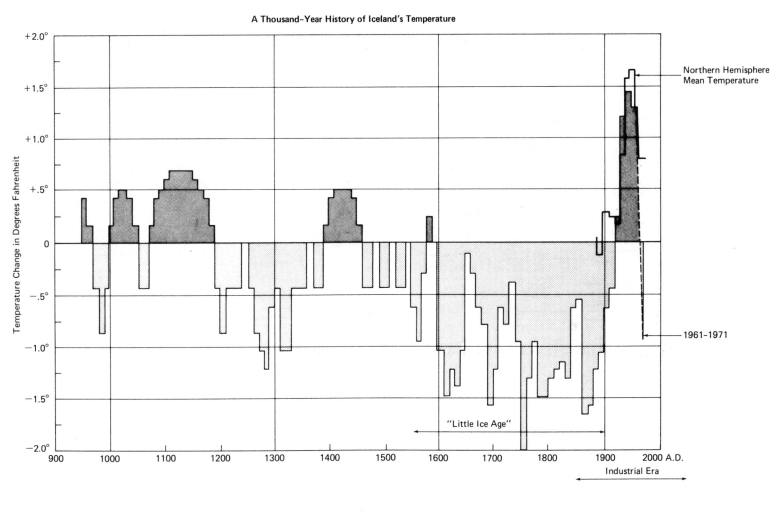

A Thousand-Year History of Iceland's Temperature

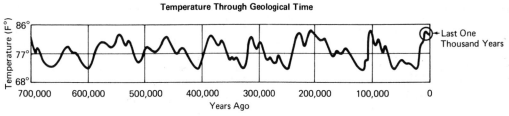

Temperature Through Geological Time

Source: Tom Alexander, "Ominous Changes in the World's Weather," *Fortune,* February 1974. Parios Studio for Fortune Magazine, "A Fifty-Year Balmy Spell Comes to an End," February, 1974.

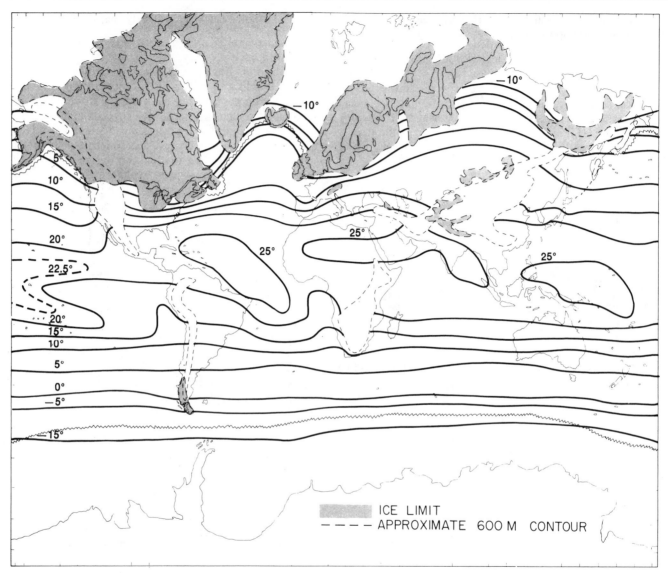

ICE LIMIT
- - - - APPROXIMATE 600 M CONTOUR

The average surface air temperature in July, 20,000 years ago. Ice sheets had been advancing for several thousand years, and were now at their greatest extent. The heavy lines on the map are isotherms—lines connecting locations with the same temperature. The pattern they reveal is fairly regular in the Southern Hemisphere, where expanses of ocean are largely uninterrupted by land masses. In the Northern Hemisphere, ice sheets (grey areas) reached almost to London's present location, where the average summer temperature was greatly affected: it was about 5°C., compared with 18° now. Isotherms have not been drawn in areas where the land rises above a 600-meter altitude, since air is normally cooler at increasing elevation, and this effect obscures the temperature changes associated with Ice Ages. The map is based upon available evidence on treeline and permafrost positions of 20,000 years ago, data on pollen distributions, and deductions of air temperature over the oceans from an analysis of the literature on sea water temperatures of the past. The map was created by Sharon Gould-Stewart, an M.I.T. undergraduate in Earth and Planetary Sciences, and Minoru Tinaka, an M.I.T. graduate student in Meteorology.

Source: Newell, Reginald E., "The Earth's Climatic History," *Technology Review*, edited at The Massachusetts Institute of Technology, December, 1974.

CHANGES IN ATMOSPHERIC CARBON DIOXIDE

During the past century, the combustion of fossil fuels has increased the atmospheric concentration of carbon dioxide from about 300 parts per million to about 320 parts per million. Since carbon dioxide molecules in the atmosphere absorb the infrared radiation from the earth, a net heating of the earth results. An additional increase of 30% in atmospheric carbon dioxide is projected by the year 2000. The concern over the continuing increase in atmospheric carbon dioxide is not in the absolute heating of the earth's surface—which might be only a fraction of a degree—but that past natural changes in the earth's climate—ice ages, widespread droughts, changes in the level of the ocean—were accompanied by only slight shifts in global circulation patterns and only small changes in the average temperature over much of the earth.

The change in the concentration of carbon dioxide is, however, only one of a number of changes in the global heat balance resulting from human activities. The reflectivity or absorptivity of the earth has been changed by the cutting of forests, plowing grasslands, the construction of cities and highways, and the irrigation of desert lands. Particulate emissions from industrial plants have increased the reflection of sunlight by the atmosphere (although windblown dust, atmospheric moisture and particulates from volcanic eruptions are perhaps more significant). Finally, all of the energy produced from the burning of fossil or nuclear fuels is ultimately dissipated as heat.

Projections of atmospheric carbon dioxide concentration are, in themselves, uncertain. Carbon dioxide is absorbed in the oceans (the oceans contain 60 times as much CO_2 as does the atmosphere) and an increase in atmospheric CO_2 causes an increase in photosynthetic plant growth. It is possible that these processes will result in the establishment of a new equilibrium level for atmospheric CO_2.

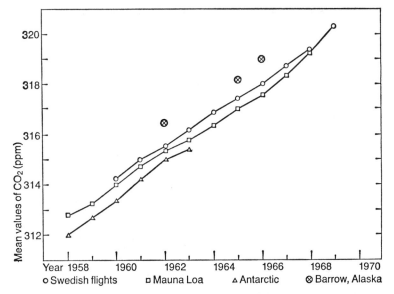

Sources: Swedish flights (Bolin and Bischof, 1969); Mauna Loa (Pales and Keeling, 1965, Bainbridge, 1970); Antarctic (Brown and Keeling, 1965); Barrow, Alaska (Kelley, 1969)

Annual mean values of CO_2.

Reprinted from *Man's Impact on the Global Environment: Assessments and Recommendations for Action* by William H. Mathews by permission of The MIT Press, Cambridge, Massachusetts, 1970.

Possible Atmospheric Carbon Dioxide Concentrations[a]

Year	Amount Added from Fossil Fuel (Mt/yr)[b]	Cumulative Amount Added over Previous Decade (Mt)	Concentration by Volume (ppm)	Total Amount in Atmosphere (Mt)	Percentage of Annual Addition Remaining in Atmosphere
1970	15,400	126,500	321	2.50×10^6	52
1980	22,800	185,000	334	2.61×10^6	52
1990	32,200	268,000	353	2.75×10^6	52
2000	45,500	378,000	379	2.95×10^6	51

Note: This table is not a projection or a prediction. These calculations have been developed to provide insight into the nature of problems that may exist over the next several decades.

[a] The data in this table are taken from a model of the carbon cycle developed at the Study by Jon Machta, Lester Machta, and Jerry Olson, which includes biospheric and oceanic uptake of CO_2 (see Appendix to the report of Work Group 2). The model calculates yearly values for atmospheric CO_2, using past fossil fuel CO_2 production data from the United Nations. It was assumed that the atmosphere and the oceans were in equilibrium in 1860 with a concentration of 290 ppm by volume. The mechanisms used for oceanic uptake were gaseous concentration difference and the chemical mixing of CO_2 into a larger pool of carbonates. The mechanism used for continental biospheric uptake was CO_2 fertilization, that is, an increase in atmospheric CO_2 concentration, causing a proportional increase in photosynthetic rate. According to Keeling (1970) the mass of CO_2 in that layer of the ocean which is in communication with the atmosphere is between 5 and 8 times the mass of CO_2 in the atmosphere. The model assumed a mixing layer in the ocean that contains 8 times the mass of CO_2 that is in the atmosphere. The fertilization coefficient was adjusted to agree roughly with the Mauna Loa measurements (See Figure 1.2).

[b] For illustrative purposes only, the numbers in the first column are based on 4 percent annual growth rate until 1980 and 3.5 percent thereafter. These numbers are slightly larger than those compiled by Work Group 7.

Note: One million metric tons equals one megaton (Mt) or 10^{12} grams (g).

Reprinted from *Man's Impact on the Global Environment. Assessment and Recommendations for Action* by William H. Mathews by permission of The MIT Press, Cambridge, Massachusetts, 1970.

CO₂ Produced by Fossil Fuel Combustion, 1950-1967
(Billions of metric tons)

Year	Coal[a]	Lignite[b]	Refined Oil Fuels[c]	Natural Gas[d]	Total
1950	3.7	0.9	1.4	0.4	6.4
1951	3.8	0.9	1.7	0.5	6.9
1952	3.8	0.9	1.8	0.5	7.0
1953	3.8	0.9	1.9	0.5	7.1
1954	3.8	0.9	2.0	0.6	7.3
1955	4.1	1.0	2.2	0.6	7.9
1956	4.4	1.1	2.4	0.7	8.6
1957	4.5	1.3	2.5	0.7	9.0
1958	4.6	1.4	2.6	0.8	9.4
1959	4.8	1.4	2.8	0.9	9.9
1960	5.0	1.4	3.1	1.0	10.5
1961	4.5	1.5	3.3	1.0	10.3
1962	4.6	1.5	3.5	1.1	10.7
1963	4.8	1.6	3.8	1.2	11.4
1964	5.0	1.7	4.2	1.3	12.2
1965	5.0	1.7	4.5	1.5	12.7
1966	5.1	1.7	4.8	1.6	13.2
1967	4.8	1.7	5.2	1.7	13.4
1980 (est.)[e]	11.1[f]		10.8	4.0	26.0

Source: Computed from Table 7.A.1

[a] Assumed carbon content, coal = 75 percent (PSAC, 1965). Note: the weight of CO_2 emissions is equal to the weight of the fuel $\times$ its percent carbon $\times$ the ratio of the molecular weight of CO_2 to the molecular weight of carbon (i.e., $\frac{44}{12} = 3.67$).

[b] Assumed carbon content, lignite = 45 percent (PSAC, 1965).

[c] Assumed carbon content, refined oil fuels = 86 percent (the figure used by PSAC, 1965, for liquid hydrocarbons).

[d] Assumed carbon content, natural gas = 70 percent (corresponding to a mixture by volume of 80 percent CH_4, 15 percent C_2H_6, and 5 percent N_2) (PSAC, 1965).

[e] The 1980 estimate was constructed by multiplying the 1965 emissions for coal (including lignite), refined oil fuels, and natural gas $\times$ growth factors (i.e., ratios of 1980 consumption to 1965 consumption for solid and liquid fuels and natural gas) derived from Darmstadter's figures in metric tons coal equivalent for 1965 and 1980.

[f] Coal and lignite combined.

Reprinted from *Man's Impact on the Global Environment: Assessment and Recommendations* by William H. Mathews by permission of The MIT Press, Cambridge, Massachusetts, 1970.

NATURAL AND MAN-MADE ENERGY DENSITIES

As illustrated here, and in 13-5 and 13-6, on a global basis the energy produced by man is very small in comparison to the energy input from the sun. On a local basis, however, energy densities in urban and industrial areas often exceed the energy from the sun and, along with other changes in the atmosphere caused by man's activities, result in changes in local climate (see 13-7).

Natural Power Densities

	W/m²
Heat balance on the earth's surface (average)	100
Latent heat density of rainfall on the continents (global av.)	75
Sensible heat density for 1°C of rainfall on the continents	0.1
Winds, waves, convections, and currents (global average)	0.7
Photosynthesis	0.075

Energy Production per Unit Area of Natural and Artificial Processes at Various Scales

	Natural Production, W/m²		Artificial Production, W/m²	
Area, m²	Event	Rate	Type of Use	Rate
5×10^{14}	Dissipation of kinetic energy	5	Man's ultimate energy production	0.8
	Solar-energy absorption by atmosphere	25		
	Solar flux at top of atmosphere	350		
10^{12}	Cyclone latent heat release (1 cm rain per day)	200	Northeastern United States ultimate production (10^8 people, 20 kw each)	2.0
	Cyclone kinetic energy production	30		
10^8	Thunderstorm		Suburban area (400 persons per km², 10 kw per capita)	4
	Kinetic energy production	100	Super energy center or city	1000
	Latent heat release (1 cm rain per 30 min)	5000	Agroindustrial complex	100
	Evaporation from lake	100		
10^4	Tornado kinetic energy production	10^4	Cooling pond (1000 kw per 2 acres)	130
			Cooling tower (10^6 kw per 10^4 m²)	10^5

Source: Parker, Frank L. "Other Ecological Impacts," in Energy, the Environment, and Human Health, © American Medical Association Congress on Environmental Health, Publishing Sciences Group, Inc., Acton, Massachusetts, 1974.

	Area (km²)	Population 10^6	EC density (W/m²)	EC per Capita, kw.	Average net radiation (W/m²)
Nordrhein-Westfalen	34,039	16.84	4.2	8.0	50
Same, industrial area only	10,296	11.27	10.2	8.9[b]	51
West Berlin	234[a]	2.3	21.3	2.0	57
Moscow	878	6.42	127	16.8[b]	42
Sheffield (1952)	48[a]	0.5	19	1.6	46
Hamburg	747	1.83	12.6[a]	5.0	55
Cincinnati	200[a]	0.54	26	9.3	99
Los Angeles County	10,000	7.0	7.5	10.3	108
Los Angeles	3,500[a]	7.0	21	10.3	108
New York, Manhattan	59	1.7	630	21.0	93
21 metropolitan areas (Washington-Boston)	87,000	33	4.4	11.2[c]	~90
Fairbanks, Alaska[a]	37	0.03	18.5	21.8	18

[a] Building area only   [b] Related to industrial production   [c] Eastern United States

Rises in energy use are predicted by various sources to be anywhere from 3.8 to 7 per cent per year. Yet the energy consumption of some urban areas is today of the same order of magnitude as the average net natural radiation of the area, as this table indicates, and for New York and Moscow it is larger. (The table covers data from 1965–1968.) The data was collected by H. Flohn and presented to the Study of Man's Impact on Climate this summer in Stockholm. (See Robert C. Cowen's Science Report in Technology Review for October/November, p. 7.)

Source: Harleman, Donald R. F., "Heat—the Ultimate Waste," Technology Review, edited at The Massachusetts Institute of Technology, December, 1971.

Man-Made Power Densities

	Consumption			
	Future		Present	
Global average	$\dfrac{20 \text{ kW/cap} \cdot 10^{10} \text{ cap}}{1.48 \times 10^{14} \text{ m}^2} = 1.35 \text{ W/m}^2$		$\dfrac{1.5 \text{ kW/cap} \cdot 3.3 \times 10^9 \text{ cap}}{1.48 \times 10^{14} \text{ m}^2} = 0.033 \text{ W/m}^2$	
F. R. Germany	$\dfrac{20 \text{ kW/cap} \cdot 6 \times 10^7 \text{ cap}}{2.5 \times 10^{11} \text{ m}^2} = 5 \text{ W/m}^2$		$\dfrac{4 \text{ kW/cap} \cdot 6 \times 10^7 \text{ cap}}{2.5 \times 10^{11} \text{ m}^2} = 1 \text{ W/m}^2$	
Industrial area (Ruhr)	$\dfrac{100 \text{ kW/cap} \cdot 10^8 \text{ cap}}{10^{10} \text{ m}^2} = 1,000 \text{ W/m}^2$		$\dfrac{18 \text{ kW/cap} \cdot 6 \times 10^6 \text{ cap}}{6.5 \times 10^9 \text{ m}^2} = 17 \text{ W/m}^2$	

Production

Large nuclear power parks 30,000 MWe → 100,000 MWth

$$\frac{7 \times 10^{10} \text{ Wth (waste)}}{3.5 \times 10^6 \text{ m}^2} = 20,000 \text{ W/m}^2$$

Source: Wolf Häfele, "A systems approach to energy," American Scientist, Vol. 62, July–August 1974.

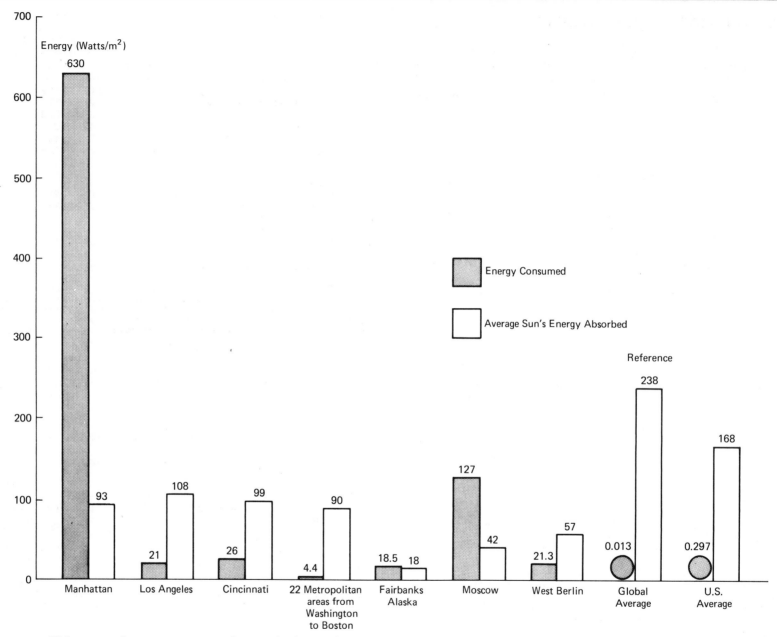

Energy (Watts/m²)

Energy Consumed

Average Sun's Energy Absorbed

While man on the average consumes only one unit of energy for every 16,700 units which are absorbed on earth from the sun, there are "islands" where the heat release from energy consumption is a significant share of the average solar input. Energy release on Manhattan Island (the extreme case) is 630 watts per square meter, more than six times the sun's average of 93 watts/m².

Reprinted from *Man's Impact on Climate* by W. H. Mathews, W. H. Kellogg, and R. D. Robinson by permission of The M.I.T. Press, Cambridge, Massachusetts, 1971.

GROWTH IN THERMAL ENERGY PRODUCTION

Thermal Waste Energy
(in units of 10^6 MW)

Geographic Location	Assumed Yearly Increase (percent)	1970	1980	2000
World	5.7	5.5	9.6	31.8
North America	4	2.2	3.4	7.5
(United States)[a]	(4)	(2.0)	(3)	(6.5)
(Canada)	(7)	(0.183)	(0.36)	(1.0)
Central America	6	0.12	0.2	0.68
South America	6	0.09	0.16	0.5
Western Europe	4	1.08	1.6	3.5
Western Asia	10	0.05	0.13	0.81
Far East	10	0.44	1.1	7.2
Oceania	8	0.069	0.145	0.64
(Australia)	(8)	(0.06)	(0.13)	(0.58)
Africa	6	0.1	0.18	0.57
East Europe	7	1.37	2.7	10.4
(Russia)	(8)	(0.98)	(2.0)	(9)

Source: United Nations, *World Energy* [a] ( ) indicates subregion of main region.
Supplies.

Reprinted from *Man's Impact on the Global Environment: Assessment and Recommendations for Action* by William H. Mathews by permission of The MIT Press, Cambridge, Massachusetts, 1970.

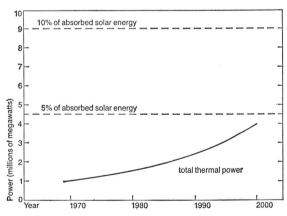

Source: Greenfield, 1970
Area: East, North Central States (Mich., Ill., Ind., Wis., Ohio); Middle Atlantic States (N.Y., N.J., Pa.) 351,028 mi³
Assumptions:
1. Electric power + waste heat = 30 percent of all power
2. Electrical capacity increases by factor of 10, 1970–2000
3. All other energy doubles, 1970–2000

Thermal power generation in a climatically significant area in the United States.

Reprinted from *Man's Impact of the Global Environment: Assessment and Recommendations for Action* by William H. Mathews by permission of The MIT Press, Cambridge, Massachusetts, 1970.

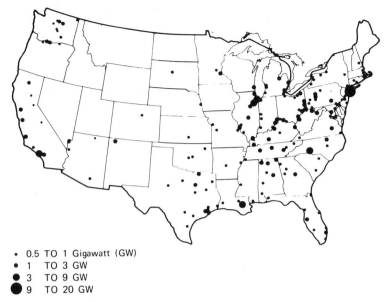

- • 0.5 TO 1 Gigawatt (GW)
- • 1 TO 3 GW
- • 3 TO 9 GW
- • 9 TO 20 GW

Electrical generation for the conterminous United States: 1970 (Source: U.S. Federal Power Commission)

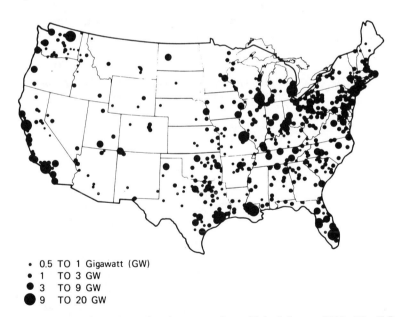

- • 0.5 TO 1 Gigawatt (GW)
- • 1 TO 3 GW
- • 3 TO 9 GW
- • 9 TO 20 GW

Possible electrical generation pattern for the conterminous United States: 2000. The U.S. Federal Power Commission originally suggested this pattern for 1990; we have assumed this same pattern for the year 2000, based on the slower growth rate expected for electrical power consumption (based on data from U.S. Federal Power Commission)

Source: Environmental Quality–1976, The Seventh Annual Report of the Council on Environmental Quality, September 1976.

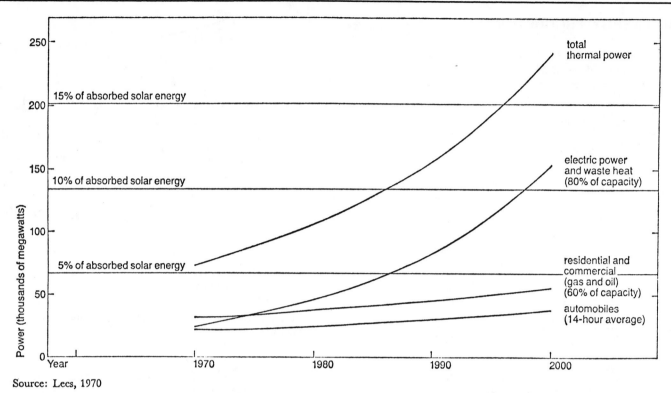

Thermal power generation in the Los Angeles basin.

Source: Lees, 1970

Reprinted from *Man's Impact on the Global Environment: Assessment and Recommendations for Action* by William H. Mathews by permission of The MIT Press, Cambridge, Massachusetts, 1970.

LIMITS OF THERMAL ENERGY PRODUCTION

At present, energy use throughout the world is about 0.250 Q (10^{18} Btu) per year. With a world population of about four billion, per capita consumption is about 60 million Btu per person per year. Assuming the world population levels off at about 10 billion people and that per capita consumption rises to about 300 million Btu per year (United States per capita consumption is now about 370 million Btu per year), the level of world thermal energy production would reach 3 Q per year. According to the chart by Luten, this level of energy production would result in a trivial increase in temperature of less than 0.1°F.

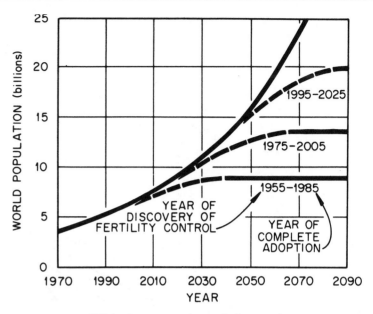

Effect of momentum in population growth.

Source: **Alvin M. Weinberg and R. Philip Hammond,** *Global Effects of Increased Use of Energy,* A/CONF.–49/P–033; Fourth U.N. International Conference on the Peaceful Uses of Atomic Energy, Geneva, Switzerland, September 1971.

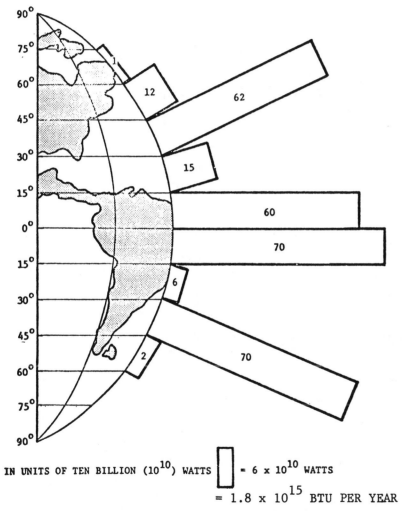

IN UNITS OF TEN BILLION (10^{10}) WATTS $\square$ = 6×10^{10} WATTS

= 1.8×10^{15} BTU PER YEAR

Estimated thermal waste in year 2100 for longitudinal section, 60°W–90°W (total world population— 10 billion; energy consumption per capita = 300×10^6 Btu per year).

Source: Solar Energy Final Task Force Report, Project Independence Blueprint, Federal Energy Administration, November 1974.

Range of Total World Energy Use

WORLD POPULATION	AVERAGE PER CAPITA ENERGY USE			
	100	300	500	1000 MMBTU/YR.
6 Billion	0.6	1.8	3	6×10^{18} BTU/YR.
10 Billion	1	3	5	10×10^{18} BTU/YR.
20 Billion	2	6	10	20×10^{18} BTU/YR.

Source: Solar Energy Final Task Force Report, Project Independence Blueprint, Federal Energy Administration, November 1974.

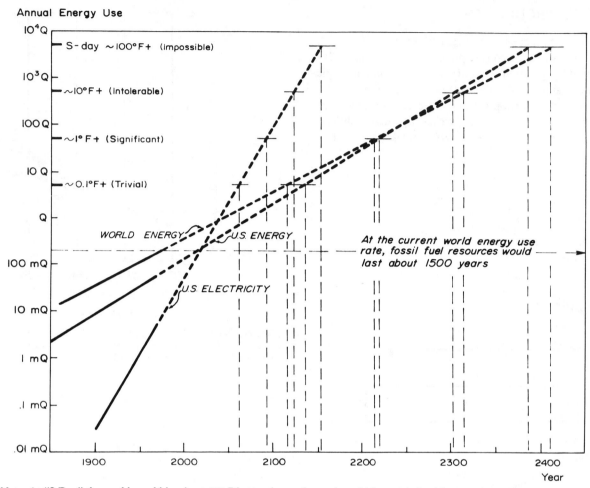

Annual Energy Use

Note: At "S Day" the world would be about 100 F hotter than today and would be uninhabitable; accordingly, the condition is unrealizable. At one tenth the S-day energy use, the earth would still be 10 F hotter and much of it uninhabitable, an intolerable condition. Note how long it is before even trivial heating occurs and how quickly it worsens.

Extrapolation of U.S. energy growth.

Source: Luten, Daniel B. "United States Requirements," in *Energy, The Environment, and Human Health,* © American Medical Association, Publishing Sciences Group, Inc., Acton, Massachusetts, 1974.

CLIMATIC CHANGES FROM URBANIZATION

A number of factors contribute to the climatic differences between urban and rural environments. The materials used to construct buildings and streets have a higher thermal conductivity than does soil and can therefore absorb heat from the sun at a faster rate. This stored heat is released at night keeping the city at a higher temperature than the surrounding countryside. Building structures in the city have a braking effect on the wind thereby reducing the amount of heat carried away. The spaces between buildings provide traps for the incoming sunlight with light being reflected from one surface to another until it is absorbed. Large quantities of heat are generated within the city by heating and air-conditioning systems. The cooling effect of evaporation of moisture from the ground is largely absent in the city since rain is immediately diverted to sewers and transported away from the city. Finally, activities in the city generate gaseous and particulate contaminents in the air that partially block incoming sunlight and prevent the outflow of heat.

These effects combine to create a different climate for the city in contrast to the countryside. The city has less ultraviolet radiation, more clouds and fog, more rain, a higher temperature, a lower relative humidity, and a lower average wind speed.

Climatic Changes Produced by Cities

Element	Comparison with Rural Environs
Contaminants	
Dust particles	10 times more
Sulfur dioxide	5 times more
Carbon dioxide	10 times more
Carbon monoxide	25 times more
Radiation	
Total on horizontal surface	15 to 20% less
Ultraviolet, winter	30% less
Ultraviolet, summer	5% less
Cloudiness	
Clouds	5 to 10% more
Fog, winter	100% more
Fog, summer	30% more
Precipitation	
Amounts	5 to 10% more
Days with 0.2 in.	10% more
Temperature	
Annual mean	1 to 1.5°F more
Winter minima	2 to 3° F more
Relative humidity	
Annual mean	6% less
Winter	2% less
Summer	8% less
Wind speed	
Annual mean	20 to 30% less
Extreme gusts	10 to 20% less
Calms	5 to 20% less

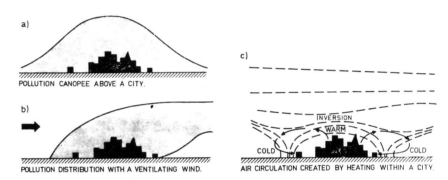

Urban heat and pollution islands. During calm weather, particles of dust and smoke accumulate over the centers of large cities. There is some circulation between the cooler suburbs and the warmer central portion of the cities. This heat and pollution is removed by ventilating winds.

Source: Bert Bolin, "Energy and Climate, Secretariat for Future Studies, Stockholm, Sweden.

NATURAL AND MAN-MADE PARTICULATE AIR POLLUTION

As indicated in the accompanying tables, on a global basis, man-made pollutants contribute about 20% of the total pollutants in the atmosphere. Man-made pollutants are, however, generated in localized urban and industrial areas and it is in these areas that measures for control are being implemented. The U.S. Environmental Protection Agency has identified five pollutants for which standards are being set. These five substances include: carbon monoxide (CO), nitrogen oxides (NO_x), photochemical oxidants (O_x), sulfur oxides (SO_x), and particulate matter (P). Particulates are solid or liquid airborne substances, the remainder are gases. Photochemical oxidants are not emitted directly, but are produced from nitrogen oxides and hydrocarbon gases through a number of atmospheric reactions. Hydrocarbon emissions (HC) are not harmful in themselves, but because they contribute to the formation of photochemical oxidants, they have become an important air quality measure.

Estimates of Particles Smaller than 20-μ Radius Emitted into or Formed in the Atmosphere

[10^6 metric tons per year]

Natural:	
Soil and rock debris	100 to 500.
Forest fires and slash-burning debris	3 to 150.
Sea salt	(300).
Volcanic debris	25 to 150.
Particles formed from gaseous emissions:	
Sulfate from H_2S	130 to 200.
Ammonium salts from NH_3	80 to 270.
Nitrate from NO_x	60 to 430
Hydrocarbons from plant exudations	75 to 200.
Subtotal	773 to 2,200.
Man-made:	
Particles (direct emissions)	10 to 90.
Particles formed from gaseous emissions	130 to 200.
Sulfate from SO_2	130 to 200.
Nitrate from NO_x	30 to 35.
Hydrocarbons	15 to 90.
Subtotal	185 to 415.
Total	958 to 2,615

[1] Includes unknown amounts of indirect man-made contributions.

Source: Summary Report of the Cornell Workshop on Energy and the Environment, sponsored by the National Science Foundation, Committee on Interior and Insular Affairs, United States Senate, May 1972.

Sources of Airborne Particles Based on Observed Atmospheric Fallout

(Million metric tons/yr)

Sources	Observed Fallout
Sea salts	1000 to 2000[a]
Mineral dusts	
Northern Hemisphere	50[b]
Southern Hemisphere	20[c]

[a]*Source:* Woodcock, A. H., 1962. This estimate indicates the probable rate of production at the sea surface of salt particles that remain airborne long enough to be transported up to local cumulus-base altitudes.

[b]The rate of dust production of a given area is a poorly understood function of numerous variables including the size of its land surface, its topographical features (mountains, forest, desert) and the use to which it is put. Nevertheless, a minimum estimate of the annual production of atmospheric dust can be made by assuming steady state and using measurements (Windom, 1969) of dust accumulations in five North American glaciers between Greenland (77°N) and Mt. Popocatepetl, Mexico (19°N) in the three major wind systems of the Northern Hemisphere. Particles accumulate at a median rate of about 0.00002 g/cm²-yr (ranging from 0.07 to 3.2×10^{-5} g/cm²-yr). (No size data for particles were given.) Assuming that fallout was uniform over the Northern Hemisphere (2.55×10^{18} cm²), the total fallout was about 5×10^{13} g/yr, or 50 million metric tons per year.

[c]Less dust is produced in the Southern Hemisphere owing to the relative scarcity of arid lands there as compared to the Northern Hemisphere (Bogdanov, 1963). We assumed that the dust production of the Southern Hemisphere was 40% of the Northern Hemisphere.

Reprinted from *Man's Impact on the Global Environment: Assessment and Recommendations for Action* by William H. Mathews by permission of The MIT Press, Cambridge, Massachusetts, 1970.

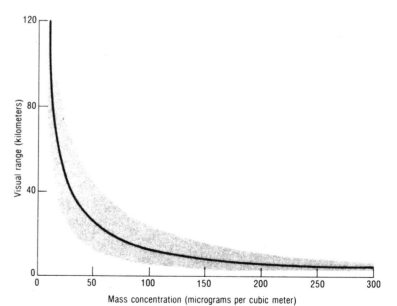

Source: U.S. Department of Health, Education, and Welfare, Public Health Service, *Air Quality Criteria for Particulate Matter* (Washington, D.C.: Government Printing Office, 1969), p. 57.

Relationship between concentration of particulates and visibility.

Source: The Sixth Annual Report of the Council on Environmental Quality, December 1975.

Selected Sources of Airborne Particles Based on Inputs
(Million metric tons per year)

Sources	U.S. Total Particles of All Sizes[a]	U.S. 5-micron-diameter Particles	U.S./World Ratio	Estimated World Release 5-micron-diameter Particles[b]
Agriculture and forestry burning				
Forest	6.09	0.9[c]	6[d]	5.4
Crop wastes and ranges	2.18	0.33[c]	10[e]	3.3
Forest cleaning	(no reliable available estimates . . . .)			
Grain processing	0.73	0.02[f]	10[g]	0.2
Mining	1.45	(no reliable available estimates . . .)		
Waste incineration	1.0	0.18[h]	( )[i]	——

[a] NAPCA, 1970.

[b] Computed from columns 2 and 3 of this table.

[c] Duprey, 1968, gives 15 percent of all particles released by all types of refuse incinerators as less than 5 microns in diameter. In the absence of other data, we have taken 15 percent as the fraction of particles less than 5 microns emitted by burning.

[d] The United States has about one-sixth of the forest acreage under exploitation and also about one-sixth of the world forest area. Therefore, to convert U.S. data on forest fire emissions to a world estimate, we have multiplied by a factor of 6.

[e] The United States has about one-ninth of the cropland, pastureland, and rangeland of the world. We have added an additional factor to account for the greater use of fire in other continents.

[f] Duprey, 1968, cites one test of particle size distribution from feed and grain mills in which 3 percent of the particles by weight determination were less than 5 microns in diameter.

[g] Using the 1967 production of wheat flour, both global and U.S. (United Nations, *Statistical Yearbook*, 1968), the United States accounts for one-tenth of world production.

[h] NAPCA, 1969. The percent (by weight) of all released particles that are less than 5 microns in diameter at two municipal incinerators have been analyzed: Los Angeles 30 percent, and Milwaukee 6 percent. The figure in the table assumes the average of these two numbers, 18 percent for the entire United States. This assumption is not necessarily correct, but the results are illustrative.

[i] No estimate is available for this figure. It might be noted that in much of the industrialized world (Europe) disposal methods are in many cases better than those in the United States, while in the underdeveloped world waste incineration is small relative to other sources of particles and compared to that in industrialized countries.

Reprinted from *Man's Impact on the Global Environment: Assessment and Recommendations for Action* by William H. Mathews by permission of The MIT Press, Cambridge, Massachusetts, 1970.

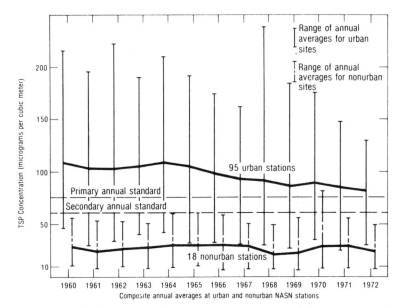

Trends in total suspended particulate concentrations, 1960 through 1972, at National Air Sampling Network (NASN) stations.

Source: The Fifth Annual Report of the Council on Environmental Quality, December, 1974.

Total Suspended Particulates: Estimated Total Nationwide Emissions
[In millions of tons per year]

	1940	1950	1960	1970
Controllable	19.2	20.8	21.0	22.3
Total [1]	44.9	33.2	29.9	25.5

[1] Includes sources which are considered "uncontrollable" by EPA's definition. These include such categories as forest, structural, and agricultural fires. Because the year-to-year variability in "uncontrollable" sources can be great, the apparent downward trend in estimated *total* emissions should be viewed with caution. Estimated total TSP emissions for 1969, for example, were 35 million tons.

Source: Environmental Protection Agency, *The National Air Monitoring Program: Air Quality and Emissions Trends* (1973), Vol. 1, pp. 1–6ff.

Source: The Fifth Annual Report of the Council on Environmental Quality, December, 1974.

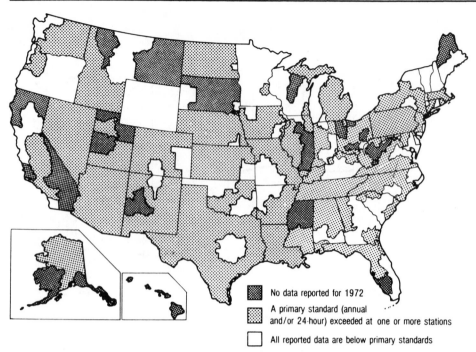

No data reported for 1972

A primary standard (annual and/or 24-hour) exceeded at one or more stations

All reported data are below primary standards

Total suspended particulate levels, 1972.

Source: The Fifth Annual Report of the Council on Environmental Quality, December, 1974.

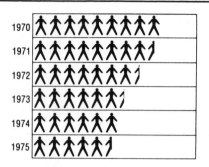

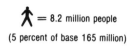

= 8.2 million people
(5 percent of base 165 million)

Source: U.S. Environmental Protection Agency, *Trends in the Quality of the Nation's Air* (Washington, D.C.: Government Printing Office, 1977), p. 5.

Trends in national population exposure to particulate matter levels above the National Primary Ambient Air Quality Standard (NAAQS), 1970–75.

Source: U.S. Environmental Protection Agency, *Trends in the Quality of the Nation's Air* (Washington, D.C.: Government Printing Office, 1977), p. 5.

SULFUR DIOXIDE AIR POLLUTION

About two-thirds of the sulfur dioxide emissions come from the combustion of fuels, primarily coal. The remaining third stems principally from industrial processes with minor contributions from transportation, solid waste disposal, and agricultural waste disposal. The geographic distribution follows, therefore, the pattern of industrial development and the movement of these emissions by the prevailing westerly winds.

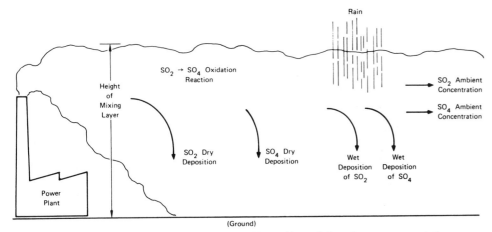

Model for emission to ambient relationship for sulfur oxide emissions from a representative power plant.

Source: *Air Quality and Stationary Source Emission Control*, A Report by the Commission on Natural Resources, prepared for the Committee on Public Works, United States Senate, March 1975.

Components of the Global Sulfur Cycle

(Values listed are in units of million (10^6) tons per year, calculated as sulfur)

Component	Land	Sea	Total	Percent of Total
Sources to atmosphere:				
Windblown sea salt ($SO_4^=$)	—	43	43	24
Biogenic H_2S, organic sulfides	—	—	88	48
Burning fossil fuels, Smelting (SO_2)	50	—	50	27
Volcanoes, etc. (H_2S, SO_2, $S°$, $SO_4^=$)	0.66	—	0.66	0.36
Total			182	
Sinks from atmosphere:				
Rain over the oceans (SO_2, $SO_4^=$)	—	72	72	39
Rain over the land (SO_2, $SO_4^=$)	85	—	85	47
Plant uptake and dry deposition (SO_2)	15	—	15	8
Plant uptake ($SO_4^=$)	10	—	10	5
Total			182	

Source: Data of Kellog et al. (1972), recalculated in terms of amounts of sulfur.
SO_2 = 50% sulfur, H_2S = 94% sulfur, $SO_4^=$ = 33% sulfur. The agreement between sources and sinks is artificial since the quantity of biogenic H_2S was calculated in order to make the values agree.
For a comparison of other estimates, see Friend (1973).
Reproduced with permission of the National Academy of Science.

SOURCE: Finklea *et al.* (1974).

Nationwide geographic variation in SO_2 emission density.

Reproduced with permission of the National Academy of Sciences.

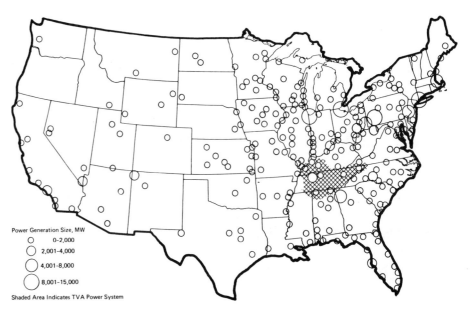

SOURCE: Finklea *et al.* (1974).

Location of major coal- and oil-fired power units, 1971.

Reproduced with permission of the National Academy of Sciences.

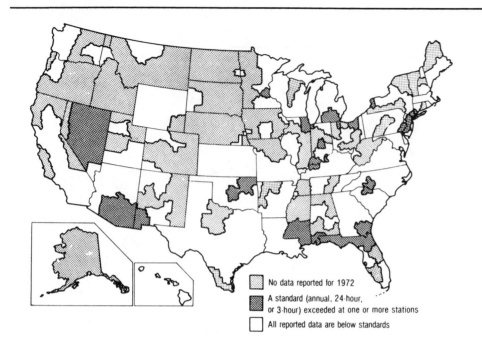

No data reported for 1972

A standard (annual, 24-hour, or 3-hour) exceeded at one or more stations

All reported data are below standards

Source: EPA, *Monitoring and Air Quality Trends Report, 1972* (1973), Figure 3–2.

Sulfur dioxide levels, 1972.

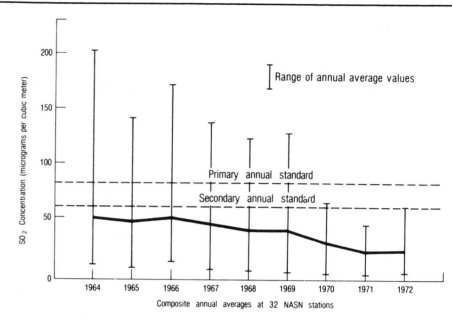

Trends in sulfur dioxide concentrations, 1964–72, National Air Sampling Network (NASN) stations.

Sulfur Dioxide: Estimated Total Nationwide Emissions

[In millions of tons per year]

	1940	1950	1960	1970
Controllable	22.2	24.3	22.6	33.3
Total	22.8	24.9	23.2	33.4

Source: Environmental Protection Agency, *The National Air Monitoring Program: Air Quality and Emissions Trends* (1973), Table 1–5.

Source: The Fifth Annual Report of the Council on Environmental Quality, December 1974.

Emission of SO$_2$ from Stationary Sources in the ECE Region (1973)

	10^3 tons
Austria	700
Belgium	998
Bulgaria	1,000
Canada[a]	7,037
Czechoslovakia	3,000
Denmark	623
Finland	549
France	3,224
Federal Republic of Germany	3,928
German Democratic Republic	4,000
Greece	—
Hungary	1,500
Ireland	330
Italy	3,169
Luxembourg	46
Netherlands	782
Norway	181
Poland	3,000
Portugal	—
Spain	1,297
Sweden	829
Switzerland	151
United Kingdom	5,605
United States of America[b]	30,126
USSR	23,500
Total	96,575

According to the data presented here, the SO$_2$ emission from stationary sources in the ECE region, in the year 1973 could be estimated at around 100 million tons. It is important to mention that more than 75% of world man-made air pollution is emitted in ECE countries.[c]

[a]ENV/WP.1/R.9 page 5.

[b]United States Statement.

[c]The secretariat's calculation according to world and ECE countries fuel consumption. (See ENV/R.25.)

Source: Introductory Report by the Secretariat, ENV/SEM.4/R.1, Economic Commission for Europe, Second Seminar on Desulphurization of Fuels and Combustion Gases, Washington, D.C. Nov. 11–20, 1975.

Mechanisms That Convert Sulfur Dioxide to Sulfates or Sulfuric Acid Aerosols

Mechanism	Overall reaction	Factors on which sulfate formation primarily depends (in addition to sulfur dioxide concentration)
Direct photooxidation	$SO_2 \xrightarrow[\text{water}]{\text{light, oxygen}} H_2SO_4$	Sunlight intensity
Indirect photooxidation	$SO_2 \xrightarrow[\text{organic oxidants, OH}]{\text{smog, water, NO}_x} H_2SO_4$	Organic oxidant concentration, OH, NO$_x$
Air oxidation in liquid droplets	$SO_2 \xrightarrow{\text{liquid water}} H_2SO_3$ $NH_3 + H_2SO_3 \xrightarrow{\text{oxygen}} NH_4^+ + SO_4^=$	Ammonia concentration
Catalyzed oxidation in liquid droplets	$SO_2 \xrightarrow[\text{heavy metal ions}]{\text{oxygen, liquid water}} SO_4^=$	Concentration of heavy metal (Fe, Mn) ions
Catalyzed oxidation on dry particles	$SO_2 \xrightarrow[\text{carbon, water}]{\text{oxygen, particulate}} H_2SO_4$	Carbon particle concentration (surface area)

Source: Environmental Protection Agency.

Source: The Sixth Annual Report of the Council on Environmental Quality, December 1975.

U.S. SOURCES OF AIR POLLUTION

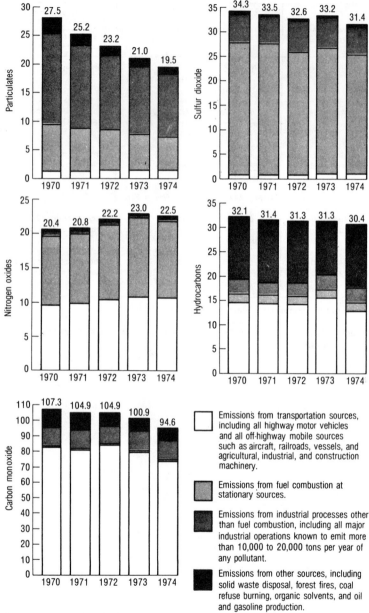

Legend:

☐ Emissions from transportation sources, including all highway motor vehicles and all off-highway mobile sources such as aircraft, railroads, vessels, and agricultural, industrial, and construction machinery.

▦ Emissions from fuel combustion at stationary sources.

◪ Emissions from industrial processes other than fuel combustion, including all major industrial operations known to emit more than 10,000 to 20,000 tons per year of any pollutant.

■ Emissions from other sources, including solid waste disposal, forest fires, coal refuse burning, organic solvents, and oil and gasoline production.

Source: Environmental Protection Agency.

Estimated emissions of major air pollutants, nationwide, 1970-74 (millions of tons per year)

Source: The Sixth Annual Report of the Council on Environmental Quality, December 1975.

Nationwide Air Pollution Emissions, by Pollutant and Source, 1970-74[1]
(In millions of tons per year)

Pollutants and sources	1970	1971	1972	1973	1974
Particulates					
Transportation [2]	1.2	1.2	1.3	1.3	1.3
Fuel combustion in stationary sources	8.3	7.5	7.1	6.4	5.9
Industrial processes [3]	15.7	14.5	13.1	11.9	11.0
Solid waste disposal	1.1	0.8	0.7	0.6	0.5
Miscellaneous [4]	1.2	1.2	1.0	0.8	0.8
Total	27.5	25.2	23.2	21.0	19.5
Sulfur oxides					
Transportation [2]	0.7	0.7	0.7	0.8	0.8
Fuel combustion in stationary sources	27.0	26.7	25.2	25.6	24.3
Industrial processes [3]	6.4	6.0	6.6	6.7	6.2
Solid waste disposal	0.1	0.0	0.0	0.0	0.0
Miscellaneous [4]	0.1	0.1	0.1	0.1	0.1
Total	34.3	33.5	32.6	33.2	31.4
Carbon monoxide					
Transportation [2]	82.3	80.9	83.4	79.3	73.5
Fuel combustion in stationary sources	1.1	1.0	1.0	1.0	0.9
Industrial processes [3]	11.8	11.6	12.0	13.0	12.7
Soid waste disposal	5.5	3.9	3.2	2.8	2.4
Miscellaneous [4]	6.6	7.5	5.3	4.8	5.1
Total	107.3	104.9	104.9	100.9	94.6
Hydrocarbons					
Transportation [2]	14.7	14.3	14.1	13.7	12.8
Fuel combustion in stationary sources	1.6	1.7	1.7	1.7	1.7
Industrial processes [3]	2.9	2.7	2.9	3.1	3.1
Solid waste disposal	1.4	1.0	0.8	0.7	0.6
Miscellaneous [4]	11.5	11.7	11.8	12.1	12.2
Total	32.1	31.4	31.3	31.3	30.4
Nitrogen oxides					
Transportation [2]	9.3	9.8	10.5	11.0	10.7
Fuel combustion in stationary sources	10.1	10.1	10.8	11.2	11.0
Industrial processes [3]	0.6	0.6	0.6	0.6	0.6
Solid waste disposal	0.3	0.2	0.2	0.1	0.1
Miscellaneous [4]	0.1	0.1	0.1	0.1	0.1
Total	20.4	20.8	22.2	23.0	22.5

[1] The data presented in this table are estimates based on common data sources, uniform estimation methods, and consistent assumptions. At the time these estimates were prepared, most 1974 and some 1973 source data were not available from the standard references. Missing items were projected on the basis of forecasts from industrial trade journals, economic indications, or extrapolations from 1967–72 data.

[2] Includes all highway motor vehicles and all off-highway mobile sources such as aircraft, railroads, vessels, and agricultural, industrial, and construction machinery.

[3] Emissions for over 80 industrial processes/products were computed, including all major operations known to emit more than 10,000–20,000 tons per year of some criteria pollutant on a national scale.

[4] Sources not appropriately classified under previous categories are included here. Among these are forest fires, coal refuse burning, organic solvents, and oil and gasoline production.

Source: Environmental Protection Agency.

Source: The Sixth Annual Report of the Council on Environmental Quality, December 1975.

Nationwide Emissions in the United States, 1968[a]
(Million short tons per year)[b]

Source	Emission Particles[c] 10⁶ short tons	Percentage	SO$_x$ 10⁶ short tons	Percentage	HC 10⁶ short tons	Percentage	Emission NO$_x$ 10⁶ short tons	Percentage	CO 10⁶ short tons	Percentage
Fuel Combustion	10.1	35.7	25.2	75.9	17.3	54.1	18.1	87.8	65.7	65.7
Transportation	1.2	4.3	0.8	2.4	16.6	51.9	8.1	39.3	63.8	63.8
Motor vehicles	0.8	2.8	0.3	0.9	15.6	48.8	7.2	34.9	59.2	59.2
Gasoline	0.5	1.8	0.2	0.6	15.2	47.5	6.6	32.0	59.0	59.0
Diesel	0.3	1.0	0.1	0.3	0.4	1.3	0.6	2.9	0.2	0.2
Aircraft[d]	N[e]	N	N	N	0.3	0.9	N	N	2.4	2.4
Railroads	0.2	0.7	0.1	0.3	0.3	0.9	0.4	1.9	0.1	0.1
Vessels	0.1	0.4	0.3	0.9	0.1	0.3	0.2	1.0	0.3	0.3
Nonhighway motor fuels use	0.1	0.4	0.1	0.3	0.3	1.0	0.3	1.5	1.8	1.8
Stationary sources	8.9	31.4	24.4	73.5	0.7	2.2	10.0	48.5	1.9	1.9
Coal	8.2	29.0	20.1	60.5	0.2	0.6	4.0	19.4	0.8	0.8
Fuel oil	0.3	1.0	4.3	13.0	0.1	0.3	1.0	4.8	0.1	0.1
Natural gas	0.2	0.7	N	N	N	N	4.8	23.3	N	N
Wood	0.2	0.7	N	N	0.4	1.3	0.2	1.0	1.0	1.0
Other Sources	18.2	64.3	8.0	24.1	14.7	45.9	2.5	12.2	34.4	34.3
Industrial processes[f]	7.5	26.5	7.3	22.0	4.6	14.4	0.2	1.0	9.7	9.6
Solid waste disposal	1.1	3.9	0.1	0.3	1.6	5.0	0.6	2.9	7.8	7.8
Miscellaneous[g]	9.6	33.9	0.6	1.8	8.5	26.5	1.7	8.3	16.9	16.9
Total	28.3	100.0	33.2	100.0	32.0	100.0	20.6	100.0	100.1	100.0

[a] Source: NAPCA, 1970. The authors of this document comment that the "numbers presented here should be representative of current emissions" but that "because of the increasing availability of more comprehensive data and emission factors, revisions will be made in emission quantities. Therefore, the numbers presented in this document are subject to change."

[b] Data were given in short tons and were not converted in the table to metric tons because of the distortion in the percentage figures that would result from rounding off the numbers in the emissions columns. Some of the overall figures are presented below in million metric tons:

[c] No estimates were given of the likely size distribution of particles. For the burning of coal, 79 percent of the particle emissions from electric generating plants and 62 percent from industrial sources were assumed to have been collected by control devices (NAPCA, 1970).

[d] Emissions are those under 3,000 feet only. Twenty percent of the fuel used in each flight is assumed to be consumed under 3,000 feet.

More detailed data can be found in Northern Research and Engineering Corp., 1969.

[e] N = Negligible.

[f] For some greater detail, see the report of Work Group 5 in this book.

[g] Includes such sources as forest fires, structural fires, coal refuse, agriculture, organic solvent evaporation, and gasoline marketing.

Source	Particles	Oxides of Sulfur	Hydrocarbons	Oxides of Nitrogen	Carbon Monoxide
Fuels	9.2	22.9	15.7	16.5	59.7
Other	16.5	7.3	13.4	2.3	31.3
Total	25.7	30.2	29.1	18.8	91.0

Reprinted from *Man's Impact on the Global Environment: Assessment and Recommendations for Action* by William H. Mathews by permission of The MIT Press, Cambridge, Massachusetts, 1970.

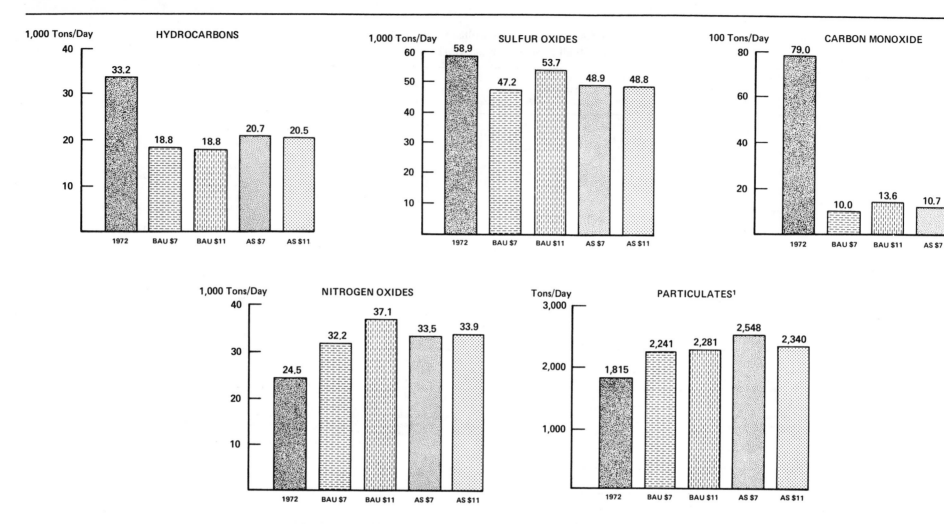

1,000 Tons/Day — HYDROCARBONS

1972	BAU $7	BAU $11	AS $7	AS $11
33.2	18.8	18.8	20.7	20.5

1,000 Tons/Day — SULFUR OXIDES

1972	BAU $7	BAU $11	AS $7	AS $11
58.9	47.2	53.7	48.9	48.8

100 Tons/Day — CARBON MONOXIDE

1972	BAU $7	BAU $11	AS $7	AS $11
79.0	10.0	13.6	10.7	11.8

1,000 Tons/Day — NITROGEN OXIDES

1972	BAU $7	BAU $11	AS $7	AS $11
24.5	32.2	37.1	33.5	33.9

Tons/Day — PARTICULATES[1]

1972	BAU $7	BAU $11	AS $7	AS $11
1,815	2,241	2,281	2,548	2,340

[1] Major sources of particulates in 1985 are coal- and oil-fired electrical generation, oil refining, and oil shale processing. It is assumed that national ambient air quality standards have been promulgated.

Projected environmental impacts, 1985 – Air.

Note: BAU and AS refer to Business as Usual and Accelerated Supply assumptions made in the Project Independence projections of the Federal Energy Administration.

Source: Energy Perspectives, U.S. Department of the Interior, February 1975.

ACID RAINFALL

Clean rainwater is slightly acidic with a pH[1] in the range of 5.5 to 5.7 due to small amounts of atmospheric carbon dioxide dissolved in the rain. During the past 20 years, rainfall in the northern hemisphere has been becoming steadily more acidic at a rate of about 2-3% per year. This increase is roughly parallel to the increase in atmospheric sulfur dioxide and the increase in the acidity of rain is attributed to the sulfuric acid formed from the oxidation of the sulfur compounds in the atmosphere.

The increasing acidity of rain falling in Sweden due to atmospheric sulfur generated in central Europe and Great Britain was reported at the United Nations conference in 1971. Measurements in the United States also indicate a steady increase in the acidity of rainfall, especially in the industrialized eastern half of the country.

There is evidence that low pH accompanying acid rain may be harmful to vegetation, aquatic environments, and possibly to soils. Reductions in forest growth have been observed in Sweden and reductions in some species of fish in a lake near a large smelting complex in Ontario, have been attributed to the acidity of the rain and snow.

[1] Acidity is measured in terms of pH which represents the logarithm of the reciprocal of the hydrogen ion concentration of a solution. A neutral solution has a pH of 7.0. As the hydrogen ion concentration increases, the solution becomes more acid and the pH falls.

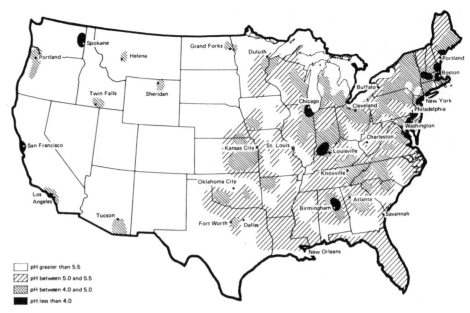

pH greater than 5.5
pH between 5.0 and 5.5
pH between 4.0 and 5.0
pH less than 4.0

Acidity of rainfall in the U.S.

Reproduced with permission of the National Academy of Sciences.

Comparison of SO₂ Emissions in Eastern North America[a] with Total Deposition of Sulfates in Precipitation
(Units: Million tons sulfur)

	1955–56	1965–66	1972–73
Estimated emissions			
Canada[b]	1.3	1.8	1.3
U.S.: Electric power plants[c]	2.9	5.6	8.9
U.S.: Other emissions[c]	3.6	3.4	3.6
Natural sources[d]	1.8	1.8	1.8
Total	9.6	12.6	15.6
Estimated deposition in precipitation	2.9	4.7	5.7[e]
Deposition as % of emissions	30%	38%	37%

[a] The area considered is that from the Mississippi River east to the Atlantic Coast and Nova Scotia (60E), north of central Alabama (33N) and south of 50N (Winnipeg to the Gulf of St. Lawrence).

[b] Canadian emissions for 1970 are given in Rennie and Halstead (1973): these were dominated by the emissions from the Sudbury smelters, which have probably been relatively constant during the period considered (Beamish and Harvey 1972, Balsillie, personal communication).

[c] U.S. emissions were estimated by interpolating from the data for 1950, 1960, and 1970 in EPA (1973), assuming that 80 of power plant emissions and 50 of other emissions were in the area considered.

[d] Natural emissions were estimated by scaling the global estimate of ref. 1 in proportion to the area under consideration (about 2% of the earth's land surface) (cf. Rodhe 1972).

[e] An increase of 3% per year is assumed for the period 1965–72.

Source: Air Quality and Stationary Source Emission Control, A Report of the Commission on Natural Resources, National Academy of Sciences, prepared for the Committee on Public Works, United States Senate, March 1975.

NITROGEN OXIDE AIR POLLUTION

Nitrogen comprises 79% of the atmosphere and, in itself is an inert, harmless gas. However, a number of compounds of nitrogen, formed by both natural and man-made processes, are reactive and have potential adverse health effects. The most important of these compounds are: NO (nitric oxide), NO_2 (nitrogen dioxide), NH_3 (ammonia), and to a lesser extent N_2O (nitrous oxide).

About 90% of the nitrogen compounds in the atmosphere are formed by natural processes—organic protein decay, bacterial oxidation, and fixation by lightning. Man-made nitrogen compounds in the atmosphere are derived largely from the combustion of fossil fuel through the oxidation of organic nitrogen compounds in the fossil fuels and by the fixation of atmospheric nitrogen gas, N_2, at the high temperatures prevailing in the combustion zone. Based on estimates of the global background level and the rate of production of nitrogen oxides, the average residence time of nitrogen oxides in the atmosphere is 3 to 4 days. This short residence time reflects the action of natural scavenging processes.

Although man-made atmospheric nitrogen compounds represent only 10% of the global inventory, concentrations in urban and industrial areas due to human activities can be much higher and produce undesirable health effects.

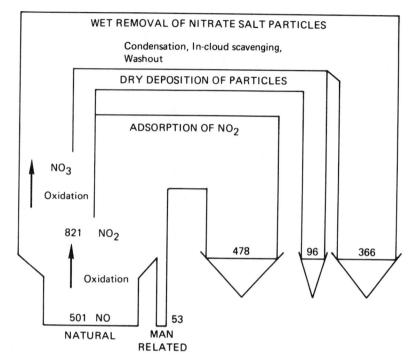

Nitrogen oxide cycle in the atmosphere (10^6 tons/year).

Source: Air Quality and Stationary Source Emission Control, A Report by the Commission on Natural Resources, prepared for the Committee on Public Works, United States Senate, March 1975.

World-Wide Urban Emissions of Nitrogen Dioxide

Fuel	Source	NO_2 Emissions (10^6 Tons)	% Total	Sub %
Total		52.9	100	
Coal			51	100
	power generation	12.2	23	47
	industrial	13.7	26	52
	domestic/commercial	1.0	2	3
Petroleum			41	100
	refinery production	0.7	1	3
	gasoline	7.5	14	34
	kerosene	1.3	2	6
	fuel oil	3.6	7	16
	residual oil	9.2	17	41
Natural gas			4	100
	power generation	0.6	1	25
	industrial	1.1	2	50
	domestic/commercial	0.4	<1	25
Other				
	incineration	0.5	<1	
	wood	0.3	<1	
	forest fires	0.5	1	

Source: Modified from Robinson and Robbins (1968).
Source: Air Quality and Stationary Source Emission Control, A Report by the Commission on Natural Resources, prepared for the Committee on Public Works, United States Senate, March 1975.

Summary of Nationwide NOx Emissions by Source Type and Fuel Use, 1972.

Source Type	Emissions (10⁶ tons/year)			% of Total Emissions		
Stationary fuel combustion	12.27			49.7		
Electric generation		5.94			24.1	
coal			3.95			16.0
oil			0.85			3.4
natural gas			1.14			4.7
Industrial fuel combustion		5.39			21.8	
coal			0.76			3.0
oil			0.41			1.6
process gas			2.58			10.5
natural gas			1.64			6.7
Commercial-institutional		0.65			2.6	
Residential		0.29			1.2	
Transportation	8.72			35.4		
gasoline		6.62			26.9	
diesel		1.90			7.7	
other		0.20			0.8	
Industrial process losses	2.88			11.7		
Solid waste disposal	0.18			0.7		
Miscellaneous	0.17			0.7		
New York point sources	0.42			1.8		
Total	24.64			100.0		

Source: Air Quality and Stationary Source Emission Control, A Report by the Commission on Natural Resources, prepared for the Committee on Public Works, United States Senate, March 1975.

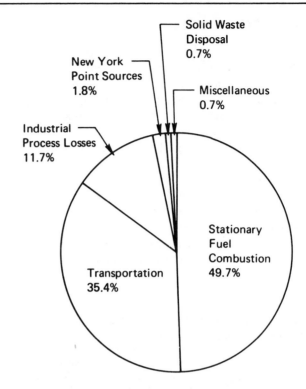

Source Type	Emissions (10⁶ ton/year)
Stationary Fuel Combustion	12.27
Transportation	8.72
Industrial Process Losses	2.88
Solid Waste Disposal	0.18
Miscellaneous	0.17
New York Point Sources	0.42
Total	24.64

Nationwide emissions of nitrogen oxides, 1972.

Source: Air Qualtiy and Stationary Source Emission Control, A Report by the Commission on Natural Resources, prepared for the Committee on Public Works, United States Senate, March 1975.

AIR POLLUTION FROM VEHICLES

The fuels used in the engines of vehicles are distillate fractions from petroleum and consist primarily of hydrocarbons, that is, compounds containing hydrogen and carbon. Complete combustion of these compounds produces carbon dioxide and water. In the average engine, however, combustion processes are seldom complete—carbon is oxidized to carbon monoxide (CO) rather than to carbon dioxide (CO_2), and some hydrocarbons remain unburned. In addition, the high temperatures in the combustion chambers of the engine cause some of the nitrogen in the air to be converted to nitrogen oxides. Trace elements in the fuel, such as sulfur, or additives, such as lead, are also converted to gases or particulates in the exhaust.

Amendments to the Clean Air Act in 1970 require that air-pollutant emissions from light-duty motor vehicles be reduced. A 90% reduction in carbon monoxide and hydrocarbon emissions was required by the 1975 model year, and 90% reduction in nitrogen oxides by the 1976 model year. These standards were to be met for the useful life of the vehicle—five years or 50,000 miles, whichever came first. Techniques for reducing engine emissions are discussed in 14-15.

Measured Emissions from 1976 Certified Cars Equipped with Catalysts in Comparison to the Statutory Standards[4]

Car	Forty-nine-state data [a]			California data [a]		
	Hydro-carbons	Carbon monoxide	Nitrogen oxides	Hydro-carbons	Carbon monoxide	Nitrogen oxides
Ford (302) [b]	0.7	3.0	3.0	0.6	3.5	1.0
Chevrolet (350)	0.5	8.0	2.3	0.5	3.7	1.7
Plymouth (318)	0.8	7.0	2.2	0.9	3.4	1.1
Datsun B210	1.2	9.0	2.1	0.5	4.1	1.6
Mercedes (168)	0.4	3.0	1.6	0.4	3.3	1.6
Hornet (304)	0.4	3.0	2.7	0.6	5.8	1.9
Standards	1.5	15.0	3.1	0.9	9.0	2.0

[a] Emissions data are in grams per mile and include a deterioration factor.
[b] Numbers in parethenses denote engine displacement class in model tested.

Source: *Environmental Quality–1976*, The Seventh Annual Report of the Council on Environmental Quality, September 1976.

Relative Importance of Mobile and Stationary Sources in Large Urban Areas

Region	Study Year	CO (percent)		HC (percent)		NO$_x$ (percent)	
		Mobile	Stationary	Mobile	Stationary	Mobile	Stationary
Chicago	1967	94	6	81	19	35	65
Denver	1967	93	7	78	22	48	52
Los Angeles	1966	95	5	72	28	73	27
New York	1965	96	4	84	16	38	62
Philadelphia	1967	70	30	47	53	27	73
Washington, D.C.	1966	96	4	86	14	44	56
Dallas	1967	97	3	93	7	80	20
Phoenix-Tucson	1967	94	6	87	13	71	29
Portland, Oregon	1968	72	28	64	36	79	21
Cincinnati	1967	85	15	83	17	34	66
Louisville	1967	75	25	83	17	35	65
Miami	1968	90	10	7	93	60	40
Atlanta	1968	89	11	86	14	71	29
Houston	1967	75	25	58	42	43	57
New Orleans	1968	47	53	49	51	56	44
Oklahoma City	1968	98	2	49	51	69	31
Pittsburgh	1967	80	20	70	30	29	71
St. Louis	1967	77	23	80	20	48	52
Charlotte	1968	92	8	86	14	28	72
Hartford	1967	95	5	82	18	52	48
Indianapolis	1967	85	15	86	14	52	48
Providence	1967	95	5	88	12	56	44
National Level		65	35	46	54	38	62
National Totals (millions of tons)		98	53	17	20	9	15

Source: Report of the Committee on Regulatory Effects on the Costs of Automotive Transportation, Appendix I-G, p. 5.

Source: *The Automobile and the Regulation of Its Impact on the Environment*, NSF/RA/X-74-023, Columbia University, New York, 1975.

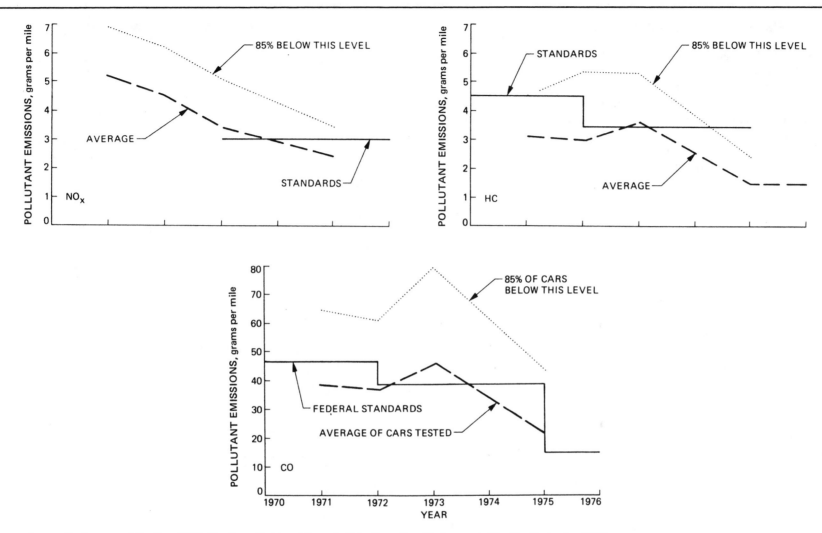

Source: *Environmental Quality–1976*, The Seventh Annual Report of the Council on Environmental Quality, September 1976.

WATER REQUIREMENTS FOR ENERGY PRODUCTION

In a study[1] by the Water Availability Task Group of the National Petroleum Council's Committee on the U.S. Energy Outlook, the Task Group concluded that there is an adequate supply of water in the United States to meet the requirements for a maximum effort to develop energy resources until 1985 except in Arizona and New Mexico. In other areas of the critical basins of the Upper Colorado River and the Upper Missouri River, major transmission facilities for interbasin water transfers would be required to meet the needs for the 1980–85 period. After 1985, water availability will likely be a constraint on the growth in the production of synthetic fuels and electricity in the Colorado and Missouri Basins.

[1] *U.S. Energy Outlook: Water Availability*, Report of the Water Availability Task Group, National Petroleum Council, 1973.

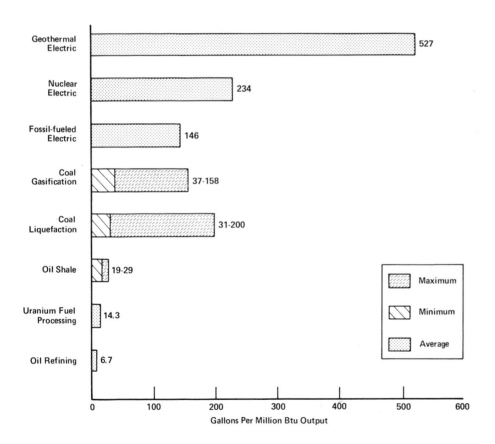

Water consumption in energy conversion and refining.

Source: U.S. Geological Circular 703, Water Demands for Expanding Energy Development, 1974.

Source: Energy Perspectives, U.S. Department of the Interior, February 1975.

PROJECTED WATER USE IN U.S. 1980-2000

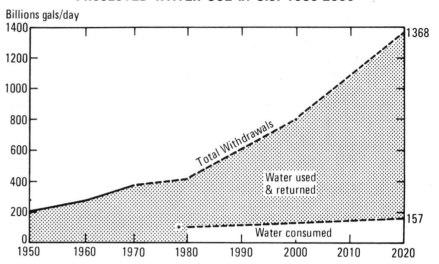

USE BY PURPOSE

	Billions gal/day withdrawn		PERCENT OF	
			Withdrawals	Consumption
Agriculture irrigation (98%)	1950: 110 / 1970: 130 / 1980: 138 / 2000: 153 / 2020: 166	1980 / 2020	31% / 12	80% / 64
Steam Electric Power	40 / 170 / 193 / 1470 / 914	1980 / 2020	44 / 67	2 / 8
Industry (Self supplied)	37 / 47 / 75 / 127 / 211	1980 / 2020	17 / 15	7 / 10
Municipal (Public supplied)	14 / 27 / 34 / 51 / 74	1980 / 2020	8 / 5	10 / 6
Rural domestic	†	1980 / 2020	0.6 / 0.2	1.7 / 1.6
			100%	100%
			both periods	

*Data on past water consumption not currently available.

Source: Natl. Commission on Water Quality. Redrawn from Murray op cit. p 10, fig. 12 and from U.S. Water Resources Council (1968). "The Nations Water Resources", U.S. Govt. Printing Office, Washington, D.C., Part I, p 8.
November 1975

Source: Staff Draft Report, National Commission on Water Quality, November 1975.

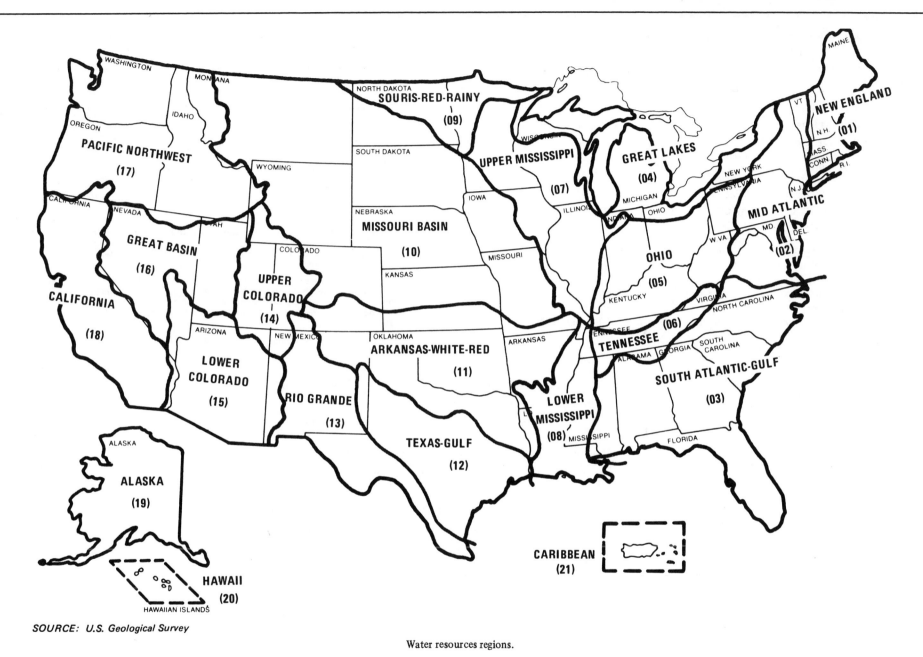

Water resources regions.

SOURCE: U.S. Geological Survey

Source: U.S. Energy Outlook: Water Availability, National Petroleum Council, 1973.

Water Consumption in Energy Conversion and Refining
(Gallons Per Million Btu Output)

PROCESS OR PRODUCT	CONSUMPTIVE USE
URANIUM FUEL	14.3
OIL REFINING	6.7
PIPELINE GAS FROM COAL:	
(A) WATER COOLING	
(90 Percent Load Factor)	72-158
(B) PARTIAL AIR COOLING	
(90 Percent Load Factor)	37-79
SYNTHETIC OIL FROM COAL	31-200
ELECTRICAL GENERATION	
(80 Percent Load Factor):	
(A) FOSSIL-FUELED PLANTS	146
(B) NUCLEAR	234
(C) GEOTHERMAL	527
OIL FROM SHALE	19-29

Source: Energy Perspectives, U.S. Department of the Interior, February 1975.

U.S. Water Use, by Category, 1940-72
[In billion gallon daily average]

Water use	1940	1950	1960	1970	1972
Total	**136.43**	**202.70**	**322.90**	**327.30**	**350.34**
Groundwater	22.56	35.19	58.17	54.27	56.27
Irrigation [1]	71.03	100.00	135.00	119.18	122.51
Public water utilities	10.10	14.10	22.00	27.03	28.34
Rural domestic [2]	3.10	4.60	4.00	4.34	4.44
Industrial and miscellaneous [3]	29.00	38.10	41.20	55.95	59.76
Steam electric utilities	23.20	45.90	98.70	120.80	135.30

[1] Total take, including delivery losses, but not including reservoir evaporation.
[2] Rural farm and nonfarm household and garden use, and water for farm stock and dairies.
[3] For 1940–1960, includes manufacturing and mineral industries, rural commercial industries, air-conditioning, resorts, hotels, motels, military and other State and Federal agencies, and other miscellaneous uses; thereafter, includes manufacturing, mining and mineral processing, ordnance, and construction.

Sources: U.S. Bureau of Competitive Assessment and Business Policy, 1940-60 based principally on committee prints of the U.S. Senate Committee on Water Resources, thereafter on *National Assessment of the Water Resources Council* (1968), as cited in U.S. Bureau of the Census, *Statistical Abstract of the United States, 1973* (1973), Table 285.

Source: The Fifth Annual Report of the Council on Environmental Quality, December 1974.

Summary of Water Supplies and Demands by Major Water Regions—1980
(Billions of Gallons Per Day)

Region	Water Supply						Water Demand		Gross Demand/ Supply Ratio (Percent)*	Net Demand/ Supply Ratio (Percent)†
	Regulated Runoff	Imports-Exports	Basin Inflows	Withdrawal Supply	In-Channel Requirement	Consumptive Use Supply	Total Withdrawal	Total Consumptive Use		
North Atlantic	145.10	—	—	145.10	1.92	143.18	44.02	2.69	30	2
South Atlantic-Gulf	169.30	—	—	169.30	19.70	149.60	41.18	3.32	24	2
Great Lakes	51.80	—	—	51.80	6.32	45.48	47.89	1.88	93	4
Ohio	108.90	—	—	108.90	8.76	100.14	41.75	1.62	38	2
Tennessee	38.60	—	—	38.60	4.15	34.45	12.25	0.57	32	2
Upper Mississippi	49.10	—	—	49.10	6.46	42.64	14.80	1.10	30	3
Lower Mississippi	38.20	—	283.73	321.93	32.84	289.09	12.62	3.01	4	1
Souris-Red-Rainy	4.30	+0.39	—	4.69	0.62	4.07	0.94	0.22	20	5
Missouri	42.20	+0.06‡	—	42.26	11.00	31.26	23.26	13.16	55	42
Arkansas-White-Red	69.90	+0.15	—	70.05	1.92	68.13	17.28	8.48	25	12
Texas-Gulf	23.90	—	—	23.90	0.78	23.12	26.38	9.42	110	41
Rio Grande	3.30	+0.11	—	3.41	0.34	3.07	8.33	4.68	244	152
Upper Colorado	12.00	-0.76	—	11.24	7.37	3.87	5.68	2.70	51	70
Lower Colorado	1.80	-4.18	8.38	6.00	1.86	4.14	8.50	4.08	142	99
Great Basin	4.00	+0.05	—	4.05	0.47	3.58	7.06	3.30	174	92
Columbia-North Pacific	189.00	—	—	189.00	14.70	174.30	41.41	13.58	22	8
California	56.00	+4.18	—	60.18	1.30	58.88	37.99	29.02	63	49
Total	**1,007.40**	**0**	**292.11**	**1,299.51**	**120.51**	**1,179.00**	**391.34**	**102.83**	**30**	**9**

Source: The Water Resources Council, *The Nation's Water Resources, The First National Assessment*, (Washington, D.C., 1968).

* Gross Demand/Supply Ratio = $\dfrac{\text{Total Withdrawal}}{\text{Withdrawal Supply}}$ (Percent)

† Net Demand/Supply Ratio = $\dfrac{\text{Total Consumptive Use}}{\text{Consumptive Use Supply}}$ (Percent)

‡ Imports of 0.45 bgd and exports of 0.39 bgd.

Source: U.S. Energy Outlook: Water Availability, National Petroleum Council, 1973.

ENVIRONMENTAL IMPACT—AQUATIC SYSTEMS

The temperature of an aquatic environment is one of the major controlling factors in determining the nature of biological systems in that environment. Extremely high or low temperatures can be lethal to biological organisms, and changes in temperature throughout the seasons both control and direct biological development. Natural bodies of water undergo thermal changes with the seasons which have determined, in various regions of the country, the nature of the ecosystems which can tolerate the local condi-

tions. The addition of warm waste water from power plants alters the temperature profile of natural bodies of water and may adversely affect the environment if temperature changes are extreme and the body of water is restricted in size.

Uncertainty over the long-term environmental effects of thermal discharges has led the Environmental Protection Agency to require that all new power plants over 25 megawatts in size employ closed-cycle cooling systems so that heated water is not discharged into natural bodies of water. Variances from this regulation can be obtained upon review of particular plant sites.

Aquatic Systems—General Considerations for Siting and Design

Characteristics	Why Important *re* Environment Interaction with Electric Generation [a]	Alternatives	
		Short term	Long term
A. PHYSICAL			
I. Temperature	a. Biological "master factor" in that it can be a lethal, controlling, and directive factor to biological systems (Gunter, 1957; Brett, 1960; Kinne, 1963, 1964; Naylor, 1965; Mihursky and Kennedy 1967; Krendel and Parker, 1969). SES thermal discharges have been reported to exceed tolerances of aquatic species (Trembley, 1965; Coutant, 1962; Churchill and Wojtalik, 1969; Mihursky, 1969).	1. Release heat to atmosphere by system other than "one pass cooling" *re* closed cooling system. 2. Obtain rapid mixing of thermal discharge to reduce temperature change to tolerable levels. 3. Reduce temperature differential across condensers. 4. When feasible and appropriate, employ accessory cooling methods such as ponding, sprays, cooling towers.	a. Regional distribution to atmosphere to prevent local overloading and indirect effects on water resource.
	b. Southern latitude species in warmer seasons are already near their upper temperature tolerance limit (Naylor, 1965). There is more biological temperature flexibility to the north.	1. Revert to seasonal closed cycle cooling systems. 2. Site SES in northerly latitudes.	a. Move to deep oceanic sites. b. Closed loop systems using cooling reservoirs; no encroachment on natural public resource. c. Operate southern SES during cooler seasons only.
	c. Low temperatures can limit metabolic activity during winter seasons in northerly latitudes.	1. Proper use of waste heat in a constructive manner during cooler seasons to stimulate biological production.	
	d. Low temperatures in winter can cause cold kills in southerly latitudes.	1. Proper use of waste heat in winter can prevent cold kills.	
	e. Faunistic displacement options exist in heating northern waters; options are lost in heating southern waters.		

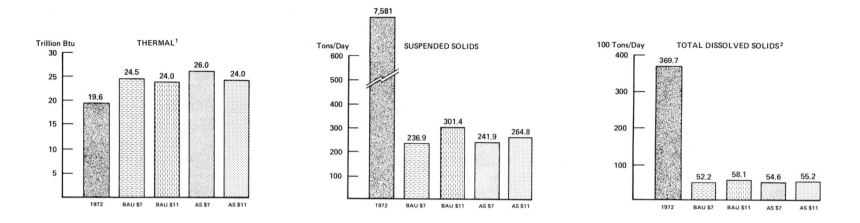

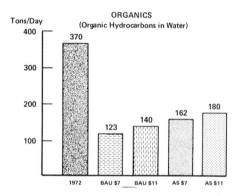

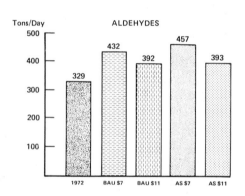

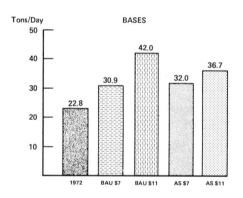

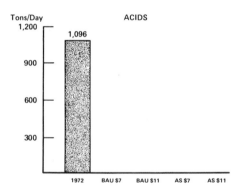

Source: Project Independence Report, Environmental Quality Analyses, Environmental Protection Agency, 1974, pages 46-49.

[1] The estimates appear to be extraordinarily low.

[2] Dissolved solids include acids, bases, phosphates, and nitrates.

Projected domestic environmental impacts, 1985—Water.

Source: *Energy Perspectives*, U.S. Department of the Interior, February 1975.

SOURCES OF OCEAN OIL POLLUTION

Estimates of natural oil seepage into the ocean range from 0.15 to 6.0 million metric tons per year, with the best estimate placed at 0.6 million metric tons per year. Estimates of oil spills from industrial activities range from 2 to 5 million metric tons per year. About half of the oil discharged into the ocean results from the operation of tankers and other ships. Less than 5% results from offshore oil production activities.

Summary of Worldwide Seepage Rates: Regional Breakdown by Oceans
(In 10^6 metric tons per year)

Ocean	Case I, P_{16}	Case II, $P_{1.0}$	Case III, $P_{0.1}$
Pacific	2.83×10^6	2.69×10^5	0.689×10^5
Atlantic	2.06×10^6	1.96×10^5	5.04×10^4
Indian	9.30×10^5	8.85×10^1	2.28×10^4
Arctic	2.14×10^5	2.30×10^5	5.20×10^3
Southern	1.88×10^1	1.74×10^3	4.51×10^2
Total	6.05×10^6	0.558×10^5	0.148×10^6

Wilson, R. D., et al, "Natural Marine Oil Seepage," *Science*, Vol. 184 pp. 857–865, May 24, 1974. © American Association for the Advancement of Sciences.

World Crude Oil Production[a] and Transport by Tanker[b]
(Millions of metric tons per year)

Year	Production	Tanker Transported
1960	1,040	——
1962	1,210	——
1964	1,420	——
1965	1,500	——
1969	1,820	1,180
1975	2,700	1,820
1980	4,000	2,700

[a] 1960 to 1965 production taken from American Petroleum Institute, 1967. 1969 estimates from Esso Research and Engineering Company (Standard Oil of New Jersey). 1975 and 1980 estimates averaged between predictions in Bachman, 1969, and *Marine Resources and Legislative and Political Arrangements for Their Development*, 1969, assuming production equal to consumption.

[b] 1969 estimate from Esso Research and Engineering Company (Standard Oil of New Jersey). Estimates for 1975 and 1980 from *Oil and Gas Journal*, 1968.

Reprinted from *Man's Impact on the Global Environment: Assessment and Recommendations for Action* by William H. Mathews by permission of The MIT Press, Cambridge, Massachusetts, 1970.

Estimates of Direct Losses into the World's Waters, 1969
(Metric tons per year)

	Loss	Percentage of Total Loss
Tankers (normal operations)[a]		
Controlled	30,000	1.4
Uncontrolled	500,000	24.0
Other ships (bilges, etc.)[b]	500,000	24.0
Offshore production (normal operations)[c]	100,000	4.8
Accidental spills[d]		
Ships	100,000	4.8
Nonships	100,000	4.8
Refineries[e]	300,000	14.4
In rivers carrying industrial automobile wastes[f]	450,000	21.6
Total[g]	2,080,000	100.0

[a] Sources: Inter-governmental Maritime Consultative Organization, 1965, and Esso Research and Engineering Company (Standard Oil of New Jersey). It is estimated that 80 percent of the world's tanker fleet use Load-on-Top to reduce discharges from cleaning and 20 percent do not. These estimates are based on this distribution and some measured values for oil content of discharge using both methods.

[b] Source: Moss, 1963. Firm data in this area are not available anywhere. Records are not kept of bilge pumping nor are estimates of oil content. This area warrants further development; surveys need to be run to produce further data.

[c] Sources: Surveys by Esso Research and Engineering Company (Standard Oil of New Jersey) and American Petroleum Institute, 1969.

[d] Sources: U.S. Coast Guard, 1969, and American Petroleum Institute, 1969. These data are not firm. Mechanisms are needed to accumulate data to improve the reliability of this estimate.

[e] Sources: American Petroleum Institute, 1968; Federal Water Pollution Control Administration, 1968; and Esso Research and Engineering Company (Standard Oil of New Jersey). Petrochemical production wastes are included.

[f] Using U.S. Department of Health, Education, and Welfare, 1963, data, we have assumed that the average concentration of hydrocarbons in the U.S. rivers entering the sea is approximately 85 parts per billion. Multiplying this figure by the estimated river runoff of 1,750 billion tons/year yields a figure of approximately 150,000 tons of hydrocarbons, which are annually carried to the oceans from the United States (Revelle, 1963). If the river runoff for the world is three times that of the United States, and if the average concentration is the same as the estimated U.S. average, then about 450,000 tons of hydrocarbons are carried into the oceans. In addition hydrocarbons are introduced into the ocean through sewage outfalls. We estimate that this amount is less than that in the rivers perhaps by a factor of 3 or 4, but this is not included in the table.

[g] Oil from pleasure craft, whose use is increasing in the United States, is not included, nor are there any measurements of natural seeps, which would be valuable for comparative purposes.

Reprinted from *Man's Impact on the Global Environment: Assessment and Recommendations for Action* by William H. Mathews by permission of The MIT Press, Cambridge, Massachusetts, 1970.

Spills in U.S. Waters: 1971–1975
(Source: U.S. Coast Guard)

Spills, by category[a]	Number of incidents					Total volume, million gallons				
	1971	1972	1973	1974	1975	1971	1972	1973	1974	1975
Type of location										
Inland waters	631	682	1,722	2,815	1,995	1,409.8	2,270	7,117	8,924	6,884
Coastal waters	7,201	7,441	9,871	9,503	6,622	6,720	14,277	15,490	6,799.8	6,993
Open waters (Great Lakes or territorial seas)	315	423	571	251	1,020	37	24	419	108	3,364
Contiguous zone (from 3 to 12 miles from coastline)	396	801	483	164	133	651	34	1,218.8	24	4
High seas	193	583	681	1,233	728	20	2,197	68.9	52	6,879
Type of pollutant										
Light oil[b]	4,320	4,290	4,104	2,657	2,677	2,822	6,578	6,415	3,181	7,185
Heavy oil[c]	1,603	2,049	2,851	5,084	2,698	2,934	1,761	4,538	12,754.8	9,565
Solvent	(d)	(d)	49	44	45	(d)	(d)	32	13	10
Waste oil	930	890	1,003	1,094	958	164	8,067	1,211	111.9	3,483.8
Other oil	462	1,151	2,976	2,774	2,414	2,673	357	2,650	728	421
Other materials (including sewage, refuse, etc.)	269	428	774	470	552	115	2,025	8,339	1,193	3,437
Unknown		945	1,123	1,551	1,843	1,194	89	15	1,128	56.8	23.8
Source										
Vessels	2,134	2,493	3,550	3,726	2,804					
Dry cargo vessels	271	402	353	377	294	418	42	650	90.9	2
Tank ships	386	453	825	973	546	1,665	2,583.9	4,494	1,434	7,637
Tank barges	828	830	718	833	679	1,197.8	3,739	1,572	2,468	3,472
Combatant vessels	261	294	246	278	208	440.8	40.9	17.9	39	17
Other vessels	388	494	1,408	1,265	1,082	180	96	1,184	253	1,322
Land vehicles	77	145	305	373	312	101	172	741	7,885	1,105
Nontransportation-related facilities, refineries	188	185	214	155	2,509	2,206	42	166	772	4,268
Pipelines	(e)	216	559	557	534	(e)	1,237	1,847	6,205	2,416
Other transportation facilities	22	68	162	3,489	174	159.9	13	151	2,695	219
All other onshore and offshore facilities[f]	3,723	3,804	3,904	799	402	2,158	10,483	6,479	1,567	122.8
Miscellaneous and unknown	2,592	3,040	4,634	4,867	3,803	310	354[g]	7,009	603	3,524
Cause										
Casualty (includes collision, grounding, and blowouts)	124	360	2,793	952	147	2,045	4,082	16,068	4,861	5,874
Rupture, leak, or structural failure	2,757	2,201		2,352	1,018	2,715	4,823[h]		7,234.9	11,486
Equipment failure (valves, pumps)	947	1,542	1,072	2,103	2,502	274	293	800	1,100	1,843
Personnel failure	829	1,287	2,204	2,707	1,865	1,035.9	940	1,127.8	3,544	962
Deliberate discharges	359	457	599	316	473	50	68	2,176	292	192
Natural phenomena	94	257	354	380	345	5.8	8,045.9	2,051	241	172
Unknown	3,536	3,827	5,506	5,206	4,078	712	551	2,090	858	3,573.9
Total spills reported	13,966	9,931	13,328	13,966	10,538	8,839	18,085	24,314.9	16,916	24,126

[a] The U.S. Coast Guard has no reason to believe that the number of discharges in 1973 was any greater than in 1971 or 1972. The increase in the number reported probably reflected public awareness of the legal requirement to report discharges.
[b] Data for 1971, 1972 include gasoline, light fuel oil, kerosene, and light crude; 1973 data include crude oil, gasoline, and other distillate fuel oil.
[c] Data for 1971, 1972 include diesel oil, heating oil, heavy fuel oil, heavy crude, and asphalt; 1973 data include diesel oil, asphalt, and residual fuel oil.
[d] Data for 1973 added the category of solvents, previously included under several other categories.
[e] Pipeline data for 1971 are included under other categories.
[f] Changes in 1973 "sources" categories make it necessary to combine some onshore and offshore production, storage, and transfer facilities in order to compare data to those for 1971 and 1972.
[g] Changes in 1973 "cause" categories make it necessary to combine the categories of "casualty" and "rupture, leak, or structural failure."
[h] Includes one 6-million-gallon sewage spill.

Source: Environmental Quality–1976, The Seventh Annual Report of the Council on Environmental Quality, September 1976.

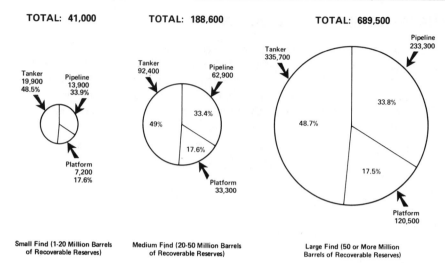

| TOTAL: 41,000 | TOTAL: 188,600 | TOTAL: 689,500 |

Small Find (1-20 Million Barrels of Recoverable Reserves) Tanker 19,900 48.5%, Pipeline 13,900 33.9%, Platform 7,200 17.6%

Medium Find (20-50 Million Barrels of Recoverable Reserves) Tanker 92,400 49%, Pipeline 62,900 33.4%, Platform 33,300 17.6%

Large Find (50 or More Million Barrels of Recoverable Reserves) Tanker 335,700 48.7%, Pipeline 233,300 33.8%, Platform 120,500 17.5%

Source: *OCS Oil and Gas: An Environmental Assessment,* Council on Environmental Quality, 1974, pages 4-34, 4-35.

Oil spilled over the life of an average OCS field, by cause (barrels).

Source: Energy Perspectives, U.S. Department of the Interior, February 1975.

TOTAL: 37 Million Barrels

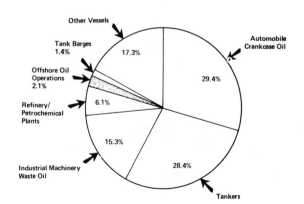

Other Vessels 17.3%
Tank Barges 1.4%
Offshore Oil Operations 2.1%
Refinery/Petrochemical Plants 6.1%
Industrial Machinery Waste Oil 15.3%
Tankers 28.4%
Automobile Crankcase Oil 29.4%

Source: Porricelli, J. D., and Keith, V. F., *Tankers and U.S. Energy Situation—an Economic and Environmental Analysis,* 1973, page 063.

Sources of ocean oil pollution, 1972.

Source: Energy Perspectives, U.S. Department of the Interior, February 1975.

LAND REQUIREMENTS—COAL AND SHALE OIL

The total land area of the United States is 2.26 billion acres. About half, or 1.06 billion acres, is classified as farmland. Between 1930 and 1971 the land utilized for the production of coal was 1.47 million acres, about one-tenth of 1% of the land used in farms. Of the land utilized by the coal industry, 68% has been reclaimed. Production of coal to meet energy requirements is expected to at least double over the next decade. Recent regulations require that essentially all the land disturbed by coal recovery be rehabilitated.

Land Utilized and Reclaimed by the Bituminous Coal Industry, 1930–1971

(Acres)

Land utilized:	
1930–71	1,470,000
1971	73,000
Land reclaimed:	
1930–71	1,000,000
1971	94,600
Percent reclaimed, 1930–71	68

Source: Energy Perspectives, U.S. Department of the Interior, February 1975.

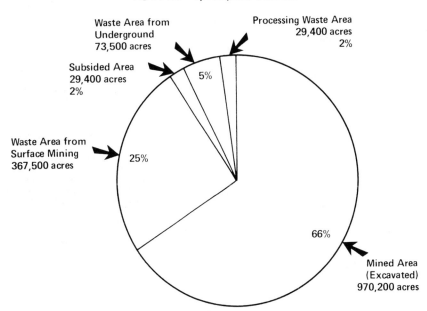

TOTAL: 1,470,000 Acres

Waste Area from Underground 73,500 acres

Processing Waste Area 29,400 acres 2%

Subsided Area 29,400 acres 2%

5%

Waste Area from Surface Mining 367,500 acres

25%

66%

Mined Area (Excavated) 970,200 acres

Source: *IC 8642, Land Utilization and Reclamation in the Mining Industry, 1930-71*, U.S. Bureau of Mines, 1974.

Cumulative land use, bituminous coal mining, 1930–71

Source: Energy Perspectives, U.S. Department of the Interior, February 1975.

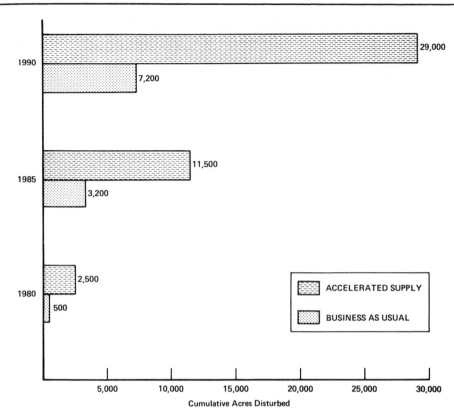

Source: *Potential Future Role of Oil Shale: Prospects and Constraints*
Project Independence Report, 1974, page 8.

Land impacts, shale oil production, 1980-90.

Source: Energy Perspectives, U.S. Department of the Interior, February 1975.

U.S. Land Use and Government Ownership, 1950-70

[In acres]

Land use	1950	1960	1970
Land Area (million)			
Total	**2,273**	**¹ 2,271**	**² 2,264**
Farm	1,162	¹ 1,124	¹ 1,064
Grazing land ³	402	¹ 319	² 288
Forest land not grazed	368	¹ 438	² 475
Other ⁴	341	¹ 390	² 437
Park Area (thousands)			
Total	**29,137**	**32,321**	**38,064**
National Parks	23,836	25,704	28,543
State parks ⁵	4,657	5,602	8,555
County and municipal parks	644	1,015	966
Federally owned land (million)			
Total ⁶	**⁷ 754**	**772**	**762**
Bureau of Land Management	NA	500	474
Forest Service	180	186	187
Park Service	23.8	25.7	28.5
Fish and Wildlife Service	⁷ 7	16	28
Department of Defense ⁸	30	31	31
Civil works ⁹	NA	15	16
Other ¹⁰	NA	7	8

¹ 1959 data.

² 1969 data.

³ Includes grasslands, arid woodlands, and other forested land grazed.

⁴ Includes urban, industrial, and residential areas, rural parks, wildlife refuges, highways, railroad rights of way, ungrazed desert, rocky barren and swamp land, tundra and other land.

⁵ Excludes state forests, wildlife refuges, and waysides not administered by state park agencies.

⁶ Excludes outlying area beginning in 1960.

⁷ 1955 data.

⁸ Army (excluding Corps of Engineers), Navy and Air Force.

⁹ Corps of Engineers and Bureau of Reclamation.

¹⁰ Atomic Energy Commission, Tennessee Valley Authority and Bureau of Indian Affairs.

Sources: U.S. National Park Services, *Areas Administered by the National Park Service*, semiannual; U.S. Bureau of Outdoor Recreation, *State Outdoor Recreation Statistics—1962* (1963);National Recreation and Park Association, Arlington, Va., *State Park Statistics, 1970, Parks and Recreation* (1971), and *Recreation and Park Yearbook*; General Services Administration, *Inventory Report on the Real Property Owned by the U.S. Throughout the World,* annual; and U.S. Department of Agriculture, Economic Research Service, *Agricultural Statistics,* annual, as cited in U.S. Bureau of the Census, *Statistical Abstract of the United States, 1973* (1973), Tables 323, 328, 329, and 983.

Source: The Fifth Annual Report of the Council on Environmental Quality, December 1974.

NATURAL AND MAN-MADE RADIATION SOURCES

Radiation from natural sources—cosmic rays, radioactivity in rocks, and radioactivity within the human body—results in an annual exposure of about 100 milliroentgen (mr) per person. Man-made sources of radiation contribute about 100 mr additionally per year per person. Over 90% of the man-made radiation is derived from diagnostic and therapeutic use of x-rays. Radiation from the normal effluents of nuclear power plants results in an exposure of about 0.01 mr/year.

Effects of radiation on body chemistry begin to be observed at radiation levels of 25,000 mr with rapid recovery after exposure. Lethal effects are observed with whole-body irradiations at levels of 300,000 to 500,000 mr. Genetic effects are often assumed to be linear with radiation dose although experimental difficulties have not made it possible to determine whether there is a threshold level of radiation below which genetic damage does not occur. The National Council on Radiation Protection has recommended that exposures of individuals in the general population be limited to 170 mr/year from sources other than natural sources.

Effects of Large Whole-Body Doses of Radiation

Dose	Effect
10,000,000 mrem	Death within hours due to damage to central nervous system.
1,200,000 mrem	Death within several days due to damage to gastrointestinal system.
600,000 mrem	Death within several weeks due to damage to blood-forming organs.
450,000 mrem	50-50 chance of death within 30 days.
100,000 mrem	Possible temporary impairment, but probable recovery.

Source: The Environmental Impact of Electrical Power Generation: Nuclear and Fossil, WASH-1261, U.S. Atomic Energy Commission, 1973.

Liquid Radioactivity Levels

	(Picocuries/liter)*
Typical nuclear power plant radioactive waste discharge	1–10
Domestic tap water	20
River water	10–100
4% beer	130
Ocean water	350
Whiskey	1200
Milk	1400
Salad oil	4900

*A curie is a unit used to measure radioactivity
 A picocurie is a trillionth of a curie
 A liter is approximately one quart

Reprinted by permission of General Electric

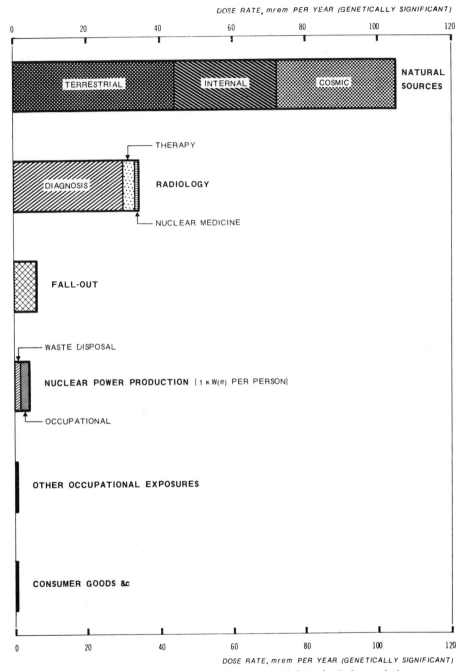

Annual genetically significant dose rate, as averaged through whole population.

Source: Pochin, Edward E. *Estimated Population Exposure from Nuclear Power Production and Other Radiation Sources*, Nuclear Energy Agency, Organization for Economic Co-operation and Development, Paris, France, Jan., 1976.

Relative Alpha Activity of Foods*

Food Stuff	Relative Activity
Brazil nuts	1400
Cereals	60
Teas	40
Liver and kidney	15
Flours	14
Peanuts and peanut butter	12
Chocolates	8
Biscuits	2
Milks (evaporated)	1–2
Fish	1–2
Cheeses and eggs	0.9
Vegetables	0.7
Meats	0.5
Fruits	0.1

*From *Proceedings of the Second United Nations International Conference on the Peaceful Uses of Atomic Energy,* September 1–13, 1958, Geneva, Switzerland, United Nations Publication, Volume 23, Experience in Radiological Protection, page 153, W. V. Mayneord.

Source: The Natural Radiation Environment, United States Atomic Energy Commission, 1968.

Dose Rates from Natural Radioactivity*

Source	Dose, mrad
External Irradiation	
Cosmic rays	28.7
Terrestrial radiation	50
Internal Radiation	
^{40}K	20
other radionuclides	1.6
Total	100

* estimated by UNSCEAR as average annual dose to gonads received by world population.

Summary of Annual Whole-Body doses per Capita in the United States from Man-Made Sources*

Radiation Source	Dose, mrem
Medical	
Diagnostic	103
Therapeutic	6
Radiopharmaceutical	2
Occupational	0.8
Environmental	
Global fallout	4
Worldwide ^{3}H and ^{85}Kr	0.05
AEC installations	0.01
Nuclear power reactors	0.002
Fuel reprocessing	0.0008
Miscellaneous	2.6
Total	114
Natural	130

* estimated for year 1971 by the U.S. Environmental Protection Agency

Source: Nuclear Power and the Environment, International Atomic Energy Agency, Vienna, Austria, 1972.

Calculation of Annual Radiation Exposure

	Common Source of Radiation	Your Annual Inventory (mrem/year)
WHERE YOU LIVE	Location: Cosmic radiation at sea level	40
	Add 1 for every 100 feet of elevation where you live	———
	House construction: Wood 35	
	Concrete 50	
	Brick 75	
	Stone 70	———
	Ground (U.S. Average)	56
WHAT YOU EAT DRINK AND BREATHE	Water and food (U.S. Average)	25
	Air (U.S. Average)	5
HOW YOU LIVE	Jet Airplanes: Number of 6000-mile flights x 4	———
	Radium Dial Wrist Watch: Add 2	———
	Television Viewing:	
	Black and white: Number of hours per day x 1	———
	Color: Number of hours per day x 2	———
	X-ray Diagnosis and Treatment	
	Limb x-ray: 420	
	Chest x-ray: 150	
	Stomach x-ray: 350	
	Colon x-ray: 450	
	Head x-ray: 50	
	Spinal x-ray: 250	
	Gastrointestinal tract x-ray: 2000	
	Dental x-ray: 20	———
HOW CLOSE YOU LIVE TO A NUCLEAR PLANT	At Site Boundary: Number of hours per day x .2	———
	One Mile Away: Number of hours per day x 0.02	———
	Five Miles Away: Number of hours per day x 0.002	———
	TOTAL	———

Compare your dose to the U.S. Average of 200 mrem/year

Source: The Environmental Impact of Electrical Power Generation: Nuclear and Fossil, WASH-1261, U.S. Atomic Energy Commission, 1973.

U.S. Population Exposure From Different Sources of Radiation, 1970[a]

Source	U.S. population at risk in 1970 Percent	U.S. population at risk in 1970 Millions	Average per capita dose to population at risk in 1970[b] (millirems)	Average per capita dose to U.S. population in 1970[b] (millirems)	Total man-rems per year in 1970 (millions)	Total estimated man-rems per year in 2000[c] (millions)
A. Natural						
1. Cosmic...............................	100%	205	45	45	9.2	14.4
2. Terrestrial:						
a. External..........................	100	205	60	60	12.3	19.3
b. Internal..........................	100	205	25	25	5.1	8.0
Subtotal (Natural)......................	100	205	130	130	26.6	41.7
B. Radioactive contamination in the environment[d]						
1. Worldwide fallout except tritium (^{3}H)..	100	205	4.0	4.0	0.82	1.6
2. Worldwide ^{3}H[e].......................	100	205	0.07	0.07	0.015	0.0084
3. Operating nuclear facilities[d]:						
a. Reactors..........................	23	47.6[f]	0.0092	0.0021	0.00043	0.056
b. Fuel reprocessing.................	0.7	1.5[g]	0.113	0.0008	0.00017	0.065
c. Worldwide ^{85}Kr	100	205	0.0004	0.0004	0.00008	0.012
d. AEC facilities....................	0.8	1.6	1.5	0.01	0.0025	0.0038
Subtotal (Nuclear facilities)[d].......				0.01	0.0032	0.137
Subtotal (Radioactive contamination)......				4.1	0.84	1.8
C. Man-made sources excluding "environmental" contamination						
1. Medical irradiation:						
a. Diagnostic x-rays						
(i) Medical......................	33	67.3[h]	230[i]	103[i,j]	18.7[k]	45[k]
(ii) Dental.......................	22	45.9[h]		1[i]	0.2	
b. Radiotherapy (external or sealed source)[l]..................	0.01	0.025[m]		5.1[n]	0.5	
c. Internally administered radioisotopes.....................	0.7	1.5		1.1	0.2	
Subtotal (Medical irradiation)........				110	19	47
2. Occupational irradiation..............	0.4	0.772	212	0.79	0.16	0.3
3. Consumer devices[o]	100	205	2.0	0.2	0.4	47
Subtotal (Man-made sources)..............				110	20	47

[a]Based on data contained in a June 1971 draft of "Estimates of Ionizing Radiation Doses in the United States 1960–2000," Special Studies Group under the auspices of EPA.
[b]Whole body dose unless otherwise noted.
[c]Based on an estimated U.S. population in 2000 equal to 321×10^6.
[d]Does not include small radioactive releases from fossil fuel power plants.
[e]Tritium is produced naturally, by nuclear explosives (fallout), and in nuclear reactors. In 1970, the vast majority of the worldwide tritium was from fallout. By 2000 approximately one-half of the estimated worldwide tritium will be reactor produced.
[f]Within 50 miles of reactor site.
[g]Within 100 km of plant site.
[h]Population at risk in 1964.
[i]The whole body dose is taken as the gonad dose.
[j]The estimated 1970 mean genetically significant dose (GSD) to entire U.S. population is 61 mrem per capita in the United States. In 2000 the projected GSD = 93 mrem per capita in the United States.
[k]Assumed to equal 1.5 times the projected GSD.
[l]Only GSD for treatment of malignant disease reported.
[m]1966 data
[n]Average dose to 93.5 million, the U.S. population under 30.
[o]Television, watches, air transport, etc.

Source: Sam H. Schurr, *Energy Research Needs*, PB-207 516, prepared by Resources for the Future, Inc. for the National Science Foundation, October 1971.

BIOLOGICAL EFFECTS—HIGH VOLTAGE ELECTRIC FIELDS

Along with public concern over the possible effects of man's activities on the environment, concern has been expressed over the possible effects of the electric and magnetic environment upon man. Human perception of a voltage field begins at a field gradient of about 10 kV/m through stimulation of the hair. Persons wearing shoes with insulated soles walking beneath high voltage power lines can just begin to perceive electrical shocks by touching grounded objects at field gradients of about 6 kV/m, a gradient found at some points under extra high voltage lines.

Approximate Magnetic Field Under Center Phase of 765 kV Line

Line Height = 49 Feet (13.7 meters)
Height of Measurement Above Ground = 1.5 Meters (5 Feet)

Current in One Phase of 765 kV Line	Magnetic Field Strength
1000 Amperes	0.155 Gauss
2000 Amperes	0.310 Gauss

60 Hz Electric Fields in the Vicinity of Electrical Applicances[a]

Appliance	Electric Field (volt/meter)
Electric range	4
Toaster	40
Electric blanket	250
Iron	60
Broiler	130
Hair dryer	40
Vaporizer	40
Refrigerator	60
Color TV	30
Stereo	90
Coffee pot	30
Vacuum cleaner	16
Clock	15
Hand mixer	50
Incandescent light bulb	2
Phonograph	40

[a]The measurements were made at a distance of 30 cm from appliances.
Source: Biological Effects of High Voltage Electric Fields, prepared by the IIT Research Institute for the Electric Power Research Institute EPRI 381-1, November 1975.

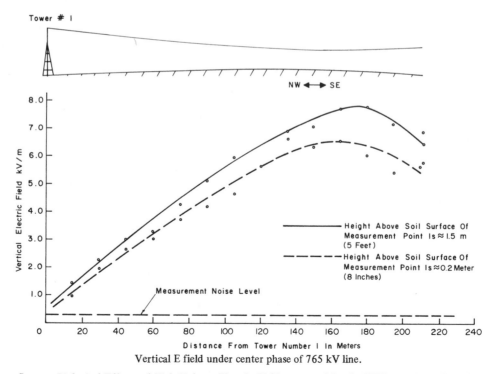

Vertical E field under center phase of 765 kV line.

Source: Biological Effects of High Voltage Electric Fields, prepared by the IIT Research Institute for the Electric Power Research Institute, EPRI 381-1, November 1975.

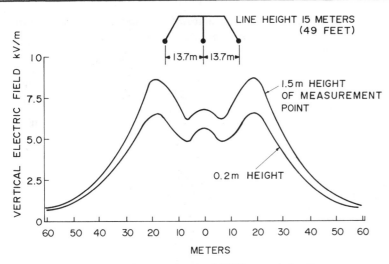

Electric field measured under 765 kV transmission line.

Source: Biological Effects of High Voltage Electric Fields, prepared by the IIT Research Institute for the Electric Power Research Institute, EPRI 381-1, November 1975.

Localized 60 Hz Magnetic Flux Densities Produced by Some Electrical Appliances

10–25 gauss	*0.01–1.0 gauss*
325 watt soldering gun	toy auto transformer
magnetic stirrer	garbage disposal
power feeder cable	clothes dryer
hair dryer	black/white television set
	vacuum cleaner
5–10 gauss	heating pad
can opener	electric toaster
140 watt soldering gun	bell transformer
fluorescent desk lamp	
kitchen range	*0.01–0.1 gauss*
electric shaver	home electric service unit
	kitchen fluorescent lamp
1–5 gauss	dishwasher
bench grinder	laundry washer
arc welder	phonograph
food mixer	calculator
power transformer	electric iron
induction motor	
color television set	*0.001–0.01 gauss*
food blender	refrigerator
electric drill	
portable heater	

Source: Biological Effects of High Voltage Electric Fields, prepared by the IIT Research Institute for the Electric Power Research Institute, EPRI 381-1, November 1975.

Summary of American and Western European Tests Using Human Subjects

Investigator	Subject	Frequency	Stress	Indicator	Results
Koeppen[45a,46]	humans	50 Hz	7.5/2.5/27 kV/m 30 minutes	EKG	No effects
Strumza[45a,49]	525 humans and controls	50 Hz	Long duration fields within 25 m of 200/400 kV line vs fields beyond 125 m for controls	Visits to and by physicians, use of medicine, medical histories	No significant difference between exposed and control groups
Johansson[47]	10 males 10 females	50 Hz	100 kV/m	Psychological test subjective responses	No statistical difference was observed in test performance, some discomfort reported by a few subjects also influenced by weather
Busby[64]	farm workers and livestock, 18 farms	60 Hz	Fields from 765 kV line	Questionnaire response	No significant effect attributed to fields from lines
Kouwenhoven[21]	human 11 American linemen	60 Hz	Fields encountered in normal line and bare-hand work	Physical examination CV, ECG, kidney, visual, auditory emotional status	No effects
R. Hauf[44]	10 humans	50 Hz	1, 15 kV/m intermittent 45 minute exposure	ECG, EEG, pulse blood pressure, reaction time	No significant changes except for small decreases in reaction time
G. Hauf[45]	6 humans (3 male)	50 Hz	1/15/20 kV/m, alternating 45 min exposures	EKG, EEG, blood pressure reaction time	No pathological changes

[a]Indicates source of data cited.
Source: Biological Effects of High Voltage Electric Fields, prepared by the IIT Research Institute for the Electric Power Research Institute, EPRI 381-1, November 1975.

Research in Western Europe and the United States has not produced evidence that human exposure to the normally encountered electric and magnetic fields from high voltage power lines has any harmful biological effect. On the other hand, studies in the USSR on workers employed in high voltage switchyards have indicated some undesirable effects at field intensities of 100 kV/m, some ten times higher than the level normally found under high voltage lines.

HEALTH EFFECTS OF AIR POLLUTANTS

Except for carbon monoxide (CO), which has a very specific effect on human metabolism, it is not possible to separate the effects of other air pollutants—sulfur dioxide, nitrogen oxides, particulates, hydrocarbons, and photochemical oxidants. All of these latter pollutants occur simultaneously in the atmosphere and all have somewhat similar effects with particular impact on those with respiratory or cardiovasculor difficulties. In addition, the measurements which have been made on the effect of air pollutants on health have often been made when other forms of environmental stress have been present—cold, damp weather; hot, humid weather; overcrowding; poor nutrition, or poor hygiene. These stresses, plus that from the smoking of tobacco, have made many individuals sensitive to any additional environmental stress.

Carbon Monoxide. Carbon monoxide (CO) is an odorless, nonirritant gas that displaces oxygen from hemoglobin in the blood, forming carboxyhemoglobin which interferes with the oxygen transport capability of the blood. Exposure to high concentrations of CO seriously impairs the function of oxygen-dependent tissues in the body—particularly the brain, heart, and skeletal muscles. Evidence indicates that health effects begin to be observed at a carboxyhemoglobin concentration in the blood of 3-5% particularly in individuals with heart or circulatory conditions (the present CO standard of 9 parts per million for 8 hours results in a carboxyhemoglobin content in the blood of about 1.5%). For those who smoke, it might be observed that the CO content in inhaled tobacco smoke is 400-450 parts per million.

Calculated Carboxyhemoglobin at Equilibrium with Inspired Carbon Monoxide Concentration, P, in Parts per Million

(Applicable to nonsmokers only)

Inspired Carbon Monoxide Concentration, ppm	Carboxyhemoglobin Concentration, %[a]	Carboxyhemoglobin Concentration from Empirical Equation, % COHb = 0.4 + (P/7)[b]
0	0.36	0.4
5	1.11	1.1
8.7	1.66	1.6
10	1.85	1.8
15	2.57	2.5
20	3.29	3.3
30	4.69	4.7
40	6.05	6.1
50	7.36	7.5

[a]Data from Coburn *et al.*[38] Assumptions used in equation of Coburn *et al.*: carbon monoxide production = 0.4 ml/hr STPD; $D_{L_{CO}}$ = 20 ml/min-torr; barometric pressure = 760 torr; alveolar ventilation = 3500 ml/min STPD; Haldane constant = 220;[150] mean pulmonary capillary pO_2 = 100 torr; fraction of unbound hemoglobin constant at 3%.

[b]P = carbon monoxide concentration, ppm. This equation applicable only up to an inspired carbon monoxide concentration of 50 ppm.

Approximate Physiologically Equivalent Altitudes (in Feet) at Equilibrium with Ambient CO Levels

Ambient CO Concentration (ppm)	Actual Altitude (ft)		
	0 (Sea Level)	5,000	10,000
	Physiologically Equivalent Altitudes with COHb		
0	0 (sea level)	5,000	10,000
25	6,000	8,300	13,000
50	10,000	12,000	15,400
100	12,300	15,300	18,000

Source: Air Quality and Automobile Emission Control, A Report by the Coordinating Committee on Air Quality Studies, prepared for the Committee on Public Works, United States Senate, Volume 2, September 1974.

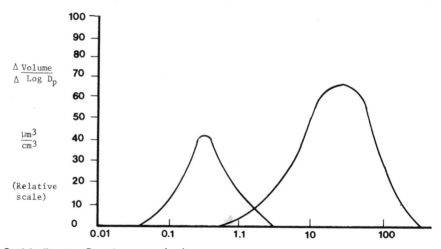

Particle diameter, D_p, micrometers (μm)

GENERAL AEROSOL PROPERTIES

0.01-2μm

1. Soluble
2. Produced by condensation (enlarge by coagulation)
3. Contain: $SO_4^=$, NO_3^-, NH_4^+ organic condensates, trace metals, (lead, arsenic, beryllium, cadmium, mercuty, etc.)

2-100 + μm

1. Insoluble
2. Produced mechanically
3. Contain: soil, tire dust, sea salt, fly ash, SiO_2, $CaCO_3$

Source: Measurements of the effects of Fossil Fuel Combustion on Atmospheric Composition (Professor Robert Charlson, University of Washington.)

Bimodal distribution of airborne particles.

Reproduced with permission of the National Academy of Sciences.

Sulfur Dioxide and Particulates. By itself, sulfur dioxide is not considered to be a hazard to health. SO_2 is highly soluble in body tissue fluids and virtually all of the SO_2 in the air is absorbed in the airways before reaching the lungs. SO_2 is also readily converted, however, into sulfuric acid or into sulfate salts which, in combination with particulate matter, can penetrate the lungs. Deposition of particles in the airways of the body depends largely upon the size of the particles inhaled. Particles larger than 15 microns in size generally settle before they enter the body. Particles larger than one micron are deposited mostly in the upper airways and at the branching sites of the tracheobronchial system. The submicron particles settle in the narrow peripheral airways and the alveoli in the lungs.

Although sulfur dioxide may not be the direct agent in causing adverse health effects, there apparently is a correlation between SO_2 concentration in the average urban or industrial atmosphere and health effects although contributions to these effects may be made by a number of other factors.

Comparison of ψ Values with Pollutant Concentrations, Descriptor Words, General Health Effects, and Cautionary Statement

Index value	Air quality level	Pollutant levels					Health effect descriptor	General health effects	Cautionary statements
		TSP (24 hour), $\mu g/m^3$	SO₂ (24 hour), $\mu g/m^3$	CO (8 hour), mg/m^3	O₃ (1 hour), $\mu g/m^3$	NO₂ (1 hour), $\mu g/m^3$			
500	Significant harm	1000	2620	57.5	1200	3750		Premature death of ill and elderly. Healthy people will experience adverse symptoms that affect their normal activity	All persons should remain indoors, keeping windows and doors closed. All persons should minimize physical exertion and avoid traffic
400	Emergency	875	2100	46.0	1000	3000	Hazardous	Premature onset of certain diseases in addition to significant aggravation of symptoms and decreased exercise tolerance in healthy persons	Elderly and persons with existing diseases should stay indoors and avoid physical exertion. General population should avoid outdoor activity
300	Warning	625	1600	34.0	800	2260	Very unhealthful	Significant aggravation of symptoms and decreased exercise tolerance in persons with heart or lung disease, with widespread symptoms in the healthy population	Elderly and persons with existing heart or lung disease should stay indoors and reduce physical activity
200	Alert	375	800	17.0	400ᶜ	1130	Unhealthful	Mild aggravation of symptoms in susceptible persons, with irritation symptoms in the healthy population	Persons with existing heart or respiratory ailments should reduce physical exertion and outdoor activity
100	NAAQS	260	365	10.0	160	ª	Moderate		
50	50% of NAAQS	75ᵇ	80ᵇ	5.0	80	ª	Good		
0		0	0	0	0	ª			

ª No index values reported at concentration levels below those specified by "Alert Level" criteria.
ᵇ Annual primary NAAQS.
ᶜ 400 $\mu g/m^3$ was used instead of the ozone Alert Level of 200 $\mu g/m^3$.

Source: Environmental Quality—1976, The Seventh Annual Report of the Council on Environmental Quality, September 1976.

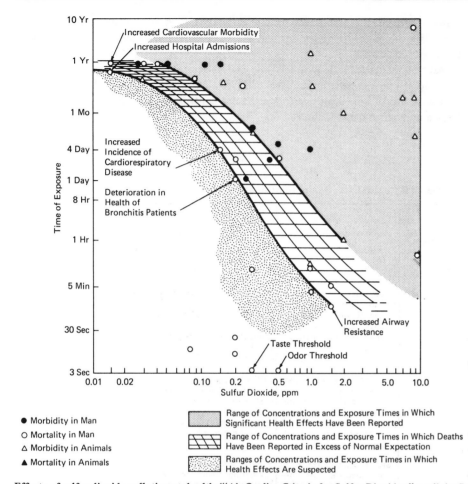

Effects of sulfur dioxide pollution on health ("Air Quality Criteria for Sulfur Dioxides," a talk by Bernard E. Conley, chief, Air Quality Criteria, National Center for Air Pollution Control)

Source: Starr, C. Greenfield, M. A. and Hausknecht, D. F. "A Comparison of Public Health Risks: Nuclear vs Oil-fired Power Plants," *Nuclear News*, October 1972.

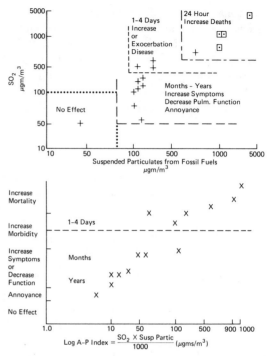

Dose response curve for sulfur dioxide and total suspended particulates.

Source: Air Quality and Automobile Emission Control, A Report by the Coordinating Committee on Air Quality Studies, prepared for the Committee on Public Works, United States Senate, Volume 2, September 1974.

Response to High Concentrations of Nitrogen Dioxide in Occupational Exposures

Nitrogen Dioxide Concentration		Clinical Effect	Time between Exposure and Termination of Effect
mg/m³	ppm		
940	500	acute pulmonary edema fatal	Within 48 hr
564	300	bronchopneumonia—fatal	2–10 days
282	150	bronchiolitis, fibrosa obliterans—fatal	3–5 weeks
94	50	bronchiolitis, focal pneumonitis—recovery	6–8 weeks
47	25	bronchitis, broncho- pneumonia—recovery	6–8 weeks

Source: Air Quality and Automobile Emission Control, A Report by the Coordinating Committee on Air Quality Studies, prepared for the Committee on Public Works, United States Senate, Volume 2, September 1974.

Nitrogen Oxides. The chief health concern with regard to the oxides of nitrogen is their capacity to combine, under the action of sunlight, with a variety of hydrocarbons to form reactive, unstable oxidants, principally ozone and peroxyactyl nitrate. These compounds are not only eye irritants but act on the peripheral airways and alveolar capillary membranes of the lungs. Exposures of healthy volunteers to high concentrations of NO_2 alone have not produced any changes in pulmonary function.

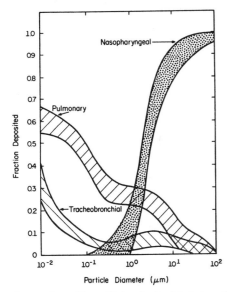

Respiratory deposition efficiencies for inhaled particles.

Natusch, D. F. S. and Wallace, J. R., "Urban Aerosol Toxicity: The Influence of Particle Size," *Science*, Vol. 186, pp. 695–699, November 22, 1974. © Association for the Advancement of Science.

Results of Epidemiological Studies

Adverse health effect	Concentration[a] at which effect was observed SO₂, μg/m³ (ppm)	Sulfates, μg/m³	Averaging time
Increased mortality	300–400 (0.11–0.15)	NA[b]	24 hr
Aggravation of symptoms in elderly	365 (0.14)	8–10	24 hr
Aggravation of asthma	180–250 (0.07–0.09)	6–10	24 hr
Decreased lung function in children	220 (0.075)	11	Annual mean
Increased acute lower respiratory disease in families	90–100 (0.034–0.037)	9	Annual mean
Increased prevalence of chronic bronchitis	95 (0.035)	14	Annual mean
Increased acute respiratory disease in families	106 (0.039)	15	Annual mean
Increased respiratory disease related illness absences in female workers	NA[b]	13	Annual mean
Primary standard	365	-	24 hr
Primary standard	80	-	Annual mean

[a]Effects levels are best judgment estimates based on a synthesis of several studies.

[b]NA = not available.

Source: Position Paper on Regulation of Atmospheric Sulfates, EPA-450/2-75-007, U.S. Environmental Protection Agency, September 1975.

Response to Low Concentrations of Nitrogen Dioxide

Effect	Nitrogen Dioxide Concentration, mg/m³ (ppm)		Time to Effect	Reference
Odor threshold	0.23	(0.12)	immediate	Henscheler[116]
	0.23	(0.12)	immediate	Shalamberidze[117]
Threshold for dark adaptation	0.14	(0.075)	not reported	Shalamberidze[117]
	0.50	(0.26)	not reported	Bondareva[9]
	1.3–3.8	(0.7–2)	20 min[a]	Suzuki[120]
Increased airway resistance	3.0–3.8	(1.6–2.0)	15 min	von Nieding[13,126,128]
	2.8	(1.5)	45 min[b]	Rokaw[130]
	3.8	(2.0)	45 min[c]	Rokaw[130]
	5.6	(3.0)	45 min[d]	Rokaw[130]
	7.5–9.4	(4–5)	40 min[e]	Abe[119]
	9.4	(5)	15 min	Stresemann[173]
	11.3–75.2	(6–40)	5 min	Nakamura[121]
	13.2–31.8	(7–17)	10 min[f]	Yokoyama[122–125]
Decreased pulmonary diffusing capacity	7.5–9.4	(4–5)	15 min	von Nieding[126]
Increased alveolar-arterial pO₂ difference	9.4	(5)	25 min[g]	von Nieding[13]
No change in sputum histamine concentration	0.9–6.6	(0.5–3)	45 min	Thomas[129]

[a]Exposure lasted 10 min. Effect on flow resistance was observed 10 min after termination of exposure.
[b]Effect was produced at this concentration when normal subjects and those with chronic respiratory disease exercised during exposure.
[c]Effect occurred at rest in subjects with chronic respiratory disease.
[d]Effect occurred at rest in normal subjects.
[e]Exposure lasted 10 min. Maximal effect on flow resistance was observed 30 min later.
[f]Also failed to find increased flow resistance over the range of nitrogen dioxide exposures from 5.1 to 30.1 mg/m³ (2.7–16 ppm).
[g]Effect occurred 10 min after termination of 15-min exposure period.

Source: Air Quality and Automobile Emission Control, A Report by the Coordinating Committee on Air Quality Studies, prepared for the Committee on Public Works, United States Senate, Volume 2, September 1974.

COMPARATIVE HEALTH EFFECTS RESULTING FROM ENERGY USE

Standards for the control of pollutants do not have a common basis in the United States. Standards for the levels of sulfur dioxide are set near the point where medically perceivable effects have been observed and at a level about 100 times above the natural background level. Standards for nitrogen oxides are set about a factor of ten below the level where medical effects are observed and about a factor of ten above natural background levels. Standards for radiation are set about a factor of 100 below the natural background level which, in turn, is a factor of 100 below the level where medically perceivable effects are observed.

Richard Wilson (see table) has prepared figures for the number of deaths related to various aspects of energy production. He has suggested that since we accept 50,000 deaths a year from the use of energy in automobiles, that we accept other energy-related activities on the basis of a similar risk-benefit comparison. He suggests that any cost outside that context be labeled the "cost of public confidence."

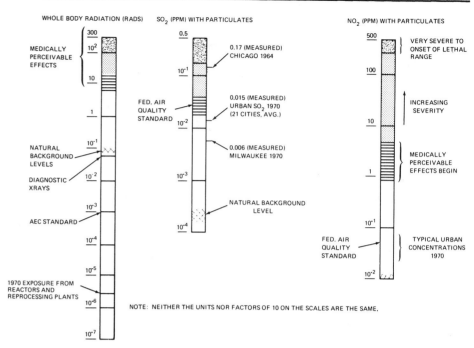

Comparison of pollutant standards, background levels, manmade exposures, and health effects.

Source: Comparative Risk-Cost-Benefit Study of Alternative Sources of Electrical Energy, WASH-1224, U.S. Atomic Energy Commission, December 1974.

Energy Production's Impact on Deaths in the U.S.
(Based on 1965 figures)

		Per Kilowatt Hour	Per Billion Megawatt Hours
Coal mining: black lung disease		10^{-9}	1000
Coal mining: accidents		6×10^{-11}	60
Petroleum refining and oil well accidents		7×10^{-12}	7
Uranium mining cancers	(no breeder)	1.5×10^{-11}	15
	(breeder)	7×10^{-14}	0.07
Uranium processing and fuel fabrication accidents	(breeder)	2×10^{-13}	0.2
(estimate)	(no breeder)	2×10^{-11}	20
Coal/gas/oil air pollution		3×10^{-9}	3000
Radiation cancers from normal operation of reactors and processing plants (0.1 mrem per year at year 2000)		3×10^{-13}	0.3
Potential reactor accidents (one per 30 years of WASH severity by year 2000)			
(a) direct deaths		3×10^{-11}	30
(b) possible extra cancer cases		3×10^{-12}	3
Gas main explosives		6×10^{-12}	3
Gas poisoning		6×10^{-12}	6
Dam failures (one per 50 years)		1×10^{-10}	100

Source: Wilson, Richard; Remarks at the Conference on Advanced Energy Systems, Edward Teller Center for Science, Technology and Political Thought, Denver, Colorado, June 1974, as reported in the *Weekly Energy Report*, July 1, 1974.

Pollutant	Health Effects	Effects on Vegetation	Effects on Aquatic and Terrestrial Organisms	Air Standards	Water Standards
Sulfur Oxides SO_X	$SO_2 \rightarrow SO_3 \rightarrow H_2SO_4$. The heart and lungs are the major target organs for SO_X. The presence of SO_X increases bronchio constriction hence aggravating asthma and emphysema and decreasing lung ventilation.	Sulfur oxides are highly toxic to vegetation; effects include: interveinal necroses, yellowing of broadleaf species and reddish discoloration of conifer needles. Acid rains may also damage vegetation or alter soil conditions.	Aquatic communities may be affected by increasing acid conditions due to acid rains. Animals are sensitive to high SO_X concentrations.	TLV* for SO_2 is 5.0 ppm. Federal Primary Ambient Air Standards for SO_2 are: 365 $\mu g/m^3$ (0.14 ppm) - 24 hr. standard; 80 $\mu g\ m^3$ (0.03 ppm) annual standard. The secondary standard is 1300 $\mu g/m^3$ (0.5 ppm).	
Nitrogen Oxides NO_X	$NO \rightarrow NO_2$ by photochemical oxidation. NO_2 is four times as toxic as NO. At high levels NO_2 causes pulmonary edema and death while at low levels the effects include emphysema, polycythemia, leucocytosis and sensitivity to infection. In addition to lung damage, liver kidney and heart damage may occur. Eye and skin irritation may also occur. NO_X exposure correlates with lung cancer induction. NO_3^- and NO_2^- in water may cause methemoglobinemia and death. NO_2^- may cause cancer.	Adverse effects on plants from NO_2 include: defoliation, chlorosis, irregular necrotic spots, tip and margin burn, high leaf gloss, inhibition of photosynthesis and growth retardation. Middle age, rapidly growing leaves are most sensitive. Nitrate, because it is an important nutrient, is considered an asset in irrigation water.	Nitrates in water are rapidly removed by aquatic plants and may result in eutrophication. Nitrogen (nitrate or ammonia) should not exceed 0.3 mg/l in lakes or 1.0 mg/l in free-flowing streams to prevent algal blooms. Nitrate ion, a minor component, is toxic to aquatic organisms but they are very resistant to nitrate. Livestock poisoning may occur from nitrite ingestion. It is recommended that the nitrate plus nitrite nitrogen not exceed 100 ppm and that NO_2-N alone not exceed 10 ppm.	TLV for NO is 25 ppm. TLV for NO_2 is 5.0 ppm. TLV for HNO_3 is 2.1 ppm. Odor perception at 0.12 ppm. Federal Primary Ambient Air Standard is 100 $\mu g/m^3$ (0.05 ppm) as an annual arithmetic mean.	National Interim Primary Drinking Water Standard for nitrate as nitrogen is 10 mg/ml. It has also been recommended that Nitrate-Nitrogen in drinking water not exceed 1.0 m mg/l .
Particulate Matter	Particles 0.5 to 5.0μ in diameter are most likely to cause disease. Chronic symptoms due to lung scarring include: difficulty breathing, chest pain, cough, decreased vital capacity and heart disease.	Excessive dusting can clog the stomates of plant leaves, preventing air and water exchange.	Suspended solids harm aquatic biota by reducing light penetration, suffocating bottom dwellers, physical abrasion and habitat destruction. This is especially serious in nursery or spawning site. The following levels of suspended solids are recommended; <25 mg/l (high protection); 26-80 mg/l (moderate); 81-40 mg/l (low); over 400 mg/l (very low protection).	TLV* for nuisance particulates is 10 mg/m³ total particulates. Primary National Ambient Air Quality Standard for Suspended Particulates is 75 $\mu g/m^3$ (annual), and 260 $\mu g/m^3$ (24 hr). The secondary standard is 60 $\mu g/m^3$ (annual) and 150 $\mu g/m^3$ (24 hr.).	
Carbon Monoxide CO	CO reacts with hemoglobin to form carboxyhemoglobin. May result in brain damage due to oxygen deprivation. Symptoms of exposure include: headache, dizziness, nausea, vomiting, systemic pain, cherry red skin color, and fatigue.	Plants are insensitive to CO levels known to affect man. At high concentrations the following symptoms are observed: leaf curling, increased aging, reduced gravity response, reduced leaf size, and feminization.	In water CO $\rightarrow CO_2$. See next section for effects.	TLV = 50 ppm. National Primary Ambient Air Standard is: 10000 $\mu g/m^3$ for a yearly average; 40000 $\mu g/m^3$ for a 24 hr. average.	

Pollutant	Health Effects	Effects on Vegetation	Effects on Aquatic and Terrestrial Organisms	Air Standards	Water Standards
Hydrogen Sulfide H_2S	H_2S is highly toxic. It is a pulmonary irritant but its major effect is paralysis of the nerves governing respiration leading to asphyxiation. Low level exposure may result in: fatigue, metallic taste, nausea, vomiting, diarrhea, pulmonary edema, eye irritation and dizziness. Chronic exposure can cause kidney, liver and/or brain damage.	At low concentrations little effect. At 20 to 40 ppm tan or white markings may appear on young, growing leaves.	H_2S is extremely toxic to aquatic organisms. A maximum level of dissociated hydrogen sulfide assumed to be safe for all aquatic organisms is 0.002 ppm.	TLV* = 10 ppm (skin) odor perception between 1 45 $\mu g/m^3$.	
Uranium	Exposure to the radiation from uranium can result in induction of leukemia, induction of neoplasms especially lung cancer, cataracts, reduced life expectancy, genetic effects, sterility and suppression of immune responses. Uranium especially accumulates in and affects lungs, bones, kidneys and liver.	Radiation affects plants in the following order of severity: tall plants (most severe), shrubs, hedges, mosses and lichens (least severe). Fields are generally more resistant to radiation effects than complex forest ecosystems.	Aquatic organisms often concentrate radioactive elements. In general the following order of sensitivity to radiation exists: large herbivorous mammals > small mammals and birds >herbivorous insects > filter feeding aquatic invertebrates > unicellular animals and plants.	For occupational exposure U^m natural = 7 x 10⁻¹¹ $\mu c/ml$. For nonoccupational exposure: U natural = 3 x 10⁻¹² $\mu c/l$. Standards also exist for ^{230}U, ^{232}U, ^{233}U, ^{234}U, ^{235}U, ^{236}U, ^{238}U and ^{240}U.	For occupational exposure: U natural = 5 x 10⁻⁴ $\mu c/$ml. For nonoccupational exposure: U natural = 2 x 10⁻⁵ $\mu c/ml$.
Carbon Dioxide CO_2	CO_2 is not ordinarily considered a toxic gas. At high concentrations it stimulates respiration and breathing becomes labored. It forms carboxyhemoglobin and deprives the brain of oxygen. Symptoms of exposure include: headache, dizziness, tinnitus, difficulty breathing, muscle tremor, fatigue, and unconsciousness.	Plants require CO_2 for photosynthesis. High CO_2 concentrations may increase the acidity of rain, secondarily affecting vegetation.	Concentrations of free CO_2 rarely exceed 20 ppm in surface waters. Fish can acclimate to concentrations as high as 60 ppm but will try to avoid even minor increases in CO_2.	TLV = 5000 ppm.	

*TLV = Threshold Limit Value. The concentration of a substance to which a worker can be exposed 8 hours per day or 40 hours per week without significant health effects or discomfort.

Source: Energy/Environment Fact Book, United States Environmental Protection Agency, EPA-600/9-77-041, March 1978.

COMPARATIVE ENVIRONMENTAL IMPACTS OF POWER PLANTS

	1,000-Mw. coal-fired plant	1,000-Mw. nuclear plant
Net fuel consumed (per year)	2.5 million tons	1 ton uranium (130 tons U_3O_8)*
Onsite fuel inventory	0.5 million tons	80 to 140 tons uranium (500 tons U_3O_8)*
Solid waste produced (per year)	25,000 tons ash	50 tons uranium (total 2,500 tons of cask shipments)
Gaseous waste produced (per year)	9 million tons CO_2, 0.1 million tons SO_2-NO_x, 0.1 million tons fly ash, 0.005 curies (radium and thorium)	3 to 300,000 curies (mostly xenon and krypton)
Liquid waste produced	—	3 to 3,000 curies H^3, 0.1 to 10 curies of other materials
Heat produced:		
To stack	300 thermal Mw.	—
To condenser	1,300 thermal Mw.	2,100 thermal Mw.
Total	1,600 thermal Mw.	2,100 thermal Mw.
Water consumed:		
If heated 10°F.	2,000 cu.ft./sec.	3,300 cu.ft./sec.
If evaporated	26 cu.ft./sec.	42 cu.ft./sec.

*The equivalent net annual requirement for U_3O_8 needed to replace the fuel consumed or to make new fuel.

In terms of the fuel consumed and the waste effluents produced, fossil-fueled and nuclear-fueled electric generating plants are almost totally different. A good site for one type of plant may be an inefficient site for the other, and the problem of coordinating these characteristics with social and ecological constraints promises nothing but more effort and frustration for the energy industry in the next decades. The table compares the fuel used and effluents produced annually by 1,000-Mw. coal and nuclear units.

Source: Lowe, William W., "Creating Power Plants: The Costs of Controlling Technology, *Technology Review*, edited at The Massachusetts Institute of Technology January, 1972.

Environmental Effects of Electrical Power Generation

Energy Source	Effect on Land	Effects on Water	Effects on Air	Biological Effects	Supply
Coal	Disturbed land Large amounts of solid waste	Acid mine drainage Increased water temperature	Sulfur oxides Nitrogen oxides Particulates Some radio-active gases	Respiratory problems from air pollutants	Large reserves
Oil	Wastes in the form of brine, pipeline construction	Oil spills Increased water temperature	Nitrogen oxides Carbon monoxide Hydrocarbons	Respiratory problems from air pollutants	Limited domestic reserves
Gas	Pipeline construction	Increased water temperature	Some oxides of nitrogen	Little known effects	Extremely limited domestic reserves
Uranium	Disposal of radioactive waste	Increased water temperature some radioactive liquids	Some radioactive gases	None detectable in normal operation	Large reserves if breeders are developed

Source: The Environmental Impact of Electrical Power Generation: Nuclear and Fossil, WASH-1261, U.S. Atomic Energy Commission, 1973.

Comparison of Costs and Impacts of Alternate Electrical Energy Production Systems[a]

Basis: 1000 MWe Power Plant, 75% CF 6.57 × 10⁹ kWhe

	Coal	Oil	Gas	LWR
Power Plant and Energy System Efficiencies				
Electrical energy (billion kWhe/year)	6.57	6.57	6.57	6.57
Power plant heat rate (Btu/kWhe)	8,900	8,830	9,110	10,850
Power plant thermal efficiencies (kwe/kw$_t$, %)	38	39	38	32
Energy system efficiency				
(kwh$_e$ consumer/kwh$_t$ input, %)	35	35	34	28
Consumption of Nonrenewable Fuel Resources				
Power plant fuel consumption (annual)	2.3 M tons	10 M barrels	64 B cubic ft	~130 M tons U[b]
Fraction of reserves consumed (annual)	0.000006	0.0001	0.0004	0.0002
Conventional Costs (mills/kWhe)[c]				
Plant	7.8	7.2	6.4	11.7
O&M	0.8	0.6	0.6	0.8
Fuel	9.8	27.4	36.0	6.0
Total	18.4	35.2	43.0	18.5
Selected Abatement Costs (mills/kWhe)[c]	4.7	2.0	0.6	0.6
Occupational Health and Safety				
Occupational health (MDL/yr)	600	U	U	480
Occupational safety				
fatalities (deaths/year)	1.1	0.17	0.08	0.1
nonfatal injuries (#/year)	46.8	13.1	5.3	6.0–7.0
total man-days lost (MDL/year)	9,250	1,725	780	900–1000
Public Health and Safety				
Public health				
routine pollutant release (MDL/year)	U	U	U	180–210
Public safety				
Transportation injuries				
fatalities (deaths/year)	0.55	U	U	0.009
nonfatal (injuries/year)	1.2	U	U	0.08
total man-days lost (MDL/year)	3,500	U	U	60
Environmental degradation				
Land				
land use, inventory (acres)	22,400	~1,600	~3,600	~1,000
land use, consumption (acres/year)	740	S	S	12
Air				
SO_2 release, w/o abatement (tons/year)	120,000	38,600	20	3,600
SO_2 release, w abatement	24,000	21,000	0	720
NO_x releases, w/o abatement (tons/year)	27,000	26,000	13,400	810
particulate releases, w/o abatement (tons/year)	270,000	26,000	518	8,000
particulate releases, w abatement	2,000	150	4	60
trace metals releases (tons/year)	0.5 Hg	1,500 V	U	S
radioactivity releases (Ci/year)	0.02	0.0005	S	250–500 T
thermal discharge, power plant stack (billion kwh$_t$/year)	1.64	1.71	2.2	0
Water				
cooling water use (billion gal/year)	263	263	263	424
process water use (billion gal/year)	1.46	1.75	1.42	0.095
radioactivity releases (Ci/year)	0	0	0	500–1000
other impacts (billion gal/year)	16.8	7.9	0	S
thermal discharge, power plant (billion kwht/year)	9	9	9	14

U ≡ Unevaluated; S ≡ Small; M ≡ Million; B ≡ billion; T ≡ Thousand.

[a]The number of digits shown is not generally indicative of precision. In many cases, several digits are retained merely for calculational purposes.

[b]About 99% of this figure is not irretrievably consumed; rather, it is available in the form of enrichment plant tails for use in breeder reactors.

[c]1980 dollars.

Source: Comparative Risk-Cost-Benefit Study of Alternative Sources, of Electrical Energy, WASH-1224, U.S. Atomic Energy Commission, December 1974.

Ash Collection and Utilization in the United States
1971 (tons)

	Fly Ash	Bottom Ash	Boiler Slag
Ash Uses:			
1. Mixed with raw material before forming cement clinker	94,529	—	83,421
2. Mixed with cement clinker or mixed with pozzolan cement	14,998	—	—
3. Partial replacement of cement in:			
(a) concrete products	160,689	32,086	69,442
(b) structural concrete	168,218	—	—
(c) dams and other mass concrete	64,769	—	—
4. Lightweight aggregate	162,257	12,645	—
5. Fill material for roads, construction sites, etc.	329,590	484,049	2,384,398
6. Stabilizer for road bases, parking areas, etc.	33,503	7,147	44,954
7. Filler in asphalt mix	133,923	2,569	74,101
8. Miscellaneous	89,613	431,203	388,219
Subtotal	1,252,093	969,701	3,044,538
Ash removed from plant site (at no cost to utility, but not covered in categories listed above)	1,698,564	492,405	346,269
Total ash utilized	2,950,657	1,462,107	3,390,808
Ash removed to disposal areas	22,219,548	7,661,375	1,117,694
Total ash collected, 1971	25,170,205	9,123,483	4,508,502
Estimated 1976 ash production	33,553,953	106,492,323	2,283,556

Source: Chemical Engineering 16 April, 1973.

Source: Introductory Report by the Secretariat, ENV/SEM.4/R.1, Economic Commission for Europe, Second Seminar on Desulphurization of Fuels and Combustion Gases, Washington, D.C., Nov. 11–20, 1975.

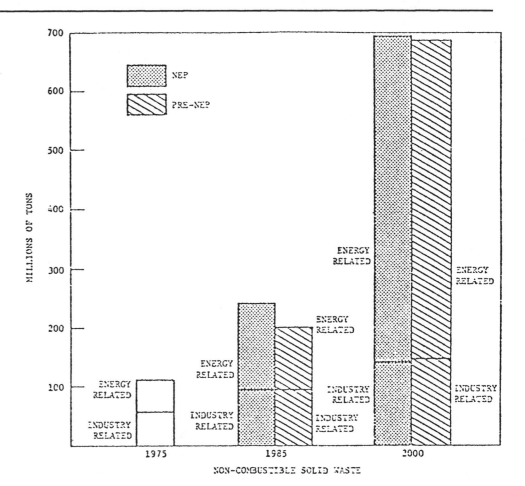

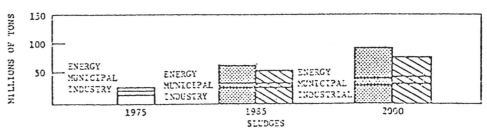

Noncombustible solid waste and sludge generation by source and with and without provisions of the National Energy Plan of 1977.

Source: Annual Environmental Analysis Report, An Environmental Analysis of Energy Technologies using the Assumptions of the National Energy Plan, Volume 1, Technical Summary, MTR-7626, A MITRE Report, prepared for the Energy Research and Development Administration, Sept. 1977.

Environmental Control of Energy

The industry's largest wet scrubber air quality control system is shown between the 700-foot stack and the plant proper at the 820-megawatt coal-fired plant of Kansas City Power & Light Co., and Kansas Gas and Electric Co., the co-owners, near La Cygne, Kansas. The scrubber was built by Babcock & Wilcox, Barberton, Ohio, at a cost of approximately $45 million. An estimated two years of additional time and $5 million to $10 million more in investment will be required for the system to reach control design capability in removing sulfur dioxide and fly ash from exhaust emissions.

THE ASSESSMENT OF RISKS

The setting of standards for the control of the environmental impacts of the production and use of energy involves a conscious or intuitive acceptance of levels of risk in order to obtain real or perceived benefits. The process is complex. The use of energy in various forms is pervasive in our society, touching every sector of the economy—agriculture, transportation, housing, communications, and industry. The environmental effects of energy production and use are, in some instances, quite obvious—as, for example, air and water pollution. Many effects, however, are more subtle and more difficult to quantify, for example, the effects on global weather of an increase in atmospheric carbon dioxide or of waste heat production. Even in instances where environmental damage can be observed, it is often not possible to identify a single causative agent or to agree on specific standards which represent a public consensus with regard to an acceptable risk level.

From a study of the risks associated with various activities of man, there appears to be a general correlation between risks that are deemed acceptable and the benefits derived from the risks taken. The higher the perceived benefits, the higher are the acceptable risks. At the same time, there is also an apparent marked difference in the acceptability of voluntary and involuntary risks. Individuals seem to be willing to accept voluntary risks—smoking, hunting, skiing, sport flying—at a level about 1000 times higher than they will accept from involuntary risks—those associated with everyday living. Perhaps not coincidentally, the risk of death due to disease falls between the acceptable levels for voluntary and involuntary risks. Perhaps the risk of death due to disease forms the intuitive base line against which individuals gauge their personal willingness to accept risks.

The setting of standards for the activities related to energy production and use is very much dependent upon the risk-benefit relationships as seen by each member of society. As a group the members of society will ultimately decide which risks are real and which benefits are worthwhile.

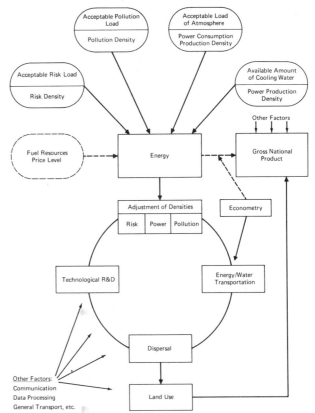

This schematic representation of an energy system depicts the relationship of energy, land use, and gross national product.

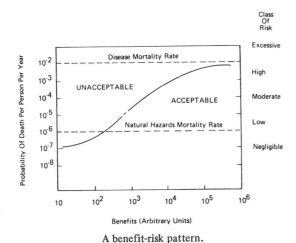

A benefit-risk pattern.

Source: *Reactor Safety Study: An Assessment of Accident Risks in U.S. Commercial Nuclear Power Plants*, WASH-1400 (NUREG-75/014), United States Nuclear Regulatory Commission, October 1975.

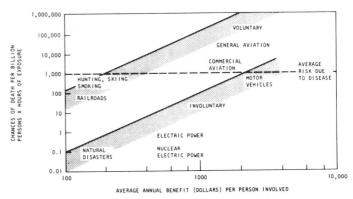

The benefit is measured by the amount of money spent on each voluntary activity, and by the contribution to an individual's income for involuntary activities.

Source: C. Starr, "Benefit-Cost Relationships in Socio-Technical Systems," *United Nations Symposium on Environmental Aspects of Nuclear Power Stations*, IAEA-SM-146/47, International Atomic Energy Agency, August 1970.

FATALITY FREQUENCY—NATURAL AND MAN-CAUSED EVENTS

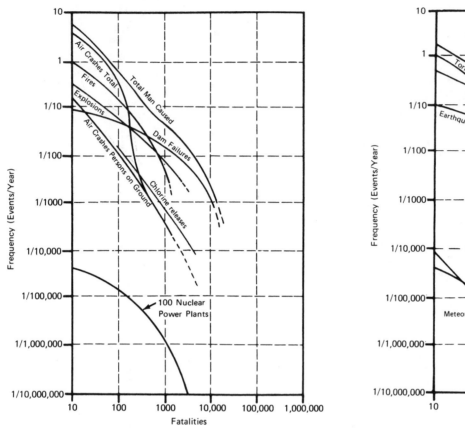

Frequency of fatalities due to man-caused events.

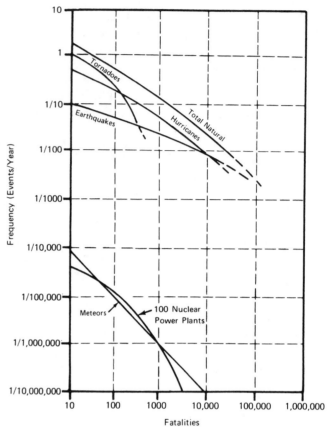

Frequency of fatalities due to natural events.

Source: Reactor Safety Study: An Assessment of Accident Risks in U.S. Commercial Nuclear Power Plants, WASH-1400 (NUREG-75/014), United States Nuclear Regulatory Commission, October 1975.

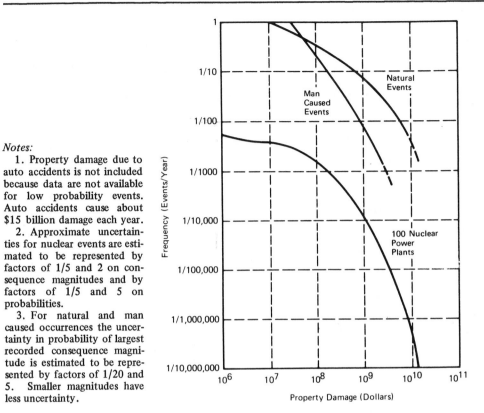

Frequency of property damage due to natural and man-caused events.

Notes:

1. Property damage due to auto accidents is not included because data are not available for low probability events. Auto accidents cause about $15 billion damage each year.

2. Approximate uncertainties for nuclear events are estimated to be represented by factors of 1/5 and 2 on consequence magnitudes and by factors of 1/5 and 5 on probabilities.

3. For natural and man caused occurrences the uncertainty in probability of largest recorded consequence magnitude is estimated to be represented by factors of 1/20 and 5. Smaller magnitudes have less uncertainty.

U.S. Fatalities—by Major Categories (1969)

Type	Fatalities/100,000 persons	Approximate Individual Risk Fatality Probability/year
Diseases	819	8×10^{-3}
Accidents	57.6	6×10^{-4}
Other Causes	76.2	8×10^{-4}
Total	951.9	1×10^{-2}

Annual Accident Fatalities and Injuries in the U.S.

Accident Type	Total United States Fatalities	Total United States Injuries	People Within 25 Miles of Nuclear Sites Fatalities	People Within 25 Miles of Nuclear Sites Injuries
Automobile	55,000	5×10^{6}	4200	375,000
Falls	20,000	1×10^{6}	1500	75,000
Fire	7,500	0.3×10^{6}	560	22,000
Other	33,000	1.6×10^{6}	2500	120,000
TOTAL	115,000	7.9×10^{6}	8760	592,000
Reactor Accidents (for 100 plants from Table 5-6, Chapter 5)	7×10^{-2}	1	3×10^{-3}	2×10^{-1}

Average Risk of Fatality by Various Causes

Accident Type	Total Number	Individual Chance per Year
Motor Vehicle	55,791	1 in 4,000
Falls	17,827	1 in 10,000
Fires and Hot Substances	7,451	1 in 25,000
Drowning	6,181	1 in 30,000
Firearms	2,309	1 in 100,000
Air Travel	1,778	1 in 100,000
Falling Objects	1,271	1 in 160,000
Electrocution	1,148	1 in 160,000
Lightning	160	1 in 2,000,000
Tornadoes	91	1 in 2,500,000
Hurricanes	93	1 in 2,500,000
All Accidents	111,992	1 in 1,600
Nuclear Reactor Accidents (100 plants)		1 in 5,000,000,000

U.S. Economic Losses From Various Causes

Source	Estimated Annual Losses (Millions of $)
Automobile Accidents (1970)	5,000
Fires (Property - 1970)	2,200
Hurricanes (1952-72 average)	500
Fires (Forest - 1970)	70
Tornadoes (1970)	50
Reactor Accidents from 100 plants (See Table 5-6, Chapter 5)	2

Source: Reactor Safety Study: An Assessment of Accident Risks in U.S. Commercial Nuclear Power Plants, WASH-1400 (NUREG-75/014), United States Nuclear Regulatory Commission, October 1975.

Source: Reactor Safety Study: An Assessment of Accident Risks in U.S. Commercial Nuclear Power Plants, WASH-1400 (NUREG-75/014), United States Nuclear Regulatory Commission, October 1975.

CONSEQUENCES OF NATURAL DISASTERS

Consequences of Major U.S. Hurricanes (1900–1972) [a]

No.	Date (month/year)	Fatalities (U.S. only)	Damage Range In Million $	No.	Date (month/year)	Fatalities (U.S. only)	Damage Range In Million $
1.	8/00	6,000	5 - 50	27.	9/33	21	0.5 - 5
2.	9/28	1,836	5 - 50	28.	10/44	18	50 - 500
3.	9/19	787	5 - 50	29.	9/56	15	5 - 50
4.	9/38	600	50 - 500	30.	9/67	15	50 - 500
5.	8/35	408	5 - 50	31.	7/70	13	50 - 500
6.	6/57	390	50 - 500	32.	7/34	11	0.5 - 5
7.	9/09	350	0.5 - 5	33.	9/55	7	50 - 500
8.	8/69	323	500 - 5000	34.	6/16	7	0.5 - 5
9.	8/15	275	5 - 50	35.	6/34	6	0.5 - 5
10.	9/15	275	5 - 50	36.	10/35	5	5 - 50
11.	9/26	243	50 - 500	37.	8/64	5	50 - 500
12.	8/55	184	500 - 5000	38.	9/41	4	5 - 50
13.	6/72	122	500 - 5000	39.	9/45	4	50 - 500
14.	10/54	95	50 - 500	40.	10/50	4	5 - 50
15.	8/65	75	500 - 5000	41.	9/48	3	5 - 50
16.	8/54	60	50 - 500	42.	8/49	2	50 - 500
17.	8/60	50	50 - 500	43.	9/49	2	5 - 50
18.	8/40	50	0.5 - 5	44.	9/50	2	0.5 - 5
19.	9/47	51	50 - 500	45.	10/35	2	0.5 - 5
20.	9/61	46	50 - 500	46.	10/47	1	0.5 - 5
21.	9/44	46	50 - 500	47-51.	Five others	0	5 - 50
22.	8/32	40	5 - 50		TOTAL 51	TOTAL 12,577	TOTAL ~ 12 Billion
23.	8/33	40	5 - 50				
24.	9/64	38	50 - 500				
25.	8/55	25	50 - 500				
26.	9/54	21	5 - 50				

(a) From "Some Devastating North Altantic Hurricanes of the 20th Century," U.S. Department of Commerce.

Source: Reactor Safety Study: An Assessment of Accident Risks in U.S. Commercial Nuclear Power Plants, WASH-1400 (NUREG-75/014), United States Nuclear Regulatory Commission, October 1975.

Dam and Levee Failures in the U.S. (1889-1972)

Year	Name/Location	Type of Structure	Lives Lost
1889	/Johnston, Pa.	Dam	~2000
1890	Walnut Grove/Prescott, Ariz.	Dam	150
1894	Mill River/Mass.	Dam	143
1900	Austin/Austin, Pa.	Dam	8
1928	St. Francis Dam/Ca.	Dam	~450
1955	/Yuba City, Ca.	Levee	~38
1963	Baldwin Hills/Los Angeles, Ca.	Reservoir	5
1972	/Buffalo Creek, W. Va.	Dam	125
1889-Present		Total	2919

Annual Rates of Fires with Large Economic Losses

Dollar Loss	Annual Frequency (approximate average)
>1 million	50
>3 million	14
>10 million	3
>20 million	1.2
>40 million [a] (4 in 8 years)	0.5

(a) Includes 3 large forest fires estimated at 40-50 million each and 1 large industrial fire at 75 million.

Consequences of Major U.S. Earthquakes (1900-1972)[a]

Date	Place	Fatalities	Damage (millions)
1906	San Francisco, California	~750	400
1925	Santa Barbara, California	13	6.5
1933	Long Beach, California	102	45
1935	Helena, Montana	4	3.5
1940	Imperial Valley, California	9	5.5
1949	Olympia, Washington	8	20
1952	Kern County, California	11	48
1954	Eureka, California	1	1
1957	San Francisco, California	0	1
1959	Hebgen Lake, Montana	28	4
1964	Anchorage, Alaska	125	310
1965	Puget Sound, Washington	6	12
1969	Santa Rosa, California	0	7
1971	San Fernando, California	58	480

Source: Reactor Safety Study: An Assessment of Accident Risks in U.S. Commercial Nuclear Power Plants, WASH-1400, (NUREG-75/014), United States Nuclear Regulatory Commission, October 1975.

Fatalities in Major Airplane Crashes Throughout the World (1960-1973)

Number of Fatalities	Number of Crashes
50 - 100	40
100 - 150	21
150 - 200	6

Early Fatalities in Major Explosions Throughout the World (1925-1971)

Fatalities	Number of Explosions	Probability of Event Per Year in U.S. With Fatalities > N	
9 - 50	30	N=8	0.47 year^{-1}
50 - 100	4	N=50	0.13 year^{-1}
100 - 200	4	N=100	0.087 year^{-1}
200 - 1000	3	N=200	0.043 year^{-1}
Largest (1100)	1	N=1000	0.01 year^{-1}

Source: Reactor Safety Study: An Assessment of Accident Risks in U.S. Commercial Nuclear Power Plants, WASH-1400 (NUREG-75/014), United States Nuclear Regulatory Commission, October 1975.

2.	Factors Tending to Decrease Average Lifetime	Decrease of Average Lifetime
	Overweight by 25%	3.6 years
	Male rather than Female	3.0 years
	Smoking 1 pack per day	7.0 years
	2 packs per day	10.0 years
	City rather than country living	5.0 years
	Actual radiation from nuclear power plants in 1970	less than 1 minute
	Estimate for the year 2000 assuming hundredfold increase in nuclear power production	less than 30 minutes

Source: Advisory Committee on the Biological Effects of Radiation, "The Effects on Populations of Exposure to Low Levels of Ionizing Radiation," Division of Medical Science, National Academy of Sciences, National Research Council (Nov. 1972).

COMPARATIVE RISKS OF NUCLEAR POWER PLANTS

Most Likely Consequences of a Core Melt Accident

	Consequences
Fatalities	<1
Injuries	<1
Latent Fatalities per year	<1
Thyroid Nodules per year	<1
Genetic Defects per year	<1
Property Damage [a]	<$1,000,000

(a) This does not include damage that might occur to the plant or costs for replacing the power generation lost by such damage.

Annual Fatalities and Injuries Expected Among the 15 Million People Living Within 25 Miles of U.S. Reactor Sites

Accident Type	Fatalities	Injuries
Automobile	4,200	375,000
Falls	1,500	75,000
Fire	560	22,000
Electrocution	90	--
Lightning	8	--
Reactors (100 plants)	2	20

Source: Reactor Safety Study: An Assessment of Accident Risks in U.S. Commercial Nuclear Power Plants, WASH-1400 (NUREG-75/014), United States Nuclear Regulatory Commission, October 1975.

Consequences of Reactor Accidents for Various Probabilities for 100 Reactors

Chance Per Year	Consequences				
	Early Fatalities	Early Illness	Total Property Damage 10^9	Decontamination Area Square Miles	Relocation Area Square Miles
One in 200 [a]	<1.0	<1.0	<0.1	<0.1	<0.1
One in 10,000	<1.0	300	0.9	2000	130
One in 100,000	110	300	3	3200	250
One in 1,000,000	900	14000	8	(b)	290
One in 10,000,000	3300	45000	14	(b)	(b)

(a) This is the predicted chance per year of core melt considering 100 reactors.
(b) No change from previously listed values.

Approximate Values of Early Illness and Latent Effects for 100 Reactors

Chance Per Year	Consequences			
	Early Illness	Latent Cancer Fatalities [b] (per yr)	Thyroid Illness [b] (per yr)	Genetic Effects [c] (per yr)
1 in 200 [a]	<1.0	<1.0	4	<1.0
1 in 10,000	300	170	1400	25
1 in 100,000	3000	460	3500	60
1 in 1,000,000	14,000	860	6000	110
1 in 10,000,000	45,000	1500	8000	170
Normal Incidence Per Year	4×10^5	17,000	8000	8000

(a) This is the predicted chance per year of core melt for 100 reactors.

(b) This rate would occur approximately in the 10 to 40 year period after a potential accident.

(c) This rate would apply to the first generation born after the accident. Subsequent generations would experience effects at decreasing rates.

Source: Reactor Safety Study: An Assessment of Accident Risks in U.S. Commercial Nuclear Power Plants, WASH-1400 (NUREG-75/014), United States Nuclear Regulatory Commission, October 1975.

Average Probability of Major Man-Caused and Natural Events

Type of Event	Probability of 100 or More Fatalities	Probability of 1000 or More Fatalities
Man-Caused		
Airplane Crash	1 in 2 years	1 in 2000 years
Fire	1 in 7 years	1 in 200 years
Explosion	1 in 16 years	1 in 120 years
Toxic Gas	1 in 100 years	1 in 1000 years
Natural		
Tornado	1 in 5 years	very small
Hurricane	1 in 5 years	1 in 25 years
Earthquake	1 in 20 years	1 in 50 years
Meteorite Impact	1 in 100,000 years	1 in 1,000,000 years
Reactors		
100 plants	1 in 100,000 years	1 in 1,000,000 years

Approximate Average Societal and Individual Risk Probabilities per Year from Potential Nuclear Plant Accidents[a]

Consequence	Societal	Individual
Early Fatalities [b]	3×10^{-3}	2×10^{-10}
Early Illness [b]	2×10^{-1}	1×10^{-8}
Latent Cancer Fatalities [c]	7×10^{-2}/yr	3×10^{-10}/yr
Thyroid Nodules [c]	7×10^{-1}/yr	3×10^{-9}/yr
Genetic Effects [d]	1×10^{-2}/yr	7×10^{-11}/yr
Property Damage ($)	2×10^{6}	——

(a) Based on 100 reactors at 68 current sites.

(b) The individual risk value is based on the 15 million people living in the general vicinity of the first 100 nuclear power plants.

(c) This value is the rate of occurrence per year for about a 30-year period following a potential accident. The individual rate is based on the total U.S. population.

(d) This value is the rate of occurrence per year for the first generation born after a potential accident; subsequent generations would experience effects at a lower rate. The individual rate is based on the total U.S. population.

Source: Reactor Safety Study: An Assessment of Accident Risks in U.S. Commercial Nuclear Power Plants, WASH-1400 (NUREG-75/014), United States Nuclear Regulatory Commission, October 1975.

NUCLEAR POWER PLANT SAFETY

Differences in the public perception of risks and benefits are nowhere more apparent than in the assessment of risks and the evaluation of benefits for nuclear power plants. The safety record for nuclear power has been good—there is no recorded case of injury to a member of the general public from commercial nuclear power activities. The risk of a major reactor accident has been calculated to be comparable to the risk of being struck by a meteor. Yet there remains a public perception of unusual risk with regard to nuclear power. Such risk perceptions are normally based on a history of experience with accidents and personal judgments of the likelihood of being affected by a particular type of accident. The lack of accident experience with nuclear power plants has made personal judgments difficult. As a result, the debate continues on the safety of nuclear power.

The safety of nuclear power plants, from the standpoint of controlling the exposure of the public to radiation, may be considered under two sets of circumstances—normal operating conditions and reactor accident conditions. The control of radioactive effluents from commercial nuclear power plants has proved to be quite effective with discharges generally a factor of 100 below permissible levels. The control of radioactivity under reactor accident conditions has not been directly demonstrated in operating commercial reactors since such accident conditions have not developed (while several accidents have occurred in experimental reactors but without injury to the public, the experience gained cannot be translated directly to the larger commercial reactors).

The most serious accident postulated for nuclear plants is one in which there is rupture in the primary coolant system and a subsequent loss of coolant accident (LOCA). If shutdown systems and the emergency core cooling system (ECCS) fail to function, the decay

Model of the floating nuclear plant and breakwater which Public Service Electric and Gas Company of New Jersey hopes to put in the Atlantic Ocean southwest of Little Egg Inlet, within the three-mile limit. It will have two units, each with a capacity of 1150 megawatts (*Courtesy Public Service Electric and Gas Company*)

Multilayered Protection from Fission Products in Nuclear Power Plants

Barrier or Layer	Effectiveness
1. Ceramic fuel pellets	Only a fraction of the gaseous and volatile fission products are released from the pellets.
2. Metal fuel tubes (cladding)	These contain the fission products released from the pellets. During the life of the fuel, less than 0.5% of the tubes may develop pinhole sized leaks through which some fission products escape.
3. Reactor vessel and piping	The 8- to 10-inch-thick steel vessel and 3- to 4-inch-thick steel piping contain the reactor cooling water. A portion of the circulating water is continuously passed through a filtering trap to keep the radioactivity low.
4. Concrete shield	Operators and equipment are protected from high levels of core radiation by concrete 7 to 10 feet thick.
5. Containment dome	The entire reactor part of the plant is enclosed to prevent release of radioactivity in case of reactor cooling water pipe leakage or rupture.
6. Exclusion area	A designated area around each plant separates the plant from the public. Entrance is restricted.
7. Plant separation distance	Plants are located at a distance from population centers.

Source: Nuclear Power and the Environment: Questions and Answers; American Nuclear Society, Hinsdale, Illinois, 1973.

heat of the radioactive fission products in the fuel can cause the fuel temperature to rise to the point where the fuel is vaporized and the fission products are dispersed outside the primary reactor system. In reactors cooled by water or sodium, this dispersal process can be assisted by the vaporization of the water or the sodium.

Protection against the release of radioactive fission products to the environment outside the reactor building is provided by a number of engineered safety features (ESF). Neutron-absorbing safety rods are inserted into the reactor core to provide rapid shutdown of the reactor to limit core heat production. An emergency core cooling system provides cooling for the core to offset the loss of primary coolant. A cooling system is provided within the containment structure to condense steam vapors and reduce pressures. Sprays and filters are provided to collect the radioactivity from the atmosphere within the containment structure, and finally, the containment structure is designed to be airtight to prevent leakage to the outer environment.

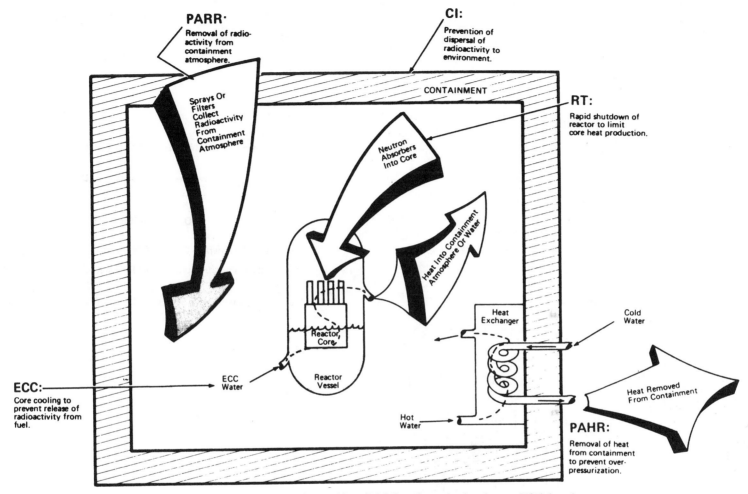

Power water reactor loss of coolant accident (LOCA) engineered safety feature (ESF) functions.

Source: Reactor Safety Study: An Assessment of Accident Risks in U.S. Commercial Nuclear Power Plants, WASH-1400 (NUREG-75/014), United States Nuclear Regulatory Commission, October 1975.

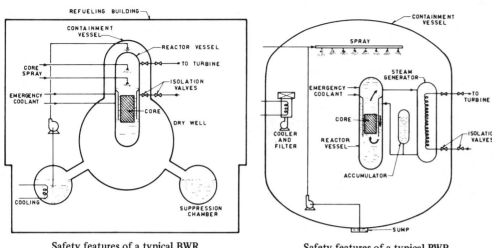

Safety features of a typical BWR.

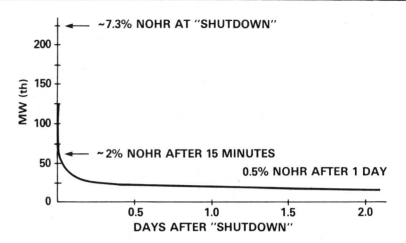

★ NOHR IS NORMAL OPERATING HEAT-RATE AT DESIGN POWER

Decay-heat generation rate vs time for LWR 1000 MWe (3077 MW thermal).

Source: Nuclear Reactor Safety, Hearings before the Joint Committee on Atomic Energy, Congress of the United States, Part 1, January 3, September 25, 26, 27, and October 1, 1973.

Safety features of a typical PWR.

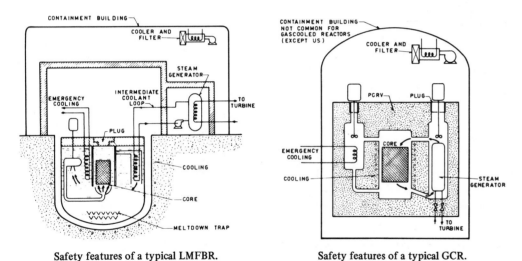

Safety features of a typical LMFBR.

Safety features of a typical GCR.

Source: Nuclear Power and the Environment, International Atomic Energy Agency, Vienna, Austria, 1972.

Dose-Limiting Recommendations

	NCRP Report #39	FRC Report #1	ICRP Publ. #9 AEC Part 20
Maximum permissible dose equivalent for occupational exposure:			
Combined whole-body occupational exposure			
Prospective annual limit	5 rems in any one year	—	5 [a]
Retrospective annual limit	10–15 rems in any one year	—	12
Long-term accumulation to age N years	$(N-18) \times 5$ rems	$(N-18) \times 5$	$(N-18) \times 5$
Bone		0.1 μg Ra^{226} (or equiv.)	30 rems
Skin	15 rems in any one year	30	30
Hands	75 rems in any one year (25/qtr)	75 (25)	75
Forearms	30 rems in any one year (10/qtr)	75	75
Other organs, tissues and organ systems	15 rems in any one year (5/qtr)	15 (5)	15
Thyroid		30 (10)	30
Fertile women (with respect to fetus)	0.5 rem in gestation period	—	1.3 rems per quarter
Dose limits for the public, or occasionally exposed individuals:			
Individual—whole-body	0.5 rem in any one year	0.5	0.5 [a]
Skin, bone, thyroid		—	3 rems in any year [b]
Hands, forearms, feet, ankles		—	7.5 rems in any year
Other single organs		—	1.5 rems in any year
Students	0.1 rem in any one year	—	—
Population dose limits:			
Genetic	0.17 rem average per year	⎱ 0.17 average of	0.17
Somatic	0.17 rem average per year	⎰ suitable sample	—

[a] Gonads and red blood marrow.
[b] 1.5 rems to thyroid of children up to 16 years of age.

Reproduced with permission of the National Academy of Sciences. A Report of the National Academy of Engineering. Published by the National Academy of Sciences.

U.S. AIR QUALITY STANDARDS

Air quality standards in the United States are set by the Environmental Protection Agency. The States, which have responsibility for implementing the standards, may themselves set standards that are more stringent. Each of the standards is subject to continuing debate. The analysis by Kurt Yeager recounts the development of standards for sulfur dioxide.

The EPA sets two standards for air quality. The *primary* ambient air quality standard represents a limit on the concentration of a pollutant in the atmosphere which must be maintained to protect the health of members of the public. The *secondary* standard represents a level which will protect the welfare of the public and includes effects on soils, water, crops, vegetation, man-made materials, animals, climate, damage to and deterioration of property, hazards to transportation, economic values, and personal comfort and well-being. The EPA may use its judgment in determining the level of air quality required to protect all aspects of the public welfare.

The EPA has the responsibility not only of setting the primary and secondary air quality standards but also of defining methods by which these standards shall be met. The EPA has adopted two methods—"emission limitations" and "the best available control technology"—to meet these standards. Both methods are subject to continuing debate since both tend to require that emissions from a particular plant be well below the primary and secondary standards.

The Significant Deterioration Issue. In 1972, the Sierra Club and other environmental groups filed suit against the Environmental Protection Agency for failure to promulgate regulations under the Clean Air Act to "protect and enhance" the environment through prevention of significant deterioration in air quality. As a result of the District Court decision, which was upheld by the U.S. Court of Appeals and the Supreme Court, EPA adopted regulations in December 1974, which are based on allowable increments of pollutant concentrations for specific categories of new major industrial sources under an area classification procedure. These area classifications include:

Class I—Applies to areas in which practically any air quality deterioration would be considered significant, thus allowing little or no major energy or industrial development.

Class II—Applies to areas in which deterioration that would normally accompany moderate, well-controlled growth would not be considered significant.

Class III—Applies to areas in which deterioration would be permitted to allow concentrated or very large-scale energy or industrial development, as long as the national secondary ambient air quality standards are not exceeded.

The EPA regulations initially designated all regions as Class II, subject to redesignation as Class I or Class III by initiative at the State and local level. As of December 1975, no formal applications had been made to reclassify regions to the Class I or Class III designations.

National Ambient Air Quality Standards[a]

Pollutant	Averaging time	Primary standards	Secondary standards
Particulate matter	Annual (geometric mean)	75 $\mu g/m^3$	60 $\mu g/m^3$
	24 hour[b]	260 $\mu g/m^3$	150 $\mu g/m^3$
Sulfur oxides	Annual (arithmetic mean)	80 $\mu g/m^3$ (0.03 ppm)	—
	24 hour[b]	365 $\mu g/m^3$ (0.14 ppm)	—
	3 hour[b]	—	1300 $\mu g/m^3$ (0.5 ppm)
Carbon monoxide	8 hour[b]	10 mg/m³ (9 ppm)	10 mg/m³ (9 ppm)
	1 hour[b]	40 mg/m³ (35 ppm)	40 mg/m³ (35 ppm)
Nitrogen dioxide	Annual (arithmetic mean)	100 $\mu g/m^3$ (0.05 ppm)	100 $\mu g/m^3$ (0.05 ppm)
Photochemical oxidants	1 hour[b]	160 $\mu g/m^3$ (0.08 ppm)	160 $\mu g/m^3$ (0.08 ppm)
Hydrocarbons (nonmethane)	3 hour (6 to 9 a.m.)	160 $\mu g/m^3$ (0.24 ppm)	160 $\mu g/m^3$ (0.24 ppm)

[a] The air quality standards and a description of the reference methods were published on April 30, 1971, in 42 C.F.R. 410, recodified to 40 C.F.R. 50 on July 1, 1975.

[b] Not to be exceeded more than once per year.

Source: Environmental Quality–1976, The Seventh Annual Report of the Council on Environmental Quality, September 1976.

New Source Standards of Performance for Fossil Fuel-Fired Steam Generators

Pollutant	Standard
Particulate matter	0.10 lb per million Btu heat input, maximum two hour average
	20% opacity (except that 40% opacity is permissible for not more than two minutes in any hour)
Sulfur dioxide	0.80 lb per million Btu heat input, maximum two hour average when liquid fossil fuel is burned
	1.2 lbs per million Btu heat input, maximum two hour average when solid fuel is burned
Nitrogen oxides	0.20 lb per million Btu heat input, maximum two hour average, expressed as NO_2, when gaseous fossil fuel is burned
	0.30 lb per million Btu heat input, maximum two hour average, expressed as NO_2, when liquid fossil fuel is burned
	0.70 lb per million Btu heat input, maximum two hour average, expressed as NO_2, when solid fossil fuel (except lignite) is burned

Source: EPA Regulations 40 CFR 60.42 to 40 CFR 60.44.
Source: Air Quality and Stationary Source Emission Control, A Report by the Commission on Natural Resources, prepared for the Committee on Public Works, United States Senate, March 1975.

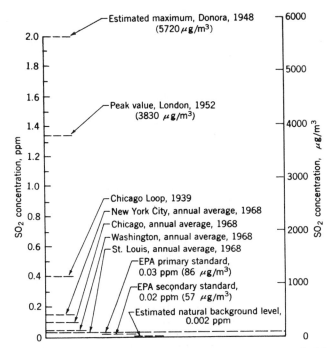

Comparison of SO$_2$ levels for disasters, averages in cities, and standards gives an insight into severity of EPA limits.

Reprinted with permission from *POWER*, September, 1974.

Automobile Emission Standards, Federal Standards for 1977 and Barth Study Recommendations

Emission	Federal Standard[a]	Barth Recommendation[b]
CO	3.4 g/mi	6.16 g/mi
HC	.41 g/mi	.14 g/mi
NO$_x$	.4 g/mi	.40 g/mi

[a]38 Federal Register 10317 (1973).
[b]D. S. Barth, et al., "Federal Motor Vehicle Emission Goals for CO, HC, and NO$_x$ Based on Desired Air Quality Levels," in The Impact of Auto Emission Standards, Report of the Staff of the Subcommittee on Air and Water Pollution to the Committee on Public Works, United States Senate, Washington, October 1973, p. 21.

Ambient Air Quality Standards, Federal Standards and Barth Study, "Desired Air Quality"

Pollutant	Federal Primary Standard[a]	"Desired Air Quality" in Barth Study
CO	9 ppm–8 hr max.	9 ppm–8 hr max.
HC	.24 ppm–3 hr max.	—
NO$_x$	.05 ppm–ann. avg.	.10 ppm–1 hr max.
Oxidants	.08 ppm–1 hr max.	.06 ppm–1 hr max.
Particulates	260 μg/m^3–24 hr max.	—
SO$_x$	.14 ppm–24 hr max.	—

[a]36 Federal Register 8187 (1971). ppm–parts per million; μg/m^3–micrograms per cubic meter.
Source: The Automobile and the Regulation of Its Impact on the Environment, NSF/RA/X-74-023, Columbia University, 1975.

Typical SO$_2$ Regulations for Existing Boilers in States Using Coal[a]

State	Regulation (or Equivalent)
Ohio	1.0 lb SO$_2$/10^6 Btu fired[b]
Tennessee	1.5 lb SO$_2$/10^6 Btu fired[b,c,f]
Kentucky	1.2 lb SO$_2$/10^6 Btu fired[d,e]
Alabama	1.2 lb SO$_2$/10^6 Btu fired[c,e]
Michigan	1.0 lb SO$_2$/10^6 Btu fired[e,f]
Pennsylvania	0.6 to 1.0 lb SO$_2$/10^6 Btu fired depending on size[e]
West Virginia	2.7 lb SO$_2$/10^6 Btu fired[e]
Illinois	1.8 lb SO$_2$/10^6 Btu fired[e]

[a]As of June 1973.
[b]Effective 1 July 1975.
[c]Under revision.
[d]Effective 1 July 1977.
[e]Less stringent in some areas of state.
[f]Regulations in different terms.
Source: Introductory Report by the Secretariat, ENV/SEM.4/R.1, Economic Commission for Europe, Second Seminar on Desulphurization of Fuels and Combustion Gases, Washington, D.C., Nov. 11–20, 1975.

The Basis for Sulfur Dioxide Standards. (quoted from Kurt E. Yeager, "Stacks vs. Scrubbers," Research Progress Report FF-3, Fossil Fuel and Advanced Systems Division, Electric Power Research Institute, July 1975.)

Should control standards be based on ambient concentrations of SO_2? Or of sulfates? Both? Neither? As background for evaluating the status and relative merits of stacks and scrubbers, it is useful to consider the existing criteria supporting sulfur oxide (SO_x) air quality standards.

The standards were developed during the late 1960s and early 1970s after scores of studies to correlate mortality and sickness rates with ambient air SO_2 and particulate concentrations. These studies indicated that the most susceptible groups —asthmatics, the aged, etc.,—experience no adverse effects when the average annual ambient concentration at ground level is below 100 $\mu g/m^3$. EPA therefore set the annual standard at 80 μg (allowing a margin for safety) and, on the basis of a statistical model, established 365 μg as the 24-hour standard. Healthy people remain unaffected by much higher concentrations of SO_2 alone.

Ironically, however, as EPA has stepped up its enforcement, further research has led scientists to regard the phenomenon of SO_x pollution as much more complicated than they once believed. This, in turn, is shaking their previous convictions about the direct relationship between SO_2 and health. In short, a regulatory process mandated by Congress has been built on an uncertain scientific foundation.

The problem with the early studies seems to be that they did not cope with the fact that when SO_2 levels in the atmosphere are low, the levels of other variables—such as temperature, humidity, or particulates—also tend to be low. When SO_2 levels are high, the others often are high, too. As a result, the studies did not identify the independent effect of SO_2 alone.

The Role of Sulfates

EPA studies now indicate that SO_2 in ambient concentrations should be viewed more as an indicator of harmful secondary chemical combinations rather than as a toxic agent in its own right. From preliminary epidemiological studies, laboratory experiments, and acid rainfall measurements, EPA in the past two years has concluded that sulfuric acid mist and certain sulfates are stronger irritants than SO_2. Still, there is less understanding of the health effects of sulfate concentrations and other complex pollutant interactions than EPA believes desirable to initiate a new air quality standard-setting process.

EPA's epidemiological studies, primarily the Community Health and Environmental Surveillance Studies (known as CHESS), have correlated asthma attacks and cardio-pulmonary dysfunctions with ambient levels of both SO_2 and suspended sulfates. There is a more significant statistical association for sulfates (measured concentrations from 8 to 15 $\mu g/m^3$) than for SO_2. *But these results are not conclusive.* There are deficiencies in the measurement method; particle size distribution and chemical characterization were not available; and the studies may have been confounded by the presence of other pollutants.

Data on comparative toxicity indicate that SO_2 alone is only a mild respiratory irritant, whereas specific sulfate and sulfuric acid aerosols are more potent. Particle size and chemical composition, as well as mass concentration, are of paramount importance. The importance of humidity has been demonstrated. Different species of sulfates, and even various particle sizes in the same sulfate, engender different responses when inhaled by animals. Since little is known about the nature of sulfates in the ambient atmosphere, it is difficult to translate these findings to human effects from ambient exposures. Furthermore, the toxicity of sulfates formed in urban atmospheres, which are a complex of other pollutants and meteorological variables, may be quite different from the toxicity of sulfates generated in a laboratory chamber.

SO₂/Sulfate Correlation Undefined

The primary mechanism for the formation of sulfuric acid and sulfate in the Midwest and Northeast, where high ambient sulfate concentrations exist, appears to the the catalytic conversion of atmospheric SO_2 and water in the presence of metallic catalysts such as vanadium, iron, etc., in fine particulate form. This conversion does not occur, however, below a pH value of 2. The buffering agent is probably ammonia. If one considers the measured ground-level concentration of ammonia in the Midwest and Northeast (5–10 $\mu g/m^3$), it becomes clear that a measured sulfate concentration exceeding the tentative EPA threshold of about 8 $\mu g/m^3$ is possible at very low SO_2 levels. Hence, a reduction in ambient SO_2 concentration may not lead to a concurrent reduction in ambient sulfate concentration.

National and regional trends show a steady decline in SO_2 levels from 1962 to 1970. The distribution of sulfate has not matched this trend; indeed, sulfate concentrations in certain nonurban areas appear to have increased. Unfortunately, 1970–1973 sulfate data will not be available until the fall of 1975. There are questions whether the trend data reflect actual national atmospheric conditions, and what may be the role of pollutant transport mechanisms.

The EPA Science Advisory Board Study Panel on Sulfates, which recently reviewed the existing EPA sulfate data base, has further amplified these limitations in its March 7, 1975, report entitled *Scientific and Technical Issues Relating to Sulfates.* It concluded that the major toxic components of "reducing" type air pollutants are in the SO_x/particulate complex (SPC). The panel also concluded that the water-soluble sulfate content of total suspended particulates (TSP) is an unreliable index of their concentration. Rational and efficient control will depend on verification of the active components in SPCs and development of a practical analytical method for their detection.

These basic conclusions were apparent at least 10 years ago, the EPA Science Advisory Board panel noted, but the preoccupation at that time with SO_2—legitimate in that it is the essential precursor, but misleading in the assumption that it is the prime toxic agent—delayed rational attack on the SPC problem. A decision now to develop an ambient air standard based on the water-soluble sulfate content of TSP—using present analytical techniques—could again divert attention from the central issue and lead to additional loss of precious time.

There are indications that copollutants (particulates, catalysts, and alkaline materials) are critical and of major significance. It is therefore inappropriate to prepare a control strategy based exclusively on SO_2 reduction when control of copollutants may be required for reducing atmospheric sulfates.

Conclusions

As a result of these considerations, the EPA Science Advisory Board sulfate panel concluded that "the siting of new power plants in *nonurban* areas should not have an important influence on either SO_2 or sulfate concentrations in *urban* areas" (emphasis added). The panel also suggested a research strategy that would focus on these central issues: (1) more accurate identification of significant toxic agents, (2) development of analytical methods for them, and (3) understanding of the kinetics of their formation and transport. These goals will probably require a technical program stretching over 5 to 10 years.

In conclusion, it is uncertain what conditions govern the rate at which sulfates and acid aerosols are formed in the atmosphere, how they behave, and how they can be controlled—or whether they, too, are only indicators of a more complex interaction among air pollutant species.

Major Air Pollutants[1]

Pollutant	Characteristics	Principal sources	Principal effects	Controls	National ambient standards [2,3] (in micrograms per cubic meter)		
Total suspended particulates (TSP)	Any solid or liquid particles dispersed in the atmosphere, such as dust, pollen, ash, soot, metals, and various chemicals; the particles are often classified according to size as settleable particles: larger than 50 microns; aerosols: smaller than 50 microns; and fine particulates: smaller than 3 microns	Natural events such as forest fires, wind erosion, volcanic eruptions; stationary combustion, especially of solid fuels; construction activities; industrial processes; atmospheric chemical reactions	Health: Directly toxic effects or aggravation of the effects of gaseous pollutants; aggravation of asthma or other respiratory or cardiorespiratory symptoms; increased cough and chest discomfort; increased mortality. Other: Soiling and deterioration of building materials and other surfaces, impairment of visibility, cloud formation, interference with plant photosynthesis	Cleaning of flue gases with inertial separators, fabric filters, scrubbers, or electrostatic precipitators; alternative means for solid waste reduction; improved control procedures for construction and industrial processes	Primary: Annual=75, 24-hour=260; Secondary: Annual=60, 24-hour=150; Alert: 24-hour=375		
Sulfur dioxide (SO₂)	A colorless gas with a pungent odor; SO₂ can oxidize to form sulfur trioxide (SO₃), which forms sulfuric acid with water	Combustion of sulfur-containing fossil fuels, smelting of sulfur-bearing metal ores, industrial processes, natural events such as volcanic eruptions	Health: Aggravation of respiratory diseases, including asthma, chronic bronchitis, and emphysema; reduced lung function; irritation of eyes and respiratory tract; increased mortality. Other: Corrosion of metals; deterioration of electrical contacts, paper, textiles, leather, finishes and coatings, and building stone; formation of acid rain; leaf injury and reduced growth in plants	Use of low-sulfur fuels; removal of sulfur from fuels before use; scrubbing of flue gases with lime or catalytic conversion	Primary: Annual=80, 24-hour=365; Alert: 24-hour=800		
Carbon monoxide (CO)	A colorless, odorless gas with a strong chemical affinity for hemoglobin in blood	Incomplete combustion of fuels and other carbon-containing substances, such as in motor vehicle exhausts; natural events such as forest fires or decomposition of organic matter	Health: Reduced tolerance for exercise, impairment of mental function, impairment of fetal development, aggravation of cardiovascular diseases. Other: Unknown	Automobile engine modifications (proper tuning, exhaust gas recirculation, redesign of combustion chamber); control of automobile exhaust gases (catalytic or thermal devices); improved design, operation, and maintenance of stationary furnaces (use of finely dispersed fuels, proper mixing with air, high combustion temperature)	Primary: 8-hour=10,000, 1-hour=40,000; Alert: 8-hour=17,000		

See footnotes at end of table.

Major Air Pollutants[1] (*Continued*)

Pollutant	Characteristics	Principal sources	Principal effects	Controls	National ambient standards [2,3] (in micrograms per cubic meter)	
Photochemical oxidants (O_x)	Colorless, gaseous compounds which can comprise photochemical smog, e.g., ozone (O_3), peroxyacetyl nitrate (PAN), aldehydes, and other compounds	Atmospheric reactions of chemical precursors under the influence of sunlight	Health: Aggravation of respiratory and cardiovascular illnesses, irritation of eyes and respiratory tract, impairment of cardiopulmonary function Other: Deterioration of rubber, textiles, and paints; impairment of visibility; leaf injury, reduced growth, and premature fruit and leaf drop in plants	Reduced emissions of nitrogen oxides, hydrocarbons, possibly sulfur oxides	Primary: Alert:	1-hour=160 1-hour=200
Nitrogen dioxide (NO_2)	A brownish-red gas with a pungent odor, often formed from oxidation of nitric oxide (NO)	Motor vehicle exhausts, high-temperature stationary combustion, atmospheric reactions	Health: Aggravation of respiratory and cardiovascular illnesses and chronic nephritis Other: Fading of paints and dyes, impairment of visibility, reduced growth and premature leaf drop in plants	Catalytic control of automobile exhaust gases, modification of automobile engines to reduce combustion temperature, scrubbing flue gases with caustic substances or urea	Primary: Alert:	Annual=100 24-hour=282 1-hour=1,130
Hydrocarbons (HC)	Organic compounds in gaseous or particulate form, e.g., methane, ethylene, and acetylene	Incomplete combustion of fuels and other carbon-containing substances, such as in motor vehicle exhausts; processing, distribution, and use of petroleum compounds such as gasoline and organic solvents; natural events such as forest fires and plant metabolism; atmospheric reactions	Health: Suspected contribution to cancer Other: Major precursors in the formation of photochemical oxidants through atmospheric reactions	Automobile engine modifications (proper tuning, crankcase ventilation, exhaust gas recirculation, redesign of combustion chamber); control of automobile exhaust gases (catalytic or thermal devices); improved design, operation, and maintenance of stationary furnaces (use of finely dispersed fuels, proper mixing with air, high combustion temperature); improved control procedures in processing and handling petroleum compounds	Primary:	3-hour=160

[1] Pollutants for which national ambient air quality standards have been established.

[2] Primary standards are intended to protect against adverse effects on human health. Secondary standards are intended to protect against adverse effects on materials, vegetation, and other environmental values.

[3] The federal episode criteria specify that meteorological conditions are such that pollutant concentrations may be expected to remain at these levels for 12 or more hours or to increase; in the case of oxidants, the situation is likely to reoccur within the next 24 hours unless control actions are taken.

Source: Based on information compiled by Enviro Control, Inc.

Source: The Sixth Annual Report of the Council on Environmental Quality, December 1975.

SULFUR DIOXIDE CONTROL ALTERNATIVES

The reduction in the emission of sulfur dioxide in the gases produced from the combustion of fuels can be achieved by a number of means: fuels with inherently low sulfur content can be used, the sulfur content of coal can be reduced to some extent by physical and chemical methods, the sulfur can be removed from coal by gasification or liquefaction processes, the sulfur can be removed by adding limestone or magnesia to the coal as it is being burned in fluidized-bed combustion systems, and the sulfur can be removed from the combustion gases in lime/limestone stack gas scrubbing systems. The Environmental Protection Agency is currently requiring the installation of stack gas cleaning systems as the best available technology for reducing the sulfur omissions from the burning of high sulfur coals.

In addition to the processes above which afford an *overall* reduction in the emission of sulfur dioxide into the atmosphere, another control strategy has been adopted in the United Kingdom to keep the concentration of atmospheric sulfur dioxide below the primary ambient standard. This intermittent control system (ICS) depends upon varying the SO_2 emission rates on the basis of existing or predicted meteorological conditions. During periods of air stagnation when stack gases are not dispersed and diluted by air movements, power plants may be fired with fuels of lower sulfur content, power levels may be reduced, or plants in particular locations may be shut down entirely.

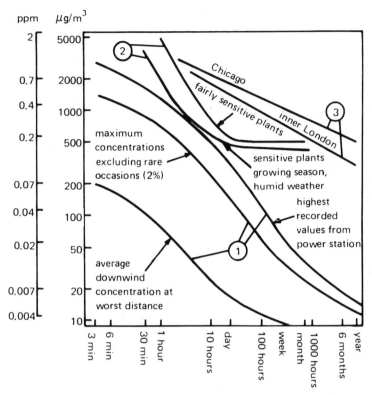

Measurements of atmospheric sulfur dioxide in the United Kingdom. Figure illustrates the dependence on sampling time of (1) maximum concentrations from a large power station, (2) concentrations which affect vegetations, and (3) maximum concentrations in cities.

Source: A. J. Clarke, D. H. Lucas, and F. F. Ross, "Tall Stacks—How Effective are They?" Central Electricity Generating Board, *Second International Clean Air Conference*, Washington, December 1970.

Summary of Control Alternatives for SO$_2$

Control Technique	Sulfur Removal Efficiency (3% S base), %	Energy Penalty, %	Incremental Cost, mills/kWh	Timing	Annual Capacity 10^6 tons coal	Likely Applicability	Other Environmental Effects	Comments
Nonregenerable (lime/limestone) scrubbing (new plants)	90	3–7	3.0	1975–77	230 (1980)	All new utilities and large industrial boilers	Waste disposal problems	Operating load factor 80%, 1000-MW power plant
Nonregenerable scrubbing (existing plants)	90	3–7	4.4–6.1	1975–77	60 (1980)	Boilers ⩾ 100 MW ⩽ 20 yr age	Waste disposal problems	Operating load factor 60%, 300-MW power plant
Regenerable scrubbing (new plants)	80–95	3–7	3.4–7.0	1980	—	New utilities and large industrial boilers	Reduced waste disposal problems	Limited data on utilization and costs
Physical coal cleaning	30–50	3–10	0.6–1.6	1974–80	—	All coal combustors	Land, air, waste disposal at cleaning site	Cleaning costs $1.50/ton
Coal liquefaction (solvent refined cleaning)	75–85	20	10	1985	110	Existing and mine mouth power plants, industrial boilers, small point sources	Sulfur and other emissions partially concentrated at central plants; increased mining necessary; unidentified air, water, and solid waste problems	Economics of scale possible
Low-Btu gas	80–98	10–40	6–1.6	1980–85 / 1980	13 / 200	New and large existing power plants		Energy penalty reduced by combined cycle application. Capital intensive
Pipeline quality	98+	40	NA	1980 / 1985	29 / 50–100	Small and area sources		Minimal energy penalty if used to replace electric heat
Fluidized bed	90+	Minimal	0–3	1985	150	New power plants and industrial boilers	Substantial reductions in NO$_x$ emissions possible	Energy efficiency can be greater than conventional power plants
Refuse combustion	20	Minimal	Minimal	1985–90	50	Coal mix, 10–20% refuse	Benefits due to solid waste reduction	
Low sulfur western coals	75–85	Minimal	0–5.3	1980	250	Western and fringe eastern boilers	Increased particulate control necessary	3% energy penalty for transportation to eastern areas
SIP compliance coal (Eastern)	35–70	Minimal	1.0–3.0	1975	95	Eastern existing boilers		Cost for 1% S or less compared to 3% S coal
Desulfurized residual oil 0.3%	85	—	2.0	1980–85	NA	Existing oil burners	Also removes V, other trace elements	

Source: Position Paper on Regulation of Atmospheric Sulfates, EPA-450/2-75-007, U.S. Environmental Protection Agency, September 1975.

SULFUR DIOXIDE CONTROL PROCESSES

Summary Description of Major Flue Gas Desulfurization Processes.

Process	Classification/ operating principles	SO$_2$ particulate efficiency	Development status	Application	Implementation[a]	Advantages	Disadvantages
Lime/limestone scrubbing	Nonregenerable process/ wet absorption in scrubber by slurry; insoluble sulfites and sulfates disposed of as waste.	Up to 90 percent SO$_2$ removal/99 percent fly ash removal by most scrubbers.	32 full-scale units (9,829 MW) in operation or planned for start-up by 1977.	Old or new power plant; coal- or oil-fired.	An additional 21 units (10,349 MW) planned for installation by 1980; 15 units (8,099 MW) considered for unspecified date; 4-5 years lead time needed for new plants; 3 years for retrofit of old plants.	Cheapest of existing processes; elimination of particulate control requirement.	Waste and water pollution problems; reheat of scrubber exit gases needed; supply and handling of large volumes of reactant may be problems.
Double alkali process	Nonregenerable process/ wet absorption in scrubber; reactants and soluble; reaction products precipitated and removed from recycled reactant solution outside of scrubber; most common reactant sodium sulfite.	High efficiency >90 percent SO$_2$ removal/high particulate removal as above.	Active area but no fullscale demonstration as yet; 2 units operating; G.M. installed a unit on a coal-fired boiler in February 1974; several sulfate removal schemes under study.	As above with potentially lower cost and greater ease of operation favoring some inroads into smaller plants.	Research-Cottrell estimates $600 million a year market by 1979; a second generation lime/limestone system; lead times as above for power plants.	Potentially cheaper and more reliable than lime/limestone system.	Similar to above and all throwaway systems.
Magnesium oxide scrubbing	Regenerable process/wet absorption by magnesium oxide slurry; fly ash removed prior to or after scrubbing; magnesium oxide regenerated by calcining with carbon; SO$_2$ by-product can be converted to sulfuric acid or sulfur.	90 percent SO$_2$ removal/particulates removal as required by prescrubber.	One full-scale unit on test at Boston Edison 150-MW oil-fired unit; Potomac power unit started in 1973 (coal-fired 100 MW).	Similar to lime/limestone but oil-fired boilers will not require particulate control upstream of scrubber.	4 additional units (850 MW) planned or under consideration; lead times as for lime/ limestone systems.	May be more reliable than lime/limestone process; no known waste disposal problems; regeneration facility need not be located at utility.	Cost of regeneration; marketing of sulfur products; reheat.
Wellman-Lord	Regenerable process/ sodium base scrubbing with sulfite to produce bisulfite; regeneration in an evaporative crystalizer. Sulfate formed either purged or removed by selective crystallization.	>90 percent SO$_2$ removal; particulate removal as above by prescrubber.	Reliably operated (>9000 hours) in Japan; full-scale demonstration scheduled at Northern Indiana Public Service coal-fired 115-MW boiler to start December 1975.	As above.	5 additional units (1800 MW) planned by 1980; lead times same as for limestone systems.	More reliable than lime/limestone system based on Japanese experience; simplicity of unit operations in regenerator; waste disposal problems reduced.	Some bleed of solution to remove undesirable reaction products a source of water pollution, otherwise as above.

[a]Twenty-six units (13,772 MW) under consideration have not yet selected a specific FGD process. Seven units (655 MW) are operating or planned using processes not summarized here.

Source: Position Paper on Regulation of Atmospheric Sulfates, EPA-450/2-75-007, U.S. Environmental Protection Agency, September 1975.

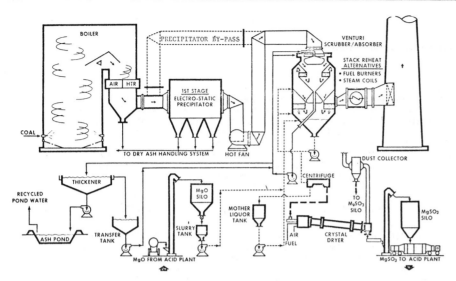

Process flow diagram for magnesia slurry SO_2 recovery system at Potomac Electric Power Company's Dickerson Station. After fly ash is removed from the flue gas by the cyclone separator and the electrostatic precipitator, the gas is passed through the scrubber where SO_2 in the flue gas is absorbed by magnesium oxide (MgO). The desulfurized flue gas then passes to the stack and the magnesium sulfite from the scrubber is transferred to an acid plant for recovery of MgO for recycle.

Source: Stern, Richard D., "Flue Gas Desulfurization—Regenerable—A Status Report on Full-Scale Regenerable Processess," ENV/SEM. 4/Cb. 4, Discussion Paper for United Nations Economic Commission for Europe Second Seminar on the Desulfurization of Fuels and Combustion Gases, Washington, D.C., November 1975.

ATMOSPHERIC FLUIDIZED-BED COMBUSTION SYSTEM (a)

PRESSURIZED FLUIDIZED-BED COMBUSTION SYSTEM (b)

Process schematics compare atmospheric and pressurized fluidized-bed combustion systems. A substantial portion of the SO_2 released during combustion is retained by the limestone fed to the combustor along with the coal. A practical process for the regeneration of the calcium sulfate needs to be developed, and better methods for the clean-up of the hot, dusty product gas are required to prevent damage to the blades of the gas turbines. Fluidized bed systems also reduce NO_x emissions since combustion temperatures in the bed can be kept relatively low at a level of 1600°F to 1700°F.

Source: Research Progress Report FF-2, Fossil Fuel and Advanced Systems Division, Electric Power Research Institute, Palo Alto, California, January 1975.

Second Generation Regenerable Flue Gas Desulfurization Processes Operated in the United States

Developer	Type	Sorbent	Regeneration Method	Intermediate	Product	Development Status
Atomics International	Wet	Na_2CO_3	Reduction (Coke)	H_2S	S	Pilot (coal-fired utility)
Bergbau-Forschung	Dry	Carbon	Thermal	SO_2	S or H_2SO_4	Prototype (coal-fired utility)
Bureau of Mines (U.S.)	Wet	Sodium Citrate	Reduction (H_2S)	None	S	Pilot (smelter)[a]
Consolidation Coal	Wet	K_2CO_3	Reduction (CO)	H_2S	S	Pilot (coal-fired utility)
EPA/TVA	Wet	NH_4OH	Acidulation	SO_2	S or H_2SO_4	Pilot (coal-fired utility)[a]
Pfizer	Wet	Sodium Citrate	Reduction (H_2S)	None	S	Pilot (coal-fired utility)[a]
Shell	Dry	CuO	Reduction (H_2)	SO_2	S or H_2SO_4	Pilot (coal-fired utility) (A full-scale 40 MW system is installed on an oil refinery in Japan)
Stauffer	Wet	Sodium Phosphate	Reduction (H_2S)	None	S	Pilot (oil-fired utility)
Stone & Webster/Ionics	Wet	NaOH	Electrolysis	SO_2	S or H_2SO_4	Pilot (coal-fired utility)
Westvaco	Dry	Carbon	Reduction (H_2S)	None	S	Pilot (oil-fired industrial)

[a]EPA and the Bureau of Mines plan to demonstrate citrate process technology on a coal-fired prototype installation under a joint-funding agreement.
Source: Stern, Richard D., "Flue gas desulfurization—Regenerable—A status report on full-scale regenerable processes," ENV/SEM.4/Cb.4, Discussion Paper for United Nations Economic Commission for Europe Second Seminar on the Desulfurization of Fuels and Combustion Gases, Washington, D.C., November 1975.

CONSTRUCTION STATUS OF FLUE-GAS DESULFURIZATION SYSTEMS

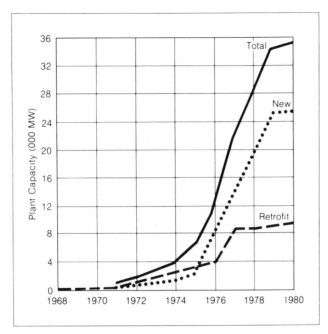

FLUE-GAS DESULFURIZATION SYSTEMS

Status	Units	Capacity (MW)
Operational	20	3,311
Under construction	19	6,340
Planned		
Contract awarded	9	3,901
Bids requested	13	6,556
Letter of intent	7	2,125
Design study	33	14,339
Totals	101	36,572

SOURCE: *Summary Report—Flue-Gas Desulfurization Systems,* Pedco-Environmental, Inc., March 1975.

Projected growth in generating capacity using scrubbers is heavily weighted toward new plant construction. Combined new and retrofitted capacity of 27,000 MW between 1975 and 1980 represents about 20 installations annually and may be limited by the capability of experienced FGD system suppliers.

Yeager, Kurt E., "Stacks vs. Scrubbers," Research Progress Report FF-3, Fossil Fuel and Advanced Systems Division, Electric Power Research Institute, Palo Alto, California, July 1975.

A Report on the Status of Scrubber Installations

Utility	Plant	Size of FGD Unit (Mw)	Process	Vendor	Start-Up
SCRUBBERS IN OPERATION OR UNDER CONSTRUCTION (by process and start-up date)					
Kansas P&L	Lawrence-4	125	Limestone	Combustion Engineering	Dec. 1968
	Lawrence-5	400	Limestone	CE	Nov. 1971
Commonwealth Ed	Will County-1	167	Limestone	Babcock & Wilcox	Feb. 1972
Kansas City P&L	Hawthorn-4	100	Limestone	CE	Aug. 1972
	Hawthorn-3	140	Limestone	CE	Nov. 1972
Key West, Fla.	Key West	37	Limestone	Zurn	Oct. 1972
Kansas City P&L	La Cygne-1	820	Limestone	B&W	June 1973
Arizona PS	Cholla-1	115	Limestone	Research-Cottrell	Dec. 1973
Southern Cal Ed	Mohave-1	160	Limestone	Universal Oil Products	Oct. 1974*
Detroit Ed	St. Clair-6	180	Limestone	Peabody Engineering	Jan. 1975
TVA	Widow's Creek-8	550	Limestone	TVA	May 1975
No. States Pwr	Sherburne-1	680	Limestone	CE	May 1976
Louisville G&E	Paddy's Run-6	65	Lime	CE	April 1973

A Report on the Status of Scrubber Installations (*Continued*)

Utility	Plant	Size of FGD Unit (Mw)	Process	Vendor	Start-Up
Southern Cal Ed	Mohave-2	160	Lime	SCE/Stearns-Roger	Nov. 1973*
Duquesne Light	Phillips	410	Lime	Chemico	March 1973
Dairyland Pwr Co-op	Alma	80	Lime	Foster Wheeler	Aug. 1971
Duquesne Light	Elrama	510	Lime	Chemico	Late 1974
Penn Power	Mansfield-1	880	Lime	Chemico	Early 1975
	Mansfield-2	800	Lime	Chemico	Early 1975
Montana Power	Colstrip-1	360	Lime	Combustion Equipment Associates	May 1975
	Colstrip-2	360	Lime	CEA	May 1975
TVA	Shawnee-10	30	Limestone/Lime	TVA	April 1972
Boston Ed	Mystic-6	150	Magnesium Oxide	Chemico	April 1972
Potomac E&P	Dickerson-3	100	Mag-Ox	Chemico	Sept. 1973
Phila. Elec.	Eddystone-1 (part)	120	Mag-Ox	United Engineers	June 1975
Illinois Power	Wood River-4	110	Catalytic Oxidation	Monsanto	Sept. 1972
No. Ind. PS	Mitchell-11	115	Wellman-Lord	Davy-Powergas/Allied Chemical	Mid-1975
PS New Mexico	San Juan-1	375	Wellman-Lord	Davy-Powergas	1977
	San Juan-2	340	Wellman-Lord	Davy-Powergas	1977
Nevada Power	Gardner-1	125	Aqueous Sodium Base; Non-Regenerable	CEA	May 1974
	Gardner-2	125	Aqueous Sodium; N-R	CEA	May 1974
	Gardner-3	125	Aqueous Sodium; N-R	CEA	1975
General Motors	Parma Chevy	32	Double Alkali	GM	April 1974

Units will be dismantled and replaced in 1977 by scrubbers listed below

SCRUBBERS IN PLANNING STAGE

Contract Awarded

Utility	Plant	Size of FGD Unit (Mw)	Process	Vendor	Start-Up
Columbus & So. Ohio	Conesville-5	375	Lime	UOP	1976
	Conesville-6	375	Lime	UOP	1976
Kansas P&L	Jeffrey-1	700	Limestone	CE	1979
	Jeffrey-2	700	Limestone	CE	1979
	Jeffrey-3	700	Limestone	CE	1979
	Jeffrey-4	700	Limestone	CE	1979
Kentucky Util	Green River-1,2,3	64	Lime	American Air Filter	1975
Louisville G&E	Cane Run-4	178	Lime	American Air Filter	1975
	Mill Creek-3	425	Lime	American Air Filter	1977
No. States Power	Sherburne-2	680	Limestone	CE	1977
PS Indiana	Gibson	650	Limestone	CE	1976
Southwestern PS	Harrington-1	343	Lime	CE	1976
Springfield, Mo.	Southwest-1	200	Limestone	UOP	1976
Texas Utilities	Martin Lake-1	793	Limestone	R-C	1976
	Martin Lake-2	793	Limestone	R-C	1977
	Martin Lake-3	793	Limestone	R-C	1978
	Martin Lake-4	793	Limestone	R-C	1979

Letter of Intent Signed

Utility	Plant	Size of FGD Unit (Mw)	Process	Vendor	Start-Up
Arizona Elec Pwr	Apache-2	200	Limestone	R-C	1979
	Apache-3	205	Limestone	R-C	1979
Arizona PS	Cholla-2	250	Limestone	R-C	1977
	Cholla-3	250	Limestone	R-C	1978

Out for Bids

Utility	Plant	Size of FGD Unit (Mw)	Process	Vendor	Start-Up
Central Ill PS	Duck Creek-1	100	Limestone	Riley Stoker/ Environeering	1976
	Newton-1	600	Limestone/Lime		1977

A Report on the Status of Scrubber Installations (*Continued*)

Utility	Plant	Size of FGD Unit (Mw)	Process	Vendor	Start-Up
Cincinnati G&E	Miami Fort-8	500	Limestone/Lime		1977
General Public Util	Homer City-3	650	Not Selected		1977
Indianapolis P&L	Petersburg-3	515	Limestone/Lime		1977
So. Carolina PS	Georgetown-2	140	Limestone/Lime		1977
Texas Utilities	Monticello-3	800	Limestone		1978
Conducting Engineering Studies					
Arizona PS	Four Corners-1	175	Lime	Chemico	1976
Basin Electric	Missouri Basin-1	550	Limestone		1979
	Missouri Basin-2	550	Limestone		1980
	Missouri Basin-3	550	Limestone		1983
Cincinnati G&E	East Bend-1	600	Not Selected		1979
Colorado Ute Elec	Craig-1	450	Limestone		1978
	Craig-2	450	Limestone		1978
	Hayden-1	202	Limestone		1978
	Hayden-2	250	Limestone		1978
New England Elec	Brayton Pt.-3	650	Not Selected		—
Salt River Project	Navajo-1	750	Limestone/Lime		1976
	Navajo-2	750	Limestone/Lime		1976
	Navajo-3	750	Limestone/Lime		1977
Southern Cal Ed	Mohave-1	640	Limestone/Lime		1977
	Mohave-2	640	Limestone/Lime		1977
So. Miss. Elec Pwr	Hattiesburg-1	223	Limestone		1979
	Hattiesburg-2	223	Limestone		1979
Initial Planning Stage					
Arizona PS	Cholla-4	372	Limestone		1979
	Four Corners-2	175	Lime	Chemico	1977
	Four Corners-3	229	Lime		1977
	Four Corners-4	800	Lime	Chemico	1977
	Four Corners-5	800	Limestone/Lime		1977
Louisville G&E	Cane Run-1	110	Lime		1980
	Cane Run-2	107	Lime		1980
	Cane Run-3	137	Lime		1980
	Cane Run-5	183	Lime		1976
	Cane Run-6	277	Lime		1977
	Mill Creek-1	330	Lime		1979
	Mill Creek-2	330	Lime		1978
	Mill Creek-4	425	Lime		1979
Phila. Elec	Eddystone-1	241	Mag-Ox	United Engineers	May 1978
	Eddystone-2	334	Mag-Ox	UE	May 1978
	Cromby-1	153	Mag-Ox	UE	May 1978

Compiled from a September 1974 EPA report, "Flue Gas Desulfurization: Installations and Operations."
Reproduced from *Electrical Week*, Sept. 30, 1974. Copyright 1974 by McGraw-Hill, Inc.

OPERATIONAL AVAILABILITY OF FLUE-GAS DESULFURIZATION SYSTEMS

Availability Data
Flue-Gas Desulfurization Demonstration Plants

Utility & Plant	Capacity (MW)	Process Type	Startup Date	Annual Availability (%)			Highest Monthly Availability		Longest Continuous Run (Days)	Comments
				1972	1973	1974	%	Month		
Kansas Power & Light										
Lawrence 4	125	Limestone injection/ wet scrubbing	10/68	See Comments			100	—	40	Since July 1973, Unit 4 has had 100% availability; the scrubbers are now cleaned manually once per week per module; the reheaters and mist eliminators are washed nightly with the wash lances when the boiler is at half load.
Lawrence 5	400	Limestone injection/ wet scrubbing	9/71	See Comments			100	—	5	Scrubber was off-line for 3 or 4 days during 1973 because of pump failures and loss of marbles caused by breaks in the pot cover welds; there is a nightly check of the modules that are out of service at half load to repair plugged nozzles; reheaters and mist eliminators are washed once per week.
Hawthorn 3	125	Limestone injection/ wet scrubbing	11/72	N.A.[a]	40	40	N.A.	—	22	Scrubbers have to be cleaned at frequent intervals.
Hawthorn 4	140	Tail-end limestone scrubbing	8/72	N.A.	40	40	N.A.	—	22	Limestone injected after air preheater to prevent plugging of boiler tubes; scrubbers have to be cleaned manually at frequent intervals.
La Cygne	820	Tail-end limestone scrubbing	6/73	—	45	45	N.A.	—	N.A.	System design calls for 86% availability of each scrubbing module (1 of 7 is off-line at any given time); however, maintenance and cleanup requirements have reduced overall availability to 45%; an 8th module is being planned to improve availability.
Commonwealth Edison										
Will County 1	165	Tail-end limestone scrubbing	2/72	29.5	27.1	69.0	99.2	12/74	N.A.	Availability data is for A module; fresh water mist eliminator wash greatly improved availability; mist eliminator wash water is discharged from system.
City of Key West										
Stock Island[b]	42	Tail-end limestone scrubbing	8/73	—	low	low	N.A.	—	5	Very little scrubber operation since startup; high liquid level in scrubber causes high pressure drop, which kicks system off-line.
Arizona Public Service										
Cholla[c]	125	Tail-end limestone scrubbing	12/73	—	—	91.5	N.A.	—	45	Utility has not yet run its independent acceptance test; open-loop operation; low-sulfur coal.
Louisville Gas & Electric										
Paddy's Run 6	70	Lime scrubbing	4/73	—	90	100	100	8/74	45	System utilizes carbide lime; peaking unit with very low load factor.

Availability Data (*Continued*)

										Comments
Duquesne Light										
Phillips	387	Lime scrubbing	8/73	—	low	<50	N.A.	—	N.A.	Only 2 of 4 scrubber trains are required and these are constantly alternated.
Southern California Edison										
Mohave 2[c]	790	Lime scrubbing	11/73	—	—	85.4	N.A.	—	N.A.	170-MW module on 790-MW boiler; inlet SO_2 concentration to scrubber is 200 ppm; operated with high availability.
TVA										
Shawnee 10	150[d]	Lime/limestone scrubbing	8/72	See Comments			N.A.	—	90	Program objective is to achieve a long-term reliability run of at least 4 to 6 months; as of 9/74 the longest run was 50 days on TCA using limestone and 90 days on venturi-spray tower using lime.
Nevada Power										
Reid Gardner 1[c]	150	Single alkali scrubbing	3/74	—	—	50	N.A.	—	10	Most of the downtime was due to a lack of soda ash; the scrubbers have not been operating long enough to get a good assessment of availability.
Reid Gardner 2[c]	125	Single alkali scrubbing	4/74	—	—	25	N.A.	—	8	See Reid Gardner 1.
Boston Edison										
Mystic 6[b]	155	Magnesia scrubbing	4/72	low	32	57	87	3/74	7	Scrubber availability improved from March through June 1974; during this period, the inability to demonstrate runs longer than 7 days was due primarily to boiler problems not related to scrubber operation; scrubber was shut down at the end of June at the conclusion of the 2-year test program.
Illinois Power										
Wood River 4	109	Catalytic oxidation	9/72	See Comments			N.A.	—	N.A.	System was started up in September 1972 but was shut down almost immediately (October) to convert the reheat burners to oil firing, which was not completed until June 1973. A successful performance test was then made, but because of the catalyst plugging hazard, the system was shut down again to install an external burner system, which was not completed until April 1974. Various other problems with standard equipment have been encountered since, with the result that only 651 hours of operating time have been logged in almost 2 years.

[a]N.A. - not available
[b]Oil-fired boiler
[c]Low-sulfur coal
[d]Only part of flue gas, equivalent to about 10 MW, is diverted through each of three different scrubbers: a venturi followed by a spray tower, a turbulent contact absorber (TCA), and a marble-bed absorber; the marble-bed absorber is no longer in use.

Source: Status of Stack Gas Control Technology, EPRI 209, prepared by Battelle Columbus Laboratories for the Electric Power Research Institute, Palo Alto, California, July, 1975.

CONTROL OF PARTICULATE EMISSIONS

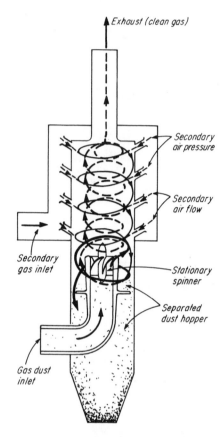

Cyclone collector (above) is desirable for capturing particulates 10 microns and smaller. As unit efficiency increases, however, operation becomes more expensive, since pressure drop increases with a decrease in diameter. The greater the drop, the greater the power cost

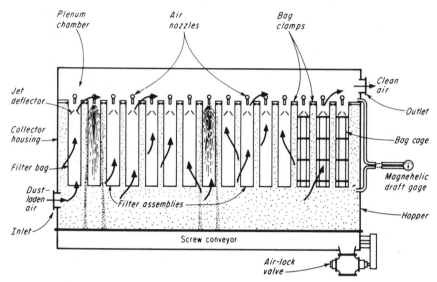

Fabric filter performs effectively to 0.1-micron particle size. Automatic shakers, or air nozzles operated periodically, keep the fabric filter surfaces clean

Comparing average values of key emission-control devices

Equipment	Efficiency, %	Pressure drop, in. H$_2$O	Gas velocity, 1000 ft/min	Installed cost, $/cfm
Cyclone collector	85-95	3.5-5.0	—	0.35-1.05
Fabric filter				
Shaker type	99	2.5	0.002-0.006	0.75-1.25
Reverse jet	99 +	3.5	0.018-0.028	1.00-1.50
Precipitator	99 +	0.9	0.2-0.5	1.00-3.50
Venturi scrubber				
Low energy	95 +	5-15	6[1]; 0.6[2]	0.75-1.50
Medium Energy	98 +	15-20	9[1]; 0.6[2]	1.00-2.00

(1) At throat   (2) Through cyclone separator

Reprinted with permission from *POWER*, April, 1974.

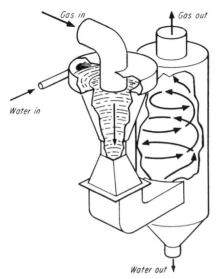

Gas in

Gas out

Water in

Water out

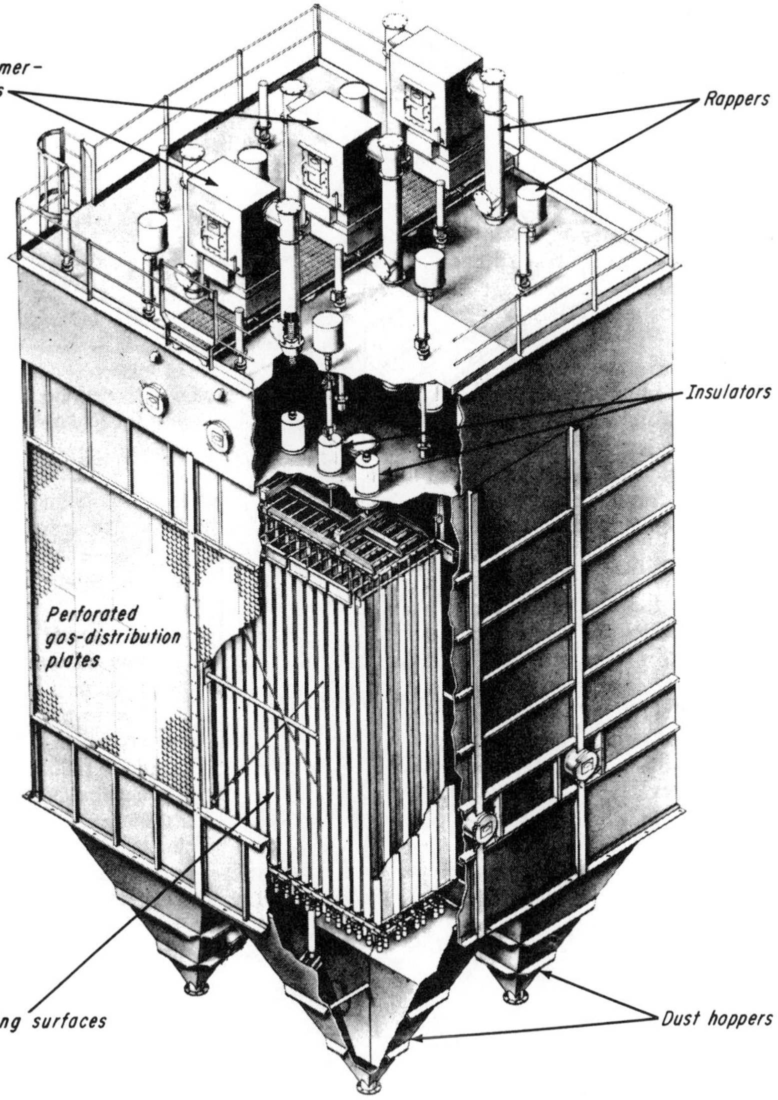

Wet-approach venturi scrubber
(above) is compact, has capability for
collecting particulates down to
submicron sizes. Low- and medium-
energy devices are usually adequate
for capturing coal dust and flyash, thus
avoiding high energy costs. Scrubber
has no limitation on temperature or
humidity of incoming gas stream.
Electrostatic precipitator (right) has
capability for capturing entire range of
emissions from coal-fired boilers—and
then some. For particulate limits in the
0.2-lb/million Btu category and lower,
it is the device to select. Although its
installed cost is relatively high,
operating cost is low because the gas
stream need not be accelerated.
Caution: Deliveries may take as long as
two years or more

Reprinted with permission from *POWER*, April, 1974.

Transformer-
rectifiers

Rappers

Insulators

Perforated
gas-distribution
plates

Collecting surfaces

Dust hoppers

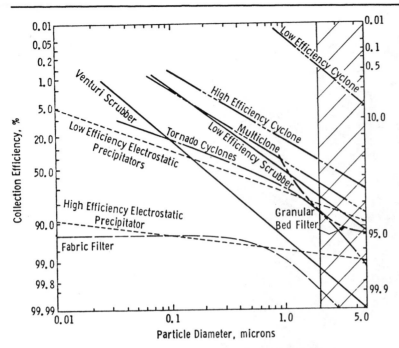

Extrapolated fractional efficiency of particulate removal devices.

Source: Clean Power Generation from Coal, Research and Development Report No. 84, Office of Coal Research, Department of the Interior, 1973.

Summary of Particulate Removal Systems

	OPERATING CONDITIONS		SIZE RELATIVE TO CYCLONES	EFFICIENCY OF DUST COLLECTION			INLET OR FACE VELOCITY	CAPITAL COST	OPERATING COST	DEVELOPER
	OPERATED °F/PSIG	PROJECTED °F/PSIG		7 MICRON	2 MICRON	<2 MICRON	CF/MIN./FT²	$/KW	MILS/KWH	
COMMERCIAL SYSTEMS										
CYCLONES	1700/HIGH	>2000/HIGH	1	99+	40	—	4000	4–6	0.2	
AERODYNE TORNADO	900/500	~1700/500	1	—	80	60	—	13	—	AERODYNE DEVELOP. CORP.
ELECTROSTATIC PRECIPITATORS	900/15	~1700/300	3 TO 8	—	—	95	200 TO 300	15–35	0.1	
BAG HOUSES	550/15	—	3 TO 6	—	—	99.9	5 TO 10	15–30	0.2	
GRAVEL BED FILTERS	>900/HIGH	>2000/300+	3	—	80	—	100	9–12		REXNORD AIR POL. CONTROL
DEVELOPMENTAL SYSTEMS										
SILICA FIBRES		1500	3 TO 6							J. P. STEVENS CO.
METAL FABRICS	800/15	>1500/HIGH	0.3 TO 2.0	—	—	99.9	100–120			BRUNSWICK CORP.
ELECTROSTATIC FILTERS	250/15	550/HIGH	0.6 TO 2.0	—	—	99.9	20–30			AMER. PREC., DUSTEX DIV.
PANEL BED	1000/15	>2000/HIGH	0.6 TO 3.0	—	—	99.9	30–45			CITY COLLEGE OF N.Y. (SQUIRES)
DUCON	600/15	>1500/HIGH	0.3 TO 2.0	—	99	—	30–90			DUCON CO.
PEBBLE BED	250/15	1500/HIGH	3	—	80	—	30–100			COMBUSTION POWER CO.
CERAMIC[1]	3450/15	—	5 TO 10	—	—	99.9	2–5			NORTON CO.
SINTERED POROUS METAL[2]	900/15	1500/HIGH	5 TO 10	—	98	—	7–10			PALL TRINITY MICRO CORP.
IMPINGEMENT (VEFL)	100/15	>1500/HIGH	2 TO 3	—	99.9	—	—			BABCOCK & WILCOX

1. $10/FT² FOR MATERIAL ONLY.   2. $60–110/FT²

Source: Purification of Hot Fuel Gases from Coal or Heavy Oil, EPRI 243-1, prepared by Stone & Webster Engineering Corp for the Electric Power Research Institute, Palo Alto, California, November, 1974.

CONTROL OF NITROGEN OXIDES

About half the NO_x emissions in the United States are produced from stationary boilers and about half are produced from the engines of vehicles. In both cases the primary source of NO_x comes from the fixation of the nitrogen in the air used to burn the fuel. There is also a small contribution from the combustion of the chemically bound nitrogen in the fuel itself.

The fixation of nitrogen from the air is strongly dependent upon the temperature and control processes are based on techniques for reducing peak temperatures in the combustion process. Among the methods for reducing combustion temperatures and NO_x emissions in power plant boilers are: injection of steam or water into the combustion zone, recirculation of flue gas into the boiler, two-stage combustion where heat is released at a slower rate, and fluidized-bed combustion where temperatures are kept low in a large, uniform combustion zone.

REFERENCE

William J. Aghassi and Paul N. Cheremisinoff, "NO_x control in central station boilers," *Power Engineering*, June 1975.

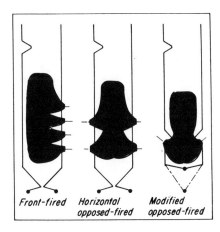

Front-fired   Horizontal opposed-fired   Modified opposed-fired

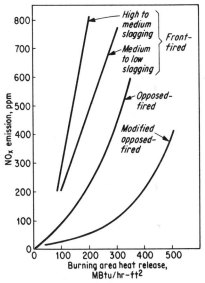

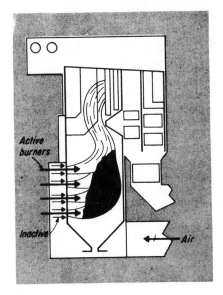

Active burners

Inactive — Air

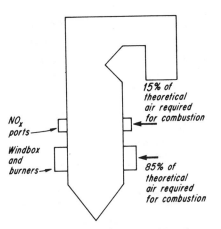

NO_x ports

Windbox and burners

15% of theoretical air required for combustion

85% of theoretical air required for combustion

Comparison of Leading NO$_x$ Control Techniques

Technique	Comments
Off-stoichiometric (biased firing, overfire airports, low excess air)	Method relies on creation of fuel-rich, reducing atmosphere. *Biased firing* allows air to go only through selected burners; *overfire* adds NO_x ports above burners to stage combustion; *low excess air* requires special burners. Method is inexpensive, effective but fireside deposits, corrosion, slagging a problem
Low combustion air temperature	Valid method but will lead to decrease in boiler efficiency unless economizer is enlarged, feedwater temperature lowered
Flue-gas recirculation through burners	Method cools combustion air and is effective, especially with pulverized coal. Expensive to apply and operate, requiring added controls. Best choice is on large, new units
Improved burner/furnace combinations	Aim is to design furnace enclosure and burner locations so sufficient time is allocated in furnace for complete combustion of coal when fired with slower and more controlled rates of heat releases. Method is inexpensive, effective
Increase furnace cooling surface	An effective method but very costly, requiring more radiant superheater surface for more diffuse burning, less turbulence

Reprinted with permission from *POWER*, April, 1974.

Improved burner/furnace combination (above) is a simple means for reducing NO_x emissions with the boiler. Two-stage firing and flue-gas recirculation (middle) are also used, although latter can be expensive and may upset combustion

Off-stoichiometric firing (above) has the advantage of requiring no additional power, and is moderately effective. Disadvantages: May result in unburned carbon, fireside deposits and corrosion

Source: "Combustion, Pollution Controls," *POWER*, April 1974. Reprinted with permission from *POWER*, April, 1974.

WASTE HEAT DISCHARGE STANDARDS

The standards for the discharge of waste heat from steam electric generating units require that all new power plants install closed-cycle cooling systems. Some existing generating units must be converted to closed-cycle cooling by the early 1980s—all the generating units of 500 MW or larger which were placed into service after January 1, 1970, and all the units 25 MW to 499 MW placed into service after January 1, 1974.

Summary of EPA's Promulgated Effluent Limitations on Heat Discharged by Steam Electric Generating Units*

All "no discharge" limitations allow for blowdown to be discharged at temperature not to exceed cold-side (after cooling) temperature, except where unit has existing closed-cycle cooling, blowdown may exceed the cold-side temperature. All limitations for existing units to be achieved by no later than July 1, 1981, except where system reliability would be seriously impacted, the compliance date can be extended to no later than July 1, 1983.

EXISTING GENERATING UNITS	
Capacity 500 Mw and greater	
Placed into service prior to January 1, 1970	NO LIMITATION
Placed into service January 1, 1970 or thereafter	NO DISCHARGE**
Capacity 25 Mw to 499 Mw	
Placed into service prior to January 1, 1974	NO LIMITATION
Placed into service January 1, 1974 or thereafter	NO DISCHARGE**
Capacity less than 25 Mw	NO LIMITATION

** Exceptions prescribed on a case-by-case basis for units in systems of less than 150 Mw capacity, units with cooling ponds or cooling lakes, units without sufficient land available, units with blowdown TDS 30,000 mg/l or greater and neighboring land within 500 ft of cooling tower(s), and units where FAA finds a hazard to commercial aviation would exist.

NEW SOURCES	NO DISCHARGE

*No effluent limitations on heat from sources other than main condenser cooling water.

Source: Development Document for Effluent Limitations Guidelines and New Source Performance Standards for the Steam Electric Power Generating Point Source Category, EPA 440/1-74029-a, October 1974, page 5.

Source: Staff Draft Report, National Commission on Water Quality, 1111 18th St., N.W., Washington, D.C., November 1975.

The Trojan nuclear power plant on the Columbia River near Prescott, Washington, 1972. (*Courtesy U.S. Environmental Protection Agency*)
Credit: EPA-Documerica Gene Daniels

Steam Electric Cooling System In-Place as of January, 1975

Region	Percent of Generating Capacity Provided by Units Using the Following Types of Cooling Systems:			
	Once Through-Navigable Waters	Closed Cycle-Towers	Impoundment-Mainly Once Through	Mixed
New England [1]	87.3%	0%	0%	12.7%
Middle Atlantic [2]	89.3%	8.9%	0%	1.8%
EastNorth Central [3]	90.4%	0.8%	4.2%	4.6%
West North Central [4]	59.5%	16.1%	8.5%	15.9%
South Atlantic [5]	79.3%	5.3%	3.4%	12.0%
East South Central [6]	79.9%	6.7%	0%	13.4%
West South Central [7]	26.5%	27.8%	25.5%	20.2%
Mountain [8]	8.7%	54.2%	29.1%	8.0%
Pacific [9]	85.5%	14.5%	0%	0%
Non-Contiguous [10]	100%	0%	0%	0%
All	72.9%	11.2%	6.7%	9.2%

(1) Conn., Maine, Mass., New Hamp., R.I., Vermont
(2) New Jersey, N.Y., Penn.
(3) Ill, Indiana, Mich., Ohio, Wisc.
(4) Iowa, Kansas, Minn., Miss., Nebraska, N. Dakota, S. Dakota
(5) Del., D.C., Florida, Georgia, Md., N.C., SC., Va., W. VA.

(6) Ala., KY, Miss., Tenn.
(7) Ark., LA, Ok., Texas
(8) Arizona, Colo., Id., MO, Nevada, New Mex., Utah, Wyo.
(9) Calif., Oregon, Wash.
(10) Alaska, Hawaii, Puerto Rico, Virgin Islands

Source: Teknekron, Inc., Capabilities and Costs of Technology for Water Pollution Control for the Steam Electric Power Industry, Interim Report to the National Commission on Water Quality, April 1975.

Source: Staff Draft Report, National Commission on Water Quality, 1111 18th St., N.W., Washington, D.C. November 1975.

POWER PLANT COOLING SYSTEM ALTERNATIVES

Qualitative Advantages and Disadvantages of Cooling System Alternatives

Cooling System	Advantages	Disadvantages
Once-Through System	1. Normally simplest, most economical.	1. Cooling water supply problem for large power plants 2. Possibility of significant ecological problems with receiving waters
Cooling Pond	1. Serves as settling basin for suspended solids 2. Relatively simple operation 3. May be beneficial for other purposes e.g., recreation.	1. Large land area required 2. Large capital cost 3. Blowdown disposal problem.
Spray Module Canal	1. Requires less land area as compared to cooling pond 2. Economical, compared to cooling pond.	1. Performance dependent upon wind conditions 2. Low altitude fogging and icing 3. Increased water loss due to drift 4. Blowdown disposal problem.
Mechanical Draft Wet Tower	1. Positive control over air supply 2. Relatively close control over cold water temperature 3. Generally low pumping head 4. Ambient relative humidity has minimal effect on tower performance 5. More packing per unit volume of tower 6. Usually lower capital cost than natural draft wet tower.	1. Operating and maintenance costs higher than natural draft tower 2. Subject to recirculation of the humid exhaust air 3. More severe localized fogging and icing problem than natural or fan-assisted wet tower 4. Blowdown disposal problem 5. Greater concentrations of salt deposition near site boundary than non-mechanical draft systems.
Natural Draft Wet Tower	1. Fewer mechanical and electrical components than mechanical draft wet tower 2. Lower operating and maintenance costs than mechanical draft wet tower 3. Better plume dispersion because of high elevation discharge.	1. Internal resistance to air flow must be kept minimal 2. Requires greater tower height to produce draft 3. Substantially larger capital cost than mechanical draft wet tower 4. Performance dependent on both wet bulb and dry bulb temperatures 5. Normal fogging and icing 6. Soil and wind loading problems caused by large size and weight of structure 7. Blowdown disposal problem.

Qualitative Advantages and Disadvantages of Cooling System Alternatives (*Continued*)

Cooling System	Advantages	Disadvantages
Fan-Assisted Natural Draft Wet Tower	1. Better control of air flow than natural draft tower 2. Smaller structure and lighter weight than natural draft tower 3. Consumes less power than mechanical draft wet tower if fans are turned off when not needed. 4. Less serious fog-ice problems than mechanical draft towers.	1. Higher capital cost than mechanical draft tower. 2. More serious fog-ice problems than natural draft towers.
Mechanical Draft Dry Tower	1. Avoids problems of fogging and icing 2. Eliminates such water problems as: availability, evaporative losses, and blowdown.	1. Substantially higher capital and operating costs than other systems 2. High maintenance costs 3. Larger volumes of air must be circulated 4. Decreased plant output and efficiency at high dry bulb temperatures 5. Larger land area required than wet towers. 6. Dry towers may have greater freeze-protection problems than wet towers.
Natural Draft Dry Tower	1. Same as mechanical draft dry tower, items 1 and 2 2. Fewer mechanical and electrical components than mechanical draft dry tower 3. Lower operating and maintenance costs than mechanical draft dry tower.	1. Higher capital cost than all other cooling systems, including mechanical draft dry tower 2. Soil and wind loading problems caused by large size and weight of structure. 3. Dry towers may have greater freeze-protection problems than wet towers.

Source: Heat Sink Design and Cost Study for Fossil and Nuclear Power Plants, WASH-1360, United States Atomic Energy Commission, December 1974.

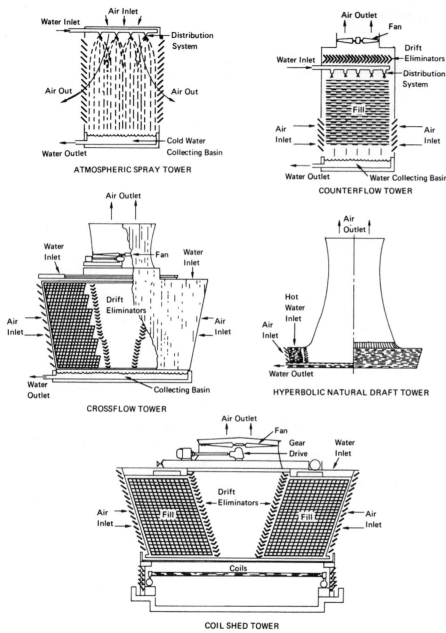

ATMOSPHERIC SPRAY TOWER

COUNTERFLOW TOWER

CROSSFLOW TOWER

HYPERBOLIC NATURAL DRAFT TOWER

COIL SHED TOWER

Schematic of cooling towers.

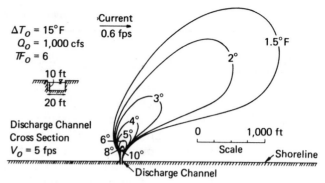

$\Delta T_o = 15°F$
$Q_o = 1,000$ cfs
$\mathbb{F}_o = 6$

Current
0.6 fps

1.5°F

Discharge Channel Cross Section
$V_o = 5$ fps

Isotherms of surface temperature rise for a surface discharge.

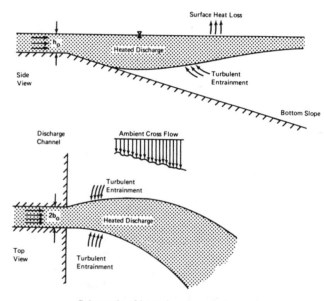

Schematic of heated surface discharge.

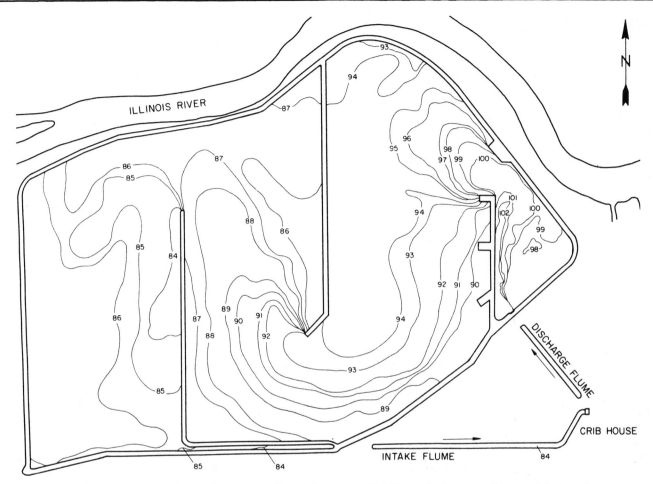

Temperature profiles in cooling pond at the Powerton Station of Commonwealth Edison. Surface area of the pond is about 3 square miles. Contour interval is 1°F.

Source: Daedalus Enterprises, Inc., P.O. Box 1869, Ann Arbor, Michigan 48106.

CONTROL OF AUTOMOTIVE EMISSIONS

Control of the atmospheric level of automotive emissions can be achieved by a number of methods. The number of vehicles on the highways can be reduced by car-pooling and increased use of mass transit. Lower fuel consumption can be achieved by establishing speed limits at which operating efficiences are higher. The most effective technique, however, has been to set standards for the emissions permitted per vehicle per mile of travel.

Reduction of the emissions from individual vehicles has been achieved by two methods —one method has been to shift to smaller vehicles which consume less fuel per mile and therefore emit less exhaust; the other method has involved modifications of the engine exhaust systems to achieve complete combustion of the fuel since both the hydrocarbon (HC) and carbon monoxide (CO) emissions are a result of incomplete combustion. Reduction in the NO_x emissions requires design modifications to reduce the flame temperatures in the engine to prevent the fixation of the nitrogen in the air used to burn the fuel. Modifications include recirculation of crankcase and exhaust gases, leaner fuel mixtures, changes in timing of spark ignition, decreased compression ratios, and catalytic exhaust recombiners.

Summary of Typical Emission Modifications

Model Year Added	Modification
1963	"Open" crankcase ventilation system
1968	a) Leaner mixture carburetor calibration
	b) Intake air pre-heat to constant temperature
	c) Intake manifold pre-heat
	d) Retarded idle timing
	e) "Closed" positive crankcase ventilation (PVC) system
1970	a) Decreased compression ratio
	b) Air injection (on some models)
	c) Modified spark advance control
1971	Fuel evaporative control system
1972	a) Anti-dieseling solenoid
	b) Fast acting choke
	c) Air injection
1973	a) Exhaust gas recirculation (EGR)
	b) Hardened valve seats
	c) Spark advance control
1974	a) Precision cams, bores, pistons
	b) Improved EGR systems

Source: The Automobile and the Regulation of Its Impact on the Environment, NSF/RA/X-74-023, Columbia University, 1975.

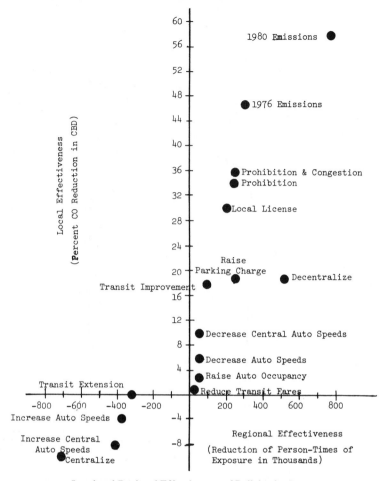

Local and Regional Effectiveness of Policies for Boston.

Source: Air Quality and Automobile Emission Control, A report by the Coordinating Committee on Air Quality Studies, prepared for the Committee on Public Works, United States Senate, Volume 4, September 1974.

Proposed Automobile Emissions Standards by Year in Which Emissions Would be Required

Proposal	Hydrocarbons			Carbon monoxide			Nitrogen oxides				
Emissions standard, grams per mile	1.5	0.9	0.41	15.0	9.0	3.4	3.1	2.0	1.5	1.0	0.4
1970 Clean Air Act	1976		1977	1976		1977	1976	1977			1978
Schedule as of January 1976	1976-77		1978	1976-77		1978	1976	1977			1978
EPA Proposal (March 5, 1975)	1977-79	1980-81	1982	1977-79	1980-81	1982		1977-81			
President's Proposal (July 28, 1975)	1976-81			1976-81			1976-81				
Senate Staff Proposal (August 8, 1975)	1976-77		1978	1976-77		1978	1976-77			1978-81[a]	1982
House Subcommittee	1976-77	1978-79	1980	1976-77	1978-79	1980		1977-79			1980[b]
Senate Committee Bill	1976-78		1979	1976-78		1979	1976	1977-79		1979[c]	
House Committee Bill	1976-79		1980	1976-79		1980	1976	1977-80			1981[d]

[a] Applies to only 50 percent of vehicles and engines in 1978 and 1979.

[b] Administrator may grant annual suspensions based on unavailability of technology or excessive fuel penalties, but interim standards are not to exceed 1.5 gpm in 1980-81 and 1.0 gpm in 1982-84.

[c] Applies to only 10 percent of vehicles in 1979 and 100 percent in 1980.

[d] Administrator would be able to grant yearly suspension of this requirement. Maximum allowable standards with such suspension would be 2.0 for 1981-82, 1.5 for 1983-84, 0.4 for 1985 and beyond.

Source: Environmental Quality–1976, The Seventh Annual Report of the Council on Environmental Quality, September 1976.

Automobile Emission Control Schedules

	1976	1977	1978	1979	1980	1981	1982
1970 Clean Air Act							
HC	1.5	0.41 ———————————————————→					
CO	15.0	3.4 ————————————————————→					
NO$_x$	3.1	2.0	0.4 ————————————————→				
As of July 1, 1977							
HC	1.5 ——→ 0.41 ——————————————→						
CO	15 ——→ 3.4 ———————————————→						
NO$_x$	3.1	2.0	0.4				
1977 Amendments							
HC	1.5 ———————————→ 0.41						
CO	15 ———————————→ 7.0				3.4 ——→		
NO$_x$	3.1	2.0			1.0 ——→		

Source: Environmental Quality, the eighth annual report of the Council on Environmental Quality, December 1977.

Energy Costs

GNP and Personal Income

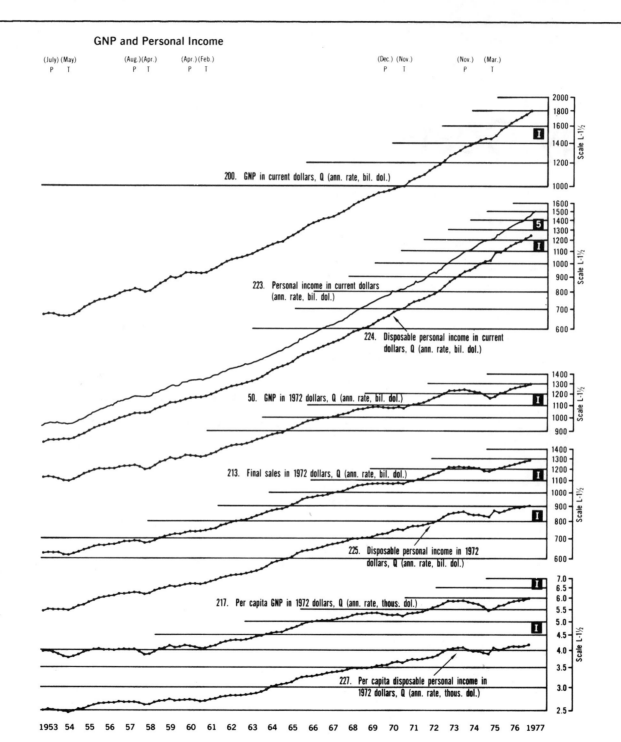

Source: Business Conditions Digest, Bureau of Economic Analysis, U.S. Department of Commerce, June 1977.

WORLD ENERGY CONSUMPTION AND GROSS NATIONAL PRODUCT

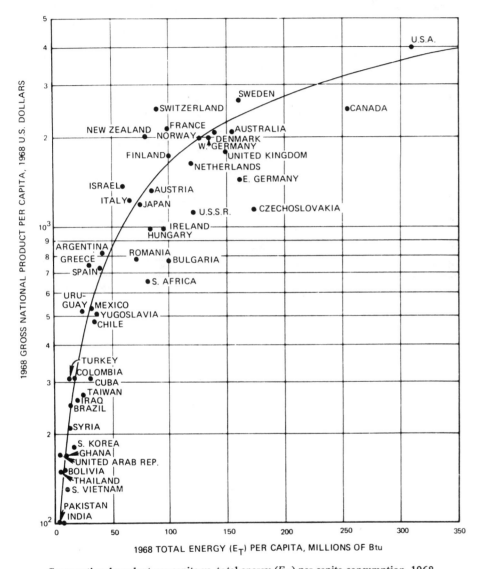

Gross national product per capita vs. total energy (E_T) per capita consumption, 1968.

Source: The U.S. Energy Problem, Volume 1: Summary, ITC Report C645 to the National Science Foundation, InterTechnologo Corporation, Warrenton, Virginia, November 1971.

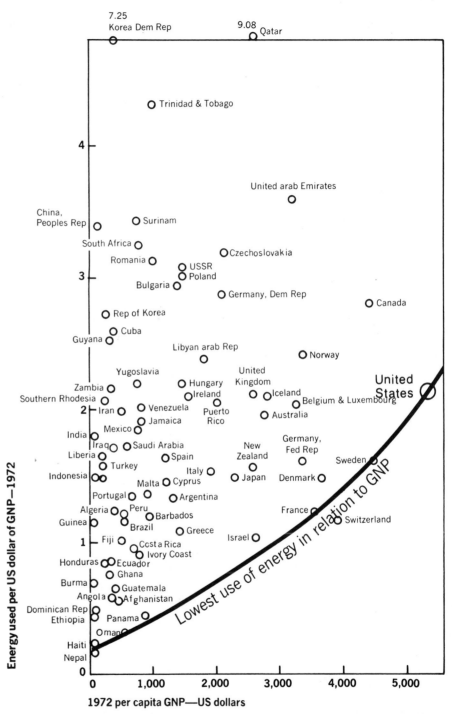

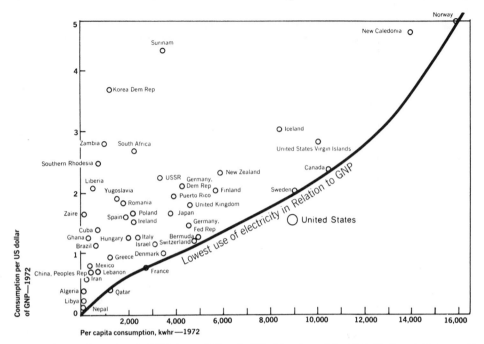

COUNTRIES	POPULATION			GROSS NATIONAL PRODUCT Constant 1967 U.S. Dollars			ELECTRICITY PRODUCTION		
	Millions			Dollars Per Capita			Kilowatt-Hours Per Capita		
	1975	1976 Est.	1977 Est.	1975	1976 Est.	1977 Est.	1975	1976 Est.	1977 Est.
AUSTRALIA	13.5	13.7	13.9	2704	2693	2777	5489	5584	5676
AUSTRIA	7.5	7.6	7.6	3253	3382	3526	4693	4908	5158
BELGIUM	9.8	9.8	9.8	3898	4051	4204	4184	4449	4673
CANADA	22.8	23.1	23.5	4092	4229	4319	11961	12844	13383
EAST GERMANY	16.9	16.9	16.9	4219	4527	4787	5006	5278	5467
FRANCE	52.9	53.2	53.5	3155	3291	3372	3355	3558	3553
INDIA	614.0	627.5	641.3	86	87	88	123	140	150
ISRAEL	3.4	3.5	3.6	706	714	722	2853	2971	3028
ITALY	55.8	56.2	56.5	1310	1374	1398	2595	2806	2904
JAPAN	110.9	112.3	113.6	2370	2504	2641	4049	4077	4273
MEXICO	59.2	61.3	63.5	358	349	345	730	783	839
POLAND	34.0	34.4	34.7	1676	1802	1876	2859	3026	3179
SPAIN	35.6	36.0	36.4	1197	1206	1217	2315	2539	2813
SWEDEN	8.2	8.2	8.2	4866	5012	5098	9659	10000	10500
U.S.S.R.	254.3	256.6	258.9	1931	2011	2075	4067	4313	4574
UNITED KINGDOM	56.1	56.1	56.2	1255	1273	1292	4852	4902	4941
VENEZUELA	12.8	13.2	13.7	1188	1220	1234	1656	1795	1971
WEST GERMANY	62.0	62.0	62.1	4519	4765	4973	4868	5208	5618
UNITED STATES	213.6	215.1	216.8	4447	4683	4884	9377	9872	10502

Source: Based on McGraw-Hill Publications' Department of Economics, June 17, 1977; U.S. Department of Commerce, Ebasco Data.

ECONOMIC DATA—OECD MEMBER COUNTRIES—1973

Economic Data—OECD Member Countries—1974

Country	Population (thousands)	Inhabitants (per square kilometer)	Crude Birth Rate (per thousand)	Annual Population Increase (1964-1974) (%)	Per Capita Gross Domestic Product (US$)	Per Capita Energy Consumption (tons of oil equivalent)	Infant Mortality (deaths in 1st year per 1,000 live births)	Access to Higher Education (percent of relevant age group)	Dwellings Completed (per 1,000 inhavitants)	Animal Protein in Diet (grams per inhabitant per day)	Telephones (per 1,000 inhabitants)	Television Sets (per 1,000 inhabitants)
Australia	13,338	2	18.4	1.8	5,880	4.39	16.1	30.5	11.3	69	355	227
Austria	7,545	90	12.9	0.4	4,370	3.05	23.5	15.7	5.6	54	246	237
Belgium	9,772	320	12.6	0.4	5,470	4.65	16.2	28.5	6.9	59	257	244
Canada	22,479	2	15.4	1.5	6,460	8.56	16.8	49.8	11.4	65	528	348
Denmark	5,045	117	14.1	0.7	6,030	3.53	12.0	35.0	9.6	66	400	282
Finland	4,688	14	13.2	0.3	4,710	4.70	10.2	23.3	15.1	67	329	263
France	52,492	96	15.2	0.8	5,060	3.39	12.0	26.4	9.5	67	217	237
Germany	62,054	250	10.1	0.7	6,200	4.25	21.1	15.8	9.7	58	287	298
Greece	8,962	68	16.1	0.5	2,150	1.42	23.9	---	17.6	45	187	58
Iceland	215	2	19.5	1.3	6,280	5.35	9.6	---	10.5	--	381	220
Ireland	3,086	44	22.3	0.7	2,180	2.40	17.8	---	8.2	62	120	176
Italy	55,361	184	15.9	0.7	2,710	2.47	22.6	27.9	3.0	49	229	208
Japan	109,710	295	18.9	1.2	4,130	3.05	10.8	23.8	13.4	32	357	229
Luxembourg	357	137	11.0	0.8	5,950	13.92	13.5	---	6.4	59	382	220
Netherlands	13,545	367	13.7	1.1	5,110	4.55	11.2	20.5	10.8	58	320	258
New Zealand	3,045	11	19.3	1.6	4,440	3.20	15.5	---	11.5	74	475	304
Norway	3,985	12	15.1	0.8	5,850	4.91	10.4	39.2	10.4	55	329	249
Portugal	8,782	96	19.2	-- 0.2	1,300	0.90	44.8	9.8	5.2	39	109	66
Spain	34,956	69	19.6	1.0	2,100	1.72	13.6	27.1	9.9	46	181	164
Sweden	8,157	18	13.5	0.6	6,880	5.45	9.6	31.1	10.5	58	594	339
Switzerland	6,443	156	12.9	0.9	6,970	3.43	12.8	---	12.7	58	560	253
Turkey	39,066	50	(39.6)	2.6	750	0;45	153.0	---	2.7	--	21	7
United Kingdom	56,068	230	13.1	0.4	3,370	3.82	16.7	21.5	4.9	55	340	309
United States	211,894	23	14.9	1.0	6,600	8.09	17.6	43.9	9.4	73	657	523

Source: The OECD Observer, Organization for Economic Cooperation and Development, Paris, France. March–April, 1976.

Energy Sources, 1972

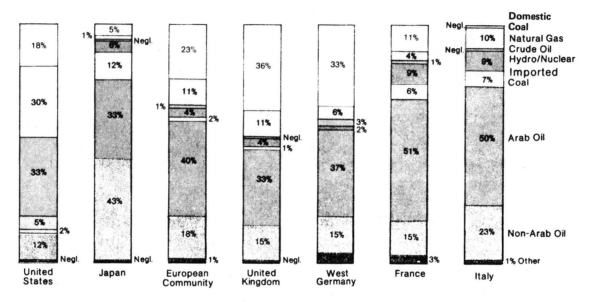

Source: Council on International Economic Policy Annual Report, February 1974.

Source: Staff Data and Materials on U.S. Trade and Balance of Payments, Committee on Finance, United States Senate, February 26, 1974.

INTERNATIONAL ECONOMIC INDICATORS

Imports of Petroleum—Annual and Quarterly

Period	United States[1]	France	F.R. Germany	Italy	Netherlands	United Kingdom	Japan	Canada
	Transaction value f.a.s.[2]				($ billion)			
		c.i.f.	c.i.f.	c.i.f.	c.i.f.	c.i.f.	c.i.f.	f.o.b.
1960	1.5	0.7	0.7	0.5	0.5	1.4	0.1	0.4
1969	2.6	1.6	2.0	1.6	0.9	2.1	2.3	0.6
1970	2.8	1.9	2.3	1.9	1.3	2.2	2.8	0.6
1971	3.6	2.4	3.2	2.4	1.8	2.9	3.6	0.7
1972	4.6	3.0	3.3	2.6	2.1	2.9	4.5	0.9
1973	8.1	3.9	5.7	3.6	3.0	4.1	6.7	1.2
1974	26.1	10.8	12.5	10.4	5.6	10.6	21.1	3.1
1975	26.6	11.0	11.8	9.3	5.8	9.3	21.0	3.5
1976	34.2	13.3	14.0	10.2		10.0	23.3	3.5
1977								
1975: I	6.9	2.9	2.6	2.1	1.4	2.5	5.3	1.0
II	5.8	2.7	3.0	2.1	1.4	2.3	4.9	1.0
III	7.0	2.5	3.0	2.3	1.5	2.1	5.1	0.8
IV	7.0	3.0	3.3	2.8	1.5	2.3	5.8	0.7
1976: I	7.9	3.3	3.1	2.2	1.7	2.7	5.6	1.0
II	7.8	3.0	3.3	2.6	1.9	2.4	5.7	1.0
III	9.1	3.3	3.8	2.5	1.7	2.4	5.9	0.8
IV	9.3	3.6	3.7	2.8		2.4	6.1	0.7
1977: I	11.5							
II								
III								
IV								

[1] Includes crude petroleum imports into the Virgin Islands, most of which are refined and sent to the United States, but are not in U.S. import statistics because the Virgin Islands are not part of the U.S. customs area, and U.S. trade with this (or any other) possession is not included in U.S. foreign trade data. [2] Prior to 1974 Customs values are shown.

Source: International Economic Indicators, U.S. Department of Commerce, June 1977.

Balance on Current Account—Annual and Quarterly

Period	United States	France	F.R. Germany	Italy	Netherlands	United Kingdom	Japan	Canada
				($ billion)				
1960	1.7	(1)	1.1	−0.7	0.3	−0.7	0.1	−1.2
1969	−2.0	−1.8	1.9	2.7	(2)	1.1	2.1	−0.8
1970	−0.4	−0.3	0.9	0.9	−0.5	1.8	2.0	1.1
1971	−4.0	0.5	0.9	2.0	−0.2	2.6	5.8	0.4
1972	−9.8	0.3	0.8	2.3	1.3	0.4	6.6	−0.4
1973	(2)	−0.7	4.3	−2.5	2.3	−1.8	−0.1	0.1
1974	−3.6	−6.0	9.7	−7.8	2.0	−7.9	−4.7	−1.5
1975	11.7	−0.1	3.8	−0.5	1.6	−3.6	−0.6	−4.9
1976	−0.6		3.4	−2.9 p	2.4	−2.6	3.7	−4.4
1977								
Seasonally adjusted.*								
1975: I	1.4	−0.1	2.2	−0.5	0.1	−1.3	0.4	−1.3
II	3.9	0.8	0.9	0.1	0.6	−0.7	0.5	−1.2
III	3.2	0.1	0.5	0.9	0.2	−1.2	−0.8	−1.1
IV	3.1	−0.5	0.3	−1.1	0.7	−0.5	−0.6	−1.3
1976: I	0.3	−1.3	1.2	−1.4	0.8	−0.2	1.0	−1.3
II	0.6	−0.4	0.7	−1.3	0.2	−0.8	1.4	−1.2
III	−0.7	−2.4	0.8	0.7	0.5	−0.9	0.2	−0.7
IV	−0.8		0.6	−0.8 p	0.9	−0.6	0.7	−1.1
1977: I						−0.5	0.9	
II								
III								
IV								

* Except for Italy and Netherlands. [1] Comparable data not available. [2] Less than $0.05 billion.

Source: International Economic Indicators, U.S. Department of Commerce, June 1977.

Unit Labor Costs and Productivity in Manufacturing—Annual

Period	United States	France	F.R. Germany	Italy	Netherlands	United Kingdom	Japan	Canada
Unit Labor Costs in U.S. Dollars (1967=100)								
1960	**97.7**	81.7	78.1	76.5	65.4	85.7	82.5	106.3
1969	**108.7**	98.8	103.1	104.3	102.7	92.8	107.2	101.9
1970	**116.5**	98.9	124.6	119.2	108.4	106.0	113.3	111.5
1971	**117.6**	105.5	141.8	135.6	120.3	117.9	130.7	115.7
1972	**118.1**	120.8	162.2	152.2	138.9	126.1	160.1	121.9
1973	**123.2**	148.1	208.3	172.5	174.1	134.2	194.3	126.6
1974	**140.9**	162.0	235.3	183.4	202.8	157.2	233.4	145.4
1975	**156.4**	224.1	267.5	243.2	249.8	199.6	272.0	160.3
1976	**157.9**P							180.8P
1977								
Unit Labor Costs in National Currencies (1967 = 100)								
1960	**97.7**	81.4	81.8	76.1	68.4	84.0	82.0	95.6
1969	**108.7**	104.0	101.4	104.8	103.3	106.8	106.1	101.7
1970	**116.5**	111.1	114.0	119.8	108.8	121.7	112.1	107.9
1971	**117.6**	118.2	123.7	134.3	116.5	132.6	125.4	108.3
1972	**118.1**	123.9	130.0	142.4	123.7	138.7	134.0	111.9
1973	**123.2**	133.6	138.4	160.8	134.4	150.6	145.4	117.4
1974	**140.9**	158.3	152.4	191.2	151.1	184.8	187.8	131.8
1975	**156.4**	195.0	164.7	254.2	175.0	247.1	222.8	151.1
1976	**157.9**P							165.2P
1977								
Output per Man-Hour (1967=100)								
1960	**78.8**	68.7	66.4	65.1	68.1	76.8	52.6	75.5
1969	**104.9**	115.4	113.8	112.2	120.5	108.4	130.0	113.3
1970	**104.5**	121.2	116.6	117.8	132.2	109.1	146.5	115.2
1971	**110.3**	127.5	122.5	123.5	140.6	114.3	151.7	122.9
1972	**116.0**	135.9	130.3	132.9	152.0	121.2	163.9	127.4
1973	**119.4**	142.2	138.6	147.8	163.9	128.1	184.3	132.2
1974	**114.7**	146.1	145.6	155.6	173.3	127.9	187.5	132.3
1975	**114.9**	139.8	150.4	151.0	169.9	123.9	181.7	134.4
1976	**122.4**P							138.2P
1977								

Source: International Economic Indicators, U.S. Department of Commerce, June 1977.

GNP Growth Rates at Constant 1972 Prices—Annual and Quarterly per Capita GNP Growth Rates at 1972 Prices—Annual

Period	United States	France[1]	F.R. Germany	Italy[1]	Netherlands	United Kingdom[1]	Japan	Canada
GNP Growth Rates (% change from preceding year)								
1960	**2.3**	7.4	17.5	6.4	8.4	4.7	13.4	2.9
1969	**2.6**	7.6	8.2	5.7	6.8	1.4	10.8	5.3
1970	**−0.3**	6.0	5.8	5.0	6.6	2.3	10.9	2.5
1971	**3.0**	5.8	3.0	1.6	4.2	2.5	7.3	6.5
1972	**5.7**	5.9	3.4	3.1	4.1	2.6	9.1	5.9
1973	**5.5**	6.1	5.1	6.9	6.3	6.0	9.8	7.2
1974	**−1.7**	2.9	0.5	3.9	2.5	0.1	−1.3	3.2
1975	**−1.8**	−2.4	−3.2	−3.5	−2.2	−1.6	2.4	0.6
1976	**6.1**		5.6	5.6		1.3	6.3	4.6
1977								
Seasonally adjusted (% change from preceding quarter at annual rates)								
1975: I	**−9.9**	−6.0	−10.5	−2.2	(2)	−1.1	−1.2	−1.4
II	**5.6**	5.0	0.6	−4.5	(2)	−9.4	3.5	6.3
III	**11.4**	−1.5	2.5	−1.2	(2)	−3.2	5.7	6.3
IV	**3.3**	13.9	12.5	10.9	(2)	5.6	4.4	0.6
1976: I	**9.2**	12.9	6.9	9.7	(2)	8.4	13.3	13.0
II	**4.5**	5.4	3.2	5.9	(2)	−9.3	5.5	1.6
III	**3.9**		1.1	0.8	(2)	6.3	1.4	2.0
IV	**2.6**		7.3	8.0	(2)	5.7	2.3	−2.5
1977: I	**5.2**							
II								
III								
IV								
Per Capita GNP Growth Rates (% change from preceding year)								
1960	**0.7**	6.4	16.4	5.8	7.1	3.9	12.7	0.7
1969	**1.6**	6.8	7.2	5.0	5.6	1.0	9.5	1.6
1970	**−1.4**	5.1	4.8	4.3	5.3	2.1	9.6	1.1
1971	**1.9**	4.8	1.9	0.9	3.0	2.2	6.2	5.1
1972	**4.8**	4.9	2.7	2.4	3.0	2.3	6.7	4.7
1973	**4.7**	5.2	4.6	6.0	5.4	5.7	8.4	5.8
1974	**−2.4**	2.2	0.4	3.0	1.7	0.1	−2.5	1.6
1975	**−2.6**	−2.9	−2.8	−4.2	−3.0	−1.6	1.2	−0.9
1976	**5.3**							
1977								

[1] Gross domestic product.   [2] Not available.
Note: Growth rates are calculated from unrounded figures.

Source: International Economic Indicators, U.S. Department of Commerce, June 1977.

Wholesale Prices, All Commodities

Country	Annual percentage changes				Quarterly percentage charges				
	1973-74	1974-75	1975-76	1/76 1/77*	IV/75 1/76	I-II	1976 II-III	II-IV	IV/76 1/77*
United States	18.9	9.2	4.6	5.9	0.4	1.5	1.2	1.0	2.2
France	23.6	1.0	10.3	13.9[1]	2.6	4.6	3.8	2.2	
Germany, F.R.	14.9	7.0	5.4	2.3	3.2	2.0	0.3	-0.4	0.4
Italy	40.8	8.6	22.9	31.1[1]	6.6	11.6	4.0	5.9	
Netherlands	14.0	6.9	15.8	10.5[1]	7.2	5.9	-1.9	-0.8	0.4
Japan	31.3	3.0	5.5	4.7	2.0	1.6	1.9	0.8	
Canada	22.4	6.5	4.4	6.4	0.4	1.6	0.3	0.6	3.8

Wholesale Prices, All Commodities—Annual and Quarterly

Period	United States	France	F.R. Germany	Italy	Netherlands	United Kingdom	Japan	Canada
				(1967 = 100)				
1960	94.9	88.0	95.2	86.6	85.2	(1)	93.9	87.4
1969	106.5	110.1	96.6	104.3	101.2	(1)	103.0	106.9
1970	110.4	118.1	102.0	111.9	107.5	(1)	106.7	108.4
1971	113.9	122.6	106.6	115.7	108.5	(1)	105.9	109.8
1972	119.1	129.7	110.6	120.4	112.4	(1)	106.7	117.5
1973	134.7	146.2	119.7	140.8	126.6	(1)	123.7	142.7
1974	160.1	180.7	137.5	198.2	144.3	(1)	162.4	174.7
1975	174.9	182.5	147.1	215.2	154.2	(1)	167.3	186.0
1976	182.9	201.3	155.0	264.4	178.5	(1)	176.5	194.1
1977								
1975: I	171.2	183.6	146.2	212.7	150.7	(1)	166.6	185.6
II	173.0	180.6	146.9	212.9	150.3	(1)	166.3	183.1
III	176.7	180.8	147.0	214.8	154.6	(1)	167.2	188.4
IV	178.6	185.0	148.0	220.2	161.3	(1)	169.1	190.5
1976: I	179.4	189.8	152.7	234.7	172.9	(1)	172.5	191.2
II	182.1	198.6	155.7	261.9	183.1	(1)	175.2	194.3
III	184.2	206.1	156.1	272.5	179.7	(1)	178.5	194.8
IV	186.0	210.7	155.5	288.7	178.2	(1)	179.9	196.0
1977: I	190.0		156.2			(1)	180.6	203.4[2]
II								
III								
IV								

Source: International Economic Indicators, U.S. Department of Commerce, June 1977.

Prices since 1661 (1661 = 100)*

Figures for each decade are along a horizontal line; eg that for 1667 = 88, for 1944 = 195.

	0	1	2	3	4	5	6	7	8	9
166-		100	103	101	96	96	92	88	88	84
167-	85	84	81	80	86	92	88	81	82	87
168-	85	82	82	80	81	83	84	74	74	73
169-	75	76	75	78	87	87	89	90	95	98
170-	85	74	73	70	73	66	75	65	68	79
171-	90	100	75	72	76	77	73	70	69	72
172-	75	74	68	66	70	72	75	71	73	77
173-	70	65	66	63	65	66	64	69	67	66
174-	74	80	73	70	62	63	69	67	70	71
175-	70	67	69	67	67	68	68	81	78	74
176-	73	70	70	74	75	78	79	81	80	73
177-	74	79	87	88	86	84	84	80	87	82
178-	81	85	86	95	93	89	88	87	90	87
179-	92	90	90	95	101	109	114	110	110	118
180-	157	169	129	115	119	138	136	138	151	157
181-	153	152	175	180	155	141	127	140	144	128
182-	115	105	101	104	106	115	102	102	97	95
183-	95	97	95	93	97	97	107	102	103	113
184-	111	105	96	91	94	95	95	100	87	82
185-	82	79	82	97	108	108	108	110	96	100
186-	104	100	104	105	103	102	104	102	100	93
187-	95	100	111	110	105	102	100	95	88	85
188-	89	86	88	88	82	76	72	70	73	73
189-	76	75	71	71	64	62	63	64	68	73
190-	79	75	75	75	72	75	81	84	76	80
191-	84	86	91	91	91	116	146	193	207	222
192-	270	167	141	139	150	146	136	131	129	124
193-	104	89	86	85	103	103	106	110	113	113
194-	152	205	195	177	195	191	191	205	219	227
195-	234	251	269	273	276	287	301	312	319	319
196-	322	333	347	354	365	379	396	404	425	446
197-	474	513	545	595						

Linked index. Main sources: Mitchell and Deane, Abstract of British Historical Statistics, and Department of Employment, British Labour Statistics Historical Abstract 1886–1968.
Basic series: Schumpeter-Gilboy price index 1661–1697 (1697 = 100) and 1696–1823 (1701 = 100); Rousseaux price indexes 1800–1923 (1865 to 1885 = 100); Sauerbeck-Statist price indexes 1846–1938 (1867 to 1877 = 100); DE index of the internal purchasing power of the pound 1914–1968 (1963 = 100).
Series rebased on 1661 = 100 using the multipliers: 1697 = 100, 0.9174; 1701 = 100, 0.7399; 1865–1885 = 100, 0.8679; 1867–1877 = 100, 1.0761; 1963 = 100, 3.5417.

Source: The Economist, London, July 13, 1974.

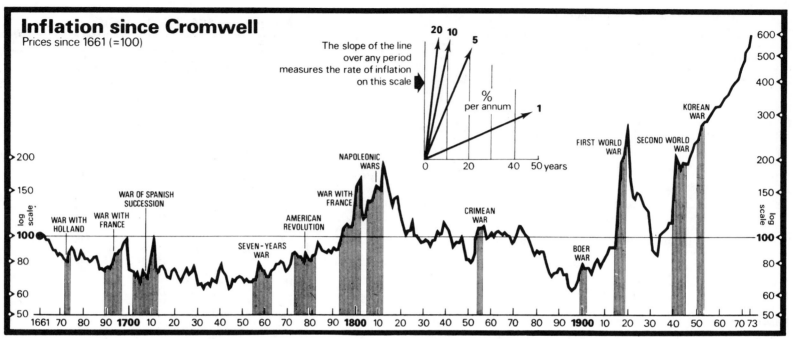

Inflation since Cromwell
Prices since 1661 (=100)

The slope of the line over any period measures the rate of inflation on this scale

% per annum

0   20   40   50 years

WAR WITH HOLLAND
WAR WITH FRANCE
WAR OF SPANISH SUCCESSION
SEVEN-YEARS WAR
AMERICAN REVOLUTION
WAR WITH FRANCE
NAPOLEONIC WARS
CRIMEAN WAR
BOER WAR
FIRST WORLD WAR
SECOND WORLD WAR
KOREAN WAR

1661 70 80 90 1700 10 20 30 40 50 60 70 80 90 1800 10 20 30 40 50 60 70 80 90 1900 10 20 30 40 50 60 70 73

Source: The Economist, London, July 13, 1974.

GROSS NATIONAL PRODUCT								COUNTRIES	INDUSTRIAL PRODUCTION				ELECTRICITY PRODUCTION			
Billions of Current Dollars				Billions of Constant 1967 Dollars					1967 = 100				Billions of Kwh			
1975	1976	1977 Forecast	1978 Forecast	1975	1976	1977 Forecast	1978 Forecast		1975	1976	1977 Forecast	1978 Forecast	1975	1976	1977 Forecast	1978 Forecast
66.7	78.1	91.5	105.1	36.5	36.9	38.6	40.3	AUSTRALIA	128	131	134	137	74.1	76.5	78.9	81.2
39.9	44.5	49.3	54.2	24.4	25.7	26.8	27.8	AUSTRIA	155	168	180	191	35.2	37.3	39.2	40.8
64.4	72.8	81.4	91.0	38.2	39.7	41.2	42.7	BELGIUM	130	140	150	162	41.0	43.6	45.8	48.1
154.0	176.4	196.1	218.9	93.3	97.7	101.5	106.4	CANADA	139	145	149	158	272.7	296.7	314.5	333.4
71.3	76.5	80.9	85.4	71.3	76.5	80.9	85.4	EAST GERMANY	164	174	181	190	84.6	89.2	92.4	96.1
284.1	326.6	368.7	412.0	166.9	175.1	180.4	184.9	FRANCE	141	155	170	184	177.5	189.3	190.1	195.8
83.0	95.3	110.5	127.1	53.0	54.3	56.7	59.9	INDIA	140	151	162	175	75.7	87.7	96.5	106.1
7.6	10.4	14.0	18.9	2.4	2.5	2.6	2.7	ISRAEL	226	232	246	258	9.7	10.4	10.9	11.4
128.1	159.3	191.2	224.3	73.1	77.2	79.0	80.6	ITALY	126	141	149	155	144.8	157.7	164.1	170.6
524.2	594.7	675.7	766.4	262.8	281.2	300.0	321.1	JAPAN	172	193	206	222	449.0	457.9	485.4	514.5
44.4	56.1	73.0	89.6	21.2	21.4	21.9	22.9	MEXICO	178	182	185	192	43.2	48.0	53.3	59.7
69.4	80.0	89.0	98.1	57.0	62.0	65.1	68.3	POLAND	212	223	232	241	97.2	104.1	110.3	115.8
81.5	98.1	120.1	160.8	42.6	43.4	44.3	45.6	SPAIN	194	203	207	215	82.4	91.4	102.4	112.6
65.7	75.0	81.9	92.3	39.9	41.1	41.8	42.8	SWEDEN	138	137	136	143	79.2	82.0	86.1	93.0
491.0	516.0	537.2	564.1	491.0	516.0	537.2	564.1	U.S.S.R.	183	191	202	214	1,034.3	1,106.7	1,184.2	1,278.8
147.9	171.1	198.5	223.0	70.4	71.4	72.6	73.3	UNITED KINGDOM	121	122	125	127	272.2	275.0	277.7	280.5
25.4	28.9	34.0	40.8	15.2	16.1	16.9	18.4	VENEZUELA	127	137	149	162	21.2	23.7	27.0	31.2
443.3	482.4	524.3	570.9	280.2	295.4	308.8	319.8	WEST GERMANY	140	150	158	164	301.8	322.9	348.9	368.0
1,528.8	1,706.5	1,900.0	2,095.0	949.9	1,007.3	1,058.9	1,063.2	UNITED STATES	118	130	138	147	2,003.0	2,123.4	2,276.9	2,419.5

Source: Based on McGraw-Hill Publications' Department of Economics, June 17, 1977; U.S. Department of Commerce, Ebasco Data.   EBASCO

U.S. ENERGY CONSUMPTION AND ECONOMIC OUTPUT

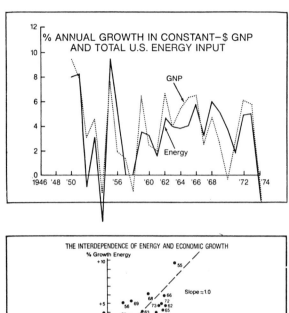

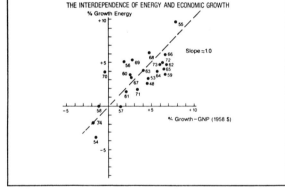

Source: John W. Simpson, Statement before the Ways and Means Committee, U.S. House of Representatives, March 5, 1975.

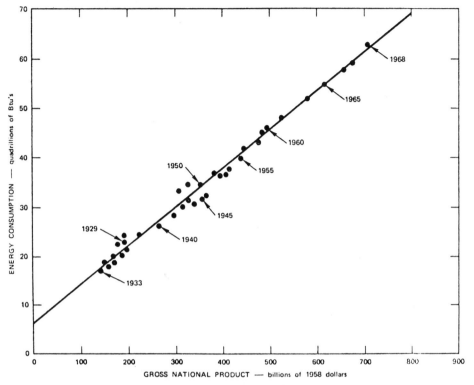

NOTE: Redrawn from publication "The Outlook For Fossil Fuels" by Warren B. Davis, Director-Economics, Gulf Oil Corporation

Source: Support of Energy Program Planning, Stanford Research Institute, sponsored by the Advanced Research Projects Agency, September 1972.

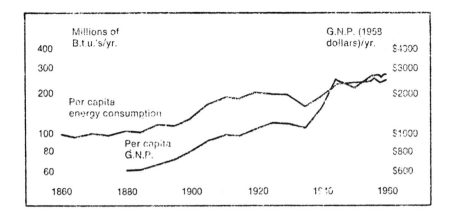

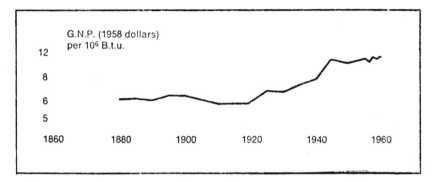

Per capita energy consumption in the United States stayed constant for the first 35 years of the record beginning in 1850, then began the relatively steady rise which led to doubling between 1885 and 1920, fell precipitously during the great depression, and has had a generally upward trend since 1932. The ratio of G.N.P. to energy consumption is shown by the single lower curve, which is in fact a measure of our generally increasing efficiency of energy use in terms of G.N.P. achieved for a given expenditure of energy.

Source: "Energy Technology to the Year 2000," A Special Symposium, *Technology Review*, edited at the Massachusetts Institute of Technology, October–November 1971.

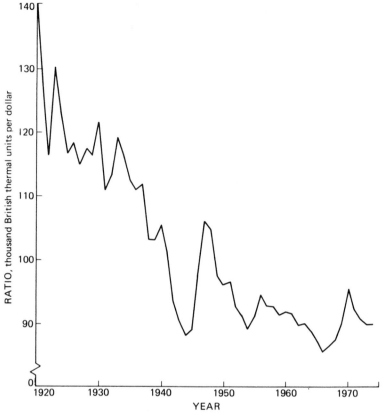

Source: Environmental Quality–1976, The Seventh Annual Report of the Council on Environmental Quality, September 1976.

PRICE TRENDS IN THE UNITED STATES

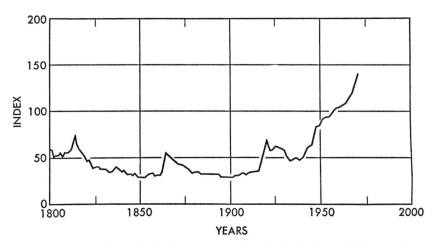

Consumer price index, 1800–1971 (U.S. Bureau of Labor Statistics) [1957–1959 = 100].

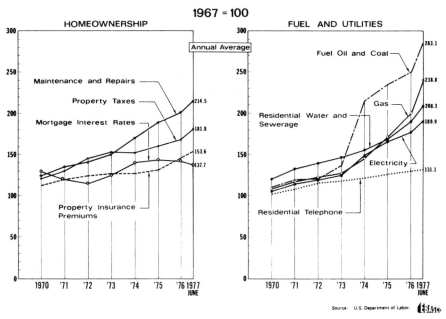

Selected components of the Consumer Price Index.

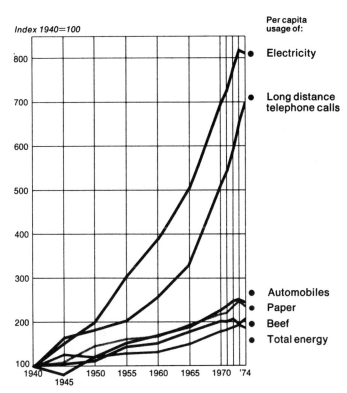

U.S. increase in per capita consumption.
Tom Cardamone for *Fortune Magazine*, April, 1975.

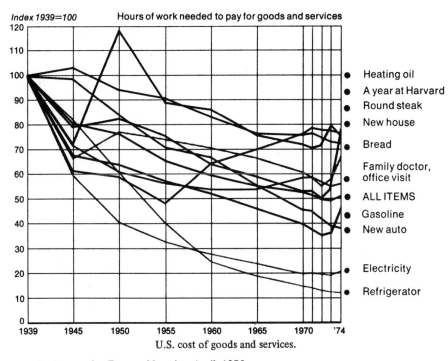

U.S. cost of goods and services.
Tom Cardamone for *Fortune Magazine*, April, 1975.

Price Movements

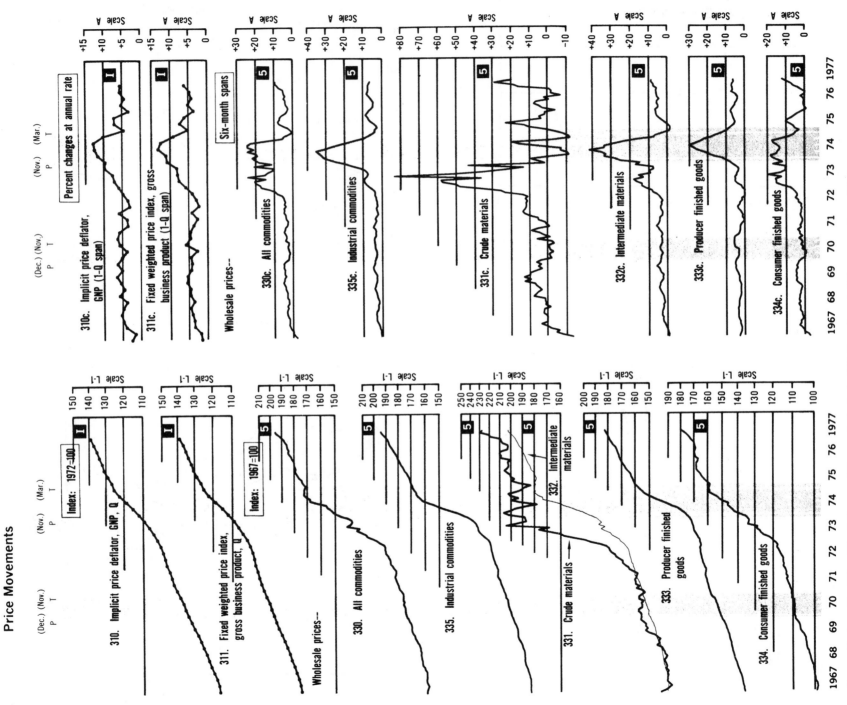

Source: Business Conditions Digest, Bureau of Economic Analysis, U.S. Department of Commerce, June, 1977.

TRENDS IN U.S. ELECTRIC POWER COSTS

Average Costs for Operating Home Appliances. Central air conditioner, 30,000 BTU's, in a 1,500 square foot home (about three bedrooms), set at 75 degrees: $165.00 for the season from May to September.

Room air conditioner, 6,000 BTUs, 75 degrees, average use: $45.00 from May to September.

Washing machine, automatic, 20 hours per month: 40 cents.

Clothes dryer, 16 hours per month: $3.20.

Water heater, 50 gallon, standard: $14.08. Quick-recovery (with extra heating element): $16.20.

Electric toothbrush or electric knife: 4 cents per month.

Dishwasher, 25 hours per month: $1.20.

Oven and range, about 85 hours per month: $4.00. With self-cleaning oven: $4.16.

Refrigerator-freezer, frost-free, 14 cubic feet: $6.40 per month. Non-frost-free: $3.32.

Attic fan: about $3.50 per month. (Varies greatly with size of house and the weather.)

Electric lawn mower, six hours per month: 24 cents.

Circulating pump for 30′ by 15′ swimming pool, about $30 per month.

Microwave oven: 15 hours per month: 88 cents.

Television, 100 hours per month, color solid state: 80 cents. Color tube type: 1.20. Black and white solid state: 22 cents. Black and white tube type: 64 cents.

Toaster oven, 10 hours per month: 60 cents.

Trash compactor, average use per month: 4 cents.

Waste disposer, 7 hours per month: 12 cents.

Source: Potomac Electric Power Company, Washington, D.C., September, 1975.

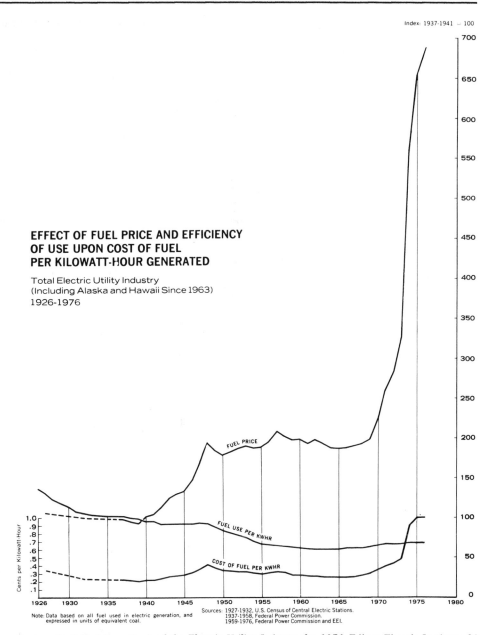

Index: 1937-1941 = 100

EFFECT OF FUEL PRICE AND EFFICIENCY OF USE UPON COST OF FUEL PER KILOWATT-HOUR GENERATED

Total Electric Utility Industry
(Including Alaska and Hawaii Since 1963)
1926-1976

Note: Data based on all fuel used in electric generation, and expressed in units of equivalent coal.

Sources: 1927-1932, U.S. Census of Central Electric Stations.
1937-1958, Federal Power Commission.
1959-1976, Federal Power Commission and EEI.

Source: Statistical Year Book of the Electric Utility Industry for 1976, Edison Electric Institute, 90 Park Avenue, New York, N.Y. 10016, October 1977.

Average Revenues per Kilowatt-Hour Sold—Total Electric Utility Industry
cents per kilowatt-hour

Year	Residential	Commercial and Industrial		Street and Highway Lighting	Other Public Authorities	Railroads and Railways	Total Ultimate Customers
		Small Light and Power	Large Light and Power				
1976	3.45¢	3.46¢	2.07¢	5.03¢	2.44¢	3.78¢	2.89¢
1975	3.21	3.23	1.92	4.74	2.24	3.72	2.70
1974	2.83	2.85	1.55	4.31	1.92	3.43	2.30
1973	2.38	2.30	1.17	3.83	1.47	2.20	1.86
1972	2.29	2.22	1.09	3.69	1.37	1.92	1.77
1971	2.19	2.12	1.03	3.53	1.29	1.72	1.69
1970	2.10	2.01	0.95	3.38	1.20	1.49	1.59
1969	2.09	1.99	0.91	3.25	1.16	1.39	1.54
1968	2.12	2.00	0.90	3.18	1.18	1.35	1.55
1967	2.17	2.04	0.90	3.12	1.19	1.33	1.56
1966	2.20	2.06	0.89	3.05	1.24	1.31	1.56
1965	2.25	2.13	0.90	2.99	1.34	1.31	1.59
1964	2.31	2.19	0.91	2.99	1.35	1.31	1.62
1963	2.37	2.28	0.93	3.05	1.36	1.32	1.65
1962	2.41	2.37	0.96	3.04	1.39	1.34	1.68
1961	2.45	2.35	0.97	3.04	1.41	1.34	1.69
1960	2.47	2.46	0.97	3.18	1.49	1.31	1.69
1959	2.51	2.38	0.96	3.08	1.47	1.28	1.69
1958	2.54	2.43	0.97	3.08	1.49	1.25	1.71
1957	2.56	2.44	0.94	3.05	1.47	1.22	1.67

Based on sales data in Table 19 S and revenue data in Table 33 S.

ULTIMATE CUSTOMER

Total Electric Utility Industry
(Including Alaska and Hawaii Since 1960)
Average Use per Customer and
Average Revenue per Kilowatt-Hour · 1926-1976

Based on Tables 44 S and 45 S, pages 52 and 53.

Source: Statistical Year Book of the Electric Utility Industry for 1976, Edison Electric Institute, 90 Park Avenue, New York, N.Y. 10016, October 1977.

Consumer Price Index—All-City Average*
1955-1976
Index Numbers: 1967 = 100

Date	All Items	Electricity	Gas and Electricity	Total Housing	Food	Apparel
December, 1976	174.3	182.4	201.4	181.6	181.7	151.8
December, 1975	166.3	171.4	179.0	172.2	180.7	145.2
December, 1974	155.4	157.5	156.7	159.9	169.7	141.9
December, 1973	138.5	129.0	131.0	140.6	151.3	130.5
December, 1972	127.3	120.2	122.5	131.2	126.0	125.0
December, 1971	123.1	116.0	118.2	126.8	120.3	121.8
December, 1970	119.1	109.9	110.7	122.6	115.3	119.2
December, 1969	112.9	104.2	104.8	114.2	112.8	114.7
December, 1968	106.4	101.0	101.4	107.0	105.2	109.0
December, 1967	101.6	100.5	100.2	101.5	100.9	102.5
December, 1966	98.6	99.2	99.4	98.9	99.7	98.5
December, 1965	95.4	99.1	99.5	95.7	96.0	94.8
December, 1964	93.6	99.3	99.8	94.3	92.8	93.5
December, 1963	92.5	100.0	99.6	93.5	91.5	93.1
December, 1962	91.0	100.3	99.6	92.0	89.8	91.1
December, 1961	89.9	100.1	99.4	91.3	88.5	90.8
December, 1960	89.3	100.0	99.3	90.8	89.3	90.5
December, 1959	88.0	99.6	97.0	89.5	86.6	89.4
December, 1958	86.7	97.6	93.5	87.9	87.3	88.0
December, 1957	85.2	96.3	90.3	87.1	85.4	88.1
December, 1956	82.7	95.5	88.6	84.8	83.1	87.5
December, 1955	80.4	95.4	88.1	82.9	80.6	85.7

*Alaska and Hawaii included since 1963.
Source: U. S. Bureau of Labor Statistics.

Residential Electricity and Consumer Price Index—Total Electric Utility Industry*

Year	Average Revenue per Residential Kwhr Used	100 Kwhr	250 Kwhr	500 Kwhr	750 Kwhr	1,000 Kwhr	Electricity	Consumer Price Index‡
			Average Rate per Kwhr Based on Average of Typical Bills in Cities** Price per Kwhr for Monthly Use of: (Federal Power Commission)				U.S. Bureau of Labor Statistics Index Numbers Based on 1967=100†	
1976	3.45¢						177.6	170.5
1975	3.21	6.15¢	4.82¢	3.85¢	3.57¢	3.49¢	167.0	161.2
1974	2.83	5.89	4.60	3.59	3.30	3.23	147.5	147.7
1973	2.38	4.99	3.79	2.82	2.55	2.49	124.9	133.1
1972	2.29	4.65	3.47	2.51	2.26	2.19	118.9	125.3
1971	2.19	4.51	3.34	2.40	2.15	2.07	113.2	121.3
1970	2.10	4.25	3.14	2.23	2.00	1.92	106.2	116.3
1969	2.09	4.09	3.00	2.10	1.90	1.83	102.8	109.8
1968	2.12	4.05	2.96	2.06	1.86	1.80	100.9	104.2
1967	2.17	4.03	2.95	2.07	1.89	1.83	100.0	100.0
1966	2.20	4.03	2.95	2.07	1.89	1.83	99.1	97.2
1965	2.25	4.00	2.94	2.07	1.89	1.83	99.1	94.5
1964	2.31	4.02	2.95	2.08	1.91	1.86	99.6	92.9
1963	2.37	4.03	2.97	2.12	1.93	1.89	100.1	91.7
1962	2.41	4.06	2.99	2.13	1.95	...	100.1	90.6
1961	2.45	4.06	2.99	2.13	...	...	100.1	89.6
1960	2.47	4.05	2.98	2.13	...	...	99.8	88.7
1959	2.51	4.04	2.98	2.12	...	...	98.5	87.3
1958	2.54	3.98	2.94	2.10	...	...	97.1	86.6
1957	2.56	3.93	2.92	2.09	...	...	95.9	84.3
1956	2.61	3.89	2.89	2.08	...	...	95.5	81.4
1955	2.65	3.88	2.88	2.07	...	...	95.2	80.2

* Alaska and Hawaii included since 1964.
** As determined by the Federal Power Commission in its "Typical Electric Bills." The figures cover rate schedules as of December 31 in all cities with populations of 2,500 or more.
† "Index numbers" are percentages of the base year 1967. For a description of the components and characteristics of this Index, see BLS Handbook of Methods, Bulletin #1711 for 1971.
‡ The "Consumer Price Index" measures the average change in prices of goods and services purchased by urban wage-earner and clerical-worker families. Electricity is one of these components. The Index does not reflect the over-all effect of the increased use of electricity by Residential customers—for which see "Average Revenue per Residential Kwhr Used," column 2 of this table; also see Tables 44 S, 45 S and 60 S of this publication.

Source: Statistical Year Book of the Electric Utility Industry for 1976, Edison Electric Institute, 90 Park Avenue, New York, N.Y. 10016, October 1977.

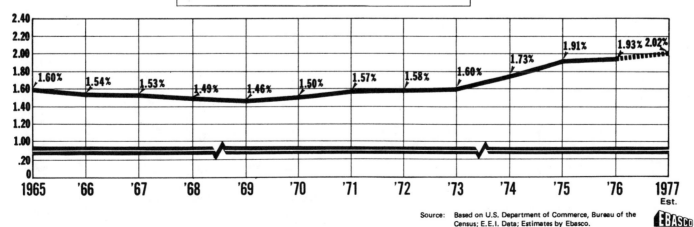

RESIDENTIAL ELECTRICITY-AVERAGE ANNUAL BILL AS A PERCENT OF MEDIAN FAMILY INCOME

Source: Based on U.S. Department of Commerce, Bureau of the Census; E.E.I. Data; Estimates by Ebasco.

TRENDS IN CRUDE OIL PRICES

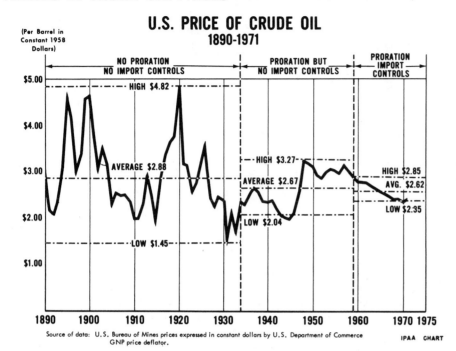

U.S. PRICE OF CRUDE OIL
1890-1971

(Per Barrel in Constant 1958 Dollars)

Source of data: U.S. Bureau of Mines prices expressed in constant dollars by U.S. Department of Commerce GNP price deflator.

IPAA CHART

Source: Tom B. Medders, Jr., Independent Petroleum Association of America, Statement in *Trends in Oil and Gas Exploration,* Hearings before the Committee on Interior and Insular Affairs, United States Senate, Part I, August 8 and 9, 1972.

Prices

Year	Crude oil at well (per barrel)			Motor gasoline retail (cents per gallon)		
	Current dollars	Constant 2958 dollars	Constant 1972 dollars	Excise tax	Taxes	Total
1953	2.68	3.03	4.42	21.28	7.41	28.69
1954	2.78	3.10	4.52	21.56	7.48	29.04
1955	2.77	3.05	4.45	21.42	7.65	29.07
1956	2.79	2.97	4.33	21.57	8.36	29.93
1957	3.09	3.17	4.63	22.11	8.85	30.96
1958	3.01	3.01	4.39	21.47	8.91	30.38
1959	2.90	2.85	4.16	21.18	9.31	30.49
1960	2.88	2.79	4.07	20.99	10.14	31.13
1961	2.89	2.76	4.03	20.53	10.23	30.76
1962	2.90	2.74	4.00	20.36	10.28	30.64
1963	2.89	2.70	3.93	20.11	10.31	30.42
1964	2.88	2.65	3.86	19.98	10.37	30.35
1965	2.86	2.58	3.76	20.71	10.46	31.17
1966	2.88	2.53	3.69	21.57	10.51	32.08
1967	2.91	2.47	3.61	22.55	10.60	33.15
1968	2.94	2.40	3.51	22.93	10.78	33.71
1969	3.09	2.41	3.51	23.85	10.99	34.84
1970	3.18	2.35	3.43	24.55	11.14	35.69
1971	3.39	2.39	3.49	25.24	11.24	36.48
1972	3.39	2.32	3.39	24.46	11.67	36.13

Source: Fiscal Policy and the Energy Crisis, Hearings before the Subcommittee on Energy of the Committee on Finance, United States Senate, Part 3, January 23 and 24, 1974.

The Dramatic Jump in Middle East Crude Prices

Period	Posted price [1]	Arab tax	Total cost (production cost plus tax)
1960–65	$1.80	$0.82	$0.92
1966–67	1.80	.85	.95
1968–69	1.80	.88	.98
Jan. 1 to Nov. 14, 1970	1.80	.91	1.01
Nov. 15, 1970 to Feb. 14, 1971	1.80	.99	1.10
Feb. 15 to May 31, 1971	2.18	1.26	1.37
June 1, 1971 to Jan. 19, 1972	2.28	1.32	1.43
Jan. 20, 1972 to Jan 1, 1973	2.48	1.44	1.55
Jan. 1 to Mar. 31, 1973	2.59	1.51	1.62
Apr. 1 to May 31, 1973	2.75	1.61	1.71
June 1973	2.90	1.70	1.80
July 1973	2.95	1.74	1.84
August 1973	3.07	1.80	1.90
Oct. 1 to Oct. 15, 1973	3.01	1.77	1.87
Oct. 16 to Dec. 31, 1973	5.12	3.05	3.15
Jan. 1, 1974	11.56	7.00	7.10

[1] Posted price is the base on which OPEC countries figure their tax; it bears little relationship to the producers' selling price.

Source: International Crude Oil & Product Prices.

Source: Fiscal Policy and the Energy Crisis, Hearings before the Subcommittee on Energy of the Committee on Finance, United States Senate, Part 4, January 25, 28, and 29, 1974.

Oil Prices, Oil Imports (Crude and Product) Total U.S. Oil Products Consumption, U.S. Crude Production January 1972 through March 1975 (Selected Months 1972, 1973)

[In thousands of barrels of oil per day]

Year and month	Imports, price per barrel [1]	Domestic, uncontrolled well head price	Imports, crude	Imports, refined products	Total oil imports	Total U.S. consumption oil products	U.S. crude production
1972: January	$2.48		2,046	2,721	4,767	16,735	9,114
1973:							
August	3.07		3,593	2,913	6,906	17,414	9,161
October	5.12		3,740	2,785	6,525	17,095	9,172
November	5.18		3,452	3,412	6,864	18,434	9,144
December	5.04		2,891	3,055	5,746	17,429	9,041
1974:							
January	9.59	$9.82	2,382	2,973	5,355	17,270	8,907
February	12.45	9.87	2,248	2,973	5,221	17,371	9,156
March	12.73	9.88	2,462	2,753	5,215	16,045	8,950
April	12.72	9.88	3,267	2,703	5,970	15,919	8,952
May	13.02	9.88	3,908	2,454	6,362	15,740	8,903
June	13.06	9.95	3,925	2,218	6,143	16,191	8,777
July	12.75	9.95	4,091	2,140	6,231	15,853	8,698
August	12.68	9.98	3,924	2,281	6,205	15,803	8,717
September	12.53	10.10	3,758	2,180	5,938	16,318	8,622
October	12.44	10.75	3,936	2,361	6,297	17,121	8,651
November	12.53	10.90	3,997	2,581	6,578	17,129	8,458
December	12.82	11.08	3,964	2,623	6,587	17,588	8,471
1975:							
January	12.77	11.28	3,934	2,324	6,298	17,425	8,577
February	[1] 13.05	11.39	4,061	2,136	6,197	17,338	8,496
March	13.28	11.47	3,834	1,938	5,772	16,635	8,394
April	13.28	[2] 11.57					
May	[2] 13.11						
June							

[1] Beginning in 1974, includes fees and transportation to refineries.   [2] Preliminary.

Source: FEA publications.

Source: Highlights of Energy Legislation in the 94th Congress, (Through June 30, 1975), A Background Paper prepared by the Congressional Research Service at the request of the Committee on Interior and Insular Affairs, United States Senate, 1975.

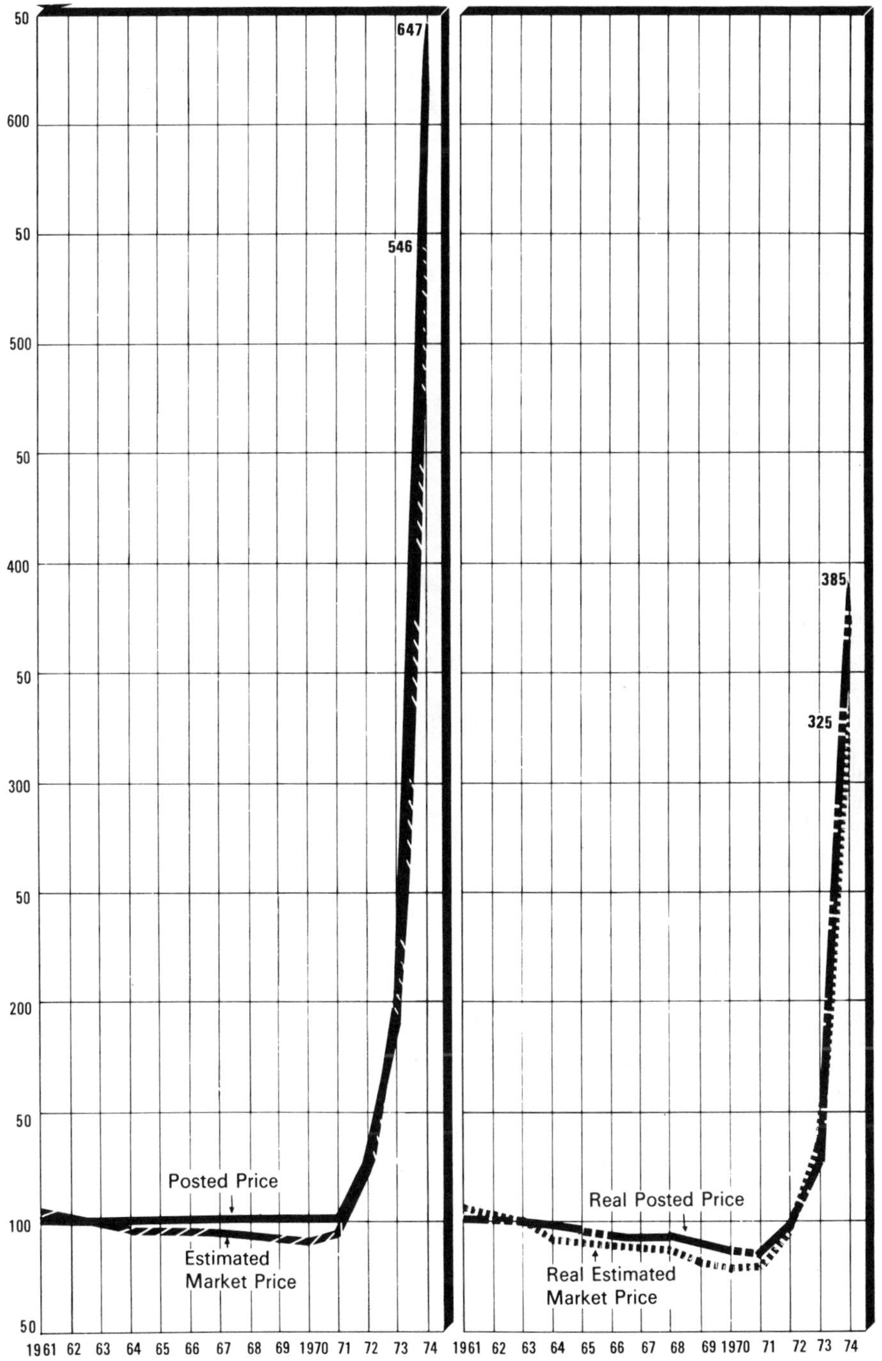
50
600
50
500
50
400
50
300
50
200
50
100
50
647
546
Posted Price
Estimated
Market Price
385
325
Real Posted Price
Real Estimated
Market Price
1961 62 63 64 65 66 67 68 69 1970 71 72 73 74
1961 62 63 64 65 66 67 68 69 1970 71 72 73 74

OPEC–Oil prices (1963 = 100).

Source: Energy Prospects to 1985, Volume 1, Organisation for Economic Co-operation and Development, Paris, France, 1974.

TRENDS IN FUEL PRICES

U.S. Fossil Fuel Prices[1] in Cents per Million Btu, 1950–75
[In both current and 1972 prices]

Year	Crude oil		Natural gas liquids		Natural gas		Bituminous coal		Anthracite coal		Composite[2]		GNP implicit price deflators[3]
	Current	Deflated	Current	Deflated	Current	Deflated	Current	Deflated	Current	Deflated	Current	Deflated	
1950	43.3	80.7	55.0	102.5	6.0	11.2	18.5	34.5	35.0	65.2	26.3	49.0	53.64
1951	43.6	76.1	59.0	103.0	7.1	12.4	18.8	32.8	37.4	65.3	26.9	47.0	57.27
1952	43.6	75.2	59.9	103.3	7.5	12.9	18.7	32.2	36.9	63.6	27.3	47.1	58.00
1953	46.2	78.5	59.8	101.6	8.9	15.1	18.8	31.9	38.1	64.7	28.7	48.7	58.88
1954	47.9	80.2	55.0	92.1	9.8	16.4	17.3	29.0	33.5	56.1	29.1	48.8	59.69
1955	47.8	78.4	52.4	85.9	9.8	16.1	17.2	28.2	30.9	50.7	28.5	46.7	60.98
1956	48.1	76.5	56.7	90.1	10.4	16.5	18.4	29.3	32.2	51.2	29.2	46.4	62.90
1957	53.3	82.0	54.8	84.3	10.9	16.8	19.4	29.8	35.4	54.4	31.3	48.1	65.02
1958	51.9	78.6	55.7	84.3	11.6	17.6	18.6	28.2	35.0	53.0	24.8	37.5	66.06
1959	50.0	74.1	56.2	83.2	12.5	18.5	18.2	27.0	32.4	48.0	29.9	44.3	67.52
1960	49.7	72.4	56.7	82.6	13.6	19.8	17.9	26.1	30.8	44.9	29.8	43.4	68.67
1961	49.8	71.9	51.4	74.2	14.6	21.1	17.5	25.3	31.7	45.8	28.0	40.4	69.28
1962	50.0	70.9	50.9	72.1	15.0	21.3	17.1	24.2	31.3	44.4	29.9	42.4	70.55
1963	49.8	69.6	47.4	66.2	15.4	21.5	16.8	23.5	33.1	46.2	29.4	41.1	71.59
1964	49.7	68.4	46.7	64.2	15.0	20.6	17.0	23.4	34.1	46.9	28.9	39.7	72.71
1965	49.3	66.3	49.0	65.9	15.1	20.3	16.9	22.7	33.5	45.1	28.6	38.5	74.32
1966	49.7	64.7	53.3	69.4	15.2	19.8	17.3	22.5	31.8	41.4	29.1	37.9	76.76
1967	50.3	63.7	54.5	69.0	15.5	19.6	17.6	22.3	32.1	40.6	29.8	37.7	79.02
1968	50.7	61.4	48.6	58.9	15.8	19.1	17.8	21.6	34.6	41.9	29.8	36.1	82.57
1969	53.3	61.5	45.2	52.1	16.2	18.7	19.0	21.9	39.0	45.0	30.8	35.5	86.72
1970	54.8	60.0	50.0	54.7	16.6	18.2	23.9	26.2	42.6	46.6	30.4	33.3	91.36
1971	58.4	60.8	54.6	56.9	17.7	18.4	29.1	30.3	46.7	48.6	35.7	37.2	96.02
1972	58.4	58.4	56.2	56.2	18.0	18.0	31.9	31.9	47.2	47.2	36.5	36.5	100.0
1973	67.0	63.3	72.9	68.9	21.0	19.8	35.5	33.5	52.0	49.1	41.9	39.6	105.92
1974	116.2	100.0	124.2	106.9	29.5	25.4	65.6	56.5	86.1	74.1	75.7	65.0	116.20
1975[4]	137.9	109.1	155.8	123.3	42.4	33.6	78.1	61.8	98.4	77.9	87.9	69.6	126.35

[1] All fuel prices taken as close to the point of production as possible.

[2] Weighted by relative importance of individual fuels in demand for fossil fuels.

[3] GNP implicit price deflators are used in computing the deflated prices shown. 1975 estimated. 1972 = 100.

[4] Estimated.

Source: Energy Perspectives 2, U.S. Department of the Interior, June 1976.

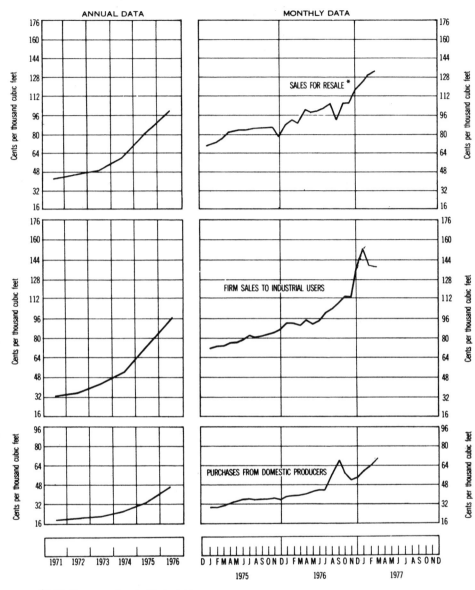

ANNUAL DATA   MONTHLY DATA

SALES FOR RESALE *

FIRM SALES TO INDUSTRIAL USERS

PURCHASES FROM DOMESTIC PRODUCERS

1971 1972 1973 1974 1975 1976

D J FMAM J J ASOND J FMAM J J ASOND J F MAM J J AS OND
1975   1976   1977

* To intrastate gas companies under FPC rate schedules.

Average prices of gas bought and sold by natural gas interstate pipeline companies.

Source: *FPC News*, Federal Power Commission, Volume 10, No. 36, Sept. 9, 1977.

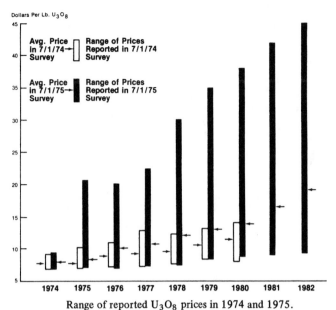

Range of reported U_3O_8 prices in 1974 and 1975.

Source: *National Energy Outlook*, FEA-N-75/713, Federal Energy Administration, February 1976.

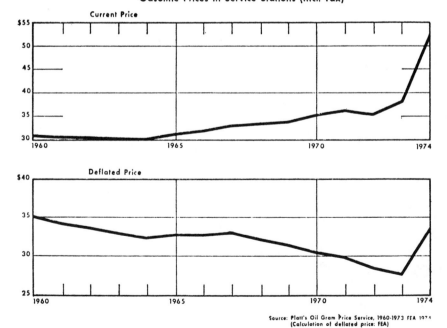

Gasoline Prices in Service Stations (Incl. Tax)

Current Price

Deflated Price

Source: Platt's Oil Gram Price Service, 1960-1973 FEA 1974
(Calculation of deflated price: FEA)

Source: *Energy Statistics*, Committee on Finance, United States Senate, July 4, 1975.

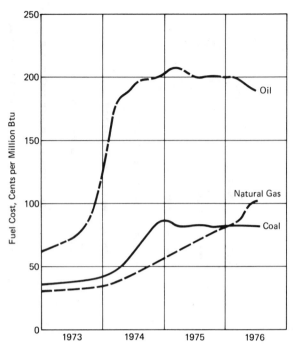

Cost of fossil fuels delivered to U.S. steam-electric utility plants, 25 MW or greater.

Source: Adapted from *FPC News*, Volume 10, Number 20, Federal Power Commission, May 20, 1977.

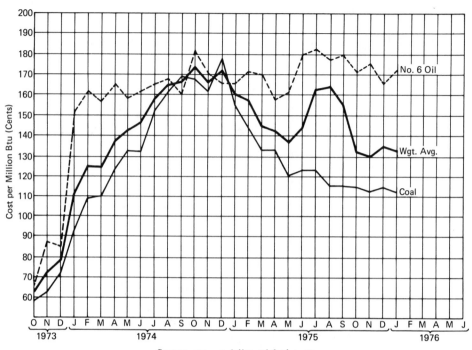

System average delivered fuel costs.

Source: Potomac Electric Power Company, Washington, D.C.

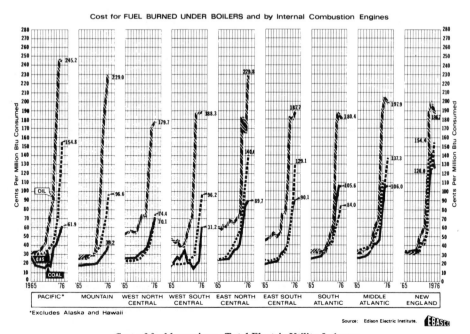

Cost of fuel by regions—Total Electric Utility Industry.

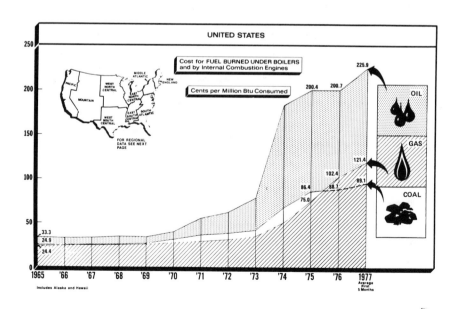

Cost of fuel for electric generation—Total Electric Utility Industry.

COSTS OF SYNTHETIC FUELS FROM COAL

Costs of Synthetic Fuels from Coal[1] , 1973

PRODUCT	HEAT VALUE		PLANT OUTPUT		TOTAL PLANT INVESTMENT (Millions of Dollars)		PRODUCT COSTS[2] (Dollars Per Million Btu)	
	Btu per cf	Btu per Bbl	cf/day	Bbls/day	EASTERN COAL	WESTERN COAL	EASTERN COAL	WESTERN COAL
LIQUID BOILER FUEL		6,400,000		65,000	427-540	450-590	1.98-2.20	1.60-1.80
DISTILLATE MIX		6,400,000		65,000	595-750	620-815	2.26-2.45	1.75-2.00
PIPELINE GAS:								
CURRENT TECHNOLOGY (Lurgi)	930		250x10⁶		---	475-625	---	2.53-2.83
ADVANCED TECHNOLOGY	930		250x10⁶		380-480	380-500	2.70-2.98	2.25-2.50
LOW-BTU UTILITY GAS	130		1700x10⁶		70-90	70-95	1.15-1.27	.80-.90
COMBINED OIL/GAS	930	6,400,000	225x10⁶	40,000	450-600	475-665	1.85-2.18	1.50-1.70

[1] Synthetic liquids are desulfurized
[2] Based on a 15 percent discounted cash flow.

Source: Energy Perspectives, U.S. Department of the Interior, February 1975.

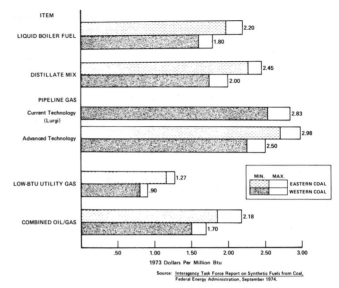

Costs of synthetic fuels from coal, 1973.

Source: Energy Perspectives, U.S. Department of the Interior, February 1975.

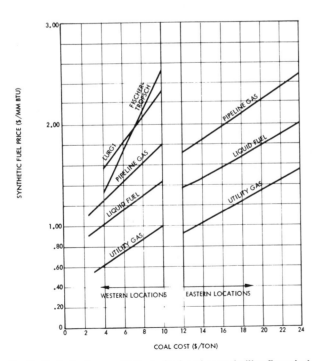

Synthetic fuel price sensitivity to feed coal costs (utility financing).

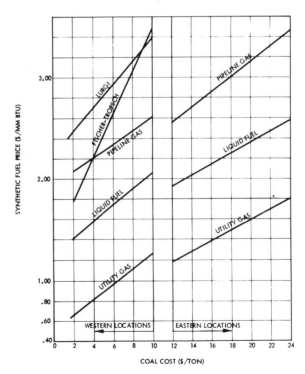

Synthetic fuel price sensitivity to feed coal costs (15% DCF).

Source: Synthetic Fuels From Coal, Task Force Report, Project Independence Blueprint, Federal Energy Administration, November 1974.

Estimated Synthetic Fuel Plant Investment[1] and Operating Costs[2]
(In Late 1973 Dollars)

	Low Btu Utility Gas		Pipeline Gas		Liquid Fuel		Lurgi[4]	Fischer-Tropsch[6]
	East	West	East	West	East	West	West	West
Coal Consumption (MM tons/yr)	3.564	4.831	5.132	6.930	8.237	11.167	9.742	18.092
Total Capital Requirement	83.825	83.825	380.000	380.000	456.200[5]	456.200[5]	427.097	715.00
Working Capital	7.620	7.620	34.200	34.200	41.470	41.470	38.830	65.00
By-product Credits[3]	2.700	2.700	10.380	10.380	0	0	13.020	0
Total Net Operating Costs	52.745	29.550	87.730	53.370	151.780	96.660	64.790	142.910
Annual Depreciation	3.810	3.8.0	17.100	17.100	20.740	20.740	19.410	32.500
Total Plant Investment	76.205	76.205	342.000	342.000	414.730[5]	414.730[5]	388.270	650.000
Raw Materials								
Coal	42.770	19.325	61.586	27.720	98.842	44.670	28.248	72.370
Process Water	0.300	0.400	2.450	1.958	4.572	3.802	1.140	0.355
Cooling Water	0.397	0.336						1.550
Electric Power	0.991	0.991	0[7]	0[7]	13.750	13.750	0[7]	5.697
Catalysts and Chemicals	0.285	0.285	1.275	1.275	6.660	6.660	1.120	8.860
Labor	3.153	3.153	10.390	10.390	11.050	11.050	13.230	12.900
Supplies								
Operating	2.456	2.456	2.732	2.732	Included in catalysts		2.910	15.900
Maintenance	1.350	1.350	4.693	4.693	and chemicals		2.020	(Included above)
Indirect Costs								
Taxes, Insurance	2.058	2.058	8.750	8.750	11.200	11.200	10.480	17.550
Indirect Payroll and G and A	1.892	1.892	6.234	6.234	5.537	5.537	7.940	7.390
Total Gross Operating Costs	55.441	32.246	98.108	63.753	151.783	96.660	77.813	142.910
Total Output, 10^{12} Btu/yr	67.267	67.267	76.477	76.477	153.183	153.183	76.088	137.800
Fuel Unit Price, $/$10^6$ Btu								
Utility	0.93	0.59	1.75	1.30	1.37	1.01	1.44	1.70
12% DCF	1.11	0.77	2.46	2.01	1.81	1.45	2.25	2.47
15% DCF	1.18	0.82	2.73	2.28	1.97	1.61	2.55	2.76
20% DCF	1.39	1.04	3.56	3.11	2.48	2.12	3.49	3.75

Notes:
1. Investment costs do not include the coal mine, feed coal transportation to the plant or product transportation
2. Operating cost estimates per the FPC Synthetic Gas-Coal method
3. No by-product credit given to elemental sulfur
4. Lurgi plant above differs from ElPaso Natural Gas Co. design in having coal-fired boilers rather than low-BTU gas-fired boilers; this results in lower capital investment.
5. No East/West difference shown because change in capital investment is function of specific coal used and will vary from location to location.
6. Fischer-Tropsch plant listed is a dual-product facility, producing gasoline and pipeline gas.
7. Electric power is plant-produced.

Source: Synthetic Fuels From Coal, Task Force Report, Project Independence Blueprint, Federal Energy Administration, November 1974.

COSTS OF ALTERNATE FUELS

Process	Capital cost, in millions of 1973 dollars
SNG from coal, old technology	$400
SNG from coal, new technology	$300 to 350
SNG from oil shale	$350
Syncrude from coal	$350
Syncrude from oil shale	$450
Methanol from coal	$350

Capital cost of synthetic fuel plants. The figures shown are for the construction of plants that will produce each day fuel with a total heating value of 250 × 109 B.t.u. Estimating costs for large plants using unproven technology is very difficult, yet these estimates indicate that the costs of building different types of synthesizing plants will be roughly equivalent. (The cost shown for the oil-shale processing plant includes an investment in mining and in waste-disposal facilities; the plant alone would cost perhaps $300 million).

	SNG from coal, using old technology	SNG from coal, using new technology	Syncrude from coal	Syncrude from oil shale	Methanol from coal
Capital, at 15 per cent per year	59	44	51	37	51
Operating costs	22	16	22	22	44
Fuel costs	48	44	37	37	48
Total cost	129	104	110	96	143
Cost per million B.t.u. of product	$1.56	$1.26	$1.33	$1.17	$1.73
Cost per barrel (oil equivalent)	$9.05	$7.30	$7.70	$6.80	$10.00

The annual operating cost, in millions of 1973 dollars, of Various Synthesizing Plants, each producing daily a product with a total heating value of 250×10^9 Btu. While the predictions are necessarily imprecise, it appears that syncrude from oil shale will be less expensive to produce than the other products shown. (The capital and fuel costs for the oil-shale plant reflect the costs of mining, crushing, and handling raw shale, and disposing of spent shale. The cost of coal is calculated using a price of 32¢ per million Btu.)

Source: "Energy Self-Sufficiency: An Economic Evaluation," *Technology Review*, edited at the Massachusetts Institute of Technology, May, 1974.

Estimated Costs for Conversion of Natural Gas to Methanol

Starting Energy Form	Product Energy Form	Methane Costs (¢/MM BTU)	Capital Requirements ($/MM BTU/Day)	Plant Thermal Efficiency (Percent)	Energy-Related Costs (¢/MM BTU)	Operating Costs (¢/MM BTU)	Capital-Related Costs (¢/MM BTU)	Total Cost of Product (¢/MM BTU)
Methane	Methanol	10	400	55	18	15	24	57
Methane	Methanol	20	400	55	36	15	24	75
Methane	Methanol	40	400	55	73	15	24	112
Methane	Methanol	60	400	55	108	15	24	143
Methane	Methanol	80	400	55	146	15	24	185
Methane	Methanol	100	400	55	182	15	24	221

Source: *U.S. Energy Outlook: New Energy Forms*, National Petroleum Council 1973.

Comparison of Liquid and Gaseous Fuels from Crude Oil

Starting Energy Form	Product Energy Form	Crude Oil Cost ($/Bbl)	Crude Oil Cost (¢/MM BTU)	Capital Requirements ($/MM BTU/day)	Thermal Efficiency (Percent)	Energy Related Costs (¢/MM BTU)	Operating Costs (¢/MM BTU)	Capital-Related Costs (¢/MM BTU)	Total Refinery Gate Energy Costs (¢/MM BTU)
High-S Crude Oil	Fuel Oil	4	69	250	92	75	6	15	96
High-S Crude Oil	Methane	4	69	520	80	86	8	32	126
High-S Crude Oil	Fuel Oil	6	103	250	92	112	6	15	133
High-S Crude Oil	Methane	6	103	520	80	129	8	32	169
High-S Crude Oil	Fuel Oil	8	138	250	92	150	6	15	171
High-S Crude Oil	Methane	8	138	520	80	173	8	32	213

Source: *U.S. Energy Outlook: New Energy Forms*, National Petroleum Council, 1973.

Economics—Hydrogen by Electrolysis

Power Costs (Mills/KWH)	Capital Costs ($/MM BTU Hydrogen per Day)	Electrolysis Annual Load Factor (Percent)	Cell Energy Requirements, (KWH/MM BTU of Hydrogen)	Total Capital Factor (Percent)	Capital Costs (¢/MM BTU Hydrogen)	Energy Costs (¢/MM BTU Hydrogen)	Operating Costs (¢/MM BTU Hydrogen)	Credit for Oxygen, ($/Ton)	Selling Price of Electrolytic Hydrogen (¢/MM BTU)
10	1,480	100	426	20	81	426	8	None	515
10	1,480	100	426	15	61	426	8	None	495
10	1,480	100	426	20	81	426	8	9	456
10	1,480	100	426	15	61	426	8	9	436
2	1,480	40	426	20	207	85	8	None	300
2	1,480	40	426	15	155	85	8	None	248
2	1,480	40	426	20	207	85	8	9	241
2	1,480	40	426	15	155	85	8	9	189
2	1,200	40	414	20	164	83	8	None	255
2	1,200	40	414	15	125	83	8	None	216
2	1,200	40	414	20	164	83	8	11	183
2	1,200	40	414	15	125	83	8	11	144
2	1,000	40	403	20	137	81	8	None	226
2	1,000	40	403	15	103	81	8	None	192
2	1,000	40	403	20	137	81	8	11	154
2	1,000	40	403	15	103	81	8	11	120

Source: *U.S. Energy Outlook: New Energy Forms*, National Petroleum Council, 1973.

ENERGY TRANSPORT AND STORAGE COSTS

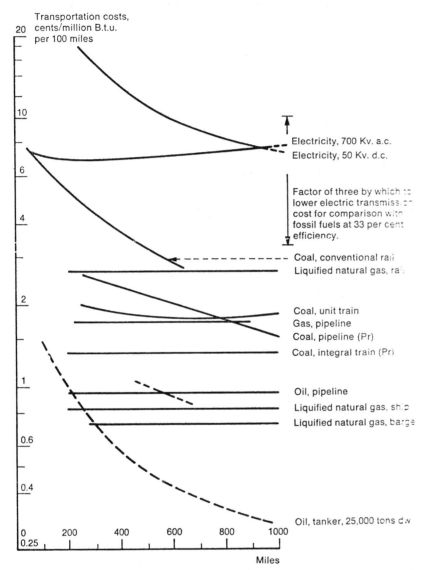

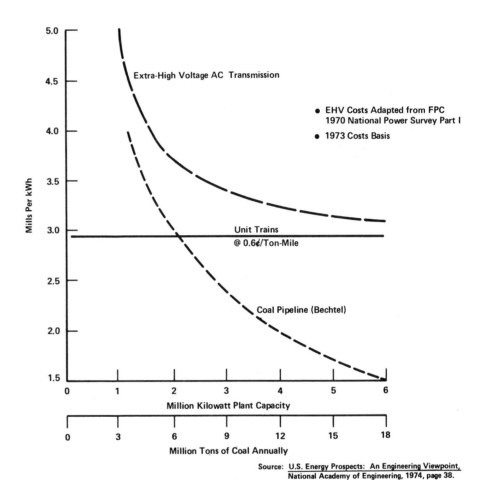

U.S. coal transportation costs, by mode (for a 1,000-mile transport distance).

Source: Energy Perspectives, U.S. Department of the Interior, February 1975.

Though the chart above proposes to show comparative costs of transporting fuel and energy in different forms and by different means, it is in fact so simplified as to be only vaguely correct. For the actual transportation cost depends critically on the efficiency of the particular facility being considered as well as on the particular place and time when it is being used; for these reasons—and because parallel data are difficult or impossible to acquire—a critical issue in developing our energy strategies is only vaguely suggested by this presentation.

Source: "Energy Technology to the Year 2000," A Special Symposium, *Technology Review*, edited at the Massachusetts Institute of Technology, October/November, 1971.

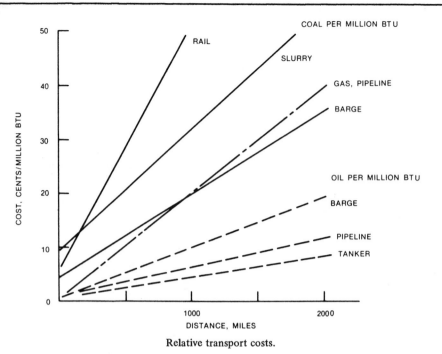

Relative transport costs.

Source: *Project Independence*, Project Independence Report, Federal Energy Administration, November 1974.

Relative Transportation Costs

Energy Form	Transportation Method	Capital-Related Costs (¢/MM BTU)	Energy-Related Costs (¢/MM BTU)	Other Operating Costs (¢/MM BTU)	Total Costs per 1,000 Miles (¢/MM BTU)
Pre-1972 Situation					
Methane	36-42 inch Pipeline	16	2	2	20
Petroleum Liquids	24-30 inch Pipeline	5	0.5	1	6.5
Bituminous Coal	Unit Train	—	—	—	20
1976-2000 Situation					
Substitute Methane	36-42 inch Pipeline-vapor	16	10	2	28
Hydrogen	36-42 inch Pipeline-vapor	20	13	3	36
Methane	Pipeline-LNG	8	3	3	14
Methanol	Pipeline	10	1	2	13
Petroleum Liquids	Pipeline	5	1	1	7

Source: *U.S. Energy Outlook: New Energy Forms*, National Petroleum Council, 1973.

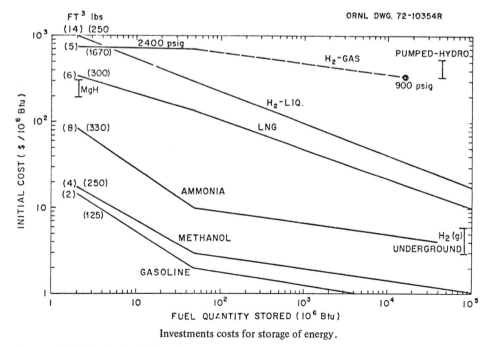

Investments costs for storage of energy.

Source: Oak Ridge National Laboratory.

COSTS OF GEOTHERMAL ENERGY

Selected Comparative Cost Data for Geothermal Energy

Geothermal Field	Geothermal Production	Local Average, Other Fuel
Electricity, U.S. mills/kwh		
Namafjall, Iceland	2.5–3.5	—
Larderello, Italy	4.8–6.0	~7.5
Matsukawa, Japan	4.6	~6.0
Cerro Prieto, Mexico	4.1–4.9	~8.0
Pauzhetsk, U.S.S.R.	7.2	~10.0
The Geysers, United States	5.0	7.0
Space heating, U.S.$/Gcal energy		
Reykjavik, Iceland	4.0	6.7
Szeged, Hungary	3.0	11.0
Refrigeration, U.S.$/Gcal energy		
Rotorua, New Zealand	0.12	2.40
Drying diatomite, U.S.$/ton		
Namafjall, Iceland	~2	~12

"Worldwide Status of Geothermal Resources Development," by James B. Koenig in Geothermal Energy: Resources, Production, Stimulation, edited by Paul Kruger and Carel Otte, with permission of the publishers, Stanford University Press. © 1973 by the Board of Trustees of the Leland Stanford Junior University.

Comparison of Estimated Costs of Producing Electricity

Oil Price ($/BBL)	Electricity (Mills/kWh)
4	22.4
7	27.7
11	34.8

GEOTHERMAL

	Geysers	Other
Busbar price of electricity	11.4	18.03

$$BPE = \frac{cc \times FCR}{c \times 8.76} + E + O\&M$$

where:

cc = capital cost $/kW	364.26	561.63–861.63
FCR = annual fixed charge rate	.15	.15
c = capacity factor	.60	.60
E = energy cost	0	0
O&M = operations and maintenance costs, mills/kWh	1	2

Source: Geothermal Energy Task Force Report, Project Independence, Federal Energy Administration, November 1974.

Estimated Capital Costs for Geothermal Power Plants

Type	Installed Capacity (MW)	Plant Cost $/kW	Geothermal Energy Cost ($/kW)	Total Capital Cost
Geyser[a]	200	140	224.63[c]	364.26
Brine	200	300–600[b]	261.63	561.63–861.63
Hot dry rock	200	300–600	261.63	561.63–861.63[d]

[a]Based on Pacific Gas and Electric Company data.

[b]Plant cost estimates for using energy from hot water range between 300 and 600 per installed kilowatt.

[c]Based on estimates made by B. Greider, Chevron Oil Company.

[d]Until further data are available we have assumed the same costs for hot dry rock that were developed for brine.

Steam winning:	
Wells necessary to produce steam for 110,000-kw. plant (16 wells at $150,000)	$2,400,000
Steam transmission lines at $10/kw.	1,100,000
Total	3,500,000
Overhead charges:	
Annual fixed charges at 14%	$ 490,000
Royalty payment to landowners	175,600
Total annual	665,600
Heat cost:	
Steam winning cost in mills/kwh.	0.80
Exploration, mills/kwh.	0.33
Disposal charge, mills/kwh.	0.02
Total energy cost, mills/kwh.	1.15
Conversion cost:	
Fixed charges, mills/kwh.	1.95
Operating cost, mills/kwh.	0.25
Total power cost, mills/kwh.	3.35

This exhibit shows how the various costs associated with developing, building, and operating a geothermal plant contribute to the total cost of power from it, in the case of a utility producing its own steam from its own wells. Exploration charges are based on spending $2 million to find a field capable of supporting a 110,000-kw. installation.

Source: Bowen, Richard G. and Groh, Edward A. "Geothermal—Earth's Primordial Energy," Technology Review, edited at The Massachusetts Institute of Technology, October/November, 1971.

Electric Power Costs:

	Geothermal	Nuclear	Hydropower	Coal
Plant investment, $/kw.	$110.00	$225.00	$250.00	$150.00
Fixed charges, 14%/year/kw.	15.40	31.50	35.00	21.00
Fixed charges, mills/kwh.	1.95	4.00	6.10	4.36
Operating costs, mills/kwh.	0.25	0.50	0.10	0.25
Energy costs, mills/kwh.	2.66	2.00	——	3.00
Total costs:				
Variable load factor, mills/kwh.	4.86	6.50	6.20	7.61
90% load factor, mills/kwh.	4.86	6.50	4.55	5.92

The price of power generated from geothermal sources is competitive with that by other methods. The plant investment figures given at the right are based (except for hydropower) on the average for new tax-paying, privately financed plants ordered in the U.S. in 1970. Oil-fired plants have essentially the same energy cost (3 mills/kwh.) as coal-fired; the energy cost shown here for geothermal (2.66 mills/kwh.) is that paid by Pacific Gas and Electric Co. at The Geysers. The variable load factors assumed in the total cost computation are 55 per cent for coal and oil, 65 per cent for hydropower, and 90 percent for nuclear and geothermal.

Source: Bowen, Richard G. and Groh, Edward A. "Geothermal—Earth's Primordial Energy," *Technology Review*, edited at The Massachusetts Institute of Technology, October/November, 1971.

Comparison of Power Plants Using Alternative Energy Sources

	GEOTHERMAL	FOSSIL	NUCLEAR
Present Maximum Unit Size	Small (110 MWe)	Very large (1200 MWe)	Very large (1200 MWe)
Fuel Preparation and Handling	Expensive but simple	Complex, expensive but unsophisticated	Very elaborate, sophisticated and expensive
Heat Production	Naturally occurring	Within combustion boiler	Within reactor vessel
Electric Generation	Similar	Similar	Similar
Cooling system #	Similar	Similar	Similar
Waste Disposal	Very simple, minimum risk, inexpensive	Moderately complex and expensive	Elaborate, sophisticated expensive
Safety Complexity	Very minor, inexpensive	Moderate	Very severe, costly
Flexibility of Plant location with respect to demand source	No flexibility, resource site dependent	High flexibility, cooling-water dependent	Fair flexibility, cooling-water dependent
Capital Cost ($/installed KW)	$200 (Dry Steam) $400 (Binary Systems)	$500-700 (Oil Fired)	$700-1000
Price of Electricity at Busbar (Mills/KWH)	11.0 (Dry Steam) 14.5 (Binary System, 300°F) 12.5 (Binary System, 500°F)	27.7 *	17.0 **

\# Complexity varies with size and location; geothermal can provide own cooling water.
* Assumes $7/BBL oil.
** Includes 2 mills for fuel.

Source: Geothermal Energy Task Force Report, Project Independence, Project Independence Blueprint, Federal Energy Administration, November 1974.

COSTS OF SOLAR SPACE HEATING

Collector Area Required for Single-Family Residences

Region	50% Solar Dependency			80% Solar Dependency		
	HO	HP	HC	HO	HP	HC
Northeast (Wilmington)	440	400	680	1,120	1,120	1,380
Southeast (Atlanta)	240	160	560	820	(2,000)	1,260
Gulf Coast (Mobile)	120	100*	580	240	(---)	1,200
Great Lakes (Madison)	1,200	480	860	(---)	(---)	(---)
West (Santa Maria)	160	100*	100*	220	120	160

Key: Solar collector area (ft^2) required for:

HO   —   Heating Only

HP   —   Solar-Assisted Heat Pump

HC   —   Heating and Cooling (Absorption)

*   —   Required for domestic hot water only

( ) —   Collector area exceeds available roof area

Source: Solar Heating and Cooling of Buildings, Phase O, NSF-RA-N-74-023A, Westinghouse Electric Corporation, prepared for the National Science Foundation, May 1974.

Costs of Space Heating (1970 Prices) in Dollars Per Million BTU Useful Delivery.

Location	Optimized solar heating cost in 25,000 BTU/degree-day house, capital charges @ 6%, 20 years		Electric heating, usage 30,000 kwh/year	Fuel heating, fuel cost only	
	Collector @ $2/ft.^{2}	Collector @ $4/ft.^{2}		Gas	Oil
Santa Maria	1.10	1.59	4.28[1]	1.52	1.91
Albuquerque	1.60	2.32	4.63	0.95	2.44
Phoenix	2.05	3.09	5.07	0.85	1.89
Omaha	2.45	2.98	3.25[3]	1.12	1.56
Boston	2.50	3.02	5.25	1.85	2.08
Charleston	2.55	3.56	4.22	1.03	1.83
Seattle-Tacoma	2.60	3.82	2.29[2,3]	1.96	2.36
Miami	4.05	4.64	4.87	3.01	2.04

Notes: [1] Electric power costs are for Santa Barbara. Electric power data for Santa Maria were not available.

[2] Electric power costs are for Seattle.

[3] Publicly owned utility.

Solar heat costs are from optimal design systems yielding least cost heat.

Electric power heat costs are from U.S. Federal Power Commission, All Electric Homes, Table 2 (1970). Conventional heat fuel costs are derived from prices per million BTU reported in P. Balestra, The Demand for Natural Gas in the United States, Tables 1.2 and 1.3 (North Holland Publishing Co., 1967). The 1962 costs were updated to 1970 by use of national price indexes on gas (121.1 in 1970 versus 112.8 in 1962) and on fuel oil (119.22 in 1970 versus 101.2 in 1962) as adjustment factors on each fuel price in each state. Bureau of Labor Statistics fuel prices indexes obtained from Gas Facts. Fuel prices were converted to fuel costs by dividing by the following national average heat (combustion) efficiencies: gas, 75%; oil, 75%. Heat efficiencies are from American Society of Heating, Refrigerating and Air Conditioning Engineers, Guide and Data Book 692-694 (1963 ed.).

All solar heat costs based on amortizing entire solar system capital costs in 20 years at 6 percent interest. Capital investment based on current prices of solar water heaters at $4 per sq. ft. plus current costs of other components, and on anticipated near-term solar collector price of $2 per sq. ft.

Source: Solar Energy as a National Energy Resource, NSF/NASA Solar Energy Panel, National Science Foundation, December 1972.

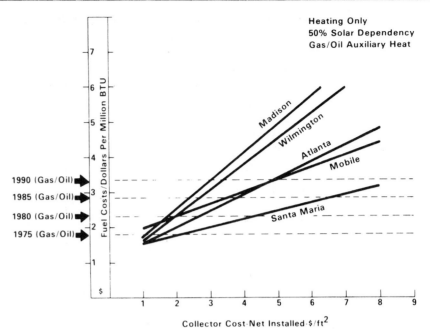

Economic feasibility of solar systems—heating only.

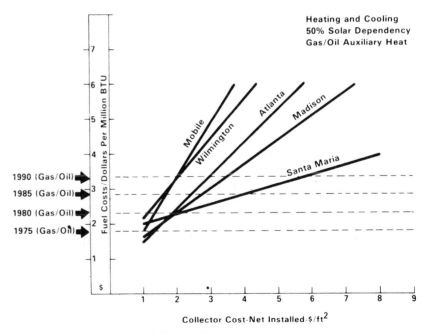

Economic feasibility of solar systems—heating and cooling.

Source: Solar Heating and Cooling of Buildings, Phase O, NSF-RA-N-74-023A, Westinghouse Electric Corporation, prepared for the National Science Foundation, May 1974.

COSTS OF OCEAN THERMAL ENERGY CONVERSION POWER PLANTS

Comparison of Several OTEC Plant Cost Estimates

	CMU[1]			U Mass[2]		APL[3]	TRW[4]	LMC[5]
	Low U=4000	Med 1000	High 400	Low	High	~450	~450	~450
ΔT	40°F	40°F	40°F	32°F	32°F	39°F	40°F	34°F
Working Fluid	NH₃	NH₃	NH₃	Propane	Propane	NH₃	NH₃	NH₃
Cost/kw excluding operating cost	436	656	1187	387	712	357	1812	2594
Cost excluding hull, CWP, moor, deploymt	346	559	1086	329	610	262	1102	1901
Cost/kw adjusted to 34°F	519	839	1630	283	524	370	1654	1901
Cost/kw adjusted to 34°F and $9/ft²	571	978	1995	No data		1540	1654	1901

[1] Abrahim Lavi, Final Report: Solar Sea Power Project, January 1975, NSF/RANN/SE/GI-39114/PR/74/6, pp. 108-113.

[2] J. G. McGowan et al., Variations in Heat Exchanger Design for Ocean Thermal Difference Power Plants, August 1974, NSF/RANN/SE/GI-34979/TR/74/4, p. 5.

[3] G. L. Dugger et al., Tropical Ocean Thermal Power Plants and Potential Products, AIAA Paper No. 75-617, Presented at AIAA/AAS Solar Energy for Earth Conference, Los Angeles, April 21-24, 1975, pp. 23-26.

[4] Robert H. Douglass et al., Ocean Thermal Energy Conversion Research on an Engineering Evaluation and Test Program, TRW Systems Group, Quarterly Report 26555-6003-TU-00, February 1975, section 5, pp. 1-5.

[5] Lloyd C. Trimble et al., Ocean Thermal Energy Conversion (OTEC) Power Plant Technical and Economic Feasibility, Vol. I, Technical Report, Ocean Systems Lockheed Missles and Space Company, Inc., NSF/RANN/SE/GI-C937/FR/75/1, April 1975, section 2, pp. 87-90.

Source: Lavi, Abrahim, "Solar Sea Power Plants, Cost and Economics," *Proceedings, Third Workshop on Ocean Thermal Energy Conversion (OTEC)*, Houston, Texas, May 8-10, 1975, APL/JHU SR 75-2, The Johns Hopkins University, Applied Physics Laboratory, August, 1975.

Projected Electric Energy Cost Comparison from Conventional Power Plants

	OIL	COAL	NUCLEAR 80% Cap.	NUCLEAR 55% Cap.[3]
Capital Cost	11.6	12.9	16.8	24.4
Fuel Cost	29.0[1]	17.3[2]	5.4	5.4
Operating & Maintenance	2.	3.	2.5	2.5
Total mills/kwh	42.6	33.2	24.7	32.3

[1] Based on $12.20/barrel in 1980.

[2] Basis of coal price not given, estimated coal price at $37/ton.

[3] See "Not Man Apart: Refuting Nuclear Boosters", published by Friends of the Earth, Mid-April 1975 issue, pp. 12-13.

Note: Comparative electrical generating costs in the first half of 1975 were: nuclear, 11.41 miles/kwh; oil, 32.73 mills/kwh; coal, 14.71 mills/kwh (see page 15-19).

Source: Lavi, Abrahim, "Solar Sea Power Plants, Cost and Economics," *Proceedings, Third Workshop on Ocean Thermal Energy Conversion (OTEC)*, Houston, Texas, May 8-10, 1975, APL/JHU SR 75-2, The Johns Hopkins University, Applied Physics Laboratory, August, 1975.

Source	(0)	(1) TRW	(2) Aluminium	(3) Optimized (2)	(4) Theoretical	(5) Optimized (4)
Overall Heat Transfer Coeff. BTU/hr ft² °F		450	1000	1000	4000	4000
Capital Cost $/kw		2100	1130	1040	731	690
Return Rate	11%	$231.2	$124.3	$114.4	$ 80.4	$ 75.9
Depreciation/Amortization	88%	18.5	9.9	9.2	6.4	6.1
Administrative	1%	21.	11.3	10.4	7.3	6.9
Taxes (local)	3%	63.	33.9	31.2	21.9	20.7
Insurance on Capital Cost	2%	42.	22.6	20.8	14.6	13.8
Protection		1.5				
Crew		6.7				
Logistics		2.6				
Maintenance	(Marine and power plant at 3/4% of construction)	15.4				
Miscellaneous		2.8				
Overhead		4.2	33.2	33.2	33.2	33.2
TOTAL		$408.9	$235.2	$219.2	$163.8	$156.6
Mills/kwh						
80% capacity		58.36	33.57	31.28	23.38	22.35
90% capacity		51.88	29.84	27.81	20.78	19.87

Comparison of busbar energy costs using different SSPP designs. Differences in estimated power costs are related cost of materials chosen for heat exchangers (tantalum or aluminum) which are deemed suitable for operation in the marine environment and to the assumed heat transfer coefficients based on estimates of the extent of fouling of surfaces.

Source: Lavi, Abrahim, "Solar Sea Power Plants, Cost and Economics," *Proceedings, Third Workshop on Ocean Thermal Energy Conversion* (*OTEC*), Houston, Texas, May 8-10, 1975, APL/JHU SR 75-2, The Johns Hopkins University, Applied Physics Laboratory, August, 1975.

COSTS OF BIOMASS ENERGY

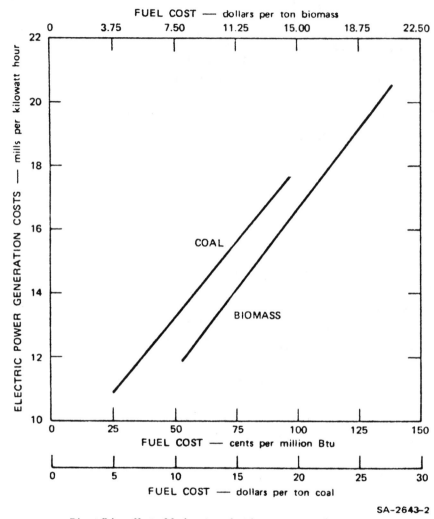

Direct firing effect of fuel cost on electric power generation costs.

SA-2643-2

Source: Effective Utilization of Solar Energy to Produce Clean Fuel, PB-233 956, Stanford Research Institute, prepared for the National Science Foundation, June 1974.

Comparative Costs of Electric Power Generation from Direct Firing of Biomass and Coal.

A 1000 megawatt station operating at an annual load factor of 80% would require a land area of about 245 square miles to provide the biomass at an average yield of 15,070 tons per day (30 tons per acre-year). Yield is based on plantation cultivation in regions with a growing season of 240 days or longer and with irrigation requirements of 700 million gallons per day (2150 acre feet/day).

	Biomass at 65¢/MM Btu	Coal[a] at 35¢/MM Btu
Power plant capital investment (millions of dollars)	$245	$290[b]
Electric generation costs (mills/kWh)		
Capital and related	5.6	6.6
Maintenance and operational	1.0	1.0
SO_x removal variable	—	0.9
Fuel cost	6.5	3.3
Total	13.1	11.8

[a]HR = 9400 Btu/kWh for coal.
[b]Includes SO_2 at $50/kW.

Source: Effective Utilization of Solar Energy to Produce Clean Fuel. Stanford Research Institute, prepared for the National Science Foundation, PB-233 956, June 1974.

Heat Available from Fired-Dried Bagasse

Heat in Flue Gas	Btu/100 Pounds Fuel
Carbon dioxide	8,870
Nitrogen	31,760
Oxygen (excess)	1,720
Water	
Combustion product	52,190
Moisture in air	990
Moisture in fuel	17,320
Total	112,850
Heat value of fuel	700,000
Theoretical efficiency (%)	83.9%

Source: Effective Utilization of Solar Energy to Produce Clean Fuel, PB-233 956, Stanford Research Institute, prepared for the National Science Foundation, June 1974.

COSTS OF WIND POWER

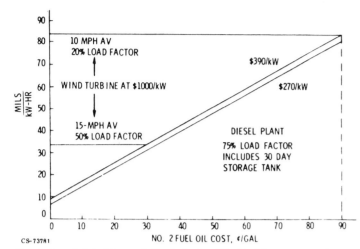

Comparison of costs for wind turbine and diesel plants. Wind turbine power costs are competitive in locations where a 50% load factor can be achieved at average wind velocities of 15 miles per hour. Limiting factor is the need for stand-by generation capacity during calm periods since energy storage costs are not included in calculations. Commercially available wind turbine systems are priced near $1500 per kilowatt.

Source: Wind Energy Developments in the 20th Century, Lewis Research Center, National Aeronautics and Space Administration, 1975.

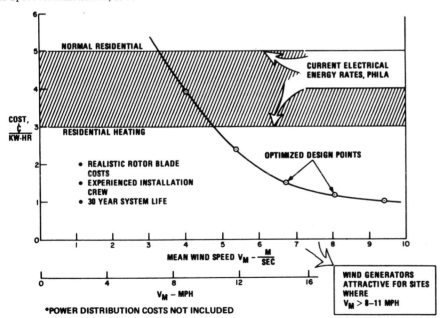

Effect of median wind velocity on electrical energy generation cost.

Source: Killen, Robert "G.E. Systems Studies of Large-Scale WECS," in *Proceedings of the Second Workshop on Wind Energy Conversion Systems*, Washington, D.C., June 9–11, 1975, NSF-RA-N-75-050, The MITRE Corporation.

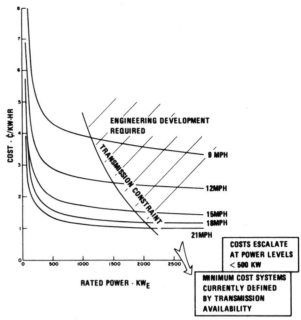

Effect of Median Wind Speed and Rated Power on Estimated Energy Cost (Power Distribution Costs Not Included). Estimate is based on production of 100 wind turbine units with rotor blade diameters approaching 200 feet. Transmission constraint refers to mechanical transmission between rotor and electrical generator.

Source: Killen, Robert "G.E. Systems Studies of Large-Scale WECS," in *Proceedings of the Second Workshop on Wind Energy Conversion Systems*, Washington, D.C. June 9–11, 1975, NSF-RA-N-75-050, The MITRE Corporation.

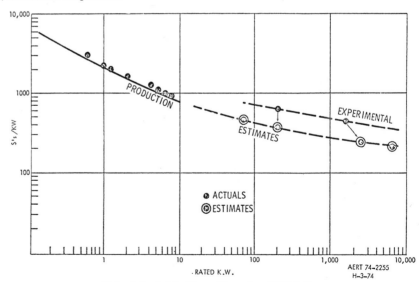

Expected wind turbine selling price, 1973 dollars.

Source: Solar Energy Task Force Report, Project Independence, Federal Energy Administration, November 1974.

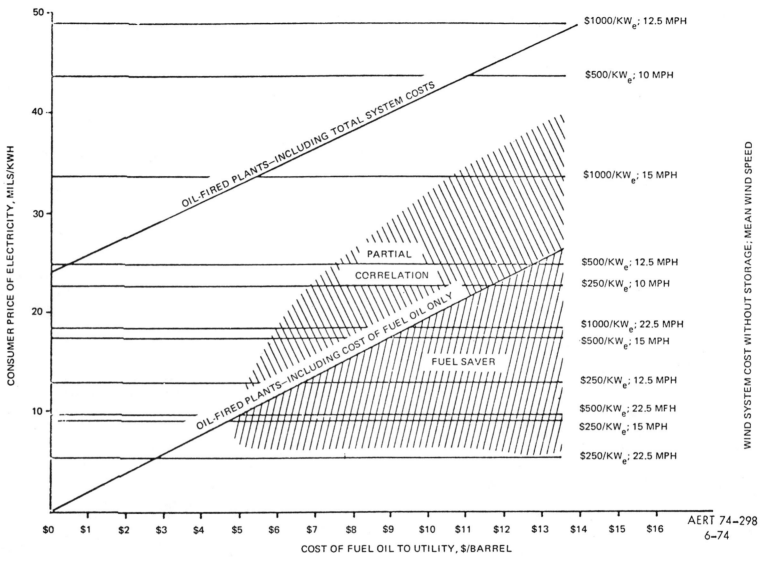

Breakeven costs, wind turbines versus price of oil.

Source: Solar Energy Task Force Report, Project Independence, Federal Energy Administration, November 1974.

ENERGY UNIT COSTS AND CAPITAL CALCULATIONS

Unit Costs and Capital Calculations

Unit capital investment requirements for each form of energy production were estimated or obtained from literature sources and are given in Table C-1. Table C-2 supplements this table for oil and gas and gives the total cost of oil and gas production. Notes and references for Table C-1 follow.

Table C-1. Approximate unit investment costs (1974 dollars)

Fuel	Unit size	Unit rate	Capital cost per unit	% depreciable
Coal				
Mines				
Strip	—	—	$10 per ton - yr. initial	
			$4 per ton - yr. deferred	95
Deep	—	—	$14 per ton - yr. initial	
			$21 per ton - yr. deferred	78
Trains	12,000 tons/train	300,000 tons/yr.	$880,000 per train	100
Slurry pipelines	three-24 in. diam. pipelines	10^{15} Btu/yr.	$740x10^6/10^{15}$ Btu x 1000 mi	100
Oil				
Drill rigs	—	—	$1 x 10^6/rig	100
Wells (only)[a]	4600 ft. avg.	—	$21.80 per ft. (including dry holes)	30
Refineries	—	150,000 bbl/da	$2000 per bbl/da	100
Pipelines	24 in. diam. pipeline	0.4 x 10^{15} Btu/yr.	$170x10^3 per mi at 0.2x10^6 bbl/da	100
Tankers	300,000 DWT = 2.1x10^6 bbl	30,000 bbl/da (Persian Gulf)	$1400 per bbl/da=$4.2x10^7 per ship	100
Super ports	—	—	$600 x $10^6	100
Gas				
Drill rigs	—	—	1 x $10^6 per rig	100
Wells (only)[a]	6600 ft. avg.	—	$30.50 per ft. (including dry holes)	27
Synthetic fuels				
Coal liquefaction	40,000 bbl/da	2.5 x 10^{11} Btu/da	$480 x $10^6	100
Methanol	15,000 tons/da	2.5 x 10^{11} Btu/da	$360 x $10^6	100
Oil shale	50,000 bbl/da	2.5 x 10^{11} Btu/da	$300 x $10^6	100
Solar				
Home heating	750 ft.^{2}	7.7 x $10^6 Btu/yr.	$4500	–
Home cooling	750 ft.^{2}	—	$1500 additional	–
Uranium				
Mining	—	—	$750 per ton U output/yr.	–
Milling	—	—	$15,000 per ton U output/yr.	–
Enrichment	10^7 separative work units (SWU/yr.)	—	$150 per SWU-yr.	100
Fuel processing	—	—	$55 x $10^6 per 1000 tons U/yr.	100
Electricity				
Generation[b]				
Fossil-fired	1000 MW	—	$240 per kW	100
Hydro	1000 MW	—	$260 per kW	100
Pumped hydro	1000 MW	—	$260 per kW	100
Nuclear	1000 MW	—	$360 per kW[c]	100
Waste conversion	—	—	$575 per kW	100
Geothermal (dry steam)	—	—	$190 per kW	100
Peaking plants	—	—	$150 per kW	–
Transmission	(13% of industry investment)	—	—	–
Distribution	(27% of industry investment)	—	—	–
Miscellaneous	(5% of industry investment)	—	—	–
Sulfur cleanup				
Low-Btu gas	150 x 10^6Mcf/da	1.5 x 10^{11} Btu/da	$200 x $10^6	100
Solvent refined coal	10,000 tons/da	3.0 x 10^{11} Btu/da	$300 x $10^6	100
Stack gas cleanup	750-MW plant	—	$50 per kW	100

[a]See Table C-2 for total costs of oil and gas production.
[b]On the average, generation equals 55% of total industry investments.
[c]In addition, $40 per kW for first core loading was added in the last year of construction.

Source: An Assessment of U.S. Energy Options for Project Independence, UCRL-51638, Lawrence Livermore Laboratory, University of California, September 1, 1974.

References and Notes for Table C-1 Unit Costs and Capital Calculations

COAL

Mining. Ref.: *Coal Availability*, NPC (1972).

For surface mines —	1970 dollars	1974 dollars
Initial capital	8.20 per ton	10.41≈$10
Deferred	3.10	3.93≈$4
Total	11.30	14.35≈$14

Underground mines —		
Initial capital	11.39 per ton	14.47≈$14
Deferred	16.62	21.10≈$21
Total	28.01	35.57≈$35

Assume future production will be 75% strip; 25% mined.

Strip is 95% tangible; deep is 70% tangible.
(Ref.: private communication, National Coal Association (1974).

Assume deferred capital is spread evenly over the following 30 yr:
Strip = 4/30 = $0.13 per ton-yr.
Deep = 21/30 = $0.70 per ton-yr.
Then, tangible costs:
Strip = 0.95 x 0.75 x (10.41+0.13) = $7.51
Deep = 0.70 x 0.25 x $14.47+0.70) = $2.65
Total = $10.16 per ton-yr.

Intangible costs: strip = $0.40 per ton-yr and deep = $1.14 per ton-yr which totals $1.54 per ton-yr.
Total = $11.70 per ton-yr.

To replace exhausted mines, additional capacity equal to 3% of the total capacity was added each year, in addition to expanded capacity.

Table C-2. Petroleum industry capital investment[a] for exploration and production (1974 costs[b])

Investments	% tangible	% intangible	% of total cost
Lease acquisitions	0	19.6	19.6
Producing wells and platforms	10.4	24.3	34.7
Lease equipment[c]	18.3	0	18.3
Miscellaneous tangible	1.3	0	1.3
Gas separation plants	1.3	0	1.3
Geological and biophysical	1.3	0	6.5
Lease rentals	0	6.5	2.1
Dry wells	0	16.2	16.2
	31.3	68.7	100.0
Ratio to total well cost[d]	0.62	1.35	

[a]Based on NPC Case 1, 1970-1985 total investment.
[b]1970 costs escalated 1.7 for well costs and 1.27 for all others.
[c]Present-day investments would have a lower percentage of lease equipment. This is largely a secondary/tertiary recovery cost in the above figures.
[d]Total well cost includes producing wells and platforms and dry wells.

Trains. Ref.: *Coal Availability (NPC* $1972) Table J-8, p. 179. 1970 prices inflated from 1970 to 1974 at 1.29.

Slurry Pipelines. LLL estimate — based on costs of oil pipelines for transporting equivalent volumes of fluid. See oil pipelines below.

OIL

Drill Rigs. Based on conversations with several different contractors, approximate drill rig costs were estimated as:
Onshore rigs (90% of total rigs):
Less than 10,000-ft capability ~ $350,000
15-20,000-ft capability ~ $600,000 } in 1972
Offshore rigs (10% of rigs):
15-20,000-ft capability $ 600,000 (in 1972)
Plus platform $1,000,000 (in 1974)

Inflate 1972-74 x 1.15
Onshore (inflated) $ 400,000 (say $600,000)
$ 690,000
Offshore (inflated) $ 690,000
Platform $1,000,000
$1,700,000
Average total cost = 0.9 x 600 + 0.1 x 1.7 = $710,000.

In summary, assume a conservative figure of $1 million each.

Source: An Assessment of the U.S. Energy Options for Project Independence, UCRL-51638, Lawrence Livermore Laboratory, University of California, September 1, 1974.

Oil Wells. Ref.: *Oil and Gas Availability*, NPC, for average well depths (4300 ft in 1970 and increasing at 40 ft/yr). 1970 prices for drilling and oil well casing were separately escalated using cost indexes reported in *OGJ* (March 18, 1974), p. 28. Overall effect is to escalate 1970 costs of oil (and gas) wells by multiplying by 1.7 to obtain 1974 costs. The cost listed includes the dry hole allowance for 42% of total holes drilled.

Refineries. Ref.: E. K. Grigsby, E. W. Mills, and D. C. Collins, "What will Future Refineries Cost?" *Hydrocarbon Processing* (May, 1973).

Oil Pipelines. Ref.: J. P. O'Donnell, "Pipeline Economics," *OGJ* (Aug. 14, 1972). For 24 in. diam. line—in 1972 costs ranging from $127,000 per mile to $257,000 per mile with an average of $146,000 per mile. Inflate by multiplying by 1.15 to 1974 dollars = $170,000 per mile.

Tankers. Ref.: See Refineries above.

Superports. Ref.: "Monobuoy System Proposed for U.S. First Superport" *World Dredging and Marine Construction* (Feb. 1974).

GAS

Drill Rigs. (see Oil Drill Rigs above).

Gas wells. (see Oil Wells above). Average depth from NPC *Oil and Gas Availability* is 5900-ft avg, increasing at 90 ft/yr. 1974 cost of $30.50 per ft includes a dry hole allowance for 42% of total holes drilled.

Pipelines. For total costs of gas transmission, the NPC Case 1 figures were adopted and adjusted for inflation (1.27 × 1970 costs) and for the shorter period (1974–1985). Costs given by NPC for LNG transport and storage were not included.

Gas distribution and storage costs were derived from American Gas Association data. Ref.: *Gas Facts*, American Gas Association (1971).

Storage, distribution, and miscellaneous costs averaged 1.25 times the cost of gas transmission for the years 1960 to 1971. The NPC figures for gas transport in the lower 48 states for 1974–1985 were multiplied by 1.25 to obtain gas storage and distribution costs.

SYNTHETIC FUELS

Coal Liquefaction. Ref.: H. C. Hottel and J. B. Howard, *New Energy Technology: Some Facts and Assessments* (MIT Press, Boston, 1971) p. 170. Escalated to 1974 dollars.

Methanol. Ref.: *Project Independence; An Economic Evaluation*, MIT Energy Laboratory Policy Study Group (1974).

Oil Shale—Surface Retorting. Ref.: *Oil Shale Task Group Report*, NPC (1972). Escalated to 1974 dollars.

SOLAR HEATING AND COOLING

LLL estimate.

URANIUM

Mining. Ref.: *Mining Engineers Handbook*, I. A. Given, Ed. (AIMMPE, N.Y., 1973). Gives complete costs for mining and milling of uranium; costs are presumed to be for the mid-1960's. Costs as given were escalated to 1974 costs by multiplying by 1.8.

Enrichment. Ref.: *Uranium Enrichment, Guidelines for Assessment and Planning*. Western Interstate Nuclear Board (1973).

Fuel Reprocessing. Ref.: *Resource Needs for Nuclear Power Growth*, (Atomic Industrial Forum, New York, 1973).

ELECTRICITY

Fossil-Fired Generation. Ref.: J.F. Kaufmann, *Cost Study of Electrical Generation Facilities*, Office of Planning and Analysis, AEC, Washington, D.C. (1973). A 3-yr construction interval was assumed; the referenced cost of $210 per kw (in current dollars) was spread equally over this period and then inflated to 1974 dollars.

Hydroelectric and Pumped Storage. Ref.: Federal Power Commission, Rept. FPC-S-232 (1971) Table 4.

Nuclear. Ref.: See Fossil-Fired Generation above. A 5-yr construction interval was assumed; the $300 per kw cost given in the reference in current dollars was spread equally over this period and then inflated to 1974 dollars.

Waste Conversion. LLL estimate, based on costs of 20 to 40 MW fossil-fired power plants.

Geothermal. LLL estimate, primarily based on costs for generation from dry steam at The Geysers, Cal.

Peaking Plants. Ref.: *1970 National Power Survey*, FPC. We used 1968 costs of gas turbines, $85 per kw; and of diesel plants, $125 per kw. Inflate by multiplying 1.4 to 1974 dollars. Take mean costs as about $150 per kw.

Transmission, Distribution, and Miscellaneous. Capital investment trends for period 1948-1972 were graphed and a judgment was made as to the future trend for ratios of expenditure for these categories to expenditure for generation.

Low Btu Gasification. Ref.: H.C. Hottel and J.B. Howard, *New Energy Technology, Some Facts and Assessments*, (MIT Press, Boston, 1971).

Solvent-Refined Coal. Ref.: "Big SRC Pilot Plant Under Construction," *OGJ* (Nov. 6, 1972). Escalated to 1974 dollars.

Stack-Gas Cleanup. Ref.: *Final Report of Sulfur Oxide Control Technology Assessment Panel on Projected Utilization of Stack-Gas Cleanup Systems by Steam-Electric Plants* EPA, Rept. APTD-1569 (1973). Mean value of $50 per kw taken from Table IV-1, p. 45.

Source: An Assessment of U.S. Energy Options for Project Independence, UCRL-51638, Lawrence Livermore Laboratory, University of California, September 1, 1974.

ENERGY CAPITAL REQUIREMENTS

Energy Facility Cumulative Capital Requirements, 1975–85
(Billions of 1973 Dollars)

ITEM	NATIONAL PETROLEUM COUNCIL	ARTHUR LITTLE CO.	FEA ACCELERATED SUPPLY
OIL AND GAS (Including Refining)	133	122	98.4
COAL	8	6	11.9
SYNTHETIC FUELS	10	6	.6
NUCLEAR	7	84	138.5
ELECTRIC POWER PLANTS (Excluding Nuclear)	137	43	60.3
ELECTRIC TRANSMISSION	42	90	116.2
TRANSPORTATION	43	43	25.5
OTHER	–	8	2.2
TOTAL	380	402	453.6

[1] Based upon an energy conservation scenario.

Source: *Energy Perspectives*, U.S. Department of the Interior, February 1975.

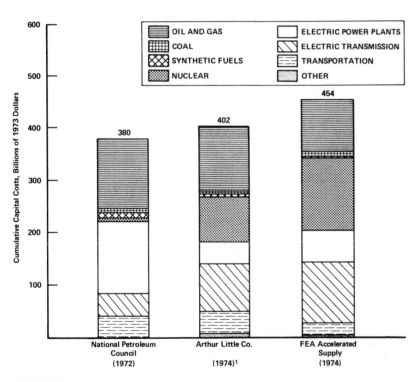

Energy facility cumulative capital requirements, 1975–85.

Source: *Energy Perspectives*, U.S. Department of the Interior, February 1975.

[1] Estimate is based upon an energy conservation scenario.   Source: Project Independence Report, 1974, page 282.

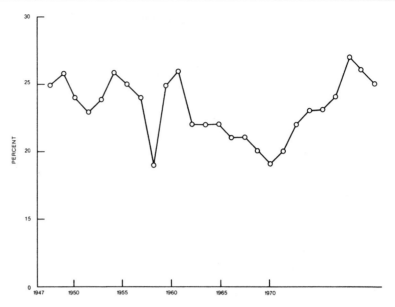

Energy's annual share of business investment.

Source: *Project Independence*, Project Independence Report, Federal Energy Administration, November 1974.

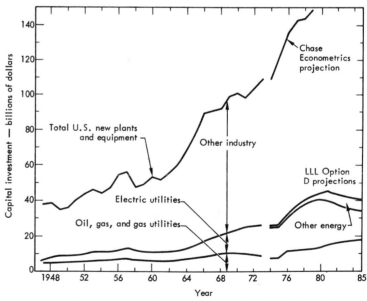

Historical trends in capital investment in the energy industry since 1947 shown in 1974 constant dollars (inflated with the wholesale price index for industrial commodities).

Source: *An Assessment of U.S. Energy Options for Project Independence*, UCRL-51638, Lawrence Livermore Laboratory, September 1, 1974.

Construction Expenditures—Investor-Owned Electric Utilities (Excluding Alaska and Hawaii) by Type of Electric Utility Plant
Millions of Dollars

Year	Total	Production	Transmission	Distribution	Other	Year	Total	Production	Transmission	Distribution	Other
1976	$16 979	$11 512	$1 758	$2 844	$865	1963	$3 319	$1 165	$644	$1 323	$187
1975	15 090	9 828	1 734	2 817	711	1962	3 154	1 078	609	1 305	162
1974	16 350	10 145	2 060	3 360	785	1961	3 256	1 267	579	1 265	145
1973	14 907	8 775	2 047	3 371	714	1960	3 331	1 342	537	1 300	152
1972	13 385	7 931	1 748	3 073	633	1959	3 383	1 519	554	1 163	147
1971	11 894	6 702	1 806	2 774	612						
1970	10 145	5 429	1 680	2 614	422	1958	3 764	1 879	608	1 125	152
1969	8 294	3 992	1 554	2 421	327	1957	3 679	1 647	594	1 270	168
1968	7 140	3 189	1 503	2 135	313	1956	2 910	1 029	455	1 274	152
1967	6 120	2 553	1 323	1 977	267	1955	2 719	1 064	434	1 093	128
1966	4 932	1 789	1 137	1 769	237	1954	2 835	1 280	464	993	98
1965	4 027	1 300	940	1 585	202	1953	2 876	1 391	442	938	105
1964	3 551	1 114	824	1 424	189	1952	2 599	1 251	379	879	90

In these figures an attempt has been made to eliminate wherever possible capital expenditures made for the purchase of existing properties rather than for actual new construction.

Construction expenditures are, in general, the gross amounts spent for construction of all kinds, including the acquisition of real estate and all necessary equipment. The figures include money spent for replacements, additions, and betterments (but not for maintenance of existing plant), as well as for new construction.

Source: Estimated by Edison Electric Institute on basis of statistics collected by *Electrical World*.

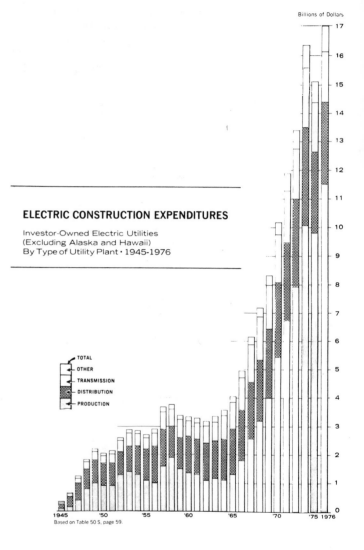

ELECTRIC CONSTRUCTION EXPENDITURES

Investor-Owned Electric Utilities
(Excluding Alaska and Hawaii)
By Type of Utility Plant · 1945-1976

Source: Statistical Year Book of the Electric Utility Industry for 1977, Edison Electric Institute, 90 Park Avenue, New York.

COMPARATIVE COSTS OF POWER-GENERATING PLANTS

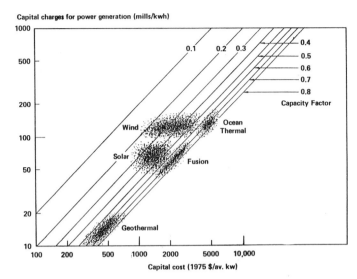

Capital costs & capital charges for new electric power generation options.

Source: Dr. Chauncey Starr, "New Electric Power Generation Options for the 21st Century," presented at Energy Forum IV, sponsored by the U.S. National Committee–World Energy Conference, Washington, D.C., June 21, 1976.

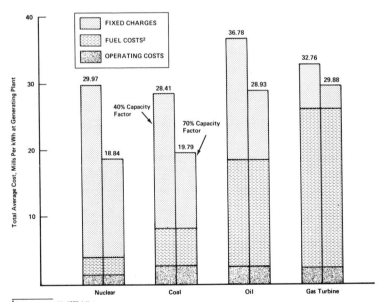

Comparative costs of power-generating plants[1].

[1]Costs are expressed in 1973 dollars.
[2]Fuel costs are $14.52 a ton for coal, $10.59 per barrel for residual oil, and $11.77 for distillate.

Source: Project Independence Report, 1974, page 286.

Source: Energy Perspectives, U.S. Department of the Interior, February 1975.

Comparison of Conventional Costs and Some Evaluated Nonconventional Costs of Alternate Energy Systems

Basis: Annual operation of one 1000 MWe Power Plant and Supporting Fuel Cycle (6.57 billion kWhe). 1980 dollars.

		Coal	Oil	Gas	LWR
Conventional costs	(10^6/yr)				
capital plant		51	47	42	77
fuel cycle		64	180	237	39
O&M		5.4	4.0	3.7	5.2
rounded totals		120	231	283	121
Abatement costs	(10^6/yr)				
cooling towers		3.6	5.9	7.1	2.4
sulfur/SO_2 removal		25.9	4.9	NA	NA
strip-mined land reclamation		0.1	NA	NA	S
near-zero radwaste		NA	NA	NA	1.2–1.8
rounded totals		30	11	7	3–4
Conventional & abatement	(10^6/yr)	150	242	290	125
Abatement component	(%)	20	5	2	3
Safety	(10^6/yr)				
occupational[a]		0.46	0.086	0.039	0.05
public[b]		0.18	U	U	0.003
subtotal		0.64	>0.086	>0.039	0.053
Health	(10^6/yr)				
occupational		0.03[c]	U	U	0.024[d]
public		U	U	U	0.01[e]
subtotal		>0.03	U	U	0.034
Total human health & accident costs	(10^6/yr)	>0.67	>0.086	>0.039	0.087
Environmental effects	(10^6/yr)				
water base		0.4	0.4	0.4	0.6
air base		0.8	0.6	0.1	S
land base		0.2	S	S	S
subtotal		1.4	1.0	0.5	0.6
Total human & env. effects	(10^6/yr)	2.1	1.1	0.5	0.7
Percent of conv.	(%)	3	1	0.5	0.9

Key: U = unevaluated; NA = not applicable; S = small.
[a]Conventional injuries in routine industrial accidents, including fatal and nonfatal injuries; 1 death = 6,000 MDL = $300,000.
[b]Conventional injuries in accidents in transportation of fuels; 1 death = 6,000 MDL = $300,000.
[c]Coal workers' pneumoconiosis (CWP).
[d]Radiological health effects, including lung cancers among uranium miners.
[e]Radiological health effects from routine emissions.
Source: Comparative Risk-Cost-Benefit Study of Alternative Sources of Electrical Energy, WASH-1224, United States Atomic Energy Commission, December 1974.

Cost Comparisons of Alternative Power-Generating Plants, 1975 Dollars
[Mills per kWh at generating plant]

Item	Type of power-generating plant		
	Nuclear	Coal	Oil
Fuel cost[1]	4.0	6.8	18.3
Operating and maintenance expenses:[2]	1.5	3.0	2.8
Fixed charges:[3]			
Capacity factor = 0.7	15.0	13.0	6.0
Capacity factor = 0.4	27.0	22.0	11.0
Total costs:			
Capacity factor = 0.7	20.5	22.8	27.1
Capacity factor = 0.4	32.5	30.8	32.1

[1] $14.81 per ton for coal; $11.21 per barrel for residual oil.
[2] Includes environmental costs.
[3] Fixed charges based on estimates of capital investments by Ebasco Services, Inc., *Fossil and Nuclear 1000 MW Central Station Power Plants Investment Estimates,* EPRI/TPS-75-601, Sept. 1975. Stack gas cleanup system investment costs included in coal-fixed powerplant costs.

Source: Energy Perspectives 2, U.S. Department of the Interior, June 1976.

Estimated Cost of Electric Energy Generation in Different Types of Power Plants, July 1974[e]

	Gas-Fired[a,c]	Coal-Fired[a]	Hydroelectric	Nuclear	Geothermal[b] (Dry Steam)	Oil-Fired (September 1973)	Oil-Fired[a] (February 1974)
Unit investment cost of plant, dollars per kilowatt	125	225	390	400	100	185	185
Annual fixed charge, percent of investment	17	17	17	17	17	17	17
Kilowatt-hours generated per year per kilowatt capacity	7,000	7,000	7,000	7,000	7,000	7,000	7,000
Heat rate, Btu/kWh	9,100	8,800	—	10,500	22,000	8,900	8,900
Cost of fuel:							
cents per million Btu	35.7	49.0 (est.)	—	17.5	14.3	79.7	150.0 (est.)
dollars per unit	0.37 (mcf)	10.86 (est.) (ton)	—	—	0.35 (ton)	4.89 (bbl)	9.20 (est.) (bbl)
Quality of fuel, Btu/unit	1,022 (cu ft)	11,084[d] (lb)	—	—	1,210 (lb)	145,816 (gal)	145,996[d] (gal)
Cost of electricity, mills/kWh							
plant investment	3.0	6.2	9.5	9.7	2.4	4.5	4.5
operation and maintenance	0.6	0.8	0.1	0.4	0.5	0.7	0.7
fuel	3.6	4.7	—	1.9	3.1	8.4	15.8
Total	7.2	11.7	9.6	12.0	6.0	13.6	21.0

References: FPC Form 423
FPC Form 1
Finney, J. P., "Design and operation of the Geysers power plant," in Paul Kruger and Carel Otto (eds.), *Geothermal Energy: Resources, Production and Stimulation* (Stanford, Calif.: Stanford University Press, 1973), pp. 145–161.
Stone, R., testimony before Senate Subcommittee on Water and Power Resources in Washington, D.C., June 13, 1973, pp. 107–116.

[a] Fuel costs are preliminary averages for February 1974 computed from FPC Form 423, Monthly Report on Cost and Quality of Fuels.
[b] All numbers for Geothermal are reported by Pacific Gas & Electric on FPC Form 1 for 1973 operations at The Geysers Steam Plant.
[c] FPC policy presently discourages construction of gas-fired steam-electric plants.
[d] Average energy contents of fuels reported on FPC Form 423 for November 1973.
[e] Personal communication, Dr. Charles F. Reusch, Federal Power Commission, May 20, 1974.
Source: A Technology Assessment of Geothermal Energy Resource Development, prepared for the National Science Foundation by The Futures Group, April 15, 1975.

U.S. Electrical Generating Costs in First Half, 1975

UTILITY	NUCLEAR 1ST QUARTER Cost (1) (mills per Kwh)	NUCLEAR 1ST QUARTER Net Kwh (Millions)	NUCLEAR 2ND QUARTER Cost (1) (mills per Kwh)	NUCLEAR 2ND QUARTER Net Kwh (Millions)	OIL 1ST QUARTER Cost (1) (mills per Kwh)	OIL 1ST QUARTER Net Kwh (Millions)	OIL 2ND QUARTER Cost (1) (mills per Kwh)	OIL 2ND QUARTER Net Kwh (Millions)	COAL 1ST QUARTER Cost (1) (mills per Kwh)	COAL 1ST QUARTER Net Kwh (Millions)	COAL 2ND QUARTER Cost (1) (mills per Kwh)	COAL 2ND QUARTER Net Kwh (Millions)
(2)Baltimore G&E	* (*)	*	— (2.69)	715	— (23.39)	1,483	— (23.74)	1,055	— (8.34)	1,271	— (9.09)	1,164
Boston Edison	18.74 (1.95)	713	18.75 (1.96)	793	32.22 (21.63)	1,599	31.36 (20.77)	1,415	* (*)	*	* (*)	*
Carolina P&L	5.70 (2.00)	1,435	11.24 (2.11)	617	* (*)	*	* (*)	*	21.81 (14.43)	4,050	17.55 (11.60)	4,916
Con. Ed. of NY	13.58 (2.48)	936	11.20 (3.94)	1,508	35.40 (21.79)	5,235	35.19 (21.66)	5,468	* *	*	* *	*
Consumers Power	— (2.30)	22	— (6.80)	910	— (27.90)	703	— (28.10)	855	— (9.90)	3,397	— (9.00)	3,404
Dairyland Coop.	15.85 (2.97)	94	37.14 (3.16)	34	90.89 (22.63)	2	57.42 (13.59)	2	10.93 (7.35)	889	16.48 (8.34)	517
Duke Power	12.65 (2.23)	2,127	7.76 (1.95)	3,960	— (18.72)	24	— (28.62)	3	15.48 (10.82)	7,824	17.59 (11.49)	6,067
(3)Florida P&L	— (1.79)	2,694	— (1.92)	1,446	— (13.31)	5,125	— (15.76)	7,759	* (*)	*	* (*)	*
Jersey Central P&L	13.96 (2.04)	1,607	15.77 (2.05)	623	54.16 (25.79)	715	66.23 (26.69)	509	14.82 (6.24)	431	14.27 (6.68)	451
Maine Yankee	13.60 (4.43)	1,175	18.58 (4.84)	475	* (*)	*	* (*)	*	* (*)	*	* (*)	*
Metro. Edison	14.81 (2.61)	844	18.81 (2.76)	526	— (—)	57	— (—)	26	21.03 (13.46)	1,216	19.11 (13.02)	1,292
Niagara Mohawk	12.86 (3.56)	909	10.98 (3.59)	1,151	25.94 (17.83)	1,447	26.00 (17.01)	1,306	21.90 (14.06)	1,816	24.12 (13.74)	1,600
Northeast Util.	9.63 (3.93)	1,992	9.96 (4.11)	3,806	30.78 (23.08)	2,907	30.45 (22.14)	5,247	* (*)	*	* (*)	*
(4)Omaha Pub. Pow.	19.20 (1.92)	276	19.59 (3.05)	316	* (*)	*	* (*)	*	10.67 (7.05)	866	11.32 (6.88)	759
Pacific G&E	— (2.26)	94	— (2.25)	75	— (20.50)	3,664	— (24.30)	853	* (*)	*	* (*)	*
Phila. Elec.	12.76 (2.75)	1,252	12.73 (2.65)	1,274	43.21 (24.02)	1,323	46.32 (24.14)	1,042	17.49 (9.68)	1,919	18.36 (9.76)	1,867
Rochester G&E	9.40 (2.30)	772	— (2.50)	221	* (*)	*	* (*)	*	20.60 (14.20)	290	19.90 (14.00)	340
(5)Sacramento M.U.D.	* (*)	*	— (3.14)	1,338	* (*)	*	* (*)	*	* (*)	*	* (*)	*
(6)So. Cal. Edison	— (1.20)	2,538	— (1.20)	2,490	— (22.90)	24,053	— (24.30)	25,678	— (3.00)	8,698	— (2.80)	8,171
(7)TVA	3.78 (1.52)	1,929	— (—)	—	* (*)	*	* (*)	*	9.43 (6.76)	16,157	10.46 (8.23)	17,843
Vermont Yankee	15.42 (3.31)	935	15.96 (3.39)	885	* (*)	*	* (*)	*	* (*)	*	* (*)	*
Va. Elec. & P. Co.	11.61 (2.74)	2,263	12.47 (2.92)	2,115	24.42 (19.30)	3,857	27.96 (19.83)	3,521	22.65 (12.28)	1,752	21.75 (11.72)	1,801
Wisc. Elec. Power	9.32 (2.75)	1,430	7.64 (2.76)	1,812	* (*)	*	* (*)	*	15.95 (8.64)	2,622	18.56 (9.09)	1,997
Wisc. Pub. Serv.	16.28 (2.44)	362	18.10 (2.52)	311	— (22.79)	—	— (22.13)	—	19.85 (12.27)	770	22.63 (11.44)	641
Yankee Atomic	8.47 (2.68)	372	10.03 (2.67)	370	* (*)	*	* (*)	*	* (*)	*	* (*)	*
(8)Weighted Average												
Quarters:	11.36		11.46		32.43		33.01		14.50		14.92	
First Half:	11.41				32.73				14.71			

(1)Total cost of each kilowatt hour (kwh) is in mills and includes amortized capital allocations. Figures in parentheses is cost of fuel alone.

(2)Nuclear contribution did not begin until May 8.

(3)One unit was being refueled in second quarter.

(4)Fort Calhoun Unit 1 was out of service for refueling and inspection from February 7 through May 9.

(5)Rancho Seco Unit 1 did not go into service until April 18.

(6)Southern California Edison figures represent the 12-month period through June.

(7)Browns Ferry Units 1 and 2 were not operating in second quarter.

(8)Weighted average for each period in each fuel category is derived by dividing the total net kwh produced into the total of net kwh times mills per kwh at each utility.

* Not applicable.

Note: Figures for Commonwealth Edison were not ready in time for this survey, but will be included in the survey to be completed after the third quarter.

Source: "U.S. Electrical Generation Costs in First Half, 1975, Atomic Industrial Forum, Inc., New York, N.Y.

COST ESCALATION—NUCLEAR POWER STATIONS

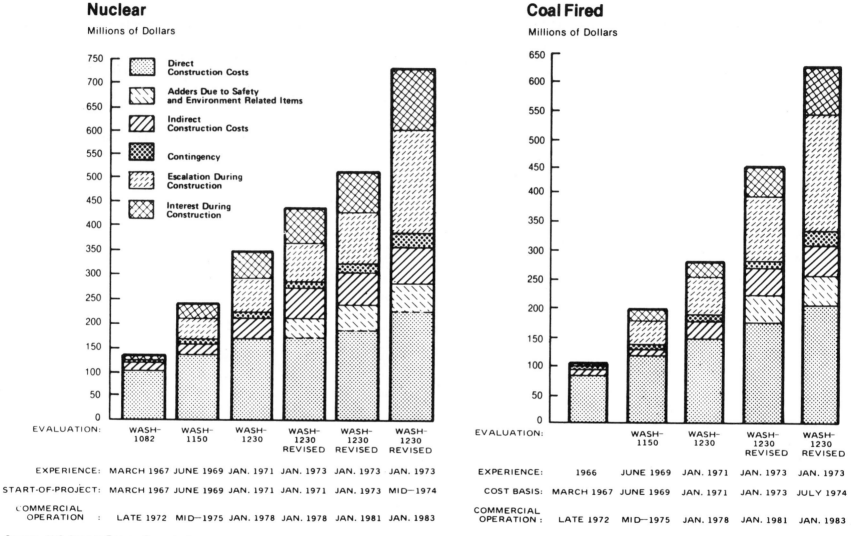

Nuclear

Millions of Dollars

Legend:
- Direct Construction Costs
- Adders Due to Safety and Environment Related Items
- Indirect Construction Costs
- Contingency
- Escalation During Construction
- Interest During Construction

EVALUATION:	WASH-1082	WASH-1150	WASH-1230	WASH-1230 REVISED	WASH-1230 REVISED	WASH-1230 REVISED
EXPERIENCE:	MARCH 1967	JUNE 1969	JAN. 1971	JAN. 1973	JAN. 1973	JAN. 1973
START-OF-PROJECT:	MARCH 1967	JUNE 1969	JAN. 1971	JAN. 1971	JAN. 1973	MID—1974
COMMERCIAL OPERATION :	LATE 1972	MID—1975	JAN. 1978	JAN. 1978	JAN. 1981	JAN. 1983

Source: U.S. Atomic Energy Commission.

Coal Fired

Millions of Dollars

EVALUATION:	WASH-1150	WASH-1230	WASH-1230 REVISED	WASH-1230 REVISED
EXPERIENCE:	1966	JUNE 1969	JAN. 1971	JAN. 1973	JAN. 1973
COST BASIS:	MARCH 1967	JUNE 1969	JAN. 1971	JAN. 1973	JULY 1974
COMMERCIAL OPERATION :	LATE 1972	MID—1975	JAN. 1978	JAN. 1981	JAN. 1983

Comparison of plant cost estimates (total investment for 1000 MWe plants).

Source: As given in *National Energy Outlook*, FEA-N-75/713, Federal Energy Administration, February 1976.

Nuclear Plant Costs and Schedules—1974, Updated Cost Ranges and Schedules for Nuclear Plants as of 3-31-74

Year Contract Let	No. of Plants	Cost Range-$/Kwe Low	High	Average Cost-$/Kwe	Percent Increase	Average Size-Mwe	Schedule Range-Years	Average Schedule —Years
1965	6*	$124	$330	$199	(+68%**)	645	5 to 8	6.3
1966	20	112	482	260	+30	821	4 to 9	6.7
1967	30	109	652	354	+36	852	5 to 13	7.5
1968	14	197	720	413	+17	925	6 to 13	8.7
1969	7	187	530	395	−4	1030	6 to 13	9.2
1970	14	240	577	370	−7	1020	6 to 10	8.4
1971	20	296	572	475	+28	983	6 to 12	8.7
1972***	36	290	645	458	−4	1105	7 to 11	8.5
1973	38	313	650	456	—	1120	6 to 13	8.8
1974 (3 mo.)	8	425	655	558	+22	1215	8 to 12	9.4

*Fort St. Vrain not included
**% increase over original estimate of $119. See Table 2, line 1.
***Demo 1 excluded.
The number of plants in Col. 2 does not correspond exactly to the numbers in Table 2 because of cancellations and rescheduling.
Method of calculation for this table followed the procedure used for Table 2.

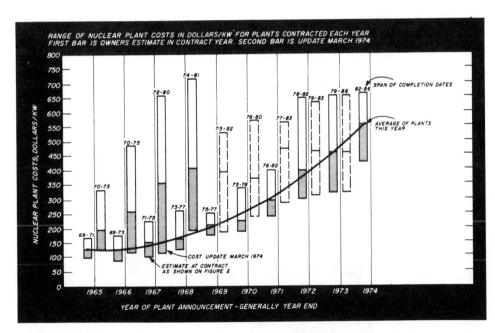

Source: Olds, F. C., "Power Plant Capital Costs Going Out of Sight," *Power Engineering,* August, 1974.

Fuel Cycle Costs

	Contribution to Total Fuel Cycle Cost of Light Water Reactor
Uranium	34%
Conversion	4
Enrichment	41
Fabrication	21
Reprocessing	8
Pu credit	−8

In the case of the boiling water reactor (L.W.R.), using 1970 figures, the cost of uranium ore represents less than 35% of total fuel cycle costs; the cost of enrichment was of roughly equal significance. As the author points out, the elasticities of fuel cycle costs to changes in ore or enrichment cost are small.

Source: Bupp, Irvin C and Derian, Jean-Claude; "The Breeder Reactor in the U.S.: A New Economic Analysis," *Technology Review,* edited at The Masschusetts Institute of Technology, July/August, 1974.

Cost Trends in Nuclear Power Generation

	1967 Plant on Operation by 1970[a]	1970 Plant on Operation by 1974–75[b]	Early 1973 Plant on Operation by 1981[c]	Late 1973 Plant on Operation by 1983[d]
Capital	2.58	6.31	11.30	16.52
Operation and maintenance	0.28	0.38	0.64	1.66
Fuel	1.63	1.87	2.30	3.11
Total cost of electricity	4.49	8.56	14.24	21.29

All figures in mills/kWh.

[a]U.S. Atomic Energy Commission, WASH-1098, November 1967.
[b]Manson Benedict, Electric Power from Nuclear Fission, Bulletin of the Atomic Scientists, Sept. 1971.
[c]M.I.T. Center for Policy Alternatives.
[d]A study of base load alternatives for the Northeast Utility System, Arthur D. Little, Inc., July 1973.

Fuel Cycle Cost (in constant 1967 dollars)	
1967:	1.63 mills/kWh
1970:	1.62 mills/kWh
1973:	2.18 mills/kWh

Cost trends of L.W.R. power generation between 1967 and 1973 are all sharply up, except those of fuel cycle, which have remained low chiefly as a result of economics of scale. When current dollars are converted to constant dollars, the fuel cost increase is flattened even more.

U.S. POLLUTION ABATEMENT COSTS

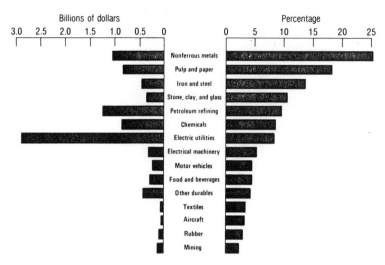

Source: Based on U.S. Department of Commerce, Bureau of Economic Analysis, *Survey of Current Business,* Vol. 55, No. 7, July 1975, p. 16.

BEA estimates of pollution abatement investments as a percentage of total plant and equipment expenditures for selected industries, 1973 plus 1974.

Source: The Sixth Annual Report of the Council on Environmental Quality, December 1975.

Comparison of Automobile Retail Price Increase Estimates
for Automobile Pollution Control
(In current dollars)

Model year	Environmental Protection Agency	National Academy of Sciences	Bureau of Labor Statistics	Industry
1968	5.40	6.65	16.00	—
1969	5.40	6.65	16.00	—
1970	35.00	39.38	21.50	—
1971	35.00	39.38	40.50	—
1972	35.00	39.38	47.50	—
1973	87.00	54.53	75.20	80.00
1974	87.00	54.53	76.60	80.00
1975	200.00	159.00	195.80	208.00

Sources: Environmental Protection Agency, *The Cost of Clean Air*, annual (Washington, D.C.: Government Printing Office, 1974); National Academy of Sciences, "Report by the Committee on Motor Vehicle Emission," prepared for the Environmental Protection Agency under contract no. 68-01-0402, February 1973; U.S. Department of Labor, Bureau of Labor Statistics, press releases, 1968-74; letter from J. J. Ricardo, President, Chrysler Corp., to Senator P. V. Domenici, May 24, 1975; letter from L. A. Iacocca, President, Ford Motor Co., to Senator P. V. Domenici, May 13, 1975; E. E. Estes, President, General Motors Corp., written testimony, Senate Committee on Public Works, Subcommittee on Environmental Pollution, 94th Cong., 1st sess., May 15, 1975, p. 13.

Source: The Sixth Annual Report of the Council on Environmental Quality, December 1975.

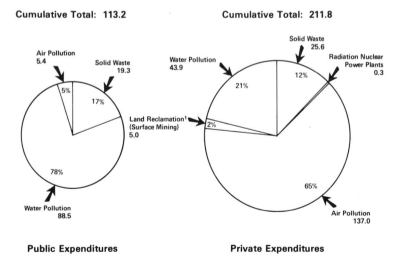

Cumulative Total: 113.2   **Cumulative Total: 211.8**

Public Expenditures   **Private Expenditures**

Source: <u>Fifth Annual Report</u>, Council on Environmental Quality, 1974.

Projected U.S. pollution control expenditures, 1973–1982 (billions of 1973 dollars).

Source: Energy Perspectives, U.S. Department of the Interior, February 1975.

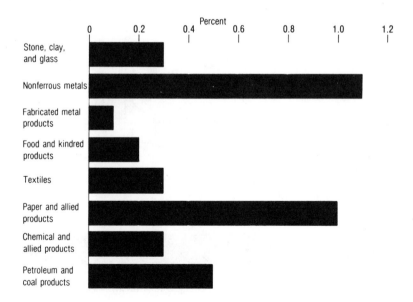

Source: Based on U.S. Department of Commerce, Bureau of Economic Analysis, *Survey of Current Business,* Vol. 55, July 1975, p. S–5, and CEQ cost estimates.

Direct abatement costs as a percentage of value of shipments for selected industries, 1974.

Source: The Sixth Annual Report of the Council on Environmental Quality, December 1975.

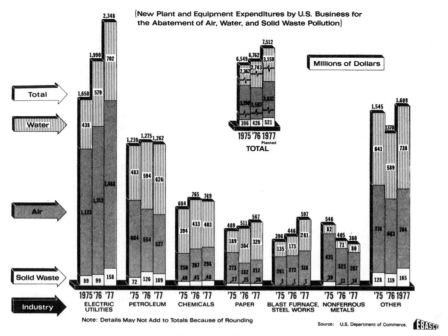

Expenditures for pollution abatement.

Estimated Incremental Pollution Control Expenditures (in billions of 1975 dollars): 1975-1984[a]

Pollutant/source	1975			1984			Cumulative (1975-1984)			
	Operation and maintenance costs [b]	Capital costs [b]	Total annual costs [c]	Operation and maintenance costs [b]	Capital costs [b]	Total annual costs [c]	Capital investment	Operation and maintenance costs [b]	Capital costs [b]	Total annual costs [c]
Air pollution										
Public	0.1	0.1	0.2	0.6	0.2	0.8	1.9	4.2	1.8	6.0
Private										
Mobile	3.4	1.5	4.9	1.3	4.4	5.7	29.1	20.9	31.7	52.6
Industrial	1.2	1.5	2.7	3.1	3.6	6.7	15.1	21.7	25.4	47.1
Utilities	0.6	0.6	1.2	2.7	2.7	5.4	15.6	15.4	15.4	30.8
Subtotal	**5.3**	**3.7**	**9.0**	**7.7**	**10.9**	**18.6**	**61.7**	**62.2**	**74.3**	**136.5**
Water pollution										
Public										
Federal	0.2	<0.05	0.2	0.2	0.1	0.3	2.0	2.1	0.4	2.5
State and local	1.1	0.2	1.3	4.8	2.3	7.1	24.2	28.6	11.2	39.8
Private										
Industrial	1.3	1.2	2.5	6.0	4.7	10.7	36.5	30.9	26.1	57.0
Utilities	0.4	0.3	0.7	0.9	0.5	1.4	2.3	7.2	4.3	11.5
Subtotal	**3.0**	**1.7**	**4.7**	**11.9**	**7.6**	**19.5**	**65.0**	**68.8**	**42.0**	**110.8**
Radiation										
Nuclear powerplants	<0.05	<0.05	<0.05	<0.05	<0.05	<0.05	0.2	0.1	<0.05	0.2
Solid waste										
Public	0.2	0.1	0.3	0.3	0.1	0.4	1.1	2.6	0.8	3.4
Private	0.3	0.1	0.4	0.7	0.2	0.9	0.8	4.3	0.2	4.5
Subtotal	**0.5**	**0.2**	**0.7**	**1.0**	**0.3**	**1.3**	**1.9**	**6.9**	**1.0**	**7.9**
Land reclamation										
Surface mining [d]										
Noise	NA	NA	NA	0.3	0.3	0.6	1.7	1.8	1.6	3.4
Total	**8.8**	**5.6**	**14.4**	**20.9**	**19.1**	**40.0**	**130.5**	**139.8**	**118.9**	**258.8**

[a] Incremental costs are expenditures made pursuant to federal environmental legislation beyond those that would have been made in the absence of this legislation.
[b] Interest and depreciation.
[c] Operation and maintenance plus capital costs.
[d] Not included in this year's estimate.

NA = not available.

Source: Environmental Quality—1976, The Seventh Annual Report of the Council on Environmental Quality, September 1976.

COSTS OF SULFUR DIOXIDE CONTROL

Installation Costs
Flue-Gas Desulfurization System
(Two 800-MW Boilers)

Scrubber System

Scrubber contract	$32,000,000
Foundations, electrical, I&C, linings, oil storage, control room, erection	32,200,000
Sludge ponds with lining	2,500,000
Lime dock, handling and storage facilities	11,700,000
Chimney with lining	7,600,000
Subtotal	$86,000,000
Distributables	18,300,000
Contingency	3,000,000
Escalation	5,000,000
Engineering	1,500,000
Owner's cost (including allowances for funds during construction)	15,100,000
Total Scrubber System	$128,900,000

Off-Site Waste Disposal

Disposal contract	$59,500,000
Land, right-of-way, electrical, calcilox harbor, grits conveyor, pipe rack	2,800,000
Subtotal	$62,300,000
Contingency	6,000,000
Escalation	5,300,000
Engineering	4,000,000
Owner's costs (including allowances for funds during construction)	6,700,000
Total Off-Site Waste Disposal	$84,300,000

SOURCE: *Air Quality and Stationary Source Emission Control*, Serial 94-4, prepared by the National Academy of Engineering for the U.S. Senate Committee on Public Works, March 1975; data for the Bruce Mansfield Plant, December 1974.

Cost Calculation for New Plant

Without flue gas desulfurization:	
capital cost: $500/kW × 0.17 ÷ 7000 hours	12.1 mills/kWh
fuel cost: heat rate of 8982 × $1/MM Btu of fuel	8.9
other operating costs:	0.5
Cost of power generation without scrubber	21.6 mills/kWh
Added cost of lime scrubbing process:	
capital cost: $100/kW × 0.17 ÷ 7000 hours	2.4 mills/kWh
operating cost: labor, chemicals, etc.	0.5
sludge disposal	0.3
Subtotal	3.2 mills/kWh
energy loss of 6% of plant output needed for scrubber operation:	0.6
capacity derating of plant resulting from 6% loss in output capacity:	0.7
Subtotal	1.3
added cost of lime scrubbing process	4.5 mills/kWh
Cost of power generation with scrubber:	26.2 mills/kWh

Cost Calculation for Retrofit to Existing Plant, Rural Location

Without flue gas desulfurization:	
capital cost $250/kW × 0.14 ÷ 6000 hours	5.8 mills/kWh
fuel costs: heat rate of 10,342 × $1/MM Btu of fuel	10.3
other operating costs	1.0
Cost of power generation without scrubber	17.2 mills/kWh
Added cost of lime scrubbing process:	
capital cost of $125/kW × 0.17 ÷ 6000 hours	3.5 mills/kWh
operating cost: labor, chemicals, etc.	0.6
sludge disposal	0.5
Subtotal	4.6 mills/kWh
energy loss, 6% of plant output needed for scrubber operation	0.7
capacity derating of 6% (replacement at $500/kW, 17% fixed charge)	0.8
Subtotal	1.5
added cost of lime scrubbing process	6.1 mills/kWh
Cost of power generation with scrubber:	23.3 mills/kWh

New Plant: Sensitivity Calculations on Added Cost from Scrubber

Low case:	
$60/kW; 0.15 mills/kWh for sludge disposal	2.1 mills/kWh
5% energy loss and capacity derating	1.1
Added cost of scrubber, low case:	3.2 mills/kWh
High case:	
$130/kW; 1.0 mills/kWh for sludge disposal	4.7 mills/kWh
7% energy loss and capacity derating	1.5
Added cost of scrubber, high case:	6.2 mills/kWh

Source: Air Quality and Stationary Source Emission Control, A Report by the Commission on Natural Resources, Prepared for the Committee on Public Works, United States Senate, Serial No. 94-4, March 1975.

COSTS OF POWER PLANT COOLING SYSTEMS

Cost of Condenser Cooling Itemized, Compared
(Thousands of Dollars)

	Natural Waterway	Cooling Towers Mech. Draft	Cooling Towers Nat. Draft
STRUCTURES AND IMPROVEMENTS			
Ground improvements	13	42	28
River screen house	3,345	—	—
Forebay for screen house or make-up pump house at river	65	10	10
Discharge structure at river	25	—	—
Make-up pump house at river	—	525	525
Mechanical draft cooling tower basin	—	1,380	—
Flume for mechanical draft cooling towers	—	220	—
Natural draft cooling tower basin	—	—	Included with tower
Pump house for cooling towers	—	680	680
Circulating water piping earthwork	280	225	195
Blowdown piping earthwork	—	5	8
Blowdown piping outfall structure at river	—	7	7
Blowdown retention basin	—	—	—
Make-up piping earthwork	—	3	9
TOTAL—Structures and Improvements	3,728	3,097	1,462
MECHANICAL			
Cooling towers	—	3,584	11,400
Circulating water pumps and motor drives	965	965	1,160
Make-up pumps and motor drives	—	108	108
Screen wash pumps and motor drives	10	6	6
Traveling screens	480	28	28
pH control system	—	30	30
Chlorination system	60	35	35
Pipes, valves and fittings			
a. Circulating water piping	2,324	1,950	1,700
b. Make-up piping c. Blowdown piping	—	57	112
d. Chemical feed piping	30	30	30
Instruments and controls	10	30	25
Miscellaneous accessories	15	15	15
TOTAL—Mechanical	3,894	6,838	14,649
ELECTRICAL			
Structures and improvements	128	128	74
Accessory electrical equipment	261	500	304
Electrical installation	161	237	138
TOTAL—Electrical	550	865	516
Sub-Total—Structures, Mechanical and Electrical	8,172	10,800	16,627
Contingency—5%	409	540	831
Total estimated construction cost	8,581	11,340	17,458
Estimated construction cost/kW	7.46	9.86	15.18

Source: *Nuclear Industry*, The Monthly Magazine of the Atomic Industrial Forum, Inc., New York, N.Y., January 1975.

Unit and Total Costs of Cooling Devices

Cooling Device	Area Required (Normalized)	Water Required (gal/kW/yr)	Additional Unit Costs ($/kW)	Cost to Year 2000 ($ billions)
Open cycle	—	500,000	—	—
Pond	500–1,000	—	—	—
Mechanical draft				
1. Wet	1–2	4,000–8,000	7	11
2. Dry	—	70	27	—
Natural draft				
1. Wet	1	4,000–8,000	11	16
2. Dry	—	70	25	60

Economic Comparison of Cooling Towers

	Induced Draft Wet Tower	Natural Draft Wet Tower	Natural Draft Dry Tower	Induced Draft Dry Tower	Mechanical Draft Fin Tube Exchanger
Circulating water flow GPM	91,000	95,000	115,000	160,000	95,000
Cooling range, °F	21	21	17	12	20
Design approach, °F	20	20	38	33	25
Wet bath temperature, °F	65	62	—	—	—
Dry bath temperature, °F	—	—	95	95	95
Land area, square feet	18,600	32,000	69,000	22,500	60,000
Total investment cost—millions of dollars	28,715	29,580	31,905	32,305	30,980
Backpressure, in. Hg (44% station load)	1.3	1.4	1.5	1.9	1.5
Capability loss, $1,000	105	69	448	430	292
Annual costs plus capability loss, $1,000	6,453	6,476	7,217	7,283	6,979

Economic Comparison of Cooling Methods

Heat Rejection Method	Temperature Rise (°F)	Inlet Temperature (°F)	Condenser Back Pressure	Capacity ($/kW)
1. Run of river cooling	17	55	1.2	5
2. Bay/lake cooling	—	—	1.2	6
3. Natural draft cooling towers— run of river makeup	28	70	1.5, 2.16	7.5
4. Natural draft cooling towers— makeup reservoir	—	—	1.5, 2.16	11
5. Cooling pond	—	65	—	10
6. Dry cooling towers	—	—	—	22

Source: *Engineering for Resolution of the Energy-Environment Dilemma*, reproduced with permission of the National Academy of Sciences, A Report of the National Academy of Engineering. Published by the National Academy of Sciences, 1972.

Summary of Costs and Economic Penalties for Optimized Cooling Systems (Middletown) Interfaced with 1000 MWe Fossil Power Plant and the Plant Outputs and Heat Rates at Optimum Cooling System Design Conditions

Cooling System / Item	Once-Through	Mech. Wet	Fan Wet	Nat. Wet	Pond	Spray Canal	Mech. Dry	Nat. Dry
Total Capital Investment (Equipment, Material & Labor, and Indirect Charges) (10^6)	11.27	16.04	20.65	20.05	28.63	17.84	25.50	28.16
Total Economic Penalties Assigned to the Optimized Cooling System (10^6)	2.74	13.03	10.94	11.82	13.56	13.94	58.49	51.79
Total Evaluated Cost (Total Capital Cost plus Total Economic Penalties) (10^6)	14.01	29.07	31.59	31.87	42.19	31.78	83.99	79.95
Gross Plant Output at Optimum Cooling System Design Conditions, MWe	1043	1020	1026	1014	1000	1020	938	932
Gross Plant Heat Rate, Btu/kWh	7365	7531	7487	7576	7682	7531	8189	8242

Source: Heat Sink Design and Cost Study for Fossil and Nuclear Power Plants, WASH-1360, United States Atomic Energy Commission, December 1974.

COSTS OF POLLUTION

Summary of Selected Studies on Health Damage Costs

Study	Pollutants	Diseases studied	Effects observed	Estimated cost of effects for U.S. in millions of dollars per year	Method of estimation
Ridker (1967)	General air pollution	Lung cancer, chronic bronchitis, acute bronchitis, emphysema, asthma, pneumonia	Higher urban mortality rates for these diseases	360–400	Total costs of these diseases computed to include lost earnings, early burial, treatment, and absenteeism; based on urban–rural mortality differentials, 18–20 percent of costs attributed to air pollution
Lave and Seskin (1970)	Particulates, sulfates	Bronchitis, lung cancer, other cancer, pneumonia, infant mortality	Increased mortality	2,080 (mortality and morbidity)	Multiple regression analysis across U.S. and British cities; various other variables used to explain mortality rates, including temperature–humidity, age, and race
Justus, Williams, and Clement (1973)	Particulates, sulfur dioxide, nitrogen dioxide, oxidants, carbon monoxide	Respiratory disease (nonrespiratory diseases assumed to be unaffected by pollution)	Mortality and morbidity	62–311	Respiratory disease costs taken from Ridker (1967); based on literature search, 95–99 percent of costs assigned to nonpollution factors such as smoking
Jaksch and Stoevener (1974)	Airborne suspended particulates (other pollutants not measured)		Statistically significant increase in number of hospital outpatients during and immediately after high pollution days	(1)	Multiple regression analysis; 14 other variables included, such as age, sex, and temperature–humidity; overall explanatory power (R^2) of equations is low
Singley et al. (1974)	Bacteria in drinking water	Salmonellosis, shigellosis, hepatitis	Bacteria count in water supply has statistically significant effect on reported cases	145	Multiple regression analysis across counties of reported diseases that could be caused by water; other explanatory variables include population density, age distribution, and income; dollar cost assigned to each case—i.e., $184 for one case of salmonellosis
Abel, Tihansky, and Walsh (1975)	Diseases carried in drinking water	Various	Mortality and morbidity	644	Tabulated reported cases attributed to water, estimated number of unreported cases, assigned a dollar value to each case, i.e., $1,250 for one case of salmonellosis
Page, Harris, and Epstein (1975)	Chlorinated hydrocarbons in drinking water	Cancer	Statistically significant relationship between use of chlorinated Mississippi water and various cancers	(1)	Multiple regression analysis across Louisiana parishes; other explanatory variables include median family income and fraction of work force in petroleum and coal products industries

[1] No national estimate.

Sources: R. G. Ridker, *Economic Costs of Air Pollution: Studies in Measurement* (New York: Frederick A. Praeger, 1967); L. B. Lave and E. P. Seskin, "Air Pollution and Human Health," *Science*, 169 (August 21, 1970) pp. 723–33; C. G. Justus, J. R. Williams, and J. D. Clement, *Economic Costs of Air Pollution Damage*, prepared for Southern Services, Inc., Birmingham, Alabama, by Science Technology and Research, Inc., Atlanta, May 1973; John A. Jaksch and Herbert H. Stoevener, *Outpatient Medical Costs Related to Air Pollution in the Portland, Oregon Area*, EPA 600/5–74–017 (Washington: Government Printing Office, 1974); J. Edward Singley et al., "A Benefit/Cost Evaluation of Drinking Water Hygiene Programs," prepared for the Environmental Protection Agency (preliminary, 1974); Fred H. Abel, Dennis P. Tihansky, and Richard G. Walsh, "National Benefits of Water Pollution Control," Environmental Protection Agency (1975); Talbot Page, Robert H. Harris, and Samuel P. Epstein, "Relation between Cancer Mortality and Drinking Water in Louisiana" (mimeo, 1975).

Source: The Sixth Annual Report of the Council on Environmental Quality, December 1975.

Summary of Selected Studies on Vegetation Damage Costs

Study	Location of study	Pollutants	Type of vegetation	Damage costs, in millions of dollars per year	Method
Lacasse and Weidensaul (1970)	Pennsylvania	(In descending order of importance) Oxidants, sulfur oxides, lead, hydrogen chloride, particulates, herbicides, ethylene	(In descending order of importance) Vegetables, fruits, agronomic crops, lawns, shrubs, woody ornamentals, timber, commercial flowers	11.0 (includes some avoidance costs)	Field observation
Benedict, Miller, and Olson (1971)	National	Ozone, peroxyacyl nitrates, oxides of nitrogen, sulfur dioxide, fluorides	Crops and ornamentals	Crops 85.6 Ornamentals 46.2 ———— 131.8	Assumed damage functions based on some published studies
Lacasse (1971)	Pennsylvania	(In descending order of importance) Oxidants, sulfur oxides, lead, hydrogen chloride, particulates, herbicides, ethylene	(In descending order of importance) Vegetables, fruits, agronomic crops, lawns, shrubs, woody ornamentals, timber, commercial flowers	0.22 (includes some avoidance costs)	Field observation
Millecan (1971)	California (15 counties)	(In descending order of importance) Ozone, peroxyacyl nitrates, fluorides, ethylenes, sulfur dioxide, particulates	Citrus and other crops	26	Field observation
Feliciano (1972)	New Jersey	(In descending order of importance) Peroxyacyl nitrates, ammonia, fluoride, particulates	Vegetables and field crops accounted for 85 percent of damages	1.19	Field observation
Naegele, Feder, and Brandt (1972)	New England	Oxidants accounted for over 90 percent of damages	Fruits, vegetables, agronomic crops	1.1	Field observation
Pell (1973)	New Jersey	(In descending order of importance) Oxidants, hydrogen fluoride, ethylene, sulfur dioxide, anhydrous ammonia	Agronomic and ornamental plantings	0.13	Field observation

Source: T. E. Waddell, *The Economic Damages of Air Pollution*, EPA–600/5–74–012 (Washington, D.C.: Government Printing Office, 1974).

Source: The Sixth Annual Report of the Council on Environmental Quality, December 1975.

Summary of Selected Studies on the Value of Wildlife

Study	Species/location	Group receiving benefits	Unit benefit	Total benefit per year	Method
Davis (1964)	Big game/private forest in Maine	Hunters	Average benefit of $1.96 per household per day	$29,080 for 500,000-acre forest, about 9 percent of its value in logging	Onsite interviews
Brown and Hammack (1973)	Waterfowl/United States	Hunters	Marginal value of $3.10 to $4.65 per bagged waterfowl	(No estimate)	Mail–circulated questionnaire
Meyer (1974)	Salmon/Fraser River in British Columbia	Fishermen and nonusers	$632 per household per year for fishing plus $223 per household per year for preservation (value to nonusers)	$186 million for fishing plus $101 million for preservation	Household interviews
Horvath (1974)	Wildlife/Southeastern United States	Hunters, fishermen, watchers, photographers	$2,183–3,107 per household per year for all uses	$24–31 billion for all uses	Household survey

Sources: Robert K. Davis, "The Value of Big Game Hunting in a Private Forest," *Transactions of the 29th North American Wildlife and Natural Resources Conference,* March 9–11, 1964 (Washington, D.C.: Wildlife Management Institute, 1964), pp. 393–403; Gardner Mullard Brown, Jr., and Judd Hammack, "Dynamic Economic Management of Migratory Waterfowl, *Review of Economic Statistics,* 55 (February 1973), pp. 73–82; Philip A. Meyer, *Recreational and Preservation Values Associated with the Salmon of the Fraser River,* Environment Canada, Fisheries and Marine Service, Info. Report Series PAC/N–74–1 (Ottawa); Joseph C. Horvath, "Economic Survey of Wildlife Recreation," Georgia State College, Environmental Research Group, 1974.

Source: The Sixth Annual Report of the Council on Environmental Quality, December 1975.

Summary of Selected Studies on Materials Damage and Avoidance Costs

Study	Pollutant	Affected material	Damaging effect	Avoidance effect	Estimated damage cost, in millions of dollars per year	Estimated avoidance cost, in millions of dollars per year	Method
Mueller and Stickney (1970)	Ozone	Rubber products	Early replacement	Pollution-resistant materials	226	170	Literature search, questionnaire
Robbins (1970)	H_2S, SO_2, organic gases, particulates	Electrical contacts	Corrosion, fouling	Plating, air conditioning	15	50	Literature search and consultation with manufacturers
Salmon (1970)	Not specified	Miscellaneous	Accelerated deterioration	(1)	3,800	(1)	Literature search, interviews with manufacturers and users
Salvin (1970)	NO_x and ozone	Textile fibers and dyes	Fading and yellowing	Research, quality control, more expensive dyes and textiles	2 206		Consultations with industry
ITT (1971)	Not specified	Electrical components	Component failure	Air filtering, etc.	13.2	2.4	Literature search, interviews with manufacturers and users
Fink, Buttner, and Boyd (1971)	Sulfur oxides and other pollutants	Susceptible metal products	Accelerated corrosion	Extra protection and maintenance	2 1,450		Engineering-economic
Spence and Haynie (1972)	Particulates, sulfur oxides	Paints	Loss of service life	(1)	704	(1)	Literature survey
Gillette (1973)	Sulfur oxides	Metals and paints	Accelerated corrosion	Extra protection and maintenance	2 400		Engineering-economic

1 No estimate made of avoidance cost.

2 Includes both damage and avoidance costs.

Source: T. E. Waddell, *The Economic Damages of Air Pollution*, EPA–600/5–74–012 (Washington, D.C.: Government Printing Office, 1974).

Source: The Sixth Annual Report of the Council on Environmental Quality, December 1975.

Energy Futures

IN THIS FIGURE, DOMESTIC OIL INCLUDES CRUDE AND NATURAL GAS LIQUIDS

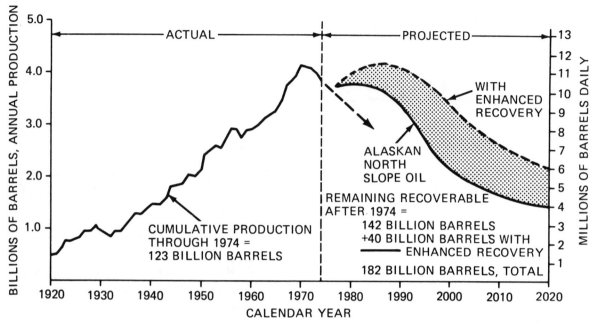

Projected domestic oil production.

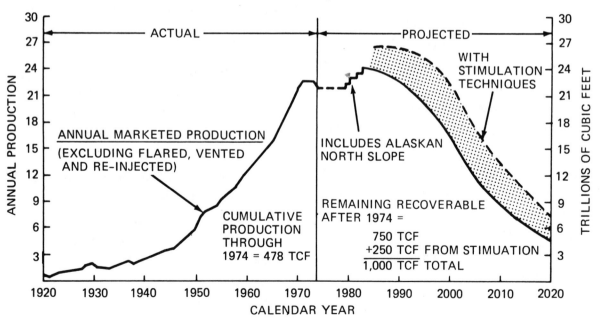

Projected domestic natural gas production.

Source: *A National Plan for Energy Research, Development & Demonstration: Creating Energy Choices for the Future, Volume 1: The Plan, ERDA-48.* Energy Research and Development Administration, June 1975.

ENERGY IN THE FUTURE

The nature of the world energy problem hinges on several factors: the increasing world population, the increasing per capita energy consumption, the depletion of the world fossil fuel energy resource base, and the need to develop substitute energy forms. This problem can be considered as encompassing three phases:

Near-Term Phase. 1975–90; characterized by shortages of petroleum and gas in certain countries of the world and the need for governments to administer these shortages and to begin development of alternate energy sources.

Transition Phase. 1985–2020; characterized by the increasing use of coal and coal derivatives in place of oil and natural gas and the use of nuclear energy for the generation of electricity.

Terminal Phase. 2020–; characterized by gradual transition to zero population growth, a leveling off of per capita energy consumption, and reliance on energy derived from nuclear fission and fusion, and from renewable resources such as solar, wind, geothermal, etc.

Two estimates of world energy supply and demand are given in the accompanying tables which indicate the uncertainties regarding the extent of these resources but also indicate the eventual exhaustion of fossil fuel resources (additional data on energy resources appear in Chapters 2 and 3). While there is little argument that the fossil fuel resources will be depleted at some time in the future, the extent of these resources, in terms of meeting near-term needs, presents a particularly difficult problem in planning the development of alternate energy sources. Most of the remainder of this chapter is devoted to resumes of various plans that have been suggested for meeting the energy problem in the United States.

One projection of the long-range future of energy consumption and the sources of that energy, is given in the accompanying chart prepared by Earl Cook.

Energy Content of World Supply of Fossil Fuel
(In Units of $Q \equiv 10^{18}$ Btu)

	Known Recoverable[a]	Undiscovered and or Marginal[a]	Eventually Recoverable[b]	%[b]
Coal	17.3	320	192	88.8
Crude oil	1.73	23	11.1	5.2
Natural gas	1.95	20	10.1	4.7
Natural gas liquids	0.21	3.2		
Tar-sand oil	0.23	6.3	1.7	0.8
Shale oil	0.87	77	1.1	0.5
Total	22.5 Q	450 Q	216 Q	

[a]According to V. E. McKelvey and D. C. Duncan, ref. 5.
[b]According to M. K. Hubbert, ref. 6.

Value of $Q \equiv 10^{18}$ Btu

$= 2.52 \times 10^{17}$ kcal
$= 1.05 \times 10^{21}$ joule
$= 2.93 \times 10^{14}$ kWh thermal
$= 1.22 \times 10^{10}$ MWd thermal
$= 3.35 \times 10^{7}$ MW year thermal

Source: Wolf Häfele, "A Systems Approach to Energy," *American Scientist*, Vol. 62, July–August 1974.

Energy Budget for a Steady-State Civilization

	kW(th)/capita
Present U.S. level	10.0
Adjustment for the future	6.0
Steel, aluminum and magnesium production	0.1
Recovery and recycling of scarce elements	2.0
Electrolytic hydrogen	2.5
Water by desalination (100 gal/day)	0.3
Water transport to cities	0.1
Air-conditioning to cities	0.3
Intensive food production	0.2
Sewage and waste treatment	0.5
Contingency	4.0
Total	20.0

Source: Weinberg and Hammond, ref. 3.

Energy Consumption

United States	
1970	0.007Q/year
2000	0.16 Q/year
World	
1970	0.24 Q/year
	4×10^9 people, 2 kW(th)/capita
2000	2.1 Q/year
	7×10^9 people, 10 kW(th)/capita
2050	6 Q/year
	10×10^9 people, 20 kW(th)/capita

Source: Wolf Häfele, "A Systems Approach to Energy," *American Scientist*, Vol. 62, July–August 1974.

World Energy Balance Sheet

Depletion Sources

Source	Supply[a] (Q)	Cumulative Demand, A.D. 1960–2000 (Q)	Reference Annual Demand, A.D. 2000[b] (Q/yr)	Surplus, A.D. 2000[c] (Q)	Supply, A.D. 2000[d] (yr)
Coal	20–30				
Liquid petroleum	3–6				
Gas	2–5				
Nuclear		no breakdown	breakdown depends on alternatives	no breakdown	no breakdown
light water reactors	~100				
fast breeder reactors	~10,000				
Total, excluding nuclear	25–41	10–20	0.4–0.5	5–31	10–77

Continuous Sources

Source	"Reasonable" Annual Capability by A.D. 2000 (Q/yr)	Maximum Estimated Capability (Q/yr)	Utilization of Continuous Sources (%)
Wood fuel	0.04		
Biological photosynthesis	Nil	All solar energy forms total about 3,000	
Hydropower	0.04		no breakdown
Windpower	0.001		
Solar direct conversion	0.001		
Tidal	0.00025	0.03	
Geothermal	0.00001	0.006	
Total	~0.08	>3.000	<0.003

[a]"Known recoverable" and "recoverable at reasonable costs" only. Does not include resources with costs more than twice the 1960 costs.

[b]Excludes an estimated 0.08 Q/year supplied by continuous sources.

[c]Worst case. Excludes nuclear contribution; range computed using minimum supply minus maximum demand, and vice versa.

[d]Worst case. Computed using minimum surplus and maximum annual demand, and vice versa.

Reproduced with permission of the National Academy of Sciences, *Engineering for the Benefit of Mankind*, a Report of the National Academy of Engineering. Published by the National Academy of Sciences, 1970.

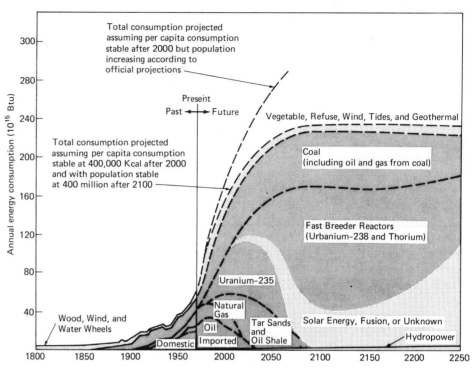

What does the future hold for U.S. energy consumption—and the sources whence it comes? This proposal projects consumption under two different population conditions—and shows for a stabilized population how energy needs might be met into the first centuries of the third millenium. Energy will surely become more expensive, but—at least through the year 2250—it will still be available. This stable-population assumption produces a plateau of about four times the present rate of energy consumption.

Source: Earl Cook, "Energy for Millenium Three," *Technology Review*, edited at The Massachusetts Institute of Technology December, 1972.

OECD—PROJECTED ENERGY REQUIREMENTS

In a study released by the Organization for Economic Cooperation and Development (OECD) on energy prospects to 1985,[1] a decrease in the rate of energy growth among member countries[2] was forecast as a result of higher oil prices. The projection of the rate of energy growth at 4.9% annually at pre-October 1973 oil prices was reduced to 4.3% per year with oil at $6 per barrel, and to 3.8% per year with oil at $9 per barrel.

[1] *Energy Prospects to 1985*, Volumes 1 and 2, Organization for Economic Co-Operation and Development, Paris 1974.

[2] Australia, Austria, Belgium, Canada, Denmark, Finland, France, Germany (FRG), Greece, Iceland, Ireland, Italy, Japan, Luxembourg, The Netherlands, New Zealand, Norway, Portugal, Spain, Sweden, Switzerland, Turkey, the United Kingdom, and the United States.

These projections on energy consumption do not include reductions which also might be made by changes in government policy or active conservation programs. Inclusion of such government actions along with price-induced reductions lead to a projection of overall energy savings in 1985 of 15–20% from 1973 levels. Projected savings include 15% in industry, 15–25% in transportation, and 20–25% in the residential-commercial sector.

The study also concludes that nuclear power will become more attractive in the future as fossil fuel prices rise, and that nuclear power also offers considerable environmental and land-use advantages. An accelerated program for the construction of nuclear power plants was suggested that would increase installed nuclear capacity within the OECD countries from 58,500 megawatts in 1974 to 700,000 megawatts in 1985 and 1,439,000 megawatts in 1990.

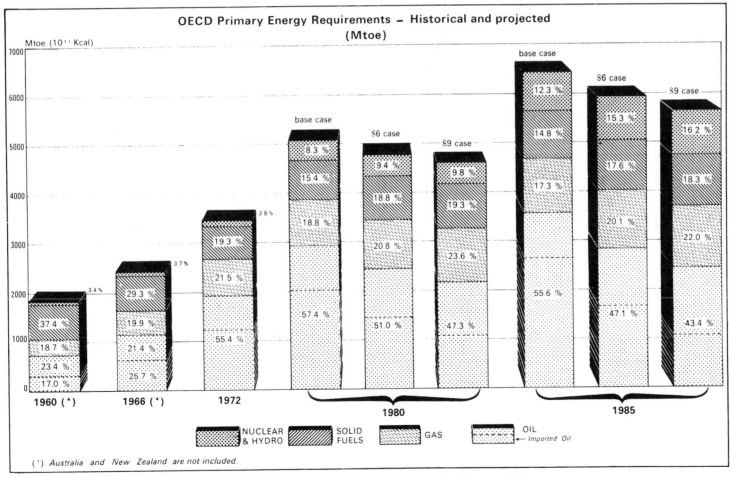

Source: Energy Prospects to 1985, Organisation for Economic Co-operation and Development, Paris, France 1974.

OECD—ENERGY SELF SUFFICIENCY

The degree to which member countries of the OECD (see 16-2) were dependent in 1972 upon imported energy supplies is shown in one of the accompanying charts, and the impact of higher oil prices on the total petroleum energy requirement (TPER) to 1985 is shown in the second chart.

Only two of the OECD countries, Canada and Australia, produce more primary energy supplies than they consume, although Canada imports coal and Australia imports oil. While the figures for total energy self-sufficiency give some indication of the dependence of a particular country on foreign energy supplies, the total self-sufficiency figures do not reflect dependence on particular fuel forms, such as oil and gas, for which other fuel forms, such as coal and nuclear, cannot be readily substituted.

Percentage of Indigenous Energy Production to Consumption in OECD Countries in 1972

	Coal (1)	Oil (1)	Gas (1)	Hydro Electricity(2)	Nuclear Electricity(2)	Total Energy Self Sufficiency(3)
Canada	77	115	178	76	3	120
Australia	171	56	100	20	–	108
USA	112	68	97	14	3	86
Norway	36	23	–	107	–	72
Netherlands	63	6	171	–	1	70
UK	98	2	97	2	11	51
Germany	115	7	64	5	3	50
New Zealand	102	3	100	72	–	42
Turkey	100	35	–	29	–	31
Austria	25	25	55	59	–	30
Iceland	–	–	–	96	–	24
France	69	1	54	30	9	23
Greece	85	–	–	20	–	21
Spain	85	–	–	49	7	21
Switzerland	–	–	–	77	14	21
Sweden	–	–	–	75	2	17
Belgium	65	–	–	2	–	16
Portugal	44	–	–	80	–	16
Italy	5	1	93	33	3	15
Japan	32	–	68	20	2	10
Finland	–	–	–	38	–	7
Ireland	8	–	–	10	–	2
Luxembourg	–	–	–	42(4)	–	2
Denmark	–	–	–	–	–	–

1) The ratio of indigenous supply of the fuel to total indigenous consumption of the fuel.

2) The ratio of electricity supplied from the energy source to total electricity consumed.

3) The ratio of total indigenous supply to total primary energy.

4) Mainly electricity from pumped storage integrated as peak load generating plant into the international grid.

This table does not include less common sources such as wood or peat which can in some cases make a significant contribution. For example in Finland these two sources provide about 18 per cent of total energy consumption.

Source: Energy Prospects to 1985, Organisation for Economic Cooperation and Development, Paris, France 1974.

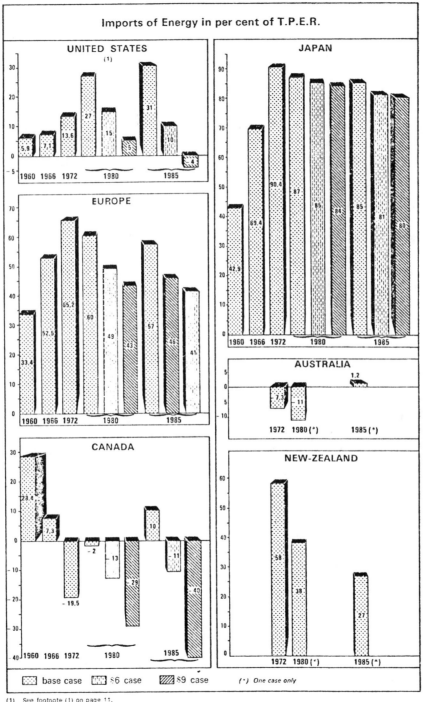

Imports of Energy in per cent of T.P.E.R.

base case $6 case $9 case   (*) One case only

(1) See footnote (1) on page 11.

Source: Energy Prospects to 1985, Organisation for Economic Co-operation and Development, Paris, France 1974.

U.S.—DEPLETION OF PETROLEUM AND GAS

In a study[1] of production of oil and gas on the outer continental shelf, the Office of Technology Assessment of the U.S. Congress outlined the uncertainties in resource estimates and the implications of these uncertainties in the formulation of energy policy as follows:

If the optimistic (resource) figures are valid, then we have plenty of time to develop alternatives in a deliberate manner, and can perhaps reasonably aim at effectively eliminating oil imports by 1985 or 1990. But if the pessimistic estimates are correct, it may be necessary not only to take very strong measures to curb demand and to accelerate the development of alternative sources of petroleum products, but also to limit production from domestic sources below the maximum efficient rate and to accept a relatively high level of imports, in order to avoid a period of extremely heavy dependence on imports toward the end of this century.

In 1973, the U.S. consumed petroleum liquids at a rate of 17.3 million barrels per day, or 6.3 billion barrels per year. Of this amount, 11.1 million barrels were produced in the U.S. and 6.2 million (35.9%) were imported. According to 1974 estimates of the American Petroleum Institute (API) and the American Gas Association (AGA), the U.S. has 46.9 billion barrels of proved and indicated reserves of oil and natural gas liquids (NGL). This amount represents only 11.6 years of production at the 1973 rate of production of 11.1 million barrels per day, or 7.4 years of production at the 1973 rate of consumption.

Of course, existing reserves cannot produce at a constant rate; instead, the rate of production declines continuously over the lifetime of a reserve. . . . If we are simply to replace both the projected decline of 4.4 million barrels per day of domestic production and the 1973 import level of 6.2 million barrels per day by 1985, without taking into account any growth in domestic consumption, we would have to provide an additional 10.6 million barrels per day of new production by 1985.

The magnitude of the oil supply problem becomes more evident if we take into account the effects of an annual rate of growth of demand for petroleum liquids of a conservative 2% per year. This is well below the 5.6% growth rate in the U.S. between

[1]*An Analysis of the Feasibility of Separating Exploration from Production of Oil and Gas on the Outer Continental Shelf*. Congress of the United States, Office of Technology Assessment, May 1975.

Estimates of Undiscovered Recoverable Oil Resources of the United States

Source	Oil and natural gas liquids (billions of barrels)		
	Onshore	Offshore	Total
1. National Petroleum Council (1972)[1]	90	64	154
2. Mobil Oil Corp. (1974)[2]	34	54	88
3. National Academy of Sciences (1975)[3]	[4]	[4]	113
4. Hubbert (1974)[5]	27	45	[6]72
5. U.S. Geological Survey (1974)[7]	136–272	64–128	200–400

[1] The National Petroleum Council estimates were for undiscovered oil-in-place, rather than for recoverable oil. The figures in the table were obtained by applying an average recovery factor of 40 percent to the oil-in-place estimates. National Petroleum Council, "U.S. Energy Outlook."
[2] Robert Gillette, "Oil and Gas Resources: Did USGS Gush Too High?", Science, July 12, 1974, p. 128, table 1.
[3] National Academy of Sciences, "Mineral Resources and the Environment," February 1975, p. 8.
[4] No breakdown given.
[5] National Academy of Sciences, op. cit., p. 89, table 2.
[6] The breakdown between onshore and offshore resources is based on rough estimates provided by Dr. Hubbert in a personal communication.
[7] National Academy of Sciences, op. cit., p. 89, table 2.

1970 and 1973 and below the USGS 1972 projection of a 3.6% annual growth rate from 1972 to 1985. . . . Over 10 years a 2% annual growth rate represents an additional demand of 3.8 million barrels per day in 1985. When added to the 10.6 million barrels per day that would be needed to replace current imports and projected declines in current output, this implies a need for 14.4 million barrels per day of new production in 1985, or additional imports of 8.2 million barrels per day.

However, the complete replacement of imports by new domestic production could create a need for greater imports by the end of this century. This can be seen by examining current estimates of remaining U.S. oil resources. The first table compares some of the most important recent estimates.

The implications of the differences in resource estimates are substantial. The Federal Energy Administration's projections of long-term oil production that are based on an estimate of about 200 billion barrels of undiscovered recoverable resources, the same as the lower limit of the USGS estimate, indicate that production will peak in the mid-to-late 1980s and will decline below current levels around 2030. In contrast, Hubbert's estimate of 72 billion barrels implies that the peak has already occurred. In fact, there has been a consistent decline in domestic production since November, 1970. . . .

Another way of looking at the long-term implications of the differences in resource estimates is to calculate the number of years of supply that the estimates represent

Years of Remaining Domestic Production of Oil and Natural Gas Liquids

Estimate	Undiscovered oil and NGL (billions of barrels)	Total remaining production of oil and NGL (billions of barrels)[1]	No growth in consumption (6.3 billion barrels/year)				2.5 percent annual growth			
			No imports		35 percent imports		No imports		35 percent imports	
			Years of production	Year of exhaustion	Years of production	Year of exhaustion	Years of production	Year of exhaustion	Years of production	Year of exhaustion
1. National Petroleum Council	154	217	34	2009	53	2028	25	2000	34	2009
2. Mobil Oil	88	151	24	1999	37	2012	19	1994	26	2001
3. National Academy of Sciences	113	176	28	2003	43	2018	21	1996	29	2004
4. Hubbert	72	135	21	1996	33	2008	17	1992	24	1999
5. USGS	200–400	263–463	42–73	2017–2048	64–113	2039–2088	28–42	2003–2117	38–54	2013–2029

[1] This is the sum of undiscovered recoverable oil and NGL plus an additional 63 billion barrels of oil and NGL estimated to be producible from known fields.

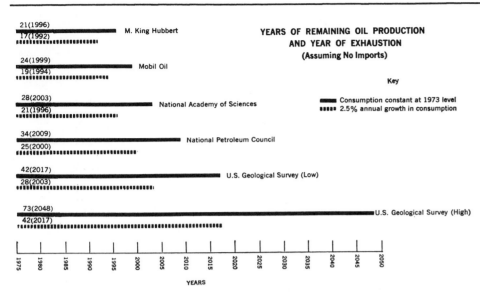

YEARS OF REMAINING OIL PRODUCTION
AND YEAR OF EXHAUSTION
(Assuming No Imports)

Key

▬▬ Consumption constant at 1973 level
▪▪▪▪▪ 2.5% annual growth in consumption

21(1996)
17(1992)   M. King Hubbert

24(1999)
19(1994)   Mobil Oil

28(2003)
21(1996)   National Academy of Sciences

34(2009)
25(2000)   National Petroleum Council

42(2017)
28(2003)   U.S. Geological Survey (Low)

73(2048)
42(2017)   U.S. Geological Survey (High)

YEARS

Years of remaining oil production and year of exhaustion (assuming no imports).

in terms of specified rates of consumption. The second table shows the results of such calculations, and the dates of exhaustion they imply, based on both the 1973 rate of consumption of 6.3 billion barrels per year and the lower rate of 4.1 billion barrels a year that would result if imports are allowed to continue at 35% of the 1973 total. . . .

Three major points are highlighted by these figures. First, the range between the most pessimistic and most optimistic estimates is considerable—33 years to exhaustion compared to 113 years, if there is no growth in consumption and imports continue to supply 35% of domestic needs. The energy policies implied by these two extremes differ enormously in terms of the need for immediate remedial actions. Second, even a relatively low 2.5% annual growth rate of consumption will substantially reduce the time to exhaustion; for example, the time implied by the NAS estimate if imports continue at present levels would be reduced from 43 years with no growth to 29 years at the 2.5% growth rate. Third, the goal of the elimination of dependence upon imported oil may be quite costly if the lower estimates are correct, since its attainment could reduce by a decade or more the already limited time available to develop acceptable ways of producing alternatives such as shale oil and coal synthetics. . . .

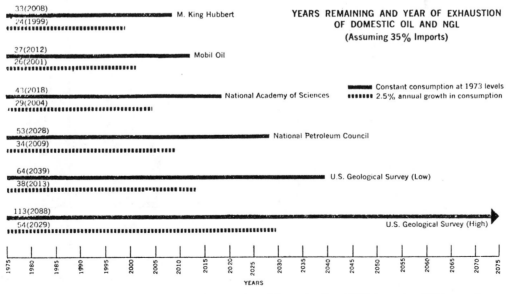

YEARS REMAINING AND YEAR OF EXHAUSTION
OF DOMESTIC OIL AND NGL
(Assuming 35% Imports)

33(2008)
24(1999)   M. King Hubbert

37(2012)
26(2001)   Mobil Oil

43(2018)
29(2004)   National Academy of Sciences

▬▬ Constant consumption at 1973 levels
▪▪▪▪▪ 2.5% annual growth in consumption

53(2028)
34(2009)   National Petroleum Council

64(2039)
38(2013)   U.S. Geological Survey (Low)

113(2088)
54(2029)   U.S. Geological Survey (High)

YEARS

Years remaining and year of exhaustion of domestic oil and NGl (assuming 35% imports).

NPC—U.S. ENERGY BALANCE

At the request of the U.S. Department of the Interior, the National Petroleum Council initiated, in January 1970, a comprehensive study of the energy outlook in the United States for the remainder of the century. The study projected energy demand, energy resources, and changes in government policy or economic conditions that might improve the national energy posture.

The Initial Appraisal, published in July, 1971, was based on the assumption that government policies and economic conditions would remain unchanged through 1985. Total energy demand was projected to increase at a rate of 4.2% per year from 1970 to 1985 and reach a level of 125 quadrillion Btu per year. Imports of energy supplies from foreign sources were projected to increase from 12 to 30% of the total supply by 1985.

In a report[2] published in December, 1972, the National Petroleum Council considered three growth trends; a low trend at 3.4% per year, an intermediate trend at 4.2% per year (the Initial Appraisal case); and a high trend at 4.4% per year. These three growth trends were further studied in four cases (see 16-6).

[1] *U.S. Energy Outlook: An Initial Appraisal 1971-1985*, National Petroleum Council, July 1971.

[2] *U.S. Energy Outlook*, A Summary Report of the National Petroleum Council, Washington, D.C., December 1972.

The following reports have been published in NPC U.S. Energy Outlook series. Reports are available from: National Petroleum Council, 1625 K Street, N.W., Washington, D.C. 20006.

Quantity	U.S. Energy Outlook—Initial Appraisal (1971)	Unit Price	Quantity	U.S. Energy Outlook—Final Reports (1972)	Unit Price
	U.S. Energy Outlook—Vol. I	$ 4.00		U.S. Energy Outlook—Summary Report	$ 6.50
	U.S. Energy Outlook—Vol. II	10.00		U.S. Energy Outlook—Full Report: casebound	17.50
	Task Group Reports			U.S. Energy Outlook—Full Report: paperback	15.00
	Energy Demand	5.00		Guide to NPC Energy Report†	1.50
	Gas Supply	6.00		**Task Group Reports**	
	New Energy Forms	10.00		Coal Availability	18.00
	Nuclear	3.00		Energy Demand (Includes Oil Demand)	15.00
	Oil Demand	8.00		Fuels for Electricity	6.00
	Oil Shale	8.00		Gas Demand	5.00
	Oil Supply	10.00		Gas Transportation	12.00
				New Energy Forms	15.00
				Nuclear Energy Availability	10.00
				Oil & Gas (O&G Supply; For. O&G Avail.)	25.00
				Oil Shale Availability	8.00
				Water Availability	7.00

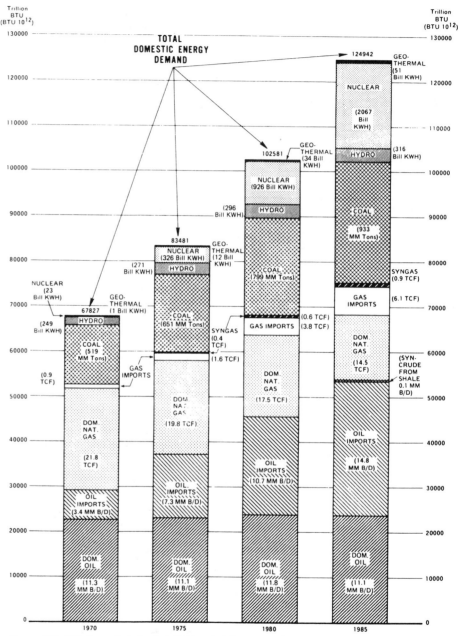

Source: NPC, "U.S. Energy Outlook: An Initial Appraisal 1971-1985," Volume One (July 1971), p. 15.

U.S. energy balance—initial appraisal.

Source: Oil Refinery Capacity, Hearing before the Committee on Interior and Insular Affairs, United States Senate, August 2, 1973, U.S. Government Printing Office, Washington, 1973.

Projections of U.S. Total Energy Demand Under Three Different Sets of Assumptions

	Growth Rate (Average Annual % Gain)			Volume (Quadrillion BTU's)	
Case	1970–1980	1981–1985	1971–1985	1980	1985
High	4.5	4.3	4.4	105.3	130.0
Intermediate (Initial Appraisal)	4.2	4.0	4.2	102.6	124.9
Low	3.5	3.3	3.4	95.7	112.5

Variant Projections of U.S. Energy Demand by Major Consuming Sector

	Demand Volume—Quadrillion BTU's						
	1970 Actual	1980 Low[†]	1980 Intermediate	1980 High[†]	1985 Low[†]	1985 Intermediate	1985 High[†]
Residential/Commercial	15.8	21.1	22.4	23.4	23.9	26.6	28.5
Industrial	20.0	24.7	26.8	27.2	27.1	30.9	31.9
Transportation	16.3	23.0	23.9	24.4	26.7	28.3	29.0
Electricity Conversion	11.6	20.7	22.8	23.5	26.7	30.2	31.4
Non-Energy	4.1	6.2	6.7	6.8	8.1	8.9	9.2
Total	67.8	95.7	102.6	105.3	112.5	124.9	130.0

	Growth Rates—Average Annual Percent Change						
	1960-1970 Historical	1970-1980 Low[†]	1970-1980 Intermediate	1970-1980 High[†]	1980-1985 Low[†]	1980-1985 Intermediate	1980-1985 High[†]
Residential/Commercial	4.0	3.0	3.6	4.0	2.5	3.5	4.0
Industrial	3.4	2.1	2.9	3.1	1.9	2.9	3.2
Transportation	4.2	3.5	3.9	4.1	3.0	3.4	3.5
Electricity Conversion	7.2	5.9	6.9	7.3	5.2	5.8	6.0
Non-Energy	3.4	4.3	5.1	5.3	5.5	5.9	6.2
Total	4.3	3.5	4.2	4.5	3.3	4.0	4.3

Electricity is allocated to each consuming sector and is converted at 3,412 BTU's per KWH and included in the total energy demand for the appropriate sector; the energy used by utilities for generation is shown in the Electricity Conversion category. The following figures show a reconciliation of electricity demands in these sectors with the total Electric Utility energy inputs, for the Intermediate Case only:

Demand Volumes—Quadrillion BTU's	1970	1980	1985
Residential/Commercial	2.8	5.7	7.8
Industrial	2.3	4.4	6.3
Transportation	—	0.1	0.1
Electricity Conversion	11.6	22.8	30.2
Total Utility Inputs	16.7	33.0	44.4

† Based on the variables deemed to be the most significant long-range determinants of energy demand.

Source: *U.S. Energy Outlook*, A Summary Report of the National Petroleum Council, Washington, D.C., December 1972.

Historical and Projected U.S. Energy Demand
(Quad. BTU)

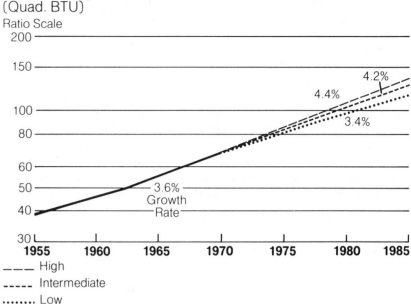

Energy Demand Trends to 2000

	Volume (Quadrillion BTU)		Growth Rate	
	1985	2000	1971-85	1985-2000
High Case	130.0	215	4.4%	3.4%
Intermediate Case	124.9	200	4.2%	3.2%
Low Case	112.5	170	3.4%	2.8%

Source: *Guide to National Petroleum Council Report on United States Energy Outlook*, National Petroleum Council, Washington, D.C., December, 1972.

NPC—U.S. ENERGY SUPPLY AND CONSUMPTION

The results of the Initial Appraisal study of the United States energy outlook by the National Petroleum Council indicated that if a growing dependence on imported energy supplies were to be avoided, significant changes in government policy and economic conditions would need to be made. The ongoing NPC studies review four cases of energy supply based on an intermediate energy growth rate of 4.2% per year resulting in a projected energy demand of 125 quadrillion Btu by 1985. These cases represented different sets of assumptions regarding government policy and economic conditions. Case I assumed an early resolution of environmental issues, a ready availability of government land for resource development, adequate economic incentives, and higher rates of discovery for energy resources. Case IV assumed that development of energy supplies would be constrained by environmental considerations, that government policies would hinder rather than help resource development, and that discovery rates for oil and gas would not improve. Cases II and III represented intermediate assumptions with regard to these issues.

U.S. Energy Supply and Consumption in 1985
(Quad. BTU)

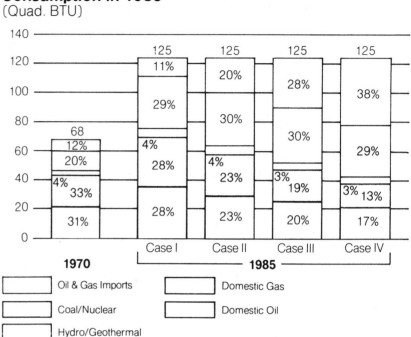

Source: Guide to National Petroleum Council Report on United States Energy Outlook, National Petroleum Council, Washington, D.C., December, 1972.

Potential Domestic Energy Supply Availability
(Data in Conventional Units)

	Units	Initial Appraisal	Case I	Case II	Case III	Case IV
1975						
Oil—Domestic Liquid Production	MMB/D	11.08	10.24	10.19	9.75	9.62
—Shale Syncrude	MMB/D	0	0	0	0	0
—Coal Syncrude	MMB/D	0	0	0	0	0
Subtotal—Oil	MMB/D	11.08	10.24	10.19	9.75	9.62
Gas—Domestic Production	TCF/yr	19.8	23.7	23.6	22.0	21.8
—Nuclear Stimulation	TCF/yr	0	0	0	0	0
—Syngas (Coal)	TCF/yr	0	0	0	0	0
Subtotal—Gas	TCF/yr	19.8	23.7	23.6	22.0	21.8
Hydroelectric	Billion KWH/yr	271	271	271	271	271
Geothermal (Capacity)	MWe	1,500	1,500	1,500	1,500	1,500
Coal	MMT/yr	621	665	621	621	603
Nuclear (Capacity)	MWe	59,000	64,000	64,000	64,000	28,000
Nuclear (U_3O_8)	MT/yr	18.4	19.1	19.1	19.1	11.5
1980						
Oil—Domestic Liquid Production	MMB/D	11.80	13.58	12.94	11.61	8.90
—Shale Syncrude	MMB/D	0	.15	.10	.10	0
—Coal Syncrude	MMB/D	0	.08	0	0	0
Subtotal—Oil	MMB/D	11.80	13.81	13.04	11.71	8.90
Gas—Domestic Production	TCF/yr	17.5	25.9	24.3	20.4	17.3
—Nuclear Stimulation	TCF/yr	0	.2	.1	.1	0
—Syngas (Coal)	TCF/yr	.2	.6	.4	.4	.2
Subtotal—Gas	TCF/yr	17.7*	26.7	24.8	20.9	17.51
Hydroelectric	Billion KWH/yr	296	296	296	296	296
Geothermal (Capacity)	MWe	4,500	10,250	5,250	4,500	2,500
Coal	MMT/yr	734	851	734	734	705
Nuclear (Capacity)	MWe	150,000	188,000	188,000	150,000	107,000
Nuclear (U_3O_8)	MT/yr	34.2	50.9	45.6	36.5	29.1
1985						
Oil—Domestic Liquid Production	MMB/D	11.08	15.46	13.89	11.83	10.38
—Shale Syncrude	MMB/D	.10	.75	.40	.40	.10
—Coal Syncrude	MMB/D	0	.68	.08	.08	0
Subtotal—Oil	MMB/D	11.18	16.89	14.37	12.31	10.48
Gas—Domestic Production	TCF/yr	14.5	30.6	26.5	20.4	15.0
—Nuclear Stimulation	TCF/yr	0	1.3	.8	.8	0
—Syngas (Coal)	TCF/yr	.5	2.5	1.3	1.3	.5
Subtotal—Gas	TCF/yr	15.0*	34.4	28.6	22.5	15.5
Hydroelectric	Billion KWH/yr	316	316	316	316	316
Geothermal (Capacity)	MWe	7,000	19,000	9,000	7,000	3,500
Coal	MMT/yr	863	1,093	863	863	819
Nuclear (Capacity)	MWe	300,000	450,000	375,000	300,000	240,000
Nuclear (U_3O_8)	MT/yr	59.3	108.5	89.2	70.7	60.4

* Does not include 0.4 TCF SNG from naphtha reported in Initial Appraisal as domestic supply.

Source: U.S. Energy Outlook, A Summary Report of the National Petroleum Council, Washington, D.C., December 1972.

Potential Domestic Energy Supply Availability
(All Data X 10^{12} BTU's/Year)

	Initial Appraisal	Case I	Case II	Case III	Case IV
1975					
Oil—Domestic Liquid Production	22,789	20,735	20,630	19,754	19,502
—Shale Syncrude	0	0	0	0	0
—Coal Syncrude	0	0	0	0	0
Subtotal—Oil	22,789	20,735	20,630	19,754	19,502
Gas—Domestic Production	20,430	24,513	24,300	22,766	22,421
—Nuclear Stimulation	0	0	0	0	0
—Syngas (Coal)	0	0	0	0	0
Subtotal—Gas	20,430*	24,513	24,300	22,766	22,421
Hydroelectric	2,840	2,990	2,990	2,990	2,990
Geothermal	120	120	120	120	120
Coal	16,310	16,650	15,554	15,554	15,100
Nuclear	3,340	4,000	4,000	4,000	1,661
Total Potential Supplies	65,829	69,008	67,594	65,184	61,794
1980					
Oil—Domestic Liquid Production	24,323	27,758	26,456	23,789	18,112
—Shale Syncrude	0	296	197	197	0
—Coal Syncrude	0	175	0	0	0
Subtotal—Oil	24,323	28,229	26,653	23,986	18,112
Gas—Domestic Production	18,030	26,746	25,043	21,041	17,906
—Nuclear Stimulation	0	206	103	103	0
—Syngas (Coal)	190	512	329	329	165
Subtotal—Gas	18,220*	27,464	25,475	21,473	18,071
Hydroelectric	3,033	3,240	3,240	3,240	3,240
Geothermal	343	782	401	343	191
Coal	19,928	21,200	18,284	18,284	17,550
Nuclear	9,490	11,349	11,349	9,787	6,788
Total Potential Supplies	75,337	92,264	85,402	77,113	63,952
1985					
Oil—Domestic Liquid Production	23,405	31,689	28,477	24,346	21,426
—Shale Syncrude	197	1,478	788	788	197
—Coal Syncrude	0	1,489	175	175	0
Subtotal—Oil	23,602	34,656	29,440	25,309	21,623
Gas—Domestic Production	14,960	31,604	27,324	21,049	15,474
—Nuclear Stimulation	0	1,341	825	825	0
—Syngas (Coal)	560	2,269	1,208	1,208	494
Subtotal—Gas	15,520*	35,214	29,357	23,082	15,968
Hydroelectric	3,118	3,320	3,320	3,320	3,320
Geothermal	514	1,395	661	514	257
Coal	23,150	27,100	21,388	21,388	20,300
Nuclear	21,500	29,810	25,249	20,220	16,126
Total Potential Supplies	87,404	131,495	109,415	93,833	77,594

*Does not include 380 trillion BTU's SNG from naphtha reported in Initial Appraisal as domestic supply.

Source: U.S. Energy Outlook, A Summary Report of the National Petroleum Council, Washington, D.C., December 1972.

ENERGY POLICY PROJECT—ENERGY CHOICES

In a study[1] of the energy choices facing the nation, the Energy Policy Project of the Ford Foundation reviewed three alternate futures, or scenarios, based on different assumptions concerning energy growth patterns and the policies and consequences involved in those scenarios.

The historical growth scenario assumes that growth in energy demand will continue in accord with past trends, that policies would not be adopted to change habits of energy use, and that there would be a vigorous effort to develop energy supplies to match rising demand for energy.

The Technical Fix scenario assumes a similar level and mix of goods and services as the Historical Growth scenario but includes a determined national effort to reduce energy demand through the use of energy-saving technologies. The study concludes that the Technical Fix Scenario would provide a quality of life and economic growth that differs little from the historical growth case. The study also concludes that the energy growth rates could be met if only one of the available domestic resources—Western coal or shale, or nuclear power, or oil and gas—was aggressively developed.

The zero energy growth scenario assumes that a further reduction in energy demand will result from a concern of society over the social and environmental costs of energy growth, the dehumanizing aspects of big centralized institutions and the adoption of the idea that "enough is best" in place of "more is better."

[1]*Exploring Energy Choices, A Preliminary Report*, Energy Policy Project of the Ford Foundation, Washington, D. C., 1974, and *A Time to Choose, America's Energy Future*, Final Report of the Energy Policy Project of the Ford Foundation, Ballinger Publishing Co., Cambridge, Massachusetts, 1974.

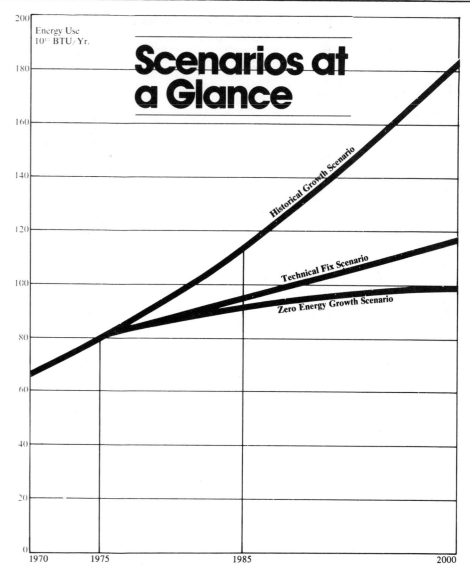

Reprinted from *Exploring Energy Choices* with permission of the Energy Policy Project of The Ford Foundation.

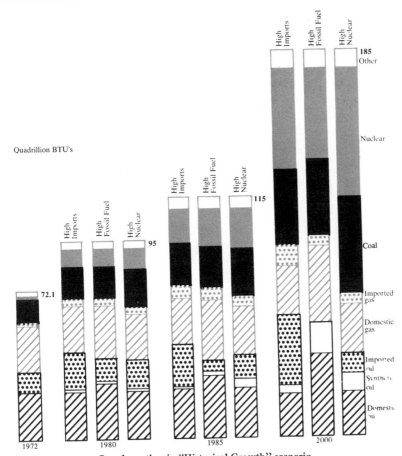

Supply option in "Historical Growth" scenario.

Reprinted from *Exploring Energy Choices* with permission of the Energy Policy Project of The Ford Foundation.

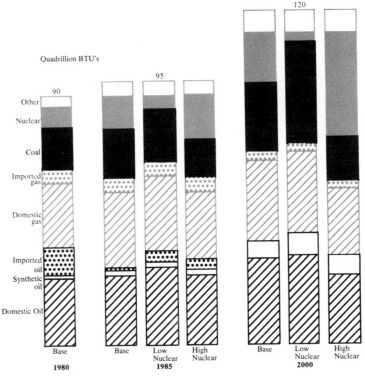

Energy supply for the technical fix scenario.

Reprinted from *Exploring Energy Choices* with permission of the Energy Policy Project of The Ford Foundation.

REFERENCES

Publications of the Energy Policy Project of the Ford Foundation, Ballinger Publishing Company, 17 Dunster Street, Harvard Square, Cambridge, Massachusetts 02138.

Energy Policy Project Staff, *A Time to Choose: America's Energy Future*, 528 pp.

Edward Berlin, Charles J. Cicchetti, and William J. Gillen, *Perspective on Power, A Study of the Regulation and Pricing of Electric Power*, 204 pp.

Donald F. Boesch, Carl H. Hershner, and Jerome H. Milgram, *Oil Spills and the Marine Environment*, 144 pp.

Gerard M. Brannon, *Energy Taxes and Subsidies*, 204 pp.

Gerard M. Brannon, *Studies in Energy Tax Policy*, 400 pp.

Charles J. Cicchetti and John L. Jurewitz, *Studies in Electric Utility Regulation*, 224 pp.

Thomas Duchesneau, *Competition in the U.S. Energy Industry*, 448 pp.

Foster Associates, *Energy Prices 1960–1973*, 292 pp.

John E. Gray, *Energy Policy—Industry Perspectives*, 144 pp.

Elias P. Gyftopoulos, Lazaros J. Lazaridis, and Thomas Widmer, *Potential Fuel Effectiveness in Industry*, 112 pp.

Jerome E. Hass, Ward J. Mitchell, and Bernell K. Stone, *Financing the Energy Industry*, 160 pp.

J. Herbert Hollomon and Michel Grenon, *U.S. Energy Research and Development Policy*, 192 pp.

Arjun Makhijani, *Energy and Agriculture in the Third World*, 168 pp.

John G. Myers et al., *Energy Consumption in Manufacturing*, 656 pp.

National Academy of Sciences and National Academy of Engineering, *Rehabilitation Potential of Western Coal Lands*, 228 pp.

Dorothy K. Newman and Dawn Day Wachtel, *The American Energy Consumer*, 384 pp.

Richard Schoen, Alan S. Hirshberg, and Jerome M. Weingart, *New Energy Technologies for Buildings*, 224 pp.

Robert Williams, Ed., *The Energy Conservation Papers*, 416 pp.

Mason Willrich, and Theodore B. Taylor, *Nuclear Theft: Risks and Safeguards*, 272 pp.

Joseph A. Yager and Eleanor B. Steinberg, *Energy and U.S. Foreign Policy*, 515 pp.

ENERGY POLICY PROJECT—CONSERVATION POTENTIAL

The focal point of the policy for the Technical Fix Scenario of the Energy Policy Project is the reduction of energy demand substantially below that of historical trends by emphasizing consumption efficiency. This reduction in demand through consumption efficiency is assumed, of course, for the Zero Growth Scenario as well but this latter scenario also assumes a further reduction in demand through changes in patterns of living.

Reductions in the use of energy in the residential-commercial sector are projected through better building design and construction, more efficient heating and cooling equipment, the use of heat pumps, the use of solar heating and cooling, and the adoption of total energy systems. Improvements in transportation efficiency are expected from the use of more efficient cars, an increase in the efficiency of aircraft utilization, and a shift in transport from less efficient to more efficient modes, i.e., a shift from air and truck to rail transport. Reductions in industrial energy use are expected from improvements in process efficiency, the use of recycled metals, and the use of combined steam-electric generation systems.

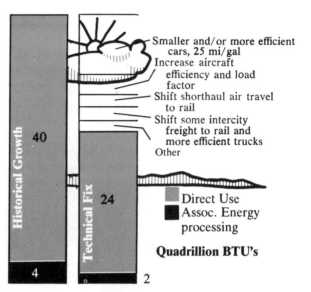

Total Transportation Energy Use in 2000

Savings in transportation.

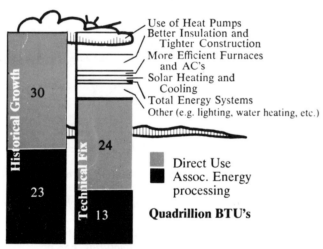

Total Residential and Commercial Energy Use in 2000

Residential and commercial savings.

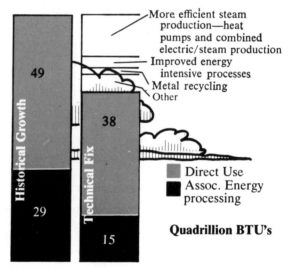

Industrial Energy Use in 2000

Industrial savings.

Reprinted from *Exploring Energy Choices* with permission of the Energy Policy Project of The Ford Foundation.

NAE ENERGY PROSPECT—SUPPLY AND DEMAND

In a study[1] prepared in 1974, the Task Force on Energy of the National Academy of Engineering, assessed the United States energy outlook for the period until 1985. The Task Force projected an unrestrained demand in 1985 based on historical growth patterns of 58 million barrels per day of oil equivalent[2] (115.5 quadrillion Btu) and 51 million

[1] *U.S. Energy Prospects, An Engineering Viewpoint*, National Academy of Engineering, Washington, D.C., 1974.

[2] Total United States energy demand is expressed in terms of barrels of oil by conversion of the energy content of other fuel forms into barrels of oil at 5,800,000 Btu per barrel.

barrels per day (101 quadrillion Btu) as the demand if reduced by conservation. They projected an increase in domestic energy supply from 37.2 million barrels per day in 1973 to 40 million barrels per day in 1985—leaving a domestic shortfall in supply of 11–18 million barrels per day, which would need to be met by imports (imports in 1973 were 6.6 million barrels per day).

The Task Force outlined a series of actions necessary to increase domestic energy supplies to match the demand in 1985 assuming conservation measures were also undertaken. These actions are outlined on pages 16-10 and 16-11.

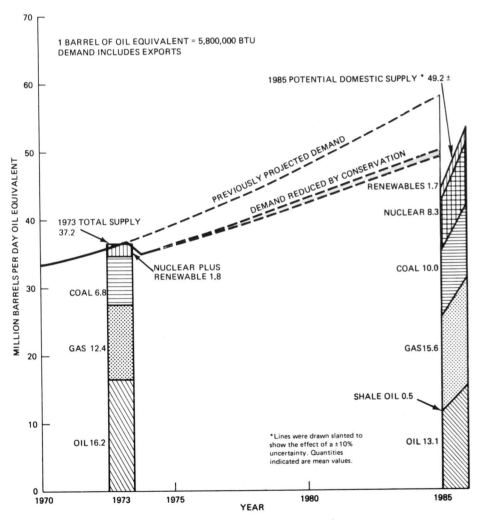

Possible U.S. energy supply and demand.

Reproduced with permission of the National Academy of Sciences. A Report of the National Academy of Engineering. Published by the National Academy of Sciences.

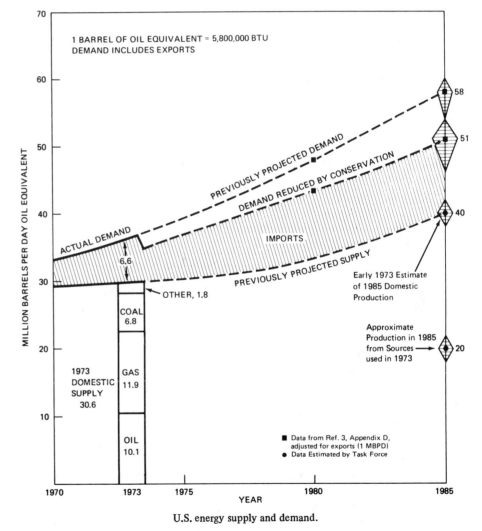

U.S. energy supply and demand.

Reproduced with permission of the National Academy of Sciences. A Report by the National Academy of Engineering. Published by the National Academy of Sciences.

NAE ENERGY PROSPECTS—RESOURCE DEVELOPMENT

The Task Force on Energy of the National Academy of Engineering has estimated that the 1985 unrestrained demand of 58 million barrels per day of petroleum could be reduced by some 8 to 9 million barrels per day by conservation measures, leaving 49 to 50 million barrels a day to be met from domestic energy supplies. To meet this reduced 1985 demand from domestic supplies would require the following actions:

Increase coal production by a factor of 2.
Increase nuclear power capacity to one-third of total electrical capacity.
Duplicate present domestic oil and gas capacity to get a net 25% increase in output.
Use new technologies to produce:

1.1 MBPD of synthetic gases from coal,
0.6 MBPD of synthetic liquids from coal,
0.5 MBPD of oil from shale.

Modest increases in hydroelectric and geothermal power.

Coal Production.

To double coal production we must, typically:

develop 140 new 2 MTPY eastern underground mines;
develop 30 new 2 MTPY eastern surface mines;
develop 100 new 5 MTPY western surface mines;
recruit and train 80,000 new eastern coal miners;
recruit and train 45,000 new western coal miners;
manufacture 140 new 100-cubic-yard shovels and draglines;
manufacture 2400 continuous mining machines.

To transport the increased production we must, typically:

construct 60 new 2 MTPY eastern rail-barge systems of 100 to 500 miles each;
construct 70 new 3 MTPY western rail-barge systems of 1000 to 1200 miles each;
construct 4 new 25 MTPY slurry pipelines of 1000 miles each;
construct 2 new 2.5 BCFD gas pipelines of 1000 miles each;
manufacture 8000 railroad locomotive units;
manufacture 150,000 gondola and hopper cars of 100-ton capacity each.

Nuclear Power.

To increase nuclear power to one-third of total electrical capacity, we must:

design and construct an average of two to three new nuclear units per month for the next 10 years;
fabricate and deliver an average of 30 reactor pressure shells, and 30 turbine-generator sets each year for 10 years;
acquire 10 to 12 approved nuclear plant sites per year for 10 years;
recruit and train 30,000 to 40,000 plant operators and maintenance personnel in the next 10 years;
increase uranium production by a factor of 6;

increase uranium enrichment capacity by 50% over existing and presently planned AEC capacity.

Oil and Gas.

To increase oil and gas production by 25% by 1985, it will be necessary to:

find and open new fields—most of these will be offshore or in Alaska;
increase the production rate from existing fields through new outpost wells, workover of existing wells, extended life of stripper wells, and increased secondary recovery projects;
accelerate tertiary recovery methods;
open up known reserve fields for production;
develop recovery methods for low-permeability sands;
drill an average of 58,000 wells each year until 1985, twice the 1973 rate.

1985 Estimated Possible Oil/Gas Production

Source	Gas BCFD	Gas MBPD Equiv.	Oil (MBPD)	Total (MBPD) Equiv.
1973 Production	67.0	11.9	10.1	22.0
Declines in 1973 fields	(39.4)	(7.0)	(6.0)	(13.0)
Production from new fields	36.6	6.5	5.5	12.0
Accelerated production and leasing	13.8	2.5	1.9	4.4
Additional known fields	0.1	--	0.2	0.2
Accelerated tertiary oil recovery methods	0.4	0.1	0.8	0.9
Gas production from low-permeability sands	3.0	0.5	--	0.5
Subtotal: Incremental production by 1985	14.5	2.6	2.4	5.0
Grand Total	81.5	14.5	12.5	27.0

Estimated 1985 U.S. Energy Demand Reduction

Category	MBPD
By Conservation	
Industrial conservation measures	1.5
Transportation	
Lower speeds, car pooling	1.0
Airplane load factors	0.3
Space heating efficiency	1.0
By Use of Energy-saving Equipment	
Smaller, more efficient cars	2.0
Other transportation savings	1.1
Better building insulation standards	1.1
Residential and commercial equipment	0.4
Industrial process efficiency	1.0
Total conservation potentials	9.4
Less 15 percent for partial overlap	8.0

Coal-Based Synthetic Fuels.

To produce 1.7 MBPD of synthetic fuels from coal by 1985, it will be necessary to:

construct 20 new 250-million CFD methane-from-coal plants;

construct 8 new 40,000 BPD methanol-from-coal plants;

construct 10 new medium Btu gas-from-coal plants;

conduct an aggressive program to develop economical liquefaction processes using hydrogenation;

construct 10 new 30,000 BPD coal liquefaction plants.

Shale Oil.

To produce 0.5 MBPD of oil from shale by 1985, it will be necessary to:

bring into production 50 new 5 MTPY shale mines and retorting plants;

lay, stabilize, and restore 5 square miles of tailings that are 40 feet deep each year;

construct 10 new 50,000 BPD shale oil extraction and upgrading plants, including pipelines to markets;

develop and convey 80,000 acre-feet per year of new water supplies in water-scarce areas.

Expansion of Coal-Based Power Plants.

To meet electrical needs with coal, we must:

reconvert 5 oil conversion units per month over 4 years;

design, construct, and bring on-line 2 new 700 Mwe coal-fired units per month for 10 years;

design and manufacture 24 new 700 Mwe turbine generators every year for 10 years;

design and erect 180 new 2.5 MTPY coal-fired furnaces;

purchase and develop 50 new power plant sites;

design, construct, and start up 10 new medium Btu gas conversion plants to supply 50 existing gas-fired units of 400 Mwe each;

provide 260,000 acre-feet of water per year.

Overall Generating Capacity (Gwe)

	1973	1985
Nuclear	21[b]	325[a]
Coal (solids)	143[b]	330
Oil/gas	175[b]	130
Gas turbine & internal combustion	38	50
Medium-Btu syngas	--	20
Hydro	54	75
Geothermal	Nil	7
Pumped storage	8	45
Total	439	982

[a] At end of year; average for year, 250 Gwe.

[b] Estimated split. Total fossil-fired steam was 318 Gwe.[20]

NAE ENERGY PROSPECTS—DEVELOPMENT CONSTRAINTS

A Perspective On Capital Requirements. The $500 to $600 billion for in-place facilities will require about another $100 billion for infrastructure support. $700 billion would average $60 billion per year until 1985. For comparison:

> current investments in the energy industry are about $30 billion per year;
>
> 1970 industrial plant and equipment investment for all purposes was about $100 billion.

Other Important Considerations. If these programs are to be carried out by 1985 a number of important problems will have to be solved.

> A balance will have to be reached between environmental and energy objectives.
>> Sulfur dioxide;
>> land reclamation;
>> offshore oil and gas operations;
>> oil refineries;
>> nuclear power plants.
> An extensive manpower training program will be needed.
>> 30,000 additional engineers;
>> 190,000 additional skilled construction workers;
>> 240,000 additional operators.
> Water must be made available.

"Guesstimated" U.S. Production Facility Capital Requirements (billion $)
(1974–1985 Incremental Production)

Oil and gas	160–200
Uranium	11–14
Coal	
Solid	18–24
Synthetics	16–22
Shale oil	3–5
Power supply	
Nuclear generation	90–110
Fossil + renewable generation	60–70
Transmission + distribution	135–165
Total (Rounded)	490–610

Reprinted with permission of the National Academy of Sciences. A Report of the National Academy of Engineering. Published by the National Academy of Sciences.

Typical 1973 Overall Project Times
(From go-ahead to production)

Type of Facility	Years
Coal-fired power plant	5–8
Surface coal mine	2–4
Underground coal mine	3–5
Uranium exploration and mine	7–10
Nuclear power plant	9–10
Hydroelectric dam	5–8
Produce oil and gas from new fields	3–10
Produce oil and gas from old fields	1–3

Reprinted with permission of the National Academy of Sciences. A Report of the National Academy of Engineering. Published by the National Academy of Sciences.

The Role of the Government. If the domestic energy supplies are to be developed to their maximum levels to minimize dependence on foreign sources, the government must take the leadership to:

> develop a long-term, comprehensive energy policy;
> reduce administrative constraints;
> reduce the risks to developers of new technologies;
> support selected high-risk, high-payoff developments;
> compile and disseminate energy data and information;
> establish energy conservation standards;
> foster an economic environment that will attract investment capital into the energy industry.

The Energy Challenge. If this nation chooses to minimize fuel imports as soon as possible:

> These are the *only* programs that will enable it to be done, and even so, imports must rise in the interim.
> There must be a cooperative effort between the government and the energy industry.
>> There is not time to create new institutions.
>> The government must establish a well-defined national energy policy and conduct licensing and regulatory actions so that large sums of private capital are invested and the public interest is protected.
>> The energy industry has the managerial and technical skills to do the job.

—Decisions must be reached *now*.

CEQ HALF AND HALF PLAN—ENERGY PROJECTIONS

In a plan[1] outlined in March 1974, Russell W. Peterson, Chairman of the Council on Environmental Quality, called for a serious long-term national program to conserve energy and meet the needs of a growing economy.

Basic Considerations.

1. The production and consumption of energy is the single major source of environmental degradation and energy development must be undertaken with full regard and protection of the environment.

2. The era of energy growth through exploitation of domestic supplies of oil and gas is over. Domestic production of oil and natural gas within the 48 states, which provide over 75% of the total United States energy, will shortly begin a sustained decline.

3. The United States must achieve a capacity for energy self-sufficiency consistent with national security. The United States should plan so that energy supplies outside the United States can be devoted to increasing the quality of life for the other 94% of the people of the globe.

4. Over the past 25 years, growth in net energy consumed per capita has averaged 1.4% per year and since the mid-1960s the per capita growth rate has exceeded 3% per year, a level which cannot be sustained. The United States must return to its historic growth rate of 1.4%.

Half and Half Plan Elements.

1. The target for gross energy consumption in the year 2000 should be 121 quadrillion Btus per year, an increase of 49 quadrillion Btus over the 1972 consumption, and representing an annual growth rate of 1.8%.

2. This target is based on a growth in net per capita energy consumption of 0.7% per year and a conservation effort to save energy at a rate of 0.7% per year through improved efficiency and elimination of waste. This program—half growth and half conservation—would provide an effective increase in usable energy of 1.4% per year.

Energy Demand Implications.

1. Per capita net energy consumption in the year 2000 would be 25% above present levels in the residential and commercial sector, 35% above present levels in the industrial sector, and 10% above present per capita levels in the transportation sector.

2. In terms of effective energy, per capita consumption would be supplemented by savings in energy through conservation—more efficient appliances, better insulation, energy-conscious building design, recycling of materials, more efficient industrial process design, smaller and more efficient cars, greater use of mass transit, and more rational land use.

[1] Russell W. Peterson, *A National Energy Conservation Program: The Half and Half Plan*, Council on Environmental Quality, March 1974.

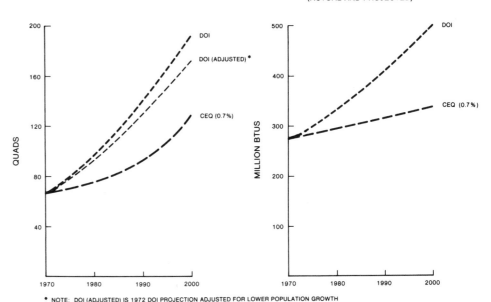

GROSS ENERGY CONSUMPTION
(ACTUAL AND PROJECTED)

NET ENERGY CONSUMPTION PER CAPITA
(ACTUAL AND PROJECTED)

* NOTE: DOI (ADJUSTED) IS 1972 DOI PROJECTION ADJUSTED FOR LOWER POPULATION GROWTH

Source: Russell W. Peterson, *A National Energy Conservation Program: The Half and Half Plan*, Council on Environmental Quality, March 1974.

Energy Supply Implications.

1. Major reliance must be placed on coal and nuclear power with coal inputs increasing from 12.6 to 33.4 quadrillion Btus between 1971 and 2000, and nuclear power from 0.4 to 35 quadrillion Btus.

2. Over 42% of the total energy inputs will be used to produce electricity.

3. Limited petroleum resources must increasingly be reserved for transportation uses.

4. Major research and development should be carried out on new energy resources such as nuclear fusion, solar and geothermal energy but, even with a major effort, these sources cannot be expected to provide more than 3% of the total needs by the year 2000.

The projections of the CEQ for gross energy consumption and net energy consumption per capita, differ from those of the Department of the Interior (DOI)[2] in several respects: the CEQ assumes a population of 250 million in the year 2000 while the DOI projects a population of 279 million; the CEQ projects an annual growth rate in gross energy consumption of 1.8% while the DOI growth rate is 3.5%; and the CEQ projects an annual growth rate in net energy consumption per capita of 0.7% in contrast to the DOI projection of 2.5%.

[2] Walter G. Dupree, Jr., and James A. West, *United States Energy Through the Year 2000*, U.S. Department of the Interior, December 1972.

CEQ HALF AND HALF PLAN—ENERGY SOURCES

In its half and half plan, the Council on Environmental Quality (CEQ) compares its projections for energy sources with those of the Department of the Interior (DOI) as follows:

Coal and Oil Shale. The CEQ and DOI projections for the consumption of coal and oil shale in the year 2000 are identical at 33.4 quadrillion Btus (1971 consumption was 12.6 quadrillion Btu). The CEQ projects that 10.0 quadrillion Btu of this energy will be used to produce electricity and that 13.4 quadrillion Btu will be used to manufacture synthetic oil and gas from coal and oil shale.

Natural Gas. The CEQ projection for the consumption of natural gas in the year 2000 is 20.0 quadrillion Btu, substantially less than the 34.0 quadrillion Btu projected by the DOI (1971 consumption was 22.7 quadrillion Btu). Of the 20.0 quadrillion Btu consumption of natural gas projected by the CEQ, 3.0 quadrillion Btu would be imported.

Petroleum. The CEQ projection for the consumption of petroleum in the year 2000 is 25.4 quadrillion Btu, about one-third of the 69.4 quadrillion Btu projected by the DOI. The CEQ projects petroleum imports in the year 2000 equivalent to 10.0 quadrillion Btu (the DOI projection included 50.2 quadrillion Btu from imports and oil shale).

Electricity. The CEQ projects energy inputs into the production of electricity of 51.2 quadrillion Btu in contrast to the DOI projection of 80.4 quadrillion Btu. The CEQ projects that natural gas and petroleum will be phased out as sources of energy for the production of electricity by the mid-1980s. In the year 2000, energy sources for electrical generation will be: nuclear power, 35.0 quadrillion Btu; coal, 10.0 quadrillion Btu; geothermal, 2 quadrillion Btu; and hydro, 4.2 quadrillion Btu.

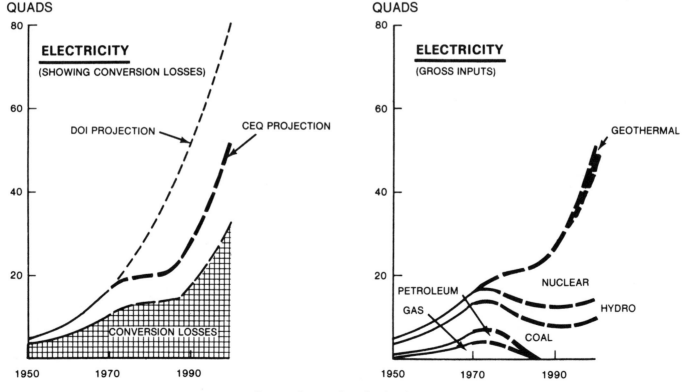

Sources of energy (actual and projected).

Source: Russell W. Peterson, *A National Energy Conservation Program: The Half and Half Plan*, Council on Environmental Quality, March 1974.

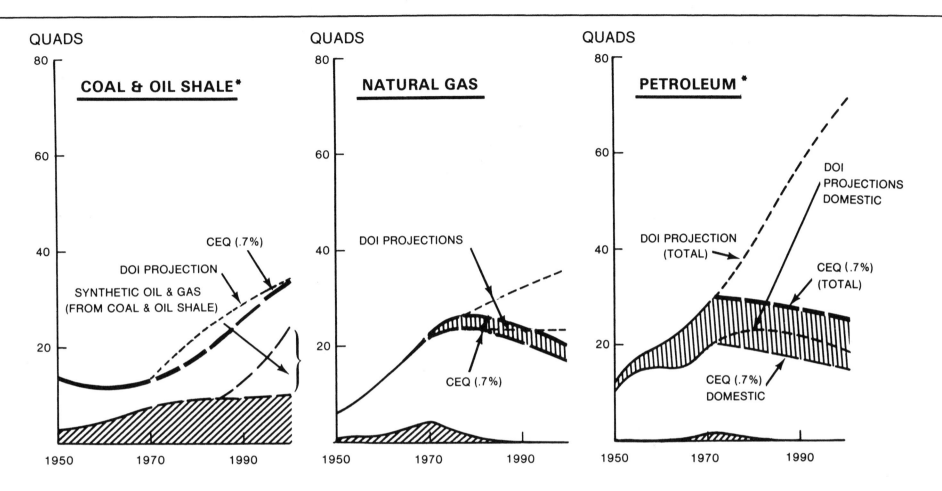

QUADS

COAL & OIL SHALE*

CEQ (.7%)

DOI PROJECTION

SYNTHETIC OIL & GAS
(FROM COAL & OIL SHALE)

1950   1970   1990

QUADS

NATURAL GAS

DOI PROJECTIONS

CEQ (.7%)

1950   1970   1990

QUADS

PETROLEUM*

DOI
PROJECTIONS
DOMESTIC

DOI PROJECTION
(TOTAL)

CEQ (.7%)
(TOTAL)

CEQ (.7%)
DOMESTIC

1950   1970   1990

* NOTE : AN UNSPECIFED AMOUNT OF OIL SHALE IS INCLUDED IN THE DOI PETROLEUM PROJECTIONS BUT IN THE CEQ COAL PROJECTIONS.

IMPORTED

USED IN PRODUCING ELECTRICITY

CEQ 0.7% PROJECTIONS

DOI PROJECTIONS (1972)

Sources of energy (actual and projected).

Source: Russell W. Peterson, *A National Energy Conservation Program: The Half and Half Plan,* Council on Environmental Quality, March 1974.

CEQ HALF AND HALF PLAN—SECTOR CONSUMPTION

The projections of the Council on Environmental Quality for energy consumption in the residential and commercial, industrial, and transportation sectors are given in the accompanying charts and tables. The assumptions upon which these projections are based are given in pages 16-12 and 16-13.

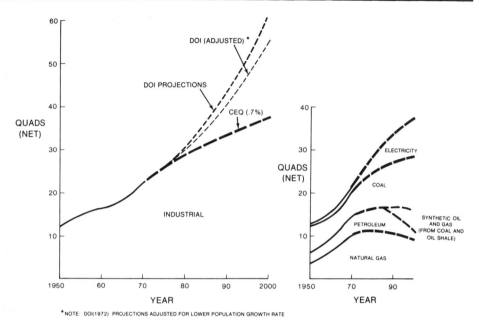

Industrial sector (actual and projected).

Source: Russell W. Peterson, *A National Energy Conservation Program: The Half and Half Plan*, Council on Environmental Quality, March 1974.

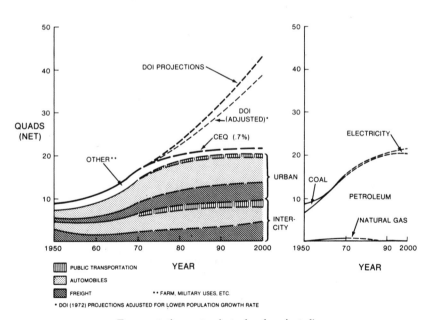

Transportation sector (actual and projected).

Source: Russell W. Peterson, *A National Energy Conservation Program: The Half and Half Plan*, Council on Environmental Quality, March 1974.

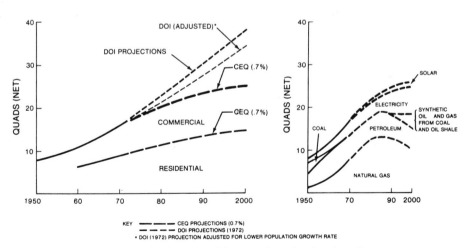

Residential and commercial sectors (actual and projected).

Source: Russell W. Peterson, *A National Energy Conservation Program: The Half and Half Plan*, Council on Environmental Quality, March 1974.

Energy Source and Consuming Sector-1971 (Actual)
(Quadrillion Btus)

	Total	(Electricity)	Residential	Commercial	Industrial	Transportation
Nuclear	0.4	(0.4)	(0.1)	(0.1)	(0.2)	—
Hydro	2.8	(2.8)	(0.9)	(0.7)	(1.2)	—
Coal						
direct	4.9	—	—	0.4	4.5	—
electricity	7.7	(7.7)	(2.5)	(1.9)	(3.3)	—
Solar	—		—	—	—	—
Geothermal	—		—	—	—	—
Natural gas						
direct[a]	18.6	—	4.2	3.2	10.4	0.8
electricity	4.1	(4.1)	(1.4)	(1.0)	(1.7)	
Petroleum[b]						
direct	28.0		3.5	2.6	5.7	16.2
electricity	2.5	(2.5)	(0.8)	(0.6)	(1.1)	—
Total gross inputs	69.0	(17.5)	13.4	10.5	28.1	17.0
Conversion losses	11.9		3.9	2.9	5.1	—
Net consumption	57.1		9.5	7.6	23.0	17.0

[a]Imports were 0.9 quads.
[b]Imports were 8.3 quads.

The Half and Half Plan
Energy Source and Consuming Sector-2000 (Projected)
(Quadrillion Btus)

	Total	(Electricity)	Residential	Commercial	Industrial	Transportation
Nuclear	35.0	(35.0)	(4.0)	(6.0)	(22.5)	(2.5)
Hydro	4.2	(4.2)	(2.0)	(1.0)	(1.2)	—
Coal						
direct	10.0	—	—	—	10.0	—
electricity	10.0	(10.0)	(3.0)	(3.0)	(4.0)	—
synthetic[a]	13.4	—	3.0	3.0	7.4	—
Solar	1.0	—	0.5	0.5	—	—
Geothermal	2.0	(2.0)	—	—	(2.0)	—
Natural Gas[b]	20.0	—	7.0	4.0	9.0	—
Petroleum[c]	25.4	—	2.0	0.4	2.4	20.6
Total gross inputs	121.0	(51.2)	21.5	17.9	58.5	23.1
Conversion losses	36.1		6.6	7.2	20.8	1.5
Net consumption	84.9		14.9	10.7	37.7	21.6

[a]Including oil from shale.
[b]Imports represent 3.0 quads.
[c]Imports represent 10.0 quads.

PROJECT INDEPENDENCE—BACKGROUND AND OIL ASSESSMENT

In November 1974, the Federal Energy Administration published their analysis of the nation's energy problem in a series of reports under the general title of *Project Independence*.[1] The information on this and following pages is taken from the Executive Summary of that study.

Introduction

—The report is an evaluation of the Nation's energy problem.
 o It assesses the "base case" situation through 1985, if current policies prevail.
 o It evaluates the impacts and implications of a wide range of major energy policy alternatives.
—The FEA study is not a "blueprint" for reaching zero imports by 1980, nor does it make specific policy recommendations.
—Rather than evaluate hundreds of alternative actions, the study contrasts the broad strategic options available to the U.S.
 o Increasing domestic supply,
 o Conserving and managing energy demand,
 o Establishing standby emergency programs.
—The strategies are evaluated in terms of their impact on:
 o Development of alternative energy sources,
 o Vulnerability to import disruptions,
 o Economic growth, inflation, and unemployment.
 o Environmental effects,
 o Regional and social impacts.
—The strategies are only illustrative and, in reality, a national energy policy will probably contain elements from each.
—The study provides the analytical and factual basis for focusing debate on the difficult choices and tradeoffs, and selecting a national energy policy.

Background

—Domestic energy demand has been growing at 4–5 percent per year.
—The U.S. was self-sufficient in energy through about 1950, but our situation has deteriorated rapidly since then.
 o Coal production is still at 1940's levels,
 o Crude oil production has been declining since 1970,
 o Natural gas consumption has been exceeding new discoveries since 1968.
—Our dependence on foreign oil has grown to 35 percent of domestic petroleum consumption in 1973 (see Figure 1).
—The world oil market is dominated by several Middle East countries.
 o They have 60 percent of world reserves (see Figure 2).
 o They produce 70 percent of world oil exports (see Figure 3).

[1]*Project Independence*, Project Independence Report, Federal Energy Administration, Washington, D. C., November 1974.

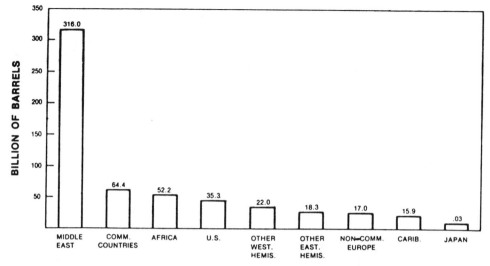

1973 crude petroleum reserves for major producing areas.

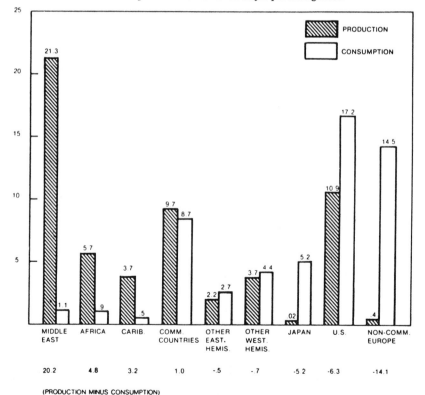

(PRODUCTION MINUS CONSUMPTION)

1973 crude petroleum production and petroleum product consumption for major producing and consuming areas [MB/D].

Source: Project Independence Executive Summary, Federal Energy Administration, November 1974.

—The 1973 embargo demonstrated our domestic vulnerability to insecure imports.
  ○ The embargo affected 14 percent of U.S. petroleum consumption.
  ○ Its economic impact was a $10–20 billion drop in GNP.
  ○ 500,000 additional people were unemployed at its peak.

World Oil Assessment

—The world oil price will largely determine U.S. energy prices and, in turn, affect both United States supply possibilities and rate of energy growth.

—World oil prices are highly uncertain and could decline to about $7 per barrel (FOB U.S.) and might fall somewhat lower.
  ∞ World supply/demand can be brought into balance at $7, but would require significant OPEC production cutbacks from the expected doubling of their capacity by 1985.
  ○ OPEC has already cut back production 10 percent in four months to eliminate the estimated 2–3 million barrel per day (MMBD) world surplus.

—Major OPEC cutbacks would be required to sustain $11 world oil prices.
  ○ In the short term, prices can be supported by moderate production cutbacks.
  ○ Much of the expected increase over 1973 OPEC production levels must be foregone by 1985 to support $11 prices.
  ○ Decisions by major oil exporters will be more political than economic because greater revenues are not needed by the key suppliers to support their economic growth.

—Foreign sources of oil have a significant probability of being insecure in the 1974–1985 time frame.

—The resolution of pressing international financial, economic, and political problems will ultimately determine world oil prices and security of supply.
  ○ The study contrasts differences in the United States' situation based on a $7 and on an $11 world oil price.
  ○ The study also estimates potential levels of world oil disruptions and their impact on the U.S.

PROJECT INDEPENDENCE—BASE CASE

Domestic Energy Through 1985: The Base Case

—If major policy initiatives are not implemented, the U.S. energy picture will be substantially different from pre-1974 trends, and is described below.

Energy Demand

—At $11 world oil prices, domestic energy demand will grow at substantially lower rates than it has in the past.
  ○ Total demand will grow at a rate of 2.7 percent per year between 1972 and 1985, compared to 4-5 percent during 1960-1970,
  ○ 1985 demand will be about 103 quadrillion Btu's (quads) as contrasted with most other forecasts in the 115-125 quads range,
  ○ Electric demand will also be below its recent high growth rates,
  ○ Petroleum demand will be about constant between 1974 and 1977 and only grow at about 1-2 percent per year thereafter.
—At $7 prices, total energy demand will grow at 3.2 percent through 1985, and petroleum consumption will be about 5 million barrels per day (MMBD) higher than at $11 levels by 1985.

Energy Supply

—Petroleum production is severely constrained in the short run and greatly affected by world oil prices in the long run (Figure 4).
  ○ Between 1974 and 1977, there is little that can prevent domestic production from declining or at best remaining constant.
  ○ By 1985, at $7 world oil prices, production will rise to 8.9 MMBD from the current 8.6 MMBD. "Lower 48" production will decline from 8.2 MMBD to 4.2 MMBD, but is offset by Alaskan and Outer Continental Shelf (OCS) production.
  ○ If oil prices remain at $11, production could reach 12.8 million barrels per day by 1985. This further increase comes mainly from the use of more expensive secondary and tertiary recovery in the lower 48 States.
—Coal production will increase significantly, but is limited by lack of markets.
  ○ By 1985, coal use will be between 1.0 and 1.1 billion tons per year depending on world oil prices.
  ○ Production could be expanded greatly by 1985, but lower electric growth, increasing nuclear capacity, and environmental restrictions limit this increase.
—Potential increases in natural gas production are limited, but continued regulation could result in significant declines.
  ○ Continued regulation at today's price will reduce production to 15.2 TCF by 1985, or 38 percent below the deregulated case.
  ○ With deregulation of gas, production will rise from 22.4 trillion cubic feet (TFC) in 1972 to 24.6 TFC by 1985. Alaska production will be 1.6 TCF of this total.
—Nuclear power is expected to grow from 4.5 percent to 30 percent of total electric power generation.

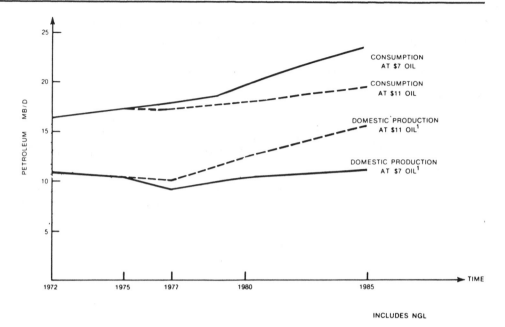

Effects of $7 VS $11 foreign oil.

Source: Project Independence Executive Summary, Federal Energy Administration, November 1974.

  ○ This forecast is lower than many others due to continued schedule deferments, construction delays and operating problems.
—Synthetic fuels will not play a major role between now and 1985.
  ○ At $7 they are marginally economic.
  ○ At $11 they are economic, but given first commercial operation in the late 1970's, their contribution by 1985 is small.
  ○ Research and development (R&D) on these technologies is important if they are to replace a growing liquid and gaseous fuels gap which may develop after 1985.
—Shale oil could reach 250,000 B/D by 1985 at $11 prices, but would be lower if expectations for $7 prices prevail.
—Geothermal, solar, and other advanced technologies are large potential sources, but will not contribute to our energy requirements until after 1985.
  ○ R&D is needed so that these important sources, which can have less environmental impact than current sources and are renewable (do not deplete existing reserves), can be useful beyond 1985.

Constraints and Barriers

—Even achieving the Base Case will require actions to alleviate potentially serious barriers.
  ○ Rather than stimulating coal use, current Clean Air Act requirements could, by mid-1975, preclude 225 million tons of coal now used in utilities.

○ The financial situation of the electric utility industry is particularly critical, and inadequate rates of return will not only reduce their internal funds, but hamper their ability to attract debt or equity financing.

○ Current manpower, equipment, and materials shortages are likely to persist in the short-term and inhibit production increases.

○ Continued problems with growth in the nuclear industry are possible, unless reliability problems, future shortages of enrichment capacity, and the waste disposal problem are resolved.

Oil Imports and Domestic Vulnerability

—Oil imports will remain level or rise in the next few years, no matter what long-term actions we take.

—Our domestic vulnerability to future disruptions is dependent on world oil prices.

○ At $7 oil and no new domestic policy actions, imports will reach 12.3 MMBD in 1985, of which 6.2 MMBD are susceptible to disruption. A one year embargo could cost the economy $205 billion.

○ At $11, imports will decline to 3.3 MMBD by 1985, and only 1.2 MMBD are susceptible to disruption, at a cost of $40 billion for a one year embargo.

Economic and Environmental Assessment

—Higher energy prices are likely in any event, but $11 world oil will magnify these price trends and have several major effects.

○ $11 oil prices, as opposed to $7, will reduce U.S. economic growth from 3.7 percent to 3.2 percent.

○ Dollar outflows for petroleum imports will be higher in the near term for $11 than for $7, but by 1980 the situation will be reversed.

○ At $11 oil prices, large regional price disparities exist with eastern oil-dependent regions at the high end of the spectrum.

○ At $7 prices, these disparities narrow and the Northeast is no longer the highest cost region.

○ Because energy costs as a percentage of total consumption are higher for lower income groups, higher energy costs will impact the poor more heavily.

—Energy production through 1985 will have mixed environmental impacts.

○ Most sources of water pollution should be below 1972 levels, due to the Federal water pollution standards.

○ Emission controls will lessen the air pollution impact of increased energy use, but some regions will still be affected significantly.

○ Surface mining will continue to increase and problems of secondary economic development in the West and Alaska are likely.

PROJECT INDEPENDENCE—ALTERNATIVE ENERGY STRATEGIES

Alternative Energy Strategies

—U.S. options to reduce vulnerability fall into three distinct categories. While each has significant impact, a national energy policy will probably combine elements from each.

1. Accelerating Domestic Supply

—Federal policies to lease the Atlantic OCS, reopen the Pacific OCS and tap the Naval Petroleum Reserves can dramatically increase domestic oil production.
- At $7 prices, domestic production by 1985 could rise from 8.9 MMBD to 12.8 MMBD.
- At $11 prices, production could reach as high as 17 MMBD, although less is needed to achieve zero imports (Figure 5).

—Shale oil production could reach one MMBD in 1985.
- Prices close to $11 would be needed for economic viability.
- Potential water and environmental constraints would have to be overcome.

—Accelerating nuclear power plant construction does not reduce imports much; in general, it replaces new coal-fired power plants.

—Accelerating synthetic fuel production would require by-passing key research steps and may not be cost-effective or practical in the 1985 time frame.

—Accelerating domestic energy production could be inhibited by several key constraints:
- In the short-term, many shortages of materials, equipment, and labor will persist.
- By 1985, however, most critical shortages will be overcome sufficiently to meet the requirements of the accelerated supply scenario.
- Availability of drilling rigs and fixed and mobile platforms will be a major constraint in reaching the projected oil levels.
- Financial and regulatory problems in the utility and railroad industries could hamper their ability to purchase needed facilities and equipment.
- Water availability will be a problem in selected regions by 1985.

2. Energy Conservation and Demand Management

—Energy conservation actions can reduce demand growth to about 2.0 percent per year between 1972 and 1985.
- To achieve reductions beyond those induced by price could require new standards on products and buildings, and/or subsidies and incentives.
- Major actions could include standards for more efficient new autos, incentives to reduce miles traveled, incentives for improved thermal efficiency in existing homes and offices and minimum thermal standards for new homes and offices.
- Petroleum demand could be reduced by 2.2 MMBD by 1985 (Figure 6).
- Electricity consumption could be reduced from 12.3 quads to about 11.0 quads in 1985, compared with 5.4 in 1972.

—Demand management can further reduce dependence on limited oil and gas supplies by actions that involve switching from petroleum and natural gas consumption to coal or coal-fired electric power.

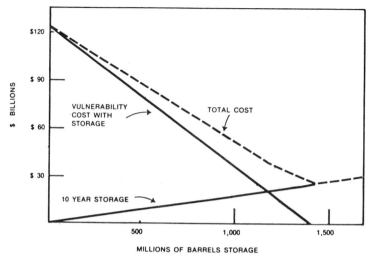

Costs of vulnerability as a function of stockpile size.

Source: Project Independence Executive Summary, Federal Energy Administration, November 1974.

- Switching existing power plants and industrial users, prohibiting new oil or gas-fired power plants and encouraging electric space heating is most important at lower oil prices, and can substitute 400 million tons of coal per year for 2.5 MMBD of petroleum and 2.5 TCF per year of natural gas.
- Implementation may be limited by environmental restrictions and financial inability of the electric utility industry to support a large electrification strategy.

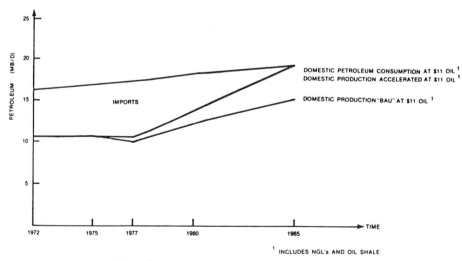

Effect of domestic production on petroleum imports.

Source: Project Independence Executive Summary, Federal Energy Administration, November 1974.

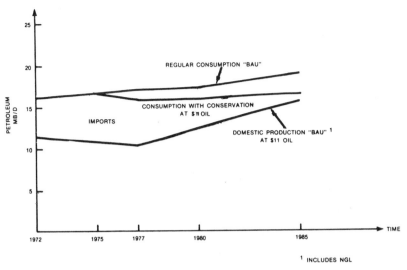

Conservation management.

Source: Project Independence Executive Summary, Federal Energy Administration, November 1974.

○ Electrification to increase coal use in the pre-1985 period must be weighed against the possibility of increasing coal use by liquefication and gasification in the post-1985 period.

3. Emergency Programs

—Standby conservation or curtailment measures can reduce vulnerability.
  ○ Depending on the level of demand in 1985, curtailment measures in response to an embargo can cut consumption by 1–3 MMBD.
  ○ At higher world oil prices curtailment is less effective because there is less "fat" in energy consumption.
  ○ They involve almost no cost when not needed and relatively small administrative costs and some economic impact when implemented.
  ○ They can be instituted in 60–90 days.
—Emergency storage is cost-effective in reducing the impact of an embargo (Figure 7).
  ○ Storage to insure against a one MMBD cutoff for one year would cost $6.3 billion over ten years.
  ○ A one MMBD interruption of oil supply for one year during that period could cost the economy $30–40 billion.
  ○ This cost effectiveness holds for any level of insecure imports, and applies if there is a one-in-five chance of one disruption in ten years.
—The International Energy Program (IEP) developed in Brussels will foster consumer nation cooperation and reduce the United States' economic impact of a supply disruption during the next several years.
  ○ It can reduce the likelihood of an import disruption.

○ It includes a formula for allocating shortages which avoids excessive bidding and divisive scramble for oil by the participants during the most vulnerable period of the next few years. If, in the 1980's, the United States achieves the low import levels which are possible at high oil prices and by pursuing aggressive strategies of accelerating supply and conservation, the IEP would still act as a deterrent to an interruption, but its utility in protecting the U.S. against the economic impact of supply disruptions would be diminished.

Comparison of Alternative Energy Strategies

Import Vulnerability

—Domestic supply and demand actions can greatly reduce U.S. vulnerability to import disruptions by 1985.
  ○ At $7 per barrel oil with all supply and demand actions implemented, 3 MMBD could still be subject to cutoff.
  ○ At $11, either all the demand actions or only a portion of the supply strategy would completely eliminate our vulnerability.
—Domestic supply and demand strategies are cheaper in economic terms than imported oil or any other emergency option.
  ○ At either $7 or $11, they have a lower present resource cost than imports, and reduce insecure imports.
—After domestic actions, standby demand curtailment is most effective in reducing vulnerability.
—Demand curtailment and storage can be designed to buffer against large levels of insecure imports.

Economic and Regional Impacts

—Accelerating domestic supply or reducing demand will mean lower energy costs for the Nation and, hence, higher economic growth.
—Reduced energy costs will benefit lower income groups.
—Increased domestic supply may result in wider regional price disparities than if no action is taken.
—The economy can absorb the increased financial costs of reducing vulnerability.
—Both supply and demand actions will have economic impacts, regionally and in key sectors of the economy.

Environmental Impacts

—The conservation strategy has the lowest environmental impact.
—A demand management strategy which substitutes coal for oil and gas will result in the greatest increase in environmental impact over the base case.
—The accelerated supply strategy has mixed environmental impacts.
  ○ Air pollution is lower due to more nuclear plants and increased oil and gas production.
  ○ Solid waste is up dramatically due to increased shale oil production.
  ○ Many virgin resource areas will be disturbed for the first time.

PROJECT INDEPENDENCE—UNCERTAINTIES AND POLICY

Major Uncertainties

—The degree to which price will dampen demand
- If demand is much less sensitive to price than is assumed, we will be much more vulnerable in 1977 and in 1985 at higher world oil prices.
- Mandatory energy conservation measures or diversification and acceleration of supply hedge against this uncertainty.

—The ultimate production potential of frontier oil areas
- Literally all the new oil production forecast comes from frontier areas in Alaska and the Atlantic OCS, or from improved tertiary recovery techniques.
- If the frontier areas do not prove productive:
  - —1985 domestic production could decline to 5 MMBD at $7 oil prices;
  - —at $11 and with accelerated supply actions, total production could still not exceed about 11 MMBD.
- Synthetic fuels, switching from petroleum and gas to coal and mandatory conservation may be necessary if frontier areas are not lucrative.

—Time required to implement domestic measures
- While lead times were taken into account, other factors could delay achievement.
- Federal inaction or local opposition.
- Materials and equipment constraints.
- Delays in private investment decisions due to price uncertainty.

Policy Implications

—Although $11 world oil prices make achievement of self-sufficiency easier, the United States is still better off economically with lower world oil prices. The implementation of a limited number of major supply or demand actions could make us self-sufficient. By 1985, we could be at zero imports at $11, and down to 5.6 MMBD of imports at $7 prices.
- Not all of these actions may be warranted, but they indicate we have significant flexibility when one considers:
  - —some projected imports in 1985 are from secure sources.
  - —some insecure imports can be insured against.

—not all of the supply and demand actions must be implemented to achieve the desired result.

—While we cannot delay all action, we can pick from those that make the most economic, environmental, and regional sense.

—Accelerating domestic supply, while economic, has some important drawbacks:
- It will adversely affect environmentally clean areas.
- It requires massive regional development in areas which may not benefit from or need increased supply.
- It is a gamble on as yet unproved reserves of oil and gas.
- It may well be constrained by key materials and equipment shortages.

—Implementing a conservation strategy has positive environmental effects and alleviates constraint problems, but:
- It requires intervention and regulation in previously free market areas.
- It results in increased nonmarket costs due to more limited individual choice and changed lifestyles.

—While cost effective, there are several important ramifications to a storage program:
- It will take a few years to implement and our vulnerability will be greatest during that period.
- It requires more imports now, which will act to sustain *cartel prices* in the near term.
- We could suffer major capital losses—$4 billion for each one billion barrels stored if the world oil price drops from $11 to $7.

—Our actions to increase domestic self-sufficiency could have an appreciable impact on world oil price.
- U.S. reduction in imports can make even $7 hard for OPEC to maintain.
- World oil price reductions could jeopardize domestic energy investments and could require price guarantees or other supports.

—Any domestic energy policy must be designed to resolve uncertainties and minimize the risk of not anticipating world oil prices correctly.
- Policy programs should include actions to reduce domestic uncertainty, such as exploring the frontier areas.
- Policies may be needed to avoid or defer major investments or actions, if they involve significant costs of being wrong, until world uncertainty is reduced.
- A flexible and dynamic approach must be balanced against the need for a stable long term policy which encourages domestic energy investment.

NATIONAL ENERGY POLICY—GOALS AND NEAR-TERM ACTIONS

In January 1975, President Gerald Ford outlined the goals and principles of a national energy policy and specific actions recommended by his administration.[1] Details of this policy are given on this and pages following.

NATIONAL ENERGY POLICY GOALS AND PRINCIPLES ANNOUNCED BY THE PRESIDENT

I. Near-Term (1975–1977): Reduce oil imports by one million b/d by the end of 1975 and two million b/d by the end of 1977, through immediate actions to reduce energy demand and increase domestic supply.

(A) With no action, imports would be about eight million b/d by the end of 1977, more than 20 percent above the 1973 pre-embargo levels.

(B) Acting to meet the 1977 goal will reduce imports below 1973 levels, assuring reduced vulnerability from an embargo and greater consumer nation cooperation.

(C) More drastic short-term reductions would have unacceptable economic impacts.

II Mid-Term (1975–1985): Eliminate vulnerability by achieving the capacity for full energy independence by 1985. This means 1985 imports of no more than three to five million barrels of oil per day, all of which can be replaced immediately from a strategic storage system and managed with emergency measures.

(A) With no action, oil imports by 1985 could be reduced to zero at prices of $11 per barrel or more—or they could go substantially higher if world oil prices are reduced (e.g., at $7 per barrel, U.S. consumption could reach 24 million b/d with imports of above 12 million, or above 50 percent of the total.)

[1] Fact sheet on the President's State of the Union Message, The White House, January 15, 1975.

Impacts of Near and Midterm Actions on Petroleum Consumption and Imports
Near-Term Program (MMB/D)

	1975	1977
Consumption if no new actions	18.0	18.3
Imports if no new actions	6.5	8.0
	Import Savings	
Less service savings by short-term actions:	1975	1977
production from Elk Hills	0.2	0.3
coal conversion	0.1	0.3
tax package	0.9	1.6
Total import savings	1.2	2.2
Remaining imports	5.3	5.8

(B) The U.S. anticipates a reduction in world oil prices over the next several years. Hence, plans and policies must be established to achieve energy independence even at lower prices—countering the normal tendency to increase imports as the price declines.

(C) Actions to meet the 1985 goal will hold imports to no more than 3–5 million b/d even at $7 per barrel prices. Protection against an embargo of the remaining imports can then be handled most economically with storage and standby emergency measures.

III. Long-Term (Beyond 1985): Within this century, the U.S. should strive to develop technology and energy resources to enable it to supply a significant share of the free world's energy needs.

(A) Other consuming nations have insufficient fossil fuel resources to reach domestic energy self-sufficiency.

(B) The U.S. can again become a world energy supplier and foster world energy price stability—much the same as the nation did prior to the 1960s when it was a major supplier of world oil.

ACTIONS ANNOUNCED BY THE PRESIDENT

I. Actions To Meet Near-Term Goals (1975–1977)

To meet the national goals, the President outlined a comprehensive program of legislative proposals to the Congress which he requested be enacted within 90 days and administrative actions that he will begin implementing immediately. The legislative package is more effective and equitable than the administrative program, but the President indicated that the seriousness of the situation demanded immediate action. These actions will reduce overall energy demand, increase domestic production, increase conversion to coal, and reduce oil imports. They include:

(A) Administrative Actions

1. Import Fee—Because of the seriousness of the problem and because time is required for Congressional action on his legislative proposals, the President is acting immediately within existing authorities to increase the import fees on crude oil and petroleum products. These new import fees would be modified upon passage of the President's legislative package.

(a) Import fees on crude oil and petroleum products under the authority of the Trade Expansion Act of 1962, as amended, will be increased by $1 effective Feb. 1, 1975; an additional $1 effective March 1; and another $1 effective April 1, for total increase of $3.00 per barrel. Currently existing fees will also remain in effect.

(b) FEA's Old Oil Entitlements program will be utilized to spread price increases on crude among all refiners and to lessen disproportionate regional effects, particularly in the Northeast.

(c) As of February, 1975, product imports will cease to be covered by FEA's Old Oil Entitlements program. In order to overcome any severe regional impacts that could be caused by large fees in import dependent areas, imported products will receive a rebate corresponding to the benefit which would have been obtained under that program. The rebate should be approximately $1.00 in February, $1.40 in March, and $1.80 per barrel in April.

(d) This import fee program would reduce imports by about 500,000 b/d. In April it would generate about $400 million per month in revenues.

2. Backup Import Control Program—The energy conservation measures and tax proposals will be supplemented by the use of Presidential power to limit oil imports as necessary to achieve the near-term goals.

3. Crude Oil Price Decontrol—To stimulate production and further cut demand, steps will be taken to remove price controls on domestic crude oil by April, 1975, subject to congressional disapproval.

4. Increase Public Education on Energy Conservation—Energy Resources Council will step up its efforts to provide information on energy conservation methods and benefits.

(B) Legislative Proposals

1. Comprehensive Tax and Decontrol Program—The President asked the Congress to pass within 90 days a comprehensive legislative package which could lead to reduction of oil imports of 900,000 b/d by 1975 and 1.6 million b/d by 1977. Average oil prices would rise about $4.00 per barrel or $.10 per gallon. The package which will raise $30 billion in revenues on an annual basis includes:

(a) **Windfall Profits Tax**—A tax on all domestic crude oil to capture the windfall profits resulting from price decontrol. The tax would take 88 percent of the windfall profits on crude oil and would phase out over several years. The tax would be retroactive to January 1, 1975.

(b) **Petroleum Excise Tax and Import Fee**—An excise tax on all domestic crude oil of $2 per barrel and a fee on imported crude oil and product imports of $2 per barrel. The new, administratively established import fee of $3 on crude oil would be reduced to $2.00, and $1.20 fee on products would be increased to $2.00 when the tax is enacted. The product import fee would keep the excise tax from encouraging foreign refining and the related loss of jobs to the U.S.

(c) **New Natural Gas Deregulation**—Remove federal interstate price regulation on new natural gas to increase domestic production and reduce demand for scarce natural gas supplies.

(d) **Natural Gas Excise Tax**—An excise tax on natural gas of 37 cents per thousand cubic feet (Mcf), which is equivalent on a Btu basis to the $2 per barrel petroleum excise tax and fee. This will discourage attempts to switch to natural gas and acts to reduce natural gas demand curtailments. Since the usual result of gas curtailments is a switch to oil, this will limit the growth of oil imports.

2. Elk Hills Naval Petroleum Reserve. The President is asking the Congress to permit production of the Elk Hills Naval Petroleum Reserve (NPR No. 1) under Navy control. Production could reach 160,000 b/d early in 1975 and 300,000 b/d by 1977. The oil produced would be used to top off Defense Department storage tanks, with the remainder sold at auction or exchanged for refined petroleum products used by the Department of Defense. Revenues would be used to finance further exploration, development and production of the Naval petroleum reserves and the strategic petroleum storage.

3. Conversion to the Use of Domestic Coal. The President is asking the Congress to amend the Clean Air Act and the Energy Supply and Environmental Coordination Act of 1974 to permit a vigorous program to make greater use of domestic coal to reduce the need for oil. This program would reduce the need for oil imports by 100,000 b/d in 1975 and 300,000 b/d in 1977. These amendments would extend FEA's authority to grant prohibition orders from 1975 to 1977, prohibit power plants early in the planning process from burning oil and gas, extend FEA enforcement authority from 1978 to 1985, and make clear that coal burning installations that had originally planned to convert from coal to oil be eligible for compliance date extensions. It would give EPA authority to extend compliance dates and eliminate restrictive regional environmental limitations. A plant could convert as long as its own emissions do not exceed ambient air quality standards.

NATIONAL ENERGY POLICY—MIDTERM SUPPLY ACTIONS

II. Actions Announced By The President To Meet Mid-Term Goals (1975–1985)

These actions are designed to meet the goal of achieving the capability for energy independence by 1985. The actions include measures to increase domestic energy production (including measures to cope with constraints and strike a balance between environmental and energy objectives), reduce energy demand, and prepare for any future emergency resulting from an embargo.

(A) Supply Actions

1. Naval Petroleum Reserve No. 4 (legislative proposal)—The President is asking the Congress to authorize the exploration, development and production of NPR-4 in Alaska to provide petroleum for the domestic economy, with 15–20 percent earmarked for military needs and strategic storage. The reserves in NPR-4 which are now largely unexplored could provide at least two million b/d of oil by 1985. Under the legislative proposal:

(a) the President would be authorized to explore, develop and produce NPR-4.

(b) The government's share of production (approximately 15–20 percent) would be used to help finance the strategic storage system and to help fulfill military petroleum requirements. Any other receipts go to the United States Treasury as miscellaneous receipts.

2. OCS Leasing (administrative)—The President reaffirmed his intention to continue an aggressive Outer Continental Shelf leasing policy, including lease sales in the Atlantic, Pacific, and Gulf of Alaska. Decisions on individual lease sales will await completion of appropriate environmental studies. Increased OCS leasing could add domestic production of 1.5 million b/d of oil and additional supplies of natural gas by 1985. There will be close cooperation with coastal states in their planning for possible increased local development. Funding for environmental studies and assistance to states for planning has been increased in FY 1975.

3. Reducing Domestic Energy Price Uncertainty (legislative proposal)—Legislation will be requested authorizing and requiring the President to use tariffs, import quotas, import price floors, or other measures to achieve domestic energy price levels necessary to reach self-sufficiency goals. This legislation would enable the President to cope with possible large-scale fluctuations in world oil prices.

4. Clean Air Act Amendments (legislative proposal)—In addition to the amendments outlined earlier for short-term goals, the President is asking for other Clean Air Act amendments needed for a balance between environmental and energy goals. These include:

(a) legislative clarification to resolve problems resulting from court decisions with respect to significant air quality deterioration in areas already meeting health and welfare standards.

(b) extension of compliance dates through 1985 to implement a new policy regarding stack gas scrubbers—to allow use of intermittent control systems in isolated power plants through 1985 and requiring other sources to achieve control as soon as possible.

Midterm Program

Consumption if no new actions	23.9 MMB/D
Imports if no new actions	12.7 MMB/D

Less savings achieved by following actions:	1985 Impact on Imports	
OCS leasing	1.5	
NPR-4 development	2.0	
Coal conversion	0.4	
Synthetic fuel commercialization	0.3	
Auto efficiency standards	1.0	
Continuation of taxes	2.1	
Appliance efficiency goals	0.1	
Insulation tax credit	0.3	
Thermal standards	0.3	
Total import savings by actions		8.0
Remaining imports		4.7
Less:		
emergency storage	3.0	
standby authorities	1.7	
Net import vulnerability		0

(c) a pause for 5 years (1977–1981 model years) for nationwide auto emission standards at the current California levels for hydrocarbons (0.9 grams per mile) and carbon monoxide (9.1 grams per mile), and at 1975 standards (3.1 grams per mile) for oxides of nitrogen (with the exception of California which has adopted the 2.0 standard). These standards for hydrocarbons (HC) and carbon monoxide (CO) are more stringent than now required nationwide for 1976 model year cars. The change from the levels now required for 1977–1981 model years in the law will have no significant impact on air quality standards, yet they will facilitate attainment of the goal of 40 percent increase in auto fuel efficiency by the 1980 model year.

(d) EPA will shortly begin comprehensive hearings on emission controls and fuel economy which will provide more detailed data for Congressional consideration.

5. Surface Mining (legislative proposal)—The President is asking the Congress to pass a surface mining bill which strikes a balance between our desires for reclamation and environmental protection and our need to increase domestic coal production substantially over the next ten years. The proposed legislation will correct the problems which led to the President's veto of a surface mining bill last year.

6. Coal Leasing (administrative)—To assure rapid production from existing leases and to make new, low sulfur coal supplies available, the President directed the Secretary of the Interior to:

(a) adopt legal diligence requirements to assure timely production from existing leases.

(b) meet with Western governors to explore regional questions on economic, environmental and social impacts associated with new federal coal leases.

(c) design a program of new coal leasing consistent with timely development and adequate return on public assets, if proper environmental safeguards can be provided.

7. Electric Utilities—The President is asking the Congress for legislation concerned with utilities. In recent months 60 percent of planned nuclear capacity and 30 percent of non-nuclear capacity additions have been postponed or cancelled by electric utilities. Financing problems are worsening and state utility commission practices have not assured recovery of costs and adequate earnings. The transition from oil and gas-fired plants to coal and nuclear has been slowed greatly—contributing to pressure for higher oil imports. Actions involve:

(a) **Uniform Investment Tax Credit (legislative)**—an increase in the investment tax credit to eliminate the gap between utilities and other industries—currently a four percent rate applies to utilities and seven percent to others.

(b) **Higher Investment Tax Credit (legislative)**—an increase in investment tax credit for all industry including utilities, for one year—to 12 percent. The 12 percent rate would be retained for two additional years for all power plants except oil and gas-fired facilities.

(c) **Preferred Stock Dividend Deductions (legislative)**—A change in tax laws applicable to all industries, including utilities, which allows deductions of preferred stock dividends for tax purposes to reduce the cost of capital and stimulate equity rather than debt financing.

(d) **Mandated Reform of State Utility Commission Processes (legislative)**—the legislation would selectively reform utility commission practices by: (1) setting a maximum limit of five months for rate or service proceedings; (2) requiring fuel adjustment pass-throughs, including taxes; (3) requiring that construction work in progress be included in a utility's rate base; (4) removing any rules prohibiting a utility from charging lower rates for electric power during off-peak hours; and (5) allowing the cost of pollution control equipment to be included in the rate base.

(e) **Energy Resources Council Study (administrative)**—review and report to the President on the entire regulatory process and financial situation relating to electric utilities and determine what further reforms or actions are needed. ERC will consult with state utility commissions, governors, public utilities and consumers.

8. Nuclear Power—To accelerate the growth of nuclear power which supplies only one percent of our energy needs, the President is proposing, in addition to actions outlined above:

(a) **Expedited Licensing and Siting (legislative)**—a Nuclear Facility Licensing Act to assure more rapid siting and licensing of nuclear plants.

(b) **1976 Budget Increase (legislative)**—an increase of $41 million in appropriations for nuclear safety, safeguards and waste management.

9. Energy Facilities Siting (legislative)—Legislation would reduce energy facility siting bottlenecks and assure sites for needed facilities with proper land use considerations:

(a) the legislation would require that states have a comprehensive and coordinated process for expeditious review and approval of energy facility applications, and state authorities which ensure that final state energy facility decisions cannot be nullified by actions of local governments.

(b) provision for owners of eligible facilities or citizens to sue states for inaction.

(c) provide no federal role in making case by case siting decisions for the states.

NATIONAL ENERGY POLICY–CONSERVATION, EMERGENCY AND LONG-TERM ACTIONS

(B) Energy Conservation Actions

The President announced a number of energy conservation measures to reduce demand, including:

1. Auto Gasoline Mileage Increases (administrative)–The Secretary of Transportation has obtained written agreements with each of the major domestic automobile manufacturers which will yield a 40 percent improvement in fuel efficiency on a weighted average for all new autos by 1980 model year. These agreements are contingent upon relaxation of Clean Air Act auto emission standards. The agreement provides for interim goals, federal monitoring, and public reporting of progress.

2. Building Thermal Standards (legislative)–The President is asking Congress for legislation to establish national mandatory thermal (heating and cooling) efficiency standards for new homes and commercial buildings which would save the equivalent of over 500,000 b/d of oil by 1985. Under this legislation:

(a) the Secretary of Housing and Urban Development shall consult with engineering, architectural, consumer, labor, industry, and government representatives to advise on development of efficiency standards.

(b) thermal standards for one and two-family dwellings will be developed and implementation would begin within one year. New minimum performance standards for energy in commercial and residential buildings would be developed and implemented as soon thereafter as practicable.

(c) standards would be implemented by state and local governments through local building codes.

(d) the President also directed the Secretary of Housing and Urban Development to include energy conservation standards in new mobile home construction and safety standards.

3. Residential Conservation Tax Credit (legislative)–The President is asking Congress for legislation to provide incentives to homeowners for making thermal efficiency improvements in existing homes. This measure, along with a stepped-up public information program, could save the equivalent of over 500,000 b/d by 1985. Under this legislation:

(a) a 15 percent tax credit retroactive to January 1, 1975 for the cost of certain improvements in thermal efficiency in residences would be provided. Tax credits would apply to the first $1,000 of expenditures and can be claimed during the next three years.

(b) improvements such as storm windows, and insulation, would qualify for the tax credit.

4. Low-Income Energy Conservation Program (legislative)–The President is proposing legislation to establish a Low-Income Energy Conservation Program to offer direct subsidies to low-income and elderly homeowners for certain energy conservation improvements such as insulation. The program is modeled upon a successful pilot program in Maine.

(a) the program would be administered by FEA, under new legislation, and the President is requesting supplemental appropriations in 1975 and $55 million in fiscal year 1976.

(b) acting through the states, federal funds would be provided to purchase materials. Volunteers or community groups could install the materials.

5. Appliance Efficiency Standards (administrative)–The President directed the Energy Resources Council to develop energy efficiency goals for major appliances and to obtain agreements within six months from the major manufacturers of these appliances to comply with the goals. The goal is a 20 percent average improvement by 1980 for all major appliances, including air conditioners, refrigerators and other home appliances. Achievement of these goals would save the equivalent of over one-half million barrels of oil per day by 1985. If agreement cannot be reached, the President will submit legislation to establish mandatory appliance efficiency standards.

6. Appliance and Auto Efficiency Labelling Act (legislative)–The President will ask the Congress to enact a mandatory labelling bill to require that energy efficiency labels be placed on new appliances and autos.

(C) Emergency Preparedness

The President announced that comprehensive energy emergency legislation will be proposed, encompassing two major components.

1. Strategic Petroleum Storage (legislative)–Development of an energy storage system of one billion barrels for domestic use and 300 million barrels for military use. The legislation will authorize the government to purchase and prepare the storage facilities (salt domes or steel tanks), while complex institutional questions are resolved and before oil for storage is actually purchased. FEA will develop the overall program in cooperation with the Department of the Interior and the Department of Defense. All engineering, planning, and environmental studies would be completed within one year. The 1.3 billion barrels will not be complete for some years, since time is required to purchase, prepare, and fill the facilities.

2. Standby and Planning Authorities (legislative)–The President is requesting a set of emergency standby authorities to be used to deal with any significant future energy shortages. These authorities would also enable the United States to fully implement the agreement on an International Energy Program between the United States and other nations signed on November 18, 1974. This legislation would include the authority to:

(a) implement energy conservation plans to reduce demand for energy;

(b) allocate petroleum products and establish price controls for allocated products;

(c) ration fuels among end users;

(d) allocate materials needed for energy production where such materials may be in short supply;

(e) increase production of domestic oil; and

(f) regulate petroleum inventories.

III. Actions Announced By The President To Meet Long-Term Goals (Beyond 1985)

The expanded research and development program on which the nation is embarked will provide the basis for increasing domestic energy supplies and maintaining energy independence. It will also make it possible in the long run for U.S. to export energy supplies and technology to others in the free world. Important elements are:

(A) Synthetic Fuels Program (administrative)—The President announced a National Synthetic Fuels Commercialization Program to ensure at least one million b/d equivalent of synthetic fuels capacity by 1985, using technologies now nearing commercial application.

1. Synthetic fuel types to be considered will include synthetic crude from oil shale and a wide range of clean solid, liquid, and gaseous fuels derived from coal.

2. The program would entail federal incentives (possibly including price guarantees, purchase agreements, capital subsidies, leasing programs, etc.), granted competitively, and would be aimed at the production of selected types of gaseous and liquid fuels from both coal and oil shale.

3. The program will rely on existing legislative authorities, including those contained in the Federal Non-Nuclear Energy Research and Development Act of 1974, but new legislative authorities will be requested if necessary.

(B) Energy Research and Development Program—In the current fiscal year, the federal government has greatly increased its funding for energy research and development programs. These federal programs are a part of a much larger national energy R&D effort and are carried out in cooperation with industry, colleges and universities and others. The President stated that his 1976 budget will continue to emphasize these accelerated programs which include research and the development of technology for energy conservation and on all forms of energy including fossil fuels, nuclear fission and fusion, solar and geothermal.

(C) Energy Research and Development Administration (ERDA)—The President has signed an executive order which activates, effective Jan. 19, 1975, the Energy Research and Development Adminstration. ERDA will bring together in a single agency the major federal energy R&D programs which will have the responsibility for leading the national effort to develop technology to assure that the U.S. will have an ample and secure supply of energy at reasonable prices. ERDA consolidates major R&D functions previously handled by the AEC, Department of the Interior, National Science Foundation and Environmental Protection Agency. ERDA will also continue the basic research, nuclear materials production and weapons programs of the AEC.

DOI ENERGY PERSPECTIVES—ENERGY CONSUMPTION, BUSINESS AS USUAL

Based on data derived from Project Independence studies, the Department of the Interior (DOI) has projected an increase in gross energy consumption at a rate of 3.2% per year between 1973 and 1985 under business as usual conditions and with oil at $7.00 per barrel. With oil at $11.00 per barrel, the growth in gross energy consumption will be 2.7% per year without conservation and 2.0% per year with conservation during the same period.

[Business as Usual, $7 Oil]

Gross Energy Consumption, 1973-90

(Quadrillion Btu)[1]

SOURCE	1973	1977		1980		1985		1990	
		WOC	WC	WOC	WC	WOC	WC	WOC	WC
PETROLEUM	34.7	37.8	35.7	41.6	38.4	47.9	42.6	55.2	49.1
NATURAL GAS	22.8	21.6	21.3	22.9	21.9	23.9	22.5	25.0	23.6
COAL	13.4	16.9	16.7	18.1	16.9	19.9	16.7	21.9	17.9
HYDROPOWER & GEOTHERMAL	2.9	3.5	3.5	4.0	4.0	4.8	4.8	5.6	5.6
NUCLEAR	.9	2.8	2.8	4.8	4.8	12.5	12.5	21.5	21.5
TOTAL	74.7	82.6	80.0	91.4	86.0	109.0	99.1	129.2	117.7
PETROLEUM IMPORTS (Million Barrels Per Day)	6.2			9.7	8.1	12.4	9.8	15.9	13.0

[1] 1 Quadrillion Btu = 500,000 barrels petroleum per day for a year
= 40 million tons of bituminous coal
= 1 trillion cubic feet of natural gas
= 100 billion kWh (based on a 10,000-Btu/kWh heat rate)

WOC = without conservation; WC = with conservation.

Source: Energy Perspectives, U.S. Department of the Interior, February 1975.

[Business as Usual, $11 Oil]

U.S. Gross Energy Consumption, 1973-90

(Quadrillion Btu)[1]

SOURCE	1973	1977		1980		1985		1990	
		WOC	WC	WOC	WC	WOC	WC	WOC	WC
PETROLEUM	34.7	34.1	32.4	34.9	32.5	38.0	33.5	41.3	36.5
NATURAL GAS	22.8	21.7	21.7	23.1	22.2	24.8	23.7	26.6	25.2
COAL	13.4	16.6	16.4	19.4	18.6	22.9	19.7	26.9	23.2
HYDROPOWER & GEOTHERMAL	2.9	3.5	3.5	4.0	4.0	4.8	4.8	5.6	5.6
NUCLEAR	.9	2.8	2.8	4.8	4.8	12.5	12.5	21.5	21.5
TOTAL	74.7	78.7	76.8	86.2	82.1	103.0	94.2	121.9	112.0
PETROLEUM IMPORTS (Million Barrels Per Day)	6.2	N/A	N/A	4.6	3.3	3.3	1.2	3.9	1.8

[1] 1 Quadrillion Btu = 500,000 barrels petroleum per day
= 40 million tons of bituminous coal
= 1 trillion cubic feet of natural gas
= 100 billion kWh (based on a 10,000-Btu/kWh heat rate)

WOC = without conservation; WC = with conservation.

Source: Energy Perspectives, U.S. Department of the Interior, February 1975.

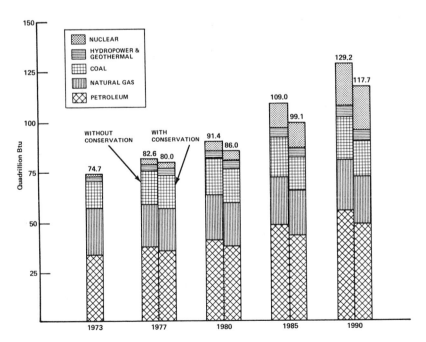

U.S. gross energy consumption, 1973-90 (business as usual, $7 oil).

Source: Energy Perspectives, U.S. Department of the Interior, February 1975.

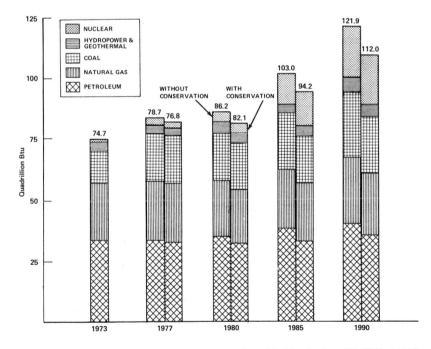

U.S. gross energy consumption, 1973-90 (business as usual, $11 oil).

Source: Energy Perspectives, U.S. Department of the Interior, February 1975.

DOI ENERGY PERSPECTIVES—ENERGY CONSUMPTION, ACCELERATED SUPPLY

Based on data from Project Independence studies, the Department of the Interior (DOI), has projected the annual rate of growth for gross energy consumption under conditions of acceleration of production of domestic supplies, at 3.2% per year with oil at $7.00 per barrel and 2.8% per year with oil at $11.00 per barrel during the years 1973–85. Assuming conservation measures are taken, these growth rates are 2.4% per year with $7.00 oil and 2.0% per year with $11.00 oil.

[Accelerated Supply, $7 Oil]

U.S. Gross Energy Consumption, 1973-90
(Quadrillion Btu)[1]

SOURCE	1973	1980		1985		1990	
		WOC	WC	WOC	WC	WOC	WC
PETROLEUM	34.7	41.1	37.9	47.6	41.3	55.1	47.9
NATURAL GAS	22.8	23.6	23.1	24.7	23.6	25.9	24.7
COAL	13.4	17.4	16.4	17.7	15.2	18.0	15.5
HYDROPOWER & GEOTHERMAL	2.9	4.0	4.0	4.8	4.8	5.6	5.6
NUCLEAR	.9	5.7	5.7	14.7	14.7	25.3	25.3
TOTAL	74.7	91.8	87.1	109.5	99.6	129.9	119.0
PETROLEUM IMPORTS (Million Barrels Per Day)	6.2	8.4	6.8	8.5	5.6	12.0	8.7

[1] 1 Quadrillion Btu = 500,000 barrels petroleum per day for a year
= 40 million tons of bituminous coal
= 1 trillion cubic feet of natural gas
= 100 billion kWh (based on a 10,000-Btu/kWh heat rate)

WOC = without conservation; WC = with conservation.

Source: Energy Perspectives, U.S. Department of the Interior, February 1975.

[Accelerated Supply, $11 Oil]

U.S. Gross Energy Consumption, 1973-90
(Quadrillion Btu)[1]

SOURCE	1973	1980		1985		1990	
		WOC	WC	WOC	WC	WOC	WC
PETROLEUM	34.7	34.2	32.2	38.1	35.7	42.4	39.6
GAS	22.8	24.2	23.5	25.5	24.6	26.9	25.9
COAL	13.4	18.9	17.3	21.1	16.4	23.5	18.4
HYDROPOWER & GEOTHERMAL	2.9	4.0	4.0	4.8	4.8	5.6	5.6
NUCLEAR	.9	5.7	5.7	14.7	14.7	25.3	25.3
TOTAL	74.7	87.0	82.7	104.2	96.2	123.7	114.8
PETROLEUM IMPORTS (Million Barrels Per Day)	6.2	2.9	2.0	0	0	0	0

[1] 1 Quadrillion Btu = 500,000 barrels petroleum per day for a year
= 40 million tons of bituminous coal
= 1 trillion cubic feet of natural gas
= 100 billion kWh (based on a 10,000-Btu/kWh heat rate)

WOC = without conservation; WC = with conservation.

Source: Energy Perspectives, U.S. Department of the Interior, February 1975.

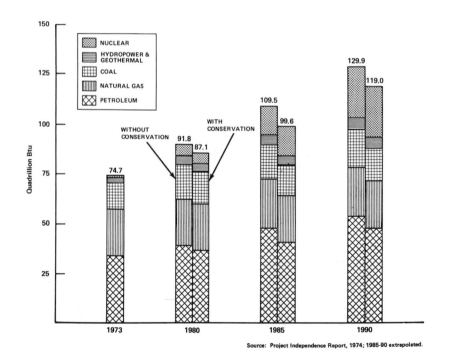

U.S. gross energy consumption, 1973–90 (accelerated supply, $7 oil).

Source: Energy Perspectives, U.S. Department of the Interior, February, 1975.

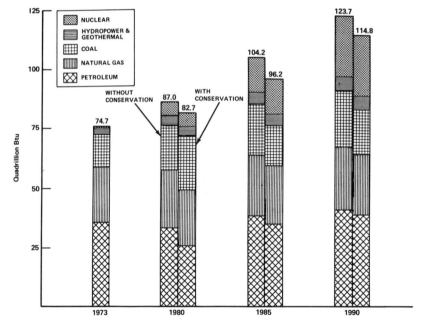

U.S. gross energy consumption, 1973–90 (accelerated supply, $11 oil).

Source: Energy Perspectives, U.S. Department of the Interior, February 1975.

DOI ENERGY PERSPECTIVES—ELECTRICAL SECTOR INPUTS

Based on Project Independence studies, the Department of the Interior (DOI) has projected inputs of energy into the electrical sector as growing at a rate of 6.0% per year between 1973 and 1990 under business as usual conditions, with oil at $7.00 per barrel, and without conservation measures. The rate of growth is projected to be 5.1% per year with conservation. With oil at $11.00 per barrel, the growth rate will be 5.9% per year without conservation and 4.8% per year with conservation.

[Business as Usual, $7 Oil]

Energy Inputs in the Electrical Sector, 1973-90

(Quadrillion Btu)[1]

SOURCE	1973	1977		1980		1985		1990	
		WOC	WC	WOC	WC	WOC	WC	WOC	WC
PETROLEUM	3.5	4.9	4.2	5.9	5.4	7.4	7.3	8.5	6.7
NATURAL GAS	3.7	2.7	2.7	2.9	2.9	2.7	2.7	2.4	2.4
COAL	8.7	11.8	11.7	12.3	11.4	13.5	10.7	15.3	9.7
HYDROPOWER & GEOTHERMAL	2.9	3.5	3.5	4.0	4.0	4.8	4.8	5.6	5.6
NUCLEAR	.9	2.8	2.8	4.8	4.8	12.5	12.5	21.5	21.5
TOTAL	19.7	25.7	24.9	29.9	28.5	40.9	38.0	53.3	45.9

[1] 1 Quadrillion Btu = 500,000 barrels petroleum per day for a year
= 40 million tons of bituminous coal
= 1 trillion cubic feet of natural gas
= 100 billion kWh (based on a 10,000-Btu/kWh heat rate)

WOC = without conservation; WC = with conservation.

Source: Energy Perspectives, U.S. Department of the Interior, February 1975.

[Business as Usual, $11 Oil]

Energy Inputs in the Electrical Sector, 1973-90

(Quadrillion Btu)[1]

SOURCE	1973	1977		1980		1985		1990	
		WOC	WC	WOC	WC	WOC	WC	WOC	WC
PETROLEUM	3.5	3.4	3.3	3.2	2.6	3.0	2.4	3.1	2.1
GAS	3.7	2.9	3.0	3.0	2.8	2.5	2.5	2.4	2.4
COAL	8.7	11.6	11.5	13.6	13.0	16.4	13.4	20.2	12.2
HYDROPOWER & GEOTHERMAL	2.9	3.5	3.5	4.0	4.0	4.8	4.8	5.6	5.6
NUCLEAR	.9	2.8	2.8	4.8	4.8	12.5	12.5	21.5	21.5
TOTAL	19.7	24.2	24.1	28.6	27.2	39.2	35.6	52.8	43.8

[1] 1 Quadrillion Btu = 500,000 barrels petroleum per day for a year
= 40 million tons of bituminous coal
= 1 trillion cubic feet of natural gas
= 100 billion kWh (based on a 10,000-Btu/kWh heat rate)

WOC = without conservation; WC = with conservation.

Source: Energy Perspectives, U.S. Department of the Interior, February 1975.

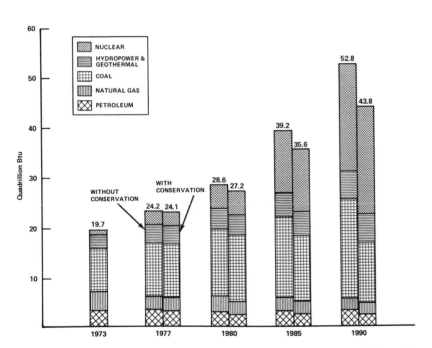

Source: Project Independence Report, 1974; 1985-90 extrapolated.

U.S. energy inputs in the electrical sector, 1973–90 (business as usual, $11 oil).

Source: Energy Perspectives, U.S. Department of the Interior, February, 1975.

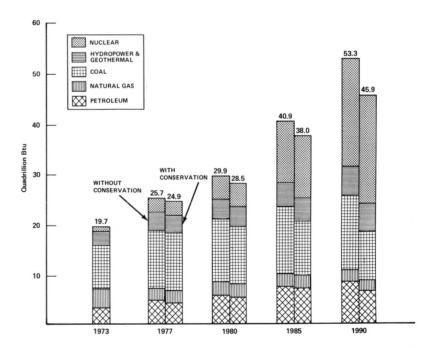

Source: Project Independence Report, 1974, 1985-90 extrapolated.

U.S. energy inputs in the electrical sector, 1973–90 (business as usual, $7 oil).

Source: Energy Perspectives, U.S. Department of the Interior, February 1975.

DOI ENERGY PERSPECTIVES—ELECTRIC POWER CAPACITY

Based on Project Independence and Atomic Energy Commission reports, the Department of the Interior has projected that the generating capacity of the electric utility industry will grow at a rate of 6.7% per year between 1973 and 1985 under business as usual conditions and with oil at $11.00 per barrel. Nuclear capacity will increase from 5 to 22% of the total capacity during that period, combustion turbines will increase from 8 to 18%. Oil-fired steam power will decline from 18 to 9% and gas-fired steam power from 14 to 5% of the total capacity.

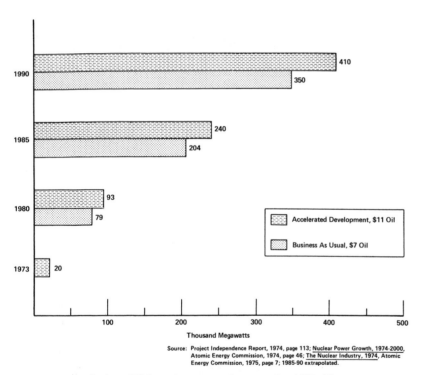

Source: Project Independence Report, 1974, page 113; Nuclear Power Growth, 1974-2000, Atomic Energy Commission, 1974, page 46; The Nuclear Industry, 1974, Atomic Energy Commission, 1975, page 7; 1985-90 extrapolated.

Projected U.S. nuclear power plant capacity, 1973-90.

Source: Energy Perspectives, U.S. Department of the Interior, February 1975.

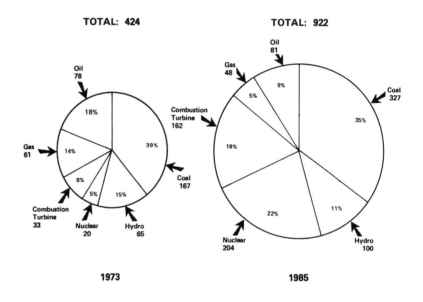

[1] 1 megawatt = 1,000 kilowatts.

Source: Project Independence Report, 1974, page 127.

Capacity of the U.S. electric utility industry, 1973 and 1985 (thousands of magawatts at year end)[1].

Source: Energy Perspectives, U.S. Department of the Interior, February 1975.

DOI ENERGY PERSPECTIVES—CRUDE OIL PRODUCTION

Based on Project Independence data, the Department of the Interior has projected that domestic production of crude oil in 1985 will be 9.2 million barrels per day under business as usual conditions and with oil at $7.00 per barrel, and 16.5 million barrels per day under accelerated supply, $11.00 oil conditions.

These crude oil production figures do not include natural gas liquids which would add 2.3 million barrels per day with oil at $7.00 per barrel and 2.9 million barrels per day with oil at $11.00 per barrel.

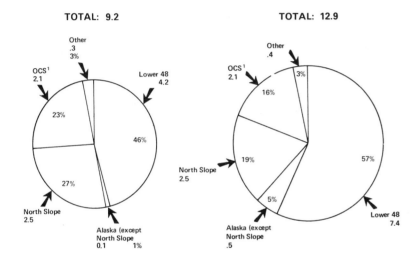

Business as Usual, $7 Oil   Business as Usual, $11 Oil

[1]OCS = Outer Continental Shelf.

Source: Project Independence Report, 1974, page 47.

Projected U.S. crude oil production, 1985 (millions of barrels per day).

Source: Energy Perspectives, U.S. Department of the Interior, February 1975.

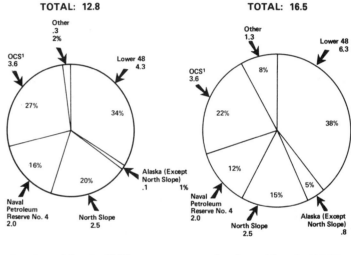

Accelerated Supply, $7 Oil   Accelerated Supply, $11 Oil

[1]OCS = Outer Continental Shelf.

Source: Project Independence Report, 1974, page 47.

Projected U.S. crude oil production, 1985 (millions of barrels per day).

Source: Energy Perspectives, U.S. Department of the Interior, February 1975.

DOI ENERGY PERSPECTIVES—NATURAL GAS PRODUCTION AND CONSUMPTION

Based on Project Independence and Bureau of Mines data, the Department of the Interior has projected that the domestic production of natural gas will increase at a rate of 0.5% per year from 22.6 trillion cubic feet per year in 1973 to 24.1 trillion cubic feet per year in 1985. Onshore production in the lower 48 states is expected to decline from 86 to 66% of the total while offshore production will increase from 14 to 30% of the total.

Consumption of natural gas during the 1973–85 period will closely match domestic production. Relative sector consumption will change, however, with consumption in the industrial sector declining from 48.2 to 41% of the total, and in the electrical sector from 16 to 10% of the total. At the same time, consumption in the household and commercial sector will increase from 33 to 44% of the total.

[Business as Usual, $11 Oil]

Projected U.S. Marketed Natural Gas Production[1]
(Trillions of Cubic Feet)

ITEM	1973	1977		1980		1985	
		$7	$11	$7	$11	$7	$11
LOWER 48 ONSHORE	19.4	16.2	16.6	15.2	15.8	14.6	15.8
LOWER 48 OCS	3.2	4.5	4.5	5.4	5.4	7.2	7.2
NORTH SLOPE	0	0	0	.8	.8	1.6	1.1
TOTAL	22.6	20.7	21.1	21.4	22.0	23.4	24.1

[1]Marketed natural gas refers to gross production less gas used for repressuring and quantities vented and flared.

Source: Energy Perspectives, U.S. Department of the Interior, February 1975.

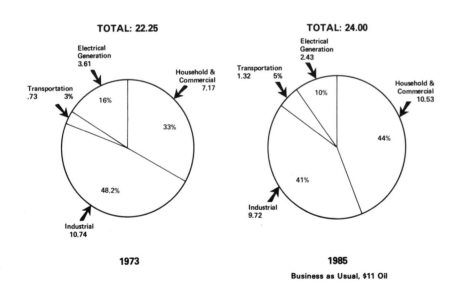

U.S. natural gas consumption by sector, 1973 and 1985 (trillions of cubic feet).

Source: Energy Perspectives, U.S. Department of the Interior, February 1975.

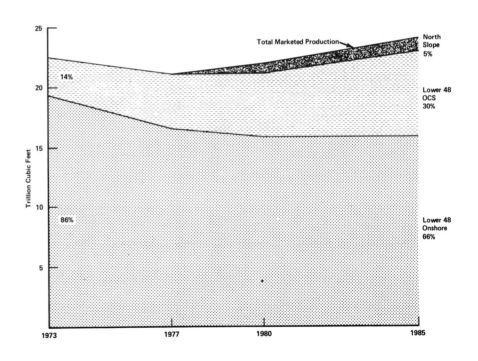

Projected U.S. marketed natural gas production.

Source: Energy Perspectives, U.S. Department of the Interior, February 1975.

DOI ENERGY PERSPECTIVES—SYNTHETIC FUELS

Based on Project Independence and Federal Energy Administration reports, the Department of the Interior has projected the production of synthetic fuels—high Btu gas and oil—from coal and oil shale during the period 1973-90. Under business as susual, $11.00 oil conditions, production of pipeline gas from coal would increase at an annual rate of 29% during the period 1985-90 and reach a level of 1.3 trillion cubic feet per year in 1990. Under accelerated supply, $11.00 oil conditions, a production level of 2.8 trillion cubic feet per year would be reached by 1990.

Under accelerated supply conditions, synthetic high Btu gas would constitute 3% of the total gas consumption in 1985, and synthetic oil production would constitute 2 to 3% of the total oil consumption.

Production of oil from shale is projected to range from 0.450 to 1.60 million barrels per day in 1990 depending upon whether business as usual or accelerated supply policies are adopted.

Projected Production of Synthetic Fuels from Coal

FUEL	TOTAL OUTPUT		NUMBER OF PLANTS	
	BAU	AS	BAU	AS
HIGH-BTU PIPELINE GAS[1] (Billion Cubic Ft./Year):				
1980	6	6	1	1
1985	360	1,300	7	20
1990	1,300	2,800	17	34
LIQUID FUEL[1] (Thousand Bbl/Day):				
1982	---	65	---	1
1985	---	500	---	8
1990	65	1,500	1	18

[1] Given a crude oil price of $11/barrel.   BAU = business as usual; AS = accelerated supply.

Source: Energy Perspectives, U.S. Department of the Interior, February 1975.

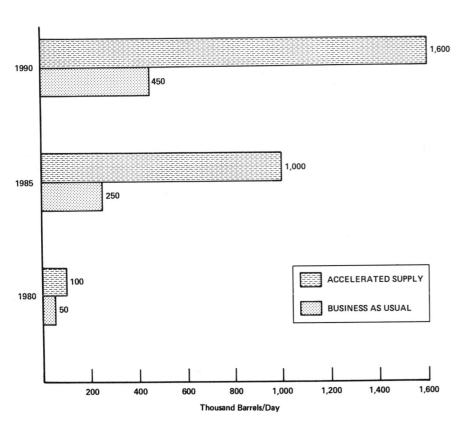

Source: **Potential Future Role of Oil Shale; Prospects and Constraints, Project Independence Report, 1974, page 36.**

U.S. shale oil production potential, 1980–90.

Source: Energy Perspectives, U.S. Department of the Interior, February 1975.

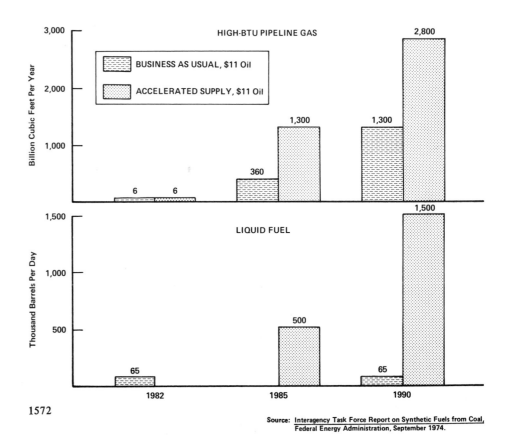

1572

Source: **Interagency Task Force Report on Synthetic Fuels from Coal, Federal Energy Administration, September 1974.**

Projected production of synthetic fuels from coal.

Source: Energy Perspectives, U.S. Department of the Interior, February 1975.

CTAB NATIONAL ENERGY PROGRAM— DEMAND-SUPPLY ESTIMATES

In a report[1] published in February 1975, the Panel on Project Independence Blueprint, Commerce Technical Advisory Board (CTAB), United States Department of Commerce, compared estimates of energy supply and demand in the year 1985 as projected by other organizations. They noted that these estimates ranged from 88 to 125 quadrillion Btu in 1985 while the CTAB panel itself selected 108 quadrillion Btu as a planning base. The Panel concluded that complete independence from imports was not likely by 1985 but that this dependence could be declining if appropriate actions were taken. Assuming these actions are taken, the Panel projected a modest increase in domestic oil and natural gas production (returning to 1972 levels), a rapid increase in coal production and nuclear energy, and some reduction in energy demand as a result of price effects and conservation measures. The recommendations for action of the CTAB Panel are given in 16-30.

[1] *CTAB Recommendations for a National Energy Program*, U.S. Department of Commerce, Commerce Technical Advisory Board, February 1975.

Domestic Reserves of Oil, Natural Gas and Coal

	Oil	Gas	Coal
Proven Recoverable with Current Technology and Economics	40 Billion Bbl.	250 TCF	434 Billion Tons
Years Remaining at Current Use Rates	7 Years	11 Years	700 Years
Total Proven Reserves	300 Billion Bbl.	500 TCF	1600 Billion Tons
Years Remaining at Current Use Rates	50 Years	22 Years	2600 Years
Ultimately Discoverable Reserves	600 Billion Bbl.	900 TCF	3200 Billion Tons
Years Remaining at Current Use Rates	100 Years	40 Years	5200 Years

Source: National Petroleum Council / Federal Energy Administration

Source: CTAB Recommendations for a National Energy Program, U.S. Department of Commerce, Commerce Technical Advisory Board, February, 1975.

Comparisons of 1985 Supply Estimates (10^{15} Btu)

	FEA SUMMARY REPORT (11/74) $7 Oil Base Case	$7 Oil Acc. Supply	$11 Oil Base Case	$11 Oil Acc. Supply	FORD FOUNDATION REPORT (9/74) Historical Growth	Technical Fix	Zero Energy Growth	NATIONAL PETROLEUM COUNCIL 12/72 Case I	Case II	Case III	Case IV	8/74 Medium Case	N.A.E.[1] 5/74	J.C.A.E.[2] 5/74	I.G.T.[3] 12/73	RANGE OF THESE STUDIES Low—High	CTAB PANEL STUDY
Domestic Oil	22.5	29.9	30.7	35.8	32	30	28	31.7	28.5	24.3	21.4	26.5	26.5	23.3	30.7	21.4— 35.8	25
Domestic Natural Gas	23.9	24.7	24.8	25.3	29	27	25	32.9	28.1	21.9	15.5	21.6	27.2	15.1	30.1	15.1— 31.6	22
Nuclear	12.5	14.7	12.5	14.7	10	8	5	36.9	37.6	37.8	16.1	14.2	17.6	19.9	21.8	5 — 21.8	16.5
Coal	19.9	17.7	22.9	20.7	25	16	14	*(see above)*			20.3	19.7	21.2	23.5	22.3	14 — 25	22.7
Synthetic Oil	0	0	0	.1	1.5	0	—	1.5	.2	.2	—	.2	1.3	—	2.8	0 — 2.8	.3
Synthetic Gas	0	0	0	.2	1.5	—	—	2.3	1.2	1.2	.5	1.8	2.3	1.5	3.6	0 — 3.6	3.0
Shale Oil	.6	.6	.6	2.1	2	1	—	1.5	.8	.8	.2	.8	1.0	.2	2.1	0 — 2.1	.5
Hydro-electric	4.8	4.8	4.8	4.8	3	3	3	3.3	3.3	3.3	3.3	3.5	3.2	3.2	4.1	3 — 4.1	3.5
Other—Geo-thermal, Solar	*(see above)*				1	—	2	1.4	.7	.5	.3	.5	.4	.4	—	0 — 2.0	.5
Imports	24.8	17.1	6.5	0	11	7	11	13.4	24.5	34.9	47.3	20.4	—	37.3	7.8	0 — 47.3	14.0
TOTAL SUPPLY	109.1	109.6	102.9	104.2	116	92	88	124.9	124.9	124.9	124.9	108.8	100.7	124.4	125.4	88 —125.4	108.0

Note: The values 36.9, 37.6, 37.8 under National Petroleum Council Cases I, II, III are bracketed spanning Nuclear and Coal. The value 4.8 under all four FEA columns is bracketed spanning Hydroelectric and Other—Geothermal, Solar.

[1] National Academy of Engineering.
[2] Joint Committee on Atomic Energy.
[3] Institute of Gas Technology.
NOTE: Due to rounding, fuel supplies may not add up to totals.

Source: CTAB Recommendations for a National Energy Program, U.S. Department of Commerce, Commerce Technical Advisory Board, February 1975.

CTAB Panel Domestic Demand—Supply Estimates for 1985

DEMAND	Range Quads	Panel's Planning Base Quads
If Historical Trends Continue	125	125
Reduction Due to Price Effects	10—15	13
Reduction Due to Additional Conservation	2— 5	4
Projected Demand	105—113	108

SUPPLY	Range of Panel Estimates (in conventional units)		Conventional Units	Panel's Planning Base (in conventional units)	Panel's Planning Base (in equivalent Quads/yr.)
Coal	1.0	— 1.4	Billion Tons/Yr.	1.2	26
Oil	11	— 13.5	Million Bbl./Day	12	25
Natural Gas	21	— 24	Trillion Cu. Ft./Yr.	22	22
Nuclear	250	— 300	Gigawatts	275	16.5
Shale Oil	0.25	— 0.5	Million Bbl./Day	0.25	0.5
Hydroelectric	58	— 65	Gigawatts	58	3.5
Geothermal	3	— 8	Gigawatts	3	0.2
Solar	0.1	— 0.2	Quads/Yr.	0.1	0.1
Solid Waste	0.1	— 0.5	Quads/Yr.	0.2	0.2
Synthetic Oil	0.1	— 0.2	Million Bbl./Yr.	0.1	*
Synthetic Gas	1.0	— 2.0	Trillion Cu. Ft./Yr.	2.0	*
Total Available Supply	85	— 107	Quads/Yr.		94

BALANCE	Demand	Supply	Imports Required	Cost of Imports at $11 per Barrel
Imports Using Panel's Planning Base	108	94	14 Quads/Yr. (7 MMBPD)	$28 Billion/Yr.
Imports Using Panel's Maximum Expected Demand and Minimum Expected Supply	113	85	28 Quads/Yr. (13 MMBPD)	$52 Billion/Yr.
Imports Using Panel's Minimum Expected Demand and Maximum Expected Supply	105	107	Balanced	—

Notes:

In addition, these demand-supply estimates have been converted in Appendix 6 to equivalent values utilizing other energy measurement units. Conversion factors are given in Appendix 4.
* Included with Coal

The planning base demand and supply levels of the panel are premised upon the immediate implementation of the recommendations contained in this report. It would be imprudent to expect to attain the planning supply levels if programs are not in place during 1975 and any delay will reduce the attainable levels.

Source: CTAB Recommendations for a National Energy Program, U.S. Department of Commerce, Commerce Technical Advisory Board, February 1975.

CTAB NATIONAL ENERGY PROGRAM—RECOMMENDATIONS

In their report[1] of February 1975, the Panel on Project Independence Blueprint of the Commerce Technical Advisory Board noted that, since the embargo of October 1973, domestic production of petroleum had declined 5%, production of coal had remained the same, production of natural gas was down 5%, and imports of petroleum were up 14%. Nuclear power generation was up 20% but, during 1974, more than half of the planned nuclear generating capacity was deferred and some capacity was canceled. The Panel stated that decisions must be made today, and actions taken now, to ensure the Nation's future economic growth with adequate supplies of domestic energy.

Panel Recommendations. To achieve the supply and demand balance that the Panel believes prudent and attainable, immediate positive steps must be taken. Not only must

[1]*CTAB Recommendations for a National Energy Program*, U.S. Department of Commerce, Commerce Technical Advisory Board, February 1975.

the Nation increase the domestic supply of and reduce the demand for energy; but also, it must transfer much of its future energy growth requirements from petroleum and natural gas to coal (through direct burning, gasification and liquefaction) and to nuclear energy. To these ends, the Panel believes that:

- The main underlying principle is the maintenance of a strong, competitive market. The competitive market needs time to absorb and adjust to the significantly higher cost of energy. Due to the urgency of the situation, certain actions may be required, in the short term, to accelerate the supply and demand responses. Mandatory actions, however, must be carefully screened for their effect on the economy.
- The role of the Government is to act as a catalyst and provide a climate for the private sector to achieve the required goals. Elements of this climate include a well-informed public, a clear definition of national energy policies, a commitment to those policies, a stable domestic economy, and the promotion of the ready flow of energy at stable prices in international commerce.

- Laws and regulations should reflect the need for efficient extraction, transportation, and utilization of energy. Increasing the supply of and decreasing the demand for energy is deterred by many existing laws and regulations enacted during a period of time when supplies of energy were readily available and inexpensive.

- The solutions to the long-term energy supply and demand problem must come from technological developments. Research and development must also play a significant role in the nearer-term, providing new techniques for more efficient extraction, transportation, and utilization of energy, and for alleviating impacts on the environment and public health.

- A system to identify supply and demand trends in their early stages is fundamentally necessary to provide a basis for prompt and sound actions. Such a system would monitor the results of higher energy prices, conservation, and measures designed to stimulate domestic energy production; and would provide cost/benefit studies of the economic and social effects of various alternative actions.

With these underlying principles in mind, the Panel has put together a comprehensive and integrated set of recommendations for a National Energy Program. The recommendations constitute a minimum program to reduce energy demand without serious effects on the rest of the economy, to increase domestic supply with due regard to health, safety, and environmental needs, and to monitor the process to assure prompt implementation of other emergency actions if they become necessary.

The Panel makes the following recommendations:

Have the Energy Resources Council coordinate governmental review of this report to obtain immediate action in appropriate areas.

Develop a Public Information Program:

- In order to elicit the full support of the Nation a comprehensive public information program should be initiated immediately. The program should provide a full appreciation and understanding of the

energy problems facing the Country and their possible solutions.

Take Appropriate Energy Conservation Measures:

- Stimulate voluntary conservation by all segments of the Nation.

- Revise transportation regulations to promote the efficient use of energy.

- Develop vehicles with improved fuel economy.

- Promote energy efficiency in residential and commercial buildings, via appropriate standards and incentives.

- Encourage and support industry to generate electric power while producing process steam; and encourage utilities to provide steam to industry from central power stations.

- Encourage all industries to participate in energy conservation programs.

- Stimulate substitution of coal and nuclear energy in the production of electricity and steam. This is necessary to make oil and natural gas available for higher priority uses.

- Study socio-economic effects of possible implementation of accelerated mandatory conservation actions as emergency alternatives.

Increase Domestic Energy Supply:

- Promote increased coal production and utilization.

—Provide an economic, regulatory and environmental climate conducive to the rapid development of coal supply and coal utilization.

—Provide Federal funds to mitigate social and environmental problems of regions impacted by coal production and conversion.

- Initiate measures for increasing oil and natural gas supplies.

—Remove price controls on petroleum and petroleum products and deregulate the price of new natural gas production.

—Pool talent and technology to maximize secondary and tertiary recovery of petroleum.

—Open Naval Petroleum Reserve No. 4 to

exploration and development on the same basis as any other Federal lands.

—Expedite approvals for transportation systems from areas of new oil and gas discovery, particularly for Alaskan natural gas.

- Stimulate nuclear development.

—Address the public concern regarding nuclear plant safety and environmental compatibility.

—Improve the licensing process for nuclear power by simplifying, and shortening, and by eliminating duplication in Federal and State procedures.

—Delineate and evaluate uranium resources and increase enrichment capability.

- Promote synthetic fuels from coal.

—Develop an integrated comprehensive program for a viable synthetics industry.

—Streamline regulatory procedures.

- Expand leasing of Government lands.

—Conduct lease acreage sales in each of the prospective frontier Outer Continental Shelf areas and in the Gulf of Mexico at the maximum rate that can be efficiently assimilated by the industry for exploration and development.

—Modify the Federal Mineral Leasing Act (1920) so that lessees can acquire sufficient resources for large-scale coal development.

—Promote adequate shale leasing acreages.

- Ensure the availability of skilled human resources.

—Expand the support by Federal and State Governments for new technical and degree-level education programs.

—Establish a program of Federal loans and grants to generate employment in labor-intensive energy enterprises.

- Study socio-economic effects of possible implementation of wartime supply provisions as emergency alternatives.

Provide a Balanced Environmental Climate:

- Modify existing Federal air quality regulations to

permit the maximum use of fossil fuels within the limits of health standards.

- Review and modify legislation controlling air and water pollution, where this legislation restricts energy supply growth.

- Streamline site certification procedures for all energy facilities.

- Establish realistic Federal environmental regulations for strip mining.

- Consider cost/benefit trade-offs for all environmental regulations.

Improve the Economic Climate for Energy Production:

- Revise tax laws pertaining to dividends, capital gains, and capital losses to attract additional capital.

- Revise tax laws to increase investment tax credit and increase allowed tax depreciation to improve corporate cash flows.

- Review and modify Federal and State rules and regulations for rate proceedings to reduce delays and increase return on equity.

- Strengthen the mechanisms by which bond and equity securities are marketed.

Strengthen and Expand Research and Development Programs:

- Have Government assume primary responsibility for funding long-term programs with high risk but potentially high payout including breeder reactors, solar energy, and fusion.

- Increase Federal funding for research and development programs related to improved coal mining, coal conversion and utilization, advanced power generation methods, radioactive waste disposal, and energy related public health effects.

- Provide Government assistance for construction and operation of at least the following advanced commercial-scale demonstration plants: coal gasification (2 plants), coal liquefaction (2 plants), and oil shale (1 plant).

- Support further development of computerized energy information systems to assist in the evaluation of future energy policy issues.

EEI ECONOMIC GROWTH—POPULATION AND GNP

In June, 1975, the Policy Committee on Economic Growth, Pricing and Energy Use of the Edison Electric Institute, published a report on "Economic Growth in the Future." This study encompassed three growth scenarios. The High Growth Scenario assumed a return to the fertility rate of 2.5 of the 1960s such that the United States population in the year 2020 would be about 350 million and would continue to grow (to over 2 billion by the year 2200). The High Growth scenario also assumed an annual increase of 4.2% in the Gross National Product, slightly below the 4.3% growth experienced during the 1960s.

The Moderate Growth scenario assumed a fertility rate at the replacement level of 2.1 births per woman with the population leveling off at 330 million. The annual increase in GNP was projected to be 3.5 to 3.7% per year, the average increase in energy demand was projected to be 3.0% per year, and the average increase in electric energy demand was projected to be 5.3-5.8% per year.

The Low Growth scenario assumed a fertility rate of 1.8 births per woman with population peaking at 270 million, an annual increase in GNP of 2.3%, and an average increase in electric energy demand of 3.7%.

The Committee concluded that the most likely scenario was that of moderate growth, representing a mid position between a policy of high growth as a continuation of historic trends with no serious social or physical resource constraints, and that of low growth which would require a national policy objective to achieve a steady-state society by imposing increasingly rigid constraints on the use of energy and materials.

U.S. population.

Source: Economic Growth in the Future, Report of the EEI Committee on Economic Growth, Pricing and Energy Use, Edison Electric Institute, 90 Park Avenue, New York, N.Y. 10016, June 1975.

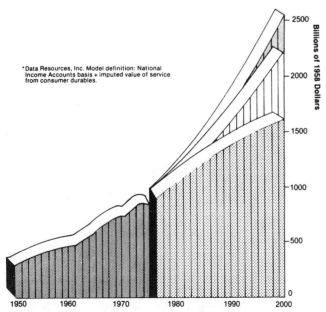

U.S. gross national product (Data Resources, Inc. Model definition: National Income Accounts basis + imputed value of service from consumer durables).

Source: Economic Growth in the Future, Report of the EEI Committee on Economic Growth, Pricing and Energy Use, Edison Electric Institute, 90 Park Avenue, New York, N.Y. 10016, June 1975.

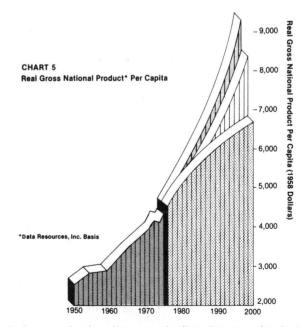

Real gross national product per capita (Data Resources, Inc. Basis).

Source: Economic Growth in the Future, Report of the EEI Committee on Economic Growth, Pricing and Energy Use, Edison Electric Institute, 90 Park Avenue, New York, N.Y. 10016, June 1975.

EEI ECONOMIC GROWTH—ENERGY SUPPLY AND DEMAND

In its study of *Economic Growth in the Future*, the Committee on Economic Growth, Pricing and Energy Use of the Edison Electric Institute, came to the following conclusions with regard to energy supply and demand.

Energy Demand. Energy demand will continue to grow at a slower pace than real GNP with a 3% per year increase over the next 25 years under the moderate economic growth case in contrast to a growth rate of 3.7% for real GNP. Growth in the consumption of electric energy will be much faster than for energy as a whole with a projection of 5.3 to 5.8% annual growth for the moderate growth case. Even under the low-growth conditions, electric energy demand would grow at a rate of 3.7% per year. Consumption of energy sources for the production of electricity will constitute between 45 and 50% of all energy consumption in the year 2000 in contrast to 27% in 1975.

Energy Supply. Fuel consumption patterns will shift over the next 25 years from dominance of oil and natural gas to an increasing use of coal and nuclear power. Oil and natural gas presently provide 77% of the nation's energy with coal, hydropower and nuclear fuel providing the remainder. The share provided by coal, hydro, and nuclear is expected to be over 50% by the end of the century. The potential contribution from geothermal, tides, winds, and trash are small, although geothermal may be locally significant. Solar and fusion power are not likely to make major contributions until well into the next century.

Energy imports under moderate growth conditions are projected to remain at about 18% of the total demand until 1980 and gradually decline to about 10% by the year 2000. A 10 to 12% level of imports is judged to be a reasonable balance between risks of interruption of imported energy and the costs of providing additional domestic supplies. In the low-growth case, domestic supplies are expected to be limited because of environmental constraints on resource development. In the low-growth scenario, imports are projected as rising to 24% of total consumption in 1980, staying at that level through 1985 and then declining to 18% by the year 2000.

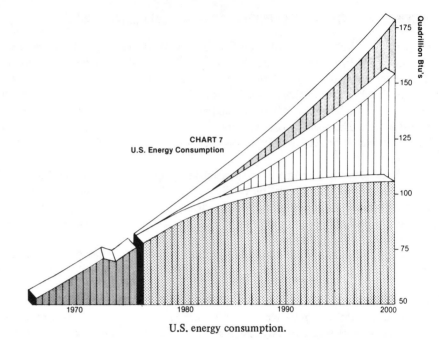

CHART 7
U.S. Energy Consumption

U.S. energy consumption.

Source: Economic Growth in the Future, Report of the EEI Committee on Economic Growth, Pricing and Energy Use, Edison Electric Institute, 90 Park Avenue, New York, N.Y. 10016, June 1975.

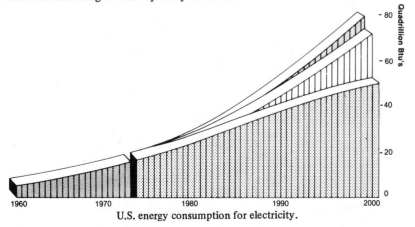

U.S. energy consumption for electricity.

Source: Economic Growth in the Future, Report of the EEI Committee on Economic Growth, Pricing and Energy Use, Edison Electric Institute, 90 Park Avenue, New York, N.Y. 10016, June 1975.

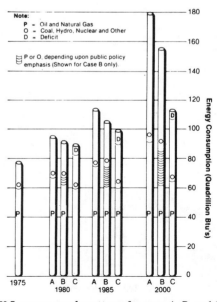

Note:
P = Oil and Natural Gas
O = Coal, Hydro, Nuclear and Other
D = Deficit

P or O, depending upon public policy emphasis (Shown for Case B only).

U.S. energy supply patterns for cases A, B, and C.

Source: Economic Growth in the Future, Report of the EEI Committee on Economic Growth, Pricing and Energy Use, Edison Electric Institute, 90 Park Avenue, New York, N.Y. 10016, June 1975.

EEI ECONOMIC GROWTH—CONCLUSIONS AND RECOMMENDATIONS

In its study of economic growth in the future, the Policy Committee on Economic Growth, Pricing and Energy Use of the Edison Electric Institute came to the following conclusions and recommendations:

- Potential limiting factors can be managed so that desirable forms of economic growth can be sustained by the United States for the foreseeable future. Under policies which provide a balance of economic and environmental needs, continued growth at moderate rates is both feasible and desirable. Under such policies an average growth rate for real GNP of 3.5 to 3.7 percent per year could be expected over the next 25 years. Even under policies intended to curtail growth and achieve a "steady state" as rapidly as possible, the transition would take 10 to 15 years and the resulting annual increase in real GNP over the next 25 years would average about 2.3 percent.

- Continuing economic growth will improve the "quality of life" as well as the standard of living. Growth is necessary to: aid in maintaining employment; improve the conditions of the poor; finance cleaning up the environment; and maintain a favorable position in international relations, particularly with regard to trade, balance of payments, and defense.

- An attempt to halt growth arbitrarily would require authoritarian measures with a commensurate loss of individual freedom.

- United States economic growth over the next 25 years is likely to be constrained more by shortages of capital than by shortages of energy or raw materials. Revisions in public policy are needed to encourage capital formation and private investment in productive facilities. Government deficits are a primary cause of inflation and serve to preempt capital funds needed for investment in productive facilities.

- Improvement in environmental quality can be made consistent with economic growth through the use of environmental controls supplemented by economic incentives. Environmental standards should be based on cost-benefit analyses. Once established, environmental regulations should be stabilized and not subjected to frequent revisions.

- With the birth rate in the United States currently below the replacement rate and declining, there is no cause for concern about population exceeding the capacity of the nation to accommodate everyone. Even if birth rates shift back slightly above the replacement rate, the nation's agricultural capacity would be more than adequate.

- Energy demand will grow at a slower rate than GNP. Under conditions of moderate economic growth and increasing environmental investment, energy growth can be expected to average 3 percent per year over the next 25 years.

- Oil and natural gas cannot provide the basis for long-term economic growth. To achieve energy independence it will be necessary to make a basic shift from oil and natural gas to coal and nuclear fuels.

- Electric energy consumption would increase at an average rate of 5.3 to 5.8 percent per year under policies assumed for a moderate-growth scenario. Electric energy will grow as a result of the shift from oil and gas to coal and nuclear fuels. As is also the case of total energy consumption, electric energy growth rates will vary significantly over time and from region to region.

- An ultimate solution in regard to supplies of non-energy mineral resources requires deep-sea mining and the ability to mine "average rock" for its varied but low-grade mineral contents. These solutions are not yet technologically feasible. Recycling possibilities lessen, but do not eliminate, the concern about this situation. Thus, mineral exploration, mining, and refining warrant increased R & D effort now to avoid or minimize later problems.

- The national energy council should be continued and strengthened to give it authority and responsibility for United States energy policy.

- Primary reliance should be placed on a combination of free-market prices and improving technology to make available increasing quantities of basic natural resources and discourage wasteful uses.

- Oil and gas wellhead prices should be deregulated as rapidly as possible without causing major disruptions in the domestic economy such as resulted from the abrupt oil price increases in 1973-74.

- Agricultural output should be maximized so that food can be sold in world markets to meet the needs of developing nations and to help overcome the United States balance-of-payments problems arising from the importing of oil and other raw materials.

- The study also contains a number of conclusions and recommendations relative to the electric utility industry. To accomplish the shift to coal and nuclear fuels, electric utilities will have to increase their construction activities. Unless steps are taken to increase the present level of internally generated funds, almost two-thirds of the capital needed to fund this construction must be raised in the nation's competitive capital markets. In the next 15 years the industry will need approximately $500 billion from the capital markets. One hundred billion or more of this will have to come through newly issued common stock. These amounts can be raised only if the industry is financially healthy. An average return on common equity of 15 percent or more will be needed to demonstrate financial health, even if the rate of inflation subsides to the 4 or 5 percent per year range. Higher returns will be needed by some companies, depending on individual corporate financial circumstances. Continued high inflation will also require higher returns on equity.

  Recommended changes in regulatory policy which will promote financial health include: (1) use of forward test years; (2) allowing tariff changes to become effective without suspension, subject to later refund; (3) increasing book depreciation rates and normalizing tax deferrals; and (4) inclusion of construction work in progress in the rate base as construction proceeds. The primary objectives in electric rate structure design should be to approximate as closely as feasible the cost to serve each customer, and to provide sufficient revenues to cover all necessary costs of doing business including adequate return on investment.

Source: Economic Growth in the Future—Executive Summary, Report of the EEI Committee on Economic Growth, Pricing and Energy Use, Edison Electric Institute, 90 Park Avenue, New York, N.Y. 10016, June 1975.

ERDA ENERGY CHOICES—SUMMARY

In its study of *A National Plan for Energy Research, Development and Demonstration*, the Energy Research and Development Administration has outlined the options available for the development of new and improved energy technologies. A partial summary of the ERDA plan follows.

A serious and continuing energy problem exists in this country.

Imports, in the form of petroleum, petroleum products, and natural gas account for 20 percent of the total domestic energy consumption, at an annual cost of over $25 billion in 1974.

This heavy reliance on imported energy has serious national security implications. Dependence on imports makes the United States vulnerable to undesirable external influences on U.S. foreign and domestic policy. Foreign powers can threaten life styles and economic stability by curtailing the supply of petroleum or effecting arbitrary and sudden price changes. The quadrupling of the world price of petroleum in the past two years has disrupted the U.S. economy and the economies of all other importing nations.

This Plan recognizes five national policy goals as a focus for energy policy:

- To maintain the security and policy independence of the Nation.
- To maintain a strong and healthy economy, providing adequate employment opportunities and allowing fulfillment of economic aspirations (especially in the less affluent parts of the population).
- To provide for future needs so that life styles remain a matter of choice and are not limited by the unavailability of energy.
- To contribute to world stability through cooperative international efforts in the energy sphere.

National Energy R, D & D Goals

> I. Expand the domestic supply of economically recoverable energy producing raw materials
>
> II. Increase the utilization of essentially inexhaustible domestic energy resources
>
> III. Efficiently transform fuel resources into more desirable forms
>
> IV. Increase the efficiency and reliability of the processes used in the energy conversion and delivery systems
>
> V. Transform consumption patterns to improve energy utilization
>
> VI. Increase end-use efficiency
>
> VII. Protect and enhance the general health, safety, welfare and environment related to energy
>
> VIII. Perform basic and supporting research and technical services related to energy

Source: A National Plan for Energy Research, Development and Demonstration: Creating Energy Choices for the Future, ERDA-48, Volume 1: The Plan, Energy Research and Development Administration, June 1975.

- To protect and improve the Nation's environmental quality by assuring that the preservation of land, water, and air resources is given high priority.

The national energy problem is best understood as one of limited choices today.

- The U.S. energy system currently relies most on the least plentiful domestic energy resources, and least on the most abundant resources.
- Over 75 percent of the Nation's energy consumption is based on petroleum and natural gas. Domestic supplies of these commodities are dwindling.
- Coal, the most abundant domestic fossil fuel, provides less than 20 percent of current energy needs.
- Uranium, the domestic energy source with the greatest energy potential, provides about 2 percent of the Nation's energy.
- Solar energy, available to all, but diffuse, provides a negligibly small percentage of current needs.

To overcome this problem and to achieve our National policy goals, the Nation must have the flexibility of a broad range of energy choices.

To generate the necessary options, the Plan is designed to facilitate the changeover from dependence on a narrow base of diminishing domestic resources to reliance on a broader range of less limited or unlimited alternatives.

To accomplish this transition a framework of national energy technology goals has been established.

All the national energy technology goals must be pursued together. Concentration on only one or a few technological avenues is not likely to solve the energy problem.

Based upon an analysis of scenarios, the status of the candidate technologies, and the extent of the resources they would use, a national ranking of R,D&D technologies has been developed to identify priorities for emphasis in the Plan.

The ranked list is presented in the table on page 669.

For the near-term (now to 1985) and beyond, the priorities are:

- To preserve and expand major domestic energy systems: coal, light water reactors (the highest nuclear priority), and gas and oil both from new sources and from enhanced recovery techniques.
- To increase the efficiency of energy used in all sectors of the economy and to extract more usable energy from waste materials.

For the mid-term (1985–2000) and beyond, priorities are:

- To accelerate the development of new processes for production of synthetic fuels from coal and for extraction of oil from shale.
- To increase the use of under-used fuel forms, such as geothermal energy, solar energy for heating and cooling, and extraction of more usable energy from waste heat. None of these technologies has a major long-term impact, but each can be quite useful in relieving mid-term shortages.

For the long-term (past 2000), priorities are:

National Ranking of R, D & D Technologies

Near-Term Major Energy Systems	Coal—Direct Utilization in Utility/Industry Nuclear—Converter Reactors Oil and Gas—Enhanced Recovery	Highest Priority Supply
New Sources of Liquids and Gases for the Mid-Term	Gaseous & Liquid Fuels from Coal Oil Shale	
"Inexhaustible" Sources for the Long-Term	Breeder Reactors Fusion Solar Electric	
Near-Term Efficiency (Conservation) Technologies	Conservation in Buildings & Consumer Products Industrial Energy Efficiency Transportation Efficiency Waste Materials to Energy	Highest Priority Demand
Under Used Mid-Term Technologies	Geothermal Solar Heating and Cooling Waste Heat Utilization	Other Important Technologies
Technologies Supporting Intensive Electrification	Electric Conversion Efficiency Electric Power Transmission and Distribution Electric Transport Energy Storage	
Technologies Being Explored for the Long-Term	Fuels from Biomass Hydrogen in Energy Systems	

Source: A National Plan for Energy Research, Development and Demonstration: Creating Energy Choices for the Future, ERDA-48, Volume 1: The Plan, Energy Research and Development Administration, June 1975.

- To pursue vigorously those candidate technologies which will permit the use of essentially inexhaustible resources:
  - —Nuclear breeders.
  - —Fusion.
  - —Solar electric energy from a variety of technological options, including wind power, thermal and photovoltaic approaches, and use of ocean thermal gradients.

    None of the above three technologies is assured of large scale application. All have unique unresolved questions in one or more areas: technical, economic, environmental or social. The benefits to be gained in achieving success in one or more of these approaches require that vigorous development efforts proceed now on all three.
- To provide the technologies to use the new sources of energy which may be distributed as electricity, hydrogen or other forms throughout all sectors of the economy.

    (As an example, long term efforts are needed to develop a full range of electric vehicle capabilities.)

Substantial effort is required now if the significant energy contributions defined above are to become available in the mid- and long-term as needed.

It should be noted that outlays for Federally supported programs may not necessarily conform to the national ranking developed here. This is because many of the technologies will be developed in the private sector and there are differences in the scope of the program effort and the extent of development required.

The above priority ranking and accompanying Plan itself reflect ERDA's determination that five major changes are needed in the nature and scope of the Nation's energy R,D&D program.

These changes, which must be made rapidly and simultaneously and many of which are already reflected in the President's program for 1976, are:

- *Emphasis on overcoming the technical problems inhibiting expansion of high leverage existing systems—notably coal and light water reactors.*

    Achieving an expansion requires the solution of several critical problems involving operational reliability and acceptable environmental impact.
- *An immediate focus on conservation efforts.*

    These efforts implement first generation existing technology, extend this technology with improved capabilities, demonstrate its viability and widely disseminate the results.

    The primary targets are automotive transportation, buildings and industrial processes.
- *Acceleration of commercial capability to extract gaseous and liquid fuels from coal and shale.*

    A two-pronged effort is needed to achieve this objective. Existing technologies must be implemented as soon as possible to gain needed experience with large scale synthetic fuel production. A Synthetic Fuels Commercialization program is now being developed to implement the President's synthetic fuels goal announced in the 1975 State of the Union Message. Also required is aggressive pursuit of parallel efforts, now underway, to develop a more efficient generation of plants with lower product costs and less environmental impact.
- *Inclusion of the solar electric approach among the "inexhaustible" resource technologies to be given high priority.*

    The technologies for producing essentially inexhaustible supplies of electric power from solar energy will be given priority comparable to fusion and the breeder reactor.
- *Increased attention to under-used new technologies that can be rapidly developed.*

    The technologies that are close to implementation and promise a significant impact for the mid-term and beyond are principally solar heating and cooling and the use of geothermal power.

To attain the national energy goals, it is necessary not only to demonstrate the technical feasibility of new energy systems but to ensure that the environmental, health, and safety aspects of these systems are socially acceptable. This will require that environmental effects assessment be initiated early in the R,D&D process and that environmental and safety controls be developed as an integral part of energy system design. Ensuring social acceptability demands vigorous program over-view and assessment, open reporting of findings and progress, and frequent public interaction on the part of the R,D&D establishment.

To assist in the development of the energy supply technologies assigned priority in the Plan, supporting technologies are also required.

Implementation of the National Plan for energy R,D&D will require coordination and cooperation among all sectors of the society.

ERDA ENERGY CHOICES—SCENARIOS

In the development of its *National Plan for Energy Research Development and Demonstration*, the Energy Research and Development Administration analyzed a number of different scenarios for meeting national energy requirements. The assumptions behind these scenarios and the conclusions drawn from them follow.

A number of strategies have been advanced to solve the energy problem. The first is to place primary national emphasis on reduction of energy waste and inefficiencies to ease supply problems. The second is to put primary emphasis on the use of the vast energy residing in the Nation's coal and oil shale resources to produce synthetic fuels that will substitute directly for diminishing supplies of oil and gas. The third is to emphasize the alteration of consumption patterns, shifting from reliance on petroleum and gas to reliance on electricity, which can be provided from all the domestically abundant energy sources.

To derive a fuller perspective, these strategies need to be contrasted with views of the future in which (a) no significant new initiatives are undertaken, (b) a key technology (such as nuclear power) is eliminated from consideration, and (c) some combination of all of the primary responses is assumed to have a high—even unrealistically high—degree of success. ERDA has examined all six strategies called scenarios in this report:

Scenario 0	No New Initiatives
Scenario I	Improved Efficiencies in End-Use
Scenario II	Synthetics from Coal and Shale
Scenario III	Intensive Electrification
Scenario IV	Limited Nuclear Power
Scenario V	Combination of All Technologies

Analysis of these scenarios focuses on drawing forth insights on the nature of:

- The energy system itself, viewed as a system.
- The role of technologies within the system.
- How the above characteristics change with time.

It should be emphasized that the scenarios are not forecasts or predictions. They are illustrations of possible strategies—"paper and pencil experiments."

The same demand for energy services was used as a basis for all scenarios. The demand assumes continuation of historical trends by use sector modified to reflect recent price increases. The appropriate technology mix of each scenario, together with the oil and gas production estimates shown in Figures 1 and 2, leads to estimates of the amounts of imported fuel needed to satisfy demand. Results are shown in Figure 5.

The import levels for the primary Scenarios I, II and III are unacceptably high in the year 2000, representing an increase over today's levels and in some cases an accelerating increase. These levels reflect, of course, the result of emphasizing only a single set of technological approaches to deal with the energy problem. The analysis suggests that:

- All scenarios, except V, are unacceptable individually: they show increasing imports. That is, only the successful development and implementation of a large

Scenario 0	No New Initiatives
Scenario I	Improved Efficiencies in End-Use
Scenario II	Synthetics from Coal and Shale
Scenario III	Intensive Electrification
Scenario IV	Limited Nuclear Power
Scenario V	Combination of All Technologies

Inputs for Scenario 0—No New Initatives

Supply Assumptions

- Oil and gas production draws on remaining recoverable domestic resources
  - According to lower estimates by the U.S. Geological Survey (1975) and the National Academy of Sciences (see Figures 2–1 and 2–2)
  - Without tertiary or other new recovery
- Coal and nuclear converter reactors continue to expand to meet electricity demand, limited by ability to construct or convert plants
- Other energy sources (e.g., geothermal, hydroelectric, and urban wastes) expand according to historic projections of existing technologies which do not reflect recognition of a serious energy problem.

Demand Assumptions

- Current consumption patterns continue with no improvement in residential, commercial, or industrial end-use and most transportation efficiencies
- A 40 percent efficiency improvement for energy use in automobiles is realized by 1980 because of a trend toward smaller autos.

Inputs for Scenario I—Improved Efficiencies in End-Use

Supply Assumptions*

- Domestic oil and gas production is increased above the base case (Scenario 0) by new enhanced recovery technologies
- Solar heating and cooling are introduced
- Geothermal heat is used for process and space heating
- Waste materials are employed as fuels or are recycled to save net energy in production.

Demand Assumptions*

- Residential and commercial sector technologies are improved with regard to
  - The structure itself in order to reduce heating and cooling requirements
  - Improved air conditioners, furnaces, and heat pumps
  - Appliances and consumer products
- Industrial process efficiency improvements are achieved in
  - Process heat and electric equipment
  - Petrochemicals
  - Primary metals
- Efficiencies of electricity transmission and distribution are increased
- Improved transportation efficiencies derived from new technologies (in contrast to efficiencies from smaller vehicles) are assumed for land and air transportation
- Waste heat (e.g., from electric generation) is employed for other low-grade uses now requiring separate energy input.

* Other assumptions are essentially those of Scenario 0.

Inputs for Scenario II—Synthetic Fuels from Coal and Shale

Supply Assumptions*

- Substantial new synthetic fuels production is introduced from
  - Coal
  - Oil Shale
  - Biomass
- Enhanced oil and gas recovery levels of Scenario I are included
- Under-used solar, geothermal, and waste sources included in Scenario 0 are not included here.

Demand Assumptions*

- No end-use efficiency improvements are assumed.

* The assumptions, unless otherwise stated, are those of the previous scenarios to ensure that comparisons are being made only of the impacts of stated energy options.

Inputs to Scenario III—Intensive Electrification

Supply Assumptions*

- Electric power is intensively generated by coal and nuclear power as in prior scenarios
- New technology energy sources are introduced as available to generate electricity
  - Breeder reactors
  - Solar electric (wind, thermal, photovoltaics and ocean thermal)
  - Fusion
  - Geothermal electric
- A minimal contribution is assumed from waste materials (as in Scenario 0).

Demand Assumptions*

- Improved electric conversion efficiencies are introduced
- Widespread use of electric autos begins
- Technologies to improve efficiency of electricity transmission and distribution are implemented

* Supply assumptions are consistent with Scenario I and demand assumptions with Scenario 0, unless otherwise stated.

Inputs to Scenario IV—Limit on Nuclear Power

Supply Assumptions

- Converter reactor energy levels are constrained to 200,000 megawatts electric
- Coal electric is at the levels in other scenarios to permit coal to be employed for synthetics
- Additional sources of electricity depend on
  - Accelerated geothermal development (more than a factor of two over Scenario III)
  - Accelerated solar development (a factor of two over Scenario III)
  - Fusion as in Scenario III
- Solar and geothermal heating are used (as in Scenarios I and III)
- Synthetic fuels are produced from coal, shale, and biomass at the level of Scenario II.

Demand Assumptions

- Industrial efficiency aspect of conservation scenario (Scenario I) is included
- Electric transmission efficiencies are not included, as electricity use grows too slowly to justify changes.

Source: A National Plan for Energy Research, Development & Demonstration: Creating Energy Choices for the Future, Volume 1: The Plan, ERDA-48, Energy Research and Development Administration, June, 1975.

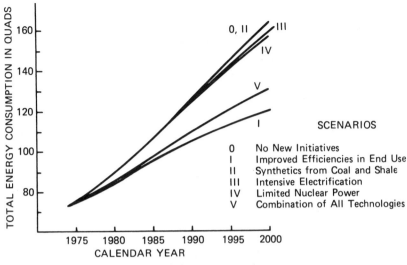

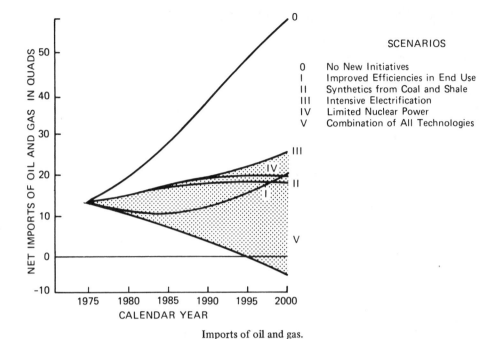

Imports of oil and gas.

Source: A National Plan for Energy Research, Development & Demonstration: Creating Energy Choices for the Future, Volume 1: The Plan, ERDA-48, Energy Research and Development Administration, June, 1975.

number of technologies in a combination of approaches can make importing fuel a matter of choice.

● Curtailment of any major existing option (such as nuclear power) places heavy demands on all the remaining options and precludes an acceptable solution (low level of imports or no imports).

The target area for R,D&D contribution is represented by the shaded area in Figure 5, bounded on the top by the curves for the primary Scenarios I through III and bounded below by the extensive technological success assumed in Scenario V.

The actual future levels of imports will depend not only on the technological results within the above spectrum but also upon:

● The actual amount of oil and gas found and produced in the United States.
● The actual life style (demand for services) either chosen for the future or forced upon the public by a continuing energy supply problem.

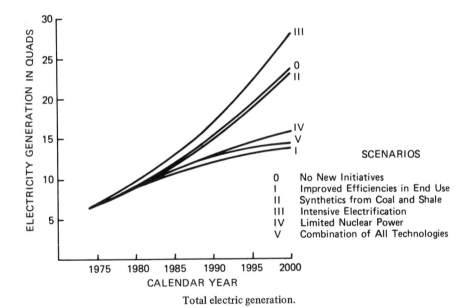

Total electric generation.

Source: A National Plan for Energy Research, Development & Demonstration: Creating Energy Choices for the Future, Volume 1: The Plan, ERDA-48, Energy Research and Development Administration, June, 1975.

Total energy consumption.

TIME SCALE FOR ENERGY DEVELOPMENT

One of the important aspects of planning for energy development to meet expected energy demand is the lead time required for the construction of facilities to extract energy resources and convert these resources into useful forms. The production and plant lead times indicated in the accompanying charts are for well-established technologies for which an industrial base exists. In the case of new technologies, these production and plant lead times may be preceded by many years of research, development and demonstration.

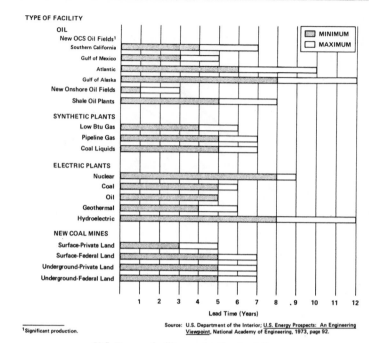

U.S. Energy facility production lead times.

Source: Energy Perspectives, U.S. Department of the Interior, February 1975.

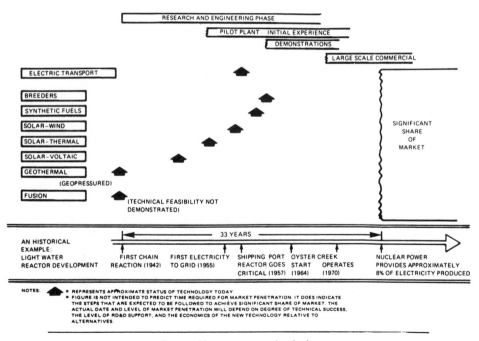

Status of longer range technologies.

Source: A National Plan for Energy Research, Development and Demonstration: Creating Energy Choices for the Future, ERDA 76-1, Energy Research and Development Administration, April 15, 1976.

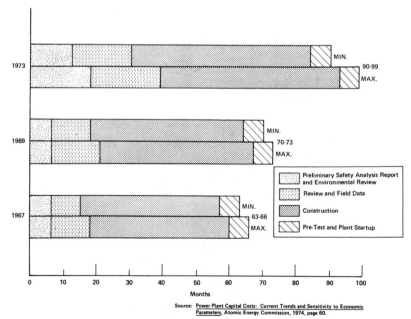

Nuclear power plant lead time trends.

Source: Energy Perspectives, U.S. Department of the Interior, February 1975.

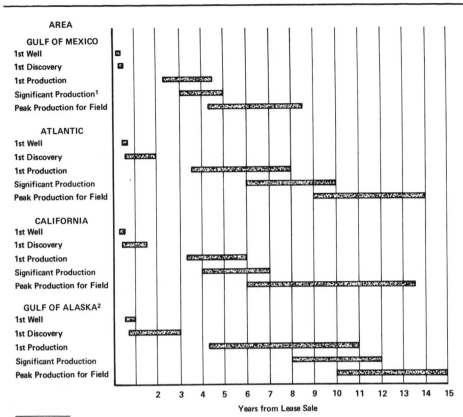

AREA

GULF OF MEXICO
1st Well
1st Discovery
1st Production
Significant Production[1]
Peak Production for Field

ATLANTIC
1st Well
1st Discovery
1st Production
Significant Production
Peak Production for Field

CALIFORNIA
1st Well
1st Discovery
1st Production
Significant Production
Peak Production for Field

GULF OF ALASKA[2]
1st Well
1st Discovery
1st Production
Significant Production
Peak Production for Field

Years from Lease Sale

[1] Significant production exists when gathering systems and pipelines are constructed and connected to OCS wells.

[2] Peak production for an entire area will take 10 to 25 years.

Source: U.S. Geological Survey, 1975.

OCS production lead times.

Source: Energy Perspectives, U.S. Department of the Interior, February 1975.

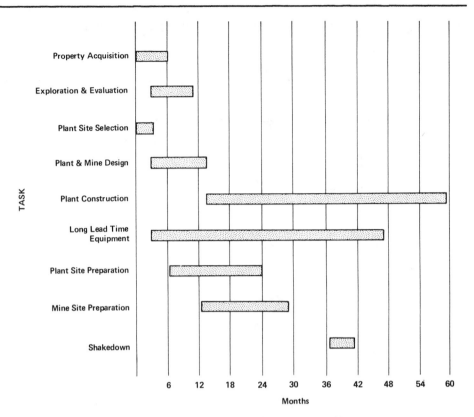

TASK

Property Acquisition
Exploration & Evaluation
Plant Site Selection
Plant & Mine Design
Plant Construction
Long Lead Time Equipment
Plant Site Preparation
Mine Site Preparation
Shakedown

Months

Source: Potential Future of Oil Shale: Prospects and Constraints, Project Independence Report, 1974, page 301.

Shale oil production lead times, concurrent activities schedule.

Source: Energy Perspectives, U.S. Department of the Interior, February 1975.

U.S. ENERGY RESEARCH AND DEVELOPMENT FUNDING

Federal Energy R&D Funding

	Agency	Fiscal year 1970	1971	1972	1973	1974
Coal:						
Resources development		30.4	49.0	73.5	94.5	119.9
Production and utilization R. & D. including gasification, liquefaction, and MHD.	DOI, OCR	13.5	18.8	30.3	43.5	52.5
	DOI, BOM	13.2	15.4	14.7	19.8	18.1
Mining health and safety research	DOI, BOM	3.7	14.8	28.5	31.2	28.3
Interior central fund (part)	DOI					21.0
Petroleum and natural gas		8.8	11.5	12.9	12.8	9.1
Petroleum extraction technology	DOI, BOM	2.7	2.7	3.2	3.1	3.1
Nuclear gas stimulation	AEC	3.7	6.1	7.1	7.2	4.0
Oil shale	DOI, BOM	2.4	2.7	2.6	2.5	2.0
Nuclear fission		283.4	295.2	358.0	412.0	475.4
Liquid metal fast breeder reactor	AEC	144.3	167.9	236.0	269.0	320.0
	TVA			.2	3.0	3.0
Other civilian nuclear power	AEC	108.5	96.6	86.8	98.0	90.5
Nuclear materials process development	AEC	30.6	30.7	35.0	42.0	61.9
Nuclear fusion		37.5	42.2	52.8	65.5	88.5
Magnetic confinement	AEC	34.3	32.2	33.3	39.6	47.3
Laser	AEC	3.2	10.0	19.5	25.9	41.2
Solar energy	NSF			1.7	4.2	12.2
Geothermal energy		.2	.2	1.4	3.4	4.1
	NSF			.7	.7	1.4
	DOI–GS	.2	.2	.7	2.5	2.5
	DOI–BOM				.2	.2
Electrical generation, transmission and storage.			1.3	2.2	4.9	4.1
	NSF		.5	1.3	2.4	.9
	DOI		.8	.9	1.0	1.0
	AEC				1.5	2.2
Control technology (stationary sources)				28.6	38.1	47.5
Air pollution control technology	EPA	19.8	17.4	24.5	29.5	21.5
SOX removal	TVA			1.1	3.0	18.0
Thermal effects	EPA	.8	.6	.7	1.0	1.0
	AEC	1.5	1.8	2.3	4.6	7.0
Miscellaneous				6.3	6.9	11.0
Systems and resource studies	NSF			4.4	5.3	5.3
Energetics research	NSF			1.9	1.6	1.7
Interior central fund (part)	DOI					4.0
Total research and development		382.4	419.2	537.4	642.3	771.8
AEC		326.1	345.3	420.0	487.8	574.1
EPA		20.6	18.0	25.2	30.5	22.5
NSF			.5	10.0	14.2	21.5
DOI		35.7	55.4	80.9	103.8	132.7
TVA				1.3	6.0	21.0

Agency codes: AEC—Atomic Energy Commission; DOI, BOM—Department of the Interior, Bureau of Mines; DOI, GS—Department of the Interior, Geological Survey; DOI, OCR—Department of the Interior, Office of Coal Research; NSF—National Science Foundation; TVA—Tennessee Valley Authority.

Source: Fiscal Policy and the Energy Crisis, Hearings before the Subcommittee on Energy, Committee on Finance, United States Senate, January 23 and 24, 1974.

Privately-Owned Electric Utilities in the United States Research and Development—1974

	1974 Amount	1974 Percent	1973 Amount	1973 Percent
WITHIN THE COMPANIES				
Power plants:				
Hydroelectric:				
Recreation, fish and wildlife	$ 790,453	0.3	$ 580,349	0.2
Other hydroelectric	342,461	0.1	378,454	0.2
Fossil-fuel steam	48,086,894	19.6	76,184,096	31.8
Internal combustion or gas	520,343	0.2	170,174	0.1
Nuclear	12,017,659	4.9	18,329,870	7.7
Unconventional generation	1,481,244	0.6	1,629,914	0.7
Siting and heat rejection	3,629,639	1.5	5,749,539	2.4
System planning, engineering and operation	5,406,500	2.2	6,352,783	2.6
Transmission:				
Overhead	4,158,240	1.7	3,645,197	1.5
Underground	2,092,613	0.9	2,858,175	1.2
Distribution	1,563,909	0.6	2,446,449	1.0
Environment	18,658,823	7.6	6,903,013	2.9
Other	19,275,352	7.8	12,376,968	5.2
Total	$ 118,024,130	48.0	$ 137,604,981	57.5
OUTSIDE THE COMPANIES				
Research support to:				
Electric Research Council or Electric Power Research Institute	$ 72,428,822	29.4	$ 47,212,203	19.8
Nuclear Power Groups	13,461,097	5.5	11,565,062	4.8
Others	42,084,634	17.1	42,840,948	17.9
Total	$ 127,974,553	52.0	$ 101,618,213	42.5
Grand total	$ 245,998,683	100.0	$ 239,223,194	100.0

Source: FPC News, Vol. 8, No. 32, August 8, 1975.

Federal Energy R&D
(Dollars in Millions)

	FY 75 BA	FY 75 BO	FY 76* BA	FY 76* BO	FY 77 BA	FY 77 BO
Direct Energy R&D						
ERDA	$1,317.0	$1,011.0	$1,657.0	$1,427.0	$2,435.0	$2,009.0
DOI	89.9	54.2	104.0	93.3	98.3	96.3
EPA	80.8	18.2	56.8	76.6	55.4	76.6
NRC	58.9	51.7	87.5	76.9	104.0	98.2
NASA	0.8	0.8	1.7	1.0	–0–	0.8
Subtotal	1,547.4	1,135.9	1,907.0	1,674.8	2,692.7	2,280.9
Supporting R&D						
ERDA	362.0	313.0	403.0	373.0	430.0	404.0
DOI	33.2	30.9	59.0	56.7	66.8	65.2
EPA	53.2	5.0	43.2	43.4	41.6	43.4
NRC	2.3	2.1	9.6	9.1	5.3	5.0
NSF	103.2	65.9	114.6	74.2	123.4	106.9
Subtotal	553.9	416.9	629.4	556.4	667.1	624.5
Total Federal Energy R&D	$2,101.3	$1,552.8	$2,536.4	$2,231.2	$3,359.8	$2,905.4

* Funds for FY 76 Transition Quarter are not included.

Source: A National Plan for Energy Research, Development and Demonstration: Creating Energy Choices for the Future, ERDA-76-1, Energy Research and Development Administration, April 15, 1976.

Estimated Industrial Expenditures for Energy R, D & D—1963 and 1973

Expenditures in Millions of Dollars

Energy Sources	1963			1973			
	In 1963 Dollars	In 1973 Dollars	Percent of Total	Percent of Total	Total	Federal Funds	Company Funds
Oil and Gas	$336	$483	(55%)	(35%)	$303	$ 7	$296
Coal	11	16	(2)	(7)	66	5	61
Other Fossil	not indicated	not indicated	—	(6)	53	0	53
Electricity	267	384	(43)	(45)	392	263	129
Geothermal, solar and all other	<1	1	(<1)	(7)	61	10	51
Total	$614	$884	(100%)	(100%)	$875	$285	$590

SOURCES: **Energy R&D and National Progress,** prepared by the Interdepartmental Energy Study under the chairmanship of the Director of the Office of Science and Technology, 1964.
Science Resources' Studies Highlights, National Science Foundation, NSF 74-319, December 4, 1974.
NOTES: 1. The yearly data are not strictly comparable because of differences in sources, definitions and data aggregation into the energy source grouping.
2. 1963 data are order-of-magnitude estimates and may include research or other commodities. The Implicit Price Deflator was used to put 1963 expenditures into 1973 dollars. "Electricity" includes nuclear fission and fusion plus generation and distribution R&D expenditures plus other electrical related projects.
3. 1973 data have been grouped using the NSF source material. "Electricity" includes nuclear fission and fusion expenditures only. Some portion of "Other Fossil" and/or "All Other" should probably be included in "Electricity". The funds are for operating expenses incurred by a company in conduct of R&D in its own facilities. Federal funds are receipts for work performed. Company funds are for internal company-sponsored R&D. Some arbitrary assignments of funding source have been made in the grouping. "Other Fossil" includes $11 million in shale and $42 million in unidentified energy sources.

Major Energy R, D & D in Other Nations

Country	Estimated Current Budget	Major Thrust
France	$700 million	Sixty percent of budget devoted to current and future nuclear technology. Also coal mining technology and oil and gas exploration, storage and transport.
West Germany	$450 million	Sixty-five percent devoted to nuclear and twenty-five percent to coal extraction and utilization.
United Kingdom	$350 million	Sixty percent devoted to nuclear power.
Canada	$220 million	Fifty percent devoted to derived fuels mainly in situ conversion of coal and tar sands. Thirty percent devoted to nuclear technology.
Japan	$200 million	Eighty percent devoted to the nuclear option.
USSR	Unknown	All aspects of technology believed receiving emphasis in a broadly based program.

Selected Examples of Other Agency Energy R, D & D Efforts*

Agency	Indicated FY 76 Energy, R,D&D Budget (millions)	Major Thrusts
Department of the Interior	$160	Oil and gas recovery, resource assessment and mining and extractive technology
National Science Foundation	$155	Basic research
Environmental Protection Agency	$140	Alleviation of environmental damage to energy systems and measurement and monitoring of health effects and pollutants
Nuclear Regulatory Commission	$90	Confirmatory nuclear safety R,D&D and studies on safeguards, safety systems and siting guides.

* A survey of current energy R,D&D in other Government agencies was conducted early in the study. The survey was initiated by letter dated March 13, 1975, from the Administrator of ERDA. The heads of twenty Federal agencies were asked for material to incorporate into the ERDA National Energy Plan. The call requested information on each agency's program activities and objectives, the strategy and rationale for chosen program approaches, and the specifics of programs used to implement the strategy. Survey results (shown in Table C-2) were used as a basis for indicating general agency interest. The nature of the budget data, difference in interpretations of requirements and definitions, etc., preclude exacting reliance on the statistical data.

Source: A National Plan for Energy Research, Development and Demonstration: Creating Energy Choices for the Future, Volume 1: The Plan, ERDA-48, Energy Research and Development Administration, June 1975.

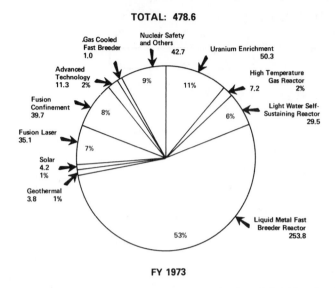

TOTAL: 478.6

FY 1973

[1] Data are based upon estimated fiscal year 1973 expenditures.

Source: *The Nation's Energy Future*, Atomic Energy Commission, 1973, page 29.

Federal funding, nuclear and renewable energy R&D, 1973[1] ($ million).

Source: Energy Perspectives, U.S. Department of the Interior, February 1975.

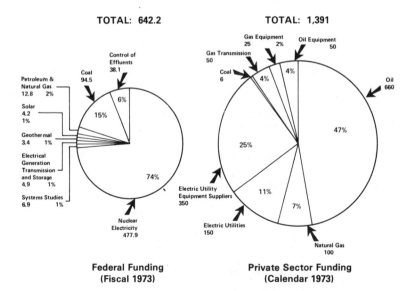

TOTAL: 642.2   **TOTAL: 1,391**

Federal Funding   **Private Sector Funding**
(Fiscal 1973)   **(Calendar 1973)**

Source: *The Nation's Energy Future*, Atomic Energy Commission, 1973, page 29;
A Time to Choose, Ford Foundation, 1974, pages 307-308.

U.S. energy R&D expenditures, federal and private sectors, 1973 ($ million).

Source: Energy Perspectives, U.S. Department of the Interior, February 1975.

FY 1977 ERDA Budget to Congress Program Total

	(In Millions)			
	FY76 Estimate		FY77 Estimate	
	B/A	B/O	B/A	B/O
Conservation R&D				
Electric energy systems and energy storage				
Electric energy systems	$19.6	$14.1	$ 24.5	$19.4
Energy storage systems	16.3	13.8	22.3	18.5
Total	35.9	27.9	46.8	37.9
End use conservation and technology to improve efficiency				
Industry conservation	4.2	2.0	12.4	9.7
Buildings conservation	12.6	8.2	21.6	18.4
Transportation energy conservation	12.9	10.4	23.7	20.4
Improved conversion efficiency	9.1	7.0	15.5	4.6
Total	38.8	27.6	73.2	53.1
Total conservation research and development	$74.7	$55.5	$120.0	$91.0
Fossil Energy Development				
Coal				
Liquefaction	$ 89.9	$ 92.9	$ 73.9	$ 79.5
Gasification				
high Btu	53.4	37.3	45.2	59.3
low Btu	24.6	36.0	33.0	40.0
Advanced power systems	10.0	7.5	22.5	12.8
Direct combustion	38.1	32.6	52.4	52.1
Advanced research and supporting technology	35.4	32.1	37.1	36.6
Demonstration plants	51.9	23.3	107.2	80.9
Magnetohydrodynamics	29.5	18.4	37.4	27.3
Other capital equipment	0	0	.2	.1
Total coal	$332.8	$280.1	$408.9	$388.6
Petroleum and natural gas				
Gas and oil extraction	$41.5	$32.9	$35.2	$30.5
Supporting research	1.8	1.6	1.8	1.8
Other capital equipment	0	0	.2	.1
Total petroleum and natural gas	43.3	34.5	37.2	32.4
In-situ technology				
Oil shale	14.0	10.1	21.5	12.5
In-Situ coal gasification	6.1	7.6	8.2	6.7
Supporting research	1.3	1.1	1.3	1.3
Other capital equipment	0	0	.1	.1
Total in-situ technology	21.4	18.8	31.1	20.6
Total fossil energy development	$397.5	$333.4	$477.2	$441.6

FY 1977 ERDA Budget to Congress Program Total

(In Millions)

	FY76 Estimate		FY77 Estimate	
	B/A	B/O	B/A	B/O
Solar Energy Development				
Direct thermal				
Applications heating and cooling	$ 35.1	$24.8	$ 45.3	$ 34.5
Agricultural and process heat	4.8	3.7	3.9	2.5
Solar electric applications	64.9	49.1	102.5	73.0
Technology support and utilization	5.4	4.6	4.0	3.0
Fuel from biomass	4.5	3.8	4.3	3.0
Total solar energy development	$114.7	$86.0	$160.0	$116.0
Geothermal energy development				
Engineering R&D	$ 10.8	$ 9.5	$ 11.7	$ 11.7
Hydrothermal technology applications	5.9	11.9	12.9	10.8
Advanced technology applications	7.1	4.6	10.6	8.5
Environmental control and institutional studies	4.0	2.5	4.8	4.8
Resource exploration and assessment	3.6	3.2	10.1	9.7
Geothermal loan guarantee program	0	0	50.0	4.4
Total geothermal energy development	$ 31.4	$31.7	$100.1	$ 49.9
Fusion Power R&D				
Magnetic fusion				
Confinement systems	$ 76.9	$ 67.3	$ 88.8	$ 79.6
Development and technology	41.5	34.6	67.5	49.0
Research	31.1	37.8	43.0	41.4
Reactor projects	17.4	5.1	91.8	44.0
Total magnetic fusion	166.9	144.8	291.1	214.0
Laser fusion	83.6	79.0	101.0	90.5
Total fusion power R&D	$250.5	$223.8	$392.1	$304.5
Fission Power Reactor Development				
Liquid metal fast breeder reactor				
Base program R&D	$313.3	$310.4	$343.3	$333.3
Clinch river breeder reactor	107.0	57.0	237.6	171.0
Reactor safety	49.4	45.8	58.4	55.6
Advanced fuels	15.1	14.8	16.2	15.5
Total LMFBR	484.8	428.0	655.5	575.4
Water cooled breeder reactor program	42.1	43.9	48.9	39.6
Gas cooled reactors				
Thermal reactor	$ 15.3	$ 14.3	$ 16.2	$ 16.3
Fast breeder reactors	6.4	6.2	8.1	7.5
Reactor safety	4.4	4.0	5.7	5.4
Total gas cooled reactors	26.1	24.5	30.0	29.2
Molten salt breeder reactor	3.7	4.1	0	0
Light water reactor technology	4.0	3.0	12.5	10.0
Supporting activities	17.4	14.8	20.1	21.3
Other capital equipment	24.4	3.4	22.7	8.8
Total fission power reactor development	$602.5	$521.7	$789.7	$684.3
Reactor safety facilities	$ 0	$ 0	$ 33.3	$ 24.7

FY 1977 ERDA Budget to Congress Program Total

(In Millions)

	FY76 Estimate		FY77 Estimate	
	B/A	B/O	B/A	B/O
Nuclear Fuel Cycle & Safeguards R&D				
Fuel cycle R&D				
Uranium resource assessment	$ 18.9	$ 14.9	$ 36.5	$ 30.1
Support of nuclear fuel cycle	36.4	32.0	61.5	54.2
Waste management (commercial)	13.6	12.2	80.8	62.8
Total fuel cycle R&D	68.9	59.1	178.8	147.1
U-235 process development	54.8	61.1	95.9	68.3
Advanced isotope separation technology	32.7	27.6	43.8	39.2
Nuclear materials security and safeguards	16.6	14.7	28.2	27.3
Total nuclear fuel cycle and safeguards	$173.0	$162.5	$346.7	$281.9
Environmental control technology	$ 13.0	$ 11.8	$ 16.1	$ 14.7
Environmental Research				
Biomedical research				
Health studies	$ 71.3	$ 68.9	$ 70.5	$ 69.1
Biological studies	40.9	39.2	41.4	41.1
Environmental studies	45.4	45.1	49.0	49.9
Physical and technological studies	18.6	16.3	19.0	18.3
Analysis and assessment	11.9	11.5	19.3	17.0
Education and training	3.6	3.3	2.3	2.6
Other capital equipment	.5	.5	.5	.5
Total environmental research	$192.2	$184.8	$202.0	$198.5
Basic Energy Sciences				
Nuclear sciences	$107.5	$ 90.9	$ 93.2	$ 94.0
Material sciences	52.9	49.7	77.7	58.0
Molecular mathematical and geosciences	48.0	46.0	53.3	50.3
Other capital equipment	2.1	1.3	2.3	2.1
Total basic energy sciences	$210.5	$187.9	$226.5	$204.4

DEPARTMENT OF ENERGY
FY 1979 BUDGET
($ IN MILLIONS)

MISSION: ENERGY SUPPLY – RESEARCH & TECHNOLOGY DEVELOPMENT

	FY78 ESTIMATE		FY79 ESTIMATE	
	BA	BO	BA	BO
COAL	$ 579.1	$ 520.1	$ 618.2	$ 609.4
PETROLEUM	74.1	62.5	79.7	73.7
GAS	31.0	26.5	25.9	30.0
SOLAR	302.7	225.2	309.2	281.4
GEOTHERMAL	106.2	81.9	129.7	118.9
MAGNETIC FUSION	325.4	279.1	334.0	327.0
FUEL CYCLE R&D	284.9	212.8	247.4	227.7
BREEDER REACTOR	517.1	610.6	367.4	445.0
NUCLEAR RESEARCH & APPLICATIONS ..	226.7	206.8	278.5	253.8
LIGHT WATER REACTOR FACILITIES ...	28.1	24.6	10.0	19.0
INTERNATIONAL SPENT FUEL STORAGE .	5.0	5.0	3.0	3.0
HYDROELECTRIC	10.0	7.5	8.0	6.0
BIOMASS	20.7	15.2	26.9	15.3
BASIC ENERGY SCIENCES	177.4	164.1	211.5	201.1
ADVANCED TECHNOLOGY & ASSESSMENT .	7.5	5.7	21.0	17.0
TOTAL ENERGY SUPPLY – R&TD	$2,695.9	$2,447.6	$2,670.4	$2,628.3

MISSION: ENERGY SUPPLY – PRODUCTION, DEMONSTRATION AND DISTRIBUTION

	FY78 ESTIMATE		FY79 ESTIMATE	
	BA	BO	BA	BO
COAL	$ 10.6	$ 2.9 1/	$ 2.9	$ 3.6 1/
NAVAL PETROLEUM AND SHALE RESERVES	152.1	83.0 1/	199.0	212.2 1/
SOLAR	87.2	83.7	64.4	69.4
URANIUM ENRICHMENT ACTIVITIES (NET) ...	471.0	409.3	201.6	233.7
URANIUM RESOURCE ASSESSMENT	64.8	54.7	95.2	78.1
MULTI-RESOURCE	0.5	0.6	1.0	0.6
GEOTHERMAL RESOURCES DEVELOPMENT FUND .	15.0	6.6	–	6.0
POWER MARKETING ADMINISTRATION	107.3	159.8	127.5	162.7
TOTAL ENERGY SUPPLY – PRODUCTION, DEMONSTRATION AND DISTRIBUTION	$908.5	$800.6	$691.6	$766.3

MISSION: CONSERVATION

	FY78 ESTIMATE		FY79 ESTIMATE	
	BA	BO	BA	BO
UTILITIES	$ 54.6	$ 43.9	$ 54.7	$ 52.2
RESIDENTIAL AND COMMERCIAL	125.5	109.6	263.2	184.0
INDUSTRIAL	29.6	21.7	49.1	36.2
TRANSPORTATION	65.0	51.0	97.5	83.8
FEDERAL	0.6	0.7	10.2	10.2
STATE/LOCAL	297.1	147.9	368.3	399.4
MULTI-SECTOR	119.5	98.7	166.5	136.4
TOTAL CONSERVATION	$ 691.9	$ 473.5	$1,009.5	$ 902.2

MISSION: BASIC SCIENCES

	FY78 ESTIMATE		FY79 ESTIMATE	
	BA	BO	BA	BO
LIFE SCIENCES RESEARCH AND BIOMEDICAL APPLICATIONS	$ 42.5	$ 41.0	$ 40.8	$ 40.0
HIGH ENERGY PHYSICS	274.2	238.4	294.7	288.1
NUCLEAR PHYSICS	80.4	83.3	90.8	86.7
TOTAL BASIC SCIENCES	$397.1	$362.7	$426.3	$414.8

MISSION: ENVIRONMENT

	FY78 ESTIMATE		FY79 ESTIMATE	
	BA	BO	BA	BO
ENVIRONMENTAL RESEARCH AND DEVELOPMENT				
Overview and Assessment	$ 51.4	$ 44.8	$ 51.1	$ 48.9
Biomedical and Environmental Research	166.4	158.0	157.5	152.1
TOTAL ENVIRONMENTAL RESEARCH AND DEVELOPMENT	$217.8	$202.8	$208.6	$201.0

COMPARISON OF THREE ENERGY PLANS

The following comparison of three energy plans—the plans of the President, the Ways and Means Committee of the House of Representatives and the Democratic Majority of the House of the 94th Congress—was prepared by the Congressional Research Service.[1]

A COMPARISON OF THREE CURRENT ENERGY PLANS

The three plans which are compared are:

1. "The President's Plan"—The summary includes some measures undertaken as administrative acts as well as the provisions of the Administration's omnibus energy bill, introduced in the Senate as S. 594 on February 5, 1975 and in the House as H.R. 2633 and H.R. 2650 on February 4, 1975. An official summary was reprinted in the *Congressional Record*, February 5, 1975, pp. S1421-26.

2. "Ways and Means"—The summary is taken from the suggested alternative proposals drawn up by eight task forces made up of the Democratic members of the House Ways and Means Committee and issued March 3, 1975. A text is reprinted in the *Congressional Record*, March 4, 1975, pp. H1370-1374. The draft proposals differ somewhat from the provisions of H.R. 5005 introduced by Mr. Ullman, March 17, 1975.

3. "Democratic Majority"—The summary is drawn from the recommendations prepared jointly by the Democratic Policy and Steering Committee of the House and Democratic Policy Committee of the Senate as the Democratic Majority's comprehensive proposals covering both energy and economic recovery. Entitled, "The Congressional Program of Economic Recovery and Energy Sufficiency," it was announced February 27, 1975 and was printed by the Senate Democratic Policy Committee for use by Members of Congress.

[1]*Highlights of Energy Legislation in the 94th Congress*, A background paper prepared by the Congressional Research Service, for the Committee on Interior and Insular Affairs, United States Senate, Serial No. 94-19 (92-109), U.S. Government Printing Office, Washington, D.C., 1975.

A Comparison of 3 Current Energy Plans

Goal and Task	President's Plan	Ways and Means	Democratic Majority
I. Reduce dependence on foreign oil	Essential	Agree—but not so fast and not through tariffs	Agree—but economic restoration is Nation's highest priority.
By what means?	Import fees, tariffs, and excise taxes: Rely on price and market to lower imports, through lowered consumption.	Quantitative controls: Import quotas, Federal Petroleum Purchasing Agency to buy all imports.	Conservation mainly: Let new National Energy Production Board decide whether import quotas or other means are needed.
How much and how fast?	1,000,000 bbl/d, end 1975; 2,000,000 bbl/d end 1976. But this still means rise in imports from 5,300,000 bbl/d, 1975 to 5,800,000 bbl/d, 1977 (administration figures).	1,000,000 bbl/d over 2 to 3 yr, an estimated 400,000 bbl/d 1976.	Believe total program will reduce "domestic consumption of imported petroleum" 500,000 bbl/d, 1st yr; 1,600,000 bbl/d, 2d yr; 5,000,000 bbl/d, 1980.
Reduce and limit imports to what levels?	3,000,000 to 5,000,000 bbl/d by 1985	To about 25 percent of expected domestic oil output by "early 1980's" [estimated 4,000,000 bbl/d based on estimated domestic oil production of 15,800,000 bbl/d in 1985].	To about 10 percent of expected domestic energy consumption by 1985. [Estimated 4,500,000 bbl/d based on plan's target of holding energy consumption to 45,000,000 bbl/d in oil equivalent in 1975.]
II. Protect against disruption of imports. By creating a system of strategic petroleum reserves.	Essential. Fully explore, develop and produce all (4) naval petroleum reserves. Use oil and revenue gained to create a national petroleum reserve stockpile of not more than: 1,000,000,000 bbl for civilian use. 3,000,000 bbl—military use. Reserve 20 percent of production NPR No. 4 for strategic reserves; balance for the public economy. (Note: It is anticipated that NPR No. 4 will be able to produce a minimum of 2,000,000 bbl/d by 1985. 80 percent of this would be 1,600,000 bbl/d for current public consumption.) Require any domestic importer or refiner to maintain stored petroleum reserves as determined by President.	Agree. Accumulate reserve stockpile by— Federal purchase of imports (outside any quota limit) and/or From naval petroleum reserves. Goal "during next few years": 50 percent of annual imports, (e.g., 6 mo. supply); at import level of 5,000,000 bbl/d, this would equal 600,000,000 bbl.	Agree. Accumulate reserve stockpile by— Federal purchase of imports and from naval petroleum. From naval petroleum reserves, OCS and market place. Goal: 3,000,000 bbl/d for 6 mo by 1980. [Equals 500,000,000 bbl] and 3,000,000 bbl/d for full year by 1985 [equals 1,000,000,000 bbl.] National Energy Production Board would "oversee establishment".

A Comparison of 3 Current Energy Plans (Continued)

Goal and Task	President's Plan	Ways and Means	Democratic Majority
Authorize standby energy authority including particularly: Allocation of fuels_____	Essential_____ To deal with future embargoes, international commitments, other emergencies. To be invoked only if President finds emergency situation exists.	Agree_____ But in addition: Mandatory allocation of oil needs to be continued in any case to insure shifts from oil and other conservation measures. Continue current legislative authority and FEA allocation system. (Public Law 93–159).	Agree. But the allocation and other standby authority should be vested in the National Energy Production Board, and we should extend allocation system now to "accomodate" reduced dependence by "managing and controlling" excessive consumption.
End-user consumer rationing_____ III. Restrain and reshape demand_____ By what means in general?	Only in emergency, e.g., embargo_____ Essential_____ Rely mainly on increased prices, jawboning for voluntary compliance, modest subsidies and relaxed auto emission requirements.	Agree_____ Agree on goal but not on method_____ Specific incentive and disincentive taxes and subsidies for specific conservation targets, plus selected mandatory conservation measures.	Agree. Agree on goal but not on method. Similar to ways and means plus extended mandatory allocation program to manage and control excessive energy consumption.
By what means specifically? Increase prices_____	(a) Import fees, tariffs on imported oil_____ (b) Excise taxes on domestic oil and gas_____ (c) Deregulate old oil_____ (d) Deregulate new gas_____ (e) Establish oil floor price_____	(a) Reject_____ (a) Reject. (b) Reject general tax—Propose specific taxes on gasoline, inefficient autos. (c) and (d) Reject as conservation measure—but—propose phased deregulation to encourage increased production. (e) Reject_____	(e) Reject—Recommend some price control to shield consumer from OPEC-inflated prices.
Improve auto fuel efficiency_____	Relax clean air standards on cars; voluntary effort by industry to improve 40 percent by 1980.	Auto efficiency tax, beginning 1977 model year of $0 to $500 for 20> —15 m.p.g. Tax credit for new auto purchase, graduated_____ "Consider" mandatory fuel standards by 1979 model year.	Mandatory fuel and efficiency standards to improve 50 percent by 1980, 100 percent by 1985. Purchase rebates for low-use cars, excise tax for high-use car, R. & D. for low-fuel, low-polluting new auto prototype.
Gasoline tax_____ Insulate, retrofit residential and commercial buildings.	None_____ Design thermal efficiency standards for new buildings, to be promulgated through States (after about 4 yr). Finance 3-yr State "winterizing" program mainly for low-income homes—$9,000,000 1975, $55,000,000 1976–77.	5 cents per gallon 1975, rising to 40 cents 1979_____ "Consider" tax credits for residential energy improvements including solar equipment, and rapid amortization commercial solar equipment. Investment credits for industrial retrofitting on large scale.	5 cents per gallon but primarily for revenue. Major loan guarantees, grant, tax credit program for insulation, etc., to upgrade 40,000,000 houses in 10 yr. Upgrade standards for new construction, especially Federal buildings.
Encourage conversion to coal_____	Relax clean air standards_____ Expand authority to require conversion to coal_____	Relax clean air standards_____ Require switch to coal "ASAP."_____ 5-yr amortization of costs of new coal equipment___	"Discourage" use of natural gas in new plants. Tax credit and loans to retrofit for coal; "suggests" current clean air standards not be increased for such plants.
More efficient electric utilities_____	Change "offpeak" pricing to encourage more balanced use. Permit increased rates submitted in this plan as method to expand supplies but effect could be restrained use.	"Encourage" experiments with utility rate design alternatives; establish rate restructuring under FPC.	Financing for improved transmission, extended lines. Better use of existing capacity, redesign rate structure.
Encourage mass transit_____	Supports mass transit (included in budget request)_	Supported as object of energy trust fund expenditures. Use highway trust fund to expand mass transit_____ Increase Federal aid to railroads_____	Added funding for public transport, railroad rehabilitation, electrification.
IV Increase domestic energy supply: Oil and natural gas: Price incentives_____	Deregulate old oil. Deregulate new gas. Establish floor price for oil.	Accept phased deregulation of oil and natural gas—over several years. Alternative approaches are being considered covering time frame of 5 to 5 yr. Reject floor prices.	Accept phased deregulation of oil and natural gas—over several years. But maintain control to separate price of domestic energy from OPEC. United States should determine its own energy prices. Reject floor price.
Tax incentives_____	Retain depletion allowance [however, windfall tax would reduce base]. Impose graduated tax (15 percent to 90 percent) which would phase out on all domestic crude to capture windfall profits. No plowback.	Eliminate depletion allowance on all crude oil as of Jan. 1, 1975 (H.R. 2166). Phase out small producers by Jan. 1, 1978. Eliminate depletion allowance for most natural gas. Graduated "windfall" profits tax plus plow back__	Eliminate depletion allowance on all foreign drilling. Retain D/A only for small producers who do not operate retail outlets. Excess profits tax, avoidable by "plowback" reinvestment. Proceeds to energy trust fund.

A Comparison of 3 Current Energy Plans

Goal and Task	President's Plan	Ways and Means	Democratic Majority
	No change in taxes which encourage United States companies to invest in oil/gas production abroad, but import fee of $1.20 on product imports to be increased to $2 per barrel after excise tax is enacted to keep excise tax from encouraging foreign refining.	"Neutralize" tax laws to remove incentives to invest abroad, several alternatives being considered.	Eliminate foreign tax subsidies so United States capital is not encouraged to invest abroad.
Alaskan oil and gas	Devote 80% NPR No. 4 output to current use (est. 1,000,000 bbl/d by 1985).	Not discussed in current proposal.	Develop NPPs rapidly to make "estimated 10,000,000,000 to 40,000,000,000 bbl" available for "storage or commercial use."
	Legislation to be submitted for constructing a natural gas pipeline from Alaska.	Not discussed in current proposal.	"Expedited consideration" of natural gas delivery system from Alaska.
Outer continental shelf	Accelerated development and production under existing law, including lease sales in Atlantic, Pacific, and Gulf of Alaska.	No mention	Revise OCS Act to accelerate "exploration consistent with public interest," public data bank, production under leases to prevent withholding, mandatory unitization.
Enlist more independent producers	No special provision but antitrust legislation to be enforced against "price fixing and bid-rigging" (In Oct. 8, 1974, address, reaffirmed Jan. 15, 1975.)	No special provision except possible delay in eliminating depletion allowance for small producers. No special antitrust provision.	Change bidding system for all Federal leases "to permit greater participation by small companies." Similar possible concession on depletion allowance. Strengthen antitrust laws to promote free enterprise and completition.
Coal Environmental considerations	Strip mining—Modified version of vetoed bill. Design new program of coal leasing.	Strip mining reclamation and mine safely programs are listed as programs which could be funded from an energy trust fund.	Enact the Strip Mining Control Act.
Tax incentives	No special provision, investment credit increased for all investors to 12% for 1 year.	5 year amortization new coal mining and coal using equipment and associated railroad expansion. 50 year amortization gradings and bores "might be be considered."	No special provision: mentions "Coal conversion incentives of major proportion" to induce shift of all electric utilities and heavy industrial boilers in 10 years but coal mining not mentioned as candidate for major loans, etc.
Capital incentives	No special provision	"Capital must be raised for opening new coal mines."	No special provisions.
Nuclear	Expedited licensing and siting (separate legislation) and 1976 budget increase of $41,000,000 for nuclear safeguards and waste management.	Tax incentives suggested to "encourage and facilitate expansion" if environmental issues are settled.	No special provisions.
Alternate fuels	National synthetic fuels commercialization program (announced Jan. 15, 1975) to produce 1,000,000 bbl/d synthetic fuels by 1985 from oil shale and coal. Rely on ERDA for continuing program in other fuels development.	Recycled oil—tax incentives to encourage. Solid wastes as fuel—tax and other incentives. Geothermal—expand definition of intangible drilling expense to include costs associated with geothermal. Solar energy—tax incentives to encourage retrofitting. Synthetic oil and gas; "Be alert" to possible encouragement measures.	National Energy Production Board to directly undertake or finance on contract or joint venture; commercial demonstration of synthetic fuels oil shale; MHD; geothermal; solar.
Research and development	To be administered by and through ERDA and its budget.	To be financed in part through an energy trust fund.	To be financed in part through an energy trust fund, similar to highway trust fund, into which 5 cents gasoline tax would be paid.
Energy management measures designed to speed up development and production of energy supplies	Reform utilities regulation to allow increased rates to finance capital expansion—readjust off-peak pricing—allow full pass-through of all costs. Expedite energy facilities planning and siting: FEA prepares a national plan; states prepare related management plans.	No special provisions.	Reform utilities rate structure and speed up FPC and State regulatory procedures, but reject any automatic pass-through on cost. Enact regional planning mechanism in which states can participate and resolve regional concerns through other agencies.

NATIONAL ENERGY OUTLOOK

In February 1976, the Federal Energy Administration published *National Energy Outlook*, a document incorporating revisions of the information presented in the *Project Independence Report* of November 1974. The findings and conclusions of the 1976 report are as follows:

FINDINGS AND CONCLUSIONS

- Last year's Project Independence Report has been revised to incorporate the latest information on energy prices, legislation, and resources.

- The events of 1975 have changed the energy outlook:

  - Energy legislation has been enacted which largely removes the oil depletion allowance, sets a new oil pricing policy, establishes conservation measures and provides standby authorities in the event of another embargo.

  - New Federal estimates of oil and gas resources are substantially lower than previous figures.

  - Further increases in the price of imported oil have occurred and the possibility of a rapid drop in price now seems remote.

  - Higher energy prices have spurred the search for oil and gas and dramatically cut the rate of growth in energy demand.

- Taking account of these changes, the FEA analysis shows that energy independence can still be achieved:

  - Over the next 10 years, the Nation can greatly expand its domestic energy production and cut the rate of growth in energy demand, and still meet its economic objectives.

  - Our dependence in the next few years would have increased, but the recently enacted legislation will hold our vulnerability to about current levels.

  - The post-1985 prospects for maintaining independence are less certain, unless technological and economic breakthroughs occur.

- If we do not establish policies to stimulate domestic energy production and cut energy use, or if regional growth restrictions, less reserves than expected, or extended price controls occur, our dependence on foreign oil could rise dramatically above today's levels.

  - The uncertainties and energy impacts associated with each of these factors is large and makes accurate forecasting difficult.

  - But unless the Nation weighs the impacts of alternative policy assumptions, it cannot choose a national energy policy which balances economic, energy, environmental, and social objectives.

- The figures which follow briefly summarize the major findings of this year's National Energy Outlook.

What Are the Roots of Our Energy Problem?

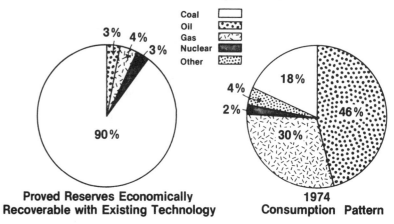

Proved Reserves Economically Recoverable with Existing Technology

1974 Consumption Pattern

- The United States has vast reserves of coal amounting to three times the energy contained in the Middle East's oil reserves, and currently accounting for more than 90 percent of U.S. proved energy reserves.

- Yet, over the last 75 years, the United States has switched from using coal for over 90 percent of its energy needs to depending on oil and gas for 75 percent of its energy.

- Thus, the Nation depends upon its least abundant energy resources to provide most of its energy needs.

- The result has been a growing dependence on imported energy, the availability and price of which are controlled by a few Middle East countries.

- Our task between now and 1985 is to find and develop more oil and gas and stimulate conservation to offset currently dwindling production of these fuels, as our economy is converted to the more abundant resources, such as coal and nuclear power.

- In the post-1985 period, our task is to develop new technologies that can dramatically expand our economically usable reserves of oil and uranium and make greater use of non-depletable resources, such as solar and geothermal energy.

How Did We Become So Vulnerable To Oil Imports?

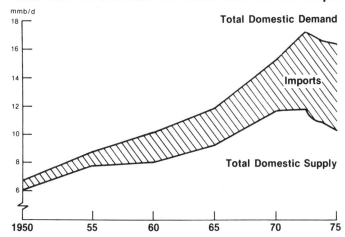

- The availability of inexpensive imported oil served as a disincentive to domestic production which peaked at 9.6 million barrels per day (MMB/D) in 1970, has been declining ever since, and now stands at 8.2 MMB/D.

- Meanwhile, petroleum demand grew at an annual rate of 4.6 percent in the 1960's and early 1970's, in response to low prices, air pollution restrictions on coal use, and the growing use of automobiles.

- The combination of declining domestic production and rising demand led to a rapid growth in imports:

  - From 1.8 million barrels per day (MMB/D), or 19 percent of consumption, in 1960.

  - To 3.4 MMB/D, or 23 percent of consumption, in 1970.

  - To a high of 6.0 MMB/D, or 37 percent of consumption, in 1975.

- The Arab oil embargo in 1973/74 demonstrated that the U. S. is vulnerable to severe supply disruptions and oil price increases.

- Today, the United States spends about $37 billion, or $125 per person, for imported oil; as compared to about $3 billion, or $15 per person, in 1970.

How Much Energy Will the Nation Consume?

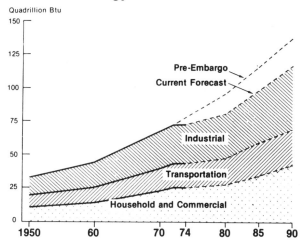

- Energy demand grew at a rate of 3.6 percent in the 20 years before the 1973 embargo, and in 1975 the United States consumed about 73 quadrillion Btu's (quads).

  - Electricity consumption grew at twice the rate of all energy demand (about 7 percent per year).

- By 1985, as a result of higher prices, energy demand will be much lower than historic growth rates even with continued economic expansion.

  - With a continuation of current oil prices (about $13 per barrel in 1975 dollars), demand will be 98.9 quads in 1985 -- a 2.8 percent growth rate.

  - Electricity will still grow about twice as fast as overall energy demand, but at reduced levels of 5.4 percent per year.

  - Consumption will gradually shift from oil and gas to coal and nuclear power.

- If world oil prices decline to $8 per barrel, or if oil and gas prices are regulated substantially below market prices, energy demand would grow faster at an annual rate of between 3.0 and 3.2 percent.

- The largest reductions in energy growth will be in the Household/Commercial and Transportation Sectors, which will respond most sharply to higher energy prices.

How Will the U.S. Meet Its Growing Energy Demands by 1985?

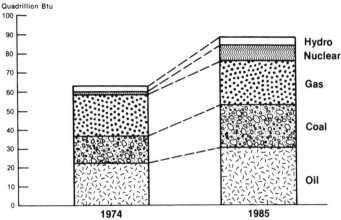

- Total domestic supply is forecast to increase by 40 percent between now and 1985, with all major fuels playing a large role:

  - Coal production could increase to over one billion tons, from current levels of 640 million tons.

  - Oil production could reach 13.9 MMB/D, if Outer Continental Shelf leasing is strongly pursued and market prices prevail.

  - Natural gas production could reach 22.3 trillion cubic feet (Tcf) if new gas prices are deregulated, but will be 17.9 Tcf under current regulations.

  - Although nuclear power has experienced significant delays, it could grow from current levels of 8.6 percent to about 26 percent of electricity generation.

  - Emerging technologies such as the conversion of coal into oil or gas, solar, and geothermal energy, will be important in the post-1985 period, but will not produce much energy in the next ten years.

- Each of these supply increases, while technically and economically feasible, requires significant growth of the energy producing sectors and will not be forthcoming unless pricing and government regulatory policies encourage it. Institutional barriers and policy uncertainty will also delay development.

- If one or more domestic energy sources do not achieve these projected levels, imports will make up the shortage because other domestic fuel sources could not compensate for the loss.

How Much Energy Can Be Saved?

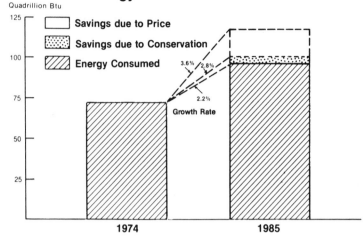

- Higher energy prices should significantly cut energy demand growth during the next ten years, reducing the growth rate to 2.8 percent from the historical rate of 3.6 percent.

- An active conservation program could further reduce energy demand by the equivalent of 3 million barrels per day, reducing the annual energy growth rate to 2.2 percent through 1985.

  - Savings could be achieved in all the major sectors: residential, commercial, industrial and transportation.

  - Actions which improve automobile efficiency and the efficiency of homes and office buildings would have the greatest impact in the next ten years.

- While conservation can reduce energy demand, it does not appear feasible to cut the growth rate to zero or to obviate the need for expanding existing supplies of energy.

What Will Oil Imports Be by 1985?

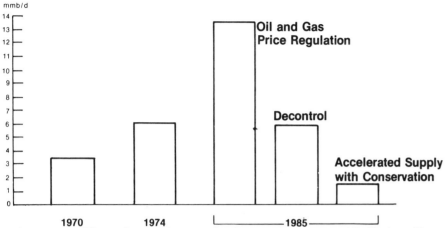

- Imports will continue to increase in the next two years, until Alaskan oil production begins.

  - There is little that can be done to cut imports through 1977, although conserving energy and increasing oil recovery in existing fields will provide some help.

- We have much greater ability to cut imports by 1985, if appropriate policy actions are taken:

  - If oil and gas prices are regulated at low levels, imports could reach 13.5 MMB/D in 1985.

  - With gradual deregulation of oil and gas prices, and a continuation of current world oil prices, imports could drop to 5.9 MMB/D, slightly below today's level.

  - A maximum effort to increase supply and cut demand could reduce imports to about 1.0 MMB/D, making the United States invulnerable by 1985.

- By 1990, however, imports could increase as domestic production from older fields again declines. This decline will need to be offset by the growing use of nuclear power, synthetic fuels, solar and other emerging technologies.

Where Will New Oil Supplies Come From?

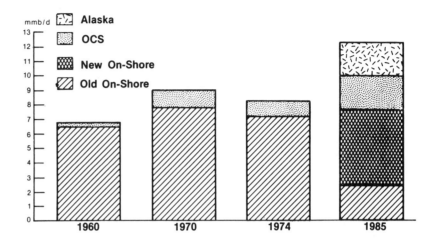

- Domestic crude oil production could increase to 12.3 MMB/D in 1985 (from 8.4 MMB/D in 1975), if today's market prices are allowed to stimulate domestic production and an aggressive OCS leasing and development program is followed.

- While total production will increase from today's levels, oil supply from existing onshore reserves could decline to 2.4 MMB/D by 1985, as older fields are depleted.

  - But, more intensive use of secondary and tertiary recovery in current fields and new discoveries onshore can keep onshore production about constant.

- If the current OCS leasing and development schedules are followed, OCS production could more than double by 1985, to about 2.3 MMB/D.

- Alaska will be the greatest new source of production, increasing to about 2.4 MMB/D by 1985.

How Quickly Will This Country Run Out of Oil?

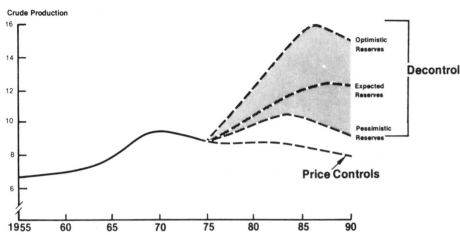

Crude Production

Is Natural Gas About to be Depleted?

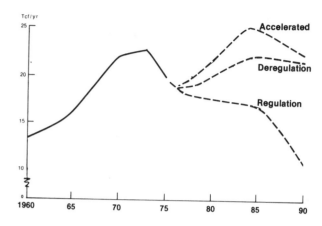

Tcf/yr

- The amount of oil we discover and produce depends on how much oil is in the ground (reserves) and whether oil prices are high enough to justify their production.

- If oil prices remain at current levels and the Federal Government's best estimate of known and expected reserves proves correct, domestic production could reach 12.3 MMB/D (or 13.9 MMB/D including natural gas liquids) by 1985 and begin to decline in the late 1980's.

- If world prices fall to $8 per barrel or domestic prices are regulated over a long period, production is never likely to be much above today's levels, and will decline again in the early 1980's. Crude production could be as low as 8.3 MMB/D in 1985 under these circumstances.

    - This decline will occur because the more expensive enhanced recovery techniques and some frontier area production, such as that from Alaska, would not be economic at lower prices.

- It is highly uncertain how much oil remains to be found and produced:

    - If reserves are much lower than expected, production could not peak much above historic levels and would decline rapidly thereafter, even at high world oil prices.

    - If reserves are much higher than expected, oil production could be maintained at least at today's levels for many years.

- Natural gas production can be expected to decrease during the next few years but then increase to 22.3 Tcf in 1985 if prices are deregulated (as compared to 20.1 Tcf in 1975).

    - Most of the new gas production will come from the Gulf of Mexico and intensive production from onshore fields.

    - If OCS leasing is accelerated and resource availability proves more favorable than expected, natural gas production could reach 25.5 Tcf in 1985.

- If present regulations continue, natural gas production could decline to 17.9 Tcf in 1985. The lower the price, the more rapid will be the decline in natural gas exploration and production.

- Because of limited reserves and uncertain supplemental gas supplies, natural gas production is likely to peak in the 1980's and then probably decline again.

- Supplemental gas supplies could play an important role in the 1980's and later:

    - Gas from Alaska could supply over 1 Tcf before 1985, if needed transportation systems are completed.

    - Liquefied natural gas could supply about 2 Tcf by 1985.

    - If financial incentives are provided, over 1 Tcf of synthetic gas from coal could be delivered by 1985, and this source could supplement dwindling supplies of naturally occurring gas in the post-1985 period.

Where Will New Coal Production Come From?

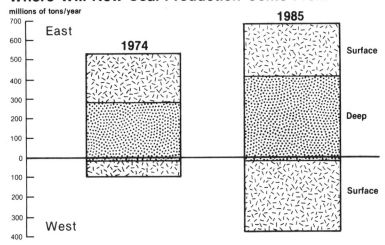

millions of tons/year

East
1974
1985
Surface
Deep
Surface
West

What Will Be the Sources of Electricity in the Future?

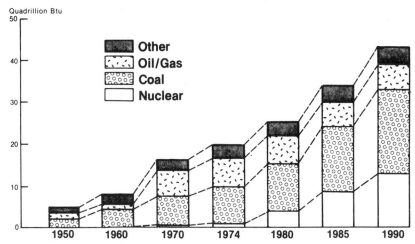

Quadrillion Btu

- Other
- Oil/Gas
- Coal
- Nuclear

1950 1960 1970 1974 1980 1985 1990

- Coal production is projected to increase from 603 million tons in 1974 to 1040 million tons in 1985.

  - Coal production will not increase this fast if long-term utility demand is uncertain and if major environmental and transportation issues are unresolved.

  - More coal could probably be produced, but will not be because its markets are limited primarily by the growth in electric power and synthetic fuels.

- The major expansion of production will occur in the Western regions, increasing from 92 million tons in 1974 to about 380 million tons in 1985.

  - Western coal production will continue to be mainly surface mining.

- Eastern mining could expand by about 30 percent.

  - Underground mining should reverse recent trends and increase more than surface mining in the East, since low-cost surface reserves are being depleted.

- Electricity could continue to grow at about twice the rate of total energy demand; but its growth rate will be reduced to about 5.4 percent, compared to its historical rate of about 7 percent.

- The use of natural gas and oil to generate baseload electricity could be phased out due to higher prices, and be replaced with less expensive nuclear and coal plants.

- Coal's use in electric generation could increase by 77 percent in the next 10 years.

  - Over 700 million tons of coal could be used to generate electricity in 1985, as compared to 392 million tons in 1974.

- Nuclear energy could represent about 26 percent of electric power generation in 1985, as compared to 8.6 percent in 1975.

  - However, nuclear power accounts for a 30 percent smaller contribution than previously projected, reflecting actual cancellations and deferments caused by reductions and uncertainty in demand growth, financial difficulties, and licensing delays.

- Electricity will represent a continually increasing share of energy in the future, rising from 28 percent in 1974 to 37 percent in 1990.

How Much Can New Technologies Contribute?

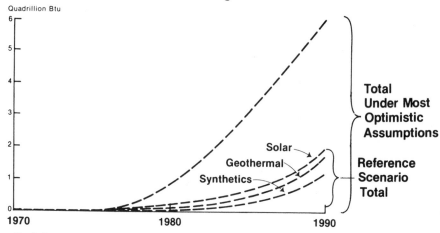

Quadrillion Btu

- Solar, geothermal and synthetic fuels will make only a small contribution to domestic energy supplies by 1985 (about 1 percent).

- The major contribution from solar, geothermal and synthetic fuels will not be felt until after 1990.

- The technology for these sources exists, but must be proven economically viable on a commercial scale. It will take several years to build the first full-size plants; hence a large industry will not be possible during the next 10 years.

- It is likely that few, if any, synthetic fuel plants will be built by 1985 without Federal financial assistance.

- Unless commercial size plants are started now and proven economic by 1985, it will not be possible for these new sources to replace dwindling supplies of oil and gas in the post-1985 period.

How Much Will Energy Supply Investment Cost?

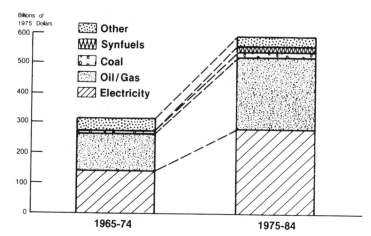

Billions of 1975 Dollars

- Energy investments in the U.S. will be about 580 billion (in 1975 dollars) in the next ten years.

  - While this investment seems large, it is about 30 percent of fixed business investment, which is energy's historical share.

  - In certain sectors, such as utilities, large demands will be placed on the capital markets.

- Oil, gas, and electric utility capital spending will almost double in the next 10 years.

- The largest portion of the energy investment will be in the electric utility sector which could account for 47 percent of the total.

- Oil and gas investment depends greatly on the pricing and policy strategies adopted and could range from about $160 to $315 billion.

- Coal investment could increase to $18 billion or only 3 percent of the total, but representing a 200 percent increase from the 1965-1974 total of $6 billion.

- Investments to increase energy efficiency could also be significant, perhaps an additional $250 billion through 1985.

  - Conservation investments are difficult to separate from non-energy investments and will be spread throughout the economy.

ERDA NATIONAL PLAN

FACT SHEET

ERDA's Revised National Plan
For Energy Research, Development and Demonstration
April 15, 1976

Dr. Robert Seamans, Administrator of the Energy Research and Development Administration, today presented to the President and submitted to the Congress a revised plan for energy research, and development. The new report, entitled *A National Plan for Energy Research, Development and Demonstration* (*ERDA 76-1*), revises, expands and updates the plan submitted by ERDA on June 30, 1975. Specifically it:

—Contains further discussion and revision of ERDA's recommendations for National energy RD&D goals and priorities.
—Summarizes ERDA's completed and ongoing analyses of the energy situation and the Nation's energy RD&D needs.
—Summarizes current and planned federal energy RD&D activities.

Background

• On June 30, 1975, ERDA submitted to the President and the Congress a report entitled *A National Plan for Energy Research, Development and Demonstration* (*ERDA-48*)
  —Within the context of the President's goals for energy independence, the 1975 plan:
    —Recommended energy RD&D goals and objectives for the Nation.
    —Examined the potential timing and contribution of major energy technology options.
    —Ranked major technologies and related them to their potential energy contribution in the near, mid and long term.
    —Discussed Federal and private sector roles in energy RD&D and described the Federal energy RD&D effort.
  —The Plan served as an important input to the development of the President's amended 1976 and 1977 budget requests for energy RD&D funding.
• On February 26, 1976, the President sent to the Congress a comprehensive message summarizing progress made in moving the Nation toward energy independence; outlining actions he has taken to achieve our goals; and listing legislation which awaits action by the Congress. The President urged the Congress to act promptly on all the proposals that are needed to achieve the Nation's energy goals, including his proposed 1977 budget of $2.9 billion (outlays) for energy research, development and demonstration—an increase of 30% over 1976.

The President's 1977 Budget Request for Energy RD&D

• The primary responsibility for bringing into use new technologies for energy conservation and expanding domestic energy production rests with the private sector. The Federal Government's responsibility is to assist the private sector in the development and market penetration of new energy technologies by: establishing an appropriate policy environment for private sector action, sharing risks with the private sector and conducting a complementary RD&D program.
• Within this framework, the President's 1977 Budget recognizes the high priority of energy RD&D by proposing a greatly expanded program at a level appropriate to the responsibilities of the Federal Government. Specifically it:
  —Accelerates energy RD&D programs directed at achieving greater long-term energy independence.
    —Expands efforts to assure the safety, reliability, and availability of commercial nuclear power plants.
  —Places greatest funding on technologies with the highest potential payoff in terms of recoverable resources (i.e., nuclear and fossil).
  —Greatly increases the Federal investment in conservation technologies.
  —Continues to expand the investigation of other technologies where they can make significant contributions to meeting the long-term energy requirements of the U.S. (i.e., solar and fusion).
  —Encourages cost-sharing with private industry (e.g., coal liquefaction demonstrations) and avoids undertaking RD&D more appropriately the responsibility of the private sector (e.g., in areas of conservation technology).
  —Supports the commercial demonstration of synthetic fuel production from coal, oil shale, and other domestic resources by providing loan guarantees beginning in FY 1976.
• Details of the Budget requests for energy RD&D are summarized in Table 1.

ERDA's Report (ERDA 76-1) on Revisions to its National Plan for Energy RD&D

• The ERDA report being released today summarizes ERDA's current views on the energy technologies the Nation will need to achieve longer-term energy independence. The report refines and extends the scope of its June 30, 1975 Plan. Specifically it:
  —Stresses the paramount role of the private sector in the development and commercialization of new energy technologies.
  —Develops in more detail the Federal program and strategy to accelerate the market penetration of energy technologies with near-term potential.
  —Underscores the need for greater integration of socioeconomic and environmental factors into technology development decisions at regional levels as well as the national level.
  —Reviews changes in the energy situation which have occurred since publication of the previous plan.
  —Presents new analytic results that have had an impact on shaping the current revision of the plan.
  —Outlines an overall approach to energy RD&D program planning and resource allocation within ERDA.
  —Outlines ERDA's planning goals for the coming year which are to:
    —Increase effort on benefit cost studies;
    —Examine the priorities of specific programs, and
    —Identify and analyze energy RD&D activities in the private sector.

The Strategy of Technology Introduction

Time of Impact	Strategic Element	Technology	Impact in Year 2000 (Quads)
Near-term (now to 1985 and beyond)	Increase efficiency of energy use and convert waste to energy	Conservation in Buildings and Consumer Products	7.1
		Industrial Energy Efficiency	8.0
		Transportation Efficiency	9.0
		Waste Materials to Energy	4.9
	Preserve and expand oil, gas, coal, and nuclear	Coal-Direct Utilization in Utility/Industry	24.5
		Nuclear-Converter Reactors	28.0
		Oil & Gas Enhanced Recovery	13.6
Mid-Term (1985-2000 and beyond)	Accelerate development of synthetic fuels from coal and shale	Gaseous and liquid fuels from Coal	14.0
		Oil Shale	7.3
	Increase use of under-used (limited application) fuel forms and extract more usable energy from waste heat	Geothermal	3.1–5.6
		Solar Heating & Cooling	5.9
		Waste Heat Utilization	4.9
Long-Term (past 2000)	Develop the technologies necessary to use the essentially inexhaustible fuel resources	Breeder Reactors	3.1
		Fusion	—
		Solar Electric	2.1–4.2
	Develop the technology necessary to change the existing distribution systems to accommodate the distribution of new energy sources.	Electric Conversion Efficiency	2.6
		Electric Power Transmission & Distribution	1.4
		Electric Transport	1.3
		Energy Storage	—
		Hydrogen in Energy Supplies	—
		Fuels from Biomass	1.4

Source: A National Plan for Energy Research, Development and Demonstration: Creating Energy Choices for the Future, ERDA 76-1, Energy Research and Development Administration, April 15, 1976.

- The major change reflected in this first annual revision of ERDA's plan is the singling out of conservation (energy efficiency) technologies for increased attention, ranking them with several supply technologies as being of the highest priority for national action. The report emphasizes the fact that the primary responsibility for developing technologies to improve energy efficiency rests with the private sector. It also points out that the President's FY 1977 Budget provides a substantial increase for energy conservation RD&D–from $55 million in FY 1976 to $91 million in FY 1977, or 64%–to encourage and stimulate further industrial and other private sector activity.
- In preparing this report (ERDA 76-1) on revisions to the Plan, ERDA undertook new studies and analyses as well as reviewed and considered inputs from a wide variety of concerned and interested parties.

- Although just released today, the analyses and conclusions contained in this report (ERDA 76-1) were an important input to the process of developing the President's 1977 budget requests for energy and research and development.

Source: A National Plan for Energy Research, Development and Demonstration: Creating Energy Choices for the Future, ERDA 76-1, Energy Research and Development Administration, April 15, 1976.

Federal Energy R&D Budget Exclusive of ERDA
(Authority in Millions)

Energy RD&D Programs	FY 75	FY 76*	FY 77
Nuclear Fuel Cycle and Safeguards	6.6	14.4	25.0
Conservation	4.1	5.7	1.2
Geothermal	10.9	12.2	5.9
Fusion	-0-	-0-	-0-
Fission	57.6	78.7	85.0
Solar	0.8	1.7	-0-
Fossil	69.6	80.0	84.7
Environmental Control Tech.	80.8	57.3	55.9
Subtotal	230.4	250.0	257.7
Supporting Research			
Basic Energy Sciences	84.6	95.8	103.6
Environmental Research	107.3	130.6	133.5
Subtotal	191.9	226.4	237.1
Total Non-ERDA Energy RD&D	422.3	476.4	494.8

* 76 dollars do not include transition quarter.

Federal Energy R&D Budget Exclusive of ERDA
(Outlays in Millions)

Direct Energy R&D	FY 75	FY 76†	FY 77	FY 76 to FY 1977 percent change*
Nuclear Fuel Cycle and Safeguards	$ 120	$ 163	$ 282	73
Conservation	21	55	91	64
Geothermal	21	32	50	57
Fusion	151	224	304	36
Fission	538	522	709	36
Solar	15	86	116	35
Fossil	138	333	442	33
Environmental Control Tech.	7	12	15	24
Subtotal	$1,011	$1,427	$2,009	
Supporting Research				
Basic Energy Sciences	$ 165	$ 188	$ 205	9
Environmental Research	148	185	199	7
Subtotal	$ 313	$ 373	$ 404	
Total ERDA Energy RD&D	$1,324	$1,800	$2,413	

† Funds for FY 76 Transition Quarter are not included.
° Percentage change calculated prior to rounding outlays.

Numbers of Major Facilities Potentially Required by Year 2000

Type of Energy Facilities	Number of Equivalent Facilities Potentially Required in 2000	Existing in 1975
Fossil Power Plants (1000 MWe equivalent)	500–800	400
Nuclear Power Plants (1000 MWe equivalent)	450–800	37
Geothermal Power Plants (1000 MWe equivalent)	40–100	0.5
Solar Electric Power Plants (1000 MWe equivalent)	50–100	0
Coal Mines (one million tons/year equivalent)	1600–2370	640
Coal Liquefaction or Gasification Plants (50,000 b/d oil equivalent)	70–140	0
Oil from Shale Plants (50,000 b/d oil equivalent)	40–80	0
Buildings (millions) Solar Heated and Cooled	13–22	negligible
Heat Pumps (millions)	13–26	0.2
Electric Automobiles (millions)	10–20	negligible

Notes
- The list is non-additive because different technologies serve the equivalent energy markets in different scenarios.
- Source for the date is the ERDA–48 scenarios.
- MWe = megawatts electric
- b/d = barrels per day

ERDA Energy Related Budget FY 75–76–77
(Authority in Millions)

Energy RD&D Programs	FY 75	FY 76†	FY 77	FY 76 to FY 77 percent change*
Nuclear Fuel Cycle and Safeguards	$ 118	$ 173	$ 347	101
Conservation	36	75	120	60
Geothermal	28	31	100**	223
Fusion	183	250	392	57
Fission	567	602	823	37
Solar	42	115	160	39
Fossil	335	398	477	20
Environmental Control Tech.	8	13	16	23
Subtotal	$1,317	$1,657	$2,435	
Supporting Research				
Basic Energy Sciences	191	210	227	8
Environmental Research	171	193	203	5
Subtotal	362	403	430	
Total ERDA Energy RD&D	$1,679	$2,060	$2,865	

† 76 dollars do not include transition quarter.
° Percentage change calculated prior to rounding authority.
** Includes $50 Million for Geothermal Loan Guarantee Program.

ERDA Energy R&D Budget
(Outlays in Millions)

Direct Energy R&D	FY 75	FY 76†	FY 77
Nuclear Fuel Cycle and Safeguards	6.3	13.7	23.9
Conservation	1.3	7.0	1.9
Geothermal	10.7	10.1	5.8
Fusion	-0-	-0-	-0-
Fission	50.6	68.7	80.2
Solar	0.8	1.0	0.8
Fossil	37.0	70.2	82.2
Environmental Control Tech.	18.2	77.1	77.1
Subtotal	124.9	247.8	271.9
Supporting Research			
Basic Energy Sciences	55.5	62.8	89.7
Environmental Research	48.4	120.6	130.8
Subtotal	103.9	183.4	220.5
Total Non-ERDA Energy RD&D	228.8	431.2	492.4

† Funds for FY 76 Transitional Quarter are not included.

Source: A National Plan for Energy Research, Development and Demonstration: Creating Energy Choices for the Future, ERDA 76-1, Energy Research and Development Administration, April 15, 1976.

Technologies Now Available for Pursuing Major Energy Technology Goals

The last column of this table presents data from ERDA-48. It represents the maximum impact of the technology in any scenario measured in terms of additional oil which would have to be marketed if the technology were not implemented. Basis for the calculation is explained in Appendix B of ERDA-48. These data are being reexamined, and changes will be made when analysis is completed. In a number of cases, revised projections of impacts will be lower.

Technology	Term of Impact*	Direct Substitution For Oil & Gas**	RD&D Status	Impact in Year 2000 in Quads
GOAL I: Expand the Domestic Supply of Economically Recoverable Energy Producing Raw Materials				
Oil and Gas—Enhanced Recovery	Near	Yes	Pilot	13.6
Oil Shale	Mid	Yes	Study/Pilot	7.3
Geothermal	Mid	No	Lab/Pilot	3.1-5.6
GOAL II: Increase the Use of Essentially Inexhaustible Domestic Energy Resources				
Solar Electric	Long	No	Lab	2.1-4.2
Breeder Reactors	Long	No	Pilot/Demo	3.1
Fusion	Long	No	Lab	—
GOAL III: Efficiently Transform Fuel Resources Into More Desirable Forms				
Coal—Direct Utilization Utility/Industry	Near	Yes	Pilot/Demo	24.5
Waste Materials to Energy	Near	Yes	Comm	4.9
Gaseous & Liquid Fuels from Coal	Mid	Yes	Pilot/Demo	14.0
Fuels from Biomass	Long	Yes	Lab	1.4
GOAL IV: Increase the Efficiency and Reliability of the Processes Used in the Energy Conversion and Delivery Systems				
Nuclear Converter Reactors	Near	No	Demo/Comm	28.0
Electric Conversion Efficiency	Mid	No	Lab	2.6
Energy Storage	Mid	No	Lab	—
Electric Power Transmission and Distribution	Long	No	Lab	1.4
GOAL V: Transform Consumption Patterns to Improve Energy Utilization				
Solar Heat & Cooling	Mid	Yes	Pilot/Demo	5.9
Waste Heat Utilization	Mid	Yes	Study/Demo	4.9
Electric Transport	Long	Yes	Study/Lab	1.3
Hydrogen in Energy Systems	Long	Yes	Study	—
GOAL VI: Increase End-Use Efficiency				
Transportation Efficiency	Near	Yes	Study/Lab	9.0
Industrial Energy Efficiency	Near	Yes	Study/Comm	8.0
Conservation in Buildings and Consumer Products	Near	Yes	Study/Comm	7.1

* Near—now through 1985
Mid—1985 through 2000
Long—Post-2000
** Assumes no change in end-use device.

OECD WORLD ENERGY OUTLOOK

In 1977, the Organization for Ecomonic Co-operation and Development published the following summary[1] of its findings on the world energy situation.

EXECUTIVE SUMMARY

PURPOSE AND ORIGIN OF THE REPORT

This report analyses projected trends in energy needs and supplies and the implied oil import demand of OECD and other major global regions for the next 15 years. It also calls for realising the opportunities still available to improve the energy supply-demand balance and thus reduce oil imports. While the scope is global and covers all energy, the focus is on OECD's needs and opportunities, and on oil as the most versatile and crucial fuel in the mid-term.

The report is a revision and extension of OECD's *Energy Prospects to 1985*, which when published at the beginning of 1975 was one of the first studies to take account of the significant increases in energy prices. The earlier study, begun in 1973, was prepared on the basis of only a year's experience with higher prices and even shorter experience with the severity of the economic recession of 1974-75. Further experience has produced a lowering of expectations about early response of supply and demand of energy, and also realisation that medium-term economic growth rates may be lower than expected prior to 1974. But great uncertainty still prevails in the continuing disequilibrium in both energy markets and the world economy. This report is but one of many re-assessments that will be required as the world energy market evolves towards a new equilibrium.

SCOPE AND BASIC ASSUMPTIONS

The findings of this study arise from several scenarios or cases. There are three different economic growth rate cases. The central growth rate case is combined with two different energy policy scenarios. The other two growth cases are briefly considered to demonstrate the importance of the growth parameter. Throughout there is but one price assumption for the reference Saudi crude: $11.51 per barrel in 1975 dollars[1].

While this study parallels *Energy Prospects to 1985* in many aspects, recent developments have called for some modification of scope, focus and technique.

[1] *World Energy Outlook*, A Reassessment of Long Term Energy Developments and Related Policies, Organisation for Economic Co-operation and Development, Paris 1977.

Projected World Oil Trade
Mtoe (Mb/d)
Net Imports (+)/Exports (—)

	1974		1980		1985 Reference Case		1985 Accelerated Policy Case	
Canada	−9	(−0.2)	38	(0.8)	54	(1.1)	35	(0.7)
United States	290	(5.9)	458	(9.3)	477	(9.7)	211	(4.3)
OECD Europe	708	(14.2)	624	(12.4)	738	(14.7)	554	(11.0)
Japan	263	(5.2)	348	(6.9)	441	(8.7)	382	(7.6)
Australia/New Zealand	14	(0.3)	30	(0.6)	40	(0.8)	35	(0.7)
Total OECD	1,266	(25.3)	1,498	(30.0)	1,750	(35.0)	1,218	(24.4)
Centrally Planned Europe	−41	(−0.8)	−25	(−0.5)	20	(0.4)	20	(0.4)
Centrally Planned Asia	−7	(−0.1)	−25	(−0.5)	−60	(−1.2)	−60	(−1.2)
Oil Importing Developing Countries	161	(3.2)	148	(2.9)	149	(3.0)	149	(3.0)
Other Countries	39	(0.8)	57	(1.1)	61	(1.2)	61	(1.2)
Non-OPEC Oil Exporting Developing Countries	−46	(−0.9)	−150	(−3.0)	−189	(−3.8)	−189	(−3.8)
Residual[1]	72	(1.4)	25	(0.5)	25	(0.5)	25	(0.5)
Total net import demand	1,444	(28.9)	1,528	(30.6)	1,756	(35.1)	1,224	(24.5)
OPEC Consumption[2]	97	(1.9)	146	(2.9)	208	(4.2)	204	(4.1)
OPEC Production	1,541	(30.8)	1,674	(33.5)	1,964	(39.3)	1,428	(28.6)

1. Change in stocks at sea and statistical difference.
2. Includes bunkers.

This study therefore:

— draws upon extensive re-assessments completed by nearly all major energy consuming and some energy producing countries after the oil price increases of 1973-74, but at the same time relies less upon member states' official forecasts than the first study;

— highlights the uncertainty in growth by putting three growth cases;

— presents an alternative or accelerated policy case to demonstrate the potential for changing projected results with stronger policies;

— encompasses the world, not simply the OECD as in the original study;

— looks at horizons beyond 1985; and

[1] After the OPEC Ministerial Conference in Doha, Qatar, on 15-17 December, 1976, it was announced that Saudi Arabia and the United Arab Emirates had decided to raise their crude oil prices by 5 per cent as of 1 January, 1977, and the other OPEC member countries (Algeria, Ecuador, Gabon, Indonesia, Iran, Iraq, Kuwait, Libya, Nigeria, Venezuela and Qatar) had decided to increase the price of "marker" crude oil from $11.51 per barrel to $12.70 per barrel, an increase of about 10 per cent, as of 1 January, 1977, and to $13.30, a further increase of about 5 per cent, as of 1 July, 1977. At the same time, the Saudi Arabian Oil Minister indicated that Saudi Arabia would lift its ceiling on crude oil production and would review its position in mid-1977 in the light of Middle East peace developments and the results of the Conference on International Economic Co-operation. Under these circumstances, it is impossible to determine at the present time what the effective price of crude oil will be. Consequently, the price assumption which underlies the projections of energy demand and supply given in the *World Energy Outlook* has been retained unchanged.

— explores the potential savings that might accrue from adoption of various conservation practices.

KEY CONCLUSIONS AND IMPLICATIONS

Continuance of present policies governing supply expansion and conservation will lead down the path of higher demand for oil imports by industrial countries. OECD net oil import needs in 1985 are put at 35 million barrels per day (Mb/d) or 1 750 million tons of oil equivalent (Mtoe). This level is 9.7 Mb/d (484 Mtoe) greater than in 1974. Demands by the rest of the world (excluding OPEC countries) may be roughly balanced by their projected supplies. Hence demand for exports from the present members of OPEC could be 35 Mb/d (1 750 Mtoe) and when OPEC consumption is included, total required OPEC production could be 39.3 Mb/d (1 964 Mtoe).

By 1985, the present members of OPEC may have expanded their rated production capacity to about 45 Mb/d and other exporters to some 5.5 Mb/d but then as now the margin between the rated capacity and the exporters' desired level of output may be quite large. This could produce a close and uncertain balance between crude liftings and the demand for exports. The exporters' desired level of output may not be sufficient to accommodate possibly greater oil demand generated by spurts of economic growth in the mid-1980's that exceed the growth of industrial countries assumed here. Similarly, that level of output could be pressed if the developing world's import needs are larger than indicated by a recent staff study of the World Bank.

The future desired level of production by oil exporting countries will be determined by a complex set of forces working on each exporting country and is difficult to predict, except, of course, for those countries whose present reserves can no longer sustain past levels of output. But even in the case of those countries with a greater measure of discretion in setting output levels, there is cause for concern. The large volume of revenues raises questions about the degree of absorption of the revenues in the domestic economies to spur economic growth, and faster growth has implications for societal change. On a deeper, philosophical level are other important forces: the preferences of some societies of oil exporting countries to hold physical assets such as oil, or to defer receipt of current income to the future, or simply to see the opportunities of future generations in a different light than seen in industrial societies. These preferences may also be reinforced by apprehensions about depletion. As a consequence, all the oil projected to be sought at today's price may not be made available by exporting countries, in which event large increase in the real price of oil would occur as a result of unilateral price action by producers. This would necessitate rapid and costly adjustments by energy consumers throughout the world which could threaten maintenance of moderate economic growth which is so vital to the employment of the rising world population, not to say improvement of real standards of living of the peoples of the developing countries. This risk, on the supply side, is compounded by the increased vulnerability of industrial states, resulting from higher import demand, to a deliberate reduction of oil supplies to achieve a political objective of some oil exporters.

Notwithstanding the doubts about exporters satisfying the projected demand of 35 Mb/d (1 750 Mtoe) in the reference case, OECD's net import needs could rise even higher than projected if there were any slackening in the present commitment to supply expansion and conservation embodied in current policies (some of which await implementation). For example, replenishing of depleted reserves to maintain present oil production levels in some OECD countries would require doubling the historical finding rates. Holding oil imports of OECD down to the projected level of 1985 will require (assuming a 5 percent annual depletion of reserves) the finding and developing of reserves sufficient to support an output of about 10 Mb/d (500 Mtoe) of crude oil and natural gas liquids merely to sustain the 1974 production level, plus additional reserves to support a further 5 Mb/d (250 Mtoe) to increase OECD output to the projected 1985 level. This reference scenario implies strenuous effort, in the form of considerable capital mobilization and reduced energy consumption. The lesson of the past three years is that it is an achievement that cannot be taken for granted.

BASIC POLICY ALTERNATIVES

Faced with the economic and political consequences of increasing import demand, OECD countries have but one realistic alternative to minimize inherent risks of enlarged imports of oil: to take positive action to expand energy supply —particularly the "conventional" sources of oil, natural gas, coal and nuclear energy; to realise greater energy savings through more effective conservation measures; and to build larger stockpiles to buffer the effect of any deliberate supply reductions. Failure to do so, and instead deferring the acceptance of the consumer and taxpayer cost of these actions until the dangers are more apparent and the time available to make adjustment is shorter, courts the danger of considerable economic dislocation, slower growth and higher unemployment.

In regard to the "conventional" energy sources, policy action is required promptly—that is, within the next year or so—in order to make any appreciable addition to supplies by the early 1980's. Equally prompt action is needed in regard to the "non-conventional" sources if the necessary technologies are to be developed on a commercial basis by the 1990's.

The type of policies needed to bring about increased production include elimination of price controls on crude oil and natural gas; accelerated granting of exploration leases and production licences, principally in Alaska, the North American Continental Shelf, and some sectors of the North Sea; moderation of environmental requirements of new energy development, particularly coal and nuclear power; incentive measures for application of secondary and tertiary oil recovery techniques; and active promotion of coal utilisation by removal of demand constraints. Coal and nuclear fission should be substituted for oil wherever possible. Government support programmes for project development in solar energy, coal conversion and extraction techniques from shale oil and tar sands should also be strengthened.

On the demand or conservation side, significant potential exists for further reducing future energy demand per unit of aggregate output. A number of measures could be taken now. Pricing energy at world levels would have some immediate conservation effect through market forces. Firm and outspoken political

Energy Production, Consumption and Imports in the OECD to 1985 - Reference and Accelerated Policy Scenarios
Mtoe

	Production				Net Imports[1]				Total Energy Requirements[2]			
			1985				1985				1985	
	1974	1980	Reference	Accelerated	1974	1980	Reference	Accelerated	1974	1980	Reference	Accelerated
Solid Fuel	634.2	723.5	863.7	870.7	34.4	52.8	60.2	54.4	685.6	776.3	923.4	925.1
Oil and NGL	634.8	810.0	887.2	1,007.7	1,266.1	1,497.7	1,750.3	1,217.6	1,780.9	2,195.1	2,513.8	2,137.4
Natural Gas	695.6	694.1	773.2	871.1	12.5	56.5	112.2	122.8	704.2	750.6	885.4	993.9
Nuclear	58.0	216.3	464.3	522.3					58.0	216.3	464.3	522.3
Hydro and Other	237.3	264.5	307.2	307.2					237.3	264.5	307.2	307.2
Total	2,259.9	2,708.0	3,295.1	3,579.0	1,313.1	1,607.9	1,922.7	1,394.8	3,466.1	4,203.7	5,094.1	4,885.4

Energy Projections for the OECD for 1985 - Low Growth, Reference and High Growth Scenarios
Mtoe

	Production			Net Imports			Total Energy Requirements[2]		
	Low Growth	Reference	High Growth	Low Growth	Reference	High Growth	Low Growth	Reference	High Growth
Solid Fuel	838.7	863.2	877.4	39.9	60.2	67.4	878.6	923.4	944.8
Oil and NGL	887.2	887.2	887.2	1,593.3	1,750.3	1,938.0	2,366.8	2,513.8	2,689.5
Natural Gas	772.2	773.2	773.2	93.5	112.2	131.1	865.7	885.4	904.3
Nuclear	441.7	464.3	484.7				441.7	464.3	484.7
Hydro and Other	291.9	307.2	320.1				291.9	307.2	320.1
Total	3,231.7	3,295.1	3,342.8	1,726.7	1,922.7	2,136.5	4,844.7	5,094.1	5,343.6

1. Columns do not sum to total for 1974 and 1980. since net imports of electricity are not included in the column breakdown but are included in the total.
2. The rows do not sum to total. For 1980 and 1985 the differences are due to the omission of marine bunkers from total energy requirements. For 1974, the differences are due to marine bunkers and stock changes.

commitment by national leaders to the need for energy conservation, together with selected mandatory measures such as speed limits, could heighten public awareness and create the psychological climate necessary to reinforce market effects. Measures of a more structural nature could be implemented now for growing effect in future years, such as automobile and other transportation efficiency standards, industry conservation monitoring and insulation standards and building codes. Other structural measures in the transportation, industry and residential/commercial sectors could enhance even further the long-term prospect for achieving large energy savings.

The accelerated policy alternative described in this study reflects these additional policy options which could be taken in the fields of conservation and increased production of indigenous sources. Taken together, they represent a potential to reduce OECD countries' net oil imports by 1985 by as much as 10.6 Mb/d (532 Mtoe)—a saving of 30 percent under the level of imports that may be required in the absence of prompt policy action.

The total savings would result from increasing energy supply by 290 Mtoe and lowering demand for energy by 242 Mtoe[2]. Although more than half this potential saving could be achieved in North America alone, every region could make an important contribution. The thrust of the policy options on supply is that faster development will come mostly through relaxation of policies that impede the operation of market forces. On demand, however, the policy options call for more active government programmes to bolster the dampening effect of price on demand through dissemination of information on cost-savings or by tax or other inducements.

The purpose of more vigorous policy on the part of OECD countries to arrest the current trends toward supply-demand imbalances would be, of course, to moderate the general economic and political consequences of growing oil

2. Included in the savings of energy demand is 36 Mtoe in reduced demand for bunkers which is not a "conservation" measure in itself.

imports. As the more detailed examination of regions and countries shows, the time lags—especially in creating new production—are long and require that new policies be implemented shortly if they are to have any influence on oil import levels by 1985.

FACTORS INFLUENCING OECD ENERGY DEMAND TO 1985

The total energy needs in 1985 of OECD projected in the reference case—5 094 Mtoe—are lower than in the comparable price case in *Energy Prospects to 1985*[3] primarily because the recent severe economic recession coupled with inflation has produced expectations of slower growth rates (4.0 per cent annually, 1972 to 1985, as contrasted to 4.9 percent, 1972 to 1985).

Total energy needs are seen as increasing more slowly than economic growth rates—at 3.6 percent annually from 1974 to 1985 for an implied energy/GDP growth elasticity of 0.84, which compares with 0.99 prevailing between 1960 and 1974. It has been difficult to identify separately the effect upon energy demand in the past two years produced alternatively by: recession, higher energy prices, milder weather and non-price conservation—although all observers put the likely level of effect of these forces in that descending order.

The slower growth rate case assumes an average growth rate one-half percentage point lower than in the central cases, throughout the period to 1985, which produces an OECD energy demand of 4 845 Mtoe or nearly 5 percent lower than in the reference case. The higher growth rate case, a half percentage point higher, produces an OECD energy demand of 5 344 Mtoe by 1985, or about 5 percent greater than in the reference case. Expressed differently, in terms of the effect upon oil imports, a full one percent increase in GDP in 1985 would increase the demand for oil imports 2 percent or 700 000 b/d by 1985.

The accelerated policy case projects a reduction of demand through implementation of conservation programmes amounting to 242 Mtoe in 1985. More than one-third is assumed achievable in the United States alone, in transportation (higher fuel efficiency standards in cars), the household sector (greater home insulation and appliance labelling), and the industrial sector (monitoring of industry conservation practices). There is also considerable potential for savings from automobile fuel efficiency, insulation and information programmes in the European and Pacific regions.

SUPPLY EXPANSION WITHIN OECD BY 1985

The current estimates of indigenous supply expansion within the OECD region are lower for all sectors than the estimates given in *Energy Prospects to 1985*.

Continued reduction in oil production in the United States; technical delays in the North Slope and portions of the North Sea; postponements in issuing licences to build pipelines in North America; and delays in granting leases to explore new regions or issuing licences to produce proved fields of some sections of the North Sea and Outer Continental Shelf of North America, have resulted in lower

production estimates than in *Energy Prospects to 1985*—887 Mtoe as contrasted to the earlier estimate of 1 312 Mtoe—despite new discoveries and mark-up of official estimates of reserves in some areas.

Natural gas production prospects for 1985 are also lower than those of the earlier study, due largely to the continuance of price controls upon natural gas in the United States, the slower-than-expected pace of approving a gas pipeline system from Alaska and the Canadian north, and the slowing of projected production in the Netherlands. The estimate of 773 Mtoe by 1985 is less than the 1 016 Mtoe projected in *Energy Prospects to 1985*.

Coal output for 1985, projected at 863 Mtoe, is also down from the 1 025 Mtoe of the earlier study.

It is nuclear power, however, where the greatest short-fall from earlier expectations has occurred. The projection of OECD nuclear capacity in 1985 is now put at only 325 gigawatts with an output of 464 Mtoe—considerably less than the 748 Mtoe foreseen in the comparable price case of the earlier study, and below government targets set prior to the setting of higher oil prices at the end of 1973. The continual mark-down of nuclear expansion results from resistance to nuclear energy from conservationists and environmentalists, the escalation of capital costs primarily to meet higher security standards, and lower expectations of generating load requirements (in turn due to slower economic growth).

OECD ENERGY PROSPECTS TO 1990 AND BEYOND

Before considering the energy prospects for the rest of the world, it is imperative to look beyond 1985 even though extrapolation of trends becomes less reliable. Some order of magnitude estimates of the possible course of oil import requirements are instructive for today's policy formulation. The justification for more vigorous policy does not lie only in reducing dependence in 1985; it lies even more strongly in preventing the consequences of possible "policy drift" from emerging beyond 1985.

By 1990, OECD countries will need to achieve a very significant expansion of their nuclear power programmes in order to displace fossil fuels in electricity generation sufficiently to hold net oil imports below the high of 40.5 Mb/d (2 025 Mtoe) projected in the reference case; they would have to do even better to achieve the reduction implied in the accelerated policy case. Alternatively, large new reserves of oil would need to have been proved, developed and licensed for production in the OECD region. While economic growth and electricity growth might slow below the already reduced rates projected to 1985, this would imply economic and social costs for industrial states. Through 1995, continued large-scale expansion of nuclear power would be required to supply simply the growth increment in energy needs.

THE REST OF THE WORLD TO 1985

The countries outside the OECD comprise the developing world and from an energy viewpoint are considered in this study as belonging to the broad groupings: the centrally planned economies of the USSR, Eastern Europe and Asia; the many developing countries that are not members of OPEC; and the

---

3. The figures quoted from *Energy Prospects* have been converted by the new conversion factors now being used at the OECD and as a result differ somewhat from the numbers in *Energy Prospects*.

OPEC countries. Obviously the latter will continue to play a crucial role in the world energy market as suppliers of oil and natural gas. But the centrally planned economies and the non-OPEC developing countries, including new or re-emerging oil exporters, while subject to considerable uncertainty because of either state secrecy or scanty information about both their energy needs and resources, as well as their energy and development policies, could soon become a deciding factor in determining whether the potential demand for oil moving in international trade will be satisfied.

The Eastern European trading bloc has expanded energy trade both with industrial states of OECD, especially in natural gas exports, and with OPEC countries. There should continue to be a modest increase in this trade but the net exports of energy from the countries of this bloc are projected to decline. China is expected to find increasing internal needs for its own rapidly expanding oil production, and since it is not compelled by large foreign debt to export oil it may make available no more than 1.2 Mb/d (60 Mtoe) for exports in 1985 —but the exportable surplus could vary a half million barrels daily in either direction as the result of a change in domestic oil needs.

The estimates of the energy needs for the remainder of the developing countries (outside OPEC) were derived by the staff of the World Bank and were largely adopted here. These countries have widely disparate needs and opportunities in energy, but as a group they are projected to go from their aggregate net oil imports of 2.3 Mb/d (115 Mtoe) in 1974 to net oil exports of 0.8 Mb/d (40 Mtoe) in 1985—in large part because of the growing oil exports of Mexico, Egypt and Malaysia. The aggregate level, of course, masks the plight of the many low-income, energy-poor countries which will not share directly in the projected reduction in import needs. If, for instance, the three named countries ultimately became members of OPEC, the remaining developing countries outside OPEC might see net oil import requirements rise to as much as 1.3 Mb/d (65 Mtoe) in 1985. The section dealing with the prospects of the developing countries, however, highlights some of the frailties of these projections that are known to all who have attempted to make them.

Finally, estimates are offered for the oil needs of OPEC, which are put at about 4.2 Mb/d (208 Mtoe) (including bunkers) in 1985. The supplies of crude oil offered by OPEC for export are assumed to match the potential world demand for OPEC oil even though, as mentioned earlier, the gap between expanded productive capacity and demand will narrow and the gap between demand and the exporters' combined desired level of output will be even narrower. It is also clear that the discretion in setting the combined level of output will become more concentrated in the governments of one or two member states of OPEC as world demand increases under assumed constant prices and the reserves of less important exporters are depleted without significant replenishment by reserve additions.

The costly, abrupt adjustments that might be necessary to realign demand to a set level of supply from OPEC are, of course, the very risks that would be avoided by a transformation and revitalization of the energy policies of many of the industrial states of the OECD, such as the policy options contained in the accelerated policy case.

*
**

A further elaboration of these findings appears in the regional summaries and detailed country assessments in the main body of the report. The tables of energy balances and the technical annexes will further assist the reader to understand the basis for the findings.

THE NATIONAL ENERGY PLAN

On April 29, 1977, the Executive Office of the President of the United States presented the following overview[1] of a national energy plan for the United States:

Overview

The diagnosis of the U.S. energy crisis is quite simple: demand for energy is increasing, while supplies of oil and natural gas are diminishing. Unless the U.S. makes a timely adjustment before world oil becomes very scarce and very expensive in the 1980's, the nation's economic security and the American way of life will be gravely endangered. The steps the U.S. must take now are small compared to the drastic measures that will be needed if the U.S. does nothing until it is too late.

How did this crisis come about?

Partly it came about through lack of foresight. Americans have become accustomed to abundant, cheap energy. During the decades of the 1950's and 1960's, the real price of energy in the U.S. fell 28 percent. And from 1950 until the quadrupling of world oil prices in 1973–1974, U.S. consumption of energy increased at an average annual rate of 3.5 percent. As a result of the availability of cheap energy, the U.S. developed a stock of capital goods—such as homes, cars, and factory equipment—that uses energy inefficiently.

The Nature of the Problem

The most critical increase in demand has been for oil, the most versatile and widely used energy resource. To meet that growing demand, the U.S. has turned increasingly to imports. In January and February of 1977, the U.S. imported about 9 million barrels of oil per day, half of total domestic oil consumption. By 1985, U.S. oil consumption could equal 12 to 16 million barrels per day.

U.S. domestic oil production has been declining since 1970. New production from Alaska, the deep Outer Continental Shelf, and new recovery methods should reverse the decline, but will be unable to satisfy the projected growth in U.S. demand. Other major additions to domestic oil supply are unlikely.

The principal oil-exporting countries will not be able to satisfy all the increases in demand expected to occur in the U.S. and other coun-

[1] *The National Energy Plan*, Executive Office of the President, Energy Policy and Planning, April 29, 1977.

tries throughout the 1980's. In 1976, the 13 OPEC countries exported 29 million barrels of oil per day. If world demand continues to grow at the rates of recent years, by 1985 it could reach or exceed 50 million barrels per day. However, many OPEC countries cannot significantly expand production; and, in some, production will actually decline. Thus, as a practical matter, overall OPEC production could approach the expected level of world demand only if Saudi Arabia greatly increased its oil production. Even if Saudi Arabia did so, the highest levels of OPEC production probably would be inadequate to meet increasing world demand beyond the late 1980's or early 1990's.

There are physical and economic limits on the world's supply of oil. A widely used geological estimate of total recoverable world oil resources, past and present, is about 2 trillion barrels. More than 360 billion barrels have already been consumed. Current proved crude reserves are 600 billion barrels. World consumption of oil has grown at an average annual rate of 6.6 percent since 1940, and it grew by as much as 8 percent annually during the 1960's.

If it could be assumed that world demand for oil would grow at an annual rate of only 3 percent, and if it were possible (which it is not) that production would keep pace with that rate of growth, the world's presently estimated recoverable oil resources would be exhausted before 2020. At a conjectural growth rate of 5 percent, those resources would be exhausted by 2010. Despite some uncertainty about the exact size of recoverable world oil resources, and about the rate of increase of productive capacity, this fundamental fact is clear: *within about four generations, the bulk of the world's supply of oil, created over hundreds of millions of years, will have been substantially consumed.*

Of course, actual physical exhaustion of oil resources will not occur. Even today, well over half the oil in existing fields is being left in the ground because additional recovery would be too expensive. As production by conventional methods declines and oil becomes more scarce, its price will rise and more expensive recovery methods and novel technologies will be used to produce additional oil. As this process continues, the price of oil will become prohibitive for most energy uses. Eventually the nations of the world will have to seek substitutes for oil as an energy source, and oil will have to be reserved for petrochemical and other uses in which it has maximum value.

The world now consumes about 20 billion barrels of oil per year. To maintain even that rate of consumption and keep reserves intact, *the world would have to discover another Kuwait or Iran roughly every three years, or another Texas or Alaska roughly every six months.* Although some large discoveries will be made, a continuous series of such finds is unlikely. Indeed, recent experience suggests that, compared to world oil consumption, future discoveries will be small or moderate in size, will occur in frontier areas, and will yield oil only at very high cost. Obviously, continued *high rates of growth* of oil consumption simply cannot be sustained.

Natural gas supplies are also limited. In the U.S., natural gas constitutes only 4 percent of conventional energy reserves, but supplies 27 percent of energy consumption. Gas consumption grew about 5.7 percent per year between 1960 and 1970. From 1970 to 1974, however, consumption dropped 1.3 percent. The demand for gas is considerably higher than the amount that can be supplied. Hence, gas is rationed by prohibitions on hook-ups for new homes in many areas.

Gas is not only in short supply, but its allocation across the country is distorted, and its distribution among end-uses is unsatisfactory. Federal regulation of the wellhead price of natural gas in interstate commerce has discouraged its distribution from gas producing States to other States, and has encouraged consumption of this premium fuel for less essential uses. Industry and utilities currently consume almost 60 percent of U.S. natural gas, despite the fact that other fuels could be used in a majority of cases.

During the 1973–75 period, only 19 percent of new gas reserve additions were made available to the interstate market, and much of that gas was from the Federal domain. Since the price of intrastate gas is not regulated, there are strong economic incentives to sell gas within the producing States. *The existing distinction between intrastate and interstate sales has given intrastate users first claim to natural gas.*

Strategies and Objectives

The U.S. has three overriding energy objectives:
—as an immediate objective that will become even more important in the future, to reduce dependence on foreign oil and vulnerability to supply interruptions;
—in the medium term, to keep U.S. imports sufficiently low to weather the period when world oil production approaches its capacity limitation; and
—in the long term, to have renewable and essentially inexhaustible sources of energy for sustained economic growth.

The U.S. and the world are at the early stage of an energy transition. Previous energy transitions in the U.S. were stimulated by new technologies, such as the development of the railroad and the mass production of automobiles, which fostered the use of coal and oil, respectively. The latest transition springs from the need to adjust to scarcity and higher prices.

To make the new transition, the U.S. should adhere to basic principles that establish a sound context for energy policy and provide its main guidelines. The energy crisis must be addressed comprehensively by the Government and by a public that understands its seriousness and is willing to make necessary sacrifices. Economic growth with high levels of employment and production must be maintained. National policies for the protection of the environment must be continued. Above all, the U.S. must solve its energy problems in a manner that is fair to all regions, sectors and income groups.

The salient features of the National Energy Plan are:
—conservation and fuel efficiency;
—rational pricing and production policies;
—reasonable certainty and stability in Government policies;
—substitution of abundant energy resources for those in short supply; and
—development of nonconventional technologies for the future.

Conservation and fuel efficiency are the cornerstone of the proposed National Energy Plan. Conservation is cheaper than production of new supplies, and is the most effective means for protection of the environment. It can contribute to international stability by moderating the growing pressure on world oil resources. Conservation and improved efficiency can lead to quick results. For example, a significant percentage of poorly insulated homes in the United States could be brought up to strict fuel-efficiency standards in less time than it now takes to design, build, and license one nuclear powerplant.

Although conservation measures are inexpensive and clean compared with energy production and use, they do sometimes involve sacrifice and are not always easy to implement. If automobiles are to be made lighter and less powerful, the American people must accept sacrifices in comfort and horsepower. If industry is required to make energy-saving investments and to pay taxes for the use of scarce resources, there will be some increases in the cost of consumer products. These sacrifices, however, need not result in major changes in the American way of life or in reduced standards of living. Automobile fuel efficiency can be greatly improved through better design and use of materials, as well as by producing lighter and less powerful cars, without inhibiting Americans' ability to travel. With improved energy efficiency, the impact of rising energy prices can be significantly moderated.

Energy conservation, properly implemented, is fully compatible with economic growth, the development of new industries, and the creation of new jobs for American workers. Energy consumption need not be reduced in absolute terms; what is necessary is a slowing down in its rate of growth. By making adjustments in energy consumption now, the U.S. can avoid a possibly severe economic recession in the mid 1980's.

The U.S. has a clear choice. If a conservation program begins now, it can be carried out in a rational and orderly manner over a period of years. It can be moderate in scope, and can apply primarily to capital goods, such as homes and automobiles. If, however, conservation is delayed until world oil production approaches its capacity limitation, it will have to be carried out hastily under emergency conditions.

It will be sudden, and drastic in scope; and because there will not be time to wait for incremental changes in capital stock, conservation measures will have to cut much more deeply into patterns of behavior, disrupt the flow of goods and services, and reduce standards of living.

Pricing policies should encourage proper responses in both the consumption and the production of energy, without creating any windfall profits. *If users pay yesterday's prices for tomorrow's energy, U.S. resources will be rapidly exhausted. If producers were to receive tomorrow's prices for yesterday's discoveries, there would be an inequitable transfer of income from the American people to the producers, whose profits would be excessive and would bear little relation to actual economic contribution.*

Currently, Federal pricing policy encourages overconsumption of the scarcest fuels by artificially holding down prices. If, for example, the cost of expensive foreign oil is averaged with cheaper domestic oil, consumers overuse oil, and oil imports are subsidized and encouraged. Consumers are thus misled into believing that they can continue to obtain additional quantities of oil at less than its replacement cost.

Artificially low prices for some energy sources also distort interfuel competition. The artificially low price of natural gas, for example, has encouraged its use by industry and electric utilities, which could use coal, and in many areas has made gas unavailable for new households, which could make better use of its premium qualities.

These misguided Government policies must be changed. But neither Government policy nor market incentives can improve on nature and create additional oil or gas in the ground. From a long-term perspective, prices are an important influence on production and use. As long as energy consumers are misled into believing they can obtain energy cheaply, they will consume energy at a rate the U.S. cannot afford to sustain. Their continued overuse will make the nation's inevitable transition more drastic and difficult.

A national energy policy should encourage production. The energy industries need adequate incentives to develop *new* resources and are entitled to sufficient profits for exploration for *new* discoveries. But they should not be allowed to reap large windfall profits as a result of circumstances unrelated to the marketplace or their risk-taking.

The fourfold increase in world oil prices in 1973–74 and the policies of the oil-exporting countries should not be permitted to create unjustified profits for domestic producers at consumer's expense. By raising the world price of oil, the oil-exporting countries have increased the value of American oil in existing wells. That increase in value has not resulted from free market forces or from any risk-taking by U.S. producers. *National energy policy should capture the increase in oil value for the American people.* The distribution of the proceeds of higher prices among domestic producers and consumers must be equitable and economically efficient if the United States is to spread the cost fairly across the population and achieve its energy goals.

The pricing of oil and natural gas should reflect the economic fact that the true value of a depleting resource is the cost of replacing it. An effective pricing system would provide the price incentives that producers of oil and natural gas need by focusing on harder to find new supplies. The system should also moderate the adjustment that households will have to make to rising fuel costs. It should end the distortions of the intrastate-interstate distinction for new natural gas, which is a national resource. It should also promote conservation by raising the ultimate price of products made by energy-intensive processes.

Reasonable certainty and stability in Government policies are needed to enable consumers and producers of energy to make investment decisions. A comprehensive national energy plan should resolve a wide range of uncertainties that have impeded the orderly development of energy policy and projects. Some uncertainties are inherent in a market economy, and Government should not shelter industry from the normal risks of doing business. But Government should provide business and the public with a clear and consistent statement of its own policies, rules, and intentions so that intelligent private investment decisions can be made.

Resources in plentiful supply should be used more widely as part of a process of moderating use of those in short supply. Although coal comprises 90 percent of United States total fossil fuel reserves, the United States meets only 18 percent of its energy needs from coal. Seventy-five percent of energy needs are met by oil and natural gas although they account for less than 8 percent of U.S. reserves. This imbalance between reserves and consumption should be corrected by shifting industrial and utility consumption from oil and gas to coal and other abundant energy sources.

As industrial firms and utilities reduce their use of oil and gas, they will have to turn to coal and other fuels. The choices now for electric utilities are basically coal and nuclear power. Expanding future use of coal will depend in large part on the introduction of new technologies that permit it to be burned in an environmentally acceptable manner, in both power plants and factories. Efforts should also be made to develop and perfect processes for making gas from coal.

Light-water nuclear reactors, subject to strict regulation, can assist in meeting the United States energy deficit. The 63 nuclear plants operating today provide approximately 10 percent of U.S. electricity, about 3 percent of total energy output. That contribution could be significantly increased. The currently projected growth rate of nuclear energy is substantially below prior expectations due mainly to the recent drop in demand for electricity, labor problems, equipment delays, health and safety problems, lack of a publicly accepted waste disposal program, and concern over nuclear proliferation. The Government should ensure that risks from nuclear power are kept as low as humanly possible, and should also establish the framework for resolving problems and removing unnecessary delays in the nuclear licensing process.

To the extent that electricity is substituted for oil and gas, the total amounts of energy used in the country will be somewhat larger due to the inherent inefficiency of electricity generation and distribution. But conserving scarce oil and natural gas is far more important than saving coal.

Finally, *the use of nonconventional sources of energy must be vigorously expanded.* Relatively clean and inexhaustible sources of energy offer a hopeful prospect of supplementing conventional energy sources in this century and becoming major sources of energy in the next. Some of these nonconventional technologies permit decentralized production, and thus provide alternatives to large, central systems. Traditional forecasts of energy use assume that nonconventional resources, such as solar and geothermal energy, will play only a minor role in the United States energy future. Unless positive and creative actions are taken by Government and the private sector, these forecasts will become self-fulfilling prophecies. Other technologies that increase the efficiency of energy use should also be encouraged, such as cogeneration, the simultaneous production of industrial process steam and electricity.

A national energy plan cannot anticipate technological miracles. Even so, nonconventional technologies are not mere curiosities. Steady technological progress is likely, breakthroughs are possible, and the estimated potential of nonconventional energy sources can be expected to improve. Some nonconventional technologies are already being used, and with encouragement their use will grow. Because nonconventional energy sources have great promise, the Government should take all reasonable steps to foster and develop them.

The National Energy Plan is based on this conceptual approach. It contains a practical blend of economic incentives and disincentives as well as some regulatory measures. It strives to keep Government intrusion into the lives of American citizens to a minimum. It would return the fiscal surpluses of higher energy taxes to the American people.

Finally, the Plan sets forth goals for 1985 which, although ambitious, can be achieved with the willing cooperation of the American people. These goals are:
— reduce the annual growth of total energy demand to below 2 percent;
— reduce gasoline consumption 10 percent below its current level;
— reduce oil imports from a potential level of 16 million barrels per day to 6 million, roughly one-eighth of total energy consumption;
— establish a Strategic Petroleum Reserve of 1 billion barrels;
— increase coal production by two-thirds, to more than 1 billion tons per year;
— bring 90 percent of existing American homes and all new buildings up to minimum energy efficiency standards; and
— use solar energy in more than 2½ million homes.

The Plan would reverse the recent trend of ever-rising oil imports and ever-increasing American dependence on uncertain foreign sources of supply. It would prepare the United States for the time when the world faces a limitation on oil production capacity and consequent skyrocketing oil prices. It would achieve substantial energy savings through conservation and increased fuel efficiency, with minimal disruption to the economy, and would stimulate the use of coal in a manner consistent with environmental protection.

The United States is at a turning point. It can choose, through piecemeal programs and policies, to continue the current state of drift. That course would require no hard decisions, no immediate sacrifices, and no adjustment to the new energy realities. That course may, for the moment, seem attractive. But, with each passing day, the United States falls farther behind in solving its energy problems. Consequently, its economic and foreign policy position weakens, its options dwindle, and the ultimate transition to scarce oil supplies and much higher oil prices becomes more difficult. If the United States faces up to the energy problem now and adopts the National Energy Plan, it will have the precious opportunity to make effective use of time and resources before world oil production reaches its capacity limitation.

The energy crisis presents a challenge to the American people. If they respond with understanding, maturity, imagination, and their traditional ingenuity, the challenge will be met. Even the "sacrifices" involved in conservation will have their immediate rewards in lower fuel bills and the sense of accomplishment that comes with achieving higher efficiency. By preparing now for the energy situation of the 1980's, the U.S. will not merely avoid a future time of adversity. It will ensure that the coming years will be among the most creative and constructive in American history.

Summary of the National Energy Plan

Conservation

In the transportation sector, the Plan proposes the following major initiatives to reduce demand:

—a graduated excise tax on new automobiles with fuel efficiency below the fleet average levels required under current legislation; the taxes would be returned through rebates on automobiles that meet or do better than the required fleet averages and through rebates on all electric automobiles;

—a standby gasoline tax, to take effect if total national gasoline consumption exceeds stated annual targets; the tax would begin at 5 cents per gallon, and could rise to 50 cents per gallon in 10 years if targets were repeatedly exceeded by large or increasing amounts; the tax would decrease if a target were met; taxes collected would be returned to the public through the income tax system and transfer payment programs; States would be compensated for lost gasoline tax revenues through sources such as the Highway Trust Fund;

—fuel efficiency standards and a graduated excise tax and rebate system for light-duty trucks;

—removal of the Federal excise tax on intercity buses;

—increase in excise tax for general aviation fuel, and elimination of the existing Federal excise tax preference for motorboat fuel;

—improvement in the fuel efficiency of the Federal automobile fleet, and initiation of a vanpooling program for Federal employees.

To reduce waste of energy in existing buildings, the Plan proposes a major program containing the following elements:

—a tax credit of 25 percent of the first $800 and 15 percent of the next $1,400 spent on approved residential conservation measures;

—a requirement that regulated utilities offer their residential customers a "turnkey" insulation service, with payment to be made through monthly bills; other fuel suppliers would be encouraged to offer a similar service;

—facilitating residential conservation loans through opening of a secondary market for such loans;

—increased funding for the current weatherization program for low-income households;

—a rural home conservation loan program;

—a 10 percent tax credit (in addition to the existing investment tax credit) for business investments in approved conservation measures;

—a Federal grant program to assist public and non-profit schools and hospitals to insulate their buildings;

—inclusion of conservation measures for State and local government buildings in the Local Public Works Program.

The development of mandatory energy efficiency standards for new buildings will be accelerated. In addition, the Federal Government will undertake a major program to increase the efficiency of its own buildings.

The Plan proposes the establishment of mandatory minimum energy efficiency standards for major appliances, such as furnaces, air conditioners, water heaters, and refrigerators.

The Plan proposes to remove major institutional barriers to cogeneration, the simultaneous production of process steam and electricity by industrial firms or utilities, and to provide an additional 10 percent tax credit for investment in cogeneration equipment. Encouragement will also be given to district heating, and the Energy Research and Development Administration (ERDA) will undertake a study to determine the feasibility of a district heating demonstration program at its own facilities.

To promote further industrial conservation and improvements in industrial fuel efficiency, an additional 10 percent tax credit for energy-saving investments would be available for certain types of equipment (including equipment for use of solar energy) as well as conservation retrofits of buildings.

The Plan also contains a program for utility reform, with the following elements:

—a phasing out of promotional, declining block, and other electric utility rates that do not reflect cost incidence; declining block rates for natural gas would also be phased out;

—a requirement that electric utilities either offer daily off-peak rates to customers willing to pay metering costs or provide a direct load management system;

—a requirement that electric utilities offer customers interruptible service at reduced rates;

Fuel Balances by Sector
[Millions of barrels of oil equivalent per day]

	1976	1985 without Plan	1985 with Plan	1985 Plan plus additional conservation
Demand	37. 0	48. 3	46. 4	45. 2
Residential and commercial:				
Oil	3. 5	3. 2	2. 7	------------
Natural gas	3. 9	3. 8	4. 1	------------
Electricity	6. 3	9. 1	8. 4	------------
Coal	. 1	(1)	(1)	------------
Total [2]	13. 8	16. 1	15. 2	------------
Industry:				
Oil	3. 2	7. 0	4. 0	------------
Natural gas	4. 4	4. 5	4. 5	------------
Electricity	4. 2	7. 2	7. 1	------------
Coal	1. 9	2. 7	5. 0	------------
Total [2]	13. 7	21. 4	20. 6	------------
Transportation:				
Oil	9. 2	10. 6	10. 2	------------
Natural gas	. 3	. 2	. 3	------------
Total [2]	9. 5	10. 8	10. 5	------------
Electricity:[3]				
Oil	1. 6	2. 0	1. 3	------------
Natural gas	1. 5	. 9	. 5	------------
Coal	4. 9	8. 2	8. 3	------------
Nuclear	1. 0	3. 6	3. 8	------------
Other	1. 5	1. 6	1. 6	------------
Total [2]	10. 5	16. 3	15. 5	------------
Supply	37. 0	48. 5	46. 4	45. 2
Domestic:				
Crude oil [4]	9. 7	10. 4	10. 6	------------
Natural gas	9. 5	8. 2	8. 8	------------
Coal	7. 9	12. 2	14. 5	------------
Nuclear	1. 0	3. 7	3. 8	------------
Other	1. 5	1. 7	1. 7	------------
Refinery gain	. 4	. 9	. 6	------------
Total [2]	30. 0	37. 1	40. 0	------------
Imports/exports (−):				
Oil	7. 3	11. 5	7. 0	5. 8
Natural gas	. 5	1. 2	. 6	------------
Coal	−. 8	−1. 2	−1. 2	------------
Total [2]	7. 0	11. 5	6. 4	5. 2

[1] Less than 0.05 million barrels of oil equivalent per day.
[2] Detail may not add due to rounding.
[3] Included in previous sectoral totals.
[4] Includes natural gas liquids.

—a prohibition of master metering in most new structures;

—a prohibition of discrimination by electric utilities against solar and other renewable energy sources;

—Federal authority to require additional reforms of gas utility rates;

—Federal Power Commission (FPC) authority to require interconnections and power pooling between utilities even if they are not now subject to FPC jurisdiction, and to require wheeling.

Oil and Natural Gas

Government policy should provide for prices that encourage development of new fields and a more rational pattern of distribution; but it should also prevent windfall profits. It should promote conservation by confronting oil and gas users with more realistic prices, particularly for those sectors of the economy where changes can be made without hardship. To promote these ends, the Plan proposes a new system for pricing oil and natural gas.

The proposal for oil pricing contains the following major elements:

—price controls would be extended;

—newly discovered oil would be allowed to rise over a 3 year period to the 1977 world price, adjusted to keep pace with the domestic price level; thereafter, the price of newly discovered oil would be adjusted for domestic price increases;

—the incentive price for "new oil" would be applicable to oil produced from an onshore well more than 2½ miles from an existing well, or from a well more than 1,000 feet deeper than any existing well within a 2½ mile radius; the incentive price would be applicable to oil from Federal offshore leases issued after April 20, 1977;

—the current $5.25 and $11.28 price ceilings for previously discovered oil would be allowed to rise at the rate of domestic price increases;

—stripper wells and incremental tertiary recovery from old fields would receive the world price;

—all domestic oil would become subject in three stages to a crude oil equalization tax equal to the difference between its controlled domestic price and the world oil price; the tax would increase with the world price, except that authority would exist to discontinue an increase if the world price rose significantly faster than the general level of domestic prices;

—net revenues from the tax would be entirely returned to the economy: residential consumers of fuel oil would receive a dollar-for-dollar rebate, and the remaining funds would be returned to individuals through the income tax system and transfer payment programs;

—once the wellhead tax is fully in effect, the entitlements program would be terminated, along with certain related activities, but would be retained on a standby basis.

The proposal for natural gas pricing contains the following major provisions:

—all new gas sold anywhere in the country from new reservoirs would be subject to a price limitation at the Btu equivalent of the average refiner acquisition cost (before tax) of all domestic crude oil;

—that price limitation would be approximately $1.75 per thousand cubic feet (Mcf) at the beginning of 1978; the interstate-intrastate distinction would disappear for new gas;

—new gas would be defined by the same standards used to define new oil;

—currently flowing natural gas would be guaranteed price certainty at current levels, with adjustments to reflect domestic price increases;

—authority would exist to establish higher incentive pricing levels for specific categories of high-cost gas, for example, from deep drilling, geopressurized zones and tight formations;

—gas made available at the expiration of existing interstate contracts or by production from existing reservoirs in excess of contracted volumes would qualify for a price no higher than the current $1.42 per Mcf ceiling; gas made available under the same circumstances from existing intrastate production would qualify for the same price as new gas;

—the cost of the more expensive new gas would be allocated initially to industrial rather than residential or commercial users;

—Federal jurisdiction would be extended to certain synthetic natural gas facilities;

—taxes would be levied on industrial and utility users of oil and natural gas to encourage conservation and conversion to coal or other energy sources.

The Plan contains the following additional proposals for oil and natural gas:

—to encourage full development of the oil resources of Alaska, Alaskan oil from existing wells would be subject to the $11.28 upper tier wellhead price and would be treated as uncontrolled oil for purposes of the entitlements program; new Alaskan oil finds would be subject to the new oil wellhead price;

—production from Elk Hills Naval Petroleum Reserve would be limited to a ready reserve level at least until the west-to-east

Balances by Fuel [1]
[Millions of barrels of oil equivalent per day]

	1976	1985 without Plan	1985 with Plan	1985 Plan plus additional conservation
Oil:				
Consumption	17. 4	22. 8 [2]	18. 2	17. 0
Domestic supply [3]	9. 7	10. 4	10. 6	10. 6
Refinery gain	. 4	. 9	. 6	. 6
Imports	7. 3	11. 5	7. 0	5. 8
Natural gas:				
Consumption	10. 0	9. 4	9. 4	
Domestic supply	9. 5	8. 2	8. 8	
Imports	. 5	1. 2	. 6	
Coal:				
Consumption	6. 8	10. 9	13. 3	
Domestic supply	7. 9	12. 2	14. 5	
Exports	. 8	1. 2	1. 2	

[1] Detail may not add up to total due to rounding.
[2] Assuming compliance with automobile efficiency standards under current law, and reduced driving as a result of higher gasoline prices. Without these assumptions, consumption would be 25 million barrels per day.
[3] Includes natural gas liquids.

transportation systems for moving the surplus Alaskan oil are in place or until California refineries have completed a major retrofit program to enable more Alaskan oil to be used in California;

—the Outer Continental Shelf Lands Act would be amended to require a more flexible leasing program using bidding systems that enhance competition, to assure a fair return to the public, and to assure full development of the OCS resources;

—shale oil will be entitled to the world oil price;

—the guidelines established by the Energy Resources Council in the previous administration would be replaced by a more flexible policy: projects for importation of liquified natural gas (LNG) should be analyzed on a case-by-case basis with respect to the reliability of the selling country, the degree of American dependence the project would create, the safety conditions associated with any specific installation and all costs involved; imported LNG would not be concentrated in any one region; new LNG tanker docks would be prohibited in densely populated areas;

—Federal programs for development of gas from geopressurized zones and Devonian shale would be expanded;

—the Administration hopes to eliminate gasoline price controls and allocation regulations next fall; to maintain competition among marketers, it supports legislation similar to the pending "dealer day in court" bill;

—as part of the extension of oil and natural gas price controls, the Administration would urge that independent producers receive the same tax treatment of intangible drilling costs as their corporate competitors;

—a Presidential Commission will study and make recommendations concerning the national energy transportation system.

To provide relative invulnerability from another interruption of foreign oil supply, the Strategic Petroleum Reserve will be expanded to 1 billion barrels; efforts will be made to diversify sources of oil imports; contingency plans will be transmitted to the Congress; and development of additional contingency plans will be accelerated.

Coal

Conversion by industry and utilities to coal and other fuels would be encouraged by taxes on the use of oil and natural gas.

The Plan also contains a strong regulatory program that would prohibit all new utility and industrial boilers from burning oil or natural gas, except under extraordinary conditions. Authority would also exist to prohibit the burning of oil or gas in new facilities other than boilers. Existing facilities with coal-burning capability would generally be prohibited from burning oil and gas. Permits would be required for any conversion to oil or gas rather than to coal. By 1990, virtually no utilities would be permitted to burn natural gas.

While promoting greater use of coal, the Administration will seek to achieve continued improvement in environmental quality. A strong, but consistent and certain, environmental policy can provide the confidence industry needs to make investments in energy facilities. The Administration's policy would:

—require installation of the best available control technology in all new coal-fired plants, including those that burn low sulfur coal;

—protect areas where the air is still clean from significant deterioration;

—encourage States to classify lands to protect against significant deterioration within 3 years after enactment of Clean Air Act amendments;

—require Governors to announce intent to change the classification of allowable air quality for a given area within 120 days

after an application is made to construct a new source in that area;

—require States to approve or disapprove the application within 1 year thereafter.

Further study is needed of the Environmental Protection Agency's policies allowing offsetting pollution trade-offs for new installations. A committee will study the health effects of increased coal production and use, and the environmental constraints on coal mining and on the construction of new coal-burning facilities. A study will also be made of the long-term effects of carbon dioxide from coal and other hydrocarbons on the atmosphere.

The Administration supports uniform national strip mining legislation.

An expansion is proposed for the Government's coal research and development program. The highest immediate priority is development of more effective and economic methods to meet air pollution control standards. The program will include research on:

—air pollution control systems;

—fluidized bed combustion systems;

—coal cleaning systems;

—solvent refined coal processes;

—low Btu gasification processes;

—advanced high Btu gasification processes;

—synthetic liquids technology;

—coal mining technology.

Nuclear Power

It is the President's policy to defer any U.S. commitment to advanced nuclear technologies that are based on the use of plutonium while the United States seeks a better approach to the next generation of nuclear power than is provided by plutonium recycle and the plutonium breeder. The U.S. will defer indefinitely commercial reprocessing and recycling of plutonium. The President has proposed to reduce the funding for the existing breeder program, and to redirect it toward evaluation of alternative breeders, advanced converter reactors, and other fuel cycles, with emphasis on nonproliferation and safety concerns. He has also called for cancellation of construction of the Clinch River Breeder Reactor Demonstration Project and all component construction, licensing, and commercialization efforts.

To encourage other nations to pause in their development of plutonium-based technology, the United States should seek to restore confidence in its willingness and ability to supply enrichment services. The United States will reopen the order books for U.S. uranium enrichment services, and will expand its enrichment capacity by building an energy-efficient centrifuge plant. The President is also proposing legislation to guarantee the delivery of enrichment services to any country that shares U.S. nonproliferation objectives and accepts conditions consistent with those objectives.

To resolve uncertainties about the extent of domestic uranium resources, ERDA will reorient its National Uranium Resources Evaluation Program to improve uranium resource assessment. The program will also include an assessment of thorium resources.

The United States has the option of relying on light-water reactors to provide nuclear power to meet a share of its energy deficit. To enhance the safe use of light-water reactors:

—the Nuclear Regulatory Commission (NRC) has already increased the required number of guards at nuclear plants and the requirements for the training that guards receive;

—the President is requesting that the NRC expand its audit and inspection staff to increase the number of unannounced inspections and to assign one permanent Federal inspector to each nuclear power plant;

—the President is requesting that the Commission make mandatory the current voluntary reporting of minor mishaps and component failures at operating reactors;

—the President is requesting that the NRC develop firm siting criteria with clear guidelines to prevent siting of nuclear plants in densely populated locations, in valuable natural areas, or in potentially hazardous regions.

The President has directed that a study be made of the entire nuclear licensing process. He has proposed that reasonable and objective criteria be established for licensing and that plants which are based on a standard design not require extensive individual licensing.

To ensure that adequate waste storage facilities are available by 1985, ERDA's waste management program has been expanded to include development of techniques for long-term storage of spent fuel. Also, a task force will review ERDA's waste management program. Moreover, improved methods of storing spent fuel will enable most utilities at least to double their current storage capacity without constructing new facilities.

Hydroelectric Power

The Department of Defense (Corps of Engineers), together with other responsible agencies, will report on the potential for installation of additional hydroelectric generating capacity at existing dams throughout the country.

Nonconventional Resources

America's hope for long-term economic growth beyond the year 2000 rests in large measure on renewable and essentially inexhaustible sources of energy. The Federal Government should aggressively promote the development of technologies to use these resources.

Solar Energy

Solar hot water and space heating technology is now being used and is ready for widespread commercialization. To stimulate the development of a large solar market, a tax credit is proposed. The credit would start at 40 percent of the first $1,000 and 25 percent of the next $6,400 paid for qualifying solar equipment. The credit would decline in stages to 25 percent of the first $1,000 and 15 percent of the next $6,400. The credit would be supported by a joint Federal-State program of standards development, certification, training, information gathering, and public education. Solar equipment used by business and industry would be eligible for an additional 10 percent investment tax credit for energy conservation measures.

Geothermal Energy

Geothermal energy is a significant potential energy source. The tax deduction for intangible drilling costs now available for oil and gas drilling would be extended to geothermal drilling.

Research, Development and Demonstration

An effective Federal research, development and demonstration program is indispensable for the production of new energy sources. The Federal Government should support many research options in their early stages, but continue support into the later stages only for those that meet technical, economic, national security, health, safety, and environmental criteria. Research and development should be accompanied by preparation for commercialization so that successful projects can rapidly be put to practical use.

Additional research, development and demonstration initiatives are proposed, with emphasis on small, dispersed and environmentally sound energy systems.

An Office of Small-Scale Technologies would be established to fund small, innovative energy research and development projects. The office would enable individual inventors and small businesses to contribute to the national energy research and development effort.

Information

A three-part energy information program is proposed. A Petroleum Production and Reserve Information System would provide the Federal Government with detailed, audited data on petroleum reserve estimates and production levels. A Petroleum Company Financial Data System would require all large companies and a sample of small firms engaged in crude oil or natural gas production to submit detailed financial information to the Federal Government. Data required from integrated companies would permit evaluation of the performance of their various segments by providing vertical accountability. An Emergency Management Information System would provide the Federal and State governments with information needed to respond to energy emergencies.

Competition

Effective competition in the energy industries is a matter of vital concern. The Under Secretary for policy and evaluation in the proposed Department of Energy would be responsible for making certain that policies and programs of the Department promote competition. Although at this time it does not appear necessary to proceed with new legislation for either horizontal or vertical divestiture of the major oil companies, their performance will be monitored. The proposed information program would greatly assist that effort.

A present anomaly in the availability of the tax deduction for intangible drilling costs within the oil industry would be removed as part of the program for extending oil and natural gas price controls.

Emergency Assistance for Low-Income Persons

Existing emergency assistance programs are deficient in assisting low-income persons to meet sharp, temporary increases in energy costs due to shortages or severe winters. A redesigned program will be completed promptly and submitted to the Congress.

WAES ENERGY GLOBAL PROSPECTS: 1985-2000

In May 1977, the Workshop on Alternative Energy Strategies, a group representing 15 countries, published a report[1] on the global prospects for energy in the period 1985-2000. The conclusions of this group are as follows:

Conclusions

After two years of study we conclude that world oil production is likely to level off—perhaps as early as 1985—and that alternative fuels will have to meet growing energy demand. Large investments and long lead times are required to produce these fuels on a scale large enough to fill the prospective shortage of oil, the fuel that now furnishes most of the world's energy. The task for the world* will be to manage a transition from dependence on oil to greater reliance on other fossil fuels, nuclear energy and, later, renewable energy systems.

Our major conclusions are as follows:

1. The supply of oil will fail to meet increasing demand before the year 2000, most probably between 1985 and 1995, even if energy prices rise 50% above current levels in real terms. Additional constraints on oil production will hasten this shortage, thereby reducing the time available for action on alternatives.

2. Demand for energy will continue to grow even if governments adopt vigorous policies to conserve energy. This growth must increasingly be satisfied by energy resources other than oil, which will be progressively reserved for uses that only oil can satisfy.

3. The continued growth of energy demand requires that energy resources be developed with the utmost vigor. The change from

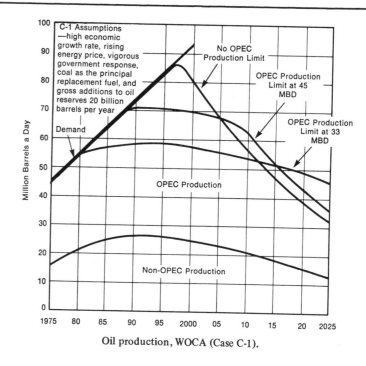

Oil production, WOCA (Case C-1).

a world economy dominated by oil must start *now*. The alternatives require 5 to 15 years to develop, and the need for replacement fuels will increase rapidly as the last decade of the century is approached.

4. Electricity from nuclear power is capable of making an important contribution to the global energy supply although worldwide acceptance of it on a sufficiently large scale has yet to be established. Fusion power will not be significant before the year 2000.

5. Coal has the potential to contribute substantially to future energy supplies. Coal reserves are abundant, but taking advantage of them requires an active program of development by both producers and consumers.

6. Natural gas reserves are large enough to meet projected demand provided the incentives are sufficient to encourage the development of extensive and costly intercontinental gas transportation systems.

---

* This study of energy has focused on the World Outside Communist Areas (shortened in this report to the acronym WOCA) which is often called the non-Communist world. The U.S.S.R. and China are major energy producers and suppliers, although their trade in fuels with countries in WOCA has to date been relatively small. In the WAES projections we have assumed that this situation would continue to the end of the century.

[1]*Energy: Global Prospects 1985-2000*, Report of the Workshop on Alternative Energy Strategies, a Project Sponsored by the Massachusetts Institute of Technology, McGraw-Hill Book Company, New York, 1977.

7. Although the resource base of other fossil fuels such as oil sands, heavy oil, and oil shale is very large, they are likely to supply only small amounts of energy before the year 2000.

8. Other than hydroelectric power, renewable resources of energy —e.g., solar, wind-power, wave-power—are unlikely to contribute significant quantities of additional energy during this century at the global level, although they could be of importance in particular areas. They are likely to become increasingly important in the 21st century.

9. Energy efficiency improvements, beyond the substantial energy conservation assumptions already built into our analysis, can further reduce energy demand and narrow the prospective gaps between energy demand and supply. Policies for achieving energy conservation should continue to be key elements of all future energy strategies.

10. The critical interdependence of nations in the energy field requires an unprecedented degree of international collaboration in the future. In addition it requires the will to mobilize finance, labor, research and ingenuity with a common purpose never before attained in time of peace; and it requires it now.

Failure to recognize the importance and validity of these findings and to take appropriate and timely action will almost certainly result in a world different from the one on which these projections have been based. Failure to act could lead to substantially higher energy prices as the supply/demand imbalance becomes more apparent—with the depressant effects on the economies of the world and the consequent frustration of the aspirations of the less-developed countries. The major political and social difficulties that might arise could cause energy to become a focus for confrontation and conflict.

In addition, the longer the world delays facing this issue the more serious the outcome will be. Even with prompt action the margin between success and failure in the 1985-2000 period is slim. Time has become one of the most precious of our resources. Recognizing the importance of time and the need to respond can help us through the period of transition that lies ahead.

Part I is a summary of the analysis on which these conclusions are based. Part II is a more detailed analysis of demand and conservation, oil, gas, coal, nuclear and other fuels, and finally the system we used to combine our estimates of future supply and demand in order to see where and when prospective shortages might appear. Much greater detail will be found in three Technical Vol-

Summary of Oil Balance in the Year 2000

Economic Growth:	High	High	Low	Low
Energy Price (1985-2000):	Rising	Rising	Constant	Constant
Principal Replacement Fuel:	Coal	Nuclear	Coal	Nuclear
WAES Scenario Case:	C-1	C-2	D-7	D-8
Major Importer's Desired Imports	(all numbers are in MBD)			
North America*	10.4	10.7	15.8	15.8
Western Europe	16.5	16.4	13.2	12.5
Japan	15.2	14.4	8.2	7.9
Non-OPEC Rest of WOCA	11.2	9.5	9.6	9.0
International Bunkers**	5.4	5.4	4.5	4.5
Total Desired Imports	58.7	56.4	51.3	49.7
Major Exporters' Potential Exports				
OPEC***	38.7	37.2	35.2	34.5
Prospective (shortage) or surplus	(20.0)	(19.2)	(16.1)	(15.2)
as a percentage of total WOCA potential oil production	27%	26%	28%	26%

* Takes account of domestic production of oil shale and oil sands in addition to conventional oil production.
** International bunkers represent the oil used in international shipping.
*** OPEC potential exports equal OPEC potential production minus OPEC internal demand.

North America: Canada, U.S.A.

Western Europe:
	WAES-Europe	plus	non-WAES Europe	
	Denmark	The Netherlands	Austria	Luxembourg
	Finland	Norway	Belgium	Portugal
	France	Sweden	Greece	Spain
	F.R.G.	U.K.	Iceland	Switzerland
	Italy		Ireland	

OPEC (Organization of Petroleum Exporting Countries): Algeria, Ecuador, Gabon, Indonesia, Iran, Iraq, Kuwait, Libya, Nigeria, Qatar, Saudi Arabia, United Arab Emirates (Abu Dhabi, Dubai and Sharjah), Venezuela.

Non-OPEC Rest of WOCA: all other countries outside Communist areas.

WOCA: World Outside Communist Areas.

* *Energy Demand Studies: Major Consuming Countries* (November 1976, MIT Press), *Energy Supply to the Year 2000: Global and National Studies*, and *Energy Supply-Demand Integrations to the Year 2000: Global and National Studies* (June 1977, MIT Press).

PROJECT INTERDEPENDENCE

In June 1977, the Congressional Research Service of the U.S. Library of Congress summarized its findings[1] on the U.S. and world energy outlook through 1990 as follows:

PROJECT INTERDEPENDENCE: U.S. AND WORLD ENERGY OUTLOOK THROUGH 1990

MAJOR FINDINGS

DEMAND

1. Assuming long-term real GNP growth in the United States of 3.5 percent per year for the period 1976–90, CRS base case projects 2.9 percent average annual energy growth rates. The energy demand forecast assumes no supply constraints but higher prices for oil and gas compared with the present.

2. Imposition of a gradually escalating Btu tax could result in bringing average annual energy growth rates down considerably, but the price to be paid for lower energy growth will be dependent on the ability of the economy to substitute labor, capital and other materials for energy. If there is enough flexibility in the economy, GNP growth rates and employment will only be marginally affected; if, on the other hand, there is little flexibility in the economy, the Btu tax could result in substantially lower GNP growth rates and higher unemployment. This is currently a matter of some controversy.

DOMESTIC SUPPLY

3. Under optimistic political and economic assumptions, domestic oil production could rise from 10.0 million barrels per day (b/d) in 1976 to 10.4 million b/d in 1980, 10.9 million b/d in 1985, and 11.4 million b/d by 1990. These supply figures assume—among other things—decontrol of oil and an aggressive OCS development plan. It seems more probable that the political and economic climate will be less favorable and finding rates lower than projected in the base case. This could result in production of 9.5 million b/d in 1980 and 1985 and 10.1 million b/d in 1990.

[1] *Project Interdependence: U.S. and World Energy Outlook Through 1990*, A summary report prepared by the Congressional Research Service, Library of Congress, for the Committee on Energy and Natural Resources, U.S. Senate, and the Subcommittee on Energy and Power, Committee on Interstate and Foreign Commerce, U.S. House of Representatives, Publication No. 95-31, June 1977.

TABLE 1.—U.S. ENERGY DEMAND AND SUPPLY: 1977-90, PROJECTIONS

[In million barrels per day oil equivalent]

	1977	1980	1985	1990
Supply:				
Oil and NGL	9.6	10.4	10.9	11.4
Natural gas	9.2	8.5	8.3	8.3
Shale oil			.08	.2
Nuclear power	1.5	2.3	3.4	5.0
Coal (supply for domestic)	6.7	6.8	7.7	10.5
Hydro/solar/geo	1.3	1.7	1.9	2.4
Total U.S. supply	28.3	29.7	32.3	37.8
Demand	36.9	40.4	45.1	51.9
Imports:				
Natural gas	.5	.7	1.0	1.2
Crude oil and products	8.1	10.0	11.8	12.9
Oil for oil storage program	.1	.4	.3	

[In conventional units]

	1977	1980	1985	1990
Supply:				
Oil and NGL (million b/d)	9.6	10.4	10.9	11.4
Natural gas (TCF/yr)	18.6	17.4	16.9	16.9
Shale oil (million b/d)			.08	.2
Nuclear power (gigawatts)	54	81	131	175
Coal (million short tons/yr utilization)	630	695	850	1.135
Hydro/solar/geo (gigawatts)	2.8	3.5	4.0	5.0
Imports:				
Natural Gas (TCF/yr)	.99	1.5	2.1	2.4
Crude oil and products (million b/d)	8.1	10.0	11.8	12.9
Oil storage (million b/d)	.1	.4	.3	

4. Under similar optimistic political and economic assumptions, natural gas production is projected to decline only gradually from 19.7 trillion cubic feet (TCF) in 1976 to 17.4 TCF in 1980, and 16.9 TCF each in 1985 and 1990. These supply figures assume early decontrol of new natural gas; an aggressive OCS development plan; a completed natural gas pipeline for Alaskan gas by 1985; very high-finding rates in frontier areas between 1976 and 1990. Under less optimistic, but currently more probable political and economic assumptions and continuing decline in gas findings in the lower 48 States, production is likely to decline faster to 16.3 TCF in 1980, 15.7 TCF in 1985, and 15.5 TCF in 1990.

5. Of all potential sources of synthetic fuels, only synthetic gas from coal is projected to make a substantial contribution to supply in 1985 (0.5 TCF) and 1990 (1.5 TCF). Potentially promising nonconventional sources of methane—such as methane occluded in coal, geopressure zones, natural gas from Devonian shale and tight formation—are not expected to make a substantial contribution to U.S. energy supply between 1976 and 1990. Shale oil is projected to supply 75,000 b/d in 1985 and 200,000 b/d by 1990.

6. While coal reserves are huge, production is likely to be limited to 775 million tons in 1980, 940 million tons in 1985, and 1,225 million tons by 1990. Of these figures, about 80 million tons is projected for exports in 1980, and 90 million tons may be exported by 1985 and 1990. In terms of Btu's, production increases are lower due to the higher proportion of low-Btu western coal, as compared with current production from primarily higher Btu eastern coal. Coal production and utilization is subject to numerous constraints, and the domestic production and utili-

zation figures used in the base case of this study assume that some impediments to higher coal use will be removed soon.

7. During the period of our forecast, the main constraint to expansion of nuclear power capacity is the rate of licensing new nuclear powerplants. The base case assumes that licensing will be speeded up from a recent low of about one plant every 8 weeks to about one plant every 5 weeks.

8. All other domestic energy sources: hydropower has been projected to increase slowly; geothermal is not likely to gain a role of significance during the next 15 years; and, solar energy is projected to supply about 1 percent of total U.S. energy demand in 1985 and 2 percent by 1990. The projected solar energy contribution assumes substantial government support for solar energy in the form of tax incentives.

CAPITAL

9. Uncertainties about energy policy have delayed investment decisions in many sectors of the energy industry. Capital requirements to meet projected energy supply are substantial, but can be met assuming prices will not be regulated to the extent of preventing a fair return on investment. Some form of government support in the form of price floors or loan guarantees might be needed to encourage development of high-risk energy sources such as shale oil, tar sands, synthetic gas, and other alternative energy resources.

OIL AND NATURAL GAS IMPORTS

10. Under base case demand and supply conditions laid down in this report, natural gas imports are expected to grow from 0.9 TCF in 1976 to 1.5 TCF in 1980, 2.1 TCF in 1985, and 2.4 TCF by 1990. Oil imports—crude oil and oil products—are projected to increase from 7.2 million b/d in 1976 to 10 million b/d in 1980, 11.8 million b/d in 1985, and 12.9 million b/d by 1990. The strategic oil storage program is projected to add another 0.4 million b/d in 1980 and 0.3 million b/d in 1985.

11. Additional oil imports will come primarily from Arab countries in the Middle East and North Africa. Total U.S. imports from the Middle East and North Africa are projected to grow from about 2.5 million b/d in 1976 to 6.6 million b/d in 1980, 7.4 million b/d in 1985, and 7.9 million b/d by 1990. The adverse effects of another oil embargo on the economy at current and projected Arab oil import levels could be anywhere between very substantial and disastrous.

12. The implications of a dependence on the Middle East and North Africa in 1985 equal to total oil imports in 1976 are manifold and include: (a) potential oil supply interruptions as a result of war in the Middle East, or confrontation with the Soviet Union or, acts of sabotage by terrorist groups; (b) unknown effects of the accumulation of vast monetary reserves in Saudi Arabia and a few other Arab Gulf countries on the world economy; (c) potential conservation policies of Saudi Arabia or other Arab Gulf countries with limited capital absorptive capacity, resulting in escalating oil prices; (d) reduced leverage in the exercise of U.S. foreign policy in the region; (e) possible friction within the western alliance systems related to potential competition for available Middle East oil supplies.

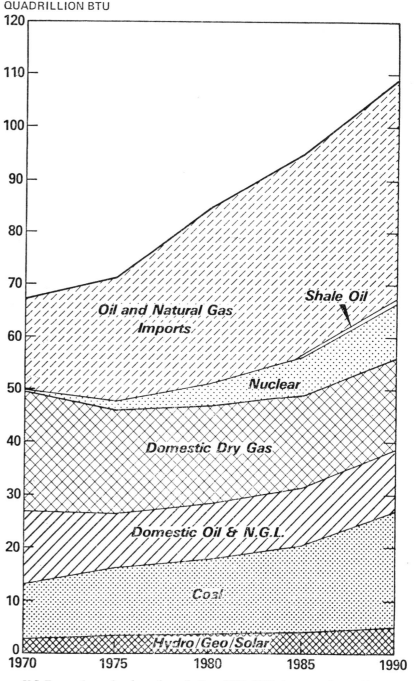

U.S. Energy demand and supply projections 1970–1990—base case (in quads).

13. World oil productive capacity can meet total world demand for oil throughout most of the decade of the 1980's, if the Soviet bloc does not become a major importer of oil. Sometime between the late 1980's and middle 1990's productive capacity of world oil is likely to peak, followed by declining world oil production (conventional oil sources only).

14. In spite of the production capacity there is no guarantee that producer states—in particular Saudi Arabia and a few smaller producers in the Arab Gulf region with low capital absorptive capacity—will continue to increase oil production. For political and economic reasons world oil demand could outpace supply as early as the early 1980's or as late as the early or middle 1990's. A few producer states—in particular Saudi Arabia—are likely to be in a position to determine when world oil production levels will peak. This action will be followed by subsequent world oil price increases.

15. Economic incentives by themselves are not likely to encourage the Saudis and other Arab Gulf producers with limited capital absorptive capacity to continue to increase oil production. A key incentive to a general atmosphere of cooperation between the industrial world and OAPEC is early progress toward a comprehensive peace settlement in the Arab-Israeli dispute. Other incentives to encourage further growth in OAPEC oil exports include: (*a*) progress in the "North-South" dialogs; (*b*) favorable foreign investment policy in the industrial world, encouraging surplus nations to produce more oil and invest their surpluses in the rest of the world; (*c*) assistance in modernization of developing oil-producing states, leading toward further interdependence between consumer and producer states.

16. There seems no practical way to avoid development and implementation of energy policies in the industrial world—spearheaded by the United States—leading to an aggressive conservation ethic and a major program of energy resource development. If the industrial nations can show that they are willing to make adequate sacrifices at home, leading to reduced long-term reliance on imported oil, the Gulf States may be convinced to fill the energy gap in the transition period. A considerable degree of cooperation and understanding of interdependence can only be avoided at great cost.

17. The Soviet bloc is not expected to become a major net exporter or importer of energy throughout the 1980's. The Peoples Republic of China is projected to export 0.5 million b/d in 1980 and 1 million b/d by 1985.

Glossary

absorbed dose. When ionizing radiation passes through matter, some of its energy is imparted to the matter. The amount absorbed per unit mass of irradiated material is called the absorbed dose, and is measured in rems and rads.

absorber. Any material that absorbs or diminishes the intensity of ionizing radiation. Neutron absorbers, like boron, hafnium and cadmium are used in control rods for reactors. Concrete and steel absorb gamma rays and neutrons in reactor shields. A thin sheet of paper or metal will absorb or attenuate alpha particles and all except the most energetic beta particles.

absorption. The process by which the number of particles or photons entering a body of matter is reduced by interaction of the particles or radiation with the matter; similarly, the reduction of the energy of particles or photons while traversing a body of matter.

acre-foot. A quantity of water covering 1 acre to a depth of 1 foot. Contains 43,560 gallons (U.S.). One acre-foot of water can satisfy the municipal and industrial energy demands of four people for 1 year.

activation. The process of making a material radioactive by bombardment with neutrons, protons, or other nuclear particles or photons.

acute radiation sickness syndrome. An acute organic disorder that follows exposure to relatively severe doses of ionizing radiation. It is characterized by nausea, vomiting, diarrhea, blood cell changes, and in later stages by hemorrhage and loss of hair.

AEC. The U.S. Atomic Energy Commission. The AEC was abolished January 19, 1975 and its activities transferred to the Energy Research and Development Administration (ERDA) and the Nuclear Regulatory Commission (NRC).

AGA. American Gas Association. The trade association of the private gas industry.

air sampling. The collection and analysis of samples of air to measure its radioactivity or to detect the presence of radioactive substances, particulate matter or chemical pollutants.

alpha particle. (Symbol α) A positively charged particle emitted by certain radioactive materials. It is made up of two neutrons and two protons bound together. Hence it is identical with the nucleus of a helium atom. It is the least penetrating of the three common types of decay radiation.

alternating current (AC). An electric current whose direction is reversed at regular intervals. Electric power in the United States alternates with a frequency of 60 hertz, or cycles per second. Many countries use a frequency of 50 hertz.

ampere. A unit of measure for an electric current; the amount of current which flows in a circuit in which the electromotive force is 1 volt and the resistance is 1 ohm.

animal waste conversion. The process for obtaining oil from animal wastes. A Bureau of Mines experiment has obtained 80 gallons of oil per ton from cow manure. In comparison, average oil shale yields 25 gallons of oil per ton of ore.

anthracite. A hard, black, lustrous coal that burns efficiently and is therefore valued for its heating quality.

API. American Petroleum Institute. A trade association of the American petroleum industry.

associated-dissolved gas. Associated gas is free natural gas in immediate contact, but not in solution, with crude oil in the reservoir; dissolved gas is natural gas in solution in crude oil in the reservoir.

ATGAS. A process for coal gasification being developed for the Department of the Interior by Applied Technology Inc. The primary feature of the process is dissolving of coal in a bath of molten iron.

atom. A particle of matter whose nucleus is indivisible by chemical means. It is the fundamental building block of the chemical elements.

atomic energy. The energy released by a nuclear reaction or by radioactive decay. (See radioactivity, fission, fusion, nuclear reactors.)

Atomic Energy Commission. (See AEC.)

atomic mass. (See atomic weight.)

atomic mass unit. (Abbreviation amu.) One-twelfth the mass of a neutral atom of the most abundant isotope of carbon, carbon-12.

atomic number. (Symbol Z.) The number of protons in the nucleus of an atom, and also its positive charge. Each chemical has its characteristic atomic number, and the numbers of the known elements form a complete series from 1 (hydrogen) to 105.

atomic reactor. A nuclear reactor.

atomic weight. The basis of the scale of atomic weights is the oxygen atom, which has been arbitrarily assigned an atomic weight of 16. The unit of the scale is 1/16 the weight of oxygen-16, or roughly the mass of the proton or neutron. The atomic weight of an element, therefore, is approximately equivalent to the total number of protons and neutrons in its nucleus.

autoradiograph. A photographic record of radiation from radioactive material in an object, made by placing the object very close to a photographic film or emulsion. The process is called autoradiography. It is used, for instance, to locate radioactive atoms or tracers in metallic or biological samples.

backup. Reserve generating capacity of a power system.

background radiation. The radiation in man's natural environment, including cosmic rays and radiation from the naturally radioactive elements, both outside and inside the bodies of humans and animals. It is also called natural radiation.

backscatter. When radiation of any kind strikes matter (gas, solid, or liquid) some of it may be reflected or scattered back in the general direction of the source. An understanding or exact measurement of the amount of backscatter is important when beta particles are being counted in an ionization chamber, in medical treatment with radiation, or in the use of industrial radioisotopic thickness gauges.

barrel (bbl). A liquid measure of oil, usually crude oil, equal to 42 American gallons or about 306 pounds. One barrel equals 5.6 cubic feet or 0.159 cubic meters. For crude oil 1 bbl is about 0.136 metric tons, 0.134 long tons, and 0.150 short tons. The energy values of petroleum products per barrel are: crude petroleum 5.6 million Btu/bbl; residual fuel oil−6.29; distillate fuel oil−5.83; gasoline−5.25; jet fuel−5.67; jet fuel (naphtha-type)−5.36; kerosene−5.67; petroleum coke−6.02 and asphalt−6.64.

barrier shield. A wall or enclosure shielding the operator from an area where radioactive material is being used or processed by remote control equipment.

base load. The minimum load of a utility (electric or gas) over a given period of time.

base load station (gas). A station which is normally operated to take all or part of the base load of a system and which, consequently, operates at a high load factor.

Bbl/d. Barrels per day.

Bbls. Barrels.

bender process. A continuous, fixed-bed chemical treating process using a lead sulfide catalyst for sweetening light distillates. The process converts mercaptans to disulfides by oxidation.

benzene. C_6H_6. A colorless liquid hydrocarbon, made from coal tar and by catalytic reforming of naphthenes. It is used in the manufacture of phenol, styrene, nylon, detergents, aniline, phthalic anhydride and other compounds; as a solvent; and as a component of high-octane gasoline.

beta particle. (Symbol β^-) An elementary particle emitted from a nucleus during radioactive decay, with a single electrical charge and a mass equal to 1/1837 that of a proton. A negatively charged beta particle is identical to an electron. A positively charged beta particle is called a positron. Beta radiation may cause skin burns, and beta-emitters are harmful if they enter the body. Beta particles are easily stopped by a thin sheet of metal.

BeV. Symbol for a billion (10^9) electron volts. (See electron volt.)

bi-gas. A process for coal gasification being developed by the Department of Energy and the American Gas Association.

binding energy. The binding energy of a nucleus is the minimum energy required to dissociate it into its component neutrons and protons.

biological dose. The radiation dose absorbed in biological material. Measured in rems.

biological half-life. The time required for a biological system, such as a human or animal, to eliminate by natural processes half the amount of a substance (such as radioactive material) that has entered it.

biological shield. A mass of absorbing material placed around a reactor or radioactive source to reduce the radiation to a level safe for humans.

bitumen. A general name for various solid and semisolid hydrocarbons; a native substance of dark color, comparatively hard and nonvolatile, composed principally of hydrocarbon.

bituminous coal. Soft coal; coal that is high in carbonaceous and volatile matter. When volatile matter is removed from bituminous coal by heating in the absence of air, the coal becomes coke.

blended fuel oil. A mixture of residual and distillate fuel oils.

blending naphtha. A distillate used to thin heavy stocks to facilitate processing, e.g., to thin lubricating oil in dewaxing processes.

blending stock. Any of the stocks used to make commercial gasoline. These include: natural gasoline, straight-run gasoline, cracked gasoline, polymer gasoline, alkylate, and aromatics.

body burden. The amount of radioactive material present in the body of man or animals.

boiling water reactor. A nuclear reactor in which water, used as both coolant and moderator, is allowed to boil in the core. The resulting steam generally is used directly to drive a turbine.

bone seeker. A radioisotope that tends to accumulate in the bones when it is introduced into the body. An example is Sr-90, which behaves chemically like calcium.

bottoming cycle. A means to increase the thermal efficiency of a steam electric generating system by converting some waste heat from the condenser into electricity rather than discharging all of it to the environment.

breeder reactor. A nuclear reactor designed to produce more fuel than it consumes. The new fissionable materials are created by neutron capture in the fertile materials U-238 or Th-232. There are three types of breeder reactors: the liquid metal fast breeder (LMFBR); the gas cooled fast breeder (GCBR); and the molten salt breeder (MSBR).

breeding ratio. The number of new fission atoms produced in a breeder reactor per fissionable atom consumed in the reactor.

British Thermal Unit (Btu). The quantity of heat necessary to raise the temperature of one pound of water 1°F. One Btu equals 252 calories, 778 foot-pounds, 1055 joules, or 0.293 watt-hours.

bulk plant. A wholesale distributing unit for petroleum products, often having facilities on railroad sidings. It may have tank storage for light oils and a warehouse with storage for products sold in barrels and packages.

Bunker "C" fuel oil. A heavy residual fuel oil used by ships, industry, and for large-scale heating installations. In industry it is often referred to as No. 6 fuel.

burnable poison. A neutron absorber (or poison) such as boron, which when incorporated in the fuel or fuel cladding of a nuclear reactor, gradually "burns up" (is changed into nonabsorbing material) under neutron irradiation. This process compensates for the loss of reactivity in a reactor which occurs as fuel is consumed and poisonous fission products accumulate.

burn-up (nuclear). A measure of the consumption of nuclear fuel in a nuclear reactor. Fuel burn-up may be expressed in terms of total energy extracted from the fuel during its stay in the reactor, or in terms of percentage of the fuel consumed over that period. For the former, the units usually are megawatt-days of heat per metric ton. (MWD/tonne). A 1% burn-up is about 9000 MWD/tonne.

Bureau of Mines (BoM). A bureau of the Department of Interior established in 1910. The bureau ". . . conducts research and administers regulatory programs necessary for performance of the governmental function to stimulate the private sector toward the production of an appropriate and substantial share of the national mineral and fuel needs in a manner that best protects the public interest."

busbar. An electrical conductor in the form of rigid bars located in a switchyard or power plants, serving as a common connection for two or more electrical circuits.

by-products (residuals). Secondary products which have commercial value and are obtained from the processing of raw material. They may be the residues of the gas production process, such as coke, tar, and ammonia, or they may be the result of further processing of such residues, such as ammonium sulphate.

by-product material (*nuclear*). Any artificial radioactive material obtained during the production or use of source material or fissionable material. It includes fission products and radioisotopes produced in nuclear reactors, but not radioactive materials occurring in nature or those made with accelerators such as cyclotrons.

caking coal. Coal which softens and agglomerates on heating and, after volatile matter has been driven off at high temperatures, produces a hard gray cellular mass of coke.

calorie. A unit of heat energy equal to the amount of heat that will raise the temperature of 1 gram of water 1°C. The calorie is used when temperature is measured on the Centigrade scale, while the British thermal unit is used when the measurement is on the Fahrenheit scale. One calorie equals 3.97×10^3 Btu, 4.18 joules, 1.10×10^3 watt-hours.

calorie (*large calorie*). The amount of heat required to change the temperature of 1 kilogram of water 1°C. The large calorie is used to designate nutritional values.

calorific value. The heat liberated by the combustion of a unit quantity of a fuel.

carbon oxides. Compounds of carbon and oxygen produced when the carbon of fossil fuels combines with oxygen during burning. The two most common such oxides are carbon monoxide, a very poisonous gas, and carbon dioxide.

cash bonus payment. A cash consideration paid by the lessee for the execution of an oil or gas lease by a landowner. The bonus is usually computed on a per acre basis.

cask. A heavily shielded container used to store and/or ship radioactive materials.

catalytic cracking. The conversion of high-boiling hydrocarbons into lower boiling substances by means of a catalyst. Feedstocks may range from naphtha cuts to reduced crude oils.

catalytic reforming. The rearranging of hydrocarbon molecules in a gasoline feedstock to produce other hydrocarbons having a higher antiknock quality.

cathode rays. A stream of electrons emitted by the cathode, or negative electrode, of a gas-discharge tube or by a hot filament in a vacuum tube, such as a television tube.

ccf. One hundred cubic feet. A unit of measure used for small amounts of natural gas.

CEQ. Council on Environmental Quality.

cf. Cubic feet.

cfh. Cubic feet per hour.

cfm. Cubic feet per minute.

cfs. Cubic feet per second.

chain reaction. A nuclear reaction that is self-sustaining. In a fission chain reaction, a fissionable nucleus absorbs a neutron and fissions, releasing additional neutrons. These in turn can be absorbed by other fissionable nuclei, releasing still more neutrons. A fission chain reaction is self-sustaining when the number of neutrons released equals or exceeds the number of neutrons lost by absorption in nonfissionable material or by escape from the system.

charged particle. An ion; an elementary particle that carries a positive or negative electric charge.

char-oil process (*COED*). A process being developed by the Department of Energy for low-temperature distillation of coal carbonization products. The process is designed to produce clean liquids, gases and char for fuel, with the product balance depending upon economic factors.

chromosome. A biological structure within a cell which is responsible for inherited characteristics.

cladding. The outer jacket of nuclear fuel elements. It prevents corrosion of the fuel by the coolant and the release of fission products into the coolant. Aluminum or its alloys, stainless steel and zirconium alloys are common cladding materials.

closed-cycle reactor system. A reactor design in which the primary heat of fission is transferred outside the reactor core by means of a coolant circulating in a completely closed system that includes a heat exchanger.

CO_2-acceptor process. A process being developed by the Department of Energy and the American Gas Association to gasify lignite. Commercial application could provide a clean synthetic gas or low-sulfur solid fuel for power generation.

coal augering. A surface mining method used to extract coal from seams exposed by cuts in hilly terrain. The mining machines consist of large single and double augers which drill horizontally into the seams to extract the coal.

coal classification systems. In all countries the basis for classification is content of volatile matter. Anthracite is 10% volatile; lean coal, semianthracite or dry-steam coal is 10–13% volatile; variously designated coal is 14–20% volatile; coking coal is 20–30% volatile.

coal gas. An artificial gaseous fuel produced by heating coal in the absence of oxygen. This fuel, known as town gas, has a heating value of about 450 Btu per standard cubic foot. Use of coal gas in the United States ceased when inexpensive natural gas became widely available.

coal gasification. The conversion of coal to a gas suitable for use as a fuel. Cf. Hygas, CO^2-acceptor, Bi-gas, methanation, Lurgi, ATGAS processes.

coal liquefaction. The conversion of coal into liquid hydrocarbon and related compounds by hydrogenation.

coal oil. Oil obtained by the destructive distillation of bituminous coal.

coal preparation. A collective term for physical and mechanical processes applied to coal to make it suitable for a particular use.

coal preparation plant. A plant for the cleaning and sizing of the raw coal before it is loaded into railway cars or trucks.

coal slurry pipeline. A pipeline which transports coal in pulverized form suspended in water.

coal tar. A gummy, black substance produced as a by-product when bituminous coal is distilled.

coastwise shipping. Goods shipped from one United States port to another along the same coastal region.

COED. A Department of Energy project for development of liquid fuel from coal char.

coke. A porous, solid residue resulting from the incomplete combustion of coal heated in a closed chamber, or oven, with a limited supply of air. Coke is largely carbon and is a desirable fuel in certain metallurgical industries.

coking coal. The most important of the bituminous coals, which burns with a long yellow flame and creates an intense heat.

combination utility. A utility which supplies more than one service (gas, electricity, water, traction, etc.).

combined-cycle plant. A plant which achieves higher efficiency by employing two thermodynamic cycles in tandem. For example, heat rejected from a gas-fired turbine is used to generate steam to operate a steam turbine.

containment (*nuclear*). A gas-tight shell or other enclosure around a nuclear reactor designed to contain radioactive vapors and gases that might otherwise be released to the atmosphere in a reactor accident.

contamination. The presence of unwanted radioactive matter on objects or materials.

continental shelf. The extension of the continental land mass into the oceans, under relatively

control rod. Rod, plate, or tube containing a strong neutron absorbing material (hafnium, boron, etc.) used to control the power of a nuclear reactor. A control rod absorbs neutrons, preventing them from causing further fissions.

conventional gas. Natural gas as contrasted to synthetic gas.

conventional oil. Crude oil and condensate as contrasted with synthetic oil from shale or coal.

conversion. The chemical processing of uranium ore concentrates into uranium hexafluoride gas.

conversion factors: The energy content of most fuels can vary depending on their source and composition. The following energy equivalents are among those commonly used. Coal:

> Anthracite = 25.4 million Btu/ton.
> Bituminous = 26.2 million Btu/ton.
> Subbituminous = 19.0 million Btu/ton.
> Lignite = 13.4 million Btu/ton.

The average heating value of bituminous coal and lignite used in electricity generation and in industry in 1969 in the United States was 24.7 million Btu/ton.
Petroleum:

> Crude petroleum—5.60 million Btu/bbl (42 gal).
> Residual fuel oil—6.29 million Btu/bbl.
> Distillate fuel oil—5.83 million Btu/bbl.
> Gasoline (including aviation)—5.25 million Btu/bbl.
> Jet fuel (kerosene type)—5.67 million Btu/bbl.
> Jet fuel (naphtha-type)—5.36 million Btu/bbl.
> Kerosene—5.67 million Btu/bbl.
> Asphalt and road oil—6.64 million Btu/bbl.

Natural Gas:

> Dry—1031 Btu/cu ft at STP.
> Wet—1103 Btu/cu ft at STP.
> Liquids (avg)—4.1 million Btu/bbl.
> Fissionable material—74 million Btu/gm U-235 fissioned.

converter reactor. A nuclear reactor that produces some new fissionable material but less than the nuclear material it consumes. Light water reactors and high temperature gas-cooled reactors are converters.

coolant. A substance circulated through a nuclear reactor to remove or transfer heat. Common coolants include water, air, carbon dioxide, helium, and liquid sodium.

cooling pond. An artificial pond used to receive and dissipate waste heat, usually from a steam-electric power plant. Approximately an acre of pond surface is needed per megawatt of electric output for a modern steam-electric power plant.

cooling tower, wet. A unit or structure, usually built of wood, for the cooling of water by evaporation.

cooling tower, dry. A heat exchanger in which the heat in water is transferred to the air. The radiator of an automobile performs in this manner.

core. The central part of a nuclear reactor which contains the nuclear fuel.

counter. A general designation applied to radiation detection instruments or survey meters that detect and measure radiation.

cosmic rays. Radiation of many sorts that stem directly or indirectly from sources outside the earth's atmosphere. Cosmic radiation is part of the natural background radiation, and some of its constituents have extremely high energies.

cps. Cycles per second.

cracking. Processing that breaks down and rearranges the molecular structure of hydrocarbon chains. In thermal cracking, high temperature and high pressure are applied; in catalytic cracking, temperature and pressure are applied in the presence of a catalyst.

cracking plant. An oil refinery.

cracking still. The combined equipment—furnace, reaction chamber, fractionator—for the thermal conversion of heavier charging stock to gasoline.

critical mass. The smallest mass of fissionable material that will support a self-sustaining chain reaction under stated conditions.

criticality. The state of a nuclear reactor when a chain reaction has been achieved.

crude. Oil in its natural state, before refining or processing.

crude naphtha. Light distillate made in the fractionation of crude oil.

crude oil. A mixture of hydrocarbons as present in natural underground reservoirs.

cryogenics. The study and production of very low temperatures and their associated phenomena.

cryogenic techniques. Techniques involving extremely low temperatures used to keep certain fuels in liquid form; i.e., liquefied by hydrogen. CF. super conductivity.

cubic foot (cu ft). The most common unit of measurement of gas volume. It is the amount of gas required to fill a volume of one cubic foot under stated conditions of temperature, pressure, and water vapor. One cubic foot equals 28317. 01 cubic centimeters; 1728 cubic inches; 7.48 gallons (U.S.); and 28.31 liters. One cubic foot/second equals 1.98 acre-feet/day; 448.8 gallons/minute; and 0.646 million gallons/day.

cubic meter. A measure of volume in the metric system. One cubic meter equals 8.6 barrels (U.S. liquids); 35.3 cubic feet; 1.3 cubic yards; 264.1 gallons (U.S.); and 999.9 liters.

cubic yard. A measure of volume. One cubic yard equals 27 cubic feet and 0.76 cubic meters.

curie. (Abbreviation Ci.) The basic unit to describe the intensity of radioactivity in a sample of material. The curie is equal to 37 billion disintegrations per second, which is approximately the rate of decay of 1 gram of radium. A curie is also a quantity of any nuclide having 1 curie of radioactivity. Named by Marie and Pierre Curie, who discovered radium in 1898. Common fractions are:

> Megacurie: One million curies (MCi).
> Microcurie: One millionth of a curie.
> Millicurie: One thousandth of a curie (mCi).

CUP. Cascade uprating program.

daughter. A nuclide formed by the radioactive decay of another nuclide, which in this context is called the parent (see radioactive series).

dc. Direct current.

dead-weight tons (dwt). The total lifting capacity of a ship expressed in long tons (2240 lbs). For example, the oil tanker Universe Ireland is listed as 312,000 dwt which means it can carry 312,000 tons of oil or about 1.9 million barrels.

decay, radioactive. The process whereby atoms of radioactive substances experience transformation into atoms of other elements with attendant emission of penetrating radiations (gamma rays) and some nuclear particles. Each radioactive substance has a unique decay rate which may range from a fraction of a second to hundreds of years or more.

decay cooling. The storage of irradiated fuel elements to allow for radioactive decay of short-lived radioisotopes prior to initiating fuel reprocessing.

decay heat. The heat produced by radioactive decay of radioactive fission products in a nuclear core.

decay product. A nuclide resulting from the radioactive disintegration of a radionuclide, formed either directly or as the result of successive transformations in a radioactive series. A decay product may be either radioactive or stable.

decontamination. The removal of radioactive contaminents from surfaces or equipment, as by cleaning or washing with chemicals.

deep mining. The exploration of coal or mineral deposits at depths exceeding about 1000 feet.

degree day, cooling. A measure of the need for air-conditioning (cooling based on temperature and humidity). Although cooling degree days are published for many weather stations, a specific procedure has not been generally accepted.

degree day heating. A measure of heating requirements, based on the extent to which the daily mean temperature falls below a reference temperature usually 65°F.

deliverability. The volume of gas a well, field, pipeline, or distribution system can supply in a given period of time.

demand. The demand on a system or any of its parts over an interval of time, determined by dividing the total energy supplied by the number of units of time in the interval.

depleted uranium. Uranium containing less U-235 than the 0.71% found in nature. Depleted uranium is a by-product of the enrichment process.

depletion allowance. A tax allowance extended to the owner of exhaustible resources based on an estimate of the permanent reduction in value caused by the removal of the resource.

design basis accident. The design basis accident is the most serious reactor accident that can be hypothesized from an adverse combination of equipment malfunction, operating errors, and other foreseeable causes.

desulfurization. The process by which sulfur and sulfur compounds are removed from gaseous or liquid hydrocarbon mixtures.

detector. Material or device that is sensitive to radiation and can produce a response signal suitable for measurement or analysis. A radiation detection instrument.

deuterium. (Symbol ^{2}H or D.) An isotope of hydrogen whose nucleus contains one neutron and one proton and is therefore about twice as heavy as the nucleus of normal hydrogen, which is a single proton. Deuterium is often referred to as heavy hydrogen; it occurs in nature as 1 atom to 6500 atoms of normal hydrogen. It is nonradioactive. (See heavy water.)

diesel engine. An internal combustion engine in which combustion is caused by the heat generated by the compression of an internal combustion engine in which combustion is caused by the heat generated by the compression of an air-fuel mixture.

diesel fuel. Fuel used for internal combustion in diesel engines; usually that fraction which distills after kerosene; similar to gas oil.

direct current (dc). Electricity that flows continuously in one direction as contrasted with alternating current.

direct energy conversion. The generation of electricity from an energy source in a manner that does not include transference of energy to a working fluid. Direct conversion methods usually produce direct current. Some methods include thermoelectric conversion, thermionic conversion, and magnetohydrodynamic conversion.

distillate fuel oil. Any fuel oil, topped crude oil, or other petroleum oil, derived by refining or processing crude oil, which has a boiling range from 550° to 1200°F.

Doppler effect. A shift in the measured frequency of a wave pattern caused by movement of either the receiving device or the wave source. For example, a moving receiver will intercept more or fewer waves per unit time depending on whether it is moving toward or away from the source of the waves. By analogy, in a reactor, since fission cross sections depend on relative velocity of the neutrons and the uranium atoms (neutron movement can be considered wave motion), vibration of the uranium atoms in a fuel element due to the increased operating temperature leads to a Doppler effect. This Doppler effect can vary the reactivity of the reactor.

dose. The amount of ionizing radiation energy absorbed per unit mass of irradiated material. Measured in reps, rems, and rads.

dose rate. The radiation dose delivered per unit time.

dosimeter. A device that measures radiation dose, such as a film badge or ionization chamber.

doubling time. In the long-term operation of a breeder reactor system, the time required to achieve a net doubling of the inventory of fissionable material present in the system, expressed in years.

drift. Water lost from an evaporative cooling tower as liquid droplets entrained in the exhaust air.

drift mine. A coal mine which is entered directly through a horizontal opening.

dry hole. A drilled well which does not yield gas and/or oil in quantities or condition to support commercial production.

ecology. The science dealing with the relationship of all living things with each other and with their environment.

ecosystem. A complex of the community of living things and the environment forming a functioning whole in nature.

efficiency, thermal. A percentage indicating the available heat input that is converted to useful purposes. It is applied, generally, to combustion equipment. E = Btu output/Btu input.

EHV. Extra-high voltage.

electron. (Symbol e$^-$.) An elementary particle with a unit negative charge and a mass 1/1837 that of the proton. Electrons surround the positively charged nucleus and determine the chemical properties of the atom.

electron volt. (Abbreviation ev or eV.) The amount of kinetic energy gained by an electron when it is accelerated through an electric potential of 1 volt. It is equivalent to 1.603×10^{-12} erg.

element. One of the 105 known chemical substances that cannot be divided into simpler substances by chemical means. Examples are hydrogen, lead, and uranium.

emergency core cooling system (ECCS). The system which provides an emergency supply of cooling water to the core of a nuclear reactor in the event that the heat transfer fluid normally used is suddenly lost.

energy. The capability of doing work. There are several forms of energy, including kinetic, potential, thermal, and electromagnetic. One form of energy may be changed to another, such as burning coal to produce steam to drive a turbine which produces electricity. Except for some hydroelectric and nuclear power, most of the world's energy comes from energy in the form of fossil fuels, which are burned to produce heat.

Energy Research and Development Administration (ERDA). The creation of the Energy Research and Development Administration (ERDA) in January 1975 brought together the key elements of the Federal Government's energy research and development programs which had been conducted as major thrusts in several different agencies. These key elements are: the nuclear fission and fusion programs of the Atomic Energy Commission, together with the fossil, solar, geothermal, energy conservation programs of the Department of the Interior, the National Science Foundation, and the Environmental Protection Agency. ERDA will also conduct the energy production, physical and life sciences research, and weapons programs formerly of AEC. The functions of ERDA were incorporated into the Department of Energy in October 1977.

enriched uranium. Uranium in which the amount of U-235 present has been artificially increased above the 0.71% found in nature. Uranium enriched between 3 and 6% is a common fuel for civil nuclear power stations. Uranium enriched to 90% or more is used for nuclear propulsion of warships and submarines, and in atomic bombs.

enriching, (gas). Increasing the heat content of gas by mixing it with a gas of higher Btu content.

enrichment. A process by which the proportion of the fissionable uranium isotope (U-235) is increased above the 0.7% contained in natural uranium.

enthalpy. The heat content per unit mass, expressed in Btu per pound.

environment. The total surroundings of an organism which act upon it.

environmental impact statements. Analytical statements which present the environmental costs and benefits of a proposed Federal action. Required by the National Environmental Policy Act, sec. 102 (2) (c).

EPA. Environmental Protection Agency. A Federal agency created in 1970 to provide coordinated and effective governmental action for protection of the environment by the systematic abatement and control of pollution through integration of research, monitoring, standard setting, and enforcement activities.

ERDA. (see Energy Research and Development Administration).

excess reactivity. Reactivity over and above that needed to achieve criticality. Excess reactivity is built into a reactor (by using extra fuel) in order to compensate for fuel burn-up and the accumulation of fission-product poisons during operation.

exchange gas. Gas that is received from (or delivered to) another party in exchange for gas delivered to (or received from) such other party.

exclusion area. An area immediately surrounding a nuclear reactor where human habitation is controlled to assure safety in the event of an accident.

excursion. A sudden, rapid rise in the power level of a reactor. Excursions are usually quickly suppressed by the negative temperature coefficient of the reactor and/or by automatic control rods.

exponential growth. Growth which occurs at a rate which is a constant percentage of the whole.

fallout. Radioactive material that settles to earth after a nuclear explosion. Fallout takes two forms. The first, called "local fallout," consists of the denser particles injected into the atmosphere that descend to earth within 24 hours near the site of the detonation and in an area extending downwind for some distance (often hundreds of miles), depending on meteorological conditions and the yield of the detonation. The other form, called "worldwide fallout," consists of lighter particles which ascend into the upper troposphere and stratosphere and are distributed over a wide area of the earth by atmospheric circulation. They then are brought to earth, mainly by rain or snow, over periods ranging from months to years.

fast breeder reactor. A nuclear reactor that operates with neutrons at the fast speed of their initial emission from the fission process, and that produces more fissionable material than it consumes.

fast neutron. A neutron with kinetic energy greater than approximately 1,000,000 electron volts.

fast reactor. A nuclear reactor in which the fission chain reaction is sustained primarily by fast neutrons. Fast reactors contain no moderator and inherently require enriched fuel. They are of interest because of favorable neutron economy which makes them suitable for breeding.

fast flux testing facility (FFTF). A major Department of Energy experimental facility to be completed in 1979. When completed, it will provide test conditions for breeder fuels comparable to those expected in commercial fast breeder reactors.

fertile material. A material, not itself fissionable by thermal neutrons, which can be converted into a fissionable material by irradiation in a nuclear reactor. The two basic fertile materials are U-238 and Th-232. When these fertile materials capture neutrons, they become fissionable Pu-239 and U-233, respectively.

FFTF. Fast flux testing facility.

film badge. A light-tight package of photographic film, worn like a badge by workers in nuclear industry or research, used to measure exposure to ionizing radiation. The absorbed dose can be calculated by the degree of film darkening caused by the irradiation.

finished products. Petroleum oils, or a mixture or combination of such oils, which are to be used without further processing.

firedamp. A highly explosive mixture of methane and air found in coal seams. It is frequently the cause of explosions in coal mines.

fireflooding. A method to increase recovery of oil from existing fields. Cf. *in situ* combustion.

fissile material. An element which can be converted into a fissionable element by neutron irradiation. The primary fissile materials are U-238 and Th-232.

fission. The splitting of a heavy nucleus into two approximately equal parts (which are nuclei of lighter elements), accompanied by the release of a relatively large amount of energy and generally one or more neutrons. Fission can occur spontaneously, but usually is caused by nuclear absorption of gamma rays, neutrons, or other particles.

fission fragments. The two or more nuclei which are formed by the fission of a nucleus. Also referred to as primary fission products. They are of medium atomic weight, and are radioactive.

fission products. The nuclei formed by the fission of heavy elements, plus nuclides formed by the fission fragments radioactive decay. Fission products are intensely radioactive.

fissionable material. Any isotope fissionable by neutrons. The three basic fissionable isotopes are U-235, Pu-239, and U-233.

flare gas. Natural gas burned in flares at an oil field; waste gas.

flue gas. Gas from the combustion of fuel.

fluidized bed. A bed of solid particles held in suspension by a flow of gas or liquid.

flux (neutron). A measure of the intensity of neutron radiation. It is the number of neutrons passing through one square centimeter in one second. Expressed as $n \times v$, where n = the number of neutrons per cubic centimeter and v = their velocity in centimeters per second.

fly ash. Small particles of air-borne ash produced by the burning of fuels.

food chain. The pathway by which any material (such as radioactive material from fall-out) passes from the first absorbing organism through plants and animals to humans.

fuel. Any substance that can be burned to produce heat. The energy content of common fuels are as follows;

1 barrel (Bbl) of crude oil equals 5,800,000 Btu.
1 cubic foot (cu ft) of natural gas equals 1035 Btu.
1 ton of coal equals 14,000,000 to 26,000,000 Btu.
Two trillion Btus per year are about equal to 1000 barrels of crude oil per day.

fuel (nuclear). Fissionable material used to produce energy in a reactor.

fuel energy conversion factors. Coal:

Anthracite (Penn.)—25.4 million Btu/ton.
Bituminous-26.2 million Btu/ton.
Subbituminous-19.0 million Btu/ton.
Lignite—13.4 million Btu/ton.

Petroleum:

Crude—5.6 million Btu/bbl.
Residual fuel oil—6.29 million Btu/bbl.
Distillate fuel oil—5.83 million Btu/bbl.
Gasoline—5.25 million Btu/bbl.
Jet fuel (kerosene-type)—5.67 Btu/bbl.
Kerosene—5.67 million Btu/bbl.
Petroleum coke—6.02 million Btu/bbl.

fuel cycle. The series of steps involved in supplying fuel for nuclear power reactors. It includes mining, refining of uranium, fabrication of fuel elements, their use in a nuclear reactor, chemical processing to recover remaining fissionable material, re-enrichment of the fuel, refabrication into new fuel elements, and waste storage.

fuel cell. A device in which hydrogen is fed to an electrode where it is catalytically converted to hydrogen ions, releasing electrons to flow through an external circuit. The electrons flowing through the external circuit constitute an electric current. Upon return to the fuel cell the electrons react to form oxygen ions which, in turn, unite with hydrogen ions to form water. The process is, essentially, the reverse of electrolysis.

fuel depot. A bulk storage installation composed of storage tanks and related facilities such as docks, loading racks, and pumping units.

fuel fabrication. The manufacture and assembly of reactor fuel elements containing fissionable and fertile nuclear material.

fuel oils. Fuel oils are the petroleum fractions with a higher boiling range than kerosene. They are generally classified as distillates or residuals. Distillates (Nos. 1, 2, and 4) are the lighter oils used primarily for central heating of homes, small apartment houses, commercial buildings, and for transportation. Residuals (Nos. 5 and 6), often called bunker oils, are heavier, high viscosity oils which usually need to be heated before they can be pumped. They are used in industry, large commercial buildings, and for the generation of electricity.

fuel rate. The amount of fuel needed to generate one kilowatt-hour of electricity. In 1969 the rates were 0.88 pounds of coal, 0.076 gallons of oil, or 10.4 cubic feet of natural gas.

fuel reprocessing. The processing of reactor fuel to recover the unused, residual fissionable materials.

fusion. The process in which two atomic nuclei combine to form a larger nucleus whose mass is less than the aggregate mass of the original nuclei. The lost mass appears as energy.

gallon. A unit of measure. A United States gallon contains 231 cu in, 0.133 cu ft, or 3.785 liters. It is 0.83 times the Imperial gallon. One United States gallon of water weighs 8.3 lb.

gamma rays. High-energy, short-wavelength electromagnetic radiation emitted by nuclei. Energies of gamma rays are usually between 0.010 and 10 MeV. Gamma radiation usually accompanies alpha and beta emissions and always accompanies fission. Gamma rays are very penetrating and are best attenuated by dense materials.

gas, associated. Free natural gas in immediate contact, but not in solution, with crude oil in the reservoir.

gas cap. A layer of gas on top of oil in an underground structure or reservoir.

gas, casinghead. Unprocessed natural gas produced from a reservoir containing oil. Sometimes called Bradenhead Gas.

gas centrifuge process. A method of enrichment in which heavier uranium atoms are partially separated from lighter ones by centrifugal force.

gas-cooled fast breeder reactor (GCFBR). A fast breeder reactor which is cooled by a gas, usually helium, under pressure.

gaseous diffusion. A method of isotopic separation based on the phenomenon that gas molecules of different masses will diffuse through a porous barrier at different rates. This method has been used in a number of countries, including the United States, to produce uranium enriched in U-235 either for civilian reactor fuel or for nuclear weapons.

gas, illuminating. A gas containing relatively large amounts of unsaturated and/or heavy hydrocarbon gases which burn with a luminous flame.

gas impurities. Undesirable matter in gas, such as dust, water vapor, hydrogen sulfide, tar, and ammonia.

gas, liquefied. Liquefied natural gas (LNG) boils at a temperature of -259°F. In volume, it occupies 1/600 of the gas in vapor state.

gas, manufactured. A gas obtained by destructive distillation of coal, by thermal decomposition of oil, or by the reaction of steam passing through a bed of heated coal or coke. Examples are coal gas, coke oven gas, producer gas, blast furnace gas, blue (water) gas, and carbureted water gas.

gas, natural. A naturally occurring mixture of hydrocarbons. Gases found in porous geologic formations beneath the earth's surface, often in association with petroleum. The principal constituent is methane.

gas, nonassociated. Free natural gas not in contact with, or dissolved in, crude oil in the reservoir.

gas, sour. Gas found in its natural state, containing amounts of compounds of sulfur which make it impractical to use, without purifying, because of its corrosive effect on piping and equipment.

gas, sweet. Gas found in its natural state and containing such small amounts of compounds of sulfur that it can be used without purifying with no deleterious effect on piping and equipment.

gas turbine. A prime mover in which gas, under pressure or formed by combustion, is directed against a series of turbine blades; the energy in the expanding gas is converted into mechanical energy supplying power at the shaft.

gas, wet. Wet natural gas is unprocessed or partially processed natural gas, produced from strata containing condensable hydrocarbons. The term is subject to varying legal definition as specified by certain state statutes.

gasoline. A refined petroleum distillate, including naphtha, jet fuel or other petroleum oil, having a boiling range at atmospheric pressure from 80° to 400°F.

GCBR. Gas-cooled fast breeder reactor.

Geiger-Muller counter. A radiation detection and measuring instrument. It consists of a gas-filled tube containing electrodes. When ionizing radiation passes through the tube, a short, intense pulse of current passes from the negative electrode to the positive electrode and is measured or counted. The number of pulses per second measures the intensity of radiation. It is sometimes called simply a Geiger counter, or a G–M counter.

genetic effects of radiation. Radiation effects that can be transferred from parent to offspring. Any radiation-caused change in the genetic material of sex cells.

genetically significant dose. A population-averaged dose which estimates the potential genetic effects of radiation on future generations. It takes into consideration the number of people in various age groups, the average dose to the reproductive organs to which people in these groups are exposed, and their expected number of future children.

Geological Survey. A bureau of the Department of the Interior established in 1879. The objectives of the Survey are to "perform surveys, investigations, and research covering topography, geology, and the mineral and water resources; enforce departmental regulations, applicable to oil, gas, and other mining leases, permits, licenses, development contracts, and gas storage contracts; and publish and disseminate data relative to the foregoing activities.

geothermal; geothermic. Of or relating to the heat of the earth's interior.

geothermal gradient. The change in temperature of the earth with depth. The mean rate of increase in temperature with depth in areas that are not adjacent to volcanic regions is about 1°F in about 55 feet, corresponding to about 100°F per mile of depth.

gigawatt (GW). 1,000,000 kilowatts, 1,000 megawatts.

gpm. Gallons per minute.

gps. Gallons per second.

graphite (reactor grade). A very pure form of carbon used as a moderator in nuclear reactors.

gross national product (GNP). The total market value of the goods and services produced by the nation before the deduction of depreciation charges and other allowances for capital consumption. A widely used measure of economic activity.

half-life. The time in which half of the atoms in a radioactive substance disintegrate. Half-lives vary from millionths of a second to billions of years.

half-life, biological. The time required for a biological system to eliminate, by natural processes, half the amount of a substance which has entered it.

half-thickness. The thickness of any given absorber that will reduce the intensity of a beam of radiation to one–half its initial value.

health physics. The science concerned with recognition, evaluation, and control of health hazards from ionizing radiation.

heat exchanger. Any device that transfers heat from one fluid to another or to the environment.

heat pump. A device which transfers heat from a colder to a hotter reservoir by the expenditure of mechanical energy. A heat pump is a reversed refrigeration apparatus. Heat pumps are more efficient for residential heating than electrical resistive heating.

heat sink. Anything that absorbs heat; usually part of the environment, such as the air, a river, or outer space.

heating value. The amount of heat produced by the complete combustion of a unit quantity of fuel.

hertz. Cycles per second. United States electrical supply has a frequency of 60 hertz.

heterogeneous reactor. A nuclear reactor in which the fuel is separate from the moderator and is arranged in discrete bodies, such as fuel elements. Most reactors are heterogeneous.

heavy water. (Symbol D_2O.) Water containing the heavy isotope of hydrogen (deuterium). Heavy water is used as a moderator in some reactors because it slows down neutrons effectively and also has a low cross section for absorption of neutrons than ordinary (light) water.

heavy water moderated reactor. A reactor that uses heavy water as its moderator. Heavy water is an excellent moderator and thus permits the use of natural uranium as a fuel.

high-temperature gas-cooled reactor. (See HTGCR.)

holder, gas. A gas-tight receptacle or container in which gas is stored for future use. There are two general ways of storing gas: (1) at approximately constant pressure in which case the volume changes, and (2) in containers of constant volume in which case the quantity of gas varies with the pressure.

homogeneous reactor. A reactor in which the fuel is mixed with the moderator or coolant. Example: fused-salt reactor.

hopper car. A railway car for coal, gravel, etc., shaped like a hopper, with an opening to discharge the contents.

horsepower (Hp). A standard unit of power equal to 746 watts in the United States. One horsepower equals 2545 Btu/hour or 550 foot-pounds/second.

horsepower hour. One horsepower expended for one hour. One horsepower hour equals 1,980,000 foot-pounds, 0.745 kilowatt-hours or 2454 Btu.

Hp. Horsepower.

HTGCR. High temperature gas-cooled reactor. A nuclear reactor cooled with helium.

hydraulic fracturing. A general term, for which there are numerous trade or service names, for the fracturing of rock in an oil or gas reservoir by pumping a fluid under high pressure into the well. The purpose is to produce artificial openings in the rock in order to increase permeability.

hydrocarbon fuels. Fuels that contain an organic chemical compound of hydrogen and carbon.

hydrocracking. A process combining cracking or pyrolysis with hydrogenation. Feedstocks can include crude oil, petroleum tars, and asphalts.

hydroelectric plant. A plant in which the kinetic energy of falling water is used to turn a turbine generator producing electricity.

hydroforming. A process in which naphthas are passed over a catalyst at elevated temperatures and moderate pressures, in the presence of added hydrogen or hydrogen-contained gases, to form high-octane motor fuel or aromatics.

hydrotreating. The removal of sulfur from low-octane gasoline feedstock by replacement with hydrogen.

induced radioactivity. Radioactivity that is created when substances are bombarded with neutrons as from a nuclear explosion, in a reactor, or with charged particles and photons produced by accelerators.

injection (gas injection, water injection). Introduction of gas or water into an oil well in order to increase pressure within the well to force oil to the surface.

in situ. In the natural or original position; applied to a rock, soil, or fossil when occuring in the situation in which it was originally formed or deposited.

in situ combustion. An experimental means of recovery of oil of low gravity and high viscosity. The oil in the oil-bearing strata is ignited and the combustion supported by air injection. The heat breaks the oil down to coke and light oils and as the combustion front advances, the light oils move ahead of the fire and into the bore of a producing well. Also known as fireflooding.

in situ recovery. Refers to methods to extract the fuel component of a deposit without removing the deposit from its bed.

intangible drilling costs. Expense items that are written off in the year incurred for tax purposes.

ion. An atom or molecule that has lost or gained one or more electrons creating an ion which is electrically charged.

ion exchange. A chemical process in which ions from two different molecules are exchanged.

ionization. The process of adding one or more electrons to, or removing one or more electrons from atoms or molecules, thereby creating ions. High temperatures, electrical discharges, or nuclear radiations can cause ionization.

ionization chamber. An instrument that measures radiation by measuring the electrical current which is created due to ionization of gas in a chamber as a result of the radiation.

ionization event. An occurrence in which an ion or group of ions is produced; for example, by passage of a charged particle through matter.

ionized gas. A gas that is capable of carrying an electric current.

ionizing radiation. Any radiation capable of displacing electrons from atoms or molecules, thereby producing ions. Examples: alpha, beta, and gamma radiation; shortwave ultraviolet light. Ionizing radiation may produce severe skin or tissue damage.

irradiation. Exposure to radiation, as in a nuclear reactor.

isotopes. Atoms with the same atomic number but different atomic weights. Thus, $^{12}_{6}C$, $^{13}_{6}C$, and $^{14}_{6}C$ are isotopes of the element carbon, the subscripts denoting their common atomic numbers, the superscripts denoting the varying atomic weights.

isotope separation. The process of separating isotopes from one another, or changing their relative concentration, as by gaseous diffusion or electromagnetic separation.

isotopic enrichment. A process by which the relative concentrations of the isotopes of a given element are altered, thus producing a form of the element which has been enriched in one particular isotope and depleted in its other isotopic forms.

JCAE. Joint Committee on Atomic Energy, U.S. Congress. The JCAE was created in 1947 to handle the legislative matters related to the civilian and military uses of atomic energy. The committee was abolished in 1977 and its functions transferred to a number of committees of the Senate and the House of Representatives.

joule. A unit of energy or work which is equivalent to one watt per second or 7.737 footpounds.

kerosene. The petroleum fraction containing hydrocarbons that are slightly heavier than those found in gasoline and naphtha, with a boiling range between 180° and 300°C. Used today as fuel for gas turbines and jet engines.

kilogram (kg). The unit of weight in the metric system, equal to 1000 grams or 2.2 lb.

kiloton (kt). A measure of explosive force which originated in the early nuclear weapons program. One kiloton represents the energy of 10^{12} calories, or 3.9×10^{9} Btu, or 4×10^{12} joules.

kilovolt (kV). 1000 volts.

kilovolt-ampere (kVA). A term indicating the energy in an electrical circuit. It is the product of voltage and current.

kilowatt (kW). 1000 watts. A unit of power equal to 1000 watts, or to energy consumption at a rate of 1000 joules per second. It is usually used for electrical power. An electric motor rated at one horsepower uses electrical energy at a rate of about $\frac{3}{4}$ kilowatt.

kilowatt-hour (kWh). A unit of work or energy equal to that expended by one kilowatt in one hour. It is equivalent to 3413 Btu of heat energy.

kinetic energy. The energy of motion; the ability of an object to do work because of its motion.

kV. Kilovolt.

kVA. Kilovolt-ampere.

kt. Kiloton.

laser-induced fusion. A process in which the high temperature required to initiate fusion is produced by bombarding frozen pellets of deuterium and tritium with intense bursts of radiation from one or more lasers.

lethal dose. A dose of ionizing radiation sufficient to cause death. Median lethal dose (MLD or LD-50) is the dose required to kill within a specific period of time (usually 30 days) half of the individuals in a large group of organisms similarly exposed. The LD-50/30 for man is about 400,000 to 450,000 mrem.

light-water reactor (LWR). Nuclear reactor in which water is the primary coolant and moderator. There are two commercial light-water reactor-types—the boiling water reactor (BWR) and the pressurized water reactor (PWR).

lignite. A low-grade cool intermediate between peat and bituminous coal.

liquefied natural gas (LNG). Natural gas that has been cooled to about −160°C for storage or shipment as a liquid. Liquefaction greatly reduces the volume of a gas and, thus, the cost of shipping and storage is reduced.

liquefied petroleum gas (LPG). Consists of propanes and butanes recovered from natural gas and in petroleum refining. Its energy content ranges between 2000 and 3500 Btu per standard cubic foot. Sometimes called "bottled gas." Used widely as a fuel for internal combustion engines when pollution must be minimized and as a substitute for natural gas in areas not served by pipelines.

liquid metal fast breeder (LMFBR). A nuclear breeder reactor cooled by molten sodium in which fission is caused by fast neutrons.

liquids, natural gas. Liquid hydrocarbon mixtures which are gaseous at reservoir temperatures and pressures but are recoverable by condensation and absorption. Natural gasoline and liquefied petroleum gases fall in this category.

liter. The primary standard of capacity in the metric system, equal to the volume of one kilogram of pure water at maximum density, at approximately 4°C, and under normal atmospheric pressure. One liter = 0.264 gallons (U.S.), 1.05 quarts (U.S.) or 2.11 pints (U.S.).

lithium. Element No. 3 (symbol Li; atomic weight 6.94). As found in nature, lithium consists of a mixture of two stable isotopes—lithium-6 (7.5%) and lithium-7 (92.4%). Lithium-6 is of interest as a possible fuel, or source thereof, for the generation of power from a controlled thermonuclear reaction.

LNG. Liquified natural gas.

load. The amount of power needed to be delivered to a given point on an electric system.

longwall mining. A method of working coal seams that originated in England in the 17th century. The seam is removed in a continuous operation from a long working face, or wall. The space from which the coal has been removed is either allowed to collapse or is filled with stone and debris.

low population zone. An area of low population density sometimes required around a nuclear installation to assure, with reasonable probability, that effective protective measures can be taken in the event of a serious accident.

LPG. Liquid petroleum gas.

Lurgi process. A commercial process for coal gasification originated in Germany. This process has had limited application in the United States.

LWR. Light-water reactor.

magnetohydrodynamics (MHD). A branch of physics that deals with phenomena arising from the motion of electrically conducting fluids in the presence of electric and magnetic fields. In open-cycle MHD generators, the working fluid is exhausted to the atmosphere. In closed-cycle MHD, the working fluid is continuously recirculated through a closed loop.

magnetic bottle. A magnetic field used to confine or contain a plasma in controlled fusion (thermonuclear) experiments.

magnetic mirror. A magnetic field used in controlled fusion experiments to reflect charged particles back into the central region of a magnetic bottle.

marketable natural gas. Raw gas from which certain hydrocarbon and nonhydrocarbon compounds have been removed or partially removed by processing. Marketable natural gas is often referred to as pipeline gas, residue gas, or sales gas.

mass. The quantity of matter in a body. Often used as a synonym for weight, which, strictly speaking, is a measure of the gravitational force exerted on a body by the earth.

mass-energy equation. The statement developed by Albert Einstein, German-born physicist, that the mass of a body is a measure of its energy content, as an extension of his 1905 special theory of relativity. The statement was subsequently verified experimentally by measurements of mass and energy in nuclear reactions. The equation, usually given as $E = mc^2$, shows that when the energy of a body changes by an amount E (no matter what form the energy takes), the mass, m, of the body will change by an amount equal to E/c^2. The factor c^2, the square of the speed of light in a vacuum, may be regarded as the conversion factor relating units of mass and energy. The equation predicted the possibility of releasing enormous amounts of energy by the conversion of mass to energy. It is also called the Einstein equation.

matter. The substance of which a physical object is composed. All materials in the universe have the same inner nature, that is, they are composed of atoms, arranged in different (and often complex) ways; the specific atoms and the specific arrangements identify the various materials.

maximum credible accident. The most serious reactor accident that can reasonably be imagined from any adverse combination of equipment malfunction, operating errors, and other foreseeable causes. The term is used to analyze the safety characteristics of a reactor. Reactors are designed to be safe even if a maximum credible accident should occur.

maximum permissible concentration. The concentration of radioactive material in air, water, and foodstuffs which competent authorities have established as the maximum that would not create undue risk to human health.

maximum permissible dose. That dose of ionizing radiation established by competent authorities as an amount below which there is no reasonable expectation of risk to human health. (See radiation protection guide.)

Mcf. One thousand cubic feet.

Mcfd. One thousand cubic feet per day.

Mcfh. One thousand cubic feet per hour.

mean life. The average time during which an atom, an excited nucleus, a radionuclide, or a particle exists in a particular form.

median lethal dose. (See lethal dose.)

mega-. A prefix that multiplies a basic unit by 1,000,000.

megawatt (MW). 1000 kilowatts, 1 million watts.

megawatt-day per ton (MWd/t). A unit that expresses the burn-up of nuclear fuel in a reactor; specifically the number of megawatt-days of heat output per metric ton of fuel in the reactor.

metallurgical coal. Coal with strong or moderately strong coking properties that contains no more than 8.0% ash and 1.25% sulfur, as mined or after conventional cleaning.

methane (CH$_4$). The lightest in the paraffin series of hydrocarbons. It is colorless, odorless, and flammable. It forms the major portion of marsh gas and natural gas.

methyl alcohol (CH$_3$OH). A poisonous liquid also known as methanol, which is the lowest member of the alcohol series. Also known as wood alcohol, since its principal source is the destructive distillation of wood.

metric ton. 1000 kilograms, equal to 2204.6 lbs.

Mev. One million (10^6) electron volts. Also written as MeV.

middle distillate. One of the distillates obtained between kerosene and lubricating oil fractions in the refining process. These include light fuel oils and diesel fuel.

milli-. A prefix that multiplies a basic unit by 1/1000.

mine. An opening or excavation in the earth for the purpose of extracting minerals; a pit or excavation in the earth from which metallic ores or other mineral substances are taken by digging.

mine-mouth plant. A steam-electric plant or coal gasification plant built close to a coal mine and usually associated with delivery of output via transmission lines or pipelines over long distances as contrasted with plants located nearer load centers and at some distance from sources of fuel supply.

Mmcf. Million cubic feet.

moderator. A material such as ordinary water, heavy water, or graphite, used in a reactor to slow down high-velocity neutrons, thus increasing the likelihood of further fission.

molecule. A group of atoms held together by chemical forces. The atoms in the molecule may be identical, as in H_2, S_2, and S_8, or different, as in H_2O and CO_2. A molecule is the smallest unit of a compound which can exist by itself and retain all its chemical properties. (Compare atom, ion.)

molten-iron process. A process for the gasification of coal in which the coal is reacted with air or oxygen in a molten-iron bath. The product gases are essentially methane, carbon monoxide, and hydrogen which with methanation, can be made into pipeline-quality gas.

molten salt breeder reactor (MSBR). A breeder reactor in which the fuel is in the form of a molten salt of plutonium or uranium. It offers several technical advantages, but poses severe, unresolved engineering problems. AEC support for MSBR research terminated in June 1973.

Mtce. Million tons of coal equivalent. A comparative unit of energy content widely used in the oil industry. 1 Mtce = 4.48 million bbl oil = 25.19 trillion cubic feet natural gas.

mutation. A permanent chromosomal change which results in an alteration of the characteristics of offsprings.

naphtha. A petroleum fraction with a boiling point ranging from 125° to 240°C. Its principle uses are in solvents, paint thinners, and as a raw material for the production of organic chemicals. It is also used as a raw material for the production of synthetic natural gas.

natural gas. A gaseous fossil fuel usually found associated with oil. As drawn from the well, the gas consists of 60 to 80% methane, 5 to 9% ethane, 3 to 18% propane, and 2 to 14% heavier hydrocarbons. Nonhydrocarbons such as nitrogen, carbon dioxide, and hydrogen sulfide are sometimes present. Propane and the heavier hydrocarbons are usually removed and sold as liquefied petroleum gas. Pipeline natural gas is principally methane and has an energy content of 980 to 1050 Btu per standard cubic foot.

natural gas liquids. Propane, butanes, and pentanes that are subject to recovery from raw gas by processing in field separators, scrubbers, gas processing and reprocessing plants. The propane and butane components are often referred to as liquefied petroleum gases or LPG.

natural gas products. Liquids including natural gasoline, which are recovered by processes of absorption, adsorption, compression, or refrigeration from natural gas.

natural gasoline. A mixture of liquid hydrocarbons extracted from natural gas and stabilized to obtain a liquid product suitable for blending with refinery gasoline.

natural radiation or natural radioactivity. Background radiation.

natural uranium. Uranium as found in nature. It contains 0.7% U-235, 99.3% U-238, and a trace of U-234. It is also called normal uranium.

neutron. (Symbol n.) An uncharged elementary particle with a mass slightly greater than that of the proton, and found in the nucleus of every atom heavier than hydrogen-1. Neutrons sustain the fission chain reaction in a nuclear reactor.

neutron capture. The process in which an atomic nucleus absorbs or captures a neutron.

nitrogen oxides (NO_x). Products of combustion of fossil fuels through the combustion of nitrogen compounds in the fuel or through fixation of atmospheric nitrogen.

normal uranium. See natural uranium.

nonassociated gas. Free natural gas not in contact with, nor dissolved in, crude oil in a reservoir.

nuclear fuel cycle. The various steps which involve the production, processing, use, and reprocessing of nuclear fuels.

nuclear power plant. Any device, machine, or assembly that converts nuclear energy into some form of useful power, such as mechanical or electrical power.

nuclear reactor. A device in which a fission chain reaction can be initiated, maintained, and controlled. Its essential component is a core, containing fissionable fuel, moderator, reflector, shielding, coolant, and control mechanisms.

nucleon. A constituent of an atomic nucleus, that is, a proton or a neutron.

nucleus. The small, positively charged core of an atom. It is only about 1/10,000 the diameter of the atom, but contains nearly all the atom's mass. All nuclei contain both protons and neutrons, except the nucleus of ordinary hydrogen, which consists of a single proton.

nuclide. A general term applicable to all atomic forms of the elements. The term is often erroneously used as a synonym for isotope, which properly has a more limited definition. Whereas isotopes are the various forms of a single element (hence are a family of nuclides) and all have the same atomic number and number of protons, nuclides comprise all the isotopic forms of all the elements. Nuclides are distinguished by their atomic mass and energy state.

OAPEC. Organization of Arab Petroleum Exporting Countries. It was founded in 1968 for cooperation in economic and petroleum affairs. Original members were Saudi Arabia, Kuwait, and Libya. In 1970, Abu Dhabi, Algeria, Bahrain, Dubai, and Qatar joined.

OCS. Outer continental shelf.

Office of Coal Research. (OCR). A bureau established in the Department of the Interior in 1960 to develop new and more efficient methods of mining, preparing, and utilizing coal; this office was transferred to the Energy Research and Development Administration in January 1975 and to the Department of Energy in October 1977.

Offshore Windpower System (OSWS). A proposed system to generate electricity by wind turbines mounted on offshore platforms.

OPEC. Organization of Petroleum Exporting Countries. Founded in 1960 to unify and coordinate petroleum policies of the members. The members and the dates of membership are: Abu Dhabi (1967); Algeria (1969); Indonesia (1962); Iran (1960); Iraq (1960); Kuwait (1960); Libya (1962); Nigeria (1971); Qatar (1961); Saudi Arabia (1960); and Venezuela (1960). OPEC headquarters are in Vienna, Austria.

oil shale. An expression used to cover a range of materials containing organic matter (kerogen) which can be converted into crude shale oil, gas, and carbonaceous residue by heating.

organic-cooled. A nuclear reactor that uses waxlike organic chemicals, such as mixtures of polyphenyls and terphenyls, as coolant and usually also as moderator.

original-oil-in-place. The estimated number of barrels of crude oil in known reservoirs prior to any production, usually expressed as "stock tank" barrels, or the volume that goes into a stock tank after the shrinkage that results when dissolved gas is separated from the oil.

outage. The period in which a generating unit, transmission line, or other facility, is out of service.

overburden. Material of any nature, consolidated or unconsolidated, that overlies a deposit or useful materials, ores, or coal, especially those deposits that are mined from the surface by open cuts.

overpressure. The transient pressure over and above atmospheric pressure resulting from a blast wave from a nuclear explosion.

parent. A radionuclide that upon decay or disintegration yields a specific nuclide (the daughter), either directly or as a later member of a radioactive series.

particulates. Small particles of solid material produced by burning of fuels.

permissible dose. (See maximum permissible dose.)

personnel monitoring. Determination by either physical or biological measurement of the amount of ionizing radiation to which an individual has been exposed, such as by measuring the darkening of a film badge or performing a radon breath analysis.

petroleum. An oily flammable bituminous liquid that may vary from almost colorless to black. A complex mixture of hydrocarbons with small amounts of other substances. Prepared for use as gasoline, naphtha, or other products by various refining processes.

petroleum naphtha. A genetic term applied to refined, partially refined, or unrefined petroleum products and liquid products of natural gas. The naphthas used for specific purposes, such as cleaning, manufacture of rubber, paints, varnishes, etc., are made to have high volatility.

petroleum spirits. A refined petroleum distillate with volatility, flash point, and other properties making it suitable as a thinner and solvent in paints, varnishes, and similar products.

petroleum tar. A viscous black or dark brown product obtained in petroleum refining which will yield a substantial quantity of solid residue when partly evaporated or fractionally distilled.

photon. A discrete quantity of electromagnetic energy. Photons have momentum but no mass or electrical charge.

photosynthesis. The process in which sunlight falling on green plants causes carbon dioxide and water to be converted into more complex organic materials such as glucose.

photovoltaic cell. A type of semiconductor device in which the absorption of light brings about a separation of electric charges. Such a separation produces a voltage which can be used to set up an electric current in an external circuit, thus, directly converting radiant energy to electrical energy. Materials which can be used in the construction of photovoltaic cells are silicon, cadmium sulfide, and gallium arsenide.

physical half-life. (See half-life.)

pig. A heavy shielding container (usually lead) used to ship or store radioactive materials.

pile. Old term for nuclear reactor. This name was used because the first reactor was built by piling up graphite and natural uranium.

pilot plant. A small-scale industrial process unit operated to test the application of a chemical or other manufacturing process under conditions that will yield information useful in the design and operation of full-scale manufacturing equipment. The pilot unit serves to disclose the special problems to be solved in adapting a successful laboratory method to commercial-sized units.

plowshare. The Atomic Energy Commission program of research and development on peaceful uses of nuclear explosives. The possible uses include large-scale excavation, such as for canals and harbors, crushing ore bodies, and producing heavy transuranic isotopes. The term is based on a biblical reference, Isiah 2:4.

plutonium. (Symbol Pu.) A heavy, radioactive, man-made metallic element with atomic number 94. Its most important isotope is fissionable Pu-239, produced by neutron irradiation of U-238. It is used for reactor fuel and in weapons.

pollution. The addition of any undesirable agent to an ecosystem.

pool reactor. A reactor in which the fuel elements are suspended in a pool of water that serves as the reflector, moderator, and coolant. Popularly called a swimming pool reactor, it is usually used for research and training.

population density. The number of persons per unit area (usually per square mile) who inhabit an area.

positron. A subatomic particle with the mass of an electron but having a positive charge of the same magnitude as the electron's negative charge.

power. The rate at which work is done, therefore, the rate at which energy is transferred. Power is measured in units of work per unit of time. Typical units are the watt and the horsepower.

power density. Rate of heat generated per unit volume of a nuclear reactor core.

power reactor. A nuclear reactor designed for use in a power plant, as distinguished from reactors used primarily for research or for producing radiation or fissionable materials.

ppm. Parts per million.

pressure vessel. A strong-walled container housing the core of most types of power reactors; it usually also contains moderator, reflector, thermal shield, and control rods.

pressurized water reactor. A power reactor in which heat is transferred from the core to a heat exchanger by water kept under high pressure to achieve high temperature without boiling in the primary system. Steam is generated in a secondary circuit. Many reactors producing electric power are pressurized water reactors.

primary fission products. Fission fragments.

probable reserves. A realistic assessment of the reserves that will be recovered from known oil or gas fields based on the estimated ultimate size and reservoir characteristics of such fields.

production reactor. A reactor designed primarily for large-scale production of plutonium by neutron irradiation of U-238.

prompt criticality. The state of a reactor when the fission chain reaction is sustained solely by prompt neutrons, that is, without the help of delayed neutrons.

protection. Provisions to reduce exposure of persons to radiation. For example, protective barriers to reduce external radiation or measures to prevent inhalation of radioactive materials.

proved reserves. The estimated quantity of crude oil, natural gas liquids, or sulfur which analysis or geological and engineering data demonstrate with reasonable certainty to be recoverable from known oil or gas fields under existing economic and operating conditions.

psi. Pounds per square inch.

psia. Pounds per square inch absolute. A measure of pressure that includes atmospheric pressure.

pumped hydroelectric storage. A means for the large-scale storage of electrical energy. Excess electricity produced during periods of low demand is used to pump water up into an elevated reservoir. When demand is high the water is released to operate a hydroelectric generator.

Q unit. One quintillion Btu (1×10^{18} Btu). A very large unit of energy. 1Q = 38.46 billion tons of coal, 172.4 billion bbls of oil, 968.9 trillion cu ft natural gas.

quality factor. The factor by which a radiation dose is multiplied to obtain a quantity that expresses, on a common scale for all ionizing radiations, the irradiation incurred by exposed persons. It is used because some types of radiation such as alpha particles are more biologically damaging than other types.

rad. (Acronym for radiation absorbed dose.) The basic unit of absorbed dose of radiation. A dose of one rad means the absorption of 100 ergs of radiation energy per gram of absorbing material.

radiation. The emission and propagation of energy through matter or space by means of electromagnetic disturbances which display both wavelike and particlelike behavior; in this context the particles are known as photons. The term has been extended to include streams of fast-moving particles (alpha and beta particles, free neutrons, cosmic radiation, etc.).

radiation area. Any accessible area in which the level of radiation is such that a major portion of an individual's body could receive in any one hour a dose in excess of 5 millirem, or in any five consecutive days a dose in excess of 150 millirem.

radiation burn. Radiation damage to the skin.

radiation damage. A general term for the harmful effects of radiation on matter.

radiation detection instruments. Devices that detect and record the characteristics of ionizing radiation.

radiation monitoring. Continuous or periodic determination of the amount of radiation present in a given area.

radiation protection. Legislation and regulations to protect the public and laboratory or industrial workers against radiation. Also measures to reduce exposure to radiation.

radiation protection guide. The officially determined radiation doses which should not be exceeded without careful consideration of the reasons for doing so. These are equivalent to the older term maximum permissible dose.

radiation shielding. Reduction of radiation by interposing a shield of absorbing material between any radioactive source and a person, laboratory area, or radiation-sensitive device.

radiation source. Usually a man-made sealed source of radioactivity used in teletherapy, radiography, as a power source for batteries, or in various types of industrial gauges. Machines such as accelerators and natural radionuclides may also be considered sources.

radiation standards. Exposure standards, permissible concentrations, rules for safe handling, regulations for transportation, regulations for industrial control of radiation, and control of radiation by legislative means. (See radiation protection, radiation protection guide.)

radiation sterilization. Use of radiation to cause a plant or animal to become sterile, that is, incapable of reproduction. Also the use of radiation to kill all forms of life (especially bacteria) in food, surgical sutures, etc.

radiation warning symbol. An officially prescribed symbol (a magenta trefoil on a yellow background) which should be displayed when radiation hazard exists.

radioactive. Exhibiting radioactivity or pertaining to radioactivity.

radioactive contamination. Deposition of radioactive material in any place where it may harm persons, spoil experiments, or make products or equipment unsuitable or unsafe for some specific use. The presence of unwanted radioactive material found on the walls of vessels in used-fuel processing plants, or radioactive material that has leaked into a reactor coolant. Often referred to only as contamination.

radioactive dating. A technique for measuring the age of an object or sample of material by determining the ratios of various radioisotopes or products of radioactive decay it contains. For example, the ratio of C-14 to C-12 reveals the approximate age of bones, pieces of wood, or other archaeological specimens that contain carbon extracted from the air at the time of their origin.

radioactive isotope. A radioisotope.

radioactive series. A succession of nuclides, each of which transforms by radioactive disintegration into the next until a stable nuclide results. The first member is called the parent, the intermediate members are called daughters, and the final stable member is called the end product.

radioactive waste. (See waste, radioactive.)

radioactivity. The spontaneous change of atomic nuclei from one nuclear species to another. Such changes are always accompanied by the emission of corpuscular or electromagnetic radiation or both. These emissions are called alpha, beta, and gamma radiations. Alpha radiations are positively charged helium nuclei (2 protons and 2 neutrons) having velocities from 5 to 7% that of light. Beta radiations are negatively and positively charged electrons which may have velocities up to about 99.9% of the speed of light. Gamma radiations are electromagnetic, thus, they are of the same nature as X-rays and light.

radioactivity concentration guide. The concentration of radioactivity in air and water, which through continuous consumption of air or water for 50 years, or until equilibrium is reached, yields "radiation protection guide" exposures. This Federal Radiation Council term replaces the former "maximum permissible concentration."

radiobiology. The study of the scientific principles, mechanisms, and effects of the interaction of ionizing radiation with living matter.

radioecology. The body of knowledge and the study of the effects of radiation on species of plants and animals in natural communities.

radioisotope. A radioactive isotope. An unstable isotope of an element that decays or disintegrates spontaneously, emitting radiation. More than 1300 natural and artificial radioisotopes have been identified.

radioisotopic generator. A small power generator that converts the heat released during radioactive decay directly into electricity. These generators generally produce only a few watts of electricity and use thermoelectric or thermionic converters. Some also function as electrostatic converters to produce a small voltage. Sometimes called an atomic battery.

radiology. The science which deals with the use of all forms of ionizing radiation in the diagnosis and treatment of disease.

radiomutation. A permanent, transmissible change in form, quality, or other characteristic of a cell or offspring from the characteristics of its parent, due to radiation exposure. (See genetic effects of radiation, mutation.)

radioresistance. A relative resistance of cells, tissue, organs, or organisms to the injurious action of radiation. (Compare radiosensitivity.)

radiosensitivity. A relative susceptibility of cells, tissues, organs, or organisms to the injurious action of radiation. (Compare radioresistance.)

radium. (Symbol Ra.) A radioactive metallic element with atomic number 88. As found in nature, the most common isotope has an atomic weight of 226. It occurs in minute quantities associated with uranium in pitchblende, carnotite, and other minerals.

radon. (Symbol Rn.) A radioactive element, one of the heaviest gases known. Its atomic number is 86, and its atomic weight is 222. It is a daughter of radium in the uranium radioactive series.

raw gas. Natural gas, in its natural state, existing in or produced from a gas or oil field.

raw materials. Ores and crude concentrates of uranium and thorium.

reactivity. A measure of the departure of a nuclear reactor from criticality. It is equal to the multiplication factor minus one and is thus zero precisely at criticality. If there is excess reactivity (positive reactivity), the reactor power will rise. Negative reactivity will result in a decreasing power level.

reactor. (See nuclear reactor.)

recoverable reserves. Minerals expected to be recovered by present-day techniques and under present economic conditions.

recycling. The reuse of fissionable material, after it has been recovered by chemical processing from spent or depleted reactor fuel, reenriched and then refabricated into new fuel elements.

reduced crude. A residual product remaining after the removal, by distillation or other means, of an appreciable quantity of the more volatile components of crude oil.

refine. To cleanse or purify by removing undesired components; to process a material to make it usable.

refinery. A process in which crude oil is distilled into usable substances. Components of low molecular weight vaporize first. Typical crude fractions are: ether, methane, ethane, propane, butane, kerosene, fuel oil, lubricants, jelly paraffin, asphalt, and tar.

refinery gas. Any form or mixture of gas gathered in a refinery from the various stills.

refining. The separation of crude oil into component parts, and the manufacture of products needed for the market. Important processes in refining are distillation, cracking, chemical treating, and solvent extraction.

reflector. A layer of material immediately surrounding a reactor core which scatters back or reflects into the core many neutrons that would otherwise escape. The returned neutrons can then cause more fissions and improve the neutron economy of the reactor. Common reflector materials are graphite, beryllium, and natural uranium.

reforming. The thermal or catalytic conversion of naphtha into more volatile products of higher octane number.

refrigeration ton. A unit of cooling capacity. In commercial usage, 12,000 Btu per hour or 200 Btu per minute of heat removal. Originally, the amount of heat required to melt a ton of ice in 24 hours.

regulating rod. A reactor control rod used for making frequent fine adjustment in reactivity.

relative biological effectiveness (*RBE*). A factor used to compare the biological effectiveness of different types of ionizing radiation. It is the inverse ratio of the amount of absorbed radiation, required to produce a given effect to a standard or reference radiation required to produce the same effect.

rem. (Acronym for roentgen equivalent man.) The unit of dose of any ionizing radiation which produces the same biological effect as a unit of absorbed dose or ordinary X-rays. The RBE dose (in rems) = RBE × absorbed dose (in rads).

remaining reserves. Those quantities of crude oil, natural gas liquids, natural gas, and sulfur estimated as proved or probable reserves after deducting those quantities produced up to the respective date of the estimate.

rep. (Acronym for roentgen equivalent physical.) An obsolete unit of absorbed dose of any ionizing radiation, with a magnitude of 93 ergs per gram. It has been superseded by the rad.

reprocessing. Chemical recovery of unburned uranium and plutonium and certain fission products from spent nuclear fuel elements.

residual fuel oil. A heavy fraction from the distillation of crude oil, often used as a fuel for power plants.

retort. A vessel used for the distillation of volatile materials, as in the separation of some metals and the destructive distillation of coal; also a long semicylinder, now usually of fire clay or silica, for the manufacture of coal gas.

roentgen. (Abbreviation r.) A unit of exposure to ionizing radiation. It is that amount of gamma or X-rays required to produce ions carrying 1 electrostatic unit of electrical charge (either positive or negative) in 1 cubic centimeter of dry air under standard conditions. Named after Wilhelm Roentgen, German scientist who discovered X-rays in 1885.

roentgen equivalent, man. (See rem.)

roentgen rays. X-rays.

royalty bidding. Competitive bidding for leases in which the lease is awarded to the company offering to pay the land owner the largest share of the net proceeds of production.

Rio Blanco. Name of an AEC-industry experiment to stimulate production of natural gas by use of multiple nuclear explosions and to test the economic feasibility of future utilization of nuclear stimulation of an entire gas field. The test was made on May 16, 1973 near Meeker, in Rio Blanco County, Colorado.

Rulison. Name of an AEC-industry experiment to stimulate production of natural gas by use of a nuclear explosive to fracture impermeable rocks. Conducted in 1969.

safety rod. A standby control rod used to shut down a nuclear reactor rapidly in emergencies.

scaler. An electronic instrument for rapid counting of radiation-induced pulses from Geiger counters or other radiation detectors. It permits rapid counting by reducing by a definite scaling factor the number of pulses entering the counter.

Scf. Standard cubic feet.

Scfd. Standard cubic feet per day.

Scram. The sudden shutdown of a nuclear reactor, usually by rapid insertion of the safety rods. Emergencies or deviations from normal reactor operation cause the reactor operator or automatic control equipment to scram the reactor.

secondary recovery. Oil and gas obtained by the injection of air, gas, or water into a production formation.

separative work. A measure of the work required to separate U_{235} and U_{238} isotopes in the gaseous diffusion process.

shield (*shielding*). A body of material used to reduce the passage of radiation.

smog. A mixture of smoke and fog. A fog made heavier and usually darker by smoke and chemical fumes.

smoke. Suspension of small particles in a gas.

solar energy. The energy transmitted from the sun, which is in the form of electromagnetic radiation. Although the Earth receives only one-half of one billionth of the total solar energy output, this amounts to about 5300×10^{18} Btu annually.

solar furnace. An optical device with large mirrors that focuses the rays from the sun upon a small focal point to produce very high temperatures.

solar power. Useful power derived from solar energy. Both steam and hot-air engines have been operated from solar energy. Solar steam engines were built in California, Arizona, and Egypt between 1900 and 1914. Use of these engines was discontinued because of competition from the gasoline engine and the electric motor.

solvent refined coal. A process to treat coal to remove ash, sulfur, and other impurities. The end product contains about 16,000 Btu per pound, has an ash content of 0.1%, and a sulfur content of about 0.5%. The product is solid at room temperature, but can be liquefied by use of relatively low heat.

somatic effects of radiation. Effects of radiation limited to the exposed individual, as distinguished from genetic effects, which also affect subsequent unexposed generations. Large radiation doses can be fatal. Smaller doses may make the individual noticeably ill, may merely produce temporary changes in blood cell levels detectable only in the laboratory, or may produce no detectable effects whatever. Also called physiological effects of radiation. (Compare genetic effects of radiation.)

source material. As defined in the Atomic Energy Act of 1954, any material except special nuclear material, which contains 0.03% or more of uranium, thorium, or any combination of the two.

SNG. Synthetic natural gas.

SO_2. Sulfur dioxide.

spent (*depleted*) *fuel*. Nuclear reactor fuel that has been irradiated (used) to the extent that it can no longer effectively sustain a chain reaction.

spill. The accidental release of radioactive material.

stable. Incapable of spontaneous change. Not radioactive.

stable isotope. An isotope that does not undergo radioactive decay.

stack gas desulfurization. Treatment of stack gases to remove sulfur compounds.

standard cubic foot (*SCF*). The amount of gas contained in a volume of 1 cubic foot under a pressure of 1 atmosphere (14.73 pounds per square inch) at a temperature of $60°$F.

steam-electric plant. A plant in which the prime movers (turbines) connected to the generators are driven by steam.

strip mining. The mining of coal by surface methods as distinguished from the surface mining of ores which is commonly designated as open-pit mining.

stripper well. A nearly depleted oil well which produces less than 10 barrels/day.

stripping. Removal of the lightest fractions from a mixture.

subcritical assembly. A reactor consisting of a mass of fissionable material and moderator which cannot sustain a chain reaction. Used primarily for educational purposes.

subcritical mass. An amount of fissionable material insufficient in quantity or of improper geometry to sustain a fission chain reaction.

subsidence. A sinking or lowering of strata, including the surface, due to underground excavations, often coal mines.

sulfur oxides. Compounds composed of sulfur and oxygen. Produced by the burning of sulfur in coal, oil and gas. Harmful to the health of man, plants, and animals, and may cause damage to materials at sufficiently high concentrations.

supertanker. A very large oil tanker. The definition changes with advancing marine technology. In the late 1940s, 45,000 dwt tankers were considered super tankers; in the 1950s, 100,000 dwt was a super tanker; now common usage is 500,000 dwt.

supercritical reactor. A reactor in which the power level is increasing.

superheating. The heating of a vapor, particularly steam, to a temperature much higher than the boiling point at the prevailing pressure. Superheated steam is used in power plants to improve efficiency and to reduce condensation in the turbines.

surface mining. The mining of coal from outcroppings or by the removal of overburden from a seam of coal, as opposed to underground mining. Also called strip mining; placer mining; opencast; opencut mining; open-pit mining.

survey meter. Any portable radiation detection instrument especially adapted for surveying or inspecting an area to establish the existence and amount of radioactive material present.

sweetening. The process by which petroleum products are improved in odor and color by oxidizing or removing the sulfur-containing and unsaturated compounds.

SWU. Separative work unit.

syncrude. Synthetic crude oil derived from coal or oil shale.

syngas. Synthetic natural gas (SNG).

synthetic natural gas (SNG). A gaseous fuel manufactured from naptha or coal. It contains 95 to 98% methane, and has an energy content of 980 to 1035 Btu per standard cubic foot, about that of natural gas.

Tcf. Trillion cubic feet. A unit of measure commonly used for natural gas. 1 Tcf = 39.3 million tons of coal, $18\frac{1}{2}$ million bbl oil.

tar sands. Hydrocarbon-bearing deposits distinguished by the high viscosity of the hydrocarbon, which is not recoverable by ordinary oil production methods.

temperature coefficient. The change in reactor reactivity occurring when the operating temperature changes. The coefficient is said to be positive when an increase in temperature increases reactivity, negative when an increase in temperature decreases reactivity. Negative temperature coefficients are desirable because they help to prevent power excursions.

tertiary recovery. Methods other than fluid injection to augment oil recovery.

thermal breeder reactor. A breeder reactor in which the fission chain reaction is sustained by thermal neutrons.

thermal efficiency. In power plants, the ratio of heat converted to useful energy to the heat input.

thermal pollution. The addition of heat to a body of water, such as a lake or stream, such that the temperature is raised to an undesirable level. This heat may change the ecological balance, making it impossible for some types of life to survive, or it may favor the survival of other organisms, such as algae.

thermal power plant. Any electric power plant which operates by generating heat and converting the heat to electricity.

thermal reactor. A nuclear reactor in which the fission process is propagated mainly by thermal neutrons, i.e., by neutrons that have been slowed down until they are in thermal equilibrium with the atoms of the moderator.

thermal shield. A layer of high-density material placed around a reactor core to absorb neutrons and protect the reactor vessel.

thermionic conversion. A conversion device in which electrical energy is produced directly from heat energy. Theoretical efficiencies range from 15 to 33% with actual performance of 5 to 16%.

thermodynamics. The science and study of the relationships between heat and mechanical work. First law: Energy can neither be created nor destroyed; Second law: Heat cannot pass from a colder to a warmer body without the additional expenditure of mechanical energy.

thermoelectric conversion. Conversion of heat energy into electricity based on the Seebeck effect in which a difference in temperature between junctions comprised of two dissimilar metals produces an electric current.

thermonuclear reaction. A reaction in which very high temperatures allow the fusion of two light nuclei to form the nucleus of a heavier atom, releasing a large amount of energy.

threshold dose. The minimum dose of radiation that will produce a detectable biological effect.

thorium (Th). A naturally radioactive element with atomic number 90 and, as found in nature, an atomic weight of approximately 232. The fertile Th-232 isotope is abundant and can be transmuted to fissionable U-233 by neutron irradiation.

ton. A unit of weight equal to 2000 pounds in the United States, Canada and the Union of South Africa, and 2240 pounds in Great Britain. The American ton is often called the short ton, while the British ton is called the long ton. The metric ton, or 1000 kilograms, equals 2204.62 pounds.

topping. The distillation of crude oil to remove light fractions only.

topping cycle. A means to increase thermal efficiency of a steam-electric power plant by increasing temperatures and interposing a device, such as a gas turbine, between the heat source and the conventional steam-turbine generator to convert some of the additional heat energy into electricity.

total energy. Use of packaged energy systems of high efficiency, in which gas-fired turbines or engines produce electrical energy and in which the exhaust heat is used for heating or cooling.

tracer, isotopic. An isotope of an element, a small amount of which may be incorporated into a sample of material (the carrier) in order to follow (trace) the course of that element through a chemical, biological or physical process. The tracer may be radioactive, in which case observations are made by measuring the radioactivity. If the tracer is nonradioactive, mass spectrometers or neutron activation analysis may be employed to determine isotopic composition. Tracers also are called labels or tags, and materials are said to be labeled or tagged when radioactive tracers are incorporated in them.

tritium. A form of hydrogen consisting of 1 proton and 2 neutrons. Tritium is radioactive with a half-life of 12.4 years and spontaneously changes to helium-3 with the emission of a negative beta particle. Tritium is expected to be used as a fuel in fusion power plants.

turbine. A rotary engine made with a series of curved blades mounted on a shaft. The shaft is rotated by a current of fluid such as water, steam, or gas impinging on the blades.

UHV. Ultrahigh voltage transmission.

ultimate recoverable reserves. The total quantity of crude oil, natural gas, natural gas liquids, or sulfur estimated to be ultimately producible from an oil or gas field as determined by an analysis of current engineering data. This includes any quantities already produced up to the date of the estimate.

ultrahigh voltage transmission (UHV). Transmission of electricity at voltages higher than 800 kV.

underground coal gasification. The *in situ* process for producing synthetic gas from coal in underground deposits by burning of the coal in place.

unit train. A system for delivering coal efficiently in which a string of cars, with distinctive markings, and loaded to "full visible capacity," is operated without service frills or stops.

uranium (U). A radioactive element with the atomic number 92 and an average atomic weight of approximately 238. The two principal natural isotopes are U-235 (0.7% of natural uranium) which is fissionable (capable of being split and thereby releasing energy) and U-238 (99.3% of natural uranium) which is fertile (having the property of being convertible to a fissionable material). Natural uranium also includes a minute amount of U-234.

USGS. The United States Geological Survey.

uranium hexafluoride (UF$_6$). A compound of uranium and fluorine. UF_6 gas is the process fluid in the gaseous diffusion enrichment process.

uranium tetrafluoride (UF$_4$). A solid green compound; an intermediate product in the production of uranium hexafluoride gas, which in turn is used in the separation of U-235 in gaseous diffusion plants. Called green salt.

uranium trioxide (UO$_3$). An intermediate product in the refining of uranium. Called orange oxide.

volt. A unit of electrical force equal to that amount of electromotive force that will cause a steady current of one ampere to flow through a resistance of one ohm.

voltage. The amount of electromotive force, measured in volts, that exists between two points.

Wagon Wheel. A proposed industrial experiment with nuclear explosives that would use sequential firing of several charges to stimulate a natural gas field.

wastes, radioactive. Equipment and materials from nuclear operations which are radioactive and for which there is no further use. Wastes are generally classified as high-level (having radioactivity concentrations of hundreds to thousands of curies per gallon or cubic foot), low level (in the range of 1 microcurie per gallon or cubic foot), or intermediate.

waterflooding. A secondary oil recovery operation in which water is injected into a petroleum reservoir to bring more oil to the surface.

water gas. A mixture of gases produced by forcing steam through a very hot coke or coal. It is a mixture of carbon monoxide and hydrogen with small amounts of nitrogen and carbon dioxide and is sometimes used as a fuel for heating and cooking.

watt. The rate of energy transfer equivalent to one ampere under an electrical pressure of one volt. One watt equals 1/746 horsepower, or one joule per second.

watt-hour. The amount of energy needed to power a one-watt device for one hour.

wheeling. Transmission of electricity by a utility over its lines for another utility; also includes the receipt from and delivery to another system of like amounts but not necessarily the same energy.

wild cat. A well drilled in an area which has not produced gas or oil previously; usually exploratory and often without geophysical investigation. On the average, one of nine or ten wildcat wells strike oil or gas deposits.

work. Work is done whenever a force is exerted through a distance. The amount of work done is the product of force and distance. The metric unit of work is the joule (J). When a force of 1 newton is exerted through a distance of 1 meter, 1 joule of work is done.

X-ray. A penetrating form of electromagnetic radiation emitted either when the inner orbital electrons of an excited atom return to their normal state (these are characteristic X-rays), or when a metal target is bombarded with high-speed electrons (these are bremsstrahlung). X-rays are always nonnuclear in origin.

Energy Conversion Factors

CONVERSION FACTORS

Conversion Factors

To Convert	Into	Multiply by
acres	ft^2	43,560
acres	m^2	4,047
acres	hectares	0.4047
British thermal unit (Btu)	cal	252
Btu	joules (J)	1,055
Btu	kilowatt-hours (kWh)	2.93×10^{-4}
Btu	megawatt-years (MW yr)	3.34×10^{-11}
Btu/ft^2	langleys (cal/cm^2)	0.271
calories (cal)	Btu	3.97×10^{-3}
cal	ft lb	3.09
cal	J	4.18
cal/minute (cal/min)	watts (W)	0.0698
centimeters (cm)	inches (in.)	0.394
feet (ft)	meters	0.305
foot-pounds (ft lb)	cal	0.324
ft lb	J	1.36
ft lb	kilogram-meters (kg m)	0.138
ft lb	kWh	3.77×10^{-7}
gallons (gal)	liters	3.79
gal	lb	0.00220
gigawatts (GW)	W	1×10^9
hectares	acres	2.47
horsepower (hp)	kW	0.745
J	Btu	9.48×10^{-4}
J	cal	0.239
J	ft lb	0.738
J	W hr	2.78×10^{-4}
1000 cal (Kcal)	Btu	3.97
Kcal/min	kW	0.0698
kg m	ft lb	7.23
kg	lb	2.20
kg	tons	0.00110

Conversion Factors (Continued)

To Convert	Into	Multiply by
kilowatts (kW)	hp	1.34
kWh	Btu	3,413
kWh	ft lb	2.66×10^6
kWh	W hr	1×10^3
kW	Kcal/min	14.3
langleys	Btu/ft^2	3.69
langleys/min	W/sc cm	0.0698
megawatt-hours	Btu	3.41×10^6
MW yr	Btu	2.99×10^{10}
miles	meters	1,609
cm^2	ft^2	0.00108
cm^2	in.^{2}	0.155
ft^2	in.^{2}	144
ft^2	m^2	0.0929
in.^{2}	cm^2	6.45
m^2	ft^2	10.8
m^2	mile^2	3.86×10^{-7}
mile^2	acres	640
mile^2	ft^2	2.79×10^7
mile^2	m^2	2.59×10^6
tons (short)	kg	907
tons (short)	lb	2000
tons (metric)	tons (short)	1.1025
tons (short)	tons (metric)	0.907
W hr	J	3600
°F	°C	subtract 32 and multiply by 0.555
°C	°F	multiply by 1.8 and add 32
cal/cm^2-sec-°C	Btu/ft^2-hr-°F	7380
Btu/ft^2-hr-°F	cal/cm^2-sec-°C	1.35×10^{-4}

UNITS AND CONVERSION FACTORS

Units and Conversion Factors

Weight

1 kilogram (kg) = 1000 grams (g) = 1,000,000 milligrams (mg) = 1,000,000,000 micrograms (μg)

1 kg = 2.205 pounds (lb); 1 g = 0.035 ounce (oz)

1 lb = 453.6 g = 0.4536 kg

1 metric ton = 2205 lb = 1000 kg

1 short ton = 2000 lb = 907.2 kg

1 megaton = 1,000,000 tons

Length

0.001 kilometer (km) = 1 meter (m) = 100 centimeters (cm) = 1,000 millimeters (mm) = 1,000,000 microns (μ)

1 km = 0.6214 statute mile; 1 m = 39.37 inches (in.) = 3.281 feet (ft); 1 cm = 0.3937 in.

1 mile = 1.609 km; 1 ft = 0.3048 m; 1 in. = 2.54 cm

Area

1 hectare = 10,000 square meters (m^2)

1 hectare = 2.47 acres = 0.003861 square mile

1 square mile = 640 acres = 259 hectares

Volume and Cubic Measure

1 cubic meter (m^3) = 1,000,000 cubic centimeters (cm^3)

1 m^3 = 35.31 cubic feet (ft^3); 1 cm^3 = 0.061 cubic inch ($in.^3$)

1 liter (l) = 1000 cm^3

1 l = 61.02 $in.^3$ = 0.2642 gallon (gal)

1 ft^3 = 0.02832 m^3 = 28.32 l; 1 gal = 231 in^3 = 3.785 l

Energy and Work

1 British thermal unit (Btu) = 252 calories (cal) = 0.0002931 kilowatt-hour = 1055 joules (kWh)

1 joule = 1 watt-sec = 10^7 erg = 0.73 ft lb

1 gm-cal = 4.184 joules

1 kWh (kilowatt-hour) = 3.6 X 10^6 joules = 3412 Btu

1 watt-yr = 3.15 X 10^7 joules = 3 X 10^4 Btu

1 hp = 550 ft lb/sec = 746 watt

Power

1 megawatt (MW) = 1000 kilowatts (kW) = 1,000,000 watts (W) = 3,413,000 Btu/hour (Btu/hr) = 1341 horsepower (hp)

Pressure

1 atmosphere (atm) = 76 cm mercury = 14.70 lb/$in.^2$ = 1,013 millibars (mb)

Temperature scales

	Absolute Zero	Ice Point (water)	Steam Point (water)
Degrees Fahrenheit ($^\circ$F)	−459.7	32	212
Degrees Celsius or Centigrade ($^\circ$C)	−273.15	0	100
Degrees Kelvin ($^\circ$K)	0	273.15	373.15
Degrees Réaumur ($^\circ$R)		0	80

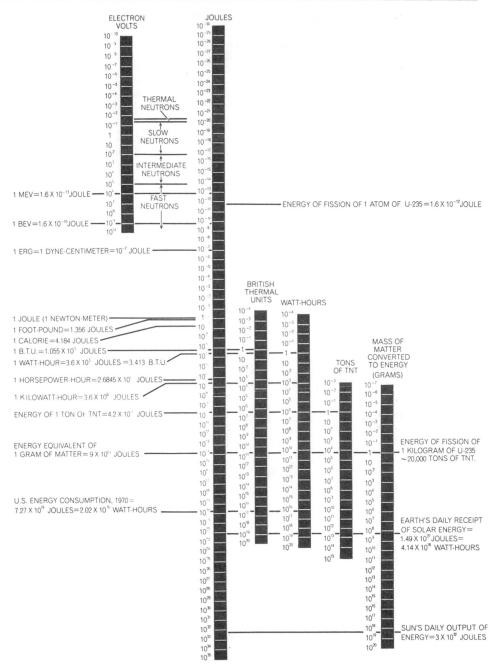

From "Energy and Power," by Chauncy Starr. Copyright © 1971 by Scientific American Inc. All rights reserved.

BTU EQUIVALENTS OF FUELS

Btu Equivalents of Common Fuels[1]

Fuel	Common Measure	Btus
Crude oil	barrel (bbl.)—42 gallons	5,800,000
Natural gas	1000 cubic feet	1,035,000
Natural gas	therm	100,000
Coal (bituminous)	ton (short ton −907.2 kg)	26,000,000
Coal (lignite)	ton	14,000,000
Gasoline	gallon	124,000
Electricity	kilowatt-hour (kWh)	3,412[2]

[1] A Btu is the amount of heat required to raise the temperature of 1 lb of water 1°F.

[2] Because of conversion losses in the generation of electric power from heat, about 10,000 Btu are required to produce one kilowatt-hour.

Energy Consumption Factors in Btus

8000 Btu = energy/day needed to sustain 1 man = 2,000,000 calories

50×10^3 Btu = daily per capita consumption by primitive agricultural man

1.0×10^6 Btu = daily per capita consumption by 1973 American

70×10^6 Btu = annual consumption by American family car

80×10^6 Btu = annual consumption for heating typical American house

75×10^{15} Btu = United States gross consumption in 1973 = 2.5×10^{12} watt-yr.

250×10^{15} Btu = world annual gross consumption (1973)

Equivalent Factors for Energy Consumption in Quad Units (10^5 Btu)

1 quad = 10^{15} Btu = 10^{-3} Q-units

$= 33 \times 10^9$ watt-yr.

$= 1.0 \times 10^{18}$ joules

$= 250 \times 10^{15}$ calories

$= 3.0 \times 10^{12}$ cubic feet of hydrogen

$= 1.0 \times 10^{12}$ cubic feet of natural gas = 1 TCF (trillion cubic feet)

$= 300 \times 10^9$ kWh—(thermal) $= 100 \times 10^9$ kWh (electrical)

$= 8.0 \times 10^9$ gallons of gasoline

$= 7.0 \times 10^9$ gallons of crude oil

$= 170 \times 10^6$ barrels of crude oil[a]

$= 50 \times 10^6$ short tons subbituminous coal

$= 40 \times 10^6$ short tons bituminous coal

$= 4.0 \times 10^6$ pounds of uranium (light-water reactor LWR)

$= 50 \times 10^3$ pounds of uranium (in a nuclear breeder with 60% efficiency)

[a] One million bbl per day corresponds to 2.12 quad per year.

Heat Content of Fuels in Btus

	Btu	Per Unit
Coal		
Anthracite (Pa.)	25,400,000	ton
Bituminous	26,200,000	ton
Blast furnace gas	100	ft³
Briquettes and package fuels	28,000,000	ton
Coke	24,800,000	ton
Coke-breeze	20,000,000	ton
Coke-oven gas	550	ft³
Coal tar	150,000	gal.
Coke-oven and manufactured gas products, light oils	5,460,000	bbl.
Natural gas (dry)	1,035	ft³
Natural gas liquids (average)	4,011,000	bbl.
Butane	4,284,000	bbl.
Propane	3,843,000	bbl.
Petroleum:		
Asphalt	6,640,000	bbl.
Coke	6,024,000	bbl.
Crude oil	5,800,000	bbl.
Diesel	5,806,000	bbl.
Distillate fuel oil	5,825,000	bbl.
Gasoline, aviation	5,048,000	bbl.
Gasoline, motor fuel	5,253,000	bbl.
Jet fuel:		
Commercial	5,670,000	bbl.
Military	5,355,000	bbl.
Kerosene	5,670,000	bbl.
Lubricants	6,060,000	bbl.
Miscellaneous oils	5,588,000	bbl.
Refinery still gas	5,600,000	bbl.
Heavy fuel oil	6,287,000	bbl.
Road oils	6,640,000	bbl.
Wax	5,570,000	bbl.
Shale oil	5,800,000	bbl.
Uranium:		
Total contained energy	60,000,000,000,000	short ton U_3O_8
Energy available with present technology	500,000,000,000	short ton U_3O_8

Sources: The U.S. Energy Problem, Volume II (NTIS), 1972, p. A-82.
The U.S. Energy Dilemma (Illinois State Geological Survey), 1973, p. 28.

Energy Unit Conversion Chart*

CUBIC FEET NATURAL GAS** (CF)	BARRELS OIL (bbl)	SHORT TONS BITUMINOUS COAL (T)	BRITISH THERMAL UNITS (Btu)	KILOWATT HOURS ELECTRICITY (kWhr)
–	–	–	1	0.000293
1	0.00018	0.00004	1000	0.293
3.41	0.00061	0.00014	3413	1
1000 (1 MCF)	0.18	0.04	1 MILLION	293
3413	0.61	0.14	3.41 MILLION	1000 (1 MWhr)
5600	1	0.22	5.6 MILLION	1640
25,000	4.46	1	25 MILLION	7325
1 MILLION (1 MMCF)	180	40	1 BILLION	293,000
3.41 MILLION	610	140	3.41 BILLION	1 MILLION (1 GWhr)
1 BILLION (1 BCF)	180,000	40,000	1 TRILLION	293 MILLION
1 TRILLION (1 TCF)	180 MILLION	40 MILLION	1 QUADRILLION (QUAD) (Q)	293 BILLION

* Based on the following nominal fuel heating values:
1 Cubic Foot Natural Gas = 1000 Btu
1 Barrel Crude Oil = 5.6 Million Btu
1 Pound Bituminous Coal = 12,500 Btu

** Substitute Natural Gas (SNG) and Liquefied Natural Gas (LNG) will have approximately the same heating value.

IGT
EDUCATION · RESEARCH

Institute of Gas Technology, 3424 South State Street, IIT Center Chicago, Ill. 60616.

CONVERSION FACTORS FOR OIL

Energy Equivalents of Oil

One 42 gallon barrel of oil

5.8×10^6 Btu
5.6×10^3 cubic feet of natural gas
1.70 thermal megawatt-hours
0.58 electrical megawatt-hours[1]
0.232 tons bituminous coal
0.42 tons lignite
6.119×10^9 joules

One million barrels of oil per day

2.1×10^{15} Btu/year
2.1 quad/year
0.365 billion barrels/year
232,000 tons bituminous coal/day
420,000 tons lignite/day
5.6 trillion cubic feet (TCF) natural gas/day
580 million kilowatt hours of electricity/day[1]

[1] At a heat rate of 10,000 Btu per kWh of electricity

Approximate Calorific Equivalents

One million tonnes of oil equals approximately:		Heat units and Other fuels expressed in terms of million tonnes of oil.		
Heat Units				million tonnes of oil
39	million million Btu	10	million million Btu approximates to	0·26
395	million therms	100	million therms approximates to	0·25
10 000	Teracalories	10 000	Teracalories approximates to	1·00
Solid Fuels				
1·5	million tonnes of coal	1	million tonnes of coal approximates to	0·67
4·9	million tonnes of lignite	1	million tonnes of lignite approximates to	0·20
3·3	million tonnes of peat	1	million tonnes of peat approximates to	0·30
Natural Gas (1 cub. ft = 1 000 Btu) (1 cub. metre = 9 000 kcal)				
1·167	thousand million cub. metres	1	thousand million cub. metres approximates to	0·86
41·2	thousand million cub. ft.	10	thousand million cub. ft. approximates to	0·24
113	million cub. ft./day for a year	100	million cub. ft./day for a year approximates to	0·88
Town Gas (1 cub. ft. = 470 Btu) (1 cub. metre = 4 200 kcal)				
2·5	thousand million cub. metres	1	thousand million cub. metres approximates to	0·40
88·3	thousand million cub. ft.	10	thousand million cub. ft. approximates to	0·11
242	million cub. ft./day for a year	100	million cub. ft./day for a year approximates to	0·41
Electricity (1 kWh = 3 412 Btu) (1 kWh = 860 kcal)				
12	thousand million kWh	10	thousand million kWh approximates to	0·82
One million tonnes of oil produces about 4,000 million units (kWh) of electricity in a modern power station.				

NOTE

In previous years, the BP Statistical Review has employed a conversion for other energy sources into oil equivalent of 10 500 kcal/kg. This year a value of 10 000 kcal/kg has been employed which downrates the heating value of oil, thus slightly increasing the oil equivalent value of other fuels.

Source: The British Petroleum Company Limited, Britannic House, Moor Lane, London EC2Y 9BU.

Approximate Conversion Factors for Crude Oil*

FROM \ INTO	Tonnes (Metric Tons)	Long Tons	Short Tons	Barrels	Kilolitres (cub. metres)	1 000 Gallons (Imp.)	1 000 Gallons (U.S.)
	MULTIPLY BY						
Tonnes (Metric Tons)	1	0·984	1·102	7·33	1·16	0·256	0·308
Long Tons	1·016	1	1·120	7·45	1·18	0·261	0·313
Short Tons	0·907	0·893	1	6·65	1·05	0·233	0·279
Barrels	0·136	0·134	0·150	1	0·159	0·035	0·042
Kilolitres (cub. metres)	0·863	0·849	0·951	6·29	1	0·220	0·264
1 000 Gallons (Imp.)	3·91	3·83	4·29	28·6	4·55	1	1·201
1 000 Gallons (U.S.)	3·25	3·19	3·58	23·8	3·79	0·833	1

TO CONVERT	FROM			
	Barrels to Tonnes	Tonnes to Barrels	Barrels/Day to Tonnes/Year	Tonnes/Year to Barrels/Day
	MULTIPLY BY			
Crude Oil *	0·136	7·33	49·8	0·0201
Motor Spirit	0·118	8·45	43·2	0·0232
Kerosine	0·128	7·80	46·8	0·0214
Gas/Diesel	0·133	7·50	48·7	0·0205
Fuel Oil	0·149	6·70	54·5	0·0184

*Based on world average gravity (excluding Natural Gas Liquids)

Source: The British Petroleum Company Limited, Britannic House, Moor Lane, London EC2Y 9BU.

ELECTRICAL UNITS

Power. The basic unit of power is the watt. For an electrical appliance the power rating is found by multiplying the voltage by the current (in amperes). Thus, a 125 volt appliance drawing 10 amperes has a power rating of 1250 watts. The kilowatt is simply 1000 watts and a megawatt is a million watts (10^6 watts). The horsepower is equivalent to 746 watts.

We can summarize these facts as follows:

$$\text{number of watts} = \text{voltage} \times \text{current}$$
$$1 \text{ kilowatt} = 1000 \text{ watts} = 10^3 \text{ watts}$$
$$1 \text{ megawatt} = 1{,}000{,}000 \text{ watts} = 10^6 \text{ watts}$$
$$1 \text{ horsepower} = 1 \text{ hp} = 746 \text{ watts}$$

Energy. The basic unit for energy is the kilowatt-hour (kWh), which is the energy used when a device rated at 1000 watts operates for an hour (or a 100 watt appliance operates for 10 hours). The table below gives various equivalents.

Unit or Process	Number of kWh
Calorie (diet)	$0.0012\ (1.2 \times 10^{-3})$
Calorie (ordinary)	1.2×10^{-6}
Btu	2.9×10^{-4}
One hour manual labor	0.06
Combustion of 1 gallon of gasoline	38.3
Heat one gallon of water $1°$F	0.0024

METRIC TERMINOLOGY

The following terms are frequency used in reports employing the metric system of units.

1 billion	1000 million = 10^9
1 trillion	1000 billion = 10^{12}
1 kcal	1 kilocalorie = 3.968 Btu (British thermal units)
1 therm	10^5 Btu = 25,200 kcal
1 thermie	1000 kcal
1 metric ton	1000 kilograms = 0.985 ton = 1.1023 short ton
1 toe	1 ton of oil equivalent = 10^7 kcal
1 Mtoe	1 million metric tons of oil equivalent = 10^{13} kcal
1 tce	1 ton of coal equivalent = 0.7 toe
1 Mtce	1 million tons of coal equivalent = 0.7 Mtoe
1 cubic foot	0.0283 cubic meter
1 Mcf	1 thousand cubic feet = 28.3 cubic meters
1 TCF	1 trillion cubic feet = 28.3×10^9 cubic meters
\$1 per Mcf	is approximately equivalent to \$40 per toe
f.o.b.	free on board
c.i.f.	cost including insurance and freight
GNP	Gross National Product
GDP	Gross Domestic Product

Source: Energy Prospects to 1985, Organization for Economic Co-operation and Development, Paris, France, 1974.

GROWTH RATE UNITS

Multiplication Factors for Energy Consumption at Various Growth Rates
(Reference Year 1974)

Growth Rate per Year (%)	1974	1979	1984	1999
0.5	1.00	1.025	1.051	1.133
1.0	1.00	1.051	1.105	1.284
1.5	1.00	1.078	1.162	1.455
2.0	1.00	1.105	1.221	1.650
2.5	1.00	1.133	1.284	1.868
3.0	1.00	1.162	1.350	2.117
4.0	1.00	1.217	1.480	2.666
5.0	1.00	1.276	1.629	3.386

For example: if energy consumption is 75 quad in 1974, then it will be (75×2.117) in 1999 if the growth rate is 3% per year.

SUPPLY–DEMAND ESTIMATES AND CONVERSION FACTORS

Conversion Factors

In an effort to make this study consistent with other major energy studies, the actual measurement units (tons of coal, barrels of oil, cubic feet of gas, etc.), have been converted to Btu's (British Thermal Units) and expressed as Quads (1x10¹⁵ Btu's). The Panel recognizes the need for an international unit to be used throughout the world to measure energy. Those which could be considered would include kilocalories (amount of heat required to raise 1,000 grams of water one degree centigrade: 1 kilocalorie = 3.97 Btu's) and Joules (the meter-kilogram-second unit of work or energy equal to the work done by a force of one newton when its point of application moves through a distance of one meter: equivalent to one watt-second: one Btu = 1055 Joules).

Difficulties arise in achieving consistency in conversions since energy is expressed in entirely different measuring units. For example,

1. Petroleum products have different Btu values depending on which product is chosen. Residual fuel oil generally has a heating value of about 6,300,000 Btu's/barrel. For this study, however, crude oil is used and expressed in barrels per day. Each barrel of crude oil equates to 5,800,000 Btu's. One million barrels of crude oil per day equals 2.12 Quads per year.

2. Although natural gas has a nominal heating value of 1032 Btu's per standard cubic foot (SCF), low and intermediate Btu gas has heating values in the 150 to 500 Btu/SCF range.

This report equates gas to 1000 Btu's per standard cubic foot. One trillion cubic foot/year of gas equals one Quad/year.

3. The conversion of coal presents similar problems. In general, the heating value of western coal is less than that of eastern coal—making the heating value of coal in the U.S. vary from about 13,000,000 to 26,000,000 Btu's per short ton. This study uses eastern coal at an average value of 24,500,000 Btu/short ton and western coal at an

average value of 17,000,000 Btu/short ton. In 1985, this study assumes that the use of western coal will increase to 40 percent.

4. Throughout the text of this report, electrical generating capacity (normally expressed in gigawatts or megawatts) is represented by an equivalent number of Quads/year of primary energy, namely one gigawatt equals .06 Quads/year.

This conversion was made using a capacity factor of slightly less than 69 percent and a heat rate of 10,000 Btu's/kilowatt-hour[1] These somewhat arbitrary conversion factors were used in this report to convert non-fossil as well as all fossil power generation.

In summary:

Fuel	Common Units	Btu's
Oil	Barrel (Bbl)	5.8×10^6
Gas	Standard Cubic Foot (SCF)	1,000
Coal	Short Ton (T)	$13\text{-}26 \times 10^6$ (Average of 21.6×10^6)
Electricity	(Net) Kilowatt Hour	3,413

Or, another way of viewing these data is:

1 Quad = 46 million short tons of coal
1 Quad = 175 million barrels of crude oil
1 Quad = 1 trillion standard cubic feet (scf) of gas

Since the Panel promoted the development of all fuels to cover the Nation's energy demands, its findings and recommendations are expressed in various units (tons of coal, barrels of oil, cubic feet of gas, gigawatts, etc.). To be consistent with other major energy studies, the Panel chose the Btu (British Thermal Unit) as the common unit of energy. Due to the quantity of annual energy requirements in the U.S., the Quad (1x10¹⁵ Btu) was used throughout this report.

Since a Btu or a Quad is not the most familiar or understandable unit to the American public, this table expresses the Panel's prudent planning bases in other equivalent units. The numbers used to

convert Quads to other units are shown in Appendix 4.

The Panel's 1985 demand planning base is 108 Quads. This is equivalent to:

1. 114 x 10¹⁸ Joules; or
2. 51 million barrels of oil per day; or
3. 5 billion tons of coal per year; or
4. 108 trillion cubic feet of gas per year; or
5. 31.6 trillion kilowatt-hours of electricity per year; or
6. $206 billion of fuels (at $11 per barrel of oil).

Supply/Demand Estimates (Equivalent Units)

Fuel	Units	Actual Units	Quads[1]	Equivalent in Joules[1] (10¹⁸)	Oil Equivalent (MMBPD)	Coal Equivalent (MMMTPY)	Gas Equivalent (TCFY)	Electric Equivalent (Trillion KWH)	$Billion[1] (at $11/BBL Oil Equivalent) Yr.
Coal	Billion Tons/Yr	1.2	26	27.4	12.4	1.20	26	7.6	49
Oil	MMBPD	12	25	26.4	12.0	1.16	25	7.3	48
Gas	TCFY	22	22	23.2	10.4	1.02	22	6.5	42
Nuclear	GWe	275	16.5	17.4	7.8	.76	16.5	4.9	31
Synthetics	(in coal)	-	-	-	-	-	-	-	-
Shale Oil	MMBPD	.25	.5	.5	.3	.02	.5	.1	1
Hydroelectric	GWe	58	3.5	3.7	1.7	.16	3.5	1.0	7
Geothermal*	GWe	6	.4	.4	.2	.02	.4	.1	1
Solar	Quads	.1	.1	.1	-	-	.1	-	-
Imports	Quads	14.0	14	14.8	6.6	.66	14	4.1	27
Totals			108	113.9	51.4	5.0	108	31.6	206

*Includes solid waste
[1] Per year

Source: CTAB Recommendations for a National Energy Program, U.S. Department of Commerce, Commerce Technical Advisory Board, February 1975.

[1] 1 Gigawatt (10⁶ KW) x 24 Hr/day x 365 days/year x 10,000 Btu/kwh x 69% x 1 Quad/10¹⁵ Btu = .0604 Quads (use .06)

Source: CTAB Recommendations for a National Energy Program, U.S. Department of Commerce, Commerce Technical Advisory Board, February 1975.

ENERGY USE MEASURES AND CONVERSION FACTORS

Gross Measures of Energy Used in the United States

- The two most commonly used gross measures of U.S. energy supply and demand are quadrillions of Btu (Quads or Btu x 10^{15}) and millions of barrels daily, oil equivalent.
- Two quadrillion Btu's per year are roughly equivalent to one million barrels daily (MMB/D).
- By moving straight across Fig. A–1, it is possible to match equivalent values in each of the commonly used measures of annual production or distribution for the major fuels and power sources.
- Uranium is entered as U$_3$O$_8$ in units of 10,000 short tons utilized in present thermal reactors.

- Solar energy is entered in units of 1,000 square kilometers of collector surface assuming 50% efficiency and the average amount of sunshine per year in the United States.

Conversion Factors

1 Quad = 180 million barrels of petroleum*
42 million tons of bituminous coal*
0.98 trillion cubic feet of natural gas*
293 billion kilowatt hours of electricity

* These values vary with the quality of fuel actually extracted and represent an average of recent production.

It is also of interest to note that 1 million barrels of oil per day is approximately equal to 2 Quads per year.

Source: A National Plan for Energy Research, Development and Demonstration: Creating Energy Choices for the Future, Volume 1: The Plan, ERDA-48, Energy Research and Development Administration, June 1975.

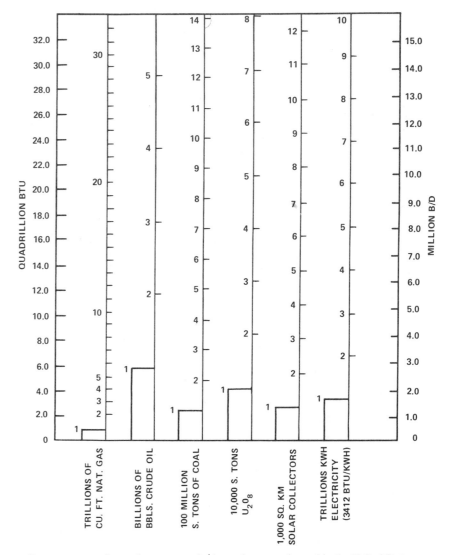

Gross measures of annual energy supply/demand commonly used in the United States.

Source: A National Plan for Energy Research, Development and Demonstration: Creating Energy Choices for the Future, Volume 1: The Plan, ERDA-48, Energy Research and Development Administration, June 1975.

Index